T0190058

The series Lecture Notes in Computer Science (LNCS), including its subseries Lecture Notes in Artificial Intelligence (LNAI) and Lecture Notes in Bioinformatics (LNBI), has established itself as a medium for the publication of new developments in computer science and information technology research, teaching, and education.

LNCS enjoys close cooperation with the computer science R & D community, the series counts many renowned academics among its volume editors and paper authors, and collaborates with prestigious societies. Its mission is to serve this international community by providing an invaluable service, mainly focused on the publication of conference and workshop proceedings and postproceedings. LNCS commenced publication in 1973.

Lecture Notes in Computer Science 14221

Hayit Greenspan · Anant Madabhushi ·
Parvin Mousavi · Septimiu Salcudean ·
James Duncan · Tanveer Syeda-Mahmood ·
Russell Taylor
Editors

Medical Image Computing and Computer Assisted Intervention – MICCAI 2023

26th International Conference
Vancouver, BC, Canada, October 8–12, 2023
Proceedings, Part II

Springer

Editors
Hayit Greenspan
Icahn School of Medicine, Mount Sinai,
NYC, NY, USA

Tel Aviv University
Tel Aviv, Israel

Parvin Mousavi
Queen's University
Kingston, ON, Canada

James Duncan
Yale University
New Haven, CT, USA

Russell Taylor
Johns Hopkins University
Baltimore, MD, USA

Anant Madabhushi
Emory University
Atlanta, GA, USA

Septimiu Salcudean
The University of British Columbia
Vancouver, BC, Canada

Tanveer Syeda-Mahmood
IBM Research
San Jose, CA, USA

ISSN 0302-9743 ISSN 1611-3349 (electronic)
Lecture Notes in Computer Science
ISBN 978-3-031-43894-3 ISBN 978-3-031-43895-0 (eBook)
https://doi.org/10.1007/978-3-031-43895-0

This Springer imprint is published by the registered company Springer Nature Switzerland AG
The registered company address is: Gewerbestrasse 11, 6330 Cham, Switzerland

Paper in this product is recyclable.

Preface

We are pleased to present the proceedings for the 26th International Conference on Medical Image Computing and Computer-Assisted Intervention (MICCAI). After several difficult years of virtual conferences, this edition was held in a mainly in-person format with a hybrid component at the Vancouver Convention Centre, in Vancouver, BC, Canada October 8–12, 2023. The conference featured 33 physical workshops, 15 online workshops, 15 tutorials, and 29 challenges held on October 8 and October 12. Co-located with the conference was also the 3rd Conference on Clinical Translation on Medical Image Computing and Computer-Assisted Intervention (CLINICCAI) on October 10.

MICCAI 2023 received the largest number of submissions so far, with an approximately 30% increase compared to 2022. We received 2365 full submissions of which 2250 were subjected to full review. To keep the acceptance ratios around 32% as in previous years, there was a corresponding increase in accepted papers leading to 730 papers accepted, with 68 orals and the remaining presented in poster form. These papers comprise ten volumes of Lecture Notes in Computer Science (LNCS) proceedings as follows:

- Part I, LNCS Volume 14220: Machine Learning with Limited Supervision and Machine Learning – Transfer Learning
- Part II, LNCS Volume 14221: Machine Learning – Learning Strategies and Machine Learning – Explainability, Bias, and Uncertainty I
- Part III, LNCS Volume 14222: Machine Learning – Explainability, Bias, and Uncertainty II and Image Segmentation I
- Part IV, LNCS Volume 14223: Image Segmentation II
- Part V, LNCS Volume 14224: Computer-Aided Diagnosis I
- Part VI, LNCS Volume 14225: Computer-Aided Diagnosis II and Computational Pathology
- Part VII, LNCS Volume 14226: Clinical Applications – Abdomen, Clinical Applications – Breast, Clinical Applications – Cardiac, Clinical Applications – Dermatology, Clinical Applications – Fetal Imaging, Clinical Applications – Lung, Clinical Applications – Musculoskeletal, Clinical Applications – Oncology, Clinical Applications – Ophthalmology, and Clinical Applications – Vascular
- Part VIII, LNCS Volume 14227: Clinical Applications – Neuroimaging and Microscopy
- Part IX, LNCS Volume 14228: Image-Guided Intervention, Surgical Planning, and Data Science
- Part X, LNCS Volume 14229: Image Reconstruction and Image Registration

The papers for the proceedings were selected after a rigorous double-blind peer-review process. The MICCAI 2023 Program Committee consisted of 133 area chairs and over 1600 reviewers, with representation from several countries across all major continents. It also maintained a gender balance with 31% of scientists who self-identified as women. With an increase in the number of area chairs and reviewers, the reviewer load on the experts was reduced this year, keeping to 16–18 papers per area chair and about 4–6 papers per reviewer. Based on the double-blinded reviews, area chairs' recommendations, and program chairs' global adjustments, 308 papers (14%) were provisionally accepted, 1196 papers (53%) were provisionally rejected, and 746 papers (33%) proceeded to the rebuttal stage. As in previous years, Microsoft's Conference Management Toolkit (CMT) was used for paper management and organizing the overall review process. Similarly, the Toronto paper matching system (TPMS) was employed to ensure knowledgeable experts were assigned to review appropriate papers. Area chairs and reviewers were selected following public calls to the community, and were vetted by the program chairs.

Among the new features this year was the emphasis on clinical translation, moving Medical Image Computing (MIC) and Computer-Assisted Interventions (CAI) research from theory to practice by featuring two clinical translational sessions reflecting the real-world impact of the field in the clinical workflows and clinical evaluations. For the first time, clinicians were appointed as Clinical Chairs to select papers for the clinical translational sessions. The philosophy behind the dedicated clinical translational sessions was to maintain the high scientific and technical standard of MICCAI papers in terms of methodology development, while at the same time showcasing the strong focus on clinical applications. This was an opportunity to expose the MICCAI community to the clinical challenges and for ideation of novel solutions to address these unmet needs. Consequently, during paper submission, in addition to MIC and CAI a new category of "Clinical Applications" was introduced for authors to self-declare.

MICCAI 2023 for the first time in its history also featured dual parallel tracks that allowed the conference to keep the same proportion of oral presentations as in previous years, despite the 30% increase in submitted and accepted papers.

We also introduced two new sessions this year focusing on young and emerging scientists through their Ph.D. thesis presentations, and another with experienced researchers commenting on the state of the field through a fireside chat format.

The organization of the final program by grouping the papers into topics and sessions was aided by the latest advancements in generative AI models. Specifically, Open AI's GPT-4 large language model was used to group the papers into initial topics which were then manually curated and organized. This resulted in fresh titles for sessions that are more reflective of the technical advancements of our field.

Although not reflected in the proceedings, the conference also benefited from keynote talks from experts in their respective fields including Turing Award winner Yann LeCun and leading experts Jocelyne Troccaz and Mihaela van der Schaar.

We extend our sincere gratitude to everyone who contributed to the success of MICCAI 2023 and the quality of its proceedings. In particular, we would like to express our profound thanks to the MICCAI Submission System Manager Kitty Wong whose meticulous support throughout the paper submission, review, program planning, and proceeding preparation process was invaluable. We are especially appreciative of the effort and dedication of our Satellite Events Chair, Bennett Landman, who tirelessly coordinated the organization of over 90 satellite events consisting of workshops, challenges and tutorials. Our workshop chairs Hongzhi Wang, Alistair Young, tutorial chairs Islem

Rekik, Guoyan Zheng, and challenge chairs, Lena Maier-Hein, Jayashree Kalpathy-Kramer, Alexander Seitel, worked hard to assemble a strong program for the satellite events. Special mention this year also goes to our first-time Clinical Chairs, Drs. Curtis Langlotz, Charles Kahn, and Masaru Ishii who helped us select papers for the clinical sessions and organized the clinical sessions.

We acknowledge the contributions of our Keynote Chairs, William Wells and Alejandro Frangi, who secured our keynote speakers. Our publication chairs, Kevin Zhou and Ron Summers, helped in our efforts to get the MICCAI papers indexed in PubMed. It was a challenging year for fundraising for the conference due to the recovery of the economy after the COVID pandemic. Despite this situation, our industrial sponsorship chairs, Mohammad Yaqub, Le Lu and Yanwu Xu, along with Dekon's Mehmet Eldegez, worked tirelessly to secure sponsors in innovative ways, for which we are grateful.

An active body of the MICCAI Student Board led by Camila Gonzalez and our 2023 student representatives Nathaniel Braman and Vaishnavi Subramanian helped put together student-run networking and social events including a novel Ph.D. thesis 3-minute madness event to spotlight new graduates for their careers. Similarly, Women in MICCAI chairs Xiaoxiao Li and Jayanthi Sivaswamy and RISE chairs, Islem Rekik, Pingkun Yan, and Andrea Lara further strengthened the quality of our technical program through their organized events. Local arrangements logistics including the recruiting of University of British Columbia students and invitation letters to attendees, was ably looked after by our local arrangement chairs Purang Abolmaesumi and Mehdi Moradi. They also helped coordinate the visits to the local sites in Vancouver both during the selection of the site and organization of our local activities during the conference. Our Young Investigator chairs Marius Linguraru, Archana Venkataraman, Antonio Porras Perez put forward the startup village and helped secure funding from NIH for early career scientist participation in the conference. Our communications chair, Ehsan Adeli, and Diana Cunningham were active in making the conference visible on social media platforms and circulating the newsletters. Niharika D'Souza was our cross-committee liaison providing note-taking support for all our meetings. We are grateful to all these organization committee members for their active contributions that made the conference successful.

We would like to thank the MICCAI society chair, Caroline Essert, and the MICCAI board for their approvals, support and feedback, which provided clarity on various aspects of running the conference. Behind the scenes, we acknowledge the contributions of the MICCAI secretariat personnel, Janette Wallace, and Johanne Langford, who kept a close eye on logistics and budgets, and Diana Cunningham and Anna Van Vliet for including our conference announcements in a timely manner in the MICCAI society newsletters. This year, when the existing virtual platform provider indicated that they would discontinue their service, a new virtual platform provider Conference Catalysts was chosen after due diligence by John Baxter. John also handled the setup and coordination with CMT and consultation with program chairs on features, for which we are very grateful. The physical organization of the conference at the site, budget financials, fund-raising, and the smooth running of events would not have been possible without our Professional Conference Organization team from Dekon Congress & Tourism led by Mehmet Eldegez. The model of having a PCO run the conference, which we used at

MICCAI, significantly reduces the work of general chairs for which we are particularly grateful.

Finally, we are especially grateful to all members of the Program Committee for their diligent work in the reviewer assignments and final paper selection, as well as the reviewers for their support during the entire process. Lastly, and most importantly, we thank all authors, co-authors, students/postdocs, and supervisors for submitting and presenting their high-quality work, which played a pivotal role in making MICCAI 2023 a resounding success.

With a successful MICCAI 2023, we now look forward to seeing you next year in Marrakesh, Morocco when MICCAI 2024 goes to the African continent for the first time.

October 2023

Tanveer Syeda-Mahmood
James Duncan
Russ Taylor
General Chairs

Hayit Greenspan
Anant Madabhushi
Parvin Mousavi
Septimiu Salcudean
Program Chairs

Organization

General Chairs

Tanveer Syeda-Mahmood	IBM Research, USA
James Duncan	Yale University, USA
Russ Taylor	Johns Hopkins University, USA

Program Committee Chairs

Hayit Greenspan	Tel-Aviv University, Israel and Icahn School of Medicine at Mount Sinai, USA
Anant Madabhushi	Emory University, USA
Parvin Mousavi	Queen's University, Canada
Septimiu Salcudean	University of British Columbia, Canada

Satellite Events Chair

Bennett Landman	Vanderbilt University, USA

Workshop Chairs

Hongzhi Wang	IBM Research, USA
Alistair Young	King's College, London, UK

Challenges Chairs

Jayashree Kalpathy-Kramer	Harvard University, USA
Alexander Seitel	German Cancer Research Center, Germany
Lena Maier-Hein	German Cancer Research Center, Germany

Tutorial Chairs

Islem Rekik	Imperial College London, UK
Guoyan Zheng	Shanghai Jiao Tong University, China

Clinical Chairs

Curtis Langlotz	Stanford University, USA
Charles Kahn	University of Pennsylvania, USA
Masaru Ishii	Johns Hopkins University, USA

Local Arrangements Chairs

Purang Abolmaesumi	University of British Columbia, Canada
Mehdi Moradi	McMaster University, Canada

Keynote Chairs

William Wells	Harvard University, USA
Alejandro Frangi	University of Manchester, UK

Industrial Sponsorship Chairs

Mohammad Yaqub	MBZ University of Artificial Intelligence, Abu Dhabi
Le Lu	DAMO Academy, Alibaba Group, USA
Yanwu Xu	Baidu, China

Communication Chair

Ehsan Adeli	Stanford University, USA

Publication Chairs

Ron Summers — National Institutes of Health, USA
Kevin Zhou — University of Science and Technology of China, China

Young Investigator Chairs

Marius Linguraru — Children's National Institute, USA
Archana Venkataraman — Boston University, USA
Antonio Porras — University of Colorado Anschutz Medical Campus, USA

Student Activities Chairs

Nathaniel Braman — Picture Health, USA
Vaishnavi Subramanian — EPFL, France

Women in MICCAI Chairs

Jayanthi Sivaswamy — IIIT, Hyderabad, India
Xiaoxiao Li — University of British Columbia, Canada

RISE Committee Chairs

Islem Rekik — Imperial College London, UK
Pingkun Yan — Rensselaer Polytechnic Institute, USA
Andrea Lara — Universidad Galileo, Guatemala

Submission Platform Manager

Kitty Wong — The MICCAI Society, Canada

Virtual Platform Manager

John Baxter	INSERM, Université de Rennes 1, France

Cross-Committee Liaison

Niharika D'Souza	IBM Research, USA

Program Committee

Sahar Ahmad	University of North Carolina at Chapel Hill, USA
Shadi Albarqouni	University of Bonn and Helmholtz Munich, Germany
Angelica Aviles-Rivero	University of Cambridge, UK
Shekoofeh Azizi	Google, Google Brain, USA
Ulas Bagci	Northwestern University, USA
Wenjia Bai	Imperial College London, UK
Sophia Bano	University College London, UK
Kayhan Batmanghelich	University of Pittsburgh and Boston University, USA
Ismail Ben Ayed	ETS Montreal, Canada
Katharina Breininger	Friedrich-Alexander-Universität Erlangen-Nürnberg, Germany
Weidong Cai	University of Sydney, Australia
Geng Chen	Northwestern Polytechnical University, China
Hao Chen	Hong Kong University of Science and Technology, China
Jun Cheng	Institute for Infocomm Research, A*STAR, Singapore
Li Cheng	University of Alberta, Canada
Albert C. S. Chung	University of Exeter, UK
Toby Collins	Ircad, France
Adrian Dalca	Massachusetts Institute of Technology and Harvard Medical School, USA
Jose Dolz	ETS Montreal, Canada
Qi Dou	Chinese University of Hong Kong, China
Nicha Dvornek	Yale University, USA
Shireen Elhabian	University of Utah, USA
Sandy Engelhardt	Heidelberg University Hospital, Germany
Ruogu Fang	University of Florida, USA

Reviewers

Alaa Eldin Abdelaal
John Abel
Kumar Abhishek
Shahira Abousamra
Mazdak Abulnaga
Burak Acar
Abdoljalil Addeh
Ehsan Adeli
Sukesh Adiga Vasudeva
Seyed-Ahmad Ahmadi
Euijoon Ahn
Faranak Akbarifar
Alireza Akhondi-asl
Saad Ullah Akram
Daniel Alexander
Hanan Alghamdi
Hassan Alhajj
Omar Al-Kadi
Max Allan
Andre Altmann
Pablo Alvarez
Charlems Alvarez-Jimenez
Jennifer Alvén
Lidia Al-Zogbi
Kimberly Amador
Tamaz Amiranashvili
Amine Amyar
Wangpeng An
Vincent Andrearczyk
Manon Ansart
Sameer Antani
Jacob Antunes
Michel Antunes
Guilherme Aresta
Mohammad Ali Armin
Kasra Arnavaz
Corey Arnold
Janan Arslan
Marius Arvinte
Muhammad Asad
John Ashburner
Md Ashikuzzaman
Shahab Aslani
Mehdi Astaraki
Angélica Atehortúa
Benjamin Aubert
Marc Aubreville
Paolo Avesani
Sana Ayromlou
Reza Azad
Mohammad Farid Azampour
Qinle Ba
Meritxell Bach Cuadra
Hyeon-Min Bae
Matheus Baffa
Cagla Bahadir
Fan Bai
Jun Bai
Long Bai
Pradeep Bajracharya
Shafa Balaram
Yaël Balbastre
Yutong Ban
Abhirup Banerjee
Soumyanil Banerjee
Sreya Banerjee
Shunxing Bao
Omri Bar
Adrian Barbu
Joao Barreto
Adrian Basarab
Berke Basaran
Michael Baumgartner
Siming Bayer
Roza Bayrak
Aicha BenTaieb
Guy Ben-Yosef
Sutanu Bera
Cosmin Bercea
Jorge Bernal
Jose Bernal
Gabriel Bernardino
Riddhish Bhalodia
Jignesh Bhatt
Indrani Bhattacharya
Binod Bhattarai
Lei Bi
Qi Bi
Cheng Bian
Gui-Bin Bian
Carlo Biffi
Alexander Bigalke
Benjamin Billot
Manuel Birlo
Ryoma Bise
Daniel Blezek
Stefano Blumberg
Sebastian Bodenstedt
Federico Bolelli
Bhushan Borotikar
Ilaria Boscolo Galazzo
Alexandre Bousse
Nicolas Boutry
Joseph Boyd
Behzad Bozorgtabar
Nadia Brancati
Clara Brémond Martin
Stéphanie Bricq
Christopher Bridge
Coleman Broaddus
Rupert Brooks
Tom Brosch
Mikael Brudfors
Ninon Burgos
Nikolay Burlutskiy
Michal Byra
Ryan Cabeen
Mariano Cabezas
Hongmin Cai
Tongan Cai
Zongyou Cai
Liane Canas
Bing Cao
Guogang Cao
Weiguo Cao
Xu Cao
Yankun Cao
Zhenjie Cao

Jaime Cardoso
M. Jorge Cardoso
Owen Carmichael
Jacob Carse
Adrià Casamitjana
Alessandro Casella
Angela Castillo
Kate Cevora
Krishna Chaitanya
Satrajit Chakrabarty
Yi Hao Chan
Shekhar Chandra
Ming-Ching Chang
Peng Chang
Qi Chang
Yuchou Chang
Hanqing Chao
Simon Chatelin
Soumick Chatterjee
Sudhanya Chatterjee
Muhammad Faizyab Ali Chaudhary
Antong Chen
Bingzhi Chen
Chen Chen
Cheng Chen
Chengkuan Chen
Eric Chen
Fang Chen
Haomin Chen
Jianan Chen
Jianxu Chen
Jiazhou Chen
Jie Chen
Jintai Chen
Jun Chen
Junxiang Chen
Junyu Chen
Li Chen
Liyun Chen
Nenglun Chen
Pingjun Chen
Pingyi Chen
Qi Chen
Qiang Chen
Runnan Chen
Shengcong Chen
Sihao Chen
Tingting Chen
Wenting Chen
Xi Chen
Xiang Chen
Xiaoran Chen
Xin Chen
Xiongchao Chen
Yanxi Chen
Yixiong Chen
Yixuan Chen
Yuanyuan Chen
Yuqian Chen
Zhaolin Chen
Zhen Chen
Zhenghao Chen
Zhennong Chen
Zhihao Chen
Zhineng Chen
Zhixiang Chen
Chang-Chieh Cheng
Jiale Cheng
Jianhong Cheng
Jun Cheng
Xuelian Cheng
Yupeng Cheng
Mark Chiew
Philip Chikontwe
Eleni Chiou
Jungchan Cho
Jang-Hwan Choi
Min-Kook Choi
Wookjin Choi
Jaegul Choo
Yu-Cheng Chou
Daan Christiaens
Argyrios Christodoulidis
Stergios Christodoulidis
Kai-Cheng Chuang
Hyungjin Chung
Matthew Clarkson
Michaël Clément
Dana Cobzas
Jaume Coll-Font
Olivier Colliot
Runmin Cong
Yulai Cong
Laura Connolly
William Consagra
Pierre-Henri Conze
Tim Cootes
Teresa Correia
Baris Coskunuzer
Alex Crimi
Can Cui
Hejie Cui
Hui Cui
Lei Cui
Wenhui Cui
Tolga Cukur
Tobias Czempiel
Javid Dadashkarimi
Haixing Dai
Tingting Dan
Kang Dang
Salman Ul Hassan Dar
Eleonora D'Arnese
Dhritiman Das
Neda Davoudi
Tareen Dawood
Sandro De Zanet
Farah Deeba
Charles Delahunt
Herve Delingette
Ugur Demir
Liang-Jian Deng
Ruining Deng
Wenlong Deng
Felix Denzinger
Adrien Depeursinge
Mohammad Mahdi Derakhshani
Hrishikesh Deshpande
Adrien Desjardins
Christian Desrosiers
Blake Dewey
Neel Dey
Rohan Dhamdhere

Maged Goubran
Sobhan Goudarzi
Mark Graham
Alejandro Granados
Mara Graziani
Thomas Grenier
Radu Grosu
Michal Grzeszczyk
Feng Gu
Pengfei Gu
Qiangqiang Gu
Ran Gu
Shi Gu
Wenhao Gu
Xianfeng Gu
Yiwen Gu
Zaiwang Gu
Hao Guan
Jayavardhana Gubbi
Houssem-Eddine Gueziri
Dazhou Guo
Hengtao Guo
Jixiang Guo
Jun Guo
Pengfei Guo
Wenzhangzhi Guo
Xiaoqing Guo
Xueqi Guo
Yi Guo
Vikash Gupta
Praveen Gurunath Bharathi
Prashnna Gyawali
Sung Min Ha
Mohamad Habes
Ilker Hacihaliloglu
Stathis Hadjidemetriou
Fatemeh Haghighi
Justin Haldar
Noura Hamze
Liang Han
Luyi Han
Seungjae Han
Tianyu Han
Zhongyi Han
Jonny Hancox
Lasse Hansen
Degan Hao
Huaying Hao
Jinkui Hao
Nazim Haouchine
Michael Hardisty
Stefan Harrer
Jeffry Hartanto
Charles Hatt
Huiguang He
Kelei He
Qi He
Shenghua He
Xinwei He
Stefan Heldmann
Nicholas Heller
Edward Henderson
Alessa Hering
Monica Hernandez
Kilian Hett
Amogh Hiremath
David Ho
Malte Hoffmann
Matthew Holden
Qingqi Hong
Yoonmi Hong
Mohammad Reza Hosseinzadeh Taher
William Hsu
Chuanfei Hu
Dan Hu
Kai Hu
Rongyao Hu
Shishuai Hu
Xiaoling Hu
Xinrong Hu
Yan Hu
Yang Hu
Chaoqin Huang
Junzhou Huang
Ling Huang
Luojie Huang
Qinwen Huang
Sharon Xiaolei Huang
Weijian Huang
Xiaoyang Huang
Yi-Jie Huang
Yongsong Huang
Yongxiang Huang
Yuhao Huang
Zhe Huang
Zhi-An Huang
Ziyi Huang
Arnaud Huaulmé
Henkjan Huisman
Alex Hung
Jiayu Huo
Andreas Husch
Mohammad Arafat Hussain
Sarfaraz Hussein
Jana Hutter
Khoi Huynh
Ilknur Icke
Kay Igwe
Abdullah Al Zubaer Imran
Muhammad Imran
Samra Irshad
Nahid Ul Islam
Koichi Ito
Hayato Itoh
Yuji Iwahori
Krithika Iyer
Mohammad Jafari
Srikrishna Jaganathan
Hassan Jahanandish
Andras Jakab
Amir Jamaludin
Amoon Jamzad
Ananya Jana
Se-In Jang
Pierre Jannin
Vincent Jaouen
Uditha Jarayathne
Ronnachai Jaroensri
Guillaume Jaume
Syed Ashar Javed
Rachid Jennane
Debesh Jha
Ge-Peng Ji

Luping Ji
Zexuan Ji
Zhanghexuan Ji
Haozhe Jia
Hongchao Jiang
Jue Jiang
Meirui Jiang
Tingting Jiang
Xiajun Jiang
Zekun Jiang
Zhifan Jiang
Ziyu Jiang
Jianbo Jiao
Zhicheng Jiao
Chen Jin
Dakai Jin
Qiangguo Jin
Qiuye Jin
Weina Jin
Baoyu Jing
Bin Jing
Yaqub Jonmohamadi
Lie Ju
Yohan Jun
Dinkar Juyal
Manjunath K N
Ali Kafaei Zad Tehrani
John Kalafut
Niveditha Kalavakonda
Megha Kalia
Anil Kamat
Qingbo Kang
Po-Yu Kao
Anuradha Kar
Neerav Karani
Turkay Kart
Satyananda Kashyap
Alexander Katzmann
Lisa Kausch
Maxime Kayser
Salome Kazeminia
Wenchi Ke
Youngwook Kee
Matthias Keicher
Erwan Kerrien
Afifa Khaled
Nadieh Khalili
Farzad Khalvati
Bidur Khanal
Bishesh Khanal
Pulkit Khandelwal
Maksim Kholiavchenko
Ron Kikinis
Benjamin Killeen
Daeseung Kim
Heejong Kim
Jaeil Kim
Jinhee Kim
Jinman Kim
Junsik Kim
Minkyung Kim
Namkug Kim
Sangwook Kim
Tae Soo Kim
Younghoon Kim
Young-Min Kim
Andrew King
Miranda Kirby
Gabriel Kiss
Andreas Kist
Yoshiro Kitamura
Stefan Klein
Tobias Klinder
Kazuma Kobayashi
Lisa Koch
Satoshi Kondo
Fanwei Kong
Tomasz Konopczynski
Ender Konukoglu
Aishik Konwer
Thijs Kooi
Ivica Kopriva
Avinash Kori
Kivanc Kose
Suraj Kothawade
Anna Kreshuk
AnithaPriya Krishnan
Florian Kromp
Frithjof Kruggel
Thomas Kuestner
Levin Kuhlmann
Abhay Kumar
Kuldeep Kumar
Sayantan Kumar
Manuela Kunz
Holger Kunze
Tahsin Kurc
Anvar Kurmukov
Yoshihiro Kuroda
Yusuke Kurose
Hyuksool Kwon
Aymen Laadhari
Jorma Laaksonen
Dmitrii Lachinov
Alain Lalande
Rodney LaLonde
Bennett Landman
Daniel Lang
Carole Lartizien
Shlomi Laufer
Max-Heinrich Laves
William Le
Loic Le Folgoc
Christian Ledig
Eung-Joo Lee
Ho Hin Lee
Hyekyoung Lee
John Lee
Kisuk Lee
Kyungsu Lee
Soochahn Lee
Woonghee Lee
Étienne Léger
Wen Hui Lei
Yiming Lei
George Leifman
Rogers Jeffrey Leo John
Juan Leon
Bo Li
Caizi Li
Chao Li
Chen Li
Cheng Li
Chenxin Li
Chnegyin Li

Dawei Li
Fuhai Li
Gang Li
Guang Li
Hao Li
Haofeng Li
Haojia Li
Heng Li
Hongming Li
Hongwei Li
Huiqi Li
Jian Li
Jieyu Li
Kang Li
Lin Li
Mengzhang Li
Ming Li
Qing Li
Quanzheng Li
Shaohua Li
Shulong Li
Tengfei Li
Weijian Li
Wen Li
Xiaomeng Li
Xingyu Li
Xinhui Li
Xuelu Li
Xueshen Li
Yamin Li
Yang Li
Yi Li
Yuemeng Li
Yunxiang Li
Zeju Li
Zhaoshuo Li
Zhe Li
Zhen Li
Zhenqiang Li
Zhiyuan Li
Zhjin Li
Zi Li
Hao Liang
Libin Liang
Peixian Liang
Yuan Liang
Yudong Liang
Haofu Liao
Hongen Liao
Wei Liao
Zehui Liao
Gilbert Lim
Hongxiang Lin
Li Lin
Manxi Lin
Mingquan Lin
Tiancheng Lin
Yi Lin
Zudi Lin
Claudia Lindner
Simone Lionetti
Chi Liu
Chuanbin Liu
Daochang Liu
Dongnan Liu
Feihong Liu
Fenglin Liu
Han Liu
Huiye Liu
Jiang Liu
Jie Liu
Jinduo Liu
Jing Liu
Jingya Liu
Jundong Liu
Lihao Liu
Mengting Liu
Mingyuan Liu
Peirong Liu
Peng Liu
Qin Liu
Quan Liu
Rui Liu
Shengfeng Liu
Shuangjun Liu
Sidong Liu
Siyuan Liu
Weide Liu
Xiao Liu
Xiaoyu Liu
Xingtong Liu
Xinwen Liu
Xinyang Liu
Xinyu Liu
Yan Liu
Yi Liu
Yihao Liu
Yikang Liu
Yilin Liu
Yilong Liu
Yiqiao Liu
Yong Liu
Yuhang Liu
Zelong Liu
Zhe Liu
Zhiyuan Liu
Zuozhu Liu
Lisette Lockhart
Andrea Loddo
Nicolas Loménie
Yonghao Long
Daniel Lopes
Ange Lou
Brian Lovell
Nicolas Loy Rodas
Charles Lu
Chun-Shien Lu
Donghuan Lu
Guangming Lu
Huanxiang Lu
Jingpei Lu
Yao Lu
Oeslle Lucena
Jie Luo
Luyang Luo
Ma Luo
Mingyuan Luo
Wenhan Luo
Xiangde Luo
Xinzhe Luo
Jinxin Lv
Tianxu Lv
Fei Lyu
Ilwoo Lyu
Mengye Lyu

Qing Lyu
Yanjun Lyu
Yuanyuan Lyu
Benteng Ma
Chunwei Ma
Hehuan Ma
Jun Ma
Junbo Ma
Wenao Ma
Yuhui Ma
Pedro Macias Gordaliza
Anant Madabhushi
Derek Magee
S. Sara Mahdavi
Andreas Maier
Klaus H. Maier-Hein
Sokratis Makrogiannis
Danial Maleki
Michail Mamalakis
Zhehua Mao
Jan Margeta
Brett Marinelli
Zdravko Marinov
Viktoria Markova
Carsten Marr
Yassine Marrakchi
Anne Martel
Martin Maška
Tejas Sudharshan Mathai
Petr Matula
Dimitrios Mavroeidis
Evangelos Mazomenos
Amarachi Mbakwe
Adam McCarthy
Stephen McKenna
Raghav Mehta
Xueyan Mei
Felix Meissen
Felix Meister
Afaque Memon
Mingyuan Meng
Qingjie Meng
Xiangzhu Meng
Yanda Meng
Zhu Meng
Martin Menten
Odyssée Merveille
Mikhail Milchenko
Leo Milecki
Fausto Milletari
Hyun-Seok Min
Zhe Min
Song Ming
Duy Minh Ho Nguyen
Deepak Mishra
Suraj Mishra
Virendra Mishra
Tadashi Miyamoto
Sara Moccia
Marc Modat
Omid Mohareri
Tony C. W. Mok
Javier Montoya
Rodrigo Moreno
Stefano Moriconi
Lia Morra
Ana Mota
Lei Mou
Dana Moukheiber
Lama Moukheiber
Daniel Moyer
Pritam Mukherjee
Anirban Mukhopadhyay
Henning Müller
Ana Murillo
Gowtham Krishnan Murugesan
Ahmed Naglah
Karthik Nandakumar
Venkatesh Narasimhamurthy
Raja Narayan
Dominik Narnhofer
Vishwesh Nath
Rodrigo Nava
Abdullah Nazib
Ahmed Nebli
Peter Neher
Amin Nejatbakhsh
Trong-Thuan Nguyen
Truong Nguyen
Dong Ni
Haomiao Ni
Xiuyan Ni
Hannes Nickisch
Weizhi Nie
Aditya Nigam
Lipeng Ning
Xia Ning
Kazuya Nishimura
Chuang Niu
Sijie Niu
Vincent Noblet
Narges Norouzi
Alexey Novikov
Jorge Novo
Gilberto Ochoa-Ruiz
Masahiro Oda
Benjamin Odry
Hugo Oliveira
Sara Oliveira
Arnau Oliver
Jimena Olveres
John Onofrey
Marcos Ortega
Mauricio Alberto Ortega-Ruíz
Yusuf Osmanlioglu
Chubin Ou
Cheng Ouyang
Jiahong Ouyang
Xi Ouyang
Cristina Oyarzun Laura
Utku Ozbulak
Ece Ozkan
Ege Özsoy
Batu Ozturkler
Harshith Padigela
Johannes Paetzold
José Blas Pagador Carrasco
Daniel Pak
Sourabh Palande
Chengwei Pan
Jiazhen Pan

Jin Pan
Yongsheng Pan
Egor Panfilov
Jiaxuan Pang
Joao Papa
Constantin Pape
Bartlomiej Papiez
Nripesh Parajuli
Hyunjin Park
Akash Parvatikar
Tiziano Passerini
Diego Patiño Cortés
Mayank Patwari
Angshuman Paul
Rasmus Paulsen
Yuchen Pei
Yuru Pei
Tao Peng
Wei Peng
Yige Peng
Yunsong Peng
Matteo Pennisi
Antonio Pepe
Oscar Perdomo
Sérgio Pereira
Jose-Antonio Pérez-Carrasco
Mehran Pesteie
Terry Peters
Eike Petersen
Jens Petersen
Micha Pfeiffer
Dzung Pham
Hieu Pham
Ashish Phophalia
Tomasz Pieciak
Antonio Pinheiro
Pramod Pisharady
Theodoros Pissas
Szymon Płotka
Kilian Pohl
Sebastian Pölsterl
Alison Pouch
Tim Prangemeier
Prateek Prasanna
Raphael Prevost
Juan Prieto
Federica Proietto Salanitri
Sergi Pujades
Elodie Puybareau
Talha Qaiser
Buyue Qian
Mengyun Qiao
Yuchuan Qiao
Zhi Qiao
Chenchen Qin
Fangbo Qin
Wenjian Qin
Yulei Qin
Jie Qiu
Jielin Qiu
Peijie Qiu
Shi Qiu
Wu Qiu
Liangqiong Qu
Linhao Qu
Quan Quan
Tran Minh Quan
Sandro Queirós
Prashanth R
Febrian Rachmadi
Daniel Racoceanu
Mehdi Rahim
Jagath Rajapakse
Kashif Rajpoot
Keerthi Ram
Dhanesh Ramachandram
João Ramalhinho
Xuming Ran
Aneesh Rangnekar
Hatem Rashwan
Keerthi Sravan Ravi
Daniele Ravì
Sadhana Ravikumar
Harish Raviprakash
Surreerat Reaungamornrat
Samuel Remedios
Mengwei Ren
Sucheng Ren
Elton Rexhepaj
Mauricio Reyes
Constantino Reyes-Aldasoro
Abel Reyes-Angulo
Hadrien Reynaud
Razieh Rezaei
Anne-Marie Rickmann
Laurent Risser
Dominik Rivoir
Emma Robinson
Robert Robinson
Jessica Rodgers
Ranga Rodrigo
Rafael Rodrigues
Robert Rohling
Margherita Rosnati
Łukasz Roszkowiak
Holger Roth
José Rouco
Dan Ruan
Jiacheng Ruan
Daniel Rueckert
Danny Ruijters
Kanghyun Ryu
Ario Sadafi
Numan Saeed
Monjoy Saha
Pramit Saha
Farhang Sahba
Pranjal Sahu
Simone Saitta
Md Sirajus Salekin
Abbas Samani
Pedro Sanchez
Luis Sanchez Giraldo
Yudi Sang
Gerard Sanroma-Guell
Rodrigo Santa Cruz
Alice Santilli
Rachana Sathish
Olivier Saut
Mattia Savardi
Nico Scherf
Alexander Schlaefer
Jerome Schmid

Adam Schmidt
Julia Schnabel
Lawrence Schobs
Julian Schön
Peter Schueffler
Andreas Schuh
Christina Schwarz-Gsaxner
Michaël Sdika
Suman Sedai
Lalithkumar Seenivasan
Matthias Seibold
Sourya Sengupta
Lama Seoud
Ana Sequeira
Sharmishtaa Seshamani
Ahmed Shaffie
Jay Shah
Keyur Shah
Ahmed Shahin
Mohammad Abuzar Shaikh
S. Shailja
Hongming Shan
Wei Shao
Mostafa Sharifzadeh
Anuja Sharma
Gregory Sharp
Hailan Shen
Li Shen
Linlin Shen
Mali Shen
Mingren Shen
Yiqing Shen
Zhengyang Shen
Jun Shi
Xiaoshuang Shi
Yiyu Shi
Yonggang Shi
Hoo-Chang Shin
Jitae Shin
Keewon Shin
Boris Shirokikh
Suzanne Shontz
Yucheng Shu
Hanna Siebert
Alberto Signoroni
Wilson Silva
Julio Silva-Rodríguez
Margarida Silveira
Walter Simson
Praveer Singh
Vivek Singh
Nitin Singhal
Elena Sizikova
Gregory Slabaugh
Dane Smith
Kevin Smith
Tiffany So
Rajath Soans
Roger Soberanis-Mukul
Hessam Sokooti
Jingwei Song
Weinan Song
Xinhang Song
Xinrui Song
Mazen Soufi
Georgia Sovatzidi
Bella Specktor Fadida
William Speier
Ziga Spiclin
Dominik Spinczyk
Jon Sporring
Pradeeba Sridar
Chetan L. Srinidhi
Abhishek Srivastava
Lawrence Staib
Marc Stamminger
Justin Strait
Hai Su
Ruisheng Su
Zhe Su
Vaishnavi Subramanian
Gérard Subsol
Carole Sudre
Dong Sui
Heung-Il Suk
Shipra Suman
He Sun
Hongfu Sun
Jian Sun
Li Sun
Liyan Sun
Shanlin Sun
Kyung Sung
Yannick Suter
Swapna T. R.
Amir Tahmasebi
Pablo Tahoces
Sirine Taleb
Bingyao Tan
Chaowei Tan
Wenjun Tan
Hao Tang
Siyi Tang
Xiaoying Tang
Yucheng Tang
Zihao Tang
Michael Tanzer
Austin Tapp
Elias Tappeiner
Mickael Tardy
Giacomo Tarroni
Athena Taymourtash
Kaveri Thakoor
Elina Thibeau-Sutre
Paul Thienphrapa
Sarina Thomas
Stephen Thompson
Karl Thurnhofer-Hemsi
Cristiana Tiago
Lin Tian
Lixia Tian
Yapeng Tian
Yu Tian
Yun Tian
Aleksei Tiulpin
Hamid Tizhoosh
Minh Nguyen Nhat To
Matthew Toews
Maryam Toloubidokhti
Minh Tran
Quoc-Huy Trinh
Jocelyne Troccaz
Roger Trullo

Chialing Tsai
Apostolia Tsirikoglou
Puxun Tu
Samyakh Tukra
Sudhakar Tummala
Georgios Tziritas
Vladimír Ulman
Tamas Ungi
Régis Vaillant
Jeya Maria Jose Valanarasu
Vanya Valindria
Juan Miguel Valverde
Fons van der Sommen
Maureen van Eijnatten
Tom van Sonsbeek
Gijs van Tulder
Yogatheesan Varatharajah
Madhurima Vardhan
Thomas Varsavsky
Hooman Vaseli
Serge Vasylechko
S. Swaroop Vedula
Sanketh Vedula
Gonzalo Vegas Sanchez-Ferrero
Matthew Velazquez
Archana Venkataraman
Sulaiman Vesal
Mitko Veta
Barbara Villarini
Athanasios Vlontzos
Wolf-Dieter Vogl
Ingmar Voigt
Sandrine Voros
Vibashan VS
Trinh Thi Le Vuong
An Wang
Bo Wang
Ce Wang
Changmiao Wang
Ching-Wei Wang
Dadong Wang
Dong Wang
Fakai Wang
Guotai Wang
Haifeng Wang
Haoran Wang
Hong Wang
Hongxiao Wang
Hongyu Wang
Jiacheng Wang
Jing Wang
Jue Wang
Kang Wang
Ke Wang
Lei Wang
Li Wang
Liansheng Wang
Lin Wang
Ling Wang
Linwei Wang
Manning Wang
Mingliang Wang
Puyang Wang
Qiuli Wang
Renzhen Wang
Ruixuan Wang
Shaoyu Wang
Sheng Wang
Shujun Wang
Shuo Wang
Shuqiang Wang
Tao Wang
Tianchen Wang
Tianyu Wang
Wenzhe Wang
Xi Wang
Xiangdong Wang
Xiaoqing Wang
Xiaosong Wang
Yan Wang
Yangang Wang
Yaping Wang
Yi Wang
Yirui Wang
Yixin Wang
Zeyi Wang
Zhao Wang
Zichen Wang
Ziqin Wang
Ziyi Wang
Zuhui Wang
Dong Wei
Donglai Wei
Hao Wei
Jia Wei
Leihao Wei
Ruofeng Wei
Shuwen Wei
Martin Weigert
Wolfgang Wein
Michael Wels
Cédric Wemmert
Thomas Wendler
Markus Wenzel
Rhydian Windsor
Adam Wittek
Marek Wodzinski
Ivo Wolf
Julia Wolleb
Ka-Chun Wong
Jonghye Woo
Chongruo Wu
Chunpeng Wu
Fuping Wu
Huaqian Wu
Ji Wu
Jiangjie Wu
Jiong Wu
Junde Wu
Linshan Wu
Qing Wu
Weiwen Wu
Wenjun Wu
Xiyin Wu
Yawen Wu
Ye Wu
Yicheng Wu
Yongfei Wu
Zhengwang Wu
Pengcheng Xi
Chao Xia
Siyu Xia
Wenjun Xia
Lei Xiang

Tiange Xiang
Deqiang Xiao
Li Xiao
Xiaojiao Xiao
Yiming Xiao
Zeyu Xiao
Hongtao Xie
Huidong Xie
Jianyang Xie
Long Xie
Weidi Xie
Fangxu Xing
Shuwei Xing
Xiaodan Xing
Xiaohan Xing
Haoyi Xiong
Yujian Xiong
Di Xu
Feng Xu
Haozheng Xu
Hongming Xu
Jiangchang Xu
Jiaqi Xu
Junshen Xu
Kele Xu
Lijian Xu
Min Xu
Moucheng Xu
Rui Xu
Xiaowei Xu
Xuanang Xu
Yanwu Xu
Yanyu Xu
Yongchao Xu
Yunqiu Xu
Zhe Xu
Zhoubing Xu
Ziyue Xu
Kai Xuan
Cheng Xue
Jie Xue
Tengfei Xue
Wufeng Xue
Yuan Xue
Zhong Xue
Ts Faridah Yahya
Chaochao Yan
Jiangpeng Yan
Ming Yan
Qingsen Yan
Xiangyi Yan
Yuguang Yan
Zengqiang Yan
Baoyao Yang
Carl Yang
Changchun Yang
Chen Yang
Feng Yang
Fengting Yang
Ge Yang
Guanyu Yang
Heran Yang
Huijuan Yang
Jiancheng Yang
Jiewen Yang
Peng Yang
Qi Yang
Qiushi Yang
Wei Yang
Xin Yang
Xuan Yang
Yan Yang
Yanwu Yang
Yifan Yang
Yingyu Yang
Zhicheng Yang
Zhijian Yang
Jiangchao Yao
Jiawen Yao
Lanhong Yao
Linlin Yao
Qingsong Yao
Tianyuan Yao
Xiaohui Yao
Zhao Yao
Dong Hye Ye
Menglong Ye
Yousef Yeganeh
Jirong Yi
Xin Yi
Chong Yin
Pengshuai Yin
Yi Yin
Zhaozheng Yin
Chunwei Ying
Youngjin Yoo
Jihun Yoon
Chenyu You
Hanchao Yu
Heng Yu
Jinhua Yu
Jinze Yu
Ke Yu
Qi Yu
Qian Yu
Thomas Yu
Weimin Yu
Yang Yu
Chenxi Yuan
Kun Yuan
Wu Yuan
Yixuan Yuan
Paul Yushkevich
Fatemeh Zabihollahy
Samira Zare
Ramy Zeineldin
Dong Zeng
Qi Zeng
Tianyi Zeng
Wei Zeng
Kilian Zepf
Kun Zhan
Bokai Zhang
Daoqiang Zhang
Dong Zhang
Fa Zhang
Hang Zhang
Hanxiao Zhang
Hao Zhang
Haopeng Zhang
Haoyue Zhang
Hongrun Zhang
Jiadong Zhang
Jiajin Zhang
Jianpeng Zhang

Jiawei Zhang
Jingqing Zhang
Jingyang Zhang
Jinwei Zhang
Jiong Zhang
Jiping Zhang
Ke Zhang
Lefei Zhang
Lei Zhang
Li Zhang
Lichi Zhang
Lu Zhang
Minghui Zhang
Molin Zhang
Ning Zhang
Rongzhao Zhang
Ruipeng Zhang
Ruisi Zhang
Shichuan Zhang
Shihao Zhang
Shuai Zhang
Tuo Zhang
Wei Zhang
Weihang Zhang
Wen Zhang
Wenhua Zhang
Wenqiang Zhang
Xiaodan Zhang
Xiaoran Zhang
Xin Zhang
Xukun Zhang
Xuzhe Zhang
Ya Zhang
Yanbo Zhang
Yanfu Zhang
Yao Zhang
Yi Zhang
Yifan Zhang
Yixiao Zhang
Yongqin Zhang
You Zhang
Youshan Zhang
Yu Zhang
Yubo Zhang
Yue Zhang
Yuhan Zhang
Yulun Zhang
Yundong Zhang
Yunlong Zhang
Yuyao Zhang
Zheng Zhang
Zhenxi Zhang
Ziqi Zhang
Can Zhao
Chongyue Zhao
Fenqiang Zhao
Gangming Zhao
He Zhao
Jianfeng Zhao
Jun Zhao
Li Zhao
Liang Zhao
Lin Zhao
Mengliu Zhao
Mingbo Zhao
Qingyu Zhao
Shang Zhao
Shijie Zhao
Tengda Zhao
Tianyi Zhao
Wei Zhao
Yidong Zhao
Yiyuan Zhao
Yu Zhao
Zhihe Zhao
Ziyuan Zhao
Haiyong Zheng
Hao Zheng
Jiannan Zheng
Kang Zheng
Meng Zheng
Sisi Zheng
Tianshu Zheng
Yalin Zheng
Yefeng Zheng
Yinqiang Zheng
Yushan Zheng
Aoxiao Zhong
Jia-Xing Zhong
Tao Zhong
Zichun Zhong
Hong-Yu Zhou
Houliang Zhou
Huiyu Zhou
Kang Zhou
Qin Zhou
Ran Zhou
S. Kevin Zhou
Tianfei Zhou
Wei Zhou
Xiao-Hu Zhou
Xiao-Yun Zhou
Yi Zhou
Youjia Zhou
Yukun Zhou
Zongwei Zhou
Chenglu Zhu
Dongxiao Zhu
Heqin Zhu
Jiayi Zhu
Meilu Zhu
Wei Zhu
Wenhui Zhu
Xiaofeng Zhu
Xin Zhu
Yonghua Zhu
Yongpei Zhu
Yuemin Zhu
Yan Zhuang
David Zimmerer
Yongshuo Zong
Ke Zou
Yukai Zou
Lianrui Zuo
Gerald Zwettler

Outstanding Area Chairs

Mingxia Liu	University of North Carolina at Chapel Hill, USA
Matthias Wilms	University of Calgary, Canada
Veronika Zimmer	Technical University Munich, Germany

Outstanding Reviewers

Kimberly Amador	University of Calgary, Canada
Angela Castillo	Universidad de los Andes, Colombia
Chen Chen	Imperial College London, UK
Laura Connolly	Queen's University, Canada
Pierre-Henri Conze	IMT Atlantique, France
Niharika D'Souza	IBM Research, USA
Michael Götz	University Hospital Ulm, Germany
Meirui Jiang	Chinese University of Hong Kong, China
Manuela Kunz	National Research Council Canada, Canada
Zdravko Marinov	Karlsruhe Institute of Technology, Germany
Sérgio Pereira	Lunit, South Korea
Lalithkumar Seenivasan	National University of Singapore, Singapore

Honorable Mentions (Reviewers)

Kumar Abhishek	Simon Fraser University, Canada
Guilherme Aresta	Medical University of Vienna, Austria
Shahab Aslani	University College London, UK
Marc Aubreville	Technische Hochschule Ingolstadt, Germany
Yaël Balbastre	Massachusetts General Hospital, USA
Omri Bar	Theator, Israel
Aicha Ben Taieb	Simon Fraser University, Canada
Cosmin Bercea	Technical University Munich and Helmholtz AI and Helmholtz Center Munich, Germany
Benjamin Billot	Massachusetts Institute of Technology, USA
Michal Byra	RIKEN Center for Brain Science, Japan
Mariano Cabezas	University of Sydney, Australia
Alessandro Casella	Italian Institute of Technology and Politecnico di Milano, Italy
Junyu Chen	Johns Hopkins University, USA
Argyrios Christodoulidis	Pfizer, Greece
Olivier Colliot	CNRS, France

Wolf-Dieter Vogl	RetInSight GmbH, Austria
Vibashan VS	Johns Hopkins University, USA
Lin Wang	Harbin Engineering University, China
Yan Wang	Sichuan University, China
Rhydian Windsor	University of Oxford, UK
Ivo Wolf	University of Applied Sciences Mannheim, Germany
Linshan Wu	Hunan University, China
Xin Yang	Chinese University of Hong Kong, China

Contents – Part II

Machine Learning – Learning Strategies

Machine Learning – Learning Strategies

OpenAL: An Efficient Deep Active Learning Framework for Open-Set Pathology Image Classification

Linhao Qu[1,2], Yingfan Ma[1,2], Zhiwei Yang[2,3], Manning Wang[1,2](✉), and Zhijian Song[1,2](✉)

[1] Digital Medical Research Center, School of Basic Medical Science, Fudan University, Shanghai 200032, China
{lhqu20,mnwang,zjsong}@fudan.edu.cn
[2] Shanghai Key Lab of Medical Image Computing and Computer Assisted Intervention, Shanghai 200032, China
[3] Academy for Engineering and Technology, Fudan University, Shanghai 200433, China

Abstract. Active learning (AL) is an effective approach to select the most informative samples to label so as to reduce the annotation cost. Existing AL methods typically work under the closed-set assumption, i.e., all classes existing in the unlabeled sample pool need to be classified by the target model. However, in some practical clinical tasks, the unlabeled pool may contain not only the target classes that need to be fine-grainedly classified, but also non-target classes that are irrelevant to the clinical tasks. Existing AL methods cannot work well in this scenario because they tend to select a large number of non-target samples. In this paper, we formulate this scenario as an open-set AL problem and propose an efficient framework, OpenAL, to address the challenge of querying samples from an unlabeled pool with both target class and non-target class samples. Experiments on fine-grained classification of pathology images show that OpenAL can significantly improve the query quality of target class samples and achieve higher performance than current state-of-the-art AL methods. Code is available at https://github.com/miccaiif/OpenAL.

Keywords: Active learning · Openset · Pathology image classification

1 Introduction

Deep learning techniques have achieved unprecedented success in the field of medical image classification, but this is largely due to large amount of annotated data [5,18,20]. However, obtaining large amounts of high-quality annotated data is usually expensive and time-consuming, especially in the field of pathology image processing [5,12–14,18]. Therefore, a very important issue is how to obtain the highest model performance with a limited annotation budget.

L. Qu and Y. Ma—Contributed equally.

H. Greenspan et al. (Eds.): MICCAI 2023, LNCS 14221, pp. 3–13, 2023.
https://doi.org/10.1007/978-3-031-43895-0_1

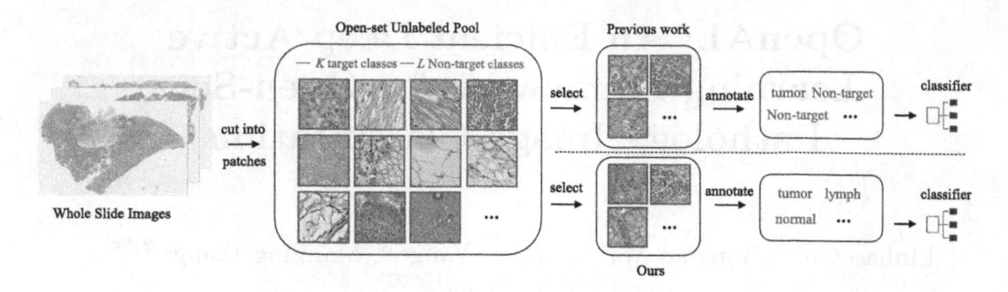

Fig. 1. Description of the open-set AL scenario for pathology image classification. The unlabeled sample pool contains K target categories (red-boxed images) and L non-target categories (blue-boxed images). Existing AL methods cannot accurately distinguish whether the samples are from the target classes or not, thus querying a large number of non-target samples and wasting the annotation budget, while our method can accurately query samples from the target categories. (Color figure online)

Active learning (AL) is an effective approach to address this issue from a data selection perspective, which selects the most informative samples from an unlabeled sample pool for experts to label and improves the performance of the trained model with reduced labeling cost [1,2,9,10,16,17,19]. However, existing AL methods usually work under the closed-set assumption, i.e., all classes existing in the unlabeled sample pool need to be classified by the target model, which does not meet the needs of some real-world scenarios [11]. Figure 1 shows an AL scenario for pathology image classification in an open world, which is very common in clinical practice. In this scenario, the Whole Slide Images (WSIs) are cut into many small patches that compose the unlabeled sample pool, where each patch may belong to tumor, lymph, normal tissue, fat, stroma, debris, background, and many other categories. However, it is not necessary to perform fine-grained annotation and classification for all categories in clinical applications. For example, in the cell classification task, only patches of tumor, lymphatic and normal cells need to be labeled and classified by the target model. Since the non-target patches are not necessary for training the classifier, labeling them would waste a large amount of budget. We call this scenario in which the unlabeled pool consists of both target class and non-target class samples open-set AL problem. Most existing AL algorithms can only work in the closed-set setting. Even worse, in the open-set setting, they even query more non-target samples because these samples tend to have greater uncertainty compared to the target class samples [11]. Therefore, for real-world open-set pathology image classification scenarios, an AL method that can accurately query the most informative samples from the target classes is urgently needed.

Recently, Ning et al. [11] proposed the first AL algorithm for open-set annotation in the field of natural images. They first trained a network to detect target class samples using a small number of initially labeled samples, and then modeled the maximum activation value (MAV) distribution of each sample using a Gaussian mixture model [15] (GMM) to actively select the most deterministic target class samples for labeling. Although promising performance is achieved, their detection of target class samples is based on the activation layer values of the detection network which has limited accuracy and high uncertainty with small initial training samples.

In this paper, we propose a novel AL framework under an open-set scenario, and denote it as OpenAL, which cannot only query as many target class samples as possible but also query the most informative samples from the target classes. OpenAL adopts an iterative query paradigm and uses a two-stage sample selection strategy in each query. In the first stage, we do not rely on a detection network to select target class samples and instead, we propose a feature-based target sample selection strategy. Specifically, we first train a feature extractor using all samples in a self-supervised learning manner, and map all samples to the feature space. There are three types of samples in the feature space, the unlabeled samples, the target class samples labeled in previous iterations, and the non-target class samples queried in previous iterations but not being labeled. Then we select the unlabeled samples that are close to the target class samples and far from the non-target class samples to form a candidate set. In the second stage, we select the most informative samples from the candidate set by utilizing a model-based informative sample selection strategy. In this stage, we measure the uncertainty of all unlabeled samples in the candidate set using the classifier trained with the target class samples labeled in previous iterations, and select the samples with the highest model uncertainty as the final selected samples in this round of query. After the second stage, the queried samples are sent for annotation, which includes distinguishing target and non-target class samples and giving a fine-grained label to every target class sample. After that, we train the classifier again using all the fine-grained labeled target class samples.

We conducted two experiments with different matching ratios (ratio of the number of target class samples to the total number of samples) on a public 9-class colorectal cancer pathology image dataset. The experimental results demonstrate that OpenAL can significantly improve the query quality of target class samples and obtain higher performance with equivalent labeling cost compared with the current state-of-the-art AL methods. To the best of our knowledge, this is the first open-set AL work in the field of pathology image analysis.

2 Method

We consider the AL task for pathology image classification in an open-set scenario. The unlabeled sample pool P_U consists of K classes of target samples and L classes of non-target samples (usually, $K < L$). N iterative queries are performed to query a fixed number of samples in each iteration, and the objective is to select as many target class samples as possible from P_U in each query, while selecting as many informative samples as possible in the target class samples. Each queried sample is given to experts for labeling, and the experts will give fine-grained category labels for target class samples, while only giving a "non-target class samples" label for non-target class samples.

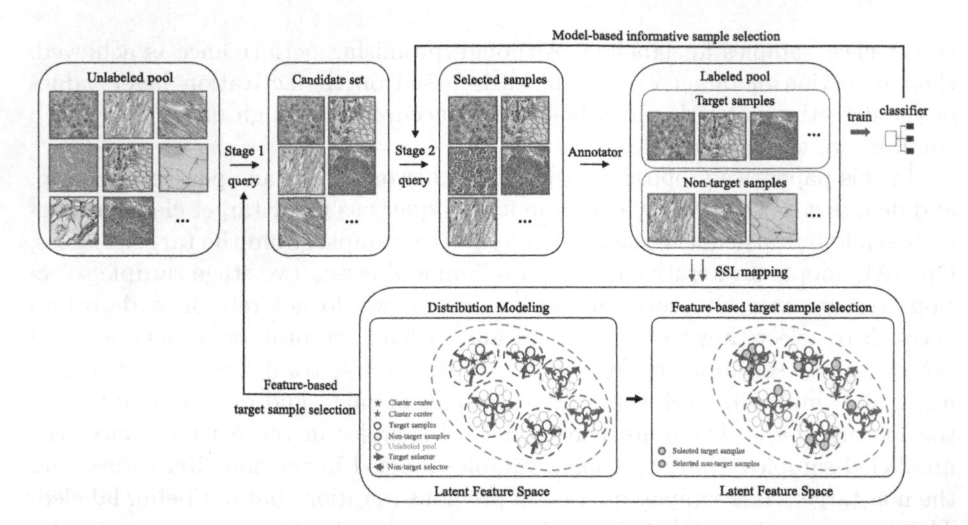

Fig. 2. Workflow of OpenAL.

2.1 Framework Overview

Figure 2 illustrates the workflow of the proposed method, OpenAL. OpenAL performs a total of N iterative queries, and each query is divided into two stages. In Stage 1, OpenAL uses a feature-based target sample selection (FTSS) strategy to query the target class samples from the unlabeled sample pool to form a candidate set. Specifically, we first train a feature extractor with all samples by self-supervised learning, and map all samples to the feature space. Then we model the distribution of all unlabeled samples, all labeled target class samples from previous iterations, and all non-target class samples queried in previous iterations in the feature space, and select the unlabeled samples that are close to the target class samples and far from the non-target class samples. In Stage 2, OpenAL adopts a model-based informative sample selection (MISS) strategy. Specifically, we measure the uncertainty of all unlabeled samples in the candidate set using the classifier trained in the last iteration, and select the samples with the highest model uncertainty as the final selected samples, which are sent to experts for annotation. After obtaining new labeled samples, we train the classifier using all fine-grained labeled target class samples with cross-entropy as the loss function. The FTSS strategy is described in Sect. 2.2, and the MISS strategy is described in Sect. 2.3.

2.2 Feature-Based Target Sample Selection

Self-supervised Feature Representation. First, we use all samples to train a feature extractor by self-supervised learning and map all samples to the latent feature space. Here, we adopt DINO [3,4] as the self-supervised network because of its outstanding performance.

Sample Scoring and Selection in the Feature Space. Then we define a scoring function on the base of the distribution of unlabeled samples, labeled target class samples and non-target class samples queried in previous iterations. Every unlabeled sample in the current iteration is given a score, and a smaller score indicates that the sample is more likely to come from the target classes. The scoring function is defined in Eq. 1.

$$s_i = s_{t_i} - s_{w_i} \tag{1}$$

where s_i denotes the score of the unlabeled sample x_i^U. s_{t_i} measures the distance between x_i^U and the distribution of features derived from all the labeled target class samples. The smaller s_{t_i} is, the closer x_i^U is to the known sample distribution of the target classes, and the more likely x_i^U is from a target class. Similarly, s_{w_i} measures the distance between x_i^U and the distribution of features derived from all the queried non-target class samples. The smaller s_{w_i} is, the closer x_i^U is from the known distribution of non-target class samples, and the less likely x_i^U is from the target class. After scoring all the unlabeled samples, we select the top $\varepsilon\%$ samples with the smallest scores to form the candidate set. In this paper, we empirically take twice the current iterative labeling budget (number of samples submitted to experts for labeling) as the sample number of the candidate set. Below, we give the definitions of s_{t_i} and s_{w_i}.

Distance-Based Feature Distribution Modeling. We propose a category and Mahalanobis distance-based feature distribution modeling approach for calculating s_{t_i} and s_{w_i}. The definitions of these two values are slightly different, and we first present the calculation of s_{t_i}, followed by that of s_{w_i}.

For all labeled target class samples from previous iterations, their fine-grained labels are known, so we represent these samples as different clusters in the feature space according to their true class labels, where a cluster is denoted as $C_t^L (t = 1, \ldots, K)$. Next, we calculate the score s_{t_i} for z_i^U using the Mahalanobis distance (MD) according to Eq. 2. MD is widely used to measure the distance between a point and a distribution because it takes into account the mean and variance of the distribution, which is very suitable for our scenario.

$$s_{t_i} = \text{Nom}\left(\min_t \left(D\left(z_i^U, C_t^L\right)\right)\right) = \text{Nom}\left(\min_t \left(z_i^U - \mu_t\right)^T \Sigma_t^{-1} \left(z_i^U - \mu_t\right)\right) \tag{2}$$

$$\text{Nom}(X) = \frac{X - X_{\min}}{X_{\max} - X_{\min}} \tag{3}$$

where $D(\cdot)$ denotes the MD function, μ_t and Σ_t are the mean and covariance of the samples in the target class t, and $\text{Nom}(\cdot)$ is the normalization function. It can be seen that s_{t_i} is essentially the minimum distance of the unlabeled sample x_i^U to each target class cluster.

For all the queried non-target class samples from previous iterations, since they do not have fine-grained labels, we first use the K-means algorithm to cluster their features into w classes, where a cluster is denoted as C_w^L $(w = 1, \ldots, W)$.

W is set to 9 in this paper. Next, we calculate the score s_{w_i} for z_i^U using the MD according to Eq. 4.

$$s_{w_i} = \text{Nom}\left(\min_{w}\left(D\left(z_i^U, C_w^L\right)\right)\right) = \text{Nom}\left(\min_{w}\left(z_i^U - \mu_w\right)^T \Sigma_t^{-1}\left(z_i^U - \mu_w\right)\right) \quad (4)$$

where μ_w and Σ_w are the mean and covariance of the non-target class sample features in the wth cluster. It can be seen that s_{w_i} is essentially the minimum distance of z_i^U to each cluster of known non-target class samples.

The within-cluster selection and dynamic cluster changes between rounds significantly enhance the diversity of the selected samples and reduce redundancy.

2.3 Model-Based Informative Sample Selection

To select the most informative samples from the candidate set, we utilize the model-based informative sample selection strategy in Stage 2. We measure the uncertainty of all unlabeled samples in the candidate set using the classifier trained in the last iteration and select the samples with the highest model uncertainty as the final selected samples. The entropy of the model output is a simple and effective way to measure sample uncertainty [7,8]. Therefore, we calculate the entropy of the model for the samples in the candidate set and select 50% of them with the highest entropy as the final samples in the current iteration.

3 Experiments

3.1 Dataset, Settings, Metrics and Competitors

To validate the effectiveness of OpenAL, we conducted two experiments with different matching ratios (the ratio of the number of samples in the target class to the total number of samples) on a 9-class public colorectal cancer pathology image classification dataset (NCT-CRC-HE-100K) [6]. The dataset contains a total of 100,000 patches of pathology images with fine-grained labeling, with nine categories including Adipose (ADI 10%), background (BACK 11%), debris (DEB 11%), lymphocytes (LYM 12%), mucus (MUC 9%), smooth muscle (MUS 14%), normal colon mucosa (NORM 9%), cancer-associated stroma (STR 10%), and colorectal adenocarcinoma epithelium (TUM, 14%). To construct the open-set datasets, we selected three classes, TUM, LYM and NORM, as the target classes and the remaining classes as the non-target classes. We selected these target classes to simulate a possible scenario for pathological cell classification in clinical practice. Technically, target classes can be randomly chosen. In the two experiments, we set the matching ratio to 33% (3 target classes, 6 non-target classes), and 42% (3 target classes, 4 non-target classes), respectively.

Metrics. Following [11], we use three metrics, precision, recall and accuracy to compare the performance of each AL method. We use precision and recall to measure the performance of different methods in target class sample selection.

As defined in Eq. 5, precision is the proportion of the target class samples among the total samples queried in each query and recall is the ratio of the number of the queried target class samples to the number of all the target class samples in the unlabeled sample pool.

$$\text{precision}_m = \frac{k_m}{k_m + l_m}, \text{recall}_m = \frac{\sum_{j=0}^{m} k_m}{n_{\text{target}}} \tag{5}$$

where k_m denotes the number of target class samples queried in the mth query, l_m denotes the number of non-target class samples queried in the mth query, and n_{target} denotes the number of target class samples in the original unlabeled sample pool. Obviously, the higher the precision and recall are, the more target class samples are queried, and the more effective the trained target class classifier will be. We measure the final performance of each AL method using the accuracy of the final classifier on the test set of target class samples.

Competitors. We compare the proposed OpenAL to random sampling and five AL methods, LfOSA [11], Uncertainty [7,8], Certainty [7,8], Coreset [17] and RA [20], of which only LfOSA [11] is designed for open-set AL. For all AL methods, we randomly selected 1% of the samples to label and used them as the initial labeled set for model initialization. It is worth noting that the initial labeled samples contain target class samples as well as non-target class samples, but the non-target class samples are not fine-grained labeled. After each query round, we train a ResNet18 model of 100 epochs, using SGD as the optimizer with momentum of 0.9, weight decay of 5e-4, initial learning rate of 0.01, and batchsize of 128. The annotation budget for each query is 5% of all samples, and the length of the candidate set is twice the budget for each query. For each method, we ran four experiments and recorded the average results for four randomly selected seeds.

3.2 Performance Comparison

Figure 3 A and B show the precision, recall and model accuracy of all comparing methods at 33% and 42% matching ratios, respectively. It can be seen that OpenAL outperforms the other methods in almost all metrics and all query numbers regardless of the matching ratio. Particularly, OpenAL significantly outperforms LfOSA [11], which is specifically designed for open-set AL. The inferior performance of the AL methods based on the closed-set assumption is due to the fact that they are unable to accurately identify more target class samples, thus wasting a large amount of annotation budget. Although LfOSA [11] utilizes a dedicated network for target class sample detection, the performance of the detection network is not stable when the number of training samples is small, thus limiting its performance. In contrast, our method uses a novel feature-based target sample selection strategy and achieves the best performance.

Upon analysis, our OpenAL is capable of effectively maintaining the balance of sample numbers across different classes during active learning. We visualize the cumulative sampling ratios of OpenAL for the target classes in each round

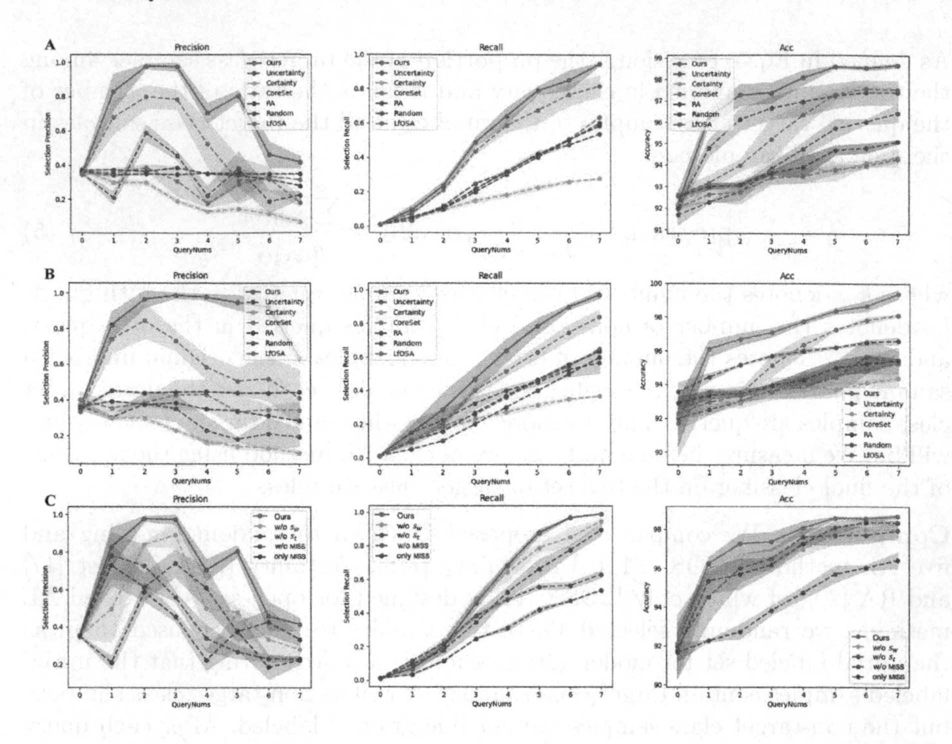

Fig. 3. A. Selection and model performance results under a 33% matching ratio. B. Selection and model performance results under a 42% matching ratio. C. Ablation Study of OpenAL under a 33% matching ratio.

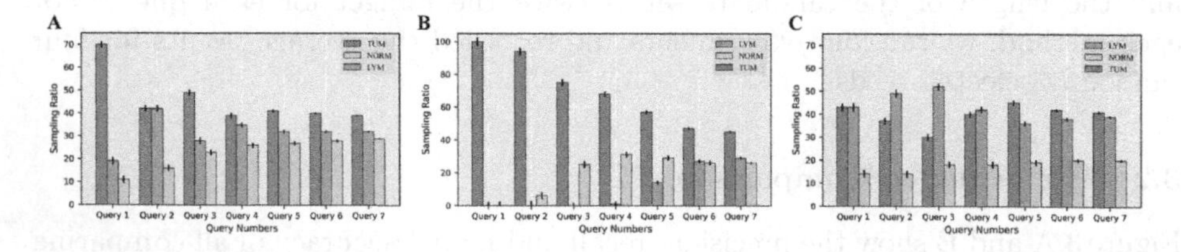

Fig. 4. A. Cumulative sampling ratios of our OpenAL for the target classes LYM, NORM, and TUM across QueryNums 1-7 on the original dataset (under 33% matching ratio). B. Cumulative sampling ratios of LfOSA on the original dataset (under 33% matching ratio). C. Cumulative sampling ratios of our OpenAL on a newly-constructed more imbalanced setting for the target classes LYM (6000 samples), NORM (3000 samples), and TUM (9000 samples).

on the original dataset with a 33% matching ratio, as shown in Fig. 4A. Additionally, we visualize the cumulative sampling ratios of the LfOSA method on the same setting in Fig. 4B. It can be observed that in the first 4 rounds, LYM samples are either not selected or selected very few times. This severe sample imbalance weakens the performance of LfOSA compared to random selection initially. Conversely, our method selects target class samples with a more balanced distribution. Furthermore, we constructed a more imbalanced setting for the target classes LYM (6000 samples), NORM (3000 samples), and TUM (9000 samples), yet the cumulative sampling ratios of our method for these three target classes remain fairly balanced, as shown in Fig. 4C.

3.3 Ablation Study

To further validate the effectiveness of each component of OpenAL, we conducted an ablation test at a matching ratio of 33%. Figure 3C shows the results, where w/o s_w indicates that the distance score of non-target class samples is not used in the scoring of Feature-based Target Sample Selection (FTSS), w/o s_t indicates that the distance score of target class samples is not used, w/o MISS means no Model-based Informative Sample Selection is used, i.e., the length of the candidate set is directly set to the annotation budget in each query, and only MISS means no FTSS strategy is used, but only uncertainty is used to select samples.

It can be seen that the distance modeling of both the target class samples and the non-target class samples is essential in the FTSS strategy, and missing either one results in a decrease in performance. Although the MISS strategy does not significantly facilitate the selection of target class samples, it can effectively help select the most informative samples among the samples in the candidate set, thus further improving the model performance with a limited labeling budget. In contrast, when the samples are selected based on uncertainty alone, the performance decreases significantly due to the inability to accurately select the target class samples. The above experiments demonstrate the effectiveness of each component of OpenAL.

4 Conclusion

In this paper, we present a new open-set scenario of active learning for pathology image classification, which is more practical in real-world applications. We propose a novel AL framework for this open-set scenario, OpenAL, which addresses the challenge of accurately querying the most informative target class samples in an unlabeled sample pool containing a large number of non-target samples. OpenAL significantly outperforms state-of-the-art AL methods on real pathology image classification tasks. More importantly, in clinical applications, on one hand, OpenAL can be used to query informative target class samples for experts to label, thus enabling better training of target class classifiers under limited budgets. On the other hand, when applying the classifier for future testing, it is also possible to use the feature-based target sample selection strategy in the OpenAL framework to achieve an open-set classifier. Therefore, this framework can be applied to both datasets containing only target class samples and datasets also containing a large number of non-target class samples during testing.

Acknowledgments. This work was supported by National Natural Science Foundation of China under Grant 82072021.

References

1. Bai, F., Xing, X., Shen, Y., Ma, H., Meng, M.Q.H.: Discrepancy-based active learning for weakly supervised bleeding segmentation in wireless capsule endoscopy images. In: Wang, L., Dou, Q., Fletcher, P.T., Speidel, S., Li, S. (eds.) MICCAI 2022. LNCS, vol. 13438, pp. 24–34. Springer, Cham (2022). https://doi.org/10.1007/978-3-031-16452-1_3
2. Balaram, S., Nguyen, C.M., Kassim, A., Krishnaswamy, P.: Consistency-based semi-supervised evidential active learning for diagnostic radiograph classification. In: Wang, L., Dou, Q., Fletcher, P.T., Speidel, S., Li, S. (eds.) MICCAI 2022. LNCS, vol. 13431, pp. 675–685. Springer, Cham (2022). https://doi.org/10.1007/978-3-031-16431-6_64
3. Caron, M., Touvron, H., Misra, I., Jégou, H., Mairal, J., Bojanowski, P., Joulin, A.: Emerging properties in self-supervised vision transformers. In: Proceedings of the IEEE/CVF International Conference on Computer Vision (ICCV), pp. 9650–9660 (2021)
4. Chen, R.J., Krishnan, R.G.: Self-supervised vision transformers learn visual concepts in histopathology. arXiv preprint arXiv:2203.00585 (2022)
5. Cheplygina, V., de Bruijne, M., Pluim, J.P.: Not-so-supervised: a survey of semi-supervised, multi-instance, and transfer learning in medical image analysis. Med. Image Anal. **54**, 280–296 (2019)
6. Kather, J.N., et al.: Multi-class texture analysis in colorectal cancer histology. Sci. Rep. **6**(1), 1–11 (2016)
7. Lewis, D.D.: A sequential algorithm for training text classifiers: corrigendum and additional data. In: ACM SIGIR Forum, vol. 29, pp. 13–19. ACM, New York (1995)
8. Luo, W., Schwing, A., Urtasun, R.: Latent structured active learning. In: Advances in Neural Information Processing Systems (NeurIPS), vol. 26 (2013)
9. Mahapatra, D., Bozorgtabar, B., Thiran, J.-P., Reyes, M.: Efficient active learning for image classification and segmentation using a sample selection and conditional generative adversarial network. In: Frangi, A.F., Schnabel, J.A., Davatzikos, C., Alberola-López, C., Fichtinger, G. (eds.) MICCAI 2018. LNCS, vol. 11071, pp. 580–588. Springer, Cham (2018). https://doi.org/10.1007/978-3-030-00934-2_65
10. Nath, V., Yang, D., Roth, H.R., Xu, D.: Warm start active learning with proxy labels and selection via semi-supervised fine-tuning. In: Wang, L., Dou, Q., Fletcher, P.T., Speidel, S., Li, S. (eds.) MICCAI 2022. LNCS, vol. 13438, pp. 297–308. Springer, Cham (2022). https://doi.org/10.1007/978-3-031-16452-1_29
11. Ning, K.P., Zhao, X., Li, Y., Huang, S.J.: Active learning for open-set annotation. In: Proceedings of the IEEE/CVF Conference on Computer Vision and Pattern Recognition (CVPR), pp. 41–49 (2022)
12. Qu, L., Liu, S., Liu, X., Wang, M., Song, Z.: Towards label-efficient automatic diagnosis and analysis: a comprehensive survey of advanced deep learning-based weakly-supervised, semi-supervised and self-supervised techniques in histopathological image analysis. Phys. Med. Biol. (2022)
13. Qu, L., Luo, X., Liu, S., Wang, M., Song, Z.: Dgmil: Distribution guided multiple instance learning for whole slide image classification. In: Wang, L., Dou, Q., Fletcher, P.T., Speidel, S., Li, S. (eds.) MICCAI 2022. LNCS, vol. 13432, pp. 24–34. Springer, Cham (2022). https://doi.org/10.1007/978-3-031-16434-7_3
14. Qu, L., Wang, M., Song, Z., et al.: Bi-directional weakly supervised knowledge distillation for whole slide image classification. In: Advances in Neural Information Processing Systems (NeurIPS), vol. 35, pp. 15368–15381 (2022)

15. Reynolds, D.A., et al.: Gaussian mixture models. Encyclopedia Biometrics **741**(659–663) (2009)
16. Sadafi, A., et al.: Multiclass deep active learning for detecting red blood cell subtypes in brightfield microscopy. In: Shen, D., et al. (eds.) MICCAI 2019. LNCS, vol. 11764, pp. 685–693. Springer, Cham (2019). https://doi.org/10.1007/978-3-030-32239-7_76
17. Sener, O., Savarese, S.: Active learning for convolutional neural networks: a core-set approach. arXiv preprint arXiv:1708.00489 (2017)
18. Srinidhi, C.L., Ciga, O., Martel, A.L.: Deep neural network models for computational histopathology: a survey. Med. Image Anal. **67**, 101813 (2021)
19. Tran, T., Do, T.T., Reid, I., Carneiro, G.: Bayesian generative active deep learning. In: International Conference on Machine Learning (ICML), pp. 6295–6304. PMLR (2019)
20. Zheng, H., et al.: Biomedical image segmentation via representative annotation. In: Proceedings of the AAAI Conference on Artificial Intelligence (AAAI), vol. 33, pp. 5901–5908 (2019)

SLPT: Selective Labeling Meets Prompt Tuning on Label-Limited Lesion Segmentation

Fan Bai[1,2,3], Ke Yan[2,3], Xiaoyu Bai[2,3], Xinyu Mao[1], Xiaoli Yin[4], Jingren Zhou[2,3], Yu Shi[4], Le Lu[2], and Max Q.-H. Meng[1,5](✉)

[1] Department of Electronic Engineering, The Chinese University of Hong Kong, Shatin, Hong Kong, China
[2] DAMO Academy, Alibaba Group, Hangzhou, China
[3] Hupan Lab, Hangzhou 310023, China
[4] Department of Radiology, Shengjing Hospital of China Medical University, Shenyang 110004, China
[5] Department of Electronic and Electrical Engineering, Southern University of Science and Technology, Shenzhen, China
mengqh@sustech.edu.cn

Abstract. Medical image analysis using deep learning is often challenged by limited labeled data and high annotation costs. Fine-tuning the entire network in label-limited scenarios can lead to overfitting and suboptimal performance. Recently, prompt tuning has emerged as a more promising technique that introduces a few additional tunable parameters as prompts to a task-agnostic pre-trained model, and updates only these parameters using supervision from limited labeled data while keeping the pre-trained model unchanged. However, previous work has overlooked the importance of selective labeling in downstream tasks, which aims to select the most valuable downstream samples for annotation to achieve the best performance with minimum annotation cost. To address this, we propose a framework that combines selective labeling with prompt tuning (SLPT) to boost performance in limited labels. Specifically, we introduce a feature-aware prompt updater to guide prompt tuning and a TandEm Selective LAbeling (TESLA) strategy. TESLA includes unsupervised diversity selection and supervised selection using prompt-based uncertainty. In addition, we propose a diversified visual prompt tuning strategy to provide multi-prompt-based discrepant predictions for TESLA. We evaluate our method on liver tumor segmentation and achieve state-of-the-art performance, outperforming traditional fine-tuning with only 6% of tunable parameters, also achieving 94% of full-data performance by labeling only 5% of the data.

Keywords: Active Learning · Prompt Tuning · Segmentation

Supplementary Information The online version contains supplementary material available at https://doi.org/10.1007/978-3-031-43895-0_2.

H. Greenspan et al. (Eds.): MICCAI 2023, LNCS 14221, pp. 14–24, 2023.
https://doi.org/10.1007/978-3-031-43895-0_2

1 Introduction

Deep learning has achieved promising performance in computer-aided diagnosis [1,12,14,24], but it relies on large-scale labeled data to train, which is challenging in medical imaging due to label scarcity and high annotation cost [3,25]. Specifically, expert annotations are required for medical data, which can be costly and time-consuming, especially in tasks such as 3D image segmentation.

Transferring pre-trained models to downstream tasks is an effective solution for addressing the label-limited problem [8], but fine-tuning the full network with small downstream data is prone to overfitting [16]. Recently, prompt tuning [5,18] is emerging from natural language processing (NLP), which introduces additional tunable prompt parameters to the pre-trained model and updates only prompt parameters using supervision signals obtained from a few downstream training samples while keeping the entire pre-trained unchanged. By tuning only a few parameters, prompt tuning makes better use of pre-trained knowledge. It avoids driving the entire model with few downstream data, which enables it to outperform traditional fine-tuning in limited labeled data. Building on the recent success of prompt tuning in NLP [5], instead of designing text prompts and Transformer models, we explore visual prompts on Convolutional Neural Networks (CNNs) and the potential to address data limitations in medical imaging.

However, previous prompt tuning research [18,28], whether on language or visual models, has focused solely on the model-centric approach. For instance, CoOp [29] models a prompt's context using a set of learnable vectors and optimizes it on a few downstream data, without discussing what kind of samples are more suitable for learning prompts. VPT [13] explores prompt tuning with a vision Transformer, and SPM [17] attempts to handle downstream segmentation tasks through prompt tuning on CNNs, which are also model-centric. However, in downstream tasks with limited labeled data, selective labeling as a data-centric method is crucial for determining which samples are valuable for learning, similar to Active Learning (AL) [23]. In AL, given the initial labeled data, the model actively selects a subset of valuable samples for labeling and improves performance with minimum annotation effort. Nevertheless, directly combining prompt tuning with AL presents several problems. First, unlike the task-specific models trained with initial data in AL, the task-agnostic pre-trained model (e.g., trained by related but not identical supervised or self-supervised task) is employed for data selection with prompt tuning. Second, in prompt tuning, the pre-trained model is frozen, which may render some AL methods inapplicable, such as those previously based on backbone gradient [9] and feature [19]. Third, merging prompt tuning with AL takes work. Their interplay must be considered. However, previous AL methods [27] did not consider the existence of prompts or use prompts to estimate sample value.

Therefore, this paper proposes the first framework for selective labeling and prompt tuning (SLPT), combining model-centric and data-centric methods to improve performance in medical label-limited scenarios. We make three main contributions: (1) We design a novel feature-aware prompt updater embedded in the pre-trained model to guide prompt tuning in deep layers. (2) We propose

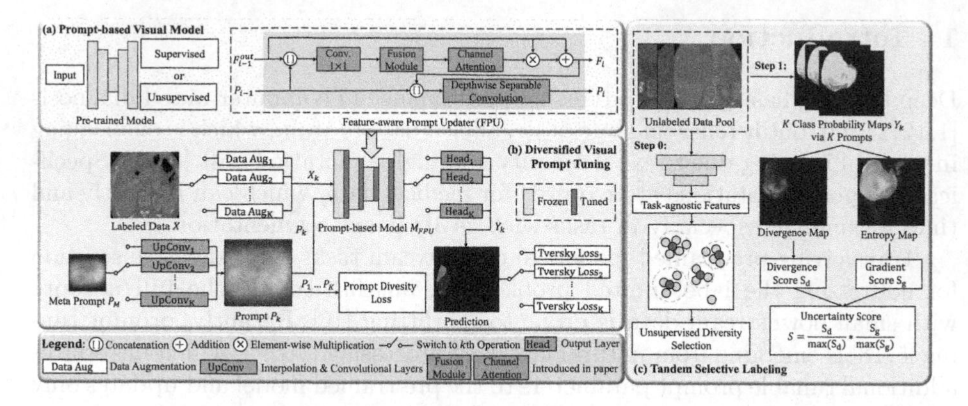

Fig. 1. Workflow of SLPT: **(1)** Create an initial label set via the pre-trained model for unsupervised diversity selection (subplot c step 0). **(2)** Insert a feature-aware prompt updater (subplot a) into the pre-trained model for prompt tuning with initial labels. **(3)** Use diversified visual prompt tuning (subplot b) to obtain prompt-based discrepant predictions. **(4)** Select valuable data by prompt-based uncertainty (subplot c step 1) and update the prompt-based model accordingly. Note: The orange modules are tunable for prompt tuning, while the gray ones are frozen. Please zoom in for details.

a diversified visual prompt tuning mechanism that provides multi-prompt-based discrepant predictions for selective labeling. (3) We introduce the TESLA strategy which includes both unsupervised diversity selection via task-agnostic features and supervised selection considering prompt-based uncertainty. The results show that SLPT outperforms fine-tuning with just 6% of tunable parameters and achieves 94% of full-data performance by selecting only 5% of labeled data.

2 Methodology

Given a task-agnostic pre-trained model and unlabeled data for an initial medical task, we propose SLPT to improve model performance. SLPT consists of three components, as illustrated in Fig. 1: (a) a prompt-based visual model, (b) diversified visual prompt tuning, and (c) tandem selective labeling. Specifically, with SLPT, we can select valuable data to label and tune the model via prompts, which helps the model overcome label-limited medical scenarios.

2.1 Prompt-Based Visual Model

The pre-trained model, learned by supervised or unsupervised training, is a powerful tool for improving performance on label-limited downstream tasks. Fine-tuning a large pre-trained model with limited data may be suboptimal and prone to overfitting [16]. To overcome this issue, we draw inspiration from NLP [18] and explore prompt tuning on visual models. In order to facilitate prompt tuning on the model's deep layers, we introduce the Feature-aware Prompt Updater (FPU). FPUs are inserted into the network to update deep prompts and features. In Fig. 1(a), an FPU receives two inputs, feature map F_{i-1}^{out} and prompt P_{i-1}, of the same shape, and updates to F_i and P_i through two parallel branches. In the feature branch, F_{i-1}^{out} and P_{i-1} are concatenated and fed into a 1x1 convolution and fusion module. The fusion module utilizes ASPP [7] to extract multi-scale contexts. Then a SE [11] module for channel attention enhances context by channel. Finally, the attention output and F_{i-1}^{out} are element-wise multiplied and added to obtain the updated feature F_i. In the prompt branch, the updated feature F_i is concatenated with the previous prompt P_{i-1}, and a parameter-efficient depth-separable convolution is employed to generate the updated prompt P_i.

To incorporate FPU into a pre-trained model, we consider the model comprising N modular M_i $(i = 1, ..., N)$ and a head output layer. After each M_i, we insert an FPU_i. Given the input F_{i-1}^{in} and prompt P_{i-1}, we have the output feature F_i, updated prompt P_i and prediction Y as follows:

$$F_{i-1}^{out} = M_i(F_{i-1}^{in}), \quad F_i, P_i = \mathrm{FPU}_i(F_{i-1}^{out}, P_{i-1}), \quad Y = \mathrm{Head}(F_N) \tag{1}$$

where input $X = F_0$, FPU and Head are tuned while M_i is not tunable.

2.2 Diversified Visual Prompt Tuning

Inspired by multi-prompt learning [18] in NLP, we investigate using multiple visual prompts to evaluate prompt-based uncertainty. However, initializing and optimizing K prompts directly can significantly increase parameters and may not ensure prompt diversity. To address these challenges, we propose a diversified visual prompt tuning approach. As shown in Fig. 1(b), our method generates K prompts $P_k \in \mathbb{R}^{1\times D\times H\times W}$ from a meta prompt $P_M \in \mathbb{R}^{1\times \frac{D}{2}\times \frac{H}{2}\times \frac{W}{2}}$ through K different upsampling and convolution operations $UpConv_k$. P_M is initialized from the statistical probability map of the foreground category, similar to [17]. Specifically, we set the foreground to 1 and the background to 0 in the ground-truth mask, and then average all masks and downsample to $1 \times \frac{D}{2} \times \frac{H}{2} \times \frac{W}{2}$. To enhance prompt diversity, we introduce a prompt diversity loss L_{div} that regularizes the cosine similarity between the generated prompts and maximizes their diversity. This loss is formulated as follows:

$$L_{div} = \sum_{k_1=1}^{K-1} \sum_{k_2=k_1+1}^{K} \frac{P_{k_1} \cdot P_{k_2}}{||P_{k_1}||_2 \cdot ||Pk_2||_2} \tag{2}$$

where P_{k_1} and P_{k_2} represent the k_1-th and k_2-th generated prompts, respectively, and $||\cdot||_2$ denotes the L2 norm. By incorporating the prompt diversity loss, we aim to generate a set of diverse prompts for our visual model.

In NLP, using multiple prompts can produce discrepant predictions [2] that help estimate prompt-based uncertainty. Drawing inspiration, we propose a visual prompt tuning approach that associates diverse prompts with discrepant predictions. To achieve this, we design K different data augmentation, heads, and losses based on corresponding K prompts. By varying hyperparameters, we can achieve different data augmentation strengths, increasing the model's diversity and generalization. Different predictions Y_k are generated by K heads, each supervised with a Tversky loss [21] $TL_k = \frac{TP}{TP+\alpha_k FP+\beta_k FN}$, where TP, FP, and FN represent true positive, false positive, and false negative, respectively. To obtain diverse predictions with false positives and negatives, we use different α_k and β_k values in TL_k. The process is formulated as follows:

$$P_k = UpConv_k(P_M), \quad X_k = DA_k(X), \quad Y_k = Head_k(M_{FPU}(X_k, P_k)) \tag{3}$$

$$L = \sum_{k=1}^{K} (\lambda_1 * TL_k(Y_k, {}^Y) + \lambda_2 * CE(Y_k, {}^Y)) + \lambda_3 * L_{div} \tag{4}$$

where $k = 1, ..., K$, M_{FPU} is the pre-trained model with FPU, CE is the cross-entropy loss, and $\lambda_1 = \lambda_2 = \lambda_3 = 1$ weight each loss component. ^{Y} represents the ground truth and L is the total loss.

2.3 Tandem Selective Labeling

Previous studies overlook the critical issue of data selection for downstream tasks, especially when available labels are limited. To address this challenge, we propose a novel strategy called TESLA. TESLA consists of two tandem steps: unsupervised diversity selection and supervised uncertainty selection. The first step aims to maximize the diversity of the selected data, while the second step aims to select the most uncertain samples based on diverse prompts.

Step 0: Unsupervised Diversity Selection. Since we do not have any labels in the initial and our pre-trained model is task-agnostic, we select diverse samples to cover the entire dataset. To achieve this, we leverage the pre-trained model to obtain feature representations for all unlabeled data. Although these features are task-independent, they capture the underlying relationships, with similar samples having closer feature distances. We apply the k-center method from Coreset [22], which identifies the B samples that best represent the diversity of the data based on these features. These selected samples are then annotated and serve as the initial dataset for downstream tasks.

Step 1: Supervised Uncertainty Selection. After prompt tuning with the initial dataset, we obtain a task-specific model that can be used to evaluate data value under supervised training. Since only prompt-related parameters can be tuned while others are frozen, we assess prompt-based uncertainty via diverse prompts, considering inter-prompts uncertainty and intra-prompts uncertainty. In the former, we compute the multi-prompt-based divergence map D, given K probability predictions Y_k through K diverse prompts P_k, as follows:

$$D = \sum_{k=1}^{K} \mathrm{KL}(Y_k || Y_{\mathrm{mean}}), \quad Y_{\mathrm{mean}} = \frac{1}{K} \sum_{k=1}^{K} Y_k \tag{5}$$

where KL refers to the KL divergence [15]. Then, we have the divergence score $S_d = \mathrm{Mean}(D)$, which reflects inter-prompts uncertainty.

In the latter, we evaluate intra-prompts uncertainty by computing the mean prediction of the prompts and propose to estimate prompt-based gradients as the model's performance depends on the update of prompt parameters θ_p. However, for these unlabeled samples, computing their supervised loss and gradient directly is not feasible. Therefore, we use the entropy of the model's predictions as a proxy for loss. Specifically, we calculate the entropy-based prompt gradient score S_g for each unlabeled sample as follows:

$$S_g = \sum_{\theta_p} ||\nabla_{\theta_p}(-\sum Y_{mean} * \log Y_{mean})||_2 \tag{6}$$

To avoid manual weight adjustment, we employ multiplication instead of addition. We calculate our uncertainty score S as follows:

$$S = \frac{S_d}{\max(S_d)} \times \frac{S_g}{\max(S_g)} \tag{7}$$

where $\max(\cdot)$ finds the maximum value. We sort the unlabeled data by their corresponding S values in ascending order and select the top B data to annotate.

3 Experiments and Results

3.1 Experimental Settings

Datasets and Pre-trained Model. We conducted experiments on automating liver tumor segmentation in contrast-enhanced CT scans, a crucial task in liver cancer diagnosis and surgical planning [1]. Although there are publicly available liver tumor datasets [1,24], they only contain major tumor types and differ in image characteristics and label distribution from our hospital's data. Deploying a model trained from public data to our hospital directly will be problematic. Collecting large-scale data from our hospital and training a new model will be expensive. Therefore, we can use the model trained from them as a starting point and use SLPT to adapt it to our hospital with minimum cost. We collected a dataset from our in-house hospital comprising 941 CT scans with eight categories: hepatocellular carcinoma, cholangioma, metastasis, hepatoblastoma, hemangioma, focal nodular hyperplasia, cyst, and others. It covers both major and rare tumor types. Our objective is to segment all types of lesions accurately. We utilized a pre-trained model for liver segmentation using supervised learning on two public datasets [24] with no data overlap with our downstream task. The nnUNet [12] was used to preprocess and sample the data into $24 \times 256 \times 256$ patches for training. To evaluate the performance, we employed a 5-fold cross-validation (752 for selection, 189 for test).

Metrics. We evaluated lesion segmentation performance using pixel-wise and lesion-wise metrics. For pixel-wise evaluation, we used the Dice per case, a commonly used metric [1]. For lesion-wise evaluation, we first do connected component analysis to predicted and ground truth masks to extract lesion instances, and then compute precision and recall per case [20]. A predicted lesion is regarded as a TP if its overlap with ground truth is higher than 0.2 in Dice.

Table 1. Evaluation of different tunings on the lesion segmentation with limited data (40 class-balanced patients). Prec. and Rec. denote precision and recall.

Method	Tuning Type	Trainable Parameters	Pixel-wise			Lesion-wise		Mean
			Dice	Prec	Rec	Prec	Rec	
Fine-tuning	All	44.81M	64.43	**87.69**	59.86	50.84	54.14	63.39
Learn-from-Scratch		44.81M	54.15	73.33	50.25	45.84	45.78	53.87
Encoder-tuning	Part	19.48M	65.61	82.00	61.96	29.36	41.10	56.00
Decoder-tuning		23.64M	67.87	77.96	**70.56**	30.82	35.92	56.63
Head-tuning		0.10M	56.73	74.45	55.57	23.29	29.74	47.96
SPM [17]	Prompt	3.15M	68.60	83.07	69.02	62.15	55.19	67.61
Ours		**2.71M**	**68.76**	79.63	69.76	**64.63**	**61.18**	**68.79**

Competing Approaches. In the prompt tuning experiment, we compared our method with three types of tuning: full parameter update (Fine-tuning, Learn-from-Scratch), partial parameter update (Head-tuning, Encoder-tuning, Decoder-tuning), and prompt update (SPM [17]). In the unsupervised diversity selection experiment, we compared our method with random sampling. In the supervised uncertainty selection experiment, we compared our method with random sampling, diversity sampling (Coreset [22], CoreCGN [6]), and uncertainty sampling (Entropy, MC Dropout [10], Ensemble [4], UncertainGCN [6], Ent-gn [26]). Unlike Ensemble, our method was on multi-prompt-based heads. Furthermore, unlike Ent-gn, which computed the entropy-based gradient from a single prediction, we calculated a stable entropy from the muti-prompt-based mean predictions and solely considered the prompt gradient.

Training Setup. We conducted the experiments using the Pytorch framework on a single NVIDIA Tesla V100 GPU. The nnUNet [12] framework was used for 3D lesion segmentation with training 500 epochs at an initial learning rate of 0.01. We integrated 13 FPUs behind each upsampling or downsampling of nnUNet, adding only 2.7M parameters. During training, we set $k = 3$ and employed diverse data augmentation techniques such as scale, elastic, rotation, and mirror. Three sets of TL parameters is ($\alpha_{1,2,3} = 0.5,0.7,0.3$, $\beta_{1,2,3} = 0.5,0.3,0.7$). To ensure fairness and eliminate model ensemble effects, we only used the model's prediction with $k = 1$ during testing. We used fixed random seeds and 5-fold cross-validation for all segmentation experiments.

3.2 Results

Evaluation of Prompt Tuning. Since we aim to evaluate the efficacy of prompt tuning on limited labeled data in Table 1, we create a sub-dataset of approximately 5% (40/752) from the original dataset. Specifically, we calculate the class probability distribution vector for each sample based on the pixel class in the mask and use CoreSet with these vectors to select 40 class-balanced samples. Using this sub-dataset, we evaluated various tuning methods for limited

Table 2. Comparison of data selection methods for label-limited lesion segmentation. Step 0: unsupervised diversity selection. Step 1: supervised uncertainty selection. The labeling budget for each step is 20 patients. Step $+\infty$ refers to fully labeled 752 data.

Step	Method	Pixel-wise			Lesion-wise		Mean
		Dice	Prec	Rec	Prec	Rec	
0	Random	65.58	80.00	65.21	23.46	39.94	54.84
	Ours	68.20	78.97	69.15	32.51	34.67	56.70
1	Random	66.67	79.95	70.67	41.45	39.45	59.64
	Entropy	66.39	80.85	66.96	37.40	39.47	58.21
	MC Dropout [10]	69.23	79.61	69.48	30.43	36.29	57.01
	Ensemble [4]	69.79	80.25	69.54	**64.38**	58.34	68.46
	CoreSet [22]	70.72	79.34	72.03	46.03	51.24	63.87
	CoreGCN [6]	70.91	77.56	72.37	51.73	49.88	64.49
	UncertainGCN [6]	71.44	75.07	**75.62**	72.83	44.99	67.99
	Ent-gn [26]	70.54	79.91	71.42	61.12	56.37	67.87
	Ours (w/o S_d)	69.54	81.97	68.59	60.47	59.82	68.08
	Ours (w/o S_g)	71.01	80.68	69.83	59.42	58.78	67.94
	Ours	**72.07**	**82.07**	72.37	61.21	**61.90**	**69.92**
$+\infty$	Fine-tuning with Full Labeled Data	77.44	85.44	77.15	62.78	68.56	74.27

medical lesion diagnosis data. The results are summarized in Table 1. Fine-tuning all parameters served as the strongest baseline, but our method, which utilizes only 6% tunable parameters, outperformed it by 5.4%. Although SPM also outperforms fine-tuning, our methods outperform SPM by 1.18% and save 0.44M tunable parameters with more efficient FPU. In cases of limited data, fine-tuning tends to overfit on a larger number of parameters, while prompt tuning does not. The pre-trained model is crucial for downstream tasks with limited data, as it improves performance by 9.52% compared to Learn-from-Scratch. Among the three partial tuning methods, the number of tuning parameters positively correlates with the model's performance, but they are challenging to surpass fine-tuning.

Evaluation of Selective Labeling. We conducted steps 0 (unsupervised selection) and 1 (supervised selection) from the unlabeled 752 data and compared our approach with other competing methods, as shown in Table 2. In step 0, without any labeled data, our diversity selection outperformed the random baseline by 1.86%. Building upon the 20 data points selected by our method in step 0, we proceeded to step 1, where we compared our method with eight other data selection strategies in supervised mode. As a result, our approach outperformed other methods because of prompt-based uncertainty, such as Ent-gn and Ensemble, by 2.05% and 1.46%, respectively. Our approach outperformed Coreset by 6.05% and CoreGCN by 5.43%. We also outperformed UncertainGCN by 1.93%. MC Dropout and Entropy underperformed in our prompt tuning, likely due to the difficulty of learning such uncertain data with only a few prompt parameters. Notably, our method outperformed random sampling by 10.28%. These results demonstrate the effectiveness of our data selection approach in practical tasks.

Ablation Studies. We conducted ablation studies on S_d and S_g in TESLA. As shown in Table 2, the complete TESLA achieved the best performance, outperforming the version without S_d by 1.84% and the version without S_g by 1.98%. It shows that each component plays a critical role in improving performance.

4 Conclusions

We proposed a pipeline called SLPT that enhances model performance in label-limited scenarios. With only 6% of tunable prompt parameters, SLPT outperforms fine-tuning due to the feature-aware prompt updater. Moreover, we presented a diversified visual prompt tuning and a TESLA strategy that combines unsupervised and supervised selection to build annotated datasets for downstream tasks. SLPT pipeline is a promising solution for practical medical tasks with limited data, providing good performance, few tunable parameters, and low labeling costs. Future work can explore the potential of SLPT in other domains.

Acknowledgements. The work was supported by Alibaba Research Intern Program. Fan Bai and Max Q.-H. Meng were supported by National Key R&D program of China with Grant No. 2019YFB1312400, Hong Kong RGC CRF grant C4063-18G, and Hong Kong Health and Medical Research Fund (HMRF) under Grant 06171066. Xiaoli Yin and Yu Shi were supported by National Natural Science Foundation of China (82071885).

References

1. Bilic, P., et al.: The liver tumor segmentation benchmark (LiTS). Med. Image Anal. **84**, 102680 (2023)
2. Allingham, J.U., et al.: A simple zero-shot prompt weighting technique to improve prompt ensembling in text-image models. arXiv preprint arXiv:2302.06235 (2023)
3. Bai, F., Xing, X., Shen, Y., Ma, H., Meng, M.Q.H.: Discrepancy-based active learning for weakly supervised bleeding segmentation in wireless capsule endoscopy images. In: Wang, L., Dou, Q., Fletcher, P.T., Speidel, S., Li, S. (eds.) MICCAI 2022. LNCS, vol. 13438, pp. 24–34. Springer, Cham (2022). https://doi.org/10.1007/978-3-031-16452-1_3
4. Beluch, W.H., Genewein, T., Nürnberger, A., Köhler, J.M.: The power of ensembles for active learning in image classification. In: Proceedings of the IEEE Conference on Computer Vision and Pattern Recognition, pp. 9368–9377 (2018)

5. Brown, T.B., et al.: Language models are few-shot learners. In: Advances in Neural Information Processing Systems, pp. 1876–1901 (2020)
6. Caramalau, R., Bhattarai, B., Kim, T.K.: Sequential graph convolutional network for active learning. In: Proceedings of the IEEE/CVF Conference on Computer Vision and Pattern Recognition, pp. 9583–9592 (2021)
7. Chen, L.C., Papandreou, G., Schroff, F., Adam, H.: Rethinking atrous convolution for semantic image segmentation. arXiv preprint arXiv:1706.05587 (2017)
8. Cheplygina, V., de Bruijne, M., Pluim, J.P.: Not-so-supervised: a survey of semi-supervised, multi-instance, and transfer learning in medical image analysis. Med. Image Anal. **54**, 280–296 (2019)
9. Dai, C., et al.: Suggestive annotation of brain tumour images with gradient-guided sampling. In: Martel, A.L., et al. (eds.) MICCAI 2020. LNCS, vol. 12264, pp. 156–165. Springer, Cham (2020). https://doi.org/10.1007/978-3-030-59719-1_16
10. Gal, Y., Ghahramani, Z.: Dropout as a Bayesian approximation: representing model uncertainty in deep learning. In: International Conference on Machine Learning, pp. 1050–1059. PMLR (2016)
11. Hu, J., Shen, L., Sun, G.: Squeeze-and-excitation networks. In: Proceedings of the IEEE Conference on Computer Vision and Pattern Recognition, pp. 7132–7141 (2018)
12. Isensee, F., Jaeger, P.F., Kohl, S.A., Petersen, J., Maier-Hein, K.H.: nnU-Net: a self-configuring method for deep learning-based biomedical image segmentation. Nat. Methods **18**(2), 203–211 (2021)
13. Jia, M., et al.: Visual prompt tuning. In: Avidan, S., Brostow, G., Cissé, M., Farinella, G.M., Hassner, T. (eds.) ECCV 2022. LNCS, vol. 13693, pp. 709–727. Springer, Cham (2022). https://doi.org/10.1007/978-3-031-19827-4_41
14. Kim, M., et al.: Deep learning in medical imaging. Neurospine **16**(4), 657 (2019)
15. Kullback, S., Leibler, R.A.: On information and sufficiency. Ann. Math. Stat. **22**(1), 79–86 (1951)
16. Kumar, A., Raghunathan, A., Jones, R., Ma, T., Liang, P.: Fine-tuning can distort pretrained features and underperform out-of-distribution. arXiv preprint arXiv:2202.10054 (2022)
17. Liu, L., Yu, B.X., Chang, J., Tian, Q., Chen, C.W.: Prompt-matched semantic segmentation. arXiv preprint arXiv:2208.10159 (2022)
18. Liu, P., Yuan, W., Fu, J., Jiang, Z., Hayashi, H., Neubig, G.: Pre-train, prompt, and predict: a systematic survey of prompting methods in natural language processing. ACM Comput. Surv. **55**(9), 1–35 (2023)
19. Parvaneh, A., Abbasnejad, E., Teney, D., Haffari, G.R., Van Den Hengel, A., Shi, J.Q.: Active learning by feature mixing. In: Proceedings of the IEEE/CVF Conference on Computer Vision and Pattern Recognition, pp. 12237–12246 (2022)
20. Powers, D.M.: Evaluation: from precision, recall and F-measure to ROC, informedness, markedness and correlation. arXiv preprint arXiv:2010.16061 (2020)
21. Salehi, S.S.M., Erdogmus, D., Gholipour, A.: Tversky loss function for image segmentation using 3D fully convolutional deep networks. In: Wang, Q., Shi, Y., Suk, H.-I., Suzuki, K. (eds.) MLMI 2017. LNCS, vol. 10541, pp. 379–387. Springer, Cham (2017). https://doi.org/10.1007/978-3-319-67389-9_44
22. Sener, O., Savarese, S.: Active learning for convolutional neural networks: a core-set approach. arXiv preprint arXiv:1708.00489 (2017)
23. Settles, B.: Active learning literature survey (2009)
24. Simpson, A.L., et al.: A large annotated medical image dataset for the development and evaluation of segmentation algorithms. arXiv preprint arXiv:1902.09063 (2019)

25. Tajbakhsh, N., Jeyaseelan, L., Li, Q., Chiang, J.N., Wu, Z., Ding, X.: Embracing imperfect datasets: a review of deep learning solutions for medical image segmentation. Med. Image Anal. **63**, 101693 (2020)
26. Wang, T., et al.: Boosting active learning via improving test performance. In: Proceedings of the AAAI Conference on Artificial Intelligence, vol. 36, pp. 8566–8574 (2022)
27. Zhan, X., Wang, Q., Huang, K.H., Xiong, H., Dou, D., Chan, A.B.: A comparative survey of deep active learning. arXiv preprint arXiv:2203.13450 (2022)
28. Zhao, T., et al.: Prompt design for text classification with transformer-based models. In: Proceedings of the 2021 Conference on Empirical Methods in Natural Language Processing, pp. 2709–2722 (2021)
29. Zhou, K., Yang, J., Loy, C.C., Liu, Z.: Learning to prompt for vision-language models. Int. J. Comput. Vision **130**(9), 2337–2348 (2022)

COLosSAL: A Benchmark for Cold-Start Active Learning for 3D Medical Image Segmentation

Han Liu[1(✉)], Hao Li[1], Xing Yao[1], Yubo Fan[1], Dewei Hu[1], Benoit M. Dawant[1], Vishwesh Nath[2], Zhoubing Xu[3], and Ipek Oguz[1]

[1] Vanderbilt University, Nashville, USA
han.liu@vanderbilt.edu
[2] NVIDIA, Nashville, USA
[3] Siemens Healthineers, Princeton, USA

Abstract. Medical image segmentation is a critical task in medical image analysis. In recent years, deep learning based approaches have shown exceptional performance when trained on a fully-annotated dataset. However, data annotation is often a significant bottleneck, especially for 3D medical images. Active learning (AL) is a promising solution for efficient annotation but requires an initial set of labeled samples to start active selection. When the entire data pool is unlabeled, how do we select the samples to annotate as our initial set? This is also known as the cold-start AL, which permits only one chance to request annotations from experts without access to previously annotated data. Cold-start AL is highly relevant in many practical scenarios but has been under-explored, especially for 3D medical segmentation tasks requiring substantial annotation effort. In this paper, we present a benchmark named COLosSAL by evaluating six cold-start AL strategies on five 3D medical image segmentation tasks from the public Medical Segmentation Decathlon collection. We perform a thorough performance analysis and explore important open questions for cold-start AL, such as the impact of budget on different strategies. Our results show that cold-start AL is still an unsolved problem for 3D segmentation tasks but some important trends have been observed. The code repository, data partitions, and baseline results for the complete benchmark are publicly available at https://github.com/MedICL-VU/COLosSAL.

Keywords: Efficient Annotation · Active Learning · Cold Start · Image Segmentation

The original version of this chapter was revised: In the header of the paper, the second and third affiliation listed wrong locations. The correction to this chapter is available at https://doi.org/10.1007/978-3-031-43895-0_74

Supplementary Information The online version contains supplementary material available at https://doi.org/10.1007/978-3-031-43895-0_3.

H. Greenspan et al. (Eds.): MICCAI 2023, LNCS 14221, pp. 25–34, 2023.
https://doi.org/10.1007/978-3-031-43895-0_3

1 Introduction

Segmentation is among the most common medical image analysis tasks and is critical to a wide variety of clinical applications. To date, data-driven deep learning (DL) methods have shown prominent segmentation performance when trained on fully-annotated datasets [8]. However, data annotation is a significant bottleneck for dataset creation. First, annotation process is tedious, laborious and time-consuming, especially for 3D medical images where dense annotation with voxel-level accuracy is required. Second, medical images typically need to be annotated by medical experts whose time is limited and expensive, making the annotations even more difficult and costly to obtain. Active learning (AL) is a promising solution to improve annotation efficiency by iteratively selecting the most *important* data to annotate with the goal of reducing the total number of annotated samples required. However, most deep AL methods require an initial set of labeled samples to start the active selection. When the entire data pool is unlabeled, which samples should one select as the initial set? This problem is known as ***cold-start active learning***, a low-budget paradigm of AL that permits only one chance to request annotations from experts without access to any previously annotated data.

Cold-start AL is highly relevant to many practical scenarios. First, cold-start AL aims to study the general question of constructing a training set for an organ that has not been labeled in public datasets. This is a very common scenario (whenever a dataset is collected for a new application), especially when iterative AL is not an option. Second, even if iterative AL is possible, a better initial set has been found to lead to noticeable improvement for the subsequent AL cycles [4,25]. Third, in low-budget scenarios, cold-start AL can achieve one-shot selection of the most informative data without several cycles of annotation. This can lead to an appealing *'less is more'* outcome by optimizing the available budget and also alleviating the issue of having human experts on standby for traditional iterative AL.

Despite its importance, very little effort has been made to address the cold-start problem, especially in medical imaging settings. The existing cold-start AL techniques are mainly based on the two principles of the traditional AL strategies: (1) *Uncertainty sampling* [5,11,15,18], where the most uncertain samples are selected to maximize the added value of the new annotations. (2) *Diversity sampling* [7,10,19,22], where samples from diverse regions of the data distribution are selected to avoid redundancy. In the medical domain, diversity-based cold-start strategies have been recently explored on 2D classification/segmentation tasks [4,24,25]. The effectiveness of these approaches on 3D medical image segmentation remains unknown, especially since 3D models are often patch-based while 2D models can use the entire image. A recent study on 3D medical segmentation shows the feasibility to use the uncertainty estimated from a proxy task to rank the importance of the unlabeled data in the cold-start scenario [14]. However, it fails to compare against the diversity-based approaches, and the proposed proxy task is only limited to CT images, making the effectiveness of this strategy unclear on other 3D imaging modalities. Consequently, no comprehensive cold-start AL baselines currently exist for 3D medical image segmentation, creating additional challenges for this promising research direction.

In this paper, we introduce the COLosSAL benchmark, the first cold-start active learning benchmark for 3D medical image segmentation by evaluating on six popular cold-start AL strategies. Specifically, we aim to answer three important open questions: (1) compared to random selection, how effective are the uncertainty-based and diversity-based cold-start strategies for 3D segmentation tasks? (2) what is the impact of allowing a larger budget on the compared strategies? (3) can these strategies work better if the local ROI of the target organ is known as prior? We train and validate our models on five 3D medical image segmentation tasks from the publicly available Medical Segmentation Decathlon (MSD) dataset [1], which covers two of the most common 3D image modalities and the segmentation tasks for both healthy tissue and tumor/pathology.

Our contributions are summarized as follows:

- We offer the first cold-start AL benchmark for 3D medical image segmentation. We make our code repository, data partitions, and baseline results publicly available to facilitate future cold-start AL research.
- We explore the impact of the budget and the extent of the 3D ROI on the cold-start AL strategies.
- Our major findings are: (1) TypiClust [7], a diversity-based approach, is a more robust cold-start selection strategy for 3D segmentation tasks. (2) Most evaluated strategies become more effective when more budget is allowed, especially diversity-based ones. (3) Cold-start AL strategies that focus on the uncertainty/diversity from a local ROI cannot outperform their global counterparts. (4) Almost no cold-start AL strategy is very effective for the segmentation tasks that include tumors.

2 COLosSAL Benchmark Definition

Formally, given an unlabeled data pool of size N, cold-start AL aims to select the optimal m samples ($m \ll N$) *without* access to any prior segmentation labels. Specifically, the optimal samples are defined as the subset of 3D volumes that can lead to the best validation performance when training a standard 3D segmentation network. In this study, we use $m = 5$ for low-budget scenarios.

2.1 3D Medical Image Datasets

We use the Medical Segmentation Decathlon (MSD) collection [1] to define our benchmark, due to its public accessibility and the standardized datasets spanning across two common 3D image modalities, i.e., CT and MRI. We select five tasks from the collection appropriate for the 3D segmentation tasks, namely tasks 2-Heart, 3-Liver, 4-Hippocampus, 7-Pancreas, and 9-Spleen. Liver and Pancreas tasks include both organ and tumor segmentation, while the other tasks focus on organs only. The selected tasks thus include different organs with different disease status, representing a good coverage of real-world 3D medical image segmentation tasks. For each dataset, we split the data into training and validation sets for AL development. The training and validation sets contain 16/4 (heart), 105/26 (hippocampus), 208/52 (liver), 225/56 (pancreas), and 25/7 (spleen) subjects. The training set is considered as the unlabeled data pool for sample selection, and the validation set is kept consistent for all experiments to evaluate the performance of the selected samples by different AL schemes.

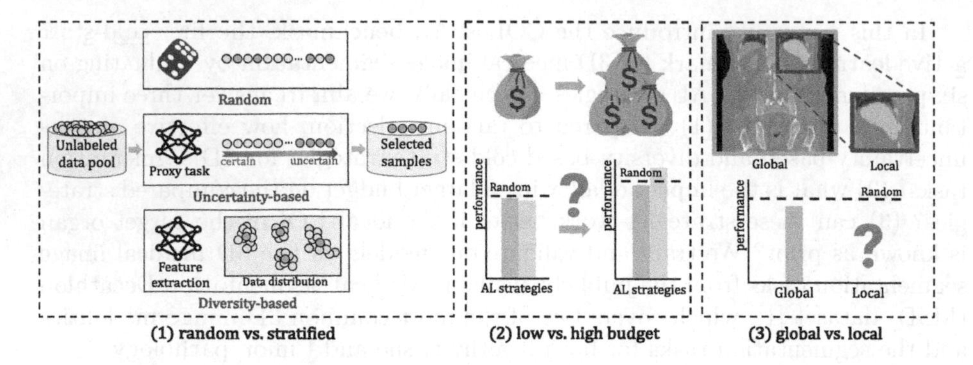

Fig. 1. Illustration of our three cold-start AL scenarios. We evaluate **(1)** uncertainty and diversity based selection strategies against random selection in a low-budget regime, **(2)** the effect of budget on performance, and **(3)** the usefulness of a local ROI for selection strategies.

2.2 Cold-Start AL Scenarios

In this study, we investigate the cold-start AL strategies for 3D segmentation tasks in three scenarios, as illustrated in Fig. 1.

1. With a low budget of 5 volumes (except for Heart, where 3 volumes are used because of the smaller dataset and easier segmentation task), we assess the performance of the uncertainty-based and diversity-based approaches against the random selection.
2. Next, we explore the impact of budgets for different cold-start AL schemes by allowing a higher budget, as previous work shows inconsistent effectiveness of AL schemes in different budget regimes [7].
3. Finally, we explore whether the cold-start AL strategies can benefit from using the uncertainty/diversity from only the local ROI of the target organ, rather than the entire volume. This strategy may be helpful for 3D tasks especially for small organs, whose uncertainty/diversity can be outweighted by the irrelevant structures in the entire volume, but needs to be validated.

Evaluation Metrics. To evaluate the segmentation performance, we use the Dice similarity coefficient and 95% Hausdorff distance (HD95), which measures the overlap between the segmentation result and ground truth, and the quality of segmentation boundaries by computing the 95th percentile of the distances between the segmentation and the ground truth boundary points, respectively.

2.3 Baseline Cold-Start Active Learners

We provide the implementation for the baseline approaches: random selection, two variants of an uncertainty-based approach named ProxyRank [14], and three diversity-based methods, namely ALPS [22], CALR [10], and TypiClust [7].

Random Selection. As suggested by prior works [3,4,7,12,20,26], random selection is a strong competitor in the cold-start setting, since it is independent and identically distributed (i.i.d.) to the entire data pool. We shuffle the entire training list with a random seed and select the first m samples. In our experiments, random selection is conducted 15 times and the mean Dice score is reported.

Uncertainty-Based Selection. Many traditional AL methods use uncertainty sampling, where the most uncertain samples are selected using the uncertainty of the network trained on an initial labeled set. Without such an initial labeled set, it is not straightforward to capture uncertainty in the cold-start setting.

Recently, Nath *et al.* [14] proposed a proxy task and then utilized uncertainty generated from the proxy task to rank the unlabeled data. By selecting the most uncertain samples, this strategy has shown superior performance to random selection. Specifically, pseudo labels were generated by thresholding the CT images with an organ-dependent Hounsfield Unit (HU) intensity window. These pseudo labels carry coarse information for the target organ, though they also include other unrelated structures. The uncertainty generated by this proxy task is assumed to represent the uncertainty of the actual segmentation task.

However, this approach [14] was limited to CT images. Here, we extend this strategy to MR images. For each MR image, we apply a sequence of transformations to convert it to a noisy binary mask: (1) z-score normalization, (2) intensity clipping to the $[1^{st}, 99^{th}]$ percentile of the intensity values, (3) intensity normalization to $[0, 1]$ and (4) Otsu thresholding [16]. We visually verify that the binary pseudo label includes the coarse boundary of the target organ.

As in [14], we compute the model uncertainty for each unlabeled data using Monte Carlo dropout [6]: with dropout enabled during inference, multiple predictions are generated with stochastic dropout configurations. Entropy [13] and Variance [21] are used as uncertainty measures to create two variants of this proxy ranking method, denoted as **ProxyRank-Ent** and **ProxyRank-Var**. The overall uncertainty score of an unlabeled image is computed as the mean across all voxels. Finally, we rank all unlabeled data with the overall uncertainty scores and select the most uncertain m samples.

Diversity-Based Selection. Unlike uncertainty-based methods which require a warm start, diversity-based methods can be used in the cold-start setting. Generally, diversity-based approaches consist of two stages. First, a feature extraction network is trained using unsupervised/self-supervised tasks to represent each unlabeled data as a latent feature. Second, clustering algorithms are used to select the most diverse samples in latent space to reduce data redundancy. The major challenge of benchmarking the diversity-based methods for 3D tasks is to have a feature extraction network for 3D volumes. To address this issue, we train a 3D auto-encoder on the unlabeled training data using a self-supervised task, i.e., image reconstruction. Specifically, we represent each unlabeled 3D volume as a latent feature by extracting the bottleneck feature maps, followed by an adaptive average pooling for dimension reduction [24].

Afterwards, we adapt the diversity-based approaches to our 3D tasks by using the same clustering strategies as proposed in the original works, but replacing the feature extraction network with our 3D version. In our benchmark, we evaluate the clustering strategies from three state-of-the-art diversity-based methods.

1. **ALPS** [22]: k-MEANS is used to cluster the latent features with the number of clusters equal to the query number m. For each cluster, the sample that is the closest to the cluster center is selected.
2. **CALR** [10]: This approach is based on the maximum density sampling, where the sample with the most information is considered the one that can optimally represent the distribution of a cluster. A bottom-up hierarchical clustering algorithm termed BIRCH [23] is used and the number of clusters is set as the query number m. For each cluster, the information density for each sample within the cluster is computed and the sample with the highest information density is selected. The information density is expressed as $I(x) = \frac{1}{|X_c|}\sum_{x' \in X_c} sim(x, x')$, where $X_c = \{x_1, x_2, ...x_j\}$ is the feature set in a cluster and cosine similarity is used as $sim(\cdot)$.
3. **TypiClust** [7]: This approach also uses the points density in each cluster to select a diverse set of typical examples. k-MEANS clustering is used, followed by selecting the most typical data from each cluster, which is similar to the ALPS strategy but less sensitive to outliers. The typicality is calculated as the inverse of the average Euclidean distance of x to its K nearest neighbors KNN(x), expressed as: Typicality$(x) = (\frac{1}{K}\sum_{x_i \in \text{KNN}(x)} ||x - x_i||_2)^{-1}$. K is set as 20 in the original paper but that is too high for our application. Instead, we use all the samples from the same cluster to calculate typicality.

2.4 Implementation Details

In our benchmark, we use the 3D U-Net as the network architecture. For uncertainty estimation, 20 Monte Carlo simulations are used with a dropout rate of 0.2. As in [14], a dropout layer is added at the end of every level of the U-Net for both encoder and decoder. The performance of different AL strategies is evaluated by training a 3D patch-based segmentation network using the selected data, which is an important distinction from the earlier 2D variants in the literature. The only difference between different experiments is the selected data. For CT pre-processing, image intensity is clipped to $[-1024, 1024]$ HU and rescaled to $[0, 1]$. For MRI pre-processing, we sequentially apply z-score normalization, intensity clipping to $[1^{\text{st}}, 99^{\text{th}}]$ percentile and rescaling to $[0, 1]$. During training, we randomly crop a 3D patch with a patch size of $128 \times 128 \times 128$ (except for hippocampus, where we use $32 \times 32 \times 32$) with the center voxel of the patch being foreground and background at a ratio of $2:1$. Stochastic gradient descent algorithm with a Nesterov momentum ($\mu = 0.99$) is used as the optimizer and L_{DiceCE} is used as the segmentation loss. An initial learning rate is set as 0.01 and decayed with a polynomial policy as in [9]. For each experiment, we train our model using 30k iterations and validate the performance every 200 iterations. A variety of augmentation techniques as in [9] are applied to achieve optimal

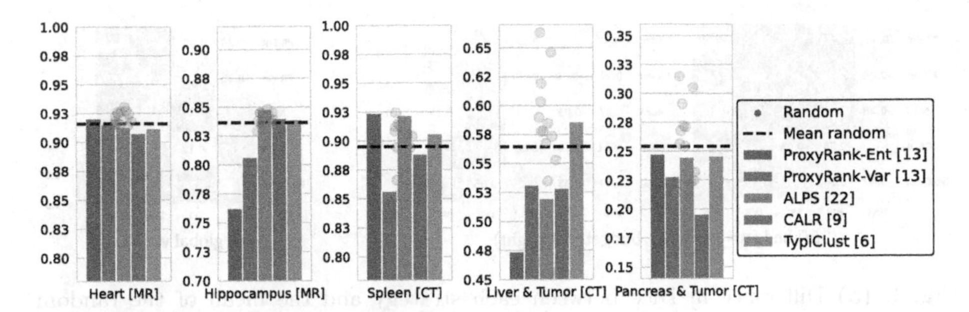

Fig. 2. Cold-start AL strategies in a low-budget regime ($m = 5$). TypiClust (orange) is comparable or superior to mean random selection, and consistently outperforms the poor random selection samples. Comprehensive tables are provided in Supp. Materials. (Color figure online)

performance for all compared methods. All the networks are implemented in PyTorch [17] and MONAI [2]. Our experiments are conducted with the deterministic training mode in MONAI with a fixed random seed=0. We use a 24G NVIDIA GeForce RTX 3090 GPU.

For the global vs. local experiments, the local ROIs are created by extracting the 3D bounding box from the ground truth mask and expanding it by five voxels along each direction. We note that although no ground truth masks are accessible in the cold-start AL setting, this analysis is still valuable to determine the usefulness of local ROIs. It is only worth exploring automatic generation of these local ROIs if the gold-standard ROIs show promising results.

3 Experimental Results

Impact of Selection Strategies. In Fig. 2, with a fixed budget of 5 samples (except for Heart, where 3 samples are used), we compare the uncertainty-based and diversity-based strategies against the random selection on five different segmentation tasks. Note that the selections made by each of our evaluated AL strategies are deterministic. For random selection, we visualize the individual Dice scores (red dots) of all 15 runs as well as their mean (dashed line). HD95 results (Supp. Tab. 1) follow the same trends.

Our results explain why random selection remains a strong competitor for 3D segmentation tasks in cold-start scenarios, as no strategy evaluated in our benchmark *consistently* outperforms the random selection *average* performance.

However, we observe that TypiClust (shown as orange) achieves comparable or superior performance compared to random selection across all tasks in our benchmark, whereas other approaches can significantly under-perform on certain tasks, especially challenging ones like the liver dataset. Hence, **TypiClust stands out as a more robust cold-start selection strategy**, which can achieve at least a comparable (sometimes better) performance against the mean of random selection. We further note that TypiClust largely mitigates the

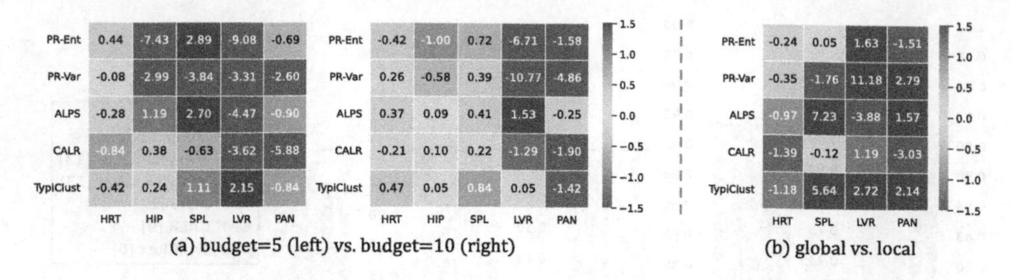

Fig. 3. (a) Difference in Dice between each strategy and the mean of the random selection (warm colors: better than random). Cold-start AL strategies are more effective under the higher budget. **(b)** Global vs. local ROI performance (warm colors: global better than local). The local ROI does not yield a consistently better performance. Comprehensive tables are provided in Supp. Materials.

risk of 'unlucky' random selection as it *consistently* performs better than the low-performing random samples (red dots below the dashed line).

Impact of Different Budgets. In Fig. 3(a), we compare AL strategies under the budgets of $m = 5$ vs. $m = 10$ (3 vs. 5 for Hearts). We visualize the performance under each budget using a heatmap, where each element in the matrix is the difference of Dice scores between the evaluated strategy and the mean of random selection under that budget. A positive value (warm color) means that the AL strategy is more effective than random selection. We observe an increasing amount of warm elements in the higher-budget regime, indicating that **most cold-start AL strategies become more effective when more budget is allowed.** This is especially true for the diversity-based strategies (three bottom rows), suggesting that when a slightly higher budget is available, the diversity of the selected samples is important. HD95 results (Supp. Tab. 1) are similar.

Impact of Different ROIs. In Fig. 3(b), with a fixed budget of $m = 5$ volumes, we compare the AL strategies when uncertainty/diversity is extracted from the entire volume (global) vs. a local ROI (local). Each element in this heatmap is the Dice difference of the AL strategy between global and local; warm color means global is better than local. The hippocampus images in MSD are already cropped to the ROI, and thus are excluded from this comparison. We observe different trends across different methods and tasks. Overall, we can observe more warm elements in the heatmap, indicating that **using only the local uncertainty or diversity for cold-start AL cannot consistently outperform the global counterparts**, even with ideal ROI generated from ground truth. HD95 results (Supp. Tab. 2) follow the same trends.

Limitations. For the segmentation tasks that include tumors (4th and 5th columns on Fig. 3(a)), we find that almost no AL strategy is very effective, especially the uncertainty-based approaches. The uncertainty-based methods heavily rely on the uncertainty estimated by the network trained on the proxy tasks, which likely makes the uncertainty of tumors difficult to capture. It may be necessary to allocate more budget or design better proxy tasks to make cold-start AL methods effective for such challenging tasks. Lastly, empirical exploration of cold-start AL on iterative AL is beyond the scope of this study and merits its own dedicated study in future.

4 Conclusion

In this paper, we presented the COLosSAL benchmark for cold-start AL strategies on 3D medical image segmentation using the public MSD dataset. Comprehensive experiments were performed to answer three important open questions for cold-start AL. While cold-start AL remains an unsolved problem for 3D segmentation, important trends emerge from our results; for example, diversity-based strategies tend to benefit more from a larger budget. Among the compared methods, TypiClust [7] stands out as the most robust option for cold-start AL in medical image segmentation tasks. We believe our findings and the open-source benchmark will facilitate future cold-start AL studies, such as the exploration of different uncertainty estimation/feature extraction methods and evaluation on multi-modality datasets.

Acknowledgements. This work was supported in part by the National Institutes of Health grants R01HD109739 and T32EB021937, as well as National Science Foundation grant 2220401. This work was also supported by the Advanced Computing Center for Research and Education (ACCRE) of Vanderbilt University.

References

1. Antonelli, M.: The medical segmentation decathlon. Nat. Commun. **13**(1), 4128 (2022)
2. Cardoso, M.J., et al.: Monai: an open-source framework for deep learning in healthcare. arXiv preprint arXiv:2211.02701 (2022)
3. Chandra, A.L., Desai, S.V., Devaguptapu, C., Balasubramanian, V.N.: On initial pools for deep active learning. In: NeurIPS 2020 Workshop on Pre-registration in Machine Learning, pp. 14–32. PMLR (2021)
4. Chen, L., et al.: Making your first choice: to address cold start problem in medical active learning. In: Medical Imaging with Deep Learning (2023)
5. Gaillochet, M., Desrosiers, C., Lombaert, H.: TAAL: test-time augmentation for active learning in medical image segmentation. In: Nguyen, H.V., Huang, S.X., Xue, Y. (eds.) DALI 2022. LNCS, vol. 13567, pp. 43–53. Springer, Cham (2022). https://doi.org/10.1007/978-3-031-17027-0_5
6. Gal, Y., Ghahramani, Z.: Dropout as a Bayesian approximation: representing model uncertainty in deep learning. In: International Conference on Machine Learning, pp. 1050–1059. PMLR (2016)
7. Hacohen, G., Dekel, A., Weinshall, D.: Active learning on a budget: opposite strategies suit high and low budgets. arXiv preprint arXiv:2202.02794 (2022)
8. Hesamian, M.H., Jia, W., He, X., Kennedy, P.: Deep learning techniques for medical image segmentation: achievements and challenges. J. Digit. Imaging **32**, 582–596 (2019)

9. Isensee, F., Jaeger, P.F., Kohl, S.A., Petersen, J., Maier-Hein, K.H.: nnU-Net: a self-configuring method for deep learning-based biomedical image segmentation. Nat. Methods **18**(2), 203–211 (2021)
10. Jin, Q., Yuan, M., Li, S., Wang, H., Wang, M., Song, Z.: Cold-start active learning for image classification. Inf. Sci. **616**, 16–36 (2022)
11. Lewis, D.D., Catlett, J.: Heterogeneous uncertainty sampling for supervised learning. In: Machine Learning Proceedings 1994, pp. 148–156. Elsevier (1994)
12. Mittal, S., Tatarchenko, M., Çiçek, Ö., Brox, T.: Parting with illusions about deep active learning. arXiv preprint arXiv:1912.05361 (2019)
13. Nath, V., Yang, D., Landman, B.A., Xu, D., Roth, H.R.: Diminishing uncertainty within the training pool: active learning for medical image segmentation. IEEE Trans. Med. Imaging **40**(10), 2534–2547 (2020)
14. Nath, V., Yang, D., Roth, H.R., Xu, D.: Warm start active learning with proxy labels and selection via semi-supervised fine-tuning. In: Wang, L., Dou, Q., Fletcher, P.T., Speidel, S., Li, S. (eds.) MICCAI 2022. LNCS, vol. 13438, pp. 297–308. Springer, Cham (2022). https://doi.org/10.1007/978-3-031-16452-1_29
15. Nguyen, V.L., Shaker, M.H., Hüllermeier, E.: How to measure uncertainty in uncertainty sampling for active learning. Mach. Learn. **111**(1), 89–122 (2022)
16. Otsu, N.: A threshold selection method from gray-level histograms. IEEE Trans. Syst. Man Cybern. **9**(1), 62–66 (1979)
17. Paszke, A., et al.: Pytorch: an imperative style, high-performance deep learning library. In: Advances in Neural Information Processing Systems, vol. 32 (2019)
18. Ranganathan, H., Venkateswara, H., Chakraborty, S., Panchanathan, S.: Deep active learning for image classification. In: 2017 IEEE International Conference on Image Processing (ICIP), pp. 3934–3938. IEEE (2017)
19. Sener, O., Savarese, S.: Active learning for convolutional neural networks: a core-set approach. arXiv preprint arXiv:1708.00489 (2017)
20. Siméoni, O., Budnik, M., Avrithis, Y., Gravier, G.: Rethinking deep active learning: using unlabeled data at model training. In: 2020 25th International Conference on Pattern Recognition (ICPR), pp. 1220–1227. IEEE (2021)
21. Yang, L., Zhang, Y., Chen, J., Zhang, S., Chen, D.Z.: Suggestive annotation: a deep active learning framework for biomedical image segmentation. In: Descoteaux, M., Maier-Hein, L., Franz, A., Jannin, P., Collins, D.L., Duchesne, S. (eds.) MICCAI 2017. LNCS, vol. 10435, pp. 399–407. Springer, Cham (2017). https://doi.org/10.1007/978-3-319-66179-7_46
22. Yuan, M., Lin, H.T., Boyd-Graber, J.: Cold-start active learning through self-supervised language modeling. arXiv preprint arXiv:2010.09535 (2020)
23. Zhang, T., Ramakrishnan, R., Livny, M.: Birch: an efficient data clustering method for very large databases. ACM SIGMOD Rec. **25**(2), 103–114 (1996)
24. Zhao, Z., Lu, W., Zeng, Z., Xu, K., Veeravalli, B., Guan, C.: Self-supervised assisted active learning for skin lesion segmentation. In: 2022 44th Annual International Conference of the IEEE Engineering in Medicine & Biology Society (EMBC), pp. 5043–5046. IEEE (2022)
25. Zheng, H., et al.: Biomedical image segmentation via representative annotation. In: Proceedings of the AAAI Conference on Artificial Intelligence, vol. 33, pp. 5901–5908 (2019)
26. Zhu, Y., et al.: Addressing the item cold-start problem by attribute-driven active learning. IEEE Trans. Knowl. Data Eng. **32**(4), 631–644 (2019)

Continual Learning for Abdominal Multi-organ and Tumor Segmentation

Yixiao Zhang, Xinyi Li, Huimiao Chen, Alan L. Yuille, Yaoyao Liu(✉), and Zongwei Zhou(✉)

Johns Hopkins University, Baltimore, USA
{yliu538,zzhou82}@jh.edu
https://github.com/MrGiovanni/ContinualLearning

Abstract. The ability to dynamically extend a model to new data and classes is critical for multiple organ and tumor segmentation. However, due to privacy regulations, accessing previous data and annotations can be problematic in the medical domain. This poses a significant barrier to preserving the high segmentation accuracy of the old classes when learning from new classes because of the catastrophic forgetting problem. In this paper, we first empirically demonstrate that simply using high-quality pseudo labels can fairly mitigate this problem in the setting of organ segmentation. Furthermore, we put forward an innovative architecture designed specifically for continuous organ and tumor segmentation, which incurs minimal computational overhead. Our proposed design involves replacing the conventional output layer with a suite of lightweight, class-specific heads, thereby offering the flexibility to accommodate newly emerging classes. These heads enable independent predictions for newly introduced and previously learned classes, effectively minimizing the impact of new classes on old ones during the course of continual learning. We further propose incorporating Contrastive Language-Image Pretraining (CLIP) embeddings into the organ-specific heads. These embeddings encapsulate the semantic information of each class, informed by extensive image-text co-training. The proposed method is evaluated on both in-house and public abdominal CT datasets under organ and tumor segmentation tasks. Empirical results suggest that the proposed design improves the segmentation performance of a baseline model on newly-introduced and previously-learned classes along the learning trajectory.

Keywords: Continual Learning · Incremental Learning · Multi-Organ Segmentation · Tumor Segmentation

1 Introduction

Humans inherently learn in an incremental manner, acquiring new concepts over time without forgetting previous ones. In contrast, deep learning models suffer

Supplementary Information The online version contains supplementary material available at https://doi.org/10.1007/978-3-031-43895-0_4.

H. Greenspan et al. (Eds.): MICCAI 2023, LNCS 14221, pp. 35–45, 2023.
https://doi.org/10.1007/978-3-031-43895-0_4

from catastrophic forgetting [10], where learning from new data can override previously acquired knowledge. In this context, the class-incremental continual learning problem was formalized by Rebuffi et al. [23], where new classes are observed in different stages, restricting the model from accessing previous data.

The medical domain faces a similar problem: the ability to dynamically extend a model to new classes is critical for multiple organ and tumor segmentation, wherein the key obstacle lies in mitigating 'forgetting.' A typical strategy involves retaining some previous data. For instance, Liu et al. [13] introduced a memory module to store the prototypical representation of different organ categories. However, such methods, reliant on an account of data and annotations, may face practical constraints as privacy regulations could make accessing prior data and annotations difficult [9]. An alternative strategy is to use pseudo labels generated by previously trained models on new data. Ozdemir et al. [18,19] extended the distillation loss to medical image segmentation. A concurrent study of ours [7] mainly focused on architectural extension, addressing the forgetting problem by freezing the encoder and decoder and adding additional decoders when learning new classes. While these strategies have been alleviating the forgetting problem, they led to tremendous memory costs for model parameters.

Therefore, we identify two main open questions that must be addressed when designing a multi-organ and tumor segmentation framework. **Q1:** Can we relieve the forgetting problem without needing previous data and annotations? **Q2:** Can we design a new model architecture that allows us to share more parameters among different continual learning steps?

To tackle the above questions, in this paper, we propose a novel continual multi-organ and tumor segmentation method that overcomes the forgetting problem with little memory and computation overhead. **First**, inspired by knowledge distillation methods in continual learning [11,14,15,17], we propose to generate soft pseudo annotations for the old classes on newly-arrived data. This enables us to recall old knowledge without saving the old data. We observe that with this simple strategy, we are able to maintain a reasonable performance for the old classes. **Second**, we propose image-aware segmentation heads for each class on top of the shared encoder and decoder. These heads allow the use of a single backbone and easy extension to new classes while bringing little computational cost. Inspired by Liu et al. [12], we adopt the text embedding generated by Contrastive Language-Image Pre-training (CLIP) [22]. CLIP is a large-scale image-text co-training model that is able to encode high-level visual semantics into text embeddings. This information will be an advantage for training new classes with the class names known in advance.

We focus on organ/tumor segmentation because it is one of the most critical tasks in medical imaging [6,21,27,28], and continual learning in semantic segmentation is under-explored in the medical domain. We evaluate our continual learning method using three datasets: BTCV [8], LiTS [1] and JHH [25] (a private dataset at Johns Hopkins Hospital)[1]. On the public datasets, the learning trajectory is to first segment 13 organs in the BTCV dataset, then learn to

[1] The JHH dataset has 200 abdominal CT scans with per-voxel annotations for 13 organs, three gastrointestinal tracts, and four cardiovascular system structures.

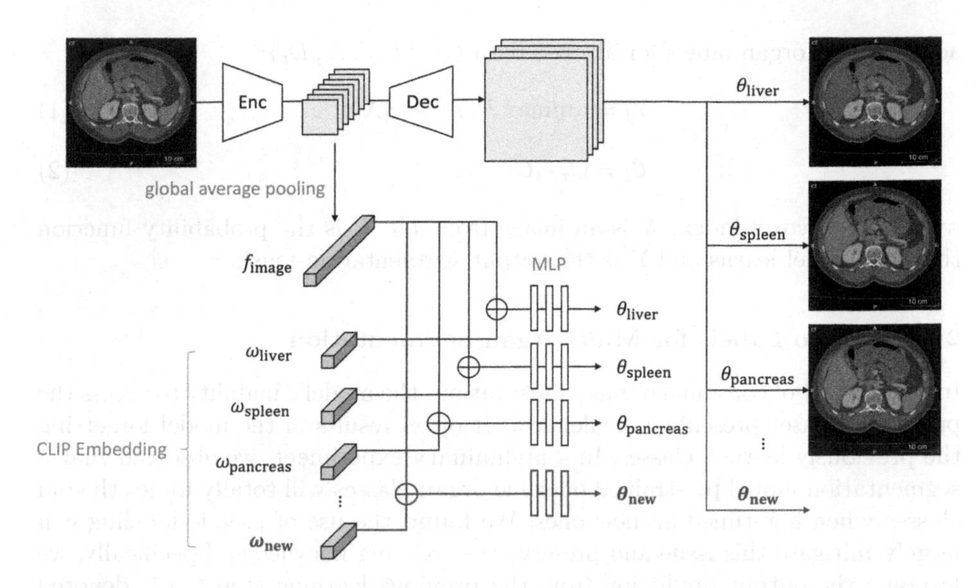

Fig. 1. An overview of the proposed method. An encoder (Enc) processes the input image to extract its features, which are then reduced to a feature vector (f_{image}) by a global average pooling layer. This feature vector is subsequently concatenated with a CLIP embedding (ω_{class}), calculated using the pre-trained CLIP model. Through a series of Multi-Layer Perceptron (MLP) layers, we derive class-specific parameters of convolution kernels (θ_{class}). These kernels, when applied to the decoder (Dec) feature, yield the mask for the respective class.

segment liver tumors in the LiTS dataset. On the private dataset, the learning trajectory is to first segment 13 organs, followed by continual segmentation of three gastrointestinal tracts and four cardiovascular system structures. In our study, we review and compare three popular continual learning baselines that apply knowledge distillation to predictions [11], features [17], and multi-scale pooled features [3], respectively. The extensive results demonstrate that the proposed method outperforms existing methods, achieving superior performance in both keeping the knowledge of old classes and learning the new ones while maintaining high memory efficiency.

2 Methodology

We formulate the continual organ segmentation as follows: given a sequence of partially annotated datasets $\{D_1, D_2, \ldots, D_n\}$ each with organ classes $\{C_1, C_2, \ldots, C_n\}$, we learn a single multi-organ segmentation model sequentially using one dataset at a time. When training on the i-th dataset D_t, the previous datasets $\{D_1, \ldots, D_{t-1}\}$ are not available. The model is required to predict the accumulated organ labels for all seen datasets $\{D_1, \ldots, D_t\}$:

$$\hat{Y}_j = \underset{c \in \mathcal{C}_t}{\operatorname{argmax}} P(Y_j = c|X) \tag{1}$$

$$\mathcal{C}_t = \cup_{\tau \leq t} C_\tau \tag{2}$$

where j is a voxel index, X is an image from D_t, P is the probability function that the model learns and $\hat{Y}$ is the output segmentation mask.

2.1 Pseudo Labels for Multi-organ Segmentation

In the context of continual organ segmentation, the model's inability to access the previous dataset presents a challenge as it often results in the model forgetting the previously learned classes. In a preliminary experiment, we observed that a segmentation model pre-trained on some organ classes will totally forget the old classes when fine-tuned on new ones. We found the use of pseudo-labeling can largely mitigate this issue and preserve the existing knowledge. Specifically, we leverage the output prediction from the previous learning step $t-1$, denoted as $\hat{Y}_{t-1}$, which includes the old classes $\mathcal{C}_{t-1}$, as the pseudo label for the current step's old classes. For new classes, we still use the ground truth label. Formally, the label $\tilde{L_t^c}$ for class c in current learning step t can be expressed as:

$$\tilde{L_t^c} = \begin{cases} L_t^c & \text{if } c \in \mathcal{C}_t - \mathcal{C}_{t-1} \\ \hat{Y}_{t-1}^c & \text{if } c \in \mathcal{C}_{t-1} \end{cases} \tag{3}$$

where L_t^c represents the ground truth label for class c in step t obtained from dataset D_t. By utilizing this approach, we aim to maintain the original knowledge and prevent the model from forgetting the previously learned information while learning the new classes. The following proposed model is trained only with pseudo labeling of old classes without any other distillation or regularization.

2.2 The Proposed Multi-organ Segmentation Model

In the following, we introduce the proposed multi-organ segmentation model for continual learning. Figure 1 illustrates the overall framework of the proposed model architecture. It has an encoder-decoder backbone, a set of image-aware organ-specific output heads, and text-driven head parameter generation.

Backbone Model: For continual learning, ideally, the model should be able to learn a sufficiently general representation that would easily adapt to new classes. We use Swin UNETR [4] as our backbone since it exhibits strong performance in self-supervised pre-training and the ability to transfer to various medical image segmentation tasks. Swin UNETR has Swin Transformer [16] as the encoder and several deconvolution layers as the decoder.

Image-Aware Organ-Specific Heads: The vanilla Swin UNETR has a Softmax layer as the output layer that predicts the probabilities of each class. We propose to replace the output layer with multiple image-aware organ-specific heads. We first use a global average pooling (GAP) layer on the last encoder features to obtain a global feature f of the current image X. Then for each organ class k, a multilayer perceptron (MLP) module is learned to map the global image feature to a set of parameters θ_k:

$$\theta_k = \mathrm{MLP}_k(\mathrm{GAP}(E(X))), \tag{4}$$

where $E(X)$ denotes the encoder feature of image X. An output head for organ class k is a sequence of convolution layers that use parameters θ_k as convolution kernel parameters. These convolution layers are applied to the decoder features, which output the segmentation prediction for organ class k:

$$P(Y_j^k = 1|X, \theta_k) = \sigma(\mathrm{Conv}(D(E(X)); \theta_k)), \tag{5}$$

where E is the encoder, D is the decoder, σ is the Sigmoid non-linear layer and $P(Y_j^k = 1)$ denotes the predicted probability that pixel j belongs to the organ class k. The predictions for each class are optimized by Binary Cross Entropy loss. The separate heads allow independent probability prediction for newly introduced and previously learned classes, therefore minimizing the impact of new classes on old ones during continual learning. Moreover, this design allows multi-label prediction for cases where a pixel belongs to more than one class (e.g., a tumor on an organ).

Text Driven Head Parameter Generation: We further equip the segmentation heads with semantic information about each organ class. With the widespread success of large-scale vision-language models, there have been many efforts that apply these models to the medical domain [2,5,26]. It is suggested that vision-language models could be used for zero-shot learning in the medical domain and recognize novel classes with well-designed prompts [20]. We propose to use CLIP [22] to generate text embeddings for the target organ names. Specifically, we produce the organ name embedding by the pre-trained CLIP text encoder and a medical prompt (e.g., "a computerized tomography of a [CLS]", where [CLS] is an organ class name). Then we use the text embeddings ω together with the global image feature f to generate parameters for the organ segmentation heads:

$$\theta_k = \mathrm{MLP}_k([\mathrm{GAP}(E(X)), \omega_k]), \tag{6}$$

where ω_k is the text embedding for organ class k. CLIP embeddings carry high-level semantic meanings and have the ability to connect correlated concepts. Therefore, it guides the MLP module to generate better convolution parameters for each organ class. More importantly, the fixed-length CLIP embedding allows us to adapt the pre-trained model to open-vocabulary segmentation and extend to novel classes.

Difference from Universal Model [12]***:*** For the purpose of continual learning, we improve the original design of Universal Model in the MLP module.

Unlike Liu et al. [12], who utilized a single MLP to manage multiple classes, we allocate an individual and independent MLP to each class. This design significantly mitigates interference among different classes.

2.3 Computational Complexity Analysis

Another key contribution of our work is the reduction of computational complexity in continual segmentation. We compare our proposed model's FLOPs (floating-point operations per second) with the baseline model, Swin UNETR [4]. Our model's FLOPs are just slightly higher than Swin UNETR's, with 661.6 GFLOPs and 659.4 GFLOPs, respectively. This is because we used lightweight output convolution heads with a small number of channels. Ji et al. [7] proposed a state-of-the-art architecture for medical continual semantic segmentation, which uses a pre-trained and then frozen encoder coupled with incrementally added decoders in each learning step. However, subsequent continual learning steps using this architecture introduce massive computational complexity. For example, Swin UNETR's decoder alone has 466.08 GFLOPs, meaning that every new learning step adds an additional 466.08 GFLOPs. In contrast, our model only needs to add a few image-aware organ-specific heads for new classes of the new task, with each head consuming only 0.12 GFLOPs. As a result, the computational complexity of our model nearly remains constant in continual learning for segmentation, while that of the architecture of Ji et al. [7] increases linearly to the number of steps.

3 Experiment and Result

Datasets: We empirically evaluate the proposed model under two data settings: in one setting, both training and continual learning are conducted on the in-house JHH dataset. It has multiple classes annotated, which can be categorized into three groups: the abdominal organs (in which seven classes are learned in step 1: spleen, right kidney, left kidney, gall bladder, liver, postcava, pancreas), the gastrointestinal tract (in which three classes are learned in step 2: stomach, colon, intestine), and other organs (in which four classes are learned in step 3: aorta, portal vein and splenic vein, celiac truck, superior mesenteric artery). The categorization is in accordance with TotalSegmentator [24]. In the other setting, we first train on the BTCV dataset and then do continual learning on the LiTS dataset. The BTCV dataset contains 47 abdominal CT images delineating 13 organs. The LiTS dataset contains 130 contrast-enhanced abdominal CT scans for liver and liver tumor segmentation. We use 13 classes (spleen, right kidney, left kidney, gall bladder, esophagus, liver, stomach, aorta, inferior vena cava, portal vein and splenic vein, pancreas, right adrenal gland, left adrenal gland) from BTCV in step 1 learning and the live tumor from LiTS in step 2 learning.

Baselines and Metrics: For a fair comparison, all the compared methods use the same Swin UNETR [4] as the backbone, which is the state-of-the-art model in a bunch of medical image segmentation tasks. We compare with three

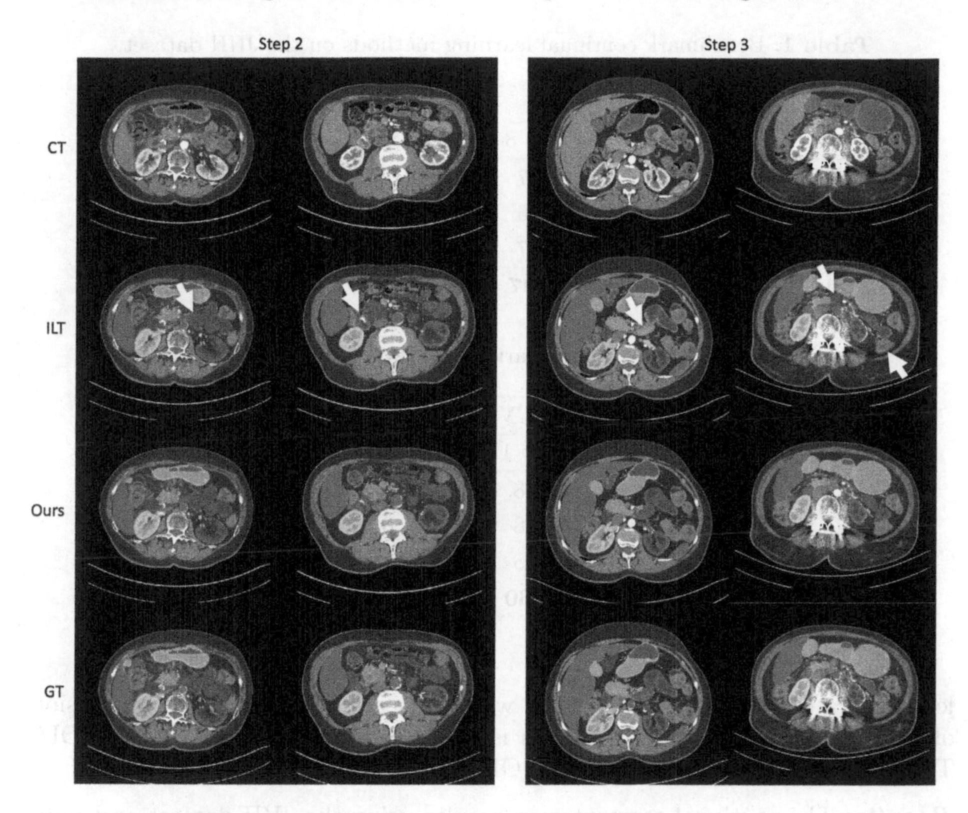

Fig. 2. The visualization comparison between our model and the baseline model Swin UNETR in continual learning steps 2 and 3 on the JHH dataset.

popular continual learning baseline methods that apply knowledge distillation, including LwF [11], ILT [17] and PLOP [3]. We compare the proposed method with different baseline models using the commonly used Dice score (DSC) metric (the Sørensen-Dice coefficient). In each learning step, we report the average DSC for the classes that are used at the current step as well as the previous steps (e.g., in step 2 of the JHH dataset, we report the average dice of the gastrointestinal tracts and the abdominal organs). The dice score at old classes reveals a model's ability to retain its previous knowledge, and the score for the current step classes indicates the model's ability to acquire new knowledge under the regularization of old ones.

Implementation Details: The proposed model architecture is trained on new classes with pseudo labeling of old classes. No other distillation techniques are used. We use a lightweight design for the image-aware organ-specific heads. Each head consists of three convolution layers. The number of kernels in the first two layers is 8, and in the last layer is 1. All the compared models are trained using the AdamW optimizer for 100 epochs with a cosine learning rate scheduler. We use a batch size of 2 and a patch size of $96 \times 96 \times 96$ for the training. The initial

Table 1. Benchmark continual learning methods on the JHH dataset.

Method	JHH_organ (7)			JHH_gastro (3)		JHH_cardiac (4)
	Step 1	Step 2	Step 3	Step 2	Step 3	Step 3
LwF [11]	**0.891**	0.777	0.767	0.530	0.486	0.360
ILT [17]	**0.891**	0.775	0.776	0.653	0.480	0.484
PLOP [3]	**0.891**	0.780	0.777	0.427	0.464	0.318
Ours	0.887	**0.783**	**0.787**	**0.695**	**0.692**	**0.636**

Table 2. Benchmark continual learning methods on the public datasets.

Method	BTCV (13)		LiTS (1)
	Step 1	Step 2	Step 2
LwF [11]	0.828	0.770	0.456
ILT [17]	0.828	0.786	0.335
PLOP [3]	0.828	0.799	0.362
Ours	**0.860**	**0.817**	**0.466**

learning rate is set as $1e^{-4}$, and the weight decay is set as $1e^{-5}$. The version of MONAI[2] used in our experiments is 1.1.0. Models are trained on NVIDIA TITAN RTX and Quadro RTX 8000 GPUs.

Results: The continual segmentation results using the JHH dataset and public datasets are shown in Tables 1 and 2, respectively. Notably, by simply using the pseudo labeling technique (LwF), we are able to achieve reasonably good performance in remembering the old classes (Dice of 0.777 in step 2 and 0.767 in step 3 for abdominal organs in the JHH dataset; Dice of 0.770 in step 2 for BTCV organs). Class-wise DSC scores are in Appendix Tables 4–7. All the compared methods use prediction-level or feature-level distillation as regularization. Among them, the proposed method achieves the highest performance in most learning steps. Specifically, the proposed method exhibits the least forgetting in old classes and a far better ability to adapt to new data and new classes.

To evaluate the proposed model designs, we also conduct the ablation study on the JHH dataset, shown in Table 3. Specifically, we ablate the performance improvement introduced by the organ-specific segmentation heads as well as the CLIP text embeddings. The first line in Table 3 shows the performance of the baseline Swin UNETR model learned with pseudo labeling (LwF). The second row introduces the organ-specific segmentation heads, but uses one-hot embeddings rather than the CLIP text embeddings for each organ. The third row gives the performance of the full method. The results show that by adapting the model to use organ-specific heads as segmentation outputs, we are able to achieve improvement of a large margin (e.g., 0.144 in step 2 and 0.179 in step

[2] https://monai.io.

Table 3. Ablation study on the JHH dataset.

Method	JHH_organ (7)			JHH_gastro (3)		JHH_cardiac (4)
	Step 1	Step 2	Step 3	Step 2	Step 3	Step 3
LwF [11]	0.891	0.777	0.767	0.530	0.486	0.360
Ours_1-hot	0.882	0.767	0.777	0.674	0.665	0.452
Ours_CLIP	**0.887**	**0.783**	**0.787**	**0.695**	**0.692**	**0.636**

3 for gastrointestinal tracts). With the application of CLIP text embeddings, we are able to further improves the performance (e.g., by a margin of 0.019 in step 2 and 0.027 in step 3 for gastrointestinal tracts). This study validates the effectiveness of the proposed organ-specific segmentation heads and the CLIP text embeddings in the continual organ segmentation task.

Finally, we show the qualitative segmentation results of the proposed method together with the best baseline method ILT on the JHH dataset. We show the results of learning steps 2 and 3 in Fig. 2, one case per column and two cases for each step. The visualization demonstrates that the proposed method successfully segments the correct organs while the best baseline method fails throughout the continual learning process.

4 Conclusion

In this paper, we propose a method for continual multiple organ and tumor segmentation in 3D abdominal CT images. We first empirically verified the effectiveness of high-quality pseudo labels in retaining previous knowledge. Then, we propose a new model design that uses organ-specific heads for segmentation, which allows easy extension to new classes and brings little computational cost in the meantime. The segmentation heads are further strengthened by utilizing the CLIP text embeddings that encode the semantics of organ or tumor classes. Numerical results on an in-house dataset and two public datasets demonstrate that the proposed method outperforms the continual learning baseline methods in the challenging multiple organ and tumor segmentation tasks.

Acknowledgements. This work was supported by the Lustgarten Foundation for Pancreatic Cancer Research and partially by the Patrick J. McGovern Foundation Award. We appreciate the effort of the MONAI Team to provide open-source code for the community.

References

1. Bilic, P., et al.: The liver tumor segmentation benchmark (LiTS). arXiv preprint arXiv:1901.04056 (2019)
2. Chen, Z., et al.: Multi-modal masked autoencoders for medical vision-and-language pre-training. In: Wang, L., Dou, Q., Fletcher, P.T., Speidel, S., Li, S. (eds.) MICCAI 2022. LNCS, vol. 13435, pp. 679–689. Springer, Cham (2022). https://doi.org/10.1007/978-3-031-16443-9_65
3. Douillard, A., Chen, Y., Dapogny, A., Cord, M.: Plop: learning without forgetting for continual semantic segmentation. In: Proceedings of the IEEE/CVF Conference on Computer Vision and Pattern Recognition, pp. 4040–4050 (2021)
4. Hatamizadeh, A., Nath, V., Tang, Y., Yang, D., Roth, H.R., Xu, D.: Swin UNETR: swin transformers for semantic segmentation of brain tumors in MRI images. In: Crimi, A., Bakas, S. (eds.) BrainLes 2021. LNCS, vol. 12962, pp. 272–284. Springer, Cham (2022). https://doi.org/10.1007/978-3-031-08999-2_22
5. Huang, S.C., Shen, L., Lungren, M.P., Yeung, S.: Gloria: a multimodal global-local representation learning framework for label-efficient medical image recognition. In: Proceedings of the IEEE/CVF International Conference on Computer Vision, pp. 3942–3951 (2021)
6. Isensee, F., Jaeger, P.F., Kohl, S.A., Petersen, J., Maier-Hein, K.H.: nnU-Net: a self-configuring method for deep learning-based biomedical image segmentation. Nat. Methods **18**(2), 203–211 (2021)
7. Ji, Z., et al.: Continual segment: towards a single, unified and accessible continual segmentation model of 143 whole-body organs in CT scans. arXiv preprint arXiv:2302.00162 (2023)
8. Landman, B., Xu, Z., Igelsias, J., Styner, M., Langerak, T., Klein, A.: MICCAI multi-atlas labeling beyond the cranial vault-workshop and challenge. In: Proceedings of MICCAI Multi-Atlas Labeling Beyond Cranial Vault-Workshop Challenge, vol. 5, p. 12 (2015)
9. Langlotz, C.P., et al.: A roadmap for foundational research on artificial intelligence in medical imaging: from the 2018 NIH/RSNA/ACR/the academy workshop. Radiology **291**(3), 781–791 (2019)
10. Lewandowsky, S., Li, S.C.: Catastrophic interference in neural networks: causes, solutions, and data. In: Interference and Inhibition in Cognition, pp. 329–361. Elsevier (1995)
11. Li, Z., Hoiem, D.: Learning without forgetting. IEEE Trans. Pattern Anal. Mach. Intell. **40**(12), 2935–2947 (2017)
12. Liu, J., et al.: Clip-driven universal model for organ segmentation and tumor detection. In: Proceedings of the IEEE International Conference on Computer Vision (2023)
13. Liu, P., et al.: Learning incrementally to segment multiple organs in a CT image. In: Wang, L., Dou, Q., Fletcher, P.T., Speidel, S., Li, S. (eds.) MICCAI 2022. LNCS, vol. 13434, pp. 714–724. Springer, Cham (2022). https://doi.org/10.1007/978-3-031-16440-8_68
14. Liu, Y., Li, Y., Schiele, B., Sun, Q.: Online hyperparameter optimization for class-incremental learning. In: Thirty-Seventh AAAI Conference on Artificial Intelligence (2023)
15. Liu, Y., Schiele, B., Sun, Q.: RMM: reinforced memory management for class-incremental learning. In: Advances in Neural Information Processing Systems 34: Annual Conference on Neural Information Processing Systems 2021, NeurIPS 2021, 6–14 December 2021, pp. 3478–3490 (2021)

16. Liu, Z., et al.: Swin transformer: hierarchical vision transformer using shifted windows. In: Proceedings of the IEEE/CVF International Conference on Computer Vision, pp. 10012–10022 (2021)
17. Michieli, U., Zanuttigh, P.: Incremental learning techniques for semantic segmentation. In: Proceedings of the IEEE/CVF International Conference on Computer Vision Workshops (2019)
18. Ozdemir, F., Fuernstahl, P., Goksel, O.: Learn the new, keep the old: extending pretrained models with new anatomy and images. In: Frangi, A.F., Schnabel, J.A., Davatzikos, C., Alberola-López, C., Fichtinger, G. (eds.) MICCAI 2018. LNCS, vol. 11073, pp. 361–369. Springer, Cham (2018). https://doi.org/10.1007/978-3-030-00937-3_42
19. Ozdemir, F., Goksel, O.: Extending pretrained segmentation networks with additional anatomical structures. Int. J. Comput. Assist. Radiol. Surg. **14**(7), 1187–1195 (2019). https://doi.org/10.1007/s11548-019-01984-4
20. Qin, Z., Yi, H., Lao, Q., Li, K.: Medical image understanding with pretrained vision language models: a comprehensive study. arXiv preprint arXiv:2209.15517 (2022)
21. Qu, C., et al.: Annotating 8,000 abdominal CT volumes for multi-organ segmentation in three weeks. arXiv preprint arXiv:2305.09666 (2023)
22. Radford, A., et al.: Learning transferable visual models from natural language supervision. In: International Conference on Machine Learning, pp. 8748–8763. PMLR (2021)
23. Rebuffi, S.A., Kolesnikov, A., Sperl, G., Lampert, C.H.: ICARL: incremental classifier and representation learning. In: Proceedings of the IEEE Conference on Computer Vision and Pattern Recognition, pp. 2001–2010 (2017)
24. Wasserthal, J., Meyer, M., Breit, H.C., Cyriac, J., Yang, S., Segeroth, M.: Totalsegmentator: robust segmentation of 104 anatomical structures in CT images. arXiv preprint arXiv:2208.05868 (2022)
25. Xia, Y., et al.: The felix project: deep networks to detect pancreatic neoplasms. medRxiv (2022)
26. Zhang, Y., Jiang, H., Miura, Y., Manning, C.D., Langlotz, C.P.: Contrastive learning of medical visual representations from paired images and text. In: Machine Learning for Healthcare Conference, pp. 2–25. PMLR (2022)
27. Zhou, Z.: Towards annotation-efficient deep learning for computer-aided diagnosis. Ph.D. thesis, Arizona State University (2021)
28. Zhou, Z., Gotway, M.B., Liang, J.: Interpreting medical images. In: Cohen, T.A., Patel, V.L., Shortliffe, E.H. (eds.) Intelligent Systems in Medicine and Health, pp. 343–371. Springer, Cham (2022). https://doi.org/10.1007/978-3-031-09108-7_12

Incremental Learning for Heterogeneous Structure Segmentation in Brain Tumor MRI

Xiaofeng Liu[1(✉)], Helen A. Shih[2], Fangxu Xing[1], Emiliano Santarnecchi[1], Georges El Fakhri[1], and Jonghye Woo[1]

[1] Gordon Center for Medical Imaging, Department of Radiology, Massachusetts General Hospital and Harvard Medical School, Boston, MA 02114, USA
xliu61@mgh.harvard.edu

[2] Department of Radiation Oncology, Massachusetts General Hospital and Harvard Medical School, Boston, MA 02114, USA

Abstract. Deep learning (DL) models for segmenting various anatomical structures have achieved great success via a static DL model that is trained in a single source domain. Yet, the static DL model is likely to perform poorly in a continually evolving environment, requiring appropriate model updates. In an incremental learning setting, we would expect that well-trained static models are updated, following continually evolving target domain data—e.g., additional lesions or structures of interest—collected from different sites, without catastrophic forgetting. This, however, poses challenges, due to distribution shifts, additional structures not seen during the initial model training, and the absence of training data in a source domain. To address these challenges, in this work, we seek to progressively evolve an "off-the-shelf" trained segmentation model to diverse datasets with additional anatomical categories in a unified manner. Specifically, we first propose a divergence-aware dual-flow module with balanced rigidity and plasticity branches to decouple old and new tasks, which is guided by continuous batch renormalization. Then, a complementary pseudo-label training scheme with self-entropy regularized momentum MixUp decay is developed for adaptive network optimization. We evaluated our framework on a brain tumor segmentation task with continually changing target domains—i.e., new MRI scanners/modalities with incremental structures. Our framework was able to well retain the discriminability of previously learned structures, hence enabling the realistic life-long segmentation model extension along with the widespread accumulation of big medical data.

1 Introduction

Accurate segmentation of a variety of anatomical structures is a crucial prerequisite for subsequent diagnosis or treatment [28]. While recent advances in data-driven deep learning (DL) have achieved superior segmentation performance [29], the segmentation task is often constrained by the availability of costly pixel-wise labeled training datasets. In addition, even if static DL models are trained with

H. Greenspan et al. (Eds.): MICCAI 2023, LNCS 14221, pp. 46–56, 2023.
https://doi.org/10.1007/978-3-031-43895-0_5

extraordinarily large amounts of training datasets in a supervised learning manner [29], there exists a need for a segmentor to update a trained model with new data alongside incremental anatomical structures [24].

In real-world scenarios, clinical databases are often sequentially constructed from various clinical sites with varying imaging protocols [19–21,23]. As well, labeled anatomical structures are incrementally increased with additional lesions or new structures of interest, depending on study goals or clinical needs [18,27]. Furthermore, access to previously used data for training can be restricted, due to data privacy protocols [17,18]. Therefore, efficiently utilizing heterogeneous structure-incremental (HSI) learning is highly desired for clinical practice to develop a DL model that can be generalized well for different types of input data and varying structures involved. Straightforwardly fine-tuning DL models with either new structures [30] or heterogeneous data [17] in the absence of the data used for the initial model training, unfortunately, can easily overwrite previously learned knowledge, i.e., catastrophic forgetting [14,17,30].

At present, satisfactory methods applied in the realistic HSI setting are largely unavailable. *First*, recent structure-incremental works cannot deal with domain shift. Early attempts [27] simply used exemplar data in the previous stage. [5,18,30,33] combined a trained model prediction and a new class mask as a pseudo-label. However, predictions from the old model under a domain shift are likely to be unreliable [38]. The widely used pooled feature statistics consistency [5,30] is also not applicable for heterogeneous data, since the statistics are domain-specific [2]. In addition, a few works [13,25,34] proposed to increase the capacity of networks to avoid directly overwriting parameters that are entangled with old and new knowledge. However, the solutions cannot be domain adaptive. *Second*, from the perspective of continuous domain adaptation with the consistent class label, old exemplars have been used for the application of prostate MRI segmentation [32]. While Li et al. [17] further proposed to recover the missing old stage data with an additional generative model, hallucinating realistic data, given only the trained model itself, is a highly challenging task [31] and may lead to sensitive information leakage [35]. *Third*, while, for natural image classification, Kundu et al. [16] updated the model for class-incremental unsupervised domain adaption, its class prototype is not applicable for segmentation.

In this work, we propose a unified HSI segmentor evolving framework with a divergence-aware decoupled dual-flow (D^{3}F) module, which is adaptively optimized via HSI pseudo-label distillation using a momentum MixUp decay (MMD) scheme. To explicitly avoid the overwriting of previously learned parameters, our D^{3}F follows a "divide-and-conquer" strategy to balance the old and new tasks with a fixed rigidity branch and a compensated learnable plasticity branch, which is guided by our novel divergence-aware continuous batch renormalization (cBRN). The complementary knowledge can be flexibly integrated with the model re-parameterization [4]. Our additional parameters are constant in training, and 0 in testing. Then, the flexible D^{3}F module is trained following the knowledge distillation with novel HSI pseudo-labels. Specifically, inspired by the self-knowledge distillation [15] and self-training [38] that utilize the previous prediction for better generalization, we adaptively construct the HSI pseudo-label with an MMD scheme to smoothly adjust the contribution of potential noisy old model predictions on heterogeneous data and progressively learned new model predictions along with the training. In addition, unsupervised self-entropy minimization is added to further enhance performance.

Our main contributions can be summarized as follow:

- To our knowledge, this is the first attempt at realistic HSI segmentation with both incremental structures of interest and diverse domains.
- We propose a divergence-aware decoupled dual-flow module guided by our novel continuous batch renormalization (cBRN) for alleviating the catastrophic forgetting under domain shift scenarios.
- The adaptively constructed HSI pseudo-label with self-training is developed for efficient HSI knowledge distillation.

We evaluated our framework on anatomical structure segmentation tasks from different types of MRI data collected from multiple sites. Our HSI scheme demonstrated superior performance in segmenting all structures with diverse data distributions, surpassing conventional class-incremental methods without considering data shift, by a large margin.

2 Methodology

For the segmentation model under incremental structures of interest and domain shift scenarios, we are given an off-the-shelf segmentor $f_{\theta^0}: \mathcal{X}^0 \rightarrow \mathcal{Y}^0$ parameterized with θ^0, which has been trained with the data $\{x_n^0, y_n^0\}_{n=1}^{N^0}$ in an initial source domain $\mathcal{D}^0 = \{\mathcal{X}^0, \mathcal{Y}^0\}$, where $x_n^0 \in \mathbb{R}^{H\times W}$ and $y_n^0 \in \mathbb{R}^{H\times W}$ are the paired image slice and its segmentation mask with the height of H and width of W, respectively. There are T consecutive evolving stages with heterogeneous target domains $\mathcal{D}^t = \{\mathcal{X}^t, \mathcal{S}^t\}_{t=1}^T$, each with the paired slice set $\{x_n^t\}_{n=1}^{N^t} \in \mathcal{X}^t$ and the current stage label set $\{s_n^t\}_{n=1}^{N^t} \in \mathcal{S}^t$, where $x_n^t, s_n^t \in \mathbb{R}^{H\times W}$. Due to heterogeneous domain shifts, $\mathcal{X}^t$ from different sites or modalities follows diverse distributions across all T stages. Due to incremental anatomical structures, the overall label space, across the previous t stages, $\mathcal{Y}^t$ is expanded from $\mathcal{Y}^{t-1}$ with the additional annotated structures $\mathcal{S}^t$ in stage t, i.e., $\mathcal{Y}^t = \mathcal{Y}^{t-1} \cup \mathcal{S}^t = \mathcal{Y}^0 \cup \mathcal{S}^1 \cdots \cup \mathcal{S}^t$. We are targeting to learn $f_{\theta^T}: \{\mathcal{X}^t\}_{t=1}^T \rightarrow \mathcal{Y}^T$ that performs well on all $\{\mathcal{X}^t\}_{t=1}^T$ for delineating all of the structures $\mathcal{Y}^T$ seen in T stages.

2.1 cBRN Guided Divergence-Aware Decoupled Dual-Flow

To alleviate the forgetting through parameter overwriting, caused by both new structures and data shift, we propose a D^{3}F module for flexible decoupling and integration of old and new knowledge.

Specifically, we duplicate the convolution in each layer initialized with the previous model $f_{\theta^{t-1}}$ to form two branches as in [13,25,34]. The first *rigidity* branch $f_{\theta^t}^r$ is fixed at the stage t to keep the old knowledge we have learned. In contrast, the extended *plasticity* branch $f_{\theta^t}^p$ is expected to be adaptively updated

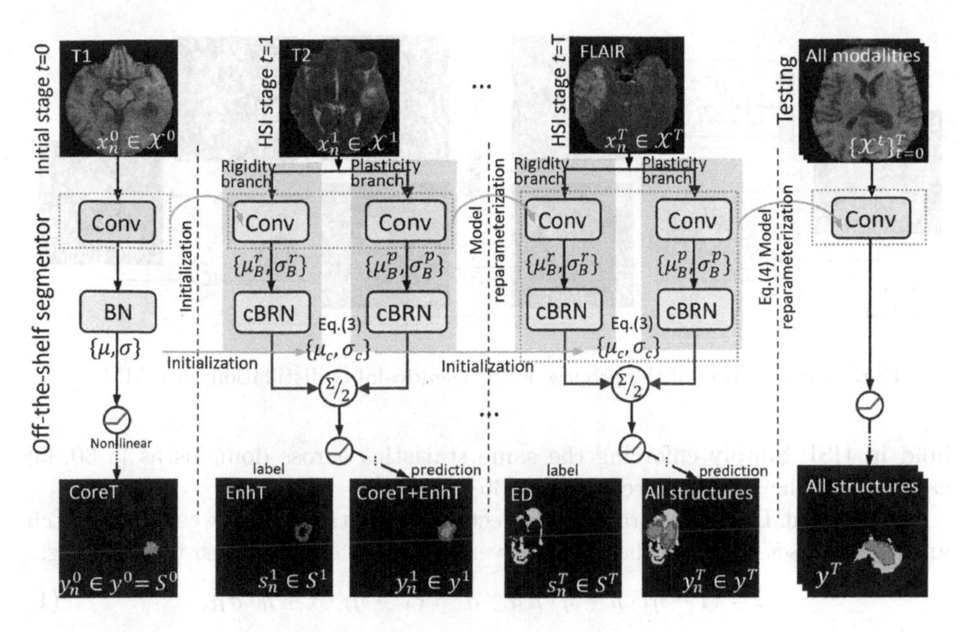

Fig. 1. Illustration of one layer in our proposed divergence-aware decoupled dual-flow module guided with cBRN for our cross-MR-modality HSI task, i.e., subject-independent (CoreT with T1) → (EnhT with T2) → (ED with FLAIR). Notably, we do not require the dual-flow or cBRN, for the initial segmentor.

to learn the new task in $\mathcal{D}^t$. At the end of current training stage t, we can flexibly integrate the convolutions in two branches, i.e., $\{W_t^r, b_t^r\}$ and $\{W_t^p, b_t^p\}$ to $\{W_{t+1}^r = \frac{W_t^r+W_t^p}{2}, b_{t+1}^r = \frac{b_t^r+b_t^p}{2}\}$ with the model re-parameterization [4]. In fact, the dual-flow model can be regarded as an implicit ensemble scheme [9] to integrate multiple sub-modules with a different focus. In addition, as demonstrated in [6], the fixed modules will regularize the learnable modules to act as the fixed one. Thus, the plasticity modules can also be implicitly encouraged to keep the previous knowledge along with its HSI learning.

However, under the domain shift, it can be sub-optimal to directly average the parameters, since $f_{\theta^t}^r$ may not perform well to predict $\mathcal{Y}^{t-1}$ on $\mathcal{X}^t$. It has been demonstrated that batch statistics adaptation plays an important role in domain generalizable model training [22]. Therefore, we propose a continual batch renormalization (cBRN) to mitigate the feature statistics divergence between each training batch at a specific stage and the life-long global data distribution.

Of note, as a default block in the modern convolutional neural networks (CNN) [8,37], batch normalization (BN) [11] normalizes the input feature of each CNN channel $z \in \mathbb{R}^{H_c \times W_c}$ with its batch-wise statistics, e.g., mean μ_B and standard deviation σ_B, and learnable scaling and shifting factors $\{\gamma, \beta\}$ as $\tilde{z}_i = \frac{z_i - \mu_B}{\sigma_B} \cdot \gamma + \beta$, where i indexes the spatial position in $\mathbb{R}^{H_c \times W_c}$. BN assumes that the same mini-batch training and testing distribution [10], which does not

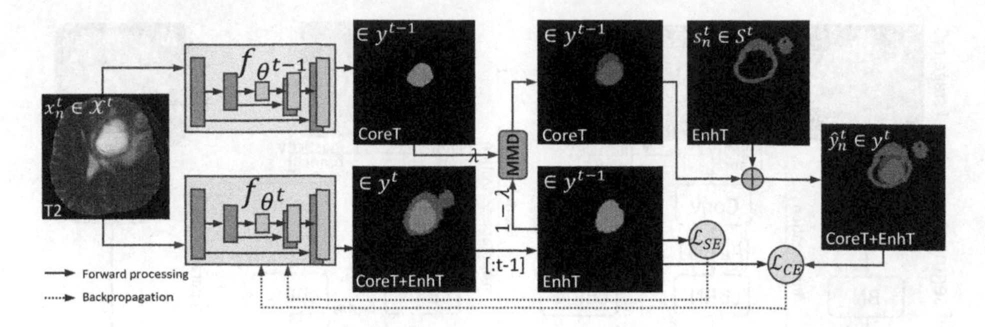

Fig. 2. Illustration of the proposed HSI pseudo-label distillation with MMD

hold in HSI. Simply enforcing the same statistics across domains as [5,30,33] can weaken the model expressiveness [36].

The recent BRN [10] proposes to rectify the data shift between each batch and the dataset by using the moving average μ and σ along with the training:

$$\mu = (1-\eta)\cdot\mu + \eta\cdot\mu_B, \quad \sigma = (1-\eta)\cdot\sigma + \eta\cdot\sigma_B, \tag{1}$$

where $\eta \in [0,1]$ is applied to balance the global statistics and the current batch. In addition, $\gamma = \frac{\sigma_B}{\sigma}$ and $\beta = \frac{\mu_B-\mu}{\sigma}$ are used in both training and testing. Therefore, BRN renormalizes $\tilde{z}_i = \frac{z_i-\mu}{\sigma}$ to highlight the dependency on the global statistics $\{\mu,\sigma\}$ in training for a more generalizable model, while limited to the static learning.

In this work, we further explore the potential of BRN in the continuously evolving HSI task to be general for all of domains involved. Specifically, we extend BRN to cBRN across multiple consecutive stages by updating $\{\mu_c,\sigma_c\}$ along with all stages of training, which is transferred as shown in Fig. 1. The conventional BN also inherits $\{\mu,\sigma\}$ for testing, while not being used in training [11]. At the stage t, μ_c and σ_c are succeeded from $t-1$ stage, and are updated with the current batch-wise $\{\mu_B^r,\sigma_B^r\}$ and $\{\mu_B^p,\sigma_B^p\}$ in rigidity and plasticity branches:

$$\mu_c = (1-\eta)\cdot\mu_c + \eta\cdot\frac{1}{2}\{\mu_B^r+\mu_B^p\}, \quad \sigma_c = (1-\eta)\cdot\sigma_c + \eta\cdot\frac{1}{2}\{\sigma_B^r+\sigma_B^p\}. \tag{2}$$

For testing, the two branches in final model f_{θ^T} can be merged for the lightweight implementation:

$$\tilde{z} = \frac{W_T^r z + b_T^r + \mu_c}{2\sigma_c} + \frac{W_T^p z + b_T^p + \mu_c}{2\sigma_c} = \frac{W_T^r + W_T^p}{2\sigma_c}z + \frac{b_T^r + b_T^p + 2\mu_c}{2\sigma_c} = \hat{W}z + \hat{b}. \tag{3}$$

Therefore, f_θ^T does not introduce additional parameters for deployment (Fig. 2).

2.2 HSI Pseudo-label Distillation with Momentum MixUp Decay

The training of our developed f_{θ^t} with D^{3}F is supervised with the previous model $f_{\theta^{t-1}}$ and current stage data $\{x_n^t, s_n^t\}_{n=1}^{N^t}$. In conventional class incremental learning, the knowledge distillation [31] is widely used to construct the combined label $y_n^t \in \mathbb{R}^{H\times W}$ by adding s_n^t and the prediction of $f_{\theta^{t-1}}(x_n^t)$. Then, f_{θ^t} can be optimized by the training pairs of $\{x_n^t, y_n^t\}_{n=1}^{N^t}$. However, with heterogeneous data in different stages, $f_{\theta^{t-1}}(x_n^t)$ can be highly unreliable. Simply using it as ground truth cannot guide the correct knowledge transfer.

In this work, we construct a complementary pseudo-label $\hat{y}_n^t \in \mathbb{R}^{H\times W}$ with a MixUp decay scheme to adaptively exploit the knowledge in the old segmentor for the progressively learned new segmentor. In the initial training epochs, $f_{\theta^{t-1}}$ could be a more reliable supervision signal, while we would expect f_{θ^t} can learn to perform better on predicting $\mathcal{Y}^{t-1}$. Of note, even with the rigidity branch, the integrated network can be largely distracted by the plasticity branch in the initial epochs. Therefore, we propose to dynamically adjust their importance in constructing pseudo-label along with the training progress. Specifically, we MixUp the predictions of $f_{\theta^{t-1}}$ and f_{θ^t} w.r.t. $\mathcal{Y}^{t-1}$, i.e., $f_{\theta^t}(\cdot)[:t-1]$, and control their pixel-wise proportion for the pseudo-label $\hat{y}_n^t$ with MMD:

$$\hat{y}_{n:i}^t = \{\lambda f_{\theta^{t-1}}(x_{n:i}^t) + (1-\lambda) f_{\theta^t}(x_{n:i}^t)[:t-1]\} \cup s_{n:i}^t, \quad \lambda = \lambda^0 \exp(-I), \tag{4}$$

where i indexes each pixel, and λ is the adaptation momentum factor with the exponential decay of iteration I. λ^0 is the initial weight of $f_{\theta^{t-1}}(x_{n:i}^t)$, which is empirically set to 1 to constrain $\lambda \in (0,1]$. Therefore, the weight of old model prediction can be smoothly decreased along with the training, and $f_{\theta^t}(x_{n:i}^t)$ gradually represents the target data for the old classes in $[:t-1]$. Of note, we have ground-truth of new structure $s_{n:i}^t$ under HSI scenarios [5,18,30,33]. We calculate the cross-entropy loss $\mathcal{L}_{CE}$ with the pseudo-label $\hat{y}_{n:i}^t$ as self-training [15,38].

In addition to the old knowledge inherited in $f_{\theta^{t-1}}$, we propose to explore unsupervised learning protocols to stabilize the initial training. We adopt the widely used self-entropy (SE) minimization [7] as a simple add-on training objective. Specifically, we have the slice-level segmentation SE, which is the averaged entropy of the pixel-wise softmax prediction as $\mathcal{L}_{SE} = \mathbb{E}_i\{-f_{\theta^t}(x_{n:i}^t)\log f_{\theta^t}(x_{n:i}^t)\}$. In training, the overall optimization loss is formulated as follows:

$$\mathcal{L} = \mathcal{L}_{CE}(\hat{y}_{n:i}^t, f_{\theta^t}(x_{n:i}^t)) + \alpha \mathcal{L}_{SE}(f_{\theta^t}(x_{n:i}^t)), \quad \alpha = \frac{I_{max} - I}{I_{max}} \alpha^0, \tag{5}$$

where α is used to balance our HSI distillation and SE minimization terms, and I_{max} is the scheduled iteration. Of note, strictly minimizing the SE can result in a trivial solution of always predicting a one-hot distribution [7], and a linear decreasing of α is usually applied, where λ^0 and α^0 are reset in each stage.

3 Experiments and Results

We carried out two evaluation settings using the BraTS2018 database [1], including cross-subset (relatively small domain shift) and cross-modality (relatively large domain shift) tasks. The BraTS2018 database is a continually evolving database [1] with a total of 285 glioblastoma or low-grade gliomas subjects,

Table 1. Numerical comparisons and ablation studies of the cross-subset brain tumor HSI segmentation task

Method	Data shift consideration	Dice similarity coefficient (DSC) [%] ↑				Hausdorff distance (HD)[mm] ↓			
		Mean	CoreT	EnhT	ED	Mean	CoreT	EnhT	ED
PLOP [5]	×	59.83 ± 0.131	45.50	57.39	76.59	19.2 ± 0.14	22.0	19.8	15.9
MargExcIL [18]	×	60.49 ± 0.127	48.37	56.28	76.81	18.9 ± 0.11	21.4	19.8	15.5
UCD [30]	×	61.84 ± 0.129	49.23	58.81	77.48	19.0 ± 0.15	21.8	19.4	15.7
HSI-MMD	√	66.87 ± 0.126	59.42	61.26	79.93	16.8 ± 0.13	18.5	17.8	14.2
HSI-D^{3}F	√	67.18 ± 0.118	60.18	63.09	78.26	16.7 ± 0.14	18.0	17.5	14.5
HSI-cBRN	√	68.07 ± 0.121	61.52	63.45	79.25	16.3 ± 0.14	17.8	17.3	13.8
HSI	√	**69.44 ± 0.119**	**63.79**	**64.71**	**79.81**	**15.7 ± 0.12**	**16.7**	**16.9**	**13.6**
Joint Static	√(upper bound)	73.98 ± 0.117	71.14	68.35	82.46	15.0 ± 0.13	15.7	16.2	13.2

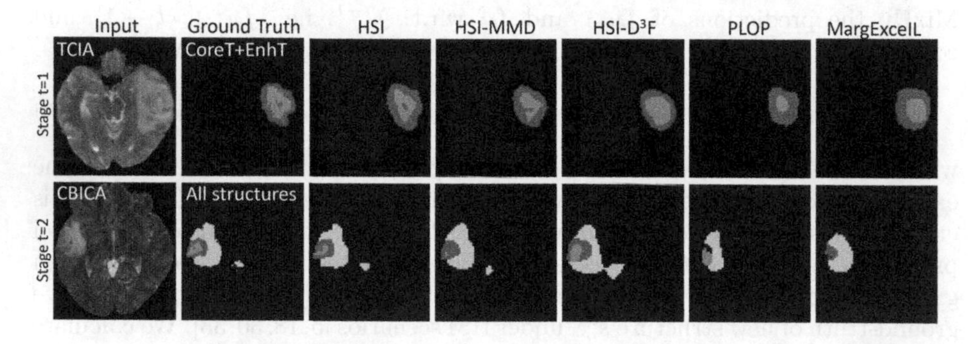

Fig. 3. Segmentation examples in $t = 1$ and $t = 2$ in the cross-subset brain tumor HSI segmentation task.

comprising three consecutive subsets, i.e., 30 subjects from BraTS2013 [26], 167 subjects from TCIA [3], and 88 subjects from CBICA [1]. Notably, these three subsets were collected from different clinical sites, vendors, or populations [1]. Each subject has T1, T1ce, T2, and FLAIR MRI volumes with voxel-wise labels for the tumor core (CoreT), the enhancing tumor (EnhT), and the edema (ED).

We incrementally learned CoreT, EnhT, and ED structures throughout three consecutive stages, each following different data distributions. We used subject-independent 7/1/2 split for training, validation, and testing. For a fair comparison, we adopted the ResNet-based 2D nnU-Net backbone with BN as in [12] for all of the methods and all stages used in this work.

3.1 Cross-Subset Structure Incremental Evolving

In our cross-subset setting, three structures were sequentially learned across three stages: (CoreT with BraTS2013) → (EnhT with TCIA) → (ED with CBICA). Of note, we used a CoreT segmentator trained with BraTS2013 as our off-the-shelf segmentor in $t = 0$. Testing involved all subsets and anatomical structures. We compared our framework with the three typical structure-incremental (SI-only) segmentation methods, e.g., PLOP [5], MargExcIL [18],

Table 2. Numerical comparisons and ablation studies of the cross-modality brain tumor HSI segmentation task

Method	Data shift consideration	Dice similarity coefficient (DSC) [%] ↑				Hausdorff distance (HD)[mm] ↓			
		Mean	CoreT	EnhT	ED	Mean	CoreT	EnhT	ED
PLOP [5]	×	39.58 ± 0.231	13.84	38.93	65.98	30.7 ± 0.26	48.1	25.4	18.7
MargExcIL [18]	×	42.84 ± 0.189	19.56	41.56	67.40	29.1 ± 0.28	46.7	22.1	18.6
UCD [30]	×	44.67 ± 0.214	21.39	45.28	67.35	29.4 ± 0.32	46.2	23.6	18.4
HSI-MMD	√	59.81 ± 0.207	51.63	53.82	73.97	19.4 ± 0.26	21.6	20.5	16.2
HSI-D^{3}F	√	60.81 ± 0.195	53.87	55.42	73.15	19.2 ± 0.21	21.4	19.9	16.2
HSI-cBRN	√	61.87 ± 0.180	54.90	56.62	74.08	18.5 ± 0.25	20.1	19.5	16.0
HSI	√	**64.15 ± 0.205**	**58.11**	**59.51**	**74.83**	**17.7 ± 0.29**	**18.9**	**18.6**	**15.8**
Joint Static	√ (upper bound)	70.64 ± 0.184	67.48	65.75	78.68	16.7 ± 0.26	17.2	17.8	15.1

and UCD [30], which cannot address the heterogeneous data across stages. As tabulated in Table 1, PLOP [5] with additional feature statistic constraints has lower performance than MargExcIL [18], since the feature statistic consistency was not held in HSI scenarios. Of note, the domain-incremental methods [17,32] cannot handle the changing output space. Our proposed HSI framework outperformed SI-only methods [5,18,30] with respect to both DSC and HD, by a large margin. For the anatomical structure CoreT learned in $t = 0$, the difference between our HSI and these SI-only methods was larger than 10% DSC, which indicates the data shift related forgetting lead to a more severe performance drop in the early stages. We set $\eta = 0.01$ and $\alpha^0 = 10$ according to the sensitivity study in the supplementary material.

For the ablation study, we denote HSI-D^{3}F as our HSI without the D^{3}F module, simply fine-tuning the model parameters. HSI-cBRN used dual-flow to avoid direct overwriting, while the model was not guided by cBRN for more generalized prediction on heterogeneous data. As shown in Table 1, both the dual-flow and cBRN improve the performance. Notably, the dual-flow model with flexible re-parameterization was able to alleviate the overwriting, while our cBRN was developed to deal with heterogeneous data. In addition, HSI-MMD indicates our HSI without the momentum MixUp decay in pseudo-label construction, i.e., simply regarding the prediction of $f_{\theta^{t-1}}(x^t)$ is ground truth for $\mathcal{Y}^{t-1}$. However, $f_{\theta^{t-1}}(x^t)$ can be quite noisy, due to the low quantification performance of early stage structures, which can be aggravated in the case of the long-term evolving scenario. Of note, the pseudo-label construction is necessary as in [5,18,30]. We also provide the qualitative comparison with SI-only methods and ablation studies in Fig. 3.

3.2 Cross-Modality Structure Incremental Evolving

In our cross-modality setting, three structures were sequentially learned across three stages: (CoreT with T1) → (EnhT with T2) → (ED with T2 FLAIR). Of note, we used the CoreT segmentator trained with T1 modality as our off-the-shelf segmentor in $t = 0$. Testing involved all MRI modalities and all structures. With the hyperparameter validation, we empirically set $\eta = 0.01$ and $\alpha^0 = 10$.

In Table 2, we provide quantitative evaluation results. We can see that our HSI framework outperformed SI-only methods [5,18,30] consistently. The improvement can be even larger, compared with the cross-subset task, since we have much more diverse input data in the cross-modality setting. Catastrophic forgetting can be severe, when we use SI-only method for predicting early stage structures, e.g., CoreT. We also provide the ablation study with respect to D^3F, cBRN, and MMD in Table 2. The inferior performance of HSI-D^3F/cBRN/MMD demonstrates the effectiveness of these modules for mitigating domain shifts.

4 Conclusion

This work proposed an HSI framework under a clinically meaningful scenario, in which clinical databases are sequentially constructed from different sites/imaging protocols with new labels. To alleviate the catastrophic forgetting alongside continuously varying structures and data shifts, our HSI resorted to a D^3F module for learning and integrating old and new knowledge nimbly. In doing so, we were able to achieve divergence awareness with our cBRN-guided model adaptation for all the data involved. Our framework was optimized with a self-entropy regularized HSI pseudo-label distillation scheme with MMD to efficiently utilize the previous model in different types of MRI data. Our framework demonstrated superior segmentation performance in learning new anatomical structures from cross-subset/modality MRI data. It was experimentally shown that a large improvement in learning anatomic structures was observed.

Acknowledgements. This work is supported by NIH R01DC018511, R01DE027989, and P41EB022544. The authors would like to thank Dr. Jonghyun Choi for his valuable insights and helpful discussions.

References

1. Bakas, S., et al.: Identifying the best machine learning algorithms for brain tumor segmentation, progression assessment, and overall survival prediction in the brats challenge. arXiv:1811.02629 (2018)
2. Chang, W.G., You, T., Seo, S., Kwak, S., Han, B.: Domain-specific batch normalization for unsupervised domain adaptation. In: CVPR, pp. 7354–7362 (2019)
3. Clark, K., et al.: The cancer imaging archive (TCIA): maintaining and operating a public information repository. J. Digit. Imaging **26**(6), 1045–1057 (2013)
4. Ding, X., Zhang, X., Ma, N., Han, J., Ding, G., Sun, J.: RepVGG: making VGG-style convnets great again. In: CVPR, pp. 13733–13742 (2021)
5. Douillard, A., Chen, Y., Dapogny, A., Cord, M.: PLOP: learning without forgetting for continual semantic segmentation. In: CVPR, pp. 4040–4050 (2021)
6. Fu, S., Li, Z., Liu, Z., Yang, X.: Interactive knowledge distillation for image classification. Neurocomputing **449**, 411–421 (2021)

7. Grandvalet, Y., Bengio, Y.: Semi-supervised learning by entropy minimization. In: NeurIPS (2005)
8. He, K., Zhang, X., Ren, S., Sun, J.: Deep residual learning for image recognition. In: CVPR (2016)
9. Huang, G., Sun, Yu., Liu, Z., Sedra, D., Weinberger, K.Q.: Deep networks with stochastic depth. In: Leibe, B., Matas, J., Sebe, N., Welling, M. (eds.) ECCV 2016. LNCS, vol. 9908, pp. 646–661. Springer, Cham (2016). https://doi.org/10.1007/978-3-319-46493-0_39
10. Ioffe, S.: Batch renormalization: towards reducing minibatch dependence in batch-normalized models. In: NeurIPS, vol. 30 (2017)
11. Ioffe, S., Szegedy, C.: Batch normalization: accelerating deep network training by reducing internal covariate shift. In: ICML, pp. 448–456. PMLR (2015)
12. Isensee, F., Jaeger, P.F., Kohl, S.A., Petersen, J., Maier-Hein, K.H.: nnU-Net: a self-configuring method for deep learning-based biomedical image segmentation. Nat. Methods **18**(2), 203–211 (2021)
13. Kanakis, M., Bruggemann, D., Saha, S., Georgoulis, S., Obukhov, A., Van Gool, L.: Reparameterizing convolutions for incremental multi-task learning without task interference. In: Vedaldi, A., Bischof, H., Brox, T., Frahm, J.-M. (eds.) ECCV 2020. LNCS, vol. 12365, pp. 689–707. Springer, Cham (2020). https://doi.org/10.1007/978-3-030-58565-5_41
14. Kim, D., Bae, J., Jo, Y., Choi, J.: Incremental learning with maximum entropy regularization: Rethinking forgetting and intransigence. arXiv:1902.00829 (2019)
15. Kim, K., Ji, B., Yoon, D., Hwang, S.: Self-knowledge distillation: a simple way for better generalization. arXiv:2006.12000 (2020)
16. Kundu, J.N., Venkatesh, R.M., Venkat, N., Revanur, A., Babu, R.V.: Class-incremental domain adaptation. In: Vedaldi, A., Bischof, H., Brox, T., Frahm, J.-M. (eds.) ECCV 2020. LNCS, vol. 12358, pp. 53–69. Springer, Cham (2020). https://doi.org/10.1007/978-3-030-58601-0_4
17. Li, K., Yu, L., Heng, P.A.: Domain-incremental cardiac image segmentation with style-oriented replay and domain-sensitive feature whitening. TMI **42**(3), 570–581 (2022)
18. Liu, P., et al.: Learning incrementally to segment multiple organs in a CT image. In: Wang, L., Dou, Q., Fletcher, P.T., Speidel, S., Li, S. (eds.) MICCAI 2022. LNCS, vol. 13434, pp. 714–724. Springer, Cham (2022). https://doi.org/10.1007/978-3-031-16440-8_68
19. Liu, X., et al.: Attentive continuous generative self-training for unsupervised domain adaptive medical image translation. Med. Image Anal. (2023)
20. Liu, X., Xing, F., El Fakhri, G., Woo, J.: Memory consistent unsupervised off-the-shelf model adaptation for source-relaxed medical image segmentation. Med. Image Anal. **83**, 102641 (2023)
21. Liu, X., et al.: Act: Semi-supervised domain-adaptive medical image segmentation with asymmetric co-training. In: Wang, L., Dou, Q., Fletcher, P.T., Speidel, S., Li, S. (eds.) MICCAI 2022. LNCS, vol. 13435, pp. 66–76. Springer, Cham (2022). https://doi.org/10.1007/978-3-031-16443-9_7

22. Liu, X., Xing, F., Yang, C., El Fakhri, G., Woo, J.: Adapting off-the-shelf source segmenter for target medical image segmentation. In: de Bruijne, M., et al. (eds.) MICCAI 2021. LNCS, vol. 12902, pp. 549–559. Springer, Cham (2021). https://doi.org/10.1007/978-3-030-87196-3_51
23. Liu, X., et al.: Subtype-aware dynamic unsupervised domain adaptation. IEEE TNNLS (2022)
24. Liu, X., et al.: Deep unsupervised domain adaptation: a review of recent advances and perspectives. APSIPA Trans. Signal Inf. Process. **11**(1) (2022)
25. Liu, Y., Schiele, B., Sun, Q.: Adaptive aggregation networks for class-incremental learning. In: CVPR, pp. 2544–2553 (2021)
26. Menze, B.H., et al.: The multimodal brain tumor image segmentation benchmark (BRATS). TMI **34**(10), 1993–2024 (2014)
27. Ozdemir, F., Fuernstahl, P., Goksel, O.: Learn the new, keep the old: extending pretrained models with new anatomy and images. In: Frangi, A.F., Schnabel, J.A., Davatzikos, C., Alberola-López, C., Fichtinger, G. (eds.) MICCAI 2018. LNCS, vol. 11073, pp. 361–369. Springer, Cham (2018). https://doi.org/10.1007/978-3-030-00937-3_42
28. Shusharina, N., Söderberg, J., Edmunds, D., Löfman, F., Shih, H., Bortfeld, T.: Automated delineation of the clinical target volume using anatomically constrained 3D expansion of the gross tumor volume. Radiot. Oncol. **146**, 37–43 (2020)
29. Tajbakhsh, N., Jeyaseelan, L., Li, Q., Chiang, J.N., Wu, Z., Ding, X.: Embracing imperfect datasets: a review of deep learning solutions for medical image segmentation. Med. Image Anal. **63**, 101693 (2020)
30. Yang, G., et al.: Uncertainty-aware contrastive distillation for incremental semantic segmentation. IEEE Trans. Pattern Anal. Mach. Intell. **45**(2), 2567–2581 (2022)
31. Yin, H., et al.: Dreaming to distill: Data-free knowledge transfer via deepinversion. In: CVPR, pp. 8715–8724 (2020)
32. You, C., et al.: Incremental learning meets transfer learning: application to multi-site prostate MRI segmentation. arXiv:2206.01369 (2022)
33. Yu, L., Liu, X., Van de Weijer, J.: Self-training for class-incremental semantic segmentation. IEEE Trans. Neural Netw. Learn. Syst. (2022)
34. Zhang, C.B., Xiao, J.W., Liu, X., Chen, Y.C., Cheng, M.M.: Representation compensation networks for continual semantic segmentation. In: CVPR (2022)
35. Zhang, H., Zhang, Y., Jia, K., Zhang, L.: Unsupervised domain adaptation of black-box source models. arXiv:2101.02839 (2021)
36. Zhang, J., Qi, L., Shi, Y., Gao, Y.: Generalizable semantic segmentation via model-agnostic learning and target-specific normalization. arXiv:2003.12296 (2020)
37. Zhou, X.Y., Yang, G.Z.: Normalization in training u-net for 2-D biomedical semantic segmentation. IEEE Robot. Autom. Lett. **4**(2), 1792–1799 (2019)
38. Zou, Y., Yu, Z., Liu, X., Kumar, B., Wang, J.: Confidence regularized self-training. In: ICCV, pp. 5982–5991 (2019)

PLD-AL: Pseudo-label Divergence-Based Active Learning in Carotid Intima-Media Segmentation for Ultrasound Images

Yucheng Tang[1,2], Yipeng Hu[3], Jing Li[2], Hu Lin[4], Xiang Xu[1], Ke Huang[4(✉)], and Hongxiang Lin[2(✉)]

[1] School of Mathematical Sciences, Zhejiang University, Hangzhou, China
[2] Zhejiang Lab, Hangzhou, China
hxlin@zhejianglab.edu.cn
[3] Centre for Medical Image Computing and Wellcome/EPSRC Centre for Interventional & Surgical Sciences, University College London, London, UK
[4] Department of Endocrinology, Children's Hospital, Zhejiang University School of Medicine, Hangzhou, China
kehuang@zju.edu.cn

Abstract. Segmentation of the carotid intima-media (CIM) offers more precise morphological evidence for obesity and atherosclerotic disease compared to the method that measures its thickness and roughness during routine ultrasound scans. Although advanced deep learning technology has shown promise in enabling automatic and accurate medical image segmentation, the lack of a large quantity of high-quality CIM labels may hinder the model training process. Active learning (AL) tackles this issue by iteratively annotating the subset whose labels contribute the most to the training performance at each iteration. However, this approach substantially relies on the expert's experience, particularly when addressing ambiguous CIM boundaries that may be present in real-world ultrasound images. Our proposed approach, called pseudo-label divergence-based active learning (PLD-AL), aims to train segmentation models using a gradually enlarged and refined labeled pool. The approach has an outer and an inner loops: The outer loop calculates the Kullback-Leibler (KL) divergence of predictive pseudo-labels related to two consecutive AL iterations. It determines which portion of the unlabeled pool should be annotated by an expert. The inner loop trains two networks: The student network is fully trained on the current labeled pool, while the teacher network is weighted upon itself and the student one, ultimately refining the labeled pool. We evaluated our approach using both the Carotid Ultrasound Boundary Study dataset and an in-house dataset from Children's Hospital, Zhejiang University School of Medicine. Our results demonstrate that our approach outperforms state-of-the-art AL approaches. Furthermore, the visualization results show that our approach less overestimates the CIM area than the rest methods, especially for severely ambiguous ultrasound images at the thickness direction.

Y. Tang—This work was performed when Yucheng Tang was visiting Zhejiang Lab as an intern.

H. Greenspan et al. (Eds.): MICCAI 2023, LNCS 14221, pp. 57–67, 2023.
https://doi.org/10.1007/978-3-031-43895-0_6

Keywords: Carotid intima-media complex · active learning · image segmentation

1 Introduction

Carotid intima-media (CIM) segmentation has been widely applied in clinical practice, providing a diagnostic basis for atherosclerotic disease (one of the complications of obesity). To identify the contour of the intima-media, i.e., the structure between the lumen-intima (LI) and the media-adventitia (MA), one of the available solutions is deep learning-based medical image segmentation for CIM. Currently, this CIM segmentation approach faces the challenges of lack of large-quantity images, high-quality labels from ultrasound experts, and a mixture of clear and ambiguous CIM areas in carotid ultrasound images.

Semi-supervised learning recently applies novel frameworks to a general segmentation task [1–4]. In particular, the combination of consistency regularization and pseudo-labeling utilizes unlabeled data to partially address the lack-of-label issue [5]. A different strategy to efficiently utilize labeling effort is active learning (AL), which can iteratively select a subset of unlabeled data for annotation by experts, but still reach a model performance otherwise requiring a much larger training set. AL has been widely applied to image classification [6–8], semantic segmentation [9,10] and medical image segmentation [11,12]. These methods have effectively improved accuracy through experts' involvement. However, carotid ultrasound images are user-end protocol dependent, and with high variability in quality, real-world labels on ultrasound images generally share the same characteristics in high variability. Therefore, after testing several state-of-the-art AL methods, we would like to incorporate methodologies from semi-supervised learning designed to extract predictive information from unlabeled data, and between labeled and unlabeled data, for AL.

In this work, we propose pseudo-label divergence-based active learning (PLD-AL) to obtain accurate CIM segmentation contributing to the clinical diagnosis of obesity and atherosclerotic disease. As shown in Fig. 1, unlike the conventional AL framework that utilizes one machine learning model, PLD-AL is composed by two networks: the student network is fully trained on the current labeled pool, and the teacher network is weighted upon previous itself and the student one. We use divergence, which measures the distance between two model predictions, to select data for annotation. Furthermore, we use the teacher network to refine the labels to reduce the noise of labels and improve the effectiveness of the next network optimization stage.

Our contributions are as follows: we propose PLD-AL, which aims to train segmentation models using a gradually enlarged and refined labeled pool. First, we automatically select and annotate large divergence data between the current and previous AI models, facilitating fast convergence of the AL model to most sound data in the unlabeled pool. Second, we propose a strategy to refine the labels in the labeled pool alternatingly with the proposed label-divergence-based AL algorithm, which improves the robustness compared to the conventional AL

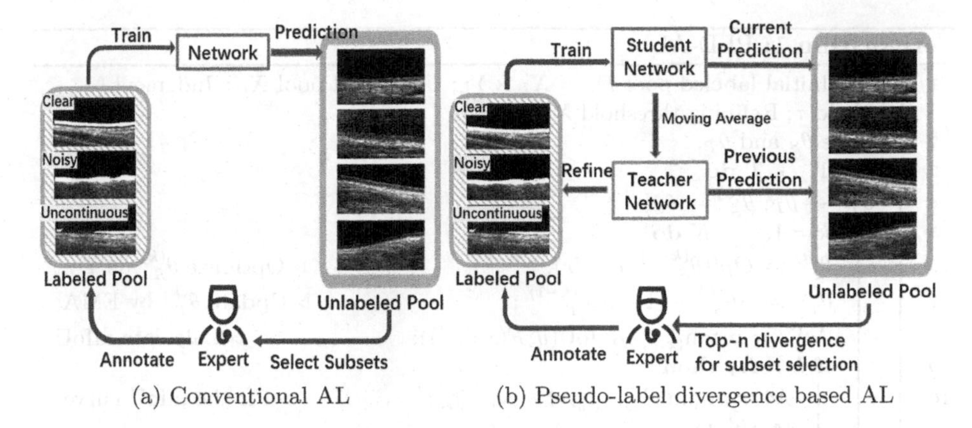

(a) Conventional AL (b) Pseudo-label divergence based AL

Fig. 1. (a) Conventional AL that trains a machine learning model to select an unlabeled subset for an expert to annotate. (b) We propose a novel AL framework to progressively annotate data by selecting top-n largest divergence between student and teacher network predictions. Additionally, such a framework can also refine the labeled data assumed to be noisy.

approach. We conducted experiments to demonstrate that our method yielded competitive performance gains over other AL methods. Finally, we applied the trained model to a real-world in-house hospital dataset with noisy labels and obtain accurate CIM segmentation results. We release our code at https://github.com/CrystalWei626/PLD_AL.

2 Method

Section 2.1 establishes mathematical formulation on the main task of CIM segmentation in our AL framework. Our proposed AL approach has two loops: in Sect. 2.2, the outer loop implements progressive annotation on the automatically selected unlabeled pool; in Sect. 2.3, the inner loop trains the neural networks on the labeled pool and subsequently refines it through a feedback routine.

2.1 Mathematical Notations and Formulation

Denote $x \in \mathbb{R}^{I\times J}$ a carotid ultrasound image and $y \in \mathbb{R}^{I\times J}$ the corresponding CIM mask. Let $D_L = X_L \times Y_L$ and X_U be the initial labeled and unlabeled pools, where X_L is the carotid ultrasound image set, and Y_L is the corresponding label set. We aim to improve generalization ability of the AI model by selecting the most informative data in X_U and delivering them to an expert for annotation.

We propose a novel AL framework: PLD-AL for CIM segmentation, as illustrated in Fig. 1 and Algorithm 1. First, AI models are trained on D_L and used to refine Y_L. Then, the AI models select data from X_U for expert to annotate, forming a new set of labeled data. Finally, we update D_L and X_U and use new D_L to train the same AI models.

Algorithm 1: PLD-AL

1 **Input:** Initial labeled pool $D_L = X_L \times Y_L$; Unlabeled pool X_U; Judgment threshold τ; Refining threshold λ;
2 Initialize θ_S and θ_T;
3 **for** $t = 1, \cdots T$ **do**
4 $\quad \theta_T^{(0)} \leftarrow \theta_T$; $\theta_S^{(0)} \leftarrow \theta_S$;
5 $\quad$ **for** $k = 1, \cdots, K$ **do**
6 $\quad\quad \theta_S^{(k)} := Opt(\theta_S^{(k-1)}; D_L, lr)$; ▷ Optimize $\theta_S^{(k)}$ on D_L
7 $\quad\quad \theta_T^{(k)} := \alpha\theta_S^{(k)} + (1-\alpha)\theta_T^{(k-1)}$; ▷ Update $\theta_T^{(k)}$ by EMA
8 $\quad\quad \mathcal{M}^{(k)} = \text{mean}_{(x,y)\in D_L} \text{IoU}(y, F(x|\theta_S^{(k)}))$; ▷ Calculate mIoU
9 $\quad\quad$ **if** $k > K_1$ **then**
10 $\quad\quad\quad \tilde{\mathcal{M}} = \text{argmin} \sum_{l=1}^{k} \|\tilde{\mathcal{M}} - \mathcal{M}^{(l)}\|_{\ell 2}^2$; ▷ Fit the mIoU curve
11 $\quad\quad\quad$ **if** $\tilde{\mathcal{M}}'(k) - \tilde{\mathcal{M}}'(k-1) < \tau$ **then**
12 $\quad\quad\quad\quad$ **for** $x \in X_L, i \in \{1, 2, ...I\}, j \in \{1, 2, ...J\}$ **do**
13 $\quad\quad\quad\quad\quad p_{ij} = F(x(i,j)|\theta_T^{(k)})$; ▷ Predict on teacher network
14 $\quad\quad\quad\quad\quad y(i,j) = \text{argmax}\{p_{ij}\}$ if $\max p_{ij} > \lambda$; ▷ Refine Y_L
15 $\quad\quad\quad\quad \theta_S \leftarrow \theta_S^{(k)}, \theta_T \leftarrow \theta_T^{(k)}$;
16 $\quad\quad\quad\quad$ **break**;
17 $\quad d(x) = \text{mean}_{i,j} \text{Div}_{KL}(x(i,j), \theta_S, \theta_T), x \in X_U$; ▷ Compute KL divergence
18 $\quad X_A = \text{arg}_{x \in X_U} \text{TOP}_n d(x)$; ▷ Select unlaleled data
19 $\quad Y_A = \{y = \text{Expert}(x) : x \in X_A\}$; ▷ Annotate by expert
20 $\quad D_A = X_A \times Y_A; D_L \leftarrow D_L \bigcup D_A; X_U \leftarrow X_U \setminus X_A$; ▷ Update D_L, X_U
21 **Output:** D_L; θ_T

In each AL iteration, we use a mean-teacher architecture as the backbone of AL. The student and the teacher networks, respectively parameterized by θ_S and θ_T, share the same neural network architecture F, which maps the carotid ultrasound image $x \in \mathbb{R}^{I \times J}$ to the extended three-dimensional CIM mask probability $p \in \mathbb{R}^{I \times J \times 2}$, whose 3rd-dimensional component $p_{ij} \in \mathbb{R}^2$ denotes the softmax probability output for binary classification at the pixel (i, j). We use the divergence between pseudo-labels generated by student and teacher networks to assist in selecting data for the expert to annotate.

2.2 Outer Loop: Divergence Based AL

The outer loop is an AL cycle that selects data for the expert to annotate according to the divergence between the predictions of the student and teacher networks. First, we initialize θ_S and θ_T. We complete the inner loop proposed in Sect. 2.3, and obtain the trained parameters for the student and teacher networks. Then, we select n data from X_U for the expert to annotate. We suggest using the Kullback-Leibler (KL) divergence to assist in selecting data, as shown in Eq. (1):

$$\mathrm{Div}_{KL}(x(i,j),\theta_S,\theta_T) = \sum_{c=1}^{2} F(x(i,j)|\theta_T) \log \frac{F(x(i,j)|\theta_T)}{F(x(i,j)|\theta_S)}. \quad (1)$$

We consider data prediction uncertainty as a decisive metric for data selection. It is deduced that the KL divergence between the output of the primary and the auxiliary models in a dual-decoder architecture can approximate the prediction uncertainty [13,14].

We compute the KL divergence scores $d(x) = \mathrm{mean}_{i,j}\mathrm{Div}_{KL}(x(i,j),\theta_S,\theta_T)$ of the data in X_U. Let X_A be the subset that contains data x in X_U corresponding to the top-n largest $d(x)$ values (denoted by $\mathrm{TOP}_n d(x)$). With this, we can next obtain the label set Y_A in terms of X_A by means of the expert's annotates and the required post-processing step; see Sect. 3.1 for details. Lastly, we add the selected dataset with its label set $X_A \times Y_A$ into D_L and delete X_A from X_U. We repeat the above steps until reaching the maximum number of AL iterations.

2.3 Inner Loop: Network Optimization and Label Refinement

The inner loop trains the neural networks by the labeled pool and refines noisy labels through a feedback routine. In the kth epoch of the inner loop, we first use the last labeled pool D_L to optimize the training parameter $\theta_S^{(k)}$ by mini-batch stochastic gradient descent. The loss function consists of a supervised loss L_{sup} between labels and predictions of the student model, and a consistency loss L_{con} between the predictions of the student and the teacher models. These can be implemented using the cross-entropy loss and the mean squared error loss, respectively. Then, we update $\theta_T^{(k)}$ by exponential moving average (EMA) with a decay rate α as Eq. (2):

$$\theta_T^{(k)} = \alpha\theta_S^{(k)} + (1-\alpha)\theta_T^{(k-1)}. \quad (2)$$

We refine noisy labels based on the idea that the fitness soars sharply at first but slows down after the model begins to fit noise [15]. We interrupt the model training before it begins to fit noise, then refine the labels utilizing the current network output. We calculate the model fitness via a series of the intersection over union (mIoU) [16] scores sampled at every training epoch. To estimate the ratio of change of the model fitness, we fit the mIoU curve $\tilde{\mathcal{M}}(k)$ via e.g., the exponential regression formed in Eq. (3) when the length of mIoU series is larger than a designated parameter $K_1 \in \mathbb{N}^+$:

$$\tilde{\mathcal{M}}(k) = a(1 - \exp\{-b \cdot k^c\}), \quad (3)$$

where a, b, and c are the fitting parameters to be determined by least squared estimate. Then we calculate the ratio of change of the model fitness γ^k via the derivative of the mIoU curve $\tilde{\mathcal{M}}'(k)$: $\gamma^k = \tilde{\mathcal{M}}'(k) - \tilde{\mathcal{M}}'(k-1)$. When training stops at this epoch k satisfying $\gamma^k < \tau$ (τ is a judgment threshold), we lastly predict the CIM mask probability $p_{ij} = F(x(i,j)|\theta_T^{(k)})$ via the teacher network for each pixel at (i,j) and update the noisy label $y(i,j)$ in Y_L if $\max p_{ij} > \lambda$ (λ is a refining threshold).

3 Experiments and Results

3.1 Experiment Settings

Implementation Details. We used the PyTorch platform (version 1.13.1) to implement our method. And we adapted the same UNet++ [17] structures as the encoding-decoding structures for the student and the teacher networks. We implemented 1000 training iterations with a total mini-batch size of 14 and initial batch size of labeled data of 2 on an Nvidia GeForce RTX 3070 GPU with 8192 MB of memory (Nvidia, Santa Clara, CA, United States). Since the number of labeled data increases after completing each AL iteration, the batch size of labeled data should increase by 2 synchronously to keep the total epoch num unchanged. We used stochastic gradient descent (SGD) as the optimizer with the parameter settings: momentum (0.9) and weight decay (0.0001). We set EMA decay rate $\alpha = \min\{1 - 1/(iter + 1), 0.99\}$, where $iter$ is the current training iteration number. 2021 regions of interest (ROI) of size 256×128 were cropped from original carotid ultrasound images for model training using template matching technique [18]. We set the number of AL iterations, fixed labeling budget, initial labeled and unlabeled data, and the test data as 5, 200,159, 1857, and 1204, respectively. During each annotation phase, experts manually marked the CIM boundaries with scatters and we subsequently generated the complete CIM masks via the Akima interpolation method [19]. θ_S and θ_T was initialized by the pre-train model[1] on ImageNet [20]. At our best practice, we chose the hyper-parameters $\lambda = 0.8$, $\tau = 0.005$ and $K_1 = 1$.

Dataset. We employed the publicly available Carotid Ultrasound Boundary Study (CUBS) dataset[2] [21] and the in-house dataset acquired at Children's Hospital, Zhejiang University School of Medicine. The CUBS dataset contains ultrasound images of the left and right carotid arteries from 1088 patients across two medical centers and three manual annotations of LI and MA boundaries by experts. According to the description of these annotations in the dataset specification, the analytic hierarchy process (AHP) [22] was adapted to weigh the three expert's annotations to obtain accurate labels for testing. We randomly performed morphological transformations (dilation and erosion) by OpenCV [23] on the accurate labels to generate noisy labels for training. The in-house dataset comes from 373 patients aged 6–12, with 2704 carotid ultrasound images. We picked 350 images with visible CIM areas and applied the model trained on CBUS to CIM segmentation. The data acquisition and the experimental protocol have been approved by the institutional review board of Children's Hospital, Zhejiang University School of Medicine.

[1] https://github.com/pprp/timm.

[2] https://data.mendeley.com/datasets/fpv535fss7/1.

Table 1. Quantitative results of performance comparison, the metrics were calculated over the test dataset and took the mean. Bold font highlights the optimal performance except for the upper limit. The asterisk $*$ denotes $p < 0.001$ compared with the rest methods.

Method	Dice (%) ↑	IoU (%) ↑	ASD (voxel) ↓	95HD (voxel) ↓	Time (S) ↓
Random	70.96 ± 8.26	57.01 ± 8.80	3.79 ± 1.05	15.95 ± 6.41	$\mathbf{118.69 \pm 6.38}$
Entropy [26]	76.62 ± 2.20	63.26 ± 2.75	2.07 ± 0.02	9.21 ± 2.09	192.24 ± 36.13
Confidence [12]	74.86 ± 0.21	61.93 ± 0.34	2.47 ± 0.89	11.15 ± 3.97	166.56 ± 2.93
CoreSet [27]	$79.92 \pm 0.31^*$	67.39 ± 0.43	$1.88 \pm 0.11^*$	$6.33 \pm 0.62^*$	199.78 ± 47.20
CDAL [28]	78.20 ± 1.61	65.15 ± 2.06	2.01 ± 0.10	7.83 ± 1.51	165.15 ± 2.74
Ours	$\mathbf{83.51 \pm 0.28^*}$	$\mathbf{72.33 \pm 0.46^*}$	$\mathbf{1.69 \pm 0.02^*}$	$\mathbf{4.72 \pm 0.11^*}$	139.06 ± 28.66
Upper Limit	84.01	73.03	1.53	4.24	213.77

Evaluation Metrics. We utilized dice coefficient (Dice) [24], intersection over union (IoU) [16], average surface distance (ASD), 95% covered Hausdorff distance (95HD) [25], and the average training time of 5 AL iterations as evaluation metrics of the CIM segmentation performance compared to the generated ground truth on the unseen test set.

3.2 Performance Comparison

We evaluated the performance of AL methods on the CIM segmentation task using the CUBS dataset.

Baselines. We compared our method to other AL methods, including AL methods with query strategy based on random selection (Random), entropy increase (Entropy) [26], prediction confidence (Confidence) [12], CoreSet [27] and predicted probability diverse contexts (CDAL) [28]. All of the backbones of these baseline methods are fully supervised models.

Furthermore, we trained a supervised model by the fully labeled pool with accurate labels yielding an upper limit of generalization ability. We compared this upper limit to the performance of all the methods.

Table 1 illustrates the quantitative results of different methods on the test dataset. It shows that our method based on the KL divergence query strategy improves the mean generalization metrics (Dice, IoU, ASD, and 95HD) compared with other AL methods. In particular, it significantly (two-tailed Wilcoxon signed-rank test with $p < 0.001$) outperforms the others in terms of any metric.

3.3 Ablation Study

We conducted ablation study on the CUBS dataset to demonstrate the importance of the label refinement module proposed in Sect. 2.3. We canceled the label refinement module and substituted the label refinement module with confidence learning (CL) for noise label correction [29].

Table 2. Quantitative results of ablation study, the metrics were calculated over the test dataset and took the mean. The abbreviations, Refine and CL, represent the label refinement module and confidence learning [29], respectively. Bold font highlights the optimal performance except for the upper limit. The asterisk $*$ denotes $p < 0.001$ compared with the rest methods.

Method	Dice (%) ↑	IoU (%) ↑	ASD (voxel) ↓	95HD (voxel) ↓	Time (S) ↓
w.o. Refine	80.17 ± 1.37	67.68 ± 1.86	1.97 ± 0.05	6.73 ± 0.99	301.93 ± 27.38
Refine→CL	81.08 ± 1.36	68.9 ± 1.84	1.86 ± 0.10	5.82 ± 0.69	689.09 ± 34.03
w/ Refine	$\mathbf{83.51 \pm 0.28}^*$	$\mathbf{72.33 \pm 0.46}^*$	$\mathbf{1.69 \pm 0.02}^*$	$\mathbf{4.72 \pm 0.11}^*$	$\mathbf{139.06 \pm 28.66}$
Upper Limit	84.01	73.03	1.53	4.24	213.77

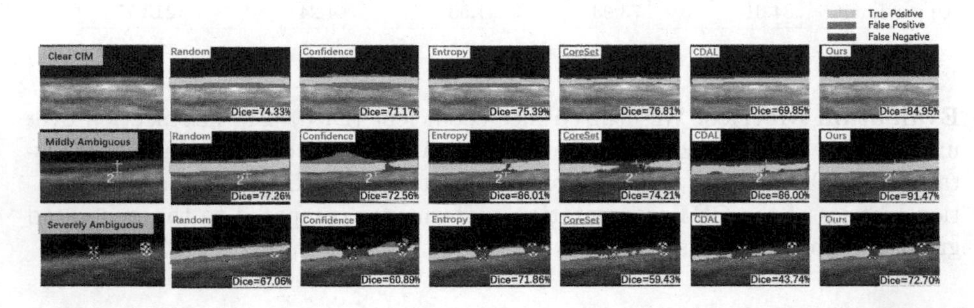

Fig. 2. Qualitative results of application study. It shows the visualization of CIM segmentation on input images with clear, mildly ambiguous, and severely ambiguous CIM areas, respectively. The images are chosen from the in-house dataset. We used the model with the best quantitative results in Sect. 3.2 to generate the masks. The green, red, and blue masks represent segmented true positive, false positive, and false negative, respectively.

Table 2 illustrates the results of ablation study experiment. Our method substantially outperforms the method without the label refinement module and slightly outperforms the method with CL. In particular, it significantly (two-tailed Wilcoxon signed-rank test with $p < 0.001$) outperforms the others in terms of all the metrics. Moreover, the training time of our method is significantly reduced compared to CL since CL needs to estimate the uncertainty during training to correct the noisy data smoothly, which leads to more computational cost.

3.4 Application on In-house Dataset

We applied the teacher network trained in Sect. 3.2 to the in-house dataset acquired at a pediatric hospital. Figure 2 visualizes three example images with different CIM area qualities (clear, mildly ambiguous, severely ambiguous). Qualitatively, the generalization ability of the model trained by our method is much better than those trained by other methods, regardless of image quality. Moreover, as shown in Fig. 2, Random over-estimates the CIM area, while CoreSet,

CDAL, and our method produces more conservative results but lost continuity in the severely ambiguous image. Quantitatively, the mean Dice, IoU, ASD, and 95HD of our method are 79.20%, 66.99%, 1.92 voxels, and 6.12 voxels, respectively, indicating a small but rational generalization loss on the in-house data.

4 Conclusion

We propose a novel AL framework PLD-AL, by training segmentation models using a gradually enlarged and refined labeled pool to obtain accurate and efficient CIM segmentation. Compared with other AL methods, it achieves competitive performance gains. Furthermore, we applied the trained model to an in-house hospital dataset and obtained accurate CIM segmentation results. In the future, we will extend our approach to subsequently calculate CIM thickness and roughness for clinical evaluation of obesity or atherosclerotic disease. We will also investigate the robustness of the proposed method in terms of inter-expert variations and noisy annotation labels. Our approach merely involves one expert in the loop, which may potentially be sensitive to the expert's experience. Multiple experts may consider minimizing inter-reader differences during human-AI interactive labeling [30].

Acknowledgement. This work was supported in part by Research Initiation Project (2021ND0PI02) and Key Research Project (2022KI0AC01) of Zhejiang Lab, National Key Research and Development Programme of China (No. 2021YFC2701902), and National Natural Science Foundation of China (No. 12071430).

References

1. Tarvainen, A., Valpola, H.: Mean teachers are better role models: weight-averaged consistency targets improve semi-supervised deep learning results. In: NeurIPS, NIPS, Long Beach (2017)
2. Xu, M.C., et al.: Bayesian pseudo labels: expectation maximization for robust and efficient semi-supervised segmentation. In: Wang, L., Dou, Q., Fletcher, P.T., Speidel, S., Li, S. (eds.) MICCAI 2022. LNCS, vol. 13435, pp. 580–590. Springer, Singapore (2022). https://doi.org/10.1007/978-3-031-16443-9_56
3. Yao, H., Hu, X., Li, X.: Enhancing pseudo label quality for semi-supervised domain-generalized medical image segmentation. In: AAAI, pp. 3099–3107. AAAI (2022)
4. Liu, F., Tian, Y., Chen, Y., Liu, Y., Belagiannis, V., Carneiro, G.: ACPL: anti-curriculum pseudo-labelling for semi-supervised medical image classification. In: CVPR, New Orleans, pp. 20697–20706. IEEE Computer Society (2022)
5. Lu, L., Yin, M., Fu, L., Yang, F.: Uncertainty-aware pseudo-label and consistency for semi-supervised medical image segmentation. Biomed. Signal Process. Control **79**(2), 104203 (2023)
6. Parvaneh, A., Abbasnejad, E., Teney, D., Haffari, G.R., Van Den Hengel, A., Shi, J.Q.: Active learning by feature mixing. In: CVPR, New Orleans, pp. 12237–12246. IEEE Computer Society (2022)
7. Sinha, S., Ebrahimi, S., Darrell, T.: Variational adversarial active learning. In: ICCV, Seoul, pp. 5972–5981. IEEE (2019)

8. Caramalau, R., Bhattarai, B., Kim, T.K.: Sequential graph convolutional network for active learning. In: CVPR, pp. 9583–9592. IEEE Computer Society (2021)
9. Casanova, A., Pinheiro, P.O., Rostamzadeh, N., Pal, C.J.: Reinforced active learning for image segmentation. arXiv preprint arXiv:2002.06583 (2020)
10. Siddiqui, Y., Valentin, J., Nießner, M.: Viewal: active learning with viewpoint entropy for semantic segmentation. In: CVPR, pp. 9433–9443. IEEE Computer Society (2020)
11. Yang, L., Zhang, Y., Chen, J., Zhang, S., Chen, D.Z.: Suggestive annotation: a deep active learning framework for biomedical image segmentation. In: Descoteaux, M., Maier-Hein, L., Franz, A., Jannin, P., Collins, D.L., Duchesne, S. (eds.) MICCAI 2017. LNCS, vol. 10435, pp. 399–407. Springer, Cham (2017). https://doi.org/10.1007/978-3-319-66179-7_46
12. Xu, Y., et al.: Partially-supervised learning for vessel segmentation in ocular images. In: de Bruijne, M., et al. (eds.) MICCAI 2021. LNCS, vol. 12901, pp. 271–281. Springer, Cham (2021). https://doi.org/10.1007/978-3-030-87193-2_26
13. Zheng, Z., Yang, Y.: Rectifying pseudo label learning via uncertainty estimation for domain adaptive semantic segmentation. Int. J. Comput. Vision **129**(4), 1106–1120 (2021)
14. Luo, X., et al.: Efficient semi-supervised gross target volume of nasopharyngeal carcinoma segmentation via uncertainty rectified pyramid consistency. In: de Bruijne, M., et al. (eds.) MICCAI 2021. LNCS, vol. 12902, pp. 318–329. Springer, Cham (2021). https://doi.org/10.1007/978-3-030-87196-3_30
15. Liu, S., Liu, K., Zhu, W., Shen, Y., Fernandez-Granda, C.: Adaptive early-learning correction for segmentation from noisy annotations. In: CVPR, New Orleans, pp. 2606–2616. IEEE Computer Society (2022)
16. Rahman, M.A., Wang, Y.: Optimizing intersection-over-union in deep neural networks for image segmentation. In: Bebis, G., et al. (eds.) ISVC 2016. LNCS, vol. 10072, pp. 234–244. Springer, Cham (2016). https://doi.org/10.1007/978-3-319-50835-1_22
17. Zhou, Z., Siddiquee, M.M.R., Tajbakhsh, N., Liang, J.: UNet++: redesigning skip connections to exploit multiscale features in image segmentation. IEEE Trans. Med. Imaging **39**(6), 1856–1867 (2019)
18. Brunelli, R.: Template Matching Techniques in Computer Vision: Theory and Practice. Wiley, Hoboken (2009)
19. Akima, H.: A method of bivariate interpolation and smooth surface fitting based on local procedures. Commun. ACM **17**(1), 18–20 (1974)
20. He, K., Girshick, R., Dollár, P.: Rethinking imagenet pre-training. In: ICCV, Seoul, pp. 4918–4927. IEEE (2019)
21. Meiburger, K.M., et al.: DATASET for "Carotid Ultrasound Boundary Study (CUBS): an open multi-center analysis of computerized intima-media thickness measurement systems and their clinical impact". Mendeley Data, V1 (2021). https://doi.org/10.17632/fpv535fss7.1
22. Sipahi, S., Timor, M.: The analytic hierarchy process and analytic network process: an overview of applications. Manag. Decis. **48**(5), 775–808 (2010)
23. Bradski, G.: The openCV library. Dr. Dobb's J. Softw. Tools Prof. Program. **25**(11), 120–123 (2000)
24. Bertels, J., et al.: Optimizing the dice score and jaccard index for medical image segmentation: theory and practice. In: Shen, D., et al. (eds.) MICCAI 2019. LNCS, vol. 11765, pp. 92–100. Springer, Cham (2019). https://doi.org/10.1007/978-3-030-32245-8_11

25. Aspert, N., Santa-Cruz, D., Ebrahimi, T.: Mesh: measuring errors between surfaces using the hausdorff distance. In: ICME, Lausanne, pp. 705–708. IEEE (2022)
26. Wang, D., Shang, Y.: A new active labeling method for deep learning. In: IJCNN, Beijing, pp. 112–119. IEEE (2014)
27. Sener, O., Savarese, S.: Active learning for convolutional neural networks: a core-set approach. arXiv preprint arXiv:1708.00489 (2017)
28. Agarwal, S., Arora, H., Anand, S., Arora, C.: Contextual diversity for active learning. In: Vedaldi, A., Bischof, H., Brox, T., Frahm, J.-M. (eds.) ECCV 2020. LNCS, vol. 12361, pp. 137–153. Springer, Cham (2020). https://doi.org/10.1007/978-3-030-58517-4_9
29. Xu, Z., et al.: Noisy labels are treasure: mean-teacher-assisted confident learning for hepatic vessel segmentation. In: de Bruijne, M., et al. (eds.) MICCAI 2021. LNCS, vol. 12901, pp. 3–13. Springer, Cham (2021). https://doi.org/10.1007/978-3-030-87193-2_1
30. Zhang, L., et al.: Learning from multiple annotators for medical image segmentation. Pattern Recognit. **138**, 109400 (2023)

Adapter Learning in Pretrained Feature Extractor for Continual Learning of Diseases

Wentao Zhang[1,4], Yujun Huang[1,4], Tong Zhang[2], Qingsong Zou[1,3], Wei-Shi Zheng[1,4], and Ruixuan Wang[1,2,4](✉)

[1] School of Computer Science and Engineering, Sun Yat-sen University, Guangzhou, China
wangruix5@mail.sysu.edu.cn
[2] Peng Cheng Laboratory, Shenzhen, China
[3] Guangdong Province Key Laboratory of Computational Science, Sun Yat-sen University, Guangzhou, China
[4] Key Laboratory of Machine Intelligence and Advanced Computing, MOE, Guangzhou, China

Abstract. Currently intelligent diagnosis systems lack the ability of continually learning to diagnose new diseases once deployed, under the condition of preserving old disease knowledge. In particular, updating an intelligent diagnosis system with training data of new diseases would cause catastrophic forgetting of old disease knowledge. To address the catastrophic forgetting issue, an **A**dapter-based **C**ontinual **L**earning framework called ACL is proposed to help effectively learn a set of new diseases at each round (or task) of continual learning, without changing the shared feature extractor. The learnable lightweight task-specific adapter(s) can be flexibly designed (e.g., two convolutional layers) and then added to the pretrained and fixed feature extractor. Together with a specially designed task-specific head which absorbs all previously learned old diseases as a single 'out-of-distribution' category, task-specific adapter(s) can help the pretrained feature extractor more effectively extract discriminative features between diseases. In addition, a simple yet effective fine-tuning is applied to collaboratively fine-tune multiple task-specific heads such that outputs from different heads are comparable and consequently the appropriate classifier head can be more accurately selected during model inference. Extensive empirical evaluations on three image datasets demonstrate the superior performance of ACL in continual learning of new diseases. The source code is available at https://github.com/GiantJun/CL_Pytorch.

Keywords: Continual learning · Adapter · Disease diagnosis

W. Zhang and Y. Huang—Authors contributed equally.

Supplementary Information The online version contains supplementary material available at https://doi.org/10.1007/978-3-031-43895-0_7.

H. Greenspan et al. (Eds.): MICCAI 2023, LNCS 14221, pp. 68–78, 2023.
https://doi.org/10.1007/978-3-031-43895-0_7

1 Introduction

Deep neural networks have shown expert-level performance in various disease diagnoses [35,36]. In practice, a deep neural network is often limited to the diagnosis of only a few diseases, partly because it is challenging to collect enough training data of all diseases even for a specific body tissue or organ. One possible solution is to enable a deployed intelligent diagnosis system to continually learn new diseases with collected new training data later. However, if old data are not accessible due to certain reasons (e.g., challenge in data sharing), current intelligent systems will suffer from catastrophic forgetting of old knowledge when learning new diseases [17].

Multiple approaches have been proposed to alleviate the catastrophic forgetting issue. One approach aims to determine part of the model parameters which are crucial to old knowledge and tries to keep these parameters unchanged during learning new knowledge [5,15,18]. Another approach aims to preserve old knowledge by making the updated model imitate the behaviour (e.g., output at certain layer) of the old model particularly with the help of knowledge distillation technique [8,17,34]. Storing a small amount of old data or synthesizing old data relevant to old knowledge and using them together with training data of new knowledge can often help significantly alleviate forgetting of old knowledge [2,3,19,22]. Although the above approaches can help the updated model keep old knowledge to some extent, they often fall into the dilemma of model plasticity (for new knowledge learning) and stability (for old knowledge preservation). In order to resolve this dilemma, new model components (e.g., neurons or layers in neural networks) can be added specifically for learning new knowledge, while old parameters are largely kept unchanged for old knowledge [20,26,30]. While this approach has shown state-of-the-art continual learning performance, it faces the problem of rapid model expansion and effective fusion of new model components into the existing ones. To alleviate the model expansion issue and meanwhile well preserve old knowledge, researchers have started to explore the usage of a pretrained and fixed feature extractor for the whole process of continual learning [14,27,28,32], where the challenge is to discriminate between different classes of knowledge with limited learnable parameters.

In this study, inspired by recent advances in transfer learning in natural language processing [7,12], we propose adding a light-weight learnable module called adapter to a pretrained and fixed convolutional neural network (CNN) for effective continual learning of new knowledge. For each round of continual learning, the CNN model will be updated to learn a set of new classes (hereinafter also called learning a new task). The learnable task-specific adapters are added between consecutive convolutional stages to help the pretrained CNN feature extractor more effectively extract discriminative features of new diseases. To the best of our knowledge, it is the first time to apply the idea of CNN adapter in the continual learning field. In addition, to keep extracted features discriminative between different tasks, a special task-specific classifier head is added when learning each new task, in which all previously learned old classes are considered as the 'out-of-distribution' (OOD) class and correspond to an additional

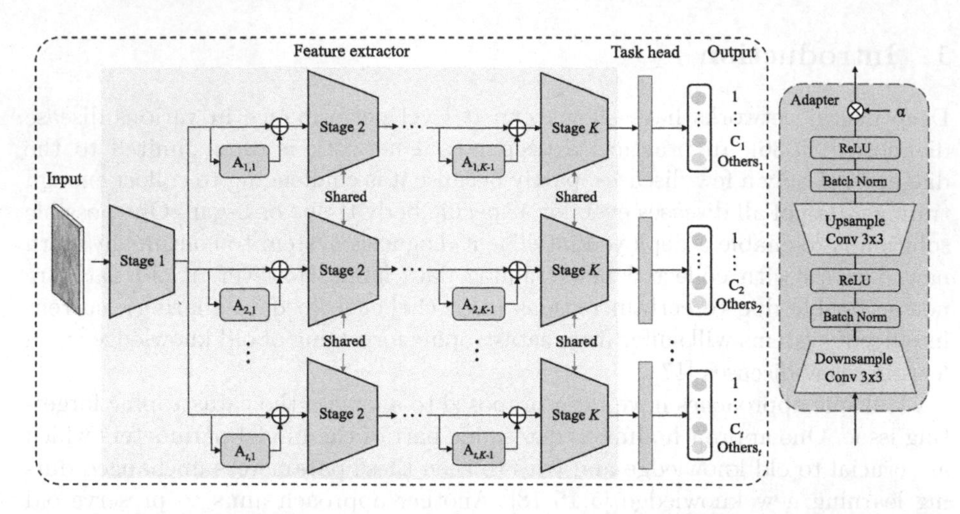

Fig. 1. The proposed framework for continual learning of new diseases. Left: task-specific adapters ($\{A_{t,1}, A_{t,1}, \ldots, A_{t,K-1}\}$ in orange) between consecutive convolutional stages are added and learned for a new task. After learning the new task-specific adapters, all tasks' classifier heads (orange rectangles) are fine-tuned with balanced training data. The pretrained feature extractor is fixed during the continual learning process. Right: the structure of each adapter, with α representing the global scaling. (Color figure online)

output neuron in each task-specific classifier head. A simple yet effective fine-tuning strategy is applied to calibrate outputs between multiple task-specific heads. Extensive empirical evaluations on three image datasets show that the proposed method outperforms existing continual learning methods by a large margin, consistently supporting the effectiveness of the proposed method.

2 Method

This study aims to improve continual learning performance of an intelligent diagnosis system. At each learning round, following previous studies [2,3] which show that rehearsal with old samples can significantly improve continual learning performance, the system will be updated based on the training data of new diseases and preserved small subset for each previously learned disease. During inference, the system is expected to accurately diagnose all learned diseases, without knowing which round (i.e., task) the class of any test input is from.

2.1 Overall Framework

We propose an **A**dapter-based **C**ontinual **L**earning framework called ACL with a multi-head training strategy. With the motivation to make full use of readily available pretrained CNN models and slow down the speed of model expansion that appears in some state-of-the-art continual learning methods (e.g.,

DER [30]), and inspired by a recently developed transfer learning strategy Delta tuning for downstream tasks in natural language process [7], we propose adding a learnable light-weight adapter between consecutive convolutional stages in a pretrained and fixed CNN model when learning new classes of diseases at each learning round (Fig. 1). Each round of continual learning as a unique task is associated with task-specific adapters and a task-specific classifier head. Model update at each round of continual learning is to find optimal parameters in the newly added task-specific adapters and classifier head. During inference, since multiple classifier heads exist, the correct head containing the class of a given input is expected to be selected. In order to establish the potential connections between tasks and further boost the continual learning performance, a two-stage multi-head learning strategy was proposed by including the idea of out-of-distribution (OOD) detection.

2.2 Task-Specific Adapters

State-of-the-art continual learning methods try to preserve old knowledge by either combining old fixed feature extractors with the newly learned feature extractor [30], or by using a shared and fixed pretrained feature extractor [32]. However, simply combining feature extractors over rounds of continual learning would rapidly expand the model, while using a shared and fixed feature extractor could largely limit model ability of learning new knowledge because only the model head can be tuned to discriminate between classes. To resolve this dilemma, we propose adding a light-weight task-specific module called adapter into a single pretrained and fixed CNN feature extractor (e.g., with the ResNet backbone), such that the model expands very slowly and the adapter-tuned feature extractor for each old task is fixed when learning a new task. In this way, old knowledge largely stored in the feature extractor and associated task-specific adapters will be well preserved when the model learns new knowledge at subsequent rounds of continual learning. Formally, suppose the pretrained CNN feature extractor contains K stages of convolutional layers (e.g., 5 stages in ResNet), and the output feature maps from the k-th stage is denoted by $\mathbf{z}_k, k \in \{1, \ldots, K-1\}$. Then, when the model learns a new task at the t-th round of continual learning, a task-specific adapter $A_{t,k}$ is added between the k-th and $(k+1)$-th stages as follows,

$$\hat{\mathbf{z}}_{t,k} = A_{t,k}(\mathbf{z}_k) + \mathbf{z}_k\,, \tag{1}$$

where the adapter-tuned output $\hat{\mathbf{z}}_{t,k}$ will be used as input to the $(k+1)$-th stage. The light-weight adapter can be flexibly designed. In this study, a simple two-layer convolution module followed by a global scaling is designed as the adapter (Fig. 1, Right). The input feature maps to the adapter are spatially downsampled by the first convolutional layer and then upsampled by the second layer. The global scaling factor α is learned together with the two convolutional layers. The proposed task-specific adapter for continual learning is inspired by Delta tuning [7] which adds learnable 2-layer perceptron(s) into a pretrained and fixed

Transformer model in natural language processing. Different from Delta tuning which is used as a transfer learning strategy to adapt a pretrained model for any individual downstream task, the proposed task-specific adapter is used as a continual learning strategy to help a model continually learn new knowledge over multiple rounds (i.e., multiple tasks) in image processing, with each round corresponding to a specific set of adapters.

Also note that the proposed adapter differs from existing adapters in CLIP-Adapter (CA) [9] and Tip-Adapter (TA) [33]. First, in structure, CA and TA use 2-layer MLP or cache model, while ours uses a 2-layer convnet with a global scaling factor. Second, the number and locations of adapters in model are different. CA and TA use adapter only at output of the last layer, while ours appears between each two consecutive CNN stages. Third, the roles of adapters are different. Existing adapters are for few-shot classification, while ours is for continual learning. It is also different from current prompt tuning. Prompts appear as part of input to the first or/and intermediate layer(s) of model, often in the form of learnable tokens for Transformer or image regions for CNNs. In contrast, our adapter appears as an embedded neural module for each two consecutive CNN stages, in the form of sub-network.

2.3 Task-Specific Head

Task-specific head is proposed to alleviate the potential feature fusion issue in current state-of-the-art methods [30] which combine task-specific features by a unified classifier head. In particular, if feature outputs from multiple task-specific feature extractors are simply fused by concatenating or averaging followed by a unified 1- or 2-layer percepton (as in [30]), discriminative feature information appearing only in those classes of a specific task could become less salient after fusion with multiple (possibly less discriminative) features from other task-specific feature extractors. As a result, current state-of-the-art methods often require storing relatively more old data to help train a discriminative unified classifier head between different classes. To avoid the possible reduction in feature discriminability, we propose not fusing features from multiple feature extractors, but a task-specific classifier head for each task. Each task-specific head consists of one fully connected layer followed by a softmax operator. Specially, for a task containing C new classes, one additional class absorbing all previously learned old classes ('others' output neuron in Fig. 1) is also included, and therefore the number of output elements from the softmax will be $C+1$. The 'others' output is used to predict the probability of the input image being from certain class of any other task rather than from the current task. In other words, each task-specific head has the ability of out-of-distribution (OOD) ability with the help of the 'others' output neuron. At the t-round of continual learning (i.e., for the t-th task learning), the task-specific adapters and the task-specific classifier head can be directly optimized, e.g., by cross-entropy loss, with the C new classes of training data and the 'others' class of all preserved old data.

However, training the task-specific classifier head without considering its relationship with existing classifier heads of previously learned tasks may cause the head selection issue during model inference. For example, a previously learned old classifier head may consider an input of latterly learned class as one of the old classes (correspondingly the 'others' output from the old head will be low). In other words, the 'others' outputs from multiple classifier heads cannot not be reliably compared (i.e., not calibrated) with one another if each classifier head is trained individually. In this case, if all classifier heads consider a test input as 'others' class with high confidence or multiple classifier heads consider a test input as one of their classes, it would become difficult to choose an appropriate classifier head for final prediction. To resolve the head selection issue, after initial training of the current task's adapters and classifier head, all the tasks' heads are fine-tuned together such that all 'others' outputs from the multiple heads are comparable. In short, at the t-th round of continual learning, the t task-specific classifier heads can be fine-tuned by minimizing the loss function L,

$$L = \frac{1}{t} \sum_{s=1}^{t} L_s^c , \tag{2}$$

where L_s^c is the cross-entropy loss for the s-th classifier head. Following the fine-tuning step in previous continual learning studies [4,30], training data of the current t-th task are sub-sampled such that training data in the fine-tuning step are balanced across all learned classes so far. Note that for each input image, multiple runs of feature extraction are performed, with each run adding adapters of a different task to the original feature extractor and extracting the feature vector for the corresponding task-specific head. Also note that in the fine-tuning step, adapters of all tasks are fixed and not tuned. Compared to training the adapters of each task with all training data of the corresponding task, fine-tuning these adapters would likely cause over-fitting of the adapters to the sub-sampled data and therefore is avoided in the fine-tuning step.

Once the multi-head classifier is fine-tuned at the t-th round of continual learning, the classifier can be applied to predict any test data as one of all the learned classes so far. First, the task head with the smallest 'others' output probability (among all t 'others' outputs) is selected, and then the class with the highest output from the selected task head is selected as the final prediction result. Although unlikely selected, the 'others' class in the selected task head is excluded for the final prediction.

3 Experiment Results

3.1 Experimental Setup

Four datasets were used to evaluate the proposed ACL (Table 1). Among them, Skin8 is imbalanced across classes and from the public challenge organized by the International Skin Imaging Collaboration (ISIC) [24]. Path16 is a subset of publicly released histopathology images collated from multiple publicly available datasets [1,6,13,25,29,35], including eleven diseases and five normal classes (see Supplementary Material for more details about dataset generation). These data

Table 1. Statistics of three datasets. '[600, 1024]': the range of image width and height.

Dataset	Classes	Train set	Test set	Number of tasks	Size
Skin8 [24]	8	3,555	705	4	[600, 1024]
Path16	16	12,808	1,607	7	224 × 224
CIFAR100 [16]	100	50,000	10,000	5, 10	32 × 32
MedMNIST [31]	36	302,002	75,659	4	28 × 28

are divided into seven tasks based on source of images, including Oral cavity (OR, 2 classes), Lymph node (LY, 2 classes), Breast (BR, 2 classes), Colon (CO, 2 classes), Lung (LU, 2 classes), Stomach (ST, 4 classes), and Colorectal polyp (CP, 2 classes). In training, each image is randomly rotated and then resized to 224 × 224 pixels.

In all experiments, publicly released CNN models which are pretrained on the Imagenet-1K dataset were used for the fixed feature extractor. During continual learning, the stochastic gradient descent optimizer was used for task-specific adapter learning, with batch size 32, weight decay 0.0005, and momentum 0.9. The initial learning rate was 0.01 and decayed by a factor of 10 at the 70th, 100th and 130th epoch, respectively. The adapters were trained for up to 200 epochs with consistently observed convergence. For fine-tuning classifier heads, the Adam optimizer was adopted, with initial learning rate 0.001 which decayed by a factor of 10 at the 55th, and 80th, respectively. The classifier heads were fine-tuned for 100 epochs with convergence observed. Unless otherwise mentioned, ResNet18 was used as the backbone, the size of memory for storing old images was 40 on Skin8, 80 on Path16, 2000 on CIFAR100 and 200 on MedMNIST.

In continual learning, the classifier sequentially learned multiple tasks, with each task a small number of new classes (e.g., 2, 10, 20). After learning each task, the mean class recall (MCR) over all classes learned so far is used to measure the classifier's performance. Note that MCR is equivalent to classification accuracy for class-balanced test set. For each experiment, the order of classes is fixed, and all methods were executed three times with different initialization. The mean and standard deviation of MCRs over three runs were reported.

3.2 Result Analysis

Effectiveness Evaluation: In this section, we compare ACL against state-of-the-art baselines, including iCaRL [19], DynaER [30], DER++ [3], WA [34], PODNet [8], and UCIR [11]. In addition, an upper-bound result (from a classifier which was trained with all classes of training data) is also reported. Similar amount of effort was taken in tuning each baseline method. As shown in Fig. 2, our method outperforms all strong baselines in almost all settings, no matter whether the classifier learn continually 2 classes each time on Skin8 (Fig. 2, first column), in two different task orders on Path16 (Fig. 2, second column), in 10 or

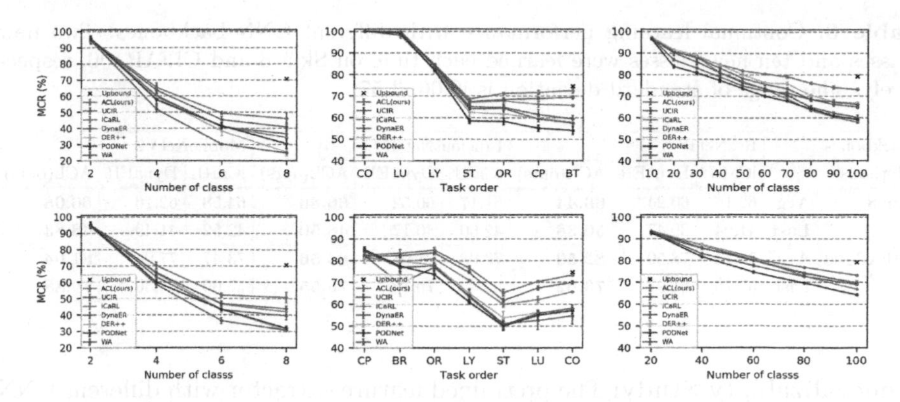

Fig. 2. Performance of continual learning on the Skin8, Path16 and CIFAR100 dataset, respectively. First column: 2 new classes each time on Skin8 respectively with memory size 16 and 40. Second column: continual learning on Path16 in different task orders. Last column: respectively learning 10 and 20 new classes each time on CIFAR100.

Table 2. Ablation study of ACL on Skin8 (with 2 new class per time) and on CIFAR100 (with 10 new classes per time). 'T.S.H.': inclusion of task-specific heads; 'Others': inclusion of the 'others' output neuron in each head. 'Avg': average of MCRs over all rounds of continual learning; 'Last': MCR at the last round.

Components				Skin8		CIFAR100	
T.S.H	Adapter	Others	Fine-tune	Avg	Last	Avg	Last
✓				$50.91_{\pm0.18}$	$27.47_{\pm0.32}$	$41.68_{\pm0.04}$	$18.64_{\pm0.14}$
✓	✓			$60.89_{\pm0.56}$	$35.4_{\pm1.20}$	$47.22_{\pm0.09}$	$21.20_{\pm0.13}$
✓	✓	✓		$60.90_{\pm1.97}$	$42.18_{\pm2.65}$	$58.72_{\pm0.07}$	$46.44_{\pm0.42}$
✓	✓	✓	✓	$\mathbf{66.44}_{\pm0.90}$	$\mathbf{50.38}_{\pm0.31}$	$\mathbf{82.50}_{\pm0.39}$	$\mathbf{73.02}_{\pm0.47}$
	✓		✓	$64.80_{\pm0.87}$	$46.77_{\pm1.58}$	$81.67_{\pm0.39}$	$70.72_{\pm0.33}$

20 classes each time on CIFAR100 (Fig. 2, last column), or in 4 different domains on MedMNIST [31] (Fig. 1 in Supplementary Material). Note that performance of most methods does not decrease (or even increase) at the last two or three learning rounds on Path16, probably because most methods perform much better on these tasks than on previous rounds of tasks.

Ablation Study: An ablation study was performed to evaluate the performance gain of each proposed component in ACL. Table 2 (first four rows) shows that the continual learning performance is gradually improved while more components are included, confirming the effectiveness of each proposed component. In addition, when fusing all the task features with a unified classifier head (Table 2, last row), the continual learning performance is clearly decreased compared to that from the proposed method (fourth row), confirming the effectiveness of task-specific classifier heads for class-incremental learning.

Table 3. Continual learning performance with different CNN backbones. Two new classes and ten new classes were learned each time on Skin 8 and CIFAR100, respectively. The range of standard deviation is [0.06, 3.57].

Backbones		ResNet18			EfficientNet-B0			MobileNetV2		
Methods		iCaRL	DynaER	ACL(ours)	iCaRL	DynaER	ACL(ours)	iCaRL	DynaER	ACL(ours)
Skin8	Avg	62.16	60.24	**66.44**	61.17	60.52	**66.86**	64.58	62.16	**66.08**
	Last	41.94	39.47	**50.38**	42.60	40.17	**48.50**	42.52	41.49	**48.83**
CIFAR100	Avg	75.74	78.59	**82.50**	73.98	81.98	**84.56**	73.47	77.04	**81.04**
	Last	57.75	64.97	**73.02**	53.40	70.13	**75.55**	55.00	64.30	**70.88**

Generalizability Study: The pretrained feature extractor with different CNN backbones were used to evaluate the generalization of ACL. As shown in Table 3, ACL consistently outperforms representative strong baselines with each CNN backbone (ResNet18 [10], EfficientNet-B0 [23] and MobileNetV2 [21]) on Skin8, supporting the generalizability of our method.

4 Conclusion

Here we propose a new adapter-based strategy for class-incremental learning of new diseases. The learnable light-weight and task-specific adapters, together with the pretrained and fixed feature extractor, can effectively learn new knowledge of diseases and meanwhile keep old knowledge from catastrophic forgetting. The task-specific heads with the special 'out-of-distribution' output neuron within each head helps keep extracted features discriminative between different tasks. Empirical evaluations on multiple medical image datasets confirm the efficacy of the proposed method. We expect such adapter-based strategy can be extended to other continual learning tasks including lesion detection and segmentation.

Acknowledgement. This work is supported in part by the Major Key Project of PCL (grant No. PCL2023AS7-1), the National Natural Science Foundation of China (grant No. 62071502 & No. 12071496), Guangdong Excellent Youth Team Program (grant No. 2023B1515040025), and the Guangdong Provincial Natural Science Fund (grant No. 2023A1515012097).

References

1. Borkowski, A.A., Bui, M.M., Thomas, L.B., Wilson, C.P., DeLand, L.A., Mastorides, S.M.: Lung and colon cancer histopathological image dataset (LC25000). arXiv preprint arXiv:1912.12142 (2019)
2. Boschini, M., et al.: Transfer without forgetting. In: Avidan, S., Brostow, G., Cissé, M., Farinella, G.M., Hassner, T. (eds.) ECCV 2022. LNCS, vol. 13683, pp. 692–709. Springer, Cham (2022). https://doi.org/10.1007/978-3-031-20050-2_40
3. Buzzega, P., Boschini, M., Porrello, A., Abati, D., Calderara, S.: Dark experience for general continual learning: a strong, simple baseline. In: NeurIPS (2020)

4. Castro, F.M., Marín-Jiménez, M.J., Guil, N., Schmid, C., Alahari, K.: End-to-end incremental learning. In: ECCV (2018)
5. Chaudhry, A., Ranzato, M., Rohrbach, M., Elhoseiny, M.: Efficient lifelong learning with a-gem. arXiv preprint arXiv:1812.00420 (2018)
6. Cruz-Roa, A., et al.: Automatic detection of invasive ductal carcinoma in whole slide images with convolutional neural networks. In: Medical Imaging 2014: Digital Pathology (2014)
7. Ding, N., et al.: Delta tuning: a comprehensive study of parameter efficient methods for pre-trained language models. arXiv preprint arXiv:2203.06904 (2022)
8. Douillard, A., Cord, M., Ollion, C., Robert, T., Valle, E.: PODNet: pooled outputs distillation for small-tasks incremental learning. In: Vedaldi, A., Bischof, H., Brox, T., Frahm, J.-M. (eds.) ECCV 2020. LNCS, vol. 12365, pp. 86–102. Springer, Cham (2020). https://doi.org/10.1007/978-3-030-58565-5_6
9. Gao, P., et al.: Clip-adapter: better vision-language models with feature adapters. arXiv preprint arXiv:2110.04544 (2021)
10. He, K., Zhang, X., Ren, S., Sun, J.: Deep residual learning for image recognition. In: CVPR (2016)
11. Hou, S., Pan, X., Loy, C.C., Wang, Z., Lin, D.: Lifelong learning via progressive distillation and retrospection. In: ECCV (2018)
12. Houlsby, N., et al.: Parameter-efficient transfer learning for NLP. In: ICML (2019)
13. Kebede, A.F.: Oral cancer dataset, version 1 (2021). https://www.kaggle.com/datasets/ashenafifasilkebede/dataset
14. Kim, G., Liu, B., Ke, Z.: A multi-head model for continual learning via out-of-distribution replay. In: Conference on Lifelong Learning Agents (2022)
15. Kirkpatrick, J., et al.: Overcoming catastrophic forgetting in neural networks. In: PNAS (2017)
16. Krizhevsky, A., Hinton, G., et al.: Learning multiple layers of features from tiny images (2009)
17. Li, Z., Hoiem, D.: Learning without forgetting. TPAMI **40**(12), 2935–2947 (2017)
18. Lopez-Paz, D., Ranzato, M.: Gradient episodic memory for continual learning. In: NeurIPS (2017)
19. Rebuffi, S.A., Kolesnikov, A., Sperl, G., Lampert, C.H.: iCaRL: incremental classifier and representation learning. In: CVPR (2017)
20. Rusu, A.A., et al.: Progressive neural networks. In: NeurIPS (2016)
21. Sandler, M., Howard, A., Zhu, M., Zhmoginov, A., Chen, L.C.: Mobilenetv 2: inverted residuals and linear bottlenecks. In: CVPR (2018)
22. Shin, H., Lee, J.K., Kim, J., Kim, J.: Continual learning with deep generative replay. In: NeurIPS (2017)
23. Tan, M., Le, Q.: Efficientnet: rethinking model scaling for convolutional neural networks. In: ICML (2019)
24. Tschandl, P., Rosendahl, C., Kittler, H.: The ham10000 dataset, a large collection of multi-source dermatoscopic images of common pigmented skin lesions. Sci. Data **5**(1), 1–9 (2018)
25. Veeling, B.S., Linmans, J., Winkens, J., Cohen, T., Welling, M.: Rotation equivariant CNNs for digital pathology. In: Frangi, A.F., Schnabel, J.A., Davatzikos, C., Alberola-López, C., Fichtinger, G. (eds.) MICCAI 2018. LNCS, vol. 11071, pp. 210–218. Springer, Cham (2018). https://doi.org/10.1007/978-3-030-00934-2_24
26. Verma, V.K., Liang, K.J., Mehta, N., Rai, P., Carin, L.: Efficient feature transformations for discriminative and generative continual learning. In: CVPR (2021)

27. Wang, Z., et al.: Dualprompt: complementary prompting for rehearsal-free continual learning. In: Avidan, S., Brostow, G., Cissé, M., Farinella, G.M., Hassner, T. (eds.) ECCV 2022. LNCS, vol. 13686, pp. 631–648. Springer, Cham (2022). https://doi.org/10.1007/978-3-031-19809-0_36
28. Wang, Z., et al.: Learning to prompt for continual learning. In: CVPR (2022)
29. Wei, J., et al.: A petri dish for histopathology image analysis. In: Artificial Intelligence in Medicine (2021)
30. Yan, S., Xie, J., He, X.: Der: Dynamically expandable representation for class incremental learning. In: CVPR (2021)
31. Yang, J., Shi, R., Ni, B.: Medmnist classification decathlon: a lightweight automl benchmark for medical image analysis. In: ISBI (2021)
32. Yang, Y., Cui, Z., Xu, J., Zhong, C., Wang, R., Zheng, W.-S.: Continual learning with Bayesian model based on a fixed pre-trained feature extractor. In: de Bruijne, M., et al. (eds.) MICCAI 2021. LNCS, vol. 12905, pp. 397–406. Springer, Cham (2021). https://doi.org/10.1007/978-3-030-87240-3_38
33. Zhang, R., et al.: Tip-adapter: training-free adaption of clip for few-shot classification. In: Avidan, S., Brostow, G., Cissé, M., Farinella, G.M., Hassner, T. (eds.) ECCV 2022. LNCS, vol. 13695, pp. 493–510. Springer, Cham (2022). https://doi.org/10.1007/978-3-031-19833-5_29
34. Zhao, B., Xiao, X., Gan, G., Zhang, B., Xia, S.T.: Maintaining discrimination and fairness in class incremental learning. In: CVPR (2020)
35. Zheng, X., et al.: A deep learning model and human-machine fusion for prediction of EBV-associated gastric cancer from histopathology. Nature Commun. **13**(1), 2790 (2022)
36. Zhou, W., et al.: Ensembled deep learning model outperforms human experts in diagnosing biliary atresia from sonographic gallbladder images. Nat. Commun. **12**(1), 1259 (2021)

EdgeAL: An Edge Estimation Based Active Learning Approach for OCT Segmentation

Md Abdul Kadir[1(✉)], Hasan Md Tusfiqur Alam[1], and Daniel Sonntag[1,2]

[1] German Research Center for Artificial Intelligence (DFKI), Saarbrücken, Germany
{abdul.kadir,hasan.alam,daniel.sonntag}@dfki.de
[2] University of Oldenburg, Oldenburg, Germany

Abstract. Active learning algorithms have become increasingly popular for training models with limited data. However, selecting data for annotation remains a challenging problem due to the limited information available on unseen data. To address this issue, we propose EdgeAL, which utilizes the edge information of unseen images as *a priori* information for measuring uncertainty. The uncertainty is quantified by analyzing the divergence and entropy in model predictions across edges. This measure is then used to select superpixels for annotation. We demonstrate the effectiveness of EdgeAL on multi-class Optical Coherence Tomography (OCT) segmentation tasks, where we achieved a 99% dice score while reducing the annotation label cost to 12%, 2.3%, and 3%, respectively, on three publicly available datasets (Duke, AROI, and UMN). The source code is available at https://github.com/Mak-Ta-Reque/EdgeAL.

Keywords: Active Learning · Deep Learning · Segmentation · OCT

1 Introduction

In recent years, Deep Learning (DL) based methods have achieved considerable success in the medical domain for tasks including disease diagnosis and clinical feature segmentation [20,28]. However, their progress is often constrained as they require large labelled datasets. Labelling medical image data is a labour-intensive and time-consuming process that needs the careful attention of clinical experts. Active learning (AL) can benefit the iterative improvement of any intelligent diagnosis system by reducing the burden of extensive annotation effort [19,25].

Ophthalmologists use the segmentation of ocular Optical Coherence Tomography (OCT) images to diagnose, and treatment of eye diseases such as Diabetic Retinopathy (DR) and Diabetic Macular Edema (DME) [6]. Here, we propose a novel Edge estimation-based Active Learning EdgeAL framework for OCT image segmentation that leverages prediction uncertainty across the boundaries

Supplementary Information The online version contains supplementary material available at https://doi.org/10.1007/978-3-031-43895-0_8.

H. Greenspan et al. (Eds.): MICCAI 2023, LNCS 14221, pp. 79–89, 2023.
https://doi.org/10.1007/978-3-031-43895-0_8

of the semantic regions of input images. The Edge information is one of the image's most salient features, and it can boost segmentation accuracy when integrated into neural model training [13]. We formulate a novel acquisition function that leverages the variance of the predicted score across the gradient surface of the input to measure uncertainty. Empirical results show that EdgeAL achieves *state-of-the-art* performance with minimal annotation samples, using a seed set as small as 2% of unlabeled data.

2 Related Work

Active learning is a cost-effective strategy that selects the most informative samples for annotation to improve model performance based on uncertainty [11], data distribution [22], expected model change [4], and other criteria [1]. A simpler way to measure uncertainty can be realized using posterior probabilities of predictions, such as selecting instances with the least confidence [9,11], or computing class entropy [14].

Some uncertainty-based approaches have been directly used with deep neural networks [24]. Gal et al. [7] propose dropout-base Monte Carlo (MC) sampling to obtain uncertainty estimation. It uses multiple forward passes with dropout at different layers to generate uncertainty during inference. Ensemble-based methods also have been widely used where the variance between the prediction outcomes from a collection of models serve as the uncertainty [18,23,27].

Many AL methods have been adopted for segmentation tasks [8,15,18]. Gorriz et al. [8] propose an AL framework Melanoma segmentation by extending Cost-Effective Active Learning (CEAL) [26] algorithm where complimentary samples of both high and low confidence are selected for annotation. Mackowiak et al. [15] use a region-based selection approach and estimate model uncertainty using MC dropout to reduce human-annotation cost. Nath et al. [18] propose an ensemble-based method where multiple AL frameworks are jointly optimized, and a *query-by-committee* approach is adopted for sample selection. These methods do not consider any prior information to estimate uncertainty. Authors in [24] propose an AL framework for multi-view datasets [17] segmentation task where model uncertainty is estimated based on *Kullback-Leibler* (KL) divergence of posterior probability distributions for a disjoint subset of prior features such as depth, and camera position.

However, viewpoint information is not always available in medical imaging. We leverage edge information as a prior for AL sampling based on previous studies where edge information has improved the performance of segmentation tasks [13]. To our knowledge, there has yet to be any exploration of using image edges as an *a priori* in active learning.

There has not been sufficient work other than [12] related to Active Learning for OCT segmentation. Their approach requires foundation models [10] to be pre-trained on large-scale datasets in similar domains, which could be infeasible to collect due to data privacy. On the other hand, our method requires a few samples (∼2%) for initial training, overcoming the limitation of the need for a large dataset.

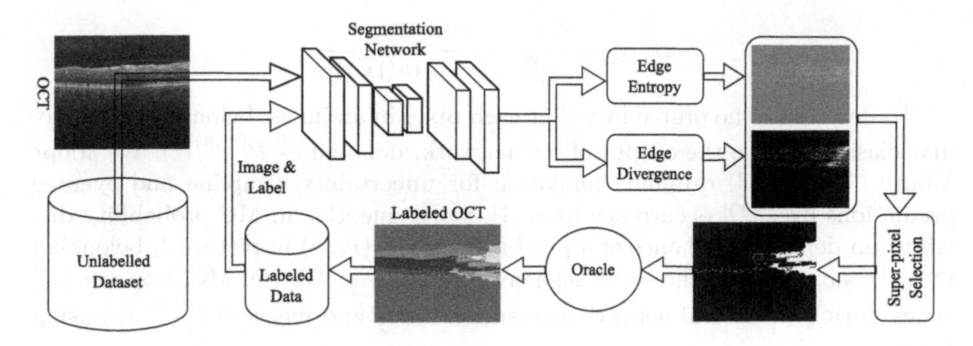

Fig. 1. The figure above illustrates the workflow of our AL framework. It first computes an OCT image's *edge entropy* and *edge divergence* maps. Later, it calculates the overlaps between superpixels based on the divergence and entropy map to recommend an annotation region.

3 Methodology

Figure 1 shows that our active learning technique consists of four major stages. First, we train the network on a subset of labeled images, usually a tiny percentage of the total collection (e.g., 2%). Following that, we compute uncertainty values for input instances and input areas. Based on this knowledge, we select superpixels to label and obtain annotations from a simulated oracle.

3.1 Segmentation Network

We trained our OCT semantic segmentation model using a randomly selected small portion of the labeled data D_s, seed set, keeping the rest for oracle imitation. We choose Y-net-gen-ffc (YN*) without pre-retrained weight initialization as our primary architecture due to its superior performance [6].

3.2 Uncertainty in Prediction

EdgeAL seeks to improve the model's performance by querying uncertain areas on unlabeled data D_u after training it on a seed set D_s. To accomplish this, we have created a novel edge-based uncertainty measurement method. We compute the *edge entropy* score and *edge divergence* score - to assess the prediction ambiguity associated with the edges. Figure 2 depicts examples of input OCT, measured *edge entropy*, and *edge kl-divergence* corresponding to the input.

Entropy Score on Edges. Analyzing the edges of raw OCT inputs yields critical information on features and texture in images. They may look noisy, but they summarize all the alterations in a picture. The Sobel operator can be used to identify edges in the input image [13]. Let us define the normalized absolute value of edges of an image I_i by S_i. $|\nabla I_i|$ is the absolute gradient.

$$S_i = \frac{|\nabla I_i| - min(|\nabla I_i|)}{max(|\nabla I_i|) - min(|\nabla I_i|)}$$

To determine the probability that each pixel in an image belongs to a particular class c, we use the output of our network, denoted as $P_i^{(m,n)}(c)$. We adopt Monte Carlo (MC) dropout simulation for uncertainty sampling and average predictions over $|D|$ occurrence from [7]. Consequently, an MC probability distribution depicts the chance of a pixel at location (m, n) in picture I_i belonging to a class c, and C is the set of segmentation classes. We run MC dropouts $|D|$ times during the neural network assessment mode and measure $P_i^{(m,n)}(c)$ using Eq. 1.

$$P_i^{(m,n)}(c) = \frac{1}{|D|} \sum_{d=1}^{D} P_{i,d}^{(m,n)}(c) \tag{1}$$

Following Zhao et al. [30], we apply contextual calibration on $P_i^{(m,n)}(c)$ by S_i to prioritize significant input surface variations. Now, S_i is linked with a probability distribution, with $\phi_i^{(m,n)}(c)$ having information about the edges of input. This formulation makes our implementation unique from other active learning methods in image segmentation.

$$\phi_i^{m,n}(c) = \frac{e^{P_i^{(m,n)}(c) \bullet S_i(m,n)}}{\sum_{k \in C} e^{P_i^{(m,n)}(k) \bullet S_i^{(m,n)}}} \tag{2}$$

We name $\phi_i^{m,n}(c)$ as contextual probability and define our *edge entropy* by following entropy formula of [14].

$$EE_i^{m,n} = -\sum_{c \in C} \phi_i^{m,n}(c) \log(\phi_i^{m,n}(c)) \tag{3}$$

Divergence Score on Edges. In areas with strong edges/gradients, *edge entropy* reflects the degree of inconsistency in the network's prediction for each input pixel. However, the degree of this uncertainty must also be measured. *KL-divergence* is used to measure the difference in inconsistency between $P_i^{(m,n)}$ and $\phi_i^{(m,n)}$ for a pixel (m, n) in an input image based on the idea of self-knowledge distillation I_i [29]. The *edge divergence* $ED_i^{m,n}$ score can be formalized using Eq. 1 and 2.

$$ED_i^{(m,n)} = D_{KL}(P_i^{(m,n)} || \phi_i^{(m,n)})$$

where $D_{KL}(P_i^{(m,n)} || \phi_i^{(m,n)})$ measures the difference between model prediction probability and contextual probability for pixels belonging to edges of the input (Fig. 2c).

3.3 Superpixel Selection

Clinical images have sparse representation, which can be beneficial for active learning annotation [15]. We use a traditional segmentation technique, SEEDS

[2], to leverage the local structure from images for finding superpixels. Annotating superpixels and regions for active learning may be more beneficial to the user than annotating the entire picture [15].

We compute mean *edge entropy* EE_i^r and mean *edge divergence* ED_i^d for a particular area r within a superpixel.

$$EE_i^r = \frac{1}{|r|} \sum_{(m,n)\in r} EE_i^{(m,n)} \tag{4}$$

$$ED_i^r = \frac{1}{|r|} \sum_{(m,n)\in r} ED_i^{(m,n)} \tag{5}$$

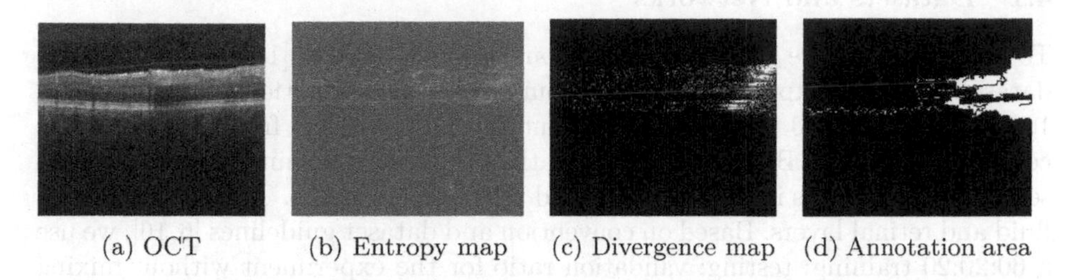

(a) OCT (b) Entropy map (c) Divergence map (d) Annotation area

Fig. 2. The figures depict an example of (a) OCT slice with corresponding (b) *edge entropy* map, (c) *edge divergence* map, (d) query regions for annotation by our EdgeAL. The figures reveal that there is less visibility of retinal layer separation lines on the right side of the OCT slice, which could explain the model's high uncertainty in that region.

Where $|r|$ is the amount of pixels in the superpixel region. We use regional entropy to find the optimal superpixel for our selection strategy and pick the one with the most significant value based on the literature [24].

$$(i, r) = \underset{(j,s)}{\arg\max} \quad EE_j^s \tag{6}$$

Following [24], we find the subset of superpixels in the dataset with a 50% overlap (r, i). Let us call it set R. We choose the superpixels with the largest *edge divergence* to determine the ultimate query (sample) for annotation.

$$(p, q) = \underset{(j,s)\in R}{\arg\max} \{ED_j^s \quad | \quad (j, s) \cap (i, r); (i, r) \in D_u)\} \tag{7}$$

After each selection, we remove the superpixels from R. The selection process runs until we have K amount of superpixels being selected from R.

After getting the selected superpixel maps, we receive the matching ground truth information for the selected superpixel regions from the oracle. The model is then freshly trained on the updated labeled dataset for the next active learning iteration.

4 Experiments and Results

This section will provide a detailed overview of the datasets and architectures employed in our experiments. Subsequently, we will present the extensive experimental results and compare them with other state-of-the-art methods to showcase the effectiveness of our approach. We compare our AL method with nine well-known strategies: softmax margin **(MAR)** [9], softmax confidence **(CONF)** [26], softmax entropy **(ENT)** [14], MC dropout entropy **(MCDR)** [7], Core-set selection **(CORESET)** [23], **(CEAL)** [8], and regional MC dropout entropy **(RMCDR)** [15], maximum representations **(MAXRPR)** [27], and random selection **(Random)**.

4.1 Datasets and Networks

To test EdgeAL, we ran experiments on Duke [3], AROI [16], and UMN [21] datasets in which experts annotated ground truth segmentations. Duke contains 100 B-scans from 10 patients, AROI contains 1136 B-scans from 24, and UMN contains 725 OCT B-scans from 29 patients. There are nine, eight, and two segmentation classes in Duke, AROI, and UMN, respectively. These classes cover fluid and retinal layers. Based on convention and dataset guidelines [6,16], we use a 60:20:20 training: testing: validation ratio for the experiment without mixing one patient's data in any of the splits. Further, we resized all the images and ground truths to 224×224 using Bilinear approximation. Moreover, we run a 5-fold cross-validation (CV) on the Duke dataset without mixing individual patient data in each fold's training, testing, and validation set. Table 1 summarizes the 5-fold CV results.

Table 1. The Table summarizes 5-fold cross-validation results (mean dice) for active learning methods and EdgeAL on the Duke dataset. EdgeAL outperforms other methods, achieving 99% performance with just 12% annotated data.

GT(%)	RMCDR	CEAL	CORESET	**EdgeAL**	MAR	MAXRPR
2%	**0.40 ± 0.05**	**0.40 ± 0.05**	0.38 ± 0.04	**0.40 ± 0.05**	**0.40 ± 0.09**	**0.41 ± 0.04**
12%	0.44 ± 0.04	0.54 ± 0.04	0.44 ± 0.05	**0.82 ± 0.03**	0.44 ± 0.03	0.54 ± 0.09
22%	0.63 ± 0.05	0.54 ± 0.04	0.62 ± 0.04	**0.83 ± 0.03**	0.58 ± 0.04	0.67 ± 0.07
33%	0.58 ± 0.07	0.55 ± 0.06	0.57 ± 0.04	**0.81 ± 0.04**	0.67 ± 0.03	0.61 ± 0.03
43%	0.70 ± 0.03	0.79 ± 0.03	0.69 ± 0.03	**0.83 ± 0.02**	0.70 ± 0.04	0.80 ± 0.04
100%	0.82 ± 0.03	0.82 ± 0.03	0.82 ± 0.03	0.82 ± 0.02	0.83 ± 0.02	0.83 ± 0.02

We run experiments using Y-net(YN) [6], U-net (UN) [10], and DeepLab-V3 (DP-V3) [24] with ResNet and MobileNet backbones [10]. We present the results in Table 2. No pre-trained weights were employed in the execution of our studies other than the ablation study presented in Table 2. We apply mixed loss of dice and cross-entropy and Adam as an optimizer, with learning rates of 0.005 and

Table 2. The table summarizes the test performance (mean dice) of various active learning algorithms on different deep learning architectures, including pre-trained weights, trained on only 12% actively selected data from the Duke dataset. The results (mean ± sd) are averaged after running two times in two random seeds. Superscript 'r' represents ResNet, 'm' represents MobileNet version 3 backbones, and '†' indicates that the networks are initialized with pre-trained weights from ImageNet [5].

Arch.	p100	EdgeAL	CEAL	CORESET	RMCDR	MAXRPR
YN^* [6]	**0.83 ± 0.02**	**0.83 ± 0.01**	0.52 ± 0.01	0.45 ± 0.02	0.44 ± 0.01	0.56 ± 0.01
YN [6]	0.82 ± 0.02	**0.81 ± 0.02**	0.48 ± 0.01	0.47 ± 0.02	0.45 ± 0.01	0.53 ± 0.01
UN [10]	0.79 ± 0.02	**0.80 ± 0.01**	0.39 ± 0.01	0.48 ± 0.02	0.63 ± 0.01	0.51 ± 0.01
$DP\text{-}V3^{r}$	0.74 ± 0.04	**0.74 ± 0.02**	0.62 ± 0.01	0.49 ± 0.01	0.57 ± 0.01	0.61 ± 0.01
$DP\text{-}V3^{m}$	0.61 ± 0.01	**0.61 ± 0.01**	0.28 ± 0.02	0.25 ± 0.01	0.59 ± 0.02	0.51 ± 0.01
$DP\text{-}V3^{r,\dagger}$	0.78 ± 0.01	**0.79 ± 0.01**	0.29 ± 0.01	0.68 ± 0.01	0.68 ± 0.01	0.73 ± 0.01
$DP\text{-}V3^{m,\dagger}$	0.78 ± 0.01	**0.79 ± 0.01**	0.18 ± 0.01	0.57 ± 0.01	**0.79 ± 0.02**	0.75 ± 0.02

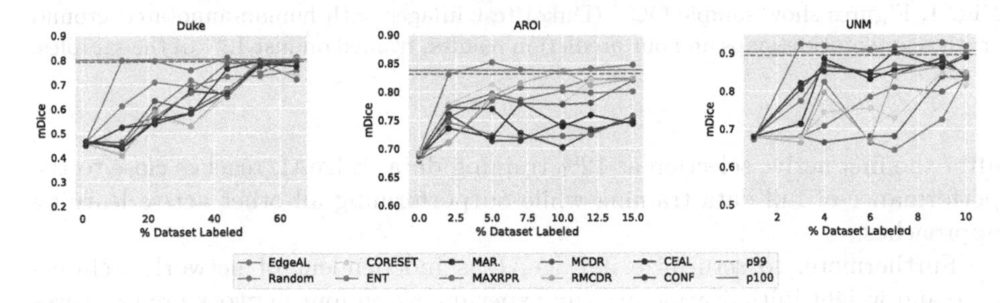

Fig. 3. EdgeAL's and other AL methods' performances (mean dice score) compared to baselines for Duke, AROI, and UNM datasets. Solid and dashed lines represent model performance and 99% of it with 100% labeled data.

weight decay of 0.0004, trained for 100 epochs with a maximum batch size of 10 across all AL iterations. We follow the hyperparameter settings and evaluation metric (dice score) of [6], which is the baseline of our experiment.

4.2 Comparisons

Figure 3 compares the performance of EdgeAL with other contemporary active learning algorithms across three datasets. Results show EdgeAL outperforms other methods on all 3 datasets. Our method can achieve 99% of maximum model performance consistently with about 12% (∼8 samples), 2.3% (∼16 samples), and 3% (∼14 samples) labeled data on Duke, AROI, and UNM datasets. Other AL methods, CEAL, RMCDR, CORESET, and MAR, do not perform consistently in all three datasets. We used the same segmentation network YN^* and hyperparameters (described in Sect. 3) for a fair comparison.

Our 5-fold CV result in Table 1 also concludes similarly. We see that after training on a 2% seed set, all methods have similar CV performance; however,

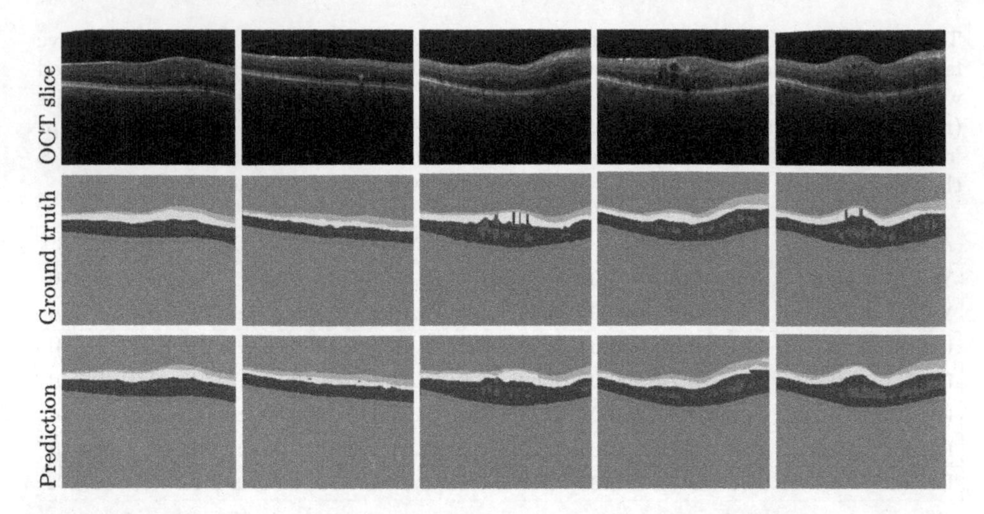

Fig. 4. Figures show sample OCT (Duke) test images with human-annotated ground truth segmentation maps and our prediction results, trained on just 12% of the samples.

after the first active selection at 12% training data, EdgeAL reaches close to the performance of full data training while outperforming all other active learning approaches.

Furthermore, to scrutinize if EdgeAL is independent of network architecture and weight initialization, we run experiments on four network architectures with default weight initialization of PyTorch (LeCun initialization)[1] and imagenet weight initialization. Table 2 presents the test performance after training on 12% of actively selected data. These results also conclude that EdgeAL's performance is independent of the architecture and weight choices, while other active learning methods (RMCDR, MAXRPR) only perform well in pre-trained models (Table 2).

5 Conclusion

EdgeAL is a novel active learning technique for OCT image segmentation, which can accomplish results similar to full training with a small amount of data by utilizing edge information to identify regions of uncertainty. Our method can reduce the labeling effort by requiring only a portion of an image to annotate and is particularly advantageous in the medical field, where labeled data can be scarce. EdgeAL's success in OCT segmentation suggests that a significant amount of data is not always required to learn data distribution in medical imaging. Edges are a fundamental image characteristic, allowing EdgeAL to be adapted for other domains without significant modifications, which leads us to future works.

[1] https://pytorch.org.

Acknowledgements. This work was partially funded by the German Federal Ministry of Education and Research (BMBF) under grant number 16SV8639 (Ophthalmo-AI) and supported by the Lower Saxony Ministry of Science and Culture and the Endowed Chair of Applied Artificial Intelligence (AAI) of the University of Oldenburg.

References

1. Bai, F., Xing, X., Shen, Y., Ma, H., et al.: Discrepancy-based active learning for weakly supervised bleeding segmentation in wireless capsule endoscopy images. In: Wang, L., Dou, Q., Fletcher, P.T., Speidel, S., Li, S. (eds.) MICCAI 2022. LNCS, vol. 13438, pp. 24–34. Springer, Cham (2022). https://doi.org/10.1007/978-3-031-16452-1_3
2. Van den Bergh, M., Boix, X., Roig, G., de Capitani, B., Van Gool, L.: SEEDS: superpixels extracted via energy-driven sampling. In: Fitzgibbon, A., Lazebnik, S., Perona, P., Sato, Y., Schmid, C. (eds.) ECCV 2012. LNCS, vol. 7578, pp. 13–26. Springer, Heidelberg (2012). https://doi.org/10.1007/978-3-642-33786-4_2
3. Chiu, S.J., Allingham, M.J., Mettu, P.S., Cousins, S.W., et al.: Kernel regression based segmentation of optical coherence tomography images with diabetic macular edema. Biomed. Opt. Express **6**(4), 1172–1194 (2015)
4. Dai, C., et al.: Suggestive annotation of brain tumour images with gradient-guided sampling. In: Martel, A.L., et al. (eds.) MICCAI 2020. LNCS, vol. 12264, pp. 156–165. Springer, Cham (2020). https://doi.org/10.1007/978-3-030-59719-1_16
5. Deng, J., Dong, W., Socher, R., Li, L.J., et al.: Imagenet: a large-scale hierarchical image database. In: 2009 IEEE Conference on Computer Vision and Pattern Recognition, pp. 248–255 (2009)
6. Farshad, A., Yeganeh, Y., Gehlbach, P., Navab, N.: Y-net: a spatiospectral dual-encoder network for medical image segmentation. In: Wang, L., Dou, Q., Fletcher, P.T., Speidel, S., Li, S. (eds.) MICCAI 2022. LNCS, vol. 13432, pp. 582–592. Springer, Cham (2022). https://doi.org/10.1007/978-3-031-16434-7_56
7. Gal, Y., Islam, R., Ghahramani, Z.: Deep Bayesian active learning with image data. In: International Conference on Machine Learning, pp. 1183–1192. PMLR (2017)
8. Gorriz, M., Carlier, A., Faure, E., Giro-i Nieto, X.: Cost-effective active learning for melanoma segmentation. arXiv preprint arXiv:1711.09168 (2017)
9. Joshi, A.J., Porikli, F., Papanikolopoulos, N.: Multi-class active learning for image classification. In: 2009 IEEE Conference on Computer Vision and Pattern Recognition, pp. 2372–2379. IEEE (2009)
10. Khan, A., Sohail, A., Zahoora, U., Qureshi, A.S.: A survey of the recent architectures of deep convolutional neural networks. Artif. Intell. Rev. **53**, 5455–5516 (2020)
11. Lee, B., Paeng, K.: A robust and effective approach towards accurate metastasis detection and pN-stage classification in breast cancer. In: Frangi, A.F., Schnabel, J.A., Davatzikos, C., Alberola-López, C., Fichtinger, G. (eds.) MICCAI 2018. LNCS, vol. 11071, pp. 841–850. Springer, Cham (2018). https://doi.org/10.1007/978-3-030-00934-2_93
12. Li, X., Niu, S., Gao, X., Liu, T., Dong, J.: Unsupervised domain adaptation with self-selected active learning for cross-domain OCT image segmentation. In: Mantoro, T., Lee, M., Ayu, M.A., Wong, K.W., Hidayanto, A.N. (eds.) ICONIP 2021. LNCS, vol. 13109, pp. 585–596. Springer, Cham (2021). https://doi.org/10.1007/978-3-030-92270-2_50

13. Lu, F., Tang, C., Liu, T., Zhang, Z., et al.: Multi-attention segmentation networks combined with the sobel operator for medical images. Sensors **23**(5), 2546 (2023)
14. Luo, W., Schwing, A., Urtasun, R.: Latent structured active learning. In: Advances in Neural Information Processing Systems, vol. 26 (2013)
15. Mackowiak, R., Lenz, P., Ghori, O., Diego, F., et al.: Cereals-cost-effective region-based active learning for semantic segmentation. arXiv preprint arXiv:1810.09726 (2018)
16. Melinščak, M., Radmilovič, M., Vatavuk, Z., Lončarić, S.: AROI: annotated retinal oct images database. In: 2021 44th International Convention on Information, Communication and Electronic Technology (MIPRO), pp. 371–376 (2021)
17. Muslea, I., Minton, S., Knoblock, C.A.: Active learning with multiple views. J. Artif. Intell. Res. **27**, 203–233 (2006)
18. Nath, V., Yang, D., Landman, B.A., Xu, D., et al.: Diminishing uncertainty within the training pool: active learning for medical image segmentation. IEEE Trans. Med. Imaging **40**(10), 2534–2547 (2020)
19. Nath, V., Yang, D., Roth, H.R., Xu, D.: Warm start active learning with proxy labels and selection via semi-supervised fine-tuning. In: Wang, L., Dou, Q., Fletcher, P.T., Speidel, S., Li, S. (eds.) MICCAI 2022. LNCS, vol. 13438, pp. 297–308. Springer, Cham (2022). https://doi.org/10.1007/978-3-031-16452-1_29
20. Nguyen, D.M.H., Ezema, A., Nunnari, F., Sonntag, D.: A visually explainable learning system for skin lesion detection using multiscale input with attention U-net. In: Schmid, U., Klügl, F., Wolter, D. (eds.) KI 2020. LNCS (LNAI), vol. 12325, pp. 313–319. Springer, Cham (2020). https://doi.org/10.1007/978-3-030-58285-2_28
21. Rashno, A., Nazari, B., Koozekanani, D.D., Drayna, P.M., et al.: Fully-automated segmentation of fluid regions in exudative age-related macular degeneration subjects: kernel graph cut in neutrosophic domain. PLoS ONE **12**(10), e0186949 (2017)
22. Samrath, S., Sayna, E., Trevor, D.: Variational adversarial active learning. In: 2019 IEEE/CVF International Conference on Computer Vision (ICCV). IEEE (2019)
23. Sener, O., Savarese, S.: Active learning for convolutional neural networks: a core-set approach. arXiv preprint arXiv:1708.00489 (2017)
24. Siddiqui, Y., Valentin, J., Nießner, M.: Viewal: active learning with viewpoint entropy for semantic segmentation. In: Proceedings of the IEEE/CVF Conference on Computer Vision and Pattern Recognition, pp. 9433–9443 (2020)
25. Tusfiqur, H.M., Nguyen, D.M.H., Truong, M.T.N., Nguyen, T.A., et al.: DRG-net: interactive joint learning of multi-lesion segmentation and classification for diabetic retinopathy grading (2022). https://doi.org/10.48550/ARXIV.2212.14615
26. Wang, K., Zhang, D., Li, Y., Zhang, R., et al.: Cost-effective active learning for deep image classification. IEEE Trans. Circuits Syst. Video Technol. **27**(12), 2591–2600 (2016)
27. Yang, L., Zhang, Y., Chen, J., Zhang, S., Chen, D.Z.: Suggestive annotation: a deep active learning framework for biomedical image segmentation. In: Descoteaux, M., Maier-Hein, L., Franz, A., Jannin, P., Collins, D.L., Duchesne, S. (eds.) MICCAI 2017. LNCS, vol. 10435, pp. 399–407. Springer, Cham (2017). https://doi.org/10.1007/978-3-319-66179-7_46
28. Yuan, W., Lu, D., Wei, D., Ning, M., et al.: Multiscale unsupervised retinal edema area segmentation in oct images. In: Wang, L., Dou, Q., Fletcher, P.T., Speidel, S., Li, S. (eds.) MICCAI 2022. LNCS, vol. 13432, pp. 667–676. Springer, Cham (2022). https://doi.org/10.1007/978-3-031-16434-7_64

29. Yun, S., Park, J., Lee, K., Shin, J.: Regularizing class-wise predictions via self-knowledge distillation. In: Proceedings of the IEEE/CVF Conference on Computer Vision and Pattern Recognition (CVPR) (2020)
30. Zhao, Z., Wallace, E., Feng, S., Klein, D., et al.: Calibrate before use: Improving few-shot performance of language models. In: Meila, M., Zhang, T. (eds.) Proceedings of the 38th International Conference on Machine Learning, Proceedings of Machine Learning Research, vol. 139, pp. 12697–12706. PMLR (2021)

Adaptive Region Selection for Active Learning in Whole Slide Image Semantic Segmentation

Jingna Qiu[1(✉)], Frauke Wilm[1,2], Mathias Öttl[2], Maja Schlereth[1], Chang Liu[2], Tobias Heimann[3], Marc Aubreville[4], and Katharina Breininger[1]

[1] Department Artificial Intelligence in Biomedical Engineering, Friedrich-Alexander-Universität Erlangen-Nürnberg, Erlangen, Germany
jingna.qiu@fau.de

[2] Pattern Recognition Lab, Department of Computer Science, Friedrich-Alexander-Universität Erlangen-Nürnberg, Erlangen, Germany

[3] Digital Technology and Innovation, Siemens Healthineers, Erlangen, Germany

[4] Technische Hochschule Ingolstadt, Ingolstadt, Germany

Abstract. The process of annotating histological gigapixel-sized whole slide images (WSIs) at the pixel level for the purpose of training a supervised segmentation model is time-consuming. Region-based active learning (AL) involves training the model on a limited number of annotated image regions instead of requesting annotations of the entire images. These annotation regions are iteratively selected, with the goal of optimizing model performance while minimizing the annotated area. The standard method for region selection evaluates the informativeness of all square regions of a specified size and then selects a specific quantity of the most informative regions. We find that the efficiency of this method highly depends on the choice of AL step size (i.e., the combination of region size and the number of selected regions per WSI), and a suboptimal AL step size can result in redundant annotation requests or inflated computation costs. This paper introduces a novel technique for selecting annotation regions adaptively, mitigating the reliance on this AL hyperparameter. Specifically, we dynamically determine each region by first identifying an informative area and then detecting its optimal bounding box, as opposed to selecting regions of a uniform predefined shape and size as in the standard method. We evaluate our method using the task of breast cancer metastases segmentation on the public CAMELYON16 dataset and show that it consistently achieves higher sampling efficiency than the standard method across various AL step sizes. With only 2.6% of tissue area annotated, we achieve full annotation performance and thereby substantially reduce the costs of annotating a WSI dataset. The source code is available at https://github.com/DeepMicroscopy/AdaptiveRegionSelection.

Keywords: Active learning · Region selection · Whole slide images

---

Supplementary Information The online version contains supplementary material available at https://doi.org/10.1007/978-3-031-43895-0_9.

H. Greenspan et al. (Eds.): MICCAI 2023, LNCS 14221, pp. 90–100, 2023.
https://doi.org/10.1007/978-3-031-43895-0_9

1 Introduction

Semantic segmentation on histological whole slide images (WSIs) allows precise detection of tumor boundaries, thereby facilitating the assessment of metastases [3] and other related analytical procedures [17]. However, pixel-level annotations of gigapixel-sized WSIs (e.g. $100,000 \times 100,000$ pixels) for training a segmentation model are difficult to acquire. For instance, in the CAMELYON16 breast cancer metastases dataset [10], 49.5% of WSIs contain metastases that are smaller than 1% of the tissue, requiring a high level of expertise and long inspection time to ensure exhaustive tumor localization; whereas other WSIs have large tumor lesions and require a substantial amount of annotation time for boundary delineation [18]. Identifying potentially informative image regions (i.e., providing useful information for model training) allows requesting the minimum amount of annotations for model optimization, and a decrease in annotated area reduces both localization and delineation workloads. The challenge is to effectively select annotation regions in order to achieve full annotation performance with the least annotated area, resulting in high sampling efficiency.

We use region-based active learning (AL) [13] to progressively identify annotation regions, based on iteratively updated segmentation models. Each region selection process consists of two steps. First, the prediction of the most recently trained segmentation model is converted to a priority map that reflects informativeness of each pixel. Existing studies on WSIs made extensive use of informativeness measures that quantify model uncertainty (e.g., least confidence [8], maximum entropy [5] and highest disagreement between a set of models [19]). The enhancement of priority maps, such as highlighting easy-to-label pixels [13], edge pixels [6] or pixels with a low estimated segmentation quality [2], is also a popular area of research. Second, on the priority map, regions are selected according to a region selection method. Prior works have rarely looked into region selection methods; the majority followed the standard approach [13] where a sliding window divides the priority map into fixed-sized square regions, the selection priority of each region is calculated as the cumulative informativeness of its constituent pixels, and a number of regions with the highest priorities are then selected. In some other works, only non-overlapping or sparsely overlapped regions were considered to be candidates [8,19]. Following that, some works used additional criteria to filter the selected regions, such as finding a representative subset [5,19]. All of these works selected square regions of a manually predefined size, disregarding the actual shape and size of informative areas.

This work focuses on region selection methods, a topic that has been largely neglected in literature until now, but which we show to have a great impact on AL sampling efficiency (i.e., the annotated area required to reach the full annotation performance). We discover that the sampling efficiency of the aforementioned standard method decreases as the AL step size (i.e., the annotated area at each AL cycle, determined by the multiplication of the region size and the number of selected regions per WSI) increases. To avoid extensive AL step size tuning, we propose an adaptive region selection method with reduced reliance on this AL hyperparameter. Specifically, our method dynamically determines an annotation

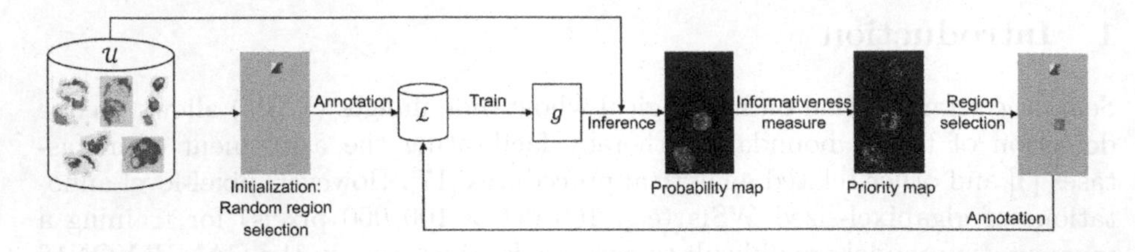

Fig. 1. Region-based AL workflow for selecting annotation regions. The exemplary selected regions are of size 8192×8192 pixels. (Image resolution: $0.25 \frac{\mu m}{px}$)

region by first identifying an informative area with connected component detection and then detecting its bounding box. We test our method using a breast cancer metastases segmentation task on the public CAMELYON16 dataset and demonstrate that determining the selected regions individually provides greater flexibility and efficiency than selecting regions with a uniform predefined shape and size, given the variability in histological tissue structures. Results show that our method consistently outperforms the standard method by providing a higher sampling efficiency, while also being more robust to AL step size choices. Additionally, our method is especially beneficial for settings where a large AL step size is desirable due to annotator availability or computational restrictions.

2 Method

2.1 Region-Based Active Learning for WSI Annotation

We are given an unlabeled pool $\mathcal{U} = \{X_1 \dots X_n\}$, where $X_i \in \mathbb{R}^{W_i \times H_i}$ denotes the ith WSI with width W_i and height H_i. Initially, X_i has no annotation; regions are iteratively selected from it and annotated across AL cycles. We denote the jth annotated rectangular region in X_i as $R_{ij} = (c_x^{ij}, c_y^{ij}, w^{ij}, h^{ij})$, where (c_x^{ij}, c_y^{ij}) are the center coordinates of the region and w^{ij}, h^{ij} are the width and height of that region, respectively. In the standard region selection method, where fixed-size square regions are selected, $w^{ij} = h^{ij} = l, \forall i, j$, where l is predefined.

Figure 1 illustrates the workflow of region-based AL for WSI annotation. The goal is to iteratively select and annotate potentially informative regions from WSIs in $\mathcal{U}$ to enrich the labeled set $\mathcal{L}$ in order to effectively update the model g. To begin, k regions (each containing at least 10% of tissue) per WSI are randomly selected and annotated to generate the initial labeled set $\mathcal{L}$. The model g is then trained on $\mathcal{L}$ and predicts on $\mathcal{U}$ to select k new regions from each WSI for the new round of annotation. The newly annotated regions are added to $\mathcal{L}$ for retraining g in the next AL cycle. The train-select-annotate process is repeated until a certain performance of g or annotation budget is reached.

The selection of k new regions from X_i is performed in two steps based on the model prediction $P_i = g(X_i)$. First, P_i is converted to a priority map M_i using a per-pixel informativeness measure. Second, k regions are selected based on M_i using a region selection method. The informativeness measure is not the focus of this study, we therefore adopt the most commonly used one that quantifies model uncertainty (details in Sect. 3.2). Next we describe the four region selection methods evaluated in this work.

2.2 Region Selection Methods

Random. This is the baseline method where k regions of size $l \times l$ are randomly selected. Each region contains at least 10% of tissue and does not overlap with other regions. ***Standard*** [13] M_i is divided into overlapping regions of a fixed size $l \times l$ using a sliding window with a stride of 1 pixel. The selection priority of each region is calculated as the summed priority of the constituent pixels, and k regions with the highest priorities are then selected. Non-maximum suppression is used to avoid selecting overlapping regions. ***Standard (non-square)*** We implement a generalized version of the standard method that allows non-square region selections by including multiple region candidates centered at each pixel with various aspect ratios. To save computation and prevent extreme shapes, such as those with a width or height of only a few pixels, we specify a set of candidates as depicted in Fig. 2. Specifically, we define a variable region width w as spanning from $\frac{1}{2}l$ to l with a stride of 256 pixels and determine the corresponding region height as $h = \frac{l^2}{w}$. ***Adaptive*** (proposed) Our method allows for selecting regions with variable aspect ratios and sizes to accommodate histological tissue variability. The k regions are selected sequentially; when selecting the jth region R_{ij} in X_i, we first set the priorities of all pixels in previously selected regions (if any) to zero. We then find the highest priority pixel (c_x^{ij}, c_y^{ij}) on M_i; a median filter with a kernel size of 3 is applied beforehand to remove outliers. Afterwards, we create a mask on M_i with an intensity threshold of τth percentile of intensities in M_i, detect the connected component containing (c_x^{ij}, c_y^{ij}), and select its bounding box. As depicted in Fig. 3, τ is determined by performing a bisection search over $[98, 100]^{th}$ percentiles, such that the bounding box size is in range $[\frac{1}{2}l \times \frac{1}{2}l, \frac{3}{2}l \times \frac{3}{2}l]$. This size range is chosen to be comparable to the other three methods, which select regions of size l^2. Note that *Standard (non-square)* can be understood as an ablation study of the proposed method *Adaptive* to examine the effect of variable region shape by maintaining constant region size.

2.3 WSI Semantic Segmentation Framework

This section describes the breast cancer metastases segmentation task we use for evaluating the AL region selection methods. The task is performed with patch-wise classification, where the WSI is partitioned into patches, each patch is classified as to whether it contains metastases, and the results are assembled. **Training.** The patch classification model $h(\mathbf{x}, \mathbf{w}) : \mathbb{R}^{d \times d} \rightarrow [0, 1]$ takes as input a patch $\mathbf{x}$ and outputs the probability $p(y = 1|\mathbf{x}, \mathbf{w})$ of containing metastases, where $\mathbf{w}$ denotes model parameters. Patches are extracted from the annotated regions at 40× magnification (0.25 $\frac{\mu m}{px}$) with $d = 256$ pixels. Following [11], a patch is labeled as positive if the center 128 × 128 pixels area contains at least

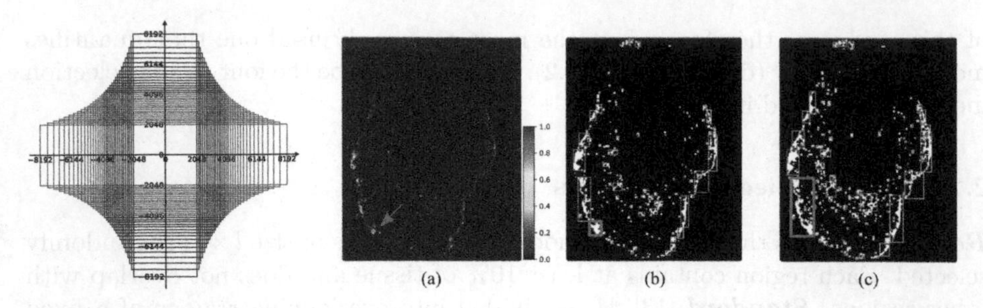

Fig. 2. *Standard (non-square)*: Region candidates for $l = 8192$ pixels.

Fig. 3. *Adaptive*: (a) Priority map M_i and the highest priority pixel (arrow). (b–c) Bisection search of τ: (b) $\tau = 99^{th}$, (c) $\tau = 98.5^{th}$.

one metastasis pixel and negative otherwise. In each training epoch, 20 patches per WSI are extracted at random positions within the annotated area; for WSIs containing annotated metastases, positive and negative patches are extracted with equal probability. A patch with less than 1% tissue content is discarded. Data augmentation includes random flip, random rotation, and stain augmentation [12]. **Inference.** X_i is divided into a grid of uniformly spaced patches (40× magnification, $d = 256$ pixels) with a stride s. The patches are predicted using the trained patch classification model and the results are stitched to a probability map $P_i \in [0, 1]^{W'_i \times H'_i}$, where each pixel represents a patch prediction. The patch extraction stride s determines the size of P_i ($W'_i = \frac{W_i}{s}, H'_i = \frac{H_i}{s}$).

3 Experiments

3.1 Dataset

We used the publicly available CAMELYON16 Challenge dataset [10], licensed under the Creative Commons CC0 license. The collection of the data was approved by the responsible ethics committee (Commissie Mensgebonden Onderzoek regio Arnhem-Nijmegen). The CAMELYON16 dataset consists of 399 Hematoxylin & Eosin (H&E)-stained WSIs of sentinel axillary lymph node sections. The training set contains 111 WSIs with and 159 WSIs without breast cancer metastases, and each WSI with metastases is accompanied by pixel-level contour annotations delineating the boundaries of the metastases. We randomly split a stratified 30% subset of the training set as the validation set for model selection. The test set contains 48 WSIs with and 80 WSIs without metastases [1].

3.2 Implementation Details

Training Schedules. We use MobileNet_v2 [15] initialized with ImageNet [14] weights as the backbone of the patch classification model. It is extended with two

[1] Test_114 is excluded due to non-exhaustive annotation, as stated by data provider.

fully-connected layers with sizes of 512 and 2, followed by a softmax activation layer. The model is trained for up to 500 epochs using cross-entropy loss and the Adam optimizer [7], and is stopped early if the validation loss stagnates for 100 consecutive epochs. Model selection is guided by the lowest validation loss. The learning rate is scheduled by the one cycle policy [16] with a maximum of 0.0005. The batch size is 32. We used Fastai v1 [4] for model training and testing. The running time of one AL cycle (select-train-test) on a single NVIDIA Geforce RTX3080 GPU (10GB) is around 7 h.

Active Learning Setups. Since the CAMELYON16 dataset is fully annotated, we perform AL by assuming all WSIs are unannotated and revealing the annotation of a region only after it is selected during the AL procedure. We divide the WSIs in $\mathcal{U}$ randomly into five stratified subsets of equal size and use them sequentially. In particular, regions are selected from WSIs in the first subset at the first AL cycle, from WSIs in the second subset at the second AL cycle, and so on. This is done because WSI inference is computationally expensive due to the large patch amount, reducing the number of predicted WSIs to one fifth helps to speed up AL cycles. We use an informativeness measure that prioritizes pixels with a predicted probability close to 0.5 (i.e., $M_i = 1 - 2|P_i - 0.5|$), following [9]. We annotate validation WSIs in the same way as the training WSIs via AL.

Evaluations. We use the CAMELYON16 challenge metric Free Response Operating Characteristic (FROC) score [1] to validate the segmentation framework. To evaluate the WSI segmentation performance directly, we use mean intersection over union (mIoU). For comparison, we follow [3] to use a threshold of 0.5 to generate the binary segmentation map and report mIoU (Tumor), which is the average mIoU over the 48 test WSIs with metastases. We evaluate the model trained at each AL cycle to track performance change across the AL procedure.

3.3 Results

Full Annotation Performance. To validate our segmentation framework, we first train on the fully-annotated data (average performance of five repetitions reported). With a patch extraction stride $s = 256$ pixels, our framework yields an FROC score of 0.760 that is equivalent to the Challenge top 2, and an mIoU (Tumor) of 0.749, which is higher than the most comparable method in [3] that achieved 0.741 with $s = 128$ pixels. With our framework, reducing s to 128 pixels improves both metastases identification and segmentation (FROC score: 0.779, mIoU (Tumor): 0.758). However, halving s results in a 4-fold increase in inference time. This makes an AL experiment, which involves multiple rounds of WSI inference, extremely costly. Therefore, we use $s = 256$ pixels for all following AL experiments to compromise between performance and computation costs. Because WSIs without metastases do not require pixel-level annotation, we exclude the 159 training and validation WSIs without metastases from all following AL experiments. This reduction leads to a slight decrease of full annotation performance (mIoU (Tumor) from 0.749 to 0.722).

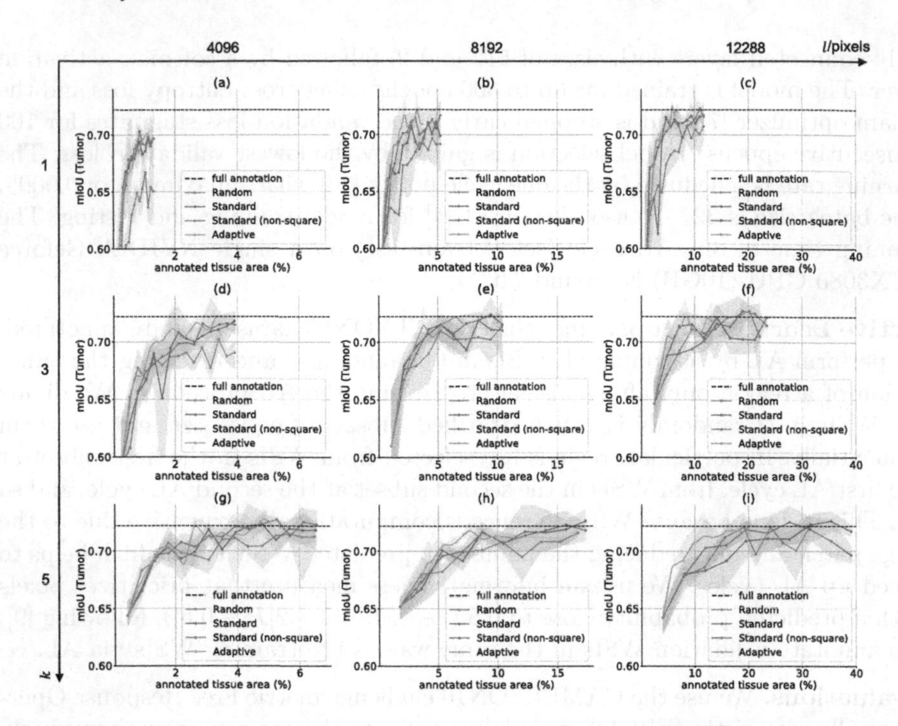

Fig. 4. mIoU (Tumor) as a function of annotated tissue area (%) for four region selection methods across various AL step sizes. Results show average and min/max (shaded) performance over three repetitions with distinct initial labeled sets. The final annotated tissue area of *Random* can be less than *Standard* as it stops sampling a WSI if no region contains more than 10% of tissue. Curves of *Adaptive* are interpolated as the annotated area differs between repetitions.

Comparison of Region Selection Methods. Figure 4 compares the sampling efficiency of the four region selection methods across various AL step sizes (i.e., the combinations of region size $l \in \{4096, 8192, 12288\}$ pixels and the number of selected regions per WSI $k \in \{1, 3, 5\}$). Experiments with large AL step sizes perform 10 AL cycles (Fig. 4 (e), (f), (h) and (i)); others perform 15 AL cycles. All experiments (except for *Random*) use uncertainty sampling.

When using region selection method *Standard*, the sampling efficiency advantage of uncertainty sampling over random sampling decreases as AL step size increases. A small AL step size minimizes the annotated tissue area for a certain high level of model performance, such as an mIoU (Tumor) of 0.7, yet requires a large number of AL cycles to achieve full annotation performance (Fig. 4 (a–d)),

Table 1. Annotated tissue area (%) required to achieve full annotation performance. The symbol "/" indicates that the full annotation performance is not achieved in the corresponding experimental setting in Fig. 4.

k	1			3			5		
l/pixels	4096	8192	12288	4096	8192	12288	4096	8192	12288
Random	/	/	/	/	/	18.1	/	/	/
Standard	/	/	/	/	9.4	14.2	4.3	17.4	31.5
Standard (non-square)	/	/	11.0	/	/	18.9	3.9	15.7	27.6
Adaptive	/	3.3	5.8	3.3	6.3	8.1	2.6	/	20.0

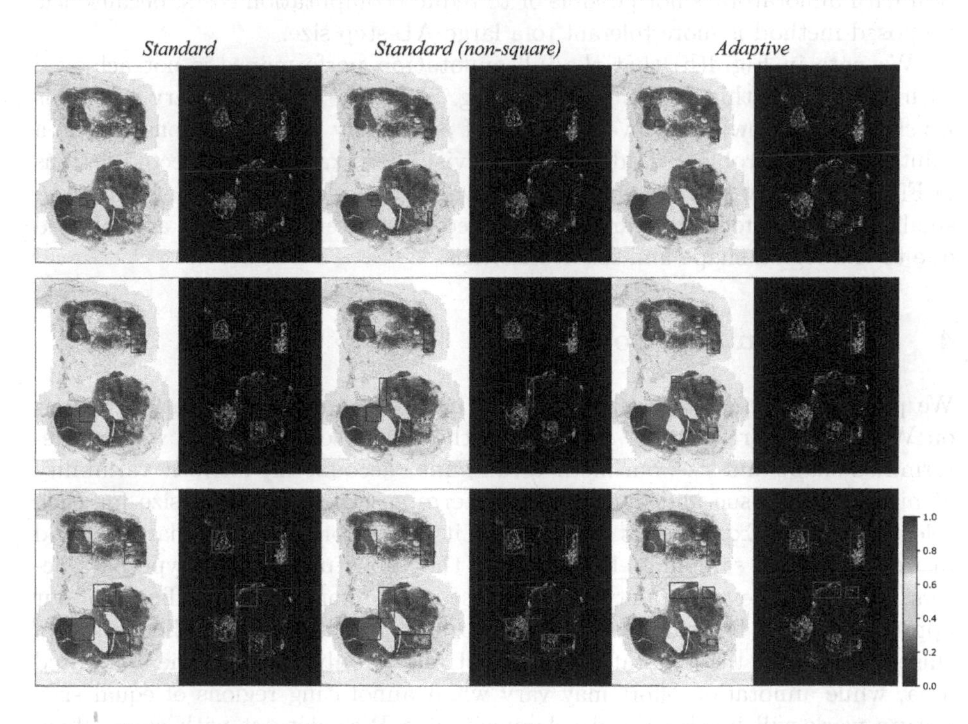

Fig. 5. Visualization of five regions selected with three region selection methods, applied to an exemplary priority map produced in a second AL cycle (regions were randomly selected in the first AL cycle, $k = 5, l = 4096$ pixels). Region sizes increase from top to bottom: $l \in \{4096, 8192, 12288\}$ pixels. Fully-annotated tumor metastases overlaid with WSI in red. (Color figure online)

resulting in high computation costs. A large AL step size allows for full annotation performance to be achieved in a small number of AL cycles, but at the expense of rapidly expanding the annotated tissue area (Fig. 4(e), (f), (h) and (i)). Enabling selected regions to have variable aspect ratios does not substantially improve the sampling efficiency, with *Standard (non-square)* outperforming *Standard* only when the AL step size is excessively large (Fig. 4(i)). However, allowing regions to be of variable size consistently improves sampling efficiency. Table 1 shows that *Adaptive* achieves full annotation performance with fewer AL cycles than *Standard* for small AL step sizes and less annotated tissue area for large AL step sizes. As a result, when region selection method *Adaptive* is used, uncertainty sampling consistently outperforms random sampling. Furthermore, Fig. 4(e–i)) shows that *Adaptive* effectively prevents the rapid expansion of annotated tissue area as AL step size increases, demonstrating greater robustness to AL step size choices than *Standard*. This is advantageous because extensive AL step size tuning to balance the annotation and computation costs can be avoided. This behavior can also be desirable in cases where frequent interaction with annotators is not possible or to reduce computation costs, because the proposed method is more tolerant to a large AL step size.

We note in Fig. 4(h) that the full annotation performance is not achieved with *Adaptive* within 15 AL cycles; in Fig. S1 in the supplementary materials we show that allowing for oversampling of previously selected regions can be a solution to this problem. Additionally, we visualize examples of selected regions in Fig. 5 and show that *Adaptive* avoids two region selection issues of *Standard*: small, isolated informative areas are missed, and irrelevant pixels are selected due to the region shape and size restrictions.

4 Discussion and Conclusion

We presented a new AL region selection method to select annotation regions on WSIs. In contrast to the standard method that selects regions with predetermined shape and size, our method takes into account the intrinsic variability of histological tissue and dynamically determines the shape and size for each selected region. Experiments showed that it outperforms the standard method in terms of both sampling efficiency and the robustness to AL hyperparameters. Although the uncertainty map was used to demonstrate the efficacy of our approach, it can be seamlessly applied to any priority maps. A limitation of this study is that the annotation cost is estimated only based on the annotated area, while annotation effort may vary when annotating regions of equal size. Future work will involve the development of a WSI dataset with comprehensive documentation of annotation time to evaluate the proposed method and an investigation of potential combination with self-supervised learning.

Acknowledgments. We thank Yixing Huang and Zhaoya Pan (FAU) for their feedback on the manuscript. We gratefully acknowledge support by d.hip campus - Bavarian aim (J.Q. and K.B.) as well as the scientific support and HPC resources provided by the Erlangen National High Performance Computing Center (NHR@FAU) of the Friedrich-Alexander-Universität Erlangen-Nürnberg (FAU). The hardware is funded by the German Research Foundation (DFG).

References

1. Bejnordi, B.E., et al.: Diagnostic assessment of deep learning algorithms for detection of lymph node metastases in women with breast cancer. JAMA **318**(22), 2199–2210 (2017)
2. Colling, P., Roese-Koerner, L., Gottschalk, H., Rottmann, M.: Metabox+: a new region based active learning method for semantic segmentation using priority maps. arXiv preprint arXiv:2010.01884 (2020)
3. Guo, Z., et al.: A fast and refined cancer regions segmentation framework in whole-slide breast pathological images. Sci. Rep. **9**(1), 1–10 (2019)
4. Howard, J., Gugger, S.: Fastai: a layered API for deep learning. Information **11**(2), 108 (2020)
5. Jin, X., An, H., Wang, J., Wen, K., Wu, Z.: Reducing the annotation cost of whole slide histology images using active learning. In: 2021 3rd International Conference on Image Processing and Machine Vision (IPMV), pp. 47–52 (2021)
6. Kasarla, T., Nagendar, G., Hegde, G.M., Balasubramanian, V., Jawahar, C.: Region-based active learning for efficient labeling in semantic segmentation. In: 2019 IEEE Winter Conference on Applications of Computer Vision (WACV), pp. 1109–1117. IEEE (2019)
7. Kingma, D.P., Ba, J.: Adam: a method for stochastic optimization. arXiv preprint arXiv:1412.6980 (2014)
8. Lai, Z., Wang, C., Oliveira, L.C., Dugger, B.N., Cheung, S.C., Chuah, C.N.: Joint semi-supervised and active learning for segmentation of gigapixel pathology images with cost-effective labeling. In: Proceedings of the IEEE/CVF International Conference on Computer Vision, pp. 591–600 (2021)
9. Lewis, D.D., Gale, W.A.: A sequential algorithm for training text classifiers. In: Croft, B.W., van Rijsbergen, C.J. (eds.) SIGIR 1994, pp. 3–12. Springer, Heidelberg (1994). https://doi.org/10.1007/978-1-4471-2099-5_1
10. Litjens, G., Bandi, P., Ehteshami Bejnordi, B., Geessink, O., Balkenhol, M., Bult, P., et al.: 1399 H&E-stained sentinel lymph node sections of breast cancer patients: the CAMELYON dataset. GigaScience **7**(6), giy065 (2018). http://gigadb.org/dataset/100439
11. Liu, Y., et al.: Detecting cancer metastases on gigapixel pathology images. arXiv preprint arXiv:1703.02442 (2017)
12. Macenko, M., Niethammer, M., Marron, J.S., et al.: A method for normalizing histology slides for quantitative analysis. In: 2009 IEEE International Symposium on Biomedical Imaging: From Nano to Macro, pp. 1107–1110. IEEE (2009)
13. Mackowiak, R., Lenz, P., Ghori, O., et al.: Cereals-cost-effective region-based active learning for semantic segmentation. arXiv preprint arXiv:1810.09726 (2018)
14. Russakovsky, O., et al.: Imagenet large scale visual recognition challenge. Int. J. Comput. Vision **115**(3), 211–252 (2015)
15. Sandler, M., Howard, A., Zhu, M., Zhmoginov, A., Chen, L.C.: Mobilenetv 2: inverted residuals and linear bottlenecks. In: Proceedings of the IEEE Conference on Computer Vision and Pattern Recognition, pp. 4510–4520 (2018)
16. Smith, L.N.: A disciplined approach to neural network hyper-parameters: part 1-learning rate, batch size, momentum, and weight decay. arXiv preprint arXiv:1803.09820 (2018)
17. Wilm, F., et al.: Pan-tumor canine cutaneous cancer histology (CATCH) dataset. Sci. Data **9**(1), 1–13 (2022)

18. Xu, Z., et al.: Clinical-realistic annotation for histopathology images with probabilistic semi-supervision: a worst-case study. In: Wang, L., Dou, Q., Fletcher, P.T., Speidel, S., Li, S. (eds.) MICCAI 2022. LNCS, vol. 13432, pp. 77–87. Springer, Cham (2022). https://doi.org/10.1007/978-3-031-16434-7_8
19. Yang, L., Zhang, Y., Chen, J., Zhang, S., Chen, D.Z.: Suggestive annotation: a deep active learning framework for biomedical image segmentation. In: Descoteaux, M., Maier-Hein, L., Franz, A., Jannin, P., Collins, D.L., Duchesne, S. (eds.) MICCAI 2017. LNCS, vol. 10435, pp. 399–407. Springer, Cham (2017). https://doi.org/10.1007/978-3-319-66179-7_46

CXR-CLIP: Toward Large Scale Chest X-ray Language-Image Pre-training

Kihyun You, Jawook Gu, Jiyeon Ham, Beomhee Park, Jiho Kim, Eun K. Hong, Woonhyuk Baek, and Byungseok Roh(✉)

Kakaobrain, Seongnam, Republic of Korea
{ukihyun,jawook.gu,jiyeon.ham,brook.park,tyler.md,amy.hong, wbaek,peter.roh}@kakaobrain.com

Abstract. A large-scale image-text pair dataset has greatly contributed to the development of vision-language pre-training (VLP) models, which enable zero-shot or few-shot classification without costly annotation. However, in the medical domain, the scarcity of data remains a significant challenge for developing a powerful VLP model. In this paper, we tackle the lack of image-text data in chest X-ray by expanding image-label pair as image-text pair via general prompt and utilizing multiple images and multiple sections in a radiologic report. We also design two contrastive losses, named ICL and TCL, for learning study-level characteristics of medical images and reports, respectively. Our model outperforms the state-of-the-art models trained under the same conditions. Also, enlarged dataset improve the discriminative power of our pre-trained model for classification, while sacrificing marginal retrieval performance. Code is available at https://github.com/kakaobrain/cxr-clip.

Keywords: Chest X-ray · Vision-Language Pre-training · Contrastive Learning

1 Introduction

Chest X-ray (CXR) plays a vital role in screening and diagnosis of thoracic diseases [19]. The effectiveness of deep-learning based computer-aided diagnosis has been demonstrated in disease detection [21]. However, one of the major challenges in training deep learning models for medical purposes is the need for extensive, high-quality clinical annotation, which is time-consuming and costly.

Recently, CLIP [22] and ALIGN [10] have shown the ability to perform vision tasks without any supervision. However, vision-language pre-training (VLP) in the CXR domain still lacks sufficient image-text datasets because many public datasets consist of image-label pairs with different class compositions. MedCLIP [26] attempted to a rule-based labler to use both image-text data and image-label data. However, it relies on the performance of the rule-based labeler and is not scalable to other diseases that the labeler cannot address.

Supplementary Information The online version contains supplementary material available at https://doi.org/10.1007/978-3-031-43895-0_10.

H. Greenspan et al. (Eds.): MICCAI 2023, LNCS 14221, pp. 101–111, 2023.
https://doi.org/10.1007/978-3-031-43895-0_10

In this paper, we propose a training method, *CXR-CLIP*, that integrates image-text data with image-label data using class-specific prompts made by radiologists. Our method does not depend on a rule-based labeler and can be applied to any image-label data. Also, inspired by DeCLIP [13], we used Multi-View Supervision (MVS) utilizing multiple images and texts in a CXR study to make more image-text pairs for efficient learning. In addition, we introduce two contrastive loss functions, named image contrastive loss (ICL) and text contrastive loss (TCL), to learn study-level characteristics of the CXR images and reports respectively.

The main contributions of this paper are summarized as follows. 1) We tackle the lack of data for VLP in CXR by generating image-text pairs from image-label datasets using prompt templates designed by radiologists and utilizing multiple images and texts in a study. 2) Two additional contrastive losses are introduced to learn discriminate features of image and text, improving image-text retrieval performances. 3) Performance of our model is validated on diverse datasets with zero-shot and few-shot settings.

2 Related Work

Data Efficient VLP. Recent studies [13,17] have proposed data-efficient VLP via joint learning with self-supervision. DeCLIP [13] suggested MVS that utilizes image and text augmentation to leverage positive pairs along with other self-supervisions. In CXR domain, GloRIA [7] aligned words in reports and sub-regions in an image for label efficiency, and BioVIL [2] combined self-supervision for label efficiency. We modify MVS as two distinct images and texts from a study and present self-supervised loss functions, ICL and TCL for efficient learning.

Self-supervision Within CXR Study. A CXR study could include several images in different views and two report sections: 'findings' and 'impression'. The impression section includes the differential diagnosis inferred from the findings section. BioVIL [2] enhanced the text encoder by matching two sections during language pre-training. MedAug [24] shows that self-supervised learning by matching images in a study is better than differently augmented images. We utilize both of multiple images and texts from a single study in VLP in an end-to-end fashion.

Leveraging Image-Label Data in VLP. MedCLIP [26] integrated unpaired images, texts, and labels using rule-based labeler [8], which is less capable of retrieving the exact report for a given image due to the effect of decoupling image-text pairs. UniCL [28] suggested using prompts to leverage image-label dataset [4], considering the samples from the same label to be a positive pair. To our knowledge, this is the first work to utilize prompting for training in CXR domain.

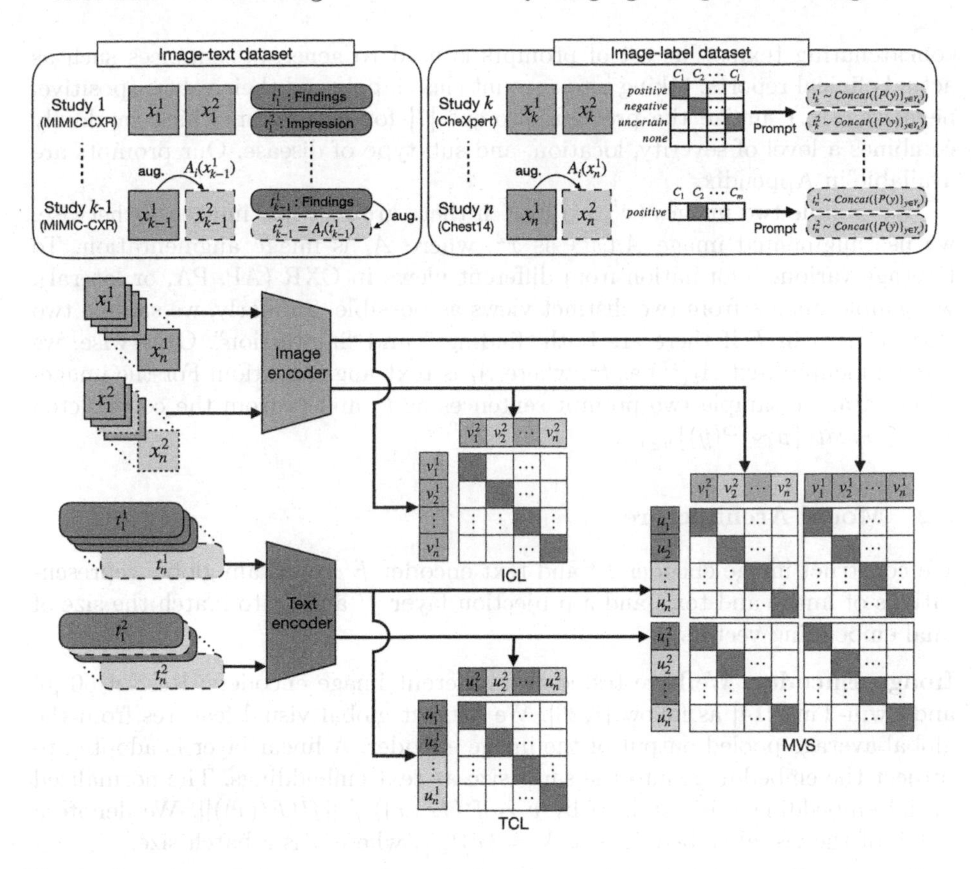

Fig. 1. Overview of the proposed method with a training batch sampling n studies, where each study has a pair of images (x^1, x^2) and a pair of text (t^1, t^2). If a study has one image or one text, data augmentation is conducted to make second examples. For the image-label data, two different prompts are generated from class labels as (t^1, t^2). Using sampled pairs, the encoders are trained with three kinds of contrastive losses (MVS, ICL, and TCL).

3 Method

CXR-CLIP samples image-text pairs from not only image-text data but also image-label data, and learns study-level characteristics with two images and two texts per study. The overview of the proposed method is illustrated in Fig. 1.

3.1 Data Sampling

We define a CXR study as $s = \{X, T\}$, where X is a set of images, and T is a set of "findings" and "impression" sections. The study of image-label dataset has a set of image labels Y instead of T. For the image-label dataset, we make prompt-based texts $T = Concat(\{p \sim P(y)\}_{y \in Y})$, where p is a sampled prompt sentence, $P(y)$ is a set of prompts given the class name and value y, and $Concat(\cdot)$ means concatenating texts. The set of prompts is used to generate sentences such as actual clinical reports, taking into account class labels and their values (positive, negative, etc.), unlike the previous prompt [7] for evaluation which randomly combines a level of severity, location, and sub-type of disease. Our prompts are available in Appendix.

We sample two images (x^1, x^2) in X if there are multiple images. Otherwise, we use augmented image $A_i(x^1)$ as x^2, where A_i is image augmentation. To leverage various information from different views in CXR (AP, PA, or lateral), we sample images from two distinct views as possible. Similarly, we sample two texts (t^1, t^2) in T if there are both "findings" and "impression". Otherwise, we use augmented text $A_t(t^1)$ as t^2, where A_t is text augmentation. For the image-label data, we sample two prompt sentences as t^1 and t^2 from the constructed $T = Concat(\{p \sim P(y)\}_{y \in Y})$.

3.2 Model Architecture

We construct image encoder E^i and text encoder E^t to obtain global representations of image and text, and a projection layer f^i and f^t to match the size of final embedding vectors.

Image Encoder. We have tested two different image encoders; ResNet-50 [6] and Swin-Tiny [14] as follow [7,26]. We extract global visual features from the global average pooled output of the image encoder. A linear layer is adopted to project the embeddings into the same size as text embeddings. The normalized visual embedding v is obtained by $v = f^i(E^i(x)) \ / \ ||f^i(E^i(x))||$. We denote a batch of the visual embeddings as $V = \{v\}_{i=1}^n$, where n is a batch size.

Text Encoder. We use BioClinicalBERT [1] model, which is the same architecture as BERT [5] but pre-trained with medical texts [11] as follow [7,26]. We use **[EOS]** token's final output as the global textual representation. Also, a linear projection layer is adopted the same as the image encoder. The normalized text embedding u is denoted as $u = f^t(E^t(t)) \ / \ ||f^t(E^t(t))||$. We denote a batch of the text embedding as $U = \{u\}_{i=1}^n$ and $(v_i,\ u_i)$ are paired.

3.3 Loss Function

In this section, we first describe CLIP loss [22] and then describe our losses (MVS, ICL, TCL) in terms of CLIP loss. The goal of CLIP loss is to pull image embedding and corresponding text embedding closer and to push unpaired image and text farther in the embedding space. The InfoNCE loss is generally adopted as a type of contrastive loss, and CLIP uses the average of two InfoNCE losses; image-to-text and text-to-image. The formula for CLIP loss is given by

$$L_{CLIP}(U,V) = -\frac{1}{2n}\Big(\sum_{u_i \in U} \log \frac{\exp(v_i^T u_i/\tau)}{\sum_{v_j \in V} \exp(u_i^T v_j/\tau)} + \sum_{v_i \in V} \log \frac{\exp(u_i^T v_i/\tau)}{\sum_{u_j \in U} \exp(v_i^T u_j/\tau)}\Big), \tag{1}$$

where τ is a learnable temperature to scale logits.

In DeCLIP [13], MVS uses four L_{CLIP} loss with all possible pairs augmented views; (x, t), $(x, A_t(t))$, $(A_i(x), t)$ and $(A_i(x), A_t(t))$. We modify DeCLIP's MVS to fit the CXR domain by the composition of the second example. DeCLIP only utilizes an augmented view of the original sample, but we sample a pair of the second image and text as described in Sect. 3.1. We denote the first and the second sets of image embeddings as U^1, U^2, and text embeddings as V^1, V^2.

$$L_{MVS} = \frac{1}{4}(L_{CLIP}(U^1, V^1)+L_{CLIP}(U^2, V^1)+L_{CLIP}(U^1, V^2)+L_{CLIP}(U^2, V^2)) \tag{2}$$

The goal of ICL and TCL is to learn modality-specific characteristics in terms of image and text respectively. We design ICL and TCL as same as CLIP loss, but the input embeddings are different. ICL only uses image embeddings; $L_{ICL} = L_{CLIP}(V^1, V^2)$ and TCL only uses text embeddings; $L_{TCL} = L_{CLIP}(U^1, U^2)$. ICL pulls image embeddings from the same study and pushes image embeddings from the different studies, so that, the image encoder can learn study-level diversity. Similarly, TCL pulls embeddings of "findings" and "impression" in the same study or diverse expressions of prompts from the same label and pushes the other studies' text embeddings, so that the text encoder can match diverse clinical expressions on the same diagnosis. Thereby, the final training objective consists of three contrastive losses balanced each component by λ_I and λ_T, formulated by $L = L_{MVS} + \lambda_I L_{ICL} + \lambda_T L_{TCL}$.

4 Experiment

4.1 Datasets

We used three pre-trained datasets and tested with various external datasets to test the generalizability of models. The statistics of the datasets used are summarized in Table 1.

MIMIC-CXR [12] consists of CXR studies, each with one or more images and free-form reports. We extracted "findings" and "impression" from the reports. We used the training split for pre-training and the test split for image-to-text retrieval.

CheXpert [8] is an image-label data with 14 classes, obtained from the impression section by its rule-based labeler, and each class is labeled as positive,

Table 1. The number of studies for each dataset and split in this paper

Data Split	Pre-training			Evaluation			
	MIMIC-CXR	CheXpert	ChestX-ray14	VinDR	RSNA	SIIM	Open-I
Train	222,628	216,478	89,696	12,000	18,678	8,422	
Valid	1,808	233	22,423	3,000	4,003	1,808	
Test	3,264	1,000		3,000	4,003	1,807	3,788

negative, uncertain, or none (not mentioned). We used the training split for pre-training with class-specific prompts. **CheXpert5x200** is a subset of CheXpert for 5-way classification, which has 200 exclusively positive images for each class. Note that only the reports of CheXpert5x200 are publicly available, but the reports of CheXpert are not. Following the previous works [7,26], we excluded CheXpert5x200 from the training set and used it for test.

ChestX-ray14 [25] consists of frontal images with binary labels for 14 diseases. Prompts are generated by sampling 3 negative classes per study. We used 20% of the original training set for validation, and the remaining 80% for pre-training.

RSNA pneumonia [23] is binary-labeled data as pneumonia or normal. We split train/valid/test set 70%, 15%, 15% of the dataset following [7] for the external classification task.

SIIM Pneumothorax[1] is also binary labeled as pneumothorax or normal. We split the train/valid/test set same ratio as RSNA pneumonia following [7] and used it for the classification task.

VinDR-CXR [18] contains 22 local labels and 6 global labels of disease, which were obtained by experienced radiologists. We split the validation set from the original training set. Of 28 classes, "other diseases" and "other lesions" classes were excluded. Then, only 18 classes having 10 or more samples within the test set were evaluated for the binary classification of each class as follow [9].

Open-I [3] is an image-text dataset. From each study, one of the report sections and one frontal-view image were sampled and used for image-to-text retrieval.

4.2 Implementation Details

We used augmentations A_i and A_t to fit medical images and reports. For A_i, we resize and crop with scale [0.8, 1.1], randomly adapt CLAHE [20], and random color jittering; brightness, hue ratios from [0.9, 1.1] and contrast, saturation [0.8, 1.2]. For A_t, to preserve clinical meaning, sentence swap and back-translation[2] from Italian to English is used. The image size and final-embedding size are set to 224 and 512 respectively as in previous work [26]. We set λ_I and λ_T to 1.0, 0.5 for balancing total loss. Two encoders were trained for 15 epochs in a mixed-precision manner, early stopped by validation loss, and optimized by AdamW [16] with an initial learning rate 5e-5 and a weight decay 1e-4. We used cosine-annealing learning-rate scheduler [15] with warm-up for 1 epoch. A training batch consists of 128 studies with 256 image-text pairs. We implemented all experiments on PyTorch with 4 NVIDIA V100 GPUs.

4.3 Comparison with State-of-the-Arts

Zero-Shot and Few-Shot Classification. Table 2 shows performance on classification tasks of our models and state-of-the-art models. To evaluate zero-

[1] https://siim.org/page/pneumothorax_challenge.
[2] https://huggingface.co/Helsinki-NLP.

Table 2. Comparison with state-of-the-art for zero-shot(ZS) or few-shot(10%) classification tasks. M, C, and C14 mean MIMIC-CXR, CheXpert, and ChestX-ray14, respectively. C* means CheXpert with reports, which are not publicly available. ResNet50 (*R*50) and SwinTiny (*SwinT*) mean the image encoder used for each model.

Model Name	Pre-train Dataset	VinDR-CXR			RSNA			SIIM			C5x200
		ZS	10%	100%	ZS	10%	100%	ZS	10%	100%	ZS-ACC
GloRIA$_{R50}$	C*	78.0	73.0	73.1	80.6	88.2	88.5	84.0	91.5	91.9	62.4*
CXR-CLIP$_{R50}$	M	78.8	82.1	82.2	83.3	88.5	89.2	85.2	88.3	90.5	56.2
CXR-CLIP$_{SwinT}$	M	78.3	84.9	85.4	81.3	88.0	88.4	85.5	86.9	88.3	54.3
MedCLIP$_{SwinT}$	M,C	82.4	84.9	85.1	81.9	88.9	89.0	89.0	90.4	90.8	59.2
CXR-CLIP$_{R50}$	M,C	**83.0**	81.4	82.1	81.7	88.5	88.9	86.4	88.4	90.7	61.7
CXR-CLIP$_{SwinT}$	M,C	82.7	86.1	86.7	**84.5**	88.1	88.8	87.9	89.6	91.2	60.1
CXR-CLIP$_{R50}$	M,C,C14	78.1	80.2	81.0	81.8	88.7	89.3	85.2	91.5	92.8	60.3
CXR-CLIP$_{SwinT}$	M,C,C14	78.9	**88.0**	**89.0**	80.1	**89.2**	**89.8**	**91.4**	**92.9**	**94.0**	**62.8**

Table 3. Comparison with state-of-the-arts for image-to-text retrieval. The notations of datasets and models are same to Table 2.

Model Name	Pre-Train Dataset	CheXpert5x200			MIMIC-CXR			Open-I			Total RSUM
		R@1	R@5	R@10	R@1	R@5	R@10	R@1	R@5	R@10	
GloRIA$_{R50}$	C*	**17.8**	**38.8**	**49.9**	7.2	20.6	30.3	1.5	4.4	6.5	177.0
CXR-CLIP$_{R50}$	M	9.4	23.0	32.6	21.4	46.0	59.2	**3.8**	8.2	**12.3**	**216.9**
CXR-CLIP$_{SwinT}$	M	8.4	21.5	30.2	**21.6**	**48.9**	**60.2**	3.6	**8.3**	11.5	214.2
MedCLIP$_{SwinT}$	M,C	2.6	3.0	3.6	1.1	1.4	5.5	0.1	0.4	0.7	18.4
CXR-CLIP$_{R50}$	M,C	5.5	19.2	27.4	20.2	45.9	58.2	3.5	8.2	12.0	200.1
CXR-CLIP$_{SwinT}$	M,C	8.5	23.0	31.6	19.6	44.2	57.1	3.1	**8.3**	11.6	207.0
CXR-CLIP$_{R50}$	M,C,C14	5.7	18.0	28.3	19.7	44.4	56.4	2.3	6.7	10.1	191.6
CXR-CLIP$_{SwinT}$	M,C,C14	7.0	20.1	29.7	20.9	46.2	58.8	2.4	6.6	9.4	201.1

shot classification fairly, we used evaluation prompts suggested from previous works [2,7,9]. The evaluation prompts are available in Appendix. We evaluate binary classification computed by Area Under ROC (AUC) and multi-class classification computed by accuracy (ACC). Our ResNet model trained with MIMIC-CXR outperforms GloRIA [7] except for CheXpert5x200, as GloRIA trained with image-text pair in CheXpert. Our SwinTiny model trained with MIMIC-CXR and CheXpert outperforms MedCLIP [26], which is the same architecture trained with the same datasets, in most of the metrics. Adding more pre-training datasets by prompting image-label datasets tends to improve performance for classifications, while the SwinTiny CXR-CLIP pre-trained with three datasets, performs the best for most of the metrics. More comparison with self-supervised models is available in Appendix.

Image-to-Text Retrieval. We evaluated image-to-text retrieval computed by $R@K$, the recall of the exact report in the top K retrieved reports for a given image. (Table 3) While GloRIA [7] uses image-text pairs in CheXpert(C*) which is not available in public, CXR-CLIP uses image-text in MIMIC-CXR. So we adapt an external image-text dataset Open-I [3] for a fair comparison. GloRIA

Table 4. Ablations and comparison with CLIP [22] and DeCLIP [13]. Our augmentations effectively preserves clinical meaning than EDA. Our full methodology (CXR-CLIP) outperforms DeCLIP.

Method	CheXpert 5x200				MIMIC-CXR			Total RSUM
	ACC	R@1	R@5	R@10	R@1	R@5	R@10	
Vanila CLIP	58.9	4.4	14.4	22.6	17.3	41.2	52.6	152.5
+ Study Level Sampling	58.7	4.6	15.1	23.2	17.8	42.5	54.2	157.4
+ Augmentations	60.6	5.7	17.0	24.9	16.1	40.2	51.5	155.4
+ MVS	61.2	5.4	17.1	24.7	16.3	40.6	53.3	157.4
+ ICL	61.6	**6.8**	**20.3**	28.6	17.5	41.6	53.2	168.0
+ TCL (CXR-CLIP)	**61.7**	6.2	18.2	**29.1**	**19.6**	**44.8**	**56.6**	**174.5**
MVS of DeCLIP (EDA)	59.5	3.2	15.5	22.9	15.8	39.1	51.5	148.0
MVS of DeCLIP (Our aug)	59.4	6.0	17.0	24.4	15.1	38.8	51.8	153.1
DeCLIP (Our aug)	59.4	5.7	16.1	24.6	18.1	44.0	55.3	163.8

has the best performance on CheXpert but our model trained with MIMIC-CXR, which has similar amounts of studies to CheXpert, outperforms on Open-I. MedCLIP almost lost the ability to retrieve image-text due to decoupling pairs of image and text during pre-training. In CXR-CLIP, adding more image-label datasets such as CheXpert and ChestX-ray14 degrades the image-text retrieval performance, possibly because the contribution of the text in original reports was diluted.

4.4 Ablations

For the ablation study, models with ResNet-50 [6] backbone were trained on MIMIC-CXR and CheXpert datasets and tested on zero-shot classification and image-to-text retrieval tasks with MIMIC-CXR and CheXpert5x200 datasets.

We conducted two ablations shown in Table 4. First, we analyzed the effect of each component of CXR-CLIP by adding the components to vanilla CLIP [22] one by one. To validate our data sampling closer, we divided the sampling method into three parts 1) study-level sampling 2) data augmentations 3) Multi-view and Multi-text sampling (MVS). Our study-level sampling strategy improves performance compared to vanilla CLIP, which uses a naive sampling method bringing an image and corresponding report. Additionally, the modified data augmentation to fit the CXR domain contributes to performance increment of classification, the similar performance on retrieval. MVS slightly improves performances in both classification and image-text retrieval. Adding more supervision (ICL and TCL) improves performance by utilizing better multi-views and multi-text inputs. However, TCL drops the performance of recalls in CheXpert5x200, TCL could be hard to optimize variation of the radiologic report and prompt not diverse as images.

In the second ablation study, CXR-CLIP was compared to DeCLIP [13] to confirm that our MVS using two image-text pairs per study is better than the MVS of DeCLIP which uses naively augmented images and texts. We show that our text augmentation outperforms DeCLIP's text augmentation named EDA [27] in terms of image-to-text recall, which implies our text augmentation preserves clinical meaning. The superiority of our MVS over DeCLIP's MVS confirms that using multiple images and texts from one study is better than using images and texts from augmented examples. Also, our full methodology (CXR-CLIP) outperforms DeCLIP, suggesting that our method efficiently learns in the CXR domain more than DeCLIP.

5 Conclusion

We presented a framework enlarging training image-text pair by using image-label datasets as image-text pair with prompts and utilizing multiple images and report sections in a study. Adding image-label datasets achieved performance gain in classification tasks including zero-shot and few-shot settings, on the other hand, lost the performance of retrieval tasks. We also proposed loss functions ICL and TCL to enhance the discriminating power within each modality, which effectively increases image-text retrieval performance. Our additional loss functions are designed to efficiently learn CXR domain knowledge along with image-text contrastive learning.

References

1. Alsentzer, E., et al.: Publicly available clinical BERT embeddings. CoRR abs/1904.03323 (2019). http://arxiv.org/abs/1904.03323
2. Boecking, B., et al.: Making the most of text semantics to improve biomedical vision-language processing. In: Avidan, S., Brostow, G., Cissé, M., Farinella, G.M., Hassner, T. (eds.) ECCV 2022. LNCS, vol. 13696, pp. 1–21. Springer, Cham (2022). https://doi.org/10.1007/978-3-031-20059-5_1
3. Demner-Fushman, D., et al.: Preparing a collection of radiology examinations for distribution and retrieval. J. Am. Med. Inform. Assoc. **23**(2), 304–310 (2016)
4. Deng, J., Dong, W., Socher, R., Li, L.J., Li, K., Fei-Fei, L.: Imagenet: a large-scale hierarchical image database. In: 2009 IEEE Conference on Computer Vision and Pattern Recognition, pp. 248–255. IEEE (2009)
5. Devlin, J., Chang, M., Lee, K., Toutanova, K.: BERT: pre-training of deep bidirectional transformers for language understanding. CoRR abs/1810.04805 (2018). http://arxiv.org/abs/1810.04805
6. He, K., Zhang, X., Ren, S., Sun, J.: Deep residual learning for image recognition. CoRR abs/1512.03385 (2015). http://arxiv.org/abs/1512.03385
7. Huang, S.C., Shen, L., Lungren, M.P., Yeung, S.: Gloria: a multimodal global-local representation learning framework for label-efficient medical image recognition. In: Proceedings of the IEEE/CVF International Conference on Computer Vision (ICCV), pp. 3942–3951 (2021)

8. Irvin, J., et al.: Chexpert: a large chest radiograph dataset with uncertainty labels and expert comparison. CoRR abs/1901.07031 (2019). http://arxiv.org/abs/1901.07031
9. Jang, J., Kyung, D., Kim, S.H., Lee, H., Bae, K., Choi, E.: Significantly improving zero-shot X-ray pathology classification via fine-tuning pre-trained image-text encoders (2022). https://arxiv.org/abs/2212.07050
10. Jia, C., et al.: Scaling up visual and vision-language representation learning with noisy text supervision. CoRR abs/2102.05918 (2021). https://arxiv.org/abs/2102.05918
11. Johnson, A., Pollard, T., Mark, R.: MIMIC-III clinical database (2020)
12. Johnson, A.E.W., Pollard, T., Mark, R., Berkowitz, S., Horng, S.: The MIMIC-CXR database (2019)
13. Li, Y., et al.: Supervision exists everywhere: a data efficient contrastive language-image pre-training paradigm. CoRR abs/2110.05208 (2021). https://arxiv.org/abs/2110.05208
14. Liu, Z., et al.: Swin transformer: hierarchical vision transformer using shifted windows. CoRR abs/2103.14030 (2021). https://arxiv.org/abs/2103.14030
15. Loshchilov, I., Hutter, F.: SGDR: stochastic gradient descent with restarts. CoRR abs/1608.03983 (2016). http://arxiv.org/abs/1608.03983
16. Loshchilov, I., Hutter, F.: Fixing weight decay regularization in adam. CoRR abs/1711.05101 (2017). http://arxiv.org/abs/1711.05101
17. Mu, N., Kirillov, A., Wagner, D.A., Xie, S.: SLIP: self-supervision meets language-image pre-training. CoRR abs/2112.12750 (2021). https://arxiv.org/abs/2112.12750
18. Nguyen, H.Q., et al.: VinDr-CXR: an open dataset of chest X-rays with radiologist's annotations. Sci. Data **9**(1), 429 (2022). https://doi.org/10.1038/s41597-022-01498-w
19. World Health Organization: Communicating radiation risks in paediatric imaging: information to support health care discussions about benefit and risk (2016)
20. Pisano, E.D., et al.: Contrast limited adaptive histogram equalization image processing to improve the detection of simulated spiculations in dense mammograms. J. Digit. Imaging **11**(4), 193 (1998). https://doi.org/10.1007/BF03178082
21. Qin, C., Yao, D., Shi, Y., Song, Z.: Computer-aided detection in chest radiography based on artificial intelligence: a survey. Biomed. Eng. Online **17**(1), 113 (2018). https://doi.org/10.1186/s12938-018-0544-y
22. Radford, A., et al.: Learning transferable visual models from natural language supervision. CoRR abs/2103.00020 (2021). https://arxiv.org/abs/2103.00020
23. Shih, G., et al.: Augmenting the national institutes of health chest radiograph dataset with expert annotations of possible pneumonia. Radiol. Artif. Intell. **1**, e180041 (2019). https://doi.org/10.1148/ryai.2019180041
24. Vu, Y.N.T., Wang, R., Balachandar, N., Liu, C., Ng, A.Y., Rajpurkar, P.: Medaug: contrastive learning leveraging patient metadata improves representations for chest x-ray interpretation. In: Jung, K., Yeung, S., Sendak, M., Sjoding, M., Ranganath, R. (eds.) Proceedings of the 6th Machine Learning for Healthcare Conference. Proceedings of Machine Learning Research, vol. 149, pp. 755–769. PMLR (2021). https://proceedings.mlr.press/v149/vu21a.html
25. Wang, X., Peng, Y., Lu, L., Lu, Z., Bagheri, M., Summers, R.M.: Chestx-ray8: hospital-scale chest X-ray database and benchmarks on weakly-supervised classification and localization of common thorax diseases. In: Proceedings of the IEEE Conference on Computer Vision and Pattern Recognition (CVPR) (2017)

26. Wang, Z., Wu, Z., Agarwal, D., Sun, J.: Medclip: contrastive learning from unpaired medical images and text (2022). https://arxiv.org/abs/2210.10163
27. Wei, J.W., Zou, K.: EDA: easy data augmentation techniques for boosting performance on text classification tasks. CoRR abs/1901.11196 (2019). http://arxiv.org/abs/1901.11196
28. Yang, J., et al.: Unified contrastive learning in image-text-label space (2022). https://arxiv.org/abs/2204.03610

VISA-FSS: A Volume-Informed Self Supervised Approach for Few-Shot 3D Segmentation

Mohammad Mozafari[1], Adeleh Bitarafan[1,2], Mohammad Farid Azampour[2], Azade Farshad[2], Mahdieh Soleymani Baghshah[1], and Nassir Navab[2](✉)

[1] Sharif University of Technology, Tehran, Iran
[2] Computer Aided Medical Procedures, Technical University of Munich, Munich, Germany
soleymani@sharif.edu

Abstract. Few-shot segmentation (FSS) models have gained popularity in medical imaging analysis due to their ability to generalize well to unseen classes with only a small amount of annotated data. A key requirement for the success of FSS models is a diverse set of annotated classes as the base training tasks. This is a difficult condition to meet in the medical domain due to the lack of annotations, especially in volumetric images. To tackle this problem, self-supervised FSS methods for 3D images have been introduced. However, existing methods often ignore intra-volume information in 3D image segmentation, which can limit their performance. To address this issue, we propose a novel self-supervised volume-aware FSS framework for 3D medical images, termed VISA-FSS. In general, VISA-FSS aims to learn continuous shape changes that exist among consecutive slices within a volumetric image to improve the performance of 3D medical segmentation. To achieve this goal, we introduce a volume-aware task generation method that utilizes consecutive slices within a 3D image to construct more varied and realistic self-supervised FSS tasks during training. In addition, to provide pseudo-labels for consecutive slices, a novel strategy is proposed that propagates pseudo-labels of a slice to its adjacent slices using flow field vectors to preserve anatomical shape continuity. In the inference time, we then introduce a volumetric segmentation strategy to fully exploit the inter-slice information within volumetric images. Comprehensive experiments on two common medical benchmarks, including abdomen CT and MRI, demonstrate the effectiveness of our model over state-of-the-art methods. Code is available at https://github.com/sharif-ml-lab/visa-fss

Keywords: Medical image segmentation · Few-shot learning · Few-shot semantic segmentation · Self-supervised learning · Supervoxels

M. Mozafari and A. Bitarafan—Equal Contribution.

Supplementary Information The online version contains supplementary material available at https://doi.org/10.1007/978-3-031-43895-0_11.

H. Greenspan et al. (Eds.): MICCAI 2023, LNCS 14221, pp. 112–122, 2023.
https://doi.org/10.1007/978-3-031-43895-0_11

1 Introduction

Automated image segmentation is a fundamental task in many medical imaging applications, such as diagnosis [24], treatment planning [6], radiation therapy planning, and tumor resection surgeries [7,12]. In the current literature, numerous fully-supervised deep learning (DL) methods have become dominant in the medical image segmentation task [5,16,18]. They can achieve their full potential when trained on large amounts of fully annotated data, which is often unavailable in the medical domain. Medical data annotation requires expert knowledge, and exhaustive labor, especially for volumetric images [17]. Moreover, supervised DL-based methods are not sufficiently generalizable to previously unseen classes. To address these limitations, few-shot segmentation (FSS) methods have been proposed [21–23,25], that segment an unseen class based on just a few annotated samples. The main FSS approaches use the idea of meta-learning [9,11,13] and apply supervised learning to train a few-shot model. However, to avoid overfitting and improve the generalization capability of FSS models, they rely on a large number of related tasks or classes. This can be challenging as it may require a large amount of annotated data, which may not always be available. Although some works on FSS techniques focus on training with fewer data [4,20,26], they require re-training before applying to unseen classes. To eliminate the need for annotated data during training and re-training on unseen classes, some recent works have proposed self-supervised FSS methods for 3D medical images which use superpixel-based pseudo-labels as supervision during training [8,19]. These methods design their self-supervised tasks (support-query pairs) by applying a predefined transformation (e.g., geometric and intensity transformation) on a support image (i.e., a random slice of a volume) to synthetically form a query one. Thus, these methods do not take into account intra-volume information and context that may be important for the accurate segmentation of volumetric images during inference.

We propose a novel volume-informed self-supervised approach for Few-Shot 3D Segmentation (VISA-FSS). Generally, VISA-FSS aims to exploit information beyond 2D image slices by learning inter-slice information and continuous shape changes that intrinsically exists among consecutive slices within a 3D image. To this end, we introduce a novel type of self-supervised tasks (see Sect. 2.2) that builds more varied and realistic self-supervised FSS tasks during training. Besides of generating synthetic queries (like [19] by applying geometric or intensity transformation on the support images), we also utilize consecutive slices within a 3D volume as support and query images. This novel type of task generation (in addition to diversifying the tasks) allows us to present a 2.5D loss function that enforces mask continuity between the prediction of adjacent queries. In addition, to provide pseudo-labels for consecutive slices, we propose the superpixel propagation strategy (SPPS). It propagates the superpixel of a support slice into query ones by using flow field vectors that exist between adjacent slices within a 3D image. We then introduce a novel strategy for volumetric segmentation during inference that also exploits inter-slice information within query volumes. It propagates a segmentation mask among consecutive slices using the

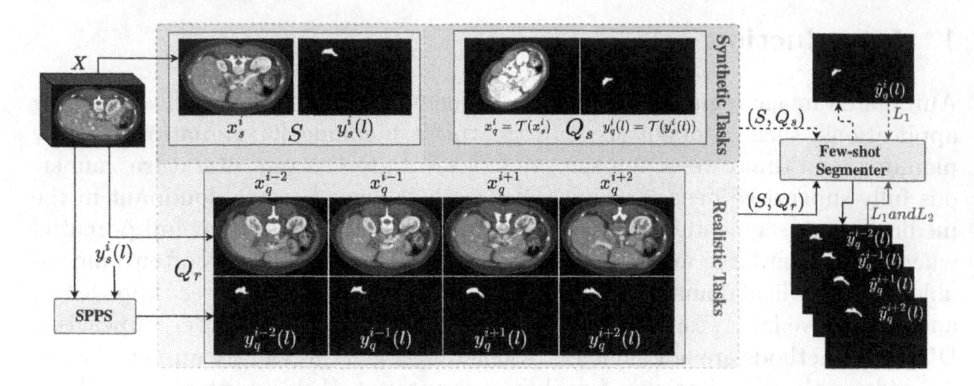

Fig. 1. An overview of the proposed VISA-FSS framework during training, where four adjacent slices of the support image, x_s^i, are taken as query images (i.e. $m = 2$). SPPS is a pseudo-label generation module for consecutive slices.

few-shot segmenter trained by VISA-FSS. Comprehensive experiments demonstrate the superiority of our method against state-of-the-art FSS approaches.

2 Methodology

In this section, we introduce our proposed VISA-FSS for 3D medical image segmentation. Our method goes beyond 2D image slices and exploits intra-volume information during training. To this end, VISA-FSS designs more varied and realistic self-supervised FSS tasks (support-query pairs) based on two types of transformations: 1) applying a predefined transformation (e.g., geometric and intensity transformation as used in [8,19]) on a random slice as support image to synthetically make a query one, 2) taking consecutive slices in a 3D volume as support and query images to learn continuous shape transformation that exists intrinsically between consecutive slices within a volumetric image (see Sect. 2.2). Moreover, the volumetric view of task generation in the second type of tasks allows us to go beyond 2D loss functions. Thus, in Sect. 2.2, we present a 2.5D loss function that enforces mask continuity between the prediction of adjacent queries during training the few-shot segmenter. In this way, the trained few-shot segmenter is able to effectively segment a new class in a query slice given a support slice, regardless of whether it is in a different volume (due to learning the first type of tasks) or in the same query volume (due to learning the second type of tasks). Finally, we propose a volumetric segmentation strategy for inference time which is elaborated upon in Sect. 2.3.

2.1 Problem Setup

In FSS, a training dataset $D_{tr} = \{(x^i, y^i(l))\}_{i=1}^{N_{tr}}, l \in L_{tr}$, and a testing dataset $D_{te} = \{(x^i, y^i(l))\}_{i=1}^{N_{te}}, l \in L_{te}$ are available, where $(x^i, y^i(l))$ denotes an image-mask pair of the binary class l. L_{tr} and L_{te} are the training and testing classes, respectively, and $L_{tr} \cap L_{te} = \varnothing$. The objective is to train a segmentation model on D_{tr} that is directly applicable to segment an unseen class $l \in L_{te}$ in a query image, $x_q \in D_{te}$, given a few support set $\{(x_s^j, y_s^j(l))\}_{j=1}^p \subset D_{te}$. Here, q and s indicate that an image or mask is from a query or support set. To simplify notation afterwards, we assume $p = 1$, which indicates the number of support images. During training, a few-shot segmenter takes a support-query pair (S, Q) as the input data, where $Q = \{(x_q^i, y_q^i(l))\} \subset D_{tr}$, and $S = \{(x_s^j, y_s^j(l))\} \subset D_{tr}$. Then, the model is trained according to the cross-entropy loss on each support-query pair as follows: $\mathcal{L}(\theta) = -\log p_\theta(y_q|x_q, S)$. In this work, we model $p_\theta(y_q|x_q, S)$ using the prototypical network introduced in [19], called ALPNet. However, the network architecture is not the main focus of this paper, since our VISA-FSS framework can be applied to any FSS network. The main idea of VISA-FSS is to learn a few-shot segmenter according to novel tasks designed in Sect. 2.2 to be effectively applicable for volumetric segmentation.

2.2 Self-supervised Task Generation

There is a large level of information in a 3D medical image over its 2D image slices, while prior FSS methods [8,19] ignore intra-volume information for creating their self-supervised tasks during training, although they are finally applied to segment volumetric images during inference. Previous approaches employ a predefined transformation (e.g., geometric and intensity) to form support-query pairs. We call these predefined transformations as synthetic transformations. On the other hand, there is continuous shape transformation that intrinsically exists between consecutive slices within a volume (we name them realistic transformation). VISA-FSS aims to, besides synthetic transformations, exploit realistic ones to learn more varied and realistic tasks. Figure 1 outlines a graphical overview of the proposed VISA-FSS framework, which involves the use of two types of self-supervised FSS tasks to train the few-shot segmenter. The two types of tasks are synthetic tasks and realistic tasks:

Synthetic Tasks. In the first type, tasks are formed the same as in [8,19]. For each slice x_s^i, its superpixels are extracted by the unsupervised algorithm [10], and its pseudo-mask is generated by randomly selecting one of its superpixels as a pseudo-organ. Thus, the support is formed as $S = (x_s^i, y_s^i(l)) \subset D_{tr}$, where l denotes the chosen superpixel. Then, after applying a random synthetic transformation $\mathcal{T}$ on S, the synthetic query will be prepared, i.e., $Q_s = (x_q^i, y_q^i(l)) = (\mathcal{T}(x_s^i), \mathcal{T}(y_s^i(l)))$. In this way, the (S, Q_s) pair is taken as the input data of the few-shot segmenter, presenting a 1-way 1-shot segmentation problem. A schematic view of a representative (S, Q_s) pair is given in the blue block of Fig. 1.

Realistic Tasks. To make the second type of task, we take $2m$ adjacent slices of the support image x_s^i, as our query images $\{x_q^j\}_{j \in N(i)}$, where $N(i) = \{i-m, ..., i-1, i+1, i+m\}$. These query images can be considered as real deformations of the support image. This encourages the few-shot segmenter to learn intra-volume information contrary to the first type of task. Importantly, pseudo-label generation for consecutive slices is the main challenge. To solve this problem, we introduce a novel strategy called SPPS that propagates the pseudo-label of the support image into query ones. Specifically, we consecutively apply flow field vectors that exist between adjacent image slices on $y_s^i(l)$ to generate pseudo-label $y_q^j(l)$ as follows: $y_q^j(l) = y_s^i(l) \circ \phi(x_s^i, x_q^{i+1}) \circ \phi(x_q^{i+1}, x_q^{i+2}) \circ ... \circ \phi(x_q^{i+m-1}, x_q^{i+m})$ for $j > m$, and $y_q^j(l) = y_s^i(l) \circ \phi(x_s^i, x_q^{i-1}) \circ \phi(x_q^{i-1}, x_q^{i-2}) \circ ... \circ \phi(x_q^{i-m+1}, x_q^{i-m})$ for $j < m$, where $\phi(x^i, x^j)$ is the flow field vector between x^i and x^j, which can be computed by deformably registering the two images using VoxelMorph [2] or Vol2Flow [3]. A schematic illustration of pseudo-label generation using SPPS is depicted in the Supplementary Materials. The yellow block in Fig. 1 demonstrates a representative (S, Q_r) pair formed using realistic tasks, where $Q_r = \{(x_q^j, y_q^j(l))\}_{j \in N(i)}$.

Loss Function. The network is trained end-to-end in two stages. In the first stage, we train the few-shot segmenter on both types of synthetic and realistic tasks using the segmentation loss employed in [19] and regularization loss defined in [25], which are based on the standard cross-entropy loss. Specifically, in each iteration, the segmentation loss L_{seg} can be followed as: $L_{seg} = \frac{-1}{HW} \sum_{h=1}^{H} \sum_{w=1}^{W} y_q^i(l)(h,w) \log(\hat{y}_q^i(l)(h,w)) + (1 - y_q^i(l)(h,w)) \log(1 - \hat{y}_q^i(l)(h,w))$, which is applied on a random query x_q^i (formed by synthetic or realistic transformations) to predict the segmentation mask $\hat{y}_q^i(l)$, where $l \in L_{tr}$. The regularization loss L_{reg} is defined to segment the class l in its corresponding support image x_s^i, as follows: $L_{reg} = \frac{-1}{HW} \sum_{h=1}^{H} \sum_{w=1}^{W} y_s^i(l)(h,w) \log(\hat{y}_s^i(l)(h,w)) + (1 - y_s^i(l)(h,w)) \log(1 - \hat{y}_s^i(l)(h,w))$.

Overall, in each iteration, the loss function during the first-stage training is $L_1 = L_{seg} + L_{reg}$.

In the second stage of training, we aim to exploit information beyond 2D image slices in a volumetric image by employing realistic tasks.

To this end, we define the 2.5D loss function, $L_{2.5D}$, which enforces mask continuity among the prediction of adjacent queries. The proposed $L_{2.5D}$ profits the Dice loss [18] to measure the similarity between the predicted mask of $2m$ adjacent slices of the support image x_s^i as follows:

$$L_{2.5D} = \frac{1}{2m-1} \sum_{j \in N(i)} (1 - Dice(\hat{y}_q^j(l), \hat{y}_q^{j+1}(l))). \tag{1}$$

Specifically, the loss function compares the predicted mask of a query slice with the predicted mask of its adjacent slice and penalizes any discontinuities between them. This helps ensure that the model produces consistent and coherent segmentation masks across multiple slices, improving the overall quality and accuracy of the segmentation. Hence, in the second-stage training, we train the network only on realistic tasks using the loss function: $L_2 = L_{seg} + L_{reg} + \lambda_1 L_{2.5D}$, where λ_1 is linearly increased from 0 to 0.5 every 1000th iteration during training. Finally, after self-supervised learning, the few-shot segmenter can be directly utilized for inference on unseen classes.

2.3 Volumetric Segmentation Strategy

During inference, the goal is to segment query volumes based on a support volume with only a sparse set of human-annotated slices, while the few-shot segmenter is trained with 2D images. To evaluate 2D segmentation on 3D volumetric images, we take inspiration from [21] and propose the volumetric segmentation propagation strategy (VSPS).

Assume, $X_s = \{x_s^1, x_s^2, ..., x_s^{n_s}\}$ and $X_q = \{x_q^1, x_q^2, ..., x_q^{n_q}\}$ denote support and query volumes, comprising of n_s and n_q consecutive slices, respectively. We follow the same setting as [8,19,21] in which slices containing semantic class l are divided into K equally-spaced groups, including $[X_s^1, X_s^2, ..., X_s^K]$ in the support, and $[X_q^1, X_q^2, ..., X_q^K]$ in the query volume, where X^k indicates the set of slices in the kth group. Suppose, in each of the k groups in the support volume, the manual annotation of the middle slice $[(x_s^c)^1, (x_s^c)^2, ..., (x_s^c)^K]$ are available as in [8,19,21]. For volumetric segmentation, previous methods [8,19,21], for each group $k \in \{1, ..., K\}$, pair the annotated center slice in the support volume with all the unannotated slices of the corresponding group in the query volume. More precisely, $((x_s^c)^k, (y_s^c)^k)$ is considered as the support for all slices in X_q^k, where $(y_s^c)^k$ is annotation of the center slice $(x_s^c)^k$. Finally, they use the 2D few-shot segmenter to find the mask of each of the query slices individually and therefore segment the whole query volume accordingly. In this work, we exploit the VSPS algorithm, which is based on two steps. In the first step, an inter-volume task is constructed to segment the center slice of each group in the query volume. More precisely, the center slice of each query group, $(x_q^c)^k$, is segmented using $((x_s^c)^k, (y_s^c)^k)$ as the support. Then, by employing the volumetric view even in the inference time, we construct intra-volume tasks to segment other slices of each group. Formally, VSPS consecutively segments each $(x_q^j)^k \in X_q^k$, starting $(x_q^c)^k$, with respect to the image-mask pair of its previous slice, i.e., $((x_q^{j-1})^k, (\hat{y}_q^{j-1})^k)$. In fact, we first find the pseudo-mask of $(x_q^c)^k$ using the 2D few-shot segmenter and consequently consider this pseudo-annotated slice as the support for all other slices in X_q^k. It is worth mentioning that our task generation strategy discussed in Sect. 2.2 is capable of handling such intra-volume tasks. Further details of the VSPS algorithm are brought in the Supplementary Materials.

3 Experiments

3.1 Experimental Setup

To unify experiment results, we follow the evaluation protocol established by [19], such as Hyper-parameters, data preprocessing techniques, evaluation metric (i.e., Dice score), and compared methods. The architecture and implementation of the network are exactly the same as developed in [19]. Moreover, during inference, a support volume with 3 annotated slices (i.e., $K = 3$) is used as a reference to segment each query volume, the same as in [19]. Also, we set $m = 3$, taking 3 adjacent slices of the support image as consecutive query images. However, the effect of this hyper-parameter is investigated in the Supplementary Materials.

Dataset. Following [8,19], we perform experiments on two common medical benchmarks, including abdominal CT image scans from MICCAI 2015 Multi-Atlas Abdomen Labeling challenge [15] and abdominal MRI image scans from ISBI 2019 Combined Healthy Abdominal Organ Segmentation Challenge [14]. In addition, in all experiments, average results are reported according to 5-fold cross-validation on four anatomical structures the same as in [8,19], including left kidney (LK), right kidney (RK), spleen, and liver.

3.2 Results and Discussion

Comparison with Existing Approaches. Table 1 compares VISA-FSS with state-of-the-art FSS methods in terms of Dice, including: Vanilla PANet [25], SE-Net [21], SSL-RPNet [23], SSL-ALPNet [19], and CRAPNet [8]. Vanilla PANet and SE-Net are baselines on natural and medical images, respectively, which utilize manual annotations for training. SSL-RPNet, SSL-ALPNet, and CRAP-Net are self-supervised methods that construct their FSS tasks using synthetic transformations (e.g., geometric and intensity) in the same way, and are only different in the network architecture. As demonstrated, VISA-FSS outperforms vanilla PANet and SE-Net without using any manual annotation in its training phase. Moreover, the performance gains of VISA-FSS compared with SSL-RPNet, SSL-ALPNet, and CRAPNet highlight the benefit of learning continuous shape transformation among consecutive slices within a 3D image for volumetric segmentation. Also, the performance of VISA-FSS was evaluated using Hausedorff Distance and Surface Dice metrics on CT and MRI datasets. On the CT dataset, VISA-FSS reduced SSLALPNet's Hausedorff Distance from 30.07 to 23.62 and improved Surface Dice from 89.31% to 90.68%. On the MRI dataset, it decreased Hausedorff Distance from 26.49 to 22.14 and improved Surface Dice from 90.16% to 91.08%. Further experiments and qualitative results are given in the Supplementary Materials, demonstrating satisfactory results on different abdominal organs.

Table 1. Comparison results of different methods (in Dice score) on abdominal images.

Method	Abdominal-CT				Mean	Abdominal-MRI				Mean
	RK	LK	Spleen	Liver		RK	LK	Spleen	Liver	
Vanilla PANet [25]	21.19	20.67	36.04	49.55	31.86	32.19	30.99	40.58	50.40	38.53
SE-Net [21]	12.51	24.42	43.66	35.42	29.00	47.96	45.78	47.30	29.02	42.51
SSL-RPNet [23]	66.73	65.14	64.01	72.99	67.22	81.96	71.46	73.55	75.99	75.74
CRAPNet [8]	74.18	74.69	70.37	75.41	73.66	86.42	81.95	74.32	76.46	79.79
SSL-ALPNet [19]	71.81	72.36	70.96	78.29	73.35	85.18	81.92	72.18	76.10	78.84
VISA-FSS (Ours)	**76.17**	**77.05**	**76.51**	**78.70**	**77.11**	**89.55**	**87.90**	**78.05**	**77.00**	**83.12**

Effect of Task Generation. To investigate the effect of realistic tasks in self-supervised FSS models, we perform an ablation study on the absence of this type of task. The experiment results are given in rows (a) and (b) of Table 2. As expected, performance gains can be observed when both synthetic and realistic tasks are employed during training. This can highlight that the use of more and diverse tasks improves the performance of FSS models.

Of note, to generate pseudo-label for consecutive slices, instead of SPPS, we can also employ supervoxel generation strategies like the popular SLIC algorithm [1]. However, we observed that by doing so the performance is 66.83 in the term of mean Dice score, under-performing SPPS (row (b) in Table 2) by about 8%. It can be inferred that contrary to SLIC, SPPS implicitly takes pseudo-label shape continuity into account due to its propagation process, which can help construct effective realistic tasks. To intuitively illustrate this issue, visual comparison of some pseudo-labels generated by SLIC and SPPS is depicted in the Supplementary Materials. In addition, to demonstrate the importance of the 2.5D loss function defined in Eq. 1 during training, we report the performance with and without $L_{2.5D}$ in Table 2 (see row (d) and (e)). We observe over 1% increase in the average Dice due to applying the 2.5D loss function.

Table 2. Ablation studies on task generation and different types of volumetric segmentation strategies on abdominal CT dataset (Results are based on Dice score).

	Training		Inference	Organs				Mean
	Tasks	Loss	Vol. Seg. Str	RK	LK	Spleen	Liver	
(a)	Syn	w.o. $L_{2.5D}$	VSS [21]	71.81	72.36	70.96	78.29	73.35
(b)	Syn. + Re	w.o. $L_{2.5D}$	VSS [21]	71.59	72.02	73.85	78.57	74.01
(c)	Syn. + Re	w.o. $L_{2.5D}$	RPS	71.13	71.72	72.68	77.69	73.31
(d)	Syn. + Re	w.o. $L_{2.5D}$	VSPS	74.67	75.14	75.00	78.74	75.88
(e)	Syn. + Re	w. $L_{2.5D}$	VSPS	76.17	77.05	76.51	78.70	77.11

Importance of the Volumetric Segmentation Strategy. To verify the influence of our proposed volumetric segmentation strategy, we compare VSPS against two different strategies: VSS and RPS. VSS (volumetric segmentation strategy) is exactly the same protocol established by [21] (explained in detail in Sect. 2.3). In addition, RPS (registration-based propagation strategy) is a ablated version of VSPS which propagates the annotation of the center slice in each query volume group into unannotated slices in the same group using registration-based models like [3] instead of using the trained few-shot segmenter. Comparison results are given in rows (b) to (d) of Table 2, demonstrating the superiority of VSPS compared with other strategies. In fact, due to learning synthetic transformations (e.g., geometric and intensity transformation) during training, VSPS, during inference, can successfully segment a new class in a query slice given a support slice from a different volume. Also, due to learning realistic transformations (e.g., intra-volume transformations), each query slice can be effectively segmented with respect to its neighbour slice.

4 Conclusion

This work introduces a novel framework called VISA-FSS, which aims to perform few-shot 3D segmentation without requiring any manual annotations during training. VISA-FSS leverages inter-slice information and continuous shape changes that exist across consecutive slices within a 3D image. During training, it uses consecutive slices within a 3D volume as support and query images, as well as support-query pairs generated by applying geometric and intensity transformations. This allows us to exploit intra-volume information and introduce a 2.5D loss function that penalizes the model for making predictions that are discontinuous among adjacent slices. Finally, during inference, a novel strategy for volumetric segmentation is introduced to employ the volumetric view even during the testing time.

References

1. Achanta, R., Shaji, A., Smith, K., Lucchi, A., Fua, P., Süsstrunk, S.: Slic superpixels. Tech. rep. (2010)
2. Balakrishnan, G., Zhao, A., Sabuncu, M.R., Guttag, J., Dalca, A.V.: An unsupervised learning model for deformable medical image registration. In: Proceedings of the IEEE Conference on Computer Vision and Pattern Recognition, pp. 9252–9260 (2018)
3. Bitarafan, A., Azampour, M.F., Bakhtari, K., Soleymani Baghshah, M., Keicher, M., Navab, N.: Vol2flow: segment 3d volumes using a sequence of registration flows. In: Medical Image Computing and Computer Assisted Intervention-MICCAI 2022: 25th International Conference, Proceedings, Part IV, pp. 609–618. Springer (2022). https://doi.org/10.1007/978-3-031-16440-8_58
4. Bitarafan, A., Nikdan, M., Baghshah, M.S.: 3d image segmentation with sparse annotation by self-training and internal registration. IEEE J. Biomed. Health Inform. **25**(7), 2665–2672 (2020)
5. Chen, T., Kornblith, S., Norouzi, M., Hinton, G.: A simple framework for contrastive learning of visual representations. In: International Conference on Machine Learning, pp. 1597–1607. PMLR (2020)
6. Chen, X., et al.: A deep learning-based auto-segmentation system for organs-at-risk on whole-body computed tomography images for radiation therapy. Radiother. Oncol. **160**, 175–184 (2021)
7. Denner, S., et al.: Spatio-temporal learning from longitudinal data for multiple sclerosis lesion segmentation. In: Crimi, A., Bakas, S. (eds.) BrainLes 2020. LNCS, vol. 12658, pp. 111–121. Springer, Cham (2021). https://doi.org/10.1007/978-3-030-72084-1_11
8. Ding, H., Sun, C., Tang, H., Cai, D., Yan, Y.: Few-shot medical image segmentation with cycle-resemblance attention. In: Proceedings of the IEEE/CVF Winter Conference on Applications of Computer Vision, pp. 2488–2497 (2023)

9. Farshad, A., Makarevich, A., Belagiannis, V., Navab, N.: Metamedseg: volumetric meta-learning for few-shot organ segmentation. In: Domain Adaptation and Representation Transfer 2022, pp. 45–55. Springer (2022). https://doi.org/10.1007/978-3-031-16852-9_5
10. Felzenszwalb, P.F., Huttenlocher, D.P.: Efficient graph-based image segmentation. Int. J. Comput. Vision **59**, 167–181 (2004)
11. Finn, C., Abbeel, P., Levine, S.: Model-agnostic meta-learning for fast adaptation of deep networks. In: International Conference on Machine Learning, pp. 1126–1135. PMLR (2017)
12. Hesamian, M.H., Jia, W., He, X., Kennedy, P.: Deep learning techniques for medical image segmentation: achievements and challenges. J. Digit. Imaging **32**, 582–596 (2019)
13. Hospedales, T., Antoniou, A., Micaelli, P., Storkey, A.: Meta-learning in neural networks: a survey. IEEE Trans. Pattern Anal. Mach. Intell. **44**(9), 5149–5169 (2021)
14. Kavur, A.E., et al.: Chaos challenge-combined (ct-mr) healthy abdominal organ segmentation. Med. Image Anal. **69**, 101950 (2021)
15. Landman, B., Xu, Z., Igelsias, J., Styner, M., Langerak, T., Klein, A.: Miccai multi-atlas labeling beyond the cranial vault-workshop and challenge. In: Proc. MICCAI Multi-Atlas Labeling Beyond Cranial Vault-Workshop Challenge. vol. 5, p. 12 (2015)
16. Li, X., Chen, H., Qi, X., Dou, Q., Fu, C.W., Heng, P.A.: H-denseunet: hybrid densely connected unet for liver and tumor segmentation from ct volumes. IEEE Trans. Med. Imaging **37**(12), 2663–2674 (2018)
17. Lutnick, B.: An integrated iterative annotation technique for easing neural network training in medical image analysis. Nat. Mach. Intell. **1**(2), 112–119 (2019)
18. Milletari, F., Navab, N., Ahmadi, S.A.: V-net: fully convolutional neural networks for volumetric medical image segmentation. In: 2016 Fourth International Conference on 3D Vision (3DV), pp. 565–571. IEEE (2016)
19. Ouyang, C., Biffi, C., Chen, C., Kart, T., Qiu, H., Rueckert, D.: Self-supervision with superpixels: training few-shot medical image segmentation without annotation. In: Vedaldi, A., Bischof, H., Brox, T., Frahm, J.-M. (eds.) ECCV 2020. LNCS, vol. 12374, pp. 762–780. Springer, Cham (2020). https://doi.org/10.1007/978-3-030-58526-6_45
20. Ouyang, C., Kamnitsas, K., Biffi, C., Duan, J., Rueckert, D.: Data efficient unsupervised domain adaptation for cross-modality image segmentation. In: Shen, D., et al. (eds.) MICCAI 2019. LNCS, vol. 11765, pp. 669–677. Springer, Cham (2019). https://doi.org/10.1007/978-3-030-32245-8_74
21. Roy, A.G., Siddiqui, S., Pölsterl, S., Navab, N., Wachinger, C.: Squeeze & excite'guided few-shot segmentation of volumetric images. Med. Image Anal. **59**, 101587 (2020)
22. Snell, J., Swersky, K., Zemel, R.: Prototypical networks for few-shot learning. In: Advances in Neural Information Processing Systems 30 (2017)
23. Tang, H., Liu, X., Sun, S., Yan, X., Xie, X.: Recurrent mask refinement for few-shot medical image segmentation. In: Proceedings of the IEEE/CVF International Conference on Computer Vision, pp. 3918–3928 (2021)
24. Tsochatzidis, L., Koutla, P., Costaridou, L., Pratikakis, I.: Integrating segmentation information into CNN for breast cancer diagnosis of mammographic masses. Comput. Methods Programs Biomed. **200**, 105913 (2021)

25. Wang, K., Liew, J.H., Zou, Y., Zhou, D., Feng, J.: Panet: few-shot image semantic segmentation with prototype alignment. In: Proceedings of the IEEE/CVF International Conference on Computer Vision, pp. 9197–9206 (2019)
26. Zhao, A., Balakrishnan, G., Durand, F., Guttag, J.V., Dalca, A.V.: Data augmentation using learned transformations for one-shot medical image segmentation. In: Proceedings of the IEEE/CVF Conference on Computer Vision and Pattern Recognition, pp. 8543–8553 (2019)

L3DMC: Lifelong Learning Using Distillation via Mixed-Curvature Space

Kaushik Roy[1,2(✉)], Peyman Moghadam[2], and Mehrtash Harandi[1]

[1] Department of Electrical and Computer Systems Engineering, Faculty of Engineering, Monash University, Melbourne, Australia
{Kaushik.Roy,Mehrtash.Harandi}@monash.edu
[2] Data61, CSIRO, Brisbane, QLD, Australia
{Kaushik.Roy,Peyman.Moghadam}@csiro.au

Abstract. The performance of a lifelong learning (L3) model degrades when it is trained on a series of tasks, as the geometrical formation of the embedding space changes while learning novel concepts sequentially. The majority of existing L3 approaches operate on a fixed-curvature (*e.g.*, zero-curvature Euclidean) space that is not necessarily suitable for modeling the complex geometric structure of data. Furthermore, the distillation strategies apply constraints directly on low-dimensional embeddings, discouraging the L3 model from learning new concepts by making the model highly stable. To address the problem, we propose a distillation strategy named L3DMC that operates on mixed-curvature spaces to preserve the already-learned knowledge by modeling and maintaining complex geometrical structures. We propose to embed the projected low dimensional embedding of fixed-curvature spaces (Euclidean and hyperbolic) to higher-dimensional Reproducing Kernel Hilbert Space (RKHS) using a positive-definite kernel function to attain rich representation. Afterward, we optimize the L3 model by minimizing the discrepancies between the new sample representation and the subspace constructed using the old representation in RKHS. L3DMC is capable of adapting new knowledge better without forgetting old knowledge as it combines the representation power of multiple fixed-curvature spaces and is performed on higher-dimensional RKHS. Thorough experiments on three benchmarks demonstrate the effectiveness of our proposed distillation strategy for medical image classification in L3 settings. Our code implementation is publicly available at https://github.com/csiro-robotics/L3DMC.

Keywords: Lifelong Learning · Class-incremental Learning · Catastrophic Forgetting · Mixed-Curvature · Knowledge Distillation · Feature Distillation

1 Introduction

Lifelong learning [31,34] is the process of sequential learning from a series of non-stationary data distributions through acquiring novel concepts while preserving already-learned knowledge. However, Deep Neural Networks (DNNs) exhibit a significant drop in performance on previously seen tasks when trained in continual learning settings.

Supplementary Information The online version contains supplementary material available at https://doi.org/10.1007/978-3-031-43895-0_12.

H. Greenspan et al. (Eds.): MICCAI 2023, LNCS 14221, pp. 123–133, 2023.
https://doi.org/10.1007/978-3-031-43895-0_12

This phenomenon is often called catastrophic forgetting [27,28,35]. Furthermore, the unavailability of sufficient training data in medical imaging poses an additional challenge in tackling catastrophic forgetting of a DNN model.

In a lifelong learning scenario, maintaining a robust embedding space and preserving geometrical structure is crucial to mitigate performance degradation and catastrophic forgetting of old tasks[23]. However, the absence of samples from prior tasks has been identified as one of the chief reasons for catastrophic forgetting. Therefore, to address the problem, a small memory buffer has been used in the literature to store a subset of samples from already seen tasks and replayed together with new samples [18,32]. Nevertheless, imbalances in data (*e.g.*, between current tasks and the classes stored in the memory) make the model biased towards the current task [18]. Knowledge distillation [4,9,16,31] has been widely used in the literature to preserve the previous knowledge while training on a novel data distribution. This approach applies constraints on updating the weight of the current model by mimicking the prediction of the old model. To maintain the old embedding structure intact in the new model, feature distillation strategies have been introduced in the literature [9,18]. For instance, LUCIR [18] emphasizes on maximizing the similarity between the orientation of old and new embedding by minimizing the cosine distance between old and new embedding. While effective, feature distillation applies strong constraints directly on the lower-dimensional embedding extracted from the old and new models, reducing the plasticity of the model. This is not ideal for adopting novel concepts while preserving old knowledge. Lower-dimensional embedding spaces, typically used for distillation, may not preserve all the latent information in the input data [19]. As a result, they may not be ideal for distillation in lifelong learning scenarios. Furthermore, DNNs often operate on zero-curvature (*i.e.*, Euclidean) spaces, which may not be suitable for modeling and distilling complex geometrical structures in non-stationary biomedical image distributions with various modalities and discrepancies in imaging protocols and medical equipment. On the contrary, hyperbolic spaces have been successfully used to model hierarchical structure in input data for different vision tasks [13,21].

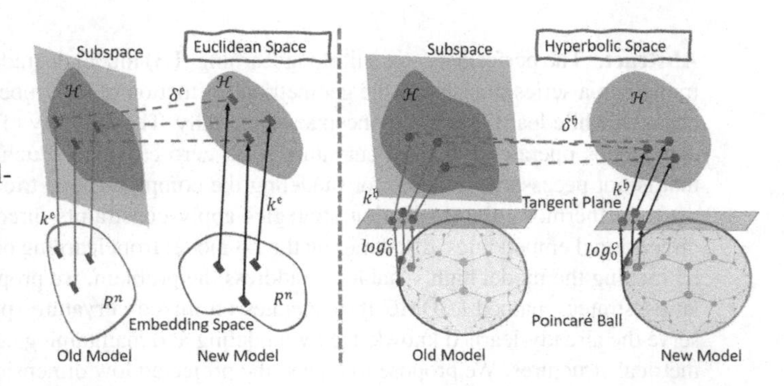

Fig. 1. Geometrical interpretation of the proposed distillation strategy, L3DMC via mixed-curvature space that optimizes the model by combining the distillation loss from Euclidean with zero-curvature (left) and hyperbolic with negative-curvature (right) space. L3DMC preserves the complex geometrical structure by minimizing the distance between new data representation and subspace induced by old representation in RKHS.

In this paper, we propose to perform distillation in a Reproducing Kernel Hilbert Space (RKHS), constructed from the embedding space of multiple fixed-curvature spacess. This approach is inspired by the ability of kernel methods to yield rich representations in higher-dimensional RKHS [10,19]. Specifically, we employ a Radial Basis Function (RBF) kernel on a mixed-curvature space that combines embeddings from hyperbolic (negative curvature), and Euclidean (zero curvature), using a decomposable Riemannian distance function as illustrated in Fig. 1. This mixed-curvature space is robust and can maintain a higher quality geometrical formation. This makes the space more suitable for knowledge distillation and tackling catastrophic forgetting in lifelong learning scenarios for medical image classification. Finally, to ensure a similar geometric structure between the old and new models in L3, we propose minimizing the distance between the new embedding and the subspace constructed using the old embedding in RKHS. Overall, our contributions in this paper are as follows:

- To the best of our knowledge, this is the first attempt to study mixed-curvature space for the continual medical image classification task.
- We propose a novel knowledge distillation strategy to maintain a similar geometric structure for continual learning by minimizing the distance between new embedding and subspace constructed using old embedding in RKHS.
- Quantitative analysis shows that our proposed distillation strategy is capable of preserving complex geometrical structure in embedding space resulting in significantly less degradation of the performance of continual learning and superior performance compared to state-of-the-art baseline methods on BloodMNIST, PathMNIST, and OrganaMNIST datasets.

2 Preliminaries

Lifelong Learning (L3). L3 consists of a series of T tasks $\mathcal{T}_t \in \{\mathcal{T}_1, \mathcal{T}_2, \cdots, \mathcal{T}_T\}$, where each task $\mathcal{T}_t$ has it's own dataset $\mathcal{D}_t = \{\mathcal{X}_t, \mathcal{Y}_t\}$. In our experiments, $\mathbf{X}_i \in \mathcal{X}_t \subset \mathcal{X}$ denotes a medical image of size $W \times H$ and $y_i \in \mathcal{Y}_t \subset \mathcal{Y}$ is its associated disease category at task t. In class-incremental L3, label space of two tasks is disjoint, hence $\mathcal{Y}_t \cap \mathcal{Y}_{t'} = \emptyset; t \neq t'$. The aim of L3 is to train a model $f : \mathcal{X} \to \mathcal{Y}$ incrementally for each task t to map the input space $\mathcal{X}_t$ to the corresponding target space $\mathcal{Y}_t$ without forgetting all previously learned tasks (*i.e.*, $1, 2, \cdots, t-1$). We assume that a fixed-size memory $\mathcal{M}$ is available to store a subset of previously seen samples to mitigate catastrophic forgetting in L3.

Mixed-Curvature Space. Mixed-Curvature space is formulated as the Cartesian product of fixed-curvature spaces and represented as $M = \times_{i=1}^{C} M_i^{d_i}$. Here, M_i can be a Euclidean (zero curvature), hyperbolic (constant negative curvature), or spherical (constant positive curvature) space. Furthermore, $\times$ denotes the Cartesian product, and d_i is the dimensionality of fixed-curvature space M_i with curvature c_i. The distance in the mixed-curvature space can be decomposed as $d_M(\mathbf{x}, \mathbf{y}) := \sum_{i=1}^{C} d_{M_i}(\mathbf{x}^i, \mathbf{y}^i)$.

Hyperbolic Poincaré Ball. Hyperbolic space is a Riemannian manifold with negative curvature. The Poincare ball with curvature $-c$, $c > 0$, $\mathbb{D}_c^n = \{\mathbf{x} \in \mathbb{R}^n : c\|\mathbf{x}\| < 1\}$ is a model of n-dimensional hyperbolic geometry. To perform vector operations on $\mathbb{H}^n$, Möbius Gyrovector space is widely used. Möbius addition between $x \in \mathbb{D}_c^n$ and $y \in \mathbb{D}_c^n$ is defined as follows

$$\mathbf{x} \oplus_c \mathbf{y} = \frac{(1 + 2c\langle \mathbf{x}, \mathbf{y} \rangle + c\|\mathbf{y}\|_2^2)\mathbf{x} + (1 - c\|\mathbf{x}\|_2^2)\mathbf{y}}{1 + 2c\langle \mathbf{x}, \mathbf{y} \rangle + c^2\|\mathbf{x}\|_2^2\|\mathbf{y}\|_2^2} \tag{1}$$

Using Möbius addition, geodesic distance between two input data points, $\mathbf{x}$ and $\mathbf{y}$ in $\mathbb{D}_c^n$ is computed using the following formula.

$$d_c(\mathbf{x}, \mathbf{y}) = \frac{2}{\sqrt{c}} \tanh^{-1}(\sqrt{c}\|(-\mathbf{x}) \oplus_c \mathbf{y}\|_2) \tag{2}$$

Tangent space of data point $\mathbf{x} \in \mathbb{D}_c^n$ is the inner product space and is defined as $T_x\mathbb{D}_c^n$ which comprises the tangent vector of all directions at $\mathbf{x}$. Mapping hyperbolic embedding to Euclidean space and vice-versa is crucial for performing operations on $\mathbb{D}^n$. Consequently, a vector $\mathbf{x} \in T_x\mathbb{D}_c^n$ is embedded onto the Poincaré ball $\mathbb{D}_c^n$ with anchor $\mathbf{x}$ using the exponential mapping function and the inverse process is done using the logarithmic mapping function $\log_{\mathbf{v}}^c$ that maps $\mathbf{x} \in \mathbb{D}_c^n$ to the tangent space of $\mathbf{v}$ as follows

$$\log_{\mathbf{v}}^c(\mathbf{x}) = \frac{2}{\sqrt{c}\lambda_{\mathbf{v}}^c} \tanh^{-1}(\sqrt{c}\| - \mathbf{v} \oplus_c \mathbf{x}\|_2) \frac{-\mathbf{v} \oplus_c \mathbf{x}}{\| - \mathbf{v} \oplus_c \mathbf{x}\|_2} \tag{3}$$

where $\lambda_{\mathbf{v}}^c$ is conformal factor that is defined as $\lambda_{\mathbf{v}}^c = \frac{2}{1 - c\|v\|^2}$. In practice, anchor $\mathbf{v}$ is set to the origin. Therefore, the exponential mapping is expressed as $\exp_0^c(\mathbf{x}) = \tanh(\sqrt{c}\|\mathbf{x}\|) \frac{\mathbf{x}}{\sqrt{c}\|\mathbf{x}\|}$.

3 Proposed Method

In our approach, we emphasize on modeling complex latent structure of medical data by combining embedding representation of zero-curvature Euclidean and negative-curvature hyperbolic space. To attain richer representational power of RKHS [17], we embed the low-dimensional fixed-curvature embedding onto higher-dimensional RKHS using the kernel method.

Definition 1. (*Positive Definite Kernel*) A function $f : \mathcal{X} \times \mathcal{X} \to \mathbb{R}$ is positive definite (pd) if and only if **1.** $k(\mathbf{x}_i, \mathbf{x}_j) = k(\mathbf{x}_j, \mathbf{x}_i)$ for any $\mathbf{x}_i, \mathbf{x}_j \in \mathcal{X}$, and **2.** for any given $n \in \mathbb{N}$, we have $\sum_i^n \sum_j^n c_i c_j k(\mathbf{x}_j, \mathbf{x}_i) \geq 0$ for any $\mathbf{x}_1, \mathbf{x}_2, \cdots, \mathbf{x}_n \in \mathcal{X}$ and $c_1, c_2, \cdots, c_n \in \mathbb{R}$. Equivalently, the Gram matrix $K_{ij} = k(\mathbf{x}_i, \mathbf{x}_j) > 0$ for any set of n samples $\mathbf{x}_1, \mathbf{x}_2, \cdots, \mathbf{x}_n \in \mathcal{X}$ should be Symmetric and Positive Definite (SPD).

Popular kernel functions (*e.g.*, the Gaussian RBF) operate on flat-curvature Euclidean spaces. In $\mathbb{R}^n$, the Gaussian RBF kernel method is defined as

$$k^{\mathfrak{e}}(\mathbf{z}_i, \mathbf{z}_j) := \exp\big(-\lambda\|\mathbf{z}_i - \mathbf{z}_j\|^2\big); \lambda > 0. \tag{4}$$

However, using the geodesic distance in a hyperbolic space along with an RBF function similar to Eq. (4) (*i.e.*, replacing $\|\mathbf{z}_i - \mathbf{z}_j\|^2$ with the geodesic distance) does not lead to a valid positive definite kernel. Theoretically, a valid RBF kernel is impossible to obtain for hyperbolic space using geodesic distance [11,12]. Therefore, we use the tangent plane of hyperbolic space and employ $\log_0^c$

$$\log_0^c(\mathbf{z}) = \frac{\mathbf{z}}{\sqrt{c}\|\mathbf{z}\|}\tanh^{-1}\big(\sqrt{c}\|\mathbf{z}\|\big)\,, \tag{5}$$

to embed hyperbolic data to RKHS via the following valid pd kernel (see [10] for the proof of positive definiteness):

$$k^{\mathfrak{h}}(\mathbf{z}_i, \mathbf{z}_j) = \exp\big(- \lambda\|\log_0^c(\mathbf{z}_i) - \log_0^c(\mathbf{z}_j)\|^2\big). \tag{6}$$

Now, in L3 setting, we have two models h^t and h^{t-1} at our hand at time t. We aim to improve h^t while ensuring the past knowledge incorporated in h^{t-1} is kept within h^t. Assume $\mathbf{Z}_t$ and $\mathbf{Z}_{t-1}$ are the extracted feature vectors for input $\mathbf{X}$ using current and old feature extractor, h^t_{feat} and h^{t-1}_{feat}, respectively. Unlike other existing distillation methods, we employ an independent 2-layer MLP for each fixed-curvature space to project extracted features to a new lower-dimensional embedding space on which we perform further operations. This has two benefits, **(i)** it relaxes the strong constraint directly applied on $\mathbf{Z}_t$ and $\mathbf{Z}_{t-1}$ and **(ii)** reduce the computation cost of performing kernel method. Since we are interested in modeling embedding structure in zero-curvature Euclidean and negative-curvature hyperbolic spaces, we have two MLP as projection modules attached to feature extractors, namely g_e and g_h.

Our Idea. Our main idea is that, for a rich and overparameterized representation, the data manifold is low-dimensional. Our algorithm makes use of RKHS, which can be intuitively thought of as a neural network with infinite width. Hence, we assume that the data manifold for the model at time $t-1$ is well-approximated by a low-dimensional hyperplane (our data manifold assumption). Let $\mathbf{Z}^{\mathfrak{e}}_{t-1} = \{\mathbf{z}^{\mathfrak{e}}_{t-1,1}, \mathbf{z}^{\mathfrak{e}}_{t-1,2}, \cdots, \mathbf{z}^{\mathfrak{e}}_{t-1,m}\}$ be the output of the Euclidean projection module for m samples at time t (*i.e.*, current model). Consider $\mathbf{z}^{\mathfrak{e}}_t$, a sample at time t from the Euclidean projection head. We propose to minimize the following distance

$$\begin{aligned}\delta^{\mathfrak{e}}(\mathbf{z}^{\mathfrak{e}}_t, \mathbf{Z}^{\mathfrak{e}}_{t-1}) &:= \Big\|\phi(\mathbf{z}^{\mathfrak{e}}_t) - \operatorname{span}\{\phi(\mathbf{z}^{\mathfrak{e}}_{t-1,i})\}_{i=1}^m\Big\|^2 \\ &= \min_{\alpha\in\mathbb{R}^m} \Big\|\phi(\mathbf{z}^{\mathfrak{e}}_t) - \sum_{i=1}^m \alpha_i\phi(\mathbf{z}^{\mathfrak{e}}_{t-1,i})\Big\|^2 .\end{aligned} \tag{7}$$

In Eq. (7), ϕ is the implicit mapping to the RKHS defined by the Gaussian RBF kernel, *i.e.* $k^{\mathfrak{e}}$. The benefit of formulation Eq. (7) is that it has a closed-form solution as

$$\delta^{\mathfrak{e}}(\mathbf{z}^{\mathfrak{e}}_t, \mathbf{Z}^{\mathfrak{e}}_{t-1}) = k(\mathbf{z}^{\mathfrak{e}}_t, \mathbf{z}^{\mathfrak{e}}_t) - k^\top_{\mathbf{zZ}}K^{-1}_{\mathbf{ZZ}}k_{\mathbf{zZ}}\,. \tag{8}$$

In Eq. (8), $K_{\mathbf{ZZ}} \in \mathbb{R}^{m\times m}$ is the Gram matrix of $\mathbf{Z}^{\mathfrak{e}}_{t-1}$, and $k_{\mathbf{zZ}}$ is an m-dimensional vector storing the kernel values between $\mathbf{z}^{\mathfrak{e}}_t$ and elements of $\mathbf{z}^{\mathfrak{e}}_{t-1}$. We provide the proof of equivalency between Eq. (7) and Eq. (8) in the supplementary material due to the lack of space. Note that we could use the same form for the hyperbolic projection module g_h to distill between the model at time t and $t-1$, albeit this time, we employ the hyperbolic kernel $k^{\mathfrak{h}}$. Putting everything together,

$$\ell_{\text{KD}}(\mathbf{Z}_t) := \mathbb{E}_{\mathbf{z}_t}\delta^{\mathfrak{e}}(\mathbf{z}^{\mathfrak{e}}_t, \mathbf{Z}^{\mathfrak{e}}_{t-1}) + \beta\mathbb{E}_{\mathbf{z}_t}\delta^{\mathfrak{h}}(\mathbf{z}^{\mathfrak{h}}_t, \mathbf{Z}^{\mathfrak{h}}_{t-1})\,. \tag{9}$$

Here, β is a hyper-parameter that controls the weight of distillation between the Euclidean and hyperbolic spaces. We can employ Eq. (9) at the batch level. Note that in our formulation, computing the inverse of an $m \times m$ matrix is required, which has a complexity of $\mathcal{O}(m^3)$. However, this needs to be done once per batch and manifold (*i.e.*, Euclidean plus hyperbolic). ℓ_{KD} is differentiable with respect to $\mathbf{Z}_t$, which enables us to update the model at time t. We train our lifelong learning model by combining distillation loss, ℓ_{KD}, together with standard cross entropy loss. Please refer to the overall steps of training lifelong learning model using our proposed distillation strategy via mixed-curvature space in Algorithm 1.

3.1 Classifier and Exemplar Selection

We employ herding based exemplar selection method that selects examples that are closest to the class prototype, following iCARL [32]. At inference time, we use exemplars from memory to compute class template and the nearest template class computed using Euclidean distance is used as the prediction of our L3 model. Assume μ_c is the class template computed by averaging the extracted features from memory exemplars belonging to class c. Then, the prediction $\hat{y}$ for a given input sample $\mathbf{X}$ is determined as $\hat{y} = \underset{c=1,\dots,t}{\arg\min} \| h^t_{feat}(\mathbf{X}) - \mu_c \|_2$.

4 Related Work

In this section, we describe Mixed-curvature space and L3 methods to tackle catastrophic forgetting.

Constant-Curvature and Mixed-Curvature Space. Constant-curvature spaces have been successfully used in the literature to realize the intrinsic geometrical orientation of data for various downstream tasks in machine learning. Flat-curvature Euclidean space is suitable to model grid data [37] while positive and negative-curvature space is better suited for capturing cyclical [2] and hierarchical [25] structure respectively. Hyperbolic representation has been used across domains ranging from image classification [26] and natural language processing [29,30] to graphs [6]. However, a constant-curvature space is limited in modeling the geometrical structure of data embedding as it is designed with a focus on particular structures [14].

Algorithm 1 Lifelong Learning using Distillation via Mixed-Curvature Space

Input: Dataset $\mathcal{D}^0, \mathcal{D}^1, ..., \mathcal{D}^T$, and Memory $\mathcal{M}$
Output: The new model at time t with parameters $\mathbf{\Theta}^t$

1: Randomly Initialize $\mathbf{\Theta}^0$; $h^0_{\mathbf{\Theta}} = h^0_{\text{feat}} \circ h^0_{\text{cls}}$
2: Train $\mathbf{\Theta}^0$ on $\mathcal{D}^0$ using ℓ_{CE}
3: **for** t in $\{1, 2, ..., T\}$ **do**
4: Initialize $\mathbf{\Theta}^t$ with $\mathbf{\Theta}^{t-1}$
5: **for** iteration 1 to max_iter **do**
6: Sample a mini batch $(\mathcal{X}_B, \mathcal{Y}_B)$ from $(\mathcal{D}^t \cup \mathcal{M})$
7: $\mathbf{Z}^t \leftarrow h^t_{\text{feat}}(\mathcal{X}_B)$
8: $\mathbf{Z}^{t-1} \leftarrow h^{t-1}_{\text{feat}}(\mathcal{X}_B)$
9: $\tilde{\mathcal{Y}}_B \leftarrow h_{\mathbf{\Theta}^t}(\mathbf{Z}^t)$
10: Compute ℓ_{CE} between $\mathcal{Y}_B$ and $\tilde{\mathcal{Y}}_B$
11: Compute ℓ_{KD} between $\mathbf{Z}^{t-1}$ and $\mathbf{Z}^t$
12: Update $\mathbf{\Theta}^t$ by minimizing the combined loss of cross-entropy ℓ_{CE} and ℓ_{KD} as in Eq. (9)
13: **end for**
14: Evaluate Model on test dataset
15: Update Memory $\mathcal{M}$ with exemplars from $\mathcal{D}^t$
16: **end for**

Kernel Methods. A Kernel is a function that measures the similarity between two input samples. The intuition behind the kernel method is to embed the low-dimensional input data into a higher, possibly infinite, dimensional RKHS space. Because of the ability to realize rich representation in RKHS, kernel methods have been studied extensively in machine learning [17].

L3 Using Regularization with Distillation. Regularization-based approaches impose constraints on updating weights of L3 model to maintain the performance on old tasks. LwF mimics the prediction of the old model into the current model but struggles to maintain consistent performance in the absence of a task identifier. Rebuff *et al.* in [32] store a subset of exemplars using a herding-based sampling strategy and apply knowledge distillation on output space like LwF [24]. Distillation strategy on feature spaces has also been studied in the literature of L3. Hou *et al.* in [18] proposes a less-forgetting constraint that controls the update of weight by minimizing the cosine angle between old and new embedding representation.

5 Experimental Details

Datasets. In our experiments, we use four datasets (*e.g.*, BloodMNIST [1], PathMNIST [20], OrganaMNIST [3]) and TissueMNIST [3] from MedMNIST collection [38] for the multi-class disease classification. BloodMNIST, PathMNIST, OrganaMNIST and TissueMNIST have 8, 9, 11, and 8 distinct classes, respectively that are split into 4 tasks with non-overlapping classes between tasks following [8]. For cross-domain continual learning experiments, we present 4 datasets sequentially to the model.

Implementation Details. We employ ResNet18 [15] as the backbone for feature extraction and a set of task-specific fully connected layers as the classifier to train all the baseline methods across datasets. To ensure fairness in comparisons, we run each experiment with the same set of hyperparameters as used in [8] for five times with a fixed set of distinct seed values, 1, 2, 3, 4, 5 and report the average value. Each model is optimized using Stochastic Gradient Decent (SGD) with a batch of 32 images for 200 epochs, having early stopping options in case of overfitting. Furthermore, we use gradient clipping by enforcing the maximum gradient value to 10 to tackle the gradient exploding problem.

Evaluation Metrics. We rely on average accuracy and average forgetting to quantitatively examine the performances of lifelong learning methods as used in previous approaches [7,32]. Average accuracy is computed by averaging the accuracy of all the previously observed and current tasks after learning a current task t and defined as: $\text{Acc}_t = \frac{1}{t}\sum_{i=1}^{t} \text{Acc}_{t,i}$, where $\text{Acc}_{t,i}$ is the accuracy of task i after learning task t. We measure the forgetting of the previous task at the end of learning the current task t using: $F_t = \frac{1}{t-1}\sum_{i=1}^{t-1} \max_{j\in\{1\ldots t-1\}} \text{Acc}_{j,i} - \text{Acc}_{t,i}$, where at task t, forgetting on task i is defined as the maximum difference value previously achieved accuracy and current accuracy on task i.

6 Results and Discussion

In our comparison, we consider two regularization-based methods (*i.e.*, EWC [22], and LwF [24]) and 5 memory-based methods (*e.g.*, EEIL [5], ER [33], Bic [36], LUCIR [18] and iCARL [32]). We employ the publicly available code[1] of [8] in our experiments to produce results for all baseline methods on BloodMNIST, PathMNIST, and OrganaMNIST datasets and report the quantitative results in Table 1. The results suggest that the performance of all methods improves with the increase in buffer size (*e.g.*, from 200 to 1000). We observe that our proposed distillation approach outperforms other baseline methods across the settings. The results suggest that the regularization-based methods, *e.g.*, EWC and LwF perform poorly in task-agnostic settings across the datasets as those methods are designed for task-aware class-incremental learning. Our proposed method outperforms experience replay, ER method by a significant margin in both evaluation metrics (*i.e.*, average accuracy and average forgetting) across datasets. For instance, our method shows around 30%, 30%, and 20% improvement in accuracy compared to ER while the second best method, iCARL, performs about 4%, 2%, and 8% worse than our method on BloodMNIST, PathMNIST, and OrganaMNIST respectively with 200

Table 1. Experimental results on BloodMNIST, PathMNIST, OrganaMNIST and TissueMNIST datasets for 4-tasks Class-Incremental setting with varying buffer size. Our proposed method outperforms other baseline methods across the settings.

Method	BloodMNIST		PathMNIST		OrganaMNIST	
	Accuracy ↑	Forgetting ↓	Accuracy ↑	Forgetting ↓	Accuracy ↑	Forgetting ↓
Upper Bound	97.98	–	93.52	–	95.22	–
Lower Bound	46.59	68.26	32.29	77.54	41.21	54.20
EWC [22]	47.60	66.22	33.34	76.39	37.88	67.62
LwF [24]	43.68	66.30	35.36	67.37	41.36	51.47
Memory Size: 200						
EEIL [5]	42.17	71.25	28.42	79.39	41.03	62.47
ER [33]	42.94	71.38	33.74	80.6	52.50	52.72
LUCIR [18]	20.76	53.80	40.00	54.72	41.70	33.06
BiC [36]	53.32	31.06	48.74	30.82	58.68	29.66
iCARL [32]	67.70	**14.52**	58.46	**-0.70**	63.02	**7.75**
Ours	**71.98**	14.62	**60.60**	21.18	**71.01**	13.88
Memory Size: 1000						
EEIL [5]	64.40	40.92	34.18	75.42	66.24	34.60
ER [33]	65.94	33.68	44.18	66.24	67.90	31.72
LUCIR [18]	20.92	28.42	53.84	30.92	54.22	23.64
BiC [36]	70.04	17.98	–	–	73.46	15.98
iCARL [32]	73.10	13.18	61.72	14.14	74.54	10.50
Ours	**77.26**	**10.9**	**67.52**	**12.5**	**76.46**	**9.08**

[1] https://github.com/mmderakhshani/LifeLonger.

Table 2. Average accuracy on the cross-domain incremental learning scenario [8] with 200 exemplars. CL3DMC outperforms baseline methods by a significant margin in both task-aware and task-agnostic settings. Best values are in bold.

Scenario	LwF	EWC	ER	EEIL	BiC	iCARL	LUCIR	L3DMC (Ours)
Task-Agnostic (Accuracy ↑)	29.45	18.44	34.54	34.54	26.79	48.87	19.05	**52.19**
Task-Aware (Accuracy ↑)	31.07	29.26	37.69	33.19	33.19	49.47	27.48	**52.83**

exemplars. Similarly, with 1000 exemplars, our proposed method shows consistent performances and outperforms iCARL by 4%, 6%, and 2% accordingly on BloodMNIST, PathMNIST, and OrganaMNIST datasets. We also observe that catastrophic forgetting decreases with the increase of exemplars. Our method shows about 2% less forgetting phenomenon across the datasets with 1000 exemplars compared to the second best method iCARL.

Table 2 presents the experimental results (*e.g.*, average accuracy) on relatively complex cross-domain incremental learning setting where datasets (BloodMNIST, PathMNIST, OrganaMNIST, and TissueMNIST) with varying modalities from different institutions are presented at each novel task. Results show an unmatched gap between regularization-based methods (*e.g.*, Lwf and EWC) and our proposed distillation method. CL3DMC outperforms ER method by around 16% on both task-aware and task-agnostic settings. Similarly, CL3DMC performs around 3% better than the second best method, iCARL.

7 Conclusion

In this paper, we propose a novel distillation strategy, L3DMC on mixed-curvature space to preserve the complex geometric structure of medical data while training a DNN model on a sequence of tasks. L3DMC aims to optimize the lifelong learning model by minimizing the distance between new embedding and old subspace generated using current and old models respectively on higher dimensional RKHS. Extensive experiments show that L3DMC outperforms state-of-the-art L3 methods on standard medical image datasets for disease classification. In future, we would like to explore the effectiveness of our proposed distillation strategy on long-task and memory-free L3 setting.

References

1. Acevedo, A., Merino, A., Alférez, S., Molina, Á., Boldú, L., Rodellar, J.: A dataset of microscopic peripheral blood cell images for development of automatic recognition systems. In: Data in brief 30 (2020)
2. Bachmann, G., Bécigneul, G., Ganea, O.: Constant curvature graph convolutional networks. In: International Conference on Machine Learning, pp. 486–496. PMLR (2020)
3. Bilic, P., et al.: The liver tumor segmentation benchmark (lits). Med. Image Anal. **84**, 102680 (2023)
4. Buzzega, P., Boschini, M., Porrello, A., Abati, D., Calderara, S.: Dark experience for general continual learning: a strong, simple baseline. In: NeurIPS (2020)

5. Castro, F.M., Marín-Jiménez, M.J., Guil, N., Schmid, C., Alahari, K.: End-to-end incremental learning. In: Ferrari, V., Hebert, M., Sminchisescu, C., Weiss, Y. (eds.) ECCV 2018. LNCS, vol. 11216, pp. 241–257. Springer, Cham (2018). https://doi.org/10.1007/978-3-030-01258-8_15
6. Chami, I., Ying, Z., Ré, C., Leskovec, J.: Hyperbolic graph convolutional neural networks. In: Advances in Neural Information Processing Systems 32 (2019)
7. Chaudhry, A., Dokania, P.K., Ajanthan, T., Torr, P.H.S.: Riemannian walk for incremental learning: understanding forgetting and intransigence. In: Ferrari, V., Hebert, M., Sminchisescu, C., Weiss, Y. (eds.) ECCV 2018. LNCS, vol. 11215, pp. 556–572. Springer, Cham (2018). https://doi.org/10.1007/978-3-030-01252-6_33
8. Derakhshani, M.M., et al.: Lifelonger: A benchmark for continual disease classification. In: Medical Image Computing and Computer Assisted Intervention-MICCAI 2022: 25th International Conference, Singapore, 18–22 September 2022, Proceedings, Part II, pp. 314–324. Springer (2022). https://doi.org/10.1007/978-3-031-16434-7_31
9. Douillard, A., Cord, M., Ollion, C., Robert, T., Valle, E.: PODNet: pooled outputs distillation for small-tasks incremental learning. In: Vedaldi, A., Bischof, H., Brox, T., Frahm, J.-M. (eds.) ECCV 2020. LNCS, vol. 12365, pp. 86–102. Springer, Cham (2020). https://doi.org/10.1007/978-3-030-58565-5_6
10. Fang, P., Harandi, M., Petersson, L.: Kernel methods in hyperbolic spaces. In: International Conference on Computer Vision, pp. 10665–10674 (2021)
11. Feragen, A., Hauberg, S.: Open problem: kernel methods on manifolds and metric spaces. what is the probability of a positive definite geodesic exponential kernel? In: Conference on Learning Theory, pp. 1647–1650. PMLR (2016)
12. Feragen, A., Lauze, F., Hauberg, S.: Geodesic exponential kernels: when curvature and linearity conflict. In: IEEE Conference on Computer Vision and Pattern Recognition, pp. 3032–3042 (2015)
13. Ganea, O., Bécigneul, G., Hofmann, T.: Hyperbolic neural networks. In: Advances In Neural Information Processing Systems 31 (2018)
14. Gu, A., Sala, F., Gunel, B., Ré, C.: Learning mixed-curvature representations in product spaces. In: International Conference on Learning Representations (2019)
15. He, K., Zhang, X., Ren, S., Sun, J.: Deep residual learning for image recognition. In: IEEE Conference on Computer Vision and Pattern Recognition, pp. 770–778 (2016)
16. Hinton, G., Vinyals, O., Dean, J., et al.: Distilling the knowledge in a neural network, vol. 2(7). arXiv preprint arXiv:1503.02531 (2015)
17. Hofmann, T., Schölkopf, B., Smola, A.J.: Kernel methods in machine learning (2008)
18. Hou, S., Pan, X., Loy, C.C., Wang, Z., Lin, D.: Learning a unified classifier incrementally via rebalancing. In: IEEE Conference on Computer Vision and Pattern Recognition, pp. 831–839 (2019)
19. Jayasumana, S., Hartley, R., Salzmann, M., Li, H., Harandi, M.: Kernel methods on riemannian manifolds with gaussian rbf kernels. IEEE Trans. Pattern Anal. Mach. Intell. **37**(12), 2464–2477 (2015)
20. Kather, J.N., et al.: Predicting survival from colorectal cancer histology slides using deep learning: a retrospective multicenter study. PLoS Med. **16**(1), e1002730 (2019)
21. Khrulkov, V., Mirvakhabova, L., Ustinova, E., Oseledets, I., Lempitsky, V.: Hyperbolic image embeddings. In: IEEE Conference on Computer Vision and Pattern Recognition, pp. 6418–6428 (2020)
22. Kirkpatrick, J., et al.: Overcoming catastrophic forgetting in neural networks. Proc. Natl. Acad. Sci. **114**(13), 3521–3526 (2017)
23. Knights, J., Moghadam, P., Ramezani, M., Sridharan, S., Fookes, C.: Incloud: incremental learning for point cloud place recognition. In: 2022 IEEE/RSJ International Conference on Intelligent Robots and Systems (IROS), pp. 8559–8566. IEEE (2022)

24. Li, Z., Hoiem, D.: Learning without forgetting. IEEE Trans. Pattern Anal. Mach. Intell. **40**(12), 2935–2947 (2017)
25. Liu, Q., Nickel, M., Kiela, D.: Hyperbolic graph neural networks. In: Advances in Neural Information Processing Systems 32 (2019)
26. Mathieu, E., Le Lan, C., Maddison, C.J., Tomioka, R., Teh, Y.W.: Continuous hierarchical representations with poincaré variational auto-encoders. In: Advances in Neural Information Processing Systems 32 (2019)
27. McCloskey, M., Cohen, N.J.: Catastrophic interference in connectionist networks: the sequential learning problem. In: Psychology of learning and motivation, vol. 24, pp. 109–165. Elsevier (1989)
28. Nguyen, C.V., Achille, A., Lam, M., Hassner, T., Mahadevan, V., Soatto, S.: Toward understanding catastrophic forgetting in continual learning. arXiv preprint arXiv:1908.01091 (2019)
29. Nickel, M., Kiela, D.: Poincaré embeddings for learning hierarchical representations. In: Advances in Neural Information Processing Systems 30 (2017)
30. Nickel, M., Kiela, D.: Learning continuous hierarchies in the lorentz model of hyperbolic geometry. In: International Conference on Machine Learning, pp. 3779–3788. PMLR (2018)
31. Parisi, G.I., Kemker, R., Part, J.L., Kanan, C., Wermter, S.: Continual lifelong learning with neural networks: a review. Neural Netw. **113**, 54–71 (2019)
32. Rebuffi, S.A., Kolesnikov, A., Sperl, G., Lampert, C.H.: icarl: incremental classifier and representation learning. In: IEEE Conference on Computer Vision and Pattern Recognition, pp. 2001–2010 (2017)
33. Riemer, M., et al.: Learning to learn without forgetting by maximizing transfer and minimizing interference. arXiv preprint arXiv:1810.11910 (2018)
34. Ring, M.B.: Child: a first step towards continual learning. Mach. Learn. **28**(1), 77–104 (1997)
35. Robins, A.: Catastrophic forgetting, rehearsal and pseudorehearsal. Connect. Sci. **7**(2), 123–146 (1995)
36. Wu, Y., et al.: Large scale incremental learning. In: IEEE Conference on Computer Vision and Pattern Recognition, pp. 374–382 (2019)
37. Wu, Z., Pan, S., Chen, F., Long, G., Zhang, C., Philip, S.Y.: A comprehensive survey on graph neural networks. IEEE Trans. Neural Netw. Learn. Syst. **32**(1), 4–24 (2020)
38. Yang, J., Shi, R., Ni, B.: Medmnist classification decathlon: a lightweight automl benchmark for medical image analysis. In: 2021 IEEE 18th International Symposium on Biomedical Imaging (ISBI), pp. 191–195. IEEE (2021)

Machine Learning – Explainability, Bias, and Uncertainty I

Weakly Supervised Medical Image Segmentation via Superpixel-Guided Scribble Walking and Class-Wise Contrastive Regularization

Meng Zhou, Zhe Xu(✉), Kang Zhou, and Raymond Kai-yu Tong(✉)

The Chinese University of Hong Kong, Hong Kong, China
jackxz@link.cuhk.edu.hk
kytong@cuhk.edu.hk

Abstract. Deep learning-based segmentation typically requires a large amount of data with dense manual delineation, which is both time-consuming and expensive to obtain for medical images. Consequently, weakly supervised learning, which attempts to utilize sparse annotations such as scribbles for effective training, has garnered considerable attention. However, such scribble-supervision inherently lacks sufficient structural information, leading to two critical challenges: (i) while achieving good performance in overall overlap metrics such as Dice score, the existing methods struggle to perform satisfactory local prediction because no desired structural priors are accessible during training; (ii) the class feature distributions are inevitably less-compact due to sparse and extremely incomplete supervision, leading to poor generalizability. To address these, in this paper, we propose the SC-Net, a new scribble-supervised approach that combines **S**uperpixel-guided scribble walking with **C**lass-wise contrastive regularization. Specifically, the framework is built upon the recent dual-decoder backbone design, where predictions from two slightly different decoders are randomly mixed to provide auxiliary pseudo-label supervision. Besides the sparse and pseudo supervision, the scribbles walk towards unlabeled pixels guided by superpixel connectivity and image content to offer as much dense supervision as possible. Then, the class-wise contrastive regularization disconnects the feature manifolds of different classes to encourage the compactness of class feature distributions. We evaluate our approach on the public cardiac dataset ACDC and demonstrate the superiority of our method compared to recent scribble-supervised and semi-supervised learning methods with similar labeling efforts.

Keywords: Weakly-supervised Learning · Segmentation · Superpixel

1 Introduction

Accurately segmenting cardiac images is crucial for diagnosing and treating cardiovascular diseases. Recently, deep learning methods have greatly advanced

M. Zhou and Z. Xu—Equal contribution.

H. Greenspan et al. (Eds.): MICCAI 2023, LNCS 14221, pp. 137–147, 2023.
https://doi.org/10.1007/978-3-031-43895-0_13

cardiac image segmentation. However, most state-of-the-art segmentation models require a large scale of training samples with pixel-wise dense annotations, which are expensive and time-consuming to obtain. Thus, researchers are active in exploring other labour-efficient forms of annotations for effective training. For example, semi-supervised learning (SSL) [14,20,22,26–29,31] is one such approach that attempts to propagate labels from the limited labeled data to the abundant unlabeled data, typically via pseudo-labeling. However, due to limited diversity in the restricted labeled set, accurately propagating labels is very challenging [15]. As another form, weakly supervised learning (WSL), i.e., our focused scenario, utilizes sparse labels such as scribbles, bounding boxes, and points for effective training, wherein scribbles have gained significant attention due to their ease of annotation and flexibility in labeling irregular objects. Yet, an intuitive challenge is that the incomplete shape of cardiac in scribble annotations inherently lacks sufficient structural information, as illustrated in Fig. 1, which easily leads to (i) poor local prediction (e.g., poor boundary prediction with high 95% Hausdorff Distance) because no structural priors are provided during training; (ii) poor generalizability due to less-compact class feature distributions learned from extremely sparse supervision. Effectively training a cardiac segmentation model using scribble annotations remains an open challenge.

Related Work. A few efforts, not limited to medical images, have been made in scribble-supervised segmentation [4,5,9,11,13,15,19,32]. For example, Tang et al. [19] introduced a probabilistic graphical model, conditional random field (CRF), to regularize the spatial relationship between neighboring pixels in an image. Kim et al. [9] proposed another regularization loss based on level-set [17] to leverage the weak supervision. S2L [11] leverages label filtering to improve the pseudo labels generated by the scribble-trained model. USTM [13] adapts an uncertainty-aware mean-teacher [31] model in semi-supervised learning to leverage the unlabeled pixels. Zhang et al. [32] adapted a positive-unlabeled learning framework into this problem assisted by a global consistency term. Luo et al. [15] proposed a dual-decoder design where predictions from two slightly different decoders are randomly mixed to provide more reliable auxiliary pseudo-label supervision. Despite its effectiveness to some extent, the aforementioned two challenges still have not received adequate attention.

In this paper, we propose SC-Net, a new scribble-supervised approach that combines **S**uperpixel-guided scribble walking with **C**lass-wise contrastive regularization. The basic framework is built upon the recent dual-decoder backbone design [15]. Besides the sparse supervision (using partial cross-entropy loss) from scribbles, predictions from two slightly different decoders are randomly mixed to provide auxiliary pseudo-label supervision. This design helps to prevent the model from memorizing its own predictions and falling into a trivial solution during optimization. Then, we tackle the aforementioned inherent challenges with two schemes. Firstly, we propose a specialized mechanism to guide the scribbles to walk towards unlabeled pixels based on superpixel connectivity and image content, in order to augment the structural priors into the labels themselves. As such, better local predictions are achieved. Secondly, we propose a class-wise

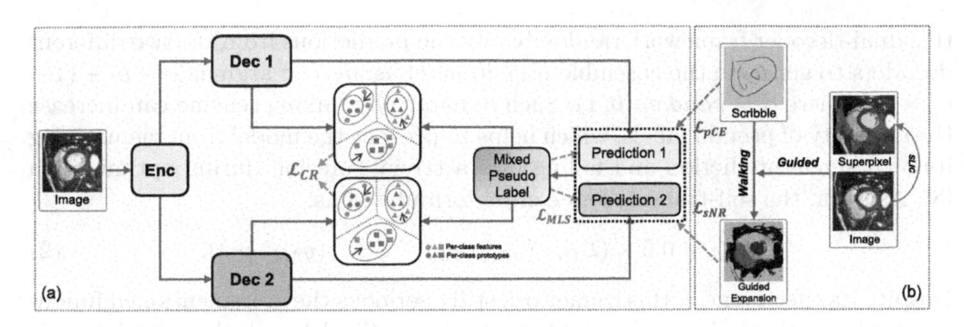

Fig. 1. Overview of SC-Net for scribble-supervised cardiac segmentation. (a) The framework consists of a shared encoder (Enc) and two independent and different decoders (Dec 1 and Dec 2). Structural priors are enriched by (b) superpixel-guided scribble walking strategy (Sect. 2.2). The class-wise contrastive regularization $\mathcal{L}_{CR}$ (Sect. 2.3) encourages the compactness of feature manifolds of different classes.

contrastive regularization term that leverages prototype contrastive learning to disconnect the feature manifolds of different classes, which addresses the issue of less-compact class feature distributions due to sparse supervision. We evaluate our approach on the public cardiac dataset ACDC and show that it achieves promising results, especially better boundary predictions, compared to recent scribble-supervised and semi-supervised methods with similar labeling efforts.

2 Methods

2.1 Preliminaries and Basic Framework

In the scribble-supervised setting, the dataset includes images and their corresponding scribble annotations. We denote an image as X with the scribble annotation $S = \{(s_r, y_r)\}$, where s_r is the pixel of scribble r, and $y_r \in \{0, 1, ..., C-1\}$ denotes the corresponding label with C possible classes at pixel s_r. As shown in Fig. 1, our framework is built upon a one-encoder-dual-decoder design [15], where the encoder (θ_{enc}) is shared and two decoders are independent and slightly different. Here, we denote the decoder 1 as θ_{Dec1} and the auxiliary decoder 2 as θ_{Dec2}. Compared to θ_{Dec1}, θ_{Dec2} introduces the dropout layer (ratio $= 0.5$) before each convolutional block to impose perturbations. In this framework, the supervised signals consist of a scribble-supervised loss and a pseudo-supervised self-training loss. For the former one, we adopt the commonly used partial cross-entropy loss for those scribble-containing pixels [11,19], formulated as:

$$\mathcal{L}_{pCE} = -0.5 \times (\sum_{c}\sum_{i \in S} \log p^c_{1(i)} + \sum_{c}\sum_{i \in S} \log p^c_{2(i)}), \tag{1}$$

where $p^c_{1(i)}$ and $p^c_{2(i)}$ are the predicted probability of pixel i belonging to class c from the two decoders θ_{Dec1} and θ_{Dec2}, respectively. For the self-training loss, this dual-decoder framework randomly mix the predictions from the two different decoders to generate the ensemble pseudo label as: $\hat{y}_{ML} = \text{argmax}[\alpha \times p_1 + (1 - \alpha) \times p_2$, where $\alpha = random(0, 1)$. Such dynamically mixing scheme can increase the diversity of pseudo labels, which helps to prevent the model from memorizing its own single prediction and falling into a trivial solution during optimization [8]. As such, the self-training loss can be formulated as:

$$\mathcal{L}_{MLS} = 0.5 \times (\mathcal{L}_{Dice}(\hat{y}_{ML}, p_1) + \mathcal{L}_{Dice}(\hat{y}_{ML}, p_2)). \quad (2)$$

Despite its effectiveness, this framework still overlooks the aforementioned fundamental limitations of sparse scribble supervision: (i) although the mixed pseudo labels provide dense supervision, they still stems from the initial sparse guidance, making it difficult to provide accurate local structural information. Thus, we propose superpixel-guided scribble walking strategy (Sect. 2.2) to enrich structural priors for the initial supervision itself. (ii) Extremely sparse supervision inevitably leads to less-compact class feature distributions, resulting in poor generalizability to unseen test data. Thus, we further propose class-wise contrastive regularization (Sect. 2.3) to enhance the compactness of class embeddings.

2.2 Superpixel-Guided Scribble Walking

In order to enhance the structural information in our initial supervision, we utilize the superpixel of the image as a guide for propagating scribble annotations to unlabeled pixels, considering that it effectively groups pixels with similar characteristics within the uniform regions of an image and helps capture the class boundaries [30]. Specifically, we employ the simple linear iterative clustering (SLIC) algorithm [1] to generate the superpixels. The algorithm works by first dividing the image into a grid of equally-sized squares, then selecting a number of seed points within each square based on the desired number K of superpixels. Next, it iteratively assigns each pixel to the nearest seed point based on its color similarity and spatial proximity (distance). This process is repeated until the clustering converges or reaches a predefined number of iterations. Finally, the algorithm updates the location of the seed points to the centroid of the corresponding superpixel, and repeats until convergence. As such, the image is coarsely segmented into K clusters. To balance accuracy and computational efficiency, the number of iterations is empirically set to 10. K is set to 150. An example of superpixel is depicted in Fig. 1.

Then, guided by the obtained superpixel, the scribbles walk towards unlabeled pixels with the following mechanisms: (i) if the superpixel cluster overlaps with a scribble s_r, the label y_r of s_r walks towards to the pixels contained in this cluster; (ii) yet, if the superpixel cluster does not overlap any scribble or overlaps more than one scribble, the pixels within this cluster are not assigned any labels. As such, we denote the set of the superpixel-guided expanded label as $\{(x_{sp}, \hat{y}_{sp})\}$, where x_{sp} represents the pixel with the corresponding label $\hat{y}_{sp} \in \{0, 1, ..., C-1\}$. An expansion example can be also found in Fig. 1. Although we use strict walking constraints to expand the labels, superpixels are primarily based on color similarity and spatial proximity to seed points.

However, magnetic resonance imaging has less color information compared to natural images, and different organs often share similar intensity, leading to some inevitable label noises. Therefore, to alleviate the negative impact of the label noises, we adopt the noise-robust Dice loss [24] to supervise the models, formulated as:

$$\mathcal{L}_{sNR} = 0.5 \times \left(\frac{\sum_i^N |p_{1(i)} - \hat{y}_{sp(i)}|^\gamma}{\sum_i^N p_{1(i)}^2 + \sum_i^N \hat{y}_{sp(i)}^2 + \epsilon} + \frac{\sum_i^N |p_{2(i)} - \hat{y}_{sp(i)}|^\gamma}{\sum_i^N p_{2(i)}^2 + \sum_i^N \hat{y}_{sp(i)}^2 + \epsilon}\right), \quad (3)$$

where N is the number of label-containing pixels. $\hat{y}_{sp}$ is converted to one-hot representation. $p_{1(i)}$ and $p_{2(i)}$ are the predicted probabilities of pixel i from θ_{Dec1} and θ_{Dec2}, respectively. Following [24], $\epsilon = 10^{-5}$ and $\gamma = 1.5$. Note that when $\gamma = 2$, this loss will degrade into the typical Dice loss.

2.3 Class-Wise Contrastive Regularization

When using extremely sparse supervision, it is difficult for the model to learn compact class feature distributions, leading to poor generalizability. To address this, we propose a class-wise contrastive regularization term that leverages prototype contrastive learning to disconnect the feature manifolds of different classes, as illustrated in Fig. 1. Specifically, using the additional non-linear projection head, we derive two sets of projected features, namely F_1 and F_2, from decoder 1 and decoder 2, respectively. Then, we filter the projected features by comparing their respective categories with that of the mixed pseudo label $\hat{y}_{ML}$ and the current predictions from the two decoders. Only features that have matching categories are retained and denoted as $\dot{F}_1^c$ and $\dot{F}_2^c$, where superscript c indicates that such feature vectors correspond to class c. Then, we use C attention modules [2] to obtain ranking scores to sort the retained features and then the top-k features are selected as the class prototypes, where the class-c prototypes are denoted as $Z_1^c = \{z_1^c\}$ and $Z_2^c = \{z_2^c\}$. Note that we extract feature prototypes in an online fashion instead of retaining cross-epoch memories as in [2], since the latter can be computationally inefficient and memory-intensive. Then, we extract the features of each category $f_1^c \in F_1$ and $f_2^c \in F_2$ using the current predictions and encourage their proximity to the corresponding prototypes z_1^c and z_2^c. We adopt the cosine similarity to measure the proximity between the class features and the class prototypes. Taking decoder 1 as example, we define its class-wise contrastive regularization loss $\mathcal{L}_{CR}^{Dec1}$ as:

$$\mathcal{L}_{CR}^{Dec1}(f_1^c, Z_1^c) = \frac{1}{C}\frac{1}{N_{z_1^c}}\frac{1}{N_{f_1^c}}\sum_{c=1}^{C}\sum_{i=1}^{N_{z_1^c}}\sum_{j=1}^{N_{f_1^c}} w_{ij}\left(1 - \frac{< z_1^{c(i)}, f_1^{c(j)} >}{||z_1^{c(i)}||_2 \cdot ||f_1^{c(j)}||_2}\right), \quad (4)$$

where w_{ij} is obtained by normalizing the learnable attention weights (detailed in [2]). $N_{z_1^c}$ or $N_{f_1^c}$ is the number of prototypes or projected features of c-th class, respectively. Similarly, we obtain such regularization loss for decoder 2, denoted as $\mathcal{L}_{CR}^{Dec2}$. As such, the overall class-wise contrastive regularization loss is formulated as:

$$\mathcal{L}_{CR} = 0.5 \times (\mathcal{L}_{CR}^{Dec1} + \mathcal{L}_{CR}^{Dec2}). \tag{5}$$

Overall, the final loss of our SC-Net is summarized as:

$$\mathcal{L} = \mathcal{L}_{pCE} + \lambda_{MLS}\mathcal{L}_{MLS} + \lambda_{sNR}\mathcal{L}_{sNR} + \lambda_{CR}\mathcal{L}_{CR}, \tag{6}$$

where λ_{MLS}, λ_{sNR} and λ_{CR} are the trade-off weights. λ_{MLS} is set to 0.5, following [15]. λ_{sNR} is set to 0.005. λ_{CR} is scheduled with an iteration-dependent ramp-up function [10] with the maximal value of 0.01 suggested by [25].

3 Experiments and Results

Dataset. We evaluate our method on the public ACDC dataset [3], which consists of 200 short-axis cine-MRI scans from 100 patients. Each patient has two annotated scans from end-diastolic (ED) and end-systolic (ES) phases, where each scan has three structure labels, including right ventricle (RV), myocardium (Myo) and left ventericle (LV). The scribbles used in this work are manually annotated by Valvano et al. [21]. Considering the large thickness in this dataset, we perform 2D segmentation rather than 3D segmentation, following [15,21].

Implementation and Evaluation Metrics. The framework is implemented with PyTorch using an NVIDIA RTX 3090 GPU. We adopt UNet [18] as the backbone with extension to dual-branch design [15]. All the 2D slices are normalized to [0, 1] and resized to 256×256 pixels. Data augmentations, including random rotation, flipping and noise injection, are applied. The SGD optimizer is utilized with the momentum of 0.9 and weight decay is 10^{-4}, the poly learning rate strategy is employed [16]. We train the segmentation model for 60,000 iterations in total with a batch size of 12. During inference, the encoder in combination with the primary decoder (Dec 1) is utilized to segment each scan slice-by-slice and stack the resulting 2D slice predictions into a 3D volume. We adopt the commonly used Dice Coefficient (DSC) and 95% Hausdorff Distance (95HD) as the evaluation metrics. Five-fold cross-validation is employed. The code will be available at https://github.com/Lemonzhoumeng/SC-Net.

Comparison Study. We compare our proposed SC-Net with recent state-of-the-art alternative methods for annotation-efficient learning. Table 1 presents the quantitative results of different methods. All methods are implemented with the same backbone to ensure fairness. According to [15,21], the cost of scribble annotation for the entire ACDC training set is similar to that of dense pixel-level annotation for 10% of the training samples. Thus, we use 10% of the training samples (8 patients) as labeled data and the remaining 90% as unlabeled data to perform semi-supervised learning (SSL). Here, we compare popular semi-supervised approaches, including AdvEnt [23], DAN [34], MT [20] and UAMT [31], as well as the supervised-only (Sup) baseline (using 10% densely labeled data only). As observed, SC-Net achieves significantly better performance than

Table 1. Quantitative results of different methods via five-fold cross-validation. Standard deviations are shown in parentheses. The best mean results are marked in **bold**.

Setting	Method	RV		Myo		LV		Mean	
		DSC	95HD	DSC	95HD	DSC	95HD	DSC	95HD
SSL	AdvEnt [23] (10% label)	0.614(0.256)	21.3(19.8)	0.758(0.149)	8.4(8.6)	0.843(0.134)	12.4(19.3)	0.737(0.179)	14.0(15.9)
	DAN [34] (10% label)	0.655(0.260)	21.2(20.4)	0.768(0.171)	9.5(11.7)	0.833(0.178)	14.9(19.5)	0.752(0.203)	15.2(17.2)
	MT [20] (10% label)	0.653(0.271)	18.6(22.0)	0.785(0.118)	11.4(17.0)	0.846(0.153)	19.0(26.7)	0.761(0.180)	16.3(21.9)
	UAMT [33] (10% label)	0.660(0.267)	22.3(22.9)	0.773(0.129)	10.3(14.8)	0.847(0.157)	17.1(23.9)	0.760(0.185)	16.6(20.5)
WSL	pCE [12] (lower bound)	0.628(0.110)	178.5(27.1)	0.602 (0.090)	176.0(21.8)	0.710(0.142)	168.1(45.7)	0.647(0.114)	174.2(31.5)
	RW [6]	0.829(0.094)	12.4(19.5)	0.708(0.093)	12.9(8.6)	0.747(0.130)	12.0(14.8)	0.759(0.106)	12.5(14.3)
	USTM [13]	0.824(0.103)	40.4(47.8)	0.739 (0.075)	133.4(42.9)	0.782(0.178)	140.4(54.6)	0.782(0.121)	104.7(48.4)
	S2L [11]	0.821(0.097)	16.8(24.4)	0.786(0.067)	65.6(45.6)	0.845(0.127)	66.5(56.5)	0.817(0.097)	49.6(42.2)
	MLoss [9]	0.807(0.089)	13.4(21.1)	0.828(0.057)	29.8(41.5)	0.868(0.074)	55.1(61.6)	0.834(0.073)	32.8(41.4)
	EM [7]	0.815(0.119)	37.9(54.5)	0.803(0.059)	56.9(53.2)	0.887(0.071)	50.4(57.9)	0.834(0.083)	48.5(55.2)
	DBMS [15]	0.861(0.087)	8.3(13.0)	0.835(0.057)	10.3(19.7)	0.899(0.062)	11.0(20.9)	0.865(0.078)	9.9(17.8)
	Ours (w/o $\mathcal{L}_{sNR}$)	0.847(0.086)	7.8(13.8)	0.823(0.091)	8.9(18.5)	0.902(0.077)	10.4(18.4)	0.858(0.093)	8.9(16.9)
	Ours (w/o $\mathcal{L}_{CR}$)	0.850(0.079)	8.3(14.3)	0.819(0.078)	9.2(17.3)	0.889(0.058)	10.7(15.7)	0.853(0.076)	9.3(16.3)
	Ours (SC-Net)	**0.862(0.071)**	**4.6(3.8)**	**0.839(0.088)**	**6.7(16.3)**	**0.915(0.083)**	**8.1(14.1)**	**0.872(0.063)**	**6.5(13.9)**
SL	Sup (10% label)	0.659(0.261)	26.8(30.4)	0.724(0.176)	16.0(21.6)	0.790(0.205)	24.5(30.4)	0.724(0.214)	22.5(27.5)
	Sup (full) (upper bound)	0.881(0.093)	6.9(10.9)	0.879 (0.039)	5.8(15.4)	0.935(0.065)	8.0(19.9)	0.898(0.066)	6.9(15.4)

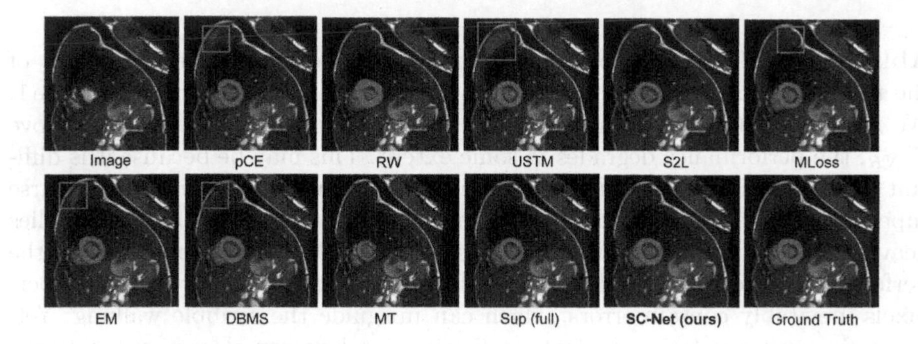

Fig. 2. Qualitative comparison of different methods.

the competing SSL methods, showing that when the annotation budget is similar, using scribble annotations can lead to better outcomes than pixel-wise annotations. Furthermore, we compare SC-Net with weakly-supervised learning (WSL) approaches on scribble annotated data, including pCE only [12] (lower bound), RW [6] (using random walker to produce additional label), USTM [13] (uncertainty-aware self-ensembling and transformation-consistent model), S2L [11] (Scribble2Label), MLoss [9] (Mumford-shah loss), EM [7] (entropy minimization) and DBMS [15] (dual-branch mixed supervision). Besides, the upper bound, i.e., supervised training with full dense annotations, is also presented. It can be observed that SC-Net achieves more promising results compared to existing methods. In comparison to DBMS, SC-Net exhibits a slight improvement in DSC, but a significant decrease in the 95HD metric ($p<0.05$). Furthermore, our method achieves slightly lower performance in DSC compared to the upper bound, but even slightly better results in 95HD. This indicates that our approach effectively addresses the inherent limitations of sparse supervision. Figure 2 presents exemplar qualitative results of our SC-Net and other methods. It can

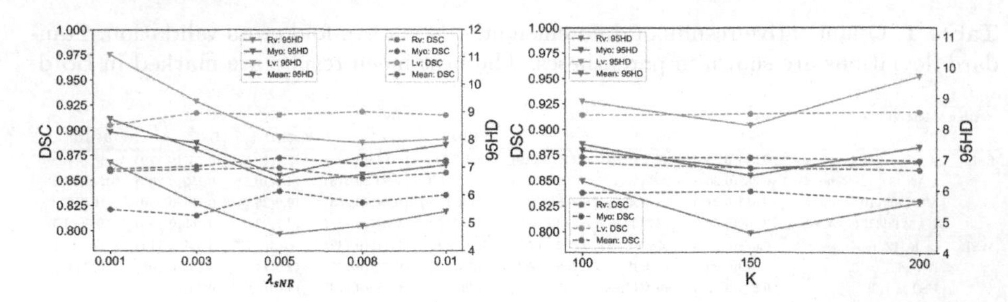

Fig. 3. Sensitivity analysis of λ_{sNR} and the cluster number K in superpixel generation.

be seen that the prediction of our SC-Net fit more accurately with the ground truth, especially in local details. Thanks to the more compact feature distributions, our method reduces false-positive predictions, as indicated in the red box.

Ablation Study. We perform an ablation study to investigate the effects of the two key components of our SC-Net. The results are also shown in Table 1. We found that the two components need to work together. When we remove $\mathcal{L}_{sNR}$, the performance degrades to some extent. This may be because it is difficult for the model to generate high-quality local pseudo-labels with only sparse supervision provided by scribbles, and class-wise contrastive regularization relies heavily on pseudo labels to separate class features. When we remove $\mathcal{L}_{CR}$, the performance also drops slightly. This is mainly because the generated superpixels inevitably contain errors, which can misguide the scribble walking. Yet, using $\mathcal{L}_{CR}$ can regularize the feature distribution between classes, reducing the impact of these errors. Meanwhile, the structure prior strengthened by superpixel guidance helps to provide higher-quality local pseudo labels to assist class-wise contrastive regularization. The two components complement each other, resulting in the best performance of our complete SC-Net.

Sensitivity Analysis. The superpixel-guided scribble walking plays an important role in our SC-Net. Thus, we conduct further assessments on the sensitivity of λ_{sNR}, which is used to weight $\mathcal{L}_{sNR}$, and the cluster number K used for superpixel generation. The results obtained by five-fold cross validation are presented in Fig. 3. As observed, increasing λ_{sNR} from 0.001 to 0.005 leads to better results in terms of both metrics. When λ_{sNR} is set to 0.01, the result exhibits only a slight decrease compared to 0.005 (0.872 vs. 0.867 in term of DSC). These observations show that our method is not so sensitive to λ_{sNR} within the empirical range. In practice, the optimal value of K depends on the characteristics of the input image, such as object complexity and texture. We find that our method is also not highly sensitive to K, but the optimal results are achieved when $K = 150$ for the cardiac MR images in our study.

4 Conclusion

In this work, we proposed the SC-Net towards effective weakly supervised medical image segmentation using scribble annotations. By combining superpixel-guided scribble walking with class-wise contrastive regularization, our approach alleviates two inherent challenges caused by sparse supervision, i.e., the lack of structural priors during training and less-compact class feature distributions. Comprehensive experiments on the public cardiac dataset ACDC demonstrated the superior performance of our method compared to recent scribble-supervised and semi-supervised methods with similar labeling efforts.

Acknowledgement. This research was supported by General Research Fund from Research Grant Council of Hong Kong (No. 14205419).

References

1. Achanta, R., Shaji, A., Smith, K., Lucchi, A., Fua, P., Süsstrunk, S.: Slic superpixels compared to state-of-the-art superpixel methods. IEEE Trans. Pattern Anal. Mach. Intell. **34**(11), 2274–2282 (2012)
2. Alonso, I., Sabater, A., Ferstl, D., Montesano, L., Murillo, A.C.: Semi-supervised semantic segmentation with pixel-level contrastive learning from a class-wise memory bank. In: Proceedings of the IEEE/CVF International Conference on Computer Vision, pp. 8219–8228 (2021)
3. Bernard, O., et al.: Deep learning techniques for automatic mri cardiac multi-structures segmentation and diagnosis: is the problem solved? IEEE Trans. Med. Imaging **37**(11), 2514–2525 (2018)
4. Can, Y.B., et al.: Learning to segment medical images with scribble-supervision alone. In: Stoyanov, D., et al. (eds.) DLMIA/ML-CDS -2018. LNCS, vol. 11045, pp. 236–244. Springer, Cham (2018). https://doi.org/10.1007/978-3-030-00889-5_27
5. Chen, Q., Hong, Y.: Scribble2d5: Weakly-supervised volumetric image segmentation via scribble annotations. In: Medical Image Computing and Computer Assisted Intervention-MICCAI 2022: 25th International Conference, Singapore, 18–22 September 2022, Proceedings, Part VIII, pp. 234–243. Springer (2022). https://doi.org/10.1007/978-3-031-16452-1_23
6. Grady, L.: Random walks for image segmentation. IEEE Trans. Pattern Anal. Mach. Intell. **28**(11), 1768–1783 (2006)
7. Grandvalet, Y., Bengio, Y.: Semi-supervised learning by entropy minimization. In: Advances in Neural Information Processing Systems 17 (2004)
8. Huo, X., et al.: Atso: asynchronous teacher-student optimization for semi-supervised image segmentation. In: CVPR, pp. 1235–1244 (2021)
9. Kim, B., Ye, J.C.: Mumford-shah loss functional for image segmentation with deep learning. IEEE Trans. Image Process. **29**, 1856–1866 (2019)
10. Laine, S., Aila, T.: Temporal ensembling for semi-supervised learning. arXiv preprint arXiv:1610.02242 (2016)
11. Lee, H., Jeong, W.-K.: Scribble2Label: scribble-supervised cell segmentation via self-generating pseudo-labels with consistency. In: Martel, A.L., et al. (eds.) MICCAI 2020. LNCS, vol. 12261, pp. 14–23. Springer, Cham (2020). https://doi.org/10.1007/978-3-030-59710-8_2

12. Lin, D., Dai, J., Jia, J., He, K., Sun, J.: Scribblesup: scribble-supervised convolutional networks for semantic segmentation. In: Proceedings of the IEEE Conference on Computer Vision and Pattern Recognition, pp. 3159–3167 (2016)
13. Liu, X.: Weakly supervised segmentation of covid19 infection with scribble annotation on ct images. Pattern Recogn. **122**, 108341 (2022)
14. Luo, X., Chen, J., Song, T., Chen, Y., Wang, G., Zhang, S.: Semi-supervised medical image segmentation through dual-task consistency. In: AAAI Conference on Artificial Intelligence (2021)
15. Luo, X., et al.: Scribble-supervised medical image segmentation via dual-branch network and dynamically mixed pseudo labels supervision. In: Medical Image Computing and Computer Assisted Intervention. pp. 528–538. Springer (2022). https://doi.org/10.1007/978-3-031-16431-6_50
16. Luo, X., et al.: Efficient semi-supervised gross target volume of nasopharyngeal carcinoma segmentation via uncertainty rectified pyramid consistency. In: de Bruijne, M., et al. (eds.) MICCAI 2021. LNCS, vol. 12902, pp. 318–329. Springer, Cham (2021). https://doi.org/10.1007/978-3-030-87196-3_30
17. Mumford, D.B., Shah, J.: Optimal approximations by piecewise smooth functions and associated variational problems. In: Communications on Pure and Applied Mathematics (1989)
18. Ronneberger, O., Fischer, P., Brox, T.: U-Net: convolutional networks for biomedical image segmentation. In: Navab, N., Hornegger, J., Wells, W.M., Frangi, A.F. (eds.) MICCAI 2015. LNCS, vol. 9351, pp. 234–241. Springer, Cham (2015). https://doi.org/10.1007/978-3-319-24574-4_28
19. Tang, M., Perazzi, F., Djelouah, A., Ayed, I.B., Schroers, C., Boykov, Y.: On regularized losses for weakly-supervised CNN segmentation. In: Ferrari, V., Hebert, M., Sminchisescu, C., Weiss, Y. (eds.) ECCV 2018. LNCS, vol. 11220, pp. 524–540. Springer, Cham (2018). https://doi.org/10.1007/978-3-030-01270-0_31
20. Tarvainen, A., Valpola, H.: Mean teachers are better role models: weight-averaged consistency targets improve semi-supervised deep learning results. In: Advances in Neural Information Processing Systems, pp. 1195–1204 (2017)
21. Valvano, G., Leo, A., Tsaftaris, S.A.: Learning to segment from scribbles using multi-scale adversarial attention gates. IEEE Trans. Med. Imaging **40**(8), 1990–2001 (2021)
22. Verma, V., et al.: Interpolation consistency training for semi-supervised learning. Neural Netw. **145**, 90–106 (2022)
23. Vu, T.H., Jain, H., Bucher, M., Cord, M., Pérez, P.: ADVENT: adversarial entropy minimization for domain adaptation in semantic segmentation. In: Proceedings of the IEEE/CVF Conference on Computer Vision and Pattern Recognition, pp. 2517–2526 (2019)
24. Wang, G., et al.: A noise-robust framework for automatic segmentation of COVID-19 pneumonia lesions from CT images. IEEE Trans. Med. Imaging **39**(8), 2653–2663 (2020)
25. Wu, Y., Wu, Z., Wu, Q., Ge, Z., Cai, J.: Exploring smoothness and class-separation for semi-supervised medical image segmentation. In: International Conference on Medical Image Computing and Computer Assisted Intervention. Springer (2022). https://doi.org/10.1007/978-3-031-16443-9_4
26. Wu, Y., Xu, M., Ge, Z., Cai, J., Zhang, L.: Semi-supervised left atrium segmentation with mutual consistency training. In: de Bruijne, M., et al. (eds.) MICCAI 2021. LNCS, vol. 12902, pp. 297–306. Springer, Cham (2021). https://doi.org/10.1007/978-3-030-87196-3_28

27. Xu, Z., et al.: Anti-interference from noisy labels: mean-teacher-assisted confident learning for medical image segmentation. IEEE Trans. Med. Imaging (2022)
28. Xu, Z., et al.: Noisy labels are treasure: mean-teacher-assisted confident learning for hepatic vessel segmentation. In: de Bruijne, M., et al. (eds.) MICCAI 2021. LNCS, vol. 12901, pp. 3–13. Springer, Cham (2021). https://doi.org/10.1007/978-3-030-87193-2_1
29. Xu, Z., et al.: All-around real label supervision: Cyclic prototype consistency learning for semi-supervised medical image segmentation. IEEE J. Biomed. Health Inform. (2022)
30. Yi, S., Ma, H., Wang, X., Hu, T., Li, X., Wang, Y.: Weakly-supervised semantic segmentation with superpixel guided local and global consistency. Pattern Recogn. **124**, 108504 (2022)
31. Yu, L., Wang, S., Li, X., Fu, C.-W., Heng, P.-A.: Uncertainty-aware self-ensembling model for semi-supervised 3d left atrium segmentation. In: Shen, D., et al. (eds.) MICCAI 2019. LNCS, vol. 11765, pp. 605–613. Springer, Cham (2019). https://doi.org/10.1007/978-3-030-32245-8_67
32. Zhang, K., Zhuang, X.: Shapepu: a new pu learning framework regularized by global consistency for scribble supervised cardiac segmentation. In: Medical Image Computing and Computer Assisted Intervention, pp. 162–172. Springer (2022). https://doi.org/10.1007/978-3-031-16452-1_16
33. Zhang, Y., Jiao, R., Liao, Q., Li, D., Zhang, J.: Uncertainty-guided mutual consistency learning for semi-supervised medical image segmentation. In: Artificial Intelligence in Medicine, p. 102476 (2022)
34. Zhang, Y., Yang, L., Chen, J., Fredericksen, M., Hughes, D.P., Chen, D.Z.: Deep adversarial networks for biomedical image segmentation utilizing unannotated images. In: Descoteaux, M., Maier-Hein, L., Franz, A., Jannin, P., Collins, D.L., Duchesne, S. (eds.) MICCAI 2017. LNCS, vol. 10435, pp. 408–416. Springer, Cham (2017). https://doi.org/10.1007/978-3-319-66179-7_47

SATTA: Semantic-Aware Test-Time Adaptation for Cross-Domain Medical Image Segmentation

Yuhan Zhang[1,2,3], Kun Huang[4], Cheng Chen[5](✉), Qiang Chen[4], and Pheng-Ann Heng[1,2]

[1] Department of Computer Science and Engineering, The Chinese University of Hong Kong, Hong Kong, China
zhangyuh@cse.cuhk.edu.hk
[2] Institute of Medical Intelligence and XR, The Chinese University of Hong Kong, Hong Kong, China
[3] Shenzhen Research Institute, The Chinese University of Hong Kong, Hong Kong, China
[4] Department of Computer Science and Engineering, Nanjing University of Science and Technology, Nanjing, China
[5] Center for Advanced Medical Computing and Analysis, Harvard Medical School and Massachusetts General Hospital, Boston, USA
cchen101@mgh.harvard.edu

Abstract. Cross-domain distribution shift is a common problem for medical image analysis because medical images from different devices usually own varied domain distributions. Test-time adaptation (TTA) is a promising solution by efficiently adapting source-domain distributions to target-domain distributions at test time with unsupervised manners, which has increasingly attracted important attention. Previous TTA methods applied to medical image segmentation tasks usually carry out a global domain adaptation for all semantic categories, but global domain adaptation would be sub-optimal as the influence of domain shift on different semantic categories may be different. To obtain improved domain adaptation results for different semantic categories, we propose Semantic-Aware Test-Time Adaptation (SATTA), which can individually update the model parameters to adapt to target-domain distributions for each semantic category. Specifically, SATTA deploys an uncertainty estimation module to measure the discrepancies of semantic categories in domain shift effectively. Then, a semantic adaptive learning rate is developed based on the estimated discrepancies to achieve a personalized degree of adaptation for each semantic category. Lastly, semantic proxy contrastive learning is proposed to individually adjust the model parameters with the semantic adaptive learning rate. Our SATTA is extensively validated on retinal fluid segmentation based on SD-OCT images. The experimental results demonstrate that SATTA consistently

Supplementary Information The online version contains supplementary material available at https://doi.org/10.1007/978-3-031-43895-0_14.

H. Greenspan et al. (Eds.): MICCAI 2023, LNCS 14221, pp. 148–158, 2023.
https://doi.org/10.1007/978-3-031-43895-0_14

improves domain adaptation performance on semantic categories over other state-of-the-art TTA methods.

Keywords: test-time adaptation · domain shift · medical image segmentation

1 Introduction

Deep learning has achieved remarkable success in medical image segmentation when the training and test data are independent and identically distributed (i.i.d) [4,13,21]. However, in many practical situations, training and test data are collected by different medical imaging devices, leading to the presence of distribution shifts. Therefore, the models trained on source-domain data perform poorly on target-domain data. An effective solution for this issue is to fine-tune the models with labeled target-domain data to adapt to the target-domain distributions [9], but it is impractical to label the target-domain data considering the high annotation cost. Existing unsupervised domain adaptation (UDA) methods [7,16,22] make full use of the labeled source-domain data and the unlabeled target-domain data in the model training, but even the target-domain data may not be available in the model training due to various practical problems.

Domain generalization (DG) methods exploit the diversity of source domains to improve the model generalization [8,14,24] when target-domain data is not available in the model training. However, it is also difficult and cost-consuming to collect multiple source-domain datasets with different domain distributions for DG. Another promising solution is test-time adaptation (TTA), which aims to gradually update the model parameters to adapt to target-domain distributions by learning from test data at test time. TTA shows greater flexibility than DG as the models could be pre-trained on single source-domain data. In TTA, a mainstream strategy is to adjust the affine parameters in BN layers for domain adaptation at test time by unsupervised loss, such as PTBN [12] and TENT [19]. Besides, auxiliary self-supervised tasks [10,17] and contrastive learning [3,20] are also considerable for TTA.

TTA methods have also been recently applied in medical image applications. Ma et al. [11] innovated distribution calibration by dynamically aggregating multiple representative classifiers via TTA to deal with arbitrary label shifts. Hu et al. [6] designed regional nuclear-norm loss and contour regularization loss for TTA on medical image segmentation tasks. Bateson et al. [1] performed inference by minimizing the entropy of predictions and a class-ratio prior, and integrated shape priors through penalty constraints for guide adaptation. Varsavsky et al. [18] introduced domain adversarial learning and consistency training in TTA for sclerosis lesion segmentation. These TTA methods have a common limitation of using a fixed learning rate for all test samples. Since test samples arrive sequentially and the scale of domain shift would change frequently, a fixed learning rate would be sub-optimal for TTA. DLTTA [23] proposed a memory bank-based discrepancy measurement for dynamic learning rate adjustment of

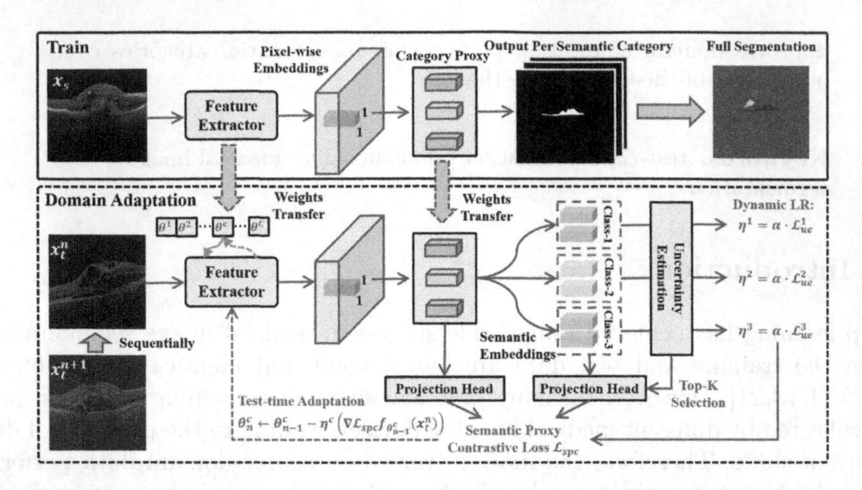

Fig. 1. Overview of SATTA for cross-domain medical image segmentation. Semantic adaptive learning rates are obtained by the uncertainty estimation module, and a semantic proxy contrastive loss is designed for individual semantic domain adaption.

TTA to effectively adapt the model to the varying domain shift. However, we find that the influence of domain shift on different semantic categories may also be different, DLTTA performed global domain adaptation for all semantic categories.

In this paper, we present **S**emantic-**A**ware **T**est-**T**ime **A**daptation (SATTA) for cross-domain medical image segmentation, aiming to perform individual domain adaptation for each semantic category at test time. SATTA first utilizes an uncertainty estimation module to effectively measure the discrepancies of different semantic categories in domain shift. Based on the estimated discrepancies, a semantic adaptive learning rate is then developed to achieve a personalized degree of adaptation for each semantic category. Lastly, a semantic proxy contrastive loss is proposed to individually adjust the model parameters with the semantic adaptive learning rate. Our SATTA is evaluated on retinal fluid segmentation based on spectral-domain optical coherence tomography (SD-OCT) images, and the experimental results show superior performance than other state-of-the-art TTA methods.

2 Methods

2.1 Test-Time Adaptation Review

Given a labeled source-domain dataset $\mathcal{S} = \{(\boldsymbol{x}_n^s, \boldsymbol{y}_n^s)\}_{n=1}^{N^s}$, model parameters θ are pre-trained on $\mathcal{S}$ by supervised risk minimization:

$$\theta^s = \arg\min_{\theta} \frac{1}{N^s} \sum_{n=1}^{N^s} \mathcal{L}_{sup}(\mathcal{F}_\theta(\boldsymbol{x}_n^s), \boldsymbol{y}_n^s) \tag{1}$$

where $\mathcal{L}_{sup}$ is the supervised loss for model optimization, such as the cross-entropy loss. However, for an unlabeled target-domain dataset $\mathcal{T} = \{(\boldsymbol{x}_n^t)\}_{n=1}^{N^t}$ that has different domain distributions with $\mathcal{S}$, the model $\mathcal{F}_{\theta^s}$ may have an obvious performance degeneration. To make the model $\mathcal{F}_{\theta^s}$ adapt to the target-domain distributions, an unsupervised TTA loss $\mathcal{L}_{tta}$ (such as rotation prediction loss [17], entropy minimization loss [19], contrastive loss [3], etc.) is designed to fine-tune model based on target-domain samples at test time:

$$\theta_n^t \leftarrow \theta_{n-1}^t - \eta(\nabla \mathcal{L}_{tta}(\mathcal{F}_{\theta_{n-1}^t}(\boldsymbol{x}_n^t))),\ n \in [1, N^t] \tag{2}$$

where η is learning rate, θ_0^t is initialized with θ^s. The final prediction on $\boldsymbol{x}_n^t$ can be given by $\boldsymbol{y}_n^t \sim \hat{\boldsymbol{y}}_n^t = \mathcal{F}_{\theta_n^t}(\boldsymbol{x}_n^t)$.

2.2 Semantic Adaptive Learning Rate

Pseudo-labeling for Semantic Aggregation. The semantic segmentation model contains a feature extractor and category predictor. For pixels $\{\boldsymbol{p}_1, \boldsymbol{p}_2, \cdots, \boldsymbol{p}_{hw}\}$ in image $\boldsymbol{x}$, feature extractor encodes them into high-dimensional pixel embeddings $\{\boldsymbol{e}_1, \boldsymbol{e}_2, \cdots, \boldsymbol{e}_{hw}\} \in \mathbb{R}^d$, and category predictor outputs the categories of the pixel embeddings. Category predictor is a projection matrix $\boldsymbol{W} \in \mathbb{R}^{d \times C} = [\boldsymbol{w}_1, \boldsymbol{w}_2, \cdots, \boldsymbol{w}_C]$ consisted of C category proxies, where each category proxy $\boldsymbol{w}_c$ can be regarded as the high-dimensional representative of the category. Category predictor measures the similarities between pixel embeddings and all category proxies. We represent the classification process of pixel $\boldsymbol{p}_i$ as:

$$\boldsymbol{y}_i \sim \hat{\boldsymbol{y}}_i \in \mathbb{R}^C = \boldsymbol{W} \cdot (f(\boldsymbol{p}_i)),\ i \in [1, hw] \tag{3}$$

where $f(\cdot)$ is the feature extractor, C is the category number, h and w are the height and width of images. Since pixel-wise labels are not available for target-domain samples at test time, we are hard to obtain the semantic information of all pixel embeddings directly. To address this problem, we assign pseudo labels to all pixel embeddings by passing them through the category predictor and then aggregate all pixel embeddings into C semantic clusters as $[\Omega_1, \Omega_2, \cdots, \Omega_C]$ according to their pseudo labels.

Semantic Uncertainty Estimation. After performing semantic aggregation by pseudo-labeling, we need to estimate the varying discrepancies of domain shift on categories. Here, we employ Monte Carlo Dropout [5] for semantic uncertainty estimation. We enable dropout at test time and perform L stochastic forward passes through the model to obtain a set of predictive outputs for pixel embedding $\boldsymbol{e}_i$:

$$u_i^l = \mathcal{M}(\boldsymbol{e}_i),\ l \in [1, L],\ i \in [1, hw] \tag{4}$$

where $\mathcal{M}(\cdot)$ is a mapping network that maps the pixel embedding $\boldsymbol{e}_i$ into an additional probability output. Then we estimate the standard deviation of the L outputs as the uncertainty score of $\boldsymbol{e}_i$:

$$s_i = std(u_i^1, u_i^2, \cdots, u_i^L),\ i \in [1, hw] \tag{5}$$

Here we take a single pixel as an example to show the uncertainty score computation, it should be noted that all pixels are performed parallel computation in the semantic segmentation model. Therefore, the computation cost does not increase with the number of pixels. For category c, its semantic uncertainty score can be calculated by:

$$\mathcal{U}^c = \frac{1}{N^{\Omega_c}} \sum_{e_i \in \Omega_c} s_i, \ c \in [1, C], \ i \in [1, hw] \tag{6}$$

where N^{Ω_c} is the number of pixel embeddings in Ω_c. $\mathcal{U}^c$ captures the unique semantic domain discrepancy over category c. Later, semantic adaptive learning rate η^c of category c for TTA is obtained directly based on the semantic domain discrepancy $\mathcal{U}^c$:

$$\eta^c = \alpha \cdot \mathcal{U}^c \tag{7}$$

where α is a scale factor. In this work, α could be set as the learning rate used for the model pre-training with the source-domain dataset. Each semantic category has its own individual learning rate in each iteration.

2.3 Semantic Proxy Contrastive Learning

General contrastive losses focus on exploring rich sample-to-sample relations, but they are hard to learn specific semantic information from samples. Proxy contrastive loss can model semantic relations by category proxies, as category proxies are more robust intuitively to noise samples [24]. Therefore, a proxy contrastive loss is more suitable for unsupervised TTA optimization with our proposed semantic adaptive learning rate.

Projection Heads. We regard each category proxy as the anchor and consider all proxy-to-sample relations. Since proxy-based methods converge very easily, we consider applying projection heads to map both pixel embeddings and category proxies to a new feature space where proxy contrastive loss is applied. Given semantic clusters $[\Omega_1, \Omega_2, \cdots, \Omega_C]$ and category proxy weights $\boldsymbol{W} = [\boldsymbol{w}_1, \boldsymbol{w}_2, \cdots, \boldsymbol{w}_C]$, We use a three-layer MLP $\mathcal{H}_1(\cdot)$ for projecting pixel embeddings and one-layer MLP $\mathcal{H}_2(\cdot)$ for projecting category proxy weights. The new pixel embedding and category proxy weight can be given by $\boldsymbol{z}_i = \mathcal{H}_1(\boldsymbol{e}_i)$ and $\boldsymbol{v}_c = \mathcal{H}_2(\boldsymbol{w}_c)$.

Top-K Selection. Pixel embeddings with high uncertainty scores contribute little to semantic proxy contrastive learning. Besides, the computation cost is huge for all pixel embeddings. To address this problem, we select K pixel embeddings with the highest confidence from each semantic cluster Ω_c. Specifically, for each semantic cluster Ω_c, we first order all pixel embeddings in it from smallest to largest according to their uncertainty scores. Then we select the first K pixel embeddings as the new semantic cluster Ω_c' for the next proxy contrastive loss by $\Omega_c' = TopK(Order(\Omega_c))$.

Semantic Proxy Contrastive Loss. For an anchor category cluster Ω'_c, we associate all pixel embeddings in it with category proxy weight $\boldsymbol{v}_c$ to form the positive pairs. We ignore the sample-to-sample positive pairs and only consider the sample-to-sample negative pairs. The semantic proxy contrastive loss for category c can be given by:

$$\mathcal{L}_{spc}(\boldsymbol{x}, \boldsymbol{W}, c) = -\frac{1}{K}\sum_{\boldsymbol{z}_i \in \Omega'_c} \log \frac{\exp(\boldsymbol{v}_c^\top \boldsymbol{z}_i \cdot \tau)}{\mathcal{Z}}$$
$$s.t.\ \mathcal{Z} = \exp(\boldsymbol{v}_c^\top \boldsymbol{z}_i \cdot \tau) + \sum_{r_0=1}^{C-1} \exp(\boldsymbol{v}_{r_0}^\top \boldsymbol{z}_i \cdot \tau) + \frac{1}{C'}\sum_{r_1=1}^{C'}\sum_{\boldsymbol{z}_j \in \Omega'_{r_1}} \exp(\boldsymbol{z}_i^\top \boldsymbol{z}_j \cdot \tau) \tag{8}$$

where $\{\boldsymbol{z}_i\}_{i=1}^K$ are obtained by $\boldsymbol{x}$ and C' is the number of categories appearing in samples.

2.4 Training and Adaptation Procedure

The overview of our SATTA is shown in Fig. 1. Given the source-domain dataset $\mathcal{S} = \{(\boldsymbol{x}_n^s, \boldsymbol{y}_n^s)\}_{n=1}^{N^s}$, the model parameters θ are pre-trained by the combination of supervised cross-entropy loss and semantic proxy contrastive loss:

$$\mathcal{L}_{total} = \mathcal{L}_{ce}(\boldsymbol{x}_n^s, \boldsymbol{y}_n^s) + \lambda \cdot \frac{1}{C}\sum_{c=1}^{C} \mathcal{L}_{spc}(\boldsymbol{x}_n^s, \boldsymbol{W}, c) \tag{9}$$

At test time, for a target-domain sample at time step n, we perform a forward pass to obtain semantic clusters and uncertainty scores and calculate the semantic adaptive learning rate of each category to serve for semantic proxy contrastive loss. For category c, the model parameters are updated to achieve desired adaptation by:

$$\theta_n^c \leftarrow \theta_{n-1}^c - \eta^c(\nabla \mathcal{L}_{spc}(f_{\theta_{n-1}^c}(\boldsymbol{x}_n^t))),\ n \in [1, N^t] \tag{10}$$

The updated model parameters are stored in a memory bank and will be loaded for the next domain adaptation of category c. If category c does not appear in the test sample by pseudo-labeling, we ignore the update of model parameters θ_{n-1}^c. We only update the parameters of the feature extractor and freeze the parameters of the category predictor.

3 Experiments

3.1 Materials

Our SATTA was evaluated on retinal fluid segmentation based on RETOUCH challenge [2], which is a representative benchmark for segmenting all of the three fluid types in SD-OCT images, including intraretinal fluid (IRF), subretinal fluid

Table 1. Quantitative comparison of different TTA methods on RETOUCH challenge using DSC metric. (Note: CD denotes cross domain.)

Methods	Domain-1			Domain-2			Domain-3		
	IRF	SRF	PED	IRF	SRF	PED	IRF	SRF	PED
U-Net [15] w/o CD	0.716	0.794	0.802	0.738	0.904	0.786	0.711	0.664	0.768
U-Net [15] w/ CD	0.637	0.751	0.667	0.587	0.733	0.624	0.536	0.512	0.565
TENT [19]	0.648	0.772	0.733	0.605	0.828	0.703	0.589	0.603	0.637
FTTA [6]	0.663	0.769	0.746	0.632	0.831	0.729	0.611	0.618	0.669
TTAS [1]	0.672	0.774	0.762	0.674	0.865	0.762	0.663	0.641	0.724
CoTTA [20]	0.668	0.776	0.755	0.683	0.852	0.754	0.670	0.634	0.716
SATTA (Ours)	**0.704**	**0.788**	**0.786**	**0.722**	**0.883**	**0.781**	**0.702**	**0.658**	**0.743**

(SRF) and pigment epithelial detachment (PED). SD-OCT images were acquired by three different vendors: Cirrus, Spectralis, and Topcon. The training set consists of 3072 (Cirrus), 1176 (Spectralis), and 3072 (Topcon) SD-OCT images, and the test set consists of 1792 (Cirrus), 686 (Spectralis) and 1792 (Topcon) SD-OCT images. We regard the SD-OCT images from three different vendors as three different domains, namely Domain-1 (Cirrus), Domain-2 (Spectralis), and Domain-3 (Topcon). We employ the dice similarity coefficient (DSC) as the quantitative segmentation metric and a higher DSC indicates a better segmentation performance.

3.2 Comparison with State-of-the-Arts

We compare our SATTA with four state-of-the-art TTA methods, including TENT [19], FTTA [6], TTAS [1] and CoTTA [20]. The four comparative methods have been reviewed in Sect. 1. To verify the TTA performance on cross-domain retinal fluid segmentation based on the RETOUCH challenge with three different domains, we train the segmentation models on two source domains and run TTA methods on the remaining target domain at test time. We carry out three times until all of three domains are tested as unseen target domains. For fair comparisons, all of TTA methods use U-Net [15] as a feature extractor and share the same experimental setting, such as initial learning rate, batch size, etc.

The quantitative comparison results are presented in Table 1. We also include "U-Net w/ CD" as the lower bound and "U-Net w/o CD" as the upper bound, where "CD" denotes cross domain. "U-Net w/o CD" denotes that the segmentation model is trained and tested on the samples from the same domain. "U-Net w/ CD" denotes that the segmentation model is trained and tested on the samples from different domains, without any domain adaptation. We observe that different TTA methods consistently improve the segmentation performance over "U-Net w/ CD". Our SATTA achieves the highest DSC than other methods. The qualitative comparison results are shown in Fig. 2. These visual results confirm

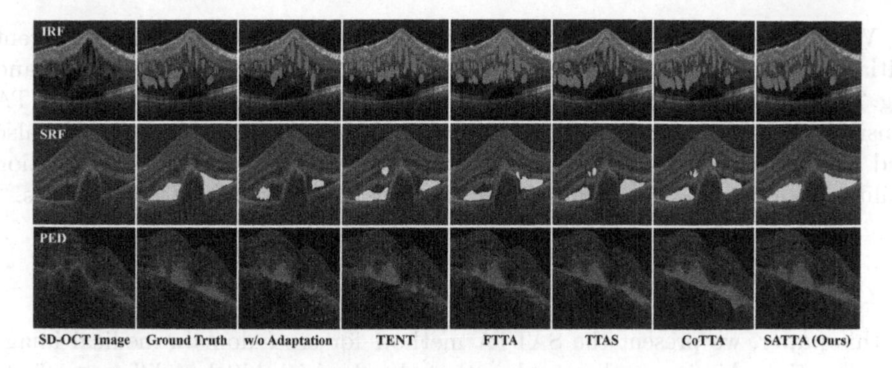

Fig. 2. Qualitative comparison of cross-domain segmentation of different TTA methods on RETOUCH challenge. The first row shows the IRF segmentation on Spectralis SD-OCT images, the second row shows the SRF segmentation on Topcon SD-OCT images, and the third row shows the PED segmentation on Cirrus SD-OCT images.

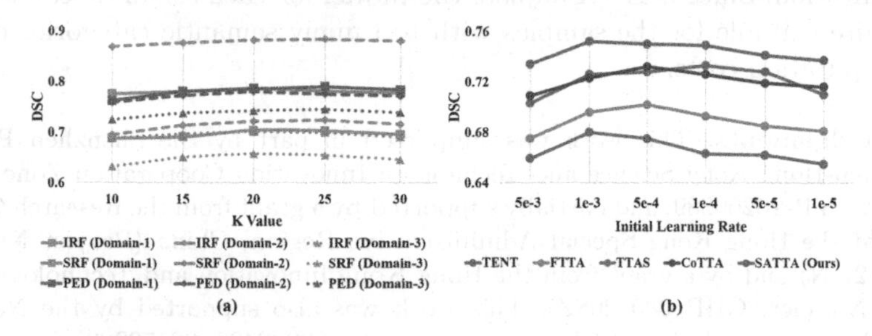

Fig. 3. (a) Ablation study with different K values in SATTA. (b) Ablation study with different initial learning rates in all TTA methods.

that a segmentation model trained only on source-domain distributions performs poorly on target-domain distributions without domain adaptation.

3.3 Ablation Study

We conduct ablation studies to analyze the key factors regarding our SATTA. We first explore the effect of K value in the Top-K selection strategy. The Top-K selection strategy aims to select pixel embeddings with high confidence scores to improve the semantic proxy contrastive learning and reduce the computation cost significantly. Figure 3(a) shows the effect of different K values on three domains for IRF, SRF, and PED. The DSC values consistently increase when rising the K value from 10 to 20, generally, peak when the K value is between 20 and 25, and consistently decrease when further rising K value. This affirms that the pixel embeddings with high confidence scores are conducive to semantic proxy contrastive learning while the pixel embeddings with low confidence scores weaken semantic proxy contrastive learning.

We also investigate the effect of the initial learning rate. We select different initial learning rates from the set {5e-3, 1e-3, 5e-4, 1e-4, 5e-5, 1e-5} for TTA, and Fig. 3(b) shows the total average DSC values of all TTA methods. Our SATTA consistently performs better than other state-of-the-art TTA methods. We also find that different initial learning rates actually affect the domain adaptation ability. Therefore, a proper initial learning rate is essential for TTA methods.

4 Conclusion

In this paper, we present the SATTA method for cross-domain medical image segmentation. Aiming at the problem that the domain shift has different effects on the semantic categories, our SATTA provides a semantic adaptive parameter optimization scheme at test time. Although our SATTA shows superior cross-domain segmentation performance than other state-of-the-art methods, it still has a limitation. Since SATTA adjusts the model for each semantic category, it is not quite suitable for the samples with too many semantic categories due to high computation costs.

Acknowledgements. This work was supported in part by the Shenzhen Portion of Shenzhen-Hong Kong Science and Technology Innovation Cooperation Zone under HZQB-KCZYB-20200089, and partially supported by a grant from the Research Grants Council of the Hong Kong Special Administrative Region, China (Project Number: T45-401/22-N) and by a grant from the Hong Kong Innovation and Technology Fund (Project Number: GHP/080/20SZ). This work was also supported by the National Natural Science Foundation of China under Grants (62202408, 62172223).

References

1. Bateson, M., Lombaert, H., Ben Ayed, I.: Test-time adaptation with shape moments for image segmentation. In: International Conference on Medical Image Computing and Computer-Assisted Intervention. pp. 736–745. Springer (2022). https://doi.org/10.1007/978-3-031-16440-8_70
2. Bogunović, H., et al.: Retouch: the retinal oct fluid detection and segmentation benchmark and challenge. IEEE Trans. Med. Imaging **38**(8), 1858–1874 (2019)
3. Chen, D., Wang, D., Darrell, T., Ebrahimi, S.: Contrastive test-time adaptation. In: Proceedings of the IEEE/CVF Conference on Computer Vision and Pattern Recognition, pp. 295–305 (2022)
4. Farshad, A., Yeganeh, Y., Gehlbach, P., Navab, N.: Y-net: a spatiospectral dual-encoder network for medical image segmentation. In: International Conference on Medical Image Computing and Computer-Assisted Intervention, pp. 582–592. Springer (2022). https://doi.org/10.1007/978-3-031-16434-7_56
5. Gal, Y., Ghahramani, Z.: Dropout as a bayesian approximation: representing model uncertainty in deep learning. In: International Conference on Machine Learning, pp. 1050–1059. PMLR (2016)
6. Hu, M., et al.: Fully Test-Time Adaptation for Image Segmentation. In: de Bruijne, M., et al. (eds.) MICCAI 2021. LNCS, vol. 12903, pp. 251–260. Springer, Cham (2021). https://doi.org/10.1007/978-3-030-87199-4_24

7. Hu, S., Liao, Z., Xia, Y.: Domain specific convolution and high frequency reconstruction based unsupervised domain adaptation for medical image segmentation. In: International Conference on Medical Image Computing and Computer-Assisted Intervention. pp. 650–659. Springer (2022). https://doi.org/10.1007/978-3-031-16449-1_62
8. Lee, S., Seong, H., Lee, S., Kim, E.: Wildnet: learning domain generalized semantic segmentation from the wild. In: Proceedings of the IEEE/CVF Conference on Computer Vision and Pattern Recognition, pp. 9936–9946 (2022)
9. Li, J., Li, X., He, D., Qu, Y.: A domain adaptation model for early gear pitting fault diagnosis based on deep transfer learning network. Proc. Instit. Mech. Eng. Part O: J. Risk Reliabi. **234**(1), 168–182 (2020)
10. Liu, Y., Kothari, P., van Delft, B., Bellot-Gurlet, B., Mordan, T., Alahi, A.: Ttt++: when does self-supervised test-time training fail or thrive? Adv. Neural. Inf. Process. Syst. **34**, 21808–21820 (2021)
11. Ma, W., Chen, C., Zheng, S., Qin, J., Zhang, H., Dou, Q.: Test-time adaptation with calibration of medical image classification nets for label distribution shift. In: International Conference on Medical Image Computing and Computer-Assisted Intervention, pp. 313–323. Springer (2022). https://doi.org/10.1007/978-3-031-16437-8_30
12. Nado, Z., Padhy, S., Sculley, D., D'Amour, A., Lakshminarayanan, B., Snoek, J.: Evaluating prediction-time batch normalization for robustness under covariate shift. arXiv preprint arXiv:2006.10963 (2020)
13. Peiris, H., Hayat, M., Chen, Z., Egan, G., Harandi, M.: A robust volumetric transformer for accurate 3d tumor segmentation. In: International Conference on Medical Image Computing and Computer-Assisted Intervention, pp. 162–172. Springer (2022). https://doi.org/10.1007/978-3-031-16443-9_16
14. Peng, D., Lei, Y., Hayat, M., Guo, Y., Li, W.: Semantic-aware domain generalized segmentation. In: Proceedings of the IEEE/CVF Conference on Computer Vision and Pattern Recognition, pp. 2594–2605 (2022)
15. Ronneberger, O., Fischer, P., Brox, T.: U-Net: convolutional networks for biomedical image segmentation. In: Navab, N., Hornegger, J., Wells, W.M., Frangi, A.F. (eds.) MICCAI 2015. LNCS, vol. 9351, pp. 234–241. Springer, Cham (2015). https://doi.org/10.1007/978-3-319-24574-4_28
16. Sun, X., Liu, Z., Zheng, S., Lin, C., Zhu, Z., Zhao, Y.: Attention-enhanced disentangled representation learning for unsupervised domain adaptation in cardiac segmentation. In: International Conference on Medical Image Computing and Computer-Assisted Intervention. pp. 745–754. Springer (2022). https://doi.org/10.1007/978-3-031-16449-1_71
17. Sun, Y., Wang, X., Liu, Z., Miller, J., Efros, A., Hardt, M.: Test-time training with self-supervision for generalization under distribution shifts. In: International Conference on Machine Learning. pp. 9229–9248. PMLR (2020)
18. Varsavsky, T., Orbes-Arteaga, M., Sudre, C.H., Graham, M.S., Nachev, P., Cardoso, M.J.: Test-time unsupervised domain adaptation. In: Martel, A.L., et al. (eds.) MICCAI 2020. LNCS, vol. 12261, pp. 428–436. Springer, Cham (2020). https://doi.org/10.1007/978-3-030-59710-8_42
19. Wang, D., Shelhamer, E., Liu, S., Olshausen, B., Darrell, T.: Tent: fully test-time adaptation by entropy minimization. arXiv preprint arXiv:2006.10726 (2020)
20. Wang, Q., Fink, O., Van Gool, L., Dai, D.: Continual test-time domain adaptation. In: Proceedings of the IEEE/CVF Conference on Computer Vision and Pattern Recognition, pp. 7201–7211 (2022)

21. Xing, Z., Yu, L., Wan, L., Han, T., Zhu, L.: Nestedformer: nested modality-aware transformer for brain tumor segmentation. In: International Conference on Medical Image Computing and Computer-Assisted Intervention. pp. 140–150. Springer (2022). https://doi.org/10.1007/978-3-031-16443-9_14
22. Xu, Z., et al.: Denoising for relaxing: unsupervised domain adaptive fundus image segmentation without source data. In: International Conference on Medical Image Computing and Computer-Assisted Intervention, pp. 214–224. Springer (2022). https://doi.org/10.1007/978-3-031-16443-9_21
23. Yang, H., et al.: Dltta: Dynamic learning rate for test-time adaptation on cross-domain medical images. arXiv preprint arXiv:2205.13723 (2022)
24. Yao, X., et al.: Pcl: proxy-based contrastive learning for domain generalization. In: Proceedings of the IEEE/CVF Conference on Computer Vision and Pattern Recognition, pp. 7097–7107 (2022)

SFusion: Self-attention Based N-to-One Multimodal Fusion Block

Zecheng Liu[1], Jia Wei[1(✉)], Rui Li[2], and Jianlong Zhou[3]

[1] School of Computer Science and Engineering, South China University of Technology, Guangzhou, China
msaiyan@mail.scut.edu.cn, csjwei@scut.edu.cn

[2] Golisano College of Computing and Information Sciences, Rochester Institute of Technology, Rochester, NY, USA
rxlics@rit.edu

[3] Data Science Institute, University of Technology Sydney, Ultimo, NSW 2007, Australia
jianlong.zhou@uts.edu.au

Abstract. People perceive the world with different senses, such as sight, hearing, smell, and touch. Processing and fusing information from multiple modalities enables Artificial Intelligence to understand the world around us more easily. However, when there are missing modalities, the number of available modalities is different in diverse situations, which leads to an N-to-One fusion problem. To solve this problem, we propose a self-attention based fusion block called SFusion. Different from preset formulations or convolution based methods, the proposed block automatically learns to fuse available modalities without synthesizing or zero-padding missing ones. Specifically, the feature representations extracted from upstream processing model are projected as tokens and fed into self-attention module to generate latent multimodal correlations. Then, a modal attention mechanism is introduced to build a shared representation, which can be applied by the downstream decision model. The proposed SFusion can be easily integrated into existing multimodal analysis networks. In this work, we apply SFusion to different backbone networks for human activity recognition and brain tumor segmentation tasks. Extensive experimental results show that the SFusion block achieves better performance than the competing fusion strategies. Our code is available at https://github.com/scut-cszcl/SFusion.

Keywords: Multimodal fusion · Missing modalities · Brain tumor segmentation · Human activity recognition

1 Introduction

People perceive the world with signals from different modalities, which often carry complementary information about varying aspects of an object or event of interest. Therefore, collecting and utilizing multimodal information is crucial for Artificial Intelligence to understand the world around us. Data collected

H. Greenspan et al. (Eds.): MICCAI 2023, LNCS 14221, pp. 159–169, 2023.
https://doi.org/10.1007/978-3-031-43895-0_15

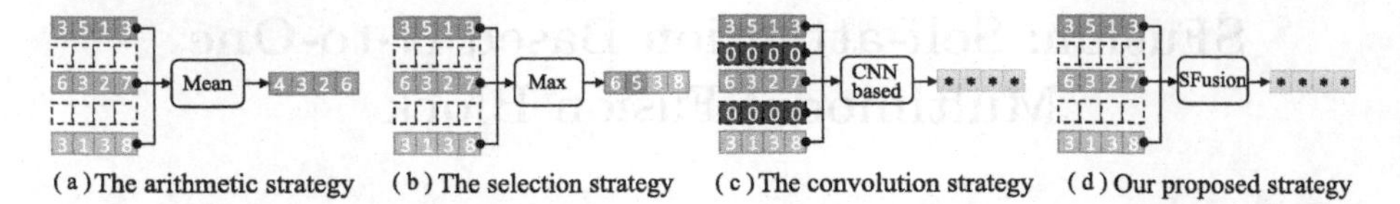

Fig. 1. Fusion strategies. $*$ denotes the value is automatically learned.

from various sensors (e.g., microphones, cameras, motion controllers) are used to identify human activity [4]. Moreover, multimodal medical images obtained from different scanning protocols (e.g., Computed Tomography, Magnetic Resonance Imaging) are employed for disease diagnosis [12]. Satisfactory performances have been achieved with these multimodal data.

In practical application, however, modality missing is a common scenario. Wirelessly connected sensors may occasionally disconnect and temporarily be unable to send any data [3]. Medical images may be missing due to artifacts and diverse patient conditions [11]. In these unexpected situations, any combinatorial subset of available modalities can be given as input. To handle this, one intuitive solution is to train a dedicated model on all possible subsets of available modalities [6,14,23]. However, these methods are ineffective and time-consuming. Another way is to predict missing modalities and perform with the completed modalities [20]. But, these approaches also require additional prediction networks for each missing situation, and the quality of the recovered data directly affects the performance, especially when there are only a few available modalities. Recently, fusing the available modalities into a shared representation received wide attention. However, it is particularly challenging due to the varying number of input modalities, which results in the N-to-One fusion problem.

Currently, existing fusion strategies to tackle this challenge can be broadly grouped into three categories: the arithmetic strategy, the selection strategy and the convolution strategy. As shown in Fig. 1(a), in the arithmetic strategy, feature representations of available modalities are merged by an arithmetic function, such as averaging, computing the first and second moments or other designed formulas [10,13,17]. For the selection strategy, as shown in Fig. 1(b), each value of fused representation is selected from the values at the corresponding position of the inputs. The selection rule can be defined as max, min or probability-based [2,8,19]. Although the above two fusion strategies are easily scalable to various data missing situations, their fusion operation is hard-coded. All available modalities contribute equally and their latent correlations are neglected. Unlike hard-coding the fusion operation, in the convolution strategy, the convolutional fusion network automatically learns how to fuse these feature representations, which is beneficial to exploiting the correlation between multiple modalities. However, as shown in Fig. 1(c), this fusion strategy needs a constant number of data to meet the requirements of the input channels in the convolutional network. Therefore, it has to simulate missing data by crudely zero-padding or replacing it with similar modalities, which inevitably introduces a bias in computation and causes performance degradation [5,18,25].

Transformer has achieved success in the field of computer vision, demonstrating that self-attention mechanism has the ability to capture the latent correlation of image tokens. However, no work has explored the effectiveness of self-attention mechanism on the N-to-One fusion, where N is variable during training, rather than fixed. Furthermore, the calculation of self-attention does not require a fixed number of tokens as input, which represents a potential for handling missing data. Therefore, we propose a self-attention based fusion block (SFusion) to tackle the problems of the above fusion strategies. As shown in Fig. 1(d), SFusion can handle any number of input data instead of fixing its number. In addition, SFusion is a learning-based fusion strategy that consists of two components: the correlation extraction (CE) module and the modal attention (MA) module. In the CE module, feature representations extracted from available modalities are projected as tokens and fed into the self-attention layers to learn multimodal correlations. Based on these correlations, a modal softmax function is proposed to generate weight maps in the MA module. Finally, it builds a shared feature representation by fusing the varying inputs with the weight maps.

The contributions of this work are:

- We propose SFusion, which is a data-dependent fusion strategy without impersonating missing modalities. It can learn the latent correlations between different modalities and builds a shared representation adaptively.
- The SFusion is not limited to specific deep learning architectures. It takes inputs from any kind of upstream processing model and serves as the input of the downstream decision model, which enables applying the SFusion to various backbone networks for different tasks.
- We provide qualitative and quantitative performance evaluations on activity recognition with the SHL [22] dataset and brain tumor segmentation with the BraTS2020 [1] dataset. The results show the superiority of SFusion over competing fusion strategies.

2 Methodology

2.1 Method Overview

For multiple modalities, let $k \in K \subseteq \{1, 2, \ldots, S\}$ index a specific modality, within the available modality set of K, where S is the number of all possible modalities. Given an input $f_k \in \mathbb{R}^{B \times C \times R_f}$, B and C denote the batch size and the number of channels, respectively. R_f represents the shape of feature representation extracted from the k-th modality of a sample data, which can be 1D (L), 2D (H×W), 3D (D×H×W) or higher-dimensional. In addition, $I = \{f_k | k \in K\}$ denotes the input set of feature representations from all the available modalities. Our goal is to learn a fusion function F that can project I into a shared feature representation f_s, denoted as $F(I) \rightarrow f_s$. To achieve the goal, we design an N-to-One fusion block, SFusion. The architecture is shown in Fig. 2, which consists of two modules: correlation extraction (CE) module and modal attention (MA) module.

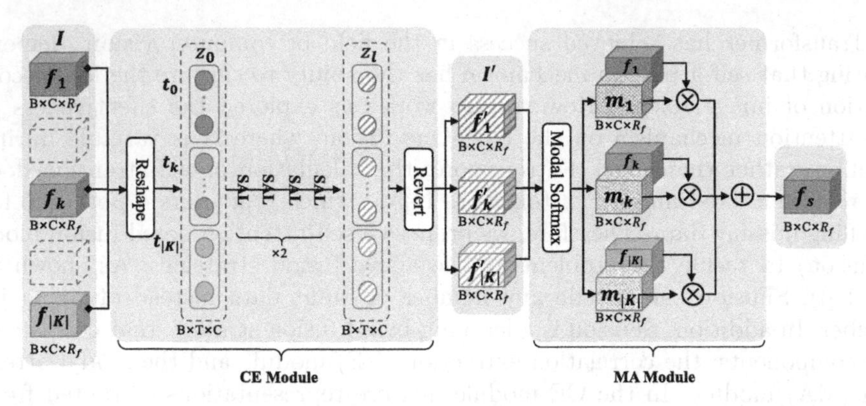

Fig. 2. The illustration of SFusion. R_f: L or H×W or D×H×W (shape of feature representation); T = L·$|K|$ or H·W·$|K|$ or D·H·W·$|K|$ (number of tokens).

2.2 Correlation Extraction

Given the feature representation $f_k \in \mathbb{R}^{B \times C \times R_f}$, we first flatten the R_f dimensions of f_k into one dimension and get a $B \times C \times R$ feature representation, where $R = L$ (1D), $R = H \times W$ (2D), $R = D \times H \times W$ (3D), etc. It can be viewed as $B \times R$ C-dimensional tokens t_k. Then, we obtain the concatenation of all the tokens $z_0 \in \mathbb{R}^{B \times T \times C}$, where $T = R \times |K|$, and $|K|$ denotes the number of available modalities.

Given z_0, the stack of eight self-attention layers (SAL) are introduced to learn the latent multimodal correlations. Each layer includes a multi-head attention (MHA) block and a fully connected feed-forward network (FFN) [21]. Layer normalization (LN) is applied before every block. The outputs of the x-th ($x \in [1, 2, \ldots, 8]$) layer can be describe as:

$$z'_x = MHA(LN(z_{x-1})) + z_{x-1} \tag{1}$$

$$z_x = FFN(LN(z'_x)) + z'_x \tag{2}$$

Therefore, we get $z_l \in \mathbb{R}^{B \times T \times C}$, which is the last SAL output. By reverting z_l to the size of $|K| \times B \times C \times R_f$, we obtain the output $I' = \{f'_k | k \in K\}$ of CE as:

$$I' = split(r(z_l)) \tag{3}$$

where $r(\cdot)$ and $split(\cdot)$ are the reshape and split operations, and I' is the set of calculated feature representations $f'_k \in \mathbb{R}^{B \times C \times R_f}$ which contains multimodal correlations and has the same size as the original input f_k.

2.3 Modal Attention

Given the calculated feature representations set I', the weight map m_k is generated with the modal attention mechanism. Feature representations extracted from different modalities are expected to have different weights for fusion at the voxel level. Therefore, we introduce a modal-wise and voxel-level softmax function to generate the weight maps from I', as shown in Fig. 3.

We denote the i-th voxel of f'_k and m_k as v_k^i and m_k^i, respectively. e is the natural logarithm. The value of weight map m_k can be defined as:

$$m_k^i = \frac{e^{v_k^i}}{\sum_{j \in K} e^{v_j^i}} \tag{4}$$

By element-wise multiplying input feature map f_k with the corresponding weight map m_k and summing all the modalities, we can obtain a fused feature map f_s as:

$$f_s = \sum_{k \in K} f_k \cdot m_k \tag{5}$$

Since the sum of $m_1^i, \dots m_{|K|}^i$ is 1, the value range of fused feature representation f_s remains stable to improve the robustness for variable input modalities. Moreover, the relative sizes of $v_1^i, \dots v_{|K|}^i$ (contain the latent multi-modal correlations learned from the CE module) are retained in the corresponding weights. In particular, when only one modality is available, all the values of the weight map are 1, which means $f_s = f_k$ ($k \in K$, $|K| = 1$). In this case, the input feature representation remains unchanged. It enables the backbone network (the upstream processing model and the downstream decision model) to enhance its capability to encode and decode information from different modalities rather than relying on a particular one. It is crucial for variable multimodal data analysis.

3 Experiments and Results

3.1 Datasets

SHL2019. The SHL (Sussex-Huawei Locomotion) Challenge 2019 [22] dataset provides data from seven sensors of a smartphone to recognize eight modes of locomotion and transportation (activities), including *still*, *walking*, *run*, *bike*, *car*, *bus*, *train*, and *subway*. The sensor data are collected from smartphones of a

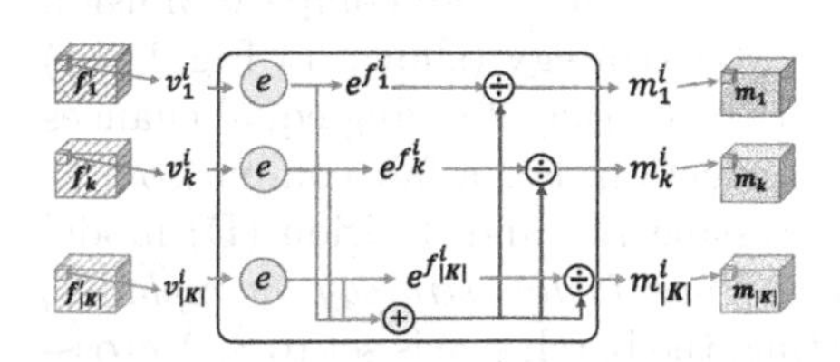

Fig. 3. The illustration of modal attention mechanism.

Table 1. Evaluation on SHL2019. *w/o* means without. † denotes results from [9].

Accuracy(%)	Bag	Hips	Torso	Hand	All
Early†	–	–	–	–	46.73
Intermediate†	–	–	–	–	63.87
Late†	–	–	–	–	63.85
Confidence†[7]	–	–	–	–	63.60
EmbraceNet† [9]	63.68	67.98	81.58	47.63	65.22
SFusion	**67.41**	**68.91**	**85.22**	**48.35**	**67.47**
SFusion *w/o* CE	56.82	63.14	74.69	46.70	60.33
SFusion *w/o* MA	65.01	67.95	83.49	47.52	65.99

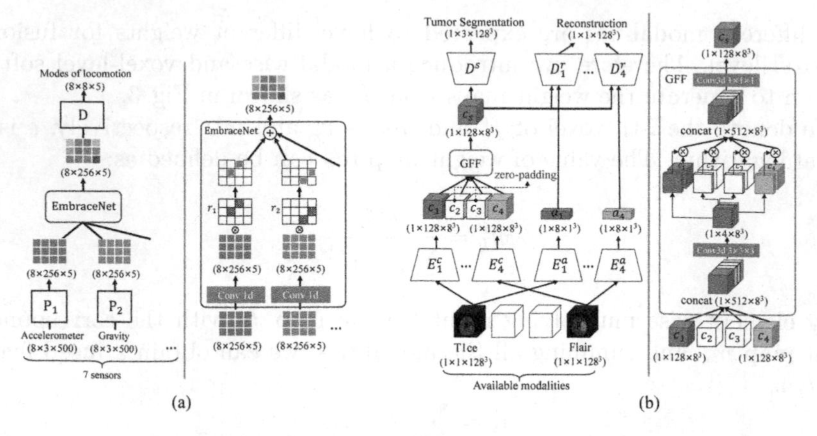

Fig. 4. (a) Activity recognition with EmbraceNet; (b) Bran tumor segmentation with GFF. (B × C × R_f) is given, where B, C and R_f denotes the batch size, channels and data shape, respectively.

person with four locations, including the bag, trousers front pocket, breast pocket and hand. Each location is called "Bag", "Hips", "Torso", and "Hand", respectively. Data acquired from the locations except the "Hand" are given in the train subset, while the validation subset provides the data of all four locations. In the test subset, only unlabeled "Hand" location data are available.

BraTS2020. The BraTS2020 [1] dataset provide four modality scans: T1ce, T1, T2, FLAIR for brain tumor segmentation. It contains 369 subjects. To better represent the clinical application tasks, there are three mutually inclusive tumor regions: the enhancing tumor (ET), the tumor core (TC), and the whole tumor (WT) [1]. We select 70% data as training data, while 10% and 20% as validation and test data respectively. To prevent overfitting, two data augmentation techniques (randomly flip the axes and rotate with a random angle in $[-10°, 10°]$) are applied during training. We apply z-score normalization [15] to the volumes individually and randomly crop $128 \times 128 \times 128$ patches as inputs to the networks.

3.2 Baseline Methods

EmbraceNet. In the experiments on activity recognition, we compare SFusion with EmbraceNet [9], which employs a selection strategy (shown in Fig. 1 (b)) by generating feature masks $(r_1, r_2, \ldots, r_7)$ with the rule of giving equal chances to all available modalities during each value selection. For a fair comparison, as shown in Fig. 4 (a), we adopt the same processing (P) and decision (D) model as used in [9]. *We obtain the performance of our fusion strategy by replacing EmbraceNet with SFusion.* Following [9] setting, the batch size is set to 8. A cross-entropy loss and the Adam optimization method [16] with $\beta_1 = 0.9$, $\beta_2 = 0.999$ are employed. The learning rate is initially set to 1×10^{-4} and reduced by a factor of 2 at every 1×10^5 steps. A total of 5×10^5 training steps are executed.

GFF. In the experiments on brain tumor segmentation, we compare SFusion with a gated feature fusion block (GFF) [5], which belongs to the convolution strategy (shown in Fig. 1(c)). As shown in Fig. 4 (b), a feature disentanglement architecture is employed. Multimodal medical images are decomposed into the modality-invariant content and the modality-specific appearance code by encoders E^c and E^a, respectively. The content codes (e.g., c_2 and c_3, shown in Fig. 4 (b)) of missing modalities are simulated with zero values. Then, all content codes are fused into a shared representation c_s by GFF. Given c_s , the tumor segmentation results are generated by the decoder D^s. For a fair comparison, we adopt the same encoders (E_i^c and E_i^a) and decoders (D^s and D_i^r) as used in [5]. *We obtain the performance of our fusion strategy by replacing GFF with SFusion and removing the zero-padding operation.* The training max_epoch is set to 200. Following [5] setting, the batch size is set to 1. Adam [16] is utilized with a learning rate of 1×10^{-4} and progressively multiplies it by (1 - epoch / max_epoch)$^{0.9}$. Losses of $\mathcal{L}_{KL}$, $\mathcal{L}_{rec}$ and $\mathcal{L}_{seg}$ are employed as [5]. During training, to simulate real missing modalities scenarios, each training patient's data is fixed to one of 15 possible missing cases. For a comprehensive evaluation, we test the performance of all 15 cases for each test patient.

Our implementations are on an NVIDIA RTX 3090(24G) with PyTorch 1.8.1.

3.3 Results

Activity recognition. We compare SFusion with the EmbraceNet [9] on SHL2019. As shown in Table 1, we also compare the results of other fusion methods, which use the same processing (P) model and decision (D) model as [9]. (1) In the early fusion method, the data of seven sensors are concatenated along their

Table 2. Dice(%) performance for MRI modalities being either absent (∘) or present (•). * denotes significant improvement provided by a Wilcoxon test (p-values < 0.05).

Modalities				WT		TC		ET	
T1ce	T1	T2	Flair	GFF	SFusion	GFF	SFusion	GFF	SFusion
•	∘	∘	∘	68.24	**69.75***	73.27	**75.63***	69.30	**71.94***
∘	•	∘	∘	64.45	**69.11***	46.93	**53.86***	23.74	**29.71***
∘	∘	•	∘	**79.78**	79.61	58.27	**61.99***	**36.13**	35.87
∘	∘	∘	•	81.82	**83.97**	50.53	**52.84**	29.50	**34.40***
•	•	∘	∘	74.99	**75.30**	75.89	**80.35***	72.09	**74.90***
•	∘	•	∘	83.93	**84.27***	79.55	**81.48***	72.87	**74.74***
•	∘	∘	•	**87.34**	87.32	79.01	**79.06**	74.89	**75.82**
∘	•	•	∘	81.76	**81.78**	59.75	**66.67***	36.50	**40.38***
∘	•	∘	•	85.86	**86.39**	61.92	**62.31**	37.52	**38.22**
∘	∘	•	•	86.99	**87.50***	61.92	**66.38***	38.94	**41.46***
•	•	•	∘	84.48	**84.59**	79.83	**82.32***	73.74	**74.78**
•	•	∘	•	88.03	**88.04**	80.50	**82.04***	74.53	**75.44**
•	∘	•	•	88.75	**89.11***	81.60	**82.06**	74.43	**74.91**
∘	•	•	•	86.84	**87.63***	65.38	**68.76***	40.90	**43.53***
•	•	•	•	88.65	**88.93**	81.29	**82.18**	**74.55**	73.76
Average				82.13	**82.89***	69.04	**71.86***	55.31	**57.32***

Table 3. Ablation experiments.

Dice(%)	*w/o* CE	*w/o* MA	SFusion
WT	82.42	82.76	**82.89**
TC	70.39	70.93	**71.86**
ET	55.65	55.56	**57.32**

Table 4. † denotes results from [24].

Dice(%)	WT	TC	ET	Overall
U-HVED† [10]	75.8	63.2	40.7	59.9
ACNet† [23]	52.5	46.9	41.8	47.1
D^2-Net† [24]	76.2	66.5	42.3	61.7
SF_FDGF	**82.1**	**69.2**	**54.9**	**68.7**

C dimension. The prediction results are obtained by inputting the concatenation into a network of P and D in series. (2) For the intermediate fusion approach, the EmbraceNet is replaced with the concatenation of feature representations along their R_f dimension. (3) In the late fusion method, an independent network of P and D in series is trained for each sensor, and then the decision is made from the averaged softmax outputs. (4) In the confidence fusion model, the EmbraceNet is replaced with the confidence calculation and fusion layers in [7]. The results of different fusion methods on the validation data are presented in Table 1. Our proposed SFusion outperforms the EmbraceNet in all four smartphone locations and improves the overall accuracy from 65.22% to 67.47%.

Brain Tumor Segmentation. The quantitative segmentation results are shown in Table 2. Compared with GFF, the network integrated with SFusion achieves better average performance over the 15 possible combinations in all three tasks. In particular, SFusion outperforms GFF for all the possible combinations in TC segmentation. Overall, SFusion achieves better Dice scores in most situations (13,15,13 situations for WT, TC and ET segmentation, respectively). In addition, we conduct the statistical significance analysis. The number of situations with significant improvement are 6, 10 and 8 for WT, TC and ET, respectively. It is provided by a Wilcoxon test (p-values < 0.05). Besides, we find no significant drop in performance caused by SFusion. In addition, we compare the SF_FDGF (where GFF is replaced by SFusion) with current state-of-the-art methods. Table 4 presents the average dice of 15 situations. For a fair comparison, we conduct experiments on BraTS2018, adopt the same data partition as [24], and cite the results in [24]. SF_FDGF achieves the best performance and verifies the effectiveness of the SFusion.

Ablation Experiments. The correlation extraction (CE) module and the modal attention (MA) module are two key components in SFusion. We evaluate the SFusion without CE and MA, respectively. SFusion without CE denotes that feature representations are directly fed into the MA module (Fig. 2). SFusion without MA means that we directly add the calculated feature representations (I') up to get the fusion result. As shown in Table 1, we can find that SFusion without CE performs worse than other methods. Compared with EmbraceNet, the improvement of SFusion without MA is inconspicuous. As shown in Table. 3, we present the averaged performance over the 15 possible combinations on BraTS2020. It shows that both the CE and MA module lead to performance improvement across all the tumor regions. Therefore, ablation experiments on two different tasks show that both CE and MA play an important role in SFusion.

4 Conclusion

In this paper, we propose a self-attention based N-to-One fusion block SFusion to tackle the problem of multimodal missing modalities fusion. As a data-dependent fusion strategy, SFusion can automatically learn the latent correlations between different modalities and builds a shared feature representation. The entire fusion process is based on available data without simulating missing modalities. In addition, SFusion has compatibility with any kind of upstream processing model and downstream decision model, making it universally applicable to different tasks. We show that it can be integrated into existing backbone networks by replacing their fusion operation or block to improve activity recognition and achieve brain tumor segmentation performance. In particular, by integrating with SFusion, SF_FDGF achieves the state-of-the-art performance. In the future, we will explore other tasks related to variable multimodal fusion with SFusion.

Acknowledgements. This work is supported in part by the Guangdong Provincial Natural Science Foundation (2023A1515011431), the Guangzhou Science and Technology Planning Project (202201010092), the National Natural Science Foundation of China (72074105), NSF-1850492 and NSF-2045804.

References

1. Bakas, S., Menze, B., Davatzikos, C., Kalpathy-Cramer, J., Farahani, K., et al.: MICCAI Brain Tumor Segmentation (BraTS) 2020 Benchmark: Prediction of Survival and Pseudoprogression (Mar 2020). https://doi.org/10.5281/zenodo.3718904
2. Chartsias, A., Joyce, T., Giuffrida, M.V., Tsaftaris, S.A.: Multimodal mr synthesis via modality-invariant latent representation. IEEE Trans. Med. Imaging **37**(3), 803–814 (2018). https://doi.org/10.1109/TMI.2017.2764326
3. Chavarriaga, R., et al.: The opportunity challenge: a benchmark database for on-body sensor-based activity recognition. Pattern Recogn. Lett. **34**(15), 2033–2042 (2013)
4. Chen, C., Jafari, R., Kehtarnavaz, N.: Utd-mhad: a multimodal dataset for human action recognition utilizing a depth camera and a wearable inertial sensor. In: 2015 IEEE International conference on image processing (ICIP), pp. 168–172. IEEE (2015)
5. Chen, C., Dou, Q., Jin, Y., Chen, H., Qin, J., Heng, P.-A.: Robust multimodal brain tumor segmentation via feature disentanglement and gated fusion. In: Shen, D., et al. (eds.) MICCAI 2019. LNCS, vol. 11766, pp. 447–456. Springer, Cham (2019). https://doi.org/10.1007/978-3-030-32248-9_50
6. Chen, C., Dou, Q., Jin, Y., Liu, Q., Heng, P.A.: Learning with privileged multimodal knowledge for unimodal segmentation. IEEE Trans. Medical Imaging (2021). https://doi.org/10.1109/TMI.2021.3119385

7. Choi, J.H., Lee, J.S.: Confidence-based deep multimodal fusion for activity recognition. In: Proceedings of the 2018 ACM International Joint Conference and 2018 International Symposium on Pervasive and Ubiquitous Computing and Wearable Computers, pp. 1548–1556 (2018)
8. Choi, J.H., Lee, J.S.: Embracenet: a robust deep learning architecture for multimodal classification. Information Fusion **51**, 259–270 (2019)
9. Choi, J.H., Lee, J.S.: Embracenet for activity: a deep multimodal fusion architecture for activity recognition. In: Adjunct Proceedings of the 2019 ACM International Joint Conference on Pervasive and Ubiquitous Computing and Proceedings of the 2019 ACM International Symposium on Wearable Computers, pp. 693–698 (2019)
10. Dorent, R., Joutard, S., Modat, M., Ourselin, S., Vercauteren, T.: Hetero-modal variational encoder-decoder for joint modality completion and segmentation. In: Shen, D., et al. (eds.) MICCAI 2019. LNCS, vol. 11765, pp. 74–82. Springer, Cham (2019). https://doi.org/10.1007/978-3-030-32245-8_9
11. Graves, M.J., Mitchell, D.G.: Body mri artifacts in clinical practice: a physicist's and radiologist's perspective. J. Magn. Reson. Imaging **38**(2), 269–287 (2013)
12. Guo, Z., Li, X., Huang, H., Guo, N., Li, Q.: Deep learning-based image segmentation on multimodal medical imaging. IEEE Trans. Radiation Plasma Med. Sci. **3**(2), 162–169 (2019)
13. Havaei, M., Guizard, N., Chapados, N., Bengio, Y.: HeMIS: hetero-modal image segmentation. In: Ourselin, S., Joskowicz, L., Sabuncu, M.R., Unal, G., Wells, W. (eds.) MICCAI 2016. LNCS, vol. 9901, pp. 469–477. Springer, Cham (2016). https://doi.org/10.1007/978-3-319-46723-8_54
14. Hu, M., et al.: Knowledge distillation from multi-modal to mono-modal segmentation networks. In: Martel, A.L., et al. (eds.) MICCAI 2020. LNCS, vol. 12261, pp. 772–781. Springer, Cham (2020). https://doi.org/10.1007/978-3-030-59710-8_75
15. Isensee, F., et al.: nnu-net: self-adapting framework for u-net-based medical image segmentation. arXiv preprint arXiv:1809.10486 (2018)
16. Kingma, D.P., Ba, J.: Adam: a method for stochastic optimization. arXiv preprint arXiv:1412.6980 (2014)
17. Lau, K., Adler, J., Sjölund, J.: A unified representation network for segmentation with missing modalities. arXiv preprint arXiv:1908.06683 (2019)
18. Ngiam, J., Khosla, A., Kim, M., Nam, J., Lee, H., Ng, A.Y.: Multimodal deep learning. In: ICML (2011)
19. Ouyang, J., Adeli, E., Pohl, K.M., Zhao, Q., Zaharchuk, G.: Representation disentanglement for multi-modal brain MRI analysis. In: Feragen, A., Sommer, S., Schnabel, J., Nielsen, M. (eds.) IPMI 2021. LNCS, vol. 12729, pp. 321–333. Springer, Cham (2021). https://doi.org/10.1007/978-3-030-78191-0_25
20. Shen, L., et al.: Multi-domain image completion for random missing input data. IEEE Trans. Med. Imaging **40**(4), 1113–1122 (2021). https://doi.org/10.1109/TMI.2020.3046444
21. Vaswani, A., et al.: Attention is all you need. In: Guyon, I., Luxburg, U.V., Bengio, S., Wallach, H., Fergus, R., Vishwanathan, S., Garnett, R. (eds.) Advances in Neural Information Processing Systems, vol. 30. Curran Associates, Inc. (2017)
22. Wang, L., Gjoreski, H., Ciliberto, M., Mekki, S., Valentin, S., Roggen, D.: Enabling reproducible research in sensor-based transportation mode recognition with the sussex-huawei dataset. IEEE Access **7**, 10870–10891 (2019)

23. Wang, Y., et al.: ACN: adversarial co-training network for brain tumor segmentation with missing modalities. In: de Bruijne, M., et al. (eds.) MICCAI 2021. LNCS, vol. 12907, pp. 410–420. Springer, Cham (2021). https://doi.org/10.1007/978-3-030-87234-2_39
24. Yang, Q., Guo, X., Chen, Z., Woo, P.Y., Yuan, Y.: D2-net: dual disentanglement network for brain tumor segmentation with missing modalities. IEEE Trans. Med. Imaging (2022)
25. Zhou, T., Canu, S., Vera, P., Ruan, S.: Latent correlation representation learning for brain tumor segmentation with missing mri modalities. IEEE Trans. Image Process. **30**, 4263–4274 (2021)

FedGrav: An Adaptive Federated Aggregation Algorithm for Multi-institutional Medical Image Segmentation

Zhifang Deng[1], Dandan Li[1], Shi Tan[2], Ying Fu[2], Xueguang Yuan[3], Xiaohong Huang[1(✉)], Yong Zhang[4], and Guangwei Zhou[5]

[1] School of Computer Science (National Pilot Software Engineering School), Beijing University of Posts and Telecommunications, Beijing, China
huangxh@bupt.edu.cn
[2] Department of Ultrasound, Peking University Third Hospital, Beijing, China
[3] School of Electronic Engineering, Beijing University of Posts and Telecommunications, Beijing, China
[4] Zhongguancun Laboratory, Beijing, China
[5] HTA Co., Ltd., Beijing, China

Abstract. With the increasingly strengthened data privacy acts and the difficult data centralization, Federated Learning (FL) has become an effective solution to collaboratively train the model while preserving each client's privacy. FedAvg is a standard aggregation algorithm that makes the proportion of the dataset size of each client an aggregation weight. However, it can't deal with non-independent and identically distributed (non-IID) data well because of its fixed aggregation weights and the neglect of data distribution. The paper presents a new aggregation strategy called FedGrav, which is designed to handle non-IID datasets and is inspired by the law of universal gravitation in physics. FedGrav can dynamically adjust the aggregation weights based on the training condition of local models throughout the entire training process, making it an effective solution for non-IID data. The model affinity is creatively proposed by considering both the differences of sample size on the client and the discrepancies among local models. It considers the client sample size as the mass of the local model and defines the model graph distance based on neural network topology. By calculating the affinity among local models, FedGrav can explore internal correlations of them and improve the aggregation weights. The proposed FedGrav has been applied to the CIFAR-10 and the MICCAI Federated Tumor Segmentation (FeTS) Challenge 2021 datasets, and the validation results show that our method outperforms the previous state-of-the-art by 1.54 mean DSC and 2.89 mean HD95. The source code will be available on Github.

Keywords: Federated Learning · Brain Tumor Segmentation · FedGrav · Model Affinity · Graph Distance

H. Greenspan et al. (Eds.): MICCAI 2023, LNCS 14221, pp. 170–180, 2023.
https://doi.org/10.1007/978-3-031-43895-0_16

1 Introduction

The demand for precise medical data analysis has led to the widespread use of deep learning methods in the medical field. However, accompanied by the promulgation of data acts and the strengthening of data privacy, it has become increasingly challenging to train models in large-scale centralized medical datasets. As one of the solutions, federated learning provides a new way out of the dilemma and attracts significant attention from researchers.

Federated learning (FL) [1,2] is a distributed machine learning paradigm in which all clients train a global model collaboratively while preserving their data locally. As a crucial core of them, the aggregation algorithm plays an important role in releasing data potential and improving global model performance. FedAvg [1], as pioneering work, was a simple and effective aggregation algorithm, which makes the proportions of local datasets size as the aggregation weights of local models. But in the real world, not only the numbers of datasets held by clients is different, but also their data distribution may be diverse, which leads to the fact that the data in the federated learning is non-Independent Identically Distribution (non-IID). The naive aggregation algorithms maybe have worse performance because of the non-IID data [3–8]. In medical image segmentation, [9] and [10] took the lead in discussing the application and safety of federated learning in brain tumor segmentation (BraTS). To solve the non-IID challenges of FL in the medical image field, FedDG [11] and FedMRCM [12] were proposed to address the domain shift issue between the source domain and the target domain, but the sharing of latent features may cause privacy concerns. Auto-FedRL [13] and Auto-FedAvg [14] were proposed to deal with the non-IID problem by using an optimization algorithm to learn super parameters and aggregate weights. IDA [15] introduced the Inverse Distance of local models and the average model of all clients to handle non-IID data. The work [16–19] proposed corresponding aggregation methods from the perspectives of clustering, frequency domain, Bayesian, and representation similarity analysis. More than this, the first computational competition on federated learning, Federated Tumor Segmentation (FeTS) Challenge[1] [20] was held to measure the performance of different aggregation algorithms on glioma segmentation [21–24]. Leon et al. [25] proposed FedCostWAvg get a notable improvement compared to FedAvg by including the cost function decreased during the last round and won the challenge. However, most of these methods improve the performance by adding other regular terms to the aggregation method, without considering all factors as a whole, which may limit the performance of the global model.

Different from the above methods, inspired by the concept of the law of universal gravitation in physics, in this paper, we propose a novel aggregation strategy, FedGrav, which unifies the differences in sample size and the discrepancies of local models among clients by defining the concept of model affinity. Specifically, we take the client sample size as the mass of the local model, and the discrepancies among the local models as their distance, which is quantified from

[1] https://fets-ai.github.io/Challenge/.

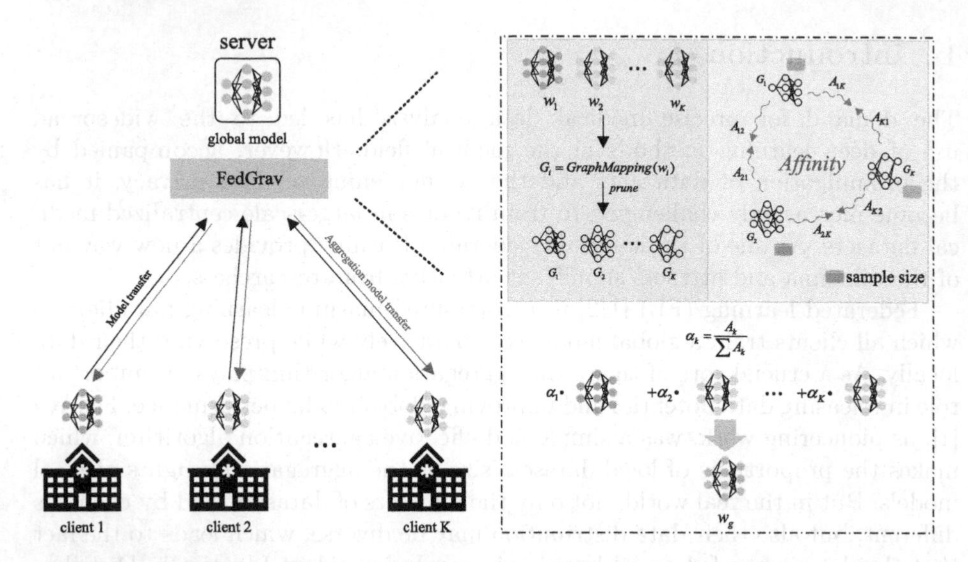

Fig. 1. Overview of the proposed FedGrav. The FedGrav defines the concept of model affinity by unifying the difference in both sample size and local model among clients to aggregates local models and explore the correlations.

the topological perspective of neural networks. Last, the formula 1 is employed to calculate the affinity and explore the internal correlation between the local models. The proposed method promotes a more effective aggregation of local models by unifying the difference between sample size and local model between clients.

The primary contributions of this paper can be summarized as: (1) We propose FedGrav, a novel aggregation strategy that unifies the difference both in sample size and local model among clients by defining the concept of model affinity; (2) We propose Model Graph Distance, a new method to quantify model differences from the perspective of neural network topology. (3) We propose an aggregation algorithm that introduces the concept of affinity and graph into federated learning, and the aggregation weights can be adjusted adaptively; (4) The superior performance is achieved by the proposed method, on the public CIFAR-10 and FeTS challenge datasets.

2 Method

2.1 Overview

Suppose K clients with private data cooperate to train a global model and share the same neural network structure, 3D-Unet [26], which is provided by the FeTS challenge and kept unchanged. For the clients, every client trains a local model w_i for local E epochs and then delivers the local model to the server. The server aggregates local models to a global model by computing the aggregation weights with the proposed FedGrav and assigns it to all clients. Specifically, given K local models, we first make graph mapping to map the network model to the topology graph, and then the graph distance is obtained after the graph pruning and comparison. For the model affinity computation, FedGrav takes the sample size of every client as the mass of the local model and combines the given graph distance to calculate the affinity between models according to the formula 1. After that, a symmetric Model Affinity Matrix $\mathbb{A} \in \mathbb{R}^{K \times K}$ is analyzed to compute aggregation weights. Last, The server aggregates local models to a global model according to the aggregation weights and assigns it to all clients. Repeat and until T rounds or other limits. An overview of the method is shown in Fig. 1.

2.2 FedGrav

Model Affinity. Inspired by the law of universal gravitation, we assume that there is similar gravitation between any two local models. We define it as model affinity in federated learning. It can be described that the affinity between two local models is proportional to the sample size of the client corresponding to the local model, and inversely proportional to the distance between two models. The equation for model affinity takes the form:

$$A_{ik} = M \frac{n_i n_k}{d_{ik}^2} \tag{1}$$

where A_{ik} is the affinity between i-th and k-th local models, n_i and n_k are the sample size of i-th and k-th client, and d_{ik} is the distance between two local models, which is quantified from the perspective of neural network topology and will be described in the following section. M is the affinity constant, it can be simplified in the subsequent analysis, so this paper will not set specific values for it. The model affinity depicts the internal correlation between two local models, which lays the foundation for accurate aggregation weights.

Graph Distance. The distance is defined to quantify model differences. The differences in local models reflect the discrepancies in the distribution of client data to a certain extent. If the differences in local models can be accurately measured, the more appropriate aggregation weights will be assigned to local models to aggregate a better global model. The key motivation is to measure the internal correlations of local models as accurately as possible. We explore the model distance from the perspective of neural network topology in this paper and define it as model graph distance. In FedGrav, the computation of graph distance goes through the following steps:

(1) Graph Mapping. Suppose the server has received local models trained by local data, and we map them into the topological graph. Inspired by [27], take the j-th convolutional layer of k-th local model with 3D-Unet structure as an example, whose kernel dimension is $3 \times 3 \times 3 \times C_{in} \times C_{out}$, it means this layer has $C_{in} \times C_{out}$ nodes with $3 \times 3 \times 3$ filter, we can obtain $C_{in} \times C_{out}$ weight matrices of size $3 \times 3 \times 3$. Thus, we get $C_{in} \times C_{out}$ nodes $W \in \mathbb{R}^{27}$. And then, we make every node W as scalar by averaging or summing, which can be formulated as:

$$w_{sum} = \sum_{d=0}^{2} \sum_{h=0}^{2} \sum_{w=0}^{2} W_{dhw}. \tag{2}$$

It can be mapped into a graph whose structure is similar to the full connection layer after the scalarization of the convolutional layer. Given a $3 \times 3 \times 3 \times C_{in} \times C_{out}$ convolutional layer, the dimensions of its input and output are C_{in} and C_{out} respectively. So, we obtain a weight matrix $W_t \in \mathbb{R}^{C_{in} \times C_{out}}$ after averaging or summing the weights of convolution kernel. We take the C_{in} and C_{out} as the number of nodes, and the weight summation w_{sum} is the edge weight.

(2) Graph Pruning. The server collects local models from clients and makes the graph mapping on them to get K graphs which have the same structure except for the edge weights. These graphs contain all the information of local models, including the part of universality and the part of characteristics of the client data. To make the graphs more distinctive, the graph pruning is conducted. In detail, we differentiated these graphs by setting an adaptive threshold δ, where the edge will be removed if the weight difference of each layer between the local models and global model in the last round is less than the threshold, otherwise, the edge will exist. It can be simplified as:

$$edge = \begin{cases} w_{kj}, & \mid w_{kj}^{t} - w_{gj}^{t-1} \mid > \delta, \\ 0, & otherwise. \end{cases} \tag{3}$$

$$\delta = Sort(\mid w_{kj}^{t} - w_{gj}^{t-1} \mid)[\lfloor \lambda \cdot C_{in} \times C_{out} \rfloor], 0 \leq \lambda < 1. \tag{4}$$

where in Eq. 3, 0 denotes the edge is removed, w_{kj}^{t} denotes edge weight of the j-th layer from the k-th graph in the t-th round, also the weight summation of the j-th layer from the k-th local model in the t-th round, w_{gj}^{t-1} is the weight summation of the j-th layer from the global model in $(t-1)$-th round. The threshold δ varies adaptively with the weights of local models, and λ is the pruning ratio which is responsible for adjusting the degree of pruning. After that we get K discriminative graphs G_i, $i \in [1, K]$.

(3) Graph Comparison. In order to measure the degree of correlation between two graphs, we measure the similarity between pairs of graphs by computing matching between their sets of embeddings, where the Pyramid Match Graph Kernel [28] is employed. We take the reciprocal of the correlation degree as the distance between them. The distance is defined as follows:

$$d_{ik} = \frac{1}{PyramidMatch(G_i, G_k)} \tag{5}$$

Aggregation Weights. According to the above process, the Affinity Matrix $\mathbb{A}$ is obtained, which reports the correlation among local models and is symmetric. The element A_{ik} in matrix $\mathbb{A}$ denotes the affinity of G_i and G_k. The elements in

Table 1. Comparisons with other state-of-the-art methods on the CIFAR-10 dataset.

Method	Accuracy (%)
FedAvg [1]	88.37 ± 0.04
FedProx [6]	87.93 ± 0.19
FedNova [29]	88.68 ± 0.26
Auto-FedAvg [14]	89.16
FedGrav	**89.35 ± 0.23**

the k-th row represent the Affinity among G_k and all graphs, so we can get the affinity of the k-th graph with all graphs, which denotes the correlations of G_k with the whole graphs. last, we normalize A_k as the aggregation weight of the k-th local model or layer.

$$\alpha_k = \frac{\sum_{i=1}^{K} A_{ik}}{\sum_{k=1}^{K}\sum_{i=1}^{K} A_{ik}} \tag{6}$$

In federated learning, clients send the updated local models back to the server each round. In round t, α_k is represented as α_k^t. The global model w_g^{t+1} is aggregated by the server:

$$w_g^t = \sum_{k=1}^{K} \alpha_k^t \cdot w_k^t \tag{7}$$

then, the server assigns the global model w_g^t to all clients. Repeat and until T rounds or other limits.

3 Experiments

3.1 Datasets and Settings

CIFAR-10. The first dataset to verify the validity of our algorithm is CIFAR-10. We partition the training set into 8 clients with heterogeneous data by sampling from a Dirichlet distribution ($\alpha = 0.5$) as in [10] to simulate the non-IID distribution, and the test set in CIFAR-10 is considered as the global test set to evaluate the performance of different algorithms. VGG-9 [30] is employed for image classification, and the other detailed settings are as follows: initial learning rate of $1e-2$; total rounds of 100; local epochs of 20; batch size of 64; SGD optimizer for clients.

MICCAI FeTS2021 Training Data. The real-world dataset used in experiments is provided by the FeTS Challenge organizer, which is the training set of the whole dataset about brain tumor segmentation. In order to evaluate the performance of FedGrav, we partition the dataset composed of 341 data samples

Table 2. Comparisons with other state-of-the-art methods on the MICCAI FeTS2021 Training dataset. D denotes DICE, H95 denotes HD95, and M denotes mean.

Method	D WT	D ET	D TC	H95 WT	H95 ET	H95 TC	M D	M H95
FedAvg [1]	90.49	73.03	69.38	4.82	33.88	39.00	77.63 ± 0.573	25.90 ± 2.731
FCW [21]	90.88	73.15	70.56	3.74	40.79	**17.16**	78.20 ± 0.749	20.56 ± 0.311
FedGrav	**91.26**	**77.21**	**70.75**	**2.80**	**27.40**	22.8	**79.74 ± 0.595**	**17.67 ± 1.692**

into training set and validation set according to the ratio of 8 : 2, and the data is unevenly distributed between 17 data clients. The segmentation network, 3D-Unet, is provided by FeTS and kept unchanged, the learning rate is $1e-4$ and the local epochs are 10. Limited by the framework and official code mechanism, the total number of rounds of training is set to 70, although the performance of the algorithm does not converge to the best.

3.2 Results

Experiment Results on the CIFAR-10. We first validate the proposed method on the CIFAR-10 dataset. Table 1 shows the quantitative results of the state-of-the-art FL methods in terms of the average accuracy, such as FedAvg [1], FedProx [6], FedNova [29], and Auto-FedAvg [14]. As can be seen from the table, the proposed FedGrav method outperforms the other competing FL aggregation methods including Auto-FedAvg, a learning-based aggregation method, which indicates the potential and superiority of FedGrav.

Experiment Results on MICCAI FeTS2021 Training Dataset. In order to verify the robustness of our method and its performance in real-world data, we conduct the experiment on the MICCAI FeTS2021 Training dataset. We evaluate the performance of our algorithm by comparing six indicators: the Dice Similarity Coefficient(DSC) and Hausdorff Distance-95th percentile(HD95) of whole tumor(WT), enhancing tumor(ET), and tumor core(TC). As is shown in Table 2, we list the average results of FedAvg, FedCostWAvg(shortened to FCW), the champion method of FeTS Challenge 2021, and the proposed FedGrav. Different from the original FedCostWAvg which changed the activation function of networks, our re-implemented version made the network unchanged to ensure a fair comparison. Through the quantitative comparison in Table 2, we can find that the proposed method FedGrav has achieved the best results in all indicators except the HD95 TC. Moreover, compared with FedCostWAvg, FedGrav has significantly improved the evaluation of segmentation performance, especially in the enhancing tumor segmentation.

The visualization results are shown in Fig. 2. It can be seen that our FedGrav achieves better segmentation results, even in the hard example, compared to FedCostWAvg and FedAvg. The results proved that the proposed method FedGrav can explore the correlations of local models better and achieved more excellent aggregation performance compared with other methods.

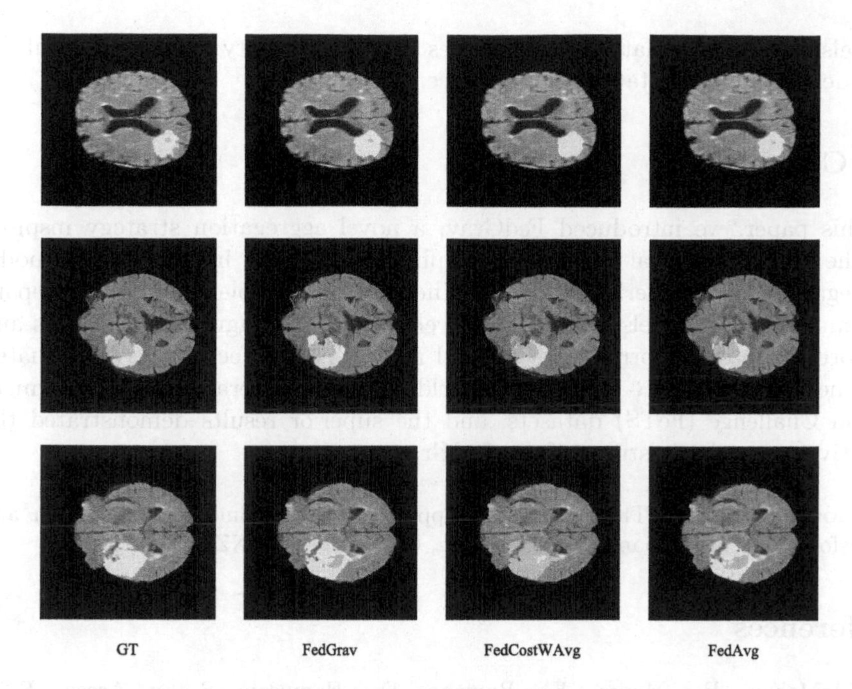

Fig. 2. The visual comparisons with previous state-of-the-art methods on the MICCAI FeTS2021 Training dataset.

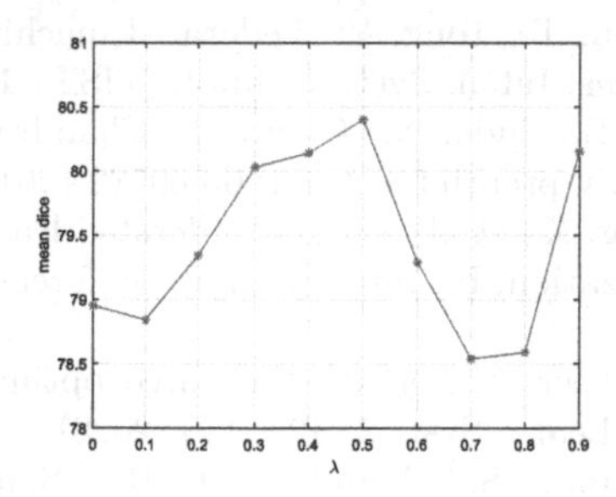

Fig. 3. Comparison of different pruning ratio λ in FedGrav on FeTS datasets.

3.3 Ablation Study

To evaluate the effectiveness and find the better configuration of FedGrav, we conduct the ablation study on the FeTS datasets, and the results are shown in Fig. 3. As we can see, the mean DSC shows a trend of rising first and then falling, because more irrelevant and redundant information will be saved in the model when pruning is not performed. The different values of λ denote the loose degree of graphs, with the gradual increase of λ, the redundant information in local models is gradually eliminated, and the unique information of each local model is preserved. While, when the pruning ratio λ increases to a certain extent, the models lack key information, which makes the model affinity inaccurate, resulting in a decline in segmentation performance.

4 Conclusion

In this paper, we introduced FedGrav, a novel aggregation strategy inspired by the law of universal gravitation in physics. FedGrav improves local model aggregation by considering both the differences in sample size and discrepancies among local models. It can adaptively adjust the aggregation weights and explore the internal correlations of local models more effectively. We evaluated our method on CIFAR-10 and real-world MICCAI Federated Tumor Segmentation Challenge (FeTS) datasets, and the superior results demonstrated the effectiveness and robustness of our FedGrav.

Acknowledgements. This work was supported by the Fund for Innovation and Transformation of Haidian District, Beijing, China(No. HDCXZHKC2021201)

References

1. McMahan, B., Moore, E., Ramage, D., Hampson, S., y Arcas, B.A.: Communication-efficient learning of deep networks from decentralized data. In: Artificial Intelligence and Statistics, pp. 1273–1282. PMLR (2017)
2. Yang, Q., Liu, Y., Chen, T., Tong, Y.: Federated machine learning: concept and applications. ACM Trans. Intell. Syst. Technol. (TIST) **10**(2), 1–19 (2019)
3. Zhao, Y., Li, M., Lai, L., Suda, N., Civin, D., Chandra, V.: Federated learning with non-IID data. arXiv preprint arXiv:1806.00582 (2018)
4. Li, X., Jiang, M., Zhang, X., et al.: FedBN: federated learning on non-IID features via local batch normalization. In: International Conference on Learning Representations (2020)
5. Li, T., Sahu, A.K., Zaheer, M., et al.: Federated optimization in heterogeneous networks. Proc. Mach. Learn. Syst. **2**, 429–450 (2020)
6. Sattler, F., Wiedemann, S., Maluller, K.-R., Samek, W.: Robust and communication-efficient federated learning from non-IID data. IEEE Trans. Neural Networks Learn. Syst. **31**, 3400–3413 (2019)
7. Karimireddy, S.P., Kale, S., Mohri, M., Reddi, S.J., Stich, S.U., Suresh, A.T.: Scaffold: stochastic controlled averaging for federated learning. ICML 2020 (2020)
8. Chen, X., Chen, T., Sun, H., Wu, Z.S., Hong, M.: Distributed training with heterogeneous data: bridging median- and mean-based algorithms. In: NeurIPS 2020 (2020)
9. Sheller, M.J., Reina, G.A., Edwards, B., Martin, J., Bakas, S.: Multi-institutional deep learning modeling without sharing patient data: a feasibility study on brain tumor segmentation. In: Crimi, A., Bakas, S., Kuijf, H., Keyvan, F., Reyes, M., van Walsum, T. (eds.) BrainLes 2018. LNCS, vol. 11383, pp. 92–104. Springer, Cham (2019). https://doi.org/10.1007/978-3-030-11723-8_9
10. Li, W., et al.: Privacy-preserving federated brain tumour segmentation. In: Suk, H.-I., Liu, M., Yan, P., Lian, C. (eds.) MLMI 2019. LNCS, vol. 11861, pp. 133–141. Springer, Cham (2019). https://doi.org/10.1007/978-3-030-32692-0_16

11. Liu, Q., Chen, C., Qin, J., et al.: Feddg: federated domain generalization on medical image segmentation via episodic learning in continuous frequency space. In: Proceedings of the IEEE/CVF Conference on Computer Vision and Pattern Recognition, pp. 1013–1023 (2021)
12. Guo, P., Wang, P., Zhou, J., Jiang, S., Patel, V.M.: Multi-institutional collaborations for improving deep learning-based magnetic resonance image reconstruction using federated learning. In: Proceedings of the IEEE/CVF Conference on Computer Vision and Pattern Recognition, pp. 2423–2432 (2021)
13. Guo, P., et al.: Auto-FedRL: federated hyperparameter optimization for multi-institutional medical image segmentation. arXiv preprint arXiv:2203.06338 (2022)
14. Xia, Y., Yang, D., Li, W., et al.: Auto-FedAvg: learnable federated averaging for multi-institutional medical image segmentation. arXiv preprint arXiv:2104.10195 (2021)
15. Yeganeh, Y., Farshad, A., Navab, N., Albarqouni, S.: Inverse distance aggregation for federated learning with non-IID data. In: Albarqouni, S., et al. (eds.) DART/DCL -2020. LNCS, vol. 12444, pp. 150–159. Springer, Cham (2020). https://doi.org/10.1007/978-3-030-60548-3_15
16. Palihawadana, C., Wiratunga, N., Wijekoon, A., et al.: FedSim: similarity guided model aggregation for Federated Learning. Neurocomputing **483**, 432–445 (2022)
17. Chen, H.Y., Chao, W.L.: FedBE: making Bayesian model ensemble applicable to federated learning. In: International Conference on Learning Representations
18. Chen, Z., Zhu, M., Yang, C., Yuan, Y.: Personalized retrogress-resilient framework for real-world medical federated learning. In: de Bruijne, M., et al. (eds.) MICCAI 2021. LNCS, vol. 12903, pp. 347–356. Springer, Cham (2021). https://doi.org/10.1007/978-3-030-87199-4_33
19. Dong, N., Voiculescu, I.: Federated contrastive learning for decentralized unlabeled medical images. In: de Bruijne, M., et al. (eds.) MICCAI 2021. LNCS, vol. 12903, pp. 378–387. Springer, Cham (2021). https://doi.org/10.1007/978-3-030-87199-4_36
20. Pati, S., et al.: The federated tumor segmentation (fets) challenge. arXiv preprint arXiv:2105.05874 (2021)
21. Bakas, S., et al.: Advancing the cancer genome atlas glioma MRI collections with expert segmentation labels and radiomic features. Sci. Data **4**(1), 1–13 (2017)
22. Reina, G.A., et al.: Open: an open-source framework for federated learning. arXiv preprint arXiv:2105.06413 (2021)
23. Sheller, M.J., et al.: Federated learning in medicine: facilitating multi-institutional collaborations without sharing patient data. Sci. Rep. **10**(1), 1–12 (2020)
24. Koer, F., et al.: Brats toolkit: translating brats brain tumor segmentation algorithms into clinical and scientific practice. Front. Neurosci. **14**, 125 (2020)
25. Mächler, L., Ezhov, I., Kofler, F., et al.: FedCostWAvg: a new averaging for better Federated Learning. In: Crimi, A., Bakas, S. (eds.) BrainLes 2021, Part II. LNCS, vol. 12963, pp. 383–391. Springer, Cham (2022). https://doi.org/10.1007/978-3-031-09002-8_34
26. Çiçek, Ö., Abdulkadir, A., Lienkamp, S.S., Brox, T., Ronneberger, O.: 3D U-Net: learning dense volumetric segmentation from sparse annotation. In: Ourselin, S., Joskowicz, L., Sabuncu, M.R., Unal, G., Wells, W. (eds.) MICCAI 2016. LNCS, vol. 9901, pp. 424–432. Springer, Cham (2016). https://doi.org/10.1007/978-3-319-46723-8_49
27. Gabrielsson, R.B.: Topological Data Analysis of Convolutional Neural Networks' Weights on Images

28. Nikolentzos, G., Meladianos, P., Vazirgiannis, M.: Matching node embeddings for graph similarity. In: Proceedings of the 31st AAAI Conference on Artificial Intelligence, pp. 2429–2435 (2017)
29. Wang, J., Liu, Q., Liang, H., Joshi, G., Poor, H.V.: Tackling the objective inconsistency problem in heterogeneous federated optimization. In: Advances in Neural Information Processing Systems, vol. 33 (2020)
30. Simonyan, K., Zisserman, A.: Very deep convolutional networks for large-scale image recognition. arXiv preprint arXiv:1409.1556 (2014)

Category-Independent Visual Explanation for Medical Deep Network Understanding

Yiming Qian[1], Liangzhi Li[6], Huazhu Fu[1], Meng Wang[1], Qingsheng Peng[2,5], Yih Chung Tham[2,3,4,5], Chingyu Cheng[2,3,4,5], Yong Liu[1], Rick Siow Mong Goh[1], and Xinxing Xu[1(✉)]

[1] Institute of High Performance Computing (IHPC), Agency for Science, Technology and Research (A*STAR), 1 Fusionopolis Way, 16-16 Connexis, Singapore 138632, Republic of Singapore
xuxinx@ihpc.a-star.edu.sg

[2] Ocular Epidemiology and Data Sciences, Singapore Eye Research Institute, Singapore, Singapore

[3] Centre for Innovation and Precision Eye Health, Yong Loo Ling School of Medicine, National University of Singapore, Singapore, Singapore

[4] Department of Ophthalmology, Yong Loo Ling School of Medicine, National University of Singapore, Singapore, Singapore

[5] Duke-NUS Medical School, National University of Singapore, Singapore, Singapore

[6] Meetyou AI Lab, Xiamen, China

Abstract. Visual explanations have the potential to improve our understanding of deep learning models and their decision-making process, which is critical for building transparent, reliable, and trustworthy AI systems. However, existing visualization methods have limitations, including their reliance on categorical labels to identify regions of interest, which may be inaccessible during model deployment and lead to incorrect diagnoses if an incorrect label is provided. To address this issue, we propose a novel category-independent visual explanation method called Hessian-CIAM. Our algorithm uses the Hessian matrix, which is the second-order derivative of the activation function, to weigh the activation weight in the last convolutional layer and generate a region of interest heatmap at inference time. We then apply an SVD-based post-process to create a smoothed version of the heatmap. By doing so, our algorithm eliminates the need for categorical labels and modifications to the deep learning model. To evaluate the effectiveness of our proposed method, we compared it to seven state-of-the-art algorithms using the Chestx-ray8 dataset. Our approach achieved a 55% higher IoU measurement than classical GradCAM and a 17% higher IoU measurement than EigenCAM. Moreover, our algorithm obtained a Judd AUC score of 0.70 on the glaucoma retinal image database, demonstrating its potential applicability in various medical applications. In summary, our category-independent visual explanation method, Hessian-CIAM, generates high-quality region of interest heatmaps that are not dependent on

Supplementary Information The online version contains supplementary material available at https://doi.org/10.1007/978-3-031-43895-0_17.

H. Greenspan et al. (Eds.): MICCAI 2023, LNCS 14221, pp. 181–191, 2023.
https://doi.org/10.1007/978-3-031-43895-0_17

categorical labels, making it a promising tool for improving our understanding of deep learning models and their decision-making process, particularly in medical applications.

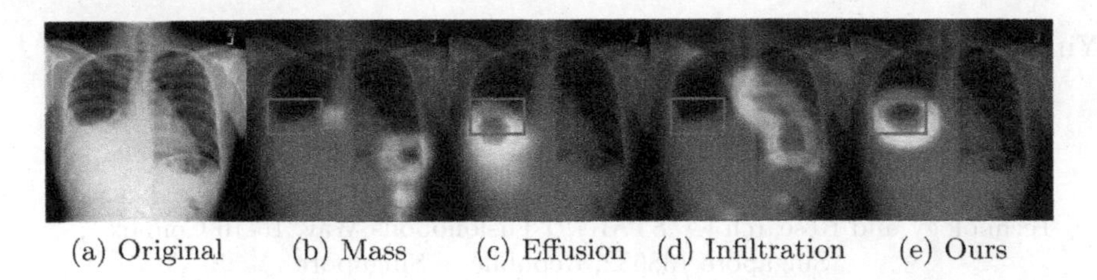

(a) Original (b) Mass (c) Effusion (d) Infiltration (e) Ours

Fig. 1. Example of GradCAM (b-d) supplied different labels vs. our method (e). Three different categorical labels lead GradCAM to generate different distinguishable heatmaps. By contrast, our Hessian-CIAM generates a stable ROI without the categorical label.

1 Introduction

Medical application is a field that has high requirements of model reliability, trustworthiness, and interpretability. According to the act proposed by the European Commission on AI system regulation [4], medical AI systems are categorized as high-risk systems. Five sets of requirements are listed: (1) high quality of data, (2) traceability, (3) transparency, (4) human oversight, (5) robustness, accuracy, and cybersecurity. These requirements impose a potential challenge for deep learning models where such a model is often used as a black-box system. To increase a model's explainability, many visualization methods are proposed to generate the region of interest (ROI) heatmap based on the output of the deep learning model [7,9]. This ROI heatmap highlights the region that deep learning algorithms focus on. This region often contains cues for researchers to investigate the algorithm's decision making process which would help doctors to gain confidence in the AI assisted products. For example, when doctors see a model make a correct prediction and at the same time highlight the right ROI, then it would help this model to gain more trust from doctors.

The state-of-art algorithms mostly focus on providing visualization during training where the categorical label is available. It becomes problematic at product deployment stage when no label is available. Without supplying the ground truth categorical labels, the false categorical labels would mislead the visualization algorithm to highlight wrong regions for cues. A sample is shown in Fig. 1. GradCAM [23] visualization is used widely on a deep learning network that classifies multiple diseases. Three different categorical labels are supplied (Fig. 1 (b-d)) which leads GradCAM to generate three distinguishable ROI heatmaps. To address this issue, we propose a method called **Hessian-C**ategory **I**ndependent **A**ctivation **M**aps (Hessian-CIAM), which utilizes the Hessian matrix as an activation weighting function to eliminate the need for categorical labels to compute the ROI heatmap. Then a polarity checking process is added to the post process which corrects the polarity error from the SVD based smoothing function. Figure 1 (e) shows the visualization from our category-independent method. We benchmark our algorithm against seven state-of-art algorithms on the Chestx-ray8 dataset which demonstrated the superior performance of our algorithm. Additionally, we demonstrate a clinical use case in glaucoma detection from retinal images which shows the flexibility of our algorithm.

2 Related Works

The visual explanation for deep networks is an essential task for researchers to interpret and debug deep networks where an ROI heat map is one of the most popular tools. This field is pioneered by Oquab et al. [18] which additional Global Max Pooling (GMP) layers are added to extract the attention region from a trained convolutional network. It is later improved by CAM [27] by attaching a Global Average Pooling (GAP) layer to the existing model. The GAP identifies the extent of the object while GMP only finds one discriminative part. One drawback of Oquab's method and CAM is the requirement of modifying the original network to output visualizations. This requirement is eliminated by a gradient-based approach GradCam [23]. In this algorithm, the activation weights from the last convolutional layer of the deep network are extracted and weighed by a gradient from the back-propagation to generate the ROI heat map. This method is later improved by GradCAM++ [3] and LayerCAM [12]. An alternative way to generate an ROI heatmap is perturbation-based methods. It removes the requirement of the gradient calculation by iteratively perturbating different parts of the activations weight [22] or image [20, 24] to identify the region on the image that has the highest impact on the prediction result. One major drawback of such an approach is its speed as it requires iteratively running the deep learning model. The gradient-based and perturbation-based methods deliver high-quality ROI heatmap when a categorical label is supplied. It is a useful visualization tool to help researchers interpret the deep network during the development stage. It becomes a different story when it comes to deployment. During the deployment, there is no such luxury of having a ground truth categorical label that is supplied to the visualization algorithm. One solution to relax this problem is using the prediction result as a target label, but this solution often generates a wrong visualization as when the deep learning algorithm outputs incorrect prediction. Muhammad [17] proposed a method to eliminate the dependence on the ground truth categorical label by directly applying SVD on the 2D activations and using its first principle component as the ROI heat map. The first principle component's polarity is bi-directional which could potentially highlight the non-interest region instead. Visualization techniques such as slot attention [15], SCOUTER [13], and SHAP [16] require modification on the original network and training to generate an ROI heatmap. It is not the main scope of our paper and we will not further discuss it here.

Medical applications have high requirements for model reliability, trustworthiness, and interpretability. The visualization tools such as GradCAM and

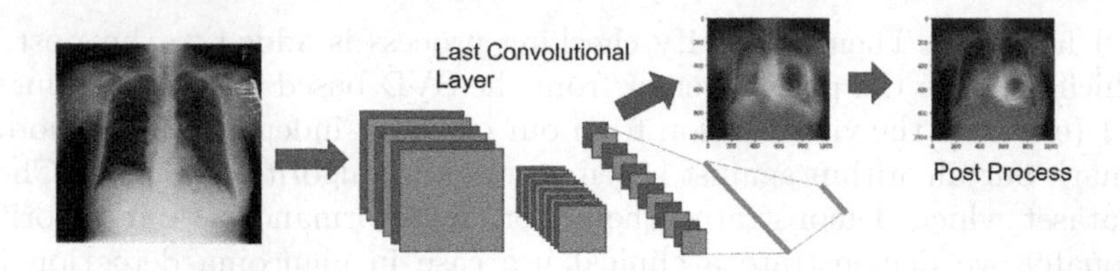

Fig. 2. Overview of our algorithm, the Hessian matrix, and activation weight from the last convolutional layer is used to create an ROI heatmap followed by a post process.

GradCAM++ are widely applied to medical applications such as retina imaging [21], X-ray [10], CT [6], MRI [26], and ultrasound [11]. However, those visualization algorithms require categorical labels to generate visual explanations. This requirement limits the usage of algorithms to the training stage where the ground truth category label is available. Generating high quality visual explanations without relying on the category label at the deployment stage remains a challenge. In this work, we propose a category-independent visual explanation method to solve this problem.

3 Method

Our algorithm generates an ROI heatmap to indicate the region on the image that the deep learning algorithms focus on when making classification decisions. Our method does not require any modification or additional training on target deep networks. The overview flow of our algorithm is illustrated in Fig. 2. Input images feed into the deep network where the activation weights from the last convolution layer are weighted by the Hessian matrix followed by a post-process to output a clean ROI heatmap.

It is well known that the Hessian matrix appears in the expansion of gradient about a point in parameter space [19], as:

$$\nabla_\omega \left(\omega + \Delta\omega\right) = \nabla_\omega(\omega) + H\Delta\omega + O(\|\Delta\omega\|^2), \tag{1}$$

where ω is a point in parameter space, $\Delta\omega$ is a perturbation of ω, $\Delta\omega$ is the gradient and H is the hessian matrix. In order to approximate the Hessian matrix H, we let $\Delta\omega = rv$, where v is the identity matrix, and r is a small number which leads the $O(r)$ term to become insignificant. So we can further simplify the equation into:

$$Hv = \frac{\nabla_\omega(\omega + rv) - \nabla_\omega(\omega)}{r} + O(r) = \frac{\nabla_\omega(\omega + rv) - \nabla_\omega(\omega)}{r}. \tag{2}$$

Our goal is to apply the Hessian matrix as a weighting function to indicate the significance of each activation function output in the CNN. So we applied an L2 normalization on the Hv, here v is an identity matrix, so we can get the normalized Hessian matrix $\hat{H} = \frac{|Hv|}{\|Hv\|_2}$. In the CNN we denote A^k as the feature activation map from the kth convolution layer. $\hat{H}^k$ denotes the normalized Hessian matrix in the kth layer. We calculate the Hadamard product between $\hat{H}^k$ and A^k, then apply ReLU to obtain the new activation map. n is the depth of the activation map. The ROI heatmap $L_H = \sum_{k=1}^{n} ReLu(\hat{H}^k \odot A^k)$.

The ROI heatmap L_H can be noisy, we follow Muhammad's approach [17] to smooth out the L_H which applies SVD on $A_H^k = ReLu(\hat{H}^k \odot A^k) = U\Sigma V^T$ where U denotes a $M \times M$ matrix. Σ denotes a diagonal matrix with size of $M \times N$. V denotes a $N \times N$ matrix. The column of U and V are the left singular vectors. The V_1 denotes the first component in V which is a weight function to create a smoothed ROI heatmap $L_{HS} = A_H^k V_1$. One drawback of Muhammad's approach [17] is the polarity of V_1 is not considered as the Eigenvectors from SVD are bidirectional. It could lead the algorithm to output non-ROI regions. To solve this problem, we revise the algorithm to calculate the correlation between the smoothed version L_{HS} and the original ROI heatmap L_H. If the correlation appears negative, we will reverse the ROI heatmap, as:

$$L_{HS} = \begin{cases} ReLu(A_H^k V_1), & \text{if } corr(A_H^k V_1, L_H) > 0, \\ ReLu(-A_H^k V_1), & \text{otherwise.} \end{cases} \tag{3}$$

4 Experiment

4.1 Experiment Setup

We conduct experiments on lung disease classification Chestx-ray8 [25] to evaluate the performance of our algorithm. The Chestx-ray8 dataset contains 100,000 x-ray images with 19 disease labels. It is a significantly imbalanced dataset with some categories having as few as 7 images. To demonstrate our visualization techniques, we simplified the dataset by selecting 6 diseases with a higher number of images. After the selection, our training set contains images from atelectasis (3135 images), effusion (2875 images), infiltration (6941 images), mass (1665 images), nodule (2036 images), and pneumothorax (1485 images). Additionally, we randomly selected 7000 images from healthy people. 20% of images in the training set were set aside as validation sets for parameter tuning. This dataset contains 881 test images with bounding boxes that indicate the location of the diseases which 644 images were in the 6 diseases we selected.

We utilize the pre-trained ResNet50 [8] as the backbone. The cross-entropy loss is used as a loss function; the learning rate is set to 0.00001; the batch size is 64. The training cycle is set to 100 epochs. Our workstation is equipped with 2 Nvidia 3090 GPU (24 GB RAM), Intel Xeon CPU (3.30 GHz), and 128 GB RAM.

4.2 Quantitative Evaluation

The algorithm is evaluated following the method proposed by Cao et al. [2]. The union of intersection (IoU) between the bounding box and ROI is measured. The

Table 1. Quantitative evaluation of visualization methods on Chestx-ray8 dataset. The IoU using prediction as the label is shown here. The value in the bracket is the IoU that uses ground truth as the label.

	IoU on different thresholds				
	0.95	0.90	0.85	0.80	0.75
Gradient based approaches					
GradCAM [23]	0.127 (0.136)	0.139 (0.151)	0.150 (0.163)	0.159 (0.175)	0.175 (0.185)
GradCAM++ [3]	0.103 (0.108)	0.117 (0.124)	0.131 (0.139)	0.144 (0.154)	0.155 (0.167)
HiResCAM [5]	0.104 (0.120)	0.112 (0.133)	0.118 (0.143)	0.124 (0.153)	0.129 (0.162)
Perturbation based approaches					
AblationCAM [22]	0.092 (0.090)	0.097 (0.094)	0.102 (0.098)	0.107 (0.103)	0.113 (0.109)
ScoreCAM [24]	0.134 (N/A)	0.141 (N/A)	0.149 (N/A)	0.158 (N/A)	0.168 (N/A)
RISE [20]	0.095 (0.097)	0.096 (0.096)	0.097 (0.097)	0.098 (0.098)	0.099 (0.099)
Category-independent approaches					
EigenCAM [17]	0.213	0.222	0.227	0.231	0.232
Ours	**0.240**	**0.253**	**0.262**	**0.267**	**0.271**

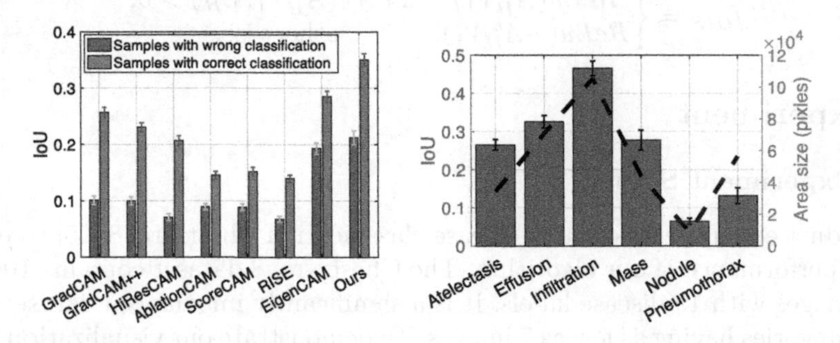

Fig. 3. (left) IoU on samples with a wrong and correct prediction on methods. (right) IoU for our method on different diseases in the bar chart (left y-axis) and the ground truth bounding box size (dashed line, right y-axis).

foreground of ROI is extracted based on applying thresholds to find the area that covers 95%, 90%, 85%, 80%, and 75% of energy from the heatmap. The gradient and perturbation-based methods require ground truth labels to generate an ROI heatmap but, at the inference time, the ground truth label is not available. To simulate the deployment scenario, we conduct two sets of evaluations. In the first evaluation, the prediction results (our ResNet model delivers 42.6% prediction accuracy) from the deep learning model are used as a label feed into the visualization methods. One drawback of this approach is the prediction result is not always reliable and the incorrect prediction could mislead the algorithm to output the wrong ROI. As a comparison, in the second evaluation, we supply ground truth labels to visualization methods. The quantitative evaluation of different visualization methods is shown in Table 1.

Three groups of visualization algorithms are evaluated in our experiment. The gradient based group contains the algorithm that relies on the gradient from the label to generate the ROI. In this group, GradCAM achieved the highest at 0.175 IoU at the 75% threshold. The pertubation based group makes small perturbations in the input image or activation weight to find the ROI that has the highest impact. In this group, the ScoreCAM achieved the highest 0.168 IoU at the 75% threshold. The category independent group contains algorithms that do not require a label to generate ROI. Our method scored the highest IoU at 0.271 IoU at the 75% threshold. When ground truth labels are supplied, the IoU for gradient based methods is improved in the range of 5% to 20%. For perturbation based methods, supplying ground truth data reduced the performance of AblationCAM and had minimal impact on RISE.

Next, we split the test set into two categories which are samples with wrong and correct predictions (shown in Fig. 3 (left)). The 75% threshold is used to calculate IoU. The samples with correct prediction consistently scored higher IoU across all visualization methods. Our method shows the highest performance in both wrong and correct prediction categories. The perturbation based methods consistently scored lower than other methods indicating this group is not suitable for X-ray image classification applications.

To further investigate the efficiency of our algorithm, we extract the IoU on each disease (Fig. 3 (right)) where a 75% threshold is applied to calculate the mean IoU. The evaluation shows our algorithm is positively correlated with the size of the ground truth bounding box. It indicates the disease with a larger infection area is easier to visualize by our algorithm.

4.3 Qualitative Evaluation

Sample images comparing our algorithm with five state-of-art algorithms are shown in Fig. 4. Our algorithm has a cleaner heatmap. The gradient methods generate a heatmap that contains a higher level of noise that covers a large area of the non-lung regions such as the shoulder. The perturbation based methods deliver the worst visualization in our evaluation. The AlbationCAM and ScoreCAM are only able to highlight the whole lung area but it does not provide any clinical value to pinpoint the disease locations. The RISE [20] method delivers multiple clusters of highlight regions that are not feasible to provide human-readable information. The last row of Fig. 4 shows the worst case in our evaluation which is a representative case to illustrate the failure mode of our algorithm. The deep learning algorithm may fail to detect the small size lesions which leads to the wrong ROI for visualization methods. More comparison is available in the supplementary material.

4.4 Clinical Application

Our algorithm has the potential to apply to many clinical applications. We conducted an additional experiment on the glaucoma retinal image database [14]

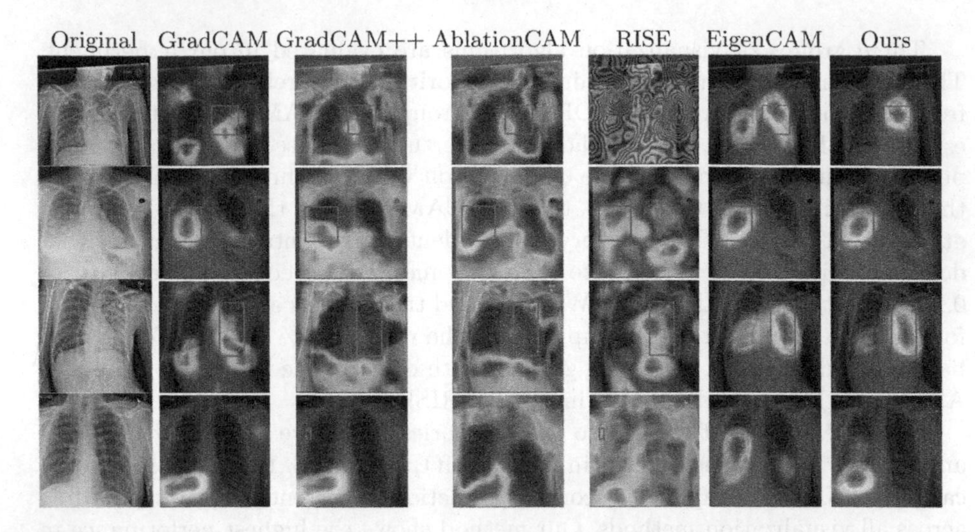

Fig. 4. Comparison of visualization methods on Chestx-ray8 dataset. The ground truth bounding box drawn by clinicians overlays on the heatmaps.

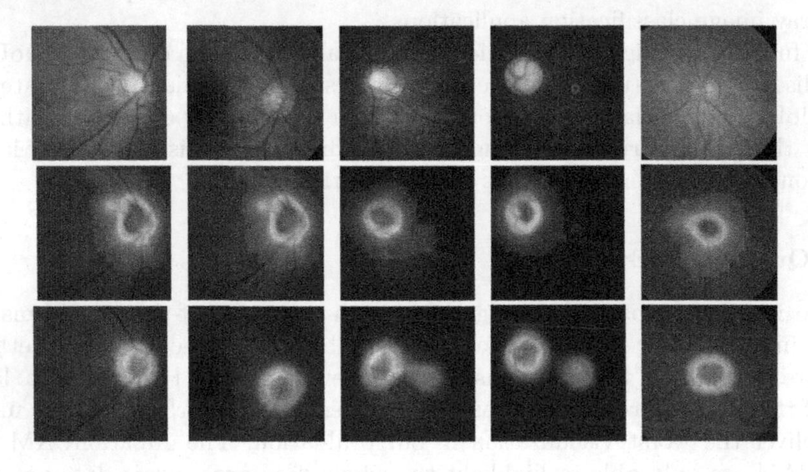

Fig. 5. Five samples of the original image (row 1), ground truth saliency map (row 2), and heatmap from our method (row 3) are shown.

with 3,144 negative and 1,712 positive glaucoma samples[1] Each sample contains a saliency map annotated by ophthalmologists by using mouse clicks to simulate the human visual attention process. Since our goal is to evaluate the explainability of our visualization algorithm, we decided to use all images to train the glaucoma classification model. We follow the work from Bylinskii et al. [1] to apply similarity (histogram intersection), cross-correlation, and Judd AUC to measure the performance of our algorithm. The 95% energy of the ROI heatmap was used

[1] the dataset is obtained from https://github.com/smilell/AG-CNN.

as a threshold to clean our heatmap. Our algorithm achieved 0.618 ± 0.0024 in similarity, 0.755 ± 0.0033 in cross-correlation, and 0.703 ± 0.0013 in Judd AUC. The complete evaluation is available in the supplementary material (Fig. 5).

5 Conclusion

In this study, we propose a novel category-independent deep learning visualization algorithm that does not rely on categorical labels to generate visualizations. Our evaluation demonstrates that our algorithm outperforms seven state-of-the-art algorithms by a significant margin on a multi-disease classification task using X-ray images. This indicates that our algorithm has the potential to enhance model explainability and facilitate its deployment in medical applications. Additionally, we demonstrate the flexibility of our algorithm by showing a clinical use case on retinal image glaucoma detection. Overall, our proposed Hessian-CIAM algorithm represents a promising tool for improving our understanding of deep learning models and enhancing their interpretability, particularly in medical applications.

Acknowledgements. This work is supported by the Agency for Science, Technology and Research (A*STAR) under its RIE2020 Health and Biomedical Sciences (HBMS) Industry Alignment Fund Pre-Positioning (IAF-PP) Grant No. H20c6a0031, the National Research Foundation, Singapore under its AI Singapore Programme (AISG Award No: AISG2-TC-2021-003), the Agency for Science, Technology and Research (A*STAR) through its AME Programmatic Funding Scheme Under Project A20H4b0141, A*STAR Central Research Fund "A Secure and Privacy Preserving AI Platform for Digital Health".

References

1. Bylinskii, Z., Judd, T., Oliva, A., Torralba, A., Durand, F.: What do different evaluation metrics tell us about saliency models? arXiv preprint arXiv:1604.03605 (2016)
2. Cao, C., et al.: Look and think twice: capturing top-down visual attention with feedback convolutional neural networks. In: 2015 IEEE International Conference on Computer Vision (ICCV), pp. 2956–2964 (2015)
3. Chattopadhay, A., Sarkar, A., Howlader, P., Balasubramanian, V.N.: Grad-CAM++: generalized gradient-based visual explanations for deep convolutional networks. In: IEEE Winter Conference on Applications of Computer Vision (WACV), pp. 839–847 (2018)
4. COMMISSION, E.: Proposal for a regulation of the European parliament and of the council (2021). https://artificialintelligenceact.eu/the-act/
5. Draelos, R.L., Carin, L.: HiResCAM: faithful location representation in visual attention for explainable 3D medical image classification. arXiv preprint arXiv:2011.08891 (2020)
6. Draelos, R.L., Carin, L.: Explainable multiple abnormality classification of chest CT volumes. Artif. Intell. Med. **132**(C), 102372 (2022)

7. Guo, Y., Liu, Y., Oerlemans, A., Lao, S., Wu, S., Lew, M.S.: Deep learning for visual understanding: A review. Neurocomputing **187**, 27–48 (2016). recent Developments on Deep Big Vision
8. He, K., Zhang, X., Ren, S., Sun, J.: Deep residual learning for image recognition. In: Proceedings of the IEEE Conference on Computer Vision and Pattern Recognition (CVPR) (2016)
9. Hohman, F., Kahng, M., Pienta, R., Chau, D.H.: Visual analytics in deep learning: an interrogative survey for the next frontiers. IEEE Trans. Visual Comput. Graphics **25**(8), 2674–2693 (2019). https://doi.org/10.1109/TVCG.2018.2843369
10. Irvin, J., et al.: Chexpert: a large chest radiograph dataset with uncertainty labels and expert comparison. In: Proceedings of the AAAI Conference on Artificial Intelligence, vol. 33, pp. 590–597 (2019)
11. Ishikawa, G., Xu, R., Ohya, J., Iwata, H.: Detecting a fetus in ultrasound images using Grad-CAM and locating the fetus in the uterus. In: ICPRAM, pp. 181–189 (2019)
12. Jiang, P.T., Zhang, C.B., Hou, Q., Cheng, M.M., Wei, Y.: Layercam: exploring hierarchical class activation maps for localization. IEEE Trans. Image Process. **30**, 5875–5888 (2021)
13. Li, L., Wang, B., Verma, M., Nakashima, Y., Kawasaki, R., Nagahara, H.: SCOUTER: slot attention-based classifier for explainable image recognition. In: Proceedings of the IEEE/CVF International Conference on Computer Vision (ICCV), pp. 1046–1055 (2021)
14. Li, L., Xu, M., Wang, X., Jiang, L., Liu, H.: Attention based glaucoma detection: a large-scale database and CNN model. In: Proceedings of the IEEE/CVF Conference on Computer Vision and Pattern Recognition (CVPR) (2019)
15. Locatello, F., et al..: Object-centric learning with slot attention. In: Advances in Neural Information Processing Systems (NIPS), vol. 33, pp. 11525–11538 (2020)
16. Lundberg, S.M., Lee, S.I.: A unified approach to interpreting model predictions. In: Guyon, I., et al. (eds.) Advances in Neural Information Processing Systems (NIPS), pp. 4765–4774. Curran Associates, Inc. (2017)
17. Muhammad, M.B., Yeasin, M.: Eigen-CAM: class activation map using principal components. In: 2020 International Joint Conference on Neural Networks (IJCNN), pp. 1–7. IEEE (2020)
18. Oquab, M., Bottou, L., Laptev, I., Sivic, J.: Is object localization for free? weakly-supervised learning with convolutional neural networks. In: Proceedings of the IEEE Conference on Computer Vision and Pattern Recognition (CVPR), pp. 685–694 (2015)
19. Pearlmutter, B.A.: Fast exact multiplication by the hessian. Neural Comput. **6**(1), 147–160 (1994)
20. Petsiuk, V., Das, A., Saenko, K.: RISE: randomized input sampling for explanation of black-box models. arXiv preprint arXiv:1806.07421 (2018)
21. Poplin, R., et al.: Prediction of cardiovascular risk factors from retinal fundus photographs via deep learning. Nature Biomed. Eng. **2**(3), 158–164 (2018)
22. Ramaswamy, H.G., et al.: Ablation-CAM: visual explanations for deep convolutional network via gradient-free localization. In: Proceedings of the IEEE/CVF Winter Conference on Applications of Computer Vision (WACV), pp. 983–991 (2020)
23. Selvaraju, R.R., Cogswell, M., Das, A., Vedantam, R., Parikh, D., Batra, D.: Grad-CAM: visual explanations from deep networks via gradient-based localization. In: Proceedings of the IEEE International Conference on Computer Vision (ICCV), pp. 618–626 (2017)

24. Wang, H., et al.: Score-CAM: score-weighted visual explanations for convolutional neural networks. In: Proceedings of the IEEE/CVF Conference on Computer Vision and Pattern Recognition Workshops (CVPRW), pp. 24–25 (2020)
25. Wang, X., Peng, Y., Lu, L., Lu, Z., Bagheri, M., Summers, R.M.: Chestx-ray8: hospital-scale chest x-ray database and benchmarks on weakly-supervised classification and localization of common thorax diseases. In: Proceedings of the IEEE Conference on Computer Vision and Pattern Recognition (CVPR), pp. 2097–2106 (2017)
26. Yang, C., Rangarajan, A., Ranka, S.: Visual explanations from deep 3D convolutional neural networks for Alzheimer's disease classification. In: AMIA Annual Symposium Proceedings, vol. 2018, p. 1571. American Medical Informatics Association (2018)
27. Zhou, B., Khosla, A., Lapedriza, A., Oliva, A., Torralba, A.: Learning deep features for discriminative localization. In: Proceedings of the IEEE Conference on Computer Vision and Pattern Recognition (CVPR), pp. 2921–2929 (2016)

Self-aware and Cross-Sample Prototypical Learning for Semi-supervised Medical Image Segmentation

Zhenxi Zhang[1], Ran Ran[2], Chunna Tian[1(✉)], Heng Zhou[1], Xin Li[3], Fan Yang[3], and Zhicheng Jiao[4]

[1] Xidian University, 2 South Taibai Road, Xi'an, Shanxi, China
zxzhang_5@stu.xidian.edu.cn , chnatian@xidian.edu.cn
[2] Cancer Center, The First Affiliated Hospital of Xi'an Jiaotong University, Xi'an, China
[3] AIQ, Abu Dhabi, United Arab Emirates
[4] Department of Diagnostic Imaging, Warren Alpert Medical School of Brown University, Providence, USA

Abstract. Consistency learning plays a crucial role in semi-supervised medical image segmentation as it enables the effective utilization of limited annotated data while leveraging the abundance of unannotated data. The effectiveness and efficiency of consistency learning are challenged by prediction diversity and training stability, which are often overlooked by existing studies. Meanwhile, the limited quantity of labeled data for training often proves inadequate for formulating intra-class compactness and inter-class discrepancy of pseudo labels. To address these issues, we propose a self-aware and cross-sample prototypical learning method (SCP-Net) to enhance the diversity of prediction in consistency learning by utilizing a broader range of semantic information derived from multiple inputs. Furthermore, we introduce a self-aware consistency learning method which exploits unlabeled data to improve the compactness of pseudo labels within each class. Moreover, a dual loss re-weighting method is integrated into the cross-sample prototypical consistency learning method to improve the reliability and stability of our model. Extensive experiments on ACDC dataset and PROMISE12 dataset validate that SCP-Net outperforms other state-of-the-art semi-supervised segmentation methods and achieves significant performance gains compared to the limited supervised training. Code is available at https://github.com/Medsemiseg/SCP-Net.

Keywords: Prototypical learning · Consistency learning · Semi-supervised segmentation

1 Introduction

With the increasing demand for accurate and efficient medical image analysis, Semi-supervised segmentation methods offer a viable solution to tackle the problems associated with scarce labeled data and mitigate the reliance on manual

---

Supplementary Information The online version contains supplementary material available at https://doi.org/10.1007/978-3-031-43895-0_18.

H. Greenspan et al. (Eds.): MICCAI 2023, LNCS 14221, pp. 192–201, 2023.
https://doi.org/10.1007/978-3-031-43895-0_18

expert annotation. It is often not feasible to annotate all images in a dataset. By exploring the information contained in the unlabeled data, semi-supervised learning [1,2] can help to improve segmentation performance compared to using only a small set of annotated examples.

Consistency constraint is a widely-used solution in semi-supervised segmentation to improve performance by making the prediction and/or intermediate features remain consistent under different perturbations. However, it's challenging to obtain universal and appropriate perturbations (e.g., augmentation [3], contexts [4], and decoders [5]) across different tasks. In addition, the efficacy of the consistency loss utilized in semi-supervised segmentation models could be weakened by minor perturbations that have no discernible effect on the predicted results. Conversely, unsuitable perturbations or unclear boundaries between structures could introduce inaccurate supervisory signals, causing a build-up of errors and leading to sub-optimal performance of the model. Recently, some unsupervised prototypical learning methods [6–10] apply the feature matching operation based on the category prototypes to generate the pseudo labels in the semi-supervised segmentation task. Then, the consistency constraint is enforced between the model's prediction and the corresponding prototypical prediction to enhance the model's performance. For example, Xu, et al. [6] propose a cyclic prototype consistency learning framework which involves a two-way flow of information between labeled and unlabeled data. Wu, et al. [7] suggest to facilitate the convergence of class-specific features towards their corresponding high-quality prototypes by promoting their alignment. Zhang, et al. [8,10] exploit the feature distances from prototypes to facilitate online correction of the pseudo label in the training course. Limited by the quantity of prototypes and insufficient feature relation learning, the only one global category prototype used in [6–8] for feature matching might omit diversity and impair the representation capability.

To put it briefly, prior research has not fully addressed the robustness and variability of prediction results in response to perturbations. To address this, unlike the global prototypes in [6,7], we propose a novel prototype generation method, namely self-aware and cross-sample class prototypes, which generates two distinct prototype predictions to enhance semantic information interaction and ensure disagreement in consistency training. We also propose to use prediction uncertainty between self-aware prototype prediction and multiple predictions to re-weight the consistency constraint loss of cross-sample prototypes. By doing so, we can reduce the adverse effects of label noise in challenging areas such as low-contrast regions or adhesive edges, resulting in a more stable consistency constraint training process. This, in turn, would lead to significantly improved model performance and accuracy. Lastly, we present SCP-Net, a parameter-free semi-supervised segmentation framework (Fig. 1) that incorporates both types of prototypical consistency constraints.

The main contributions of this paper can be summarized as: (1) We conduct an in-depth study on prototype-based semi-supervised segmentation methods and propose self-aware prototype prediction and cross-sample prototype prediction to ensure appropriate prediction diversity in consistency learning. (2) To enhance the intra-class compactness of pseudo labels, we propose a self-aware prototypical consistency learning method. (3) To boost the stability and reliability of cross-sample prototypical consistency learning, we design a dual loss re-weighting method which helps to reduce the negative effect of noisy pseudo labels. (4) Extensive experiments on ACDC and PROMISE12 datasets have demonstrated that SCP-Net effectively utilizes the unlabeled data and improves semi-supervised segmentation performance with a low annotation ratio.

2 Method

In the semi-supervised segmentation task, the training set is divided into the labeled set $\mathcal{D}_l = \{(x_k, y_k)\}_{k=1}^{N_l}$ and the unlabeled set $\mathcal{D}_u = \{x_k\}_{k=N_l+1}^{N_l+N_u}$, where $N_u \gg N_l$. Each labeled image $x_k \in \mathbb{R}^{H\times W}$ has its ground-truth mask $y_k \in \{0,1\}^{C\times H\times W}$, where H, W, and C are the height, width, and class number, respectively. Our objective is to enhance the segmentation performance of the model by extracting additional knowledge from the unlabeled dataset $\mathcal{D}_u$.

2.1 Self-cross Prototypical Prediction

The prototype in segmentation refers to the aggregated representation that captures the common characteristics of some pixel-wise features from a particular object or class. Let $p_k^c(i)$ denote the probability of pixel i belonging to class c, $f_k \in \mathbb{R}^{D\times H\times W}$ represent the feature map of sample k. The class-wise prototypes $\boldsymbol{q_k^c}$ is defined as follows:

$$\boldsymbol{q_k^c} = \frac{\sum_i p_k^c(i)\cdot f_k(i)}{\sum_i p_k^c(i)} \tag{1}$$

Let B denote the batch size. In the iterative training process, one mini-batch contains $B\times C$ prototypes for sample $k=1$ and other samples with index $j=2,3,\cdots,B$. Then, feature similarity is calculated according to the self-aware prototype $\boldsymbol{q_k^c}$ or cross-sample prototypes $\boldsymbol{q_j^c}$ to form multiple segmentation probability matrices. Specifically, $\hat{s}_{kk}^c$ is the self-aware prototypical similarity map via calculating the cosine similarity between the feature map f_k and the prototype vector $\boldsymbol{q_k^c}$ as Eq. 2:

$$\hat{s}_{kk}^c = \frac{f_k\cdot \boldsymbol{q_k^c}}{\|f_k\|\cdot\|\boldsymbol{q_k^c}\|} \tag{2}$$

Then, $softmax$ function is applied to generate the self-aware probability prediction $\hat{p}_{kk}\in\mathbb{R}^{C\times H\times W}$ based on $\hat{s}_{kk}\in\mathbb{R}^{C\times H\times W}$. Since $\boldsymbol{q_k^c}$ is aggregated in sample k itself, which can align f_k with more homologous features, ensuring the intra-class consistency of prediction. Similarly, we can obtain $B-1$ cross-sample prototypical similarity maps $\hat{s}_{kj}^c$ following Eq. 3:

$$\hat{s}_{kj}^c = \frac{f_k\cdot \boldsymbol{q_j^c}}{\|f_k\|\cdot\left\|\boldsymbol{q_j^c}\right\|} \tag{3}$$

This step ensures that features are associated and that information is exchanged in a cross-image manner. To enhance the reliability of prediction, we take the multiple similarity estimations $\hat{s}_{kj} \in \mathbb{R}^{C\times H\times W}$ into consideration and integrate them to get the cross-sample probability prediction $\hat{p}_{ko} \in \mathbb{R}^{C\times H\times W}$:

$$\hat{p}_{ko}^{c} = \frac{\sum_{j=2}^{B} e^{\hat{s}_{kj}^{c}}}{\sum_{c}\sum_{j=2}^{B} e^{\hat{s}_{kj}^{c}}} \tag{4}$$

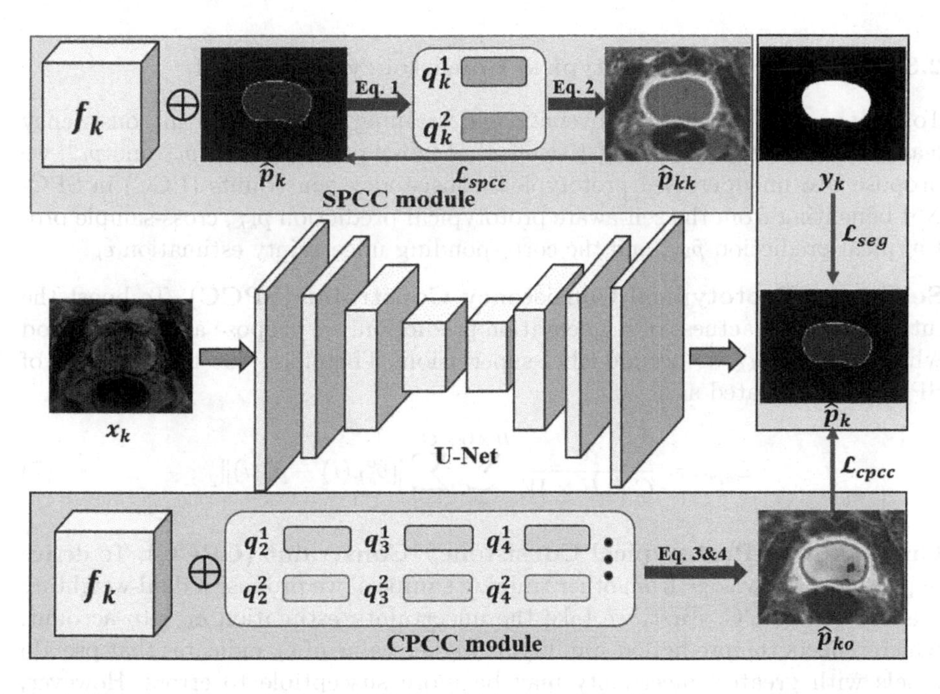

Fig. 1. The overall flowchart of SCP-Net, which consists of three parts: the supervised training with $\mathcal{L}_{seg}$, SPCC module with $\mathcal{L}_{spcc}$, CPCC module with $\mathcal{L}_{cpcc}$.

2.2 Prototypical Prediction Uncertainty

To effectively evaluate the predication consistency and training stability in semi-supervised settings, we propose a prototypical prediction uncertainty estimation method based on the similarity matrices $\hat{s}_{kk}$ and $\hat{s}_{kj}$. First, we generate B binary represented mask $\hat{m}_{kn} \in \mathbb{R}^{C\times H\times W}$ via *argmax* operation and one-hot encoding operation, where $n = 1, 2, \cdots, B$. Then, we sum all masks $\hat{m}_{kn}$ and dividing it by B to get a normalized probability $\hat{p}_{norm}$ as:

$$\hat{p}_{norm}^{c} = \frac{\sum_{n=1}^{B} \hat{m}_{kn}^{c}}{B} \tag{5}$$

And a normalized entropy is estimated from $\hat{p}_{norm}$, denoted as $e_k \in \mathbb{R}^{H\times W}$:

$$e_k = -\frac{1}{\log(C)} \sum_{c=1}^{C} \hat{p}^c_{norm} \log \hat{p}^c_{norm} \tag{6}$$

where e_k serves as the overall confidence of multiple prototypical predictions, and a higher entropy equals more prediction uncertainty. Then, we use e_k to adjust the pixel-wise weight of labeled and unlabeled samples, which will be elaborated in next subsection.

2.3 Unsupervised Prototypical Consistency Constraint

To enhance the prediction diversity and training effectiveness in consistency learning and mitigate the negative effect of noisy predictions in $\hat{p}_{kk}$ and $\hat{p}_{kj}$, we propose two unsupervised prototypical consistency constraints (PCC) in SPC-Net benefiting from the self-aware prototypical prediction $\hat{p}_{kk}$, cross-sample prototypical prediction $\hat{p}_{kj}$, and the corresponding uncertainty estimation e_k.

Self-aware Prototypical Consistency Constraint (SPCC). To boost the intra-class compactness of segmentation prediction, we propose a SPCC method which applies $\hat{p}_{kk}$ as pseudo-label supervision. Therefore, the loss function of SPCC is formulated as:

$$\mathcal{L}_{spcc} = \frac{1}{C \times H \times W} \sum_{i=1}^{H\times W} \sum_{c=1}^{C} \|\hat{p}^c_{kk}(i) - p^c_k(i)\|_2 \tag{7}$$

Cross-sample Prototypical Consistency Constraint (CPCC). To derive dependable knowledge from other training samples, we propose a dual-weighting method for CPCC. First, we take the uncertainty estimation e_k into account, which reflects the prediction stability. A higher value of e_k indicates that pseudo labels with greater uncertainty may be more susceptible to errors. However, these regions provide valuable information for segmentation performance. To reduce the influence of the suspicious pseudo labels and adjust the contribution of these crucial supervisory signals during training, we incorporate e_k in CPCC by setting a weight $w_{1ki} = 1 - e_{ki}$. Second, we introduce the self-aware probability prediction $\hat{p}_{kk}$ into the CPCC module. Specifically, we calculate the maximum value of $\hat{p}_{kk}$ along class c, termed as the self-aware confidence weight w_{2ki}:

$$w_{2ki} = \max_c \hat{p}^c_{kk}(i) \tag{8}$$

w_{2k} can further enhance the reliability of CPCC. Therefore, the optimized function of CPCC is calculated between cross-sample prototypical prediction $\hat{p}_{ko}$ and $\hat{p}_k$:

$$\mathcal{L}_{cpcc} = \frac{1}{C \times H \times W} \sum_{i=1}^{H\times W} \sum_{c=1}^{C} w_{1ki} \cdot w_{2ki} \cdot \|\hat{p}^c_{ko}(i) - p^c_k(i)\|_2 \tag{9}$$

Loss Function of SCP-Net We use the combination of cross-entropy loss $\mathcal{L}_{ce}$ and Dice loss $\mathcal{L}_{Dice}$ to supervise the training process of labeled set [11], which is defined as:

$$\mathcal{L}_{seg} = \mathcal{L}_{ce}(\hat{p}_k, y_k) + \mathcal{L}_{Dice}(\hat{p}_k, y_k) \tag{10}$$

For both labeled data and unlabeled data, we leverage $\mathcal{L}_{spcc}$ and $\mathcal{L}_{cpcc}$ to provide unsupervised consistency constraints for network training and explore the valuable unlabeled knowledge. To sum it up, the overall loss function of SCPNet is the combination of the supervised loss and the unsupervised consistency loss, which is formulated as:

$$\mathcal{L}_{total} = \sum_{k=1}^{N_l} \mathcal{L}_{seg}(\hat{p}_k, y_k) + \lambda \sum_{k=1}^{N_l+N_u} (\mathcal{L}_{spcc}(\hat{p}_k, \hat{p}_{kk}) + \mathcal{L}_{cpcc}(\hat{p}_k, \hat{p}_{ko})) \tag{11}$$

$\lambda(t) = 0.1 \cdot e^{-5(1-t/t_{max})^2}$ is a weight using a time-dependent Gaussian warming up function [12] to balance the supervised loss and unsupervised loss. t represents the current training iteration, and t_{max} is the total iterations.

3 Experiments and Results

Dataset and Evaluation Metric. We validate the effectiveness of our method on two public benchmarks, namely the Automated Cardiac Diagnosis Challenge [1] (ACDC) dataset [13] and the Prostate MR Image Segmentation challenge [2] (PROMISE12) dataset [14]. ACDC dataset contains 200 annotated short-axis cardiac cine-MRI scans from 100 subjects. All scans are randomly divided into 140 training scans, 20 validation scans, and 40 test scans following the previous work [15]. PROMISE12 dataset contains 50 T2-weighted MRI scans which are divided into 35 training cases, 5 validation cases, and 10 test cases. All 3D scans are converted into 2D slices. Then, each slice is resized to 256 × 256 and normalized to [0, 1]. To evaluate the semi-supervised segmentation performance, we use two commonly-used evaluation metrics, the Dice Similarity Coefficient (DSC) and the Average Symmetric Surface Distance (ASSD).

Implementation Details. Our method adopts U-Net [16] as the baseline. We use the stochastic gradient descent (SGD) optimizer with an initial learning rate of 0.1, and apply the "poly" learning rate policy to update the learning rate during training. The batch size is set to 24. Each batch includes 12 labeled slices and 12 unlabeled slices. To alleviate overfitting, we employ random flipping and random rotation to augment data. All comparison experiments and ablation experiments follow the same setup for a fair comparison, we use the same experimental setup for all comparison and ablation experiments. All frameworks are implemented with PyTorch and conducted on a computer with a 3.0 GHz CPU, 128 GB RAM, and four NVIDIA GeForce RTX 3090 GPUs.

[1] https://www.creatis.insa-lyon.fr/Challenge/acdc/databases.html
[2] https://promise12.grand-challenge.org

Table 1. Comparision with other methods on the ACDC test set. DSC (%) and ASSD (mm) are reported with 28 labeled scans and 112 unlabeled scans for semi-supervised training. The bold font represents the best performance.

Method	Scans Used		RV		Myo		LV		Avg	
	Labeled	Unlabeled	DSC ↑	ASSD ↓	DSC↑	ASSD↓	DSC↑	ASSD↓	DSC↑	ASSD↓
U-Net	28 (20%)	0	82.24	2.18	80.98	2.21	86.89	1.75	83.37	1.60
U-Net	140 (100%)	0	91.48	0.47	89.22	0.54	94.64	0.55	91.78	0.52
MT [12]	28	112	87.47	0.42	86.19	1.11	90.23	2.56	87.97	1.37
UAMT [17]	28	112	87.69	0.43	85.97	0.76	90.67	2.18	88.11	1.52
CCT [18]	28	112	87.97	0.45	86.07	1.30	89.60	3.38	87.88	1.71
URPC [19]	28	112	80.55	**0.39**	84.09	1.82	88.76	3.74	84.47	1.98
SSNet [7]	28	112	87.21	0.45	86.00	1.68	90.91	1.67	88.04	0.97
MC-Net [5]	28	112	82.69	0.96	84.15	1.66	88.86	3.66	85.24	2.09
SLC-Net [15]	28	112	82.19	1.93	82.57	1.21	88.97	1.25	84.58	1.47
SCP-Net (Ours)	28	112	**89.26**	0.77	**87.11**	**0.51**	**92.70**	**0.92**	**89.69**	**0.73**

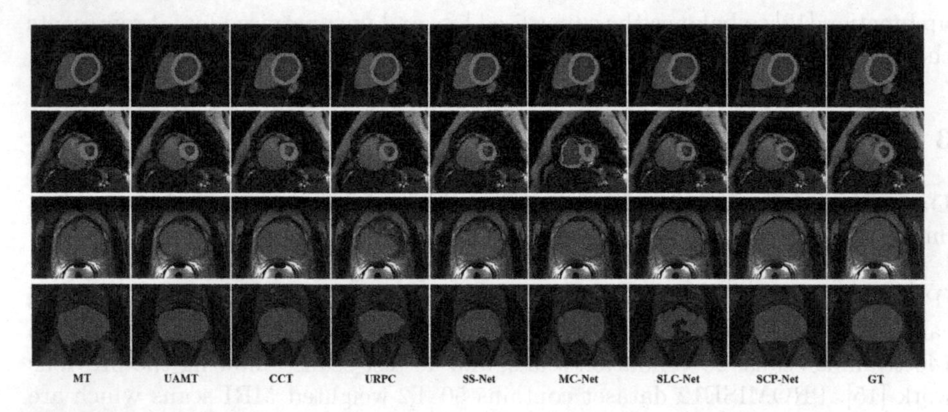

Fig. 2. Visualized segmentation results of different methods on ACDC and PROMISE12. SCP-Net better preserves anatomical morphology compared to others.

Comparison with Other Methods. To demonstrate the effectiveness of SCP-Net, we compare it with 7 state-of-the-art methods for semi-supervised segmentation and fully-supervised (100% labeled ratio) limited supervised (20% labeled ratio) baseline. The quantitative analysis results of ACDC dataset are shown in Table 1. SCP-Net significantly outperforms the limited supervised baseline by 7.02%, 6.13%, and 6.32% on DSC for RV, Myo, and LV, respectively. SCP-Net achieves comparable DSC and ASSD to the fully supervised baseline. (89.69% vs 91.78 and 0.73 vs 0.52). Compared with other methods, SCP-Net achieves the best DSC and ASSD, which is 1.58% and 0.24 higher than the second-best metric, respectively. Moreover, we visualize several segmentation examples of ACDC dataset in Fig. 2. SCP-Net yields consistent and accurate segmentation results for the RV, Myo, and LV classes according to ground truth (GT), proving that the unsupervised prototypical consistency constraints effectively extract valuable unlabeled information for segmentation performance improvement. Table 3 in supplementary material reports the quantitative result for prostate segmentation. We also perform the limited supervised and fully supervised training with 10% labeled ratio and 100% labeled ratio, respectively. SCP-Net surpasses the limited supervised baseline by 16.18% on DSC, and 10.35 on ASSD. In addition, SCP-Net gains the highest DSC of 77.06%, which is 5.63% higher than the second-best CCT. All improvements suggest that SPCC and CPCC are beneficial for exploiting unlabeled information. We also visualize some prostate segmentation examples in the last two rows of Fig. 2. We can observe that SCP-Net generates anatomically-plausible results for prostate segmentation.

Table 2. Abaliton study of the key design of SCP-Net. w means with and w/o means without.

Loss Function	Scans Used		Weight		DSC↑	ASSD ↓
	Labeled	Unlabeled	w_1	w_2		
$\mathcal{L}_{seg}$	7	0	w/o	w/o	60.88	13.87
$\mathcal{L}_{seg} + \mathcal{L}_{spcc}$	7	28	w/o	w/o	73.48	5.06
$\mathcal{L}_{seg} + \mathcal{L}_{cpcc}$	7	28	w	w	73.52	4.98
$\mathcal{L}_{seg} + \mathcal{L}_{cpcc} + \mathcal{L}_{spcc}$	7	28	w/o	w/o	74.99	4.05
$\mathcal{L}_{seg} + \mathcal{L}_{cpcc} + \mathcal{L}_{spcc}$	7	28	w	w/o	76.12	3.78
$\mathcal{L}_{seg} + \mathcal{L}_{cpcc} + \mathcal{L}_{spcc}$	7	28	w	w	77.06	3.52

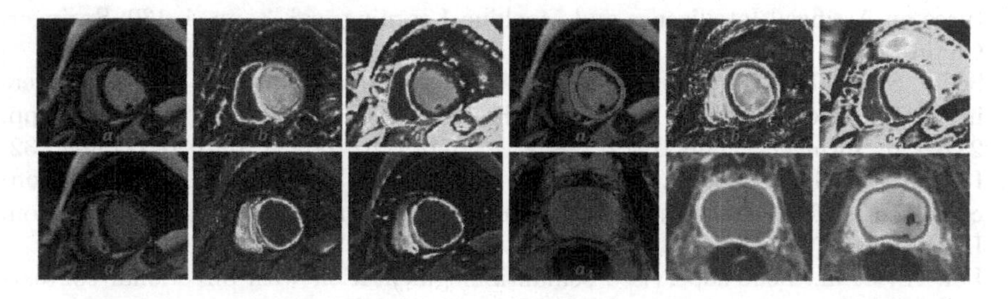

Fig. 3. Visualized results for prototypical probability predictions for RV, Myo, LV, and prostate class: (a) Ground truth, (b) Self-aware probability prediction, $\hat{p}_{kk}$, (c) Cross-sample probability prediction, $\hat{p}_{ko}$.

Ablation Study. To demonstrate the effectiveness of the key design of SCP-Net, we perform ablation study on PROMISE12 dataset by gradually adding loss components. Table 2 reports the results of ablation results. It can be observed that both the design of SPCC and CPCC promote the semi-supervised segmentation performance according to the first three rows, which demonstrates that PCC extracts valuable information from the image itself and other images, making them well-suited for semi-supervised segmentation. We also visualize the prototypical prediction $\hat{p}_{kk}$ and $\hat{p}_{ko}$ for different structures in Fig. 3. These predictions are consistent with the ground truths and show intra-class compactness and inter-class discrepancy, which validates that PCC provides effective supervision for semi-supervised segmentation. In the last three rows, the gradually improving performance verifies that the integration of prediction uncertainty w_1 and self-aware confidence w_2 in CPCC improves the reliability and stability of consistency training.

4 Conclusion

To summarize, our proposed SCP-Net, which leverages self-aware and cross-sample prototypical consistency learning, has successfully tackled the challenges of prediction diversity and training effectiveness in semi-supervised consistency learning. The intra-class compactness of pseudo label is boosted by SPCC. The dual loss re-weighting method of CPCC enhances the model's reliability. The superior segmentation performance demonstrates that SCP-Net effectively exploits the useful unlabeled information to improve segmentation performance given limited annotated data. Moving forward, our focus will be on investigating the feasibility of learning an adaptable number of prototypes that can effectively handle varying levels of category complexity. By doing so, we expect to enhance the quality of prototypical predictions and improve the overall performance.

References

1. Zhu, X., Goldberg, A.B.: Introduction to Semi-Supervised Learning. Synthesis Lectures on Artificial Intelligence and Machine Learning, vol. 3, pp. 1–130. Springer, Cham (2009)
2. Sedai, S., et al.: Uncertainty guided semi-supervised segmentation of retinal layers in OCT images. In: Shen, D., et al. (eds.) MICCAI 2019. LNCS, vol. 11764, pp. 282–290. Springer, Cham (2019). https://doi.org/10.1007/978-3-030-32239-7_32
3. Li, X., Yu, L., Chen, H., Fu, C.W., Xing, L., Heng, P.A.: Transformation-consistent self-ensembling model for semisupervised medical image segmentation. IEEE Trans. Neural Networks Learn. Syst. **32**(2), 523–534 (2020)
4. Lai, X., et al.: Semi-supervised semantic segmentation with directional context-aware consistency. In: Proceedings of the IEEE/CVF Conference on Computer Vision and Pattern Recognition, pp. 1205–1214 (2021)
5. Wu, Y., Xu, M., Ge, Z., Cai, J., Zhang, L.: Semi-supervised left atrium segmentation with mutual consistency training. In: de Bruijne, M., et al. (eds.) MICCAI 2021. LNCS, vol. 12902, pp. 297–306. Springer, Cham (2021). https://doi.org/10.1007/978-3-030-87196-3_28
6. Xu, Z., et al.: All-around real label supervision: cyclic prototype consistency learning for semi-supervised medical image segmentation. IEEE J. Biomed. Health Inform. **26**(7), 3174–3184 (2022)
7. Wu, Y., Wu, Z., Wu, Q., Ge, Z., Cai, J.: Exploring smoothness and class-separation for semi-supervised medical image segmentation. In: Wang, L., Dou, Q., Fletcher, P.T., Speidel, S., Li, S. (eds.) MICCAI 2022, Part V. LNCS, vol. 13435, pp. 34–43. Springer, Cham (2022). https://doi.org/10.1007/978-3-031-16443-9_4

8. Zhang, P., Zhang, B., Zhang, T., Chen, D., Wang, Y., Wen, F.: Prototypical pseudo label denoising and target structure learning for domain adaptive semantic segmentation. In: Proceedings of the IEEE/CVF Conference on Computer Vision and Pattern Recognition, pp. 12414–12424 (2021)
9. Zhang, Z., et al.: Model-driven self-aware self-training framework for label noise-tolerant medical image segmentation. Signal Process. **212**, 109177 (2023)
10. Zhang, Z., et al.: Dynamic prototypical feature representation learning framework for semi-supervised skin lesion segmentation. Neurocomputing **507**, 369–382 (2022)
11. Milletari, F., Navab, N., Ahmadi, S.A.: V-net: fully convolutional neural networks for volumetric medical image segmentation. In: 2016 Fourth International Conference on 3D Vision (3DV), pp. 565–571. IEEE (2016)
12. Tarvainen, A., Valpola, H.: Mean teachers are better role models: weight-averaged consistency targets improve semi-supervised deep learning results. In: Advances in Neural Information Processing Systems, vol. 30 (2017)
13. Bernard, O., et al.: Deep learning techniques for automatic MRI cardiac multi-structures segmentation and diagnosis: is the problem solved? IEEE Trans. Med. Imaging **37**(11), 2514–2525 (2018)
14. Litjens, G., et al.: Evaluation of prostate segmentation algorithms for MRI: the promise12 challenge. Med. Image Anal. **18**(2), 359–373 (2014)
15. Liu, J., Desrosiers, C., Zhou, Y.: Semi-supervised medical image segmentation using cross-model pseudo-supervision with shape awareness and local context constraints. In: Wang, L., Dou, Q., Fletcher, P.T., Speidel, S., Li, S. (eds.) MICCAI 2022, Part VIII. LNCS, vol. 13438, pp. 140–150. Springer, Cham (2022). https://doi.org/10.1007/978-3-031-16452-1_14
16. Ronneberger, O., Fischer, P., Brox, T.: U-Net: convolutional networks for biomedical image segmentation. In: Navab, N., Hornegger, J., Wells, W.M., Frangi, A.F. (eds.) MICCAI 2015, Part III. LNCS, vol. 9351, pp. 234–241. Springer, Cham (2015). https://doi.org/10.1007/978-3-319-24574-4_28
17. Yu, L., Wang, S., Li, X., Fu, C.-W., Heng, P.-A.: Uncertainty-aware self-ensembling model for semi-supervised 3D left atrium segmentation. In: Shen, D., et al. (eds.) MICCAI 2019. LNCS, vol. 11765, pp. 605–613. Springer, Cham (2019). https://doi.org/10.1007/978-3-030-32245-8_67
18. Ouali, Y., Hudelot, C., Tami, M.: Semi-supervised semantic segmentation with cross-consistency training. In: Proceedings of the IEEE/CVF Conference on Computer Vision and Pattern Recognition, pp. 12674–12684 (2020)
19. Luo, X., et al.: Efficient semi-supervised gross target volume of nasopharyngeal carcinoma segmentation via uncertainty rectified pyramid consistency. In: de Bruijne, M., et al. (eds.) MICCAI 2021. LNCS, vol. 12902, pp. 318–329. Springer, Cham (2021). https://doi.org/10.1007/978-3-030-87196-3_30

*Neuro*Explainer: Fine-Grained Attention Decoding to Uncover Cortical Development Patterns of Preterm Infants

Chenyu Xue[1], Fan Wang[2(✉)], Yuanzhuo Zhu[2], Hui Li[3], Deyu Meng[1], Dinggang Shen[4(✉)], and Chunfeng Lian[1(✉)]

[1] School of Mathematics and Statistics, Xi'an Jiaotong University, Xi'an, China
chunfeng.lian@xjtu.edu.cn

[2] Key Laboratory of Biomedical Information Engineering of Ministry of Education, School of Life Science and Technology, Xi'an Jiaotong University, Xi'an, China
fan.wang@xjtu.edu.cn

[3] Department of Neonatology, The First Affiliated Hospital of Xi'an Jiaotong University, Xi'an, China

[4] School of Biomedical Engineering, ShanghaiTech University, Shanghai, China
dgshen@shanghaitech.edu.cn

Abstract. In addition to model accuracy, current neuroimaging studies require more explainable model outputs to relate brain development, degeneration, or disorders to uncover atypical local alterations. For this purpose, existing approaches typically explicate network outputs in a post-hoc fashion. However, for neuroimaging data with high dimensional and redundant information, end-to-end learning of explanation factors can inversely assure fine-grained explainability while boosting model accuracy. Meanwhile, most methods only deal with gridded data and do not support brain cortical surface-based analysis. In this paper, we propose an *explainable geometric deep network*, the *Neuro*Explainer, with applications to uncover altered infant cortical development patterns associated with preterm birth. Given fundamental cortical attributes as network input, our *Neuro*Explainer adopts a hierarchical attention-decoding framework to learn fine-grained attention and respective discriminative representations in a spherical space to accurately recognize preterm infants from term-born infants at term-equivalent age. *Neuro*Explainer learns the hierarchical attention-decoding modules under subject-level weak supervision coupled with targeted regularizers deduced from domain knowledge regarding brain development. These prior-guided constraints implicitly maximize the explainability metrics (i.e., fidelity, sparsity, and stability) in network training, driving the learned network to output detailed explanations and accurate classifications. Experimental results on the public dHCP benchmark suggest that *Neuro*Explainer led to quantitatively reliable explanation results that are qualitatively consistent with representative neuroimaging studies. The source code will be released on https://github.com/ladderlab-xjtu/NeuroExplainer.

H. Greenspan et al. (Eds.): MICCAI 2023, LNCS 14221, pp. 202–211, 2023.
https://doi.org/10.1007/978-3-031-43895-0_19

1 Introduction

One important task for the neuroscience community is to study atypical alterations in cortices associated with brain development, degeneration, or disorders. For this aim, recent approaches, namely interpretable and explainable deep learning, rely on the training of diagnostic or predictive deep learning models [6,12] with interpretable computations and explainable results. For the aspect of preterm birth, the classification task to differentiate between preterm and term-born infants can help distinguish fine-grained differences on brain cortical surfaces, providing valuable factors for better understanding featured infantile brain development patterns related to different factors.

Although explainable deep learning methods are being actively studied in the machine learning community, they have two challenges when applying to neuroimaging data. First, existing methods typically adopt post-hoc techniques to explain a deep network [13], which is first trained for a specific classification task, and then the underlying (sparse) correlations between its input and output are analyzed offline, e.g., by backpropagating prediction gradients to the shallow layers [8]. Notably, such post-hoc approaches are established upon a common assumption that reliable explanations are the results caused by accurate predictions. This assumption could work in general applications that have large-scale training data, while cannot always hold for neuroimaging and neuroscience research, where available data are typically small-sized and much more complex (e.g., high-resolution cortical surfaces containing noisy, highly redundant, and task-irrelevant information). Second, most of these methods works on gridded data (e.g., images) [2], and does not handle 3D meshes (e.g., brain cortical surfaces) [13]. For these type of data, advanced geometric deep learning methods or mapping original meshes onto a spherical surface [14] suggested promising accuracies in multiple tasks (e.g., parcellation [14], registration [9], and longitudinal prediction [4]), yet the learned models typically lack explainability.

This paper presents an *explainable geometric deep network*, called *Neuro*Explainer, with applications to uncover altered infant cortical development patterns associated with preterm birth. *Neuro*Explainer adopts high-resolution cortical attributes as the input to develop a hierarchical attention-decoding architecture working in the sperical space. Distinct to existing post-hoc methods, the *Neuro*Explainer is constructed as an end-to-end framework, where fine-grained explanation factors can be identified in a fully learnable fashion. Our network take advantage of the explainability to boost classification for the high-dimensional neuroimaging data. Specifically, in the framework of weakly supervised discriminative localization, our *Neuro*Explainer is trained by minimizing general classification losses coupled with a set of constraints designed according to prior knowledge regarding brain development. These targeted regularizers drive the network to implicitly optimize the explainability metrics from multiple aspects (i.e., fidelity, sparsity, and stability), thus capturing fine-grained explanation factors to explicitly improve classification accuracies. Experimental results on the public dHCP benchmark suggest that our *Neuro*Explainer led to quantitatively reliable explanation results that are qualitatively consistent with

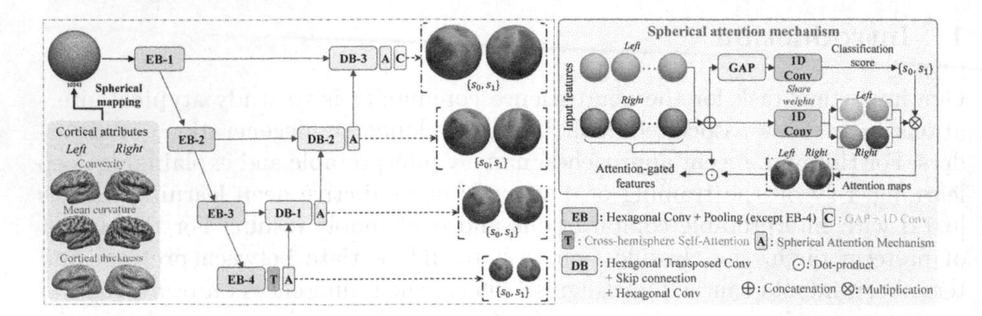

Fig. 1. The schematic diagram of our *Neuro*Explainer architecture and Spherical attention mechanism. Our *Neuro*Explainer learns to capture fine-grained by Spherical attention mechanism explanation factors to boost discriminative representation extraction.

representative neuroimaging studies, implying that it could be a practically useful AI tool for other related cortical surface-based neuroimaging studies.

2 Method

As the schematic diagram shown in Fig. 1, our *Neuro*Explainer works on the high-resolution spherical surfaces of both brain hemispheres (each with 10, 242 vertices). The inputs are fundamental vertex-wise cortical attributes, i.e., thickness, mean curvature, and convexity. The architecture has two main parts, including an encoding branch to produce initial task-related attentions on down-sampled hemispheric surfaces, and a set of attention decoding blocks to hierarchically propagate such vertex-wise attentions onto higher-resolution spheres, finally capturing fine-grained explanation factors on the input high-resolution surfaces to boost the prediction task.

2.1 Spherical Attention Encoding

The starting components of the encoding branch are four spherical convolution blocks (i.e., EB-1 to EB-4 in Fig. 1), with the learnable parameters shared across two hemispheric surfaces. Each EB adopts 1-ring hexagonal convolution [14] followed by batch normalization (BN) and ReLU activation to extract vertex-wise representations, which are then downsampled by hexagonal max pooling [14] (except in EB-4) to serve as the input of the subsequent layer. Based on the outputs from EB, we propose a learnable *spherical attention mechanism* to conduct weakly-supervised discriminative localization.

Specifically, let $\mathbf{F}^l$ and $\mathbf{F}^r \in \mathcal{R}^{162 \times M_0}$ be the vertex-wise representations (produced by EB-4) for the left and right hemispheres, respectively. We first concatenate them as a $324 \times M_0$ matrix, on which a self-attention operation [11] is applied to capturing cross-hemisphere long-range dependencies to refine the vertex-wise representations from both hemispheric surfaces, resulting in a unified feature matrix denoted as $\mathbf{F_0} = [\hat{\mathbf{F}}^l; \hat{\mathbf{F}}^r] \in \mathcal{R}^{324 \times M_0}$. As shown in Fig. 1, $\mathbf{F_0}$ is further global average pooled (GAP) across all vertices to be a holistic feature vector $f_0 \in \mathcal{R}^{1 \times M_0}$ representing the whole cerebral cortex. Both $\mathbf{F_0}$ and f_0 are then *mapped by a same vertex-wise 1D convolution* (i.e., $\mathbf{W}_0 \in \mathcal{R}^{M_0 \times 2}$, without bias) into the categorical space, denoted as $\mathbf{A}_0 = [\mathbf{A}_0^l; \mathbf{A}_0^r] \in \mathcal{R}^{324 \times 2}$ and $\mathbf{s}_0$, respectively. *Notably*, $\mathbf{s}_o$ is supervised by the one-hot code of subject's categorical label, by which $\mathbf{A}_0^l$ and $\mathbf{A}_0^r$ highlight discriminative vertices on the (down-sampled) left and right surfaces, respectively, considering that

$$\mathbf{s}_0[i] \propto \left(\mathbf{1}^T \mathbf{F_0}\right) \mathbf{W}_0[:, i] = \mathbf{1}^T \left([\hat{\mathbf{F}}^l; \hat{\mathbf{F}}^r] \mathbf{W}_0[:, i] \right) = \mathbf{1}^T \mathbf{A}_0^l[:, i] + \mathbf{1}^T \mathbf{A}_0^r[:, i], \quad (1)$$

where $\mathbf{s}_o[i]$ ($i = 0$ or 1) in our study denote the prediction scores of preterm and fullterm, respectively, and $\mathbf{1}$ is an unit vector having the same row size with the subsequent matrix. Finally, we define the hemispheric attentions as $\bar{\mathbf{A}}_0^l = \sum_{i=0}^{1} \mathbf{s}_0[i] \mathbf{A}_0^l[:, i]$ and $\bar{\mathbf{A}}_0^r = \sum_{i=0}^{1} \mathbf{s}_0[i] \mathbf{A}_0^r[:, i] \in \mathcal{R}^{324 \times 1}$, respectively, with values spatially varying and depending on the relevance to subject's category.

2.2 Hierarchically Spherical Attention Decoding

The explanation factors captured by the encoding branch are relatively coarse, as the receptive field of a cell on the downsampled surfaces (with 162 vertices after three pooling operations) is no smaller than a hexagonal region of 343 cells on the input surfaces (with 10,242 vertices). To tackle this challenge, we design a spherical attention decoding strategy to hierarchically propagate coarse attentions (from lower-resolution spheres) onto higher-resolution spheres, based on which fine-grained attentions are finally produced to improve classification.

Specifically, *Neuro*Explainer contains three consecutive decoding blocks (i.e., DB-1 to DB-3 in Fig. 1). Each DB adopts both the *attention-gated* discriminative representations from the preceding DB (except DB-1 that uses EB-4 outputs) and the local-detailed representations from the symmetric EB (at the same resolution) as the input. Let the attention-gated representations from the preceding DB be $\mathbf{F}_G^l = \left(\bar{\mathbf{A}}_{in}^l \mathbf{1}_{1 \times M_{in}}\right) \odot \hat{\mathbf{F}}_{in}^l$ and $\mathbf{F}_G^r = \left(\bar{\mathbf{A}}_{in}^r \mathbf{1}_{1 \times M_{in}}\right) \odot \hat{\mathbf{F}}_{in}^r$, respectively, where each row of $\hat{\mathbf{F}}_{in}$ has M_{in} channels, and $\odot$ denotes element-wise dot product. We first upsample $\mathbf{F}_G^l$ and $\mathbf{F}_G^r$ to the spatial resolution of the current DB, by using hexagonal transposed convolutions [14] with learnable weights shared across hemispheres. Then, the upsampled discriminative representations from each hemisphere (say $\tilde{\mathbf{F}}_G^l$ and $\tilde{\mathbf{F}}_G^r$) are channel-wisely concatenated with the local representations from the corresponding EB (say $\mathbf{F}_E^l$ and $\mathbf{F}_E^r$), followed by an 1-ring convolution to produce a unified feature matrix, such as

$$\mathbf{F}_D = [\mathcal{C}_\theta(\tilde{\mathbf{F}}_G^l \oplus \mathbf{F}_E^l); \mathcal{C}_\theta(\tilde{\mathbf{F}}_G^r \oplus \mathbf{F}_E^r)], \quad (2)$$

where $\mathcal{C}_\theta(\cdot)$ denotes 1-ring conv parameterized by θ, and $\oplus$ stands for channel concatenation. In terms of $\mathbf{F}_D$, the attention mechanism described in (1) is further applied to producing refined spherical attentions and classification scores.

Finally, as shown in Fig. 1, based on the fine-grained attentions over the input surfaces (each with 10,242 vertices), we use GAP to aggregate the attention-gated representations and apply an 1D conv to output the classification score.

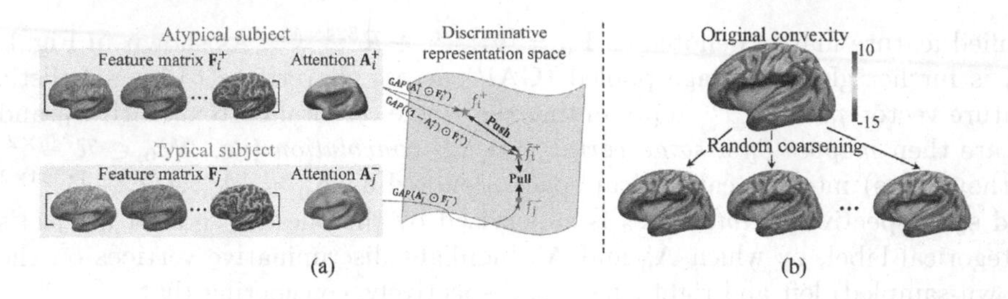

Fig. 2. Brief illustrations of (a) the explanation fidelity-aware contrastive learning strategy, and (b) explanation stability-aware data augmentation strategy.

2.3 Domain Knowledge-Guided Explanation Enhancement

To perform task-oriented learning of explanation factors, we design a set of targeted regularization strategies by considering fundamental domain knowledge regarding infant brain development. Specifically, according to existing studies, we assume that human brains in infancy have generally consistent developments, while the structural/functional discrepancies between different groups (e.g., preterm and term-born) are typically rationalized [1,10]. Accordingly, we require the preterm-altered cortical development patterns captured by our *Neuro*Explainer to be discriminative, spatially sparse, and robust, which suggests the design of the following constraints that concurrently optimize fidelity, sparsity, and stability metrics [13] in deploying an explainable deep network.

Explanation Fidelity-Aware Contrastive Learning. Given the spherical attention block at a specific resolution, we have $\mathbf{A}_i^+$ and $\mathbf{A}_j^- \in \mathcal{R}^{V\times 1}$ as the output attentions for a positive and negative subjects (i.e., preterm and fullterm infants in our study), respectively, and $\mathbf{F}_i^+$ and $\mathbf{F}_j^- \in \mathcal{R}^{V\times M}$ are the corresponding representation matrices. Based on the prior knowledge regarding infant brain development, it is reasonable to assume that $\mathbf{A}_i^+$ highlights atypically-developed cortical regions caused by preterm birth. *In contrast*, the remaining part of the cerebral cortex of a preterm infant (corresponding to $1 - \mathbf{A}_i^+$) still growths normally, i.e., looking globally similar to the cortex of a term-born infant.

Accordingly, as the illustration shown in Fig. 2(a), we design a fidelity-aware contrastive penalty to regularize the learning of the attention maps and associated representations to improve their discriminative power. Let $\boldsymbol{f}_i^+ = \mathbf{1}^T\left(\mathbf{A}_i^+\mathbf{1}_{1\times M} \odot \mathbf{F}_i^+\right)$ and $\bar{\boldsymbol{f}}_i^+ = \mathbf{1}^T\left(\{1-\mathbf{A}_i^+\}\mathbf{1}_{1\times M} \odot \mathbf{F}_i^+\right)$ be the holistic feature vector and its inverse for the ith (positive) sample, respectively. Similarly, $\boldsymbol{f}_j^- = \mathbf{1}^T\left(\mathbf{A}_j^-\mathbf{1}_{1\times M} \odot \mathbf{F}_j^-\right)$ denotes the holistic feature vector for the compared jth (negative) sample. By *pushing* $\boldsymbol{f}_i^+$ *away from both* $\bar{\boldsymbol{f}}_i^+$ *and* $\boldsymbol{f}_j^-$*, while pulling* $\bar{\boldsymbol{f}}_i^+$ *close to* $\boldsymbol{f}_j^-$, we define the respective loss as

$$\mathcal{L}_{contra} = \sum_{i\neq j}^{N} ||\bar{\boldsymbol{f}}_i^+ - \boldsymbol{f}_j^-|| + \max(m - ||\bar{\boldsymbol{f}}_i^+ - \boldsymbol{f}_i^+||, 0) + \max(m - ||\boldsymbol{f}_j^- - \boldsymbol{f}_i^+||, 0), \quad (3)$$

where i and j indicate any a pair of positive and negative cases from totally N training samples, and m is a margin setting as 1 in our implementation.

Explanation Sparsity-Aware Regularization. According to the specified prior knowledge regarding infant brain development, the attention maps produced by our *Neuro*Explainer should have two featured properties in terms of sparsity. That is, the attention map for a preterm infant (e.g., $\mathbf{A}_i^+$) should be sparse, considering that altered cortical developments are assumed to be localized. In contrast, the attention map for a healthy term-born infant (e.g., $\mathbf{A}_j^-$) should not be spatially informative, as all brain regions growth typically without abnormality. To this end, we design a straightforward entropy-based regularization to enhance results' explainability, such as

$$\mathcal{L}_{entropy} = \sum_{i \neq j}^{N} \mathbf{1}^T \left\{ \mathbf{A}_i^+ \odot \log(\mathbf{A}_i^+) - \mathbf{A}_j^- \odot \log(\mathbf{A}_j^-) \right\}, \tag{4}$$

where i and j indicate a positive and a negative cases from totally N training samples, respectively, and $\mathbf{1}$ is an unit vector to sum up the values of all vertices.

Explanation Stability-Aware Regularization. We enhance the explanation stability of our *Neuro*Explainer from two aspects. *First,* we require the spherical attention mechanisms to *robustly* decode from complex cortical-surface data fine-grained explanation factors to produce accurate predictions. To this end, we randomize the surface coarsening step by quantifying a vertex's cortical attributes (on the downsampled surface) as the average of a random subset of the vertices from the respective hexagonal region of the highest-resolution surface, such as the examples summarized in Fig. 2(b). Considering that the network is trained to produce consistently accurate predictions for all these variants with perturbations, it inversely enhances the stability of learned explanation factors.

Second, as described in Sect. 2.2, we design a cross-scale consistency regularization to refine the decoding branch. Specifically, let $\mathbf{A}_i^l$ and $\mathbf{A}_i^h$ be the spherical attentions from two different DB blocks. We simply minimize

$$\mathcal{L}_{consistent} = \sum_{i=1}^{N} \left(\mathbf{A}_i^l - \mathbf{A}_i^h\right)^2, \tag{5}$$

which encourages spherical attentions at different resolutions to be consistent.

Implementation Details. In our implementation, the feature representations produced by EB-1 to EB-4 in Fig. 1 have 32, 64, 128, and 256 channels, respectively. Correspondingly, DB-1 to DB-3, and the final classification layer have 256, 128, 64, and 32 channels, respectively. The network was trained end-to-end by minimizing the cross-entropy classification losses defined at three different spatial resolutions (overall denoted as $\mathcal{L}_{CE}$), coupled with the regularization terms introduced in Sec. 2.3, such as

$$\mathcal{L} = \mathcal{L}_{CE} + \lambda_1 \mathcal{L}_{contrast} + \lambda_2 \mathcal{L}_{entropy} + \lambda_3 \mathcal{L}_{consistent}, \tag{6}$$

where the tuning parameters were empirically set as $\lambda_1 = 0.2$, $\lambda_3 = 0.5$, and $\lambda_3 = 0.1$. The network parameters were updated by using Adam optimizer for 500 epochs, with the initial learning rate setting as 0.001 and bath size as 20.

3 Experiments

Dataset and Experimental Setup. We conducted experiments on the dHCP benchmark [5]. The structural MRIs of 700 infants scanned at term-equivalent ages (35–44 weeks postmenstrual age) were studied, including 143 preterm and 557 term-born infants. These subjects were randomly split as a training set of 500 infants (89 preterm and 411 fullterm), and a test set of the remaining 200 infants (54 preterm and 146 fullterm), where test and training sets were from different subjects. Using the data-augmentation strategy described in Sect. 2.3, the training set was augmented to have roughly 1, 250 subjects from each category for balanced network training. The input spherical surfaces contain 10, 242 vertices, and each of them has three morphological attributes, i.e., cortical thickness, mean curvature, and convexity.

Table 1. Classification results obtained by the competing geometric deep networks and different variants of our *Neuro*Explainer.

Competing Mehtods	ACC	AUC	SEN	SPE
SphericalCNN [14]	0.93	0.92	0.76	**0.98**
SphericalMoNet [9]	0.85	0.93	0.65	0.92
SubdivNet [3]	0.79	0.67	0.74	0.80
*Neuro*Explainer (**ours**)	**0.95**	**0.97**	**0.94**	0.95
w/o $\mathcal{L}_{contrast}$ (3)	0.88	0.89	0.80	0.91
w/o $\mathcal{L}_{entropy}$ (4)	0.91	0.96	0.74	0.97
w/o $\mathcal{L}_{consistent}$ (5)	0.89	0.95	0.89	0.88

Table 2. Quantitative explanation results obtained by the competing post-hoc approaches and our end-to-end *Neuro*Explainer.

Competing Methods		Fidelity	Sparsity	Stability
CAM [15] +	SphericalCNN	0.24	0.91	0.77
	SphericalMoNet	0.55	0.93	0.58
	SubdivNet	0.06	0.97	0.53
Grad-CAM [7] +	SphericalCNN	0.22	0.99	0.77
	SphericalMoNet	0.42	0.98	0.58
	SubdivNet	0.16	0.96	0.53
*Neuro*Explainer (**ours**)		**0.56**	0.73	**0.96**

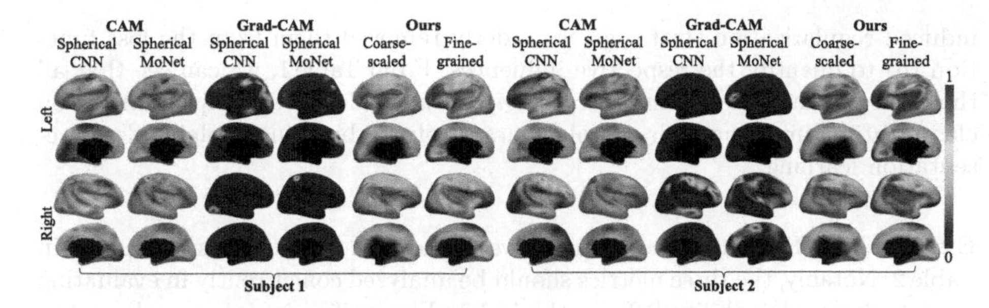

Fig. 3. Typical examples of the explanation factors captured by different methods. Higher values indicate larger links to preterm birth.

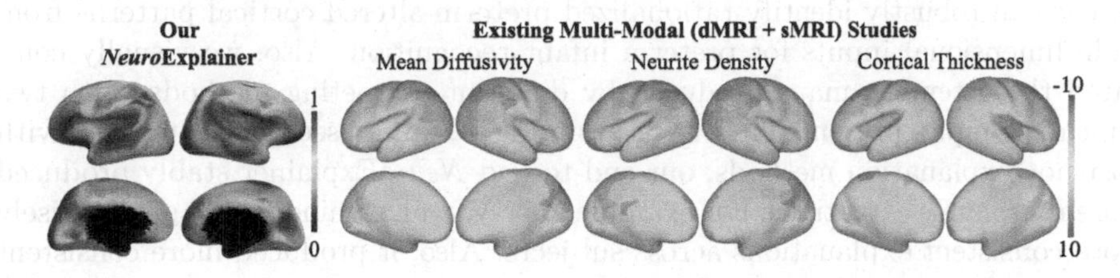

Fig. 4. Comparison of the *individualized* preterm-altered developments uncovered by *Neuro*Explainer with the *group-wise* multi-modal studies [1].

For classification, our *Neuro*Explainer was compared with three representative geometric networks, including a spherical network based on 1-ring convolution (**SphericalCNN**) [14], a MoNet reimplementation working on spherical surfaces (**SphericalMoNet**) [9], and **SubdivNet** [3] working on original meshes. The classification performance was quantified in terms of accuracy (**ACC**), area under the ROC curve (**AUC**), sensitivity (**SEN**), and specificity (**SPE**).

On the other hand, the explanation performance of our *Neuro*Explainer was compared with two representative feature-based explanation approaches, i.e., **CAM** [15] and **Grad-CAM** [7]. The explanation performance was quantitatively evaluated in terms of three metrics [13], i.e., **Fidelity**, **Sparsity**, and **Stability**. Please refer to [13] for more details regarding these metrics.

Classification Results. The classification results obtained by different competing methods are summarized in Table 1, from which we can have at least *two observations.* **1)** Our *Neuro*Explainer consistently led to better classification accuracies in terms of all metrics (especially SEN and AUC), suggesting that it can reliably identify featured development patterns associated with preterm birth to make accurate predictions in such an imbalanced learning task. **2)** These results imply that our idea to capture fine-grained explanation factors in an end-to-end fashion to boost discriminative representation extraction is beneficial for deploying an accurate classification model. **3)** To check the efficacy of the prior-induced regularization strategies, we orderly removed them from the loss function (6) to quantify the respective influences. From Table 1, we can see that all the three regularizations demonstrated *significant but different* improvements on classification, implying their complementary roles in boosting explainable representation learning.

Explanation Results. The quantitative explanation results are summarized in Table 2. Notably, the three metrics should be analyzed concurrently in evaluating a network's explainability [13], as the isolated quantification of a single metric could be biased. From Table 2, we can observe that our *Neuro*Explainer led to significantly better Fidelity and Stability, under reasonable Sparsity, suggesting that it can robustly identify rationalized preterm-altered cortical patterns from high-dimensional inputs for preterm infant recognition. Also, we visually compared the attention maps produced by different competing methods, with two typical examples presented in Fig. 3. From Fig. 3, we can see that, compared with post-hoc explanation methods, our end-to-end *Neuro*Explainer stably produced more reasonable attentions. For example, our *Neuro*Explainer led to group-wisely more consistent explanations across subjects. Also, it produced more consistent results across hemispheres, without using any related training constraints.

Finally, we compared the *individualized* preterm-altered cortical development patterns uncovered by our *Neuro*Explainer with representative *group-wise* multi-modal (dMRI and sMRI) quantitative analyses presented in [1]. As shown in Fig. 4, we can see that our observations in this paper are consistent with [1]. The discriminative cortical regions captured by our *Neuro*Explainer (using solely morphological features) are largely overlapped with the group-wise significantly different regions identified by [1] in terms of the mean diffusivity, neurite density, and cortical thickness, respectively. For example, they both highlighted some specific regions in the inferior parietal, medial occipital, and superior temporal lobe, and posterior insula, which is worth deeper evaluations in the future.

4 Conclusion

In the paper, we have proposed an geometric deep network, i.e., *Neuro*Explainer, to learn fine-grained explanation factors from complex cortical-surface data to boost discriminative representation extraction and accurate classification model construction. On the benchmark dHCP database, our *Neuro*Explainer achieved better performance than existing post-hoc approaches in terms of both explainability and prediction accuracy, in uncovering preterm-altered infant cortical development patterns. The proposed method could be a promising AI tool applied to other similar cortical surface-based neuroimage and neuroscience studies.

Funding. This work was supported in part by NSFC Grants (Nos. 62101431 & 62101430), and STI 2030-Major Projects (No. 2022ZD0209000).

References

1. Dimitrova, R., et al.: Preterm birth alters the development of cortical microstructure and morphology at term-equivalent age. Neuroimage **243**, 118488 (2021)
2. Du, M., Liu, N., Hu, X.: Techniques for interpretable machine learning. Commun. ACM **63**(1), 68–77 (2019)
3. Hu, S.M., et al.: Subdivision-based mesh convolution networks. ACM Trans. Graph. (TOG) **41**(3), 1–16 (2022)
4. Liu, P., Wu, Z., Li, G., Yap, P.-T., Shen, D.: Deep modeling of growth trajectories for longitudinal prediction of missing infant cortical surfaces. In: Chung, A.C.S., Gee, J.C., Yushkevich, P.A., Bao, S. (eds.) IPMI 2019. LNCS, vol. 11492, pp. 277–288. Springer, Cham (2019). https://doi.org/10.1007/978-3-030-20351-1_21
5. Makropoulos, A., et al.: The developing human connectome project: a minimal processing pipeline for neonatal cortical surface reconstruction. Neuroimage **173**, 88–112 (2018)
6. Ouyang, J., Zhao, Q., Adeli, E., Zaharchuk, G., Pohl, K.M.: Self-supervised learning of neighborhood embedding for longitudinal MRI. Med. Image Anal. **82**, 102571 (2022)
7. Selvaraju, R.R., Cogswell, M., Das, A., Vedantam, R., Parikh, D., Batra, D.: Grad-cam: Visual explanations from deep networks via gradient-based localization. In: Proceedings of the IEEE Conference on Computer Vision and Pattern Recognition, pp. 618–626 (2017)
8. Smilkov, D., Thorat, N., Kim, B., Viégas, F., Wattenberg, M.: Smoothgrad: removing noise by adding noise. arXiv preprint arXiv:1706.03825 (2017)
9. Suliman, M.A., Williams, L.Z., Fawaz, A., Robinson, E.C.: A deep-discrete learning framework for spherical surface registration. In: Wang, L., Dou, Q., Fletcher, P.T., Speidel, S., Li, S. (eds.) MICCAI 2022. LNCS, vol. 13436, pp. 119–129. Springer, Cham (2022). https://doi.org/10.1007/978-3-031-16446-0_12
10. Thompson, D.K., et al.: Tracking regional brain growth up to age 13 in children born term and very preterm. Nat. Commun. **11**(1), 1–11 (2020)
11. Vaswani, A., et al.: Attention is all you need. In: Advances in Neural Information Processing Systems, vol. 30 (2017)
12. Yang, Z., et al.: A deep learning framework identifies dimensional representations of Alzheimers disease from brain structure. Nat. Commun. **12**(1), 1–15 (2021)
13. Yuan, H., Yu, H., Gui, S., Ji, S.: Explainability in graph neural networks: a taxonomic survey. IEEE Trans. Pattern Anal. Mach. Intell. **45**, 5782–5799 (2022)
14. Zhao, F., et al.: Spherical U-net on cortical surfaces: methods and applications. In: Chung, A.C.S., Gee, J.C., Yushkevich, P.A., Bao, S. (eds.) IPMI 2019. LNCS, vol. 11492, pp. 855–866. Springer, Cham (2019). https://doi.org/10.1007/978-3-030-20351-1_67
15. Zhou, B., Khosla, A., Lapedriza, A., Oliva, A., Torralba, A.: Learning deep features for discriminative localization. In: Proceedings of the IEEE Conference on Computer Vision and Pattern Recognition, pp. 2921–2929 (2016)

Centroid-Aware Feature Recalibration for Cancer Grading in Pathology Images

Jaeung Lee, Keunho Byeon, and Jin Tae Kwak(✉)

School of Electrical Engineering, Korea University, Seoul, Republic of Korea
jkwak@korea.ac.kr

Abstract. Cancer grading is an essential task in pathology. The recent developments of artificial neural networks in computational pathology have shown that these methods hold great potential for improving the accuracy and quality of cancer diagnosis. However, the issues with the robustness and reliability of such methods have not been fully resolved yet. Herein, we propose a centroid-aware feature recalibration network that can conduct cancer grading in an accurate and robust manner. The proposed network maps an input pathology image into an embedding space and adjusts it by using centroids embedding vectors of different cancer grades via attention mechanism. Equipped with the recalibrated embedding vector, the proposed network classifiers the input pathology image into a pertinent class label, i.e., cancer grade. We evaluate the proposed network using colorectal cancer datasets that were collected under different environments. The experimental results confirm that the proposed network is able to conduct cancer grading in pathology images with high accuracy regardless of the environmental changes in the datasets.

Keywords: cancer grading · attention · feature calibration · pathology

1 Introduction

Globally, cancer is a leading cause of death and the burden of cancer incidence and mortality is rapidly growing [1]. In cancer diagnosis, treatment, and management, pathology-driven information plays a pivotal role. Cancer grade is, in particular, one of the major factors that determine the treatment options and life expectancy. However, the current pathology workflow is sub-optimal and low-throughput since it is, by and large, manually conducted, and the large volume of workloads can result in dysfunction or errors in cancer grading, which have an adversarial effect on patient care and safety [2]. Therefore, there is a high demand to automate and expedite the current pathology workflow and to improve the overall accuracy and robustness of cancer grading.

Recently, many computational tools have shown to be effective in analyzing pathology images [3]. These are mainly built based upon deep convolutional neural networks (DCNNs). For instance, [4] used DCCNs for prostate cancer detection and grading, [5] classified gliomas into three different cancer grades, and [6] utilized an ensemble of

H. Greenspan et al. (Eds.): MICCAI 2023, LNCS 14221, pp. 212–221, 2023.
https://doi.org/10.1007/978-3-031-43895-0_20

DCNNs for breast cancer classification. To further improve the efficiency and effectiveness of DCNNs in pathology image analysis, advanced methods that are tailored to pathology images have been proposed. For example, [7] proposed to incorporate both local and global contexts through the aggregation learning of multiple context blocks for colorectal cancer classification; [8] extracted and utilized multi-scale patterns for cancer grading in prostate and colorectal tissues; [9] proposed to re-formulate cancer classification in pathology images as both categorical and ordinal classification problems. Built based upon a shared feature extractor, a categorical classification branch, and an ordinal classification branch, it simultaneously conducts both categorical and ordinal learning for colorectal and prostate cancer grading; a hybrid method that combines DCCNs with hand-crafted features was developed for mitosis detection in breast cancer [10]. Moreover, attention mechanisms have been utilized for an improved pathology image analysis. For instance, [11] proposed a two-step framework for glioma sub-type classification in the brain, which consists of a contrastive learning framework for robust feature extractor training and a sparse-attention block for meaningful multiple instance feature aggregation. Such attention mechanisms have been usually utilized in a multiple instance learning framework or as self-attention for feature representations. To the best of our knowledge, attention mechanisms have not been used for feature representations of class centroids.

In this study, we propose a centroid-aware feature recalibration network (*CaFeNet*) for accurate and robust cancer grading in pathology images. *CaFeNet* is built based upon three major components: 1) a feature extractor, 2) a centroid update (*Cup*) module, and 3) a centroid-aware feature recalibration (*CaFe*) module. The feature extractor is utilized to obtain the feature representation of pathology images. *Cup* module obtains and updates the centroids of class labels, i.e., cancer grades. *CaFe* module adjusts the input embedding vectors with respect to the class centroids (i.e., training data distribution). Assuming that the classes are well separated in the feature space, the centroid embedding vectors can serve as reference points to represent the data distribution of the training data. This indicates that the centroid embedding vectors can be used to recalibrate the input embedding vectors of pathology images. During inference, we fix the centroid embedding vectors so that the recalibrated embedding vectors do not vary much compared to the input embedding vectors even though the data distribution substantially changes, leading to improved stability and robustness of the feature representation. In this manner, the feature representations of the input pathology images are re-calibrated and stabilized for a reliable cancer classification. The experimental results demonstrate that *CaFeNet* achieves the state-of-the-art cancer grading performance in colorectal cancer grading datasets. The source code of *CaFeNet* is available at https://github.com/colin19950703/CaFeNet.

2 Methodology

The overview of the proposed *CaFeNet* is illustrated in Fig. 1. *CaFeNet* employs a deep convolutional neural network as a feature extractor and an attention mechanism to produce robust feature representations of pathology images and conducts cancer grading with high accuracy. Algorithm 1 depicts the detailed algorithm of *CaFeNet*.

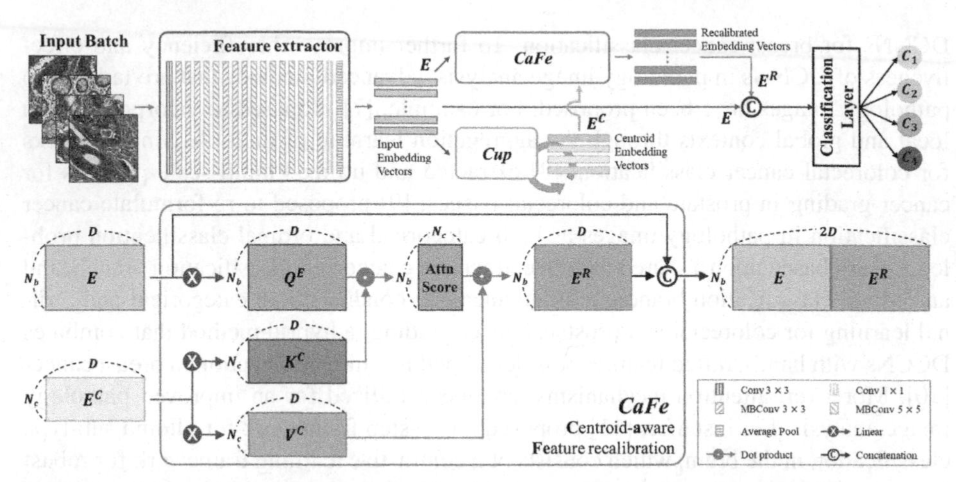

Fig. 1. Overview of *CaFeNet*. *CaFeNet* consists of a feature extractor, a *CaFe* module, a *Cup* module, and a classification layer.

2.1 Centroid-Aware Feature Recalibration

Let $\{x_i, y_i\}_{i=1}^{N}$ be a set of pairs of pathology images and ground truth labels where N is the number of pathology image-ground truth label pairs, $x_i \in \mathbb{R}^{h \times w \times c}$ is the i th pathology image, $y_i \in \{C_1, \ldots, C_M\}$ represents the corresponding ground truth label. h, w, and c denote the height, width, and the number of channels, respectively. M is the cardinality of the class labels. Given x_i, a deep neural network f maps x_i into an embedding space, producing an embedding vector $e_i \in \mathbb{R}^d$. The embedding vector e_i is fed into 1) a centroid update (*Cup*) module and 2) a centroid-aware feature recalibration (*CaFe*) module. *Cup* module obtains and updates the centroid of the class label in the embedding space $E^{\mathcal{C}} \in \mathbb{R}^{M \times D}$. *CaFe* module adjusts the embedding vector e_i in regard to the embedding vectors of the class centroids and produces a recalibrated embedding vector $e_i^{\mathcal{R}}$. e_i and $e_i^{\mathcal{R}}$ are concatenated together and is fed into a classification layer to conduct cancer grading.

Given a batch of input embedding vectors $E = \{e_i | i = 0, \ldots, N-1\}$, *Cup* module computes and updates the centroid embedding vector of each class label per epoch. Specifically, *Cup* module adds up the embedding vectors of different class labels over the iterations per epoch, computes the average embedding vectors, and updates the centroid embedding vectors $E^{\mathcal{C}} = \left\{e_j^{\mathcal{C}} | j = 0, \ldots, M-1\right\}$.

CaFe module receives a batch of embedding vectors $E = \{e_i | i = 0, \ldots, N-1\}$ and the ground truth labels $Y = \{y_i | i = 0, \ldots, N-1\}$ and a set of centroid embedding vectors $E^{\mathcal{C}} = \left\{e_j^{\mathcal{C}} | j = 0, \ldots, M-1\right\}$ and outputs a batch of recalibrated embedding vectors $E^{\mathcal{R}} = \left\{e_i^{\mathcal{R}} | i = 0, \ldots, N-1\right\}$ via an attention mechanism (Fig. 1). It first produces queries $Q^E \in \mathbb{R}^{N \times D}$ from E and keys $K^{\mathcal{C}} \in \mathbb{R}^{M \times D}$ and values $V^{\mathcal{C}} \in \mathbb{R}^{M \times D}$ from the centroid embedding vectors $E^{\mathcal{C}}$ by using a linear layer. Then, attention scores are computed via a dot product between Q^E and $K^{\mathcal{C}}$ followed by a softmax operation. Multiplying the attention scores by $V^{\mathcal{C}}$, we obtain the recalibrated feature representation

$E^{\mathcal{R}}$ for the input embedding vectors E. The process can be formulated as follows:

$$E^{\mathcal{R}} = softmax\left(Q^E K^{CT}\right)V^C. \tag{1}$$

Finally, *CaFe* concatenates E and $E^{\mathcal{R}}$ and produces them as the output.

Algorithm 1 *CaFeNet*

Inputs: A set of images with ground truth labels $X = \{X^b | b = 1, \dots, B\}$, $Y = \{Y^b | b = 1, \dots, B\}$ and current centroid embedding vectors $E^C = \{e_j^C | j = 1, \dots, N_c\}$

Outputs: Predictions $\hat{Y} = \{\hat{Y}^b | b = 1, \dots, B\}$

1: Initialize next centroids vectors $\tilde{E}^C$ for $j = 1, \dots, N_c$:
2: $\quad \tilde{e}_j^C \leftarrow 0 \in \mathbb{R}^{1 \times D}$, $S_j \leftarrow 0$
3: **for** $b = 1, \dots, B$ **do**
4: $\quad$ Take a batch of images and ground truth labels:
5: $\quad\quad X^b \leftarrow \{x_i | i = 1, \dots, N_b\}$, $x_i \in \mathbb{R}^{h \times w \times c}$
6: $\quad\quad Y^b \leftarrow \{y_i | i = 1, \dots, N_b\}$, $y_i \in \mathbb{N}$
7: $\quad$ // *CaFe* module computes recalibrated embedding vectors
8: $\quad$ Compute embedding vectors $E = \{e_i | i = 1, \dots, N_b\}$:
9: $\quad\quad E \leftarrow f(X^b) \in \mathbb{R}^{N_b \times D}$ $\quad\quad \triangleright f$ is a feature extractor
10: $\quad$ Compute query vectors from E: $Q^E \leftarrow Linear(E) \in \mathbb{R}^{N_b \times D}$
11: $\quad$ Compute key and value vectors from $E^C = \{e_j^C | j = 1, \dots, N_c\}$:
12: $\quad\quad K^C \leftarrow Linear(E^C) \in \mathbb{R}^{N_c \times D}$, $V^C \leftarrow Linear(E^C) \in \mathbb{R}^{N_c \times D}$
13: $\quad$ Compute attention scores: $AttnScore \leftarrow Q^E K^{C^T}$
14: $\quad$ Compute recalibrated vectors $E^{\mathcal{R}} = \{e_j^{\mathcal{R}} | j = 1, \dots, N_b\}$:
15: $\quad\quad E^{\mathcal{R}} \leftarrow softmax(AttnScore)V^C \in \mathbb{R}^{N_b \times D}$
16: $\quad$ Concatenate two embedding vectors: $E' \leftarrow CONCAT[E, E^{\mathcal{R}}]$
17: $\quad$ Conduct predictions: $\hat{Y}^b \leftarrow Linear(E')$
18: $\quad$ // *Cup* module adds input embedding vectors to next centroids vectors
19: $\quad$ **for** $i = 1, \dots, N_b$ **do**
20: $\quad\quad$ **for** $j = 1, \dots, N_c$ **do**
21: $\quad\quad\quad$ **if** $y_i = j$ **then**
22: $\quad\quad\quad\quad$ Add up centroids vectors: $\tilde{e}_j^C \leftarrow \tilde{e}_j^C + e_i$
23: $\quad\quad\quad\quad S_j \leftarrow S_j + 1$
24: $\quad\quad\quad$ **end if**
25: $\quad$ **end for**
26: **end for**
27: // *Cup* module updates centroids embedding vectors
28: **for** $j = 1, \dots, N_c$ **do**
29: $\quad$ Compute average centroid vectors: $\tilde{e}_j^C \leftarrow \tilde{e}_j^C / S_j$
30: **end for**
31: Update centroid vectors: $E^C \leftarrow \tilde{E}^C$

2.2 Network Architecture

We employ EfficientNet-B0 [12] as a backbone network. EfficientNet is designed to achieve the state-of-the-art accuracy on computer vision tasks while minimizing computational costs through a compound scaling method. EfficientNet-B0 is composed of one convolution layer and 16 stages of mobile inverted bottleneck blocks, of which each with a different number of layers and channels. Each mobile inverted bottleneck block comprises one pointwise convolution (1×1 convolution for the channel expansion), one depth-wise separable convolution with a kernel size of 3 or 5, and one project pointwise convolution (1×1 convolution for the channel reduction).

Table 1. Details of colorectal cancer datasets

Class	C_{Train}	$C_{Validation}$	C_{TestI}	C_{TestII}
Benign	773	374	453	27896
WD	1866	264	192	8394
MD	2997	370	738	61985
PD	1391	234	205	11895

3 Experiments and Results

3.1 Datasets

Two publicly available colorectal cancer datasets [9] were employed to evaluate the effectiveness of the proposed *CaFeNet*. Table 1 shows the details of the datasets. Both datasets provide colorectal pathology images with ground truth labels for cancer grading. The ground labels are benign (BN), well-differentiated (WD) cancer, moderately-differentiated (MD) cancer, and poorly-differentiated (PD) cancer. The first dataset includes 1600 BN, 2322 WD, 4105 MD, and 1830 PD image patches that were collected between 2006 and 2008 using an Aperio digital slide scanner (Leica Biosystems) at 40x magnification. Each image patch has a spatial size of 1024×1024 pixels. This dataset is divided into a training dataset (C_{Train}), validation dataset ($C_{Validation}$), and a test dataset (C_{TestI}). The second dataset, designated as C_{TestII}, contains 27986 BN, 8394 WD, 61985 MD, and 11985 PD image patches of size 1144×1144 pixels. These were acquired between 2016 and 2017 using a NanoZoomer digital slide scanner (Hamamatsu Photonics K.K).

3.2 Comparative Experiments

We conducted a series of comparative experiments to evaluate the effectiveness of *CaFeNet* for cancer grading, in comparison to several existing methods: 1) three DCNN-based models: ResNet [13], DenseNet [14], EfficientNet [12], 2) two metric learning-based models: triplet loss (Triplet) [15] and supervised contrastive loss (SC) [16], 3) two transformer-based models: vision transformer (ViT) [17] and swin transformer (Swin) [18], and 4) one (pathology) domain-specific model ($\mathcal{M}_{MAE-CE_o}$) [9], which demonstrates the state-of-the-art performance on the two colorectal cancer datasets under consideration. For Triplet and SC, EfficientNet was used as a backbone network. We trained *CaFeNet* and other competing networks on C_{Train} and selected the best model using$C_{Validation}$. Then, the chosen model of each network was separately applied to C_{TestI} andC_{TestII}. The results of $\mathcal{M}_{MAE-CE_o}$ were obtained from the original literature.

3.3 Implementation Details

We initialized all models using the pre-trained weights on the ImageNet dataset, and then trained them using the Adam optimizer with default parameter values (β_1= 0.9, β_2 = 0.999, ε = 1.0e-8) for 50 epochs. We employed *cosine anneal warm restart schedule* with initial learning rates of 1.0 e^{-3}, η_{min}= 1.0 e^{-3}, and T_0 = 20. After data augmentation, all patches, except for those used in ViT [17] and Swin [18] models, were resized to 512 × 512 pixels. For ViT and Swin, the patches were resized to 384 × 384 pixels. We implemented all models using the PyTorch platform and trained on a workstation equipped with two RTX 3090 GPUs. To increase the variability of the dataset during the training phase, we applied several data augmentation techniques, including affine transformation, random horizontal and vertical flip, image blurring, random Gaussian noise, dropout, random color saturation and contrast conversion, and random contrast transformations. All these techniques were implemented using the Aleju library (https://github.com/aleju/imgaug).

Table 2. Result of colorectal cancer grading on C_{TestI}.

Model	Acc (%)	Precision	Recall	F1	κ_w
ResNet [13]	87.1	0.834	0.843	0.838	0.938
DenseNet [14]	86.2	0.823	0.839	0.829	0.929
EfficientNet[12]	82.2	0.794	0.811	0.802	0.873
Triplet [15]	86.6	0.832	0.824	0.827	0.937
SC [16]	85.0	0.817	0.812	0.812	0.920
ViT [17]	85.8	0.818	0.813	0.815	0.934
Swin [18]	87.4	0.847	0.820	0.832	0.941
$\mathcal{M}_{MAE-CE_o}$[9]	87.7	–	–	0.843	0.940
CaFeNet (Ours)	87.5	0.853	0.816	0.832	0.940

3.4 Result and Discussions

We evaluated the performance of colorectal cancer grading by the proposed *CaFeNet* and other competing models using five evaluation metrics, including accuracy (Acc), precision, recall, F1-score (F1), and quadratic weighted kappa (κ_w). Table 2 demonstrates the quantitative experimental results on C_{TestI}. The results show that *CaFeNet* was one of the best performing models along with ResNet, Swin, and $\mathcal{M}_{MAE-CE_o}$. Were the best performing models. Among DCNN-based models, ResNet was superior to other DCNN-based models. Metric learning was able to improve the classification performance. EffcientNet was the worst model among them, but with the help of triplet loss (Triplet) or supervised contrastive loss (SC), the overall performance increased by ≥2.8% Acc, ≥0.023 precision, ≥0.001 recall, ≥0.010 F1, and ≥0.047 κ_w. Among the transformer-based models, Swin was one of the best performing models, but ViT showed much lower performance in all evaluation metrics.

Table 3. Result of colorectal cancer grading on C_{TestII}.

Model	Acc (%)	Precision	Recall	F1	κ_w
ResNet [13]	77.2	0.691	0.800	0.713	0.869
DenseNet [14]	78.8	0.698	0.792	0.722	0.866
EfficientNet [12]	79.3	0.701	0.802	0.727	0.870
Triplet [15]	79.1	0.702	0.815	0.730	0.886
SC [16]	79.7	0.718	0.809	0.739	0.876
ViT [17]	80.7	0.706	0.797	0.733	0.889
Swin [18]	78.6	0.690	0.785	0.712	0.873
$\mathcal{M}_{MAE-CE_o}$[9]	80.3	–	–	0.744	0.891
CaFeNet (Ours)	82.7	0.728	0.810	0.756	0.901

Moreover, we applied the same models to C_{TestII} to test the generalizability of the models. We note that C_{TestI} originated from the same set with C_{Train} and $C_{Validation}$ and C_{TestII} was obtained from different time periods and using a different slide scanner. Table 3 depicts the quantitative classification results on C_{TestII}. *CaFeNet* outperformed other competing models in all evaluation metrics except Triplet for recall. In a head-to-head comparison of the classification results between C_{TestI} and C_{TestII}, there was a consistent performance drop in the proposed *CaFeNet* and other competing models. This is ascribable to the difference between the test datasets (C_{TestI} and C_{TestII}) and the training and validation datasets (C_{Train} and $C_{Validation}$). In regard to such differences, it is striking that the proposed *CaFeNet* achieved the best performance on C_{TestII}. *CaFeNet*, ResNet, Swin, and $\mathcal{M}_{MAE-CE_o}$ were the four best performing models on C_{TestI}. However, ResNet, Swin, and $\mathcal{M}_{MAE-CE_o}$ showed a higher performance drop in all evaluation metrics. *CaFeNet* had a minimal performance drop except EfficientNet. EfficientNet, however, obtained poorer performance on both C_{TestI} and C_{TestII}. These results suggest that *CaFeNet* has the better generalizability so as to well adapt to unseen histopathology image data.

We conducted ablation experiments to investigate the effect of the *CaFe* module on cancer classification. The results are presented in Table 4. The exclusion of the *CaFe* module, i.e., EfficientNet, resulted in much worse performance than *CaFeNet*. Using only the recalibrated embedding vectors $E^{\mathcal{R}}$, a substantial drop in performance was observed. These two results indicate that the recalibrated embedding vectors complement to the input embedding vectors E. Moreover, we examined the effect of the method that merges the two embedding vectors. Using addition, instead of concatenation, there was a consistent performance drop, indicating that concatenation is the superior approach for combining the two embedding vectors together.

Table 4. Ablation study on *CaFeNet*.

Data	Model	Acc (%)	Precision	Recall	F1	κ_w
C_{Test}I	Backbone (EfficientNet)	82.2	0.794	0.811	0.802	0.873
	$E^{\mathcal{R}}$ only	77.8	0.572	0.634	0.600	0.835
	$ADD\left[E, E^{\mathcal{R}}\right]$	82.9	0.781	0.803	0.788	0.846
	$CONCAT\left[E, E^{\mathcal{R}}\right]$(Ours)	87.5	0.853	0.816	0.832	0.940
C_{Test}II	Backbone (EfficientNet)	79.3	0.701	0.802	0.727	0.870
	$E^{\mathcal{R}}$ only	56.2	0.399	0.428	0.296	−0.114
	$ADD\left[E, E^{\mathcal{R}}\right]$	75.2	0.674	0.775	0.688	0.789
	$CONCAT\left[E, E^{\mathcal{R}}\right]$(Ours)	82.7	0.728	0.810	0.756	0.901

Table 5. Model complexity of *CaFeNet* and competing models.

Model	# Params (M)	# FLOPs (M)	Training (ms/batch)	Inference (ms/batch)
ResNet [13]	23.5	21,586.0	826.1	370.9
DenseNet [14]	6.7	14,802.9	1209.9	446.2
EfficientNet [12]	4.0	141.1	609.8	306.4
Triplet [15]	4.0	141.1	623.8	315.2
SC [16]	4.0	141.1	1149.5	153.2
ViT [17]	86.2	49,391.9	682.4	228.8
Swin [18]	87.3	44,659.6	1416.14	241.8
$\mathcal{M}_{MAE-CE_o}$[9]	4.0	141.1	2344.0	519.1
CaFeNet (Ours)	8.9	155.9	622.6	302.9

In addition, we compared the model complexity of the proposed *CaFeNet* and other competing models. Table 5 demonstrates the number of parameters, floating point operations per second (FLOPs), and training and inference time (in milliseconds). The proposed *CaFeNet* was one of the models that require a relatively small number of parameters and FLOPs and a short amount of time during training and inference. DenseNet, EfficientNet, Triplet, SC, and $\mathcal{M}_{MAE-CE_o}$ contain the smaller number of parameters than that of *CaFeNet*, but these models show either the higher number of FLOPs or longer time during training and/or inference. Similar observations were made for ResNet, ViT, and Swin. These models require much larger number of parameters and FLOPs and longer training time. These results confirm that the proposed *CaFeNet* is computational efficient and it does not achieve its superior learning capability and generalizability at the expense of the model complexity.

4 Conclusions

Herein, we propose an attention mechanism-based deep neural network, called *CaFeNet*, for cancer classification in pathology images. The proposed approach proposes to improve the feature representation of deep neural networks by re-calibrating input embedding vectors via an attention mechanism in regard to the centroids of cancer grades. In the experiments on colorectal cancer datasets against several competing models, the proposed network demonstrated that it has a better learning capability as well as a generalizability in classifying pathology images into different cancer grades. However, the experiments were only conducted on two public colorectal cancer datasets from a single institute. Additional experiments need to be conducted to further verify the findings of our study. Therefore, future work will focus on validating the effectiveness of the proposed network for other types of cancers and tissues in pathology images.

Acknowledgements. This work was supported by the National Research Foundation of Korea (NRF) grant funded by the Korea government (MSIT) (No. 2021R1A2C2014557 and No. 2021R1A4A1031864).

References

1. Sung, H., et al.: Global cancer statistics 2020: GLOBOCAN estimates of incidence and mortality worldwide for 36 cancers in 185 countries. CA: Can. J. Clin. **71**, 209–249 (2021)
2. Maung, R.: Pathologists' workload and patient safety. Diagn. Histopathol. **22**, 283–287 (2016)
3. Srinidhi, C.L., Ciga, O., Martel, A.L.: Deep neural network models for computational histopathology: a survey. Med. Image Anal. **67**, 101813 (2021)
4. Arvaniti, E., et al.: Automated Gleason grading of prostate cancer tissue microarrays via deep learning. Sci. Rep. **8**, 12054 (2018)
5. Ertosun, M.G., Rubin, D.L.: Automated grading of gliomas using deep learning in digital pathology images: a modular approach with ensemble of convolutional neural networks. In: AMIA Annual Symposium Proceedings, p. 1899. American Medical Informatics Association (2015)

6. Hameed, Z., Zahia, S., Garcia-Zapirain, B., Javier Aguirre, J., Maria Vanegas, A.: Breast cancer histopathology image classification using an ensemble of deep learning models. Sensors **20**, 4373 (2020)
7. Shaban, M., et al.: Context-aware convolutional neural network for grading of colorectal cancer histology images. IEEE Trans. Med. Imaging **39**, 2395–2405 (2020)
8. Vuong, T.T., Song, B., Kim, K., Cho, Y.M., Kwak, J.T.: Multi-scale binary pattern encoding network for cancer classification in pathology images. IEEE J. Biomed. Health Inform. **26**, 1152–1163 (2021)
9. Le Vuong, T.T., Kim, K., Song, B., Kwak, J.T.: Joint categorical and ordinal learning for cancer grading in pathology images. Med. Image Anal. **73**, 102206 (2021)
10. Wang, H., et al.: Mitosis detection in breast cancer pathology images by combining handcrafted and convolutional neural network features. J. Med. Imaging **1**, 034003 (2014)
11. Lu, M., et al.: Smile: sparse-attention based multiple instance contrastive learning for glioma sub-type classification using pathological images. In: MICCAI Workshop on Computational Pathology, pp.159–169. PMLR (2021)
12. Tan, M., Le, Q.: Efficientnet: rethinking model scaling for convolutional neural networks. In: International Conference on Machine Learning, pp. 6105–6114. PMLR (2019)
13. He, K., Zhang, X., Ren, S., Sun, J.: Deep residual learning for image recognition. In: Proceedings of the IEEE conference on computer vision and pattern recognition, pp. 770–778 (2016)
14. Huang, G., Liu, Z., Van Der Maaten, L., Weinberger, K.Q.: Densely connected convolutional networks. In: Proceedings of the IEEE Conference on Computer Vision and Pattern Recognition, pp. 4700–4708 (2017)
15. Schroff, F., Kalenichenko, D., Philbin, J.: FaceNet: a unified embedding for face recognition and clustering. In: Proceedings of the IEEE Conference on Computer Vision and Pattern Recognition, pp. 815–823 (2015)
16. Khosla, P., et al.: Supervised contrastive learning. Adv. Neural. Inf. Process. Syst. **33**, 18661–18673 (2020)
17. Dosovitskiy, A., et al.: An image is worth 16x16 words: transformers for image recognition at scale. arXiv preprint arXiv:2010.11929 (2020)
18. Liu, Z., et al.: Swin transformer: hierarchical vision transformer using shifted windows. In: Proceedings of the IEEE/CVF International Conference on Computer Vision, pp. 10012–10022 (2021)
19. McInnes, L., Healy, J., Melville, J.: Umap: uniform manifold approximation and projection for dimension reduction. arXiv preprint arXiv:1802.03426 (2018)

Federated Uncertainty-Aware Aggregation for Fundus Diabetic Retinopathy Staging

Meng Wang[1], Lianyu Wang[2], Xinxing Xu[1], Ke Zou[3], Yiming Qian[1], Rick Siow Mong Goh[1], Yong Liu[1], and Huazhu Fu[1(✉)]

[1] Institute of High Performance Computing (IHPC), Agency for Science, Technology and Research (A*STAR), 1 Fusionopolis Way, #16-16 Connexis, Singapore 138632, Republic of Singapore
hzfu@ieee.org

[2] College of Computer Science and Technology, Nanjing University of Aeronautics and Astronautics, Nanjing 211100, Jiangsu, China

[3] National Key Laboratory of Fundamental Science on Synthetic Vision and the College of Computer Science, Sichuan University, Chengdu 610065, Sichuan, China

Abstract. Deep learning models have shown promising performance in the field of diabetic retinopathy (DR) staging. However, collaboratively training a DR staging model across multiple institutions remains a challenge due to non-iid data, client reliability, and confidence evaluation of the prediction. To address these issues, we propose a novel federated uncertainty-aware aggregation paradigm (FedUAA), which considers the reliability of each client and produces a confidence estimation for the DR staging. In our FedUAA, an aggregated encoder is shared by all clients for learning a global representation of fundus images, while a novel temperature-warmed uncertainty head (TWEU) is utilized for each client for local personalized staging criteria. Our TWEU employs an evidential deep layer to produce the uncertainty score with the DR staging results for client reliability evaluation. Furthermore, we developed a novel uncertainty-aware weighting module (UAW) to dynamically adjust the weights of model aggregation based on the uncertainty score distribution of each client. In our experiments, we collect five publicly available datasets from different institutions to conduct a dataset for federated DR staging to satisfy the real non-iid condition. The experimental results demonstrate that our FedUAA achieves better DR staging performance with higher reliability compared to other federated learning methods. Our proposed FedUAA paradigm effectively addresses the challenges of collaboratively training DR staging models across multiple institutions, and provides a robust and reliable solution for the deployment of DR diagnosis models in real-world clinical scenarios.

Keywords: Federated learning · Uncertainty estimation · DR staging

M. Wang and L. Wang contributed equally.

Supplementary Information The online version contains supplementary material available at https://doi.org/10.1007/978-3-031-43895-0_21.

H. Greenspan et al. (Eds.): MICCAI 2023, LNCS 14221, pp. 222–232, 2023.
https://doi.org/10.1007/978-3-031-43895-0_21

1 Introduction

In the past decade, numerous deep learning-based methods for DR staging have been explored and achieved promising results [10,11,19,27]. However, most current studies focus on centralized learning, which necessitates data collection from multiple institutions to a central server for model training. This approach poses significant data privacy security risks. Additionally, in clinical practice, different institutions may have their own DR staging criteria [3]. Consequently, it is difficult for the previous centralized DR staging method to utilize data of varying DR staging criteria to train a unified model.

Federated learning (FL) is a collaborative learning framework that enables training a model without sharing data between institutions, thereby ensuring data privacy [15,21]. In the FL paradigm, FedAvg [24] and its variants [1,4,9,16,18,22,23] are widely used and have achieved excellent performance in various medical tasks. However, these FL methods assign each client a static weight for model aggregation, which may lead to the global model not learning sufficient knowledge from clients with large heterogeneous features and ignoring the reliability of each client. In clinical practice, the data distributions of DR datasets between institutions often vary significantly due to medical resource constraints, population distributions, collection devices, and morbidity [25,29]. This variation poses great challenges for the exploration of federated DR staging methods. Moreover, most existing DR staging methods and FL paradigms mainly focus on performance improvement and ignore the exploration of the confidence of the prediction. Therefore, it is essential to develop a new FL paradigm that can provide reliable DR staging results while maintaining higher performance. Such a paradigm would reduce data privacy risks and increase user confidence in AI-based DR staging systems deployed in real-world clinical settings.

To address the issues, we propose **a novel FL paradigm, named FedUAA**, that employs a personalized structure to handle collaborative DR staging among multiple institutions with varying DR staging criteria. We utilize uncertainty to evaluate the reliability of each client's contribution. While uncertainty is a proposed measure to evaluate the reliability of model predictions [12,14,28,30], it remains an open topic in FL research. In our work, we introduce **a temperature-warmed evidential uncertainty (TWEU)** head to enable the model to generate a final result with uncertainty evaluation without sacrificing performance. Additionally, based on client uncertainty, we developed **an uncertainty-aware weighting module (UAW)** to dynamically aggregate models according to each client's uncertainty score distribution. This can improve collaborative DR staging across multiple institutions, particularly for clients with large data heterogeneity. Finally, we construct a **dataset for federated DR staging** based on five public datasets with different staging criteria from various institutions to satisfy the real non-iid condition. The comprehensive experiments demonstrate that FedUAA provides outstanding DR staging performance with a high degree of reliability, outperforming other state-of-the-art FL approaches.

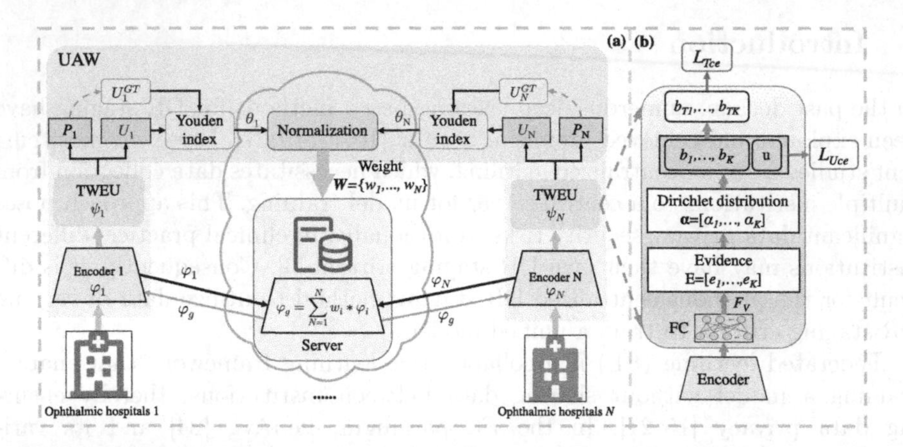

Fig. 1. The overview of FedUAA (a) with TWEU module (b). An aggregated encoder is shared by all clients for learning a global representation of fundus images, while a novel TWEU head is kept on the local client for local personalized staging criteria. Furthermore, a novel UAW module is developed to dynamically adjust the weights for model aggregation based on the reliability of each client.

2 Methodology

Figure 1 (a) shows the overview of our proposed FedUAA. During training, local clients share the encoder (φ) to the cloud server for model aggregation, while the TWEU (ψ) head is retained locally to generate DR staging results with uncertainty evaluation based on features from the encoder to satisfy local-specific DR staging criteria. The algorithm of our proposed FedUAA is detailed in **Supplementary A**. Therefore, the target of our FedUAA is:

$$\min_{\varphi\in\Phi,\psi\in\Psi} \sum_{i=1}^{N} \pounds \left(f_i \left(\varphi_i, \psi_i | X_i\right), Y_i\right), \tag{1}$$

where $\pounds$ is the total loss for optimizing the model, f_i is the model of i-th client, while X_i and Y_i are the input and label of i-th client. Different from previous personalized FL paradigms [2,4], our FedUAA dynamically adjusts the weights for model aggregation according to the reliability of each client, i.e., the client with larger distributional heterogeneity tends to have larger uncertainty distribution and should be assigned a larger weight for model aggregation to strengthen attention on the client with data heterogeneity. Besides, by introducing TWEU, our FedUAA can generate a reliable prediction with an estimated uncertainty, which makes the model more reliable without losing DR staging performance.

2.1 Temperature-Warmed Evidential Uncertainty Head

To make the model more reliable without sacrificing DR staging performance, we propose a novel temperature-warmed evidence uncertainty head (TWEU), which can directly generate DR staging results with uncertainty score based on the features from the encoder. The framework of TWEU is illustrated in Fig. 1 (b). Specifically, we take one of the client models as an example and we assume that the staging criteria of this client is K categories. Correspondingly, given a color fundus image input, we can obtain its $K+1$ non-negative mass values, whose sum is 1. This can be defined as $\sum_{i=1}^{K} b_i + u = 1$, where $b_i \geq 0$ is the probability of i-th category, while u represent the overall uncertainty score. Specifically, as shown in Fig. 1 (b), a local fully connected layer (FC) is used to learn the local DR category-related features F_V, and the *Softplus* activation function is adopted to obtain the evidence $E = [e_1, ..., e_K]$ of K staging categories based on F_V, so as to ensure that its feature value is greater than 0. Then, E is re-parameterized by Dirichlet concentration [5], as: $\boldsymbol{\alpha} = E + 1$, $i.e$, $\alpha_k = e_k + 1$ where α_k and e_k are the k-th category Dirichlet distribution parameters and evidence, respectively. Further calculating the belief masses ($\boldsymbol{b}$) and corresponding uncertainty score (u) by $b_k = \frac{e_k}{S} = \frac{\alpha_k - 1}{S}$, $u = \frac{K}{S}$, where $S = \sum_{k=1}^{K} \alpha_{i,j}^{k}$ is the Dirichlet intensities. Therefore, the probability assigned to category k is proportional to the observed evidence for category k. Conversely, if less total evidence is obtained, the greater the uncertainty score will be. As shown in Fig. 1 (b), L_{Uce} is used to guide the model optimization based on the belief masses ($\boldsymbol{b}$) and their corresponding uncertainty score (u). Finally, temperature coefficients τ is introduced to further enhance the classifier's confidence in belief masses, i.e., $b_{Ti} = \frac{e^{(b_i/\tau)}}{\sum_{i=1}^{K} e^{(b_i/\tau)}}$, where $\boldsymbol{b_T} = [b_{T1}, ..., b_{Tk}]$ is the belief masses that were temperature-warmed. As shown in Fig. 1 (b), L_{Tce} is adopted to guide the model optimization based on the temperature-warmed belief features of $\boldsymbol{b_T}$.

2.2 Uncertainty-Aware Weighting Module

Most existing FL paradigms aggregate model parameters by assigning a fixed weight to each client, resulting in limited performance on those clients with large heterogeneity in their data distributions. To address this issue, as shown in Fig. 1 (a), we propose a novel uncertainty-aware weighting (UAW) module that can dynamically adjust the weights for model aggregation based on the reliability of each client, which enables the model to better leverage the knowledge from different clients and further improve the DR staging performance. Specifically, at the end of a training epoch, each client-side model produces an uncertainty value distribution (U), and the ground truth for incorrect prediction of U^{GT} also can be calculated based on the final prediction P by,

$$u_i^{GT} = 1 - \mathbf{1}\{P_i, Y_i\}, \text{ where } \mathbf{1}\{P_i, Y_i\} = \begin{cases} 1 & \text{if } P_i = Y_i \\ 0 & \text{otherwise} \end{cases}, \tag{2}$$

where P_i and Y_i are the final prediction result and ground truth of i-th sample in local dataset. Based on U and U^{GT}, we can find the optimal uncertainty score θ, which can well reflect the reliability of the local client. To this end, we calculate the ROC curve between U and U^{GT}, and obtain all possible sensitivity ($Sens$)

and specificity ($Spes$) values corresponding to each uncertainty score (u) used as a threshold. Then, Youden index (J) [7] is adopted to obtain the optimal uncertainty score θ by:

$$\theta = \arg\max_{u} J(u), \text{ with } J(u) = Sens(u) + Spes(u) - 1. \tag{3}$$

More details about Youden index are given in **Supplementary B**. Finally, the optimal uncertainty scores $\Theta = [\theta_1, ..., \theta_N]$ of all clients are sent to the server, and a Softmax function is introduced to normalize Θ to obtain the weights for model aggregation as $w_i = e^{\theta_i} / \sum_{i=1}^{N} e^{\theta_i}$. Therefore, the weights for model aggregation are proportional to the optimal threshold of the client. Generally, local dataset with larger uncertainty distributions will have a higher optimal uncertainty score θ, indicating that it is necessary to improve the feature learning capacity of the client model to further enhance its confidence in the feature representation, and thus higher weights should be assigned during model aggregation.

3 Loss Function

As shown in Fig. 1 (b), the loss function of client model is:

$$L = L_{Uce} + L_{Tce}, \tag{4}$$

where L_{Uce} is adopted to guide the model optimization based on the features ($\boldsymbol{b}$ and u) which were parameterized by Dirichlet concentration. Given the evidence of $E = [e_1, ..., e_k]$, we can obtain Dirichlet distribution parameter $\boldsymbol{\alpha} = E + 1$, category related belief mass $\boldsymbol{b} = [b_1, ..., b_k]$ and uncertainty score of u. Therefore, the original cross-entropy loss is improved as,

$$L_{Ice} = \int \left[\sum_{k=1}^{K} -y_k \log(b_k) \right] \frac{1}{\beta(\alpha)} \prod_{k=1}^{K} b_k^{\alpha_k - 1} db = \sum_{k=1}^{K} y_k \left(\Phi(S) - \Phi(\alpha_k) \right), \tag{5}$$

where $\Phi(\cdot)$ is the digamma function, while $\beta(\alpha)$ is the multinomial beta function for the Dirichlet concentration parameter α. Meanwhile, the KL divergence function is introduced to ensure that incorrect predictions will yield less evidence:

$$L_{KL} = \log \left(\frac{\Gamma\left(\sum_{k=1}^{K} (\tilde{\alpha}_k)\right)}{\Gamma(K) \sum_{k=1}^{K} \Gamma(\tilde{\alpha}_i)} \right) + \sum_{k=1}^{K} (\tilde{\alpha}_k - 1) \left[\Phi(\tilde{\alpha}_k) - \Phi\left(\sum_{i=1}^{K} \tilde{\alpha}_k \right) \right], \tag{6}$$

where $\Gamma(\cdot)$ is the gamma function, while $\tilde{\alpha} = y + (1 - y) \odot \alpha$ represents the adjusted parameters of the Dirichlet distribution which aims to avoid penalizing the evidence of the ground-truth class to 0. In summary, the loss function L_{Uce} for the model optimization based on the features that were parameterized by Dirichlet concentration is as follows:

$$L_{Uce} = L_{Ice} + \lambda * L_{KL}, \tag{7}$$

Table 1. AUC results for different FL methods applied to DR staging.

Methods	APTOS	DDR	DRR	Messidor	IDRiD	Average
SingleSet	0.9059	0.8776	0.8072	0.7242	0.7168	0.8063
FedRep [4]	0.9372	0.8964	0.8095	0.7843	0.8047	0.8464
FedBN [23]	0.9335	0.9003	0.8274	0.7792	0.8193	0.8519
FedProx [22]	0.9418	0.8950	0.8127	0.7877	0.8049	0.8484
FedDyn [1]	0.9352	0.8778	0.8022	0.7264	0.5996	0.7882
SCAFFOLD [16]	0.9326	0.8590	0.7251	0.7288	0.6619	0.7815
FedDC [9]	0.9358	0.8858	0.7969	0.7390	0.7581	0.8236
Moon [18]	0.9436	0.8995	0.8117	0.7907	0.8115	0.8514
MDT [28]	0.9326	0.8908	0.7987	0.7919	0.7965	0.8421
Proposed	**0.9445**	**0.9044**	**0.8379**	**0.8012**	**0.8299**	**0.8636**

where λ is the balance factor for L_{KL}. To prevent the model from focusing too much on KL divergence in the initial stage of training, causing a lack of exploration for the parameter space, we initialize λ as 0 and increase it gradually to 1 with the number of training iterations. And, seen from Sect. 2.1, Dirichlet concentration alters the original feature distribution of F_v, which may reduce the model's confidence in the category-related evidence features, thus potentially leading to a decrease in performance. Aiming at this problem, as shown in Fig. 1 (b), we introduce temperature coefficients to enhance confidence in the belief masses, and the loss function L_{Tce} to guide the model optimization based on the temperature-warmed belief features $\boldsymbol{b_T}$ is formalized as:

$$L_{Tce} = -\sum_{i=1}^{K} y_i log\left(b_{Ti}\right). \quad (8)$$

4 Experimental Results

Dataset and Implementation: We construct a database for federated DR staging based on 5 public datasets, including APTOS (3,662 samples)[1], Messidor (1,200 samples) [6], DDR (13,673 samples) [20], KaggleDR (35,126 samples) (DRR)[2], and IDRiD (516 samples) [26], where each dataset is regarded as a client, More details of datasets are given in **Supplementary C**.

We conduct experiments on the Pytorch with 3090 GPU. The SGD with a learning rate of 0.01 is utilized. The batch size is set to 32, the number of epochs is 100, and the temperature coefficient τ is empirically set to 0.05. To facilitate training, the images are resized to 256×256 before feeding to the model.

[1] https://www.kaggle.com/datasets/mariaherrerot/aptos2019.
[2] https://www.kaggle.com/competitions/diabetic-retinopathy-detection.

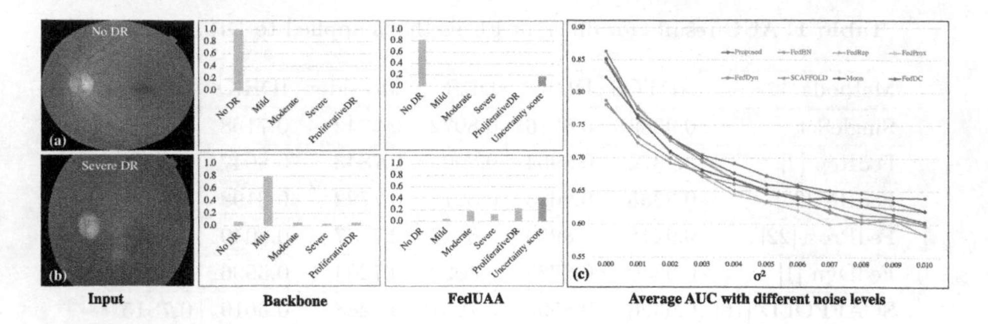

Fig. 2. (a) Instance of being correctly predicted (b) Sample with incorrect prediction result (c) Average AUC of different methods with increasing noise levels (σ^2).

Performance for DR Staging: Table 1 shows the DR staging AUC for different FL paradigms on different clients. Our FedUAA achieves the highest AUC scores on all clients, with a 1.48% improvement in average AUC compared to FedBN [23], which achieved the highest average AUC score among the compared methods. Meanwhile, most FL based approaches achieve higher DR staging performance than SingleSet, suggesting that collaborative training across multiple institutions can improve the performance of DR staging with high data privacy security. Moreover, as shown in Table 1, FL paradigms such as FedDyn [1] and SCAFFOLD [16] exhibit limited performance in our collaborative DR staging task due to the varying staging criteria across different clients, as well as significant differences in label distribution and domain features. These results indicate that our FedUAA is more effective than other FL methods for collaborative DR staging tasks. Furthermore, although all FL methods achieve comparable performance on APTOS and DDR clients with distinct features, our FedUAA approach significantly improves performance on clients with small data volumes or large heterogeneity distribution, such as DRR, Messidor, and IDRiD, by 1.27%, 1.33%, and 1.29% over suboptimal results, respectively, which further demonstrates the effectiveness of our core idea of adaptively adjusting aggregation weights based on the reliability of each client. In addition, we also conduct experiments demonstrate the statistical significance of performance improvement. As shown in Supplementary D, most average p-values are smaller than 0.05. These experimental results further prove the effectiveness of our proposed FedUAA.

Reliability Analysis: Providing reliable evaluation for final predictions is crucial for AI models to be deployed in clinical practice. As illustrated in Fig. 2 (b), the model without introducing uncertainty (Backbone) assigns high probability values for incorrect staging results without any alert messages, which is also a significant cause of low user confidence in the deployment of AI models to medical practices. Interestingly, our FedUAA can evaluate the reliability of the final decision through the uncertainty score. For example, for the data with obvious features (Fig. 2 (a)), our FedUAA produces a correct prediction result with a low uncertainty score, indicating that the decision is reliable. Conversely, even

Table 2. AUC results for different FL paradigms applied to DR staging.

Strategy	BC	EU	TWEU	UAW	APTOS	DDR	DRR	Messidor	IDRiD	Average
SingleSet	✓	✗	✗	✗	0.9059	0.8776	0.8072	0.7242	0.7168	0.8063
	✓	✓	✗	✗	0.9286	0.8589	0.8001	0.7404	0.6928	0.8042
	✓	✗	✓	✗	0.9414	0.8912	0.8279	0.7309	0.7616	0.8306
FL	✓	✗	✗	✗	0.9335	0.9003	0.8274	0.7792	0.8193	0.8519
	✓	✓	✗	✗	0.9330	0.8572	0.7938	0.7860	0.7783	0.8297
	✓	✗	✓	✗	0.9445	0.8998	0.8229	0.8002	0.8231	0.8581
	✓	✗	✓	✓	**0.9445**	**0.9044**	**0.8379**	**0.8012**	**0.8299**	**0.8636**

if our FedUAA gives an incorrect decision for the data with ambiguous features (Fig. 2 (b)), it can indicate that the diagnosis result may be unreliable by assigning a higher uncertainty score, thus suggesting that the subject should seek a double-check from an ophthalmologist to avoid mis-diagnosis. Furthermore, as shown in Fig. 2 (c), we degraded the quality of the input image by adding different levels of Gaussian noise σ^2 to further verify the robustness of FedUAA. Seen from Fig. 2 (c), the performance of all methods decreases as the level of added noise increases, however, our FedUAA still maintains a higher performance than other comparison methods, demonstrating the robustness of our FedUAA.

Ablation Study: We also conduct ablation experiments to verify the effectiveness of the components in our FedUAA. In this paper, the pre-trained ResNet50 [13] is adopted as our backbone (BC) for SingleSet DR staging, while employing FedBN [23] as the FL BC. Furthermore, most ensemble-based [17] and MC-dropout-based [8] uncertainty methods are challenging to extend to our federated DR staging task across multiple institutions with different staging criteria. Therefore, we compare our proposed method with the commonly used evidential based uncertainty approach (EU (L_{Uce})) [12].

For training model with SingleSet, as shown in Table 2, since Dirichlet concentration alters the original feature distribution of the backbone [12], resulting in a decrease in the model's confidence in category-related evidence, consequently, a decrease in performance when directly introducing EU (BC+EU (L_{Uce})) for DR staging. In contrast, our proposed BC+TWEU (L_{Uce}+L_{Tce}) achieves superior performance compared to BC and BC+EU (L_{Uce}), demonstrating that TWEU (L_{Uce}+L_{Tce}) enables the model to generate a reliable final decision without sacrificing performance. For training model with FL, as shown in Table 2, BC+FL outperforms SingleSet, indicating that introducing FL can effectively improve the performance for DR staging while maintaining high data privacy security. Besides, FL+EU (L_{Uce}) and FL+TWEU (L_{Uce}+L_{Tce}) also obtain a similar conclusion as in SingleSet, further proving the effectiveness of TWEU. Meanwhile, the performance of our FedUAA (FL+TWEU (L_{Uce}+L_{Tce})+UAW) achieves higher performance than FL+TWEU (L_{Uce}+L_{Tce}) and FL backbone, especially for clients with large data distribution heterogeneity such as DRR,

Messidor, and IDRiD. These results show that our proposed UAW can further improve the performance of FL in collaborative DR staging tasks.

5 Conclusion

In this paper, focusing on the challenges in the collaborative DR staging between institutions with different DR staging criteria, we propose a novel FedUAA by combining the FL with evidential uncertainty theory. Compared to other FL methods, our FedUAA can produce reliable and robust DR staging results with uncertainty evaluation, and further enhance the collaborative DR staging performance by dynamically aggregating knowledge from different clients based on their reliability. Comprehensive experimental results show that our FedUAA addresses the challenges in collaborative DR staging across multiple institutions, and achieves a robust and reliable DR staging performance.

Acknowledgements. This work was supported by the National Research Foundation, Singapore under its AI Singapore Programme (AISG Award No: AISG2-TC-2021-003), the Agency for Science, Technology and Research (A*STAR) through its AME Programmatic Funding Scheme Under Project A20H4b0141, A*STAR Central Research Fund "A Secure and Privacy Preserving AI Platform for Digital Health", and A*STAR Career Development Fund (C222812010).

References

1. Acar, D.A.E., Zhao, Y., Navarro, R.M., Mattina, M., Whatmough, P.N., Saligrama, V.: Federated learning based on dynamic regularization. arXiv preprint arXiv:2111.04263 (2021)
2. Arivazhagan, M.G., Aggarwal, V., Singh, A.K., Choudhary, S.: Federated learning with personalization layers. arXiv preprint arXiv:1912.00818 (2019)
3. Asiri, N., Hussain, M., Al Adel, F., Alzaidi, N.: Deep learning based computer-aided diagnosis systems for diabetic retinopathy: a survey. Artif. Intell. Med. **99**, 101701 (2019)
4. Collins, L., Hassani, H., Mokhtari, A., Shakkottai, S.: Exploiting shared representations for personalized federated learning. In: International Conference on Machine Learning, pp. 2089–2099. PMLR (2021)
5. Connor, R.J., Mosimann, J.E.: Concepts of independence for proportions with a generalization of the dirichlet distribution. J. Am. Stat. Assoc. **64**(325), 194–206 (1969)
6. Decencière, E., Zhang, X., Cazuguel, G., et al.: Feedback on a publicly distributed image database: the Messidor database. Image Anal. Stereol. **33**(3), 231–234 (2014)
7. Fluss, R., Faraggi, D., Reiser, B.: Estimation of the Youden Index and its associated cutoff point. Biometrical J.: J. Math. Methods Biosci. **47**(4), 458–472 (2005)
8. Gal, Y., Ghahramani, Z.: Dropout as a Bayesian approximation: representing model uncertainty in deep learning. In: International Conference on Machine Learning, pp. 1050–1059. PMLR (2016)
9. Gao, L., Fu, H., Li, L., Chen, Y., Xu, M., Xu, C.Z.: FEDDC: federated learning with non-IID data via local drift decoupling and correction. In: CVPR, pp. 10112–10121 (2022)

10. Gulshan, V., Peng, L., Coram, M., et al.: Development and validation of a deep learning algorithm for detection of diabetic retinopathy in retinal fundus photographs. JAMA **316**(22), 2402 (2016)
11. Gunasekeran, D.V., Ting, D.S., Tan, G.S., Wong, T.Y.: Artificial intelligence for diabetic retinopathy screening, prediction and management. Curr. Opin. Ophthalmol. **31**(5), 357–365 (2020)
12. Han, Z., Zhang, C., Fu, H., Zhou, J.T.: Trusted multi-view classification. arXiv preprint arXiv:2102.02051 (2021)
13. He, K., Zhang, X., Ren, S., Sun, J.: Deep residual learning for image recognition. In: Proceedings of the IEEE Conference on Computer Vision and Pattern Recognition, pp. 770–778 (2016)
14. Huang, L., Denoeux, T., Vera, P., Ruan, S.: Evidence fusion with contextual discounting for multi-modality medical image segmentation. In: Wang, L., Dou, Q., Fletcher, P.T., Speidel, S., Li, S. (eds.) MICCAI 2022. LNCS, vol. 13435, pp. 401–411. Springer, Cham (2022). https://doi.org/10.1007/978-3-031-16443-9_39
15. Kairouz, P., McMahan, H.B., Avent, B., et al.: Advances and open problems in federated learning. Found. Trends® Mach. Learn. **14**(1–2), 1–210 (2021). https://doi.org/10.1561/2200000083
16. Karimireddy, S.P., Kale, S., Mohri, M., Reddi, S., Stich, S., Suresh, A.T.: Scaffold: stochastic controlled averaging for federated learning. In: International Conference on Machine Learning, pp. 5132–5143. PMLR (2020)
17. Lakshminarayanan, B., Pritzel, A., Blundell, C.: Simple and scalable predictive uncertainty estimation using deep ensembles. In: Advances in Neural Information Processing Systems, vol. 30 (2017)
18. Li, Q., He, B., Song, D.: Model-contrastive federated learning. In: Proceedings of the IEEE/CVF Conference on Computer Vision and Pattern Recognition, pp. 10713–10722 (2021)
19. Li, T., et al.: Applications of deep learning in fundus images: a review. Med. Image Anal. **69**, 101971 (2021)
20. Li, T., Gao, Y., Wang, K., Guo, S., Liu, H., Kang, H.: Diagnostic assessment of deep learning algorithms for diabetic retinopathy screening. Inf. Sci. **501**, 511–522 (2019)
21. Li, T., Sahu, A.K., Talwalkar, A., Smith, V.: Federated learning: challenges, methods, and future directions. IEEE Signal Process. Mag. **37**(3), 50–60 (2020)
22. Li, T., Sahu, A.K., Zaheer, M., Sanjabi, M., Talwalkar, A., Smith, V.: Federated optimization in heterogeneous networks. Proc. Mach. Learn. Syst. **2**, 429–450 (2020)
23. Li, X., Jiang, M., Zhang, X., Kamp, M., Dou, Q.: FEDBN: federated learning on non-IID features via local batch normalization. arXiv preprint arXiv:2102.07623 (2021)
24. McMahan, B., Moore, E., Ramage, D., Hampson, S., y Arcas, B.A.: Communication-efficient learning of deep networks from decentralized data. In: Artificial Intelligence and Statistics, pp. 1273–1282. PMLR (2017)
25. Nguyen, T.X., et al.: Federated learning in ocular imaging: current progress and future direction. Diagnostics **12**(11), 2835 (2022)
26. Porwal, P., et al.: Indian diabetic retinopathy image dataset (IDRID): a database for diabetic retinopathy screening research. Data **3**(3), 25 (2018)
27. Ting, D.S.W., Cheung, C.Y.L., Lim, G., et al.: Development and validation of a deep learning system for diabetic retinopathy and related eye diseases using retinal images from multiethnic populations with diabetes. JAMA **318**(22), 2211 (2017)

28. Yu, Y., Bates, S., Ma, Y., Jordan, M.: Robust calibration with multi-domain temperature scaling. Adv. Neural. Inf. Process. Syst. **35**, 27510–27523 (2022)
29. Zhou, Y., Bai, S., Zhou, T., Zhang, Y., Fu, H.: Delving into local features for open-set domain adaptation in fundus image analysis. In: Wang, L., Dou, Q., Fletcher, P.T., Speidel, S., Li, S. (eds.) MICCAI 2022. LNCS, vol. 13437, pp. 682–692. Springer, Cham (2022). https://doi.org/10.1007/978-3-031-16449-1_65
30. Zou, K., Yuan, X., Shen, X., Wang, M., Fu, H.: TBraTS: trusted brain tumor segmentation. In: Wang, L., Dou, Q., Fletcher, P.T., Speidel, S., Li, S. (eds.) MICCAI 2022. LNCS, vol. 13438, pp. 503–513. Springer, Cham (2022). https://doi.org/10.1007/978-3-031-16452-1_48

Few Shot Medical Image Segmentation with Cross Attention Transformer

Yi Lin, Yufan Chen, Kwang-Ting Cheng, and Hao Chen(✉)

The Hong Kong University of Science and Technology, Hong Kong, China
jhc@cse.ust.hk

Abstract. Medical image segmentation has made significant progress in recent years. Deep learning-based methods are recognized as data-hungry techniques, requiring large amounts of data with manual annotations. However, manual annotation is expensive in the field of medical image analysis, which requires domain-specific expertise. To address this challenge, few-shot learning has the potential to learn new classes from only a few examples. In this work, we propose a novel framework for few-shot medical image segme ntation, termed CAT-Net, based on cross masked attention Transformer. Our proposed network mines the correlations between the support image and query image, limiting them to focus only on useful foreground information and boosting the representation capacity of both the support prototype and query features. We further design an iterative refinement framework that refines the query image segmentation iteratively and promotes the support feature in turn. We validated the proposed method on three public datasets: Abd-CT, Abd-MRI, and Card-MRI. Experimental results demonstrate the superior performance of our method compared to state-of-the-art methods and the effectiveness of each component. Source code: https://github.com/hust-linyi/CAT-Net.

Keywords: Few Shot · Cross Attention · Iterative Refinement

1 Introduction

Automatic segmentation of medical images is a fundamental step for a variety of medical image analysis tasks, such as diagnosis, treatment planning, and disease monitoring [1,2]. The emergence of deep learning (DL) has enabled the development of many medical image segmentation methods, which have achieved remarkable success [3,4,10,12,32]. Most of the existing methods follow a fully-supervised learning paradigm, which requires a considerable amount of labeled data for training. However, the manual annotation of medical images is time-consuming and labor-intensive, limiting the application of DL in medical image

Y. Lin and Y. Chen—Equal contribution.

Supplementary Information The online version contains supplementary material available at https://doi.org/10.1007/978-3-031-43895-0_22.

H. Greenspan et al. (Eds.): MICCAI 2023, LNCS 14221, pp. 233–243, 2023.
https://doi.org/10.1007/978-3-031-43895-0_22

segmentation. Specifically for the 3D volumetric medical images (*e.g.*, CT, MRI), the manual annotation is even more challenging which requires the annotators to go through hundreds of 2D slices for each 3D scan.

To address the challenge of manual annotation, various label-efficient techniques have been explored, such as self-supervised learning [15], semi-supervised learning [30,31], and weakly-supervised learning [11]. Despite leveraging information from unlabeled or weakly-labeled data, these techniques still require a substantial amount of training data [16,21], which may not be practical for novel classes with limited examples in the medical domain. This limitation encourages the few-shot learning paradigm [6,22,24,28] to be applied to medical image segmentation. Specifically, the few-shot learning paradigm aims to learn a model from a small number of labeled data (denoted as *support*) and then apply it to a new task (denoted as *query*) with only a few labeled data without any retraining. Considering the hundreds of organs and countless diseases in the human body, FSL brings great potential to the various medical image segmentation tasks where a new task can be easily investigated in a data-efficient manner.

Most few-shot segmentation methods follow the learning-to-learn paradigm, which aims to learn a meta-learner to predict the segmentation of query images based on the knowledge of support images and their respective segmentation labels. The success of this paradigm depends on how effectively the knowledge can be transferred from the support prototype to the query images. Existing few-shot segmentation methods mainly focus on the following two aspects: (1) how to learn the meta-learner [14,17,26]; and (2) how to better transfer the knowledge from the support images to the query images [5,13,18,23,25,27]. Despite prototype-based methods having shown success, they typically ignore the interaction between support and query features during training. In this paper, as shown in Fig. 1(a), we propose **CAT-Net**, a **C**ross **A**ttention **T**ransformer network for few-shot medical image segmentation, which aims to fully capture intrinsic classes details while eliminating useless pixel information and learn an interdependence between the support and query features. Different from the existing FSS methods that only focus on the single direction of knowledge transfer (*i.e.*, from the support features to the query features), the proposed CAT-Net can boost the mutual interactions between the support and query features, benefiting the segmentation performance of both the support and query images. Additionally, we propose an iterative training framework that feed the prior query segmentation into the attention transformer to effectively enhance and refine the features as well as the segmentation. Three publicly available datasets are adopted to evaluate our CAT-Net, *i.e.*, Abd-CT [9], Abd-MRI [8], and Card-CT [33]. Extensive experiments validate the effectiveness of each component in our CAT-Net, and demonstrate its state-of-the-art performance.

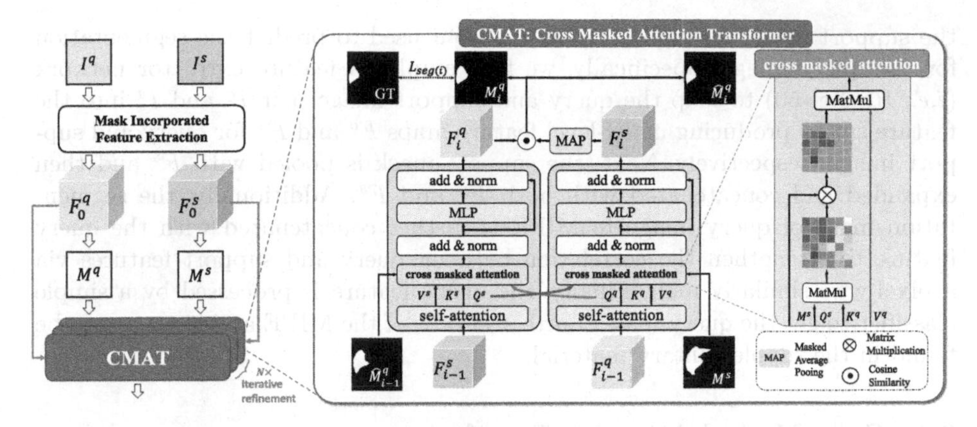

Fig. 1. (a) Overview of the CAT-NET; (b) The architecture of CMAT module.

2 Method

2.1 Problem Definition

Few-shot segmentation (FSS) aims to segment novel classes by just a few samples with densely-annotated samples. In FSS, the dataset is divided into the training set $\mathbb{D}_{\text{train}}$, containing the base classes $\mathbb{C}_{\text{train}}$, and the test set $\mathbb{D}_{\text{test}}$, containing the novel classes $\mathbb{C}_{\text{test}}$, where $\mathbb{C}_{\text{train}} \cap \mathbb{C}_{\text{test}} = \emptyset$. To obtain the segmentation model for FSS, the commonly used episode training approach is employed [29]. Each trainig/testing episode (S_i, Q_i) instantiates a N-way K-shot segmentation learning task. Specifically, the support set S_i contains K samples of N classes, while the query set Q_i contains one sample from the same class. The FSS model is trained with episodes to predict the novel class for the query image, guided by the support set. During inference, the model is evaluated directly on $\mathbb{D}_{\text{test}}$ without any re-training. In this paper, we follow the established practice in medical FSS [7,15,20] that consider the **1**-way **1**-shot task.

2.2 Network Overview

The Overview of our CAT-Net is illustrated in Fig. 1(a). It consists of three main components: 1) a mask incorporated feature extraction (MIFE) sub-net that extracts initial query and support features as well as query mask; 2) a cross masked attention Transformer (CMAT) module in which the query and support features boost each other and thus refined the query prediction; and 3) an iterative refinement framework that sequentially applies the CMAT modules to continually promote the segmentation performance. The whole framework can be trained in an end-to-end fashion.

2.3 Mask Incorporated Feature Extraction

The Mask Incorporate Feature Extraction (MIFE) sub-net takes query and support images as input and generates their respective features, integrated with the support mask. A simple classifier is then used to predict the segmentation for the query image. Specifically, we first employ a feature extractor network (*i.e.*, ResNet-50) to map the query and support image pair I^q and I^s into the feature space, producing multi-level feature maps F^q and F^s for query and support image, respectively. Next, the support mask is pooled with F^s and then expanded and concatenated with both F^q and F^s. Additionally, the segmentation mask of query image in MIFE is further concatenated with the query feature to strengthen the correlation between query and support features via a pixel-wise similarly map. Finally, the query feature is processed by a simple classifier to get the query mask. Further details of the MIFE architecture can be found in the supplementary material.

2.4 Cross Masked Attention Transformer

As shown in Fig. 1(b), the cross masked attention Transformer (CMAT) module comprises three main components: 1) a self-attention module for extracting global information from query and support features; 2) a cross masked attention module for transferring foreground information between query and support features while eliminating redundant background information, and 3) a prototypical segmentation module for generating the final prediction of the query image.

Self-Attention Module. To capture the global context information of every pixel in the query feature F_0^q and support features F_0^s, the initial features are first flattened into 1D sequences and fed into two identical self-attention modules. Each self-attention module consists of a multi-head attention (MHA) layer and a multi-perceptron (MLP) layer. Given an input sequence S, the MHA layer first projects the sequence into three sequences K, Q, and V with different weights. The attention matrix A is then calculated as:

$$A(Q, K) = \frac{QK^T}{\sqrt{d}} \tag{1}$$

where d is the dimension of the input sequence. The attention matrix is then normalized by a softmax function and multiplied by the value sequence V to get the output sequence O:

$$O = \text{softmax}(A)V \tag{2}$$

The MLP layer is a simple 1×1 convolution layer that maps the output sequence O to the same dimension as the input sequence S. Finally, the output sequence O is added to the input sequence S and normalized using layer normalization (LN) to obtain the final output sequence X. The output feature sequence of the self-attention alignment encoder is represented by $X^q \in \mathbb{R}^{HW \times D}$ and $X^s \in \mathbb{R}^{HW \times D}$ for query and support features, respectively.

Cross Masked Attention Module. We utilize cross masked attention to incorporate query features and support features with respect to their foreground information by constraining the attention region in attention matrix with support and query masks. Specifically, given the query feature X^q and support features X^s from the aforementioned self-attention module, we first project the input sequence into three sequences K, Q, and V using different weights, resulting in K^q, Q^q, V^q, and K^s, Q^s, V^s, respectively. Taking the support features as an example, the cross attention matrix is calculated by:

$$\mathrm{A}(K^q, Q^s) = \frac{(K^q)^T Q^s}{\sqrt{d}} \tag{3}$$

We expand and flatten the binary query mask M^q to limit the foreground region in attention map. The masked cross attention (MCA) map is computed as:

$$\mathrm{MCA}(K^q, Q^s, V^q, M^s) = M^s \cdot V^q(\mathrm{softmax}(A(K^q, Q^s))) \tag{4}$$

Similar to self-attention, the support feature is processed by MLP and LN layer to get the final enhanced query features F_1^s. Similarly, the enhanced query feature F_1^q is obtained with foreground information from the query feature.

Prototypical Segmentation Module. Once the enhanced query and support features are obtained, the prototypical segmentation is used to obtain the final prediction. First, a prototype of class c is built by masked average pooling of the support feature F_1^s as follows:

$$p_c = \frac{1}{K} \sum_{k=1}^{K} \frac{\sum_{k,x,y} F_{i,(k,x,y)}^s m_{(k,x,y,c)}^s}{\sum_{x,y} m_{(k,x,y,c)}^s} \tag{5}$$

where K is the number of support images, and $m_{(k,x,y,c)}^s$ is a binary mask that indicates whether pixel at the location (x, y) in support feature k belongs to class c. Next, we use the non-parametirc metric learning method to perform segmentation. The prototype network calculates the distance between the query feature vector and the prototype $P = \{P_c | c \in C\}$. Softmax function is applied to produce probabilistic outputs for all classes, generating the query segmentation:

$$\hat{M}_{i,(x,y)}^q = \mathrm{softmax}\big(\alpha \cos(F_{i,(x,y)}^q, p_c) \cdot \mathrm{softmax}(\alpha \cos(F_{i,(x,y)}^q, p_c))\big) \tag{6}$$

where $\cos(\cdot)$ denotes cosine distance, α is a scaling factor that helps gradients to back-propagate in training. In our work, α is set to 20, same as in [29].

Additionally, we design a double threshold strategy to obtain query segmentation. Specifically, we set the first threshold τ to 0.5 to obtain the binary query mask M^q, which is used to calculate the Dice loss and update the model. Then, the second threshold $\hat{\tau}$ is set to 0.4 to obtain the dilated query mask $\hat{M}^q$, which is used to generate the enhanced query feature F_2^q in the next iteration. The second threshold $\hat{\tau}$ is set lower than the first threshold τ to prevent some foreground pixels from being mistakenly discarded. The query segmentation mask M^q and dilated mask $\hat{M}^q$ are represented by:

$$M_i^q = \begin{cases} 1, & M_{i,(x,y)}^q > \tau \\ 0, & M_{i,(x,y)}^q < \tau \end{cases} \qquad \hat{M}_i^q = \begin{cases} 1, & M_{i,(x,y)}^q > \hat{\tau} \\ 0, & M_{i,(x,y)}^q < \hat{\tau} \end{cases} \tag{7}$$

2.5 Iterative Refinement Framework

As explained above, the CMAT module is designed to refine the query and support features, as well as the query segmentation mask. Thus, it's natural to iteratively apply this sub-net to get the enhanced features and refine the mask, resulting in a boosted segmentation result. The result after the i-th iteration is represented by:

$$(F_i^s, F_i^q, M_i^q, \hat{M}_i^q) = \text{CMAT}(F_{i-1}^s, F_{i-1}^q, \hat{M}_{i-1}^q, M^s) \tag{8}$$

The subdivision of each step can be specifically expressed as:

$$(F_i^s, F_i^q) = \text{CMA}(F_{i-1}^s, F_{i-1}^q, \hat{M}_{i-1}^q, M^s) \tag{9}$$

$$(M_i^q, \hat{M}_i^q) = \text{Proto}(F_i^s, F_i^q, M^s, \tau, \hat{\tau}) \tag{10}$$

where $\text{CMA}(\cdot)$ indicates the self-attention and cross masked attention module, and $\text{Proto}(\cdot)$ represents the prototypical segmentation module.

3 Experiment

3.1 Dataset and Evaluation Metrics

We evaluate the proposed method on three public datasets, *i.e.*, Abd-CT [9], Abd-MRI [8], and Card-MRI [33]. Abd-CT contains 30 abdominal CT scans with annotations of left and right kidney (LK and RK), spleen (Spl), liver (Liv). Abd-MRI contains 20 abdominal MRI scans with annotations of the same organs as Abd-CT. Card-MRI includes 35 cardiac MRI scans with annotations of left ventricular blood pool (LV-B), left ventricular myocardium (LV-M), and right ventricle (RV). We use the Dice score as the evaluation metric following [15,20].

To ensure a fair comparison, all the experiments are conducted under the 1-way 1-shot scenario using 5-fold cross-validation. We follow [15] to remove all slices containing test classes during training to ensure that the test classes are all unseen during validation. In each fold, we follow [7,15,20] that takes the last patient as the support image and the remaining patients as the query (setting I). We further propose a new validation setting (setting II) that takes every image in each fold as a support image alternately and the other images as the query. The averaged result of each fold is reported. It could evaluate the generalization ability of the model by reducing the affect of support image selection.

3.2 Implementation Details

The proposed method is implemented using PyTorch. Each 3D scan is sliced into 2D slices and reshaped into 256×256 pixels. Common 3D image pre-processing techniques, such as intensity normalization and resampling, are applied to the training data. We apply episode training with $20k$ iterations. SGD optimizer is adopted with a learning rate of 0.001 and a batch size of 1. Each episode training takes approximately 4 h using a single NVIDIA RTX 3090 GPU.

Table 1. Comparison with state-of-the-art methods in Dice coefficient (%) on Abd-CT and Abd-MRI, and Card-MRI datasets under setting I & II.

		Abd-CT [8]					Abd-MRI [9]					Card-MRI [33]			
	Methods	LK	RK	Spl.	Liv.	Avg.	LK	RK	Spl.	Liv.	Avg.	LV-B	LV-M	RV	Avg.
Setting I	SE-Net [19]	32.83	14.84	0.23	0.27	11.91	62.11	61.32	51.80	27.43	50.66	58.04	25.18	12.86	32.03
	PA-Net [29]	37.58	34.69	43.73	61.71	44.42	47.71	47.95	58.73	64.99	54.85	**70.43**	46.79	69.52	62.25
	ALP-Net [15]	63.34	54.82	60.25	73.65	63.02	73.63	78.39	67.02	73.05	73.02	61.89	87.54	76.71	75.38
	AD-Net [7]	**63.84**	56.98	61.84	73.95	64.15	71.89	76.02	65.84	76.03	72.70	65.47	88.36	78.35	77.39
	Q-Net [20]	63.26	58.37	63.36	74.36	64.83	**74.05**	77.52	67.43	78.71	74.43	66.87	89.63	79.25	78.58
	Ours	63.36	**60.05**	**67.65**	**75.31**	**66.59**	74.01	**78.90**	**68.83**	**78.98**	**75.18**	66.85	**90.54**	**79.71**	**79.03**
Setting II	ALP-Net [15]	65.99	59.49	65.02	73.50	66.05	70.17	77.05	67.71	72.45	71.85	61.61	87.13	77.35	75.36
	AD-Net [7]	67.35	59.88	64.35	76.78	67.09	72.26	76.57	**67.89**	73.96	72.67	65.08	86.26	76.50	75.95
	Q-Net [20]	66.25	62.36	**67.35**	77.33	68.32	73.96	81.07	65.39	72.36	73.20	66.35	88.40	79.37	78.04
	Ours	**68.82**	**64.56**	66.02	**80.51**	**70.88**	**75.31**	**83.23**	67.31	**75.02**	**75.22**	**67.21**	**90.54**	**80.34**	**79.36**

3.3 Comparison with State-of-the-Art Methods

We compare the proposed CAT-Net with state-of-the-art (SOTA) methods, including SE-Net [19], PANet [29], ALP-Net [15], and AD-Net [7], and Q-Net [20]. PANet [29] are the typical prototypical FSS method in the natural image domain, SE-Net [19], ALP-Net [15], AD-Net [7], and Q-Net [20] are the most representative work in medical FSS task. Experiment results presented in Table 1 demonstrate that the proposed method outperforms SOTAs on all three datasets under both setting I and setting II. Under setting I, the proposed CAT-Net achieves 66.59% Dice on Abd-CT, 75.18% Dice on Abd-MRI, and 79.03% Dice on Card-MRI in Dice, outperforming SOTAs by 1.76%, 0.75%, and 0.45%, respectively. Under setting II, CAT-Net achieves 70.88% Dice on Abd-CT, 75.22% Dice on Abd-MRI, and 79.36% Dice on Card-MRI, outperforming SOTAs by 2.56%, 2.02% and 1.32%, respectively. The consistent superiority of our method to SOTAs on three datasets and under two evaluation settings indicates the effectiveness and generalization ability of the proposed CAT-Net. In addition, the qualitative results in Fig. 2 demonstrate that the proposed method

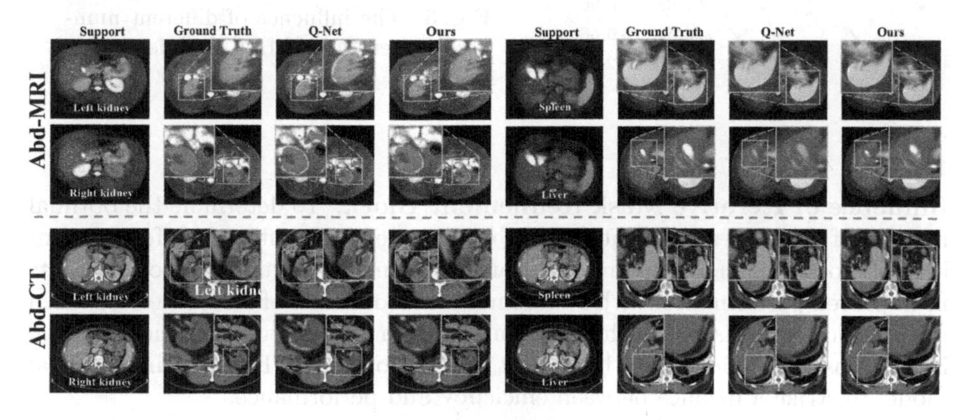

Fig. 2. Qualitative results of our method on Abd-CT and Abd-MRI.

is able to generate more accurate and detailed segmentation results compared to SOTAs.

3.4 Ablation Study

We conduct an ablation study to investigate the effectiveness of each component in CAT-Net. All ablation studies are conducted on Abd-MRI under setting II.

Effectiveness of CMAT Block: To demonstrate the importance of our proposed CAT-Net in narrowing the information gap between the query and supporting images and obtaining enhanced features, we conducted an ablation study. Specifically, we compared the results of learning foreground information only from the support ($S \rightarrow Q$) or query image ($Q \rightarrow S$) and obtaining a single enhanced feature instead of two ($S \leftrightarrow Q$). It can be observed that using the enhanced query feature ($S \rightarrow Q$) achieves 66.72% in Dice, outperforming only using the enhanced support feature ($Q \rightarrow S$) by 0.74%. With our CMAT block, the mutual boosted support and query feature ($S \leftrightarrow Q$) could improve the Dice by 1.90%. Moreover, the iteration refinement framework consistently promotes the above three variations by 0.96%, 0.56%, and 2.26% in Dice, respectively (Table 2).

Table 2. Effectiveness of each component. $S \rightarrow Q$ and $Q \rightarrow S$ denote one branch CAT-Net to enhance support or query feature, respectively. $S \leftrightarrow Q$ indicates applying cross attention to both S and Q.

$S \rightarrow Q$	$Q \rightarrow S$	$S \leftrightarrow Q$	Iter	Dice	Improve
✓				66.72	-
	✓			65.98	−0.74
		✓		68.62	+1.90
✓			✓	67.68	+0.96
	✓		✓	66.54	+0.56
		✓	✓	**70.88**	+2.26

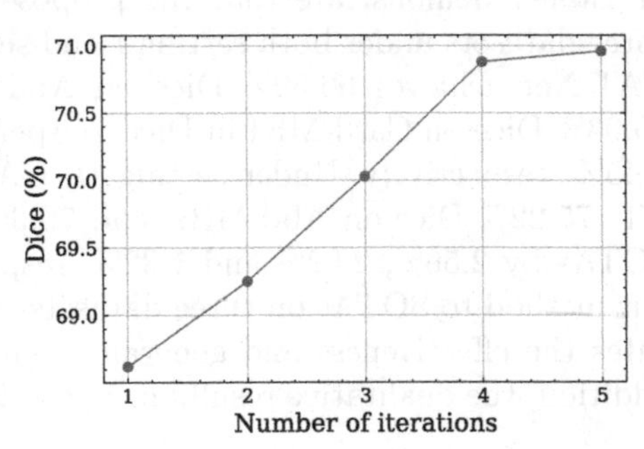

Fig. 3. The influence of different numbers of iteration CMAT modules.

Influence of Iterative Mask Refinement Block: To determine the optimal number of iterative refinement CMAT block, we experiment with different numbers of blocks. In Fig. 3, we observe that increasing the number of blocks results in improved performance, with a maximum improvement of 2.26% in Dice when using 5 blocks. Considering the performance gain between using 4 and 5 CMAT blocks was insignificant, we hence opt to use four CMAT blocks in our final model to strike a balance between efficiency and performance.

4 Conclusion

In this paper, we propose CAT-Net, Cross Attention Transformer network for few-shot medical image segmentation. Our CAT-Net enables mutual interaction between the query and support features by the cross masked attention module, enhancing the representation abilities for both of them. Additionally, the proposed CMAT module can be iteratively applied to continually boost the segmentation performance. Experimental results demonstrated the effectiveness of each module and the superior performance of our model to the SOTA methods. In the future, we plan to extend our CAT-Net from 2D to 3D networks, explore the application of our model to other medical image segmentation tasks, as well as the extension of our model to other clinical applications, such as rare diseases and malformed organs, where data and annotations are scarce and costly.

Acknowledgement. This work was supported by the Shenzhen Science and Technology Innovation Committee Fund (Project No. SGDX20210823103201011) and Hong Kong Innovation and Technology Fund (Project No. ITS/028/21FP).

References

1. Che, H., Chen, S., Chen, H.: Image quality-aware diagnosis via meta-knowledge co-embedding. In: Proceedings of the IEEE/CVF Conference on Computer Vision and Pattern Recognition, pp. 19819–19829 (2023)
2. Che, H., Cheng, Y., Jin, H., Chen, H.: Towards generalizable diabetic retinopathy grading in unseen domains. arXiv preprint arXiv:2307.04378 (2023)
3. Che, H., Jin, H., Chen, H.: Learning robust representation for joint grading of ophthalmic diseases via adaptive curriculum and feature disentanglement. In: Wang, L., Dou, Q., Fletcher, P.T., Speidel, S., Li, S. (eds.) Medical Image Computing and Computer Assisted Intervention – MICCAI 2022. MICCAI 2022. Lecture Notes in Computer Science, vol. 13433, pp. 523–533. Springer, Cham (2022). https://doi.org/10.1007/978-3-031-16437-8_50
4. Chen, J., et al.: TransUNet: Transformers make strong encoders for medical image segmentation. arXiv preprint arXiv:2102.04306 (2021)
5. Fan, Q., Pei, W., Tai, Y.W., Tang, C.K.: Self-support few-shot semantic segmentation. In: Avidan, S., Brostow, G., Cissé, M., Farinella, G.M., Hassner, T. (eds.) Computer Vision – ECCV 2022. ECCV 2022. Lecture Notes in Computer Science, vol. 13679, pp. 701–719. Springer, Cham (2022). https://doi.org/10.1007/978-3-031-19800-7_41
6. Garcia, V., Bruna, J.: Few-shot learning with graph neural networks. In: International Conference on Learning Representations (ICLR) (2018)
7. Hansen, S., Gautam, S., Jenssen, R., Kampffmeyer, M.: Anomaly detection-inspired few-shot medical image segmentation through self-supervision with supervoxels. Med. Image Anal. **78**, 102385 (2022)
8. Kavur, A.E., et al.: Chaos challenge-combined (CT-MR) healthy abdominal organ segmentation. Med. Image Anal. **69**, 101950 (2021)
9. Landman, B., Xu, Z., Igelsias, J., Styner, M., Langerak, T., Klein, A.: MICCAI multi-atlas labeling beyond the cranial vault-workshop and challenge. In: Proceedings of MICCAI Multi-Atlas Labeling Beyond Cranial Vault-Workshop Challenge, vol. 5, p. 12 (2015)

10. Lin, Y., Liu, L., Ma, K., Zheng, Y.: Seg4Reg+: consistency learning between spine segmentation and cobb angle regression. In: de Bruijne, M., et al. (eds.) MICCAI 2021. LNCS, vol. 12905, pp. 490–499. Springer, Cham (2021). https://doi.org/10.1007/978-3-030-87240-3_47
11. Lin, Y., et al.: Label propagation for annotation-efficient nuclei segmentation from pathology images. arXiv preprint arXiv:2202.08195 (2022)
12. Lin, Y., Zhang, D., Fang, X., Chen, Y., Cheng, K.T., Chen, H.: Rethinking boundary detection in deep learning models for medical image segmentation. In: Frangi, A., de Bruijne, M., Wassermann, D., Navab, N. (eds.) Information Processing in Medical Imaging. IPMI 2023. Lecture Notes in Computer Science, vol. 13939, pp. 730–742. Springer, Cham (2023). https://doi.org/10.1007/978-3-031-34048-2_56
13. Liu, Y., Liu, N., Yao, X., Han, J.: Intermediate prototype mining transformer for few-shot semantic segmentation. In: Advances in Neural Information Processing Systems (NeurIPS) (2022)
14. Luo, X., Tian, Z., Zhang, T., Yu, B., Tang, Y.Y., Jia, J.: PFENet++: boosting few-shot semantic segmentation with the noise-filtered context-aware prior mask. arXiv preprint arXiv:2109.13788 (2021)
15. Ouyang, C., Biffi, C., Chen, C., Kart, T., Qiu, H., Rueckert, D.: Self-supervised learning for few-shot medical image segmentation. IEEE Trans. Med. Imaging **41**(7), 1837–1848 (2022)
16. Pan, W., et al.: Human-Machine Interactive Tissue Prototype Learning for Label-Efficient Histopathology Image Segmentation. In: Frangi, A., de Bruijne, M., Wassermann, D., Navab, N. (eds.) Information Processing in Medical Imaging. IPMI 2023. Lecture Notes in Computer Science, vol. 13939, pp. 679–691. Springer, Cham (2023). https://doi.org/10.1007/978-3-031-34048-2_5
17. Pandey, P., Vardhan, A., Chasmai, M., Sur, T., Lall, B.: Adversarially robust prototypical few-shot segmentation with neural-ODEs. In: Wang, L., Dou, Q., Fletcher, P.T., Speidel, S., Li, S. (eds.) Medical Image Computing and Computer Assisted Intervention – MICCAI 2022. MICCAI 2022. Lecture Notes in Computer Science, vol. 13438, pp. 77–87. Springer, Cham (2022). https://doi.org/10.1007/978-3-031-16452-1_8
18. Peng, B., et al.: Hierarchical dense correlation distillation for few-shot segmentation. In: Proceedings of the IEEE/CVF Conference on Computer Vision and Pattern Recognition, pp. 23641–23651 (2023)
19. Roy, A.G., Siddiqui, S., Pölsterl, S., Navab, N., Wachinger, C.: 'squeeze & excite' guided few-shot segmentation of volumetric images. Med. Image Anal. **59**, 101587 (2020)
20. Shen, Q., Li, Y., Jin, J., Liu, B.: Q-Net: query-informed few-shot medical image segmentation. arXiv preprint arXiv:2208.11451 (2022)
21. Siam, M., Oreshkin, B.N., Jagersand, M.: AMP: adaptive masked proxies for few-shot segmentation. In: Proceedings of the IEEE/CVF International Conference on Computer Vision (ICCV), pp. 5249–5258 (2019)
22. Snell, J., Swersky, K., Zemel, R.: Prototypical networks for few-shot learning. In: Advances in Neural Information Processing Systems (NeurIPS), vol. 30 (2017)
23. Sun, L., et al.: Few-shot medical image segmentation using a global correlation network with discriminative embedding. Comput. Biol. Med. **140**, 105067 (2022)
24. Sung, F., Yang, Y., Zhang, L., Xiang, T., Torr, P.H., Hospedales, T.M.: Learning to compare: Relation network for few-shot learning. In: Proceedings of the IEEE Conference on Computer Vision and Pattern Recognition (CVPR), pp. 1199–1208 (2018)

25. Tang, H., Liu, X., Sun, S., Yan, X., Xie, X.: Recurrent mask refinement for few-shot medical image segmentation. In: Proceedings of the IEEE/CVF International Conference on Computer Vision (ICCV), pp. 3918–3928 (2021)
26. Tian, Z., et al.: Generalized few-shot semantic segmentation. In: Proceedings of the IEEE/CVF Conference on Computer Vision and Pattern Recognition (CVPR) (2022)
27. Tian, Z., Zhao, H., Shu, M., Yang, Z., Li, R., Jia, J.: Prior guided feature enrichment network for few-shot segmentation. In: IEEE Transactions on Pattern Analysis and Machine Intelligence (2020)
28. Vinyals, O., Blundell, C., Lillicrap, T., Wierstra, D., et al.: Matching networks for one shot learning. In: Advances in Neural Information Processing Systems (NeurIPS), vol. 29 (2016)
29. Wang, K., Liew, J.H., Zou, Y., Zhou, D., Feng, J.: PANet: few-shot image semantic segmentation with prototype alignment. In: Proceedings of the IEEE/CVF International Conference on Computer Vision (ICCV), pp. 9197–9206 (2019)
30. Xu, Z., et al.: All-around real label supervision: cyclic prototype consistency learning for semi-supervised medical image segmentation. IEEE J. Biomed. Health Inf. **26**(7), 3174–3184 (2022)
31. Yang, X., Lin, Y., Wang, Z., Li, X., Cheng, K.T.: Bi-modality medical image synthesis using semi-supervised sequential generative adversarial networks. IEEE J. Biomed. Health Inform. **24**(3), 855–865 (2019)
32. Zhang, D., et al.: Deep learning for medical image segmentation: tricks, challenges and future directions. arXiv preprint arXiv:2209.10307 (2022)
33. Zhuang, X.: Multivariate mixture model for myocardial segmentation combining multi-source images. IEEE Trans. Pattern Anal. Mach. Intell. **41**(12), 2933–2946 (2018)

ECL: Class-Enhancement Contrastive Learning for Long-Tailed Skin Lesion Classification

Yilan Zhang, Jianqi Chen, Ke Wang, and Fengying Xie(✉)

Image Processing Center, School of Astronautics, Beihang University, Beijing 100191, China
xfy_73@buaa.edu.cn

Abstract. Skin image datasets often suffer from imbalanced data distribution, exacerbating the difficulty of computer-aided skin disease diagnosis. Some recent works exploit supervised contrastive learning (SCL) for this long-tailed challenge. Despite achieving significant performance, these SCL-based methods focus more on head classes, yet ignoring the utilization of information in tail classes. In this paper, we propose class-**E**nhancement **C**ontrastive **L**earning (ECL), which enriches the information of minority classes and treats different classes equally. For information enhancement, we design a hybrid-proxy model to generate class-dependent proxies and propose a cycle update strategy for parameters optimization. A balanced-hybrid-proxy loss is designed to exploit relations between samples and proxies with different classes treated equally. Taking both "imbalanced data" and "imbalanced diagnosis difficulty" into account, we further present a balanced-weighted cross-entropy loss following curriculum learning schedule. Experimental results on the classification of imbalanced skin lesion data have demonstrated the superiority and effectiveness of our method. The codes can be publicly available from https://github.com/zylbuaa/ECL.git.

Keywords: Contrastive learning · Dermoscopic image · Long-tailed classification

1 Introduction

Skin cancer is one of the most common cancers all over the world. Serious skin diseases such as melanoma can be life-threatening, making early detection and treatment essential [3]. As computer-aided diagnosis matures, recent advances with deep learning techniques such as CNNs have significantly improved the performance of skin lesion classification [7,8]. However, as data-hungry approaches, deep learning models require large balanced and high-quality datasets to meet the

Supplementary Information The online version contains supplementary material available at https://doi.org/10.1007/978-3-031-43895-0_23.

H. Greenspan et al. (Eds.): MICCAI 2023, LNCS 14221, pp. 244–254, 2023.
https://doi.org/10.1007/978-3-031-43895-0_23

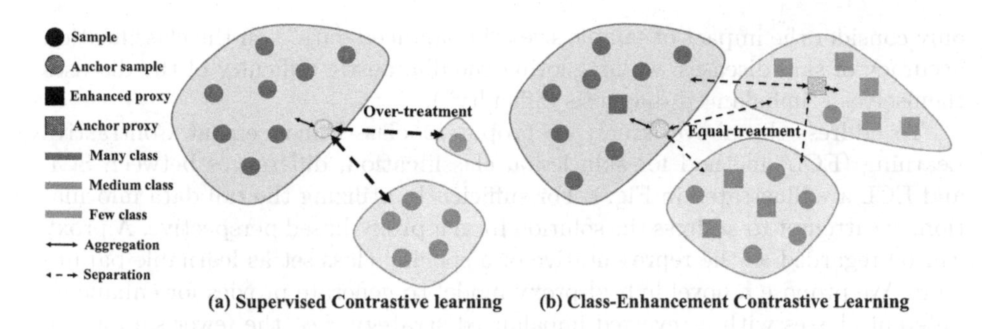

Fig. 1. Comparison between SCL (a) and ECL (b). In SCL, head classes are over-treated leading to optimization concentrating on head classes. By contrast, ECL utilizes the proxies to enhance the learning of tail classes and treats all classes equally according to balanced contrastive theory [24]. Moreover, the enriched relations in samples and proxies are helped for better representations.

accuracy and robustness requirements in applications, which is hard to suffice due to the long-tailed occurrence of diseases in the real-world. Long-tailed problem is usually caused by differences in incidence rate and difficulties in data collection. Some diseases are common while others are rare, making it difficult to collect balanced data [13]. This will cause the head classes to account for the majority of the samples and the tail classes only have small portions. Thus, existing public skin datasets usually suffer from imbalanced problems which then results in class bias of classifier, for example, poor model performance especially on tail lesion types.

To tackle the challenge of learning unbiased classifiers with imbalanced data, many previous works focus on three main ideas, including re-sampling data [1, 18], re-weighting loss [2,15,22] and re-balancing training strategies [10,23]. Re-sampling methods over-sample tail classes or under-sample head classes, re-weighting methods adjust the weights of losses on class-level or instance-level, and re-balancing methods decouple the representation learning and classifier learning into two stages or assign the weights between features from different sampling branches [21]. Despite the great results achieved, these methods either manually interfere with the original data distribution or improve the accuracy of minority classes at the cost of reducing that of majority classes [12,13].

Recently, contrastive learning (CL) methods pose great potential for representation learning when trained on imbalanced data [4,14]. Among them, supervised contrastive learning (SCL) [11] aggregates semantically similar samples and separates different classes by training in pairs, leading to impressive success in long-tailed classification of both natural and medical images [16]. However, there still remain some defects: (1) Current SCL-based methods utilize the information of minority classes insufficiently. Since tail classes are sampled with low probability, each training mini-batch inherits the long-tail distribution, making parameter updates less dependent on tail classes. (2) SCL loss focuses more on optimizing the head classes with much larger gradients than tail classes, which means tail classes are all pushed farther away from heads [24]. (3) Most methods only consider the impact of sample size ("imbalanced data") on the classification accuracy of skin diseases, while ignoring the diagnostic difficulty of the diseases themselves ("imbalanced diagnosis difficulty").

To address the above issues, we propose a class-Enhancement Contrastive Learning (ECL) method for skin lesion classification, differences between SCL and ECL are illustrated in Fig. 1. For sufficiently utilizing the tail data information, we attempt to address the solution from a proxy-based perspective. A proxy can be regarded as the representative of a specific class set as learnable parameters. We propose a novel hybrid-proxy model to generate proxies for enhancing different classes with a reversed imbalanced strategy, *i.e.*, the fewer samples in a class, the more proxies the class has. These learnable proxies are optimized with a cycle update strategy that captures original data distribution to mitigate the quality degradation caused by the lack of minority samples in a mini-batch. Furthermore, we propose a balanced-hybrid-proxy loss, besides introducing balanced contrastive learning (BCL) [24]. The new loss treats all classes equally and utilizes sample-to-sample, proxy-to-sample and proxy-to-proxy relations to improve representation learning. Moreover, we design a balanced-weighted cross-entropy loss which follows a curriculum learning schedule by considering both imbalanced data and diagnosis difficulty.

Our contributions can be summarized as follows: (1) We propose an ECL framework for long-tailed skin lesion classification. Information of classes are enhanced by the designed hybrid-proxy model with a cycle update strategy. (2) We present a balanced-hybrid-proxy loss to balance the optimization of each class and leverage relations among samples and proxies. (3) A new balanced-weighted cross-entropy loss is designed for an unbiased classifier, which considers both "imbalanced data" and "imbalanced diagnosis difficulty". (4) Experimental results demonstrate that the proposed framework outperforms other state-of-the-art methods on two imbalanced dermoscopic image datasets and the ablation study shows the effectiveness of each element.

2 Methods

The overall end-to-end framework of ECL is presented in Fig. 2. The network consists of two parallel branches: a contrastive learning (CL) branch for representative learning and a classifier learning branch. The two branches take in different augmentations $T^i, i \in \{1, 2\}$ from input images X and the backbone is shared between branches to learn the features $\tilde{X}^i, i \in \{1, 2\}$. We use a fully connected layer as a logistic projection for classification $g(\cdot) : \tilde{\mathcal{X}} \rightarrow \tilde{\mathcal{Y}}$ and a one-hidden layer MLP $h(\cdot) : \tilde{\mathcal{X}} \rightarrow \mathcal{Z} \in \mathbb{R}^d$ as a sample embedding head where d denotes the dimension. $\mathcal{L}_2$-normalization is applied to $\mathcal{Z}$ by using inner product as distance measurement in CL. Both the class-dependent proxies generated by hybrid-proxy model and the embeddings of samples are used to calculate balanced-weighted cross-entropy loss, thus capturing the rich relations of samples and proxies. For better representation, we design a cycle update strategy to optimize the proxies' parameters in hybrid-proxy model, together with a curriculum learning schedule for achieving unbiased classifiers. The details are introduced as follows.

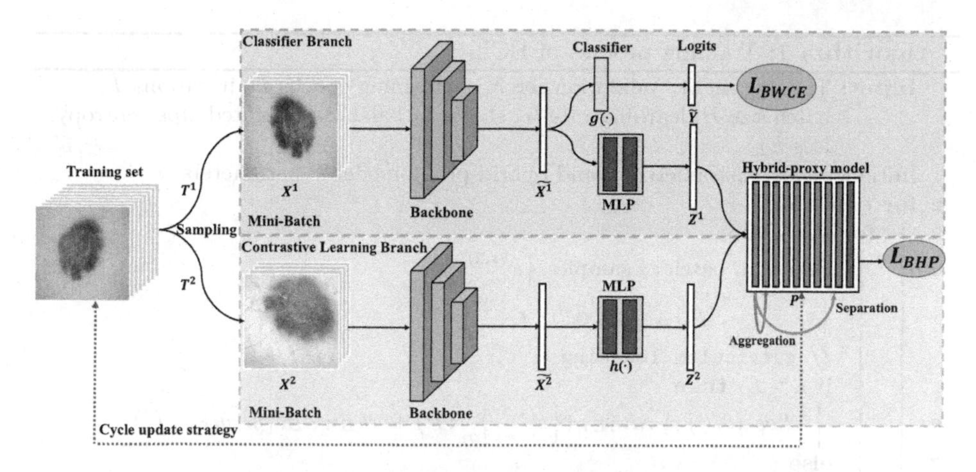

Fig. 2. Overall framework of the proposed ECL. ECL has two branches for classifier learning (guided by balanced-weighted cross-entropy loss L_{BWCE}) and contrastive learning (guided by balanced-hybrid-proxy loss L_{BHP}). Proxies in hybrid-proxy model are generated by a reserve imbalanced way (see Sect. 2.1) to strengthen the information of minority classes in a mini-batch.

2.1 Hybrid-Proxy Model

The proposed hybrid-proxy model consists of a set of class-dependent proxies $\mathcal{P} = \{p_k^c | k \in \{1, 2, ..., N_c^p\}, \ c \in \{1, 2, ..., C\}\}$, C is the class number, $p_k^c \in \mathbb{R}^d$ is the k-th proxy vector of class c, and N_c^p is the proxy number in this class. Since samples in a mini-batch follow imbalanced data distribution, these proxies are designed to be generated in a reversed imbalanced way by giving more representative proxies of tail classes for enhancing the information of minority samples. Let us denote the sample number of class c as N_c and the maximum in all classes as N_{max}. The proxy number N_c^p can be obtained by calculating the imbalanced factor $\frac{N_{max}}{N_c}$ of each class:

$$N_c^p = \begin{cases} 1 & N_c = N_{max} \\ \lfloor \frac{N_{max}}{10 N_c} \rfloor + 2 & N_c \neq N_{max} \end{cases} \tag{1}$$

In this way, the tail classes have more proxies while head classes have less, thus alleviating the imbalanced problem in a mini-batch.

As we know, a gradient descent algorithm will generally be executed to update the parameters after training a mini-batch of samples. However, when dealing with an imbalanced dataset, tail samples in a batch contribute little to the update of their corresponding proxies due to the low probability of being sampled. So how to get better representative proxies? Here we propose a cycle update strategy for the optimization of the parameters. Specifically, we introduce the gradient accumulation method into the training process to update proxies asynchronously. The proxies are updated only after a finished epoch that all data has been processed by the framework with the gradients accumulated. With such

Algorithm 1: Training process of ECL.

Input: Training set X, validation set X_{val}, training epochs E, iterations T, batch size B, learning rate lr, stages in balanced-weighted cross-entropy loss E_2

1 Initialize *model* parameters θ and hybrid-proxy model $\mathcal{P}$ parameters ϕ
2 **for** e *in* E **do**
3   **for** t *in* T **do**
4   Getting a batch of samples $\left\{x_i^{(1,2)}, y_i\right\}_B$
          $\left\{z_i^{(1,2)}\right\}_B, \{\tilde{y_i}\}_B = model(\left\{x_i^{(1,2)}\right\}_B)$
          `// curriculum learning`
5   **if** $e > E_2$ **then**
6             $Loss(\theta, \phi) = \lambda L_{BHP}(\left\{z_i^{(1,2)}\right\}_B, \mathcal{P}) + \mu L_{BWCE}(\{y_i, \tilde{y_i}\}_B, f^e)$
7   **else**
8             $Loss(\theta, \phi) = \lambda L_{BHP}(\left\{z_i^{(1,2)}\right\}_B, \mathcal{P}) + \mu L_{BWCE}(\{y_i, \tilde{y_i}\}_B)$
9         $grad_\theta^t = \nabla_\theta Loss(\theta), grad_\phi^t = \nabla_\phi Loss(\phi)$ `// calculate gradients`
10        $\theta \leftarrow \theta - lr * grad_\theta^t$`// update parameters` θ `of` *model*
11    $\phi \leftarrow \phi - \sum_t^T lr * grad_\phi^t$ `// update parameters` ϕ `of` $\mathcal{P}$
12   **if** $e > E_2$ **then**
13        $f^e = Validate(model, X_{val})$

a strategy, tail proxies can be optimized in a view of whole data distribution, thus playing better roles in class information enhancement. Algorithm 1 presents the details of the training process.

2.2 Balanced-Hybrid-Proxy Loss

To tackle the problem that SCL loss pays more attention on head classes, we introduce BCL and propose balanced-hybrid-proxy loss to treat classes equally. Given a batch of samples $\mathcal{B} = \left\{(x_i^{(1,2)}, y_i)\right\}_B$, let $\mathcal{Z} = \left\{z_i^{(1,2)}\right\}_B = \left\{z_1^1, z_2^2, ..., z_B^1, z_B^2\right\}$ be the feature embeddings in a batch and B denotes the batch size. For an anchor sample $z_i \in \mathcal{Z}$ in class c, we unify the positive image set as $z^+ = \{z_j | y_j = y_i = c, j \neq i\}$. Also for an anchor proxy p_i^c, we unify all positive proxies as p^+. The proposed balanced-hybrid-proxy loss pulls points (both samples and proxies) in the same class together, while pushes apart samples from different classes in embedding space by using dot product as a similarity measure, which can be formulated as follows:

$$L_{BHP} = -\frac{1}{2B + \sum_{c \in C} N_c^p} \sum_{s_i \in \{\mathcal{Z} \cup \mathcal{P}\}} \frac{1}{2B_c + N_c^p - 1} \sum_{s_j \in \{z^+ \cup p^+\}} log \frac{exp(s_i \cdot s_j / \tau)}{E} \tag{2}$$

$$E = \sum_{c \in C} \frac{1}{2B_c + N_c^p - 1} \sum_{s_k \in \{\mathcal{Z}_c \cup \mathcal{P}_c\}} exp(s_i \cdot s_k/\tau) \quad (3)$$

where B_c means the sample number of class c in a batch, τ is the temperature parameter. In addition, we further define $\mathcal{Z}_c$ and $\mathcal{P}_c$ as a subset with the label c of $\mathcal{Z}$ and $\mathcal{P}$ respectively. The average operation in the denominator of balanced-hybrid-proxy loss can effectively reduce the gradients of the head classes, making an equal contribution to optimizing each class. Note that our loss differs from BCL as we enrich the learning of relations between samples and proxies. Sample-to-sample, proxy-to-sample and proxy-to-proxy relations in the proposed loss have the potential to promote network's representation learning. Moreover, as the skin datasets are often small, richer relations can effectively help form a high-quality distribution in the embedding space and improve the separation of features.

2.3 Balanced-Weighted Cross-Entropy Loss

Taking both "imbalanced data" and "imbalanced diagnosis difficulty" into consideration, we design a curriculum schedule and propose balanced-weighted cross-entropy loss to train an unbiased classifier. The training phase are divided into three stages. We first train a general classifier, then in the second stage we assign larger weight to tail classes for "imbalanced data". In the last stage, we utilize the results on the validation set as the diagnosis difficulty indicator of skin disease types to update the weights for "imbalanced diagnosis difficulty". The loss is given by:

$$L_{BWCE} = -\frac{1}{B} \sum_{i=1}^{B} w_i CE(\tilde{y}_i, y_i) \quad (4)$$

$$w_i = \begin{cases} 1 & e < E_1 \\ \left(\frac{C/N_c}{\sum_{c \in C} 1/N_c}\right)^{\frac{e-E_1}{E_2-E_1}} & E_1 < e < E_2 \\ \left(\frac{C/f_c^e}{\sum_{c \in C} 1/f_c^e}\right)^{\frac{e-E_2}{E-E_2}} & E_2 < e < E \end{cases} \quad (5)$$

where w denotes the weight and $\tilde{y}$ denotes the network prediction. We assume f_c^e is the evaluation result of class c on validation set after epoch e and we use f1-score in our experiments. The network is trained for E epochs, E_1 and E_2 are hyperparameters for stages. The final loss is given by $Loss = \lambda L_{BHP} + \mu L_{BWCE}$ where λ and μ are the hyperparameters which control the impact of losses.

3 Experiment

3.1 Dataset and Implementation Details

Dataset and Evaluation Metrics. We evaluate the ECL on two publicly available dermoscopic datasets ISIC2018 [5,19] and ISIC2019 [5,6,19]. The 2018

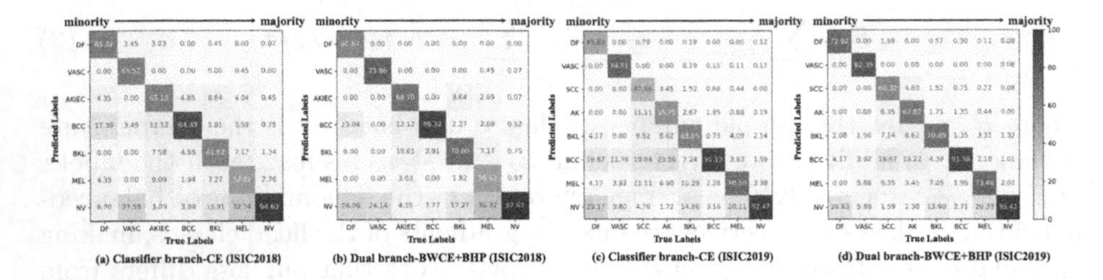

Fig. 3. The results of confusion matrix illustrate that ECL obtains great performance on most classes especially for minority classes.

dataset consists of 10015 images in 7 classes while a larger 2019 dataset provides 25331 images in 8 classes. The imbalanced factors $\alpha = \frac{N_{max}}{N_{min}}$ of the two datasets are all > 50 (ISIC2018 58.30 and ISIC2019 53.87), which means that skin lesion classification suffers a serious imbalanced problem. We randomly divide the samples into the training, validation and test sets as 3:1:1.

We adopt five metrics for evaluation: accuracy (Acc), average precision (Pre), average sensitivity (Sen), macro f1-score (F1) and macro area under curve (AUC). Acc and F1 are considered as the most important metrics in this task.

Implementation Details. The proposed algorithm is implemented in Python with Pytorch library and runs on a PC equipped with an NVIDIA A100 GPU. We use ResNet50 [9] as backbone and the embedding dimension d is set to 128. We use SGD as the optimizer with the weight decay 1e-4. The initial learning rate is set to 0.002 and decayed by cosine schedule. We train the network for 100 epochs with a batch size of 64. The hyperparameters E_1, E_2, τ, λ, and μ are set to 20, 50, 0.01, 1, and 2 respectively. We use the default data augmentation strategy on ImageNet in [9] as T_1 for classification branch. And for CL branch, we add random grayscale, rotation, and vertical flip in T_1 as T_2 to enrich the data representations. Meanwhile, we only conduct the resize operation to ensure input size $224 \times 224 \times 3$ during testing process. The models with the highest Acc on validation set are chosen for testing. We conduct experiments in 3 independent runs and report the standard deviations in the supplementary material.

3.2 Experimental Results

Quantitative Results. To evaluate the performance of our ECL, we compare our method with 10 advanced methods. Among them, focal loss [15], LDAM-DRW [2], logit adjust [17], and MWNL [22] are the re-weighting loss methods. BBN [23] is the methods based on re-balancing training strategy while Hybrid-SC [20], SCL [11,16], BCL [24], TSC [14] and ours are the CL-based methods. Moreover, MWNL and SCL have been verified to perform well in the skin disease classification task. To ensure fairness, we re-train all methods by rerun their released codes on our divided datasets with the same experimental settings.

Table 1. Comparison results on ISIC2018 and ISIC2019 datasets.

Methods	ISIC2018					ISIC2019				
	Acc	Sen	Pre	F1	AUC	Acc	Sen	Pre	F1	AUC
CE	83.89	69.56	73.62	70.34	94.81	82.41	67.02	77.32	70.90	95.37
Focal Loss	84.19	68.78	76.69	71.38	94.76	82.05	64.55	75.93	68.84	94.82
LDAM-DRW	84.20	71.74	74.65	71.98	95.22	82.29	68.08	74.61	70.84	95.65
Logit Adjust	84.15	71.54	71.78	70.77	95.55	81.93	68.94	69.12	68.64	95.17
MWNL	84.90	73.90	76.94	74.92	**96.79**	84.10	74.83	75.81	75.08	96.61
BBN	85.57	**74.96**	72.40	72.79	93.72	83.43	71.78	78.37	74.42	95.10
Hybrid-SC	86.30	73.93	75.84	74.34	96.33	84.69	70.90	76.87	73.27	96.67
SCL	86.13	70.40	80.88	74.27	96.56	84.60	70.90	81.66	75.07	96.21
BCL	84.92	72.87	71.15	71.57	95.61	83.47	73.52	74.17	73.50	95.95
TSC	85.94	73.35	77.77	74.94	95.83	84.75	71.89	79.81	75.13	95.84
Ours	**87.20**	73.01	**83.44**	**76.76**	96.55	**86.11**	**76.57**	**83.22**	**79.46**	**96.78**

We also confirmed that all models have converged and choose the best eval checkpoints. The results are shown in Table 1. It can be seen that ECL has a significant advantage with the highest level in most metrics on two datasets. Noticeably, our ECL outperforms other imbalanced methods by great gains, e.g., 2.56% in Pre on ISIC2018 compared with SCL and 4.33% in F1 on ISIC2019 dataset compared with TSC. Furthermore, we draw the confusion matrixes after normalization in Fig. 3, which illustrate that ECL has significantly improved most of the categories, from minority to majority.

Table 2. Ablation study on ISIC2019 dataset.

Methods(ISIC2019)	Proxies	Acc	Sen	Pre	F1	AUC
Classifier branch-CE	HPM	82.41	67.02	77.32	70.90	95.37
Classifier branch-BWCE	HPM	82.69	67.95	77.32	71.65	95.37
Dual branch-CE+BHP	HPM	85.49	73.35	81.61	76.76	96.52
Dual branch-BWCE+BHP	2 proxies per-class	85.52	74.03	81.46	77.22	96.53
Dual branch-BWCE+BHP	3 proxies per-class	85.36	73.49	83.00	77.53	96.74
Dual branch-BWCE+BHP	4 proxies per-class	85.79	74.09	82.03	77.42	96.53
Dual branch-BWCE+BHP	w/o cycle stategy	85.65	73.48	83.00	77.40	96.65
	HPM	**86.11**	**76.57**	**83.22**	**79.46**	**96.78**

Ablation Study. To further verify the effectiveness of the designs in ECL, we conduct a detailed ablation study shown in Table 2 (the results on ISIC2018 are shown in supplementary material Table S2). First, we directly move the contrastive learning (CL) branch and replaced the balanced-weighted cross-entropy

(BWCE) loss with cross-entropy (CE) loss. We can see from the results that adding CL branch can significantly improve the network's data representation ability with better performance than only adopting a classifier branch. And our BWCE loss can help in learning a more unbiased classifier with an improvement of 2.7% in F1 compared to CE in dual branch setting. Then we train the ECL w/o cycle update strategy. The overall performance of the network has declined compared with training w/ the strategy, indicating that this strategy can better enhance proxies learning through the whole data distribution. In the end, we also set the proxies' number of different classes equal to explore whether the classification ability of the network is improved due to the increase in the number of proxies. With more proxies, metrics fluctuate and do not increase significantly. However, the result of using proxies generated by reversed balanced way in hybrid-proxy model (HPM) outperforms equal proxies in nearly all metrics, which proves that giving more proxies to tail classes can effectively enhance and enrich the information.

4 Conclusion

In this work, we present a class-enhancement contrastive learning framework, named ECL, for long-tailed skin lesion classification. Hybrid-proxy model and balanced-hybrid-proxy loss are proposed to tackle the problem that SCL-based methods pay less attention to the learning of tail classes. Class-dependent proxies are generated in hybrid-proxy model to enhance information of tail classes, where rich relations between samples and proxies are utilized to improve representation learning of the network. Furthermore, balanced-weighted cross-entropy loss is designed to help train an unbiased classifier by considering both "imbalanced data" and "imbalanced diagnosis difficulty". Extensive experiments on ISIC2018 and ISIC2019 datasets have demonstrated the effectiveness and superiority of ECL over other compared methods.

References

1. Ando, S., Huang, C.Y.: Deep over-sampling framework for classifying imbalanced data. In: Ceci, M., Hollmén, J., Todorovski, L., Vens, C., Džeroski, S. (eds.) ECML PKDD 2017. LNCS (LNAI), vol. 10534, pp. 770–785. Springer, Cham (2017). https://doi.org/10.1007/978-3-319-71249-9_46
2. Cao, K., Wei, C., Gaidon, A., Arechiga, N., Ma, T.: Learning imbalanced datasets with label-distribution-aware margin loss. In: Advances in Neural Information Processing Systems, vol. 32 (2019)
3. Cassidy, B., Kendrick, C., Brodzicki, A., Jaworek-Korjakowska, J., Yap, M.H.: Analysis of the ISIC image datasets: usage, benchmarks and recommendations. Med. Image Anal. **75**, 102305 (2022)
4. Chen, J., Chen, K., Chen, H., Li, W., Zou, Z., Shi, Z.: Contrastive learning for fine-grained ship classification in remote sensing images. IEEE Trans. Geosci. Remote Sens. **60**, 1–16 (2022)

5. Codella, N.C., et al.: Skin lesion analysis toward melanoma detection: a challenge at the 2017 international symposium on biomedical imaging (ISBI), hosted by the international skin imaging collaboration (ISIC). In: 2018 IEEE 15th International Symposium on Biomedical Imaging (ISBI 2018), pp. 168–172. IEEE (2018)
6. Combalia, M., et al.: Bcn20000: dermoscopic lesions in the wild. arXiv preprint arXiv:1908.02288 (2019)
7. Esteva, A., et al.: Dermatologist-level classification of skin cancer with deep neural networks. Nature **542**(7639), 115–118 (2017)
8. Hasan, M.K., Ahamad, M.A., Yap, C.H., Yang, G.: A survey, review, and future trends of skin lesion segmentation and classification. Comput. Biol. Med. **155**, 106624 (2023)
9. He, K., Zhang, X., Ren, S., Sun, J.: Deep residual learning for image recognition. In: Proceedings of the IEEE Conference on Computer Vision and Pattern Recognition, pp. 770–778 (2016)
10. Kang, B., et al.: Decoupling representation and classifier for long-tailed recognition. In: International Conference on Learning Representations (2019)
11. Khosla, P., et al.: Supervised contrastive learning. Adv. Neural. Inf. Process. Syst. **33**, 18661–18673 (2020)
12. Lango, M., Stefanowski, J.: What makes multi-class imbalanced problems difficult? An experimental study. Expert Syst. Appl. **199**, 116962 (2022)
13. Li, J., et al.: Flat-aware cross-stage distilled framework for imbalanced medical image classification. In: Wang, L., Dou, Q., Fletcher, P.T., Speidel, S., Li, S. (eds.) MICCAI 2022, Part III. LNCS, vol. 13433, pp. 217–226. Springer, Cham (2022). https://doi.org/10.1007/978-3-031-16437-8_21
14. Li, T., et al.: Targeted supervised contrastive learning for long-tailed recognition. In: Proceedings of the IEEE/CVF Conference on Computer Vision and Pattern Recognition, pp. 6918–6928 (2022)
15. Lin, T.Y., Goyal, P., Girshick, R., He, K., Dollár, P.: Focal loss for dense object detection. In: Proceedings of the IEEE International Conference on Computer Vision, pp. 2980–2988 (2017)
16. Marrakchi, Y., Makansi, O., Brox, T.: Fighting class imbalance with contrastive learning. In: de Bruijne, M., et al. (eds.) MICCAI 2021, Part III. LNCS, vol. 12903, pp. 466–476. Springer, Cham (2021). https://doi.org/10.1007/978-3-030-87199-4_44
17. Menon, A.K., Jayasumana, S., Rawat, A.S., Jain, H., Veit, A., Kumar, S.: Long-tail learning via logit adjustment. In: International Conference on Learning Representations (2020)
18. Pouyanfar, S., et al.: Dynamic sampling in convolutional neural networks for imbalanced data classification. In: 2018 IEEE Conference on Multimedia Information Processing and Retrieval (MIPR), pp. 112–117. IEEE (2018)
19. Tschandl, P., Rosendahl, C., Kittler, H.: The ham10000 dataset, a large collection of multi-source dermatoscopic images of common pigmented skin lesions. Sci. Data **5**(1), 1–9 (2018)
20. Wang, P., Han, K., Wei, X.S., Zhang, L., Wang, L.: Contrastive learning based hybrid networks for long-tailed image classification. In: Proceedings of the IEEE/CVF Conference on Computer Vision and Pattern Recognition, pp. 943–952 (2021)
21. Yang, L., Jiang, H., Song, Q., Guo, J.: A survey on long-tailed visual recognition. Int. J. Comput. Vision **130**(7), 1837–1872 (2022)
22. Yao, P., et al.: Single model deep learning on imbalanced small datasets for skin lesion classification. IEEE Trans. Med. Imaging **41**(5), 1242–1254 (2021)

23. Zhou, B., Cui, Q., Wei, X.S., Chen, Z.M.: BBN: bilateral-branch network with cumulative learning for long-tailed visual recognition. In: Proceedings of the IEEE/CVF Conference on Computer Vision and Pattern Recognition, pp. 9719–9728 (2020)
24. Zhu, J., Wang, Z., Chen, J., Chen, Y.P.P., Jiang, Y.G.: Balanced contrastive learning for long-tailed visual recognition. In: Proceedings of the IEEE/CVF Conference on Computer Vision and Pattern Recognition, pp. 6908–6917 (2022)

Learning Transferable Object-Centric Diffeomorphic Transformations for Data Augmentation in Medical Image Segmentation

Nilesh Kumar[1(✉)], Prashnna K. Gyawali[2], Sandesh Ghimire[3], and Linwei Wang[1]

[1] Rochester Institute of Technology, Rochester, NY, USA
nk4856@rit.edu
[2] West Virginia University, Morgantown, USA
[3] Qualcomm Inc, San Diego, USA

Abstract. Obtaining labelled data in medical image segmentation is challenging due to the need for pixel-level annotations by experts. Recent works have shown that augmenting the object of interest with deformable transformations can help mitigate this challenge. However, these transformations have been learned globally for the image, limiting their transferability across datasets or applicability in problems where image alignment is difficult. While object-centric augmentations provide a great opportunity to overcome these issues, existing works are only focused on position and random transformations without considering shape variations of the objects. To this end, we propose a novel object-centric data augmentation model that is able to learn the shape variations for the objects of interest and augment the object in place without modifying the rest of the image. We demonstrated its effectiveness in improving kidney tumour segmentation when leveraging shape variations learned both from within the same dataset and transferred from external datasets.

Keywords: Data Augmentation · Diffeomorphic transformations · Image Segmentation

1 Introduction

A must-have ingredient for training a deep neural network (DNN) is a large number of labelled data that is not always available in real-world applications. This challenge of data annotation becomes even worse for medical image segmentation tasks that require pixel-level annotation by experts. Data augmentation (DA) is a recognized approach to tackle this challenge. Common DA strategies create new samples by using predefined transformations such as rotation, translation, and colour jitter to existing data, where the performance gains heavily relies on the choice of augmentation operations and parameters [1].

H. Greenspan et al. (Eds.): MICCAI 2023, LNCS 14221, pp. 255–265, 2023.
https://doi.org/10.1007/978-3-031-43895-0_24

To mitigate this reliance, recent efforts have focused on learning optimal augmentation operations for a given task and dataset [3,8,11,15]. However, transformations learned from these methods are typically still limited to a predefined set of simple operations such as rotation, translation, and scaling. In the meantime, another direction of research has emerged that provides an alternative way of learning more expressive augmentations based on deformation-based transformations commonly used in image registration [6,12,16]. Instead of pre-specifying a list of operations such as rotation and scaling [3], these deformation-based transformations can describe more general spatial transformations. Moreover, they are perfectly suited for modelling an object's shape changes [16] that are crucial for image segmentation tasks. It thus provides an excellent candidate for learning shape variations of an object from the data, and via which to enable shape-based augmentations for medical image segmentation tasks. [12,16].

However, to date, all existing approaches to learning deformable registration-based DA assume a perfect alignment of image pairs to learn the transformations. In other words, the deformation-based transformations are learned globally for the image. This assumption is restrictive and associated with several challenges. First, the learning of a global image-level transformation requires image alignment that may be non-trivial in many scenarios, such as the alignment of tumours that can appear at different locations of an image, or alignment of images from different modalities. The learning of transformations itself is also complicated by the presence of other objects in the image and is best suited when the object of interest is always in the same (and often centre) location in all the images, *i.e.*, images are globally aligned *a priori* [16]. Second, the application of the learned global transformations for DA is also restricted to images similar (and aligned) to those in training. It thus will be challenging to transfer the learned shape variations to even the same objects across different locations, orientations, or sizes in the image, let alone transferring across dataset (*e.g.*, to transfer the learned shape variations of an organ from one image modality to another).

Intuitively, object-centric transformations and augmentations have the potential to overcome the challenges associated with global image-level transformations. Recently, an object-centric augmentation method termed as TumorCP [13] showed that a simple object-level augmentation, via copy-pasting a tumour from one location to another, can yield impressive performance gains. However, the diversity of samples generated by TumorCP is limited to pasting tumours on different backgrounds with random distortions without further learned shape-based augmentation.

Similarly, other existing works on object-level augmentation of lesions have mostly focused on position, orientation, and random transformations of the lesion on different backgrounds [14,17]. To date, no existing works have considered shape-based object-centric augmentations. Enriching object-centric DA with learned shape variations – a factor critical to object segmentation – can result in more diverse samples and thereby improve DNN training for medical image segmentation.

In this paper, we present a novel approach for learning and transferring object-centric deformations for DA in medical image segmentation tasks. As illustrated in Fig. 1, this is achieved with two key elements:

- A generative model of object-centric deformations – constrained to C1 diffeomorphism for better DNN training – to describe shape variability learned from paired patches of objects of interest. This allows the learning to focus on the shape variations of an object regardless of its positions and sizes in the image, thus bypassing the requirement for image alignment.
- An online augmentation strategy to sample transformations from the generative model and to augment the objects of interest in place without distorting the surrounding content in the image. This allows us to add shape diversity to the objects of interest in an image regardless of their positions or sizes, eventually facilitating transferring the learned variations across datasets.

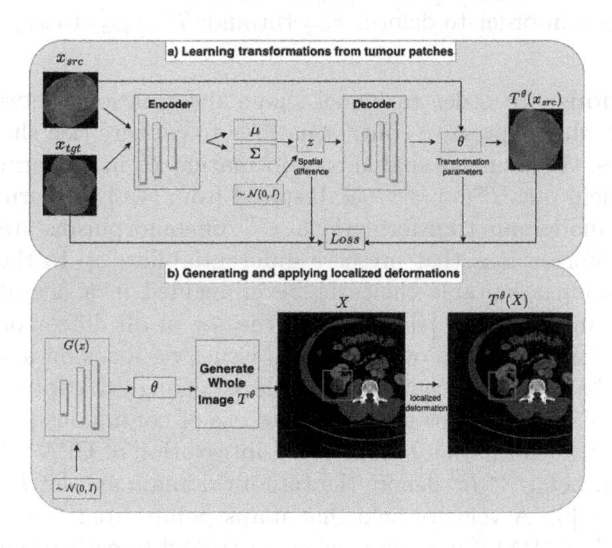

Fig. 1. Overview of the presented approach. a) Learning a generative model describing object-centric shape variations as diffeomorphic transformations. b) Sampling transformations from the learned generative model and deforming an object in place (highlighted with a red square) without distorting the surrounding content in the image. (Color figure online)

We demonstrated the effectiveness of the presented object-centric diffeomorphic augmentation in kidney tumour segmentation, including using shape variations of kidney tumours learned from the same dataset (KiTS [7]), as well as transferring those learned from a larger liver tumour dataset (LiTS [2]). Experimental results showed that it can enrich the augmentation diversity of other techniques such as TumorCP [13], and improve kidney tumour segmentation [7] using shape variations learned either within or outside the same training data.

2 Methods

We focus on DA for tumour segmentation because tumours can occur at different locations of an organ with substantially different orientations and sizes. It thus presents a challenging scenario where global image-level deformable transformations cannot apply. mentation approach comprises as outlined in Below we describe the two key methodological elements.

2.1 Object-Centric Diffeomorphism as a Generative Model

The goal of this element is to learn to generate diffeomorphic transformation parameters θ that describe shape variations – in the form of deformable transformations T^{θ} – that are present within training instances of tumour x. To realize this, we train a generative model $G(.)$ for θ such that, when given two instances of tumours (x_{src}, x_{tgt}), it is asked to generate θ from the encoded latent representations z in order to deform x_{src} through $T^{\theta}(x_{src})$ to x_{tgt}.

Transformations: In order to model shape deformations between x_{src} and x_{tgt}, we need highly expressive transformations to capture rich shape variations in tumour pairs. We assume a spatial transformation T^{θ} in the form of pixel-wise displacement field u as $T^{\theta}(x) = x + u$. Inspired from [4,6], we turn to C^1 diffeomorphisms to model our transformations. C^1 diffeomorphisms are smooth and invertible transformations that preserve differentiability up to the first derivative, making them a suitable choice to be embedded in a neural network for gradient-based optimization [4]. However, the set of all diffeomorphisms is an infinitely large Lie group. To overcome this issue, we focus on a specific finite-dimensional subset of the Lie group that is large enough to capture the relevant variations in the tumours. For this, we make use of continuous piecewise-affine-based (CPAB) transformation based on the integration of CPA velocity field v^{θ} proposed in [5]. Let $\Omega \subset R^2$ denote the tumour domain and let P be triangular tesselation of Ω [6]. A velocity field that maps points from Ω to R^2 is said to be piecewise affine (PA) if it is affine when restricted to each triangle of P. The set V of v^{θ} which are zero on the boundary of Ω can be shown to be finite-dimensional linear space [5]. The dimensionality d of V is a result of how fine P is tessellated. It can be shown that V is parameterized by θ, *i.e.*, any instance of V is a linear combination of d orthonormal CPA fields with weights θ [5]. A spatial transformation T^{θ} can be derived by integrating a velocity field v^{θ} [5] as:

$$u^{\theta}(x,t) = x + \int_0^t v^{\theta}(u^{\theta}(x,t))dt \tag{1}$$

where the integration can be done via a specialized solver [5]. The solver chosen produces faster and more accurate results than a generic ODE solver. Specifically, the cost for this solver is $O(C1)+O(C2 \text{ x Number of integration steps})$, where $C1$ is matrix exponential for the number of cells an image is divided into and $C2$ is the dimensionality of an image. The transformations T^{θ} thus can be described by a generative model of θ. We also experimented with squaring and scaling layers for integration but that resulted in texture loss when learning transformations.

Generative Modeling: The data generation process can be described as:

$$p(z) \sim \mathcal{N}(0, I), \quad \theta \sim p_\phi(\theta|z), \quad x_{tgt} \sim p(x_{tgt}|\theta, x_{src}) \tag{2}$$

$$p(x_{tgt}, z|x_{src}) = p(z) \int_\theta p(x_{tgt}|\theta, x_{src}) p_\phi(\theta|z) d\theta = p(z) p_\phi(x_{tgt}|z, x_{src}) \tag{3}$$

where z is the latent variable assumed to follow an isotropic Gaussian prior, $p_\phi(\theta|z)$ is modeled by a neural network parameterized by ϕ, and $p(x_{tgt}|\theta, x_{src})$ follows the deformable transformation as described in Equation (1).

We define variational approximations of the posterior density as $q_\psi(z|x_{src}, x_{tgt})$, modeled by a convolutional neural network that expects two inputs x_{src} and x_{tgt}. Passing a tuple of x_{src} and x_{tgt} as the input helps the latent representations to learn the spatial difference between two tumour samples. Alternatively, the generative model as described can be considered as a conditional model where both the generative and inference model is conditioned on the source tumour sample x_{src}.

Variational Inference: The parameters ψ and ϕ are optimized by the modified evidence lower bound (ELBO) of the log-likelihood $\log p(x_{tgt}|x_{src})$:

$$\begin{aligned} \log p(x_{tgt}|x_{src}) \geq \mathcal{L}_{ELBO} = E_{q_\psi(z|x_{src}, x_{tgt})} p_\phi(x_{tgt}|z, x_{src}) \\ -\beta D_{KL}(q_\psi(z|x_{src}, x_{tgt}) || p(z)) \end{aligned} \tag{4}$$

where the first term in the ELBO takes the form of similarity loss: L_2 norm on the difference between x_{tgt} and $\hat{x}_{src} = T^\theta(x_{src})$ synthesized using the θ from $G(z)$.

The second KL term constrains our approximated posterior $q_\psi(z|x_{src}, x_{tgt})$ to be closer to the isotropic Gaussian prior $p(z)$, and its contribution to the overall loss is scaled by the hyperparameter β. To further ensure that $\hat{x}_{src}$ looks realistic, we discourage $G(z)$ from generating overly-expressive transformations by adding a regularization term over the L_2 norm of the displacement field u with a tunable hyperparameter λ_{reg}. The final objective function becomes:

$$\mathcal{L} = \mathcal{L}_{ELBO} + \lambda_{reg} * ||u||_2 \tag{5}$$

Object-Centric Learning: To learn object-centric spatial transformations, x_{src} and x_{tgt} are in the forms of image patches that solely contain tumours. Given an image and its corresponding tumour segmentation mask (X, Y), we first extract a bounding box around the tumour by applying skimage.measure.regionprops from the scikit-image package to Y. We then use this bounding box to carve out the tumour x from the image X, masking out all the regions within the bounding box that do not belong to the tumour. All the tumour patches are then resized to the same scale, such that tumours of different sizes can be described by the same tesselation resolution. When pairing tumour patches, we pair each tumour with its K nearest neighbour tumours based on their Euclidean distance – this again avoids learning overly expressive transformation when attempting to deform between significantly different tumour shapes.

2.2 Online Augmentations with Generative Models

The goal of this element is to sample random object-centric transformations of T^θ from $G(z)$, to generate diverse augmentations of different instances of tumours in place. However, if we only transform the tumour and keep the rest of the image identical, the transformed tumour may appear unrealistic and out of place. To ensure that the entire transformed image appears smooth, we use a hybrid strategy to construct a deformation field for the entire image X that combines tumour-specific deformations with an identity transform for the rest of the image. Specifically, we fill a small region around the tumour with displacements of diminishing magnitudes, achieved by propagating the deformations from the boundaries of the deformation fields from $G(z)$ to their neighbours with reduced magnitudes. Repeating this process ensures that the change at the boundaries is smooth and that the transformed region appears naturally as part of the image.

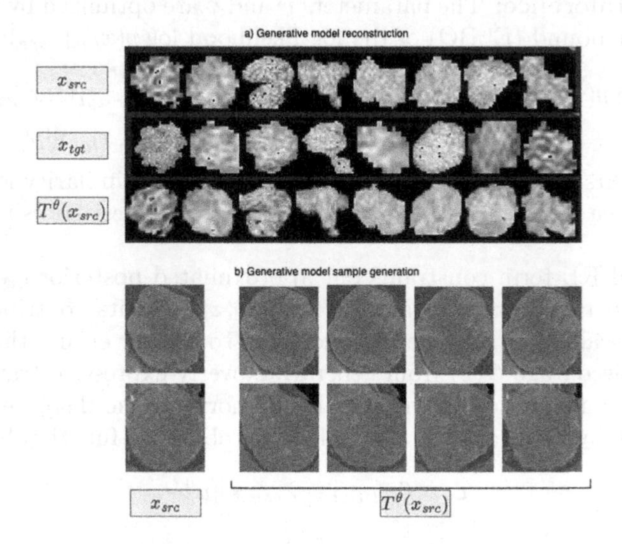

Fig. 2. Visual examples of the generative model in (a) reconstructing x_{tgt} given pairs of x_{scr} and x_{tgt}, and (b) generating deformed samples given a single x_{src}.

3 Experiments and Results

We used two publicly available datasets, LiTS [2] and KiTS [7], for our experiments. LiTS [7] contains liver and liver tumour segmentation masks for 200 scans in total, 130 train and 70 test. Similarly, KiTS [2] has kidney and kidney tumour segmentation masks for 300 scans, 168 train, 42 validation, and 90 test. We trained our generative model $G(z)$ on KiTS and LiTS separately to learn spatial variations in tumour shapes. We then used either of the learned transformation to augment kidney tumor segmentation tasks on subsets of KiTS data with varying sizes. Code link: https://github.com/nileshkumar0726/Learning_Transformations

3.1 Generative Model Implementation, Training, and Evaluation

Data: We prepared data for the generative model $G(z)$ training by first carving out tumour regions from individual slices of 3D scans using tumour segmentation masks. All tumour patches were resized to 30×30, and each was paired with eight of its closest neighbours. We trained $G(z)$ with different sizes of data for individual experiments presented in Table 1, ranging from using 11000 pairs from LiTS to only 3000 pairs when using only 25% samples from KiTS.

Model: The encoder of $G(z)$ consisted of five convolutional layers and three fully connected layers, with a latent dimension of 12 for z. The decoder consisted of five fully connected layers to output the parameters θ for T^{θ}. We trained the $G(z)$ for a total of 400 epochs and a batch size of 16. We also implemented early stopping if the validation loss does not improve for 20 epochs. We used Adam optimizer [10] with a learning rate of 1e-4.

We trained separate $G(z)$'s from KiTS and LiTS, respectively. We set $\beta = 0.001$ for both models but needed a high λ_{reg} of 0.009 for the LiTS model compared to 0.004 for KiTS model. The tumours in the LiTS have higher intensity differences, which may explain why a higher value of λ_{reg} was needed to ensure that transformed tumours did not become unrealistic.

Results: We evaluated $G(z)$ with two criteria. First, the model needs to be able to reconstruct x_{tgt} by generating θ to transform x_{src}. Second, the model needs to be able to generate diverse transformed tumour samples for a given tumour sample. Figure 2 presents visual examples of the reconstruction and generation results achieved by $G(z)$. It can be observed that the reconstruction is successful in most cases, except when x_{src} and x_{tgt} were too different. This was necessary to ensure that $T^{\theta}(x_{src})$ did not produce unrealistic examples. The averaged L2-loss of transformed $\hat{x}_{src}$ was 1.23 on the validation pairs. We also visually inspected validation samples after training to make sure that the deformed tumours were similar to the original tumours in appearance. The generated examples of tumours from a single source, as shown in Fig. 2(b), demonstrated that the generations were diverse yet realistic.

Table 1. KiTS segmentation results in terms of DICE score. The baseline model already includes standard data augmentations. Within-data augmentations used transformations learned from KiTS using the same % of training data for segmentation tasks. Cross-data augmentations used transformations learned from LiTS. TumorCP was also always performed within data.

% data for training	Augmentations	Mean Dice (std)* ↑
25%	Baseline	0.467 (0.014)
	Random Wrapping	0.535 (0.003)
	TumorCP	0.568 (0.014)
	Diffeo (within-data | cross-data)	0.497 (0.006) | 0.505 (0.002)
	TumorCP + Diffeo (within-data | cross-data)	**0.581 (0.012)** | 0.576 (0.015)
50%	Baseline	0.608 (0.017)
	Random Wrapping	0.6675 (0.0091)
	TumorCP	0.669 (0.011)
	Diffeo (within-data | cross-data)	0.640 (0.002) | 0.639 (0.014)
	TumorCP + Diffeo (within-data | cross-data)	0.689 (0.013) | **0.702 (0.016)**
75%	Baseline	0.656 (0.027)
	Random Wrapping	0.6774 (0.0036)
	TumorCP	0.690 (0.003
	Diffeo (within-data | cross-data)	0.662 (0.006) | 0.655 (0.001)
	TumorCP+ Diffeo (within-data | cross-data)	0.707 (0.001) | **0.718 (0.007)**
100%	Baseline	0.680 (0.025)
	Random Wrapping	0.6698 (0.016)
	TumorCP	0.702 (0.005)
	Diffeo (within-data | cross-data)	0.687 (0.014) | 0.688 (0.028)
	TumorCP + Diffeo (within-data | cross-data)	0.709 (0.004) | **0.713 (0.019)**

3.2 Deformation-Based da for Kidney Tumour Segmentation

Data: We then used $G(z)$ to generate deformation-based augmentations to increase the size and diversity of training samples for kidney tumour segmentation on KiTS. To assess the effect of augmentations on different sizes of labelled data, we considered training using 25%, 50%, 75%, and 100% of the KiTS training set. We considered two DA scenarios: augment with transformations learned from KiTS (within-data augmentation) versus from LiTS (cross-data augmentation).

Models: For the base segmentation network, we adopted nnU-net [9] as it contains state of the art (SOTA) pipeline for medical image segmentation on most datasets. To make the segmentation pipeline compatible with $G(z)$, we used the 2D segmentation module of nnU-net. For baselines, we considered 1) default augmentations such as rotation, scaling, and random crop in nnU-net as well as 2) TumorCP, all modified for 2D segmentation. Note that our goal is not to achieve SOTA results on KiTS, but to test the relative efficacy of the presented DA strategies in comparison with existing object-centric DA methods.

Results: We use Sørensen-Dice Coefficient (Dice) to measure segmentation network performance. Dice measures the overlap between prediction and ground truth. As summarized in Table 1, when combined with TumorCP, the presented augmentations were able to generate statistically significant (paired t-test, $p \leq 0.05$) improvements in all cases compared to TumorCP alone. This demonstrated the benefit of enriching simple copy-and-paste DA with shape variations. Interestingly, cross-data transferring of the learned augmentations (from LiTS) outperformed the within-data augmentation in the majority of the cases. Which we believe is because of two factors. Firstly, learning of the within-data augmentations is limited to the percentage of the training set used for segmentation. The number of objects to learn transformations from is thus greater in cross-data augmentation settings. Secondly, the transformations present in cross-data are completely unseen in the segmentation training network which helps in generating more diverse samples. Note that, as the transformations are learned as variations in object shapes, they can be transferred easily across datasets

Surprisingly, the improvements achieved by the presented augmentation strategy were the most prominent when the segmentation was trained on 50% and 75% of the KiTS training set. This is contrary to the expectation that DA would be most beneficial when the labelled training set is small. This may be because smaller sample sizes do not provide sufficient initial tumor samples for shape transformations. This may also explain why the combination of TumorCP boosted the performance of our augmentation strategy, as the oversampling nature of TumorCP provided more tumour samples for the presented strategy to transform to further enrich the training set. It is also worth noting that in contrast to prior literature, random wrapping of objects does not come close to the learned augmentations. We speculate that while unrealistic transformations work for whole images, they may be problematic when only augmenting specific local objects in an image.

4 Discussion and Conclusions

In this work, we presented a novel diffeomorphism-based object-centric augmentation that can be learned and used to augment the objects of interest regardless of their position and size in an image. As demonstrated by the experimental results, this allows us to not only introduce new variations to unfixed objects like tumours in an image but also transfer the knowledge of shape variations across datasets. An immediate next step will be to extend the presented approach to learn and transfer 3D transformations for 3D segmentation tasks, and to enrich the shape-based transformation with appearance-based transformations. In the long term, it would be interesting to explore ways to transfer knowledge about more general forms of variations across datasets.

Acknowledgments. This work is supported by the National Institute of Nursing Research (NINR) of the National Institutes of Health (NIH) under Award Number R01NR018301.

References

1. Alexey, D., Fischer, P., Tobias, J., Springenberg, M.R., Brox, T.: Discriminative unsupervised feature learning with exemplar convolutional neural networks. IEEE Trans. Pattern Anal. Mach. Intell. **99** (2015)
2. Bilic, P., et al.: The liver tumor segmentation benchmark (LITS). Med. Image Anal. **84**, 102680 (2023). https://doi.org/10.1016/j.media.2022.102680, https://www.sciencedirect.com/science/article/pii/S1361841522003085
3. Cubuk, E.D., Zoph, B., Mane, D., Vasudevan, V., Le, Q.V.: Autoaugment: learning augmentation strategies from data. In: Proceedings of the IEEE Conference on Computer Vision and Pattern Recognition, pp. 113–123 (2019)
4. Detlefsen, N.S., Freifeld, O., Hauberg, S.: Deep diffeomorphic transformer networks. In: 2018 IEEE/CVF Conference on Computer Vision and Pattern Recognition, pp. 4403–4412 (2018). https://doi.org/10.1109/CVPR.2018.00463
5. Freifeld, O., Hauberg, S., Batmanghelich, K., Fisher, J.W.: Highly-expressive spaces of well-behaved transformations: Keeping it simple. In: 2015 IEEE International Conference on Computer Vision (ICCV), pp. 2911–2919 (2015). https://doi.org/10.1109/ICCV.2015.333
6. Hauberg, S., Freifeld, O., Larsen, A.B.L., Fisher, J., Hansen, L.: Dreaming more data: class-dependent distributions over diffeomorphisms for learned data augmentation. In: Artificial Intelligence and Statistics, pp. 342–350 (2016)
7. Heller, N., et al.: The kits19 challenge data: 300 kidney tumor cases with clinical context, CT semantic segmentations, and surgical outcomes (2019). https://doi.org/10.48550/ARXIV.1904.00445, https://arxiv.org/abs/1904.00445
8. Ho, D., Liang, E., Stoica, I., Abbeel, P., Chen, X.: Population based augmentation: efficient learning of augmentation policy schedules. arXiv preprint arXiv:1905.05393 (2019)
9. Isensee, F., Jaeger, P.F., Kohl, S.A., Petersen, J., Maier-Hein, K.H.: nnU-Net: a self-configuring method for deep learning-based biomedical image segmentation. Nat. Methods **18**(2), 203–211 (2021)
10. Kingma, D.P., Ba, J.: Adam: a method for stochastic optimization (2014). https://doi.org/10.48550/ARXIV.1412.6980, https://arxiv.org/abs/1412.6980
11. Lim, S., Kim, I., Kim, T., Kim, C., Kim, S.: Fast autoaugment. In: Advances in Neural Information Processing Systems, pp. 6662–6672 (2019)
12. Shen, Z., Xu, Z., Olut, S., Niethammer, M.: Anatomical data augmentation via fluid-based image registration. In: Martel, A.L., et al. (eds.) MICCAI 2020. LNCS, vol. 12263, pp. 318–328. Springer, Cham (2020). https://doi.org/10.1007/978-3-030-59716-0_31
13. Yang, J., Zhang, Y., Liang, Y., Zhang, Y., He, L., He, Z.: `TumorCP`: a simple but effective object-level data augmentation for tumor segmentation. In: de Bruijne, M., et al. (eds.) MICCAI 2021. LNCS, vol. 12901, pp. 579–588. Springer, Cham (2021). https://doi.org/10.1007/978-3-030-87193-2_55
14. Zhang, X., et al.: CarveMix: a simple data augmentation method for brain lesion segmentation. In: de Bruijne, M., et al. (eds.) MICCAI 2021. LNCS, vol. 12901, pp. 196–205. Springer, Cham (2021). https://doi.org/10.1007/978-3-030-87193-2_19
15. Zhang, X., Wang, Q., Zhang, J., Zhong, Z.: Adversarial autoaugment. In: International Conference on Learning Representations (2020). https://openreview.net/forum?id=ByxdUySKvS

16. Zhao, A., Balakrishnan, G., Durand, F., Guttag, J.V., Dalca, A.V.: Data augmentation using learned transformations for one-shot medical image segmentation. In: Proceedings of the IEEE/CVF Conference on Computer Vision and Pattern Recognition (CVPR), June 2019
17. Zhu, Q., Wang, Y., Yin, L., Yang, J., Liao, F., Li, S.: Selfmix: a self-adaptive data augmentation method for lesion segmentation. In: Wang, L., Dou, Q., Fletcher, P.T., Speidel, S., Li, S. (eds.) Medical Image Computing and Computer Assisted Intervention - MICCAI 2022, pp. 683–692. Springer Nature Switzerland, Cham (2022). https://doi.org/10.1007/978-3-031-16440-8_65

Efficient Subclass Segmentation in Medical Images

Linrui Dai[1], Wenhui Lei[1,2], and Xiaofan Zhang[1,2](✉)

[1] Shanghai Jiao Tong University, Shanghai, China
{o.o111,wenhui.lei,xiaofan.zhang}@sjtu.edu.cn
[2] Shanghai Artificial Intelligence Laboratory, Shanghai, China

Abstract. As research interests in medical image analysis become increasingly fine-grained, the cost for extensive annotation also rises. One feasible way to reduce the cost is to annotate with coarse-grained superclass labels while using limited fine-grained annotations as a complement. In this way, fine-grained data learning is assisted by ample coarse annotations. Recent studies in classification tasks have adopted this method to achieve satisfactory results. However, there is a lack of research on efficient learning of fine-grained subclasses in semantic segmentation tasks. In this paper, we propose a novel approach that leverages the hierarchical structure of categories to design network architecture. Meanwhile, a task-driven data generation method is presented to make it easier for the network to recognize different subclass categories. Specifically, we introduce a Prior Concatenation module that enhances confidence in subclass segmentation by concatenating predicted logits from the superclass classifier, a Separate Normalization module that stretches the intra-class distance within the same superclass to facilitate subclass segmentation, and a HierarchicalMix model that generates high-quality pseudo labels for unlabeled samples by fusing only similar superclass regions from labeled and unlabeled images. Our experiments on the BraTS2021 and ACDC datasets demonstrate that our approach achieves comparable accuracy to a model trained with full subclass annotations, with limited subclass annotations and sufficient superclass annotations. Our approach offers a promising solution for efficient fine-grained subclass segmentation in medical images. Our code is publicly available here.

Keywords: Automatic Segmentation · Deep Learning

1 Introduction

In recent years, the use of deep learning for automatic medical image segmentation has led to many successful results based on large amounts of annotated

---

Supplementary Information The online version contains supplementary material available at https://doi.org/10.1007/978-3-031-43895-0_25.

H. Greenspan et al. (Eds.): MICCAI 2023, LNCS 14221, pp. 266–275, 2023.
https://doi.org/10.1007/978-3-031-43895-0_25

training data. However, the trend towards segmenting medical images into finer-grained classes (denoted as *subclasses*) using deep neural networks has resulted in an increased demand for finely annotated training data [4,11,21]. This process requires a higher level of domain expertise, making it both time-consuming and demanding. As annotating coarse-grained (denoted as *superclasses*) classes is generally easier than subclasses, one way to reduce the annotation cost is to collect a large number of superclasses annotations and then labeling only a small number of samples in subclasses. Moreover, in some cases, a dataset may have already been annotated with superclass labels, but the research focus has shifted towards finer-grained categories [9,24]. In such cases, re-annotating an entire dataset may not be as cost-effective as annotating only a small amount of data with subclass labels.

Here, the primary challenge is to effectively leverage superclass annotations to facilitate the learning of fine-grained subclasses. To solve this problem, several works have proposed approaches for recognizing new subclasses with limited subclass annotations while utilizing the abundant superclass annotations in classification tasks [6,8,18,25]. In general, they assume the subclasses are not known during the training stage and typically involve pre-training a base model on superclasses to automatically group samples of the same superclass into several clusters while adapting them to finer subclasses during test time.

However, to the best of our knowledge, there has been no work specifically exploring learning subclasses with limited subclass and full superclass annotations in semantic segmentation task. Previous label-efficient learning methods, such as semi-supervised learning [7,17,26], few-shot learning [10,15,19] and weakly supervised learning [13,27], focus on either utilize unlabeled data or enhance the model's generalization ability or use weaker annotations for training. However, they do not take into account the existence of superclasses annotations, making them less competitive in our setting.

In this study, we focus on the problem of efficient subclass segmentation in medical images, whose goal is to segment subclasses under the supervision of limited subclass and sufficient superclass annotations. Unlike previous works such as [6,8,18,25], we assume that the target subclasses and their corresponding limited annotations are available during the training process, which is more in line with practical medical scenarios.

Our main approach is to utilize the hierarchical structure of categories to design network architectures and data generation methods that make it easier for the network to distinguish between subclass categories. Specifically, we propose 1) a **Prior Concatenation** module that concatenates predicted logits from the superclass classifier to the input feature map before subclass segmentation, serving as prior knowledge to enable the network to focus on recognizing subclass categories within the current predicted superclass; 2) a **Separate Normalization** module that aims to stretch the intra-class distance within the same superclass, facilitating subclass segmentation; 3) a **HierarchicalMix** module inspired by GuidedMix [23], which for the first time suggests fusing similar labeled and unlabeled image pairs to generate high-quality pseudo labels for the unlabeled

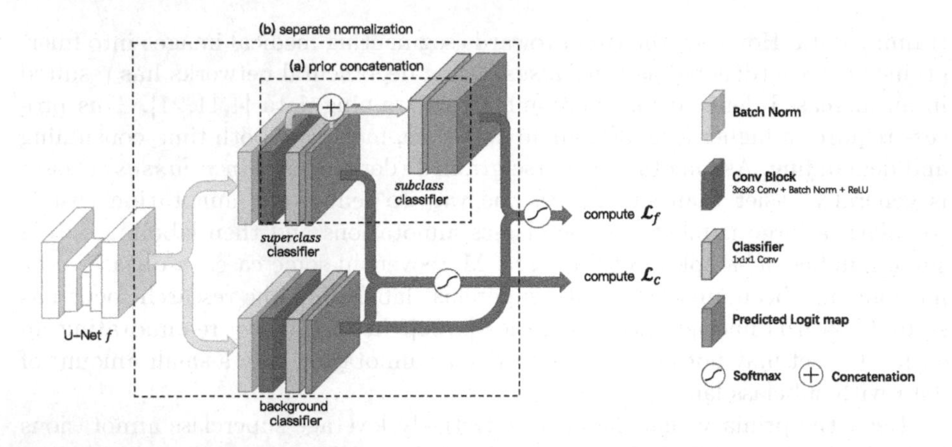

Fig. 1. Proposed network architecture, $\mathcal{L}_c$ and $\mathcal{L}_f$ stand for the superclass loss and subclass loss respectively.

samples. However, GuidedMix selects image pairs based on their similarity and fuses entire images. In contrast, our approach is more targeted. We mix a certain superclass region from an image with subclass annotation to the corresponding superclass region in an unlabeled image without subclass annotation, avoiding confusion between different superclass regions. This allows the model to focus on distinguishing subclasses within the same superclass. Our experiments on the Brats 2021 [3] and ACDC [5] datasets demonstrate that our model, with sufficient superclass and very limited subclass annotations, achieves comparable accuracy to a model trained with full subclass annotations.

2 Method

Problem Definition. We start by considering a set of R coarse classes, denoted by $\mathcal{Y}_c = \{Y_1, ..., Y_R\}$, such as background and brain tumor, and a set of N training images, annotated with $\mathcal{Y}_c$, denoted by $\mathcal{D}_c = \{(x^l, y^l)|y_i^l \in \mathcal{Y}_c\}_{l=1}^N$. Each pixel i in image x^l is assigned a superclass label y_i^l. To learn a finer segmentation model, we introduce a set of fine subclass $K = \sum_{i=1}^R k_i$ in coarse classes, denoted by $\mathcal{Y}_f = \{Y_{1,1}, ..., Y_{1,k_1}, ..., Y_{R,1}, ..., Y_{R,k_R}\}$, such as background, enhancing tumor, tumor core, and whole tumor. We assume that only a small subset of n training images have pixel-wise subclass labels $z \in \mathcal{Y}_f$ denoted by $\mathcal{D}_f = \{(x^l, z^l)|z_i^l \in \mathcal{Y}_f\}_{l=1}^n$. Our goal is to train a segmentation network $f(x^l)$ that can accurately predict the subclass labels for each pixel in the image x^l, even when $n \ll N$. **Without specification, we consider $R = 2$ (background and foreground) and extend the foreground class to multi subclass in this work.**

Prior Concatenation. One direct way to leverage the superclass and subclass annotations simultaneously is using two $1 \times 1 \times 1$ convolution layers as superclass and subclass classification heads for the features extracted from the network. The superclassification and subclassification heads are individually trained by superclass $P_c(x^l)$ labels and subclass labels $P_f(x^l)$. With enough superclass labels, the feature maps corresponding to different superclasses should be well separated. However, this coerces the subclassification head to discriminate among K subclasses under the mere guidance from few subclass annotations, making it prone to overfitting.

Another common method to incorporate the information from superclass annotations into the subclassification head is negative learning [14]. This technique penalizes the prediction of pixels being in the wrong superclass label, effectively using the superclass labels as a guiding principle for the subclassification head. However, in our experiments, we found that this method may lead to lower overall performance, possibly due to unstable training gradients resulting from the uncertainty of the subclass labels.

To make use of superclass labels without affecting the training of the subclass classification head, we propose a simple yet effective method called **Prior Concatenation (PC)**: as shown in Fig. 1 (a), we concatenate predicted superclass logit scores $S_c(x^l)$ onto the feature maps $F(x^l)$ and then perform subclass segmentation. The intuition behind this operation is that by concatenating the predicted superclass probabilities with feature maps, the network is able to leverage the prior knowledge of the superclass distribution and focus more on learning the fine-grained features for better discrimination among subclasses.

Separate Normalization. Intuitively, given sufficient superclass labels in supervised learning, the superclassification head tends to reduce feature distance among samples within the same superclass, which conflicts with the goal of increasing the distance between subclasses within the same superclass. To alleviate this issue, we aim to enhance the internal diversity of the distribution within the same superclass while preserving the discriminative features among superclasses.

To achieve this, we propose **Separate Normalization(SN)** to separately process feature maps belonging to hierarchical foreground and background divided by superclass labels. As a superclass and the subclasses within share the same background, the original conflict between classifiers is transferred to finding the optimal transformations that separate foreground from background, enabling the network to extract class-specific features while keeping the features inside different superclasses well-separated.

Our framework is shown in Fig. 1 (b). First, we use Batch Norm layers [12] to perform separate affine transformations on the original feature map. The transformed feature maps, each representing a semantic foreground and background, are then passed through a convolution block for feature extraction before further classification. The classification process is coherent with the semantic meaning of each branch. Namely, the foreground branch includes a superclassifier and a subclassifier that classifies the superclass and subclass foreground, while the background branch is dedicated solely to classify background pixels. Finally, two

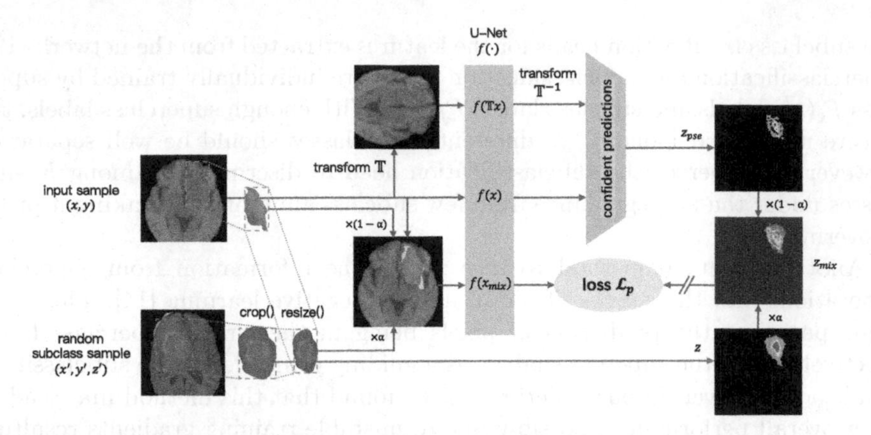

Fig. 2. The framework of *HierarchicalMix*. This process is adopted at training time to pair each coarsely labeled image x with its mixed image x_{mix} and pseudo subclass label z. "//" represents the cut of gradient backpropagation.

separate network branches are jointly supervised by segmentation loss on super- and subclass labels. The aforementioned prior concatenation continues to take effect by concatenating predicted superclass logits on the inputs of subclassifier.

HierarchicalMix. Given the scarcity of subclass labels, we intend to maximally exploit the existent subclass supervision to guide the segmentation of coarsely labeled samples. Inspired by GuidedMix [23], which provides consistent knowledge transfer between similar labeled and unlabeled images with pseudo labeling, we propose **HierarchicalMix(HM)** to generate robust pseudo supervision. Nevertheless, GuidedMix relies on image distance to select similar images and performs a whole-image mixup, which loses focus on the semantic meaning of each region within an image. We address this limitation by exploiting the additional superclass information for a more targeted mixup. This information allows us to fuse only the semantic foreground regions, realizing a more precise transfer of foreground knowledge. A detailed pipeline of HierarchicalMix is described below.

As shown in Fig. 2, for each sample (x, y) in the dataset that does not have subclass labels, we pair it with a randomly chosen fine-labeled sample (x', y', z'). First, we perform an random rotation and flipping $\mathbb{T}$ on (x, y) and feed both the original sample and the transformed sample $\mathbb{T}x$ into the segmentation network f. An indirect segmentation of x is obtained by performing the inverse transformation $\mathbb{T}^{-1}$ on the segmentation result of $\mathbb{T}x$. A transform-invariant pseudo subclass label map z_{pse} is generated according to the following scheme: Pixel (i, j) in z_{pse} is assigned a valid subclass label index $(z_{pse})_{i,j} = f(x)_{i,j}$ only when $f(x)_{i,j}$ agrees with $[\mathbb{T}^{-1}f(\mathbb{T}x)]_{i,j}$ with a high confidence τ as well as $f(x)_{i,j}$ and $x_{i,j}$ both belong to the same superclass label.

Next, we adopt image mixup by cropping the bounding box of foreground pixels in x', resizing it to match the size of foreground in x, and linearly overlaying them by a factor of α on x. This semantically mixed image x_{mix} has subclass labels $z = \text{resize}(\alpha \cdot z')$ from the fine-labeled image x'. Then, we pass it through the network to obtain a segmentation result $f(x_{mix})$. This segmentation result is supervised by the superposition of the pseudo label map z_{pse} and subclass labels z, with weighting factor α: $\mathcal{L}_p = \mathcal{L}(f(x_{mix}), \alpha \cdot z + (1-\alpha) \cdot z_{pse})$.

The intuition behind this framework is to simultaneously leverage the information from both unlabeled and labeled data by incorporating a more robust supervision from transform-invariant pseudo labels. While mixing up only the semantic foreground provides a way of exchanging knowledge between similar foreground objects while lifting the confirmation bias in pseudo labeling [1].

3 Experiments

Dataset and Preprocessing. We conduct all experiments on two public datasets. The first one is the **ACDC**[1] dataset [5], which contains 200 MRI images with segmentation labels for left ventricle cavity (LV), right ventricle cavity (RV), and myocardium (MYO). Due to the large inter-slice spacing, we use 2D segmentation as in [2]. We adopt the processed data and the same data division in [16], which uses 140 scans for training, 20 scans for validation and 40 scans for evaluation. During inference, predictions are made on each individual slice and then assembled into a 3D volume. The second is the **BraTS2021**[2] dataset [3], which consists of 1251 mpMRI scans with an isotropic 1 mm^3 resolution. Each scan includes four modalities (FLAIR, T1, T1ce, and T2), and is annotated for necrotic tumor core (TC), peritumoral edematous/invaded tissue (PE), and the GD-enhancing tumor (ET). We randomly split the dataset into 876, 125, and 250 cases for training, validation, and testing, respectively. For both datasets, image intensities are normalized to values in [0, 1] and the foreground superclass is defined as the union of all foreground subclasses for both datasets.

Implementation Details and Evaluation Metrics. To augment the data during training, we randomly cropped the images with a patch size of 256×256 for the ACDC dataset and $96 \times 96 \times 96$ for the BraTS2021 dataset. The model loss $\mathcal{L}$ is set by adding the losses from Cross Entropy Loss and Dice Loss. The weighing factor α in HierarchicalMix section is chosen to be 0.5, while τ linearly decreases from 1 to 0.4 during the training process.

We trained the model for 40,000 iterations using SGD optimizer with a 0.9 momentum and a linearly decreasing learning rate that starts at 0.01 and ends with 0. We used a batch size of 24 for the ACDC dataset and 4 for the BraTS2021 dataset, where half of the samples are labeled with subclasses and the other half

[1] https://www.creatis.insa-lyon.fr/Challenge/acdc/databases.html.

[2] http://braintumorsegmentation.org/.

only labeled with superclasses. More details can be found in the supplementary materials. To evaluate the segmentation performance, we used two widely-used metrics: the Dice coefficient (DSC) and 95% Hausdorff Distance (HD_{95}). The confidence factor τ mentioned in HierarchicalMix starts at 1 and linearly decays to 0.4 throughout the training process, along with a weighting factor α sampled according to the uniform distribution on $[0.5, 1]$.

Performance Comparison with Other Methods. To evaluate the effectiveness of our proposed method, we firstly trained two **U-Net** models [20] to serve as upper and lower bounds of performance. The first U-Net was trained on the complete subclass dataset $\{(x^l, y^l, z^l)\}_{l=1}^{N}$, while the second was trained on its subset $\{(x^l, y^l, z^l)\}_{l=1}^{n}$. Then, we compared our method with the following four methods, all of which were trained using n subclass labels and N superclass labels: **Modified U-Net (Mod)**: This method adds an additional superclass classifier alongside the subclass classifier in the U-Net. **Negative Learning (NL)**: This method incorporates superclass information into the loss module by introducing a separate negative learning loss in the original U-Net. This additional loss penalizes pixels that are not segmented as the correct superclass. **Cross Pseudo Supervision (CPS)** [7]: This method simulates pseudo supervision by utilizing the segmentation results from two models with different parameter initializations, and adapts their original network to the Modified U-Net architecture. **Uncertainty Aware Mean Teacher (UAMT)** [26]: This method modifies the classical mean teacher architecture [22] by adapting the teacher model to learn from only reliable targets while ignoring the rest, and also adapts the original network to the Modified U-Net architecture.

Table 1. Mean Dice Score (%, left) and HD_{95} (mm, right) of different methods on ACDC and BraTS2021 datasets. Sup. and Sub. separately represents the number of data with superclass and subclass annotations in the experiments. '_' means the result of our proposal is significantly better than the closet competitive result (p-value $<$ 0.05). The standard deviations of each metric are recorded in the supplementary materials.

Method	ACDC						BraTS2021					
	Sup.	Sub.	RV	MYO	LV	Avg.	Sup.	Sub.	TC	PE	ET	Avg.
U-Net	0	3	36.6, 61.5	51.6, 20.7	57.9, 26.2	48.7, 36.2	0	10	57.5, 16.6	68.8, 22.9	74.7, 12.4	67.0, 17.3
U-Net	0	140	90.6, 1.88	89.0, 3.59	94.6, 3.60	91.4, 3.02	0	876	75.8, 4.86	82.2, 5.87	83.6, 2.48	80.6, 4.40
Mod	140	3	83.1, 11.1	80.7, 6.12	83.1, 14.7	82.3, 10.6	876	10	60.3, 7.69	76.2, 7.70	80.2, 4.97	72.3, 6.79
NL [14]	140	3	61.0, 18.8	68.6, 13.7	81.5, 19.5	70.4, 17.3	876	10	59.5, 10.5	75.2, 8.35	76.8, 6.34	70.5, 8.40
CPS [7]	140	3	80.2, 9.54	80.3, 3.17	86.3, **4.21**	82.3, 5.64	876	10	62.9, 7.02	78.3, 7.08	80.8, 4.91	74.0, 6.24
UAMT [26]	140	3	79.4, 7.81	77.7, 5.87	85.5, 8.16	80.9, 7.28	876	10	60.8, 9.84	78.4, 7.11	80.1, 4.24	73.3, 7.06
Ours	140	3	**87.2**, **1.84**	**84.6**, **2.70**	**90.1**, 4.44	**87.3**, **2.99**	876	10	**65.5**, **6.90**	**79.9**, **6.38**	**80.8**, **3.59**	**75.4**, **5.62**

The quantitative results presented in Table 1 reveal that all methods that utilize additional superclass annotations outperformed the baseline method, which involved training a U-Net using only limited subclass labels. However, the methods that were specifically designed to utilize superclass information or explore the intrinsic structure of the subclass data, such as NL, CPS, and UAMT, did not consistently outperform the simple Modified U-Net. In fact, these methods sometimes performed worse than the simple Modified U-Net, indicating the difficulty of utilizing superclass information effectively. In contrast, our proposed method achieved the best performance among all compared methods on both the ACDC and BraTS2021 datasets. Specifically, our method attained an average Dice score of 87.3% for ACDC and 75.4% for BraTS2021, outperforming the closest competitor by 5.0% and 1.4%, respectively.

Table 2. Mean Dice Score (%, left) and HD_{95} (mm, right) of ablation studies on ACDC and BraTS2021 datasets (*mixup* and *pseudo* in HM column separately stands for using solely image mixup and pseudo-labeling to achieve better data utilization).

HM	PC	SN	ACDC						BraTS2021					
			Sup.	Sub.	RV	MYO	LV	Avg.	Sup.	Sub.	TC	PE	ET	Avg.
			140	3	83.1, 11.1	80.7, 6.12	83.1, 14.7	82.3, 10.6	876	10	60.3, 7.69	76.2, 7.70	80.2, 4.97	72.3, 6.79
✓			140	3	85.9, 2.55	83.6, 3.70	89.8, 5.15	86.5, 3.80	876	10	65.0, 8.00	77.0, 7.47	80.6, 3.74	74.2, 6.40
	✓		140	3	80.0, 8.06	80.4, 6.63	87.9, 5.07	82.8, 6.58	876	10	61.6, 7.00	77.3, 6.89	80.4, 6.01	73.1, 6.63
		✓	140	3	79.0, 3.32	81.2, 3.69	88.6, 4.43	82.9, 3.82	876	10	63.5, 9.03	78.9, 6.29	80.2, 4.45	74.2, 6.59
✓	✓		140	3	85.1, 1.86	81.4, 4.29	87.3, 5.55	84.6, 3.90	876	10	65.1, 7.93	78.4, 6.86	78.3, 3.97	73.9, 6.25
✓		✓	140	3	**87.6**, 2.81	83.8, **2.06**	89.9, 2.87	87.1, **2.58**	876	10	65.7, 7.56	79.6, 6.68	**81.4**, 4.25	75.5, 6.16
	✓	✓	140	3	84.7, 5.26	84.1, 2.53	89.3, **2.79**	86.0, 3.53	876	10	64.4, 7.96	79.5, 6.41	79.5, 5.07	74.4, 6.48
mixup	✓	✓	140	3	82.9, 5.42	80.6, 4.18	86.8, 6.06	83.5, 5.22	876	10	**66.2**, 6.90	79.6, 6.26	80.9, 4.19	**75.6**, 5.79
pseudo	✓	✓	140	3	78.8, 12.2	80.1, 7.66	84.3, 7.71	81.1, 9.20	876	10	62.4, 11.1	77.9, 6.55	80.0, 7.09	73.5, 8.24
✓	✓	✓	140	3	87.2, **1.84**	**84.6**, 2.70	**90.1**, 4.44	**87.3**, 2.99	876	10	65.5, **6.90**	**79.9**, **6.38**	80.8, **3.59**	75.4, **5.62**
✓	✓	✓	140	6	86.6, 1.20	84.7, 1.87	90.9, 4.23	87.4, 2.44	876	20	70.7, 7.45	81.2, 6.08	82.2, 3.58	78.0, 5.70
✓	✓	✓	140	9	86.1, 1.78	85.7, 1.92	90.8, 4.15	87.6, 2.62	876	30	71.4, 6.15	81.4, 5.84	82.5, 3.25	78.5, 5.08
UNet			0	140	90.6, 1.88	89.0, 3.59	94.6, 3.60	91.4, 3.02	0	876	75.8, 4.86	82.2, 5.87	83.6, 2.48	80.6, 4.40

Ablation Studies. In this study, we performed comprehensive ablation studies to analyze the contributions of each component and the performance of our method under different numbers of images with subclass annotations. The performance of each component is individually evaluated, and is listed in Table 2.

Each component has demonstrated its effectiveness in comparison to the naive modified U-Net method. Moreover, models that incorporate more components generally outperform those with fewer components. The effectiveness of the proposed HierarchicalMix is evident from the comparisons made with models that use only image mixup or pseudo-labeling for data augmentation, while the addition of Separate Normalization consistently improves the model performance. Furthermore, our method was competitive with a fully supervised baseline, achieving comparable results with only 6.5% and 3.4% subclass annotations on ACDC and BraTS2021.

4 Conclusion

In this work, we proposed an innovative approach to address the problem of efficient subclass segmentation in medical images, where limited subclass annotations and sufficient superclass annotations are available. To the best of our knowledge, this is the first work specifically focusing on this problem. Our approach leverages the hierarchical structure of categories to design network architectures and data generation methods that enable the network to distinguish between subclass categories more easily. Specifically, we introduced a Prior Concatenation module that enhances confidence in subclass segmentation by concatenating predicted logits from the superclass classifier, a Separate Normalization module that stretches the intra-class distance within the same superclass to facilitate subclass segmentation, and a HierarchicalMix model that generates high-quality pseudo labels for unlabeled samples by fusing only similar superclass regions from labeled and unlabeled images. Our experiments on the ACDC and BraTS2021 datasets demonstrated that our proposed approach outperformed other compared methods in improving the segmentation accuracy. Overall, our proposed method provides a promising solution for efficient fine-grained subclass segmentation in medical images.

References

1. Arazo, E., Ortego, D., Albert, P., O'Connor, N., McGuinness, K.: Pseudo-labeling and confirmation bias in deep semi-supervised learning, pp. 1–8 (2020)
2. Bai, W., et al.: Semi-supervised learning for network-based cardiac MR image segmentation. In: Descoteaux, M., Maier-Hein, L., Franz, A., Jannin, P., Collins, D.L., Duchesne, S. (eds.) MICCAI 2017. LNCS, vol. 10434, pp. 253–260. Springer, Cham (2017). https://doi.org/10.1007/978-3-319-66185-8_29
3. Baid, U., et al.: The RSNA-ASNR-MICCAI BraTS 2021 benchmark on brain tumor segmentation and radiogenomic classification. arXiv preprint: arXiv:2107.02314 (2021)
4. Bakas, S., et al.: Identifying the best machine learning algorithms for brain tumor segmentation, progression assessment, and overall survival prediction in the brats challenge. arXiv preprint: arXiv:1811.02629 (2018)
5. Bernard, O., et al.: Deep learning techniques for automatic MRI cardiac multi-structures segmentation and diagnosis: is the problem solved? IEEE Trans. Med. Imaging **37**(11), 2514–2525 (2018)
6. Bukchin, G., et al.: Fine-grained angular contrastive learning with coarse labels. In: Proceedings of the IEEE/CVF Conference on Computer Vision and Pattern Recognition, pp. 8730–8740 (2021)
7. Chen, X., Yuan, Y., Zeng, G., Wang, J.: Semi-supervised semantic segmentation with cross pseudo supervision. In: Proceedings of the IEEE/CVF Conference on Computer Vision and Pattern Recognition, pp. 2613–2622 (2021)
8. Fotakis, D., Kalavasis, A., Kontonis, V., Tzamos, C.: Efficient algorithms for learning from coarse labels. In: Conference on Learning Theory, pp. 2060–2079. PMLR (2021)
9. Guo, S., Wang, L., Chen, Q., Wang, L., Zhang, J., Zhu, Y.: Multimodal MRI image decision fusion-based network for glioma classification. Front. Oncol. **12**, 819673 (2022)
10. Hansen, S., Gautam, S., Jenssen, R., Kampffmeyer, M.: Anomaly detection-inspired few-shot medical image segmentation through self-supervision with supervoxels. Med. Image Anal. **78**, 102385 (2022)

11. He, K., et al.: Synergistic learning of lung lobe segmentation and hierarchical multi-instance classification for automated severity assessment of COVID-19 in CT images. Pattern Recogn. **113**, 107828 (2021)
12. Ioffe, S., Szegedy, C.: Batch normalization: Accelerating deep network training by reducing internal covariate shift. In: International Conference on Machine Learning, pp. 448–456. PMLR (2015)
13. Kervadec, H., Dolz, J., Wang, S., Granger, E., Ayed, I.B.: Bounding boxes for weakly supervised segmentation: global constraints get close to full supervision. In: Medical Imaging with Deep Learning, pp. 365–381. PMLR (2020)
14. Kim, Y., Yim, J., Yun, J., Kim, J.: NLNL: negative learning for noisy labels. In: Proceedings of the IEEE/CVF International Conference on Computer Vision, pp. 101–110 (2019)
15. Lei, W., et al.: One-shot weakly-supervised segmentation in medical images. arXiv preprint: arXiv:2111.10773 (2021)
16. Luo, X.: SSL4MIS (2020). https://github.com/HiLab-git/SSL4MIS
17. Luo, X., et al.: Semi-supervised medical image segmentation via uncertainty rectified pyramid consistency. Med. Image Anal. **80**, 102517 (2022)
18. Ni, J., et al.: Superclass-conditional gaussian mixture model for learning fine-grained embeddings. In: International Conference on Learning Representations (2021)
19. Ouyang, C., Biffi, C., Chen, C., Kart, T., Qiu, H., Rueckert, D.: Self-supervision with superpixels: training few-shot medical image segmentation without annotation. In: Vedaldi, A., Bischof, H., Brox, T., Frahm, J.-M. (eds.) ECCV 2020. LNCS, vol. 12374, pp. 762–780. Springer, Cham (2020). https://doi.org/10.1007/978-3-030-58526-6_45
20. Ronneberger, O., Fischer, P., Brox, T.: U-Net: convolutional networks for biomedical image segmentation. In: Navab, N., Hornegger, J., Wells, W.M., Frangi, A.F. (eds.) MICCAI 2015. LNCS, vol. 9351, pp. 234–241. Springer, Cham (2015). https://doi.org/10.1007/978-3-319-24574-4_28
21. Sekuboyina, A., et al.: Verse: a vertebrae labelling and segmentation benchmark for multi-detector CT images. Med. Image Anal. **73**, 102166 (2021)
22. Tarvainen, A., Valpola, H.: Mean teachers are better role models: weight-averaged consistency targets improve semi-supervised deep learning results. In: Advances in Neural Information Processing Systems, vol. 30 (2017)
23. Tu, P., Huang, Y., Zheng, F., He, Z., Cao, L., Shao, L.: GuidedMix-Net: semi-supervised semantic segmentation by using labeled images as reference. In: Proceedings of the AAAI Conference on Artificial Intelligence, vol. 36, pp. 2379–2387 (2022)
24. Wen, J., et al.: Multi-scale semi-supervised clustering of brain images: deriving disease subtypes. Med. Image Anal. **75**, 102304 (2022)
25. Yang, J., Yang, H., Chen, L.: Towards cross-granularity few-shot learning: coarse-to-fine pseudo-labeling with visual-semantic meta-embedding. In: Proceedings of the 29th ACM International Conference on Multimedia, pp. 3005–3014 (2021)
26. Yu, L., Wang, S., Li, X., Fu, C.-W., Heng, P.-A.: Uncertainty-aware self-ensembling model for semi-supervised 3D left atrium segmentation. In: Shen, D., et al. (eds.) MICCAI 2019. LNCS, vol. 11765, pp. 605–613. Springer, Cham (2019). https://doi.org/10.1007/978-3-030-32245-8_67
27. Zhang, K., Zhuang, X.: CycleMix: a holistic strategy for medical image segmentation from scribble supervision. In: Proceedings of the IEEE/CVF Conference on Computer Vision and Pattern Recognition, pp. 11656–11665 (2022)

Class Specific Feature Disentanglement and Text Embeddings for Multi-label Generalized Zero Shot CXR Classification

Dwarikanath Mahapatra[1,2](✉), Antonio Jose Jimeno Yepes[3], Shiba Kuanar[4], Sudipta Roy[5], Behzad Bozorgtabar[6,7], Mauricio Reyes[8], and Zongyuan Ge[9]

[1] Inception Institute of AI (IIAI), Abu Dhabi, UAE
dwarikanath.mahapatra@inceptioniai.org
[2] Faculty of Engineering, Monash University, Melbourne, Australia
[3] Unstructured Technologies, Sacramento, USA
[4] Mayo Clinic, Rochester, USA
[5] Jio Institute, Navi Mumbai, India
[6] École Polytechnique Fédérale de Lausanne (EPFL), Lausanne, Switzerland
[7] Lausanne University Hospital (CHUV), Lausanne, Switzerland
[8] University of Bern, Bern, Switzerland
[9] AIM for Health Lab, Monash University, Melbourne, Victoria, Australia

Abstract. Robustness of medical image classification models is limited by its exposure to the candidate disease classes. Generalized zero shot learning (GZSL) aims at correctly predicting seen and unseen classes and most current GZSL approaches have focused on the single label case. It is common for chest x-rays to be labelled with multiple disease classes. We propose a novel multi-label GZSL approach using: 1) class specific feature disentanglement and 2) semantic relationship between disease labels distilled from BERT models pre-trained on biomedical literature. We learn a dictionary from distilled text embeddings, and leverage them to synthesize feature vectors that are representative of multi-label samples. Compared to existing methods, our approach does not require class attribute vectors, which are an essential part of GZSL methods for natural images but are not available for medical images. Our approach outperforms state of the art GZSL methods for chest xray images.

Keywords: Multi-label · GZSL · Text Embeddings · chest x-rays · feature synthesis

1 Introduction

Deep learning methods provide state-of-the-art (SOTA) performance for a variety of medical image analysis tasks such as diabetic retinopathy grading [7], and

---

Supplementary Information The online version contains supplementary material available at https://doi.org/10.1007/978-3-031-43895-0_26.

H. Greenspan et al. (Eds.): MICCAI 2023, LNCS 14221, pp. 276–286, 2023.
https://doi.org/10.1007/978-3-031-43895-0_26

chest X-ray diagnosis [10], to name a few. SOTA fully supervised methods have access to all classes as part of the training data whereas most real world clinical applications do not provide access to all classes which leads to unseen classes being wrongly diagnosed as one of the seen classes. Zero-Shot Learning (ZSL) aims to classify unseen test data by learning their plausible representations from seen class features, and in Generalized Zero-Shot Learning (GZSL) the model should accurately classify both seen and unseen classes during test time.

Previous works on GZSL in medical images have focused on the single class scenario where an image is assigned a single disease class [18,21]. However, chest X-ray images have multiple labels and single-label methods do not work well in this setting. Hence we propose a multi-label GZSL approach that takes into account the semantic relationship between the multiple disease labels and learns a highly discriminative feature representation.

GZSL for natural images [6,12,14,22] have the advantage of providing attribute vectors for all classes that enables a model to correlate between attribute vectors and corresponding feature representations of the seen classes. Defining unambiguous attribute vectors for medical images requires deep clinical expertise and time. This is more challenging for the multi-label scenario, where many disease conditions have similar appearances and textures. For example, in lung X-ray diagnosis, many conditions frequently co-occur with labels such as *Atelectasis, Effusion, and Infiltration.* An effective class attribute vector should be able to precisely identify individual labels and differentiate them from other co-occurring disease labels, which is very challenging to define. To overcome the above challenges, we make the following contributions:

1. We propose a novel feature disentanglement method where a given image is decomposed into class-specific and class agnostic features. This enables better feature learning of different classes and subsequently contributes to better feature synthesis in the multi-label scenario.
2. We use text embedding similarities to learn the semantic relationships between different labels. This contributes to more accurate learning of multi-label interactions at a global scale and guide feature generation to synthesize feature vectors that are realistic and preserve the multi-label relationship between disease labels.
3. We solve the GZSL classification problem in terms of cluster assignment. Class specific feature disentanglement performs better for multi-label classification [11] and we use this concept to synthesize unseen class features and subsequently perform classification.

Prior Work: GZSL's objective is to recognize images from known and unknown classes. Many works have shown promising results using GANs [23,26], and Intra-Class Compactness Enhancement [12]. Recent works on multi-label zero-shot learning (ML-ZSL) use information propagation [14], attention mechanisms [9] and co-occurrence statistics with weighted combinations of seen classes [19]. ZSL in medical image analysis is a much less explored topic with limited applications such as registration [13], segmentation [1], gleason grading [16] and artifact reduction [4]. [21] used multi-modal images and medical reports for GZSL of

Table 1. Cosine similarity of the labels' BioBERT embeddings

	Atl.	Card.	Cons.	Edema	Eff.	Emph.	Fibr.	Hernia	Inf.	Mass	No Find	Nodule	PT	Pne.	Pneu.
Atelectasis	1.00	0.84	0.93	0.92	0.66	0.99	0.77	0.99	0.93	0.93	0.49	0.70	0.79	0.99	0.89
Cardiomegaly	0.84	1.00	0.97	0.97	0.93	0.88	0.98	0.83	0.95	0.97	0.81	0.96	0.98	0.87	0.60
Consolidation	0.93	0.97	1.00	0.99	0.84	0.95	0.93	0.92	0.99	0.99	0.69	0.88	0.93	0.94	0.72
Edema	0.92	0.97	0.99	1.00	0.86	0.95	0.93	0.91	0.99	0.99	0.70	0.89	0.94	0.94	0.71
Effusion	0.66	0.93	0.84	0.86	1.00	0.71	0.96	0.65	0.84	0.85	0.91	0.98	0.95	0.70	0.40
Emphysema	0.99	0.88	0.95	0.95	0.71	1.00	0.82	0.99	0.95	0.95	0.54	0.75	0.83	0.99	0.86
Fibrosis	0.77	0.98	0.93	0.93	0.96	0.82	1.00	0.76	0.91	0.93	0.87	0.98	0.99	0.80	0.52
Hernia	0.99	0.83	0.92	0.91	0.65	0.99	0.76	1.00	0.92	0.91	0.48	0.70	0.78	0.99	0.91
Infiltration	0.93	0.95	0.99	0.99	0.84	0.95	0.91	0.92	1.00	0.99	0.68	0.87	0.92	0.95	0.73
Mass	0.93	0.97	0.99	0.99	0.85	0.95	0.93	0.91	0.99	1.00	0.70	0.88	0.94	0.95	0.72
No Finding	0.49	0.81	0.69	0.70	0.91	0.54	0.87	0.48	0.68	0.70	1.00	0.91	0.85	0.53	0.23
Nodule	0.70	0.96	0.88	0.89	0.98	0.75	0.98	0.70	0.87	0.88	0.91	1.00	0.97	0.74	0.45
Pleural_Thickening	0.79	0.98	0.93	0.94	0.95	0.83	0.99	0.78	0.92	0.94	0.85	0.97	1.00	0.82	0.54
Pneumonia	0.99	0.87	0.94	0.94	0.70	0.99	0.80	0.99	0.95	0.95	0.53	0.74	0.82	1.00	0.87
Pneumothorax	0.89	0.60	0.72	0.71	0.40	0.86	0.52	0.91	0.73	0.72	0.23	0.45	0.54	0.87	1.00

chest xray (CXR) images while [17,18] used saliency maps and GANs for GZSL using only CXRs.Recently, language models pre-trained on large corpora have also been considered for GZSL of CXRs [8]. However all the above works operate in the single label setting, while we solve the multi-label problem.

2 Method

Method Overview: Given training data with seen classes we: 1) create a dictionary from the text embedddings; 2) disentangle the image into class specific and class agnostic features; 3) use class specific features to generate features of seen and unseen classes using the Mixup approach [28]; 4) for a given test image apply feature disentanglement and feature similarity analysis to identify the different class labels in the image.

Embeddings: We generate embeddings of image class labels using BioBERT [15], a BERT [5]-like pre-trained model. BioBERT [15] is pre-trained on biomedical literature, more specifically the model available from Huggingface[1], which is a base and cased model. We consider a pooled set that produces a single 768 dimension vector for a label. We then calculate the cosine similarity between each of the labels and represent it as a matrix, which we refer to as $Dict_{Text}$ - dictionary for text embeddings, shown in Table 1.

2.1 Feature Disentanglement

Our feature disentanglement method is inspired from [20] which decomposes the feature space into shape and texture for domain adaptation applications. We decompose the feature space of the seen class samples into 'class-specific'

[1] https://huggingface.co/dmis-lab/biobert-v1.1.

and 'class-agnostic' features. Class specific features encode information specific to the particular class, and have low similarity between different classes. Class agnostic features have high similarity across all classes, and have minimal semantic overlap with class specific features. The disentangled features allow for greater accuracy in identifying the multiple labels in a sample. Figure 1 (a) shows the architecture of our feature disentanglement network (FDN) consisting of L encoder-decoder networks corresponding to the L classes in the training data. The encoders and decoders (generators) are denoted, respectively, as $E_l(\cdot)$ and $G_l(\cdot)$). Similar to a classic autoencoder, the encoder, E_n, produces a latent code z_i for image $x_i \sim p$. Furthermore, we divide the latent code, z_i, into two vectors: class specific component, $z_i^{spec_l}$ for class l, and a class agnostic component, $z_i^{agn_l}$. This is achieved by having two heads instead of one (as in conventional architectures). Both vectors are combined and fed to the decoder, G_n, which reconstructs the original input. The disentanglement network is trained using the following loss:

$$\mathcal{L}_{Disent} = \mathcal{L}_{Rec} + \lambda_1 \mathcal{L}_{spec} + \lambda_2 \mathcal{L}_{agn} + \lambda_3 \mathcal{L}_{agn-spec} \tag{1}$$

Reconstruction Loss: $\mathcal{L}_{Rec}$, is the commonly used image reconstruction loss: $\mathcal{L}_{Rec} = \sum_l \mathbb{E}_{x_i \sim p_l}\left[||x_i^l - G_l(E_l(x_i^l))||\right]$. It is a sum of the reconstruction losses from the class specific autoencoders. We train different autoencoders for images of each class in order to obtain class specific features and refer to them as 'Class-specific autoencoders'.

Class Specific Loss: For given class l the class specific component $z_i^{spec_l}$ will have high similarity with samples from the same class and low similarity with the $z_i^{spec_k}$ of other classes k. These two conditions are incorporated as follows:

$$\mathcal{L}_{spec} = \sum_{i,j}\left(\sum_l\left((1 - \langle z_i^{spec_l}, z_j^{spec_l}\rangle) + \sum_{k \neq l}\langle z_i^{spec_l}, z_j^{spec_k}\rangle\right)\right) \tag{2}$$

where $\langle . \rangle$ denotes cosine similarity. The sum is calculated for all classes indexed by $\sum_l$ and over all samples indexed by i, j.

Class Agnostic Loss: Class agnostic features of different classes have similar semantic content and have high cosine similarity. L_{agn} is defined as

$$\mathcal{L}_{agn} = \sum_{i,j}\sum_l\sum_{k \neq l}(1 - \langle z_i^{agn_l}, z_j^{agn_k}\rangle) \tag{3}$$

We want class specific and class agnostic features of same-class samples to be mutually complementary and have minimal overlap in semantic content, i.e.,

$$\mathcal{L}_{agn-spec} = \sum_l \langle z_i^{agn_l}, z_j^{spec_l}\rangle \tag{4}$$

Since the above loss terms are minimized it helps us achieve our stated objectives.

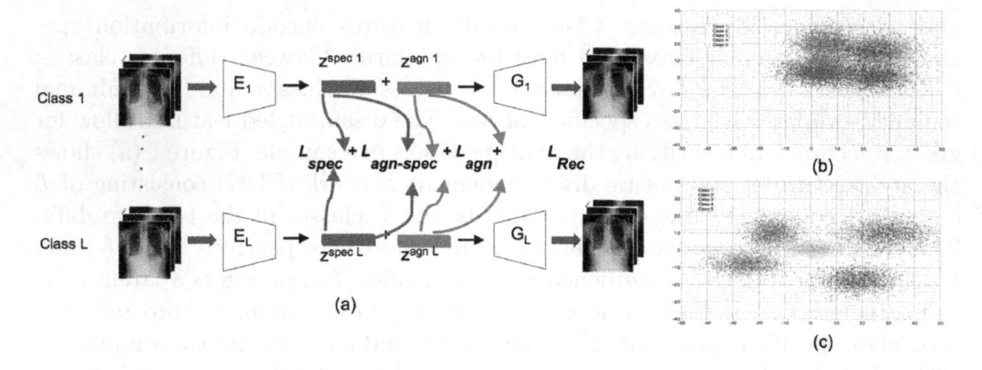

Fig. 1. (a) Architecture of class specific feature disentanglement network. Given training images from different classes of the same domain we disentangle features into class specific and class agnostic using autoencoders. T-sne results comparison between original image features and feature disentanglement output: (b) Original image features; (c) Class specific features. The classes in the tsne plot correspond to Atelectasis, Consolidation, Effusion, Infiltration and Nodule, as per the standard classes used for CheXpert.

Figure 1 (b) shows the t-sne plots of image features (taken from the fully connected layer of a multi-label DenseNet-121 image classifier) while Fig. 1 (c) shows the plot using class specific features. Plots of original features show overlapping clusters which makes it challenging to have good classification. Clusters obtained using class specific features are well separated with minimal overlap between different clusters. This clearly demonstrates the efficacy of our feature disentanglement method. The features are taken from images belonging to 5 classes from the NIH dataset. We chose 5 classes to clearly demonstrate the output and avoid cluttering.

Feature Generation Network: After disentangling the different seen class samples into their class specific components we create a distribution of each seen class feature. We generate synthetic class specific features of unseen classes using the following approach inspired by Mixup [28]:

$$\boldsymbol{z}^{spec_U} = \sum_l \Lambda_l \boldsymbol{z}_l^{spec_S}; \hat{y} = \sum \boldsymbol{y}_l \quad (5)$$

where $\boldsymbol{z}_k^{spec_U}$ is the class specific synthetic vector for unseen classes $k(\neq l)$, $\boldsymbol{z}_l^{spec_S}$ is a feature sampled from the distribution of seen class l, Λ_l is a random number drawn from a beta distribution. $\hat{y}$ is a one-hot encoded vector and is a sum of the one-hot label vectors of individual classes. Hence we do not need a weight when combining the label vectors. The weights Λ_l are such that $\sum_l \Lambda_l = 1$.

Generating unseen class features through Mixup without additional constraints can generate unrealistic features. We use the dictionary of text embeddings to guide the feature generation process. As synthetic features of the seen and unseen classes are generated we cluster them using the online self supervised learning based SwAV method [3] and calculate the centroids of each cluster. The semantic similarity of the centroid clusters should be such that their cosine similarity values are close to those obtained in Table 1, i.e., we define a loss:

$$\mathcal{L}_{ML-Cluster} = \frac{1}{N^2}\sum_i\sum_j Dict_{Text}(i,j) - Cent_{All}(i,j) \tag{6}$$

where $Cent_{All}$ refers to the changing matrix of cluster centroid similarities for all seen and unseen classes. N is the total number of classes. The final loss term for **clustering** all class samples is $\mathcal{L}_{Clust} = \mathcal{L}(x_s, x_t) + \lambda_4 \mathcal{L}_{ML-Cluster}$ where $\mathcal{L}(x_s, x_t)$ is the SwAV loss function defined in [3]. We add only those synthetic samples to classifier training data that reduce $\mathcal{L}_{Clust}$. This formulation ensures that the cluster output is well separated semantically and the cluster centroids follow the semantic relationship between all classes in Table 1.

Training, Inference and Implementation: For a given test image we use the pre-trained L class specific autoencoders to get the class specific features. An input 256×256 image is passed through the Encoder having 3 convolution layers ($64, 32, 32$ 3×3 filters) each followed by max pooling. The Decoder is symmetric to the Encoder. z^{agn} and z^{spec} are 256-dimension vectors. We then calculate the cosine similarity of the class specific features with the corresponding class centroids. If the cosine similarity is above 0.5 then the sample is assigned to the class. Following standard practice for GZSL, average class accuracies are calculated for the seen (Acc_S) and unseen (Acc_U) classes, and also the harmonic mean $H = \frac{2 \times Acc_U \times Acc_S}{Acc_U + Acc_S}$.

3 Experimental Results

Dataset Description. We demonstrate our method's effectiveness on the following chest xray datasets for multi-label classification tasks: **1.NIH Chest X-ray** Dataset [24]: having 112,120 expert-annotated frontal-view X-rays from 30,805 unique patients and has 14 disease labels. Original images were resized to 224×224. Hyperparameter values are $\lambda_1 = 1.1, \lambda_2 = 0.7, \lambda_3 = 0.9, \lambda_4 = 1.2$. **2. CheXpert** Dataset [10]: consisting of 224,316 chest radiographs of 65,240 patients labeled for the presence of 14 common chest conditions. Original images were resized to 224×224. Hyperparameter values are $\lambda_1 = 1.2, \lambda_2 = 0.8, \lambda_3 = 1.1, \lambda_4 = 1.1$. **3. PadChest** Dataset [2]: consisting of 160,868 from 67,625 patients. Hyperparameter values are $\lambda_1 = 1.3, \lambda_2 = 0.9, \lambda_3 = 0.9, \lambda_4 = 1.3$. A 70/10/20 split at patient level was done to get training, validation and test sets for both datasets.

Comparison Methods: We compare our method's performance with multiple GZSL methods - single label and multi-label techniques - employing different feature generation approaches such as CVAE or GANs. Our method is denoted as ML-GZSL (**M**ulti **L**abel **GZSL**). Our benchmark is a fully supervised learning (FSL) based method of [27] which is the top ranked method for [10], where the ranking is based on AUC. It builds upon a DenseNet-121 trained for multi-label classification.

3.1 Generalized Zero Shot Learning Results

Classification results for medical images in Table 2 show our proposed method significantly outperforms all competing GZSL methods. Note that we use the cluster centroids in place of attribute vectors for these feature synthesis methods. This significant difference in performance can be explained by the fact that the complex architectures that worked for natural images will not be equally effective for medical images which have less information. Absence of attribute vectors for medical images is another contributing factor. The class attributes provide a rich source of information about natural images which can be leveraged using existing architectures. Since those are not available for medical images these methods do not perform equally well. Different combinations of 7 seen and unseen classes are taken, and for each combination we run our model 5 times and the final reported numbers are the average of different combinations.

Table 2. GZSL Results For chest xray Images in Multi-Label setting: Average per-class classification accuracy (%) and harmonic mean accuracy (H) of generalized zero-shot learning when test samples are from seen or unseen classes. Results demonstrate the superior performance of our proposed method.

Method	NIH X-ray				CheXpert				PadChest			
	S	U	H	p	S	U	H	p	S	U	H	p
	Single Label GZSL Methods											
f-VAEGAN [26]	82.9	80.0	81.4	0.002	88.5	87.6	88.0	0.001	81.0	78.4	79.7	0.001
SDGN [25]	84.4	81.1	82.7	0.003	89.8	88.3	89.0	0.003	82.3	80.0	81.1	0.004
Feng [6]	84.7	81.4	83.0	0.0012	90.2	88.6	89.4	0.0017	82.5	80.2	81.3	0.0021
Kong [12]	84.8	81.2	82.9	0.0031	90.0	88.7	89.3	0.0034	82.7	80.5	81.6	0.0029
Su [22]	84.5	81.4	82.9	0.004	90.3	88.6	89.4	0.0045	82.3	79.8	81.03	0.0041
	Multi Label GZSL Methods											
Hayat [8]	79.1	69.2	73.8	0.005	81.2	79.8	80.5	0.0056	77.3	68.1	72.4	0.006
Lee [14]	85.1	81.3	83.1	0.008	87.4	85.7	86.5	0.0075	82.9	78.4	80.6	0.008
Huynh [9]	84.7	80.8	82.7	0.0065	86.9	85.1	86.0	0.0071	82.5	77.3	79.8	0.0073
	Proposed Method And Benchmarks											
ML-GZSL	86.2	85.0	85.6	–	90.8	90.2	90.5	–	88.2	86.1	87.1	–
FSL(Multi Label)	86.0	85.1	85.5	0.061	90.8	90.5	90.6	0.068	88.4	86.5	87.4	0.058
Mahapatra [18]	84.3	83.2	83.7	0.014	88.9	88.5	88.7	0.01	86.2	84.1	85.1	0.02

ML-GZSL's performance is almost equal to that of the benchmark fully supervised method FSL. Although GZSL methods generally perform inferior to FSL methods, our use of class specific features significantly improves performance. Additionally, the use of semantic relation between text embeddings significantly improves the performance due to better feature synthesis. The average accuracy is obtained by first calculating True Positive, False Positive, True Negative,

False Negative values and using these values to get the global accuracy. Furthermore the AUC(and F1) values for CheXpert data are as follows: FSL-93.0(91.7), ML-GZSL- 92.8(91.6), [18]-91.9(90.0), [8]-84.3(82.4).

3.2 Ablation Studies

Table 3 shows results for ablation studies. We exclude each of the three loss terms related to feature disentanglement - $\mathcal{L}_{agn}$,$\mathcal{L}_{spec}$ and $\mathcal{L}_{agn-spec}$- and report the results as ML-GZSL$_{w\mathcal{L}_{agn}}$, ML-GZSL$_{w\mathcal{L}_{spec}}$, and ML-GZSL$_{w\mathcal{L}_{agn-spec}}$. We also compare with the results of using image features obtained from a CNN based feature extractor (ResNet50 trained on Imagenet), which we denote as 'pre-train'. We observe that the class specific features has the greatest influence on the results and excluding it, ML-GZSL$_{w\mathcal{L}_{spec}}$ results in significant performance degradation compared to ML-GZSL. ML-GZSL$_{w\mathcal{L}_{agn-spec}}$ and ML-GZSL$_{w\mathcal{L}_{agn}}$ also show significantly lower performance. These results highlight the importance of the class specific features and at the same time illustrate class agnostic features have an important influence on the method's performance.

We also investigate the influence of $\mathcal{L}_{ML-Cluster}$ (Eq. 6) in the clustering process. The numbers in Table 3 show that ML-GZSL$_{w\mathcal{L}_{ML-Cluster}}$ (which is essentially the original SwAV algorithm) performs much worse. This proves the significant contribution of the text embedding dictionary in our multi-label GZSL framework.

Table 3. Ablation Results: Average per-class classification accuracy (%) and harmonic mean accuracy (H) of generalized zero-shot learning when test samples are from seen (Setting S) or unseen (Setting U) classes.

Method	NIH X-ray				CheXpert				PadChest			
	S	U	H	p	S	U	H	p	S	U	H	p
	Our Proposed Method											
ML-GZSL	86.2	85.0	85.6	-	90.8	90.2	90.5	-	88.2	86.1	87.1	-
	Feature Disentanglement Effects											
$w\mathcal{L}_{agn-spec}$	83.8	81.9	82.8	0.012	88.6	86.3	87.4	0.009	85.5	82.0	83.7	0.014
pre-train	83.4	82.0	82.7	0.017	88.2	85.3	86.7	0.009	85.1	81.7	83.4	0.011
$w\mathcal{L}_{agn}$	84.5	82.1	83.3	0.008	89.1	86.9	88.0	0.0094	86.5	83.4	84.9	0.011
$w\mathcal{L}_{spec}$	84.0	82.2	83.1	0.02	88.8	86.2	87.5	0.018	86.1	83.0	84.5	0.014
	Effect of Text Dictionary											
$w\mathcal{L}_{ML-Cluster}$	82.6	80.7	81.6	0.009	87.0	84.5	85.7	0.011	84.2	80.8	82.5	0.015

Hyperparameter Selection: The λ's were varied between $[0.4-1.5]$ in steps of 0.05 and the performance on a separate test set of 10,000 images were monitored. We optimize Eq. 1 by setting $\lambda_2 = \lambda 3 = \lambda_4 = 1$, and select the optimum value of λ_1. After fixing λ_1 we determine optimal λ_2, and subsequently λ_3, λ_4.

Realism of Synthetic Features. We reconstruct the xray images from the synthetic feature vectors using the feature disentanglement autoencoders' decoder part. We select 1000 such synthetic images from 14 classes of the NIH dataset and ask two trained radiologists, having 12 and 14 years experience in examining chest xray images for abnormalities, to identify whether the images are realistic or not. Each radiologist was blinded to the other's answers.

Results for ML-GZSL show one radiologist (*RAD* 1) identified 912/1000 (91.2%) images as realistic while *RAD* 2 identified 919 (91.9%) generated images as realistic. Both of them had a high agreement with 890 common images (89.0%) identified as realistic. Considering both *RAD* 1 and *RAD* 2 feedback, a total of 941 (94.1%) unique images were identified as realistic and 59/1000 (5.9%) images were not identified as realistic by any of the experts. ML-GZSL showed the highest agreement between *RAD* 1 and *RAD* 2.

4 Conclusion

Our experiments demonstrate that our approach of multi label GZSL is more accurate than using conventional approaches that solve the single-label scenario. We propose a novel feature disentanglement approach that obtains class specific and class agnostic features from the training images. Additionally, the relationship between text embeddings of disease labels is used to create a dictionary that guides clustering and feature synthesis. Classification results on multiple publicly available chest xray datasets demonstrate the improved performance obtained by using class specific features. The synthetic features obtained by our method are realistic since a major percentage of the corresponding reconstructed images are validated as realistic by trained clinicians.

References

1. Bian, C., Yuan, C., Ma, K., Yu, S., Wei, D., Zheng, Y.: Domain adaptation meets zero-shot learning: an annotation-efficient approach to multi-modality medical image segmentation. IEEE Trans. Med. Imaging **41**(5), 1043–1056 (2022)
2. Bustos, A., Pertusa, A., Salinas, J.M., de la Iglesia-Vayá, M.: PadChest: A large chest x-ray image dataset with multi-label annotated reports. Med. Image Anal. **66**, 101797 (2020)
3. Caron, M., Misra, I., Mairal, J., Goyal, P., Bojanowski, P., Joulin, A.: Unsupervised learning of visual features by contrasting cluster assignments. In: Larochelle, H., Ranzato, M., Hadsell, R., Balcan, M.F., Lin, H. (eds.) Advances in Neural Information Processing Systems, vol. 33, pp. 9912–9924. Curran Associates, Inc. (2020). https://proceedings.neurips.cc/paper/2020/file/70feb62b69f16e0238f741fab228fec2-Paper.pdf
4. Chen, Y., et al.: Zero-shot medical image artifact reduction. In: 2020 IEEE 17th International Symposium on Biomedical Imaging (ISBI), pp. 862–866 (2020). https://doi.org/10.1109/ISBI45749.2020.9098566
5. Devlin, J., Chang, M.W., Lee, K., Toutanova, K.: BERT: pre-training of deep bidirectional transformers for language understanding. arXiv preprint: arXiv:1810.04805 (2018)

6. Feng, Y., Huang, X., Yang, P., Yu, J., Sang, J.: Non-generative generalized zero-shot learning via task-correlated disentanglement and controllable samples synthesis. In: 2022 IEEE/CVF Conference on Computer Vision and Pattern Recognition (CVPR), pp. 9336–9345 (2022)
7. Gulshan, V., et al.: Development and validation of a deep learning algorithm for detection of diabetic retinopathy in retinal fundus photographs. JAMA **316**(22), 2402–2410 (2016). https://doi.org/10.1001/jama.2016.17216
8. Hayat, N., Lashen, H., Shamout, F.: Multi-label generalized zero shot learning for the classification of disease in chest radiographs. In: Proceeding of the Machine Learning for Healthcare Conference, pp. 461–477 (2021)
9. Huynh, D., Elhamifar, E.: A shared multi-attention framework for multi-label zero-shot learning. In: 2020 IEEE/CVF Conference on Computer Vision and Pattern Recognition (CVPR), pp. 8773–8783 (2020). https://doi.org/10.1109/CVPR42600.2020.00880
10. Irvin, J., et al.: CheXpert: a large chest radiograph dataset with uncertainty labels and expert comparison. arXiv preprint: arXiv:1901.07031 (2017)
11. Jia, J., He, F., Gao, N., Chen, X., Huang, K.: Learning disentangled label representations for multi-label classification (2022). https://doi.org/10.48550/arXiv.2212.01461
12. Kong, X., et al.: En-compactness: self-distillation embedding and contrastive generation for generalized zero-shot learning. In: 2022 IEEE/CVF Conference on Computer Vision and Pattern Recognition (CVPR), pp. 9296–9305 (2022). https://doi.org/10.1109/CVPR52688.2022.00909
13. Kori, A., Krishnamurthi, G.: Zero shot learning for multi-modal real time image registration. arXiv preprint: arXiv:1908.06213 (2019)
14. Lee, C.W., Fang, W., Yeh, C.K., Wang, Y.C.F.: Multi-label zero-shot learning with structured knowledge graphs. In: 2018 IEEE/CVF Conference on Computer Vision and Pattern Recognition, pp. 1576–1585 (2018). https://doi.org/10.1109/CVPR.2018.00170
15. Lee, J., et al.: BioBERT: a pre-trained biomedical language representation model for biomedical text mining. Bioinformatics **36**(4), 1234–1240 (2020)
16. Mahapatra, D., Bozorgtabar, B., Kuanar, S., Ge, Z.: Self-supervised multimodal generalized zero shot learning for Gleason grading. In: Albarqouni, S., et al. (eds.) DART/FAIR -2021. LNCS, vol. 12968, pp. 46–56. Springer, Cham (2021). https://doi.org/10.1007/978-3-030-87722-4_5
17. Mahapatra, D., Bozorgtabar, B., Ge, Z.: Medical image classification using generalized zero shot learning. In: Proceedings of the IEEE/CVF International Conference on Computer Vision (ICCV) Workshops, pp. 3344–3353 (2021)
18. Mahapatra, D., Ge, Z., Reyes, M.: Self-supervised generalized zero shot learning for medical image classification using novel interpretable saliency maps. IEEE Trans. Med. Imaging **41**(9), 2443–2456 (2022). https://doi.org/10.1109/TMI.2022.3163232
19. Mensink, T., Gavves, E., Snoek, C.G.: COSTA: co-occurrence statistics for zero-shot classification. In: 2014 IEEE Conference on Computer Vision and Pattern Recognition, pp. 2441–2448 (2014). https://doi.org/10.1109/CVPR.2014.313
20. Park, T., et al.: Swapping autoencoder for deep image manipulation. In: Advances in Neural Information Processing Systems (2020)
21. Paul, A., et al.: Generalized zero-shot chest x-ray diagnosis through trait-guided multi-view semantic embedding with self-training. IEEE Trans. Med. Imaging **40**, 2642–2655 (2021). https://doi.org/10.1109/TMI.2021.3054817

22. Su, H., Li, J., Chen, Z., Zhu, L., Lu, K.: Distinguishing unseen from seen for generalized zero-shot learning. In: 2022 IEEE/CVF Conference on Computer Vision and Pattern Recognition (CVPR), pp. 7875–7884 (2022). https://doi.org/10.1109/CVPR52688.2022.00773
23. Verma, V., Arora, G., Mishra, A., Rai, P.: Generalized zero-shot learning via synthesized examples. In: The IEEE Conference on Computer Vision and Pattern Recognition (CVPR), pp. 4281–4289 (2018)
24. Wang, X., Peng, Y., Lu, L., Lu, Z., Bagheri, M., Summers, R.: ChestX-ray8: hospital-scale chest x-ray database and benchmarks on weakly-supervised classification and localization of common thorax diseases. In: Proceedings of the CVPR (2017)
25. Wu, J., Zhang, T., Zha, Z.J., Luo, J., Zhang, Y., Wu, F.: Self-supervised domain-aware generative network for generalized zero-shot learning. In: The IEEE Conference on Computer Vision and Pattern Recognition (CVPR), pp. 12767–12776 (2020)
26. Xian, Y., Sharma, S., Schiele, B., Akata, Z.: F-VAEGAN-D2: a feature generating framework for any-shot learning. In: The IEEE Conference on Computer Vision and Pattern Recognition (CVPR), pp. 10275–10284 (2019)
27. Yuan, Z., Yan, Y., Sonka, M., Yang, T.: Large-scale robust deep AUC maximization: A new surrogate loss and empirical studies on medical image classification. In: 2021 IEEE/CVF International Conference on Computer Vision (ICCV), pp. 3020–3029 (2021)
28. Zhang, H., Cisse, M., Dauphin, Y.N., Lopez-Paz, D.: Mixup: beyond empirical risk minimization. In: International Conference on Learning Representations (2018). https://openreview.net/forum?id=r1Ddp1-Rb

Prediction of Cognitive Scores by Joint Use of Movie-Watching fMRI Connectivity and Eye Tracking via Attention-CensNet

Jiaxing Gao[1], Lin Zhao[2], Tianyang Zhong[1], Changhe Li[1], Zhibin He[1], Yaonai Wei[1], Shu Zhang[3], Lei Guo[1], Tianming Liu[2], Junwei Han[1], and Tuo Zhang[1](✉)

[1] School of Automation, Northwestern Polytechnical University, Xi'an, China
tuozhang@nwpu.edu.cn

[2] Cortical Architecture Imaging and Discovery Lab, Department of Computer Science and Bioimaging Research Center, The University of Georgia, Athens, GA, USA

[3] School of Computer Science, Northwestern Polytechnical University, Xi'an, China

Abstract. Brain functional connectivity under the naturalistic paradigm has been demonstrated to be better at predicting individual behaviors than other brain states, such as rest and task. Nevertheless, the state-of-the-art methods are difficult to achieve desirable results from movie-watching paradigm fMRI(mfMRI) induced brain functional connectivity, especially when the datasets are small, because it is difficult to quantify how much useful dynamic information can be extracted from a single mfMRI modality to describe the state of the brain. Eye tracking, becoming popular due to its portability and less expense, can provide abundant behavioral features related to the output of human's cognition, and thus might supplement the mfMRI in observing subjects' subconscious behaviors. However, there are very few works on how to effectively integrate the multimodal information to strengthen the performance by unified framework. To this end, an effective fusion approach with mfMRI and eye tracking, based on Convolution with Edge-Node Switching in Graph Neural Networks (CensNet), is proposed in this article, with subjects taken as nodes, mfMRI derived functional connectivity as node feature, different eye tracking features used to compute similarity between subjects to construct heterogeneous graph edges. By taking multiple graphs as different channels, we introduce squeeze-and-excitation attention module to CensNet (A-CensNet) to integrate graph embeddings from multiple channels into one. The experiments demonstrate the proposed model outperforms the one using single modality, single channel and state-of-the-art methods. The results suggest that brain functional activities and eye behaviors might complement each other in interpreting trait-like phenotypes. Our code will make public later.

Keywords: Functional Connectivity · Naturalistic Stimulus · Eye Movement · CensNet · Attention

H. Greenspan et al. (Eds.): MICCAI 2023, LNCS 14221, pp. 287–296, 2023.
https://doi.org/10.1007/978-3-031-43895-0_27

1 Introduction

There is a growing interest in leveraging brain imaging data to predict non-brain-imaging phenotypes in individual participants, since brain functional activity could intrinsically serve as an "objective" observer of a subject given that the emergence of behavior and cognition were widely attributed to the orchestration of local and remote cortical areas by means of a densely connected brain network [1]. In this context, the movie-watching paradigm has been widely demonstrated to provide richer life-like scenarios [2–5], better subject compliance in contrast to rest and task states, and were suggested to be better at predicting emotional/cognitive reaction [6–9] that could be more easily and saliently evoked by naturalistic input load, making movie watching the upper bound of current paradigms.

However, in recent works [8, 10, 11], the individual trait prediction from movie-watching paradigm fMRI (mfMRI) induced brain activity/functional connectivity can only achieve an accuracy around 0.40 (Pearson/Spearman's r). The accuracy drops even dramatically when dataset size is small. Regarding the reality that it is a challenging task to increase mfMRI data size, at least two strategies can be intuitively proposed to improve the performance on the basis of limited number of subjects: 1) Incorporation of other data modalities, such as behavior. Conceptually, a joint use of what a subject "thinks" and what the subject "reacts" to a stimulus could help to increase the accuracy of prediction of "who" he/she is. Eye movement behaviors [12], as one example, have been related to subjects' cognitive and phenotypical measures [12–14], and might supplement the fMRI derived brain activities in monitoring subjects' attention and task compliance and observing subjects' subconscious traits [15]; 2) Increase the number of different video clips watched by the same group of subjects.

Integration of multimodal data has been realized by using a graph to present the relation between subjects, where subjects are defined as nodes, node feature is fMRI connectivity, and edge is estimated by thresholding the similarity of behaviors, including eye movement [16]. This graph convolution networks could realize an embedding of a cohort of subjects' brain activity features according to their behavior similarity, and estimate a mapping of these embedded features to cognitive scores, and further propagate the mapping to other nodes as a prediction of their scores. However, behavior in this model only provides the topology of the graph but are not fully involved in the process of embeddings. Also, different definitions of edges, such as eye trajectory and pupil size in this work, yield a set of graphs with different topologies on the same set of nodes. Therefore, we aim to solve the following two technique problems: 1) how to integrate the different edge features to node ones and learn graph embedding for both node and edge; and 2) how to integrate the embeddings of heterogeneous graphs with the same set of nodes to fulfill classification or regression. Based on Convolution with Edge-Node Switching graph neural network (CensNet) [17], we proposed Attention-CensNet (A-CensNet for short), where subjects are nodes, with the mfMRI derived functional connectivity as nodal features. Eye tracking derived gaze trajectory and temporal pupil size variation were respectively used to measure the similarity between subjects and to construct a set of heterogenous edges. Each of these heterogenous graphs was taken as an independent channel, where CensNet was used to alternatively learn both node embeddings and edge embeddings. Then, Squeeze-and-Excitation attention module (SENet)

[18] was used to integrate the node-edge embeddings from multiple channels into one hybrid graph on which the final round of node embedding was performed. Note that mfMRI and eye tracking data from the same cohort exposed to different movies inputs also yield additional channels in this work.

In the following sections, we firstly introduce the dataset and preprocessing steps. Basics of CensNet and SENet, and proposal of A-CensNet are introduced followed by its application to our task. Comparative and ablation studies regarding to prediction accuracy (AUC) were presented in the Results to demonstrate the effectiveness of multiple channel integration strategy, and the better performance of A-CensNet than others.

2 Materials and Methods

2.1 Dataset

In the Human Connectome Project (HCP) 7T release [19], movie-watching fMRI and resting-state fMRI data were acquired on a 7 T Siemens Magnetom scanner [20]. Among the four scan sessions (MOVIE 1~4), MOVIE 2&3 were used as a testbed in this work. Important imaging parameters are as follows: TR = 1000 ms, TE = 22.2 ms, flip angle = 45 deg, FOV = 208 × 208 mm, matrix = 130 × 130, spatial resolution = 1.6mm^3, number of slices = 85, multiband factor = 5. During MOVIE 2 runs, 4 video clips (818 time points, TRs), separated by five 20s rest sessions (100 time points in total), were presented to subjects. In MOVIE 3 runs, 5 video clips (789 time points, TRs), separated by five 20s rest sessions (120 time points in total), were presented to subjects. Eye tracking data were acquired during MOVIE runs using an EyeLink S1000 system with a sampling rate of 1000 Hz. HCP provides many phenotypic measures from a variety of domains. As suggested by [8], we focus on measures in the cognition domain in this work. After a quality control of data modalities, 81 subjects are selected.

2.2 Preprocessing

FMRI data have been preprocessed by the minimal preprocessing pipeline for the Human Connectome Project [20, 21]. The signals were mapped to the grayordinate system, which includes 64k vertices on the reconstructed cortical surface plus 30k subcortical voxels for an individual. The within-subject cross-modal registration and cross-subject registration are adopted to warp the grayordinate vertices and voxels to the same space, such that the associated fMRI signals have cross-subject correspondence.

In this study, subcortical regions are not our major interest and not included. Destrieux atlas [22] is applied to cortical surfaces to yield 75 cortical areas on each hemisphere. We compute the mean fMRI signal averaged over vertices within a cortical area, and construct a 150-by-150 functional connectivity matrix, by means of Pearson correlation between these average signals (blue panel in Fig. 1(a)). We zero the negative correlation and 90% of the lowest positive correlation. The upper triangular matrix is converted to a vector and used as the functional feature.

For eye tracking data, time stamps are used to extract the effective data points and synchronize the eye behavior features across subjects. Blink session is not considered.

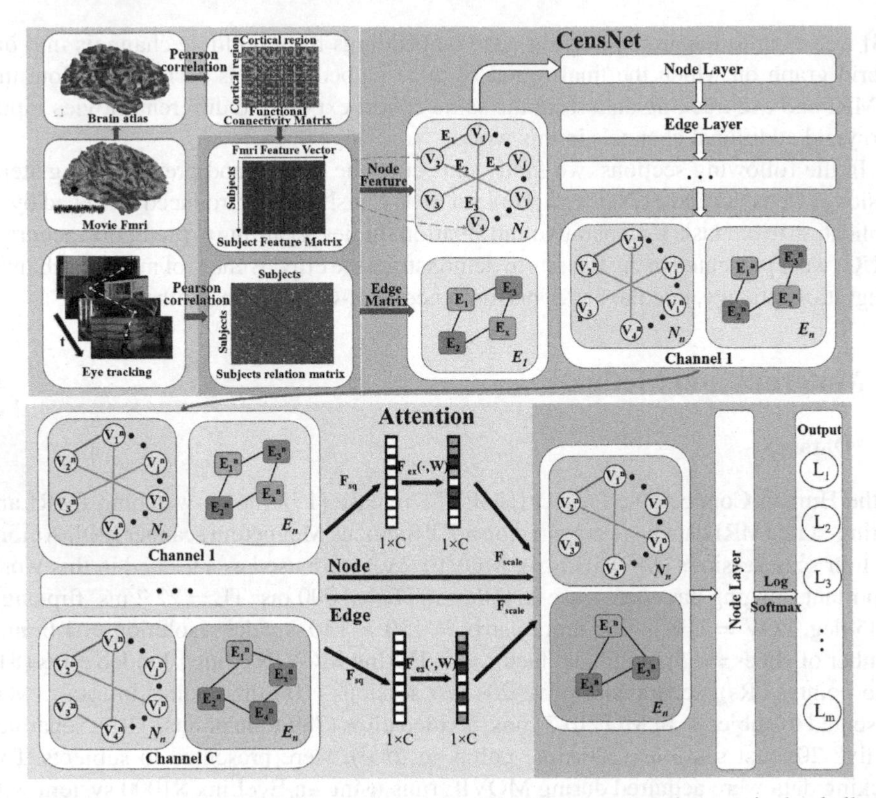

Fig. 1. The preprocessings and flowchart of A-CensNet. Data preprocessings are in the shallower blue panel. Construction of the graph is in the darker blue panel. Node feature generation is highlighted by red arrows and edge generation blue arrows. The graph (white box) is presented by its node-center version (yellow box) and edge-center version (green box) and was fed to CensNet in each channel. Gray frame highlights a typical CensNet flow. Attention module (SENet) is highlighted by carneose panel. Note that SENet was inserted in the middle of a CensNet procedure, before the last round of node convolution. (Color figure online)

Since many phenotypic measures within a domain could be correlated with one another, we perform principal components analysis (PCA) to measures classified in "cognition" domains [8]. The first principal component is used to classify the subjects to four groups by their scores. Subject number balance among groups is considered. Each group is assigned a label $l \in \mathcal{L}$.

2.3 Classification of Population via Attention-CensNet

Basics of CensNet. Supposing we have a dataset of M subjects, our objective is to assign each subject a cognitive group label l. We construct a graph $\mathcal{G} = \{\mathcal{V}, \mathcal{E}, A\}$ to represent the entire cohort as shown in Fig. 1(a), where $v \in \mathcal{V}$ is a node of the graph, the subjects in this work. Edges $\mathcal{E}$ s as well as the adjacent matrix A encode the similarity between subjects. The implementation of CensNet was detailed elsewhere [17], and we provide a summarized version as follows:

For spectral graph convolution, normalized graph Laplacian of a graph $\mathcal{G} = \{\mathcal{V}, \mathcal{E}, A\}$ is computed: $\mathcal{L} = I_N - D^{-1/2}AD^{-1/2}$ where I_N is the identity matrix and D is the diagonal degree matrix. One of the important steps is the layer-wise propagation rule based on an approximated graph spectral kernel as follows:

$$H^{l+1} = \sigma(\tilde{D}^{-1/2}\tilde{A}\tilde{D}^{-1/2}H^l W^l) \tag{1}$$

where $\tilde{A} = A + I_N$ and $\tilde{D}$ is the degree matrix, H^l and W^l are the hidden feature matrix and learnable weight of the lth layer.

On this basis, the CensNet is proposed to have both node and edge convolution layers. For convenience, the graph (white box) in Fig. 1 (a) consists of a node-center version (yellow box) and an edge-center one (green box). In the node convolution layer, the embedding of nodes in the white box is updated, while the edge adjacency matrix and edge features in the green box are intact. Then, a similar update is implemented in the edge layer. Such a node-edge switching is implemented by the following equations:

$$H_v^{l+1} = \sigma(T\Phi(H_e^l \mathrm{P}_e)T^T \odot \tilde{A}_v H_v^l W_v^l) \tag{2}$$

$$H_e^{l+1} = \sigma(T^T\Phi(H_v^l \mathrm{P}_v)T \odot \tilde{A}_e H_e^l W_e^l) \tag{3}$$

where $\tilde{A}_v = \tilde{D}_v^{-\frac{1}{2}}(A_v + I_{Nv})\tilde{D}_v^{-\frac{1}{2}}$, $\tilde{A}_e = \tilde{D}_e^{-\frac{1}{2}}(A_e + I_{Ne})\tilde{D}_e^{-\frac{1}{2}}$, $T \in \mathbb{R}^{Nv \times Ne}$ is a binary matrix that indicates whether an edge connects a node. P_e is a learnable weight vector, Φ denotes the diagonalization operation. $\odot$ denotes the element-wise product.

The loss function is defined as:

$$\mathcal{L}(\Theta) = -\sum\nolimits_{l \in Y_L} \sum\nolimits_{f=1}^{F} Y_{lf} \, log M_{lf} \tag{4}$$

where Y_L is the subset of nodes with labels, M is the softmax results of the last node layer where node feature map has F dimensions.

Squeeze-and-Excitation Attention Block. The SENet [18] is introduced to integrate multiple graphs that share the same nodes but have different node features and edges. Note that a typical CensNet includes 1) the node-and-edge switching embedding plus 2) an additional round of node embedding (for node classification). The SENet is inserted between 1) and 2) (lower panel in Fig. 1(a)). The process (node as same as edge) is expressed by:

$$z_{node} = \boldsymbol{F}_{sq}(\mathcal{F}_{node}) = \frac{1}{M \times H_n} \sum\nolimits_{i=1}^{M} \sum\nolimits_{j=1}^{H_n} \mathcal{F}_{node}(i,j) \tag{5}$$

$$s_{node} = \boldsymbol{F}_{ex}(z_{node}, W) = \sigma(g(z_{node}, W)) = \sigma(W_2\delta(W_1 z_{node})) \tag{6}$$

It is important to note that previous works [9] adopted regression scheme to predict cognitive scores, and the prediction accuracy was quantified by Pearson/Spearman correlation. However, the small dataset size significantly reduces the prediction accuracy [9], especially when sample size is below 100. We followed the suggestion in [16] to use the classification strategy instead and the accuracy of class label prediction was quantified by the AUC. Accordingly, the regression layer of some of state-of-the-art methods [8, 9, 16, 23] are replaced by classification one. We set the dimension of the final output to 4 (the number of classes) and pass it through logsoftmax function. The loss function and the AUC calculation is the same as that used in A-CensNet, while keeping the other layers by their default configuration.

3 Results

3.1 Implementation Details

It was demonstrated in He et al., 2020 that the behavior score prediction accuracy via regression drops dramatically when subject number is below 100 (no more than $r = 0.1$ via Spearman correlation). Since we only have 81 subjects, a low regression accuracy cannot be a trustworthy to be used to compare with state-of-the-arts. As a compromise solution, we adopted classification scheme to demonstrate the effectiveness of our proposed framework.

In our application, subjects are divided to four cognitive groups (around 20 subjects in each one, a total of 81 subjects). Then, we randomly split the dataset to training, validation and testing sets, respectively. We randomly selected 10 subjects from each group, a total of 40 subjects (about 50% of the total) to form the training set. Among the remaining 41 people, we randomly selected 20 people to be the verification group and 21 people to be the testing group (each accounting for 25% of the total). Such a random division of training/validation/testing subsets was repeated 100 times, independently. The results (AUCs) were the average of 100 independent replicates.

We experiment on preserving {10%, 15%, 20%} top graph and edges by their weights, and find that preserving 10% node feature and 10% edges yields the best prediction performance. We try different settings of learning rate from {0.05, 0.01, 0.005, 0.001}, dropout {0.2, 0.3, 0.4, 0.5}, hidden {16, 32, 64, 128, 512, 1024}, and found that the best performance is yielded by learning rate to 0.005, dropout to 0.2, hidden to 1024. We implement the A-CensNet structure by adding the Attention mechanism based on CensNet, which empirically produce the best performance. AUC is adopted for each independent replication experiment to evaluate the prediction performance.

3.2 Ablation Study

The prediction accuracy measured by AUC of 100 repeated experiments are reported in Table 1. As a comparison, attention module is removed, as shown in the "Attention-No" rows in Table 1. In our method, the attention block is added after the node-edge updates in each channel (see Fig. 1(a)). The results are reported in "Attention-Middle" section. We also move the attention module ahead of CensNet. The results are reported in "Attention-Before" section. It is noted that data from different datasets are taken as different channels for attention module.

Table 1. Ablation Studies. The prediction accuracy is measured by AUC. Graph structure is represented by {Node, Edge}. "+" denotes a temporal concatenation of two features. C: channel. Trj: eye movement trajectory. Ppl: pupil size variation. The index after mfMRI, Trj and Ppl indicates which movie dataset it comes from. Red: the highest AUC. Blue: the second-highest one.

<table>
<tr><th>Models</th><th colspan="4">Graph Structure</th></tr>
<tr><td rowspan="4">Attention No
(CensNet)</td><td>{mfMRI2, Trj2}
50.36±0.71</td><td>{mfMRI2, Ppl2}
52.91±0.73</td><td>{mfMRI3, Trj3}
51.49±0.75</td><td>{mfMRI3, Ppl3}
51.83±0.85</td></tr>
<tr><td colspan="4">{mfMRI2+mfMRI3, Trj2+Trj3}
51.94±0.81</td></tr>
<tr><td colspan="4">{mfMRI2+mfMRI3, Ppl2+Ppl3}
45.45±0.43</td></tr>
<tr></tr>
<tr><td rowspan="5">Attention
Middle</td><td colspan="2">C1: {mfMRI2, Trj2}
C2: {mfMRI2, Ppl2}
50.66±0.67</td><td colspan="2">C1: {mfMRI3, Trj3}
C2: {mfMRI3, Ppl3}
51.79±0.67</td></tr>
<tr><td colspan="4">C1: {mfMRI2, Trj2} C2: {mfMRI3, Trj3}
49.35±0.69</td></tr>
<tr><td colspan="4">C1: {mfMRI2, Ppl2} C2: {mfMRI3, Ppl3}
50.38±0.62</td></tr>
<tr><td colspan="4">C1: {mfMRI2+mfMRI2, Trj2+Trj3}
C2: {mfMRI2+mfMRI2, Ppl2+Ppl3}
46.21±0.55</td></tr>
<tr><td colspan="4">C1: {mfMRI2, Trj2} C2: {mfMRI2, Ppl2}
C3: {mfMRI3, Trj3} C4: {mfMRI3, Ppl3}
54.63±0.65</td></tr>
<tr><td rowspan="2">Attention
Before</td><td colspan="2">C1: {mfMRI2, Trj2}
C2: {mfMRI2, Ppl2}
49.00±0.52</td><td colspan="2">C1: {mfMRI3, Trj3}
C2: {mfMRI3, Ppl3}
48.49±0.78</td></tr>
<tr><td colspan="4">C1: {mfMRI2, Trj2} C2: {mfMRI2, Ppl2}
C3: {mfMRI3, Trj3} C4: {mfMRI3, Ppl3}
50.24±0.67</td></tr>
</table>

To evaluate whether multiple stimulus loads to the same subjects enhance the performance, we only use two channels (trajectory and pupil variation as two sets of edges) within a single movie. The results are in the first row in "Attention-Top" and "Attention-Middle" sections. We use the format {Node, Edge} to describe the graph structure.

In the non-attention algorithms, CensNet on movie2 dataset ({mfMRI2, Pupil2}) yields the best performance. Concatenation of node feature or edge feature from two movie datasets even decreases the accuracy. When attention module is added (Attention-Middle), two-channel models (gray rows) do not significantly increase the accuracy. Within one movie dataset, when node feature is fixed and eye trajectory and pupil size variation are taken as two channels, the performance is not better than that on single-channel model on movie3 dataset. When two movie datasets are considered (carneose&green rows), they are integrated by means of channels when models with attention are used. In contrast, in non-attention models, the two movie datasets are integrated by means of feature concatenation. It is seen that the channel integration outperforms feature concatenation when pupil size is used as edges (green). But this does not hold when eye trajectory was used as edges (carneose). When both trajectory and pupil size are both considered as two channels, but features from two dataset are concatenated (white rows), the prediction performance is worst (46.21 $\pm$ 0.55) in all attention models. Finally, when both trajectory and pupil size from two dataset are used as four channels (blue row) in our algorithm, the best performance was yielded.

These comparisons suggest that integration of edge features by channel attention does not always improves the performance than single use of an edge (gray rows), while integration of datasets by channel attention significantly outperforms the integration by feature concatenation. This observation still holds for "Attention-Before" models. But their accuracy is not as high as that via "Attention-Middle" models (see the cross-comparison between blue rows and gray rows), suggesting that a node-edge embedding could yield latent features more sensitive to individual variations, such that a channel attention works better at this deep feature space than being applied immediately after the original shallow features ("Attention-Before"). Finally, in addition to the 4-group classification, we evenly divided the subjects into 6 and 8 groups, respectively. AUCs via our model are 56.34 $\pm$ 0.65 and 56.95 $\pm$ 0.48, demonstrating the robustness of the algorithm to class numbers.

3.3 Comparison with State-of-the-Arts

We compare our results with the ones by state-of-the-art methods listed in Table 2. The results via linear model and GCN were applied on Movie2 and have been reported in [16], and is thus listed for reference. Note that the linear, FNN and BrainNetCNN are not GNN-related methods. They have no edges but only rely on mfMRI features.

Table 2. Comparison with state-of-the-arts. Red font highlights the highest AUC and blue font highlights the second-highest one. "+" denotes a temporal concatenation of two features.

Models	Data		AUC
	Edge	Node	
Linear [8]	-	mfMRI2	41.94±0.81
GCN [16]	Trj2	mfMRI2	48.51±0.94
CensNet [17]		mfMRI2	49.75±0.65
CensNet [17]	Trj2	mfMRI2	50.36±0.71
	Ppl2		52.91±0.73
FNN [9]	-	mfMRI2	50.81±0.66
		mfMRI3	49.45±0.85
		mfMRI2+3	50.54±1.01
BrainNet CNN [23]	-	mfMRI2	53.42±0.63
		mfMRI3	49.81±0.78
		mfMRI2+3	50.66±0.82
A-CensNet	All	All	54.63±0.65

All nonlinear deep neural networks yield significant improvement in contrast to the linear method (41.94 $\pm$ 0.81). A concatenation of mfMRI features (mfMRI2 +3 rows) does not improve the prediction accuracy in contrast to that on a single dataset, suggesting the importance of the strategy selection for concatenating features from multiple datasets. Within a single movie dataset (unshaded rows), models integrating mfMRI and eye tracking outperform the models (FNN and BrainNetCNN) that used single modality (mfMRI). After integrating mfMRI and eye behavior from multiple datasets, our results outperform all the-state-of-arts. These results demonstrate the effectiveness of integration of brain activity and eye behavior to one framework in cognition prediction. Also, given the limited subjects, multiple loads of stimuli integrated via attention modules could significantly improve the prediction performance.

4 Conclusion

We propose A-CensNet to predict subjects' cognitive scores, with subjects taken as nodes, mfMRI derived functional connectivity as node feature, different eye tracking features are used to compute similarity between subjects to construct heterogeneous graph edges. These graphs from different dataset are all taken as different channels. The proposed model integrates graph embeddings from multiple channels into one. This model outperforms the one using single modality, single channel and state-of-the-art methods. Our results suggest that the brain functional activity patterns and the behavior patterns might complement each other in interpreting trait-like phenotypes, and might provide new clues to studies of diseases with cognitive abnormality [24].

Acknowledgement. This work was supported by the National Natural Science Foundation of China (31971288, U1801265, 61936007, 62276050, 61976045, U20B2065, U1801265 and 61936007); the National Key R&D Program of China under Grant 2020AAA0105701; High-level researcher start-up projects (Grant No. 06100-22GH0202178); Innovation Foundation for Doctor Dissertation of Northwestern Polytechnical University CX2022052.

References

1. Thiebaut de Schotten, M., Forkel, S.J.: The emergent properties of the connected brain. Science, **378**(6619), 505–510 (2022)
2. Li, J., et al.: Global signal regression strengthens association between resting-state functional connectivity and behavior. Neuroimage **196**, 126–141 (2019)
3. Huijbers, W., Van Dijk, K.R.A., Boenniger, M.M., Stirnberg, R., Breteler, M.M.: Less head motion during MRI under task than resting-state conditions. Neuroimage **147**, 111–120 (2017)
4. Barch, D.M., et al.: Function in the human connectome: task-fMRI and individual differences in behavior. Neuroimage **80**, 169–189 (2013)
5. Sonkusare, S., Breakspear, M., Guo, C.: Naturalistic stimuli in neuroscience: critically acclaimed. Trends Cogn. Sci. **23**(8), 699–714 (2019)
6. Hasson, U., Nir, Y., Levy, I., Fuhrmann, G., Malach, R.: Intersubject synchronization of cortical activity during natural vision. Science **303**(5664), 1634–1640 (2004)
7. Finn, E.S., Scheinost, D., Finn, D.M., Shen, X., Papademetris, X., Constable, R.T.: Can brain state be manipulated to emphasize individual differences in functional con-nectivity? Neuroimage **160**, 140–151 (2017)
8. Finn, E.S., Bandettini, P.A.: Movie-watching outperforms rest for functional connectivity-based prediction of behavior. Neuroimage **235**, 117963 (2021)
9. He, T., et al.: Deep neural networks and kernel regression achieve comparable accuracies for functional connectivity prediction of behavior and de-mographics. Neuroimage **206**, 116276 (2020)

10. Gal, S., Coldham, Y., Bernstein-Eliav, M.: Act natural: functional connectivity from naturalistic stimuli fMRI outperforms resting-state in predicting brain activity. bioRxiv (2021)
11. He, T., et al.: Meta-matching as a simple framework to translate phenotypic predictive models from big to small data. Nature Neurosci. **25**, 1–10 (2022)
12. Lim, J.Z., Mountstephens, J., Teo, J.: Emotion recognition using eye-tracking: taxonomy, review and current challenges. Sensors **20**(8), 2384 (2020)
13. Hess, E.H., Polt, J.M.: Pupil size as related to interest value of visual stimuli. Science **132**(3423), 349–350 (1960)
14. Lohse, G.L., Johnson, E.J.: A comparison of two process tracing methods for choice tasks. Organ. Behav. Hum. Decis. Process. **68**(1), 28–43 (1996)
15. Son, J., et al.: Evaluating fMRI-based estimation of eye gaze during naturalistic viewing. Cereb. Cortex **30**(3), 1171–1184 (2020)
16. Gao, J., et al.: Prediction of cognitive scores by movie-watching FMRI connectivity and eye movement via spectral graph convolutions. In: 2022 IEEE 19th International Symposium on Biomedical Imaging (ISBI), pp. 1–5. IEEE (2022)
17. Jiang, X., Ji, P., Li, S.: CensNet: convolution with edge-node switching in graph neural networks. In: IJCAI, pp. 2656–2662 (2019)
18. Hu, J., Shen, L., Sun, G.: Squeeze-and-excitation networks. In: Proceedings of the IEEE Conference on Computer Vision and Pattern Recognition, pp. 7132–7141 (2018)
19. Elam, J.: https://www.humanconnectome.org/study/hcp-young-adult/article/first-release-of-7t-mr-image-data. Accessed 20 June 2016
20. Griffanti, L., et al.: ICA-based artefact removal and accelerated fMRI ac-quisition for improved resting state network imaging. Neuroimage **95**, 232–247 (2014)
21. Glasser, M.F., et al., Wu-Minn HCP Consortium: The minimal preprocessing pipelines for the human connectome project. Neuroimage, **80**, 105–124 (2013)
22. Destrieux, C., Fischl, B., Dale, A., Halgren, E.: Automatic parcellation of human cortical gyri and sulci using standard anatomical nomenclature. Neuroimage **53**(1), 1–15 (2010)
23. Kawahara, J., et al.: BrainNetCNN: Convolutional neural networks for brain networks; towards predicting neurodevelopment. Neuroimage **146**, 1038–1049 (2017)
24. Tye, C., et al.: Neurophysiological responses to faces and gaze direction differentiate children with ASD, ADHD and ASD + ADHD. Dev. Cogn. Neurosci. **5**, 71–85 (2013)

Partial Vessels Annotation-Based Coronary Artery Segmentation with Self-training and Prototype Learning

Zheng Zhang[1], Xiaolei Zhang[2], Yaolei Qi[1], and Guanyu Yang[1,3,4](✉)

[1] Key Laboratory of New Generation Artificial Intelligence Technology and Its Interdisciplinary Applications (Southeast University), Ministry of Education, Nanjing, China
yang.list@seu.edu.cn

[2] Department of Diagnostic Radiology, Jinling Hospital, Medical School of Nanjing University, Nanjing, China

[3] Jiangsu Provincial Joint International Research Laboratory of Medical Information Processing, Southeast University, Nanjing, China

[4] Centre de Recherche en Information Biomédicale sino-français (CRIBs), Strasbourg, France

Abstract. Coronary artery segmentation on coronary-computed tomography angiography (CCTA) images is crucial for clinical use. Due to the expertise-required and labor-intensive annotation process, there is a growing demand for the relevant label-efficient learning algorithms. To this end, we propose partial vessels annotation (PVA) based on the challenges of coronary artery segmentation and clinical diagnostic characteristics. Further, we propose a progressive weakly supervised learning framework to achieve accurate segmentation under PVA. First, our proposed framework learns the local features of vessels to propagate the knowledge to unlabeled regions. Subsequently, it learns the global structure by utilizing the propagated knowledge, and corrects the errors introduced in the propagation process. Finally, it leverages the similarity between feature embeddings and the feature prototype to enhance testing outputs. Experiments on clinical data reveals that our proposed framework outperforms the competing methods under PVA (24.29% vessels), and achieves comparable performance in trunk continuity with the baseline model using full annotation (100% vessels).

Keywords: Coronary artery segmentation · Label-efficient learning · Weakly supervised learning

Z. Zhang and X. Zhang—Contributed equally to this work.

Supplementary Information The online version contains supplementary material available at https://doi.org/10.1007/978-3-031-43895-0_28.

H. Greenspan et al. (Eds.): MICCAI 2023, LNCS 14221, pp. 297–306, 2023.
https://doi.org/10.1007/978-3-031-43895-0_28

1 Introduction

Coronary artery segmentation is crucial for clinical coronary artery disease diagnosis and treatment [4]. Coronary-computed tomography angiography (CCTA), as a non-invasive technique, has been certified and recommended as established technology in the cardiological clinical arena [15]. Thus, automatic coronary artery segmentation on CCTA images has become increasingly sought after as a means to enhance diagnostic efficiency for clinicians. In recent years, the performance of deep learning-based methods have surpassed that of conventional machine learning approaches (e.g. region growing) in coronary artery segmentation [4]. Nevertheless, most of these deep learning-based methods highly depend on accurately labeled datasets, which need labor-intensive annotations. Therefore, there is a growing demand for relevant label-efficient learning algorithms for automatic coronary artery segmentation on CCTA images.

Label-efficient learning algorithms have garnered considerable interest and research efforts in natural and medical image processing [5,6,16], while research on label-efficient coronary artery segmentation for CCTA images is slightly lagging behind. Although numerous label-efficient algorithms for coronary artery segmentation in X-ray angiograms have been proposed [19,20], only a few researches focus on CCTA images. Qi et al. [13] proposed an elabrately designed EE-Net to achieve commendable performance with limited labels. Zheng et al. [22] transformed nnU-Net into semi-supervised segmentation field as the generator of Gan, having achieved satisfactory performance on CCTA images. Most of these researches use incomplete supervision, which labels a subset of data. However, other types of weak supervision (e.g. inexact supervision), which are widely used in natural image segmentation [16], are seldom applied to coronary artery segmentation on CCTA images.

Different types of supervision are utilized according to the specific tasks. The application of various types of weak supervision are inhibited in coronary artery segmentation on CCTA images by the following challenges. 1) Difficult labeling (Fig. 1(a)). The target regions are scattered, while manual annotation is drawn slice by slice on the planes along the vessels. Also, boundaries of branches and peripheral vessels are blurred. These make the annotating process time-consuming and expertise-required. 2) Complex topology (Fig. 1(b)). Coronary artery shows complex and slender structures, diameter of which ranges from 2 mm to 5 mm. The tree-like structure varies individually. Based on these challenges and the insight that vessels share local feature (Fig. 1(b)), we propose partial vessels annotation and our framework as following.

Given the above, we propose partial vessels annotation (PVA) (Fig. 1(c)) for CCTA images. While PVA is a form of partial annotation (PA) which has been adopted by a number of researches [2,7,12,18], our proposed PVA differs from the commonly used PA methods. More specifically, PVA labels vessels continuously from the proximal end to the distal end, while the labeled regions of PA are typically randomly selected. Thus, our proposed PVA has two merits. 1) PVA balances efficiency and informativity. Compared with full annotation, PVA only requires clinicians to label vessels within restricted regions in adjacent slices,

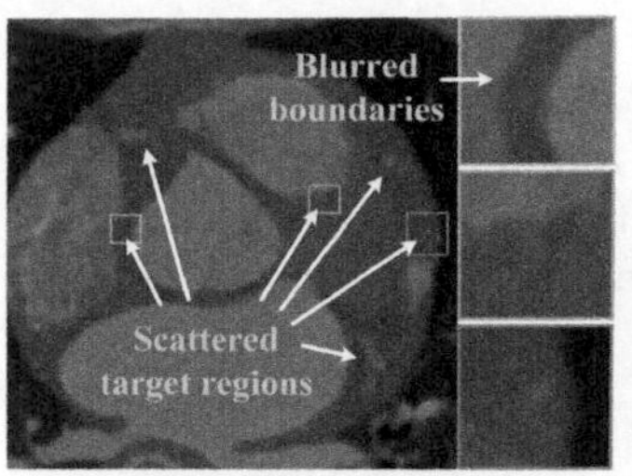

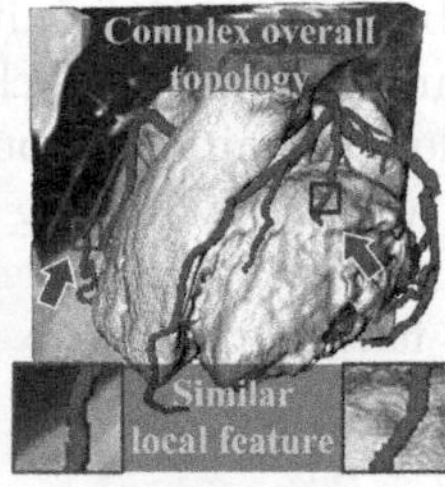

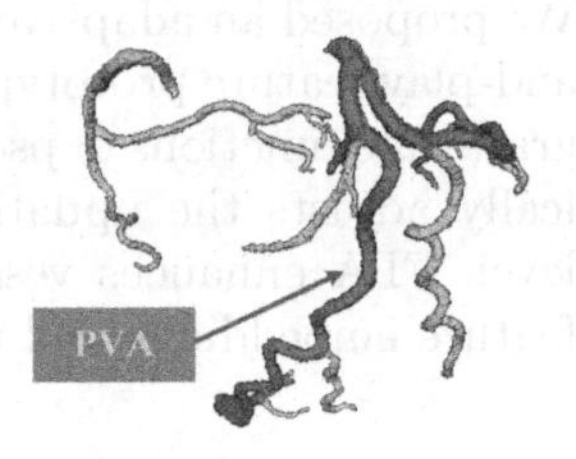

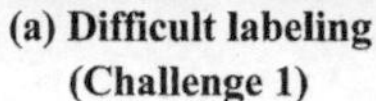

(a) Difficult labeling (Challenge 1) **(b) Complex topology (Challenge 2)** **(c) Partial Vessels Annotation (Proposed annotation)**

Fig. 1. Motivation. (a) and (b) shows the two challenges of coronary artery segmentation, while (c) shows our proposed partial vessels annotation (PVA) according to the challenges. a) Coronary artery has blurred boundaries and scattered target regions. b) Coronary artery has complex overall topology but similar local feature. c) Partial vessels annotation (red) labels less regions than full annotation (overall). (Color figure online)

rather than all scattered target regions in each individual slice. Compared with PA, PVA keep labeled vessels continuous to preserve local topology information. 2) PVA provides flexibility for clinicians. Given that clinical diagnosis places greater emphasis on the trunks rather than the branches, PVA allows clinicians to focus their labeling efforts on vessels of particular interest. Therefore, our proposed PVA is well-suited for clinical use.

In this paper, we further propose a progressive weakly supervised learning framework for PVA. Our proposed framework, using PVA (only 24.29% vessels labeled), achieved better performance than the competing weakly supervised methods, and comparable performance in trunk continuity with the full annotation (100% vessels labeled) supervised baseline model. The framework works in two stages, which are local feature extraction (LFE) stage and global structure reconstruction (GSR) stage. 1) LFE stage extracts the local features of coronary artery from the limited labeled vessels in PVA, and then propagates the knowledge to unlabeled regions. 2) GSR stage leverages prediction consistency during the iterative self-training process to correct the errors, which are introduced inevitably by the label propagation process. The code of our method is available at https://github.com/ZhangZ7112/PVA-CAS.

To summarize, the contributions of our work are three-fold:

- To the best of our knowledge, we proposed partial vessels annotation for coronary artery segmentation for the first time, which is in accord with clinical use. First, it balances efficiency and informativity. Second, it provides flexibility for clinicians to annotate where they pay more attention.
- We proposed a progressive weakly supervised learning framework for partial vessels annotation-based coronary artery segmentation. It only required 24.29% labeled vessels, but achieved comparable performance in trunk continuity with the baseline model using full annotation. Thus, it shows great potential to lower the label cost for relevant clinical and research use.

- We proposed an adaptive label propagation unit (LPU) and a learnable plug-and-play feature prototype analysis (FPA) block in our framework. LPU integrates the functions of pseudo label initialization and updating, which dynamically adjusts the updating weights according to the calculated confidence level. FPA enhances vessel continuity by leveraging the similarity between feature embeddings and the feature prototype.

2 Method

As shown in Fig. 2, our proposed framework for partial vessels annotation (PVA) works in two stages. 1) The LFE stage(Sect. 2.1) extracts and learns vessel features from PVA locally. After the learning process, it infers on the training set to propagate the learned knowledge to unlabeled regions, outputs of which are integrated with PVA labels to initialize pseudo labels. 2) The GSR stage (Sect. 2.2) utilizes pseudo labels to conduct self-training, and leverages prediction consistency to improve the pseudo labels. In our proposed framework, we also designed an adaptive label propagation unit (LPU) and a learnable plug-and-play feature prototype analysis (FPA) block. LPU initialize and update the pseudo labels; FPA block learns before testing and improves the final output during testing.

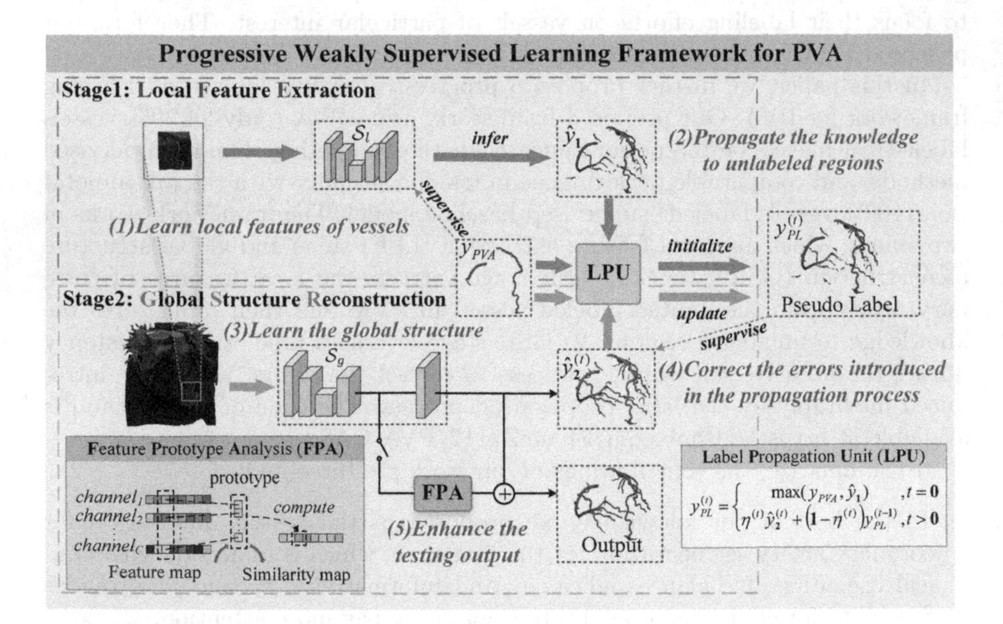

Fig. 2. Two-stage framework. LFE stage: 1) $\mathcal{S}_l$ learns local features from the labeled vessels in PVA labels. 2) $\mathcal{S}_l$ propagates the knowledge to unlabeled regions and LPU initializes the pseudo labels. **GSR** stage: 3) $\mathcal{S}_g$ learns the global structure from the pseudo labels. 4) LPU updates the pseudo labels if a quality control test is passed. 5) FPA improves the testing output after the iterative self-training process of (3) and (4).

2.1 Local Feature Extraction Stage

In LFE stage, our hypothesis is that the small areas surrounding the labeled regions hold valid information. Based on this, a light segmentation model $\mathcal{S}_l$ is trained to learn vessel features locally, with small patches centering around the labeled regions as input and output. In this manner, the negative impact of inaccurate supervision information in unlabeled regions is also reduced.

Pseudo Label Initialization in LPU. After training, $\mathcal{S}_l$ propagates the learned knowledge of local feature to unlabeled regions. For each image of shape $H \times W \times D$, the corresponding output logit $\hat{y}_1 \in [0,1]^{H \times W \times D}$ of $\mathcal{S}_l$ provides a complete estimate of the distribution of vessels, albeit with some approximation. Meanwhile, the PVA label $y_{PVA} \in \{0,1\}^{H \times W \times D}$ provides accurate information on the distribution of vessels, but only to a limited extent. Therefore, LPU integrate $\hat{y}_1$ and y_{PVA} to initialize the pseudo label y_{PL} (Eq. 1), which will be utilized in GSR stage and updated during iterative self-training.

$$\underset{\forall (h,w,d) \in \mathbb{R}^{H \times W \times D}}{y_{PL}^{(t=0)}(h,w,d)} = \begin{cases} 1, & y_{PVA}(h,w,d) = 1, \\ \hat{y}_1(h,w,d), & \text{otherwise.} \end{cases} \quad (1)$$

2.2 Global Structure Reconstruction Stage

The GSR stage mainly consists of three parts: 1) The segmentation model $\mathcal{S}_g$ to learn the global tree-like structure; 2) LPU to improve pseudo labels; 3) FPA block to improve segmentation results at testing.

Through initialization (Eq. 1), the initial pseudo label $y_{PL}^{(t=0)}$ contains the information of both PVA labels and the knowledge of local features in $\mathcal{S}_l$. Therefore, at the beginning of this stage, $\mathcal{S}_g$ learns from $y_{PL}^{(t=0)}$ to warm up. After this, logits of $\mathcal{S}_g$ are utilized to update the pseudo labels during iterative self-training.

Pseudo Label Updating in LPU. The principle of this process is that more reliable logit influences more the distribution of the corresponding pseudo label. Based on this principle, first we calculate the confidence degree $\eta^{(t)} \in [0,1]$ for $\hat{y}_2^{(t)}$. Defined by Eq. 2, $\eta^{(t)}$ numerically equals to the average of the logits in labeled regions. This definition makes sense since the expected logit equals to ones in vessel regions and zeros in background regions. The closer $\hat{y}_2^{(t)}$ gets to the expected logit, the higher $\eta^{(t)}$ (confidence degree) will be.

$$\eta^{(t)} = \frac{\sum\limits_h \sum\limits_w \sum\limits_d y_{PVA}(h,w,d) \cdot \hat{y}_2^{(t)}(h,w,d)}{\sum\limits_h \sum\limits_w \sum\limits_d y_{PVA}(h,w,d)} \quad (2)$$

Then, a quality control test is performed to avoid negative optimization as far as possible. As the confidence degree $\eta^{(t)}$ assesses the quality of predictions, which means low-confidence predictions are more likely to generate errors, our quality control test rejects low-confidence predictions to reduce the risk of error accumulation. If $\eta^{(t)}$ is higher than all elements in the set $\{\eta^{(i)}\}_{i=1}^{t-1}$, the current logit is trustworthy to pass the test to improve the pseudo label. Then, $y_{PL}^{(t)}$ is updated by the exponentially weighted moving average (EWMA) of the logits and the pseudo labels (Eq. 3). This process is similar to prediction ensemble [11], which hase been adopted to filter pseudo labels [9]. However, different from their methods, where the factor $\eta^{(t)}$ is a fixed hyperparameter coefficient and the pseudo labels are updated each or every several epoches, $\eta^{(t)}$ in our method is adaptive. Our EWMA gradually diminishes the negative influence of existing errors through the weighted average of predictions across multiple phases.

$$y_{PL}^{(t)} = \begin{cases} \eta^{(t)}\hat{y}_2^{(t)} + (1-\eta^{(t)})y_{PL}^{(t-1)}, & \eta^{(t)} = max\{\{\eta^{(i)}\}_{i=1}^{t}\} \\ y_{PL}^{(t-1)}, & \text{otherwise.} \end{cases} \tag{3}$$

Feature Prototype Analysis Block. Inspired by [21], which generates class feature prototype ρ_c (Eq. 4) from the embeddings z_i^l of labeled points in class c, we inherit the idea but further transform the mechanism into the proposed learnable plug-and-play block, FPA block. Experimental experience finds that the output of FPA block has good continuity, for which the FPA output are utilized to enhance the continuity of convolution output at testing.

$$\rho_c = \frac{1}{|\mathcal{I}_c|} \sum_{z_i^l \in \mathcal{I}_c} z_i^l \tag{4}$$

In the penultimate layer of the network, which is followed by a $1 \times 1 \times 1$ convolutional layer to output logits, we parallelly put the feature map $Z \in \mathcal{R}^{C \times H \times W \times D}$ into FPA. The output similarity map $O \in \mathcal{R}^{1 \times H \times W \times D}$ is calculated by Eq. 5, where $Z(h,w,d) \in \mathcal{R}^C$ denotes the feature embeddings of voxel (h,w,d), and $\rho_\theta \in \mathcal{R}^C$ the kernel parameters of FPA.

$$O(h,w,d) = exp(-\|Z(h,w,d) - \rho_\theta\|^2) \tag{5}$$

The learning process of FPA block is before testing, during which the whole model except FPA gets frozen. To reduce the additional overhead, ρ_θ is initialized by one-time calculated ρ_c and fine-tuned with loss $\mathcal{L}_{fpa}$ (Eq. 6), where only labeled voxels will take effect in updating the kernel.

$$\mathcal{L}_{fpa} = \frac{\sum_h \sum_w \sum_d y_{PVA}(h,w,d) \cdot log(O(h,w,d))}{\sum_h \sum_w \sum_d y_{PVA}(h,w,d)} \tag{6}$$

3 Experiments and Results

3.1 Dataset and Evaluation Metrics

Experiments are implemented on a clinical dataset, which includes 108 subjects of CCTA volumes (2:1 for training and testing). The volumes share the size of

$512 \times 512 \times D$, with D ranging from 261 to 608. PVA labels of the training set are annotated by clinicians, where only 24.29% vessels are labeled.

The metrics used to quantify the results include both integrity and continuity assessment indicators. Integrity assessment indicators are Mean Dice Coefficient (Dice), Relevant Dice Coefficient (RDice) [13], Overlap (OV) [8]; continuity assessment indicators are Overlap util First Error (OF) [14] on the three main trunks (LAD, LCX and RCA).

3.2 Implementation Details

3D U-Net [3] is set as our baseline model. Experiments were implemented using Pytorch on GeForce RTX 2080Ti. Adam optimizer was used to train the models with an initial learning rate of 10^{-4}. The patch sizes were set as $128 \times 128 \times 128$ and $512 \times 512 \times 256$ respectively for $\mathcal{S}_l$ and $\mathcal{S}_g$. When testing, sliding windows were used with a half-window width step to cover the entire volume.

3.3 Comparative Test

To verify the effectiveness of our proposed method, it is compared with both classic segmentation models (3D U-Net [3], HRNet [17], Transunet [1]) and partial annotation-related weakly supervised frameworks (EWPA [12], DMPLS [10]).

Table 1. Quantitative results of different methods under partial vessels annotation (PVA, 24.29% vessels labeled) or full annotation (FA, 100% vessels labeled).

Label	Method	Dice(%)↑	RDice(%)↑	OV(%)↑	OF↑		
					LAD	LCX	RCA
PVA	3D U-Net [3]	$60.60_{\pm 7.09}$	$69.45_{\pm 7.82}$	$62.24_{\pm 6.43}$	$0.647_{\pm 0.335}$	$0.752_{\pm 0.266}$	$0.747_{\pm 0.360}$
	HRNet [17]	$48.72_{\pm 7.16}$	$52.31_{\pm 7.96}$	$37.81_{\pm 6.61}$	$0.490_{\pm 0.297}$	$0.672_{\pm 0.301}$	$0.717_{\pm 0.356}$
	Transunet [1]	$63.08_{\pm 6.42}$	$71.97_{\pm 7.38}$	$61.21_{\pm 6.40}$	$0.669_{\pm 0.274}$	$0.762_{\pm 0.243}$	$0.728_{\pm 0.362}$
	EWPA [12]	$55.41_{\pm 6.15}$	$61.54_{\pm 6.83}$	$60.48_{\pm 5.17}$	$0.659_{\pm 0.334}$	$0.759_{\pm 0.286}$	$0.749_{\pm 0.364}$
	DMPLS [10]	$59.12_{\pm 7.69}$	$65.81_{\pm 8.15}$	$59.99_{\pm 5.80}$	$0.711_{\pm 0.292}$	$0.775_{\pm 0.284}$	$0.711_{\pm 0.358}$
	Ours	$\mathbf{71.45}_{\pm 6.07}$	$\mathbf{83.14}_{\pm 6.72}$	$\mathbf{75.40}_{\pm 6.15}$	$\mathbf{0.895}_{\pm 0.226}$	$\mathbf{0.915}_{\pm 0.190}$	$\mathbf{0.879}_{\pm 0.274}$
FA	3D U-Net	$83.14_{\pm 3.52}$	$90.91_{\pm 4.18}$	$89.00_{\pm 4.75}$	$0.913_{\pm 0.231}$	$0.843_{\pm 0.301}$	$0.873_{\pm 0.265}$

The quantitative results of different methods are summarized in Table 1, which shows that our proposed method outperforms the competing methods under PVA. The competing frameworks (EWPA and DMPLS) had achieved the best results in their respective tasks under partial annotation, but our proposed method achieved better results for PVA-based coronary artery segmentation. It is worth mentioning that the performance in trunk continuity (measured by the indicator OF) of our proposed method using PVA (24.29% vessels labeled) is comparable to that of the baseline model using full annotation (100% vessels labeled).

The qualitative visual results verify that our proposed method outperforms the competing methods under PVA. Three cases are shown in Fig. 3. All the cases show that the segmentation results of our method have good overall topology integrity, especially on trunk continuity.

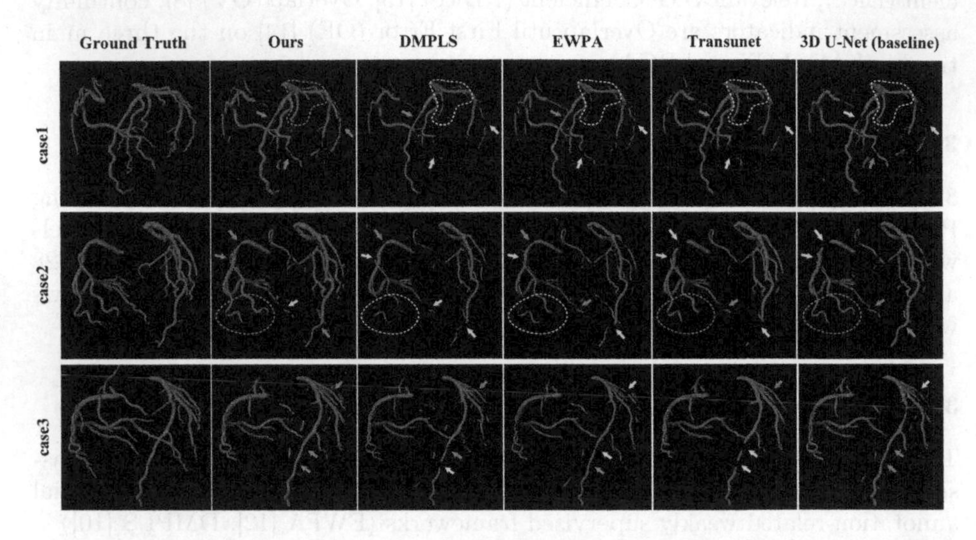

Fig. 3. Visual comparison of the segmentation results under PVA. Green symbols (arrows and dotted frames) indicate higher-quality regions than yellow symbols. (Color figure online)

3.4 Ablation Study

Ablation experiments were conducted to verify the importance of the components in our proposed framework (summarized in Table 2). The performance improvement verifies the effectiveness of pseudo label initialization (PLI) and updating (PLU) mechanisms in the label propagation unit (LPU). PLI integrates the information of PVA labels with the propagated knowledge, and PLU improves the pseudo labels during self-training. With the help of FPA block, the segmentation results gain further improvement, especially on the continuity of trunks.

Table 2. Quantitative results of ablation analysis of different components.

$\mathcal{S}_l$	LPU		$\mathcal{S}_g$	FPA	Dice(%)↑	RDice(%)↑	OV(%)↑	OF↑		
	PLI	PLU						LAD	LCX	RCA
✓					$60.60_{\pm 7.09}$	$69.45_{\pm 7.82}$	$62.24_{\pm 6.43}$	$0.647_{\pm 0.335}$	$0.752_{\pm 0.266}$	$0.747_{\pm 0.360}$
✓	✓		✓		$64.23_{\pm 6.44}$	$73.81_{\pm 6.89}$	$66.19_{\pm 5.63}$	$0.751_{\pm 0.328}$	$0.813_{\pm 0.231}$	$0.784_{\pm 0.349}$
✓	✓	✓	✓		$71.43_{\pm 7.20}$	$81.70_{\pm 6.92}$	$72.13_{\pm 5.94}$	$0.873_{\pm 0.227}$	$0.860_{\pm 0.223}$	$0.808_{\pm 0.334}$
✓	✓	✓	✓	✓	$\mathbf{71.45}_{\pm 6.07}$	$\mathbf{83.14}_{\pm 6.72}$	$\mathbf{75.40}_{\pm 6.15}$	$\mathbf{0.895}_{\pm 0.226}$	$\mathbf{0.915}_{\pm 0.190}$	$\mathbf{0.879}_{\pm 0.274}$

4 Conclusion

In this paper, we proposed partial vessels annotation (PVA) for coronary artery segmentation on CCTA images. The proposed PVA is convenient for clinical use for the two merits, providing flexibility as well as balancing efficiency and informativity. Under PVA, we proposed a progressive weakly supervised learning framework, which outperforms the competing methods and shows comparable performance in trunk continuity with the full annotation supervised baseline model. In our framework, we also designed an adaptive label propagation unit (LPU) and a learnable plug-and-play feature prototype analysis(FPA) block. LPU integrates the functions of pseudo label initialization and updating, and FPA improves vessel continuity by leveraging the similarity between feature embeddings and the feature prototype. To conclude, our proposed framework under PVA shows great potential for accurate coronary artery segmentation while requiring significantly less annotation effort.

Acknowledgements.. This research was supported by the Intergovernmental Cooperation Project of the National Key Research and Development Program of China(2022YFE0116700). We thank the Big Data Computing Center of Southeast University for providing the facility support.

References

1. Chen, J., et al.: TransUNet: transformers make strong encoders for medical image segmentation. arXiv preprint: arXiv:2102.04306 (2021)
2. Cheng, H.T., et al.: Self-similarity student for partial label histopathology image segmentation. In: Vedaldi, A., Bischof, H., Brox, T., Frahm, J.-M. (eds.) ECCV 2020. LNCS, vol. 12370, pp. 117–132. Springer, Cham (2020). https://doi.org/10.1007/978-3-030-58595-2_8
3. Çiçek, Ö., Abdulkadir, A., Lienkamp, S.S., Brox, T., Ronneberger, O.: 3D U-Net: learning dense volumetric segmentation from sparse annotation. In: Ourselin, S., Joskowicz, L., Sabuncu, M.R., Unal, G., Wells, W. (eds.) MICCAI 2016. LNCS, vol. 9901, pp. 424–432. Springer, Cham (2016). https://doi.org/10.1007/978-3-319-46723-8_49
4. Gharleghi, R., Chen, N., Sowmya, A., Beier, S.: Towards automated coronary artery segmentation: a systematic review. Comput. Methods Programs Biomed., 107015 (2022)
5. He, Y., et al.: Learning better registration to learn better few-shot medical image segmentation: Authenticity, diversity, and robustness. IEEE Trans. Neural Netw. Learn. Syst. (2022)
6. He, Y., et al.: Dense biased networks with deep priori anatomy and hard region adaptation: semi-supervised learning for fine renal artery segmentation. Med. Image Anal. **63**, 101722 (2020)
7. Ho, D.J., et al.: Deep multi-magnification networks for multi-class breast cancer image segmentation. Comput. Med. Imaging Graph. **88**, 101866 (2021)
8. Kirişli, H., et al.: Standardized evaluation framework for evaluating coronary artery stenosis detection, stenosis quantification and lumen segmentation algorithms in computed tomography angiography. Med. Image Anal. **17**(8), 859–876 (2013)

9. Lee, H., Jeong, W.-K.: Scribble2Label: scribble-supervised cell segmentation via self-generating pseudo-labels with consistency. In: Martel, A.L., et al. (eds.) MICCAI 2020. LNCS, vol. 12261, pp. 14–23. Springer, Cham (2020). https://doi.org/10.1007/978-3-030-59710-8_2
10. Luo, X., et al.: Scribble-supervised medical image segmentation via dual-branch network and dynamically mixed pseudo labels supervision. In: Wang, L., Dou, Q., Fletcher, P.T., Speidel, S., Li, S. (eds.) MICCAI 2022. Lecture Notes in Computer Science, vol. 13431. Springer, Cham (2022). https://doi.org/10.1007/978-3-031-16431-6_50
11. Nguyen, D.T., Mummadi, C.K., Ngo, T.P.N., Nguyen, T.H.P., Beggel, L., Brox, T.: Self: Learning to filter noisy labels with self-ensembling. arXiv preprint: arXiv:1910.01842 (2019)
12. Peng, L., et al.: Semi-supervised learning for semantic segmentation of emphysema with partial annotations. IEEE J. Biomed. Health Inform. **24**(8), 2327–2336 (2019)
13. Qi, Y., et al.: Examinee-examiner network: weakly supervised accurate coronary lumen segmentation using centerline constraint. IEEE Trans. Image Process. **30**, 9429–9441 (2021)
14. Schaap, M., et al.: Standardized evaluation methodology and reference database for evaluating coronary artery centerline extraction algorithms. Med. Image Anal. **13**(5), 701–714 (2009)
15. Serruys, P.W., et al.: Coronary computed tomographic angiography for complete assessment of coronary artery disease: JACC state-of-the-art review. J. Am. Coll. Cardiol. **78**(7), 713–736 (2021)
16. Shen, W., et al.: A survey on label-efficient deep image segmentation: Bridging the gap between weak supervision and dense prediction. IEEE Trans. Pattern Anal. Mach. Intell., 1–20 (2023). https://doi.org/10.1109/TPAMI.2023.3246102
17. Sun, K., Xiao, B., Liu, D., Wang, J.: Deep high-resolution representation learning for human pose estimation. In: Proceedings of the IEEE/CVF Conference on Computer Vision and Pattern Recognition, pp. 5693–5703 (2019)
18. Zhai, S., Wang, G., Luo, X., Yue, Q., Li, K., Zhang, S.: PA-Seg: learning from point annotations for 3D medical image segmentation using contextual regularization and cross knowledge distillation. IEEE Trans. Med. Imaging **42**, 2235–2246 (2023). https://doi.org/10.1109/TMI.2023.3245068
19. Zhang, J., Gu, R., Wang, G., Xie, H., Gu, L.: SS-CADA: a semi-supervised cross-anatomy domain adaptation for coronary artery segmentation. In: 2021 IEEE 18th International Symposium on Biomedical Imaging (ISBI), pp. 1227–1231. IEEE (2021)
20. Zhang, J., et al.: Weakly supervised vessel segmentation in x-ray angiograms by self-paced learning from noisy labels with suggestive annotation. Neurocomputing **417**, 114–127 (2020)
21. Zhang, Y., Li, Z., Xie, Y., Qu, Y., Li, C., Mei, T.: Weakly supervised semantic segmentation for large-scale point cloud. In: Proceedings of the AAAI Conference on Artificial Intelligence, vol. 35, pp. 3421–3429 (2021)
22. Zheng, Y., Wang, B., Hong, Q.: UGAN: semi-supervised medical image segmentation using generative adversarial network. In: 2022 15th International Congress on Image and Signal Processing, BioMedical Engineering and Informatics (CISP-BMEI), pp. 1–6. IEEE (2022)

FairAdaBN: Mitigating Unfairness with Adaptive Batch Normalization and Its Application to Dermatological Disease Classification

Zikang Xu[1,2], Shang Zhao[1,2], Quan Quan[3], Qingsong Yao[3], and S. Kevin Zhou[1,2,3(✉)]

[1] School of Biomedical Engineering, Division of Life Sciences and Medicine, University of Science and Technology of China, Hefei 230026, Anhui, People's Republic of China
skevinzhou@ustc.edu.cn

[2] Suzhou Institute for Advanced Research, University of Science and Technology of China, Suzhou 215123, Jiangsu, People's Republic of China

[3] Key Lab of Intelligent Information Processing of Chinese Academy of Sciences (CAS), Institute of Computing Technology, CAS, Beijing 100190, China

Abstract. Deep learning is becoming increasingly ubiquitous in medical research and applications while involving sensitive information and even critical diagnosis decisions. Researchers observe a significant performance disparity among subgroups with different demographic attributes, which is called **model unfairness**, and put lots of effort into carefully designing elegant architectures to address unfairness, which poses heavy training burden, brings poor generalization, and reveals the trade-off between model performance and fairness. To tackle these issues, we propose **FairAdaBN** by making batch normalization adaptive to sensitive attributes. This simple but effective design can be adapted to several classification backbones that are originally unaware of fairness. Additionally, we derive a novel loss function that restrains statistical parity between subgroups on mini-batches, encouraging the model to converge with considerable fairness. In order to evaluate the trade-off between model performance and fairness, we propose a new metric, named Fairness-Accuracy Trade-off Efficiency (FATE), to compute normalized fairness improvement over accuracy drop. Experiments on two dermatological datasets show that our proposed method outperforms other methods on fairness criteria and FATE. Our code is available at https://github.com/XuZikang/FairAdaBN.

Keywords: Dermatological · Fairness · Batch Normalization

Supplementary Information The online version contains supplementary material available at https://doi.org/10.1007/978-3-031-43895-0_29.

H. Greenspan et al. (Eds.): MICCAI 2023, LNCS 14221, pp. 307–317, 2023.
https://doi.org/10.1007/978-3-031-43895-0_29

1 Introduction

The past years have witnessed a rapid growth of applying deep learning methods in medical imaging [31]. As the performance improves continuously, researchers also find that deep learning models attempt to distinguish illness by using features that are related to a sample's demographic attributes, especially sensitive ones, such as skin tone or gender. The biased performance due to sensitive attributes within different subgroups is defined as **unfairness** [16]. For example, Seyyed-Kalantari *et. al.* [21] find that their models trained on chest X-Ray dataset show a significant disparity of True Positive Ratio (TPR) between male and female subgroups. Similar evaluations are done on brain MRI [17], dermatology [12], and mammography [15], which shows that unfairness issues exist extensively in medical applications. If the unfairness of deep learning models is not handled properly, healthcare disparity increases, and human fundamental rights are not guaranteed. Thus, there is a pressing need on investigating unfairness mitigation to eliminate critical biased inference in deep learning models.

There are two groups of methods to tackle unfairness. The first group proceeds implicitly with *fairness through unawareness* [7] by leaving out sensitive attributes when training a single model or deriving invariant representation and ignoring them subjectively when making a decision. However, plenty of evaluations prove that this may lead to unfairness, due to the entangled correlation between sensitive attributes and other variables in the data, and *statistical difference* between features of different subgroups. The second group explicitly takes sensitive attributes into consideration when training models, for example, train independent models for unfairness mitigation [18,25] with no parameters shared between subgroups. However, this may result in degraded performance because the amount of data for model building is reduced (see Table 1).

It is natural to consider whether it is possible to inherit the advantages from both worlds, that is, learning a single model on the whole dataset yet still with explicit modeling of sensitive attributes. Therefore, we propose a framework with a powerful adapter termed **Fair Adaptive Batch Normalization (FairAdaBN)**. Specifically, FairAdaBN is designed to mitigate task disparity between subgroups captured by the neural network. It integrates the common information of different subgroups dynamically by sharing part of network parameters, and enables the differential expression of feature maps for different subgroups, by adding only a few parameters compared with backbones. Thanks to FairAdaBN, the proposed architecture can minimize statistical differences between subgroups and learn subgroup-specific features for unfairness mitigation, which improves model fairness and reserves model precision at the same time. In addition, to intensify the models' ability for balancing performance and fairness, a new loss function named **Statistical Disparity Loss** (L_{SD}), is introduced to optimize the statistical disparity in mini-batches and specify fairness constraints on network optimization. L_{SD} also enhances information transmission between subgroups, which is rare for independent models. Finally, a perfect model should have both higher precision and fairness compared to current well-fitted models. However, most of the existing unfairness mitigation methods sacrifice overall performance for building a fairer model [20,22]. Therefore, following the idea of discovering the fairness-accuracy Pareto frontier [32], we propose a novel metric for evaluating the **Fairness-Accuracy Trade-off Efficiency (FATE)**, urging researchers to pay attention to the performance and fairness simultaneously when building prediction models. We evaluate the proposed method based on its application to mitigating unfairness in dermatology diagnosis.

To sum up, our contributions are as follows:

1. A novel framework is proposed for unfairness mitigation by replacing normalization layers in backbones with FairAdaBN;
2. A loss function is proposed to minimize statistical parity between subgroups for improving fairness;
3. A new metric is derived to evaluate the model's fairness-performance trade-off efficiency. Our proposed FairAdaBN has the highest FATE_{EOpp0} (48.79×10^{-2}), which doubles the highest among other unfairness mitigation methods (Ind, $22.63\ \times10^{-2}$).
4. Experiments on two dermatological disease datasets and three backbones demonstrate the superiority of our proposed FairAdaBN framework in terms of high performance and great portability.

2 Related Work

According to [4], unfairness mitigation can be categorized into pre-processing, in-processing, and post-processing based on the instruction stage.

Pre-processing. Pre-processing methods focus on the quality of the training set, by organizing fair datasets via datasets combination [21], using generative adversarial networks [11] or sketching model [27] to generate extra images, or directly resampling the train set [18,28]. However, most methods in this category need huge effort due to the preciousness of medical data.

Post-processing. Although calibration has been widely used in unfairness mitigation in machine learning tasks, medical applications prefer to use pruning strategies. For example, Wu *et. al* [26] mitigate unfairness by pruning a pre-trained diagnosis model considering the difference of feature importance between subgroups. However, their method needs extra time except for training a precise classification model, while our FairAdaBN is a one-step method.

In-Processing. In-processing methods mainly consist of two folds. Some studies mitigate unfairness by directly adding fairness constraints to the cost functions [28], which often leads to overfitting. Another category of research mitigates unfairness by designing complex network architectures like adversarial network [14,30] or representation learning [5]. This family of methods relies heavily on the accuracy of sensitive attribute classifiers in the adversarial branch, leads to bigger models and cannot make full use of pre-trained weights. While our method does not increase the number of parameters significantly and can be applied to several common backbones for dermatology diagnosis.

3 FairAdaBN

Problem Definition. We assume a medical imaging dataset $D = \{d_1, d_2, ..., d_N\}$ with N samples, the i-th sample d_i consists of input image X_i, sensitive attributes A_i and classification ground truth label Y_i. i.e. $d_i = \{X_i, A_i, Y_i\}$. A is a binary variable (e.g., skin tone, gender), which splits the dataset into the unprivileged group, $D_{A=0}$, which has a lower average performance than the overall performance, and the privileged group, $D_{A=1}$, which has a higher average performance than the overall performance. Using accuracy as the performance metric for example, for a neural network $f_\theta(\cdot)$, our goal is to minimize the accuracy gap between $D_{A=0}$ and $D_{A=1}$ by finding a proper $\hat{\theta}$.

$$\hat{\theta} = \arg\min_{\theta} \left\| \mathbb{E}_{\{X_i,Y_i\}\sim D_{A=1}} \mathbb{I}(f_\theta(X_i) = Y_i) - \mathbb{E}_{\{X_i,Y_i\}\sim D_{A=0}} \mathbb{I}(f_\theta(X_i) = Y_i) \right\| \tag{1}$$

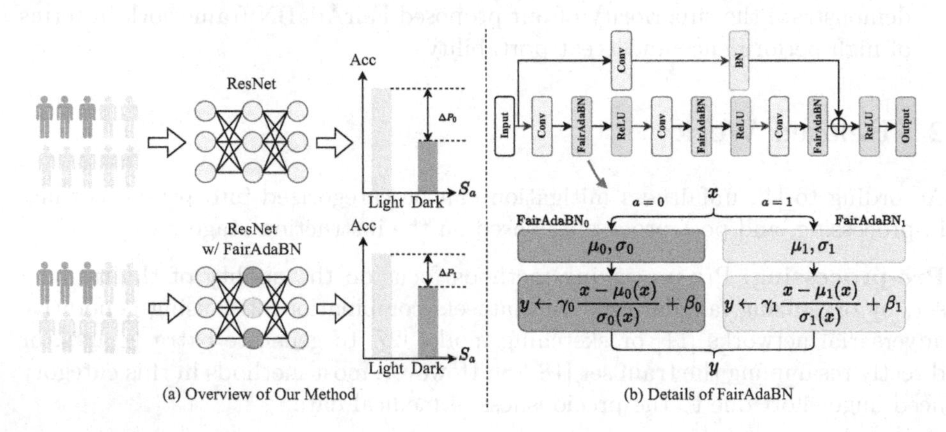

Fig. 1. Overview of Our Method. (a) Compared to ResNet (Top), ResNet w/ AdaBN (Bottom) has a smaller performance disparity (Δp) between *light* samples and *dark* samples. (b) Details of FairAdaBN (use Residual block as an example).

In this paper, we propose FairAdaBN, which replaces normalization layers in vanilla models with adaptive batch normalization layers, while sharing other layers between subgroups. The overview of our method is shown in Fig. 1.

Batch normalization (BN) is a ubiquitous network layer that normalizes mini-batch features using statistics [10]. Let $x \in \mathbb{R}^{C\times W\times H}$ denote a given layer's output feature map, where C, W, H is the number of channels, width, and height of the feature map. The BN function is defined as:

$$\mathrm{BN}(x) = \gamma \cdot \frac{x - \mu(x)}{\sigma(x)} + \beta, \tag{2}$$

where $\mu(x), \sigma(x)$ is the mean and standard deviation of the feature map computed in the mini-batch, β and γ denotes the learnable affine parameters.

We implant the attribute awareness into BN, named FairAdaBN, by parallelizing multiple normalization blocks that are carefully designed for each subgroup. Specifically, for subgroup $D_{A=a}$, its adaptive affine parameter γ_a and β_a are learnt by samples in $D_{A=a}$. Thus, the adaptive BN function for subgroup $D_{A=a}$ is given by Eq. 3.

$$\mathrm{FairAdaBN}_a(x) = \gamma_a \cdot \frac{x - \mu_a(x)}{\sigma_a(x)} + \beta_a, \tag{3}$$

where a is the index of the sensitive attribute corresponding to the current input image, $\mu_\alpha, \sigma_\alpha$ are computed across subgroups independently.

The FairAdaBN acquires subgroup-specific knowledge by learning the affine parameter γ and β. Therefore, the feature maps of subgroups can be aligned and the unfair representation between privileged and unprivileged groups can be mitigated. By applying FairAdaBN on vanilla backbones, the network can learn subgroup-agnostic feature representations by the sharing parameters of convolution layers, and subgroup-specific feature representations using respective BN parameters, resulting in lower fairness criteria. The detailed structure of FairAdaBN is shown in Fig. 1, we display the minimum unit of ResNet for simplification. Note that the normalization layer in the residual branch is not changed for faster convergence.

In this paper, we aim to retain skin lesion classification accuracy and improve model fairness simultaneously. The loss function consists of two parts: (i) the cross-entropy loss, L_{CE}, constraining the prediction precision, and (2) the statistical disparity loss L_{SD} as in Eq. 4, aiming to minimize the difference of prediction probability between subgroups and give extra limits on fairness.

$$L_{SD} = \sum_{y=1}^{N_{cg}} \left\| \mathbb{E}_{X_i \sim D_{A=0}} \mathbb{I}(f_\theta(X_i) = y) - \mathbb{E}_{X_i \sim D_{A=1}} \mathbb{I}(f_\theta(X_i) = y) \right\|^2, \tag{4}$$

where N_{cg} means the number of classification categories.

The overall loss function is given by the sum of the two parts, with a hyper-parameter α to adjust the degree of constraint on fairness. $L = L_{CE} + \alpha \cdot L_{SD}$.

4 Experiments and Results

4.1 Evaluation Metrics

Lots of fairness criteria are proposed including statistical parity [7], equalized odds [9], equal opportunity [9], counterfactual fairness [13], etc. In this paper, we use equal opportunity and equalized odds as fairness criteria. For equal opportunity, we split it into *EOpp*0 and *EOpp*1 considering the ground truth label.

$$\mathrm{EOpp0} = |P(\hat{Y} = 0 \mid Y = 0, A = 1) - P(\hat{Y} = 0 \mid Y = 0, A = 0)| \tag{5}$$

$$\text{EOpp1} = |P(\hat{Y} = 1 \mid Y = 1, A = 1) - P(\hat{Y} = 1 \mid Y = 1, A = 0)| \quad (6)$$

$$\text{EOdd} = |P(\hat{Y} = 1 \mid Y = y, A = 1) - P(\hat{Y} = 1 \mid Y = y, A = 0)|, y \in \{0, 1\} \quad (7)$$

However, these metrics only evaluate the level of fairness while do not consider the trade-off between fairness and accuracy. Therefore, inspired by [6], we propose FATE, a metric that evaluates the balance between normalized improvement of fairness and normalized drop of accuracy. The formulas of FATE on different fairness criteria are shown below:

$$\text{FATE}_{FC} = \frac{\text{ACC}_m - \text{ACC}_b}{\text{ACC}_b} - \lambda \frac{\text{FC}_m - \text{FC}_b}{\text{FC}_b}, \quad (8)$$

where FC can be one of $EOpp0, EOpp1, EOdd$. ACC denotes accuracy. The subscript m and b denote the mitigation model and baseline model, respectively. λ is a weighting factor that adjusts the requirements for fairness pre-defined by the user considering the real application, here we define $\lambda = 1.0$ for simplification. A model obtains a higher FATE if it mitigates unfairness and maintains accuracy. Note that FATE should be combined with utility metrics and fairness metrics, rather than independently.

4.2 Dataset and Network Configuration

We use two well-known dermatology datasets to evaluate the proposed method. The Fitzpatrick-17k dataset [8] contains 16,577 dermatology images in 9 diagnostic categories. The skin tone is labeled with Fitzpatrick's skin phenotype. In this paper, we regard Skin Type I to III as *light*, and Skin Type IV to VI as *dark* for simplicity, resulting in a ratio of dark : light $\approx 3 : 7$. The ISIC 2019 dataset [1,2,24] contains 25,331 images among 9 different diagnostic categories. We use gender as the sensitive attribute, where female : male $\approx 4.5 : 5.5$. Based on subgroup analysis, *dark* and *female* are treated as the privileged group, and *light* and *male* are treated as the unprivileged group.

We randomly split the dataset into train, validation, and test with a ratio of 6:2:2. The models are trained for 600 epochs and the model with the highest validation accuracy is selected for testing. The images are resized or cropped to 128×128 for both datasets. Random flipping and random rotation are used for data augmentation. The experiments are carried out on $8 \times$ NVIDIA 3090 GPUs, implemented on PyTorch, and are repeated 3 times. Pre-trained weights from ImageNet are used for all models. The networks are trained using AdamW optimizer with weight decay. The batch size and learning rate are set as 128 and 1e-4, respectively. The hyper-parameter $\alpha = 1.0$.

4.3 Results

We compare FairAdaBN with Vanilla (ResNet-152), Resampling [18], Ind (independently trained models for each subgroup) [18], GroupDRO [19], EnD [23], and CFair [29], which are commonly used for unfairness mitigation.

Results on Fitzpatrick-17k Dataset. Table 1 shows the result of these seven methods on Fitzpatrick-17k dataset. Compared to the Vanilla model, Resampling has a comparable utility, but cannot improve fairness. FairAdaBN achieves the lowest unfairness with only a small drop in accuracy. Besides, FairAdaBN has the highest FATE on all fairness criteria. This is because Ind does not share common information between subgroups, and only part of the dataset is used for training. GroupDRO and EnD rely on the discrimination of features from different subgroups, which is indistinguishable for this task. CFair is more efficient on balanced datasets, while the ratio between *light* and *dark* is skewed.

Results on ISIC 2019 Dataset. Table 1 shows the results on ISIC 2019 dataset. FairAdaBN is the fairest method among the seven methods. Resampling improves fairness sightly but does not outperform ours. GroupDRO mitigates EOpp0 while increasing unfairness on Eopp1 and Eodd. Ind and CFair cannot mitigate unfairness in ISIC 2019 dataset and EnD increases unfairness on EOpp0.

Table 1. Result on Fitzpatrick-17k and ISIC 2019 Dataset ($\text{Mean}^{\text{Std}} \times 10^{-2}$). **Best** and *Second-best* are highlighted.

Fitzpartrick-17k Dataset										
Method	Accuracy↑	Precision↑	Recall↑	F1↑	EOpp0↓	EOpp1↓	Eodd↓	E_0 ↑	E_1 ↑	E_2 ↑
Vanilla	$87.53^{0.14}$	$\mathbf{79.60^{0.33}}$	$\mathbf{80.22^{0.19}}$	$\mathbf{78.41^{0.15}}$	$1.00^{0.30}$	$10.40^{1.43}$	$10.54^{0.98}$	/	/	/
Resampling [18]†	$\underline{87.73}^{0.27}$	$\underline{79.21}^{0.40}$	$\underline{80.01}^{0.35}$	$\underline{78.27}^{0.42}$	$1.11^{0.26}$	$10.43^{1.91}$	$10.78^{2.06}$	−10.86	−0.03	−2.05
Ind [18]†	$86.33^{0.12}$	$76.11^{0.38}$	$77.48^{0.18}$	$75.20^{0.09}$	$\underline{0.78}^{0.33}$	$10.13^{0.51}$	$9.72^{0.94}$	$\underline{20.63}$	1.23	6.41
GroupDRO [19]†	$86.62^{0.19}$	$77.21^{0.62}$	$78.29^{0.52}$	$76.56^{0.56}$	$0.94^{0.34}$	$\underline{8.04}^{0.90}$	$\underline{8.23}^{1.25}$	5.07	$\underline{21.66}$	$\underline{20.91}$
EnD [23]†	$86.80^{0.52}$	$77.32^{0.60}$	$78.58^{0.53}$	$76.90^{0.66}$	$1.22^{0.31}$	$9.01^{1.60}$	$9.20^{1.59}$	−22.83	12.53	11.88
CFair [29]†	$\mathbf{87.91^{0.35}}$	$78.62^{0.49}$	$79.73^{0.37}$	$78.12^{0.38}$	$0.93^{0.28}$	$9.83^{1.65}$	$10.17^{1.57}$	10.03	12.15	10.09
FairAdaBN	$84.72^{0.40}$	$74.43^{0.22}$	$75.74^{0.33}$	$73.31^{0.48}$	$\mathbf{0.48^{0.09}}$	$\mathbf{7.67^{3.86}}$	$\mathbf{7.73^{3.95}}$	**48.79**	**23.04**	**23.45**
ISIC 2019 Dataset										
Method	Accuracy↑	Precision↑	Recall↑	F1↑	EOpp0↓	EOpp1↓	Eodd↓	E_0 ↑	E_1 ↑	E_2 ↑
Vanilla	$\underline{92.52}^{0.12}$	$\underline{82.64}^{0.31}$	$\underline{82.94}^{0.36}$	$\underline{82.60}^{0.32}$	$0.85^{0.12}$	$6.12^{1.83}$	$6.02^{1.66}$	/	/	/
Resampling [18]†	$\mathbf{92.81^{0.28}}$	$\mathbf{83.15^{0.50}}$	$\mathbf{83.42^{0.51}}$	$\mathbf{83.12^{0.52}}$	$0.86^{0.15}$	$5.65^{2.83}$	$5.76^{2.78}$	−0.80	−2.48	−5.49
Ind [18]†	$92.43^{0.11}$	$82.16^{0.15}$	$82.46^{0.12}$	$82.11^{0.08}$	$0.85^{0.11}$	$7.04^{0.96}$	$7.37^{0.77}$	−0.10	−15.13	−22.52
GroupDRO [19]†	$91.86^{0.22}$	$81.30^{0.52}$	$81.44^{0.47}$	$81.17^{0.50}$	$\underline{0.82}^{0.12}$	$6.78^{3.20}$	$6.62^{3.21}$	$\underline{2.41}$	−22.99	−22.01
EnD [23]†	$92.13^{0.08}$	$81.42^{0.48}$	$81.64^{0.35}$	$81.36^{0.38}$	$0.98^{0.09}$	$\underline{5.18}^{0.99}$	$\underline{5.10}^{1.06}$	−15.72	$\underline{14.94}$	$\underline{14.86}$
CFair [29]†	$87.39^{0.77}$	$72.39^{2.67}$	$72.60^{2.22}$	$71.28^{2.12}$	$2.83^{1.09}$	$9.21^{3.53}$	$10.80^{4.15}$	−238.49	−56.03	−84.95
FairAdaBN	$89.11^{0.09}$	$74.24^{0.13}$	$74.79^{0.18}$	$74.18^{0.14}$	$\mathbf{0.69^{0.07}}$	$\mathbf{4.85^{2.50}}$	$\mathbf{4.76^{2.73}}$	**15.14**	**17.07**	**17.24**

* E_0, E_1, E_2 denotes $\text{FATE}_{EOpp0}, \text{FATE}_{EOpp1}, \text{FATE}_{EOdd}$, respectively.
† Private implementation.

The FATE Metric. Figure 2 shows the values of FATE. According to [3], the closer the curve is to the top left corner, the smaller the fairness-accuracy

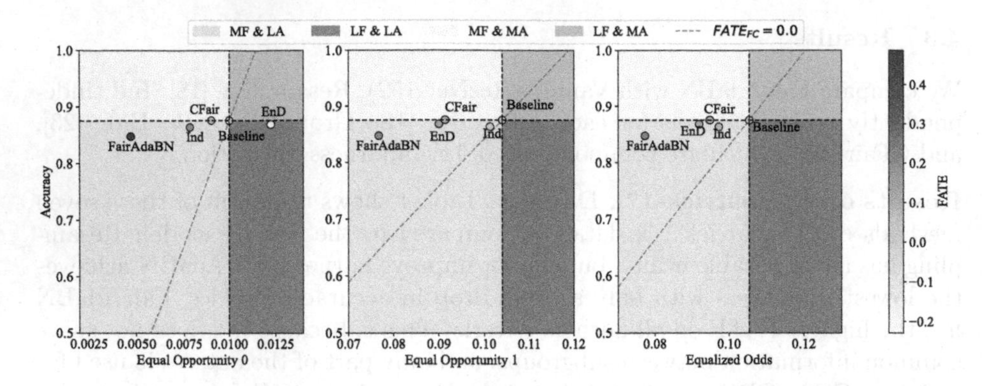

Fig. 2. FATE on different fairness criteria. The data point of baseline (fairness, accuracy) splits the space into four parts: MF: Fairer; MA: More Accurate; LF: Less Fair; LA: Less Accurate. Points on the left of the line have positive FATE, while points on the right of the line have negative FATE.

Table 2. Ablation Study ($\text{Mean}^{\text{Std}} \times 10^{-2}$). **Best** in each group are highlighted.

Method	Accuracy↑	Precision↑	Recall↑	F1↑	EOpp0↓	EOpp1↓	Eodd↓	E_0 ↑	E_1 ↑	E_2 ↑
VGG	$\mathbf{88.11^{0.51}}$	$\mathbf{79.18^{0.56}}$	$\mathbf{80.07^{0.49}}$	$\mathbf{78.55^{0.56}}$	$1.42^{0.25}$	$10.64^{2.15}$	$11.78^{2.34}$	/	/	/
VGG + FairAdaBN	$83.55^{0.24}$	$69.73^{0.83}$	$72.09^{0.41}$	$70.15^{0.69}$	$\mathbf{1.09^{0.04}}$	$\mathbf{10.58^{1.80}}$	$\mathbf{10.48^{1.97}}$	18.06	−4.61	5.86
DenseNet	$\mathbf{87.32^{0.06}}$	$\mathbf{78.12^{0.52}}$	$\mathbf{79.08^{0.38}}$	$\mathbf{77.37^{0.24}}$	$\mathbf{1.18^{0.37}}$	$10.96^{1.34}$	$11.47^{1.16}$	/	/	/
DenseNet + FairAdaBN	$80.40^{0.23}$	$65.32^{0.57}$	$69.42^{0.40}$	$65.25^{0.60}$	$1.43^{0.79}$	$\mathbf{7.70^{1.06}}$	$\mathbf{8.30^{1.58}}$	−29.11	21.82	19.71
ResNet	$\mathbf{87.53^{0.14}}$	$\mathbf{79.60^{0.33}}$	$\mathbf{80.22^{0.19}}$	$\mathbf{78.41^{0.15}}$	$1.00^{0.30}$	$10.40^{1.43}$	$10.54^{0.98}$	/	/	/
Ours w/o L_{SD}	$87.18^{0.50}$	$78.50^{0.75}$	$79.24^{0.68}$	$77.40^{0.71}$	$1.07^{0.16}$	$9.33^{0.23}$	$9.91^{0.29}$	−7.87	9.88	5.55
Ours w/o FairAdaBN	$85.02^{0.03}$	$73.76^{0.11}$	$75.67^{0.05}$	$73.63^{0.16}$	$1.39^{0.45}$	$15.30^{1.91}$	$15.05^{1.37}$	42.15	−49.94	−45.62
Ours ($\alpha = 0.1$)	$84.82^{0.79}$	$73.44^{1.11}$	$75.15^{0.98}$	$73.17^{0.95}$	$1.26^{0.18}$	$13.39^{2.98}$	$12.76^{3.28}$	−29.10	−31.85	−24.16
Ours ($\alpha = 1.0$)	$84.72^{0.40}$	$74.43^{0.22}$	$75.74^{0.33}$	$73.31^{0.48}$	$\mathbf{0.48^{0.09}}$	$\mathbf{7.67^{3.86}}$	$\mathbf{7.73^{3.95}}$	**48.79**	**23.04**	**23.45**
Ours ($\alpha = 2.0$)	$84.57^{0.38}$	$74.26^{0.22}$	$75.40^{0.11}$	$72.91^{0.87}$	$1.10^{0.60}$	$8.53^{2.79}$	$8.40^{2.75}$	−13.38	14.60	16.92

trade-off it has. The figure demonstrates that FATE has the same trend as this argument. We prefer an algorithm that obtains a higher FATE since a higher FATE denotes higher unfairness mitigation and a low drop in utility, and a negative FATE denotes that the mitigation model cannot decrease unfairness while reserving enough accuracy (not beneficial).

Limitation. Compared with other methods, FairAdaBN needs to use sensitive attributes in the test stage, which is unnecessary for EnD and CFair. Although this might be easy to acquire in real applications, improvements could be done to solve this problem.

4.4 Ablation Study

Different Backbones. Firstly, we test FairAdaBN's compatibility on different backbones, by applying FairAdaBN on VGG-19-BN and DenseNet-121. Note that the first and last BN in DenseNet are not changed. The result is shown in Table 2. The experiments are carried out on Fitzpatrick-17k dataset. The result shows that our FairAdaBN is also effective on these two backbones, except $Eopp_0$ when using DenseNet-121, showing well model compatibility. However, we also observe a larger drop in model precision compared with the baseline, which needs to be taken into consideration in future work.

Different Loss Terms. We train ResNet by only replacing BNs with FairAdaBNs (the second row of the last part), and ResNet adding L_{SD} on the total loss (the third row of the last part). The effectiveness of AdaBN is illustrated by comparing the first and second rows of the last part in Table 2. By replacing BNs with FairAdaBN, ResNet can normalize subgroup feature maps using specific affine parameters, which reduce $Eopp_1$ and $Eodd$ by 1.07×10^{-2} and 0.63×10^{-2}, respectively. Comparing the second and fourth row of the last part in Table 2, we find that by adding L_{SD}, $Eopp_0$ decreases significantly, from 1.07×10^{-2} to 0.48×10^{-2}. Besides, although adding L_{SD} on ResNet alone increases fairness criteria unexpectedly, fairness criteria decrease when using FairAdaBN and L_{SD} simultaneously. The reason could be the potential connection between FairAdaBN and L_{SD}, due to the similar form dealing with subgroups.

Hyper-parameter α. Our experiments show that $\alpha = 1.0$ has the best fairness scores and FATE compared to $\alpha = 0.1$ and $\alpha = 2.0$. Therefore we select $\alpha = 1.0$ as our final setting.

5 Conclusion

We propose FairAdaBN, a simple but effective framework for unfairness mitigation in dermatological disease classification. Extensive experiments illustrate that the proposed framework can mitigate unfairness compared to models without fair constraints, and has a higher fairness-accuracy trade-off efficiency compared with other unfairness mitigation methods. By plugging FairAdaBN into several backbones, its generalization ability is proved. However, the current study only evaluates the effectiveness of FairAdaBN on dermatology datasets, and its generalization ability on other datasets (chest X-Ray, brain MRI) or tasks (segmentation, detection), where unfairness issues also exist, needs to be evaluated in the future. We also plan to explore the unfairness mitigation effectiveness for other universal models [31].

Acknowledgement. Supported by Natural Science Foundation of China under Grant 62271465 and Open Fund Project of Guangdong Academy of Medical Sciences, China (No. YKY-KF202206).

References

1. Codella, N.C., et al.: Skin lesion analysis toward melanoma detection: a challenge at the 2017 international symposium on biomedical imaging (ISBI), hosted by the international skin imaging collaboration (ISIC). In: 2018 IEEE 15th International Symposium on Biomedical Imaging (ISBI 2018), pp. 168–172. IEEE (2018)

2. Combalia, M., et al.: BCN20000: Dermoscopic lesions in the wild. arXiv preprint: arXiv:1908.02288 (2019)
3. Creager, E., et al.: Flexibly fair representation learning by disentanglement. In: International Conference on Machine Learning, pp. 1436–1445. PMLR (2019)
4. Deho, O.B., Zhan, C., Li, J., Liu, J., Liu, L., Le Duy, T.: How do the existing fairness metrics and unfairness mitigation algorithms contribute to ethical learning analytics? Br. J. Educ. Technol. **53**, 822–843 (2022)
5. Deng, W., Zhong, Y., Dou, Q., Li, X.: On fairness of medical image classification with multiple sensitive attributes via learning orthogonal representations. arXiv preprint: arXiv:2301.01481 (2023)
6. Dhar, P., Gleason, J., Roy, A., Castillo, C.D., Chellappa, R.: PASS: protected attribute suppression system for mitigating bias in face recognition. In: Proceedings of the IEEE/CVF International Conference on Computer Vision, pp. 15087–15096 (2021)
7. Dwork, C., Hardt, M., Pitassi, T., Reingold, O., Zemel, R.: Fairness through awareness. In: Proceedings of the 3rd Innovations in Theoretical Computer Science Conference, pp. 214–226 (2012)
8. Groh, M., Harris, C., Daneshjou, R., Badri, O., Koochek, A.: Towards transparency in dermatology image datasets with skin tone annotations by experts, crowds, and an algorithm. arXiv preprint: arXiv:2207.02942 (2022)
9. Hardt, M., Price, E., Srebro, N.: Equality of opportunity in supervised learning. In: Advances in Neural Information Processing Systems, vol. 29 (2016)
10. Ioffe, S., Szegedy, C.: Batch normalization: accelerating deep network training by reducing internal covariate shift. In: International Conference on Machine Learning, pp. 448–456. PMLR (2015)
11. Joshi, N., Burlina, P.: AI fairness via domain adaptation. arXiv preprint: arXiv:2104.01109 (2021)
12. Kinyanjui, N.M., et al.: Fairness of classifiers across skin tones in dermatology. In: Martel, A.L., et al. (eds.) MICCAI 2020. LNCS, vol. 12266, pp. 320–329. Springer, Cham (2020). https://doi.org/10.1007/978-3-030-59725-2_31
13. Kusner, M.J., Loftus, J., Russell, C., Silva, R.: Counterfactual fairness. In: Advances in Neural Information Processing Systems, vol. 30 (2017)
14. Li, X., Cui, Z., Wu, Y., Gu, L., Harada, T.: Estimating and improving fairness with adversarial learning. arXiv preprint: arXiv:2103.04243 (2021)
15. Lu, C., Lemay, A., Hoebel, K., Kalpathy-Cramer, J.: Evaluating subgroup disparity using epistemic uncertainty in mammography. arXiv preprint: arXiv:2107.02716 (2021)
16. Narayanan, A.: Translation tutorial: 21 fairness definitions and their politics. In: Proc. Conf. Fairness Accountability Transp., New York, USA, vol. 1170, p. 3 (2018)
17. Petersen, E., Feragen, A., Zemsch, L.D.C., Henriksen, A., Christensen, O.E.W., Ganz, M.: Feature robustness and sex differences in medical imaging: a case study in MRI-based Alzheimer's disease detection. arXiv preprint: arXiv:2204.01737 (2022)
18. Puyol-Antón, E., et al.: Fairness in cardiac MR image analysis: an investigation of bias due to data imbalance in deep learning based segmentation. In: de Bruijne, M., et al. (eds.) MICCAI 2021. LNCS, vol. 12903, pp. 413–423. Springer, Cham (2021). https://doi.org/10.1007/978-3-030-87199-4_39
19. Sagawa, S., Koh, P.W., Hashimoto, T.B., Liang, P.: Distributionally robust neural networks for group shifts: on the importance of regularization for worst-case generalization. arXiv preprint: arXiv:1911.08731 (2019)

20. Sarhan, M.H., Navab, N., Eslami, A., Albarqouni, S.: On the fairness of privacy-preserving representations in medical applications. In: Albarqouni, S., et al. (eds.) DART/DCL -2020. LNCS, vol. 12444, pp. 140–149. Springer, Cham (2020). https://doi.org/10.1007/978-3-030-60548-3_14
21. Seyyed-Kalantari, L., Liu, G., McDermott, M., Chen, I.Y., Ghassemi, M.: CheXclusion: fairness gaps in deep chest X-ray classifiers. In: BIOCOMPUTING 2021: Proceedings of the Pacific Symposium, pp. 232–243. World Scientific (2020)
22. Suriyakumar, V.M., Papernot, N., Goldenberg, A., Ghassemi, M.: Chasing your long tails: differentially private prediction in health care settings. In: Proceedings of the 2021 ACM Conference on Fairness, Accountability, and Transparency, pp. 723–734 (2021)
23. Tartaglione, E., Barbano, C.A., Grangetto, M.: EnD: entangling and disentangling deep representations for bias correction. In: Proceedings of the IEEE/CVF Conference on Computer Vision and Pattern Recognition, pp. 13508–13517 (2021)
24. Tschandl, P., Rosendahl, C., Kittler, H.: The HAM10000 dataset, a large collection of multi-source dermatoscopic images of common pigmented skin lesions. Sci. Data **5**(1), 1–9 (2018)
25. Wang, M., Deng, W.: Mitigating bias in face recognition using skewness-aware reinforcement learning. In: Proceedings of the IEEE/CVF Conference on Computer Vision and Pattern Recognition, pp. 9322–9331 (2020)
26. Wu, Y., Zeng, D., Xu, X., Shi, Y., Hu, J.: FairPrune: achieving fairness through pruning for dermatological disease diagnosis. arXiv preprint: arXiv:2203.02110 (2022)
27. Yao, R., Cui, Z., Li, X., Gu, L.: Improving fairness in image classification via sketching. arXiv preprint: arXiv:2211.00168 (2022)
28. Zhang, H., Dullerud, N., Roth, K., Oakden-Rayner, L., Pfohl, S., Ghassemi, M.: Improving the fairness of chest x-ray classifiers. In: Conference on Health, Inference, and Learning, pp. 204–233. PMLR (2022)
29. Zhao, H., Coston, A., Adel, T., Gordon, G.J.: Conditional learning of fair representations. arXiv preprint: arXiv:1910.07162 (2019)
30. Zhao, Q., Adeli, E., Pohl, K.M.: Training confounder-free deep learning models for medical applications. Nat. Commun. **11**(1), 1–9 (2020)
31. Zhou, S.K., et al.: A review of deep learning in medical imaging: imaging traits, technology trends, case studies with progress highlights, and future promises. Proc. IEEE **109**, 820–838 (2021)
32. Zietlow, D., et al.: Leveling down in computer vision: pareto inefficiencies in fair deep classifiers. In: Proceedings of the IEEE/CVF Conference on Computer Vision and Pattern Recognition, pp. 10410–10421 (2022)

FedSoup: Improving Generalization and Personalization in Federated Learning via Selective Model Interpolation

Minghui Chen[1], Meirui Jiang[2], Qi Dou[2], Zehua Wang[1], and Xiaoxiao Li[1](✉)

[1] Department of Electrical and Computer Engineering, The University of British Columbia, Vancouver, Canada
xiaoxiao.li@ece.ubc.ca

[2] Department of Computer Science and Engineering, The Chinese University of Hong Kong, Hong Kong, China

Abstract. Cross-silo federated learning (FL) enables the development of machine learning models on datasets distributed across data centers such as hospitals and clinical research laboratories. However, recent research has found that current FL algorithms face a trade-off between local and global performance when confronted with distribution shifts. Specifically, personalized FL methods have a tendency to overfit to local data, leading to a sharp valley in the local model and inhibiting its ability to generalize to out-of-distribution data. In this paper, we propose a novel federated model soup method (*i.e.*, selective interpolation of model parameters) to optimize the trade-off between local and global performance. Specifically, during the federated training phase, each client maintains its own global model pool by monitoring the performance of the interpolated model between the local and global models. This allows us to alleviate overfitting and seek flat minima, which can significantly improve the model's generalization performance. We evaluate our method on retinal and pathological image classification tasks, and our proposed method achieves significant improvements for out-of-distribution generalization. Our code is available at https://github.com/ubc-tea/FedSoup.

1 Introduction

Federated learning (FL) has emerged as a promising methodology for harnessing the power of private medical data without necessitating centralized data governance [6,7,22,25]. However, recent study [28] has identified a significant issue in current FL algorithms, namely, the trade-off between local and global performance when encountering distribution shifts. This issue is particularly

This work is supported in part by the Natural Sciences and Engineering Research Council of Canada (NSERC), Public Safety Canada, Compute Canada and National Natural Science Foundation of China (Project No. 62201485).

Supplementary Information The online version contains supplementary material available at https://doi.org/10.1007/978-3-031-43895-0_30.

H. Greenspan et al. (Eds.): MICCAI 2023, LNCS 14221, pp. 318–328, 2023.
https://doi.org/10.1007/978-3-031-43895-0_30

prevalent in medical scenarios [12,16,17], where medical images may undergo shifts of varying degrees due to differences in imaging device vendors, parameter settings, and the patient demographics. Personalized FL (PFL) techniques are typically utilized to address the data heterogeneity problem by weighing more on in-distribution (ID) data of each client. For instance, FedRep [5] learns the entire network during local updates and keeps part of the network from global synchronization. However, they have a risk of overfitting to local data [23], especially when client local data is limited, and have poor generalizability on out-of-distribution (OOD) data. Another line of work has studied the heterogeneity issue by regularizing the updates of local model. For instance, FedProx [15] constraints local updates to be closer to the global model. An effective way to evaluate FL's generalizability is to investigate its performance on the joint global distribution following [28], which refers to testing the FL models on $\bigcup\{\mathcal{D}_i\}$, where $\mathcal{D}_i$ indicates client i's distribution[1]. Unfortunately, the existing works have not found the sweet spot between personalized (local) and consensus (global) models.

In this regard, we study a practical problem of enhancing personalization and generalization jointly in cross-silo FL for medical image classification when faced data heterogeneity. To this end, we aim to address the following two questions in FL: *What could be the causes that result in local and global trade-off?* and *How to achieve better local and global trade-off?* First, we provide a new angle to understand the trade-off. We reveal that over-personalization in FL can cause overfitting on local data and trap the model into a sharp valley of loss landscape (highly sensitive to parameter perturbation, see detailed definition in Sec. 2.2), thus limiting its generalizability. An effective strategy for avoiding sharp valleys in the loss landscape is to enforce models to obtain flat minima. In the centralized domain, weight interpolation has been explored as a means of seeking flat minima as its solution is moved closer to the centroid of the high-performing models, which corresponds to a flatter minimum [3,6,11,24]. However, research on these interpolation methods has been overlooked in FL.

With the above basis, we propose to track both local and global models during the federated training and perform model interpolation to seek the optimal balance. Our insight is drawn from the model soup method [27], which shows that averaging weights of multiple trained models with same initial parameters can enhance model generalization. However, the original model soup method requires training substantial models with varying hyper-parameters, which can be prohibitively time-consuming and costly in terms of communication during FL. Given the high communication cost and the inability to restart training from scratch in FL, we leverage global models at different time points within a single training session as the ingredients to adapt the model soup method [27] to FL.

In this paper, we propose a novel federated model soup method (FedSoup) to produce an ensembled model from local and global models that achieve better local-global trade-off. We refer the 'soup' as a combo of different federated models. Our proposed FedSoup includes two key modules. The first one is temporal model selection, which aims to select suitable models to be combined into

---
[1] In the heterogeneous setting ($\mathcal{D}_i \neq \mathcal{D}_j$), $\mathcal{D}_j$ is viewed as the OOD data for client i.

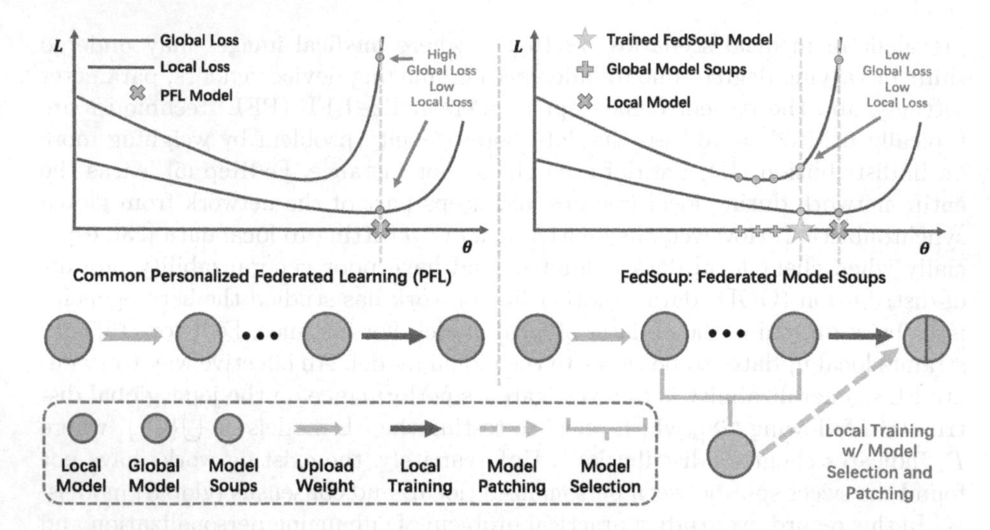

Fig. 1. The overview of our method (FedSoup) compared with common PFL methods. PFL methods typically minimize local loss but suffering high global loss. While our federated model soup method balances both local and global loss by seeking flat minima. The black dots in the figure represent ellipsis and indicate multiple rounds of model upload and model training in between. Compared to previous pFL methods, our approach introduces global model selection modules and local model interpolation with the global model (referred to as model patching).

one. The second module is Federated model patching [10], which refers to a fine-tuning technique that aims to enhance personalization without compromising the already satisfactory global performance. For the first module, temporal model selection, we utilize a greedy model selection strategy based on the local validation performance. This avoids incorporating models that could be located in a different error landscape basin than the local loss landscape (shown in Fig. 1). Consequently, each client possesses their personalized global model soups, consisting of a subset of historical global models that are selected based on their local validation sets. As for the second module, federated model patching, it introduces model patching in local client training by interpolating the local model and the global model soups into a new local model, bridging the gap between local and global domains. It promotes the personalization of the model for ID testing and also maintains good global performance for OOD generalization.

In summary, our key contributions are as follows: (i) A novel FL method called Federated Model Soups (FedSoup) is proposed to improve generalization and personalization by promoting smoothness and seeking flat minima. (ii) A new temporal model selection mechanism is designed for FL, which maintains a client-specific model soups with temporal history global model to meet personalization requirements while not incurring additional training costs. (iii) An innovative federated model patching method between local and global models is introduced in federated client training to alleviate overfitting of local limited data.

2 Method

2.1 Problem Setup

Consider a cross-silo FL setting with N clients. Let $\mathcal{D} := \{\mathcal{D}_i\}_{i=1}^N$ be a set of N training domain, each of which is a distribution over the input space $\mathcal{X}$. For each client, we have access to n training data points in the form of $(x_j^i, y_j^i)_{j=1}^n \sim \mathcal{D}_i$, where y_j^i denotes the target label for input x_j^i. We also define a set of unseen target domains $\mathcal{T} := \{\mathcal{T}_i\}_i^{N'}$ in a similar manner, where N' is the number of target domains and is typically set to one. The goal of personalization (local performance) is to find a model $f(\cdot;\theta)$ via minimizing an empirical risk $\widehat{\mathcal{E}}_{\mathcal{D}i}(\theta) := \frac{1}{n}\sum_{j=1}^n \ell(f(x^i;\theta), y^i))$ over a local client training set $\mathcal{D}_i$, where $\ell(\cdot,\cdot)$ is a loss function. On the other hands, the objective of generalization (global performance) is to minimize both population loss $\mathcal{E}_{\mathcal{D}}(\theta)$ and $\mathcal{E}_{\mathcal{T}}(\theta)$ over multiple domains by Empirical Risk Minimization (ERM) $\widehat{\mathcal{E}}_{\mathcal{D}}(\theta) := \frac{1}{Nn}\sum_{i=1}^N \sum_{j=1}^n \ell(f(x^i;\theta), y^i))$ over all training clients' training sets $\mathcal{D}$. In this work, we evaluate the local performance on local testing samples from $\mathcal{D}_i$, and evaluate the global performance on testing samples from the joint global distribution $\mathcal{D} := \{\mathcal{D}_i\}_{i=1}^N$.

2.2 Generalization and Flat Minima

In practice, ERM in deep neural networks, *i.e.*, $\arg\min_\theta \widehat{\mathcal{E}}_{\mathcal{D}}(\theta)$, can yield multiple solutions that offer comparable training loss, but vastly different levels of generalizability [3]. However, without proper regularization, models are prone to overfit the training data and the training model will fall into a sharp valley of the loss surface, which is less generalizable [4].

One common reason for failures in ERM is the presence of variations in the data distribution ($\mathcal{D}_i \neq \mathcal{D}$), which can cause a shift in the loss landscape. As illustrated in Fig. 1, the sharper the optimized minima, the more sensitive it is to shifts in the loss landscape. This results in an increase in generalization error. In cross-silo FL, each client may overfit their local training data, leading to poor global performance. This is due to the distribution shift problem, which creates conflicting objectives among the local models [23]. Therefore, when the local model converges to a sharp minima, the higher the degree of personalization (local performance) of the model, the more likely it is to have poor generalization ability (global performance).

From the domain generalization formalization in [2,3], the test loss $\mathcal{E}_{\mathcal{T}}(\theta)$ can be bounded by the robust empirical loss $\widehat{\mathcal{E}}_{\mathcal{D}}^{\epsilon}(\theta)$ as follows:

$$\mathcal{E}_{\mathcal{T}}(\theta) < \widehat{\mathcal{E}}_{\mathcal{D}}^{\epsilon}(\theta) + \frac{1}{N}\sum_{i=1}^N \sup_{A \in \mathcal{A}} |\mathcal{P}_{\mathcal{D}_i}(A) - \mathcal{P}_{\mathcal{T}}(A)| + \xi, \tag{1}$$

where the $\sup_A |\mathcal{P}_{\mathcal{D}_i}(A) - \mathcal{P}_{\mathcal{T}}(A)|$ is a divergence between domain $\mathcal{D}_i$ and $\mathcal{T}$, $\mathcal{A}$ is the set of measurable subsets under $\mathcal{D}_i$ and $\mathcal{T}$, and ξ is the confidence bound term related to the radius ϵ and the number of the training samples.

From the Equation (1), we can infer that minimizing sharpness and seeking flat minima is directly related with the generalization performance on the unseen target domain.

2.3 Our Solution: FedSoup

After analyzing the aforementioned relationship between sharpness and generalization, we expound on the distinctive challenges of seeking flat minima and mitigating the trade-off between local and global performance in FL. Consequently, we introduce two refined modules as the ingredient of our proposed FedSoup solution. FedSoup only needs to modify the training method of the client, and the algorithm implementation is shown in Algorithm 1.

Temporal Model Selection. Stochastic Weight Averaging (SWA) [11] and Sharpness-Aware Minimization (SAM) [8] are two commonly used flat minima optimizers, which seek to find parameters in wide low-loss basins. In contrast to SAM, which incurs extra computational cost to identify the worst parameter perturbation, SWA is a more succinct and effective approach for implicitly favoring the flat minima by averaging weights. The SWA algorithm is motivated by the observation that SGD often finds high-performing models in the weight space but rarely reaches the central points of the optimal set. By averaging the parameter values over iterations, the SWA algorithm moves the solution closer to the centroid of this space of points.

Nevertheless, when it comes to cross-silo FL training, the discrepancy between clients is significant, and models might lie in different basins. Merging all these models haphazardly is not effective and might hinder generalization. Recently, a selective weight averaging method called model soups [27] was introduced to enhance the generalization of fine-tuned models. The original model soups is not applicable to the FL setting, requiring high communication costs and training compute. We adapt the idea to a new approach by leveraging global models trained at different time points in one pass of FL training. Additionally, considering the heterogeneity of data distribution in cross-silo FL and the requirement of personalization, we propose a model selection strategy where each client utilizes the performance of its local validation set as a monitoring indicator. We called this module temporal model selection (see Algorithm 1 Line 7-8)

Federated Model Patching. According to previous analysis on the loss landscape, there is a loss landscape offset between different FL clients due to their domain discrepancy. Thus, simply integrating a global model can damage the model's personalization. To address this, we introduce the use of the model patching [10] (*i.e.* local and global model interpolation) during client-side local training in FL, aiming to improve model personalization and maintain the good global performance. Specifically, model patching approach forces local client not to distort global model severely and seek low-loss interpolated model between local and global, encouraging the local and global model lie in the same basin without a large barrier of linear connectivity. [19]. We called this module federated model patching. In summary, the update rule of FedSoup is implemented as follows:

$$\theta_{FedSoup} \leftarrow \frac{\theta_g^1 + \cdots + \theta_g^k + \theta_l}{k+1}, \tag{2}$$

where θ_g is global model, θ_l is local model, k is the number of selected global models. This update rule corresponds to Algorithm 1 Line 9.

Algorithm 1. FedSoup

1: **Input:** global model θ_g, local model θ_l, last epoch local model $\widehat{\theta_l^i}$, number of clients k, number of epochs n, interpolation start epoch E.
2: $\mathsf{soup} \leftarrow \{\}$
3: **for** $i = 0$ **to** n **do**
4: $\quad \theta_g \leftarrow \mathsf{Aggregation}(\widehat{\theta_l^1}, \ldots, \widehat{\theta_l^k})$
5: $\quad \theta_l \leftarrow \mathsf{ClientUpdate}(\theta_g)$
6: $\quad$ **if** $i \geq E$ **then**
7: $\quad\quad$ **if** $\mathsf{ValAcc}(\mathsf{average}(\mathsf{soup} \cup \{\theta_l\} \cup \{\theta_g\})) \geq \mathsf{ValAcc}(\mathsf{average}(\mathsf{soup} \cup \{\theta_l\}))$ **then**
8: $\quad\quad\quad \mathsf{soup} \leftarrow \{\theta_g\}$ {Module 1: Temporal Model Selection}
9: $\quad\quad \theta_l \leftarrow \mathsf{average}(\mathsf{soup} \cup \{\theta_l\})$ {Module 2: Federated Model Patching}
10: **return** θ_l

It is important to note that our proposed FedSoup algorithm requires only one carefully tuned hyper-parameter, namely the interpolation start epoch. To mitigate the risk of having empty global model soups when the start epoch is too late and to prevent potential performance degradation when the start epoch is too early, we have set the default interpolation start epoch to be 75% of the total training epochs, aligning with the default setting of SWA. Furthermore, it is worth mentioning that the modified model soup and model patching modules in our proposed FedSoup framework are interdependent. Model patching, which is a technique based on our modified model soup algorithm, provides an abundance of models to explore flatter minima and enhance performance.

3 Experiments

3.1 Experimental Setup

Dataset. We validate the effectiveness of our proposed method, FedSoup, on two medical image classification tasks. The first task involved the classification of pathology images from five different sources using Camelyon17 dataset [1], and each source is viewed as a client. The pathology experiment consists of a total of 4,600 images[2], each with a resolution of 96×96. The second task

[2] We take a random subset from the original Camelyon17 dataset to match the small data settings in FL [18].

involved retinal fundus images from four different institutions [9,21,26], and each institute is viewed as a client. The retinal fundus experiment consists of a total of 1,264 images, each with a resolution of 128×128. The objective of both datasets is to identify abnormal images from normal ones. We also maintained an equal number of samples from each client to prevent clients with more data from having a disproportionate influence on the global performance evaluation.

Table 1. Local and global performance results comparison with SOTA PFL methods on two medical image classification tasks.

Method	Pathology				Retinal Fundus			
	Local Performance		Global Performance		Local Performance		Global Performance	
	Accuracy ↑	AUC ↑	Accuracy ↑	AUC ↑	Accuracy ↑	AUC ↑	Accuracy ↑	AUC ↑
FedAvg [18]	$82.41_{(0.61)}$	$90.44_{(0.42)}$	$70.18_{(1.25)}$	$78.74_{(1.31)}$	$89.81_{(1.17)}$	$95.24_{(0.73)}$	$73.27_{(3.97)}$	$81.02_{(4.57)}$
FedProx [15]	$\mathbf{86.34}_{(0.52)}$	$\mathbf{92.78}_{(0.36)}$	$67.42_{(1.32)}$	$77.18_{(1.32)}$	$90.14_{(0.20)}$	$95.47_{(0.63)}$	$70.46_{(2.36)}$	$76.84_{(2.93)}$
MOON [14]	$85.71_{(0.42)}$	$91.98_{(0.34)}$	$70.61_{(1.54)}$	$79.27_{(1.43)}$	$89.14_{(0.77)}$	$94.91_{(0.98)}$	$77.49_{(3.15)}$	$84.89_{(2.70)}$
FedBN [17]	$82.32_{(0.87)}$	$90.07_{(0.77)}$	$65.16_{(0.69)}$	$71.61_{(0.97)}$	$89.70_{(1.45)}$	$95.49_{(0.56)}$	$75.11_{(0.52)}$	$82.70_{(1.01)}$
FedFomo [30]	$80.99_{(1.07)}$	$86.51_{(0.99)}$	$61.00_{(0.59)}$	$61.69_{(0.78)}$	$89.70_{(0.00)}$	$94.88_{(0.45)}$	$59.90_{(2.00)}$	$63.82_{(1.14)}$
FedRep [5]	$82.50_{(0.53)}$	$89.77_{(0.43)}$	$66.87_{(0.60)}$	$72.34_{(0.75)}$	$89.38_{(0.89)}$	$95.16_{(0.83)}$	$71.81_{(1.79)}$	$79.09_{(2.44)}$
FedBABU [20]	$85.18_{(0.33)}$	$92.39_{(0.29)}$	$69.56_{(1.40)}$	$77.26_{(1.48)}$	$90.25_{(1.39)}$	$95.10_{(0.33)}$	$77.65_{(0.24)}$	$85.09_{(0.21)}$
FedSoup	$85.71_{(0.37)}$	$92.47_{(0.31)}$	$\mathbf{72.87}_{(1.35)}$	$\mathbf{81.45}_{(1.40)}$	$\mathbf{90.92}_{(0.50)}$	$\mathbf{96.00}_{(0.43)}$	$\mathbf{78.64}_{(0.90)}$	$\mathbf{86.24}_{(0.86)}$

Evaluation Setup. For each client, we take 75% of the data as the training set. To assess the generalization ability and personalization of our model, we have constructed both local and global testing sets. Following [28], in our experimental setting, we first create a held-out global testing set by randomly sampling an equal number of images per source/institute, so its distribution is different from either of the client. The local testing dataset for each FL client is the remaining sample from the same source as its training set. The number of local testing set per client is approximately the same as that of the held-out global testing set. For the pathology dataset, as each subject can have multiple samples, we have ensured that data from the same subject only appeared in either the training or testing set. To align with the cross-validation setting for subsequent out-of-domain evaluations, we conducted a five-fold leave-one-client-data cross-validation, with three repetitions using different random seeds in each fold. The results of the repeated experiments without hold-out client data are provided in the appendix. For PFL methods, we report the performance by averaging the results of each personalized models.

Models and Training Hyper-Paramters. We employ the ResNet-18 architecture as the backbone model. Our approach initiates local-global interpolation at the 75% training phase, consistent with the default hyper-parameter setting of SWA. We utilize the Adam optimizer with learning rate of 1e−3, momentum coefficients of 0.9 and 0.99 and set the batch size to 16. We set the local training epoch to 1 and perform a total of 1,000 communication rounds.

3.2 Comparison with State-of-the-Art Methods

We compare our method with seven common FL and state-of-the-art PFL methods. Results in Table 1 demonstrate that our FedSoup method achieves competitive performance on local evaluation sets while significantly improving generalization with respect to global performance. Furthermore, FedSoup exhibits greater stability with lower variance of performance across multiple experiments.

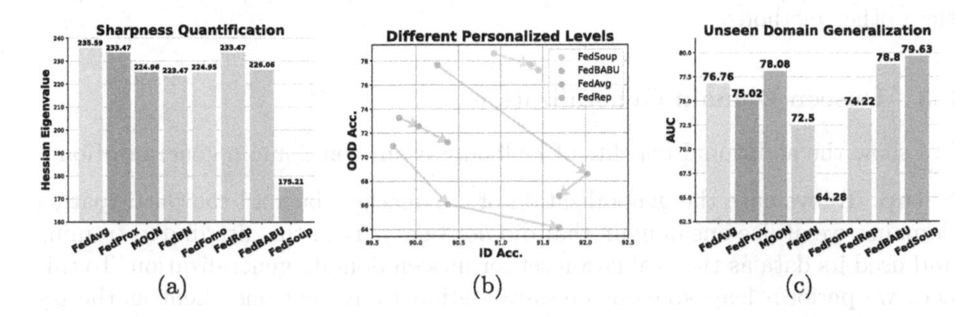

Fig. 2. Analysis of our approach: (a) sharpness quantification on the retina dataset, (b) local and global trade-off under different personalized levels (fine-tuning epochs), (c) unseen domain generalization on the pathology dataset.

Comparing the performance improvement of FedSoup across different datasets, we observed that FedSoup had a more substantial effect on the smaller retinal fundus dataset compared to the larger pathology dataset. In terms of performance gap compared to the second-best methods (FedBABU on Retina and FedProx on Pathology), our approach demonstrates a larger advantage on the Retina dataset. This observation suggests that our proposed method can mitigate the negative impact of local overfitting caused by small local datasets, thus improving the model's generalization ability.

Sharpness Quantification. The sharpness measure used in this study calculates the median of the dominating Hessian eigenvalue across all training set batches using the Power Iteration algorithm [29]. This metric indicates the maximum curvature of the loss landscape, which is often used in the literature on flat minima [13] to reflect the sharpness. The median of the dominating Hessian eigenvalue of all clients in the retinal fundus dataset was measured in this part. Based on the Fig. 2(a) presented, it is evident that the proposed method leads to flatter minima as compared to the other methods.

Trade-off at Different Personalized Levels. Following the evaluation in [28], we conducted an experiment comparing the local and globalperformance of different models at various levels of personalization using the retinal fundus datasets. We control the personalization level for the PFL methods by varying the number of iterations that the model undergoes fine-tuning using only the local training set after federated training. As we increase the number of fine-tuning iterations, we consider the level of personalization to be higher. We choose to show model performance after fine-tuning iterations 1, 7, and 15. The results in Fig. 2(b) indicated that existing FL methods often have a trade-off between local and global performance. As the number of fine-tuning iterations increases, local performance typically improves but global performance decreases. Compared to other methods, our appraoch maintains high local performance while also preventing a significant drop in global performance, which remains much higher than other methods.

3.3 Unseen Domain Generalization

We show the additional benefits of FedSoup on unseen domain generalization.

Setup. To evaluate the generalization of our method beyond the participating domains, we utilize one domain that did not take part in the distributed training and used its data as the evaluation set for unseen domain generalization. To this end, we perform leave-one-out cross-validation by having one client as the to-be-evaluated unseen set each time. To ensure a reliable results of unseen domain generalization, we conducted experiments on the Camelyon17 dataset, which has a larger number of samples.

Results. Overall, our proposed method demonstrates an advantage in terms of unseen domain generalization capabilities (see Fig. 2(c)). In comparison to FedAvg, Our approach resulted in a 2.87-point increase in the AUC index for generalization to unseen domains on the pathology dataset.

4 Conclusion

In this paper, we demonstrate the trade-off between personalization and generalization in the current FL methods for medical image classification. To optimize this trade-off and achieve flat minima, we propose the novel FedSoup method. By maintaining personalized global model pools in each client and interpolating weights between local and global models, our proposed method enhances both generalization and personalization. FedSoup outperforms other PFL methods in terms of both generalization and personalization, without incurring any additional inference or memory costs.

References

1. Bándi, P., et al.: From detection of individual metastases to classification of lymph node status at the patient level: the CAMELYON17 challenge. IEEE Trans. Med. Imaging **38**(2), 550–560 (2019)
2. Ben-David, S., Blitzer, J., Crammer, K., Kulesza, A., Pereira, F., Vaughan, J.W.: A theory of learning from different domains. Mach. Learn. **79**(1–2), 151–175 (2010)

3. Cha, J., et al.: SWAD: domain generalization by seeking flat minima. In: NeurIPS, pp. 22405–22418 (2021)
4. Chaudhari, P., et al.: Entropy-SGD: biasing gradient descent into wide valleys. In: ICLR (Poster). OpenReview.net (2017)
5. Collins, L., Hassani, H., Mokhtari, A., Shakkottai, S.: Exploiting shared representations for personalized federated learning. In: ICML. Proceedings of Machine Learning Research, vol. 139, pp. 2089–2099. PMLR (2021)
6. Dayan, I., et al.: Federated learning for predicting clinical outcomes in patients with covid-19. Nat. Med. **27**(10), 1735–1743 (2021)
7. Dou, Q., So, T.Y., Jiang, M., Liu, Q., Vardhanabhuti, V., Kaissis, G., et al.: Federated deep learning for detecting covid-19 lung abnormalities in CT: a privacy-preserving multinational validation study. NPJ Digit. Med. **4**(1), 1–11 (2021)
8. Foret, P., Kleiner, A., Mobahi, H., Neyshabur, B.: Sharpness-aware minimization for efficiently improving generalization. In: ICLR. OpenReview.net (2021)
9. Fumero, F., Alayón, S., Sánchez, J.L., Sigut, J.F., González-Hernández, M.: RIM-ONE: an open retinal image database for optic nerve evaluation. In: CBMS, pp. 1–6. IEEE Computer Society (2011)
10. Ilharco, G., et al.: Patching open-vocabulary models by interpolating weights. CoRR abs/2208.05592 (2022)
11. Izmailov, P., Podoprikhin, D., Garipov, T., Vetrov, D.P., Wilson, A.G.: Averaging weights leads to wider optima and better generalization. In: UAI, pp. 876–885. AUAI Press (2018)
12. Jiang, M., Yang, H., Li, X., Liu, Q., Heng, PA., Dou, Q.: Dynamic bank learning for semi-supervised federated image diagnosis with class imbalance. In: Wang, L., Dou, Q., Fletcher, P.T., Speidel, S., Li, S. (eds.) Medical Image Computing and Computer Assisted Intervention - MICCAI 2022. MICCAI 2022. LNCS, vol. 13433, pp. 196–206. Springer, Cham (2022). https://doi.org/10.1007/978-3-031-16437-8_19
13. Kaddour, J., Liu, L., Silva, R., Kusner, M.J.: Questions for flat-minima optimization of modern neural networks. CoRR abs/2202.00661 (2022)
14. Li, Q., He, B., Song, D.: Model-contrastive federated learning. In: CVPR, pp. 10713–10722. Computer Vision Foundation/IEEE (2021)
15. Li, T., Sahu, A.K., Zaheer, M., Sanjabi, M., Talwalkar, A., Smith, V.: Federated optimization in heterogeneous networks. In: MLSys. mlsys.org (2020)
16. Li, X., Gu, Y., Dvornek, N., Staib, L.H., Ventola, P., Duncan, J.S.: Multi-site FMRI analysis using privacy-preserving federated learning and domain adaptation: abide results. Med. Image Anal. **65**, 101765 (2020)
17. Li, X., Jiang, M., Zhang, X., Kamp, M., Dou, Q.: FedBN: federated learning on non-IID features via local batch normalization. In: ICLR. OpenReview.net (2021)
18. McMahan, B., Moore, E., Ramage, D., Hampson, S., y Arcas, B.A.: Communication-efficient learning of deep networks from decentralized data. In: AISTATS. Proceedings of Machine Learning Research, vol. 54, pp. 1273–1282. PMLR (2017)
19. Mirzadeh, S., Farajtabar, M., Görür, D., Pascanu, R., Ghasemzadeh, H.: Linear mode connectivity in multitask and continual learning. In: ICLR. OpenReview.net (2021)
20. Oh, J., Kim, S., Yun, S.: Fedbabu: Towards enhanced representation for federated image classification. CoRR abs/2106.06042 (2021)
21. Orlando, J.I., et al.: REFUGE challenge: a unified framework for evaluating automated methods for glaucoma assessment from fundus photographs. Medical Image Anal. **59**, 101570 (2020)

22. Pati, S., et al.: Federated learning enables big data for rare cancer boundary detection. Nat. Commun. **13**(1), 7346 (2022)
23. Qu, Z., Li, X., Duan, R., Liu, Y., Tang, B., Lu, Z.: Generalized federated learning via sharpness aware minimization. In: ICML. Proceedings of Machine Learning Research, vol. 162, pp. 18250–18280. PMLR (2022)
24. Ramé, A., Ahuja, K., Zhang, J., Cord, M., Bottou, L., Lopez-Paz, D.: Recycling diverse models for out-of-distribution generalization. CoRR abs/2212.10445 (2022)
25. Rieke, N., et al.: The future of digital health with federated learning. NPJ Digit. Med. **3**(1), 1–7 (2020)
26. Sivaswamy, J., Krishnadas, S., Chakravarty, A., Joshi, G., Tabish, A.S., et al.: A comprehensive retinal image dataset for the assessment of glaucoma from the optic nerve head analysis. JSM Biomed. Imaging Data Papers **2**(1), 1004 (2015)
27. Wortsman, M., et al.: Model soups: averaging weights of multiple fine-tuned models improves accuracy without increasing inference time. In: ICML. Proceedings of Machine Learning Research, vol. 162, pp. 23965–23998. PMLR (2022)
28. Wu, S., et al.: Motley: benchmarking heterogeneity and personalization in federated learning. CoRR abs/2206.09262 (2022)
29. Yao, Z., Gholami, A., Keutzer, K., Mahoney, M.W.: PyHessian: neural networks through the lens of the hessian. In: IEEE BigData, pp. 581–590. IEEE (2020)
30. Zhang, M., Sapra, K., Fidler, S., Yeung, S., Alvarez, J.M.: Personalized federated learning with first order model optimization. In: ICLR. OpenReview.net (2021)

TransLiver: A Hybrid Transformer Model for Multi-phase Liver Lesion Classification

Xierui Wang[1], Hanning Ying[2], Xiaoyin Xu[3], Xiujun Cai[2], and Min Zhang[1](✉)

[1] College of Computer Science and Technology, Zhejiang University, Hangzhou, China
min_zhang@zju.edu.cn
[2] Sir Run Run Shaw Hospital (SRRSH), affiliated with the Zhejiang University School of Medicine, Hangzhou, China
[3] Brigham and Women's Hospital, Harvard Medical School, Boston, USA

Abstract. Early diagnosis of focal liver lesions (FLLs) can decrease the fatality rate of liver cancer, which remains a big challenge. We designed a deep learning approach based on CT to assess and differentiate FLLs. To achieve high accuracy, CTs in different phases are integrated to provide more information than single-phase images. While most of the related studies use convolutional neural networks, we exploit the Transformer for multi-phase liver lesion classification. We propose a hybrid model called TransLiver, which has a transformer backbone and complementary convolutional modules. Specifically, we connect modified transformer blocks with convolutional encoder and down-samplers. For multi-phase fusion, we utilize cross phase tokens to reinforce the phases communication. In addition, we introduce a pre-processing unit to resolve realistic annotation issues. Extensive experiments are conducted, in which we achieve an overall accuracy of 90.9% on an in-house dataset of four CT phases and seven liver lesion classes. The results also show distinct advantages in comparison to state-of-art approaches in classification. The code is available at https://github.com/sherrydoge/TransLiver.

Keywords: Focal liver lesion · Multi-phase fusion · Transformer

1 Introduction

Liver cancer is one of the most deadly cancers and has the second highest fatality rate [17]. Focal liver lesions (FLLs) are the most common lesions found in liver cancer, yet FLLs are challenging to diagnose because they can be either benign lesions, such as focal nodular hyperplasia (FNH), hepatic abscess (HA), hepatic

X. Wang and H. Ying—Equal contribution.

Supplementary Information The online version contains supplementary material available at https://doi.org/10.1007/978-3-031-43895-0_31.

H. Greenspan et al. (Eds.): MICCAI 2023, LNCS 14221, pp. 329–338, 2023.
https://doi.org/10.1007/978-3-031-43895-0_31

hemangioma (HH), and hepatic cyst (HC) or malignant tumors, such as intrahepatic cholangiocarcinoma (ICC), hepatic metastases (HM), and hepatocellular carcinoma (HCC). Accurate early diagnosis of FLLs is thus critical to increasing the 5-year survival rate, a task that remains challenging as of today. Dynamic contrast-enhanced CT is a common technique for liver cancer diagnosis, where four different phases of imaging, namely, non-contrast (NC), arterial (ART), portal venous (PV), and delayed (DL) provide complementary information about the liver. Different types of FLLs acquired in the four phases are shown in Fig. 1.

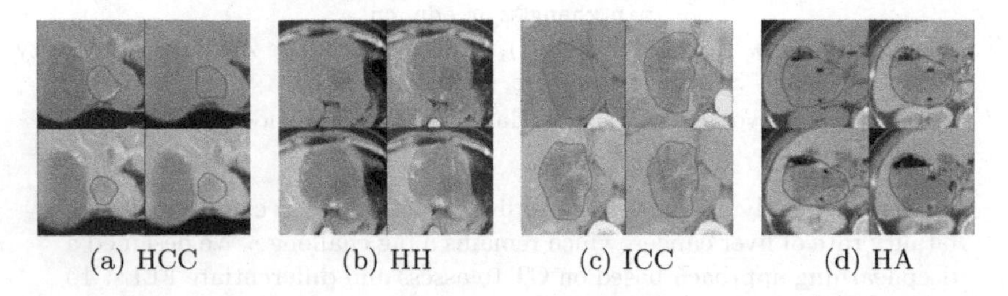

Fig. 1. Representative types of FLLs shown in different CT phases, where contours show the annotated lesion boundaries. In each image, the phase sequence from left to right and top to bottom is NC, ART, PV, and DL, respectively.

With the development of deep learning, computer-aided liver lesion diagnosis has attracted much attention [5,8,16] in recent years. Romero et al. [16] presented an end-to-end framework based on Inception-V3 and InceptionResNet-V2 to discriminate liver lesions between cysts and malignant tumors. Heker et al. [8] combined liver segmentation and classification using transfer learning and joint learning to increase the performance of CNN. As a manner to elevate the accuracy of CNNs, Frid-Adar et al. [5] designed a GAN-based network to generate synthetic liver lesion images, improving the classification performance based on CNN. It is reported in many studies [9,18] that using multi-phase data, like most professionals do in practice, can help the network get a more accurate result, which also acts in liver lesion classification [15,23,24]. Yasaka et al. [24] proposed multi-channel CNN to extract features from multi-phase liver CT by concatenation. Roboh et al. [15] proposed an algorithm based on CNNs to handle 3D context in liver CTs and utilized clinical context to assist the classification. Xu et al. [23] constructed a knowledge-guided framework to integrate liver lesion features from three phases using self-attention and fused them with a cross-feature interaction module and a cross-lesion correlation module.

A single-phase lesion annotation means the annotation of both lesion position and its class. In hospitals, collected multi-phase CTs are normally grouped by patients rather than lesions, which makes single-phase lesion annotation insufficient for feature fusion learning. However, the number of lesions inside a single patient can vary from one to dozens and they can be of different types in realistic cases. Multi-phase CTs are also not co-registered in most cases, therefore, it is necessary to make sure the lesions extracted from different phases are somehow aligned for feature fusion, which is called as multi-phase lesion annotation.

Moreover, while most works have attached much importance to liver lesion segmentation [2], its outcome is usually organized at a single-phase level. Additional effort will be needed when consolidating segmentation and multi-phase classification.

Self-attention based transformers [19] have shown strong capability in natural language processing tasks. Meanwhile, vision transformers (ViT) [4] have been shown to replace CNN with a transformer encoder in computer vision tasks and can achieve obvious advantages on large-scale datasets. To the best of our knowledge, we find no study using ViT backbone network in liver lesion classification. The reason for this is twofold. First, pure ViT has several limitations itself [6], including ignoring local information within each patch, extracting only single-scale features, and lacking inductive bias. Second, no complete open liver lesion classification datasets exist. Most relevant studies are based on private datasets, which tend to be small in size and cause overfitting in learning models.

In this paper, we construct a hybrid framework with ViT backbone for liver lesion classification, **TransLiver**. We design a pre-processing unit to reduce the annotation cost, where we obtain lesion area on multi-phase CTs from annotations marked on a single phase. To alleviate the limitations of pure transformers, we propose a multi-stage pyramid structure and add convolutional layers to the original transformer encoder. We use additional cross phase tokens at the last stage to complete a multi-phase fusion, which can focus on cross-phase communication and improve the fusion effectiveness as compared with conventional modes. While most multi-phase liver lesion classification studies use datasets with no more than three phases (without DL phase for its difficulty of collection) or no more than six lesion classes, we validate the whole framework on an in-house dataset with four phases of abdominal CT and seven classes of liver lesions. Considering the disproportion of axial lesion slice number and the relatively small scale of the dataset, we adopt a 2-D network in classification part instead of 3-D in pre-processing part and achieve a 90.9% accuracy.

2 Method

Figure 2 illustrates the overall architecture of TransLiver, where activation layers and batch normalization layers are omitted. Multi-phase liver lesion CTs are converted from single-phase annotation to multi-phase annotation by a pre-processing unit including a registration network and a lesion matcher.

For each phase, a convolutional encoder extracts preliminary lesion features on axial slices. As the backbone of the whole framework, transformer encoder employs a 4-stage pyramid structure extracting multi-scale features, with each stage connected by a convolutional down-sampler. There are two types of transformer blocks, single-phase liver transformer block (SPLTB) and multi-phase liver transformer block (MPLTB). The former is phase-specific, while the latter is in charge of multi-phase fusion. Extracted features from different phases are averaged and classified by two successive fully connected networks. A voting strategy about slices is applied to decide the classification results.

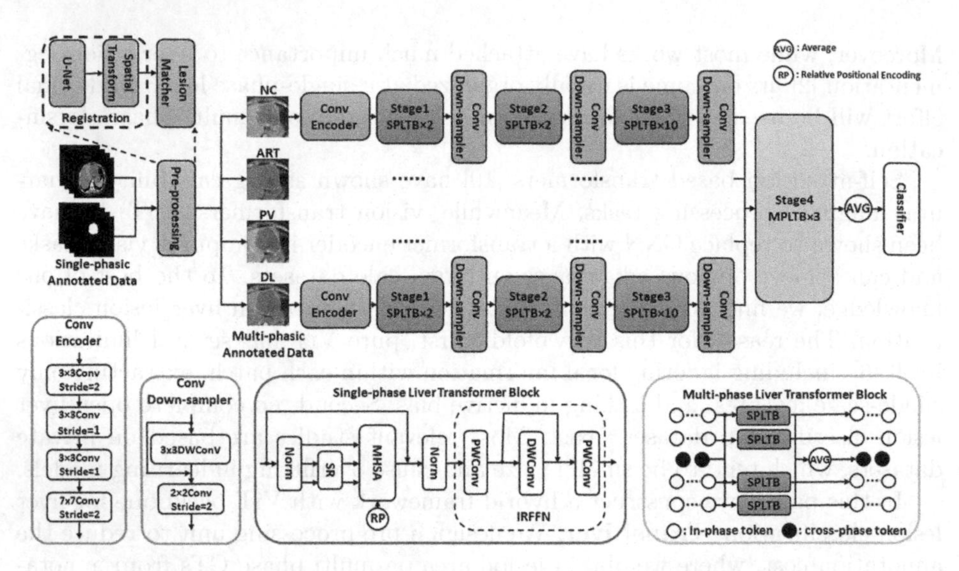

Fig. 2. The overall architecture of the proposed TransLiver model.

2.1 Pre-processing Unit

The single-phase annotated lesion has the position and class labels in all phases but they are not aligned, so we could have difficulty finding out which lesions in different phases are the same with 2 or more lesions in one patient. To reduce errors caused by unregistered data and address the situation that one patient has multiple lesions of different types, we pre-process the multi-phase liver CTs registered and grouped by lesions.

The registration network is based on Voxelmorph [1], with a U-Net learning registration field and moving data transformed by the field. We also use auxiliary Dice loss function between fixed image lesion masks and moved image lesion masks to help the registration field learning. In [1], the network needs to specify an atlas image, otherwise, pairs of images will be registered to each other. But in our work, we need to register the original data in a cross-phase form. We choose an atlas phase ART as suggested by clinicians and other phases of CTs are registered to the ART phase of every patient.

After registration, a lesion matcher finds the same lesions in different phases. We generate a minimum circumscribed cuboid with padding as the lesion window for each lesion to keep the surrounding information. The windows are then converted to 0–1 masks to calculate Dice coefficient. Lesions with the maximal window Dice coefficient that is no less than a set threshold are considered the same. Only lesions completely found in all phases will be used in the following classification network.

2.2 Convolutional Encoder and Convolutional Down-Sampler

In pure vision transformer, input images are converted to tokens by patch embedding and added with positional encoding to keep the positional information. Patch embedding consists of a linear connected layer or a convolutional layer, which does not enable to construct local relation [13]. Absolute positional encoding destroys the translation variance [10] that keeps the rotation and shift operations from altering the final results [6].

So, we construct a convolutional encoder without absolute positional encoding to replace the original embedding layer. For an input image $X \in \mathbb{R}^{B \times H \times W \times 1}$, B is the batch size, and $H \times W$ is the size of the input. The module contains four convolutional layers playing different roles. The first layer, Conv_1, with a kernel size of 3, stride of 2, and output channels of 32, reduces the size to $\frac{H}{2} \times \frac{W}{2}$. Next two layers, Conv_2 and Conv_3, each with a kernel size of 3, stride of 1, and the same output channel as Conv_1, extract local information. Conv_1, Conv_2, and Conv_3 each is followed by a GeLU activation layer and a batch normalization. Considering the design of PVTv2 [21], an overlapped convolutional layer, Conv_4, with a kernel size of 7, stride of 2, and output channels of 64, is used to strengthen the connection among patches. It is followed by layer normalization. The output Z is then reshaped from $\mathbb{R}^{B \times \frac{H}{4} \times \frac{W}{4} \times 64}$ to $\mathbb{R}^{B \times \frac{H \times W}{16} \times 64}$ to finish the tokenization of transformer.

We add convolutional down-samplers between stages of transformer encoder so that they can produce hierarchical representation like CNN structure. Each convolutional down-sampler contains a residual structure with a 3×3 depthwise convolution to increase the locality of our model. We also utilize a convolutional layer with a kernel size of 2 and stride of 2, which halves the image resolution and doubles the number of channels.

2.3 Single-Phase Liver Transformer Block

Vision Transformers can get excellent performance on large-scale datasets such as ImageNet [4], but they are also prone to overfit on small datasets such as private hospital datasets. We adopt the spatial reduction structure proposed in PVT [20] to largely reduce the computational overhead by reducing the size of K and V using depthwise convolution. Following [6,12], a learnable relative positional encoding is added here to replace the absolute positional encoding. The self-attention module can be written as:

$$\text{Attn}(Q, K, V) = \text{Softmax}\left(\frac{Q \times \text{SR}(K)}{\sqrt{d_h}} + P\right) \text{SR}(V) \quad (1)$$

where Q, K, V are the same with original ViT, d_h is the head dimension, and P is the relative positional encoding. Spatial reduction SR consists of a $k \times k$ depthwise convolution with a stride of k and a batch normalization, where k is the spatial reduction ratio set in each stage.

Feed forward network (FFN) is designed for a better capacity of representation. We use the module designed in [6] IRFFN (Inverted Residual FFN) with three convolutions instead of two linear layers in the initial vision transformer. The first and third convolutions are pointwise for dimension translation, which has a similar effect to the original linear layers. The second convolution with a shortcut connection extracts local information in a higher dimension and improves the gradient propagation ability across layers [6]. The structure also has two GeLU activation layers between convolutional layers and three batch normalizations after the GeLUs and the last convolutional layer for better performance.

2.4 Multi-phase Liver Transformer Block

Single-phase liver transformer block (SPLTB) is phase-specific, which means the model parameters of each phase are independent. It can fully extract features in different phases before fusion. Inspired by [14], in stage 4, we design a multi-phase liver transformer block (MPLTB) for communication between phases. MPLTB introduced some new parameters that are not in the original transformer. These parameters are randomly initialized, concatenated with the corresponding phase tokens respectively, and updated in phase-specific SPLTB. Then, they are separated and averaged for the next layer. The whole module is defined as:

$$\begin{aligned} \text{Concat}\left(X_i^{l+1}, t_i^l\right) &= \text{SPLTB}\left(\text{Concat}\left(X_i^l, t^l\right)\right) \\ t^{l+1} &= \text{Avg}\left(t_i^l\right) \end{aligned} \tag{2}$$

where X_i^l is phase tokens of the ith phase and the lth layer and t^l is cross phase tokens of the lth layer. Because of the phase-specific SPLTB, t_i^l represents the corresponding cross phase tokens output of the ith phase in the lth layer. Cross phase tokens need negligible extra cost and can force the information to concentrate inside the tokens [14]. Compared to the direct fusion of input images or output features like average and concatenation, cross phase tokens can also reduce fusion granularity to sufficiently explore the relationship among phases. It is worth noting that these tokens will be removed provisionally when reshaping the phase tokens in SPLTB for right execution. The fusion is conducted in deep layers because the semantic concepts are learned in higher layers which benefits the cross phase connection.

3 Experiments

3.1 Liver Lesion Classification

Dataset. The employed single-phase annotated dataset is collected from Sir Run Run Shaw Hospital (SRRSH), affiliated with the Zhejiang University School of Medicine, and has received the ethics approval of IRB. The collection process can be found in supplementary materials. The size of each CT slice is decreased to 224×224 using cubic interpolation. After the pre-processing unit with window Dice threshold of 0.3, we screen 761 lesions from 444 patients with four phases of CTs, seven types of lesions (13.2% of HCC, 5.3% of HM, 11.3% of ICC, 22.6% of HH, 31.1% of HC, 8.7% of FNH, and 7.8% of HA), and totally 4820 slices. To handle the imbalance of dataset, we randomly select 586 lesions as the training and validation set with no more than 700 axial slices in each lesion type, and the rest 175 lesions constitute the test set. Lesions from the same patient are either assigned to the training and validation set or the test set, but not both.

Implementations. The training and validation set is randomly divided with a 4:1 ratio. The data is augmented by flip, rotation, crop, shift, and scale. We initialize the backbone network using pre-trained weights of CMT-S [6]. Our models are implemented by Pytorch1.12.1 and Timm0.6.13 [22]. Then, they are trained on four NVIDIA Tesla A100 GPUs for 200 epochs using cross-entropy loss function with label smoothing and SGD optimizer with learning rate warmup and cosine annealing. The batch size is 32 and the learning rate is 1e-3. We measured performance by precision (Pre.), sensitivity (Sen.), specificity (Spe.), F1-score (F1), area under the curve (AUC), and accuracy (Acc.).

Results. We first compare the class-wise accuracy of our model against other advanced methods applying different architectures in multi-phase liver lesion classification with more than four lesion types [3,11,15,23,24]. TransLiver gets the highest overall accuracy of 90.9% classifying the most lesion types of seven (HCC 90.9%, HM 62.5%, ICC 73.7%, HH 91.7%, HC 100.0%, FNH 100.0%, and HA 93.3%). In the results of our method, HM has a relatively low performance of 62.5%, mainly due to its low proportion in our dataset. The details can be found in supplementary materials.

Because the sources of data are different among the methods compared above and to the best of our knowledge, no relevant study based on transformers was found, we further train some SOTA normal classification models on our dataset. Considering the fairness, all the models below are initialized with pre-trained weights and adopt 2-D structures using the same slice-level classification strategy. For completeness, we concatenate the multi-phase features to execute the fusion. As illustrated in Table 1, our proposed TransLiver model gets better performance than other models in all metrics. Behind our model, CMT-S achieves the best performance, indicating the effect of convolutional structures in transformer.

Table 1. Performance of TransLiver and other SOTA classification methods.

Method	Pre.	Sen.	Spe.	F1	AUC	Acc.
ResNet-18 [7]	71.7	72.6	96.1	71.2	92.6	77.1
ViT-S [4]	79.6	79.4	97.2	78.6	92.9	82.9
Swin-S [12]	77.7	78.1	97.1	77.3	93.8	82.3
CMT-S [6]	80.5	80.5	97.6	80.0	94.1	85.7
TransLiver	**88.7**	**87.4**	**98.5**	**87.3**	**95.1**	**90.9**

3.2 Ablation Study

To verify the improvement of our modules, we conduct three baseline experiments for comparison. Here convolutional encoder, convolutional down-sampler, and SPLTB as a whole is called c-SPLTB. Baseline 0 does not use c-SPLTB or cross phase tokens in MPLTB but replaces them with pure vision transformer and output feature concatenation respectively. Baseline 1 adds the c-SPLTB and Baseline 2 adds the cross phase tokens. A 3-D version of Baseline 2 utilizing 3-D patch embedding is also studied in Baseline 3 to validate the advantage of our 2-D model. The result shown in Fig. 3 demonstrates that our design choice is appropriate. It is worth mentioning that the 2-D structure is prone to redundancy between axial slices and ignores the relation between slices compared with the 3-D structure but gets observably higher accuracy. We suppose the reason is twofold. Most of lesions in our dataset having few slices weakens the redundancy between slices in 2-D pipeline, while the number of slices is still obviously larger than the number of lesions, alleviating the overfitting issue. Furthermore, vision transformers are mostly pretrained in 2-D images, causing poor performance when transferring to 3-D pipeline.

We also evaluate the model performance under different phase combinations by cutting the branch of certain phases. It shows that information from various phases can significantly influence the classification performance. A missing phase can cause an accuracy drop of about 10% and complete four-phase model outperforms single-phase model by nearly 20%. Figure 4 contains average results of phase number and details with all phase combinations can be found in supplementary materials.

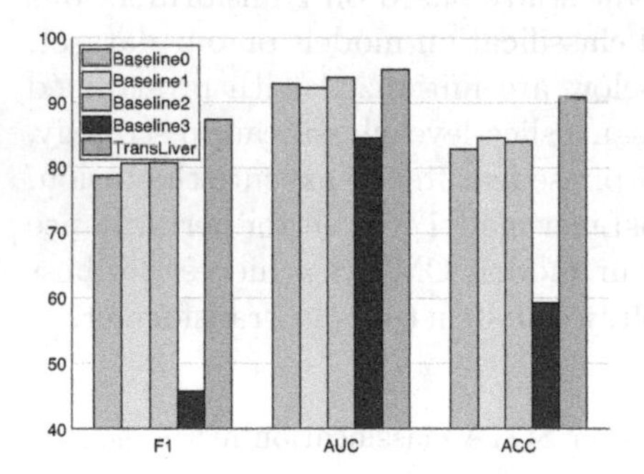

Fig. 3. Ablation study of modules.

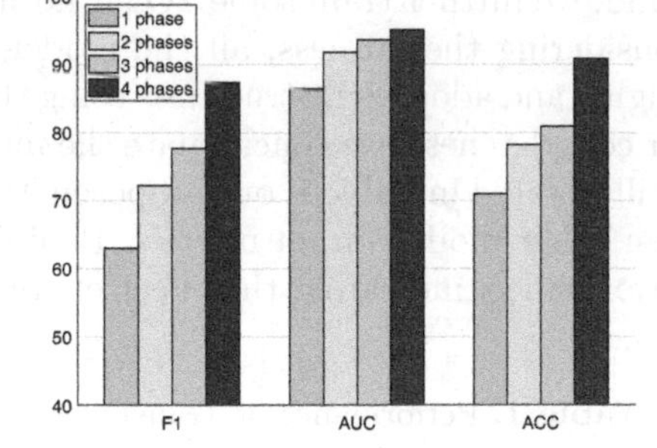

Fig. 4. Ablation study of phase number.

4 Conclusion

We have presented a hybrid architecture for multi-phase liver lesion classification in this paper. The lesion features are extracted by transformer backbone with several auxiliary convolutional modules. Then, we fuse the features from different phases through cross phase tokens to enhance their information exchange.

To handle the issues in realistic cases, we design a pre-processing unit to acquire multi-phase annotated lesions from single-phase annotated ones. We report performance of an overall 90.9% classification accuracy on a four-phase seven-class dataset through quantitative experiments and show obvious improvement compared with SOTA classification methods. In future work, we will extend classification to instance segmentation and provide an end-to-end effective model for liver lesion diagnosis.

Acknowledgements. This work was supported by the National Natural Science Foundation of China (grant numbers 62202426) and Ministry of Science and Technology of China (grant numbers 2022AAA010502).

References

1. Balakrishnan, G., Zhao, A., Sabuncu, M.R., Guttag, J., Dalca, A.V.: Voxelmorph: a learning framework for deformable medical image registration. IEEE Trans. Med. Imaging **38**(8), 1788–1800 (2019)
2. Bilic, P., et al.: The liver tumor segmentation benchmark (LiTS). Med. Image Anal. **84**, 102680 (2023)
3. Chen, X., et al.: A cascade attention network for liver lesion classification in weakly-labeled multi-phase CT images. In: Wang, Q., et al. (eds.) DART/MIL3ID -2019. LNCS, vol. 11795, pp. 129–138. Springer, Cham (2019). https://doi.org/10.1007/978-3-030-33391-1_15
4. Dosovitskiy, A., et al.: An image is worth 16×16 words: transformers for image recognition at scale. In: 9th International Conference on Learning Representations, ICLR 2021, Virtual Event, Austria, 3–7 May 2021. OpenReview.net (2021)
5. Frid-Adar, M., Klang, E., Amitai, M., Goldberger, J., Greenspan, H.: Synthetic data augmentation using GAN for improved liver lesion classification. In: 2018 IEEE 15th International Symposium on Biomedical Imaging (ISBI 2018), pp. 289–293. IEEE (2018)
6. Guo, J., et al.: CMT: Convolutional neural networks meet vision transformers. In: Proceedings of the IEEE/CVF Conference on Computer Vision and Pattern Recognition, pp. 12175–12185 (2022)
7. He, K., Zhang, X., Ren, S., Sun, J.: Deep residual learning for image recognition. In: Proceedings of the IEEE Conference on Computer Vision and Pattern Recognition, pp. 770–778 (2016)
8. Heker, M., Greenspan, H.: Joint liver lesion segmentation and classification via transfer learning. arXiv preprint arXiv:2004.12352 (2020)
9. Isen, J., et al.: Non-parametric combination of multimodal MRI for lesion detection in focal epilepsy. NeuroImage Clin. **32**, 102837 (2021)
10. Kayhan, O.S., Gemert, J.C.: On translation invariance in CNNs: convolutional layers can exploit absolute spatial location. In: Proceedings of the IEEE/CVF Conference on Computer Vision and Pattern Recognition, pp. 14274–14285 (2020)
11. Liang, D., et al.: Combining convolutional and recurrent neural networks for classification of focal liver lesions in multi-phase CT images. In: Frangi, A.F., Schnabel, J.A., Davatzikos, C., Alberola-López, C., Fichtinger, G. (eds.) MICCAI 2018. LNCS, vol. 11071, pp. 666–675. Springer, Cham (2018). https://doi.org/10.1007/978-3-030-00934-2_74

12. Liu, Z., et al.: Swin transformer: Hierarchical vision transformer using shifted windows. In: Proceedings of the IEEE/CVF International Conference on Computer Vision, pp. 10012–10022 (2021)
13. Lowe, D.G.: Object recognition from local scale-invariant features. In: Proceedings of the Seventh IEEE International Conference on Computer Vision, vol. 2, pp. 1150–1157. IEEE (1999)
14. Nagrani, A., Yang, S., Arnab, A., Jansen, A., Schmid, C., Sun, C.: Attention bottlenecks for multimodal fusion. Adv. Neural. Inf. Process. Syst. **34**, 14200–14213 (2021)
15. Raboh, M., Levanony, D., Dufort, P., Sitek, A.: Context in medical imaging: the case of focal liver lesion classification. In: Medical Imaging 2022: Image Processing, vol. 12032, pp. 165–172. SPIE (2022)
16. Romero, F.P., et al.: End-to-end discriminative deep network for liver lesion classification. In: 2019 IEEE 16th International Symposium on Biomedical Imaging (ISBI 2019), pp. 1243–1246. IEEE (2019)
17. Siegel, R.L., Miller, K.D., Fuchs, H.E., Jemal, A.: Cancer statistics, 2022. CA: a cancer J. Clin. **72**(1), 7–33 (2022)
18. Subramanian, V., Syeda-Mahmood, T., Do, M.N.: Multimodal fusion using sparse CCA for breast cancer survival prediction. In: 2021 IEEE 18th International Symposium on Biomedical Imaging (ISBI), pp. 1429–1432. IEEE (2021)
19. Vaswani, A., et al.: Attention is all you need. In: Advances in Neural Information Processing Systems, vol. 30 (2017)
20. Wang, W., et al.: Pyramid vision transformer: a versatile backbone for dense prediction without convolutions. In: Proceedings of the IEEE/CVF International Conference on Computer Vision, pp. 568–578 (2021)
21. Wang, W., et al.: PVT v2: improved baselines with pyramid vision transformer. Comput. Visual Med. **8**(3), 415–424 (2022)
22. Wightman, R.: Pytorch image models. https://github.com/rwightman/pytorch-image-models (2019). https://doi.org/10.5281/zenodo.4414861
23. Xu, X., et al.: A knowledge-guided framework for fine-grained classification of liver lesions based on multi-phase ct images. IEEE J. Biomed. Health Inform. **27**(1), 386–396 (2022)
24. Yasaka, K., Akai, H., Abe, O., Kiryu, S.: Deep learning with convolutional neural network for differentiation of liver masses at dynamic contrast-enhanced CT: a preliminary study. Radiology **286**(3), 887–896 (2018)

ArSDM: Colonoscopy Images Synthesis with Adaptive Refinement Semantic Diffusion Models

Yuhao Du[1,2], Yuncheng Jiang[1,2,3], Shuangyi Tan[1,2], Xusheng Wu[6], Qi Dou[5], Zhen Li[2,3](✉), Guanbin Li[4](✉), and Xiang Wan[1,2]

[1] Shenzhen Research Institute of Big Data, Shenzhen, China
[2] SSE, CUHK-Shenzhen, Shenzhen, China
lizhen@cuhk.edu.cn
[3] FNii, CUHK-Shenzhen, Shenzhen, China
[4] School of Computer Science and Engineering, Research Institute of Sun Yat-sen University in Shenzhen, Sun Yat-sen University, Guangzhou, China
liguanbin@mail.sysu.edu.cn
[5] The Chinese University of Hong Kong, Hong Kong, China
[6] Shenzhen Health Development Research and Data Management Center, Shenzhen, China

Abstract. Colonoscopy analysis, particularly automatic polyp segmentation and detection, is essential for assisting clinical diagnosis and treatment. However, as medical image annotation is labour- and resource-intensive, the scarcity of annotated data limits the effectiveness and generalization of existing methods. Although recent research has focused on data generation and augmentation to address this issue, the quality of the generated data remains a challenge, which limits the contribution to the performance of subsequent tasks. Inspired by the superiority of diffusion models in fitting data distributions and generating high-quality data, in this paper, we propose an **A**daptive **R**efinement **S**emantic **D**iffusion **M**odel (**ArSDM**) to generate colonoscopy images that benefit the downstream tasks. Specifically, ArSDM utilizes the ground-truth segmentation mask as a prior condition during training and adjusts the diffusion loss for each input according to the polyp/background size ratio. Furthermore, ArSDM incorporates a pre-trained segmentation model to refine the training process by reducing the difference between the ground-truth mask and the prediction mask. Extensive experiments on segmentation and detection tasks demonstrate the generated data by ArSDM could significantly boost the performance of baseline methods.

Keywords: Diffusion models · Colonoscopy · Polyp segmentation · Polyp detection

Y. Du and Y. Jiang—Equal contributions.

Supplementary Information The online version contains supplementary material available at https://doi.org/10.1007/978-3-031-43895-0_32.

H. Greenspan et al. (Eds.): MICCAI 2023, LNCS 14221, pp. 339–349, 2023.
https://doi.org/10.1007/978-3-031-43895-0_32

1 Introduction

Colonoscopy is a critical tool for identifying adenomatous polyps and reducing rectal cancer mortality. Deep learning methods have shown powerful abilities in automatic colonoscopy analysis, including polyp segmentation [5,22,26,27,29] and polyp detection [19,24]. However, the scarcity of annotated data due to high manual annotation costs results in poorly trained and low generalizable models. Previous methods have relied on generative adversarial networks (GANs) [9,25] or data augmentation methods [3,13,28] to enhance learning features, but these methods yielded limited improvements in downstream tasks. Recently, diffusion models [6,15] have emerged as promising solutions to this problem, demonstrating remarkable progress in generating multiple modalities of medical data [4,10,12,21].

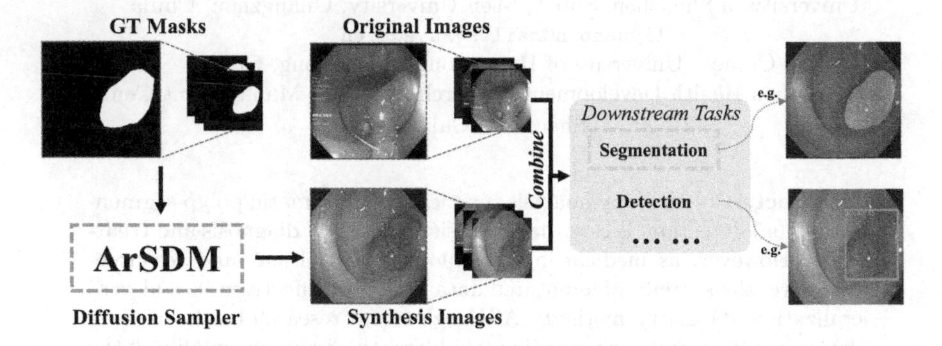

Fig. 1. Overview of the pipeline of our proposed approach, where details of **ArSDM** are described in Sect. 2.

Despite recent progress in these methods for medical image analysis, existing models face two major challenges when applied to colonoscopy image analysis. Firstly, the foreground (polyp) of colonoscopy images contains rich pathological information yet is often tiny compared with the background (intestine wall) and can be easily overwhelmed during training. Thus, naive generative models may generate realistic colonoscopy images but those images seldom contain polyp regions. In addition, in order to generate high-quality annotated samples, it is crucial to maintain the consistency between the polyp morphologies in synthesized images and the original masks, which current generative models struggle to achieve.

To tackle these issues and inspired by the remarkable success achieved by diffusion models in generating high-quality CT or MRI data [8,11,23], we creatively propose an effective adaptive refinement semantic diffusion model (ArSDM) to generate polyp-contained colonoscopy images while preserving the original annotations. The pipeline of the data generation and downstream task training is shown in Fig. 1. Specifically, we use the original segmentation masks as conditions to train a conditional diffusion model, which makes the generated samples share the same masks with the input images. Moreover, during diffusion model training, we employ an adaptive loss re-weighting method to assign loss weights for each input according to the size ratio of polyps and background, which addresses the overfitting problem for the large background. In addition, we fine-tune the diffusion model by minimizing the distance between the original ground truth masks and the prediction masks from synthesis images via a pre-trained segmentation network. Thus the refined model could generate samples better aligned with the original masks.

In summary, our contributions are three-fold: (1) **Adaptive Refinement SDM**: Based on the standard semantic diffusion model [21], we propose a novel ArSDM with the adaptive loss re-weighting and the prediction-guided sample refinement mechanisms, which is capable of generating realistic polyp-contained colonoscopy images while preserving the original annotations. To the best of our knowledge, this is the first work for adapting diffusion models to colonoscopy image synthesis. (2) **Large-Scale Colonoscopy Generation**: The proposed approach can be used to generate large-scale datasets with no/arbitrary annotations, which significantly benefits the medical image society, laying the foundation for large-scale pre-training models in automatic colonoscopy analysis. (3) **Qualitative and Quantitative Evaluation**: We conduct extensive experiments to evaluate our method on five public benchmarks for polyp segmentation and detection. The results demonstrate that our approach could help deep learning methods achieve better performances. The source code is available at https://github.com/DuYooho/ArSDM.

2 Method

Background. Denoising diffusion probabilistic models (DDPMs) [6] are classes of deep generative models, which have forward and reverse processes. The forward process is a Markov Chain that gradually adds Gaussian noise to the original data. This process can be formulated as the joint distribution $q\left(\mathbf{x}_{1:T} \mid \mathbf{x}_0\right)$:

$$q\left(\mathbf{x}_{1:T} \mid \mathbf{x}_0\right) := \prod_{t=1}^{T} q\left(\mathbf{x}_t \mid \mathbf{x}_{t-1}\right), q\left(\mathbf{x}_t \mid \mathbf{x}_{t-1}\right) := \mathcal{N}\left(\mathbf{x}_t; \sqrt{1-\beta_t}\mathbf{x}_{t-1}, \beta_t \mathbf{I}\right), \tag{1}$$

where $q\left(\mathbf{x}_0\right)$ is the original data distribution with $\mathbf{x}_0 \sim q\left(\mathbf{x}_0\right)$, $\mathbf{x}_{1:T}$ are latents with the same dimension of $\mathbf{x}_0$ and β_t is a variance schedule.

The reverse process is aiming to learn a model to reverse the forward process that reconstructs the original input data, which is defined as:

$$p_\theta\left(\mathbf{x}_{0:T}\right) := p\left(\mathbf{x}_T\right) \prod_{t=1}^{T} p_\theta\left(\mathbf{x}_{t-1} \mid \mathbf{x}_t\right), p_\theta\left(\mathbf{x}_{t-1} \mid \mathbf{x}_t\right) := \mathcal{N}\left(\mathbf{x}_{t-1}; \boldsymbol{\mu}_\theta\left(\mathbf{x}_t, t\right), \sigma_t^2 \mathbf{I}\right), \tag{2}$$

where $p\left(\mathbf{x}_T\right)$ is the noised Gaussian transition from the forward process at timestep T. In this case, we only need to use deep-learning models to represent $\boldsymbol{\mu}_\theta$ with θ as the model parameters. According to the original paper [6], the loss function can be simplified as:

$$\mathcal{L}_{\text{simple}} := \mathbb{E}_{t, \mathbf{x}_t, \boldsymbol{\epsilon} \sim \mathcal{N}(\mathbf{0}, \mathbf{I})}\left[\left\|\boldsymbol{\epsilon} - \boldsymbol{\epsilon}_\theta\left(\mathbf{x}_t, t\right)\right\|^2\right]. \tag{3}$$

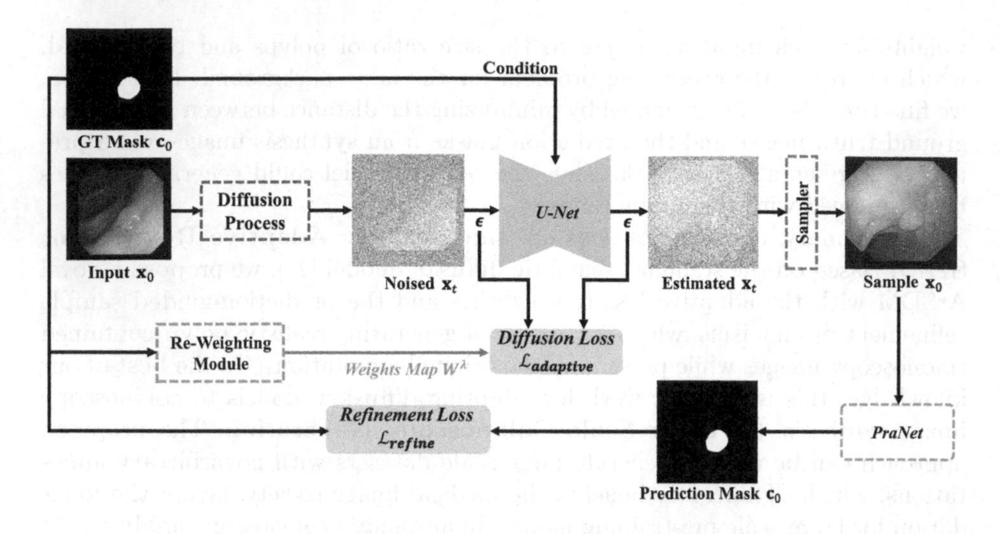

Fig. 2. The overall architecture of **ArSDM**.

Thus, instead of training the model $\boldsymbol{\mu}_\theta$ to predict $\tilde{\boldsymbol{\mu}}_t$, we can train the model $\boldsymbol{\epsilon}_\theta$ to predict $\tilde{\boldsymbol{\epsilon}}$, which is easier for parameterization and learning.

In this paper, we propose an adaptive refinement semantic diffusion model, a variant of DDPM, which has three key parts, *i.e.*, mask conditioning, adaptive loss re-weighting, and prediction-guided sample refinement. The overall illustration of our framework is shown in Fig. 2.

2.1 Mask Conditioning

Unlike the previous generative methods, our work aims to generate a synthetic image with an identical segmentation mask to the original annotation. To accomplish this, we adapt the widely used conditional U-Net architecture [21] in the reverse process, where the mask is fed as a condition. Specifically, for an input image $\mathbf{x}_0 \in \mathbb{R}^{H \times W \times C}$, $\mathbf{x}_t$ can be sampled at any timestep t with the closed form:

$$\mathbf{x}_t = \sqrt{\bar{\alpha}_t}\mathbf{x}_0 + \sqrt{1-\bar{\alpha}_t}\boldsymbol{\epsilon}, \tag{4}$$

where $\boldsymbol{\epsilon} \sim \mathcal{N}(\mathbf{0}, \mathbf{I}), \alpha_t := 1 - \beta_t$ and $\bar{\alpha}_t := \prod_{s=1}^{t} \alpha_s$. It will be fed into the encoder $\mathcal{E}$ of the U-Net, and its corresponding mask annotation $\mathbf{c}_0 \in \mathbb{R}^{H \times W}$ will be injected into the decoder $\mathcal{D}$. The model output can be formulated as:

$$\boldsymbol{\epsilon}_\theta(\mathbf{x}_t, t, \mathbf{c}_0) = \mathcal{D}(\mathcal{E}(\mathbf{x}_t), \mathbf{c}_0). \tag{5}$$

Thus, the U-Net model $\boldsymbol{\epsilon}_\theta$ in Eq. 3 becomes $\boldsymbol{\epsilon}_\theta(\mathbf{x}_t, t, \mathbf{c}_0)$, and the loss function in Eq. 3 is changed to:

$$\mathcal{L}_{\text{condition}} = \mathbb{E}_{t,\mathbf{x}_t,\mathbf{c}_0,\boldsymbol{\epsilon}\sim\mathcal{N}(\mathbf{0},\mathbf{I})}\left[\|\boldsymbol{\epsilon} - \boldsymbol{\epsilon}_\theta(\mathbf{x}_t, t, \mathbf{c}_0)\|^2\right]. \tag{6}$$

Algorithm 1: One training iteration of ArSDM

Input: $t \sim \text{Uniform}(\{1, ..., T\})$, $\mathbf{x}_0 \sim q(\mathbf{x}_0)$, $\mathbf{c}_0$, $\boldsymbol{\epsilon} \sim \mathcal{N}(\mathbf{0}, \mathbf{I})$
Output: $\tilde{\boldsymbol{\epsilon}}$, $\tilde{\mathbf{c}}_0$

1. $\mathbf{x}_t = \sqrt{\bar{\alpha}_t}\mathbf{x}_0 + \sqrt{1-\bar{\alpha}_t}\boldsymbol{\epsilon}$; $\quad \tilde{\mathbf{x}}_t = \sqrt{\bar{\alpha}_t}\mathbf{x}_0 + \sqrt{1-\bar{\alpha}_t}\boldsymbol{\epsilon}_\theta(\mathbf{x}_t, t, \mathbf{c}_0)$
2. **for** $i = t, ..., 1$ **do**
3. $\quad \mathbf{z} \sim \mathcal{N}(\mathbf{0}, \mathbf{I})$ if $i > 1$, else $\mathbf{z} = \mathbf{0}$; $\quad \tilde{\mathbf{x}}_{i-1} = \frac{1}{\sqrt{\alpha_i}}\left(\tilde{\mathbf{x}}_i - \frac{1-\alpha_i}{\sqrt{1-\bar{\alpha}_i}}\boldsymbol{\epsilon}_\theta(\tilde{\mathbf{x}}_i, i, \mathbf{c}_0)\right) + \sigma_i \mathbf{z}$
4. **end for**
5. $\tilde{\mathbf{c}}_0 = \mathcal{P}(\tilde{\mathbf{x}}_0)$
6. Take gradient descent step on $\nabla_\theta \mathcal{L}_{\text{total}}$

2.2 Adaptive Loss Re-weighting

The polyp regions in the colonoscopy images differ from the background regions, which contain more pathological information and should be adequately treated to learn a better model. However, training the diffusion models using the original loss function ignores the difference between different regions, where each pixel shares the same weights when calculating the loss. This would lead to the model generating more background-like polyps since the large background region will easily overwhelm the small foreground polyp regions during training. A simple way to alleviate this problem is to apply a weighted loss function that assigns the polyp and background regions with different weights. However, most polyps vary a lot in size and shape. Thus assigning constant weights for all polyps exacerbated the imbalance problem. In this case, to tackle this problem, we propose an adaptive loss function that vests different weights according to the size ratio of the polyp over the background. Specifically, we define a pixel-wise weights matrix $W^\lambda \in \mathbb{R}^{H \times W}$ with each entry $w^\lambda_{i,j}$ to be:

$$w^\lambda_{i,j} = \begin{cases} 1-r & , p=1 \\ r & , p=0 \end{cases}, \qquad r = \frac{\#(p=1)}{H \times W}, \tag{7}$$

where $p = 1$ means the pixel p at (h, w) belongs to the polyp region and $p = 0$ means it belongs to the background region. Thus, the loss function becomes:

$$\mathcal{L}_{\text{adaptive}} = \mathbb{E}_{t,\mathbf{x}_t,\mathbf{c}_0,\boldsymbol{\epsilon}\sim\mathcal{N}(\mathbf{0},\mathbf{I})}\left[W^\lambda \cdot \|\boldsymbol{\epsilon} - \boldsymbol{\epsilon}_\theta(\mathbf{x}_t, t, \mathbf{c}_0)\|^2\right]. \tag{8}$$

2.3 Prediction-Guided Sample Refinement

The downstream tasks of polyp segmentation and detection require rich semantic information on polyp regions to train a good model. Through extensive experiments, we found inaccurate sample images with coarse polyp boundary that is not aligned properly with the original masks may introduce large biases and noises to the datasets. The model can be confused by several conflicting training images with the same annotation. To this end, we design a refinement strategy

Table 1. Comparisons of different settings applied on three polyp segmentation baselines.

Methods	EndoScene		ClinicDB		Kvasir		ColonDB		ETIS		Overall	
	mDice	mIoU	mDice	mIoU	mDice	mIoU	mDice	mIoU	mDice	mIoU	mDice	mIoU
PraNet	87.1	79.7	89.9	84.9	89.8	84.0	70.9	64.0	62.8	56.7	74.0	67.5
+LDM	83.7	76.9	88.2	83.5	88.4	83.0	62.6	56.0	56.2	50.3	67.8	61.7
+SDM	**89.9**	**83.2**	89.2	83.7	88.4	82.6	74.2	66.5	66.4	60.3	76.4	69.6
+**Ours**	89.7	82.7	**93.3**	**88.5**	**89.9**	**84.5**	**76.1**	**68.9**	**75.5**	**68.1**	**80.0**	**73.2**
SANet	88.8	81.5	**91.6**	85.9	90.4	84.7	75.3	67.0	75.0	65.4	79.4	71.4
+LDM	72.7	60.5	88.8	82.8	88.7	82.7	64.3	55.4	58.0	49.2	68.3	59.8
+SDM	**90.2**	83.0	89.9	84.1	90.9	85.4	77.6	69.3	74.7	66.8	80.4	72.9
+**Ours**	**90.2**	**83.2**	91.4	**86.1**	**91.1**	**85.6**	**77.7**	**70.0**	**78.0**	**69.5**	**81.5**	**74.1**
PVT	**90.0**	**83.3**	93.7	88.9	**91.7**	**86.4**	80.8	72.7	78.7	70.6	83.3	76.0
+LDM	88.2	81.2	92.3	87.1	91.2	85.7	78.7	70.4	78.0	69.6	81.9	74.2
+SDM	88.8	81.7	**93.9**	**89.2**	91.2	86.1	81.3	73.5	78.7	71.1	83.4	76.3
+**Ours**	88.2	81.2	92.2	87.5	91.5	86.3	**81.7**	**73.8**	**80.6**	**72.9**	**84.0**	**76.7**

that uses the prediction of a pre-trained segmentation model on the sampled images to guide the training process and restore the proper polyp boundary information. Specifically, at each iteration of training, the output $\tilde{\epsilon} = \epsilon_\theta(\mathbf{x}_t, t, \mathbf{c}_0)$ will go into the sampler to generate sample image $\tilde{\mathbf{x}}_0$. Then, we take the sample image as the input of the segmentation model to predict the pseudo masks $\tilde{\mathbf{c}}_0$. We propose the following refinement loss based on IoU loss and binary cross entropy (BCE) loss between $\tilde{\mathbf{c}}_0$ and $\mathbf{c}_0$. The refinement loss is:

$$\begin{aligned} \mathcal{L}_{\text{refine}} &= \mathcal{L}(\mathbf{c}, \tilde{\mathbf{c}_g}) + \sum_{i=3}^{i=5} \mathcal{L}(\tilde{\mathbf{c}_i}), \\ \tilde{\mathbf{c}}_0 &= \{\tilde{\mathbf{c}_3}, \tilde{\mathbf{c}_4}, \tilde{\mathbf{c}_5}, \tilde{\mathbf{c}_g}\} = \mathcal{P}(\mathcal{S}(\tilde{\epsilon})), \end{aligned} \tag{9}$$

where $\mathcal{L} = \mathcal{L}_{IoU} + \mathcal{L}_{BCE}$ is the sum of the IoU loss and BCE loss, $\tilde{\mathbf{c}}_0$ is the collection of the three side-outputs $(\tilde{\mathbf{c}_3}, \tilde{\mathbf{c}_4}, \tilde{\mathbf{c}_5})$ and the global map $\tilde{\mathbf{c}_g}$ as described in [5]. $\mathcal{P}(\cdot)$ represents the PraNet model and $\mathcal{S}(\cdot)$ is the DDIM [16] sampler. The detailed procedure of one training iteration is shown in Algorithm 1 and the overall loss function is defined as:

$$\mathcal{L}_{\text{total}} = \mathcal{L}_{\text{adaptive}} + \mathcal{L}_{\text{refine}}. \tag{10}$$

3 Experiments

3.1 ArSDM Experimental Settings

We conducted our experiments on five public polyp segmentation datasets: EndoScene [20], CVC-ClincDB/CVC-612 [1], CVC-ColonDB [18], ETIS [14] and

Table 2. Comparisons of different settings applied on three polyp detection baselines.

Methods	EndoScene		ClinicDB		Kvasir		ColonDB		ETIS		Overall	
	AP	F1	AP	F1	AP	F1	AP	F1	AP	F1	AP	F1
Center	86.9	**91.4**	84.7	89.2	75.6	81.4	62.2	72.3	62.7	70.1	56.6	76.0
+LDM	84.1	84.4	**90.4**	89.9	81.3	81.8	73.4	74.5	65.2	71.7	62.0	76.9
+SDM	**87.8**	86.9	88.7	**91.0**	77.0	82.8	71.8	78.1	68.2	72.6	61.8	79.1
+**Ours**	85.0	89.1	86.1	90.8	**77.3**	**84.7**	**74.2**	**80.2**	**68.7**	**75.6**	**65.7**	**81.3**
Sparse	89.9	87.8	81.4	86.4	75.6	80.2	78.2	73.2	63.8	62.4	63.7	73.2
+LDM	87.4	76.3	**95.0**	**93.5**	81.5	58.8	80.0	71.0	64.4	54.3	65.3	66.3
+SDM	**94.5**	**90.5**	88.7	86.5	79.0	80.5	**81.4**	76.8	67.8	67.1	65.2	76.7
+**Ours**	92.8	86.2	92.2	90.6	**81.6**	**82.3**	80.1	**79.8**	**72.4**	**70.4**	**66.4**	**79.0**
Deform	**98.1**	**94.4**	89.7	89.9	**80.2**	74.4	**82.2**	75.5	65.3	54.7	64.5	71.8
+LDM	94.6	90.5	91.6	89.5	79.3	73.4	78.0	73.2	69.0	64.0	63.4	73.3
+SDM	96.0	90.6	90.3	91.2	82.2	78.9	80.1	75.1	67.5	66.7	65.1	75.8
+**Ours**	94.7	94.3	**92.3**	**92.0**	80.0	**80.3**	81.4	**77.3**	**74.1**	**69.3**	**67.9**	**77.9**

Kvasir [7]. Following the standard of PraNet, 1,450 image-mask pairs from Kvasir and CVC-ClinicDB are taken as the training set. The evaluations are conducted on the five datasets separately to verify the learning and generalization capability. The training image-mask pairs are padded to have the same height and width and then resized to the size of 384×384. Experiments with prediction-guided sample refinement are trained with around one-half NVIDIA A100 days, while others are trained with approximately one day for convergence. We use the DDIM sampler with a maximum timestep of 200 for sampling images.

3.2 Downstream Experimental Settings

We conduct the evaluation of our methods and the state-of-the-art counterparts on polyp segmentation and detection tasks. We generated the same number of samples as the diffusion training set using the original masks, and then combined them to create a new downstream training set. We employed PraNet [5], SANet [22], and Polyp-PVT [2] as baseline segmentation models with default settings, and evaluated them using mean Intersection over Union (IoU) and mean Dice metrics. For detection, we selected CenterNet [30], Sparse-RCNN [17], and Deformable-DETR [31] as baseline models with the same settings as the original papers, and evaluated them using Average Precision (AP) and F1-scores.

3.3 Quantitative Comparisons

The experimental results presented in Table 1 and 2 demonstrate the effectiveness of our proposed method in training better downstream models to achieve superior performance. Specifically, data generated by our approach assists the

Table 3. Ablation study of different components on polyp segmentation tasks.

Methods		PraNet		SANet	
Ada.	Ref.	mDice	mIoU	mDice	mIoU
✗	✗	76.4	69.6	80.4	72.9
✓	✗	79.1	72.4	80.5	72.8
✗	✓	78.5	71.5	81.1	73.2
✓	✓	**80.0**	**73.2**	**81.5**	**74.1**

Table 4. Ablation study of different components on polyp detection tasks.

Methods		CenterNet		Sparse.	
Ada.	Ref.	AP	F1	AP	F1
✗	✗	61.8	79.1	65.2	76.7
✓	✗	62.2	80.1	65.8	77.2
✗	✓	64.0	80.4	66.0	77.6
✓	✓	**65.7**	**81.3**	**66.4**	**79.0**

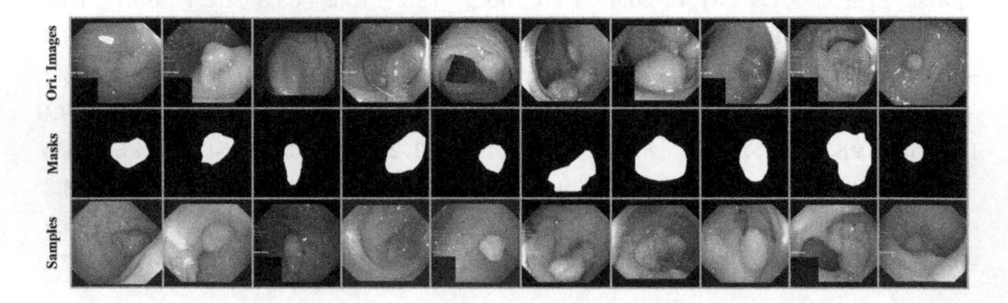

Fig. 3. Illustration of generated samples with the corresponding masks and original images for comparison reference.

significant improvements for each model in mDice and mIoU, with increases of 6.0% and 5.7% over PraNet, 2.1% and 2.7% over SANet, and 0.7% and 0.7% over Polyp-PVT. We also observe superior AP and F1-scores compared to CenterNet, Sparse-RCNN, and Deformable-DETR trained with original data, with gains of 9.1% and 5.3%, 2.7% and 5.8%, and 3.4% and 6.1%, respectively. Moreover, we conducted a comprehensive comparison with SOTA models, noting that these models were not specifically designed for colonoscopy images and may generate data that hinder the training process or lack the ability for effective improvement. Nevertheless, our experimental results confirm the superiority of our proposed method.

Ablation Study. We conducted an ablation study to assess the importance of each proposed component. Table 3 and 4 report the overall accuracies on the test set. The results demonstrate both components contribute to the accuracy improvement of baseline models, indicating their essential roles in achieving the best final performance.

3.4 Qualitative Analyses

To further investigate the generative performance of our approach, we present visualization results in Fig. 3, which displays the generated samples and their corresponding masks, alongside the original images for reference. The generated samples demonstrate differences from the original images in both the polyp regions and the backgrounds while maintaining alignment with the masks. Additionally, we sought evaluations from medical professionals to assess the authenticity of the generated samples, and non-medical professionals to locate polyps in the images, which yielded positive feedback on the quality of the generated samples.

4 Conclusion

Automatic generation of annotated data is essential for colonoscopy image analysis, where the scale of existing datasets is limited by the expertise and time required for manual annotation. In this paper, we propose an adaptive refinement semantic diffusion model (ArSDM) for generating colonoscopy images while preserving annotations by introducing innovative adaptive loss re-weighting and prediction-guided sample refinement mechanisms. To evaluate our approach comprehensively, we conduct polyp segmentation and detection experiments on five widely used datasets, where experimental results demonstrate the effectiveness of our approach, in which model performances are greatly enhanced with little synthesized data.

Acknowledgement. This work was supported in part by the Chinese Key-Area Research and Development Program of Guangdong Province (2020B0101350001), in part by the Shenzhen General Program (No. JCYJ20220530143600001), in part by the National Natural Science Foundation of China (NO. 61976250), in part by the Shenzhen-Hong Kong Joint Funding (No. SGDX20211123112401002), in part by the Shenzhen Science and Technology Program (NO. JCYJ20220818103001002, NO. JCYJ20220530141211024), and in part by the Guangdong Provincial Key Laboratory of Big Data Computing, The Chinese University of Hong Kong, Shenzhen.

References

1. Bernal, J., Sánchez, F.J., Fernández-Esparrach, G., Gil, D., Rodríguez, C., Vilariño, F.: WM-DOVA maps for accurate polyp highlighting in colonoscopy: Validation vs. saliency maps from physicians. Comput. Med. Imaging Graph. **43**, 99–111 (2015)
2. Bo, D., Wenhai, W., Deng-Ping, F., Jinpeng, L., Huazhu, F., Ling, S.: Polyp-PVT: Polyp segmentation with pyramidvision transformers (2021)
3. Chaitanya, K., et al.: Semi-supervised task-driven data augmentation for medical image segmentation. Med. Image Anal. **68**, 101934 (2021)
4. Dhariwal, P., Nichol, A.: Diffusion models beat GANs on image synthesis. Adv. Neural. Inf. Process. Syst. **34**, 8780–8794 (2021)
5. Fan, D.-P., et al.: PraNet: parallel reverse attention network for polyp segmentation. In: Martel, A.L., et al. (eds.) MICCAI 2020. LNCS, vol. 12266, pp. 263–273. Springer, Cham (2020). https://doi.org/10.1007/978-3-030-59725-2_26
6. Ho, J., Jain, A., Abbeel, P.: Denoising diffusion probabilistic models. Adv. Neural. Inf. Process. Syst. **33**, 6840–6851 (2020)
7. Jha, D., et al.: Kvasir-SEG: a segmented polyp dataset. In: Ro, Y.M., et al. (eds.) MMM 2020. LNCS, vol. 11962, pp. 451–462. Springer, Cham (2020). https://doi.org/10.1007/978-3-030-37734-2_37

8. Kim, B., Ye, J.C.: Diffusion deformable model for 4D temporal medical image generation. In: Wang, L., Dou, Q., Fletcher, P.T., Speidel, S., Li, S. (eds.) Medical Image Computing and Computer Assisted Intervention. MICCAI 2022. Lecture Notes in Computer Science, vol. 13431, pp. 539–548. Springer, Cham (2022). https://doi.org/10.1007/978-3-031-16431-6_51
9. Ma, Y., et al.: Cycle structure and illumination constrained GAN for medical image enhancement. In: Martel, A.L., et al. (eds.) MICCAI 2020. LNCS, vol. 12262, pp. 667–677. Springer, Cham (2020). https://doi.org/10.1007/978-3-030-59713-9_64
10. Park, T., Liu, M.Y., Wang, T.C., Zhu, J.Y.: Semantic image synthesis with spatially-adaptive normalization. In: Proceedings of the IEEE/CVF Conference on Computer Vision and Pattern Recognition, pp. 2337–2346 (2019)
11. Pinaya, W.H., et al.: Fast unsupervised brain anomaly detection and segmentation with diffusion models. In: Wang, L., Dou, Q., Fletcher, P.T., Speidel, S., Li, S. (eds.) Medical Image Computing and Computer Assisted Intervention. MICCAI 2022. LNCS, vol. 13438, pp. 705–714. Springer, Cham (2022). https://doi.org/10.1007/978-3-031-16452-1_67
12. Rombach, R., Blattmann, A., Lorenz, D., Esser, P., Ommer, B.: High-resolution image synthesis with latent diffusion models. In: Proceedings of the IEEE/CVF Conference on Computer Vision and Pattern Recognition, pp. 10684–10695 (2022)
13. Sandfort, V., Yan, K., Pickhardt, P.J., Summers, R.M.: Data augmentation using generative adversarial networks (cycleGAN) to improve generalizability in CT segmentation tasks. Sci. Rep. **9**(1), 16884 (2019)
14. Silva, J., Histace, A., Romain, O., Dray, X., Granado, B.: Toward embedded detection of polyps in WCE images for early diagnosis of colorectal cancer. Int. J. Comput. Assist. Radiol. Surg. **9**, 283–293 (2014)
15. Sohl-Dickstein, J., Weiss, E., Maheswaranathan, N., Ganguli, S.: Deep unsupervised learning using nonequilibrium thermodynamics. In: International Conference on Machine Learning, pp. 2256–2265. PMLR (2015)
16. Song, J., Meng, C., Ermon, S.: Denoising diffusion implicit models. In: International Conference on Learning Representations (2021)
17. Sun, P., et al.: Sparse r-cnn: End-to-end object detection with learnable proposals. In: Proceedings of the IEEE/CVF Conference on Computer Vision And Pattern Recognition, pp. 14454–14463 (2021)
18. Tajbakhsh, N., Gurudu, S.R., Liang, J.: Automated polyp detection in colonoscopy videos using shape and context information. IEEE Trans. Med. Imaging **35**(2), 630–644 (2015)
19. Tajbakhsh, N., Gurudu, S.R., Liang, J.: Automated polyp detection in colonoscopy videos using shape and context information. IEEE Trans. Med. Imaging. **35**, 630–644 (2016)
20. Vázquez, D., et al.: A benchmark for endoluminal scene segmentation of colonoscopy images. J. Healthcare Eng. **2017**, 4031790 (2017)
21. Wang, W., et al.: Semantic image synthesis via diffusion models. arXiv preprint arXiv:2207.00050 (2022)
22. Wei, J., Hu, Y., Zhang, R., Li, Z., Zhou, S.K., Cui, S.: Shallow attention network for polyp segmentation. In: de Bruijne, M., et al. (eds.) MICCAI 2021. LNCS, vol. 12901, pp. 699–708. Springer, Cham (2021). https://doi.org/10.1007/978-3-030-87193-2_66

23. Wolleb, J., Bieder, F., Sandkühler, R., Cattin, P.C.: diffusion models for medical anomaly detection. In: Wang, L., Dou, Q., Fletcher, P.T., Speidel, S., Li, S. (eds.) Medical Image Computing and Computer Assisted Intervention. MICCAI 2022. LNCS, vol. 13438, pp. 35–45. Springer, Cham (2022). https://doi.org/10.1007/978-3-031-16452-1_4
24. Wu, L., Hu, Z., Ji, Y., Luo, P., Zhang, S.: Multi-frame collaboration for effective endoscopic video polyp detection via spatial-temporal feature transformation. In: de Bruijne, M., et al. (eds.) MICCAI 2021. LNCS, vol. 12905, pp. 302–312. Springer, Cham (2021). https://doi.org/10.1007/978-3-030-87240-3_29
25. Xu, J., et al.: OfGAN: realistic rendition of synthetic colonoscopy videos. In: Martel, A.L., et al. (eds.) MICCAI 2020. LNCS, vol. 12263, pp. 732–741. Springer, Cham (2020). https://doi.org/10.1007/978-3-030-59716-0_70
26. Zhang, R., et al.: Lesion-Aware Dynamic Kernel for Polyp Segmentation. In: Wang, L., Dou, Q., Fletcher, P.T., Speidel, S., Li, S. (eds) Medical Image Computing and Computer Assisted Intervention. MICCAI 2022. LNCS, vol. 13433, pp. 99–109. Springer, Cham (2022).https://doi.org/10.1007/978-3-031-16437-8_10
27. Zhang, R., Li, G., Li, Z., Cui, S., Qian, D., Yu, Y.: Adaptive context selection for polyp segmentation. In: Martel, A.L., et al. (eds.) MICCAI 2020. LNCS, vol. 12266, pp. 253–262. Springer, Cham (2020). https://doi.org/10.1007/978-3-030-59725-2_25
28. Zhao, A., Balakrishnan, G., Durand, F., Guttag, J.V., Dalca, A.V.: Data augmentation using learned transformations for one-shot medical image segmentation. In: Proceedings of the IEEE/CVF Conference on Computer Vision and Pattern Recognition, pp. 8543–8553 (2019)
29. Zhao, X., et al.: Semi-supervised spatial temporal attention network for video polyp segmentation. In: Wang, L., Dou, Q., Fletcher, P.T., Speidel, S., Li, S. (eds.) Medical Image Computing and Computer Assisted Intervention. MICCAI 2022. LNCS, vol. 13434, pp. 456–466. Springer, Cham (2022)
30. Zhou, X., Wang, D., Krähenbühl, P.: Objects as points. arXiv preprint arXiv:1904.07850 (2019)
31. Zhu, X., Su, W., Lu, L., Li, B., Wang, X., Dai, J.: Deformable detr: Deformable transformers for end-to-end object detection. arXiv preprint arXiv:2010.04159 (2020)

FeSViBS: Federated Split Learning of Vision Transformer with Block Sampling

Faris Almalik, Naif Alkhunaizi, Ibrahim Almakky, and Karthik Nandakumar(✉)

Mohamed Bin Zayed University of Artificial Intelligence, Abu Dhabi, UAE
{faris.almalik,naif.alkhunaizi,ibrahim.almakky,
karthik.nandakumar}@mbzuai.ac.ae

Abstract. Data scarcity is a significant obstacle hindering the learning of powerful machine learning models in critical healthcare applications. Data-sharing mechanisms among multiple entities (e.g., hospitals) can accelerate model training and yield more accurate predictions. Recently, approaches such as Federated Learning (FL) and Split Learning (SL) have facilitated collaboration without the need to exchange private data. In this work, we propose a framework for medical imaging classification tasks called **Fe**derated **S**plit learning of **Vi**sion transformer with **B**lock **S**ampling **(FeSViBS)**. The FeSViBS framework builds upon the existing federated split vision transformer and introduces a *block sampling* module, which leverages intermediate features extracted by the Vision Transformer (ViT) at the server. This is achieved by sampling features (patch tokens) from an intermediate transformer block and distilling their information content into a pseudo class token before passing them back to the client. These pseudo class tokens serve as an effective feature augmentation strategy and enhances the generalizability of the learned model. We demonstrate the utility of our proposed method compared to other SL and FL approaches on three publicly available medical imaging datasets: HAM1000, BloodMNIST, and Fed-ISIC2019, under both IID and non-IID settings. Code: https://github.com/faresmalik/FeSViBS.

Keywords: Split learning · Federated learning · Vision transformer · Convolutional neural network · Augmentation · Sampling

1 Introduction

Vision Transformers (ViTs) are self-attention based neural networks that have achieved state-of-the-art performance on various medical imaging tasks [8,24,30]. Since ViTs are capable of encoding long range dependencies between input

F. Almalik and N. Alkhunaizi—Equal contribution

H. Greenspan et al. (Eds.): MICCAI 2023, LNCS 14221, pp. 350–360, 2023.
https://doi.org/10.1007/978-3-031-43895-0_33

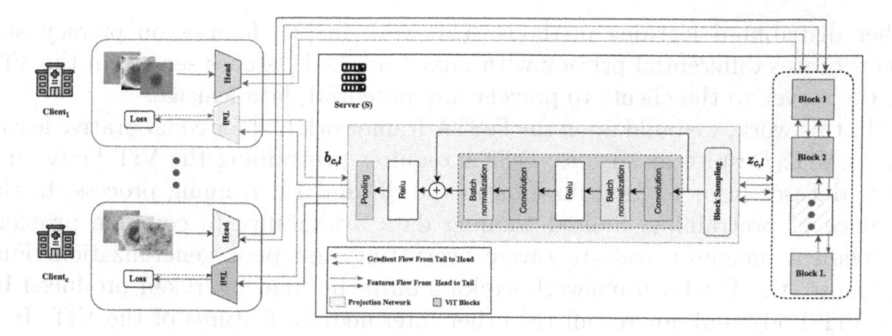

Fig. 1. FeSViBS framework. The server receives smashed representations from the clients, samples a ViT block for each client, uses a projection network to distill patch tokens into pseudo class tokens, which are sent back to the client for final prediction.

sequences [16], they are more robust against distribution shifts and are well-suited for handling heterogeneous distributions [5]. However, training ViT models typically requires significantly more data than traditional Convolutional Neural Network (CNN) models [16], which limits their application in domains such as healthcare, where data scarcity is a challenge. One way to overcome this challenge is to train such models in a collaborative and distributed manner, where large amounts of data can be leveraged from different sites without the need for sharing private data [9,11]. Federated learning and split learning are two well-known approaches for collaborative model training.

Federated Learning (FL) enables clients to collaboratively learn a global model by aggregating locally trained models [14]. Since this can be accomplished without sharing raw data, FL mitigates risks related to private data leakage. Several aggregation rules such as FedAvg [20] and FedProx [19] have been proposed for FL. However, it has been demonstrated that most FL algorithms are vulnerable to gradient inversion attacks [13], which dilute their privacy guarantees. In contrast, Split Learning (SL) divides a deep neural network into components with independently accessible parameters [10]. Since no participant in SL can access the complete model parameters, it has been claimed that SL offers better data confidentiality compared to FL. In particular, the U-shaped SL configuration, where each client has its own feature extraction head and a task-specific tail [27] can further improve client privacy, as it circumvents the need to share the data or labels. Recently, SL frameworks have been proposed for various medical applications such as tumor classification [3] and chext x-ray classification [23].

Recent studies [21,22] have demonstrated that both FL and SL can be combined to effectively train ViTs. In [22], a framework called FeSTA was proposed for medical image classification. The FeSTA framework involves a hybrid ViT architecture with U-shaped SL configuration - each client has its own CNN head and a multilayer perceptron (MLP) tail, while the shared ViT body resides on a central server. This architecture can be trained using both SL and FL in a potentially task-agnostic fashion, leading to better performance compared to other distributed learning methods. The work in [21] focuses on privacy and incorporates differential privacy with mixed masked patches sent from the ViT on the server to the clients to prevent any potential data leakage.

In this work, we build upon the FeSTA framework [22] for collaborative learning of ViT. Despite its success, FeSTA requires pretraining the ViT body on a large dataset prior to its utilization in the SL and FL training process. In the absence of pretraining, limited training data availability (a common problem in medical imaging) leads to severe overfitting and poor generalization. Furthermore, the FeSTA framework exploits only the final *cls* token produced by the ViT body and ignores all the other intermediate features of the ViT. It is well-known that intermediate features (referred to as patch tokens) also contain discriminative information that could be useful for the classification task [4].

To overcome the above limitations, we propose a framework called **Fe**derated **S**plit learning of **Vi**sion transformer with **B**lock **S**ampling **(FeSViBS)**. Our primary novelty is the introduction of a *block sampling* module, which randomly selects an intermediate transformer block for each client in each training round, extracts intermediate features, and distills these features into a pseudo cls token using a shared projection network. The proposed approach has two key benefits: (i) it effectively leverages intermediate ViT features, which are completely ignored in FeSTA, and (ii) sampling these intermediate features from different blocks, rather than relying solely on an individual block's features or the final *cls* token, serves as a feature augmentation strategy for the network, enhancing its generalization. The contributions of this work can be summarized as follows:

i. We propose the FeSViBS framework, a novel federated and split learning framework that leverages the features learned by intermediate ViT blocks to enhance the performance of the collaborative system.
ii. We introduce block sampling at the server level, which acts as a feature augmentation strategy for better generalization.

2 Methodology

We first describe the working of a typical split vision transformer before proceeding to describe FeSViBS. Each client $c \in [1, n]$ has access to local private data $(x_c, y_c) \in \{x_c^{(i)}, y_c^{(i)}\}_{i=1}^{N_c}$, where N_c is the number of training samples available at client c, x represents the input data, and y is the class label. Following [22], we assume U-shaped split learning setting, with each client having two local networks called *head* $(\mathcal{H}_{\theta_c})$ and *tail* $(\mathcal{T}_{\psi_c})$, where θ_c and ψ_c are client-specific *head* and *tail* parameters, respectively. The server consists of a ViT *body* $(\mathcal{B}_\Phi)$, which includes a stack of L transformer blocks denoted as $\mathcal{B}_{\Phi_1}, \mathcal{B}_{\Phi_2}, \cdots, \mathcal{B}_{\Phi_L}$ and $\mathcal{B}_\Phi(\cdot) = \mathcal{B}_{\Phi_L}(\cdots(\mathcal{B}_{\Phi_2}(\mathcal{B}_{\Phi_1}(\cdot))))$. Here, Φ_l represents the parameters of the lth transformer block and $\Phi = [\Phi_1, \Phi_2, \cdots, \Phi_L]$ denotes the complete set of parameters of the transformer body.

During training, the client performs a forward pass of the input data through the head to produce an embedding $h_c = \mathcal{H}_{\theta_c}(x_c) \in \mathbb{R}^{768 \times M}$ of its local data, which is typically organized as M *patch tokens* representing different patches of the input image. These embeddings (*smashed* representations) are then sent to the server. The ViT appends an additional token called the class token (*cls* $\in \mathbb{R}^{768\times1}$) and utilizes the self-attention mechanism to obtain a representation $b_c = \mathcal{B}_\Phi(h_c) \in \mathbb{R}^{768\times1}$, which is typically the *cls* token resulting from the last transformer block. This *cls* token is returned to the client for further processing. The *tail* at each client projects the received class token representation b_c into a class probability distribution to get the final prediction $\hat{y}_c = \mathcal{T}_{\psi_c}(b_c)$. This marks the end of the forward pass. Subsequently, the backpropagation starts with computing loss $\ell_c(y_c, \hat{y}_c)$, where $\ell_c(.)$ represents the client's loss function between the true labels y_c and predicted labels $\hat{y}_c$. The gradient of this loss is propagated back in the reverse order from the client's *tail*, server's *body*, to the client's *head*. We refer to this setting as Split Learning of Vision Transformer (**SLViT**), where each client optimizes the following objective in each round:

$$\min_{\theta_c,\Phi,\psi_c} \frac{1}{N_c} \sum_{i=1}^{N_c} \ell_c(y_c^{(i)}, \mathcal{T}_{\psi_c}(\mathcal{B}_\Phi(\mathcal{H}_{\theta_c}(x_c^{(i)})))) \tag{1}$$

In FeSTA [22], an additional federation step was introduced. After every few SL rounds, the local (client-specific) *heads* and *tails* are aggregated in a *unifying round* using FedAvg [20] to produce global parameters $\bar{\theta}$ and $\bar{\psi}$. Note that the above framework completely ignores all the intermediate features (*patch* tokens) extracted from various ViT blocks. In [4], it was demonstrated that these patch tokens are also discriminative and valuable for classification tasks. Hence, we aim to exploit these intermediate features to further enhance the performance.

2.1 FeSViBS Framework

The proposed FeSViBS method is illustrated in Fig. 1 and detailed in Algorithm 1. The working of the FeSViBS framework is very similar to FeSTA, except for one key difference. During the forward pass of SLViT and FeSTA, the server always returns the *cls* token from the last ViT block. In contrast, a FeSViBS server samples an intermediate block $l \in \{1, 2. \dots, L\}$ for each client c in each round and extracts the intermediate features $z_{c,l}$ from the chosen l^{th} block as follows:

$$z_{c,l} = \mathcal{B}_{\Phi_l}\Big(\mathcal{B}_{\Phi_{l-1}} \dots \mathcal{B}_{\Phi_1}(\mathcal{H}_{\theta_c}(x_c))\Big) \tag{2}$$

where $z_{c,l} \in \mathbb{R}^{768\times M}$. The server then projects the extracted intermediate features into a lower dimension using a *projection network* $\mathcal{R}$ (shared across all blocks) to obtain the final representation $b_{c,l} = \mathcal{R}_\pi(z_{c,l})$, where $b_{c,l} \in \mathbb{R}^{768\times1}$. This final representation $b_{c,l}$ can be considered as a *pseudo class token* and the role of the projection network is to distill the discriminative information contained in the intermediate features into this pseudo class token. The primary motivation for block sampling is to effectively leverage intermediate ViT features that are better at capturing local texture information (but are lost when

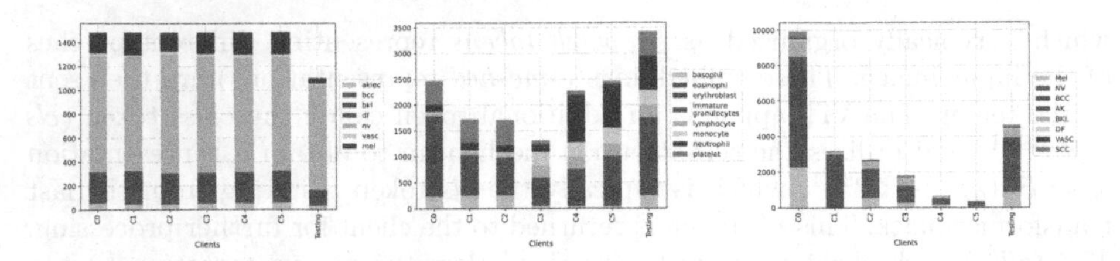

Fig. 2. Distribution of: (**left**) HAM10000, (**middle**) BloodMNIST, and (**right**) Fed-ISIC2019. Each stacked bar represents the number of samples, and each color represents each class. The last bar in each figure represents the testing set.

only the final *cls* token is used). Stochasticity in the block selection serves as a feature augmentation strategy, thereby aiding the generalization performance.

The architecture of the projection network is shown in Fig. 1 and it resembles a simple ResNet [12] block with skip connection. The pseudo class token is then sent to the client's *tail* to obtain the final prediction $\hat{y}_c = \mathcal{T}_{\psi_c}(b_{c,l})$ and complete the forward pass. Each client uses $\hat{y}_c$ along with the true labels y_c to compute the loss $\ell_c(y_c, \hat{y}_c)$. The gradients of the client's *tail* are then calculated and sent back to the server, which then carries out the back-propagation through the *projection network* and relevant blocks of the ViT *body* (only those blocks involved in the corresponding forward pass). Next, the server sends the gradients back to the client to propagate them through the *head* and end the back-propagation step. Hence, the client's optimization problem is:

$$\min_{\theta_c, \Phi_{1:l,c}, \pi, \psi_c} \frac{1}{N_c} \sum_{i=1}^{N_c} \ell_c\big(y_c^{(i)}, \mathcal{T}_{\psi_c}(b_{c,l}^{(i)})\big). \tag{3}$$

In the FeSViBS framework, the *heads* and *tails* of all the clients are assumed to have the same network architecture. Within each collaboration round, all the clients perform the forward and backward passes. While the parameters of the relevant head and tail as well as the shared projection network are updated after every backward pass, the parameters of the ViT body are updated only at the end of a collaboration round after aggregating updates from all the clients. The above protocol until this step is referred to as **SViBS**, because there is still no federation of the *heads* and *tails*. Similar to FeSTA, we also perform aggregation of the local *heads* and *tails* periodically in unifying rounds, resulting in the final FeSViBS framework. While in SViBS, the clients can initialize their heads and tails independently, FeSviBS requires a common initialization by the server and sharing of aggregated head and tail parameters after a unifying round.

3 Experimental Setup

Datasets. We conduct our experiments on three medical imaging datasets. The first dataset is HAM10000 [26], a multi-class dataset comprising of 10,015 dermoscopic images from diverse populations. HAM10000 includes 7 imbalanced

Algorithm 1. FeSViBS

Require: Local data at client c (x_c, y_c). Server initializes the body parameters (Φ), Projection Network parameters (π), client head and tail parameters $(\bar{\theta}, \bar{\psi})$

1: **for** rounds $r = 1, 2, \ldots, R$ **do**
2: **for client** $c \in [1, n]$ **do**
3: **if** $r = 1$ **or** $(r-1) \in$ Unifying Rounds **then**
4: $(\theta_c, \psi_c) \leftarrow (\bar{\theta}, \bar{\psi})$
5: **end if**
6: Client c: $h_c \leftarrow \mathcal{H}_{\theta_c}(x_c)$
7: Server:
8: Sample a ViT block (l) for client c
9: $b_{c,l} \leftarrow \mathcal{R}_\pi\Big(\mathcal{B}_{\Phi_l}\big(\mathcal{B}_{\Phi_{l-1}} \ldots \mathcal{B}_{\Phi_1}(h_c)\big)\Big)$
10: Client c:
11: Compute $\ell_c\big(y_c, \mathcal{T}_{\psi_c}(b_{c,l})\big)$ and Backprop.
12: Update (θ_c, ψ_c) with suitable optimizer
13: Server:
14: Update (π) with suitable optimizer, Compute and store $\Phi_{1:l,c}$
15: **end for**
16: Server:
17: Update body: $\Phi \leftarrow \frac{1}{n}\sum_c \Phi_{1:l,c}$
18: **if** $r \in$ Unifying Rounds **then**
19: $(\bar{\theta}, \bar{\psi}) \leftarrow (\frac{1}{n}\sum_c \theta_c, \frac{1}{n}\sum_c \psi_c)$
20: **end if**
21: **end for**

categories of pigmented lesions; we randomly perform 80%/20% split for training and testing, respectively. The second dataset [2] termed "BloodMNIST" is a multi-class dataset consisting of 17,092 blood cell images for 8 different imbalanced cell types. We followed [29] and split the dataset into 70% training, 10% validation, and 20% testing. Finally, the Fed-ISIC2019 dataset consists of 23,247 dermoscopy images for 8 different melanoma classes. This dataset was prepared by FLamby [25] from the original ISIC2019 dataset [6,7,26] and the data was collected from 6 centers, with significant differences in population characteristics and acquisition systems, representing real-world domain shifts. We use 80%/20% split for training and testing, respectively. The training samples in all datasets are divided among 6 clients, whereas the testing set is shared among them all. The distribution of each dataset is depicted in Fig. 2. Note that Fed-ISIC2019 and BloodMNIST are non-IID, whereas HAM10000 is IID.

Server's Network. For the server's body, we chose the ViT-B/16 model from timm library [28] which includes $L = 12$ transformer blocks, embedding dimension $D = 768$, 12 attention heads, and divides the input image into patches each of size 16×16 with $M = 196$ patches. We limit the block sampling to the first 6 ViT blocks. Additionally, the projection network has two convolution layers with a skip connection, which takes an input of dimension 768×196 and projects it into a lower dimension of 768.

Table 1. Average balanced accuracy for different methods. Centralized, FedAvg, FedProx, SCAFFOLD, MOON, and FeSTA have one global unified model for all clients. For local, SLViT, and SViBS, we report the standard deviation (stdev) across clients. For FeSViBS, we report stdev over stochastic sampling of ViT blocks during inference.

	Dataset		
	HAM10000	BloodMNIST	Fed-ISIC2019
Centralized	0.615	0.957	0.614
Local	0.494 ± 0.024	0.785 ± 0.017	0.290 ± 0.113
SLViT	0.540 ± 0.029	0.826 ± 0.018	0.293 ± 0.133
SViBS (ours)	0.570 ± 0.011	0.836 ± 0.014	0.330 ± 0.042
FedAvg [20]	0.564	0.894	0.476
FedProx [19]	0.568	0.892	0.472
SCAFFOLD [15]	0.290	0.880	0.330
MOON [18]	0.570	0.903	0.450
FeSTA [22]	0.638	0.929	0.430
FeSViBS (ours)	**0.682 ± 0.021**	**0.936 ± 0.002**	**0.534 ± 0.005**

Clients' Networks. Each client has two main networks: head and tail. We followed timm library's implementation of Hybrid ViTs (h-ViT) to design each client's head, which is a ResNet-50 [12] with a convolution layer added to project the features extracted by ResNet-50 to a dimension of 768×196. The tail is a linear classifier. Also, we unify the clients' networks (head and tail) every 2 rounds using FedAvg. We conduct our experiments for 200 rounds with Adam optimizer [17], a learning rate of 1×10^{-4}, and 32 batch size with a cross-entropy loss calculated at the tail. The code was implemented using PyTorch 1.10 and the models were trained using Nvidia A100 GPU with 40 GB memory.

4 Results and Analysis

Following [25], we used balanced accuracy in all experiments to evaluate the performance of the classification task across all datasets. This metric defines as the average recall on each class. In Table 1, we compare the performance of FeSViBS and SViBS frameworks with other SOTA methods. FeSViBS consistently outperforms other methods on the three datasets with both IID and non-IID settings. More specifically, for HAM10000 (IID), FeSViBS outperforms all other methods with a **4.4%** gain in performance over FeSTA and approximately **11%** over FedAvg and FedProx ($\mu = 0.006$). In the non-IID settings with both BloodMNIST and Fed-ISIC2019, FeSViBS maintains a high performance compared to other methods. Under extreme non-IID settings (Fed-ISIC2019), our approach demonstrated a performance improvement of **10.4%** compared to FeSTA and **5.8%** over FedAvg and FedProx, demonstrating the robustness of FeSViBS.

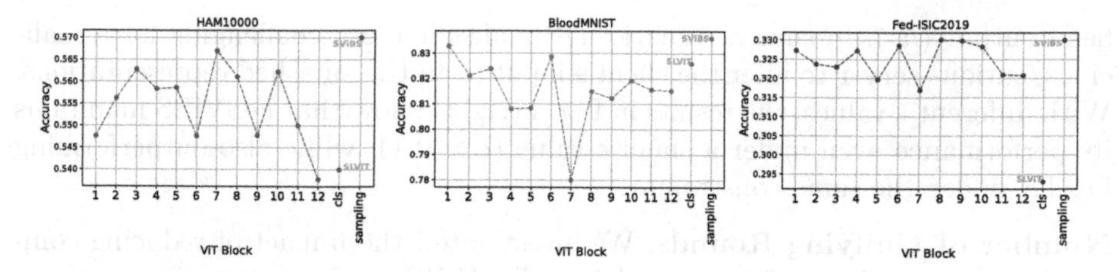

Fig. 3. Performance of each ViT block, sending *cls* token (SLViT), and SViBS. Sampling from blocks 1 to 6 (SViBS) showed better performance than individual blocks.

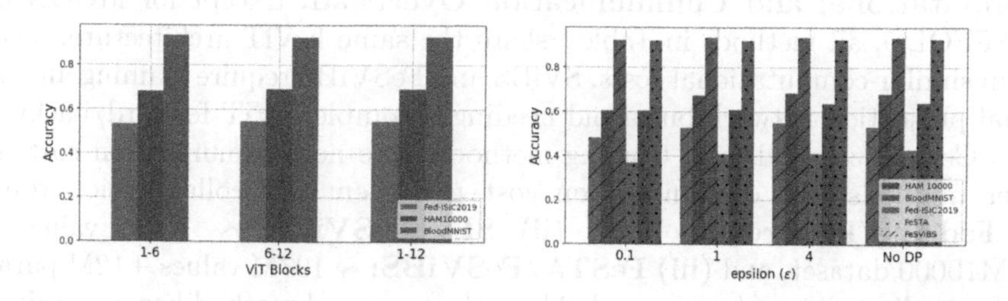

Fig. 4. FeSViBS performance with: **Left** different set of ViT blocks. **Right:** Differential Privacy with different ϵ values along with the original FeSViBS.

We investigate the impact of sampling intermediate blocks in SViBS, by analysing the individual performance of intermediate features from specific blocks during training. The results in Fig. 3 demonstrate that the majority of individual blocks outperform the vanilla split learning setting (SLViT), which is dependent on the *cls* token. On the other hand, SViBS shows dominant performance across datasets, where the sampling of ViT blocks provides augmented representations of the input images at different rounds and improves the generalizability. From Table 1, we also observe that the variance of the accuracy achieved by FeSViBS due to stochastic block sampling during inference is very low.

5 Ablation Study

Set of ViT Blocks. To study the impact of ViT blocks from which the intermediate features are sampled on the overall performance of FeSViBS, we carry out experiments choosing different sets of blocks. The results depicted in Fig. 4 (left) show consistent performance for different sets of blocks across different datasets. This indicates that implementing FeSViBS with the first 6 ViT blocks would reduce the computational cost without compromising performance.

FeSViBS with Differential Privacy. Differential Privacy (DP) [1] is a widely-used approach that aims to improve the privacy of local client's data by adding noise. We conduct experiments where we add Gaussian noise to the client's head output (h_c). In such a scenario, DP makes it more challenging for a malicious/curious server to infer the client's input from the smashed representations. With different ϵ values, the results in Fig. 4 (right) show that FeSViBS maintains its performance even under a small ϵ value ($\epsilon = 0.1$), while also outperforming FeSTA under the same constraints.

Number of Unifying Rounds. We investigated the impact of reducing communication rounds (unifying rounds) on FeSViBS performance. However, our results showed that performance was maintained even with decreasing the number of communication rounds.

Computational and Communication Overhead. Except for MOON and SCAFFOLD, all methods in Table 1 share the same h-ViT architecture, resulting in similar computational costs. SViBS and FeSViBS require training an additional projection network but avoid needing a complete ViT forward/backward pass. Centralized and local training methods have no communication cost. For other methods, the communication cost per client per collaboration round: **(i) FedAvg/FedProx:** $\sim$ 97M, **(ii) SLViT/SViBS:** $\sim$ 197M values for HAM10000 dataset, and **(iii) FeSTA/FeSViBS:** $\sim$ 197M values +12M parameters per client per unifying round. Thus, the proposed method has a marginally higher communication overload than SL and twice the communication burden as FL.

6 Conclusion and Future Directions

We proposed a novel Federated Split Learning of Vision Transformer with Block Sampling (FeSViBS), which utilizes FL, SL and sampling of ViT blocks to enhance the performance of the collaborative system. We evaluate FeSViBS framework under IID and non-IID settings on three real-world medical imaging datasets and demonstrate consistent performance. In the future, we aim to (i) extend our work and evaluate the privacy of FeSViBS under the presence of malicious clients/server, (ii) evaluate FeSViBS in the context of natural images and (iii) extend the current framework to multi-task settings.

References

1. Abadi, M., et al.: Deep learning with differential privacy. In: Proceedings of the 2016 ACM SIGSAC Conference on Computer and Communications Security, pp. 308–318 (2016)
2. Acevedo, A., Merino, A., Alférez, S., Molina, Á., Boldú, L., Rodellar, J.: A dataset of microscopic peripheral blood cell images for development of automatic recognition systems. Data Brief **30**, 1–5 (2020)
3. Ads, O.S., Alfares, M.M., Salem, M.A.M.: Multi-limb split learning for tumor classification on vertically distributed data. In: 2021 Tenth International Conference on Intelligent Computing and Information Systems (ICICIS), pp. 88–92. IEEE (2021)

4. Almalik, F., Yaqub, M., Nandakumar, K.: Self-ensembling vision transformer (SEViT) for robust medical image classification. In: Wang, L., Dou, Q., Fletcher, P.T., Speidel, S., Li, S. (eds.) Medical Image Computing and Computer Assisted Intervention. MICCAI 2022. LNCS, vol. 13433, pp. 376–386. Springer, Cham (2022). https://doi.org/10.1007/978-3-031-16437-8_36
5. Bhojanapalli, S., Chakrabarti, A., Glasner, D., Li, D., Unterthiner, T., Veit, A.: Understanding robustness of transformers for image classification. In: Proceedings of the IEEE/CVF International Conference on Computer Vision, pp. 10231–10241 (2021)
6. Codella, N.C.F., et al.: Skin lesion analysis toward melanoma detection: a challenge at the 2017 international symposium on biomedical imaging (ISBI), hosted by the international skin imaging collaboration (ISIC). In: 2018 IEEE 15th International Symposium on Biomedical Imaging (ISBI 2018), pp. 168–172 (2018). https://doi.org/10.1109/ISBI.2018.8363547
7. Combalia, M., et al.: Bcn20000: Dermoscopic lesions in the wild. arXiv:1908.02288 (2019)
8. Dai, Y., Gao, Y., Liu, F.: TransMed: transformers advance multi-modal medical image classification. Diagnostics **11**(8), 1384 (2021)
9. Dayan, I., et al.: Federated learning for predicting clinical outcomes in patients with COVID-19. Nat. Med. **27**(10), 1735–1743 (2021)
10. Gupta, O., Raskar, R.: Distributed learning of deep neural network over multiple agents. J. Netw. Comput. Appl. **116**, 1–8 (2018)
11. Ha, Y.J., Lee, G., Yoo, M., Jung, S., Yoo, S., Kim, J.: Feasibility study of multi-site split learning for privacy-preserving medical systems under data imbalance constraints in COVID-19, x-ray, and cholesterol dataset. Sci. Rep. **12**(1), 1534 (2022)
12. He, K., Zhang, X., Ren, S., Sun, J.: Deep residual learning for image recognition. In: Proceedings of the IEEE Conference on Computer Vision and Pattern Recognition, pp. 770–778 (2016)
13. Huang, Y., Gupta, S., Song, Z., Li, K., Arora, S.: Evaluating gradient inversion attacks and defenses in federated learning. Adv. Neural. Inf. Process. Syst. **34**, 7232–7241 (2021)
14. Kairouz, P., et al.: Advances and open problems in federated learning. Found. Trends ® Mach. Learn. **14**(1–2), 1–210 (2021)
15. Karimireddy, S.P., Kale, S., Mohri, M., Reddi, S., Stich, S., Suresh, A.T.: Scaffold: stochastic controlled averaging for federated learning. In: International Conference on Machine Learning, pp. 5132–5143. PMLR (2020)
16. Khan, S., Naseer, M., Hayat, M., Zamir, S.W., Khan, F.S., Shah, M.: Transformers in vision: a survey. ACM Comput. Surv. (CSUR) **54**, 1–41 (2021)
17. Kingma, D.P., Ba, J.: Adam: a method for stochastic optimization. arXiv preprint arXiv:1412.6980 (2014)
18. Li, Q., He, B., Song, D.: Model-contrastive federated learning. In: Proceedings of the IEEE/CVF Conference on Computer Vision and Pattern Recognition, pp. 10713–10722 (2021)
19. Li, T., Sahu, A.K., Zaheer, M., Sanjabi, M., Talwalkar, A., Smith, V.: Federated optimization in heterogeneous networks. Proc. Mach. Learn. Syst. **2**, 429–450 (2020)
20. McMahan, B., Moore, E., Ramage, D., Hampson, S., Arcas, B.A.: Communication-efficient learning of deep networks from decentralized data. In: Artificial Intelligence and Statistics, pp. 1273–1282. PMLR (2017)

21. Oh, S., et al.: Differentially private cutmix for split learning with vision transformer. In: First Workshop on Interpolation Regularizers and Beyond at NeurIPS 2022 (2022)
22. Park, S., Kim, G., Kim, J., Kim, B., Ye, J.: Federated split vision transformer for COVID-19 CXR diagnosis using task-agnostic training. In: 35th Conference on Neural Information Processing Systems, NeurIPS 2021, pp. 24617–24630 (2021)
23. Poirot, M.G., Vepakomma, P., Chang, K., Kalpathy-Cramer, J., Gupta, R., Raskar, R.: Split learning for collaborative deep learning in healthcare (2019). https://doi.org/10.48550/ARXIV.1912.12115, https://arxiv.org/abs/1912.12115
24. Shamshad, F., et al.: Transformers in medical imaging: a survey. arXiv preprint arXiv:2201.09873 (2022)
25. du Terrail, J.O., et al.: FLamby: datasets and benchmarks for cross-silo federated learning in realistic healthcare settings. In: Thirty-sixth Conference on Neural Information Processing Systems Datasets and Benchmarks Track (2022). https://openreview.net/forum?id=GgM5DiAb6A2
26. Tschandl, P., Rosendahl, C., Kittler, H.: The ham10000 dataset, a large collection of multi-source dermatoscopic images of common pigmented skin lesions. Sci. Data **5**(11), 180161 (2018). https://doi.org/10.1038/sdata.2018.161
27. Vepakomma, P., Gupta, O., Swedish, T., Raskar, R.: Split learning for health: Distributed deep learning without sharing raw patient data. arXiv preprint arXiv:1812.00564 (2018)
28. Wightman, R.: Pytorch image models. https://github.com/rwightman/pytorch-image-models (2019). https://doi.org/10.5281/zenodo.4414861
29. Yang, J., Shi, R., Ni, B.: MedMNIST classification decathlon: a lightweight AutoML benchmark for medical image analysis. In: IEEE 18th International Symposium on Biomedical Imaging (ISBI), pp. 191–195 (2021)
30. Zheng, S., et al.: Rethinking semantic segmentation from a sequence-to-sequence perspective with transformers. In: Proceedings of the IEEE/CVF Conference on Computer Vision and Pattern Recognition, pp. 6881–6890 (2021)

Localized Questions in Medical Visual Question Answering

Sergio Tascon-Morales(✉), Pablo Márquez-Neila, and Raphael Sznitman

University of Bern, Bern, Switzerland
{sergio.tasconmorales,pablo.marquez,raphael.sznitman}@unibe.ch

Abstract. Visual Question Answering (VQA) models aim to answer natural language questions about given images. Due to its ability to ask questions that differ from those used when training the model, medical VQA has received substantial attention in recent years. However, existing medical VQA models typically focus on answering questions that refer to an entire image rather than where the relevant content may be located in the image. Consequently, VQA models are limited in their interpretability power and the possibility to probe the model about specific image regions. This paper proposes a novel approach for medical VQA that addresses this limitation by developing a model that can answer questions about image regions while considering the context necessary to answer the questions. Our experimental results demonstrate the effectiveness of our proposed model, outperforming existing methods on three datasets. Our code and data are available at https://github.com/sergiotasconmorales/locvqa.

Keywords: VQA · Attention · Localized Questions

1 Introduction

Visual Question Answering (VQA) models are neural networks that answer natural language questions about an image [2,8,12,21]. The capability of VQA models to interpret natural language questions is of great appeal, as the range of possible questions that can be asked is vast and can differ from those used to train the models. This has led to many proposed VQA models for medical applications in recent years [7,9,14,16,23–25]. These models can enable clinicians to probe the model with nuanced questions, thus helping to build confidence in its predictions.

Recent work on medical VQA has primarily focused on building more effective model architectures [7,20,23] or developing strategies to overcome limitations in medical VQA datasets [4,15,18,19,23]. Another emerging trend is to enhance VQA performance by addressing the consistency of answers produced [22], particularly when considering entailment questions (*i.e.*, the answer to "Is the image

Supplementary Information The online version contains supplementary material available at https://doi.org/10.1007/978-3-031-43895-0_34.

H. Greenspan et al. (Eds.): MICCAI 2023, LNCS 14221, pp. 361–370, 2023.
https://doi.org/10.1007/978-3-031-43895-0_34

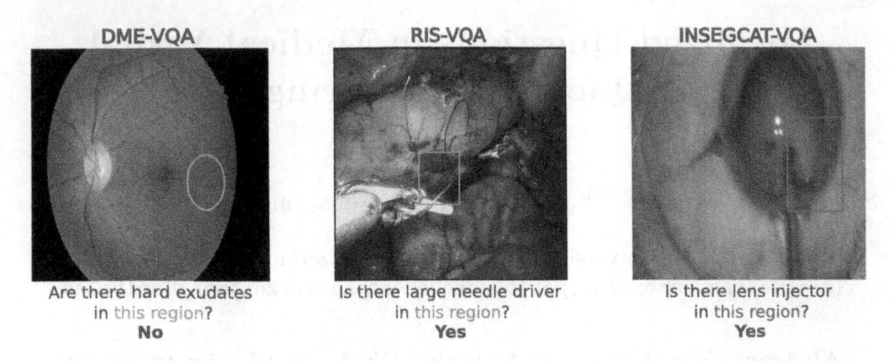

Fig. 1. Examples of localized questions. In some cases (RIS-VQA and INSEGCAT-VQA), the object mentioned in the question is only partially present in the region. We hypothesize that context can play an important role in answering such questions.

that of a healthy subject?" should be consistent with the answer to "Is there a fracture in the tibia?"). Despite these recent advances, however, most VQA models restrict to questions that consider the entire image at a time. Specifically, VQA typically uses questions that address content within an image without specifying where this content may or may not be in the image. Yet the ability to ask specific questions about regions or locations of the image would be highly beneficial to any user as it would allow fine-grained questions and model probing. For instance, Fig. 1 illustrates examples of such *localized questions* that combine content and spatial specifications. In the medical field, posing localized questions can significantly enhance the diagnostic process by providing second opinions to medical experts about suspicious regions. Additionally, this approach can improve trustworthiness by assessing the consistency between answers to both global and localized questions.

To this day, few works have addressed the ability to include location information in VQA models. In [17], localization information is posed in questions by constraining the spatial extent to a point within bounding boxes yielded by an object detector. The model then focuses its attention on objects close to this point. However, the method was developed for natural images and relies heavily on the object detector to limit the attention extent, making it difficult to scale in medical imaging applications. Alternatively, the approach from [23] answers questions about a pre-defined coarse grid of regions by directly including region information into the question (*e.g.*, "Is grasper in (0,0) to (32,32)?"). This method relies on the ability of the model to learn a spatial mapping of the image and limits the regions to be on a fixed grid. Localized questions were also considered in [22], but the region of interest was cropped before being presented to the model, assuming that the surrounding context is irrelevant for answering this type of question.

To overcome these limitations, we propose a novel VQA architecture that alleviates the mentioned issues. At its core, we hypothesize that by allowing the VQA model to access the entire images and properly encoding the region of interest, this model can be more effective at answering questions about regions. To achieve this, we propose using a multi-glimpse attention mechanism [3,22,23] restricting its focus range to the region in question, but only after the model has considered the entire image. By doing so, we preserve contextual information about the question and its region. We evaluate the effectiveness of our approach by conducting extensive experiments on three datasets and comparing our method to state-of-the-art baselines. Our results demonstrate performance improvements across all datasets.

2 Method

Our method extends a VQA model to answer localized questions. We define a *localized question* for an image $\mathbf{x}$ as a tuple $(\mathbf{q}, \mathbf{m})$, where $\mathbf{q}$ is a question, and $\mathbf{m}$ is a binary mask of the same size as $\mathbf{x}$ that identifies the region to which the question pertains. Our VQA model p_θ, depicted in Fig. 2, accepts an image and a localized question as input and produces a probability distribution over a finite set $\mathcal{A}$ of possible answers. The final answer of the model $\hat{a}$ is the element with the highest probability,

$$\hat{a} = \arg\max_{a \in \mathcal{A}} p_\theta(a \mid \mathbf{q}, \mathbf{x}, \mathbf{m}). \quad (1)$$

The model proceeds in three stages to produce its prediction: input embedding, localized attention, and final classification.

Input Embedding. The question $\mathbf{q}$ is first processed by an LSTM [11] to produce an embedding $\hat{\mathbf{q}} \in \mathbb{R}^Q$. Similarly, the image $\mathbf{x}$ is processed by a ResNet-152 [10] to produce the feature map $\hat{\mathbf{x}} \in \mathbb{R}^{C \times H \times W}$.

Localized Attention. An attention mechanism uses the embedding to determine relevant parts of the image to answer the corresponding question. Unlike previous attention methods, we include the region information that the mask defines. Our *localized attention* module (Fig. 2 right) uses both descriptors and the mask to produce multiple weighted versions of the image feature map, $\hat{\mathbf{x}}' = \text{att}(\hat{\mathbf{q}}, \hat{\mathbf{x}}, \mathbf{m})$. To do so, the module first computes an attention map $\mathbf{g} \in \mathbb{R}^{G \times H \times W}$ with G glimpses by applying unmasked attention [13,23] to the image feature map and the text descriptor. The value of the attention map at location (h, w) is computed as,

$$\mathbf{g}_{:hw} = \text{softmax}\left(\mathbf{W}^{(g)} \cdot \text{ReLU}\left(\mathbf{W}^{(x)}\hat{\mathbf{x}}_{:hw} \odot \mathbf{W}^{(q)}\hat{\mathbf{q}}\right)\right), \quad (2)$$

where the index $:hw$ indicates the feature vector at location (h, w), $\mathbf{W}^{(x)} \in \mathbb{R}^{C' \times C}$, $\mathbf{W}^{(q)} \in \mathbb{R}^{C' \times Q}$, and $\mathbf{W}^{(g)} \in \mathbb{R}^{G \times C'}$ are learnable parameters of linear transformations, and $\odot$ is the element-wise product. In practice, the transformations $\mathbf{W}^{(x)}$ and $\mathbf{W}^{(g)}$ are implemented with 1×1 convolutions and all linear transformations include a dropout layer applied to its input. The image feature maps $\hat{\mathbf{x}}$ are then weighted with the attention map and masked with $\mathbf{m}$ as,

$$\hat{\mathbf{x}}'_{cghw} = \mathbf{g}_{ghw} \cdot \hat{\mathbf{x}}_{chw} \cdot (\mathbf{m} \downarrow_{H \times W})_{hw}, \quad (3)$$

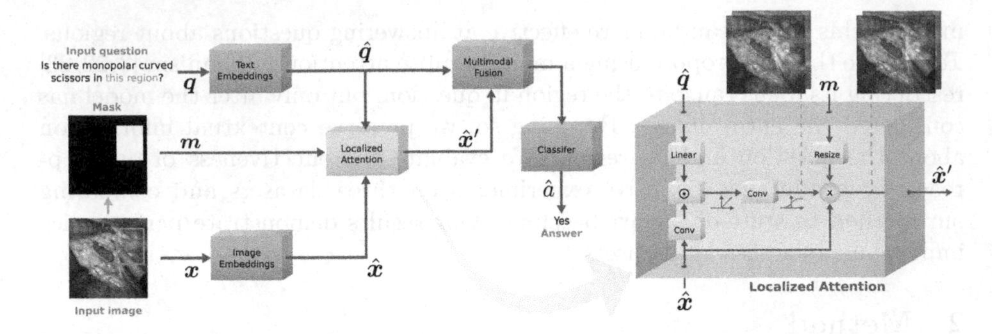

Fig. 2. Left: Proposed VQA architecture for localized questions. The Localized Attention module allows the region information to be integrated into the VQA while considering the context necessary to answer the question. **Right:** Localized Attention module.

where c and g are the indexes over feature channels and glimpses, respectively, (h, w) is the index over the spatial dimensions, and $\mathbf{m}\downarrow_{H\times W}$ denotes a binary downsampled version of $\mathbf{m}$ with the spatial size of $\hat{\mathbf{x}}$. This design allows the localized attention module to compute the attention maps using the full information available in the image, thereby incorporating context into them before being masked to constrain the answer to the specified region.

Classification. The question descriptor $\hat{\mathbf{q}}$ and the weighted feature maps $\hat{\mathbf{x}}'$ from the localized attention are vectorized and concatenated into a single vector of size $C \cdot G + Q$ and then processed by a multi-layer perceptron classifier to produce the final probabilities.

Training. The training procedure minimizes the standard cross-entropy loss over the training set updating the parameters of the LSTM encoder, localized attention module, and the final classifier. The training set consists of triplets of images, localized questions, and the corresponding ground-truth answers. As in [2], the ResNet weights are fixed with pre-trained values, and the LSTM weights are updated during training.

3 Experiments and Results

We compare our model to several baselines across three datasets and report quantitative and qualitative results. Additional results are available in the supplementary material.

3.1 Datasets

We evaluate our method on three datasets containing questions about regions which we detail here. The first dataset consists of an existing retinal fundus VQA dataset with questions about the image's regions and the entire image. The second and third datasets are generated from public segmentation datasets

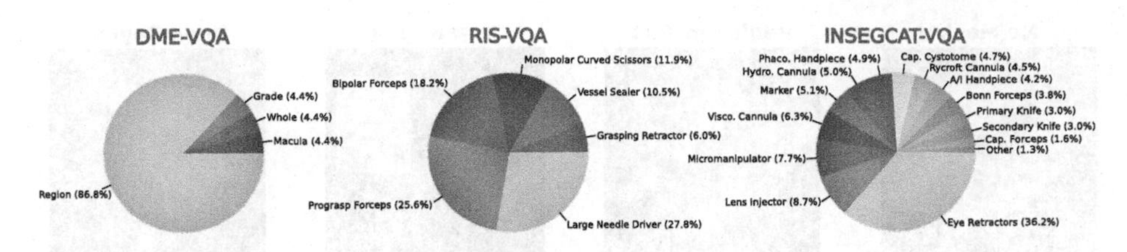

Fig. 3. Distribution by question type (DME-VQA) and by question object (RIS-VQA and INSEGCAT-VQA).

but use the method described in [23] to generate a VQA version with region questions.

DME-VQA [22]. 679 fundus images containing questions about entire images (*e.g.*, "what is the DME risk grade?") and about randomly generated circular regions (*e.g.*, "are there hard exudates in this region?"). The dataset comprises 9'779 question-answer (QA) pairs for training, 2'380 for validation, and 1'311 for testing.

RIS-VQA. Images from the 2017 Robotic Instrument Segmentation dataset [1]. We automatically generated binary questions with the structure "is there [instrument] in this region?" and corresponding masks as rectangular regions with random locations and sizes. Based on the ground-truth label maps, the binary answers were labeled "yes" if the region contained at least one pixel of the corresponding instrument and "no" otherwise. The questions were balanced to maintain the same amount of "yes" and "no" answers. 15'580 QA pairs from 1'423 images were used for training, 3'930 from 355 images for validation, and 13'052 from 1'200 images for testing.

INSEGCAT-VQA. Frames of cataract surgery videos from the InSegCat 2 dataset [5]. We followed the same procedure as in RIS-VQA to generate balanced binary questions with masks and answers. The dataset consists of 29'380 QA pairs from 3'519 images for training, 5'306 from 536 images for validation, and 4'322 from 592 images for testing.

Figure 3 shows the distribution of questions in the three datasets.

3.2 Baselines and Metrics

We compare our method to four different baselines, as shown in Fig. 4:

No mask: no information is provided about the region in the question.
Region in text [23]: region information is included as text in the question.
Crop region [22]: image is masked to show only the queried region, with the area outside the region set to zero.
Draw region: region is indicated by drawing its boundary on the input image with a distinctive color.

We evaluated the performance of our method using accuracy for the DME-VQA dataset and the area under the receiver operating characteristic (ROC) curve and Average Precision (AP) for the RIS-VQA and INSEGCAT-VQA datasets.

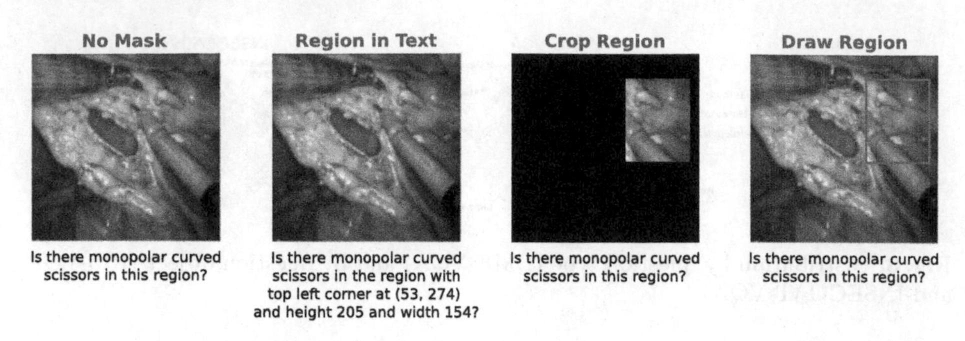

Fig. 4. Illustration of evaluated baselines for an example image.

Implementation Details: Our VQA architecture uses an LSTM [11] with an output dimension 1024 to encode the question and a word embedding size of 300. We use the ResNet-152 [10] with ImageNet weights to encode images of size 448×448, generating feature maps with 2048 channels. In the localized attention block, the visual and textual features are projected into a 512-dimensional space before being combined by element-wise multiplication. Following [6,13], the number of glimpses is set to $G = 2$ for all experiments. The classification block is a multi-layer perceptron with a hidden layer of 1024 dimensions. A dropout rate of 0.25 and ReLU activation are used in the localized attention and classifier blocks.

We train our models for 100 epochs using an early stopping condition with patience of 20 epochs. Data augmentation consists of horizontal flips. We use a batch size of 64 samples and the Adam optimizer with a learning rate of 10^{-4}, which is reduced by a factor of 0.1 when learning stagnates. Models implemented in PyTorch 1.13.1 and trained on an Nvidia RTX 3090 graphics card.

3.3 Results

Our method outperformed all considered baselines on the DME-VQA (Table 1), the RIS-VQA, and the INSEGCAT-VQA datasets (Table 2), highlighting the importance of contextual information in answering localized questions. Context proved to be particularly critical in distinguishing between objects of similar appearance, such as the bipolar and prograsp forceps in RIS-VQA, where our method led to an 8 percent point performance improvement (Table 3). In contrast, the importance of context was reduced when dealing with visually distinct objects, resulting in smaller performance gains as observed in the INSEGCAT-VQA dataset. For example, despite not incorporating contextual information, the baseline *crop region* still benefited from correlations between the location of the region and the instrument mentioned in the question (*e.g.*, the eye retractor typically appears at the top or the bottom of the image), enabling it to achieve competitive performance levels that are less than 2 percent points lower than our method (Table 2, bottom).

Table 1. Average accuracy for different methods on the DME-VQA dataset. The results shown are the average of 5 models trained with different seeds.

Method	Accuracy (%)				
	Overall	Grade	Whole	Macula	Region
No Mask	61.1 ± 0.4	80.0 ± 3.7	85.7 ± 1.2	**84.3 ± 0.5**	57.6 ± 0.4
Region in Text [23]	60.0 ± 1.5	57.9 ± 12.5	85.1 ± 1.9	83.2 ± 2.4	57.7 ± 1.0
Crop Region [22]	81.4 ± 0.3	78.7 ± 1.3	81.3 ± 1.7	82.3 ± 1.4	81.5 ± 0.3
Draw Region	83.0 ± 1.0	79.6 ± 2.5	77.0 ± 4.8	84.0 ± 1.9	83.5 ± 1.0
Ours	**84.2 ± 0.6**	**82.8 ± 0.4**	**87.0 ± 1.2**	83.0 ± 1.5	**84.2 ± 0.7**

Table 2. Average test AUC and AP for different methods on the RIS-VQA and INSEGCAT-VQA datasets. The results shown are the average over 5 seeds.

Dataset	Method	AUC	AP
RIS-VQA	No Mask	0.500 ± 0.000	0.500 ± 0.000
	Region in Text [23]	0.677 ± 0.002	0.655 ± 0.003
	Crop Region [22]	0.842 ± 0.002	0.831 ± 0.002
	Draw Region	0.835 ± 0.003	0.829 ± 0.003
	Ours	**0.885 ± 0.003**	**0.885 ± 0.003**
INSEGCAT-VQA	No Mask	0.500 ± 0.000	0.500 ± 0.000
	Region in Text [23]	0.801 ± 0.012	0.793 ± 0.014
	Crop Region [22]	0.901 ± 0.002	0.891 ± 0.003
	Draw Region	0.910 ± 0.003	0.907 ± 0.005
	Ours	**0.914 ± 0.002**	**0.915 ± 0.002**

Similar to our method, the baseline *draw region* incorporates contextual information when answering localized questions. However, we observed that drawing regions on the image can interfere with the computation of guided attention maps, leading to incorrect predictions (Fig. 5, column 4). In addition, the lack of masking of the attention maps often led the model to wrongly consider areas beyond the region of interest while answering questions (Fig. 5, column 1).

When analyzing mistakes made by our model, we observe that they tend to occur when objects or background structures in the image look similar to the object mentioned in the question (Fig. 5, column 3). Similarly, false predictions were observed when only a few pixels of the object mentioned in the question were present in the region.

Table 3. Average test AUC for different methods on the RIS-VQA dataset as a function of instrument type. Results are averaged over 5 models trained with different seeds. The corresponding table for INSEGCAT-VQA is available in the Supplementary Materials.

Method	Instrument Type					
	Large Needle Driver	Monopolar Curved Scissors	Vessel Sealer	Grasping Retractor	Prograsp Forceps	Bipolar Forceps
No Mask	0.500 ±0	0.500 ±0	0.500 ±0	0.500 ±0	0.500 ±0	0.500 ±0
Region in Text [23]	0.717 ±0.003	0.674 ±0.001	0.620 ±0.011	0.616 ±0.020	0.647 ±0.008	0.645 ±0.003
Crop Region [22]	0.913 ±0.002	0.812 ±0.003	0.752 ±0.009	0.715 ±0.015	0.773 ±0.003	0.798 ±0.004
Draw Region	0.915 ±0.003	0.777 ±0.003	0.783 ±0.004	0.709 ±0.012	0.755 ±0.004	0.805 ±0.005
Ours	**0.944 ±0.001**	**0.837 ±0.005**	**0.872 ±0.008**	**0.720 ±0.031**	**0.834 ±0.006**	**0.880 ±0.003**

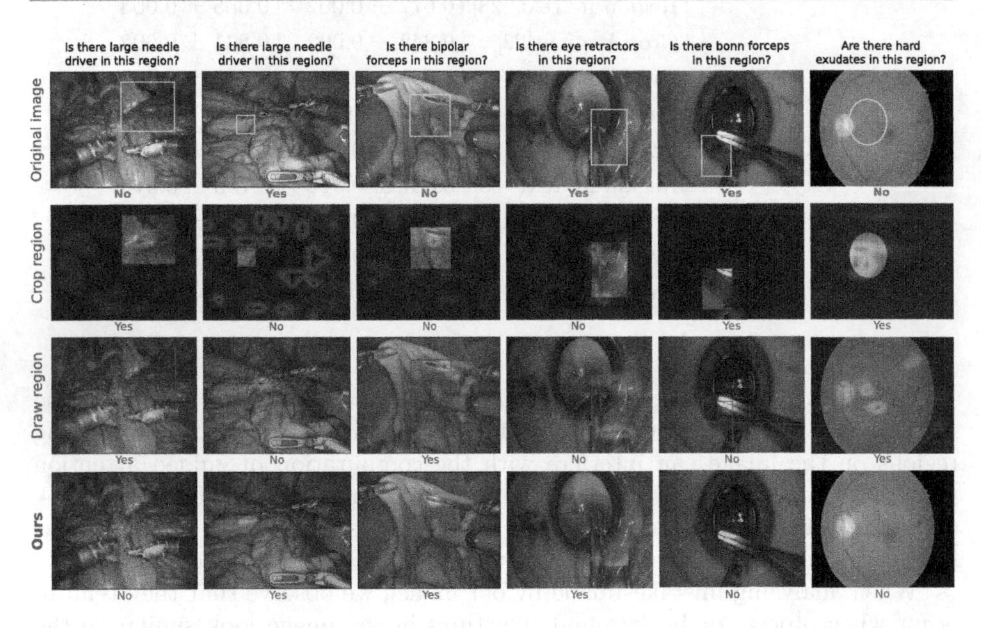

Fig. 5. Qualitative examples on the RIS-VQA dataset (columns 1–3), INSEGCAT-VQA (columns 4–5), and DME-VQA (last column). Only the strongest baselines were considered in this comparison. The first row shows the image, the region, and the ground truth answer. Other rows show the overlaid attention maps and the answers produced by each model. Wrong answers are shown in red. Additional examples are available in the Supplementary Materials.

4 Conclusions

In this paper, we proposed a novel VQA architecture to answer questions about regions. We compare the performance of our approach against several baselines and across three different datasets. By focusing the model's attention on the region after considering the evidence in the full image, we show how our method brings improvements, especially when the complete image context is required to answer the questions. Future works include studying the agreement between answers to questions about concentric regions, as well as the agreement between questions about images and regions.

Acknowledgments. This work was partially funded by the Swiss National Science Foundation through grant 191983.

References

1. Allan, M., et al.: 2017 robotic instrument segmentation challenge. arXiv preprint arXiv:1902.06426 (2019)
2. Antol, S., et al.: VQA: visual question answering. In: Proceedings of the IEEE International Conference on Computer Vision, pp. 2425–2433 (2015)
3. Ben-Younes, H., Cadene, R., Cord, M., Thome, N.: MUTAN: multimodal tucker fusion for visual question answering. In: Proceedings of the IEEE International Conference on Computer Vision, pp. 2612–2620 (2017)
4. Do, T., Nguyen, B.X., Tjiputra, E., Tran, M., Tran, Q.D., Nguyen, A.: Multiple meta-model quantifying for medical visual question answering. In: de Bruijne, M., et al. (eds.) MICCAI 2021. LNCS, vol. 12905, pp. 64–74. Springer, Cham (2021). https://doi.org/10.1007/978-3-030-87240-3_7
5. Fox, M., Taschwer, M., Schoeffmann, K.: Pixel-based tool segmentation in cataract surgery videos with mask R-CNN. In: de Herrera, A.G.S., González, A.R., Santosh, K.C., Temesgen, Z., Kane, B., Soda, P. (eds.) 33rd IEEE International Symposium on Computer-Based Medical Systems, CBMS 2020, Rochester, MN, USA, July 28–30, 2020, pp. 565–568. IEEE (2020). https://doi.org/10.1109/CBMS49503.2020.00112
6. Fukui, A., Park, D.H., Yang, D., Rohrbach, A., Darrell, T., Rohrbach, M.: Multimodal compact bilinear pooling for visual question answering and visual grounding. arXiv preprint arXiv:1606.01847 (2016)
7. Gong, H., Chen, G., Liu, S., Yu, Y., Li, G.: Cross-modal self-attention with multi-task pre-training for medical visual question answering. In: Proceedings of the 2021 International Conference on Multimedia Retrieval, pp. 456–460 (2021)
8. Goyal, Y., Khot, T., Summers-Stay, D., Batra, D., Parikh, D.: Making the v in VQA matter: elevating the role of image understanding in visual question answering. In: Proceedings of the IEEE Conference on Computer Vision and Pattern Recognition, pp. 6904–6913 (2017)
9. Hasan, S.A., Ling, Y., Farri, O., Liu, J., Lungren, M., Müller, H.: Overview of the ImageCLEF 2018 medical domain visual question answering task. In: CLEF2018 Working Notes. CEUR Workshop Proceedings, CEUR-WS.org <http://ceur-ws.org>, Avignon, France, 10-14 September 2018

10. He, K., Zhang, X., Ren, S., Sun, J.: Deep residual learning for image recognition. In: Proceedings of the IEEE Conference on Computer Vision and Pattern Recognition, pp. 770–778 (2016)
11. Hochreiter, S., Schmidhuber, J.: Long short-term memory. Neural Comput. **9**(8), 1735–1780 (1997)
12. Hudson, D.A., Manning, C.D.: GQA: a new dataset for compositional question answering over real-world images. arXiv preprint arXiv:1902.09506 3(8) (2019)
13. Kim, J.H., On, K.W., Lim, W., Kim, J., Ha, J.W., Zhang, B.T.: Hadamard product for low-rank bilinear pooling. arXiv preprint arXiv:1610.04325 (2016)
14. Liao, Z., Wu, Q., Shen, C., Van Den Hengel, A., Verjans, J.: AIML at VQA-Med 2020: knowledge inference via a skeleton-based sentence mapping approach for medical domain visual question answering (2020)
15. Liu, B., Zhan, L.M., Xu, L., Ma, L., Yang, Y., Wu, X.M.: Slake: a semantically-labeled knowledge-enhanced dataset for medical visual question answering. In: 2021 IEEE 18th International Symposium on Biomedical Imaging (ISBI), pp. 1650–1654. IEEE (2021)
16. Liu, F., Peng, Y., Rosen, M.P.: An effective deep transfer learning and information fusion framework for medical visual question answering. In: Crestani, F., et al. (eds.) CLEF 2019. LNCS, vol. 11696, pp. 238–247. Springer, Cham (2019). https://doi.org/10.1007/978-3-030-28577-7_20
17. Mani, A., Yoo, N., Hinthorn, W., Russakovsky, O.: Point and ask: incorporating pointing into visual question answering. arXiv preprint arXiv:2011.13681 (2020)
18. Nguyen, B.D., Do, T.-T., Nguyen, B.X., Do, T., Tjiputra, E., Tran, Q.D.: Overcoming data limitation in medical visual question answering. In: Shen, D., et al. (eds.) MICCAI 2019. LNCS, vol. 11767, pp. 522–530. Springer, Cham (2019). https://doi.org/10.1007/978-3-030-32251-9_57
19. Pelka, O., Koitka, S., Rückert, J., Nensa, F., Friedrich, C.M.: Radiology objects in COntext (ROCO): a multimodal image dataset. In: Stoyanov, D., et al. (eds.) LABELS/CVII/STENT -2018. LNCS, vol. 11043, pp. 180–189. Springer, Cham (2018). https://doi.org/10.1007/978-3-030-01364-6_20
20. Ren, F., Zhou, Y.: CGMVQA: a new classification and generative model for medical visual question answering. IEEE Access **8**, 50626–50636 (2020)
21. Tan, H., Bansal, M.: LXMERT: learning cross-modality encoder representations from transformers. arXiv preprint arXiv:1908.07490 (2019)
22. Tascon-Morales, S., Márquez-Neila, P., Sznitman, R.: Consistency-preserving visual question answering in medical imaging. In: Wang, L., Dou, Q., Fletcher, P.T., Speidel, S., Li, S. (eds.) Medical Image Computing and Computer Assisted Intervention. MICCAI 2022. LNCS, vol. 13438, pp. pp. 386–395. Springer, Cham (2022). https://doi.org/10.1007/978-3-031-16452-1_37
23. Vu, M.H., Löfstedt, T., Nyholm, T., Sznitman, R.: A question-centric model for visual question answering in medical imaging. IEEE Trans. Med. Imaging **39**(9), 2856–2868 (2020)
24. Yu, Y., Li, H., Shi, H., Li, L., Xiao, J.: Question-guided feature pyramid network for medical visual question answering. Expert Syst. Appl. **214**, 119148 (2023)
25. Zhan, L.M., Liu, B., Fan, L., Chen, J., Wu, X.M.: Medical visual question answering via conditional reasoning. In: Proceedings of the 28th ACM International Conference on Multimedia, pp. 2345–2354 (2020)

Reconstructing the Hemodynamic Response Function via a Bimodal Transformer

Yoni Choukroun[1(✉)], Lior Golgher[2], Pablo Blinder[3,4], and Lior Wolf[1]

[1] The School of Computer Science, Tel Aviv University, Tel Aviv-Yafo, Israel
choukroun.yoni@gmail.com

[2] The Edmond and Lily Safra Center for Brain Sciences, The Hebrew University of Jerusalem, Jerusalem, Israel

[3] Neurobiology, Biochemistry and Biophysics School, Wise Life Science Faculty, Tel Aviv University, Tel Aviv-Yafo, Israel

[4] The Sagol School for Neuroscience, Tel Aviv University, Tel Aviv-Yafo, Israel

Abstract. The relationship between blood flow and neuronal activity is widely recognized, with blood flow frequently serving as a surrogate for neuronal activity in fMRI studies. At the microscopic level, neuronal activity has been shown to influence blood flow in nearby blood vessels. This study introduces the first predictive model that addresses this issue directly at the explicit neuronal population level. Using in vivo recordings in awake mice, we employ a novel spatiotemporal bimodal transformer architecture to infer current blood flow based on both historical blood flow and ongoing spontaneous neuronal activity. Our findings indicate that incorporating neuronal activity significantly enhances the model's ability to predict blood flow values. Through analysis of the model's behavior, we propose hypotheses regarding the largely unexplored nature of the hemodynamic response to neuronal activity.

Keywords: Hemodynamic Response Function · Bimodal transformers

1 Introduction

The brain consumes copious amounts of energy to sustain its activity, resulting in a skewed energetic budget per mass compared to the rest of the body (about 25% utilized by about 3%, see [2,16] for an elaborate review of energy utilization). Given this disproportionate need, resources are allocated on a need-basis: active areas signal to the nearby blood vessel to dilate and increase blood flow, bringing a surplus of resources, to that area. This fundamental physiological process is called neurovascular coupling. It is non-trivial to model and different types of neuronal activity have been shown to elicit opposite vascular responses.

Supplementary Information The online version contains supplementary material available at https://doi.org/10.1007/978-3-031-43895-0_35.

H. Greenspan et al. (Eds.): MICCAI 2023, LNCS 14221, pp. 371–381, 2023.
https://doi.org/10.1007/978-3-031-43895-0_35

Neurovascular coupling is a cornerstone of proper brain function and also underpins the ability to observe and study the human brain in action. Imaging methods based on blood oxygenated level dependent (BOLD) approaches rely on it [13], as do methods that are based on rheological properties, such as blood volume and flow speed. Since these methods do not directly measure neuronal activity per-se, but a physiological proxy, i.e. the resulting change in vascular dynamics and oxygen levels, it is of utmost importance to know the precise transform function linking neuronal activity to the observed vascular dynamics. Given the differential response to neuronal activity (see [5] for a timely review), obtaining a cellular and population level hemodynamic response function (HRF) remains an unmet need in this field, that would finally unlock the ability to infer neuronal activity directly from blood flow dynamics [20].

The initial characterization of the hemodynamic response function (HRF) was performed at the *system* level, where *system* refers to large cortical regions encompassing tens of thousands of neurons of different types, without taking into account the fine details of different vascular compartments (see [29] for a succinct review on the original works). At this level, a canonical response function was derived from extensive work on sensory-evoked somatosensory responses. This HRF has become widely accepted and used in the interpretation of BOLD signals. This function consists of three components: an initial dip (its existence and physiological origin are much debated), a prolonged and very pronounced overshoot, followed by a shallower and much shorter undershoot. The initial dip occurs within one second of the sensory stimulus, the overshoot peaks around five seconds later, overshoot and return to baseline level occurs within 15–20 s post stimuli. It should be noted that vascular reactivity is much faster than the collective behavior described by the canonical HRF, with reports showing sensory-evoked vascular responses observed after just 300ms. Recently, more advanced imaging and analysis methods have pushed the formulation of an HRF at the single cell to single blood vessel (capillary) level, pointing to a rather narrow family of possible functions. Importantly, this work also established that the HRF derived at the microscopic level can be partially translated to macroscopic imaging approaches. Nevertheless, single neuron to single vessel responses fail to capture the more complex and varied neuronal population level responses that could be integrated across the extensive vascular network that surrounds them. Here, we exploit a unique dataset, in which neuronal and vascular responses (changes in diameter) were recorded in a volumetric fashion and with relevant temporal resolution, allowing us to establish a novel pipeline to uncover/formulate a many-to-many HRF.

Our model needs to combine neuron firing and blood vessel data and employs a multi-modal transformer. There are three types of multi-modal transformers: (i) a multi-modal Transformer where the two modalities are concatenated and separated by the `[SEP]` token [18,19], and self-attention is used, (ii) co-attention-based model modules that contextualize each modality with the other modality [21,26], and (iii) generative models containing an encoder that uses self-attention on the input and a decoder that uses both the encoded data and data from the decoder's domain as inputs [3,17,23,30–32,34]. Our model is of the third type and presents two distinctive properties: pulling from multiple time points and an attention mechanism that is modulated based on distance.

Our results show that the new transformer model can predict the state of blood vessels better than the baseline models. The utility of neuronal data in the prediction is demonstrated by an ablation study. By analyzing the learned model, we obtain insights into the link between neuronal and vascular activities.

2 Data

All procedures were approved by the Ethics Committee of Tel Aviv University for Animal Use and Welfare and followed pertinent Institutional Animal Care and Use Committee (IACUC) and local guidelines. Neuronal activity was monitored in female C57BL/6J transgenic mice expressing Thy1-GCaMP6s. Vascular dynamics were tracked using a Texas Red fluorescent dye, which was conjugated to a large polysaccharide moiety (2 mega Dalton dextran) and retro-orbitally bolus injected under brief isoflurane sedation at the beginning of the imaging day.

425 quasi-linear vascular segments and 50 putative neuronal cell bodies were manually labeled within a volume of $490 \times 500 \times 300 \mu m^3$, which was continuously imaged across two consecutive 1850-second long sessions at an imaging rate of 30.03 volumes per second [9–12]. For neuronal activity estimation, we selected a cuboid volume of interest around each neuronal cell body and summed the fluorescence within it following an axial intensity normalization corresponding to an uneven duty cycle of our varifocal lens.

For vascular diameter estimation we used the Radon transform, [1,6–9,22,24, 28] as its resilience to rotation and poor contrast are particularly useful for our application. Specifically, Gao and Drew have formerly found that thresholding the vascular intensity profile in Radon space is more resilient to noise than other thresholding methods [8]. Based on their observation, we used the time-collapsed imagery to determine a threshold in Radon space, which was then applied separately for each frame in time.

This unique ability to rapidly track neuronal and vascular interactions across a continuous brain volume bears several important advantages [9,10]. In particular, a greater proportion of the vascular ensemble that reacts to a given neuronal metabolic demand can be accounted for.

3 Method

The HRF learning problem explored in this work is defined as the prediction of current blood flow rates at different vessel segments, given the previous neuronal spikes as well as previous blood flow rates. We propose to design a parameterized deep neural network f_θ for scalar regression of blood flow rates at different vessel segments, such that at a given time t we have

$$f_\theta(S_t, F_t, X_S, X_F) \rightarrow \mathbb{R}^m \tag{1}$$

where the matrix $S_t \in \mathbb{R}^{t_s \times n}$ denotes the n neurons' spikes at the t_s previous samples, while the matrix $F_t \in \mathbb{R}^{t_v \times m}$ denotes the blood flow of the m vessel segment at the previous t_v time samples. $X_S \in \mathbb{R}^{n\times 3}$ and $X_F \in \mathbb{R}^{m\times 3}$ are the three-dimensional positions of the neurons and vessel segments, respectively.

HRF predictions should satisfy fundamental symmetries and invariance of physiological priors and of experimental bias, such as invariance to rigid spatial transformation (rotation and translation). Therefore, a positional input X_u is transformed to inter-elements Euclidean distances $D_u = \{d^u_{ij}\}_{i,j}$ where $d^u_{ij} = \|(X_u)_i - (X_u)_j\|_2$ for rigid transform invariance.

Thus, the learning problem is refined as $f_\theta : \{S_t, F_t, D_S, D_F, D_{SF}\} \rightarrow \mathbb{R}^m$, where D_S, D_F, D_{SF} represents the Euclidean distance matrix between neurons, vessel segments, and neurons to vessel segments, respectively. We do not include any further auxiliary features or prior in the input.

We model f_θ using a new variant of the Transformer family. The proposed model consists of an encoder and a decoder. The encoder embeds the neurons at *both* spatial and temporal levels. The decoder predicts vessel segment flow by utilizing *both* the past flow values and the spatial information of the vessel segments, along with the neuronal activity via the cross-attention mechanism.

Transformers. The self-attention mechanism introduced by Transformers [30] is based on a trainable associative memory with (key, value) vector pairs, where a query vector $q \in \mathbb{R}^d$ is matched against a set of k key vectors using scaled inner products, as follows

$$A(Q, K, V) = \text{Softmax}\left(\frac{QK^T}{\sqrt{d}}\right)V, \tag{2}$$

where $Q \in \mathbb{R}^{N\times d}$, $K \in \mathbb{R}^{k\times d}$ and $V \in \mathbb{R}^{k\times d}$ represent the packed N queries, k keys and values tensors respectively. Keys, queries and values are obtained using linear transformations of the sequence's elements. A multi-head self-attention layer is defined by extending the self-attention mechanism using h attention *heads*, i.e. h self-attention functions applied to the input, reprojected to values via a $dh \times D$ linear layer.

Neuronal Encoding. To obtain the initial Spatio-Temporal Encoding, for the prediction at time t, we project each neuron to a high d dimensional embedding $\phi^s_t \in \mathbb{R}^{t_s\times n\times d}$ by modulating it with its spike value such that $\phi^s_t = S_t \odot (1_d W^T)$, where $W \in \mathbb{R}^d$ denotes the neuronal encoding. The embedding is modulated by the magnitude of the spike, such that higher neuronal activities are projected farther in the embedding space.

The *temporal encoding* is defined using sinusoidal encoding [30] applied on ϕ and augmented with a learnable embedding such that $\phi^s_t \leftarrow \phi^s_t + p_t \cdot \tilde{p}$ where p_t and $\tilde{p}$ represent the sinusoidal time encoding and the learned vector, respectively. We emphasize the fact that, contrary to traditional transformers, the embedding tensor ϕ_t has an additional spatial dimension such that the tensor is three-dimensional, enabling both spatial and temporal attention.

In order to incorporate the spatial information of the neurons, we propose to insert *spatial encoding* by importing the pairwise information directly into the self-attention layer. For this, we multiply the distance relation by the similarity tensor as follows

$$A^S(Q, K, D_S) = \text{Softmax}\left(\frac{QK^T}{\sqrt{d}}\right) \odot \psi_S(D_S), \tag{3}$$

with $\odot$ denoting the Hadamard product, and $\psi_S(D_S) : \mathbb{R}^+ \to \mathbb{R}^+$ an element-wise learnable parameterized similarity function. This way, the similarity function scales the self-attention map according to the distance between the elements (in our case the neurons).

Vascular Decoding. The spatio-temporal encoding of the vascular data is similar to the embedding performed by the encoder. The information on each vascular segment is embedded in a high-dimensional vector $\phi_t^F \in \mathbb{R}^{t_v \times m \times d}$ to be further projected by the temporal encoding. The spatial geometric information is incorporated via the pairwise vascular segments' distance matrix D_F via the decoder's self-attention module A^F.

The most important element of the decoder is the cross-attention module, which incorporates neuronal information for vascular prediction. Given the final neuronal embeddings ϕ_t^s, the cross-attention module performs cross-analysis of the neuronal embeddings such that

$$A^{SF}(Q_F, K_S, D_S) = \text{Softmax}\left(\frac{Q_F K_S^T}{\sqrt{d}}\right) \odot \psi_{SF}(D_{SF}), \tag{4}$$

where Q_F and K_S represent the affine transform of ϕ_t^F and ϕ_t^s, respectively. Here also, the (non-square) cross-attention map is modulated by the neuron-vessel distance matrix D_{SF}.

The spatio-temporal map is of dimensions $A^{SF} \in \mathbb{R}^{t_v \times t_s \times h \times m \times n}$ where h denotes the number of attention heads. Thus, we perform aggregation by averaging over the neuronal time dimension, in order to remain *invariant* to the temporal neuronal embedding and to gather all past neuronal influence on blood flow rates. This way, one can observe that the proposed method is not limited to any spatial or time constraint. The model can be deployed in different spatio-temporal settings at test time, thanks to both the geometric spatial encoding and the Transformer's sequential processing ability. Finally, the output module reprojects the last time vessel embedding into the prediction space.

Architecture and Training. The initial encoding defines the model embedding dimension $d = 64$. The encoder and the decoder are defined as the concatenation of $L = 3$ layers, each composed of self-attention and feed-forward layers interleaved with normalization layers. The decoder also contains N additional cross-attention modules. The output layer is defined by a fully connected layer that projects the last vascular time embedding into the objective dimension m. An illustration of the model is given in Fig. 1.

The dimension of the feed-forward network is four times that of the embedding [30]. It is composed of GEGLU layers [25], with layer normalization set to the pre-layer norm setting, as in [15,33]. We use an eight-head self-attention module in all experiments. The geometric filtering first augments the distance using

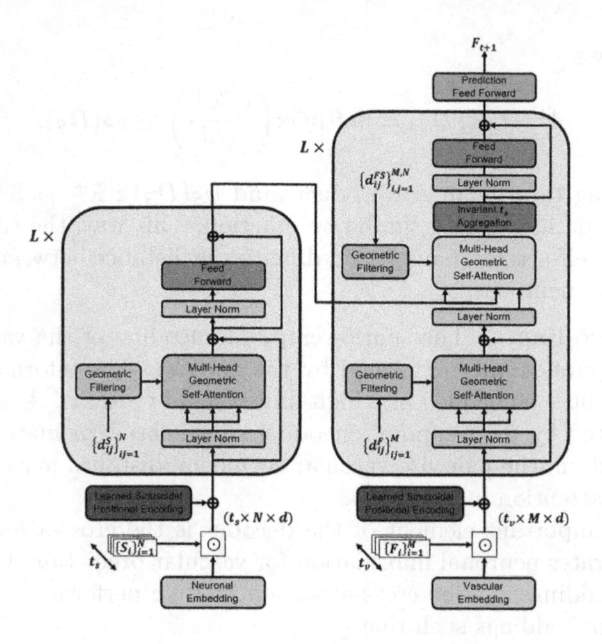

Fig. 1. Illustration of the proposed HRF Transformer architecture. The main differences from the traditional Transformers are the Geometric self-attention modules and the unified spatiotemporal analysis induced by the time aggregation module.

Fourier features [27] and the module is a fully connected neural network with two 50-dimensional hidden layers and GELU non-linearities, expanded to all the heads of the self-attention module. We provide the module with the element-wise inverse of the distance matrix instead of the regular Euclidean matrix, both in order to reduce the dynamic range and since closer elements may have a higher impact.

The training objective is the Mean Squared Error loss

$$\mathcal{L} = \mathbb{E}_t\Bigg(\sum_{j}^{m} \|f_\theta(S_t, F_t, D_S, D_F, D_{SF}) - F_{t+1}\|^2\Bigg) \tag{5}$$

The Adam optimizer [14] is used with 32 samples per minibatch, for 300 epochs. We initialized the learning rate to $5 \cdot 10^{-5}$ coupled with a cosine decay scheduler down to $1 \cdot 10^{-6}$ at the end of the training. The dataset of the first data collection session has been split by $85\%, 7.5\%$ and 7.5% for the training, validation, and testing set, respectively. Training time is approximately 20 h for time windows $t_s = t_v = 10$, on an NVIDIA RTX A600. Testing time is approximately 0.25 ms per sample.

4 Experiments

We compare the proposed method, dubbed **H**emodynamic **R**esponse **F**unction **T**ransformer (HRFT), with several popular statistical and machine-learning

Table 1. The prediction errors of the various methods on two test sets. The first is obtained in the same session for which the training samples were collected (top half of the table, and the second in a separate session 30 min later (bottom half).

Method	6 Hz		15 Hz		30.03 Hz	
	MSE	NRMSE	MSE	NRMSE	MSE	NRMSE
Persistence	24.38	0.220	12.93	0.160	6.923	0.115
Linear	13.65	0.166	9.660	0.139	5.911	0.107
RNN	13.78	0.168	11.38	0.157	10.31	0.147
HRFT-S	13.34	0.165	9.426	0.138	**5.782**	0.110
HRFT	**13.00**	**0.162**	**9.370**	**0.137**	5.783	**0.106**
Persistence	23.11	0.221	11.97	0.160	7.662	0.125
Linear	17.01	0.192	10.26	0.147	6.002	0.111
RNN	15.193	0.182	13.04	0.172	11.87	0.162
HRFT-S	14.63	0.176	9.820	0.147	5.914	**0.110**
HRFT	**14.34**	**0.173**	**9.191**	**0.143**	**5.908**	**0.110**

models: (i) naive persistence model, which predicts the previous time step's vascular input, (ii) linear regression, which concatenates all the input (blood flow and neuronal data) from all times stamps before performing the regression, and (iii) a Recurrent Neural Network composed of two stacked GRU [4] layers. All the methods are experimented with using the same inputs and the models have similar capacity (~0.5 M parameters).

In order to understand the impact of neuronal information, we also compare our method with the HRFT encoder only applied to the vascular input, referred to as HRFT-S. The only difference between this model and the full HRFT is the cross-attention module. If the neuronal input is irrelevant or the link is too weak to improve the prediction, HRFT is expected not to outperform HRFT-S, which makes the comparison pertinent.

We present both MSE and Normalized Root MSE. Because of computational constraints, we randomly subsample 55 vessel segments among the 425. We trained the model with temporal windows of size $t_s = t_v = 10$, equivalent to 300 ms according to the original data acquisition's 30.03 Hz sampling rate.

In addition to the original sampling rate, we also present results for prediction based on lower frequencies, in order to check the ability of the models to capture longer-range dependencies. We note that the error in these cases is expected to be larger, since the time gap between the last measurement and the required prediction is larger.

In order to check the generalization abilities of the methods, we test the trained models on a second dataset obtained 30 min after the sampling of the original dataset (that includes training, validation, and the first test set).

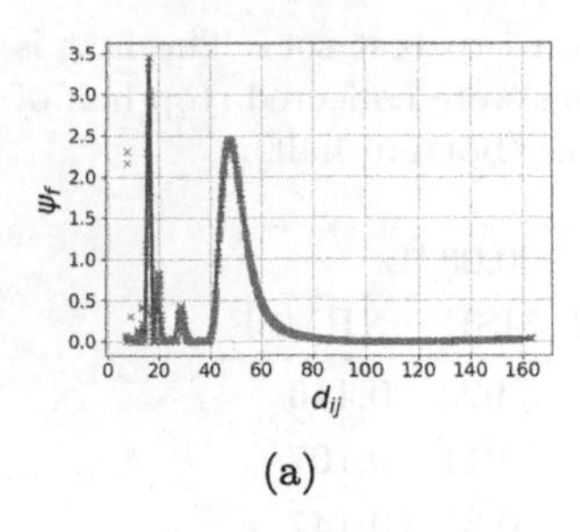

(a)

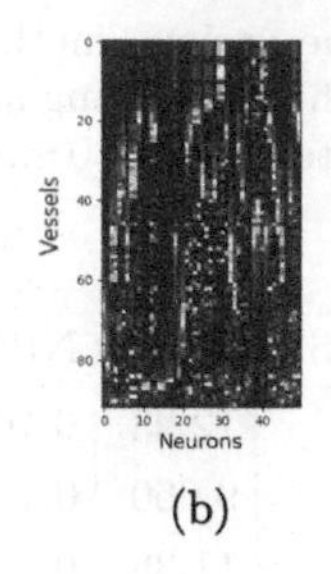

(b)

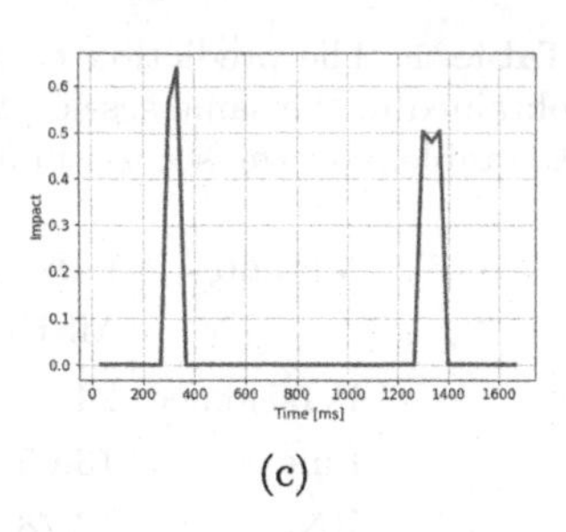

(c)

Fig. 2. (a) The learned function ψ_{SF} for the 1st cross-attention layer of the transformer. (b) The magnitude of the self-attention map between every neuron and every vessel at this layer. (c) The impact of the neurons on the vessels (saturated at 90%) for each shift in time as obtained by marginalizing over all 2nd session test samples in the 30.02 Hz dataset. More visualizations of the datasets and the learned features are provided in the Appendix.

The results are presented in Table 1. As can be seen, the HRFT method outperforms all baselines, including the HRFT-S variant, for 6 Hz and 15 Hz. At the original sampling rate, the performance of HRFT and HRFT-S is similar and better than the baselines. This is expected since at this frame rate the history of $t_v = 10$ we employ spans only 300 ms, which is at the limit of the shortest known neurovascular response reported in the literature [29]. It is reassuring that error levels for HRFT remain similar for samples taken 30min after the training set (and the first test set) were collected.

To gain insights into the HRF, we examine the HRFT model. The learned distance function ψ_{SF} of the 1st cross attention layer is depicted in Fig. 2(a) (other layers are similar). The plot shows the learned function in blue and the actual samples in red. Evidently, this prior on the attention is monotonically decreasing with the distance between the neuron and the blood vessel. Panel (b) shows the cross-attention in the same layer. We note that some neurons have little influence, and the rest of the attention is scattered relatively uniformly. Panel (c) considers the derivative of the prediction vector F_{t+1} by each of the neuron data, summed over all test samples of the 2nd session at 6 Hz, and all neurons and vessels. There are two negative peaks (contractions) that occur at 333 ms and 1333 ms, which is remarkably consistent with current knowledge [29]. There is also a dilation effect at 666 ms. The 15 Hz data with $t_v = 10$ captures shifts of 0–700 ms and the 300 ms peak is clearly visible in that model as well.

5 Conclusions and Future Work

We present the first local HRF model. While for the baseline methods, the performance is at the same level with and without neuronal data (omitted from the tables), the transformer we present supports an improved prediction capability using neuronal firing rates (ablation) and also gives rise to interesting insights regarding the behavior of the hemodynamic response function.

Limitations. Our main goal is to verify the ability to model HRF by showing that using neuronal data helps predict blood flow beyond the history of the latter. The next challenge is to scale the model in order to be able to model more vessels (without subsampling) and longer historical sequences (larger t_v, t_s). With transformers being used for very long sequences, this is a limitation of our resources and not of our method.

Acknowledgements. The authors thank David Kain for conducting the mouse surgery. This project has received funding from the ISRAEL SCIENCE FOUNDATION (grant No. 2923/20) within the Israel Precision Medicine Partnership program. It was also supported by a grant from the Tel Aviv University Center for AI and Data Science (TAD). It was also supported by the European Research Council, grant No 639416, and the Israel Science Foundation, grant No 2342/21. The contribution of the first author is part of a PhD thesis research conducted at Tel Aviv University.

References

1. Asl, M.E., Koohbanani, N.A., Frangi, A.F., Gooya, A.: Tracking and diameter estimation of retinal vessels using gaussian process and radon transform. J. Med. Imaging **4**(3), 034006 (2017)
2. Buxton, R.B.: Thermodynamic limitations on brain oxygen metabolism: physiological implications. bioRxiv, pp. 2023–01 (2023)
3. Carion, N., Massa, F., Synnaeve, G., Usunier, N., Kirillov, A., Zagoruyko, S.: End-to-end object detection with transformers. arXiv preprint arXiv:2005.12872 (2020)
4. Cho, K., van Merriënboer, B., Bahdanau, D., Bengio, Y.: On the properties of neural machine translation: encoder-decoder approaches. In: Proceedings of SSST-8, Eighth Workshop on Syntax, Semantics and Structure in Statistical Translation, pp. 103–111 (2014)
5. Drew, P.J.: Vascular and neural basis of the bold signal. Current Opin. Neurobiol. **58**, 61–69 (2019). https://doi.org/10.1016/J.CONB.2019.06.004
6. Drew, P.J., Blinder, P., Cauwenberghs, G., Shih, A.Y., Kleinfeld, D.: Rapid determination of particle velocity from space-time images using the radon transform. J. Comput. Neurosci. **29**(1), 5–11 (2010)
7. Fazlollahi, A., et al.: Efficient machine learning framework for computer-aided detection of cerebral microbleeds using the radon transform. In: 2014 IEEE 11th International Symposium on Biomedical Imaging (ISBI), pp. 113–116. IEEE (2014)
8. Gao, Y.R., Drew, P.J.: Determination of vessel cross-sectional area by thresholding in radon space. J. Cereb. Blood Flow Metab. **34**(7), 1180–1187 (2014)
9. Golgher, L.: Rapid volumetric imaging of numerous neuro-vascular interactions in awake mammalian brain. Ph.D. thesis, Sagol School of Neuroscience, Tel Aviv University (2022)
10. Gur, S., Wolf, L., Golgher, L., Blinder, P.: Microvascular dynamics from 4d microscopy using temporal segmentation. In: Pacific Symposium on Biocomputing 2020, pp. 331–342. World Scientific (2019)
11. Har-Gil, H., et al.: Pysight: plug and play photon counting for fast continuous volumetric intravital microscopy. Optica **5**(9), 1104–1112 (2018)
12. Har-Gil, H., Golgher, L., Kain, D., Blinder, P.: Versatile software and hardware combo enabling photon counting acquisition and real-time display for multiplexing, 2d and continuous 3d two-photon imaging applications. Neurophotonics **9**(3), 031920 (2022)

13. Kim, S.G., Ogawa, S.: Biophysical and physiological origins of blood oxygenation level-dependent FMRI signals. J. Cereb. Blood Flow Metab. Off. J. Int. Soc. Cereb. Blood Flow Metab. **32**, 1188–206 (2012). https://doi.org/10.1038/jcbfm.2012.23
14. Kingma, D., Ba, J.: Adam: a method for stochastic optimization. arXiv (2014)
15. Klein, G., Kim, Y., et al.: Open-source toolkit for neural machine translation. In: ACL (2017)
16. Levy, W.B., Calvert, V.G.: Communication consumes 35 times more energy than computation in the human cortex, but both costs are needed to predict synapse number. Proc. Natl. Acad. Sci. **118**(18), e2008173118 (2021)
17. Lewis, M., et al.: BART: denoising sequence-to-sequence pre-training for natural language generation, translation, and comprehension. arXiv preprint arXiv:1910.13461 (2019)
18. Li, L.H., Yatskar, M., Yin, D., Hsieh, C.J., Chang, K.W.: VisualBERT: a simple and performant baseline for vision and language. arXiv preprint arXiv:1908.03557 (2019)
19. Li, X., et al.: Oscar: object-semantics aligned pre-training for vision-language tasks. In: Vedaldi, A., Bischof, H., Brox, T., Frahm, J.-M. (eds.) ECCV 2020. LNCS, vol. 12375, pp. 121–137. Springer, Cham (2020). https://doi.org/10.1007/978-3-030-58577-8_8
20. Logothetis, N.K.: What we can do and what we cannot do with FMRI. Nature. **453**, 869–78 (2008). https://doi.org/10.1038/nature06976
21. Lu, J., Batra, D., Parikh, D., Lee, S.: VilBERT: pretraining task-agnostic visiolinguistic representations for vision-and-language tasks. In: Advances in Neural Information Processing Systems, pp. 13–23 (2019)
22. Mookiah, M.R.K., et al.: A review of machine learning methods for retinal blood vessel segmentation and artery/vein classification. Med. Image Anal. **68**, 101905 (2020)
23. Paul, M., Danelljan, M., Van Gool, L., Timofte, R.: Local memory attention for fast video semantic segmentation. arXiv preprint arXiv:2101.01715 (2021)
24. Pourreza, R., Banaee, T., Pourreza, H., Kakhki, R.D.: A radon transform based approach for extraction of blood vessels in conjunctival images. In: Gelbukh, A., Morales, E.F. (eds.) MICAI 2008. LNCS (LNAI), vol. 5317, pp. 948–956. Springer, Heidelberg (2008). https://doi.org/10.1007/978-3-540-88636-5_89
25. Shazeer, N.: GLU variants improve transformer. arXiv:2002.05202 (2020)
26. Tan, H., Bansal, M.: LXMERT: learning cross-modality encoder representations from transformers. arXiv preprint arXiv:1908.07490 (2019)
27. Tancik, M., et al.: Fourier features let networks learn high frequency functions in low dimensional domains. Adv. Neural. Inf. Process. Syst. **33**, 7537–7547 (2020)
28. Tavakoli, M., Mehdizadeh, A., Pourreza, R., Pourreza, H.R., Banaee, T., Toosi, M.B.: Radon transform technique for linear structures detection: application to vessel detection in fluorescein angiography fundus images. In: 2011 IEEE Nuclear Science Symposium Conference Record, pp. 3051–3056. IEEE (2011)
29. Uludağ, K., Blinder, P.: Linking brain vascular physiology to hemodynamic response in ultra-high field MRI. NeuroImage. **168**, 279–295 (2018). https://doi.org/10.1016/j.neuroimage.2017.02.063
30. Vaswani, A., Shazeer, N., Parmar, N., et al.: Attention is all you need. In: NeurIPS (2017)
31. Wang, H., Zhu, Y., Adam, H., Yuille, A., Chen, L.C.: Max-deeplab: end-to-end panoptic segmentation with mask transformers. arXiv preprint arXiv:2012.00759 (2020)

32. Wang, Y., et al.: End-to-end video instance segmentation with transformers. arXiv preprint arXiv:2011.14503 (2020)
33. Xiong, R., et al.: On layer normalization in the transformer architecture. arXiv:2002.04745 (2020)
34. Zhu, X., Su, W., Lu, L., Li, B., Wang, X., Dai, J.: Deformable DETR: deformable transformers for end-to-end object detection. In: ICLR (2021)

Debiasing Medical Visual Question Answering via Counterfactual Training

Chenlu Zhan[1], Peng Peng[2], Hanrong Zhang[2], Haiyue Sun[2], Chunnan Shang[2], Tao Chen[3], Hongsen Wang[3], Gaoang Wang[1,2(✉)], and Hongwei Wang[1,2(✉)]

[1] College of Computer Science and Technology, Zhejiang University, Zhejiang, China
chenlu.22@intl.zju.edu.cn
[2] ZJU-UIUC Institute, Zhejiang University, Zhejiang, China
{pengpeng,hanrong.22,haiyue.22,chunnan.22,gaoangwang}@intl.zju.edu.cn, hongweiwang@zju.edu.cn
[3] Department of Cardiology, Chinese PLA General Hospital, Beijing, China

Abstract. Medical Visual Question Answering (Med-VQA) is expected to predict a convincing answer with the given medical image and clinical question, aiming to assist clinical decision-making. While today's works have intention to rely on the superficial linguistic correlations as a shortcut, which may generate emergent dissatisfactory clinic answers. In this paper, we propose a novel DeBiasing Med-VQA model with CounterFactual training (DeBCF) to overcome language priors comprehensively. Specifically, we generate counterfactual samples by masking crucial keywords and assigning irrelevant labels, which implicitly promotes the sensitivity of the model to the semantic words and visual objects for bias-weaken. Furthermore, to explicitly prevent the cheating linguistic correlations, we formulate the language prior into counterfactual causal effects and eliminate it from the total effect on the generated answers. Additionally, we initiatively present a newly splitting bias-sensitive Med-VQA dataset, Semantically-Labeled Knowledge-Enhanced under Changing Priors (SLAKE-CP) dataset through regrouping and re-splitting the train-set and test-set of SLAKE into the different prior distribution of answers, dedicating the model to learn interpretable objects rather than overwhelmingly memorizing biases. Experimental results on two public datasets and SLAKE-CP demonstrate that the proposed DeBCF outperforms existing state-of-the-art Med-VQA models and obtains significant improvement in terms of accuracy and interpretability. To our knowledge, it's the first attempt to overcome language priors in Med-VQA and construct the bias-sensitive dataset for evaluating debiased ability.

C. Zhan and P. Peng—Contributed equally to this work.

Supplementary Information The online version contains supplementary material available at https://doi.org/10.1007/978-3-031-43895-0_36.

H. Greenspan et al. (Eds.): MICCAI 2023, LNCS 14221, pp. 382–393, 2023.
https://doi.org/10.1007/978-3-031-43895-0_36

Keywords: Medical Vision Question Answering · Language Bias · Counterfactual Sample Generation · Counterfactual Training · SLAKE-CP

1 Introduction

Medical visual question answering (Med-VQA) has attracted considerable attention in recent years. It seeks to discover the plausible answer by evaluating the visual information of a medical image and a clinical query regarding the image. The Med-VQA technology can considerably enhance the efficiency of medical professionals and fulfill the growing demand for medical resources [15, 25]. However, numerous researches have found that general VQA models are significantly influenced by superficial linguistic correlations in training set, lacking adequate visual grounding [9, 14, 26]. Since most of the existing Med-VQA datasets [12, 16] are manually spitted and annotated, the spurious over-reliant bias factor also exists in Med-VQA, as the Fig. 1 shown. Recent general VQA works dedicate to reducing the language priors through enhancing the visual information [23, 27] or data balancing [11, 13], there is bare attempt to prevent the language priors in medical domain. Current Med-VQA works [4, 5, 15, 17] devote to construct effective models and most Med-VQA datasets [12, 16] simply balance the medical images to mitigate the inherent bias. These works all neglect the cheating factors that the Med-VQA models typically resort to linguistic distributions priors, consequently ignoring the semantic clinic objects. This problem accordingly leads to disastrous results in clinic application consequences.

Therefore, we propose a novel unbiased and interpretable Med-VQA model and preliminarily construct a bias-sensitive Med-VQA dataset to address the problems mentioned above. First, with the aim of forcing the model to focus on clinic objects rather than superficial correlations, we prepare the counterfactual samples by masking clinic words with "[MASK]" tokens and meanwhile assign the irrelevant answers for implicit bias-weaken. Further, for explicitly reducing the linguist bias, we treat the language bias as the causal effect of the clinic

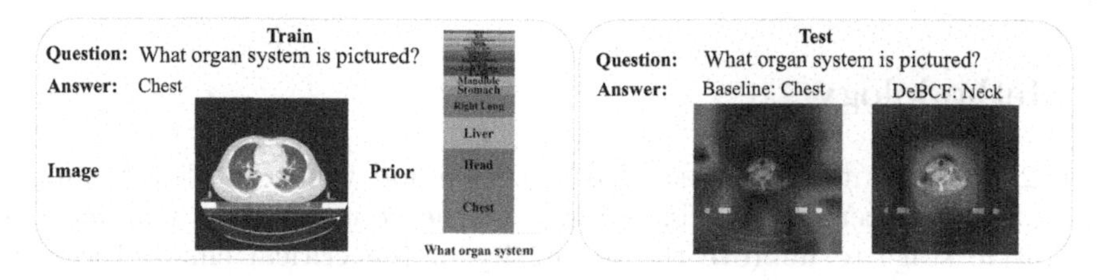

Fig. 1. The baseline generates incorrect answer "Chest" relying on the majority prior "Chest" in train-set of the publicly available SLAKE [16] dataset rather than real semantic image objects. The proposed DeBCF overcomes the language priors and generates the reliable answer with correct semantic parts.

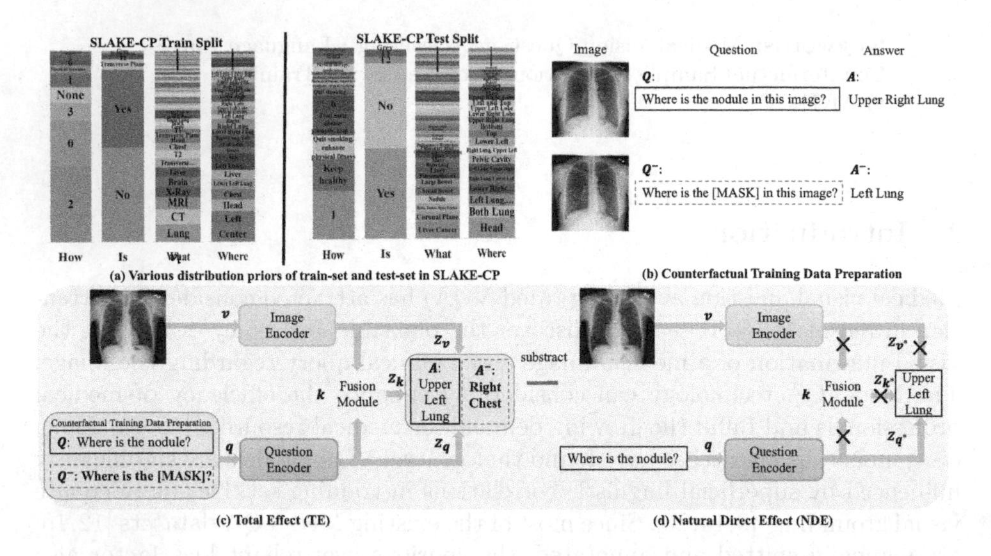

Fig. 2. (a) Various distribution priors of train-set and test-set in SLAKE-CP. (b) Counterfactual training data preparation. (c) Traditional Med-VQA causal graph with total effect. (d) Counterfactual Med-VQA causal graph with natural direct effect.

question on the generated answer and then subtract it from the total causal effect for counterfactual training. It is noted that both the original data and generated counterfactual data will be used for counterfactual training. In this way, the model may not tend to provide answers over-rely on the largest proportions of candidate answers in train-set when tested, thus concentrating on entanglement of the visual objects and language information.

Additionally, we conduct a bias-sensitive Med-VQA dataset Semantically-Labeled Knowledge-Enhanced-Changing Priors (SLAKE-CP) for evaluating the ability of disentangling the memorized linguist priors and semantic visual information. Qualitative and quantitative experimental results illustrate that our proposed model is superior to the state-of-the-art Med-VQA models on the two public benchmarks and can obtain more obvious improvements on the newly constructed SLAKE-CP.

2 Methodology

Figure 2 illustrates the proposed Med-VQA method which consists of implicit and explicit counterfactual debiased stages: the counterfactual training data preparation stage to improve the sensitivity of the critical clinic objects for implicit bias-weaken. Along with the counterfactual causal effect training stage to directly migrate the language priors.

2.1 Counterfactual Training Data Preparation

To implicitly weaken the language bias, we follow CSS [3] to prepare counterfactual training samples for improving the sensitivity of clinic objects. First, we extract the question type (e.g. "Where" in Fig. 2) of each question and calculate the importance s of the remaining words w_i in clinical question $\boldsymbol{q}$ to the label a as:

$$s(a, w_i) = S(P(a|\boldsymbol{q}, \boldsymbol{v}), w_i) := (\nabla_{w_i} P(a|\boldsymbol{q}, \boldsymbol{v}))^T \mathbf{1} \quad (1)$$

where $P(a|\boldsymbol{q}, \boldsymbol{v})$ represents the probability of predicting answer a through Med-VQA model with image $\boldsymbol{v}$ and question $\boldsymbol{q}$, ∇ is the gradient operator, S is the cosine similarity, and $\mathbf{1}$ is the all-ones vector. The top-K clinic words with the highest importance s are defined as critical words. Then, we construct counterfactual samples Q^- by replacing the critical words with "[MASK]". We also assign the Q^- with an answer A^-, and the detailed assigning procedure is as follows. We first generate the probability of predicting answer $P^+(a)$ with the question Q^+ which replaces the marginal words with "[MASK]" (all but the question type labels and the critical words), and then pick up top-N candidate answers with the highest probability as A^+. The rest answers are denoted as $A^- = \{a_i | a_i \in A, a_i \notin A^+\}$ and are assigned to Q^-.

2.2 Counterfactual Cause Effect Training Procedure

For explicitly subtracting the language priors, following [24], we introduce casual effect [21] to translate priors into quantified expressions. The causal effects can directly reflect the comparisons between the outputs with different treatments (e.g. $X = \boldsymbol{x}$ represents with-treatment and $X = \boldsymbol{x}^*$ represents the counterfactual situation where is without the treatment). The total effect (TE) of $X = \boldsymbol{x}$ on Y can be defined as two different conditions that with or without the input:

$$TE = Y_{X=\boldsymbol{x}, M(X=\boldsymbol{x})} - Y_{X=\boldsymbol{x}^*, M(X=\boldsymbol{x}^*)} \quad (2)$$

where M is the mediator between the variables X and Y. Note that the total effect can be composed of the natural direct effect (NDE) and total indirect effect(TIE). Between them, the NDE concentrates the exclusive effect of $X = \boldsymbol{x}$ on Y and prohibit the effect through M:

$$NDE = Y_{X=\boldsymbol{x}, M(X=\boldsymbol{x}^*)} - Y_{X=\boldsymbol{x}^*, M(X=\boldsymbol{x}^*)} \quad (3)$$

Thus TIE can reflect the reduction of language bias by subtracting the NDE from the TE:

$$TIE = TE - NDE = Y_{X=\boldsymbol{x}, M(X=\boldsymbol{x})} - Y_{X=\boldsymbol{x}, M(X=\boldsymbol{x}^*)} \quad (4)$$

Based on the above definition, we translate the Med-VQA task into a causal effect graph as Fig. 2 (c)(d) shown, aiming to directly formulate the language bias and subtract it. The answer set $A = \{a\}$ is caused by direct effect from medical image $V = \boldsymbol{v}$ and clinic question $Q = \boldsymbol{q}$, also the indirect effect of fusion knowledge $K(Q = \boldsymbol{q}, V = \boldsymbol{v})$ through the cross-modal fusion module. We define the notations that: $Y_{\boldsymbol{q},\boldsymbol{v},\boldsymbol{k}} = Y(Q = \boldsymbol{q}, V = \boldsymbol{v}, K = \boldsymbol{k})$. Through subtracting NDE of $Q = \boldsymbol{q}$ on A from the TE of $V = \boldsymbol{v}$, $Q = \boldsymbol{q}$ and $K = \boldsymbol{k}$ on the answer, we can explicitly capture language bias and remove it via TIE, which is defined below. In the inference stage, we choose the answer with the maximum TIE as the prediction.

$$TIE = TE - NDE = Y_{\boldsymbol{q},\boldsymbol{v},\boldsymbol{k}} - Y_{\boldsymbol{q},\boldsymbol{v}^*,\boldsymbol{k}^*} \tag{5}$$

where $\boldsymbol{k}^* = K(V = \boldsymbol{v}^*, Q = \boldsymbol{q}^*)$, $\boldsymbol{v}^*$ and $\boldsymbol{q}^*$ is the counterfactual situation where model is without $\boldsymbol{v}$, $\boldsymbol{q}$ as inputs. The $Y_{\boldsymbol{q},\boldsymbol{v},\boldsymbol{k}} = \log \sigma(Z_q + Z_v + Z_k)$, where the $Z_v = E_V(\boldsymbol{v})$, $Z_q = E_Q(\boldsymbol{q})$, $Z_k = E_F(\boldsymbol{q}, \boldsymbol{v})$ are calculated from the image encoder E_V, question encoders E_Q and the fusion module E_F respectively. The E_V, E_Q, E_F can be updated by L_{cls} [20]:

$$L_{cls}(\boldsymbol{q}, \boldsymbol{v}, a) = L_{VQA}(\boldsymbol{q}, \boldsymbol{v}, a) + L_{QA}(\boldsymbol{q}, \mathrm{a}) + L_{VA}(\boldsymbol{v}, a) \tag{6}$$

where L_{VQA}, L_{QA} and L_{VA} are corss-entropy losses over $Y_{\boldsymbol{q},\boldsymbol{v},\boldsymbol{k}}$, Z_q and Z_v.

The complete objective of our method is optimized to minimize the L_{DeBCF} which combines the L_{cls} over both the original and the counterfactual data:

$$L_{DeBCF} = \alpha L_{cls}(V, Q, A) + (1 - \alpha) L_{cls}(V, Q^-, A^-) \tag{7}$$

where α is the hyperparameter which control the ratio of counterfactual samples.

3 SLAKE-CP: Construction and Analysis

For further evaluating the debiasing ability of Med-VQA, we follow VQA-CP [1] to create a bias-sensitive Med-VQA dataset which can be called SLAKE-CP. The SLAKE-CP can be further adopted by future debiased Med-VQA researches.

Grouping. We first construct all image-question-answer samples of train-set and test-set in SLAKE [16] into a whole set together. We start by labeling each question with a question type (first few words). If the samples have the same question type and answer, then these samples can be divided into same group.

Re-Splitting. We re-split the SLAKE [16] dataset to construct disparate distribution as Fig. 2 (a) shows. In detail, we first assign 1 group to the test-set. Among the remaining groups, if there is a group with a different question type or answer from the groups in test-set, this group will be assigned to test-set otherwise to train-set, aiming to vary the prior distributions of the train and test while remaining unchanged distributions of the images. The iteration stops when the test-set approximately reaches 1/7rd of the whole set, and the remaining are added to the train-set. We ensure the newly constructed test-set and train-set cover the majority of question types (*"Is"*, *"What"*, *"Where"*, *"Which"*, *etc.*) after these procedures. Most of the data attributes of SLAKE-CP are consistent with SLAKE, such as the train-test splitting, and open-close type splitting.

Table 1. The comparison results. $*$ indicates our re-implemented result, including the mean accuracy and standard deviation by 5 runs under 5 different seeds.

Methods	SLAKE			VQA-RAD		
	Open	Closed	All	Open	Closed	All
MFB [29]	72.2	75.0	73.3	14.5	74.3	50.6
SAN [28]	74.0	79.1	76.0	31.3	69.5	54.3
BAN [10]	74.6	79.1	76.3	37.4	72.1	58.3
LPF(*) [13]	$74.8_{\pm 1.4}\%$	$77.0_{\pm 1.1}\%$	$74.9_{\pm 1.3}\%$	$41.7_{\pm 1.3}\%$	$72.1_{\pm 1.1}\%$	$60.9_{\pm 1.3}\%$
RUBi(*) [2]	$75.1_{\pm 1.2}\%$	$77.6_{\pm 1.3}\%$	$75.8_{\pm 1.3}\%$	$42.4_{\pm 1.2}\%$	$73.2_{\pm 1.0}\%$	$61.5_{\pm 1.2}\%$
GGE(*) [8]	$76.4_{\pm 1.1}\%$	$78.7_{\pm 1.2}\%$	$76.6_{\pm 1.2}\%$	$44.6_{\pm 1.4}\%$	$74.5_{\pm 1.1}\%$	$63.8_{\pm 1.1}\%$
MEVF+SAN [19]	75.3	78.4	76.5	49.2	73.9	64.1
MEVF+BAN [19]	77.8	79.8	78.6	49.2	77.2	66.1
CLIPQCR(*) [6]	$78.2_{\pm 1.3}\%$	$82.6_{\pm 1.5}\%$	$80.1_{\pm 1.3}\%$	$58.0_{\pm 1.4}\%$	$79.6_{\pm 1.1}\%$	$71.1_{\pm 1.2}\%$
CPRD+BAN [15]	79.5	83.4	81.1	52.5	77.9	67.8
Ours	$\mathbf{80.8}_{\pm 0.9}\%$	$\mathbf{84.9}_{\pm 0.7}\%$	$\mathbf{82.6}_{\pm 0.9}\%$	$\mathbf{58.6}_{\pm 1.1}\%$	$\mathbf{80.9}_{\pm 0.8}\%$	$\mathbf{71.6}_{\pm 1.0}\%$

Table 2. The additional comparison of experimental results on the SLAKE-CP dataset.

Methods	SLAKE-CP		
	Open	Closed	All
MFB(*) [29]	$10.9_{\pm 1.0}\%$	$22.1_{\pm 0.8}\%$	$21.5_{\pm 0.8}\%$
SAN(*) [28]	$11.2_{\pm 1.2}\%$	$22.7_{\pm 1.1}\%$	$23.2_{\pm 1.1}\%$
BAN(*) [10]	$11.9_{\pm 1.2}\%$	$24.4_{\pm 0.9}\%$	$24.5_{\pm 1.0}\%$
RUBi(*) [2]	$12.2_{\pm 1.3}\%$	$26.9_{\pm 1.2}\%$	$26.4_{\pm 1.3}\%$
LPF(*) [13]	$13.1_{\pm 1.4}\%$	$29.7_{\pm 1.4}\%$	$29.2_{\pm 1.3}\%$
GGE(*) [8]	$13.9_{\pm 1.1}\%$	$30.9_{\pm 1.3}\%$	$30.2_{\pm 1.3}\%$
MEVF+SAN(*) [19]	$12.6_{\pm 1.1}\%$	$29.6_{\pm 1.0}\%$	$28.7_{\pm 1.0}\%$
MEVF+BAN(*) [19]	$13.0_{\pm 1.4}\%$	$29.8_{\pm 1.2}\%$	$29.1_{\pm 1.3}\%$
CLIPQCR(*) [6]	$13.4_{\pm 1.2}\%$	$30.5_{\pm 1.1}\%$	$30.0_{\pm 1.2}\%$
CPRD+BAN(*) [15]	$13.9_{\pm 1.3}\%$	$31.2_{\pm 1.5}\%$	$30.4_{\pm 1.5}\%$
Ours	$\mathbf{18.6}_{\pm 1.1}\%$	$\mathbf{35.4}_{\pm 1.0}\%$	$\mathbf{34.2}_{\pm 1.2}\%$

4 Experiments

4.1 Datasets and Implementation Details

Datasets. SLAKE [16] is a knowledge-augmented Med-VQA dataset, consisting of 642 images and 7033 question-answer samples. The VQA-RAD [12] is a manually annotated dataset validated by clinicians, which contains 315 radiographic images and 3,515 question-answer samples. We followed the original data partition, where questions are divided into closed-ended and open-ended types.

Implementation Details. For implementation, we apply Pytorch library with 6 NVIDIA TITAN 24 GB Xp GPUs. We employ the MEVF [19] as baseline.

Table 3. Ablation results. "*CTD*": counterfactual training data preparation. "*CCE*": counterfactual cause effect training procedure.

Index	CTD	CCE	SLAKE-CP			SLAKE		
			Open	Closed	Overall	Open	Closed	Overall
1	×	×	13.0$_{\pm 0.6}$%	29.8$_{\pm 0.9}$%	29.1$_{\pm 0.7}$%	78.6$_{\pm 1.2}$%	80.5$_{\pm 1.0}$%	79.8$_{\pm 1.0}$%
2	✓	×	14.2$_{\pm 1.3}$%	31.3$_{\pm 0.9}$%	30.6$_{\pm 1.0}$%	79.4$_{\pm 1.3}$%	81.0$_{\pm 1.1}$%	80.4$_{\pm 1.1}$%
3	×	✓	16.7$_{\pm 0.8}$%	33.9$_{\pm 0.7}$%	32.9$_{\pm 0.8}$%	80.1$_{\pm 1.1}$%	81.9$_{\pm 1.3}$%	81.5$_{\pm 1.2}$%
4	✓	✓	**18.6**$_{\pm 1.1}$%	**35.4**$_{\pm 1.0}$%	**34.2**$_{\pm 1.2}$%	**80.6**$_{\pm 0.9}$%	**84.4** $_{\pm 0.7}$%	**82.6**$_{\pm 0.9}$%

Table 4. Comparisons of different top-K critical words in Sect. 2.1.

top-K	SLAKE		
	Open	Closed	All
1	**80.8**$_{\pm 1.1}$%	**84.9**$_{\pm 0.8}$%	**82.6**$_{\pm 1.0}$%
2	80.5$_{\pm 1.0}$%	84.5$_{\pm 0.8}$%	82.4$_{\pm 0.9}$%
3	79.9$_{\pm 1.2}$%	83.7$_{\pm 1.1}$%	82.1$_{\pm 1.2}$%

Table 5. Evaluations of hyperparameter α.

α	SLAKE		
	Open	Closed	All
0.3	80.1$_{\pm 1.0}$%	83.2$_{\pm 0.9}$%	81.9$_{\pm 1.1}$%
0.4	80.2$_{\pm 1.2}$%	83.5$_{\pm 1.1}$%	82.0$_{\pm 1.2}$%
0.5	80.7$_{\pm 1.2}$%	84.6$_{\pm 0.9}$%	82.5$_{\pm 1.1}$%
0.6	**80.8**$_{\pm 1.1}$%	**84.9**$_{\pm 1.0}$%	**82.6**$_{\pm 1.0}$%
0.7	80.6$_{\pm 1.3}$%	84.2$_{\pm 1.0}$%	82.3$_{\pm 1.0}$%

The vision encoders are initialized by MAML [7] and CDAE [18], and LSTM is adopted as question encoder. The BAN [10] is adopted as the fusion module E_F. The medical images are resized into 224 × 224, and questions are cut to 12 words and then embed into 300 dimensions through Golve. The proposed model is trained for 200 epochs with 64 batch size and optimized with Adam whose learning rate is $1e^{-3}$. In Sect. 2.1, we choose top-1 candidate answer as A^+ and mask top-1 critical clinic word, the hyperparameter α is set to 0.6.

4.2 Experimental Results

Comparison with State-of-the-Art Methods. We compare our DeBCF with 10 state-of-the-art Med-VQA models on the SLAKE [16] and VQA-RAD [12] public benchmarks as the Table 1 shown. The proposed model obviously outperform the existing advanced models, attaining 82.6% and 71.6% mean accuracy respectively. Specifically, the results of our proposed model have prominent improvements over the attention-based models MFB [29], SAN [28], BAN [10]. Further, the improvements over MEVF+BAN [30] and CPRD+BAN [15] which adopt the same fusion model BAN [10] as ours are 4.0%, 1.5% overall accuracy on SLAKE respectively. In particular, the proposed model conspicuously improved the overall accuracy by 2.5% and compared with the advanced CLIPQCR [6]. Moreover, our model has significant superior with other debiasing models, including RUBi [2] and LPF [13], GGE [8]. Although these works can effectively reduce language bias, they reckon without visual-linguist explicable information and contrarily weaken the inference ability. For

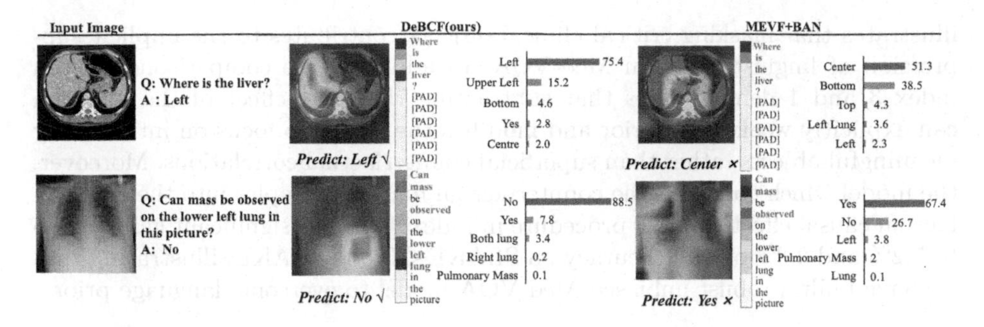

Fig. 3. Quantitative comparison analysis. The darker parts, the more contributions.

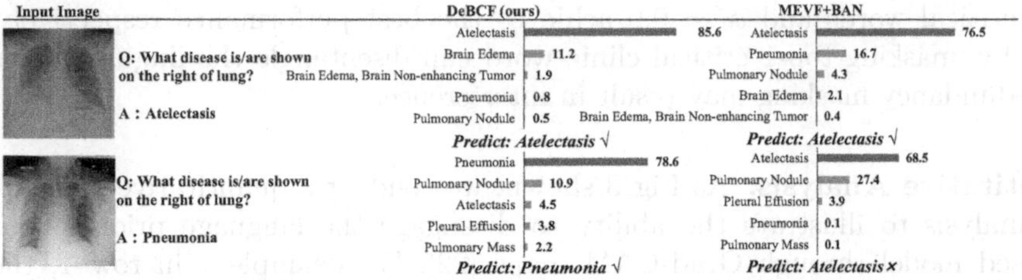

Fig. 4. The comparison analysis of sensitivity to the visual grounds.

ours, we explicitly subtract the language bias through causal effect and generate counterfactual samples to implicitly improve the sensitivity of clinical words and visual objects for inference.

Discussion of SLAKE-CP. Table 2 illustrates the superiority of the DeBCF on the newly constructed SLAKE-CP datasets which is the linguistic-bias sensitive evaluation. In particular, the DeBCF yields 34.2% mean overall accuracy on SLAKE-CP datasets. The performance of all the models has prominently dropped in the newly unbiased SLAKE-CP datasets compared with the SLAKE. It is obviously observed that the DeBCF significantly outperforms the baselines [10,19,28,29] and the debiasing methods [2,8,13]. The proposed model is also superior to the advanced models CLIPQCR [6] and CPRD+BAN [15], over-passing 4.2% and 3.8% overall accuracy. Within the bias-sensitive benchmarks, the comparisons demonstrate that our model may have the superiority to overcome the linguistic priors and force the model to generate more creditable answers rather than taking the superficial linguistic correlations as a shortcut.

Ablation Analysis. Table 3 demonstrates the ablation study which verifies the effectiveness of devised methods. We adopt MEVF+BAN [10] as the baseline in index 1. The baseline equipped with the counterfactual data preparation stage gains 1.5% and 0.6% overall accuracy on SLAKE-CP and SLAKE datasets. This illustrates that masking critical clinical objects contributes to the implicit suppression of linguistic bias in Med-VQA. In addition, the comparison between index 3 and 1 demonstrates that subtracting the cause-effect of the question can explicitly weaken the prior and modifies the model to focus on intrinsically meaningful objects rather than superficial counterfactual correlations. Moreover, the model which combines the counterfactual masking samples into the counterfactual causal effect training procedure in index 4 obtains significant gains by up to 5.2% and 2.8% overall accuracy on SLAKE-CP and SLAKE, illustrating that we have built a robust unbiased Med-VQA model to overcome language priors.

Influence of Hyperparameters. The influence results of the top-K and the hyperparameter α are conducted in Table 4 and 5, which reveal that choosing top-1 critical words and $\alpha = 0.6$ achieves the best performance respectively. Crucially, masking top-1 critical clinic word can disentangle the linguistic bias and redundancy masking may result in interference.

Quantitative Analysis. As Fig. 3 shown, we conduct a quantitative comparison analysis to illustrate the ability to disengage the language prior to our proposed model through Grad-CAM maps [22]. For example 1 in row 1, the proposed DeBCF sensitively recognizes the precise critical keywords *"Where, is, liver"* and corresponding visual image objects to predict the correct answer with the highest probability score, while the advanced model MEVF+BAN [19] is subjected to the language prior that generate the wrong answer according to the superficial context *"Where is"* and ignore the reliable visual objects. Additionally, we also conduct the comparison of sensitivity to the visual grounds in Fig. 4. Given the same question but different medical images and answers, the proposed model correctly predict the various answers while the MEVF+BAN [19] fails. The detailed comparisons illustrate the debiased ability of the proposed model to overcome language priors and ingeniously grasp the critical parts (clinic keywords and visual objects) for a precise explanation.

5 Conclusion

In this paper, we propose a novel debiasing Med-VQA model that prepares the counterfactual data by masking critical clinic words and combines it into the counterfactual training stage which subtracting the causal effect of language priors directly, aiming to migrate the linguistic-bias in Med-VQA. Additionally, we construct a linguistic-bias sensitive Med-VQA dataset SLAKE-CP by disintegrating the language priors from training. Experimental results demonstrate the superior debiasing and interpretive performance of the proposed model. It's the first attempt to construct a preliminary bias-sensitive Med-VQA dataset, which will be elaborated in our future work. The codes will be released.

Acknowledgements. This work was supported in part by Zhejiang Provincial Natural Science Foundation of China (LDT23F02023F02).

References

1. Agrawal, A., Batra, D., Parikh, D., Kembhavi, A.: Don't just assume; look and answer: overcoming priors for visual question answering. In: Proceedings of the IEEE Conference on Computer Vision and Pattern Recognition, pp. 4971–4980 (2018)
2. Cadene, R., et al.: RUBi: reducing unimodal biases for visual question answering. In: Advances in Neural Information Processing Systems, vol. 32 (2019)
3. Chen, L., Yan, X., Xiao, J., Zhang, H., Pu, S., Zhuang, Y.: Counterfactual samples synthesizing for robust visual question answering. In: Proceedings of the IEEE/CVF Conference on Computer Vision and Pattern Recognition, pp. 10800–10809 (2020)
4. Chen, Z., et al.: Multi-modal masked autoencoders for medical vision-and-language pre-training. In: Wang, L., Dou, Q., Fletcher, P.T., Speidel, S., Li, S. (eds.) MICCAI 2022. Lecture Notes in Computer Science, vol. 13435. Springer, Cham (2022). https://doi.org/10.1007/978-3-031-16443-9_65
5. Do, T., Nguyen, B.X., Tjiputra, E., Tran, M., Tran, Q.D., Nguyen, A.: Multiple meta-model quantifying for medical visual question answering. In: de Bruijne, M., et al. (eds.) MICCAI 2021. LNCS, vol. 12905, pp. 64–74. Springer, Cham (2021). https://doi.org/10.1007/978-3-030-87240-3_7
6. Eslami, S., de Melo, G., Meinel, C.: Does clip benefit visual question answering in the medical domain as much as it does in the general domain? arXiv preprint: arXiv:2112.13906 (2021)
7. Finn, C., Abbeel, P., Levine, S.: Model-agnostic meta-learning for fast adaptation of deep networks. In: Precup, D., Teh, Y.W. (eds.) Proceedings of the 34th International Conference on Machine Learning. Proceedings of Machine Learning Research, vol. 70, pp. 1126–1135. PMLR (2017)
8. Han, X., Wang, S., Su, C., Huang, Q., Tian, Q.: Greedy gradient ensemble for robust visual question answering. In: Proceedings of the IEEE/CVF International Conference on Computer vision, pp. 1584–1593 (2021)
9. Jing, C., Wu, Y., Zhang, X., Jia, Y., Wu, Q.: Overcoming language priors in VQA via decomposed linguistic representations. Proc. AAAI Conf. Artif. Intell. **34**(07), 11181–11188 (2020)
10. Kim, J.H., Jun, J., Zhang, B.T.: Bilinear attention networks. In: Advances in Neural Information Processing Systems, vol. 31 (2018)
11. KV, G., Mittal, A.: Reducing language biases in visual question answering with visually-grounded question encoder. In: Vedaldi, A., Bischof, H., Brox, T., Frahm, J.-M. (eds.) ECCV 2020. LNCS, vol. 12358, pp. 18–34. Springer, Cham (2020). https://doi.org/10.1007/978-3-030-58601-0_2
12. Lau, J.J., Gayen, S., Ben Abacha, A., Demner-Fushman, D.: A dataset of clinically generated visual questions and answers about radiology images. Sci. Data **5**, 180251 (2018). https://doi.org/10.1038/sdata.2018.251
13. Liang, Z., Hu, H., Zhu, J.: LPF: a language-prior feedback objective function for debiased visual question answering. In: Proceedings of the 44th International ACM SIGIR Conference on Research and Development in Information Retrieval, pp. 1955–1959 (2021)
14. Liang, Z., Jiang, W., Hu, H., Zhu, J.: Learning to contrast the counterfactual samples for robust visual question answering. In: Proceedings of the 2020 Conference on Empirical Methods in Natural Language Processing (EMNLP), pp. 3285–3292 (2020)

15. Liu, B., Zhan, L.-M., Wu, X.-M.: Contrastive pre-training and representation distillation for medical visual question answering based on radiology images. In: de Bruijne, M., et al. (eds.) MICCAI 2021. LNCS, vol. 12902, pp. 210–220. Springer, Cham (2021). https://doi.org/10.1007/978-3-030-87196-3_20
16. Liu, B., Zhan, L.M., Xu, L., Ma, L., Yang, Y., Wu, X.M.: SLAKE: a semantically-labeled knowledge-enhanced dataset for medical visual question answering. In: 2021 IEEE 18th International Symposium on Biomedical Imaging (ISBI), pp. 1650–1654 (2021). https://doi.org/10.1109/ISBI48211.2021.9434010
17. Liu, B., Zhan, L.M., Xu, L., Wu, X.M.: Medical visual question answering via conditional reasoning and contrastive learning. IEEE Trans. Med. Imaging **42**, 1532–1545 (2022). https://doi.org/10.1109/TMI.2022.3232411
18. Masci, J., Meier, U., Cireşan, D., Schmidhuber, J.: Stacked convolutional auto-encoders for hierarchical feature extraction. In: Honkela, T., Duch, W., Girolami, M., Kaski, S. (eds.) ICANN 2011. LNCS, vol. 6791, pp. 52–59. Springer, Heidelberg (2011). https://doi.org/10.1007/978-3-642-21735-7_7
19. Nguyen, B.D., Do, T.-T., Nguyen, B.X., Do, T., Tjiputra, E., Tran, Q.D.: Overcoming data limitation in medical visual question answering. In: Shen, D., et al. (eds.) MICCAI 2019. LNCS, vol. 11767, pp. 522–530. Springer, Cham (2019). https://doi.org/10.1007/978-3-030-32251-9_57
20. Niu, Y., Tang, K., Zhang, H., Lu, Z., Hua, X.S., Wen, J.R.: Counterfactual VQA: a cause-effect look at language bias. In: Proceedings of the IEEE/CVF Conference on Computer Vision and Pattern Recognition, pp. 12700–12710 (2021)
21. Pearl, J.: Direct and indirect effects. In: Proceedings of the 17th Conference on Uncertainty in Artificial Intelligence, 2001, pp. 411–420 (2001)
22. Selvaraju, R.R., Cogswell, M., Das, A., Vedantam, R., Parikh, D., Batra, D.: Grad-CAM: visual explanations from deep networks via gradient-based localization. In: Proceedings of the IEEE International Conference on Computer Vision, pp. 618–626 (2017)
23. Selvaraju, R.R., et al.: Taking a hint: Leveraging explanations to make vision and language models more grounded. In: Proceedings of the IEEE/CVF International Conference on Computer Vision, pp. 2591–2600 (2019)
24. Tang, K., Niu, Y., Huang, J., Shi, J., Zhang, H.: Unbiased scene graph generation from biased training. In: Proceedings of the IEEE/CVF Conference on Computer Vision and Pattern Recognition, pp. 3716–3725 (2020)
25. Tascon-Morales, S., Márquez-Neila, P., Sznitman, R.: Consistency-preserving visual question answering in medical imaging. In: Wang, L., Dou, Q., Fletcher, P.T., Speidel, S., Li, S. (eds.) Medical Image Computing and Computer Assisted Intervention - MICCAI 2022. Lecture Notes in Computer Science, vol. 13438, pp. 386–395. Springer Nature Switzerland, Cham (2022). https://doi.org/10.1007/978-3-031-16452-1_37
26. Teney, D., Abbasnedjad, E., van den Hengel, A.: Learning what makes a difference from counterfactual examples and gradient supervision. In: Vedaldi, A., Bischof, H., Brox, T., Frahm, J.-M. (eds.) ECCV 2020. LNCS, vol. 12355, pp. 580–599. Springer, Cham (2020). https://doi.org/10.1007/978-3-030-58607-2_34
27. Wu, J., Mooney, R.: Self-critical reasoning for robust visual question answering. In: Wallach, H., Larochelle, H., Beygelzimer, A., d'Alché-Buc, F., Fox, E., Garnett, R. (eds.) Advances in Neural Information Processing Systems, vol. 32 (2019)
28. Yang, Z., He, X., Gao, J., Deng, L., Smola, A.: Stacked attention networks for image question answering. In: Proceedings of the IEEE Conference on Computer Vision and Pattern Recognition, pp. 21–29 (2016)

29. Yu, Z., Yu, J., Fan, J., Tao, D.: Multi-modal factorized bilinear pooling with co-attention learning for visual question answering. In: Proceedings of the IEEE International Conference on Computer Vision, pp. 1821–1830 (2017)
30. Zhan, L.M., Liu, B., Fan, L., Chen, J., Wu, X.M.: Medical visual question answering via conditional reasoning. In: Proceedings of the 28th ACM International Conference on Multimedia, pp. 2345–2354 (2020)

Spatiotemporal Hub Identification in Brain Network by Learning Dynamic Graph Embedding on Grassmannian Manifold

Defu Yang[1,2], Hui Shen[3], Minghan Chen[4], Yitian Xue[3], Shuai Wang[1], Guorong Wu[2], and Wentao Zhu[3](✉)

[1] School of Automation, Hangzhou Dianzi University, Hangzhou, China
[2] Department of Psychiatry, University of North Carolina at Chapel Hill, Chapel Hill, USA
[3] Research Center for Augmented Intelligence, Zhejiang Lab, Hangzhou, China
wentao.zhu@zhejianglab.com
[4] Department of Computer Science, Wake Forest University, Winston-Salem, NC, USA

Abstract. Advancements in neuroimaging technology have made it possible to measure the connectivity evolution between different brain regions over time. Emerging evidence shows that some critical brain regions, known as hub nodes, play a significant role in updating brain network connectivity over time. However, current spatiotemporal hub identification is built on static network-based approaches, where hub regions are identified independently for each temporal brain network without considering their temporal consistency, and fails to align the evolution of hubs with changes in connectivity dynamics. To address this problem, we propose a novel spatiotemporal hub identification method that utilizes dynamic graph embedding to distinguish temporal hubs from peripheral nodes. Specially, to preserve the time consistency information, we put the dynamic graph embedding learning upon a smooth physics model of network-to-network evolution, which mathematically expresses as a total variation of dynamic graph embedding with respect to time. A novel Grassmannian manifold optimization scheme is further introduced to learn the embeddings accurately and capture the time-varying topology of brain network. Experimental results on real data demonstrate the highest temporal consistency in hub identification, surpassing conventional approaches.

1 Introduction

The human brain is a complex and economically organized system, consisting of interconnected regions that form a hierarchical brain network [1]. Understanding the topology of these networks is crucial for gaining insight into brain function and behavior [2]. Like many other real networks, brain network exhibits characteristics of a small-world and free-scale organization, where there are a small number of hub nodes that are densely connected to other peripheral regions [3, 4]. Recent studies show that hub nodes play a central role in adapting brain network connectivity to meet task demands [2, 5]. It has

D. Yang and H. Shen—These authors contributed equally.

H. Greenspan et al. (Eds.): MICCAI 2023, LNCS 14221, pp. 394–402, 2023.
https://doi.org/10.1007/978-3-031-43895-0_37

been also observed that most neurodegenerative and neuropsychiatric diseases are associated with alterations in dynamic functional connectivity (dFC) that occur selectively on hub nodes [6–9]. Therefore, accurate identification of hub nodes from brain networks, particularly in a dynamic scenario, is essential for understanding brain development and the neurodegenerative process.

In network science, hub nodes are often classified as provincial or connector hubs based on the information of network modules [4]. Provincial hubs have high connectivity within a single module, while connector hubs link multiple communities and play a more critical role in the overall network organization [4, 10]. With the consensus that damage to connector hubs can result in a wider disruption of the brain than damage to provincial hubs [11–13], identifying connector hubs is location-wise and function-wise more important. Multiple hub identification methods have been proposed and can be grouped into univariate-sorting-based [4, 14, 15] and multivariate-learning-based approaches [16]. The former involves selecting the top nodes ranked by certain nodal centrality, while the latter jointly identifies a set of critical nodes by learning topological features represented by low-dimensional graph embeddings.

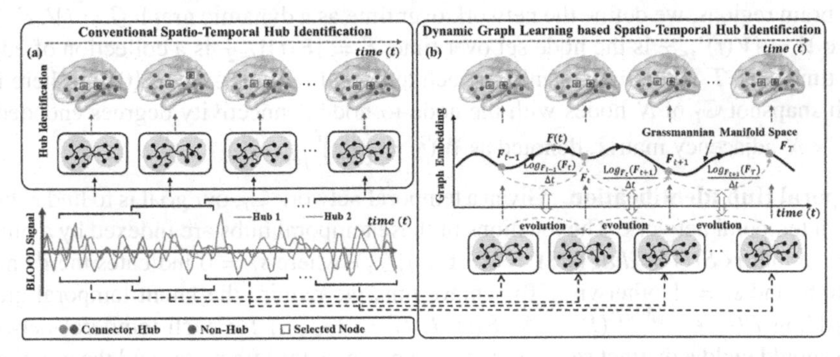

Fig. 1. Conventional (a) vs. dynamic graph embedding (b) approaches for spatiotemporal hub identification. In the proposed framework, temporal networks evolve from the previous state instead of being treated as independent static networks. Through learning a dynamic graph embedding that follows a physics-based network evolution model, the identified temporal hubs consistently align with dynamically changing connectivity.

Although various methods exist for identifying hub nodes in brain networks, there is a notable gap in effective approaches for dynamic scenarios. One common strategy is to treat each network as a separate static network and then use existing hub identification methods for each sliding window independently to obtain a set of spatiotemporal hub nodes, as illustrated in Fig. 1(a). Where a dynamic sequence of time-varying networks is generated applying a sliding window technique to BOLD signals in fMRI imaging data. While efficient, this approach lacks an in-depth exploration of how the topological evolution is linked to hub regions, and it often produces low-consistency hubs due to a low signal-to-noise ratio, as depicted in Fig. 1(a).

To address these challenges, we propose a novel learning-based spatiotemporal hub identification method. Unlike existing approaches that treat each temporal network as an independent static network, our method jointly identifies a set of temporal hub nodes using a dynamic graph embedding. Specially, due to dynamic graph embedding vectors [17–19] are the underlying representation of time-varying brain network, we can easily cast the network-to-network evolution over time as a total variation of dynamic graph embedding with respect to time, where each temporal graph embedding are not independent. Furthermore, in Fig. 1(b), as each temporal graph embedding vector is an instance residing on a Grassmannian manifold [19], we leverage the Grassmannian manifold optimization scheme to learn the dynamic graph embeddings. Overall, our spatiotemporal hub identification method ensures the temporal consistency of hub nodes over time while aligning the temporal hubs with the dynamic connectivity evolution in real time. We evaluate our proposed method on both synthetic and real brain network data, and results show that it outperforms conventional approaches.

2 Method

2.1 Dynamic Graph Embedding Learning

Dynamic Brain Network. Suppose we observe a time-varying brain network consisting of N brain regions, we define the network over time as a dynamic graph $\mathcal{G} = (\mathcal{V}, \mathcal{E}, \mathcal{T})$, where $\mathcal{V} = \{V(t)\}_{t \in \mathcal{T}}$ is the node set over time, $\mathcal{E} = \{E(t)\}_{t \in \mathcal{T}}$ is a collection of edges over time, and $\mathcal{T}$ is the time span. For each temporal point $t \in \mathcal{T} = [0, T]$, there is a graph snapshot $\mathbb{G}_t$ of N nodes with the node-to-node connectivity degrees encoded in an $N \times N$ adjacency matrix, denoted as $\boldsymbol{W}(t) = \left[w_{ij}\right]_{i,j=1}^{N}$.

Temporal Hub Identification. Given a temporal network $\mathbb{G}_t$, our goal is to find K hubs in each temporal network. The locations of these temporal hubs are indexed by a binary diagonal matrix $\boldsymbol{S}(t) = diag(\boldsymbol{s}(t)) = diag([s_i]_{i=1}^{N})$, where $s_i = 0$ indicates the ith node is a hub, and $s_i = 1$ otherwise. To achieve this, we require the latent temporal graph embedding $\boldsymbol{F}(t) \in \mathbb{R}^{N \times P} (P < N, \boldsymbol{F}(t)^{\boldsymbol{T}}\boldsymbol{F}(t) = \boldsymbol{I}_{P \times P},)$ for each temporal network $\boldsymbol{W}(t)$ should yield a distinct separation between connector hub nodes and the peripheral nodes. To link the learning of $\boldsymbol{F}(t)$ with the optimization of hub selection indicator $\boldsymbol{s}(t)$, we adopt the following objective function:

$$\arg\min_{\boldsymbol{F}(t),\boldsymbol{S}(t)} tr\Big(\boldsymbol{F}^T(t)\boldsymbol{L}_s(t)\boldsymbol{F}(t)\Big) = tr\Big(\boldsymbol{F}^T(t)(\boldsymbol{D}_s(t) - \boldsymbol{S}^T(t)\boldsymbol{W}(t)\boldsymbol{S}(t))\boldsymbol{F}(t)\Big), \quad (1)$$

where $\boldsymbol{L}_s(t) = \boldsymbol{D}_s(t) - \boldsymbol{S}^T(t)\boldsymbol{W}(t)\boldsymbol{S}(t)$ is the temporal degraded Laplacian matrix and $\boldsymbol{D}_s(t) = diag\left(\left[\sum_{j=1}^{N} s_i w_{ij} s_j\right]_{i=1}^{N}\right)$ is the diagonal matrix. The trace norm, $tr\big(\boldsymbol{F}^T(t)\boldsymbol{L}_s(t)\boldsymbol{F}(t)\big) = \sum_{i,j}^{N} s_i w_{ij} s_j \|\boldsymbol{F}_i(t) - \boldsymbol{F}_j(t)\|_2^2$, measures the smoothness of graph embedding in the context of the network topology governed by $\boldsymbol{L}_s(t)$, where $\boldsymbol{F}_i(t)$ and $\boldsymbol{F}_j(t)$ are the temporal embedding on nodes v_i and v_j. Suppose v_i is the temporal hub node and v_j is the linked non-hub node. Specifically, we want the distance term $\|\boldsymbol{F}_i(t) - \boldsymbol{F}_j(t)\|_2^2$ to be as large as possible to separate hub and peripheral nodes, while expecting that connector hubs, which have a higher connectivity degree than peripheral nodes, will have a large weight w_{ij}. Thus, identifying hub is searching for a solution that excluding all K hub nodes (let $s_i = 0$) from the objective function to minimize the trace norm.

Physics-Based Network Evolution Model. Since a temporal network evolves from the previous temporal state, the physics network-to-network evolution model can be mathematically described as follows:

$$F(t) = F(t - \Delta\tau) + \frac{\partial F(t)}{\partial t}\Delta\tau, \tag{2}$$

where $\Delta\tau$ is the time interval and $\frac{\partial F(t)}{\partial t}$ denotes the network-to-network evolution rate. In the context of neurobiological signals, the evolution of network connectivity is a smooth process rather than a mutational change. This means that as the time interval between two consecutive network states approaches zero, the 2-norm difference between two consecutive networks ($F(t)$ and $F(t - \Delta\tau)$) approaches a finite value, i.e., $\lim_{\Delta\tau\to 0} \| \frac{F(t)-F(t-\Delta\tau)}{\Delta\tau} \|_2^2 \neq \infty$. Therefore, to ensure that the network connectivity evolves smoothly over time, the evolution of $F(t)$ is subject to the minimization constraint of the integral of $F(t)$ change rate over the entire time period, i.e., $\int_0^T \| \frac{\partial F(t)}{\partial t} \|_2^2 dt$.

Spatiotemporal Hub Identification. By integrating the physics network-to-network evolution model, the spatiotemporal hub identification for the entire time series can be mathematically represented as:

$$\arg\min_{F(t),S(t)} \int_0^T tr\Big(F^T(t)\Big(D_s(t) - S^T(t)W(t)S(t)\Big)F(t)\Big)dt + \alpha \int_0^T \| \frac{\partial F(t)}{\partial t} \|_2^2 dt, \tag{3}$$

which includes two terms: the identification of temporal hub sets $S(t)$ over time using the learned dynamic graph embedding $F(t)$, and the regularization term that enforces the smoothness of $F(t)$ evolution over time. The scalar parameter α controls the trade-off between the two terms.

2.2 Optimization on Grassmannian Manifold

Equation (3) involves an integral-differential form with respect to (*w.r.t*) $F(t)$ and is hard to solve directly. To address this issue, we divide the entire time series into M segments with an interval of $\Delta t = \frac{T}{M}$, where $\lim_{M\to\infty} \frac{T}{M} = 0$. This allows us to decompose Eq. (3) into a linear discrete model based on the integration definition:

$$\arg\min_{F(t_m),S(t_m)} \sum_{m=1}^{M} \left(tr\Big(F^T(t_m)L_s(t_m)F(t_m)\Big) + \alpha \| \frac{\Delta F(t_m)}{\Delta t} \|_2^2 \right) \Delta t. \tag{4}$$

Since Eq. (4) is not a convex function, we adopt an alternative approach to jointly optimize $S(t_m)$ and $F(t_m)$ in the following.

Optimizing Dynamic Graph Embedding on Grassmannian Manifold. According to [19], each temporal graph embedding $F(t_m)$ can be represented as an instance on the Grassmannian manifold $\mathcal{I}(P, N) \in \mathbb{R}^{N\times P}$, which makes the classic Euclidean space unsuitable for measuring the variation of $\Delta F(t_m)$. In the context of physics, $\|\Delta F(t_m)\|_2^2$ signifies the evolving variation from the temporal state t_{m-1} to t_m, see Fig. 1(b). In the Grassmannian manifold space, $\|\Delta \boldsymbol{F}(t_m)\|_2^2$ can be accurately measured by the squared geodesic distance $Log_{\boldsymbol{F}(t_m)}(\boldsymbol{F}(t_{m-1})) = P - tr(\boldsymbol{F}(t_m)\boldsymbol{F}(t_m)^T\boldsymbol{F}(t_{m-1})\boldsymbol{F}(t_{m-1})^T)$. Therefore, the optimization of Eq. (4) *w.r.t* $\boldsymbol{F}(t_m)$ is as follows:

$$\boldsymbol{J} = \arg\min_{\boldsymbol{F}(t_m)} \sum_{m=1}^{M} tr\Big(\boldsymbol{F}^T(t_m)\boldsymbol{L}_s(t_m)\boldsymbol{F}(t_m) - \beta \boldsymbol{F}(t_m)\boldsymbol{F}(t_m)^T\boldsymbol{F}(t_{m-1})\boldsymbol{F}(t_{m-1})^T\Big), \quad (5)$$

where $\beta = \frac{\alpha}{\Delta t^2}$ is a scaler parameter. Following the optimization framework in [19], we calculate the Grassmannian gradient $\Delta_{\boldsymbol{F}(t_m)}$ of each temporal $\boldsymbol{F}(t_m)$ by projecting the Euclidean gradient $\frac{\partial \boldsymbol{J}(\boldsymbol{F}(t_m))}{\partial \boldsymbol{F}(t_m)} = 2(\boldsymbol{L}_s(t_m) - \sum_{i=-1, i\neq 0}^{1} \beta \boldsymbol{F}(t_{m+i})\boldsymbol{F}(t_{m+i})^T)\boldsymbol{F}(t_m)$ onto the tangent space via the orthogonal projection [19]:

$$\Delta_{\boldsymbol{F}(t_m)} = (\boldsymbol{I}_{N\times N} - \boldsymbol{F}(t_m)\boldsymbol{F}(t_m)^T)\frac{\partial \boldsymbol{J}(\boldsymbol{F}(t_m))}{\partial \boldsymbol{F}(t_m)} \quad (6)$$

Given $\Delta_{\boldsymbol{F}(t_m)}$, we update the modified $\boldsymbol{F}(t_m)$ using an exponential mapping operation [19]:

$$\boldsymbol{F}(t_m) = exp\big(-\Delta_{\boldsymbol{F}(t_m)}\big) = [\boldsymbol{F}(t_m)\boldsymbol{V}diag(\cos(\varepsilon\boldsymbol{\Sigma})) + \boldsymbol{U}diag(\sin(\varepsilon\boldsymbol{\Sigma}))]\boldsymbol{V}^T, \quad (7)$$

where $\boldsymbol{U}$, $\boldsymbol{\Sigma}$, and $\boldsymbol{V}$ are derived from the compact singular value decomposition (SVD) of $-\Delta_{\boldsymbol{F}m}$, i.e., $-\Delta_{\boldsymbol{F}m} = \boldsymbol{U}\boldsymbol{\Sigma}\boldsymbol{V}^T$. ε is a scalar parameter controlling the step size of optimization.

Optimizing Temporal Hub Node Set. After updating $\boldsymbol{F}(t_m)$, the energy function of Eq. (4) can be rearranged as Eq. (8) with $\boldsymbol{S}(t_m) = diag(\boldsymbol{s}(t_m))$:

$$\underset{\boldsymbol{s}(t_m)}{argmin} tr(\boldsymbol{s}(t_m)^T\boldsymbol{H}(t_m)\boldsymbol{s}(t_m)), \quad (8)$$

where $\boldsymbol{H}(t_m) = \left[h_{ij}\right]_{i,j}^{N}$ is an $N \times N$ matrix, and each element $h_{ij} = w_{ij}\|\boldsymbol{F}_i(t_m) - \boldsymbol{F}_j(t_m)\|_2^2$ represents the distance between the temporal graph embedding $\boldsymbol{F}(t_m)$ at the ith and j^{th} nodes. The optimal set of temporal hub nodes $\boldsymbol{s}$ at the t_m temporal point can be achieved using the convex optimization scheme proposed in [16].

3 Experiments and Results

We assess the performance of our spatiotemporal hub identification method on both simulated and real network data. Our proposed method, named Dynamic-Graph-Embedding-based method, not only learns the topological features of each temporal network but also jointly learns the evolving consistency pattern on the entire dynamic sequence. We compare our method with conventional approaches: the classic sorting-based hub identification method that uses nodal betweenness centrality [20] (referred to as Sorting-Betweenness-based method), and the multivariate-learning-based hub identification that learns the topological property independently for each temporal network (Graph-Embedding-based method [16]).

3.1 Accuracy and Robustness on Synthesized Network Data

Data Preparation. A set of synthesized time-variant networks were generated by the folling steps: (1) initialize a network with a specified number of nodes and connections; (2) set an evolution ratio (updated/total) to simulate the gradual process of network evolution. This involves keeping a fixed proportion of connections in the final state, while updating the remaining connections to generate the evolved network. Figure 2 shows toy examples of the synthesized time-varying network.

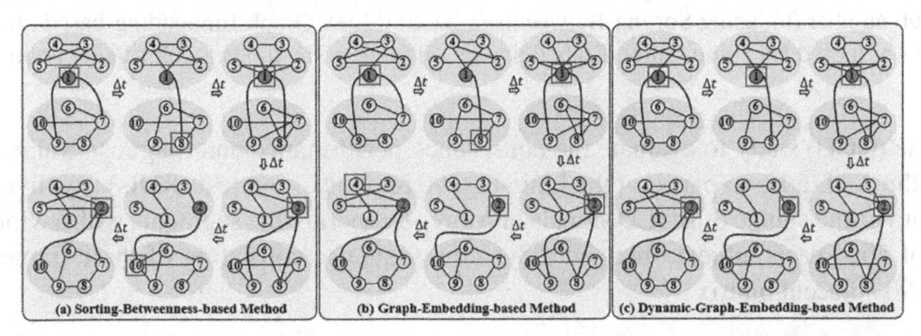

Fig. 2. Comparison of spatiotemporal hub identification using Sorting-Betweenness-based (a), Graph-Embedding-based (b), and Dynamic-Graph-Embedding-based (c) methods. The orange node represents the synthesized hub, while the other nodes are non-hubs.

Accuracy. A sequence of dynamic networks containing 6 temporal states was considered, with each temporal network containing two modules. Figure 2(a)-(c) illustrates the hubs identified for each temporal network using Sorting-Betweenness-based, Graph-Embedding-based, and Dynamic-Graph-Embedding-based methods, which are highlighted in a red box. The synthesized hubs (orange node) in the first three temporal states and the last three temporal states were labeled as node #1 and #2, respectively. It is evident that our proposed spatiotemporal hub identification approach can accurately identify the hub by incorporating the physics-based network evolution pattern.

Robustness. To further quantify the robustness of our proposed spatiotemporal hub identification method, we varied the complexity of network evolution and the network topology by changing the evolution ratio and hub number. Each time-varying network underwent 50 evolutions. The performance was evaluated by changing the evolution ratio from 40% to 70% and increasing the hub number in each temporal network while fixing other variables. We repeated the process 50 times and reported the overlap ratio between the identified temporal hubs and the ground truth across the entire dynamic sequence. Figure 3 shows that our proposed method consistently outperformed conventional methods, demonstrating its superior reliability.

3.2 Evaluation of Hub Identification on Real Brain Networks

Data Preparation. A total of 125 subjects consisting of 63 normal control (NC) and 62 obsessive-compulsive disease (OCD) were selected from an obsessive-compulsive

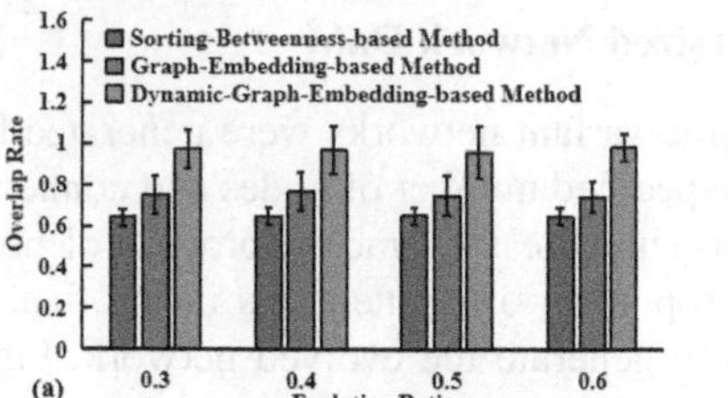

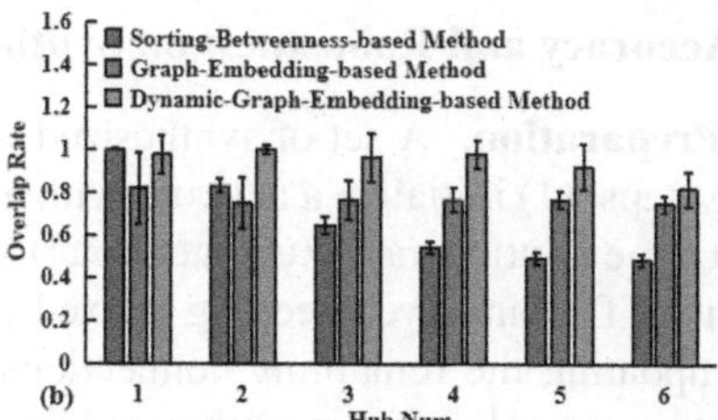

Fig. 3. Overlap ratio between ground truth and identified temporal hubs w.r.t evolution ratio (a) and hub number (b), using Sorting-Betweenness-based (blue), Graph-Embedding-based (green), and our proposed Dynamic-Graph-Embedding-based (orange) methods. (Color figure online)

disease (OCD) study to evaluate the consistency performance and diagnosis value. We used the AAL atlas to partition the brain into 90 regions of interest (ROIs). By dividing the entire time course of BOLD signals into three temporal time segments with a window size of 50 s, each subject obtained a dynamic sequence consisting of three frames of 90×90 adjacent matrix.

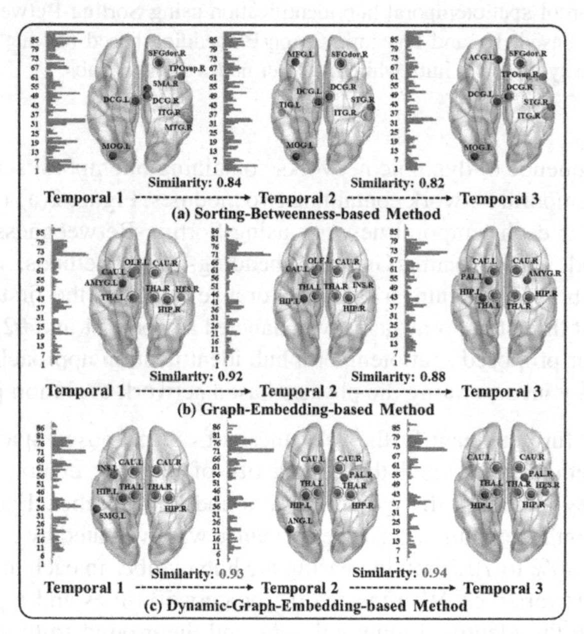

Fig. 4. Comparison of hub topology persistency across temporal states using Sorting-Betweenness-based (a), Graph-Embedding-based (b), and Dynamic-Graph-Embedding-based (c) methods. The orange circle represents the common hubs across temporal states. The left panel shows a count histogram of network nodes selected as hubs at each temporal state across subjects, while the right panel displays the most representative hub nodes voted across subjects.

Consistency of Hub Topology. We employed conventional engineering methods and our proposed spatiotemporal hub identification technique on each sliding window across subjects, as depicted in Fig. 4. The emerging evidence suggests that hubs remain stable even when switching brain states between tasks [5, 11]. Thus, it is also reasonable for us to hypothesize that hubs remain stable underlying a resting-state environment. Our proposed Dynamic-Graph-Embedding-based method yielded the highest consistency of hub locations across temporal states (6 common hubs), followed by the Graph-Embedding-based (5 common hubs) and Sorting-Betweenness-based methods (4 common hubs). To further quantify the similarity, we calculated the covariance of the count histogram between the last and current temporal states, and our method exhibited the highest similarity (Fig. 4(c)).

(a) Sorting-Betweenness-based Method

Hub	Feature1	Feature2	Feature1	Feature2	Hub
SFGdor.R	-	-	★★	★★★	STG.R
DCG.L	-	★	★	★	TPOsup.R
DCG.R	-	-	-	-	MTG.R
MOG.L	-	-	-	-	ITG.R

(b) Graph-Embedding-based Method

Hub	Feature1	Feature2	Feature1	Feature2	Hub
HIP.L	-	-	★★	★★	THA.L
HIP.R	★★★	★★	★★	★	THA.R
CAU.L	★	-	★★	★★	OLF.L
CAU.R	★	★★	★	-	AMYG.L

(c) Dynamic-Graph-Embedding-based Method

Hub	Feature1	Feature2	Feature1	Feature2	Hub
HIP.L	-	-	★★	★★	THA.L
HIP.R	★★★	★★	★★	★	THA.R
CAU.L	★	-	★★	★★	HES.R
CAU.R	★	★★	★	★	PAL.R

Feature1: Local Efficiency	Feature2: Clustering Coefficient
★★★ $p < 0.001$	★★ $0.001 \le p < 0.01$
★ $0.01 \le p \le 0.05$	- $p > 0.05$

Fig. 5. Statistical power at the hub nodes identified by Sorting-Betweenness-based (a), Graph-Embedding-based (b), and Dynamic-Graph-Embedding-based (c) methods. The significance after a two-sample t-test was indicated by red stars. (Color figure online)

Statistical Power of the Identified Hubs. As mounting evidence suggests that certain neurodegeneration and neuropsychiatric diseases selectively damage hub regions, we performed a two-sample *t*-test to assess the statistical power of the identified hub. We identified hubs for each subject at each temporal point and then voted out the eight most frequently selected nodes as the common hubs across the entire temporal series, as shown in Fig. 5. The significance was indicated by red stars. Our proposed spatiotemporal hub identification method yielded more hub nodes manifesting significant differences specific to OCD compared to the other two methods (Fig. 5(a)).

4 Conclusion

This paper introduces a novel spatiotemporal hub identification method. Our approach integrates the evolution model of network connectivity to ensure the consistency of dynamic graph embedding over time. The results on both simulated and real data are promising and suggest the great potential for investigating the role of hubs in the evolution of both task-based and resting-state-based networks.

References

1. Bullmore, E., Sporns, O.: The economy of brain network organization. Nat. Rev. Neurosci. **13**, 336 (2012)
2. Bertolero, M.A., Yeo, B.T.T., Bassett, D.S., D'Esposito, M.: A mechanistic model of connector hubs, modularity and cognition. Nat. Hum. Behav. **2**, 765–777 (2018)
3. Park, H.-J., Friston, K.: Structural and functional brain networks: from connections to cognition. Science **342**, 1238411 (2013)
4. van den Heuvel, M.P., Sporns, O.: Network hubs in the human brain. Trends Cogn. Sci. **17**, 683–696 (2013)
5. Cole, M.W., Reynolds, J.R., Power, J.D., Repovs, G., Anticevic, A., Braver, T.S.: Multi-task connectivity reveals flexible hubs for adaptive task control. Nat. Neurosci. **16**, 1348–1355 (2013)
6. Pedersen, M., Omidvarnia, A., Zalesky, A., Jackson, G.D.: On the relationship between instantaneous phase synchrony and correlation-based sliding windows for time-resolved fMRI connectivity analysis. Neuroimage **181**, 85–94 (2018)
7. Lee, W.J., et al.: Regional Aβ-tau interactions promote onset and acceleration of Alzheimer's disease tau spreading. Neuron **110**, 1932–1943, e1935 (2022)
8. Achard, S., et al.: Hubs of brain functional networks are radically reorganized in comatose patients. Proc. Natl. Acad. Sci. **109**, 20608–20613 (2012)
9. Frontzkowski, L., et al.: Earlier Alzheimer's disease onset is associated with tau pathology in brain hub regions and facilitated tau spreading. Nat. Commun. **13**, 4899 (2022)
10. Fornito, A., Zalesky, A., Breakspear, M.: The connectomics of brain disorders. Nat. Rev. Neurosci. **16**, 159 (2015)
11. Buckner, R.L., et al.: Cortical hubs revealed by intrinsic functional connectivity: mapping, assessment of stability, and relation to Alzheimer's disease. J. Neurosci. **29**, 1860–1873 (2009)
12. Gratton, C., Nomura, E.M., Pérez, F., Esposito, M.D.: Focal brain lesions to critical locations cause widespread disruption of the modular organization of the brain. J. Cogn. Neurosci. **24**, 1275–1285 (2012)
13. Tu, W., Ma, Z., Zhang, N.: Brain network reorganization after targeted attack at a hub region. Neuroimage **237**, 118219 (2021)
14. Jiao, Z., Xia, Z., Cai, M., Zou, L., Xiang, J., Wang, S.: Hub recognition for brain functional networks by using multiple-feature combination. Comput. Electr. Eng. **69**, 740–752 (2018)
15. Sporns, O.: Graph theory methods: applications in brain networks. Dialogues Clin. Neurosci. **20**, 111 (2018)
16. Yang, D., et al.: Joint hub identification for brain networks by multivariate graph inference. Med. Image Anal. **73**, 102162 (2021)
17. Newman, M.E.: Modularity and community structure in networks. Proc. Natl. Acad. Sci. **103**, 8577–8582 (2006)
18. Zhou, D., Huang, J., Schölkopf, B.: Learning with hypergraphs: clustering, classification, and embedding. In: Advances in Neural Information Processing Systems, vol. 19 (2006)
19. Cetingul, H.E., Vidal, R.: Intrinsic mean shift for clustering on Stiefel and Grassmann manifolds. In: 2009 IEEE Conference on Computer Vision and Pattern Recognition, pp. 1896–1902. IEEE (2009)
20. Tijms, B.M., et al.: Alzheimer's disease: connecting findings from graph theoretical studies of brain networks. Neurobiol. Aging **34**, 2023–2036 (2013)

Towards AI-Driven Radiology Education: A Self-supervised Segmentation-Based Framework for High-Precision Medical Image Editing

Kazuma Kobayashi[1,2](✉), Lin Gu[2,3], Ryuichiro Hataya[4,2], Mototaka Miyake[5], Yasuyuki Takamizawa[5], Sono Ito[5], Hirokazu Watanabe[5], Yukihiro Yoshida[5], Hiroki Yoshimura[6], Tatsuya Harada[3,2], and Ryuji Hamamoto[1,2]

[1] National Cancer Center Research Institute, Tokyo, Japan
kazumkob@ncc.go.jp
[2] RIKEN Center for Advanced Intelligence Project, Tokyo, Japan
[3] The University of Tokyo, Tokyo, Japan
[4] RIKEN Information R&D and Strategy Headquarters, Tokyo, Japan
[5] National Cancer Center Hospital, Tokyo, Japan
[6] Hiroshima University School of Medicine, Hiroshima, Japan

Abstract. Medical education is essential for providing the best patient care in medicine, but creating educational materials using real-world data poses many challenges. For example, the diagnosis and treatment of a disease can be affected by small but significant differences in medical images; however, collecting images to highlight such differences is often costly. Therefore, medical image editing, which allows users to create their intended disease characteristics, can be useful for education. However, existing image-editing methods typically require manually annotated labels, which are labor-intensive and often challenging to represent fine-grained anatomical elements precisely. Herein, we present a novel algorithm for editing anatomical elements using segmentation labels acquired through self-supervised learning. Our self-supervised segmentation achieves pixel-wise clustering under the constraint of invariance to photometric and geometric transformations, which are assumed not to change the clinical interpretation of anatomical elements. The user then edits the segmentation map to produce a medical image with the intended detailed findings. Evaluation by five expert physicians demonstrated that the edited images appeared natural as medical images and that the disease characteristics were accurately reproduced.

Keywords: Image editing · Self-supervised segmentation · Education

Supplementary Information The online version contains supplementary material available at https://doi.org/10.1007/978-3-031-43895-0_38.

H. Greenspan et al. (Eds.): MICCAI 2023, LNCS 14221, pp. 403–413, 2023.
https://doi.org/10.1007/978-3-031-43895-0_38

1 Introduction

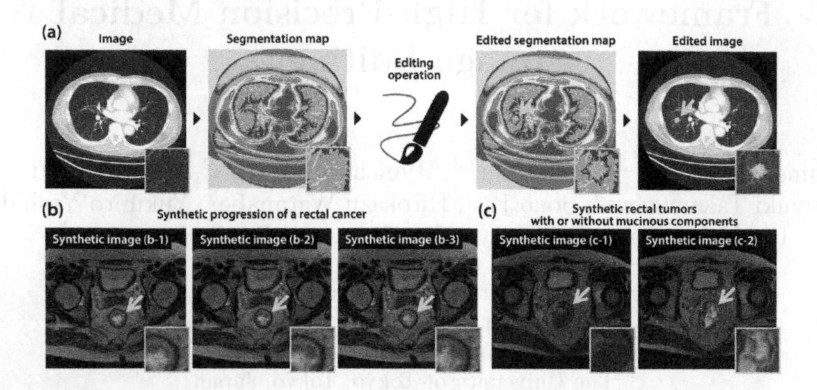

Fig. 1. Editing of anatomical elements. (a) Users can edit the segmentation map obtained from an input image to express intended fine-grained disease characteristics. A spiculated lung nodule was generated. **(b)** Synthetic disease progression showing a normal-appearing rectum (b-1), a rectal tumor extending into the submucosal layer (b-2), and the tumor extending into the muscularis propria (b-3). **(c)** A synthetic rectal tumor (c-1) and the contrasting tumor with T2 hyperintensity of extracellular mucin suspicious for mucinous adenocarcinoma (c-2).

Despite the success of artificial intelligence (AI) in aiding diagnosis, its application to medical education remains limited. Trainee physicians require several years of experience with a diverse range of clinical cases to develop sufficient skills and expertise. However, designing educational materials solely based on real-world data poses several challenges. For example, although small but significant disease characteristics (e.g., depth of cancer invasion) can sometimes alter diagnosis and treatment, collecting pairs with and without these characteristics is cumbersome. Another major challenge is longitudinal tracking of pathological progression over time (e.g., from the early stage of cancer to the advanced stage), which is difficult to understand because medical images are often snapshots. Privacy is also a concern since images of educational materials are widely distributed. Therefore, medical image editing that allows users to generate their intended disease characteristics is useful for precise medical education [3].

Image editing can synthesize low- or high-level image contents [11]. Our goal is to develop high-precision medical image editing according to the fine-grained characteristics of individual diseases, rather than at the level of disease categories. For example, even if two diseases belong to the same disease category of "lung tumor," the impression of benign or malignant will differ depending on fine-grained characteristics, such as whether the margins are "smooth" or "spiculated." In this case, our approach is to edit the tumor margins to be smooth or spiculated. These fine-grained characteristics consist of low- to mid-level image features to distinguish the substructures of organs and diseases, which we call *anatomical elements.*

Several types of image editing techniques for medical imaging have been introduced, mainly using generative adversarial networks [5] and, more recently, diffusion models [2]. Nevertheless, editing specific anatomical elements remains a challenge [1,11]. *Latent space manipulation* generates images by controlling latent feature axes [4,14], but the editable attributes are often global rather than fine-grained. *Conditional generation* can precisely edit image content by using class or segmentation labels. However, it requires manually provided labels [15] or virtual models [18], which are labor-intensive. Additionally, accurately modeling certain fine-grained characteristics, such as the textual variations of disease, can be a daunting task. *Image interpolation* [17] requires actual images with targeted content, which limits its applicability.

Here, we propose a novel framework for image editing called U3-Net that allows the generation of anatomical elements with precise conditions. The core technique is *self-supervised segmentation*, which aims to achieve pixel-wise clustering without manually annotated labels [6,7]. As shown in Fig. 1**a**, U3-Net converts an input image into a segmentation map corresponding to the anatomical elements. Once the user has completed editing, U3-Net synthesizes an image in which the targeted anatomical element has been modified. As a result, our synthesized medical images can highlight hypothetical pathological changes and significant clinical differences in a single image. For example, Fig. 1**b** shows that whether or not rectal cancer invades the muscularis propria (i.e., b-2 vs. b-3) affects cancer staging (i.e., T1 vs. T2) as well as treatment strategy (i.e., endoscopic resection vs. surgery). The distinction between mucinous and non-mucinous rectal cancers (see Fig. 1**c**) is also important to estimate the better or worse prognosis of the disease. These synthetic images can help trainees intuitively comprehend clinically significant findings and alleviate privacy concerns. Five expert physicians evaluated the edited images from a clinical perspective using two datasets: a pelvic MRI dataset and chest CT dataset.

Contributions: Our contributions are as follows:

- We propose a novel image-editing algorithm, U3-Net, to synthesize images for medical education via self-supervised segmentation.
- U3-Net can faithfully synthesize intended anatomical elements according to the editing operation on the segmentation labels.
- Evaluation by five expert physicians showed that the edited images were natural as medical images with the intended features.

2 Methodology

U3-Net consists of three neural networks: *encoder*, *decoder*, and *discriminator* (see Fig. 2). The encoder achieves self-supervised segmentation with a feature extraction (FE) module and a pixel-wise clustering (CL) module. We perform pixel-wise clustering under the constraint of invariance to photometric and

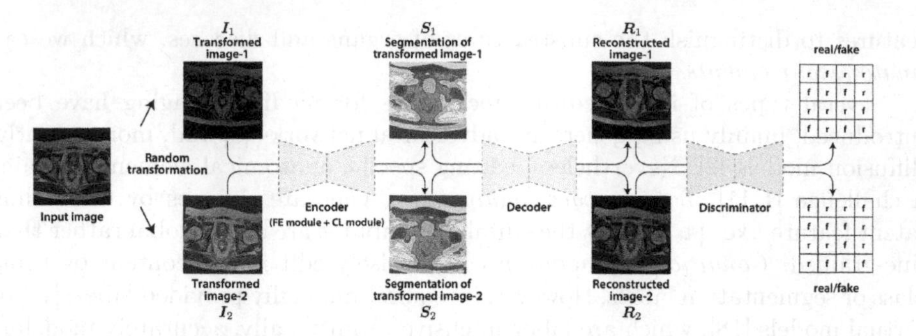

Fig. 2. Overall architecture of U3-Net. We apply two random transformations to the input image to produce images in different views, $\mathbf{I}_1$ and $\mathbf{I}_2$. The encoder converts the transformed images into quantized embedding as well as segmentation maps consisting of cluster indices, $\mathbf{S}_1$ and $\mathbf{S}_2$. Pixel-wise clustering, which should be consistent between views, is performed for the self-supervised segmentation. The decoder generates reconstructed images, $\mathbf{R}_1$ and $\mathbf{R}_2$, from the quantized embedding maps. The discriminator adversarially enhances the natural appearance by judging whether the images are real or fake on a pixel-by-pixel basis.

geometric transformations [6], with the assumption that these transformations should not change the clinical interpretation of the anatomical elements. Given a pair of differently transformed images, the FE module produces *embedding maps* corresponding to the input images. The CL module then performs K-means clustering on the embedding maps to produce two interchangeable outputs: *segmentation maps* and corresponding *quantized embedding maps.* These outputs are trained to be consistent between the two views. The decoder then estimates the corresponding images from the quantized embedding maps, while the discriminator forces the decoder to produce more realistic images.

2.1 First Training Stage for Self-supervised Segmentation

The training process for U3-Net is two-stage. First, we train the encoder and decoder (excluding the discriminator) to conduct K-class self-supervised segmentation. To achieve pixel-wise clustering that is consistent between two transformed views of the input images, we introduce four constraints: intra-cluster pull force, inter-cluster push force, cross-view consistency, and reconstruction loss.

Random Image Transformation: We consider a sequence of image transformations $[t_1, \ldots, t_n]$ specified by the type (e.g., image rotation) and magnitude (e.g., degree of rotation) of each transformation: $\mathcal{T} = t_n \circ t_{n-1} \circ \cdots \circ t_1$. Two random transformation sequences are applied to an input image $\mathbf{I} \in \mathbb{R}^{C \times H \times W}$ to produce two transformed images, $\mathcal{T}_1(\mathbf{I}) = \mathbf{I}_1$ and $\mathcal{T}_2(\mathbf{I}) = \mathbf{I}_2$. The FE module of the encoder produces two embedding maps $f(\mathbf{I}) = \mathbf{E} \in \mathbb{R}^{D \times H \times W}$, $\mathbf{E}_1$ and $\mathbf{E}_2$, which are then fed into the CL module.

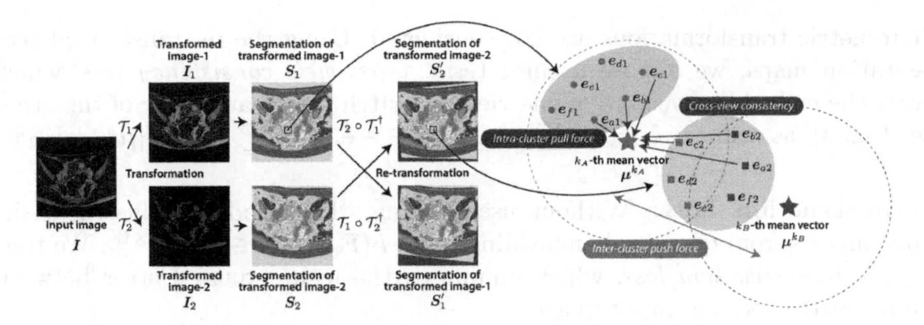

Fig. 3. Transformation-invariant pixel-wise clustering. Suppose that the majority of pixels inside the black box in $\mathbf{S}_1$ are assigned to the k_A-th cluster. The intra-cluster pull force causes the embedding vectors $\mathbf{e}_{a1}, \ldots \mathbf{e}_{f1}$ to adhere to the mean vector $\boldsymbol{\mu}^{k_A}$. From the other viewpoint, some of the same pixels, $\mathbf{e}_{a2}$, $\mathbf{e}_{b2}$, and $\mathbf{e}_{f2}$, are assigned to the k_B-th cluster, which can be assessed by re-transforming $\mathbf{S}_2$ into the coordination of $\mathbf{S}_1$. *Cross-view consistency loss* forces the embedding vectors of one view, $\mathbf{e}_{a2}, \ldots \mathbf{e}_{f2}$, to match the mean vector of the other view $\boldsymbol{\mu}^{k_A}$. The inter-cluster push force maintains the distance between the mean vectors.

Cluster Assignment and Update: In the CL module, K-means clustering in the first iteration initializes K mean vectors $\boldsymbol{\mu}^k \in \mathbb{R}^D$. Then, the embedding vector of the i-th pixel $\mathbf{e}_{i\in\{1,\ldots,H\times W\}} \in \mathbb{R}^D$ in the embedding maps, $\mathbf{E}_1$ and $\mathbf{E}_2$, is assigned to the cluster with the nearest mean vector as follows: $y_i = \mathrm{argmin}_{k\in\{1,\ldots,K\}} \|\mu^k - \mathbf{e}_i\|^2$, where y_i is the cluster index of the i-th pixel. By replacing embedding vectors with their respective mean vectors, quantized embedding maps, $\mathbf{E}_{\mathrm{q}1}$ and $\mathbf{E}_{\mathrm{q}2}$, are generated $g(\mathbf{E}) = \mathbf{E}_{\mathrm{q}} = [\boldsymbol{\mu}^{y_1}, \ldots, \boldsymbol{\mu}^{y_{H\times W}}] \in \mathbb{R}^{D\times H\times W}$. The cluster indices form the segmentation maps $\mathbf{S} = [y_1, \ldots, y_{H\times W}] \in \mathbb{R}^{H\times W}$, $\mathbf{S}_1$ and $\mathbf{S}_2$. The mean vectors $\boldsymbol{\mu}^k$ are updated by using the exponential moving average [9].

Intra-cluster Pull Force: For transformation-invariant pixel-wise clustering, we define four loss terms. The first term, *cluster loss*, forces the embedding vectors to adhere to the associated mean vector (see Fig. 3), as defined: $L_{\mathrm{cluster}} = \sum_{i\in H\times W} \|\boldsymbol{\mu}^{y_i} - \mathbf{e}_i\|^2$.

Inter-cluster Push Force: The second term, *distance loss*, pushes the distance between the mean vectors above a margin parameter m (see Fig. 3), as defined: $L_{\mathrm{dist}} = \frac{1}{K(K-1)} \sum_{k_A=1}^{K} \sum_{k_B=1, k_B\neq k_A}^{K} [2m - \|\boldsymbol{\mu}^{k_A} - \boldsymbol{\mu}^{k_B}\|]_+^2$, where k_A and k_B indicate two different cluster indices.

Cross-view Consistency: The segmentation maps from the different views, $\mathbf{S}_1$ and $\mathbf{S}_2$, should overlap after re-transforming to align the coordinates. Such a re-transform is composed of inverse and forward geometric transformations: $\mathcal{T}_2(\mathcal{T}_1^\dagger(\mathbf{S}_1)) = \mathbf{S}_1'$ and $\mathcal{T}_1(\mathcal{T}_2^\dagger(\mathbf{S}_2)) = \mathbf{S}_2'$. The inverse transformations of the photometric transformations are not considered. Using the re-transformed segmentation maps, we impose a third term, *cross-view consistency loss*, which forces the embedding vectors of one view to match the mean vector of the other (see Fig. 3), as defined: $L_{\text{cross}} = \sum_{i \in H \times W} \|\boldsymbol{\mu}^{y_{i2}} - \mathbf{e}_{i1}\|^2 + \sum_{i \in H \times W} \|\boldsymbol{\mu}^{y_{i1}} - \mathbf{e}_{i2}\|^2$.

Reconstruction Loss: Without user editing, the decoder reconstructs the input images from quantized embedding maps $h(\mathbf{E}_{\text{q}}) = \mathbf{R} \in \mathbb{R}^{C \times H \times W}$. We thus employ *reconstruction loss*, which minimizes the mean squared error between the reconstructed and input images.

Learning Objective: The weighted sum of the loss functions is set to be minimized: $L_{\text{total}} = w_{\text{cluster}} L_{\text{cluster}} + w_{\text{dist}} L_{\text{dist}} + w_{\text{cross}} L_{\text{cross}} + w_{\text{recon}} L_{\text{recon}}$.

2.2 Second Training Stage for Faithful Image Synthesis

In the second stage, we train the decoder and discriminator (excluding the encoder) to produce naturally appearing images from the quantized embedding maps. The decoder, initially optimized in the first training stage, undergoes further training to enhance its image generation capabilities. We impose adversarial learning with an extended reconstruction loss term, called *appearance loss*. The training is performed only in a single view.

Appearance Loss: Appearance loss combines *mean squared loss* L_{mse}, *focal frequency loss* L_{ffl} [8], *perceptual loss* L_{lpips} [19], and *intermediate loss* L_{int}, as follows: $L_{\text{app}} = w_{\text{mse}} L_{\text{mse}} + w_{\text{ffl}} L_{\text{ffl}} + w_{\text{lpips}} L_{\text{lpips}} + w_{\text{int}} L_{\text{int}}$, where *intermediate loss* L_{int} refers to the L2 distance of the intermediate features of the discriminator between the reconstructed and input images.

Learning Objective: We impose *generator loss* L_{gen} for the decoder to produce more faithful images by deceiving the discriminator, and *discriminator loss* L_{dis} for the discriminator to judge the real or fake of the images as the per-pixel feedback [16]. We also add *cutmix augmentation* L_{cutmix} and *consistency regularization* L_{cons} to the latter [16]. In this stage, the decoder and discriminator are trained by alternately minimizing the following competing objectives: $L_{\text{Dec}} = L_{\text{app}} + w_{\text{gen}} L_{\text{gen}}$ and $L_{\text{Dis}} = w_{\text{dis}} L_{\text{dis}} + w_{\text{cutmix}} L_{\text{cutmix}} + w_{\text{cons}} L_{\text{cons}}$.

2.3 Inference Stage for Medical Image Editing

After training, the encoder can output a segmentation map from a testing image. As shown in Fig. 1**a**, when a user edits the segmentation map $\mathbf{S} \rightarrow \mathbf{S}'$ by changing the cluster indices $y_i \rightarrow y_i'$, the quantized embedding map is subsequently updated $\mathbf{E}_{\text{q}} \rightarrow \mathbf{E}_{\text{q}}'$ by reassigning the mean vectors according to the edited indices $\boldsymbol{\mu}^{y_i} \rightarrow \boldsymbol{\mu}^{y_i'}$. Finally, the decoder converts the quantized embedding map into a synthetic image with the intended disease characteristics $h(\mathbf{E}_{\text{q}}') = \mathbf{R} \in \mathbb{R}^{C \times H \times W}$.

3 Experiments and Results

Implementation and Datasets: All neural networks were implemented in Python 3.8 using the PyTorch library 1.10.0 [12] on an NVIDIA Tesla A100 GPU running CUDA 10.2. The encoder, decoder, and discriminator were implemented based on U-Net [13] (see **Supplementary Information** for details). The *pelvic MRI dataset* with rectal cancer contained 289 image series for training and 100 image series for testing. For each image series, the min-max normalization converted the pixel values to $[-1, 1]$. The *chest CT dataset* with lung cancer contained 500 image series for training and 100 image series for testing. The CT values in the range $[-2048, 2048]$ were normalized to $[-1, 1]$. Both were in-house datasets collected from a single hospital. Every image series comprises two-dimensional (2D) consecutive slices, and we applied our algorithm on a per 2D slice basis.

Self-supervised Medical Image Segmentation: We began by optimizing the hyperparameters to achieve self-supervised segmentation. Appropriate transformations were selected from six candidate functions: t_1, `Random HorizontalFlip`, t_2, `RandomAffine`, t_3, `ColorJitter`, t_4, `RandomGaussianBlur`, t_5, `RandomPosterize`, t_6, `RandomGaussianNoise`. Because anatomical elements, including the substructures of organs and diseases, are too detailed for human annotators to segment, it was difficult to create ground-truth labels. Therefore, the training configuration was selected based on the consensus of two expert radiologists with domain knowledge. By comparing different settings on the pelvic MRI training dataset (see **Supplementary Information**), the number of segmentation classes of 10, the combination of t_1, t_2, and t_3 with moderate magnitude, the weakly imposed reconstruction loss, and a certain value of the margin parameter were considered suitable for self-supervised segmentation. In particular, we found that reconstruction loss is essential for obtaining segmentation maps corresponding to anatomical elements, although such a loss term was not included in previous studies [6,7]. A similar configuration was applied to the chest CT training dataset. The resultant segmentation maps are shown in Fig. 4**ab**. The anatomical substructures, including the histological structure of the colorectal wall and subregions within the lung, corresponded well with the segmentation maps in both the pelvic MRI and chest CT testing datasets. Because our self-supervised segmentation extracts low- to mid-level image content, a semantic object (e.g., rectum or lung cancer) typically consists of multiple segmentation classes shared with other objects (see the magnified images in Fig. 4**ab**). These anatomical elements may be too detailed for humans to annotate, demonstrating the necessity of self-supervised segmentation for high-precision medical-image editing.

Evaluation of the Synthesized Images: We measured the quality of image reconstruction using mean square error (MSE), structural similarity (SSIM), and peak signal-to-noise ratio (PSNR). The mean $\pm$ standard deviations of MSE,

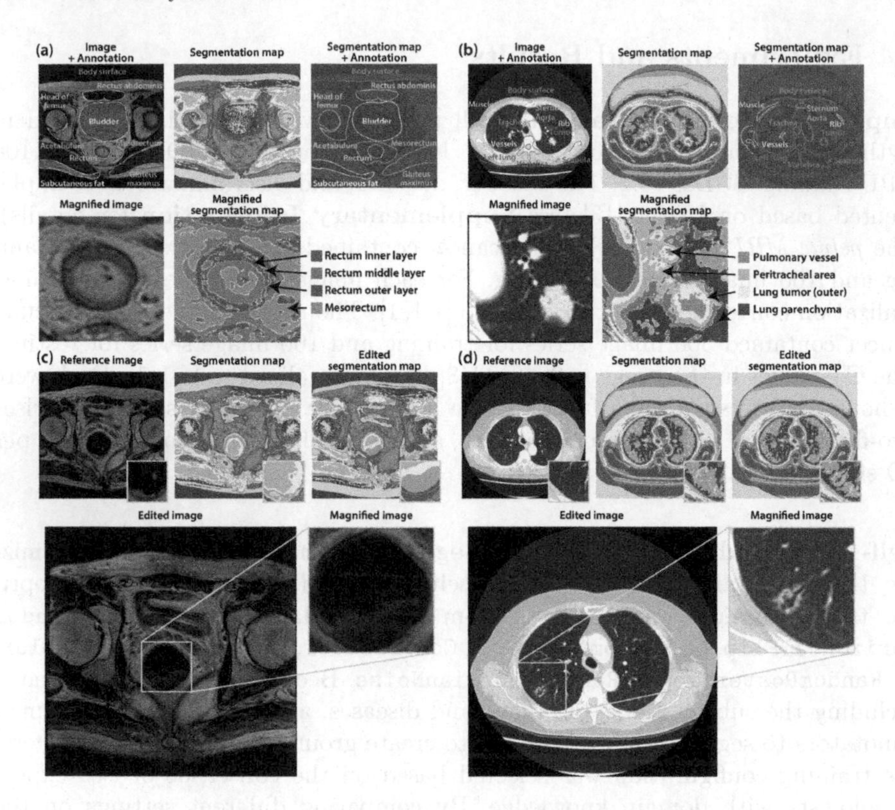

Fig. 4. Results of the image segmentation and editing. The segmentation maps were well aligned with the anatomical elements in both **(a)** the pelvic MRI and **(b)** the chest CT testing datasets. **(c)** A synthetic image generated by editing the testing image with the caption, *"Axial T2-weighted MR image shows a tumor approximately 4 cm in size on the dorsal wall of the rectum. The deepest structure of the rectal wall was intact, indicating no infiltration beyond the muscularis propria."* **(d)** A synthetic image with the caption, *"Axial CT image showing a pulmonary nodule with a length of 2–3 cm and a cavity on the dorsal side of the right upper lobe of the lung."*

SSIM, and PSNR were $1.41 \times 10^{-2} \pm 1.04 \times 10^{-2}$, $7.40 \times 10^{-1} \pm 0.57 \times 10^{-1}$, and 22.5 ± 2.7 in the pelvic MRI testing dataset and $5.03 \times 10^{-4} \pm 3.03 \times 10^{-4}$, $9.08 \times 10^{-1} \pm 0.34 \times 10^{-1}$, and 38.6 ± 1.7 in the chest CT testing dataset. Subsequently, segmentation maps from the testing images were edited to generate images with the intended characteristics (see Fig. 4**cd**). Five expert physicians (two diagnostic radiologists, two colorectal surgeons, and one thoracic surgeon) assessed them from a clinical perspective. First, we tested whether the evaluators could identify real or synthesized images from 20 images, which include ten real images and ten synthesized images. The accuracies (i.e., the ratio of images correctly identified as real or synthetic) were 0.69 ± 0.11 and 0.65 ± 0.11, for the pelvic MRI and chest CT testing datasets, respectively. Note that when the synthetic images cannot be distinguished at all, the accuracy should be 0.5. Second, we presented image captions explaining the radiological features, which also represented the editing intention for the synthetic images. We asked the evaluators to rate each presented image from A to C. ***A**: The image is natural as a medical image, and the caption is consistent with the image.* ***B**: The image is natural as a medical image, but the caption is NOT consistent with the image.* ***C**: The image is NOT natural as a medical image.* This test was conducted after informing the evaluators of the assumption that all 20 images could be synthetic, without indicating which image was real or synthetic. As a result, the ratio of synthetic images (vs. that of real images) categorized as A, B, and C were 0.80 ± 0.15 (vs. 0.78 ± 0.20), 0.02 ± 0.04 (vs. 0.08 ± 0.07), and 0.18 ± 0.11 (vs. 0.14 ± 0.13) for the pelvic MRI testing dataset, and 0.74 ± 0.28 (vs. 0.76 ± 0.30), 0.08 ± 0.09 (vs. 0.12 ± 0.15), and 0.18 ± 0.21 (vs. 0.12 ± 0.14) for the chest CT testing dataset. There were no significant differences between real and synthetic images (t-test: $p > 0.05$). Consequently, the majority of the edited images were natural-looking medical images with accurately reproduced disease features.

4 Conclusion

In this study, we propose a medical image-editing framework to edit fine-grained anatomical elements. The self-supervised segmentation extracted low- to mid-level content of medical images, which corresponded well to the clinically meaningful substructures of organs and diseases. The majority of the edited images with intended characteristics were perceived as natural medical images by several expert physicians. Our medical image editing method can be applied to medical education, which has been overlooked as an application of AI. Future challenges include improving scalability with fewer manual operations, validating segmentation maps from a more objective perspective, and comparing our proposed algorithm with existing methods, such as those based on superpixels [10].

Data use declaration and acknowledgment: The pelvic MRI and chest CT datasets were collected from the National Cancer Center Hospital. The study, data use, and data protection procedures were approved by the Ethics Committee of the National Cancer Center, Tokyo, Japan (protocol number 2016-496). Our implementation and all synthesized images will be available here: https://github.com/Kaz-K/medical-image-editing.

References

1. Chen, Y., et al.: Generative adversarial networks in medical image augmentation: a review. Comput. Biol. Med. **144**, 105382 (2022). https://doi.org/10.1016/j.compbiomed.2022.105382
2. Dhariwal, P., Nichol, A.: Diffusion Models Beat GANs on Image Synthesis. In: Ranzato, M., Beygelzimer, A., Dauphin, Y., Liang, P., Vaughan, J.W. (eds.) Advances in Neural Information Processing Systems (NeurIPS). vol. 34, pp. 8780–8794. Curran Associates, Inc. (2021). https://proceedings.neurips.cc/paper_files/paper/2021/file/49ad23d1ec9fa4bd8d77d02681df5cfa-Paper.pdf

3. Duong, M.T., et al.: Artificial intelligence for precision education in radiology. Br. J. Radiol. **92**(1103), 20190389 (2019). https://doi.org/10.1259/BJR.20190389
4. Fetty, L., et al.: Latent space manipulation for high-resolution medical image synthesis via the StyleGAN. Z. Med. Phys. **30**(4), 305–314 (2020). https://doi.org/10.1016/j.zemedi.2020.05.001
5. Goodfellow, I., et al.: Generative Adversarial Nets. In: Ghahramani, Z., Welling, M., Cortes, C., Lawrence, N., Weinberger, K. (eds.) Advances in Neural Information Processing Systems (NIPS). vol. 27 (2014). https://proceedings.neurips.cc/paper/2014/file/5ca3e9b122f61f8f06494c97b1afccf3-Paper.pdf
6. Hyun Cho, J., Mall, U., Bala, K., Hariharan, B.: PiCIE: Unsupervised Semantic Segmentation using Invariance and Equivariance in Clustering. In: IEEE/CVF Conference on Computer Vision and Pattern Recognition (CVPR), pp. 16789–16799 (2021). https://doi.org/10.1109/CVPR46437.2021.01652
7. Ji, X., Vedaldi, A., Henriques, J.: Invariant Information Clustering for Unsupervised Image Classification and Segmentation. In: IEEE/CVF International Conference on Computer Vision (ICCV), pp. 9864–9873 (2019). https://doi.org/10.1109/ICCV.2019.00996
8. Jiang, L., Dai, B., Wu, W., Loy, C.C.: Focal Frequency Loss for Image Reconstruction and Synthesis. In: IEEE/CVF International Conference on Computer Vision (ICCV), pp. 13899–13909 (2021). https://doi.org/10.1109/ICCV48922.2021.01366
9. Kaiser, L., Roy, A., Vaswani, A., Parmar, N., Bengio, S., Uszkoreit, J., Shazeer, N.: Fast Decoding in Sequence Models using Discrete Latent Variables. In: Dy, J., Krause, A. (eds.) Proceedings of the 35th International Conference on Machine Learning (ICML). Proceedings of Machine Learning Research, 80, pp. 2390–2399 (2018). https://proceedings.mlr.press/v80/kaiser18a.html
10. Li, H., Wei, D., Cao, S., Ma, K., Wang, L., Zheng, Y.: Superpixel-guided label softening for medical image segmentation. In: Martel, A.L., et al. (eds.) MICCAI 2020. LNCS, vol. 12264, pp. 227–237. Springer, Cham (2020). https://doi.org/10.1007/978-3-030-59719-1_23
11. Ling, H., Kreis, K., Li, D., Kim, S.W., Torralba, A., Fidler, S.: EditGAN: High-Precision Semantic Image Editing. In: Ranzato, M., Beygelzimer, A., Dauphin, Y., Liang, P., Vaughan, J.W. (eds.) Advances in Neural Information Processing Systems (NeurIPS) vol. 34, pp. 16331–16345 (2021). https://proceedings.neurips.cc/paper/2021/file/880610aa9f9de9ea7c545169c716f477-Paper.pdf
12. Paszke, A.,et al.: PyTorch: An Imperative Style, High-Performance Deep Learning Library. In: Wallach, H., Larochelle, H., Beygelzimer, A., d'Alché-Buc, F., Fox, E., Garnett, R. (eds.) Advances in Neural Information Processing Systems (NeurIPS), vol. 32, pp. 8024–8035 (2019). https://proceedings.neurips.cc/paper/2019/hash/bdbca288fee7f92f2bfa9f7012727740-Abstract.html
13. Ronneberger, O., Fischer, P., Brox, T.: U-Net: convolutional networks for biomedical image segmentation. In: Navab, N., Hornegger, J., Wells, W.M., Frangi, A.F. (eds.) MICCAI 2015. LNCS, vol. 9351, pp. 234–241. Springer, Cham (2015). https://doi.org/10.1007/978-3-319-24574-4_28
14. Saboo, A., Gyawali, P.K., Shukla, A., Sharma, M., Jain, N., Wang, L.: Latent-optimization based disease-aware image editing for medical image augmentation. In: 32nd British Machine Vision Conference (BMVC). p. 181 (2021). https://www.bmvc2021-virtualconference.com/assets/papers/0840.pdf

15. Sasuga, S., et al.: Image Synthesis-Based Late Stage Cancer Augmentation and Semi-supervised Segmentation for MRI Rectal Cancer Staging. In: Nguyen, H.V., Huang, S.X., Xue, Y. (eds.) Data Augmentation, Labelling, and Imperfections. pp. 1–10. Springer Nature Switzerland, Cham (2022). https://link.springer.com/chapter/10.1007/978-3-031-17027-0_1
16. Schonfeld, E., Schiele, B., Khoreva, A.: A U-Net based discriminator for generative adversarial networks. In: IEEE/CVF Conference on Computer Vision and Pattern Recognition (CVPR), pp. 8207–8216 (2020). https://doi.org/10.1109/CVPR42600.2020.00823
17. Thermos, S., Liu, X., O'Neil, A., Tsaftaris, S.A.: Controllable cardiac synthesis via disentangled anatomy arithmetic. In: de Bruijne, M., et al. (eds.) MICCAI 2021. LNCS, vol. 12903, pp. 160–170. Springer, Cham (2021). https://doi.org/10.1007/978-3-030-87199-4_15
18. Tiago, C., Snare, S.R., Šprem, J., McLeod, K.: A domain translation framework with an adversarial denoising diffusion model to generate synthetic datasets of echocardiography images. IEEE Access **11**, 17594–17602 (2023). https://doi.org/10.1109/ACCESS.2023.3246762
19. Zhang, R., Isola, P., Efros, A.A., Shechtman, E., Wang, O.: The unreasonable effectiveness of deep features as a perceptual metric. In: IEEE/CVF Conference on Computer Vision and Pattern Recognition (CVPR), pp. 586–595 (2018). https://doi.org/10.1109/CVPR.2018.00068

Rethinking Semi-Supervised Federated Learning: How to Co-train Fully-Labeled and Fully-Unlabeled Client Imaging Data

Pramit Saha(✉), Divyanshu Mishra, and J. Alison Noble

Department of Engineering Science, University of Oxford, Oxford, UK
pramit.saha@eng.ox.ac.uk

Abstract. The most challenging, yet practical, setting of semi-supervised federated learning (SSFL) is where a few clients have fully labeled data whereas the other clients have fully unlabeled data. This is particularly common in healthcare settings where collaborating partners (typically hospitals) may have images but not annotations. The bottleneck in this setting is the joint training of labeled and unlabeled clients as the objective function for each client varies based on the availability of labels. This paper investigates an alternative way for effective training with labeled and unlabeled clients in a federated setting. We propose a novel learning scheme specifically designed for SSFL which we call Isolated Federated Learning (IsoFed) that circumvents the problem by avoiding simple averaging of supervised and semi-supervised models together. In particular, our training approach consists of two parts - (a) isolated aggregation of labeled and unlabeled client models, and (b) local self-supervised pretraining of isolated global models in all clients. We evaluate our model performance on medical image datasets of four different modalities publicly available within the biomedical image classification benchmark MedMNIST. We further vary the proportion of labeled clients and the degree of heterogeneity to demonstrate the effectiveness of the proposed method under varied experimental settings.

1 Introduction

Federated Learning (FL) [10–12,27] is a distributed learning approach that allows the collaborative training of machine learning models using data from decentralized sources while preserving data privacy. However, most current FL methods have limitations, including assuming fully annotated and homogeneous data distribution among local clients. In a practical scenario, like a multi-institutional healthcare collaboration, the participating clients (*i.e.*, medical institutions and hospitals) may not have the incentive or resources to annotate their data [16]. To address this, semi-supervised federated learning (SSFL)

Supplementary Information The online version contains supplementary material available at https://doi.org/10.1007/978-3-031-43895-0_39.

H. Greenspan et al. (Eds.): MICCAI 2023, LNCS 14221, pp. 414–424, 2023.
https://doi.org/10.1007/978-3-031-43895-0_39

[4,16,28] methods have been proposed to utilize unlabeled data and integrate semi-supervised learning algorithms [2,19–21,26] into federated settings.

Based on the availability of labeled data, the existing SSFL studies can be classified into two main scenarios: (a) labels-at-client, with each client having some labeled and some unlabeled data [9,15], (b) labels-at-server, with each client possessing only unlabeled data and the server possessing some labeled data [4,7,9,28]. We argue that a more realistic SSFL scenario which is highly challenging but rarely explored in the literature is where some clients have labeled data, and others have completely unlabeled data [14,16,24].

The classic federated averaging scheme aggregates weights of all labeled and unlabeled client models trained in parallel. The labeled clients typically use cross-entropy-based loss functions while the unlabeled clients primarily use consistency regularization loss [19] or pseudo-labeling-based [1,23] semi-supervised learning schemes. This results in high gradient diversity [28] between the supervised and unsupervised models particularly in heterogeneous client settings, as these are targeted to optimize separate objective functions. As a result, the aggregated global model is weak and unable to capture a strong representation of either group of clients. This, in turn, leads to the generation of noisy targets for unlabeled clients and hence the global model fails to converge. The situation is further aggravated under non-IID data distribution conditions where the labeled client class distribution varies greatly from that of unlabeled clients. This naturally poses the following important question: *"How can we effectively co-train supervised and unsupervised models under FL setting that aim to optimize separate objective functions at their respective heterogeneous labeled data or unlabeled data clients?"*

To address this question, we present a novel SSFL algorithm which we call IsoFed that effectively improves client training by isolating the model aggregation of labeled and unlabeled client groups while still leveraging one group of models to improve another. In summary, the primary contributions of this paper are:

1. We propose IsoFed, a novel SSFL framework, that realizes isolated aggregation of labeled and unlabeled client models in the server followed by federated self-supervised pretraining of the global model in each individual site.
2. This is the first work to reformulate model aggregation for fully labeled and fully unlabeled clients under SSFL settings. To the best of our knowledge, we are the first to isolate the aggregation of labeled and unlabeled client models while switching between the two client groups.
3. This work bridges the gap between Federated Learning and Transfer Learning (TL) [22] by combining the best of both worlds for learning across sites. First, we conduct federated model aggregation among the labeled or unlabeled client groups. Next, we leverage Transfer Learning to allow knowledge transfer between the two groups. Therefore, we avoid the issue of averaging the supervised and unsupervised models with high gradient diversity in the context of SSFL while also being unaffected by catastrophic forgetting encountered in multi-domain transfer learning.

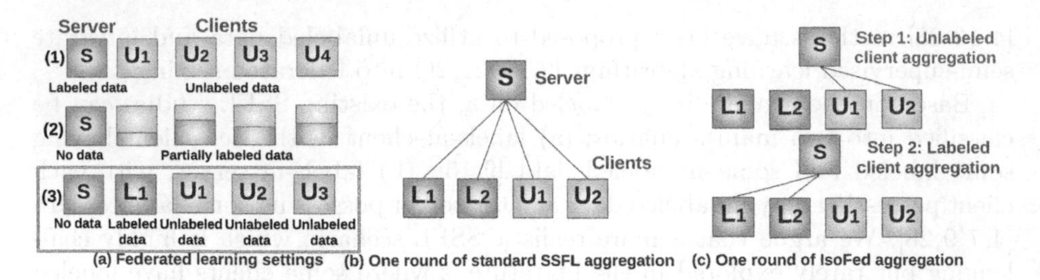

Fig. 1. Problem settings and aggregation schemes for semi-supervised federated learning. (a) Three plausible semi-supervised federated learning settings. We address the unique condition (3) with fully labeled and fully unlabeled clients. (b) One round of a standard FL aggregation scheme. (c) One round of our proposed two-step isolated aggregation scheme for labeled clients and unlabeled clients.

4. We, for the first time, extensively evaluate SSFL methods on multiple medical image benchmarks with a varying proportion of clients and degree of heterogeneity. Our results show that the proposed isolated aggregation followed by federated pretraining outperforms the state-of-the-art method, *viz.*, RSCFed [14] by **6.91%** in terms of accuracy and achieves near-supervised learning performance.

2 Methods

2.1 Problem Description

Assume a federated learning setting with m fully labeled clients denoted as $\{C_1, C_2, ..., C_m\}$ each possessing a labeled dataset $D^l = \{(X_i^l, y_i^l)\}_{i=1}^{N^l}$ and n fully unlabeled clients defined as $\{C_{m+1}, C_{m+2}, ..., C_{m+n}\}$ each possessing an unlabeled dataset $D^u = \{(X_i^u)\}_{i=1}^{N^u}$. Our objective is to learn a global model θ_{glob} via decentralized training.

2.2 Local Training

We adopt mean-teacher-based semi-supervised learning [12,14,20] to train each unlabeled client. At the beginning of each round, the global model W_{glob} is used to initialize the teacher model W_t. At the end of each communicating round, the student model W_s is returned to the server as the local model. Each batch of images undergoes two types of augmentations. The teacher model receives weakly augmented data whereas the student model receives strongly augmented data in each local iteration. In order to decrease entropy of model output, the temperature of predictions is further increased via sharpening operation [2,3,5,14] as $\hat{p}_{t,i} = Sharpen\ (p_t, \tau)_i = p_{t,i}^{\frac{1}{\tau}} / \sum_j p_{t,j}^{\frac{1}{\tau}}$ where $p_{t,i}$ and $\hat{p}_{t,i}$ denote each element in p_t before and after sharpening, respectively. τ denotes the temperature parameter. The student model is trained on the local data (D^u) via consistency regularization with the teacher model output. The consistency regularization loss is defined as $\mathcal{L}_{MSE} = \|\hat{p}_t - p_s\|_2^2$ where $\hat{p}_t$ and p_s are teacher and student predictions, respectively. $\|.\|_2^2$ denotes $L2$-norm. The student model weights are optimized via backpropagation whereas the teacher model weights are updated by exponential moving averaging (EMA) after each local iteration, as in Eq. 1:

$$W_{t+1} = \alpha W_s + (1 - \alpha) W_t \tag{1}$$

where α denotes momentum parameter. We optimize cross-entropy loss for local training on labeled clients defined as $\mathcal{L}_{CE} = -y_i \log\, p_i$, where y_i denotes labels.

2.3 Isolated Federated Aggregation

In this section, we explain the proposed isolated aggregation of labeled and unlabeled client models. Each communication round is composed of two consecutive substeps. First, the server initializes the global model W^t_{glob} and sends it to **unlabeled** clients (U_i). The global model is used to initialize the teacher model W_t in each client. At this stage, only the unlabeled clients perform local training on the global model by minimizing the consistency regularization loss. The updated semi-supervised models obtained after running the local epochs are then uploaded to the server. We adopt a dynamically weighted Federated Averaging scheme [14] to aggregate the model parameters of all unlabeled clients W_u at the server. For this, we first obtain the averaged model by performing Fed-Avg as in Eq. 2.

$$W_{avg} = \frac{\sum_{k=1}^{k=K} n_k W_k}{\sum_{k=1}^{k=K} n_k} \tag{2}$$

where K is the total number of clients. n_k is the number of samples in each client. The client models are then dynamically scaled using coefficients c_k designed as functions of the individual distances from the averaged model as denoted in Eq. 3. The global model (W_{glob}) is updated by re-aggregating the client weights scaled by new coefficients c_k. In Eq. 3, λ_c is a hyperparameter.

$$c_k = \frac{n_k \exp(-\lambda_c \frac{\|W_k - W_{avg}\|_2}{n_k})}{\sum_{k=1}^{k=K} n_k}, W_{glob} = \frac{\sum_{k=1}^{k=K} c_k W_k}{\sum_{k=1}^{k=K} c_k} \tag{3}$$

The updated global model parameters are then communicated to each **labeled** client which initializes its models using these weights and trains the local model via minimization of the standard cross-entropy loss. After a predefined number of local epochs, each labeled client uploads its local model to the server. The server then aggregates all the supervised models employing the aforementioned weighting scheme and the resultant global model W^{t+1}_{glob} is then sent to each unlabeled client at the beginning of the next round.

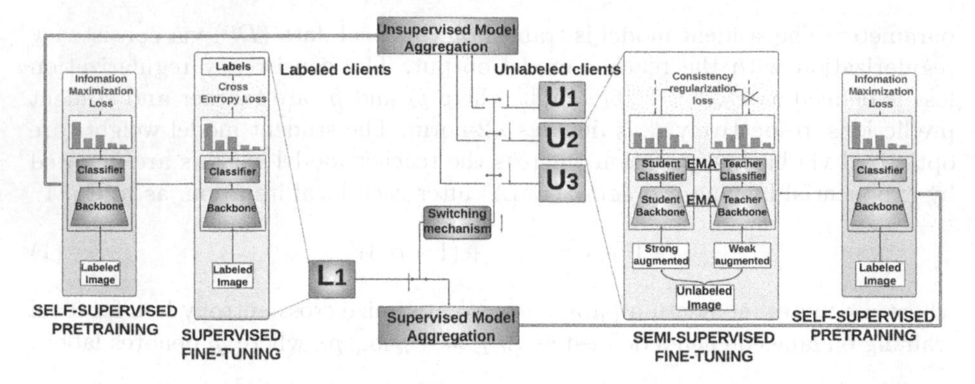

Fig. 2. Overview of our proposed methodology (IsoFed) with 1 labeled and 3 unlabeled clients. The unlabeled clients are trained using a mean-teacher-based SSL model. A switching mechanism swaps between labeled and unlabeled clients for isolated model aggregation in each round. After isolated model aggregation, an information maximization loss is used for client-adaptive pretraining to enhance the certainty and diversity of predictions of the global model for each client before actual local training.

2.4 Client-Adaptive Pretraining

Motivated by the recent success of continued pretraining in Natural Language Processing [6,8,17], we present a client-adaptive pretraining strategy as the second part of our proposed method. If we view the isolated FL from a transfer learning perspective, the global model received in one group of clients from the server can be regarded as an averaged model pretrained on the other group of clients. To improve client-specific model performance, we conduct a second phase of in-client federated pretraining on the global model before initializing it as a teacher model.

For self-supervised pretraining, we jointly learn the client-invariant features and client-specific classifier by optimizing an information-theoretic metric called information maximization (IM) loss denoted as $\mathcal{L}_{inf}$ in Eq. 4. It acts as an estimate of the expected misclassification error of the global model for each client. Optimizing the IM loss makes the global model output predictions that are individually certain but collectively diverse. With the help of a diversity preserving regularizer (first component in Eq. 4), IM avoids the trivial solution of entropy minimization where all unlabeled data collapses to the same one-hot encoding. The joint optimization is done by reducing the entropy of the output probability distribution of global model (p_i) in conjunction with maximizing the mutual information between the data distribution and the estimated output distribution yielded by the global model.

$$\mathcal{L}_{inf} = \mathbb{E}_{x \in D} \left[\left(\frac{1}{N} \sum_{i=1}^{N} p_i \right) \log \left(\frac{1}{N} \sum_{i=1}^{N} p_i \right) - \frac{1}{N} \sum_{i=1}^{N} p_i \log p_i \right] \quad (4)$$

where N is the number of classes. x denotes any instance belonging to a dataset D. The entropy minimization leads to the least number of confused predictions whereas the regularizer avoids the degenerate solution where every data sample is assigned to the same class [13,18]. The pretrained model is then initialized as the teacher model to train the local student model in each round.

3 Experiments and Results

3.1 Datasets and FL Settings

To evaluate the performance and generalisability of the proposed method, we conduct experiments on four publicly available medical image benchmark datasets with different modalities [25], *viz.*, BloodMNIST (microscopic peripheral blood cell images), PathMNIST (colon pathology), PneumoniaMNIST (chest X-ray), and OrganAMNIST (abdominal CT - axial view). Each image resolution is 28×28 pixels and is normalized before feeding it to the network. BloodMNIST contains a total of 17,092 images and is organized into 8 classes. PathMNIST has 107,180 images and has 9 types of tissues. PneumoniaMNIST is a collection of 5,856 images and the task is binary classification (diseased vs normal). OrganAMNIST is comprised of 58,850 images and the task is multi-class classification of 11 body organs. We split each training dataset between 4 clients to mimic a practical collaborative setting in healthcare. To testify the versatility of the models, we study two challenging non-IID data partition strategies with 0.5 and 0.8-Dirichlet (γ). As a result, the number of samples per class and per client widely vary from each other. Additionally, we show the impact of varying the proportion of labeled clients (75%, 50%, 25%) on model performance. See **Suppl. Sec 1** for more details.

3.2 Implementation and Training Details

For all datasets, we employ a simple CNN comprising of two 5×5 convolution layers, a 2×2 max-pooling layer, and two fully-connected layers as the feature extraction backbone followed by a two-layer MLP and a fully-connected layer as the classification network. Our model is implemented with PyTorch. We follow the settings prescribed for a training RSCFed to enable a fair comparison. See **Suppl. Sec 2** for more training details.

3.3 Results and Discussion

We use the standard metrics - accuracy, area under a ROC curve (AUC), Precision, and Recall to evaluate performance. We observe that the dynamically weighted version of Fed-Avg (discussed in Sect. 2.3) outperforms standard Fed-Avg and hence use it as a baseline in this paper instead of vanilla Fed-Avg. In order to fairly evaluate IsoFed, we compare with the following state-of-the-art SSFL benchmarks: (a) MT+wFed-Avg: a combination of Mean Teacher and dynamically weighted Fed-Avg, (b) RSCFed: Random sampling consensus-based FL [14]. Since RSCFed has already been shown to significantly outperform FedIRM [16] and Fed-Consist [24] on multiple datasets, we exclude those

Table 1. Comparison with baselines on BloodMNIST and PathMNIST. wFedAvg refers to dynamically weighted Federated averaging. UB implies Upper Bound. MT refers to Mean teacher-based SSL. Acc. and Prec. denote Accuracy and Precision. L and U denote the number of labeled and unlabeled clients respectively.

Labeling	Method	Client		Metrics (%) $\gamma = 0.8$ (less non-IID)				Metrics (%) $\gamma = 0.5$ (more non-IID)			
		L	U	Acc.	AUC	Prec.	Recall	Acc.	AUC	Prec.	Recall
Dataset 1 : BloodMNIST, Task : 8-class classification											
Fully supervised	wFed-Avg (UB)	4	0	79.57	96.61	77.65	75.70	79.45	96.80	78.28	73.31
Semi supervised	MT+wFed-Avg	3	1	77.32	96.70	74.16	73.79	70.89	95.11	73.46	65.06
	RSCFed	3	1	76.94	95.54	75.11	71.18	75.18	94.99	76.55	68.96
	IsoFed	3	1	**79.43**	**97.32**	**76.70**	**76.67**	**76.10**	**95.88**	**77.13**	**72.29**
	MT+wFed-Avg	2	2	75.88	96.56	72.85	71.94	58.29	88.35	57.85	60.46
	RSCFed	2	2	75.97	95.30	73.58	72.77	61.18	**91.50**	54.85	60.79
	IsoFed	2	2	**80.47**	**97.25**	**77.11**	**78.11**	**64.05**	90.01	**60.26**	**64.03**
	MT+wFed-Avg	1	3	75.24	95.13	72.43	70.37	52.56	89.39	57.89	55.81
	RSCFed	1	3	71.88	93.96	70.47	67.75	19.35	64.31	07.05	23.62
	IsoFed	1	3	**79.23**	**96.43**	**76.68**	**77.00**	**63.70**	**90.58**	**70.22**	**63.81**
Dataset 2 : PathMNIST, Task : 9-class classification											
Fully supervised	wFed-Avg (UB)	4	0	70.45	94.92	72.13	69.84	68.97	94.93	68.05	67.58
Semi supervised	MT+wFed-Avg	3	1	60.97	93.60	68.14	62.00	57.92	92.93	**67.20**	59.98
	RSCFed	3	1	61.55	93.71	61.00	58.95	58.33	93.59	60.68	58.73
	IsoFed	3	1	**63.10**	**94.73**	**69.25**	**64.62**	**60.23**	**93.98**	52.80	**61.66**
	MT+wFed-Avg	2	2	67.10	**95.17**	**66.41**	**66.40**	61.28	91.26	61.50	57.56
	RSCFed	2	2	64.18	93.17	60.79	58.89	58.83	90.35	58.88	55.02
	IsoFed	2	2	**70.32**	94.74	65.96	64.86	**64.00**	**93.46**	**63.88**	**61.22**
	MT+wFed-Avg	1	3	59.57	90.66	63.14	58.93	56.31	89.92	60.42	53.92
	RSCFed	1	3	64.75	**94.09**	**66.89**	**63.66**	57.42	89.43	54.96	53.53
	IsoFed	1	3	**66.48**	92.24	63.71	62.06	**64.02**	**93.99**	**66.12**	**62.39**

methods from our comparative study due to space constraints. We consider fully-supervised FL as an upper bound and report the results for both the non-IID settings on each dataset. Tables 1-2 show that overall, IsoFed outperforms RSCFed by 6.91%, 4.15%, 7.28%, and 6.71% in terms of average accuracy, AUC, Precision, and Recall respectively.

Table 1 shows our method and our baselines on 8-class classification with BloodMNIST. L and U denote the number of labeled and unlabeled clients respectively. The average accuracy for fully-supervised FL is 79.51%. Among the baselines, MT+wFed-Avg has a higher overall accuracy score of 68.36% while RSCFed has an accuracy score of 63.41%. Particularly, we find RSCFed collapses under the most extreme case of $\gamma = 0.5$ and U=3. IsoFed improves the accuracy score to 73.83% and is stable for all evaluated conditions. Table 1 further reports performance on 9-class classification with PathMNIST. The fully-supervised FL achieves an overall accuracy of 69.71%. The baselines have very similar accuracy scores of 60.53% and 60.84% respectively. IsoFed improves it to 64.69%.

Table 2 shows binary classification results on PneumoniaMNIST. The fully-supervised FL has an overall accuracy of 87.18%. MT+wFed-Avg and RSCFed

Table 2. Performance comparison of IsoFed with baselines on PneumoniaMNIST and OrganAMNIST (with ablation study). PT refers to the federated pretraining step.

Labeling	Method	Client		Metrics (%)				Metrics (%)			
		L	U	Acc.	AUC	Prec.	Recall	Acc.	AUC	Prec.	Recall
				$\gamma = 0.8$ (less non-IID)				$\gamma = 0.5$ (more non-IID)			
Dataset 3 : PneumoniaMNIST, Task : Binary classification											
Fully supervised	wFed-Avg (UB)	4	0	87.34	95.32	86.71	89.02	87.02	95.64	86.45	88.76
Semi supervised	MT+wFed-Avg	3	1	86.54	95.20	85.94	88.21	86.86	94.85	85.92	87.86
	RSCFed	3	1	86.58	95.63	**89.02**	88.68	86.70	94.50	85.75	87.65
	IsoFed	3	1	**87.10**	**95.04**	86.45	**89.00**	**89.26**	**95.80**	**88.26**	**89.44**
	MT+wFed-Avg	2	2	83.65	89.74	82.45	82.99	82.21	96.17	83.20	85.26
	RSCFed	2	2	78.37	87.36	77.31	78.76	**84.46**	**95.58**	**84.58**	**86.88**
	IsoFed	2	2	**84.70**	**90.75**	**83.56**	**84.64**	82.68	95.15	83.34	85.41
	MT+wFed-Avg	1	3	81.41	89.84	82.05	77.69	**79.97**	**94.45**	**81.28**	**83.12**
	RSCFed	1	3	78.85	86.66	77.56	76.84	62.50	50.00	31.25	50.00
	IsoFed	1	3	**85.00**	**91.68**	**83.98**	**83.95**	77.12	93.65	80.47	81.40
Dataset 4 : OrganAMNIST, Task: 11-class classification											
Fully supervised	wFed-Avg (UB)	4	0	69.72	94.41	67.44	69.60	69.50	94.63	68.12	69.60
Semi supervised	MT+wFed-Avg	3	1	68.36	93.72	68.02	69.38	66.49	93.69	67.51	68.25
	RSCFed	3	1	68.14	94.26	67.44	69.53	67.08	93.82	68.82	68.36
	IsoFed w/o PT	3	1	68.98	94.32	**68.83**	69.88	67.45	93.98	67.85	69.35
	IsoFed	3	1	**69.47**	**95.05**	68.04	**70.85**	**68.65**	**94.88**	**68.64**	**69.77**
	MT+wFed-Avg	2	2	66.28	92.77	66.12	67.63	61.71	**92.55**	**65.79**	62.66
	RSCFed	2	2	66.68	92.42	66.90	66.56	62.51	91.89	64.09	63.35
	IsoFed w/o PT	2	2	68.67	93.25	67.65	68.50	**64.37**	92.11	65.70	65.17
	IsoFed	2	2	**68.95**	**93.95**	**68.32**	**69.83**	64.08	92.45	64.56	**65.47**
	MT+wFed-Avg	1	3	57.75	90.95	61.50	55.68	50.84	87.65	60.07	48.51
	RSCFed	1	3	58.50	90.86	63.48	55.76	54.90	89.58	50.53	53.41
	IsoFed w/o PT	1	3	62.03	91.36	64.50	61.44	56.40	89.79	61.61	55.72
	IsoFed	1	3	**62.77**	**91.48**	**64.52**	**61.79**	**61.90**	**91.55**	**62.39**	**60.21**

achieve average accuracy scores of 83.44% and 79.57%. IsoFed has the best accuracy of 85.45%. Furthermore, the results of 11-class anatomy classification task on OrganAMNIST are also reported in Table 2. The upper bound accuracy is 69.61% and the baseline accuracies are 61.91% and 62.97% respectively. IsoFed achieves an overall accuracy score of 65.97%. In general, the performance of all methods decreases with γ changing from 0.8 to 0.5. It is expected as the clients become more label-skewed due to higher non-IID data partition. However, our approach is least affected by this which is reflected in its accuracy decrease by 2.19% as opposed to 4.45% and 2.94% incurred by baselines. As foreseen, performance also deteriorates with decrease in the number of labeled clients. For L:U = 3:1, 2:2, 1:3, the baseline accuracies degrade by 2.16%, 5.61%, 15.31% and 2%, 5.01%, 12.91% w.r.t. fully supervised FL setting. However, for IsoFed, the decrease in accuracy is only 0.55%, 3.09%, and 7.28%, respectively. This proves the near-supervised learning performance of the proposed training method.

The superior performance of IsoFed over the baselines and closer performance to the upper bound demonstrates better learning and generalization. This is achieved by the isolated aggregation strategy and federated pretraining on all datasets.

3.4 Ablation Study

Owing to space constraints, we show ablation experiments only on OrganAMNIST, which provides the most challenging classification task, to evaluate the impact of IsoFed components. (More results in **Suppl. Sec 2**). Table 2 demonstrates that client-adaptive pretraining improves model accuracy by 5.50% for the most extreme condition of $\gamma = 0.5$ and L:U = 1:3.

4 Conclusion

We have introduced a novel SSFL framework called IsoFed, an isolated federated learning technique, to address joint training of labeled and unlabeled clients in the context of decentralized semi-supervised learning. It opens a new research direction in learning across domains by unifying two dominant approaches - Federated Learning (among labeled or unlabeled clients) and Transfer Learning (between labeled and unlabeled clients). Our results challenge the conventional strategy of co-training fully labeled and fully unlabeled clients in SSFL. Experimental results on 4 different medical imaging datasets with varied proportion of labeled clients $(25, 50, 75\%)$ and varied non-IID distribution (0.5 & 0.8-Dirichlet) show that IsoFed achieves a considerable boost compared to current state-of-the-art SSFL method. IsoFed can be easily incorporated into other federated learning-based aggregation schemes as well as used in conjunction with any other semi-supervised learning framework in federated learning setting.

Acknowledgement. This work was supported in part by the UK EPSRC (Engineering and Physical Research Council) Programme Grant EP/T028572/1 (VisualAI), a UK EPSRC Doctoral Training Partnership award, and the InnoHK-funded Hong Kong Centre for Cerebro-cardiovascular Health Engineering (COCHE) Project 2.1 (Cardiovascular risks in early life and fetal echocardiography).

References

1. Arazo, E., Ortego, D., Albert, P., O'Connor, N.E., McGuinness, K.: Pseudo-labeling and confirmation bias in deep semi-supervised learning. In: 2020 International Joint Conference on Neural Networks (IJCNN), pp. 1–8. IEEE (2020)
2. Berthelot, D., Carlini, N., Goodfellow, I., Papernot, N., Oliver, A., Raffel, C.A.: MixMatch: a holistic approach to semi-supervised learning. In: Advances in Neural Information Processing Systems, vol. 32 (2019)
3. Chen, T., Kornblith, S., Norouzi, M., Hinton, G.: A simple framework for contrastive learning of visual representations. In: International Conference on Machine Learning, pp. 1597–1607. PMLR (2020)
4. Diao, E., Ding, J., Tarokh, V.: Semifl: communication efficient semi-supervised federated learning with unlabeled clients. arXiv preprint arXiv:2106.01432 3 (2021)

5. Goodfellow, I., Bengio, Y., Courville, A.: Deep Learning. MIT Press, Cambridge (2016)
6. Gururangan, S., et al.: Don't stop pretraining: adapt language models to domains and tasks. arXiv preprint arXiv:2004.10964 (2020)
7. He, C., Yang, Z., Mushtaq, E., Lee, S., Soltanolkotabi, M., Avestimehr, S.: SSFL: tackling label deficiency in federated learning via personalized self-supervision. arXiv preprint arXiv:2110.02470 (2021)
8. Howard, J., Ruder, S.: Universal language model fine-tuning for text classification. arXiv preprint arXiv:1801.06146 (2018)
9. Jeong, W., Yoon, J., Yang, E., Hwang, S.J.: Federated semi-supervised learning with inter-client consistency & disjoint learning. arXiv preprint arXiv:2006.12097 (2020)
10. Ji, S., Saravirta, T., Pan, S., Long, G., Walid, A.: Emerging trends in federated learning: From model fusion to federated x learning. arXiv preprint arXiv:2102.12920 (2021)
11. Kairouz, P., et al.: Advances and open problems in federated learning. Found. Trends® Mach. Learn. **14**(1–2), 1–210 (2021)
12. Li, T., Sahu, A.K., Talwalkar, A., Smith, V.: Federated learning: challenges, methods, and future directions. IEEE Sig. Process. Mag. **37**(3), 50–60 (2020)
13. Liang, J., Hu, D., Feng, J.: Do we really need to access the source data? source hypothesis transfer for unsupervised domain adaptation. In: International Conference on Machine Learning, pp. 6028–6039. PMLR (2020)
14. Liang, X., Lin, Y., Fu, H., Zhu, L., Li, X.: RSCFed: random sampling consensus federated semi-supervised learning. In: Proceedings of the IEEE/CVF Conference on Computer Vision and Pattern Recognition, pp. 10154–10163 (2022)
15. Lin, H., Lou, J., Xiong, L., Shahabi, C.: Semifed: Semi-supervised federated learning with consistency and pseudo-labeling. arXiv preprint arXiv:2108.09412 (2021)
16. Liu, Q., Yang, H., Dou, Q., Heng, P.-A.: Federated semi-supervised medical image classification via inter-client relation matching. In: de Bruijne, M., et al. (eds.) MICCAI 2021. LNCS, vol. 12903, pp. 325–335. Springer, Cham (2021). https://doi.org/10.1007/978-3-030-87199-4_31
17. Liu, Y., et al.: Roberta: a robustly optimized bert pretraining approach. arXiv preprint arXiv:1907.11692 (2019)
18. Shi, Y., Sha, F.: Information-theoretical learning of discriminative clusters for unsupervised domain adaptation. arXiv preprint arXiv:1206.6438 (2012)
19. Sohn, K., et al.: FixMatch: simplifying semi-supervised learning with consistency and confidence. Adv. Neural Inf. Process. Syst. **33**, 596–608 (2020)
20. Tarvainen, A., Valpola, H.: Mean teachers are better role models: weight-averaged consistency targets improve semi-supervised deep learning results. In: Advances in Neural Information Processing Systems, vol. 30 (2017)
21. Van Engelen, J.E., Hoos, H.H.: A survey on semi-supervised learning. Mach. Learn. **109**(2), 373–440 (2020)
22. Weiss, K., Khoshgoftaar, T.M., Wang, D.D.: A survey of transfer learning. J. Big Data **3**(1), 1–40 (2016). https://doi.org/10.1186/s40537-016-0043-6
23. Yafen, L., Yifeng, Z., Lingyi, J., Guohe, L., Wenjie, Z.: Survey on pseudo-labeling methods in deep semi-supervised learning. J. Front. Comput. Sci. Technol. **16**(6), 1279 (2022)
24. Yang, D., et al.: Federated semi-supervised learning for covid region segmentation in chest CT using multi-national data from china, Italy, japan. Med. Image Anal. **70**, 101992 (2021)

25. Yang, J., Shi, R., Wei, D., Liu, Z., Zhao, L., Ke, B., Pfister, H., Ni, B.: Medmnist v2-a large-scale lightweight benchmark for 2d and 3d biomedical image classification. Sci. Data **10**(1), 41 (2023)
26. Zhang, B., et al.: FlexMatch: boosting semi-supervised learning with curriculum pseudo labeling. Adv. Neural Inf. Process. Syst. **34**, 18408–18419 (2021)
27. Zhang, C., Xie, Y., Bai, H., Yu, B., Li, W., Gao, Y.: A survey on federated learning. Knowl.-Based Syst. **216**, 106775 (2021)
28. Zhang, Z., et al.: Improving semi-supervised federated learning by reducing the gradient diversity of models. In: 2021 IEEE International Conference on Big Data (Big Data), pp. 1214–1225. IEEE (2021)

Right for the Wrong Reason: Can Interpretable ML Techniques Detect Spurious Correlations?

Susu Sun[1](✉), Lisa M. Koch[2,3], and Christian F. Baumgartner[1]

[1] Cluster of Excellence – ML for Science, University of Tübingen, Tübingen, Germany
{susu.sun,christian.baumgartner}@uni-tuebingen.de
[2] Hertie Institute for AI in Brain Health, University of Tübingen, Tübingen, Germany
[3] Institute of Ophthalmic Research, University of Tübingen, Tübingen, Germany
lisa.koch@uni-tuebingen.de

Abstract. While deep neural network models offer unmatched classification performance, they are prone to learning spurious correlations in the data. Such dependencies on confounding information can be difficult to detect using performance metrics if the test data comes from the same distribution as the training data. Interpretable ML methods such as post-hoc explanations or inherently interpretable classifiers promise to identify faulty model reasoning. However, there is mixed evidence whether many of these techniques are actually able to do so. In this paper, we propose a rigorous evaluation strategy to assess an explanation technique's ability to correctly identify spurious correlations. Using this strategy, we evaluate five post-hoc explanation techniques and one inherently interpretable method for their ability to detect three types of artificially added confounders in a chest x-ray diagnosis task. We find that the post-hoc technique SHAP, as well as the inherently interpretable Attri-Net provide the best performance and can be used to reliably identify faulty model behavior.

Keywords: Interpretable machine learning · Confounder detection

1 Introduction

Black-box neural network classifiers offer enormous potential for computer-aided diagnosis and prediction in medical imaging applications but, unfortunately, they also have a strong tendency to learn spurious correlations in the data [11]. For the development and safe deployment of machine learning (ML) systems it is

Supplementary Information The online version contains supplementary material available at https://doi.org/10.1007/978-3-031-43895-0_40.

H. Greenspan et al. (Eds.): MICCAI 2023, LNCS 14221, pp. 425–434, 2023.
https://doi.org/10.1007/978-3-031-43895-0_40

essential to understand what information the classifiers are basing their decisions on, such that reliance on spurious correlations may be identified.

Spurious correlations arise when the training data are confounded by additional variables that are unrelated to the diagnostic information we want to predict. For instance, older patients in our training data may be more likely to present with a disease than younger patients. A classifier trained on this data may inadvertently learn to base its decision on image features related to age rather than pathology. Crucially, such faulty behavior cannot be identified using classification performance metrics such as area under the ROC curve (AUC) if the testing data contains the same confounding information as the training data, since the classifier predicts the *right* thing, but for the *wrong* reason. If undetected, however, such spurious correlations may lead to serious safety implications after deployment.

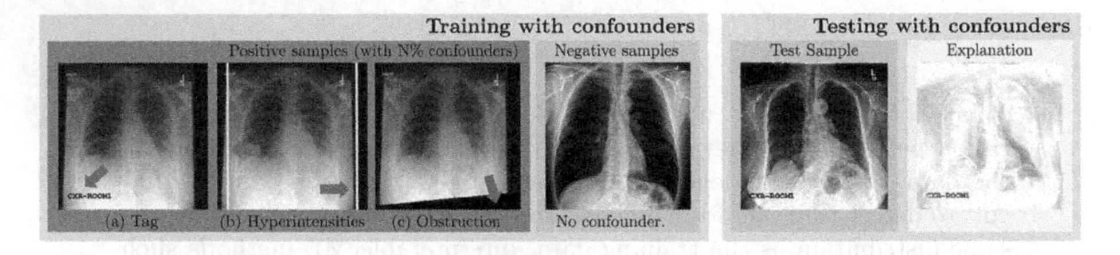

Fig. 1. Overview. We train classifiers on datasets with three types of artificially added confounders highlighted by arrows. We then evaluate the ability of explanation techniques to correctly identify reliance on these confounders (shown Attri-Net [24]).

Interpretable ML approaches may be used as a powerful tool to detect spurious correlations during development or after deployment of an ML system. Currently, the most widely used explanation modality are *visual* explanations, which highlight the pixels in the input image that are responsible for a particular decision. Common strategies include methods which leverage the gradient of the prediction with respect to the input image [7,19,22,23,25], explain the predictions by counterfactually generating an image of the opposite class [9,18,20,24], interpret the feature map of the last layer before the classification [8,10,27], or methods that build a local approximation of the decision function such as LIME [16], or SHAP [15].

The majority of visual explanation methods are *post-hoc* techniques, meaning a heuristic is applied to any trained model (e.g. a ResNet [13]) to approximately understand the decision mechanism for a given data point. However, post-hoc techniques are by definition only approximations and many techniques have been found to suffer from serious limitations [4,12,26]. *Inherently interpretable* techniques on the other hand build custom architectures that are designed to directly reveal the reasoning of the classifier to the user without the need for approximations. This class of methods does not suffer from the same limitations as post-hoc methods, and it has been argued that inherently interpretable approaches should be preferred in high-stakes applications such as medical image analysis [17]. For instance, if a classifier bases its decision on a spurious signal, an inherently interpretable classifier should by definition reveal this relationship.

Inherently interpretable visual explanation approaches are much less widely explored than post-hoc techniques, but there has recently been an increased interest in the topic. Two recently proposed methods in this category are the attribution network (Attri-Net) [24], and convolutional dynamic alignment networks (CoDA-Nets) [6]. Attri-Net first produces human-interpretable feature attribution maps for each disease category using a GAN-based counterfactual generator [24]. Then makes the final prediction with simple logistic regression classifiers based on those feature attribution maps. CoDA-Nets express neural networks as input dependent linear transformation [6]. Both approaches produce explanations on the pixel level of the input images.

Related Work on Comparing Explanation Techniques. A number of works have studied the quality of post-hoc explanation techniques. The vast majority of work focuses exclusively on gradient-based approaches (e.g. [5,21]). In their landmark study, Adebayo et al. [1] find that commonly used gradient-based explanation techniques do not pass some basic sanity checks. Arun et al. [5] extends this work to weakly supervised localisation in one of the few papers in this domain focusing on medical data. Both papers, however, do not consider other types of commonly used approaches such as counterfactual methods, or local function approximations such as LIME or SHAP.

A small number of works specifically investigate explanations' sensitivity to spurious correlations. In closely related work to ours, Adebayo et al. [3] explore a large library of post-hoc explanation techniques including LIME and SHAP, for detecting spurious image backgrounds in a bird versus dog classification task and find that many techniques are in fact able to detect the spurious background. In subsequent work, the same authors explore the usefulness of four post-hoc gradient-based explanation methods for identifying spurious correlations in hand and knee radiographs [2] and come to the conclusion that the examined methods are ineffective at identifying spurious correlations. We note that prior work is inconclusive on the usefulness of explanation techniques for identifying spurious correlations. In particular, in the medical context it is still unclear if commonly used explanation techniques are suitable for the detection of spurious correlations. Moreover, there is, to our knowledge, no evidence for the supposition that inherently interpretable techniques are better suited for this task.

Contributions. We present a rigorous evaluation of post-hoc explanations and inherently interpretable techniques for the identification of spurious correlations in a medical imaging task. Specifically, we focus on the task of diagnosing cardiomegaly from chest x-ray data with three types of synthetically generated spurious correlations (see Fig. 1). To identify whether an explanation correctly identifies a model's reliance on spurious correlations, we propose two quantitative metrics which are highly reflective of our qualitative findings. In contrast to the majority of prior work we focus on a wide range of different explanation approaches including counterfactual techniques and local function approximations, as well as post-hoc techniques and an inherently interpretable approach. Our analysis yields actionable insights which will be useful for a wide audience of ML practitioners.

2 A Framework for Evaluating Explanation Techniques

In the following, we introduce our evaluation strategy and proposed evaluation metrics, the studied confounders, as well as the evaluated explanation techniques. The strategy and evaluation metrics are generic and can also be applied to different problems. The confounders are engineered to correspond to realistic image artifacts that can appear in chest x-ray imaging[1].

2.1 Evaluation Strategy

We assume a setting in which the development data for a binary neural network based classifier contains an unknown spurious correlation with the target label. To quantitatively study this setting, we create training data with artificial spurious correlations by adding a confounding effect (e.g. a hospital tag) in a percentage of the cases with a positive label, where we vary the percentage $p \in \{0, 20, 50, 80, 100\}$. E.g., for $p = 100\%$ all of the positive images in the training set will have an artificial confounder, and for $p = 0\%$ there is no spurious signal. With increasing p the reliance on a spurious signal becomes more likely. The images with a negative label remain untouched.

In the evaluation, we consider a scenario in which the test data contain the same confounder type with the same proportion p used in the respective trainings. In this case, we can not tell if a classifier relies on the confounded features from classification performance. Our aim, therefore, is to investigate whether explanation techniques can identify that the classifier predicts the *right thing for the wrong reason.*

We perform all experiments on chest x-ray images from the widely used CheXpert dataset [14], where we focus on the binary classification task on disease cardiomegaly. We divided our dataset into a training (80%), validation (10%) test (10%) set.

2.2 Studied Confounders

We study three types of confounders inspired by real-world artefacts. Firstly, we investigate a hospital tag placed in the lower left corner of the image (see

[1] Our code can be found under https://github.com/ss-sun/right-for-the-wrong-reason.

Fig. 1a). Secondly, we add vertical lines of hyperintense signal that can be caused by foreign materials on the light path assembly (see Fig. 1b). Lastly, we consider an oblique occlusion of the image in the lower part of the image, which is an artefact that we observed for many images in the CheXpert dataset (see Fig. 1c).

2.3 Evaluation Metrics for Measuring Confounder Detection

We propose two novel metrics which reflect an explanation's ability to correctly identify spurious correlations.

Confounder Sensitivity (CS). Firstly, the explanations should be able to correctly attribute the confounder if classifier bases its decision on it. We assess this property by summing the number of true positive attributions divided by the total number of confounded pixels for each test image. We consider a pixel a true positive if it is part of the pixels affected by the confounder *and* in the top 10% attributed pixels according to a visual explanation. Thus the maximum sensitivity of 1 is obtained if all confounded pixels are in the top 10% of the attributions. Note that we do not penalise attributions outside of the confounding label as those can still also be correct. To guarantee that we only evaluate on samples for which the prediction is actually influenced by the confounder, we only include images for which the prediction with and without the confounding label is of the opposite class. To reduce computation times we use a maximum of 100 samples for each evaluation. An optimal explanation methods should obtain a CS score of 0 if the data contains $p = 0\%$ confounded data points, since in that case the spurious signal should not be attributed. For increasing p the confounder sensitivity should increase, i.e. the explanation should reflect the classifiers increasing reliance on the confounder.

Sensitivity to Prediction Changes via Explanation NCC. Secondly, the explanations should not be invariant to changes in classifier prediction. That is, if the classifier's prediction for a specific image changes when adding or removing a confounder, then the explanations should also be different. We measure this property using the average normalised cross correlation (NCC) between explanations of test images when confounders were either present or absent.Again, we only evaluate on images for which the prediction changes when adding the confounder as in these cases, we know the classifier is relying on confounders, and we evaluate a maximum of 100 samples. An optimal explanation method should obtain a high NCC score if the training data contains $p = 0\%$ confounded data points, since in that case the explanation with and without the confounder should be similar. For increasing p the NCC score should decrease to reflect the classifiers increasing reliance on the confounder.

2.4 Evaluated Explanation Methods

We evaluated five post-hoc techniques with representative examples from the approaches mentioned in the introduction: Guided Backpropgation [23] and Grad-CAM [19] (gradient-based), Gifsplanation (counterfactual), and LIME [16] and SHAP partition explainer [15] (local linear approximations). All post-hoc techniques were applied to a standard black-box ResNet50 model. We furthermore investigated the interpretable visual explanation method Attri-Net [24]. We used the default parameters for all methods. We found CoDA-Nets [6] required lengthy hyperparameter tuning for each type of experiment, and decided to exclude it in this paper.

3 Results

We first established the classifiers' performance in the presence of confounders, then compared all techniques in their ability to identify such confounders.

Classification Performance. Both investigated classifiers, the ResNet50 and the inherently interpretable Attri-Net, performed similarly in terms of classification AUC (first row of Fig. 2). For all three confounders, classification AUC consistently increased with increasing contamination p of the training dataset. This indicated that the classifiers increasingly relied on the spurious signal. For $p = 100\%$ contamination, where the confounder was present on all positive training examples, both classifiers reached almost a perfect classification AUC of 1.

Explanations. We analysed the explanations' ability to identify confounders by reporting confounder sensitivity (CS, middle row in Fig. 2) and explanation NCC (bottom row in Fig. 2). Out of the investigated methods Attri-Net and SHAP were closest to the ideal behaviour of high confounder sensitivity and low explanation NCC for $p > 0\%$. We found that SHAP performed extremely well in detecting tag confounders, but struggled with hyperintensities confounders. This can be explained by the fact that the tag confounder is relatively small and thus is more likely to be completely covered by the superpixels in SHAP. Overall, the inherently interpretable Attri-Net technique achieved the best balance. In agreement with related literature we found that gradient-based explanation methods performed poorly. In particular, Guided Backpropagation displayed similar CS-scores no matter if the classifier relies on a spurious signal ($p > 0\%$) or not ($p = 0\%$). Note that some results for $p = 100\%$ were missing because no data points fulfilled the criterion of the prediction being flipped with and without the confounders.

Figures 3, 4 & 5 contain examples explanations for the hyperintensity, tag, and edge confounder, respectively. Our qualitative analysis of the results confirms the quantitative findings, with SHAP and Attri-Net providing the most intuitive explanations. In particular, in the challenging hyperintensities scenario (see Fig. 3) AttriNet was the only method able to highlight the confounders in a human-interpretable fashion. We note that in all examples when a confounder was present, SHAP tended to highlight only the confounder, while Attri-Net also highlighted features related to Cardiomegaly. This may reflect the different decision mechanisms of the ResNet50 and the Attri-Net.

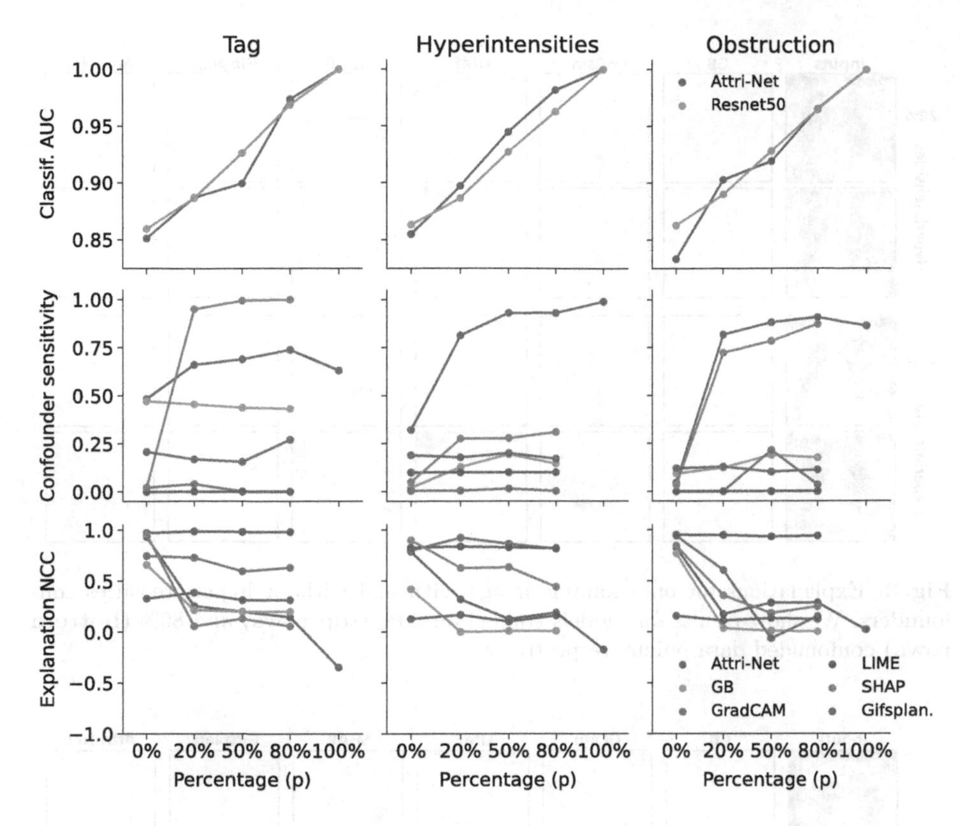

Fig. 2. (top row) Classification AUC of Attri-Net and Resnet50 on images containing hospital tags (left), hyperintensities (middle) or obstruction confounders (right column). The classifiers were trained with a varying proportion of confounders present in the positive examples in the training set (shown on the x-axes). **(bottom rows)** The explanation techniques' ability to identify confounders in terms of confounder sensitivity (middle row) and explanation NCC (bottom row, lower is better).

4 Discussion

In this paper, we proposed an evaluation strategy to assess the ability of visual explanations to correctly identify a classifier's reliance on a spurious signal. We specifically focused on the scenario where the classifier is predicting the right thing, but for the wrong reason, which is highly significant for the safe development of ML-basd diagnosis and prediction systems. Using this strategy, we assessed the performance of five post-hoc explanation techniques and one inherently interpretable technique with three realistic confounding signals. We found that the inherently interpretable Attri-Net technique, as well as the post-hoc SHAP technique performed the best, with Attri-Net yielding the most balanced performance. Both techniques are suitable for finding false reliance on a spurious signals. We also observed that the variation in the explanations' sparsity makes them perform differently in detecting spurious signals of different sizes and

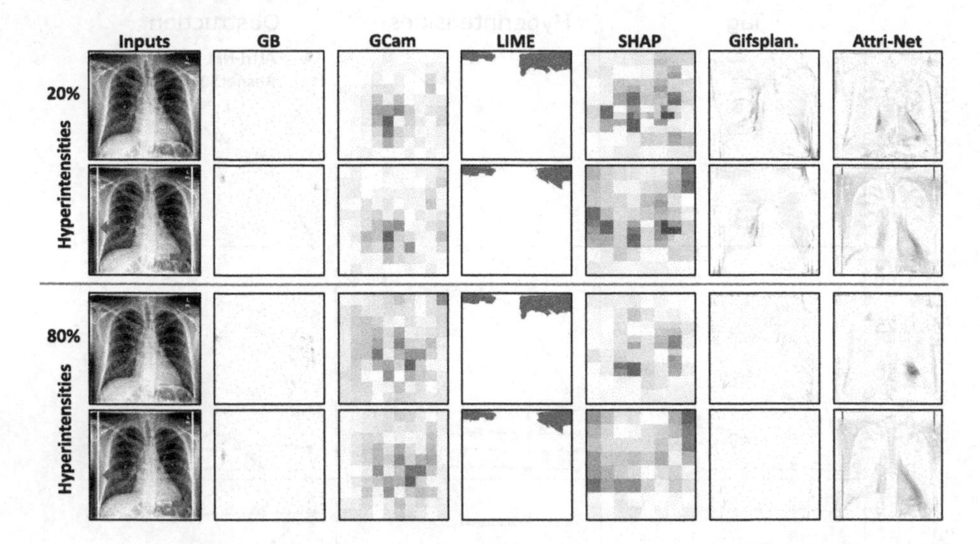

Fig. 3. Explanations for one example image with and without hyperintensities confounders. We show results for models trained on 20% (**top rows**) and 80% (**bottom rows**) confounded data points, respectively.

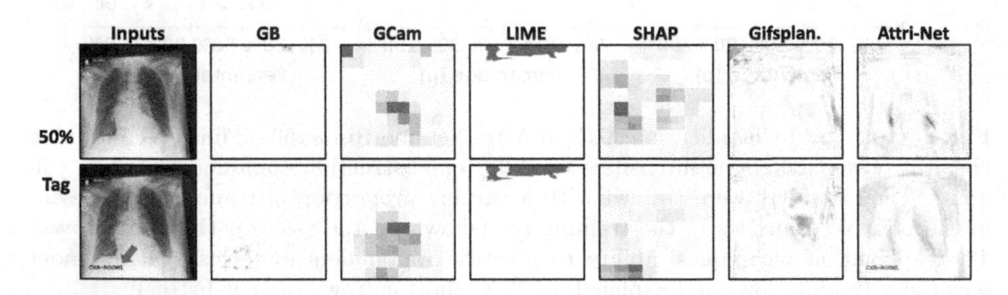

Fig. 4. Explanations for one example image with (**top**) and without (**bottom**) a tag confounder for models trained on 50% confounded data points.

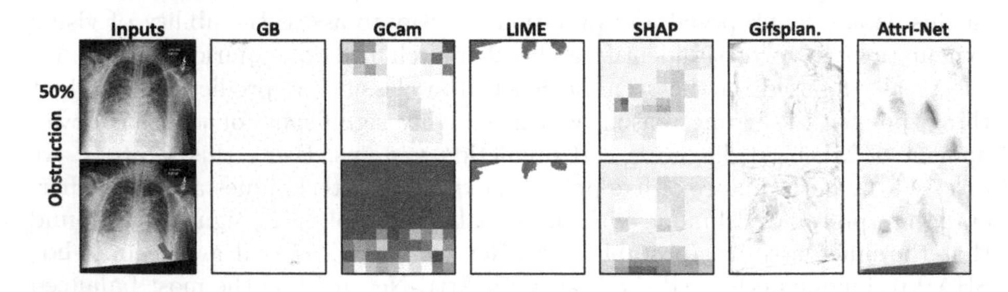

Fig. 5. Explanation for one example image with (**top**) and without (**bottom**) an obstruction confounder for models trained on 50% confounded data points.

shapes. In agreement with prior work, we found that gradient based techniques performed less robustly in our experiments.

From our experiments we draw two main conclusions. Firstly, practitioners looking to check for spurious correlations in a trained black-box model such as a ResNet should give preference to SHAP which provided the best performance out of the post-hoc techniques in our experiments. Secondly, an inherently interpretable technique, namely Attri-Net, performed the best in our experiments providing evidence to the supposition by Rudin et al. [17] that inherently interpretable techniques may provide a fruitful avenue for future work.

A major limitation of our study is the limited number of techniques we examined. Thus a primary focus of future work will be to scale our experiments to a wider range of techniques. Future work will also focus on human-in-the-loop experiments, as we believe, this will be the ultimate assessment of the usefulness of different explanation techniques.

Acknowledgements. Funded by the Deutsche Forschungsgemeinschaft (DFG, German Research Foundation) under Germany's Excellence Strategy - EXC number 2064/1 - Project number 390727645. The authors acknowledge support of the Carl Zeiss Foundation in the project "Certification and Foundations of Safe Machine Learning Systems in Healthcare" and the Hertie Foundation. The authors thank the International Max Planck Research School for Intelligent Systems (IMPRS-IS) for supporting Susu Sun, Lisa M. Koch, and Christian F. Baumgartner.

References

1. Adebayo, J., Gilmer, J., Muelly, M., Goodfellow, I., Hardt, M., Kim, B.: Sanity checks for saliency maps. In: Advances in Neural Information Processing Systems 31 (2018)
2. Adebayo, J., Muelly, M., Abelson, H., Kim, B.: Post hoc explanations may be ineffective for detecting unknown spurious correlation. In: International Conference on Learning Representations (2022)
3. Adebayo, J., Muelly, M., Liccardi, I., Kim, B.: Debugging tests for model explanations. arXiv preprint arXiv:2011.05429 (2020)
4. Alvarez-Melis, D., Jaakkola, T.S.: On the robustness of interpretability methods. arXiv preprint arXiv:1806.08049 (2018)
5. Arun, N., et al.: Assessing the trustworthiness of saliency maps for localizing abnormalities in medical imaging. Radiol.: Artif. Intell. **3**(6), e200267 (2021)
6. Bohle, M., Fritz, M., Schiele, B.: Convolutional dynamic alignment networks for interpretable classifications. In: Proceedings of the IEEE/CVF Conference on Computer Vision and Pattern Recognition, pp. 10029–10038 (2021)
7. Boreiko, V., et al.: Visual explanations for the detection of diabetic retinopathy from retinal fundus images. In: Medical Image Computing and Computer Assisted Intervention (2022)
8. Brendel, W., Bethge, M.: Approximating CNNs with bag-of-local-features models works surprisingly well on ImageNet. arXiv preprint arXiv:1904.00760 (2019)
9. Cohen, J.P., et al.: Gifsplanation via Latent Shift: a simple autoencoder approach to counterfactual generation for chest X-rays. In: Medical Imaging with Deep Learning, pp. 74–104. PMLR (2021)

10. Djoumessi, K.R., et al.: Sparse activations for interpretable disease grading. arXiv preprint arXiv:TODO (2023)
11. Geirhos, R., et al.: Shortcut learning in deep neural networks. Nat. Mach. Intell. **2**(11), 665–673 (2020)
12. Han, T., Srinivas, S., Lakkaraju, H.: Which explanation should i choose? a function approximation perspective to characterizing post hoc explanations. arXiv preprint arXiv:2206.01254 (2022)
13. He, K., Zhang, X., Ren, S., Sun, J.: Deep residual learning for image recognition. In: Proceedings of the IEEE Conference on Computer Vision and Pattern Recognition, pp. 770–778 (2016)
14. Irvin, J., et al.: Chexpert: A large chest radiograph dataset with uncertainty labels and expert comparison. In: Proceedings of the AAAI conference on artificial intelligence. vol. 33, pp. 590–597 (2019)
15. Lundberg, S.M., Lee, S.I.: A unified approach to interpreting model predictions. In: Advances in Neural Information Processing Systems 30 (2017)
16. Ribeiro, M.T., Singh, S., Guestrin, C.: why should i trust you? explaining the predictions of any classifier. In: Proceedings of the 22nd ACM SIGKDD International Conference on Knowledge Discovery and Data Mining, pp. 1135–1144 (2016)
17. Rudin, C.: Stop explaining black box machine learning models for high stakes decisions and use interpretable models instead. Nat. Mach. Intell. **1**(5), 206–215 (2019)
18. Samangouei, P., Saeedi, A., Nakagawa, L., Silberman, N.: ExplainGAN: model explanation via decision boundary crossing transformations. In: Proceedings of the European Conference on Computer Vision (ECCV), pp. 666–681 (2018)
19. Selvaraju, R.R., Cogswell, M., Das, A., Vedantam, R., Parikh, D., Batra, D.: Grad-CAM: visual explanations from deep networks via gradient-based localization. In: Proceedings of the IEEE International Conference on Computer Vision, pp. 618–626 (2017)
20. Singla, S., Pollack, B., Chen, J., Batmanghelich, K.: Explanation by progressive exaggeration. arXiv preprint arXiv:1911.00483 (2019)
21. Sixt, L., Granz, M., Landgraf, T.: When explanations lie: why many modified BP attributions fail. In: International Conference on Machine Learning, pp. 9046–9057. PMLR (2020)
22. Smilkov, D., Thorat, N., Kim, B., Viégas, F., Wattenberg, M.: SmoothGrad: removing noise by adding noise. arXiv preprint arXiv:1706.03825 (2017)
23. Springenberg, J.T., Dosovitskiy, A., Brox, T., Riedmiller, M.: Striving for simplicity: the all convolutional net. arXiv preprint arXiv:1412.6806 (2014)
24. Sun, S., Woerner, S., Maier, A., Koch, L.M., Baumgartner, C.F.: Inherently interpretable multi-label classification using class-specific counterfactuals. arXiv preprint arXiv:2303.00500 (2023)
25. Sundararajan, M., Taly, A., Yan, Q.: Axiomatic attribution for deep networks. In: International Conference on Machine Learning, pp. 3319–3328. PMLR (2017)
26. White, A., Garcez, A.D.: Measurable counterfactual local explanations for any classifier. arXiv preprint arXiv:1908.03020 (2019)
27. Zhou, B., Khosla, A., Lapedriza, A., Oliva, A., Torralba, A.: Learning deep features for discriminative localization. In: Proceedings of the IEEE Conference on Computer Vision and Precognition, pp. 2921–2929 (2016)

Interpretable Medical Image Classification Using Prototype Learning and Privileged Information

Luisa Gallée[1], Meinrad Beer[2], and Michael Götz[1,3(✉)]

[1] Experimental Radiology, University Hospital Ulm, Ulm, Germany
{luisa.gallee,michael.goetz}@uni-ulm.de
[2] Department of Diagnostic and Interventional Radiology, University Hospital Ulm, Ulm, Germany
[3] i2SouI - Innovative Imaging in Surgical Oncology Ulm, University Hospital Ulm, Ulm, Germany

Abstract. Interpretability is often an essential requirement in medical imaging. Advanced deep learning methods are required to address this need for explainability and high performance. In this work, we investigate whether additional information available during the training process can be used to create an understandable and powerful model. We propose an innovative solution called *Proto-Caps* that leverages the benefits of capsule networks, prototype learning and the use of privileged information. Evaluating the proposed solution on the LIDC-IDRI dataset shows that it combines increased interpretability with above state-of-the-art prediction performance. Compared to the explainable baseline model, our method achieves more than 6 % higher accuracy in predicting both malignancy (93.0 %) and mean characteristic features of lung nodules. Simultaneously, the model provides case-based reasoning with prototype representations that allow visual validation of radiologist-defined attributes.

Keywords: Explainable AI · Capsule Network · Prototype Learning

1 Introduction

Deep learning-based systems show remarkable predictive performance in many computer vision tasks, including medical image analysis, and are often comparable to human performance. However, the complexity of this technique makes it challenging to extract model knowledge and understand model decisions. This limitation is being addressed by the field of Explainable AI, in which significant progress has been made in recent years. An important line of research is the use of inherently explainable models, which circumvent the need for indirect, error-prone on-top explanations [14]. A common misconception is that the additional explanation comes with a decrease in performance. However, Rudin et al. [14]

H. Greenspan et al. (Eds.): MICCAI 2023, LNCS 14221, pp. 435–445, 2023.
https://doi.org/10.1007/978-3-031-43895-0_41

and others have already pointed out that this can be avoided by designing algorithms that build explainability into the core concept, rather than just adding it on top. Our work proves this once again by providing a powerful and explainable solution for medical image classification.

A promising approach for interpretability is the use of **Privileged Information**, i.e. information that is only available during training [19,20]. Besides using the additional knowledge to improve performance, it can also help to increase explainability, as has already been shown using the LIDC-IDRI dataset [3]. In addition to the malignancy of the lung nodules, which is the main goal of the prediction task, the radiologists also marked certain nodule characteristics such as sphericity, margin or spiculation. Shen et al. [16] used the attributes with a hierarchical 3D CNN approach, demonstrating the potential of using this privileged information. LaLonde et al. [11] extended this idea using capsule networks, a technique for learning individual, encapsulated representations rather than general convolutional layers [1,15]. This method was used to jointly learn the predefined attributes in the capsules and their associations with the classification target, i.e. malignancy. Explainability is enabled by providing additional attribute values that are essential to the model output. However, the predicted, possibly incorrect scores for the attributes must be trusted, which raises the question of whether there is a way to validate the predictions.

Prototype Networks are another line of research implementing the idea that the representations of images cluster around a prototypical representation for each class [17]. The goal is to find embedded prototypes (i.e. examples) that best separate the images by their classes [5]. This idea has been applied to various methods, such as unsupervised learning [13], few- and zero-shot learning [17,18,22], as well as for capsule networks [21], however without the use of privileged information. A successful approach is prototypical models with case-based reasoning, which justify their prediction by showing prototypical training examples similar to the input instance [4,12]. This idea can be used for region-wise prototypical samples [6]. However, these networks can only tell which prototypical samples resemble the query image, not why. Similar to attention models, regional explanations are learned and provided [23,24]. It is up to the user to guess which features of the image regions are relevant to the network and are exemplified by the prototypes.

Our method addresses the limitations of privileged information-based and prototype-based explanation by combining case-based visual reasoning through exemplary representation of high-level attributes to achieve explainability and high-performance. The proposed method is an image classifier that satisfies explainable-by-design with two elements: First, decisive intermediate results of a high-performance CNN are trained on human-defined attributes which are being predicted during application. Second, the model provides prototypical natural images to validate the attribute prediction. In addition to the enhanced explainability offered by the proposed approach, to our knowledge the proposed method outperforms existing studies on the LIDC-IDRI dataset.

The main contributions of our work are:

- A novel method that, for the first time to our knowledge, combines privileged information and prototype learning to provide increased explanatory power for medical classification tasks.
- A prototype network architecture based on a capsule network that leverages the benefits of both techniques.
- An explainable solution outperforming state-of-the-art explainable and non-explainable methods on the LIDC-IDRI dataset.

We provide the code with the model architecture and training algorithm of *Proto-Caps* on GitHub.

2 Methods

The idea behind our approach is to combine the potential of attribute and prototype learning for a powerful and interpretable learning system. For this, we use a capsule network of attribute capsules from which the target class is predicted. As the attribute prediction can also be susceptible to error, we use prototypes to explain the predictions made for each attribute. Based on [11], our approach, called *Proto-Caps*, consists of a backbone capsule network. The network is trained using multiple heads. An attribute head is used to ensure that each capsule represents a single attribute, a reconstruction head learns the original segmentation, and the main target prediction head learns the final classification. The model is extended by a prototype layer that provides explanations for each attribute decision. The overall architecture of *Proto-Caps* is shown in Fig. 1.

The **backbone** of our approach is a capsule network consisting of three layers: Features of the input image of size $1 \times 32 \times 32$ are extracted by a 2D convolutional layer containing 256 kernels of size 9×9. We decided not to use 3D convolutional layers, as preliminary experiments showed only marginal differences (within std. dev. of results), but required significantly more computing time. The primary capsule layer then segregates low-level features into 8 different capsules, with each capsule applying 256 kernels of size 9×9. The final dense capsule layer consists of one capsule for each attribute and extracts high-level features, overall producing eight 16-dimensional vectors. These vectors form the starting point for the different prediction branches.

The **target head**, a fully connected layer, combines the capsule encodings. The loss function for the malignancy prediction was chosen according to LaLonde et al. [11], where the distribution of radiologist malignancy annotations is optimized with the Kullback-Leibler divergence $\mathcal{L}_{mal}$ to reflect the inter-observer agreement and thus uncertainty. The **reconstruction branch** to predict the segmentation mask of the nodule consists of a simple decoder with three fully connected layers with the output filters 512, 1024, and the size of the resulting image $1 \times 32 \times 32$. The reconstruction loss $\mathcal{L}_{recon}$ implements the mean square error between the output and the binary segmentation mask. It has been shown that incorporating reconstruction learning is beneficial to performance [11].

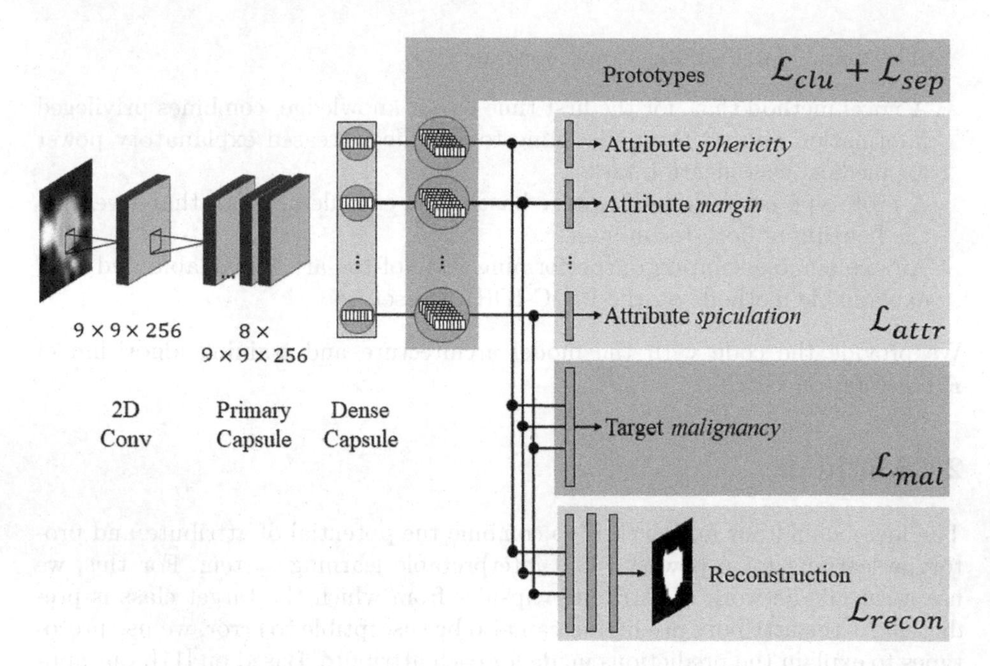

Fig. 1. Proposed Model Architecture. The backbone capsule network results in capsules representing predefined attributes. For each capsule, a set of prototypes is trained. To fit the attribute scores, the capsule vectors are fed through individual dense layers. The latent vectors of all capsules are being accumulated for a dense layer to predict a target score and for a decoder network to reconstruct the region of interest.

For the **attribute head**, we propose to use fully connected layers, instead of determining the attribute manifestation by the length of the capsule encoding, as was done previously [11]. Each capsule vector is processed by a separate linear layer to fit the respective attribute score. We formulate the attribute loss as

$$\mathcal{L}_{attr} = (1 - b) \sum_{a} \|Y_a - O_a\|^2, \tag{1}$$

where Y_a is the ground truth mean attribute score by the radiologists, O_a is the network score prediction for the a-th attribute, and b is a random binary mask allowing semi-supervised attribute learning.

Two **prototypes** are learned per possible attribute class, resulting in 8–12 prototypes per attribute (i.e. capsule). During the training, a combined loss function encourages a training sample to be close to a prototype of the correct attribute class and away from prototypes dedicated to others, similar to existing approaches [6]. Randomly initialized, the prototypes are a representative subset of the training dataset for each attribute after the training. For this, a cluster cost reduces the Euclidean distance of a sample's capsule vector O_a to the nearest prototype vector p_j of group P_{a_s} which is dedicated to its correct attribute score.

$$\mathcal{L}_{clu} = \frac{1}{A}\sum_{a}^{A} \min_{p_j \in P_{a_s}} \|O_a - p_j\|_2 . \tag{2}$$

In order to clearly distinguish between different attribute specifications, a separation loss is applied to increase the distance to the capsule prototypes that do not have the correct specification, limited by a maximum distance:

$$\mathcal{L}_{sep} = \frac{1}{A}\sum_{a}^{A} \min_{p_j \notin P_{a_s}} \max(0, \text{dist}_{max} - \|O_a - p_j\|_2). \tag{3}$$

Prototype optimization begins after 100 epochs. In addition to fitting the prototypes with the loss function, each prototype is replaced every 10 epochs by the most similar latent vector of a training sample. The original image of the training sample is stored and used for prototype visualization. During inference, the predicted attribute value is set to the ground truth attribute value of the closest prototype, ignoring the learned dense layers in the attribute head at this stage.

The overall loss function is the following weighted sum, where $\lambda_{recon} = 0.512$ was chosen according to [11], and the prototype weights were chosen empirically:

$$\mathcal{L} = \mathcal{L}_{mal} + \lambda_{recon} \cdot \mathcal{L}_{recon} + \mathcal{L}_{attr} + 0.125 \cdot (\mathcal{L}_{clu} + 0.1 \cdot \mathcal{L}_{sep}) \tag{4}$$

3 Experiments

Data. The proposed approach is evaluated using the publicly available LIDC-IDRI dataset consisting of 1018 clinical thoracic CT scans from patients with Non-Small Cell Lung Cancer (NSCLC) [2,3]. Each lung nodule with a minimum size of 3 mm was segmented and annotated with a malignancy score ranging from 1-*highly unlikely* to 5-*highly suspicious* by one to four expert raters. Nodules were also scored according to their characteristics with respect to predefined attributes, namely subtlety (difficulty of detection, 1-*extremely subtle*, 5-*obvious*), internal structure (1-*soft tissue*, 4-*air*), pattern of calcification (1-*popcorn*, 6-*absent*), sphericity (1-*linear*, 5-*round*), margin (1-*poorly defined*, 5-*sharp*), lobulation (1-*no lobulation*, 5-*marked lobulation*), spiculation (1-*no spiculation*, 5-*marked spiculation*), and texture (1-*non-solid*, 5-*solid*). The `pylidc` framework [7] is used to access and process the data. The mean attribute annotation and the mean and standard deviation of the malignancy annotations are calculated. The latter was used to fit a Gaussian distribution, which serves as the ground truth label for optimization. Samples with a mean expert malignancy score of 3-*indeterminate* or annotations from fewer than three experts were excluded in consistency with the literature [8,9,11].

Experiment Designs. To ensure comparability with previous work [8,9,11], the main metric used is Within-1-Accuracy, where a prediction within one score is considered correct. Five-fold stratified cross-validation was performed using 10 % of the training data for validation and the best run of three is reported.

The algorithm was implemented using the `PyTorch` framework version 1.13 and `CUDA` version 11.6. A learning rate of 0.5 was chosen for the prototype vectors and 0.02 for the other learnable parameters. The batch size was set to 128 and the optimizer was ADAM [10]. With a maximum of 1000 epochs, but stopping early if there was no improvement in target accuracy within 100 epochs, the experiments lasted an average of three hours on a GeForce RTX 3090 graphics card. The code is publicly available at https://github.com/XRad-Ulm/Proto-Caps.

Besides pure performance, the effect of reduced availability of attribute annotations was investigated. This was done by using attribute information only for a randomly selected fraction of the nodules during the training.

To investigate the effect of prototypes on the network performance, an ablation study was performed. Three networks were compared: **Proto-Caps** (proposed) including learning and applying prototypes during inference, **Proto-Caps**$_{\text{w/o use}}$ where prototypes are only learned but ignored for inference, and **Proto-Caps**$_{\text{w/o learn}}$ using the proposed architecture without any prototypes.

4 Results

Qualitative. Figure 2 shows examples of model output. The predicted malignancy score is justified by the closest prototypical sample of a certain attribute. The respective original image for each attribute prototype is being saved during the training process and used for visualization during inference. In case B, there are large differences between the margin and lobulation prototype and the sample. Similarly, in case C, the spiculation prediction is very different from the sample.

During application, these discrepancies between the prototypes and the sample nodule raise suspicion, and help to assess the malignancy prediction. A quantitative evaluation of the relationship between correctness in attribute and in target prediction using logistic regression analysis shows a strong relationship between both with an accuracy of 0.93/0.1.

Quantitative. Table 1 shows the results of our experiments compared to other state-of-the-art approaches, with results taken from original reports. The accuracy of the proposed method exceeds previous work in both the malignancy and almost all attribute predictions, while modelling all given attributes.

Table 2 lists the results obtained when only fractions of the training samples come with attribute information. The experiments indicate that the performance of the given approach is maintained up to a fraction of 10 %. Using no attribute annotations at all, i.e. no privileged information, achieves a similar performance, but results in a loss of explainability, as the high-level features extracted in the capsules are not understandable to humans. This result suggests that privileged information here leads to an increase in interpretability for humans by providing attribute predictions and prototypes without interfering with the model performance.

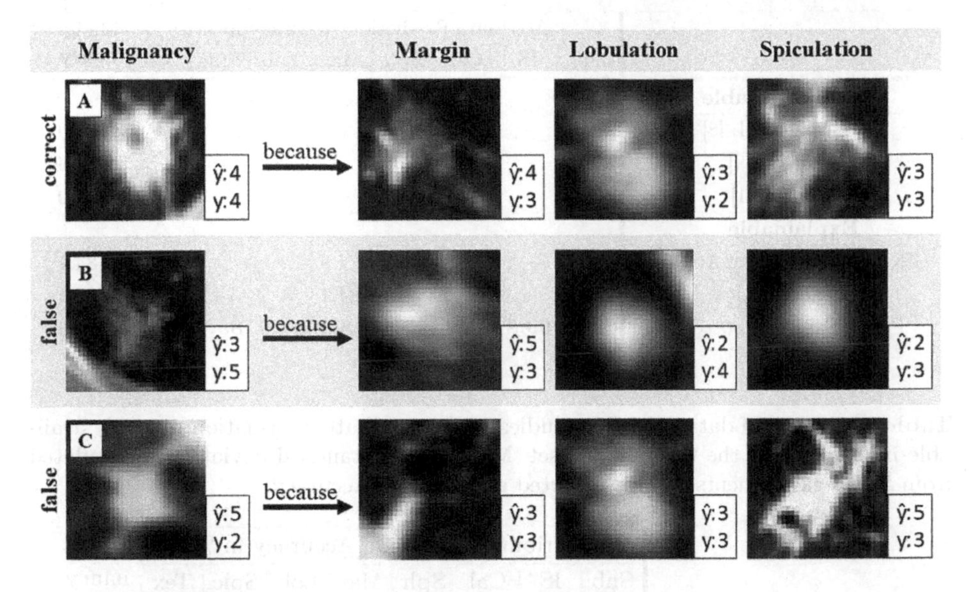

Fig. 2. One correct and two wrongly predicted examples with exemplary attribute prototypes. Prediction $\hat{y}$ and ground truth label y of malignancy and attribute respectively. Identifying false attribute predictions can help to identify misclassification in malignancy.

The ablation study shows no significant differences between the three models evaluated. For the malignancy accuracy, **Proto-Caps**$_{\text{w/o use}}$ and **Proto-Caps**$_{\text{w/o learn}}$ achieved $\mu = 93.9\,\%$ ($\sigma = 0.8$) and $\mu = 93.7\,\%$ ($\sigma = 1.1$), respectively. The average difference in attribute accuracy compared to the proposed methods is 1.7 % and 1.5 % better, respectively, and is more robust across experiments. The best result was obtained when the prototypes were learned but not used, possibly indicating that the prototypes may have a regularising effect during training, but further experiments are needed to confirm this due to the close results. To give an indication of the decoder performance, **Proto-Caps**$_{\text{w/o use}}$ achieved a dice score of 79.7 %.

Table 1. Comparison with literature values of other works, attribute scores are reported if available. Mean μ and standard deviation σ calculated from 5-fold experiments. Scores reported as Within-1-Accuracy, except for [16]. The best result is in bold.

		Attribute Prediction Accuracy in %								Malig-nancy
		Sub	IS	Cal	Sph	Mar	Lob	Spic	Tex	
Non-explainable										
3D-CNN+MTL [8]		–	–	–	–	–	–	–	–	90.0
TumorNet [9]		–	–	–	–	–	–	–	–	92.3
CapsNet [11]		–	–	–	–	–	–	–	–	77.0
Explainable										
HSCNN [16] (binary ACC.)		71.9	–	90.8	55.2	72.5	–	–	83.4	84.2
X-Caps [11]		**90.4**	–	–	85.4	84.1	70.7	75.2	93.1	86.4
Proto-Caps (proposed)	μ	89.1	**99.8**	**95.4**	**96.0**	**88.3**	**87.9**	**89.1**	**93.3**	**93.0**
	σ	5.2	0.2	1.3	2.2	3.1	0.8	1.3	1.0	1.5

Table 2. Results of data reduction studies where attribute information was only available in fractions of the training dataset. Mean μ and standard deviation σ calculated from 5-fold experiments. Scores reported as Within-1-Accuracy.

		Attribute Prediction Accuracy in %								Malig-nancy
		Sub	IS	Cal	Sph	Mar	Lob	Spic	Tex	
100 % attribute labels	μ	89.1	99.8	95.4	96.0	88.3	87.9	89.1	93.3	93.0
	σ	5.2	0.2	1.3	2.2	3.1	0.8	1.3	1.0	1.5
10 % attribute labels	μ	92.6	99.8	95.7	94.9	90.3	88.8	86.9	92.3	92.4
	σ	0.9	0.2	0.9	4.1	1.6	1.6	2.4	1.4	0.8
1 % attribute labels	μ	91.0	99.8	92.8	95.5	79.9	85.7	85.6	91.2	90.2
	σ	4.5	0.2	1.4	2.3	13.1	4.4	6.8	1.7	1.1
0 % attribute labels	μ	–	–	–	–	–	–	–	–	92.4
	σ	–	–	–	–	–	–	–	–	1.0

5 Discussion and Conclusion

We propose a new method, named *Proto-Caps*, which combines the advantages of privileged information, and prototype learning for an explainable network, achieving more than 6 % better accuracy than the state-of-the-art explainable method. As shown by qualitative results (Fig. 2), the obtained prototypes can be used to detect potential false classifications. Our method is based on capsule networks, which allow prediction based on attribute-specific prototypes. Compared to class-specific prototypes, our approach is more specific and allows better interpretation of the predictions made. In summary, *Proto-Caps* outputs prediction results for the main classification task and for predefined attributes, and provides visual validation through the prototypical samples of the attributes.

The experiments demonstrate that it outperforms state-of-the-art methods that provide less explainability. Our data reduction studies show that the proposed solution is robust to the number of annotated examples, and good results are obtained even with a 90% reduction in privileged information. This opens the door for application to other datasets by reducing the additional annotation overhead. While we did see a reduction in performance with too few labels, our results suggest that this is mainly due to inhomogeneous coverage of individual attribute values. In this respect, it would be interesting to find out how a specific selection of the annotated samples, e.g. with extremes, affects the accuracies, especially since our results show that the overall performance is robust even when the attributes are not explicitly trained, i.e. without additional privileged information. Another area of research would be to explore other types of privileged information that require less extra annotation effort, such as medical reports, to train the attribute capsules. It would also be worth investigating more sophisticated 3D-based capsule networks.

In conclusion, we believe that the approach of leveraging privileged information with comprehensible architectures and prototype learning is promising for various high-risk application domains and offers many opportunities for further research.

Acknowledgements. This research was supported by the University of Ulm (Baustein, L.SBN.0214), and the German Federal Ministry of Education and Research (BMBF) within RACOON COMBINE "NUM 2.0" (FKZ: 01KX2121). We acknowledge the National Cancer Institute and the Foundation for the National Institutes of Health for the critical role in the creation of the publicly available LIDC/IDRI Dataset.

References

1. Afshar, P., Mohammadi, A., Plataniotis, K.N.: Brain tumor type classification via capsule networks. In: 2018 25th IEEE International Conference on Image Processing (ICIP), pp. 3129–3133. IEEE (2018). https://doi.org/10.1109/ICIP.2018.8451379
2. Armato III, S.G., et al.: The lung image database consortium (LIDC) and image database resource initiative (IDRI): a completed reference database of lung nodules on CT scans. Med. Phy. **38**(2), 915–931 (2011). https://doi.org/10.1118/1.3528204, publisher: Wiley Online Library
3. Armato III, S.G., et al.: Data from LIDC-IDRI. The Cancer Imaging Archive (2015). https://doi.org/10.7937/K9/TCIA.2015.LO9QL9SX, [Data set]
4. Barnett, A.J., et al.: A case-based interpretable deep learning model for classification of mass lesions in digital mammography. Nat. Mach. Intell. **3**(12), 1061–1070 (2021). https://doi.org/10.1038/s42256-021-00423-x
5. Bien, J., Tibshirani, R.: Prototype selection for interpretable classification. Ann. Appl. Stat. **5**(4), 2403–2424 (2011). https://doi.org/10.1214/11-AOAS495
6. Chen, C., Li, O., Tao, D., Barnett, A., Rudin, C., Su, J.K.: This looks like that: deep learning for interpretable image recognition. In: Advances in Neural Information Processing Systems 32 (2019)

7. Hancock, M.C., Magnan, J.F.: Lung nodule malignancy classification using only radiologist-quantified image features as inputs to statistical learning algorithms: probing the lung image database consortium dataset with two statistical learning methods. J. Med. Imaging. **3**(4), 044504–044504 (2016). https://doi.org/10.1117/1.JMI.3.4.044504
8. Hussein, S., Cao, K., Song, Q., Bagci, U.: Risk stratification of lung nodules using 3D CNN-based multi-task learning. In: Niethammer, M., et al. (eds.) IPMI 2017. LNCS, vol. 10265, pp. 249–260. Springer, Cham (2017). https://doi.org/10.1007/978-3-319-59050-9_20
9. Hussein, S., Gillies, R., Cao, K., Song, Q., Bagci, U.: TumorNet: Lung nodule characterization using multi-view convolutional neural network with Gaussian process. In: 2017 IEEE 14th International Symposium on Biomedical Imaging (ISBI 2017), pp. 1007–1010. IEEE (2017). https://doi.org/10.1109/ISBI.2017.7950686
10. Kingma, D.P., Ba, J.: Adam: a method for stochastic optimization. arXiv preprint arXiv:1412.6980 (2014)
11. LaLonde, R., Torigian, D., Bagci, U.: Encoding visual attributes in capsules for explainable medical diagnoses. In: Martel, A.L., et al. (eds.) MICCAI 2020. LNCS, vol. 12261, pp. 294–304. Springer, Cham (2020). https://doi.org/10.1007/978-3-030-59710-8_29
12. Li, O., Liu, H., Chen, C., Rudin, C.: Deep learning for case-based reasoning through prototypes: a neural network that explains its predictions. In: Proceedings of the AAAI Conference on Artificial Intelligence, vol. 32, issue 1 (2018)
13. Pan, Y., Yao, T., Li, Y., Wang, Y., Ngo, C.W., Mei, T.: Transferrable prototypical networks for unsupervised domain adaptation. In: Proceedings of the IEEE/CVF Conference on Computer Vision and Pattern Recognition, pp. 2239–2247 (2019). https://doi.org/10.1109/CVPR.2019.00234
14. Rudin, C.: Stop explaining black box machine learning models for high stakes decisions and use interpretable models instead. Nat. Mach. Intell. **1**(5), 206–215 (2019). https://doi.org/10.1038/s42256-019-0048-x
15. Sabour, S., Frosst, N., Hinton, G.E.: Dynamic routing between capsules. In: Advances in Neural Information Processing Systems 30 (2017)
16. Shen, S., Han, S.X., Aberle, D.R., Bui, A.A., Hsu, W.: An interpretable deep hierarchical semantic convolutional neural network for lung nodule malignancy classification. Expert Syst. Appl. **128**, 84–95 (2019). https://doi.org/10.1016/j.eswa.2019.01.048, publisher: Elsevier
17. Snell, J., Swersky, K., Zemel, R.: Prototypical networks for few-shot learning. In: Advances in Neural Information Processing Systems 30 (2017)
18. Sun, S., Sun, Q., Zhou, K., Lv, T.: Hierarchical attention prototypical networks for few-shot text classification. In: Proceedings of the 2019 Conference on Empirical Methods in Natural Language Processing and the 9th International Joint Conference on Natural Language Processing (EMNLP-IJCNLP), pp. 476–485 (2019). https://doi.org/10.18653/v1/D19-1045
19. Vapnik, V., Izmailov, R., et al.: Learning using privileged information: similarity control and knowledge transfer. J. Mach. Learn. Res. **16**(1), 2023–2049 (2015)
20. Vapnik, V., Vashist, A.: A new learning paradigm: learning using privileged information. Neural Netw. **22**(5), 544–557 (2009). https://doi.org/10.1016/j.neunet.2009.06.042, advances in Neural Networks Research: IJCNN2009
21. Wang, M., Guo, Z., Li, H.: A dynamic routing CapsNet based on increment prototype clustering for overcoming catastrophic forgetting. IET Comput. Vis. **16**(1), 83–97 (2022). https://doi.org/10.1049/cvi2.12068, publisher: Wiley Online Library

22. Xu, W., Xian, Y., Wang, J., Schiele, B., Akata, Z.: Attribute prototype network for zero-shot learning. Adv. Neural. Inf. Process. Syst. **33**, 21969–21980 (2020)
23. Zheng, H., Fu, J., Mei, T., Luo, J.: Learning multi-attention convolutional neural network for fine-grained image recognition. In: Proceedings of the IEEE International Conference on Computer Vision pp. 5209–5217 (2017). https://doi.org/10.1109/ICCV.2017.557
24. Zhou, B., Khosla, A., Lapedriza, A., Oliva, A., Torralba, A.: Learning deep features for discriminative localization. In: Proceedings of the IEEE Conference on Computer Vision and Pattern Recognition, pp. 2921–2929 (2016). https://doi.org/10.1109/CVPR.2016.319

Physics-Based Decoding Improves Magnetic Resonance Fingerprinting

Juyeon Heo[1], Pingfan Song[1(✉)], Weiyang Liu[1], and Adrian Weller[1,2]

[1] University of Cambridge, Cambridge, UK
{jh2324,ps898,wl396,aw665}@cam.ac.uk
[2] The Alan Turing Institute, London, UK

Abstract. Magnetic Resonance Fingerprinting (MRF) is a promising approach for fast Quantitative Magnetic Resonance Imaging (QMRI). However, existing MRF methods suffer from slow imaging speeds and poor generalization performance on radio frequency pulse sequences generated in various scenarios. To address these issues, we propose a novel MRI physics-informed regularization for MRF. The proposed approach adopts a supervised encoder-decoder framework, where the encoder performs the main task, i.e. predicting the target tissue properties from input magnetic responses, and the decoder servers as a regularization via reconstructing the inputs from the estimated tissue properties using a Bloch-equation based MRF physics model. The physics-based decoder improves the generalization performance and uniform stability by a considerable margin in practical out-of-distribution settings. Extensive experiments verified the effectiveness of the proposed approach and achieved state-of-the-art performance on tissue property estimation.

Keywords: Magnetic Resonance Fingerprinting · Deep Neural Network · Physics-informed learning · Generalizability · Bloch equations

1 Introduction

Quantitative Magnetic Resonance Imaging (QMRI) is used to identify tissue's intrinsic physical properties, including the spin-lattice magnetic relaxation time (T1), and the spin-spin magnetic relaxation time (T2) [23]. Compared to conventional weighted (qualitative) MRI that focuses on tissue's contrast of brightness and darkness, QMRI reveals tissue's intrinsic properties with quantitative values and associated physical interpretations. Since different tissues are characterized

Supported by Turing AI Fellowship under EPSRC grant EP/V025279/1, by the Leverhulme Trust via CFI, and by Northern Ireland High Performance Computing (NI-HPC) service funded by EPSRC(EP/T022175).
J. Heo and P. Song—Co-first authors

Supplementary Information The online version contains supplementary material available at https://doi.org/10.1007/978-3-031-43895-0_42.

H. Greenspan et al. (Eds.): MICCAI 2023, LNCS 14221, pp. 446–456, 2023.
https://doi.org/10.1007/978-3-031-43895-0_42

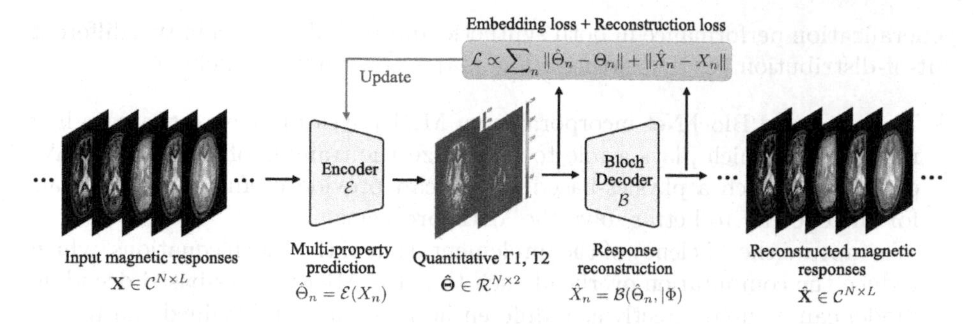

Fig. 1. BlochNet adopts a supervised encoder-decoder framework where the encoder solves an inverse problem that predicts tissue properties from input magnetic responses while the decoder leverages a Bloch equations-based MRI physics model to reconstruct the input responses from the estimated tissue properties. Such a design helps the encoder capture generalizable mapping effectively with the aid of physics-based feedback from the Bloch decoder.

by their distinct properties values, QMRI shows great potential to reduce subjectivity, with advantages in many areas including diagnosis, tissue characterization, investigation of disease pathologies, personalized medical treatment, and therapeutic assessment [1,10,28].

Despite various benefits, most QMRI approaches suffer from slow imaging speeds, and usually provide only a single intrinsic property at a time (*e.g.*, quantification of T1 alone, followed by T2 alone), resulting in low throughput. Magnetic Resonance Fingerprinting (MRF) provides an alternative QMRI framework to achieve multi-property quantification simultaneously [16]. Given a pseudo-random radio frequency (RF) pulse sequence, a distinct magnetic response, *i.e.*, fingerprint/signature, from each specific tissue is observed and then used to predict the target intrinsic tissue properties, *e.g.*, T1 and T2 values. Therefore, multi-property quantification boils down to an inverse problem that aims to infer underlying tissue properties from the observed magnetic responses.

In this work, we propose an MRI physics-regularized deep learning model for fast and robust MRF, called *BlochNet* as shown in Fig. 1. BlochNet adopts a supervised encoder-decoder framework where the encoder network solves the primary inverse problem that predicts tissue properties from input magnetic responses, while the decoder acts as a regularizer that guides the training of the encoder. In particular, the decoder leverages a Bloch equations-based MRI physics model to reconstruct the input responses from the estimated tissue properties, and compares the reconstructed inputs with the original input to provide an additional loss for the regularization purpose.

The rationale underlying the design is that domain knowledge such as well-founded physics principles bring additional useful constraints that can effectively reduce the solution space of an inverse problem. This contributes to an optimized solution, in particular, for an (ill-posed) inverse problem [11,12,15,18]. Our results verify that the proposed approach exhibits improved robustness and generalization performance in both synthetic and real MRF data in two different out-of-distribution (OOD) settings. The major contributions include:

- The proposed BlochNet incorporates an MRI physics model to the decoding mechanism, which plays a role to regularize the training of the encoder. We expect that such a physics-based design can provide useful training signals for the encoder to better solve the MRF problems.
- We improve the efficiency of the implementation of the Bloch equations, which reduce the computation overhead such that the MRI physics-based decoding model can be used directly as a differentiable module and trained end-to-end in neural networks (*e.g.*, the decoder in BlochNet).
- Compared to existing methods, BlochNet shows consistently better generalization performance across synthetic, phantom and real MRF data, and across different types of RF pulse sequences.

2 Background and Related Works

Since tissue properties lead to magnetic responses according to the MRI dynamics, quantifying tissue's properties via QMRI/MRF is a typical inverse problem, also an anti-causal task. The core idea of MRF is based on the fact that for each specific tissue, a pseudo-random pulse sequence leads to a unique magnetic response (*i.e.*, magnetization along the temporal dimension) which can serve as an identifiable signal signature, analogous to a "fingerprint" for the corresponding tissue. Once the unique identifiable magnetic responses are obtained, the estimation of tissue properties reduces to a pattern recognition problem.

Various approaches have been developed to solve the MRF problem, using either model-based techniques, *e.g.*, dictionary matching (DM), compressive sensing, or learning-based / data-driven techniques.

Model-Based Approaches. In the original MRF work [16], this task is approached via dictionary matching (DM) which finds the best matching entry in a pre-computed dictionary for each inquiry magnetic response. Accordingly, the best matching dictionary entry leads to multiple tissue properties directly via a look-up-table (LUT) operation. To alleviate the extensive computation overhead and storage burden, a number of model-based MRF approaches [7,20,25] were proposed to incorporate additional useful priors, e.g. sparsity, low rank, in order to improve reconstruction performance as well as reduce computational complexity.

Learning-Based Approaches. To address some shortcomings of model-based methods, learning-based approaches have been proposed for fast MRF by replacing the dictionary with a compact neural network. In particular, motivated by the success of deep learning in a number of tasks, there is an emerging trend [5,6,9,24] that suggests to use a trained neural network as a substitute for the MRF dictionary and LUT, so that the time-consuming dictionary matching operation can be avoided and replaced by an efficient inference through a trained network. However, despite the good performance delivered by neural networks, most of the learning-based methods were designed without taking into account the MRI physics underlying the imaging process, and may inevitably suffer from some limitations, such as degraded robustness and generalizability.

Physics-Informed Learning. Another highly related line of research is model-based learning [21,22] which provides a promising path to integrate domain knowledge with learned priors, thereby fusing the benefits of model-based methods with learning-based methods. Typical examples include algorithm unrolling [21], physics-informed neural networks (PINNs) [22], and other variants. Incorporating physics priors into the neural network design and training has demonstrated benefits in a broad range of applications [3,17], in particular for medical imaging [5,28,29]. In a similar spirit, we aim to incorporate the physics model that describes the MRI dynamics into learning-based MRF approaches so that the learned model can demonstrate improved data efficiency, robustness and generalization. Note that, in contrast to standard PINNs which are used to solve a PDE, our task is to solve an inverse PDE. Furthermore, we propose to leverage the PDE as a regularization. (More related work and comparisons are provided in the Appendix.)

3 Problem Formulation and Method

3.1 BlochNet: Regularized Networks by Physics-Based Decoding

The proposed BlochNet adopts a supervised encoder-decoder framework where the encoder solves the primary inverse problem that predicts tissue properties from input magnetic responses, while the decoder solves an auxiliary task that reconstructs the inputs from the estimated tissue properties using the Bloch equations-based MRI physics model. We highlight that a sophisticated MRI physics model is tailored and exploited as the decoder. The rationale behind such a design lies in the fact that the data generation mechanism represented by MRI physics is a useful constraint that can effectively reduce the solution

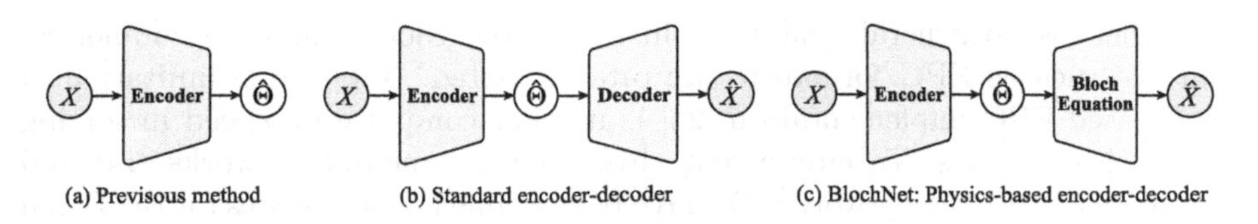

Fig. 2. Two baseline methods and our BlochNet. BlochNet exploits physics-based decoder for helping the encoder learn generalizable representation.

space of the inverse problem. Therefore, the physics-based decoder will act as a strong regularizer that can provide informative feedback and contribute to the training of a better encoder. We expect the physics prior can introduce a better and generalizable inductive bias to the encoder. Similar ideas have been explored in some other domains [11,12,18].

Specifically, in the proposed approach, the encoder uses a three-layer fully connected neural network to address the inverse problem that predicts T1, T2 tissue properties from input magnetic responses. Given an enquiry magnetic response $X_n \in \mathbb{C}^L$ for the n-th voxel where L denotes the length of each magnetic response, e.g. L = 1000 in our experiments, the encoder $\mathcal{E}$ outputs predicted tissue properties $\hat{\Theta}_n = \{\hat{T1}_n, \hat{T2}_n\} \in \mathbb{R}^p$ where $p = 2$ denotes the number of tissue properties to be predicted.

$$\hat{\Theta}_n = \mathcal{E}(X_n) \quad \forall n \in 1, \ldots, N$$

Note that, the estimation of tissue properties Θ from magnetic responses X requires long enough sequences $L > p$ to create identifiable signal evolutions that distinguish different tissues. Hence, this operation nonlinearly maps the magnetic responses from a high-dimensional manifold to a low-dimensional manifold.

In contrast, the decoder reconstructs the input magnetic responses from the estimated tissue properties $\hat{\Theta}_n$ by solving the Bloch equations [2] using our fast extended phase graph (EPG) implementation, given RF pulse sequence settings $\Phi = \{FA, TR, TE\}$ which consists of flip angles $FA \in \mathbb{C}^L$, repetition times $TR \in \mathbb{R}^L$ and echo times $TE \in \mathbb{R}^L$ across L time points.

$$\hat{X}_n = \mathcal{B}(\hat{\Theta}_n | \Phi) \quad \forall n \in 1, \ldots, N$$

where $\mathcal{B}$ denotes the decoder based on Bloch equations:

$$\frac{d\vec{M}}{dt} = \vec{M} \times \gamma \vec{B} - \begin{bmatrix} M_x/T2 \\ M_y/T2 \\ (M_z - M_0)/T1 \end{bmatrix}$$

where $\vec{M} = [M_x, M_y, M_z]^\top$ denotes the magnetization vector. $\vec{M_0}$ denotes the equilibrium magnetization; $\vec{B}$ denotes the magnetic field; and γ denotes the gyromagnetic ratio. (More details of Bloch equations are provided in the Appendix.)

3.2 Fast EPG for Solving Bloch Equations

Since there is no general analytic solution to the Bloch equations, numerical solutions such as EPG formalism are often adopted. However, a limitation of the released EPG implementation [26] is its slow computation speed in solving the Bloch equations. To circumvent this, recurrent neural networks [14] and generative adversarial networks [27] have been applied as surrogates for the Bloch equation. However, these surrogate models require a lot of training data and may generate inaccurate magnetic responses on unseen tissue properties and RF pulse settings due to complex physics dynamics and potential overfitting risks.

Instead, we adapt the EPG implementation [26] to achieve a much more efficient implementation, making it practical to use the exact MRI physics model as a decoder in the training procedure. Specifically, the adaptations involve incorporating the PyTorch jit package for efficient parallelization, using batch-wise computation for the 3 Bloch stages (including nutation+forced precession, rotation, and relaxation in Fourier domain), and handling complex values in Pytorch efficiently. The improvement leads to 500 times faster generation of magnetic responses for 1,000 sequences on CPU, making repeated EPG computations feasible during training.[1] (More details can be found in the Appendix.)

3.3 Loss Function

The loss function consists of two parts: the mean squared error (MSE) between the ground truth and the predicted tissue properties, referred to as embedding loss, and the MSE between the input and the reconstructed signatures, referred to as reconstruction loss,

$$\mathcal{L} = \frac{1}{N}\sum_{n=1}^{N}\left(\frac{1}{2}\|\hat{\Theta}_n - \Theta_n\|_2^2 + \frac{1}{2}\|\hat{X}_n - X_n\|_2^2\right).$$

4 Experiment Results

In this section, we perform an evaluation of the proposed method and conduct a comparison with other state-of-the-art MRF methods. We evaluate the generalization performance of all models across different data distributions and different RF pulse sequences. The evaluation metric is the MSE in log-scale, and therefore, the unit is the squared millisecond in log-scale. For our model, we use 3-layer encoder-decoder with varying hidden units, Adam optimizer (lr=1e-3), and maximum epochs of 100 with early stopping based on the validation set on GTX 1080 Ti. We performed ten independent trials, the results of which are presented in Table 1 and 2. The associated standard deviations are provided in the Appendix.

Table 1. Generalization performance across **different data distributions**: synthetic data for training while phantom and anatomical data for testing.

	Dictionary Matching	FC	RNN	HYDRA	Autoencoder (FC-FC)	Autoencoder (RNN-RNN)	BlochNet (FC-Bloch)
Phantom data	18.6656	0.0634	0.0457	0.1604	0.0529	0.0508	**0.0311**
Anatomical data	19.4886	0.0795	0.0574	0.3109	0.0771	0.0987	**0.0647**

[1] Our code, including the fast and end-to-end solvable EPG-Bloch code, is released on our GitHub repository at https://github.com/rmrisforbidden/CauMedical.git.

4.1 Data Settings and Baseline Methods

We exploit three types of data including synthetic data, phantom MRI data, and anatomical MRI data. In particular, synthetic data (around 80,000 samples) is used for training, while phantom data (85,645 samples) and anatomical data (7,499 samples) is for evaluation. More details about the three datasets are provided in the Appendix. We compare our approach with 6 representative state-of-the-art MRF methods, including dictionary matching (DM) [16], Fully-connected deep neural network (FC) [6], Hybrid deep learning (HYDRA) [24] as well as two auto-encoder methods with RNN encoder and RNN decoder(RNN-RNN) and FC encoder and FC decoder(FC-FC), respectively.

4.2 Experiments of Evaluating Generalization Performance

We evaluate the generalization performance of various models on two types of experiment settings: 1) across different data distributions, including synthetic, phantom and anatomical MRF data; 2) across different RF pulse sequences with different flip angles. In addition, a series of ablation studies were conducted via comparison with other methods that use different types of decoders, as shown in Table 1 and Table 2, for example, comparing BlochNet (using a physics-based decoder) with FC (using no decoder) and FC-FC (using a learned decoder) to show the effect of the encoder and the effect of MRI physics.

Generalization Across Different Data Distributions. Due to limited anatomical data with ground truth T1, T2 values, it is common practice to use a large amount of synthetic data to train models to avoid potential over-fitting, and then perform validation on anatomical data [4,7,8,13,19,20,24,25]. Following the same routine, we perform model training on synthetic MRF data, followed by model testing on phantom and anatomical data, in order to evaluate the generalization performance of trained models across different data distributions.

Table 1 includes the mean squared error (MSE) between the ground truth and predicted tissue properties for seven approaches on phantom and anatomical data (in log-scale). As shown in the table, the dictionary-matching approach gave the worst performance, because the pre-computed dictionary and LUT did not cover the OOD data samples that could be quite different from the already contained dictionary entries. Interestingly, the results show that the reconstruction loss provides benefits to between-data generalization for autoencoder models(FC-FC or RNN-RNN), in comparison with non-autoencoder models(FC or RNN), respectively, on both phantom and anatomical MRF data. Furthermore, our BlochNet outperforms all other models, indicating that reconstruction loss from the physics-based decoder has the best regularization effect that contributes to improved encoder training.

Figure 3 shows the predicted tissue properties using various models on anatomical MRF data. Each individual model shows different prediction characteristics. Specifically, HYDRA suffers from a higher loss at the rim region of the

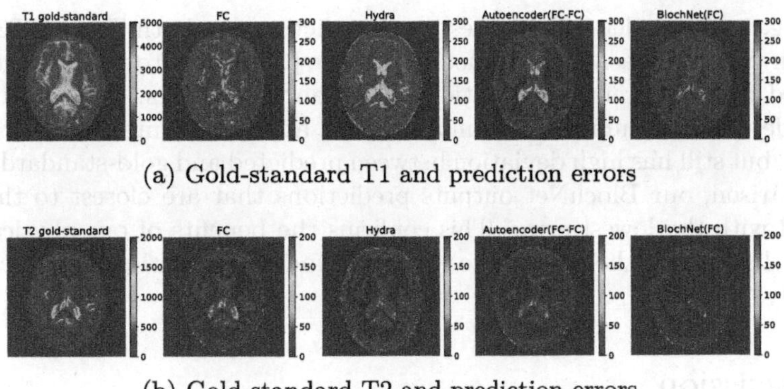

(a) Gold-standard T1 and prediction errors

(b) Gold-standard T2 and prediction errors

Fig. 3. Generalization across **different data distributions**: data in training is synthetic while in testing is anatomical MRI for four different models.

Table 2. Generalization performance across **different RF pulse sequences**: Spline5 and SplineNoisy11 in training, FISP in testing.

	Dictionary Matching	FC	RNN	HYDRA	Autoencoder (FC-FC)	Autoencoder (RNN-RNN)	BlochNet (FC-Bloch)
Phantom data	27.1956	0.8349	0.7960	0.3599	0.6968	0.6269	**0.1543**
Anatomical data	16.6150	0.7331	1.0674	0.6961	0.4021	0.9455	**0.2766**

brain, and leads to larger errors than other models. Autoencoder(FC-FC) model demonstrates better prediction of T2 values than the non-autoencoder(FC) model. The proposed BlochNet outperforms other models with the least prediction error and most stable performance across the whole range of both T1 and T2 values.

Generalization Across Different RF Pulse Sequences. In this experiment, we perform model training on one RF pulse sequence and evaluate the trained models on a different RF pulse sequence. Specifically, we adopted 3 different RF pulse sequences, including FISP [16], Spline5 [14], Spline11Noisy [14] with their flip angles shown in Figure 1 in the Appendix. More details are provided in Appendix. FISP is used exclusively in the testing stage, while Spline5 and Spline11Noisy are used exclusively in the training stage. Under such settings, the performance of our BlochNet and other six models is compared in Table 2.

In spite of degraded performance for all models under different train and test RF pulse sequences, the results clearly show the advantage of autoencoder (FC-FC or RNN-RNN) models over non-autoencoder models(FC or RNN), which confirms the benefits of incorporating a decoder to derive the reconstruction loss as additional regularization. Furthermore, the proposed BlochNet demonstrates significant gains over competing methods in such challenging cases on both phantom and anatomical MRF data.

In Fig. 4, FC model (left) makes poor predictions on both T1 and T2 values with high variance, because it cannot infer tissue properties from input signatures generated from different combinations of T1 and T2 values. Autoencoder(FC-FC) model (third column) shows more aligned and better inferences with lower variance, but still has high deviation between predicted and gold-standard values. In comparison, our BlochNet outputs predictions that are closest to the gold-standard with the lowest error. This confirms the benefits of our physics-based decoder that guides the encoder to learn the underlying anti-causal mechanism effectively.

5 Discussion

We present statistical significance tests on 10 trials for pairwise group comparisons using Tukey HSD test after the Normality Test and repeated ANOVA. For results in Table 1, corresponding to setting 1 (a relatively easy problem as the RF pulses used in training and testing are the same), our method is not always statistically better to every compared method, since all baseline methods can perform reasonably well. However, in setting 2, a much more challenging OOD setting where the RF pulses used in testing are different from those used in training and therefore can lead to different magnetic responses, our method is always statistically better (p-value < 0.001) than compared methods according to the results in Table 2. This demonstrates the robustness and generalizability of our method. As far as we are concerned, no existing approaches achieved satisfactory results in such an OOD case. While there is still ample room for improvement across all the methods, our approach took a critical step forward by incorporating physics knowledge. (More details can be found in the Appendix.)

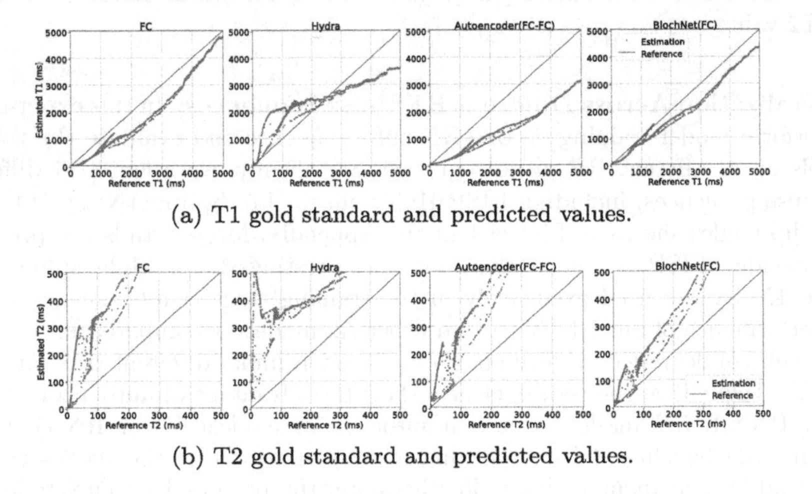

(a) T1 gold standard and predicted values.

(b) T2 gold standard and predicted values.

Fig. 4. Comparison of the generalization performance across **different RF pulse sequences**. Blue line: gold-standard. Red dots: predicted values for T1 and T2.

6 Conclusion

We propose BlochNet, a novel MRI physics-informed learning model, which consistently outperforms competing methods with better robustness and generalizability in MRF problems. In future work, we will consider k-space subsampling and incorporating spatial information for faster and more efficient QMRI/MRF.

References

1. Bipin Mehta, B., et al.: Magnetic resonance fingerprinting: a technical review. Magnetic Reson. Med. **81**(1), 25–46 (2019)
2. Bloch, F.: Nuclear induction. Phys. Rev. **70**(7–8), 460 (1946)
3. Cai, S., Wang, Z., Wang, S., Perdikaris, P., Karniadakis, G.E.: Physics-informed neural networks for heat transfer problems. J. Heat Trans. **143**(6), 4050542 (2021)
4. Cao, X., et al.: Robust sliding-window reconstruction for accelerating the acquisition of MR fingerprinting. Magnetic Reson. Med. **78**(4), 1579–1588 (2017)
5. Chen, D., Davies, M.E., Golbabaee, M.: Deep unrolling for magnetic resonance fingerprinting. In: 2022 IEEE 19th International Symposium on Biomedical Imaging (ISBI), pp. 1–4. IEEE (2022)
6. Cohen, O., Zhu, B., Rosen, M.S.: MR fingerprinting deep reconstruction network (drone). Magnetic Reson. Med. **80**(3), 885–894 (2018)
7. Davies, M., Puy, G., Vandergheynst, P., Wiaux, Y.: A compressed sensing framework for magnetic resonance fingerprinting. SIAM J. Imag. Sci. **7**(4), 2623–2656 (2014)
8. Golbabaee, M., et al.: Compressive MRI quantification using convex spatiotemporal priors and deep encoder-decoder networks. Med. Image Anal. **69**, 101945 (2021)
9. Hamilton, J.I., Currey, D., Rajagopalan, S., Seiberlich, N.: Deep learning reconstruction for cardiac magnetic resonance fingerprinting t1 and t2 mapping. Magnetic Reson. Med. **85**(4), 2127–2135 (2021)
10. Keil, V.C., et al.: A pilot study of magnetic resonance fingerprinting in Parkinson's disease. NMR Biomed. **33**(11), e4389 (2020)
11. Kilbertus, N., Parascandolo, G., Schölkopf, B.: Generalization in anti-causal learning. arXiv preprint arXiv:1812.00524 (2018)
12. Le, L., Patterson, A., White, M.: Supervised autoencoders: improving generalization performance with unsupervised regularizers. In: Advances in Neural Information Processing Systems 31 (2018)
13. Liao, C.: 3D MR fingerprinting with accelerated stack-of-spirals and hybrid sliding-window and grappa reconstruction. Neuroimage **162**, 13–22 (2017)
14. Liu, H., van der Heide, O., van den Berg, C.A., Sbrizzi, A.: Fast and accurate modeling of transient-state, gradient-spoiled sequences by recurrent neural networks. NMR Biomed. **34**(7), e4527 (2021)
15. Liu, W., Liu, Z., Paull, L., Weller, A., Schölkopf, B.: Structural causal 3D reconstruction. In: European Conference on Computer Vision (2022)
16. Ma, D., et al.: Magnetic resonance fingerprinting. Nature **495**(7440), 187 (2013)
17. Mao, Z., Jagtap, A.D., Karniadakis, G.E.: Physics-informed neural networks for high-speed flows. Comput. Methods Appl. Mech. Eng. **360**, 112789 (2020)
18. Maurer, A., Pontil, M., Romera-Paredes, B.: The benefit of multitask representation learning. J. Mach. Learn. Res. **17**(81), 1–32 (2016)

19. Mazor, G., Weizman, L., Tal, A., Eldar, Y.C.: Low rank magnetic resonance fingerprinting. In: 2016 IEEE 38th Annual International Conference of the Engineering in Medicine and Biology Society (EMBC), pp. 439–442. IEEE (2016)
20. Mazor, G., Weizman, L., Tal, A., Eldar, Y.C.: Low-rank magnetic resonance fingerprinting. Med. Phys. **45**(9), 4066–4084 (2018)
21. Monga, V., Li, Y., Eldar, Y.C.: Algorithm unrolling: interpretable, efficient deep learning for signal and image processing. IEEE Signal Process. Mag. **38**(2), 18–44 (2021)
22. Raissi, M., Perdikaris, P., Karniadakis, G.E.: Physics-informed neural networks: a deep learning framework for solving forward and inverse problems involving nonlinear partial differential equations. J. Comput. Phys. **378**, 686–707 (2019)
23. Scholand, N., Wang, X., Roeloffs, V., Rosenzweig, S., Uecker, M.: Quantitative MRI by nonlinear inversion of the Bloch equations. Magn. Reson. Med. **90**, 520–538 (2023)
24. Song, P., Eldar, Y.C., Mazor, G., Rodrigues, M.R.: Hydra: hybrid deep magnetic resonance fingerprinting. Med. Phys. **46**(11), 4951–4969 (2019)
25. Wang, Z., Li, H., Zhang, Q., Yuan, J., Wang, X.: Magnetic resonance fingerprinting with compressed sensing and distance metric learning. Neurocomputing **174**, 560–570 (2016)
26. Weigel, M.: Extended phase graphs: dephasing, RF pulses, and echoes-pure and simple. J. Magn. Reson. Imag. **41**(2), 266–295 (2015)
27. Yang, M., Jiang, Y., Ma, D., Mehta, B.B., Griswold, M.A.: Game of learning Bloch equation simulations for MR fingerprinting. arXiv preprint arXiv:2004.02270 (2020)
28. Yang, Q., et al.: Model-based synthetic data-driven learning (most-dl): application in single-shot t2 mapping with severe head motion using overlapping-echo acquisition. IEEE Trans. Med. Imag. **41**, 3167–3181 (2022)
29. Yang, Y., Sun, J., Li, H., Xu, Z.: ADMM-CSNet: a deep learning approach for image compressive sensing. IEEE Trans. Pattern Anal. Mach. Intell. **42**(3), 521–538 (2018)

Frequency Domain Adversarial Training for Robust Volumetric Medical Segmentation

Asif Hanif[1](✉), Muzammal Naseer[1], Salman Khan[1], Mubarak Shah[2], and Fahad Shahbaz Khan[1,3]

[1] Mohamed Bin Zayed University of Artificial Intelligence (MBZUAI), Abu Dhabi, UAE
{asif.hanif,muzammal.naseer,salman.khan,fahad.khan}@mbzuai.ac.ae
[2] University of Central Florida (UCF), Orlando, USA
shah@crcv.ucf.edu
[3] Linköping University, Linköping, Sweden

Abstract. It is imperative to ensure the robustness of deep learning models in critical applications such as, healthcare. While recent advances in deep learning have improved the performance of volumetric medical image segmentation models, these models cannot be deployed for real-world applications immediately due to their vulnerability to adversarial attacks. We present a 3D frequency domain adversarial attack for volumetric medical image segmentation models and demonstrate its advantages over conventional input or voxel domain attacks. Using our proposed attack, we introduce a novel frequency domain adversarial training approach for optimizing a robust model against voxel and frequency domain attacks. Moreover, we propose frequency consistency loss to regulate our frequency domain adversarial training that achieves a better tradeoff between model's performance on clean and adversarial samples. Code is available at https://github.com/asif-hanif/vafa.

Keywords: Adversarial attack · Adversarial training · Frequency domain attack · Volumetric medical segmentation

1 Introduction

Semantic segmentation of organs, anatomical structures, or anomalies in medical images (e.g. CT or MRI scans) remains one of the fundamental tasks in medical image analysis. Volumetric medical image segmentation (MIS) helps healthcare professionals to diagnose conditions more accurately, plan medical treatments, and perform image-guided procedures. Although deep neural networks (DNNs) have shown remarkable improvements in performance for different vision tasks,

---

Supplementary Information The online version contains supplementary material available at https://doi.org/10.1007/978-3-031-43895-0_43.

H. Greenspan et al. (Eds.): MICCAI 2023, LNCS 14221, pp. 457–467, 2023.
https://doi.org/10.1007/978-3-031-43895-0_43

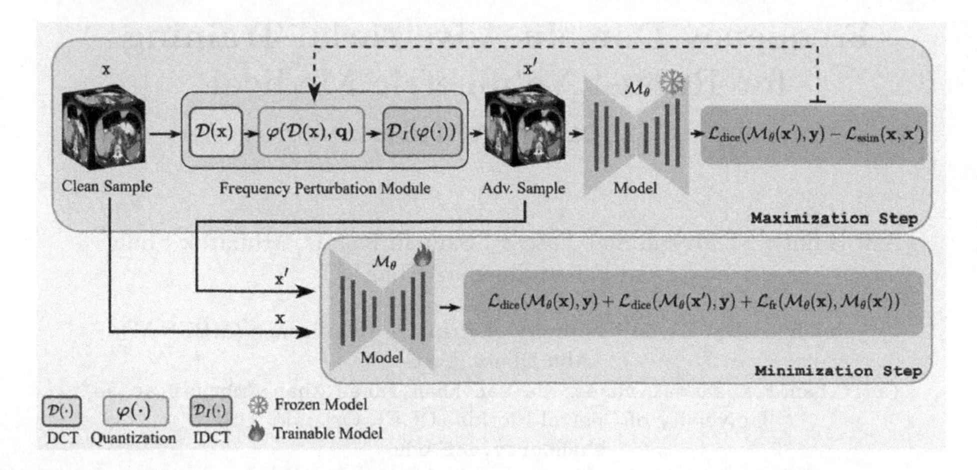

Fig. 1. Overview of Adversarial Frequency Attack and Training: A model trained on voxel-domain adversarial attacks is vulnerable to frequency-domain adversarial attacks. In our proposed adversarial training method, we generate adversarial samples by perturbing their frequency-domain representation using a novel module named "Frequency Perturbation". The model is then updated while minimizing the dice loss on clean and adversarially perturbed images. Furthermore, we propose a frequency consistency loss to improve the model performance.

including volumetric MIS, their real-world deployment is not straightforward particularly due to the vulnerabilities towards adversarial attacks [26]. An adversary can deliberately manipulate input data by crafting and adding perturbations to the input that are imperceptible to the human eye but cause the DNN to produce incorrect outputs [10]. Adversarial attacks pose a serious security threat to DNNs [1], as they can be used to cause DNNs to make incorrect predictions in a wide range of applications, including DNN-based medical imaging systems. To mitigate these threats, various techniques have been explored, including adversarial training, input data transformations, randomization, denoising auto-encoders, feature squeezing, and robust architectural changes [1]. Although significant progress has been made in adversarial defenses, however, this area is still evolving due to the development of attacks over time [3].

Ensuring the adversarial robustness of the models involved in safety-critical applications such as, medical imaging and healthcare is of paramount importance because a misdiagnosis or incorrect decision can result in life-threatening implications. Moreover, the weak robustness of deep learning-based medical imaging models will create a trust deficit among clinicians, making them reluctant to rely on the model predictions. The adversarial robustness of the medical imaging models is still an open and under-explored area [6,20]. Furthermore, most adversarial attacks and defenses have been designed for 2D natural images and little effort has been made to secure volumetric (3D) medical data [20].

In the context of 2D natural images, it has been recently observed that frequency-domain based adversarial attacks are more effective against the defenses that are primarily designed to "undo" the impact of pixel-domain adversarial noise in natural images [7]. Motivated by this observation in 2D natural images, here we explore the effectiveness of frequency-domain based adversarial attacks in the regime of volumetric medical image segmentation and aim to obtain a volumetric MIS model that is robust against adversarial attacks. To achieve this goal, we propose a *min-max* objective for adversarial training of *volumetric MIS model in frequency-domain.* For *maximization* step, we introduce **V**olumetric **A**dversarial **F**requency **A**ttack - **VAFA** (Fig. 1, Sect. 2.1) which operates in the frequency-domain of the data (unlike other prevalent voxel-domain attacks) and explicitly takes into account the 3D nature of the volumetric medical data to achieve higher fooling rate. For *minimization* step, we propose **V**olumetric **A**dversarial **F**requency-domain **T**raining - **VAFT** (Fig. 1, Sect. 2.2) to obtain a model that is robust to adversarial attacks. In VAFT, we update model parameters on clean and adversarial (obtained via VAFA) samples and further introduce a novel *frequency consistency loss* to keep frequency representation of the logits of clean and adversarial samples close to each other for a better accuracy tradeoff. In summary, our contributions are as follows:

- We propose an approach with a min-max objective for adversarial training of volumetric MIS model in the frequency domain. In the maximization step, we introduce a volumetric adversarial frequency attack (VAFA) that is specifically designed for volumetric medical data to achieve higher fooling rate. Further, we introduce a volumetric adversarial frequency-domain training (VAFT) based on a frequency consistency loss in the minimization step to produce a model that is robust to adversarial attacks.
- We conduct experiments with two different hybrid CNN-transformers based volumetric medical segmentation methods for multi-organ segmentation.

Related Work: There are three main types of popular volumetric MIS model architectures: CNN [23], Transformer [13] and hybrid [12,24]. Research has shown that medical machine learning models can be manipulated in various ways by an attacker, such as adding imperceptible perturbation to the image, rotating the image, or modifying medical text [8]. Adversarial attack studies on medical data have primarily focused on classification problems and voxel-domain adversaries. For example, Ma *et al.* [20] have used four types of pixel-domain attacks [4,10,16,21] on two-class and multi-class medical datasets. Li *et al.* [19] and Daza *et al.* [6] have focused on single-step and iterative adversarial attacks [5,10,15] on the volumetric MIS. In constant to voxel-domain adversarial attacks, our approach works in the frequency-domain.

2 Frequency Domain Adversarial Attack and Training

We aim to train a model for volumetric medical segmentation that is robust against adversarial attacks. Existing adversarial training (AT) approaches rely on min-max optimization [10,16,21] and operate in the input space. They find adversaries by adding the adversarial perturbation to the input samples by

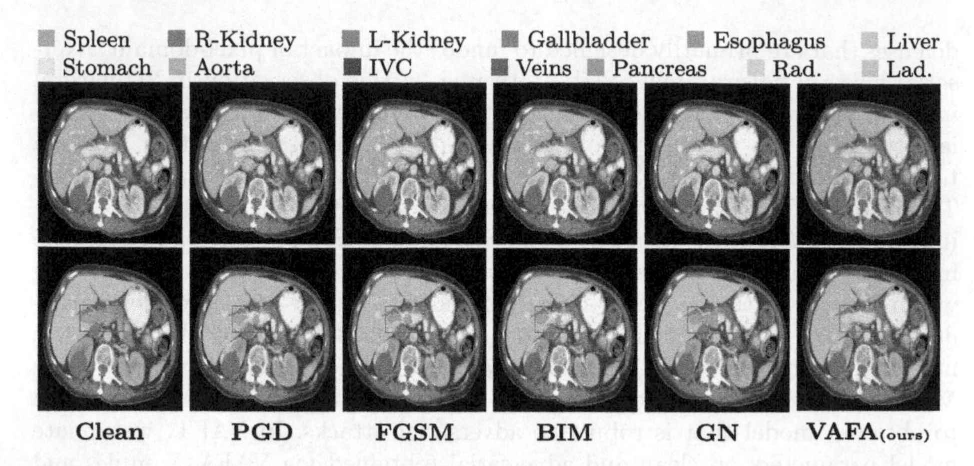

Fig. 2. Qualitative multi-organ segmentation comparison under different attacks on the UNETR [12] model. Top row shows example images and bottom row shows the corresponding segmentation masks predicted by the model under different attacks. Compared to different voxel-domain attacks (PGD [21], FGSM [10], BIM [16] and GN [14]), our attack (VAFA) achieves higher fooling rate (highlighted in red bounding box) while maintaining comparable perceptual similarity. Best viewed zoomed in.

maximizing the model loss (e.g., dice loss in segmentation). The loss function is then minimized on such adversaries to update the model parameters. In this work, we propose a frequency-domain adversarial attack that takes into account the 3D nature of the volumetric medical data and performs significantly better than the other voxel-domain as well as 2D frequency domain attacks (Table 1). Based on our attack, we then introduce a novel frequency-domain adversarial training to make the model resilient to adversarial attacks. Additionally, we observe that our approach improves/retains the performance of the robust model on clean samples when compared to the non-robust model. Our approach optimizes adversarial samples by perturbing the 3D-DCT coefficients within the frequency domain using our frequency perturbation module (Fig. 1) and adversarial guidance from the segmentation loss (Sect. 2.1). We find adversarial samples with high perceptual quality by maximizing the structural similarity between clean and adversarial samples. Using clean and adversarial samples, we propose updating the model parameters by simultaneously minimizing the segmentation loss (i.e. Dice loss) and the frequency consistency loss (Eq. 4) between the clean and adversarial outputs of the segmentation model.

3D Medical Segmentation Framework: Deep learning-based 3D medical segmentation generally uses encoder-decoder architectures [18]. The encoder produces a latent representation of the input sample. A segmentation map of the input sample is generated by the decoder using the latent feature representation. The decoder usually incorporates skip connections from the encoder to preserve spatial information [12]. Next, we describe our proposed volumetric frequency-domain adversarial attack in Sect. 2.1 and then training in Sect. 2.2.

2.1 Volumetric Adversarial Frequency Attack (VAFA)

Generally, adversarial attacks operate in the voxel domain by adding an imperceptible perturbation to the input data. In contrast, our attack perturbs the 3D-DCT coefficient to launch a frequency-domain attack for 3D medical image segmentation. Our *Frequency Perturbation Module* (FPM) transforms voxel-domain data into frequency-domain by using discrete cosine transforms (DCTs) and perturbs the DCT coefficients using a *learnable quantization*. It then takes an inverse DCT of the perturbed frequency-domain data and returns voxel-domain image. We keep the model in a "frozen" state while maximizing the dice loss [25] for segmentation and minimizing structural similarity loss [27] for perceptual quality. We represent a 3D (volumetric) single channel clean sample by $\mathrm{X} \in \mathbb{R}^{1\times H\times W\times D}$ and its ground-truth binary segmentation mask by $\mathrm{Y} \in \{0,1\}^{\mathrm{NumClass}\times H\times W\times D}$, where "NumClass" is the number of classes. We split X into n 3D patches i.e. $\mathrm{X} \mapsto \{\mathbf{x}_i\}_{i=1}^n$, where $\mathbf{x}_i \in \mathbb{R}^{h\times w\times d}$ and $h \leq H, w \leq W, d \leq D$, $h = w = d$. We apply our frequency perturbation module to each of these patches.

Frequency Perturbation Module: We apply a 3D discrete cosine transform (DCT), represented as $\mathcal{D}(\cdot)$, to each patch $\mathbf{x}_i$. The resulting DCT coefficients are then processed through a function $\varphi(\cdot)$, which performs three operations: quantization, differentiable rounding (as described in [9]), and subsequent de-quantization. $\varphi(\cdot)$ utilizes a learnable quantization table $\mathbf{q} \in \mathbb{Z}^{h\times w\times d}$ to modify the DCT coefficients, setting some of them to zero. In particular, $\varphi(\mathcal{D}(\mathbf{x}),\mathbf{q}) := \lfloor \frac{\mathcal{D}(\mathbf{x})}{\mathbf{q}} \rceil \odot \mathbf{q}$, where DCT coefficients of a patch (i.e. $\mathcal{D}(\mathbf{x})$) are element-wise divided by quantization table $\mathbf{q}$. After the division operation, the result undergoes rounding using a differentiable rounding operation [9], resulting in some values being rounded down to zero. The de-quantization step involves element-wise multiplication of $\lfloor \frac{\mathcal{D}(\mathbf{x})}{\mathbf{q}} \rceil$ with the same quantization table $\mathbf{q}$. This step allows us to reconstruct the quantized DCT coefficients. Since quantization table is in the denominator of the division operation, therefore, higher quantization table values increase the possibility of more DCT coefficients being rounded down to zero. To control the number of DCT coefficients being set to zero, we can constrain the values of the quantization table to a maximum threshold (constraint in Eq. 2). In other words, $\varphi(\cdot)$ performs a 3D adversarial lossy compression on input through a learnable quantization table. Finally, a 3D inverse DCT (IDCT) is performed on the output of $\varphi(\cdot)$ in order to obtain an adversarially perturbed voxel-domain representation, denoted by $\mathbf{x}'$. We show our frequency perturbation module in Eq. 1 as follows:

$$\mathbf{x} \mapsto \mathcal{D}(\mathbf{x}) \mapsto \underbrace{\varphi(\mathcal{D}(\mathbf{x}),\mathbf{q})}_{\substack{\text{quantization,}\\ \text{rounding and}\\ \text{de-quantization}}} \mapsto \mathcal{D}_I(\varphi(\cdot)) \mapsto \mathbf{x}' \tag{1}$$

We repeat the above mentioned sequence of transformations for all patches and then merge $\{\mathbf{x}_i'\}_{i=1}^n$ to form adversarial image $\mathrm{X}' \in \mathbb{R}^{H\times W\times D}$.

Quantization Constraint: We learn quantization table $\mathbf{q}$ by maximizing the $\mathcal{L}_{\text{dice}}$ while ensuring that $\|\mathbf{q}\|_\infty \leq q_{\max}$. Quantization threshold $q_{\max}$ controls the

Algorithm 1. Volumetric Adversarial Frequency Attack (**VAFA**)

1: Number of Steps: T, Quantization Threshold: $q_{\max}$
2: **Input:** $\mathrm{X} \in \mathbb{R}^{H\times W\times D}$, $\mathrm{Y} \in \{0,1\}^{\mathrm{NumClass}\times H\times W\times D}$ **Output:** $\mathrm{X}' \in \mathbb{R}^{H\times W\times D}$
3: **function VAFA**(X,Y)
4: $\quad \mathbf{q}_i \leftarrow \mathbf{1} \quad \forall\, i \in \{1,2,\ldots,n\}$ ▷ Initialize all *quantization tables* with ones.
5: $\quad$ **for** $t \leftarrow 1$ to T **do**
6: $\quad\quad \{\mathbf{x}_i\}_{i=1}^{n} \leftarrow \mathrm{Split}(\mathrm{X})$ ▷ Split X into 3D patches of size $(h \times w \times d)$
7: $\quad\quad \mathbf{x}_i' \leftarrow \mathcal{D}_I\big(\ \varphi(\mathcal{D}(\mathbf{x}_i), \mathbf{q}_i)\ \big) \quad \forall\, i \in \{1,2,\ldots,n\}$ ▷ Frequency Perturbation
8: $\quad\quad \mathrm{X}' \leftarrow \mathrm{Merge}(\{\mathbf{x}_i'\}_{i=1}^{n})$ ▷ Merge all adversarial patches to form X′
9: $\quad\quad \mathcal{L}(\mathrm{X},\mathrm{X}',\mathrm{Y}) = \mathcal{L}_{\mathrm{dice}}(\mathcal{M}_\theta(\mathrm{X}'),\mathrm{Y}) - \mathcal{L}_{\mathrm{ssim}}(\mathrm{X},\mathrm{X}')$
10: $\quad\quad \mathbf{q}_i \leftarrow \mathbf{q}_i + \mathrm{sign}(\nabla_{\mathbf{q}_i}\mathcal{L}) \quad \forall\, i \in \{1,2,\ldots,n\}$
11: $\quad\quad \mathbf{q}_i \leftarrow \mathrm{clip}(\mathbf{q}_i,\ \mathrm{min}{=}1,\ \mathrm{max}{=}q_{\max}) \quad \forall\, i \in \{1,2,\ldots,n\}$
12: $\quad$ **end for**
13: **end function**
14: **Return** X′

Algorithm 2. Volumetric Adversarial Frequency Training (**VAFT**)

1: Train Dataset: $\mathcal{X} = \{(\mathrm{X}_i,\mathrm{Y}_i)\}_{i=1}^{N}$, $\mathrm{X}_i \in \mathbb{R}^{H\times W\times D}$, $\mathrm{Y}_i \in \{0,1\}^{\mathrm{NumClass}\times H\times W\times D}$
2: NumSamples=N, BatchSize=B, Target Model: $\mathcal{M}_\theta$, AT Robust Model: $\mathcal{M}_{\boldsymbol{\varphi}}$
3: **for** $i \leftarrow 1$ to NumEpochs **do**
4: $\quad$ **for** $j \leftarrow 1$ to $\lfloor N/B \rfloor$ **do**
5: $\quad\quad$ Sample a mini-batch $\mathcal{B} \subseteq \mathcal{X}$ of size B
6: $\quad\quad$ $\mathrm{X}' \leftarrow$ **VAFA**(X, Y) $\forall(\mathrm{X},\mathrm{Y}) \in \mathcal{B}$ ▷ Adv. Freq. Attack on clean images.
7: $\quad\quad \mathcal{L} = \mathcal{L}_{\mathrm{dice}}(\mathcal{M}_\theta(\mathrm{X}),\mathrm{Y}) + \mathcal{L}_{\mathrm{dice}}(\mathcal{M}_\theta(\mathrm{X}'),\mathrm{Y}) + \mathcal{L}_{\mathrm{fr}}(\mathcal{M}_\theta(\mathrm{X}),\mathcal{M}_\theta(\mathrm{X}'))$
8: $\quad\quad$ Backward pass and update $\mathcal{M}_\theta$
9: $\quad$ **end for**
10: **end for**
11: $\mathcal{M}_{\boldsymbol{\varphi}} \leftarrow \mathcal{M}_\theta$ ▷ AT robust model after training completion.
12: **Return** $\mathcal{M}_{\boldsymbol{\varphi}}$

extent to which DCT coefficients are perturbed. The higher the value of $q_{\max}$, the more information is lost. The drop in perception quality of the adversarial sample and the accuracy of the model are directly proportional to the value of $q_{\max}$. To increase the perceptual quality of adversarial samples, we also minimize the structural similarity loss [27] between clean and adversarial samples, denoted by $\mathcal{L}_{\mathrm{ssim}}(\mathrm{X},\mathrm{X}')$, in optimization objective. Our attack optimizes the following objective to fool a target model $\mathcal{M}_\theta$:

$$\begin{aligned} &\underset{\mathbf{q}}{\mathrm{maximize}}\ \ \mathcal{L}_{\mathrm{dice}}(\mathcal{M}_\theta(\mathrm{X}'),\mathrm{Y}) - \mathcal{L}_{\mathrm{ssim}}(\mathrm{X},\mathrm{X}') \\ &\qquad\qquad \mathrm{s.t.}\ \ \|\mathbf{q}\|_\infty \le q_{\max}, \end{aligned} \tag{2}$$

where $\mathcal{L}_{\mathrm{ssim}}(\mathrm{X},\mathrm{X}') = 1 - \frac{1}{n}\sum_{i=1}^{n}\mathrm{SSIM}(\mathbf{x}_i,\mathbf{x}_i')$ is structural similarity loss [27]. Algorithm 1 presents our volumetric adversarial frequency attack (VAFA). An overview of the attack can be found in *maximization* step of Fig. 1.

2.2 Volumetric Adversarial Frequency Training (VAFT)

The model parameters are then updated by minimizing the segmentation loss on both clean and adversarial samples (Eq. 3). Since our attack disrupts the frequency domain to find adversaries, we develop a novel frequency consistency loss (Eq. 4) to encourage frequency domain representation of the model's output (segmentation logits) for the clean sample close to the adversarial sample. Our frequency consistency loss not only boosts the robustness of the model against adversarial attacks but also improves/retains the performance of the robust model on clean images (Sect. 3). We present our volumetric adversarial frequency training (VAFT) in Algorithm 2.

$$\underset{\theta}{\text{minimize}}\ \mathcal{L}_{\text{dice}}(\mathcal{M}_\theta(\mathrm{X}), \mathrm{Y}) + \mathcal{L}_{\text{dice}}(\mathcal{M}_\theta(\mathrm{X}'), \mathrm{Y}) + \mathcal{L}_{\text{fr}}(\mathcal{M}_\theta(\mathrm{X}), \mathcal{M}_\theta(\mathrm{X}')), \quad (3)$$

$$\mathcal{L}_{\text{fr}}(\mathcal{M}_\theta(\mathrm{X}), \mathcal{M}_\theta(\mathrm{X}')) = \|\mathcal{D}(\mathcal{M}_\theta(\mathrm{X})) - \mathcal{D}(\mathcal{M}_\theta(\mathrm{X}'))\|_1, \quad (4)$$

where $\mathrm{X}' = \mathbf{VAFA}(\mathrm{X}, \mathrm{Y})$ and $\mathcal{D}(\cdot)$ is 3D DCT function. An overview of the adversarial training can be found in *minimization* step of Fig. 1.

Figure 2 presents a qualitative results of adversarial examples under different attacks on the standard UNETR model. We highlight areas by red bounding box in Fig. 2 to show the impact of each attack on the model performance, when compared with prediction on clean sample. Our attack (VAFA) achieves higher fooling rate as compared to other voxel-domain attacks, while maintaining comparable perceptual similarity.

3 Experiments and Results

Implementation Details: We demonstrate the effectiveness of our approach using two medical segmentation models: UNETR [12], UNETR++ [24] and two datasets: Synapse (18–12 split) [17], and ACDC [2]. Using pre-trained models from open-source Github repositories by the corresponding authors, we launch different adversarial attacks and conduct adversarial training with default parameters. We use the Pytorch framework and single NVIDIA A100-SXM4-40GB GPU for our experiments. For a pixel/voxel range $[0, 255]$, we create l_∞ adversarial examples under perturbation budgets of $\epsilon \in \{4, 8\}$ for voxel-domain attacks following [7] and compare it with our attack **VAFA**. Unless otherwise specified, all attacks are run for a total of 20 optimization steps. More details about the parameters of the attacks used in different experiments can be found in Appendix. We use mean Dice Similarity Score (DSC), mean 95% Hausdorff Distance (HD95). We also report perceptual similarity between clean and adversarial sample (LPIPS) [28].

Results: For each evaluation metric, we take mean across all classes (including background) and test images. In each table (where applicable), green values show DSC and HD95 on clean images. Table 1 shows comparison of voxel-domain attacks (e.g. PGD [21], FGSM [10], BIM [16], GaussianNoise(GN) [14]) with VAFA-2D (2D DCT in FPM applied on each scan independently) and VAFA on

Table 1. Voxel vs. Freq. Attacks

Attack	DSC↓	LPIPS↑
-	74.31	-
PGD	62.67	**98.94**
FGSM	62.77	98.82
BIM	62.76	98.93
GN	74.19	97.71
VAFA-2D	61.66	98.43
VAFA	**52.54**	97.84

Table 2. Impact of $q_{\max}$ on VAFA

$q_{\max}$	DSC↓	LPIPS↑
-	74.31	-
10	65.95	**99.10**
20	56.24	98.70
30	50.96	98.33
40	49.58	97.90
60	48.83	96.60
80	**48.76**	94.50

Table 3. Impact of steps on VAFA

Steps	DSC↓	LPIPS↑
-	74.31	-
10	61.33	**98.85**
20	56.24	98.70
30	54.37	98.64
40	53.31	98.59
50	52.97	98.54
60	**52.25**	98.52

Table 4. Impact of patch size on VAFA

Size	DSC↓	LPIPS↑
-	74.31	-
4	63.48	**98.90**
8	56.24	98.70
16	41.30	98.14
32	32.40	97.49
48	28.19	97.16
96	**28.08**	96.47

Table 5. Comparison of VAFA with other voxel-domain attacks (Synapse dataset).

Models →	UNETR			UNETR++		
Attacks ↓	DSC↓	HD95↑	LPIPS↑	DSC↓	HD95↑	LPIPS↑
Clean Images	74.3	14.0	-	84.7	12.7	-
PGD ($\epsilon = 4/8$)	62.7/50.8	40.4/64.5	**98.9**/95.3	77.5/67.1	48.1/78.3	95.7/85.1
FGSM ($\epsilon = 4/8$)	62.8/53.9	34.8/48.7	98.8/94.7	73.1/67.1	37.3/43.2	94.7/82.2
BIM ($\epsilon = 4/8$)	62.8/50.7	39.9/**65.8**	98.8/95.3	77.3/66.8	46.6/78.1	**95.8**/85.3
GN ($\sigma = 4/8$)	74.2/73.9	17.0/15.4	97.7/91.1	84.7/84.3	12.3/13.4	93.3/78.2
VAFA ($q_{\max} = 20/30$)	**32.2/29.8**	**57.6**/59.9	97.5/**96.9**	**45.3/39.3**	**73.9/85.2**	94.2/**94.7**

Table 6. Performance of different attacks on adversarially trained (robust) models.

	Attacks →	UNETR					UNETR++				
	Models ↓	Clean	PGD	FGSM	BIM	VAFA	Clean	PGD	FGSM	BIM	VAFA
Synapse	$\mathcal{M}^{\text{PGD}}$	73.47	65.53	65.68	65.51	42.47	75.43	67.81	67.82	67.80	38.22
	$\mathcal{M}^{\text{FGSM}}$	72.44	64.80	66.31	64.76	39.02	81.06	73.84	74.76	73.77	37.48
	$\mathcal{M}^{\text{BIM}}$	75.12	67.78	68.32	67.78	45.97	74.80	67.58	67.46	67.57	35.72
	$\mathcal{M}^{\text{GN}}$	73.17	61.40	61.77	61.29	30.00	80.05	76.23	70.96	74.51	41.44
	$\mathcal{M}^{\text{VAFA}}$	74.67	64.83	65.49	64.73	66.31	81.88	69.09	65.40	68.90	76.47
	$\mathcal{M}^{\text{VAFA-FR}}$	**75.66**	65.90	66.79	65.83	66.33	**82.65**	70.61	67.00	70.41	78.19
ACDC	$\mathcal{M}^{\text{VAFA}}$	81.95	60.77	68.16	60.75	69.76	89.00	76.28	80.41	76.56	88.45
	$\mathcal{M}^{\text{VAFA-FR}}$	**83.44**	60.63	69.33	60.61	73.05	**91.36**	85.42	87.42	83.90	91.23

UNETR model (Synapse). VAFA achieves a higher fooling rate as compared to other attacks with comparable LPIPS. We posit that VAFA-2D on volumetric MIS data is sub-optimal and it does not take into account the 3D nature of the data and model's reliance on the 3D neighborhood of a voxel to predict its class. Further details are provided in the supplementary material. We show impacts of different parameters of VAFA e.g. quantization threshold ($q_{\max}$), steps, and patch size ($h \times w \times d$) on DSC and LPIPS in Table 2, 3 and 4 respectively. DSC and LPIPS decrease when these parameters values are increased. Table 5 shows a comparison of VAFA (patch size = $32 \times 32 \times 32$) with other voxel-domain attacks on UNETR and UNETR++ models. For adversarial training experiments, we use $q_{\max} = 20$ (for Synapse), $q_{\max} = 10$ (for ACDC) and patch-size of $32 \times 32 \times 32$ (chosen after considering the trade-off between DSC and

LPIPS from Table 4) for VAFA. For voxel-domain attacks, we use $\epsilon = 4$ (for Synapse) and $\epsilon = 2$ (for ACDC) by following the work of [11,22]. Table 6 presents a comparison of the performance (DSC) of various adversarially trained models against different attacks. $\mathcal{M}^{\text{VAFA-FR}}$, $\mathcal{M}^{\text{VAFA}}$ denote our robust models which were adversarially trained with and without frequency consistency loss ($\mathcal{L}_{\text{fr}}$, Eq. 4) respectively. In contrast to other voxel-domain robust models, our approach demonstrated robustness against both voxel and frequency-based attacks.

4 Conclusion

We present a frequency-domain based adversarial attack and training for volumetric medical image segmentation. Our attack strategy is tailored to the 3D nature of medical imaging data, allowing for a higher fooling rate than voxel-based attacks while preserving comparable perceptual similarity of adversarial samples. Based upon our proposed attack, we introduce a frequency-domain adversarial training method that enhances the robustness of the volumetric segmentation model against both voxel and frequency-domain based attacks. Our training strategy is particularly important in medical image segmentation, where the accuracy and reliability of the model are crucial for clinical decision making.

References

1. Akhtar, N., Mian, A.: Threat of adversarial attacks on deep learning in computer vision: a survey. IEEE Access **6**, 14410–14430 (2018)
2. Bernard, O., et al.: Deep learning techniques for automatic MRI cardiac multi-structures segmentation and diagnosis: is the problem solved? IEEE Trans. Med. Imaging **37**(11), 2514–2525 (2018)
3. Carlini, N., Wagner, D.: Adversarial examples are not easily detected: bypassing ten detection methods. In: Proceedings of the 10th ACM Workshop on Artificial Intelligence and Security, pp. 3–14 (2017)
4. Carlini, N., Wagner, D.: Towards evaluating the robustness of neural networks. In: 2017 IEEE Symposium on Security and Privacy (SP), pp. 39–57. IEEE (2017)
5. Croce, F., Hein, M.: Reliable evaluation of adversarial robustness with an ensemble of diverse parameter-free attacks. In: International Conference on Machine Learning, pp. 2206–2216. PMLR (2020)
6. Daza, L., Pérez, J.C., Arbeláez, P.: Towards robust general medical image segmentation. In: de Bruijne, M., et al. (eds.) MICCAI 2021. LNCS, vol. 12903, pp. 3–13. Springer, Cham (2021). https://doi.org/10.1007/978-3-030-87199-4_1
7. Duan, R., Chen, Y., Niu, D., Yang, Y., Qin, A.K., He, Y.: Advdrop: adversarial attack to DNNs by dropping information. In: Proceedings of the IEEE/CVF International Conference on Computer Vision, pp. 7506–7515 (2021)
8. Finlayson, S.G., Bowers, J.D., Ito, J., Zittrain, J.L., Beam, A.L., Kohane, I.S.: Adversarial attacks on medical machine learning. Science **363**(6433), 1287–1289 (2019)
9. Gong, R., et al.: Differentiable soft quantization: bridging full-precision and low-bit neural networks. In: Proceedings of the IEEE/CVF International Conference on Computer Vision, pp. 4852–4861 (2019)

10. Goodfellow, I.J., Shlens, J., Szegedy, C.: Explaining and harnessing adversarial examples. arXiv preprint arXiv:1412.6572 (2014)
11. Guo, C., Rana, M., Cisse, M., Van Der Maaten, L.: Countering adversarial images using input transformations. arXiv preprint arXiv:1711.00117 (2017)
12. Hatamizadeh, A., et al.: UNETR: transformers for 3D medical image segmentation. In: Proceedings of the IEEE/CVF Winter Conference on Applications of Computer Vision, pp. 574–584 (2022)
13. Karimi, D., Vasylechko, S.D., Gholipour, A.: Convolution-free medical image segmentation using transformers. In: de Bruijne, M., et al. (eds.) MICCAI 2021. LNCS, vol. 12901, pp. 78–88. Springer, Cham (2021). https://doi.org/10.1007/978-3-030-87193-2_8
14. Kim, H.: Torchattacks: a pytorch repository for adversarial attacks. arXiv preprint arXiv:2010.01950 (2020)
15. Kurakin, A., Goodfellow, I., Bengio, S.: Adversarial machine learning at scale. arXiv preprint arXiv:1611.01236 (2016)
16. Kurakin, A., Goodfellow, I.J., Bengio, S.: Adversarial examples in the physical world. In: Artificial Intelligence Safety and Security, pp. 99–112. Chapman and Hall/CRC (2018)
17. Landman, B., Xu, Z., Igelsias, J., Styner, M., Langerak, T., Klein, A.: MICCAI multi-atlas labeling beyond the cranial vault-workshop and challenge. In: Proceedings of MICCAI Multi-Atlas Labeling Beyond Cranial Vault-Workshop Challenge, vol. 5, p. 12 (2015)
18. Lei, T., Wang, R., Wan, Y., Du, X., Meng, H., Nandi, A.K.: Medical image segmentation using deep learning: a survey. arXiv preprint arXiv:2009.13120 (2020)
19. Li, Y., et al.: Volumetric medical image segmentation: a 3D deep coarse-to-fine framework and its adversarial examples. In: Lu, L., Wang, X., Carneiro, G., Yang, L. (eds.) Deep Learning and Convolutional Neural Networks for Medical Imaging and Clinical Informatics. ACVPR, pp. 69–91. Springer, Cham (2019). https://doi.org/10.1007/978-3-030-13969-8_4
20. Ma, X., et al.: Understanding adversarial attacks on deep learning based medical image analysis systems. Pattern Recogn. **110**, 107332 (2021)
21. Madry, A., Makelov, A., Schmidt, L., Tsipras, D., Vladu, A.: Towards deep learning models resistant to adversarial attacks. arXiv preprint arXiv:1706.06083 (2017)
22. Prakash, A., Moran, N., Garber, S., DiLillo, A., Storer, J.: Deflecting adversarial attacks with pixel deflection. In: Proceedings of the IEEE Conference on Computer Vision and Pattern Recognition, pp. 8571–8580 (2018)
23. Ronneberger, O., Fischer, P., Brox, T.: U-Net: convolutional networks for biomedical image segmentation. In: Navab, N., Hornegger, J., Wells, W.M., Frangi, A.F. (eds.) MICCAI 2015. LNCS, vol. 9351, pp. 234–241. Springer, Cham (2015). https://doi.org/10.1007/978-3-319-24574-4_28
24. Shaker, A., Maaz, M., Rasheed, H., Khan, S., Yang, M.H., Khan, F.S.: UNETR++: delving into efficient and accurate 3D medical image segmentation. arXiv preprint arXiv:2212.04497 (2022)
25. Sudre, C.H., Li, W., Vercauteren, T., Ourselin, S., Jorge Cardoso, M.: Generalised dice overlap as a deep learning loss function for highly unbalanced segmentations. In: Cardoso, M.J., et al. (eds.) DLMIA/ML-CDS -2017. LNCS, vol. 10553, pp. 240–248. Springer, Cham (2017). https://doi.org/10.1007/978-3-319-67558-9_28
26. Szegedy, C., et al.: Intriguing properties of neural networks. arXiv preprint arXiv:1312.6199 (2013)

27. Wang, Z., Bovik, A.C., Sheikh, H.R., Simoncelli, E.P.: Image quality assessment: from error visibility to structural similarity. IEEE Trans. Image Process. **13**(4), 600–612 (2004)
28. Zhang, R., Isola, P., Efros, A.A., Shechtman, E., Wang, O.: The unreasonable effectiveness of deep features as a perceptual metric. In: CVPR (2018)

Localized Region Contrast for Enhancing Self-supervised Learning in Medical Image Segmentation

Xiangyi Yan[1(✉)], Junayed Naushad[1,2], Chenyu You[3], Hao Tang[1], Shanlin Sun[1], Kun Han[1], Haoyu Ma[1], James S. Duncan[3], and Xiaohui Xie[1]

[1] University of California, Irvine, USA
{xiangyy4,htang6,shanlins,kunh7,haoyum3,xhx}@uci.edu
[2] University of Oxford, Oxford, UK
jnaushad@yale.edu
[3] Yale University, New Haven, USA
{chenyu.you,james.duncan}@yale.edu

Abstract. Recent advancements in self-supervised learning have demonstrated that effective visual representations can be learned from unlabeled images. This has led to increased interest in applying self-supervised learning to the medical domain, where unlabeled images are abundant and labeled images are difficult to obtain. However, most self-supervised learning approaches are modeled as image level discriminative or generative proxy tasks, which may not capture the finer level representations necessary for dense prediction tasks like multi-organ segmentation. In this paper, we propose a novel contrastive learning framework that integrates Localized Region Contrast (LRC) to enhance existing self-supervised pre-training methods for medical image segmentation. Our approach involves identifying Super-pixels by Felzenszwalb's algorithm and performing local contrastive learning using a novel contrastive sampling loss. Through extensive experiments on three multi-organ segmentation datasets, we demonstrate that integrating LRC to an existing self-supervised method in a limited annotation setting significantly improves segmentation performance. Moreover, we show that LRC can also be applied to fully-supervised pre-training methods to further boost performance.

Keywords: Self-supervised Learning · Contrastive Learning · Semantic Segmentation

1 Introduction

Multi-organ segmentation is a crucial step in medical image analysis that enables physicians to perform diagnosis, prognosis, and treatment planning. However, manual segmentation of large volume computed tomography (CT) and magnetic resonance (MR) images is time-consuming and prone to high inter-rater

H. Greenspan et al. (Eds.): MICCAI 2023, LNCS 14221, pp. 468–478, 2023.
https://doi.org/10.1007/978-3-031-43895-0_44

variability [30]. In recent years, deep convolutional neural networks (CNNs) have achieved state-of-the-art performance on a wide range of segmentation tasks for natural images [12,16]. However, in the medical domain, there is often a lack of labeled examples to optimally train a deep neural network from scratch. Since unlabeled medical images are comparatively easier to obtain in larger quantities, an alternative strategy is to perform self-supervised learning and generate pre-trained models from unlabeled datasets. Self-supervised learning involves automatically generating a supervisory signal from the data itself and learning a representation by solving a pretext task.

In computer vision, current self-supervised learning methods can be broadly divided into discriminative modeling and generative modeling. In earlier times, discriminative self-supervised pretext tasks are designed as rotation prediction [15], jigsaw solving [18], and relative patch location prediction [6], etc. Recently, contrastive learning achieves great success, whose core idea is to attract different augmented views of the same image and repulse augmented views of different images. Based on this, MoCo [11] is proposed, which greatly shrink the gap between self-supervised learning and fully-supervised learning. More advanced techniques have emerged recently [3,9]. Contrastive learning frameworks have also shown promising results in the medical domain, achieving good performance with few labeled examples [1]. Generative modeling also provides a feasible way for self-supervised pre-training [21,35]. Recently, He *et al.* propose MAE [28] and yield a nontrivial and meaningful generative self-supervisory task, by masking a high proportion of the input image. Transfer learning performance in downstream tasks outperforms supervised pre-training and shows promising scaling behavior. In medical image domain, Model Genesis [36] uses a "painting" operation to generate a new image by modifying the input image. Several self-supervised learning approaches have also achieved state-of-the-art performance in the medical domain on both classification and segmentation tasks while significantly reducing annotation cost [1,2,14,20,29,31–34]

However, most self-supervised pre-training strategies are image [11] or patch [1] level, which are not capable of capturing the detailed feature representations required for accurate medical segmentation. To address this issue, in this paper, we propose a novel contrastive learning framework that integrates Localized Region Contrast (LRC) to enhance existing self-supervised pre-training methods for medical image segmentation.

Our proposed framework leverages Felzenszwalb's algorithm [8] to formulate local regions and defines a novel contrastive sampling loss to perform localized contrastive learning. Our main contributions include

- We propose a standalone localized contrastive learning module that can be integrated into most existing pre-training strategy to boost multi-organ segmentation performance by learning localized feature representations.
- We introduce a novel localized contrastive sampling loss for dense self-supervised pre-training on local regions.

- We conduct extensive experiments on three multi-organ segmentation benchmarks and demonstrate that our method consistently outperforms current supervised and unsupervised pre-training approaches.

2 Methodology

Figure 1 illustrates our complete framework, which comprises two stages: the contrastive pre-training stage and the fine-tuning stage. Although LRC can be integrated with most of the current popular pre-training strategies, for the purpose of illustration, in this section, we demonstrate how to integrate our LRC module with the classical global (image-level) contrast pre-training strategy MoCo [11], using both its original global contrast and our localized contrastive losses during the contrastive pre-training stage. During the fine-tuning stage, we simply concatenate the local and global contrast models and fine-tune the resulting model on a small target dataset. Further details about each stage are discussed in the following subsections.

2.1 Pre-training Stage

In the pre-training stage, for each batch an image $\boldsymbol{x}_q$ is randomly chosen from B images as a query sample, and the rest of the images $\boldsymbol{x}_n \in \{\boldsymbol{x}_1, \boldsymbol{x}_2, ..., \boldsymbol{x}_B\}$ are considered as negative key samples, where $n \neq q$. To formulate a positive key sample $\boldsymbol{x}_p$, elastic transforms are performed on the query sample $\boldsymbol{x}_q$.

Global Contrast. To explore global contextual information, we train a latent encoder $\mathcal{E}_g$ following the contrastive protocol in [11]. Three sets of latent embeddings $\boldsymbol{z}_q$, $\boldsymbol{z}_p$, $\boldsymbol{z}_n$ are extracted by $\mathcal{E}_g$ from $\boldsymbol{x}_q$, $\boldsymbol{x}_p$, $\boldsymbol{x}_n$ respectively. Using dot product as a measure of similarity, a form of a contrastive loss function called InfoNCE [19] is considered: $\mathcal{L}_g = -\log \frac{\exp(\boldsymbol{z}_q \cdot \boldsymbol{z}_p / \tau_g)}{\sum_{i=1}^{B} \exp(\boldsymbol{z}_q \cdot \boldsymbol{z}_i / \tau_g)}$, where τ_g is the global temperature hyper-parameter per [27]. Note that in the global contrast branch, we only pre-train $\mathcal{E}_g$.

Local Region Contrast. Unlike global contrast, positive and negative pairs for local contrast are only generated from input image $\boldsymbol{x}_q$ and its transform $\boldsymbol{x}_p$. We differentiate local regions and formulate the positive and negative pairs by using Felzenszwalb's algorithm. For an input image $\boldsymbol{x}$, Felzenszwalb's algorithm provides K local regions $\mathcal{R} = \{\boldsymbol{r}^1, \boldsymbol{r}^2, .., \boldsymbol{r}^K\}$, where $\boldsymbol{r}^k$ is the k-th local region cluster for image $\boldsymbol{x}$. We then perform elastic transform for both the query image $\boldsymbol{x}_q$ and its local regions $\mathcal{R}_q$ so that we have the augmented image $\boldsymbol{x}_p = T_e(\boldsymbol{x}_q)$ and its local regions $\mathcal{R}_p = \{\boldsymbol{r}_p^1, \boldsymbol{r}_p^2, .., \boldsymbol{r}_p^{K_p}\}$, where $\boldsymbol{r}_p^k = T_e(\boldsymbol{r}_q^k)$. Note that $K_q = K_p$ always holds since $\mathcal{R}_p$ is a one-to-one mapping from $\mathcal{R}_q$. Following the widely used U-Net [22] model design, the query image $\boldsymbol{x}_q$ and augmented image $\boldsymbol{x}_p$ are then forwarded to a randomly initialized U-Net variant, which includes a convolutional encoder $\mathcal{E}_l$ and a convolutional decoder $\mathcal{D}_l$. We get corresponding feature maps $\boldsymbol{f}_q$ and $\boldsymbol{f}_p$ with the same spatial dimensions as $\boldsymbol{x}_q$ and $\boldsymbol{x}_p$

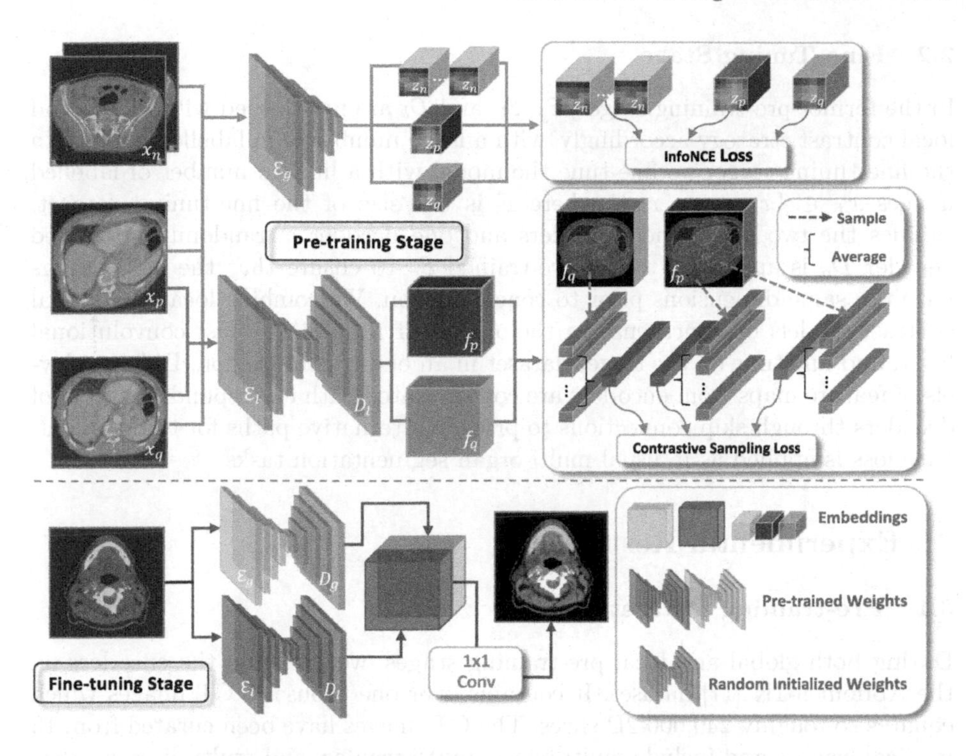

Fig. 1. Overview of our self-supervised framework with pre-training stage, consisting of global (with an example of MoCo) and our localized contrastive loss, and fine-tuning stage.

and D channels from the last convolutional layer of $\mathcal{D}_l$. Afterwards, we sample N vectors with dimension D from the local region $\boldsymbol{r}_q^k$ in $\boldsymbol{f}_q$, and formulate the sample mean $\overline{\boldsymbol{f}_q^k} = \frac{1}{N}\sum_{n=1}^{N} \boldsymbol{f}_q^{k,n}$, where $\boldsymbol{f}_q^{k,n}$ is the n-th vector sampled from feature map $\boldsymbol{f}_q$ within the k-th local region $\boldsymbol{r}_q^k$. Our sampling strategy is straightforward: we sample random points with replacement following a uniform distribution. We simply refer to this as "random sampling". Similarly, for feature map $\boldsymbol{f}_p$, its sample mean $\overline{\boldsymbol{f}_p^k}$ can be provided following the same random sampling process. Each local region pair of $\overline{\boldsymbol{f}_q^k}$ and $\overline{\boldsymbol{f}_p^k}$ is considered a positive pair. For the negative pairs, we sample both $\boldsymbol{f}_q$ and $\boldsymbol{f}_p$ from the rest of the local regions $\{\boldsymbol{r}^1, \boldsymbol{r}^2, ..., \boldsymbol{r}^{k-1}, \boldsymbol{r}^{k+1}, ..., \boldsymbol{r}^K\}$. The local contrastive loss can be defined as follows:

$$\mathcal{L}_l = -\sum_{k_1=1}^{K} \log \frac{\exp\left(\overline{\boldsymbol{f}_q^{k_1}} \cdot \overline{\boldsymbol{f}_p^{k_1}} / \tau_l\right)}{\sum_{k_2=1}^{K} \exp\left(\overline{\boldsymbol{f}_q^{k_1}} \cdot \overline{\boldsymbol{f}_q^{k_2}} / \tau_l\right) + \exp\left(\overline{\boldsymbol{f}_q^{k_1}} \cdot \overline{\boldsymbol{f}_p^{k_2}} / \tau_l\right)}$$

where τ_l is the local temperature hyper-parameter. Compared to the global contrast branch, in local contrastive learning, we pre-train both $\mathcal{E}_l$ and $\mathcal{D}_l$.

2.2 Fine-Tuning Stage

In the former pre-training stages, $\mathcal{E}_g$, $\mathcal{E}_l$, and $\mathcal{D}_l$ are pre-trained with global and local contrast strategy accordingly, with a large number of unlabelled images. In the fine-tuning stage, we fine-tune the model with a limited number of labelled images $x_f \in \{x_1, x_2, ..., x_F\}$, where F is the size of the fine-tuning dataset. Besides the two pre-trained encoders and one decoder, a randomly initialized decoder $\mathcal{D}_g$ is appended to the pre-trained $\mathcal{E}_g$ to ensure that the embeddings have the same dimensions prior to concatenation. We combine local and global contrast models by concatenating the output of $\mathcal{D}_g$ and $\mathcal{D}_l$'s last convolutional layer, and fine-tune on the target dataset in an end-to-end fashion. Different levels of feature maps from encoders are concatenated with corresponding layers of decoders through skip connections to provide alternative paths for the gradient. Dice loss is applied as in usual multi-organ segmentation tasks.

3 Experimental Results

3.1 Pre-training Dataset

During both global and local pre-training stages, we pre-train the encoders on the Abdomen-1K [17] dataset. It contains over one thousand CT images which equates to roughly 240,000 2D slices. The CT images have been curated from 12 medical centers and include multi-phase, multi-vendor, and multi-disease cases. Although segmentation masks for liver, kidney, spleen, and pancreas are provided in this dataset, we ignore these labels during pre-training since we are following the self-supervised protocol.

3.2 Fine-Tuning Datasets

During the fine-tuning stage, we perform extensive experiments on three datasets with respect to different regions of the human body.

ABD-110 is an abdomen dataset from [25] that contains 110 CT images from patients with various abdominal tumors and these CT images were taken during the treatment planning stage. We report the average DSC on 11 abdominal organs (large bowel, duodenum, spinal cord, liver, spleen, small bowel, pancreas, left kidney, right kidney, stomach and gallbladder).

Thorax-85 is a thorax dataset from [5] that contains 85 thoracic CT images. We report the average DSC on 6 thoracic organs (esophagus, trachea, spinal cord, left lung, right lung, and heart).

HaN is from [24] and contains 120 CT images covering the head and neck region. We report the average DSC on 9 organs (brainstem, mandible, optical chiasm, left optical nerve, right optical nerve, left parotid, right parotid, left submandibular gland, and right submandibular gland).

3.3 Implementation Details

All images are re-sampled to have spacing of 2.5 mm × 1.0 mm × 1.0 mm, with respect to the depth, height, and width of the 3D volume. In the pre-training stage, we apply elastic transform to formulate positive samples. In the global contrast branch, we use the SGD optimizer to pre-train a ResNet-50 [13] (for MAE [10], we use ViT-base [7].) encoder $\mathcal{E}_g$ for 200 epochs. In the local contrast branch, we use the Adam optimizer to pre-train both encoder $\mathcal{E}_l$ and decoder $\mathcal{D}_l$ for 30 epochs. The dimension of sampled vectors D is 64 since $\boldsymbol{f}_q$ and $\boldsymbol{f}_p$ have 64 channels. In the fine-tuning stage, we use the Adam optimizer to train the whole framework in an end-to-end fashion. All optimizers in both pre-training and fine-tuning stages are set to have momentum of 0.9 and weight decay of 10^{-4}.

Table 1. Comparison of our proposed pre-training strategy combining local contrast with different pre-training methods. Models are fine-tuned on three datasets where $|X_T|$ is the number of labeled CT images, and the evaluation metric is Dice score. **Bold** numbers indicate corresponding global pre-training method is enhanced by LRC.

Global Pre-training Method	ABD-110				Thorax-85				HaN			
	$\|X_T\|$=10		$\|X_T\|$=60		$\|X_T\|$=10		$\|X_T\|$=60		$\|X_T\|$=10		$\|X_T\|$=60	
w/ or wo/ LRC	w/o	w/	w/o	w/	w/o	w/	w/o	w/	w/o	w/	w/o	w/
Random init	68.8	**70.9**	76.0	**78.0**	85.9	**87.8**	89	**89.4**	50.9	**52.6**	77.8	**78.0**
Supervised Pre-training on Natural Images												
ImageNet[23]	70.9	**72.7**	77.6	**78.6**	87.2	**88.1**	89.4	**90.3**	67.5	**72.6**	77.0	**77.9**
Discriminative Self-supervised Pre-training												
Relative Loc[6]	69.1	**72.6**	76.4	**78.0**	86.1	86.0	89.4	**89.6**	55.2	**60.3**	77.9	77.9
Rotation Pred[15]	69.1	**70.2**	77.0	**78.1**	86.3	84.6	89.2	**89.7**	54.7	**59.6**	76.8	**77.3**
MoCo v1[11]	72.1	**75.3**	77.7	**79.0**	86.6	**88.6**	89.3	**90.1**	52.3	**70.6**	76.0	**77.2**
MoCo v2[3]	72.2	**75.2**	77.9	**79.6**	86.6	**87.4**	89.7	**90.0**	55.8	**68.2**	76.7	**77.5**
BYOL[9]	71.6	**74.8**	78.0	**79.0**	87.3	**88.2**	89.2	**89.5**	53.1	**61.1**	76.3	**76.6**
DenseCL[26]	71.6	**72.0**	77.2	**78.5**	84.3	**85.1**	87.5	**87.8**	59.5	**62.2**	76.7	76.7
SimSiam[4]	73.4	**76.0**	79.2	**79.5**	88.2	87.0	88.6	**89.9**	57.2	**63.2**	78.9	77.2
Generative Self-supervised Pre-training												
Models Genesis[36]	72.9	**73.2**	80.2	**80.6**	88.2	**88.4**	90.1	**91.3**	64.0	**67.2**	74.2	73.2
MAE[10]	71.5	71.2	79.5	**79.7**	86.2	**86.5**	89.3	89.0	52.8	**55.2**	77.3	**77.5**

3.4 Quantitative Results

In Table 1, we select 9 self-supervised pre-trained with 1 ImageNet supervised pre-trained networks and combine with our proposed localized region contrast (LRC). Through extensive experiments on 3 different datasets, we demonstrate LRC is capable of enhancing these pre-training algorithms in a consistent way. We use Sørensen-Dice coefficient (DSC) to measure our experimental results.

For ABD-110, LRC enhances 9 and 10 out of 10 pre-training approaches, with $|X_T| = 10$ and 60 respectively. For thorax-85, LRC enhances 7 and 9 out of 10 pre-training approaches, with $|X_T| = 10$ and 60 respectively. For HaN, LRC enhances 10 and 6 out of 10 pre-training approaches, with $|X_T| = 10$ and 60 respectively. The experiments consistently show LRC boosts the multi-organ segmentation performance of most global contrast model across all three datasets.

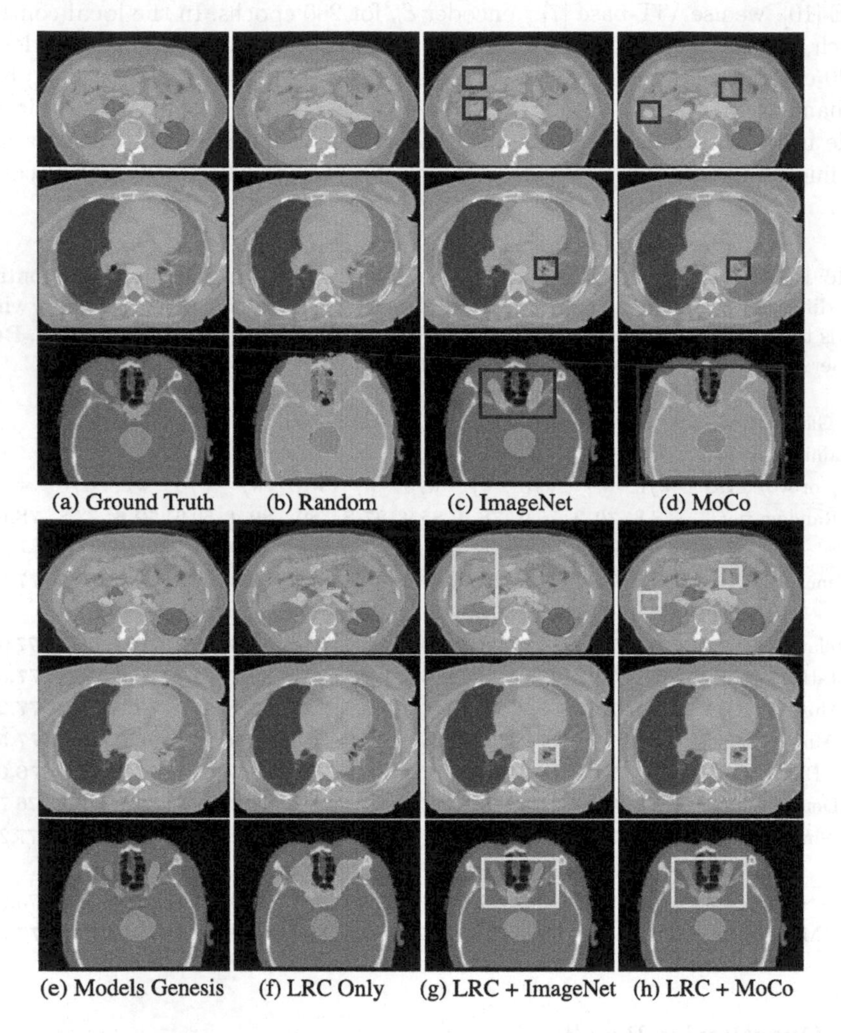

Fig. 2. Ground truth segmentation masks and predictions on a slice from each dataset. Due to limited space, we only demonstrate selected global pre-training methods. By comparing (c) with (g) and (d) with (h), our method shows significant improvement, particularly on the challenging HaN dataset.

3.5 Qualitative Results

In Fig. 2, we show segmentation results on ABD-110, Thorax-85, and HaN datasets respectively. All the results are provided by models trained with target dataset size $|X_T| = 10$. By comparing (c) with (g) and (d) with (h), our method shows significant improvement, particularly on the challenging HaN dataset. However, due to limited space, we are only capable of demonstrating selected global pre-training methods.

Table 2. Comparison of integrating LRC vs MoCo with different pre-training methods. We show LRC enhances global pre-training approaches by integrating localized features rather than simply adding additional parameters.

Global Pre-training Method	ImageNet [23]	Relative Loc [6]	Rotation Pred [15]	MAE [10]	BYOL [9]	SimSiam [4]
w/ MoCo [11]	78.4	77.1	77.3	79.5	78.2	**79.5**
w/ LRC	**78.6**	**78.0**	**78.1**	**79.7**	**79.0**	**79.5**

3.6 Visualization of Localized Regions

Figure 3 presents three pairs of localized region visualizations generated by Felzenszwalb's algorithm (with a black background) and the corresponding feature representations extracted from LRC. We use K-Means clustering to formulate these feature representations into K clusters, which are shown in purple in the figure. Our results demonstrate that LRC learns informative semantic feature representations that can be effectively clustered using a simple K-Means algorithm.

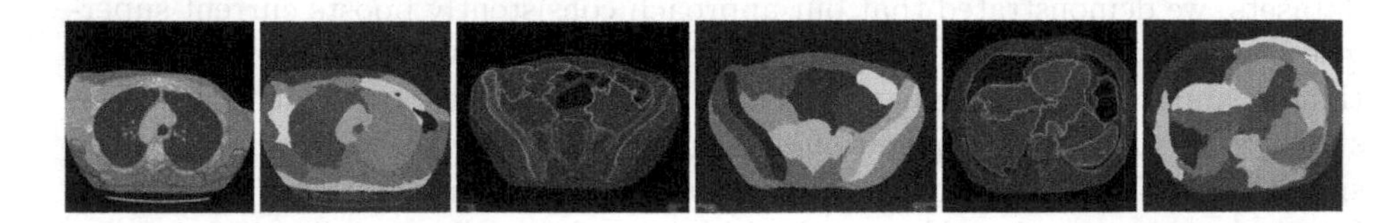

Fig. 3. Three pairs of examples comparing the local regions generated by Felzenszwalb's algorithm on the left and K-Means clustering of the embeddings from the local contrast model on the right.

3.7 Ablation Study

Effect of Additional Parameters. Additional parameters do bring performance enhancement in machine learning. However, in Table 2, we show our proposed LRC enhances the performance of general global pre-training approaches by integrating localized features rather than simply adding additional parameters. We prove this argument by adding the same amount of MoCo pre-trained network parameters, to the above global pre-trained methods. As a result, LRC outperforms MoCo under every setting. In this experiment, we use ABD-110 as fine-tuning dataset and set $|X_T| = 60$.

Number of Samples N. In Table 3, we explore the effect of different number of samples N to the contrastive sampling loss. When the sample mean $\overline{f^k}$ is only averaged from a small number of vectors, the capability of representing certain region level can be limited. In the opposite, when the number of samples N is large, the sampling bias can be high, since the number of pixels can be smaller than N. Therefore, we need a proper choice of N. With $N = 50$, our method demonstrates the best DSC score of 0.732.

Table 3. Different number of samples N largely influences the fine-tuning results. Results are provided by LRC + MoCo fine-tuned on ABD-110 with $|X_T| = 10$.

N	10	50	100
DSC	0.695	**0.732**	0.717

4 Conclusion

In this paper, we propose a contrastive learning framework, which integrates a novel localized contrastive sampling loss and enables the learning of fine-grained representations that are crucial for accurate segmentation of complex structures. Through extensive experiments on three multi-organ segmentation datasets, we demonstrated that our approach consistently boosts current supervised and unsupervised pre-training methods. LRC provides a promising direction for improving the accuracy of medical image segmentation, which is a crucial step in various clinical applications. Overall, we believe that our approach can significantly benefit the medical image analysis community and pave the way for future developments in self-supervised learning for medical applications.

References

1. Chaitanya, K., Erdil, E., Karani, N., Konukoglu, E.: Contrastive learning of global and local features for medical image segmentation with limited annotations. In: Advances in Neural Information Processing Systems, vol. 33 (2020)
2. Chaitanya, K., Erdil, E., Karani, N., Konukoglu, E.: Local contrastive loss with pseudo-label based self-training for semi-supervised medical image segmentation (2021)
3. Chen, X., Fan, H., Girshick, R., He, K.: Improved baselines with momentum contrastive learning (2020)

4. Chen, X., He, K.: Exploring simple Siamese representation learning. In: CVPR (2021)
5. Chen, X., et al.: A deep learning-based auto-segmentation system for organs-at-risk on whole-body computed tomography images for radiation therapy. Radiother. Oncol. **160**, 175–184 (2021)
6. Doersch, C., Gupta, A., Efros, A.A.: Unsupervised visual representation learning by context prediction. In: Proceedings of the IEEE International Conference on Computer Vision, pp. 1422–1430 (2015)
7. Dosovitskiy, A., et al.: An image is worth 16x16 words: transformers for image recognition at scale. In: ICLR (2021)
8. Felzenszwalb, P.F., Huttenlocher, D.P.: Efficient graph-based image segmentation. Int. J. Comput. Vision **59**(2), 167–181 (2004)
9. Grill, J.B., et al.: Bootstrap your own latent: a new approach to self-supervised learning. In: NeurIPS (2020)
10. He, K., Chen, X., Xie, S., Li, Y., Dollár, P., Girshick, R.: Masked autoencoders are scalable vision learners (2021)
11. He, K., Fan, H., Wu, Y., Xie, S., Girshick, R.: Momentum contrast for unsupervised visual representation learning. In: 2020 IEEE/CVF Conference on Computer Vision and Pattern Recognition (CVPR), pp. 9726–9735 (2020)
12. He, K., Gkioxari, G., Dollár, P., Girshick, R.: Mask R-CNN. In: 2017 IEEE International Conference on Computer Vision (ICCV), pp. 2980–2988. IEEE (2017)
13. He, K., Zhang, X., Ren, S., Sun, J.: Deep residual learning for image recognition. In: Proceedings of the IEEE Conference on Computer Vision and Pattern Recognition, pp. 770–778 (2016)
14. Hu, X., Zeng, D., Xu, X., Shi, Y.: Semi-supervised contrastive learning for label-efficient medical image segmentation (2021)
15. Komodakis, N., Gidaris, S.: Unsupervised representation learning by predicting image rotations. In: ICLR (2018)
16. Long, J., Shelhamer, E., Darrell, T.: Fully convolutional networks for semantic segmentation. In: Proceedings of the IEEE Conference on Computer Vision and Pattern Recognition, pp. 3431–3440 (2015)
17. Ma, J., et al.: Abdomenct-1k: is abdominal organ segmentation a solved problem? IEEE Trans. Pattern Anal. Mach. Intell. **44**, 6695–6714 (2021)
18. Noroozi, M., Favaro, P.: Unsupervised learning of visual representations by solving jigsaw puzzles. In: Leibe, B., Matas, J., Sebe, N., Welling, M. (eds.) ECCV 2016. LNCS, vol. 9910, pp. 69–84. Springer, Cham (2016). https://doi.org/10.1007/978-3-319-46466-4_5
19. van den Oord, A., Li, Y., Vinyals, O.: Representation learning with contrastive predictive coding (2019)
20. Ouyang, C., Biffi, C., Chen, C., Kart, T., Qiu, H., Rueckert, D.: Self-supervision with superpixels: training few-shot medical image segmentation without annotation (2020)
21. Pathak, D., Krahenbuhl, P., Donahue, J., Darrell, T., Efros, A.A.: Context encoders: feature learning by inpainting. In: Proceedings of the IEEE Conference on Computer Vision and Pattern Recognition, pp. 2536–2544 (2016)
22. Ronneberger, O., Fischer, P., Brox, T.: U-Net: convolutional networks for biomedical image segmentation. In: Navab, N., Hornegger, J., Wells, W.M., Frangi, A.F. (eds.) MICCAI 2015. LNCS, vol. 9351, pp. 234–241. Springer, Cham (2015). https://doi.org/10.1007/978-3-319-24574-4_28
23. Russakovsky, O., et al.: ImageNet large scale visual recognition challenge. Int. J. Comput. Vision **115**(3), 211–252 (2015)

24. Tang, H., et al.: Clinically applicable deep learning framework for organs at risk delineation in CT images. Nat. Mach. Intell. **1**, 1–12 (2019)
25. Tang, H., Liu, X., Sun, S., Yan, X., Xie, X.: Recurrent mask refinement for few-shot medical image segmentation. In: Proceedings of the IEEE/CVF International Conference on Computer Vision (ICCV), pp. 3918–3928 (2021)
26. Wang, X., Zhang, R., Shen, C., Kong, T., Li, L.: Dense contrastive learning for self-supervised visual pre-training. In: CVPR (2021)
27. Wu, Z., Xiong, Y., Stella, X.Y., Lin, D.: Unsupervised feature learning via non-parametric instance discrimination. In: Proceedings of the IEEE Conference on Computer Vision and Pattern Recognition (2018)
28. Xiao, T., Singh, M., Mintun, E., Darrell, T., Dollár, P., Girshick, R.: Early convolutions help transformers see better (2021)
29. Yan, X., et al.: Representation recovering for self-supervised pre-training on medical images. In: WACV, pp. 2685–2695 (2023)
30. Yan, X., Tang, H., Sun, S., Ma, H., Kong, D., Xie, X.: AFTer-UNet: axial fusion transformer U-Net for medical image segmentation (2021)
31. You, C., et al.: Rethinking semi-supervised medical image segmentation: a variance-reduction perspective. arXiv preprint: arXiv:2302.01735 (2023)
32. You, C., Dai, W., Min, Y., Staib, L., Duncan, J.S.: Bootstrapping semi-supervised medical image segmentation with anatomical-aware contrastive distillation. In: Frangi, A., de Bruijne, M., Wassermann, D., Navab, N. (eds.) IPMI 2023. Lecture Notes in Computer Science, vol. 13939. Springer, Cham (2023). https://doi.org/10.1007/978-3-031-34048-2_49
33. You, C., Zhao, R., Staib, L.H., Duncan, J.S.: Momentum contrastive voxel-wise representation learning for semi-supervised volumetric medical image segmentation. In: Wang, L., Dou, Q., Fletcher, P.T., Speidel, S., Li, S. (eds.) MICCAI 2022. Lecture Notes in Computer Science, vol. 13434. Springer, Cham (2022). https://doi.org/10.1007/978-3-031-16440-8_61
34. You, C., Zhou, Y., Zhao, R., Staib, L., Duncan, J.S.: SimCVD: simple contrastive voxel-wise representation distillation for semi-supervised medical image segmentation. IEEE Transa. Med. Imaging **41**, 2228–2237 (2022)
35. Zhang, R., Isola, P., Efros, A.A.: Colorful image colorization. In: Leibe, B., Matas, J., Sebe, N., Welling, M. (eds.) ECCV 2016. LNCS, vol. 9907, pp. 649–666. Springer, Cham (2016). https://doi.org/10.1007/978-3-319-46487-9_40
36. Zhou, Z., Sodha, V., Pang, J., Gotway, M.B., Liang, J.: Models genesis (2020). https://doi.org/10.1016/j.media.2020.101840

A Spatial-Temporal Deformable Attention Based Framework for Breast Lesion Detection in Videos

Chao Qin[1(✉)], Jiale Cao[2], Huazhu Fu[3], Rao Muhammad Anwer[1], and Fahad Shahbaz Khan[1,4]

[1] Mohamed bin Zayed University of Artificial Intelligence, Abu Dhabi, United Arab Emirates
chao.qin@mbzuai.ac.ae
[2] Tianjin University, Tianjin, China
[3] Institute of High Performance Computing, Agency for Science, Technology and Research, Singapore, Singapore
[4] Linköping University, Linköping, Sweden

Abstract. Detecting breast lesion in videos is crucial for computer-aided diagnosis. Existing video-based breast lesion detection approaches typically perform temporal feature aggregation of deep backbone features based on the self-attention operation. We argue that such a strategy struggles to effectively perform deep feature aggregation and ignores the useful local information. To tackle these issues, we propose a spatial-temporal deformable attention based framework, named STNet. Our STNet introduces a spatial-temporal deformable attention module to perform local spatial-temporal feature fusion. The spatial-temporal deformable attention module enables deep feature aggregation in each stage of both encoder and decoder. To further accelerate the detection speed, we introduce an encoder feature shuffle strategy for multi-frame prediction during inference. In our encoder feature shuffle strategy, we share the backbone and encoder features, and shuffle encoder features for decoder to generate the predictions of multiple frames. The experiments on the public breast lesion ultrasound video dataset show that our STNet obtains a state-of-the-art detection performance, while operating twice as fast inference speed. The code and model are available at https://github.com/AlfredQin/STNet.

Keywords: Breast lesion detection · Ultrasound videos · Spatial-temporal deformable attention · Multi-frame prediction

1 Introduction

Ultrasound imaging is a very effective technique for breast lesion diagnosis, which has high sensitivity. Automatically detecting breast lesions is a challenging problem with a potential to aid in improving the efficiency of radiologists in ultrasound-based breast cancer diagnosis [18,21]. Some of the challenges associated with automatic breast lesion detection include blurry boundaries and changeable sizes of breast lesions.

H. Greenspan et al. (Eds.): MICCAI 2023, LNCS 14221, pp. 479–488, 2023.
https://doi.org/10.1007/978-3-031-43895-0_45

Most existing breast lesion detection methods can be categorized into image-based [10,11,16,17,19] and video-based [1,9] breast lesion detection approaches. Image-based breast lesion detection approaches perform detection in each frame independently. Compared to image-based breast lesion detection approaches, methods based on videos are capable of utilizing temporal information for improved detection performance. For instance, Chen *et al.* [1] exploited temporal coherence for semi-supervised video-based breast lesion detection. Recently, Lin *et al.* [9] proposed a feature aggregation network, termed as CVA-Net, that executes intra-video and inter-video fusions at both video and clip levels based on attention blocks. Although the recent CVA-Net aggregates clip and video level features, we distinguish two key issues that hamper its performance. First, the self-attention based cross-frame feature fusion is a global-level operation and it operates once before the encoder-decoder, thereby ignoring the useful local information and in turn missing an effective deep feature fusion. Second, CVA-Net only performs one-frame prediction based on multiple frame inputs, which is very time-consuming.

To address the aforementioned issues, we propose a spatial-temporal deformable attention based network, named STNet, for detecting the breast lesions in ultrasound videos. Within our STNet, we introduce a spatial-temporal deformable attention module to fuse multi-scale spatial-temporal information among different frames, and further integrate it into each layer of the encoder and decoder. In this way, different from the recent CVA-Net, our proposed STNet performs both deep and local feature fusion. In addition, we introduce multi-frame prediction with encoder feature shuffle operation that shares the backbone and encoder features, and only perform multi-frame prediction in the decoder. This enables us to significantly accelerate the detection speed of the proposed approach. We conduct extensive experiments on a public breast lesion ultrasound video dataset, named BLUVD-186 [9]. The experimental results validate the efficacy of our proposed STNet that has a superior detection performance. For example, our proposed STNet achieves a mAP of 40.0% with an absolute gain of 3.9% in terms of detection accuracy, while operating at two times faster, compared to the recent CVA-Net [9].

2 Method

Here, we describe our proposed spatial-temporal deformable attention based framework, named STNet, for detecting breast lesions in the ultrasound videos. Figure 1(a) presents the overall architecture of our proposed STNet, which is built on the end-to-end detector deformable DETR [22]. Within our STNet, we introduce spatial-temporal deformable attention into the encoder and the decoder. As in CVA-Net [9], we take six frames I_{k-1}, I_k, I_{k+1}, I_{r1}, I_{r2}, I_{r3} from one ultrasound video as inputs, where there are three neighboring frames I_{k-1}, I_k, I_{k+1} and three randomly-selected frames I_{r1}, I_{r2}, I_{r3}. Given these input frames, we use the backbone, such as ResNet-50 [6], to extract deep multi-scale features F_{k-1}, F_k, F_{k+1}, F_{r1}, F_{r2}, F_{r3}. Afterwards, we introduce a

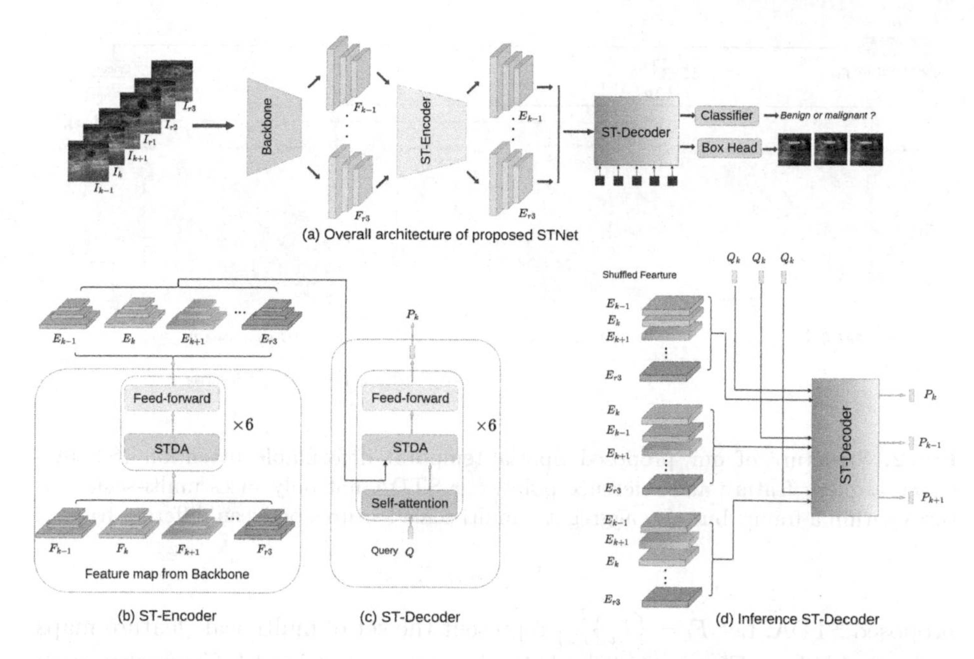

Fig. 1. **(a)** Overall architecture of the proposed STNet. The proposed STNet takes six frames as inputs and extracts multi-scale features of each frame. Afterwards, the proposed STNet utilizes a spatial-temporal deformable attention (STDA) based encoder **(b)** and decoder **(c)** for spatial-temporal multi-scale information fusion. Finally, the proposed STNet performs classification and regression. **(d)** During inference, we introduce a encoder feature shuffle strategy for multi-frame prediction.

spatial-temporal deformable attention based encoder (ST-Encoder) to perform intra-frame and inter-frame multi-scale feature fusion. Then, we introduce a spatial-temporal deformable attention based decoder (ST-Decoder) to generate output feature embeddings P_k, which are fed to a classifier and a box predictor for classification and bounding-box regression. During inference, we take three neighboring frames and three randomly-selected frames as the inputs, and simultaneously predict the results of three neighboring frames using our encoder feature shuffle strategy. As a result, our approach operates at a faster inference speed.

2.1 Spatial-Temporal Deformable Attention

Given a reference point, deformable attention [22] aggregates the features of a group of key sampling points near it. Compared to original transformer self-attention [13], deformable attention has low-complexity along with a faster convergence speed. Motivated by this, we adopt deformable attention for breast lesion detection and extend it to spatial-temporal deformable attention (STDA). Our STDA not only aggregates the features of current frame, but also aggregates the features of the rest of the frames. Figure 2 presents the structure of our

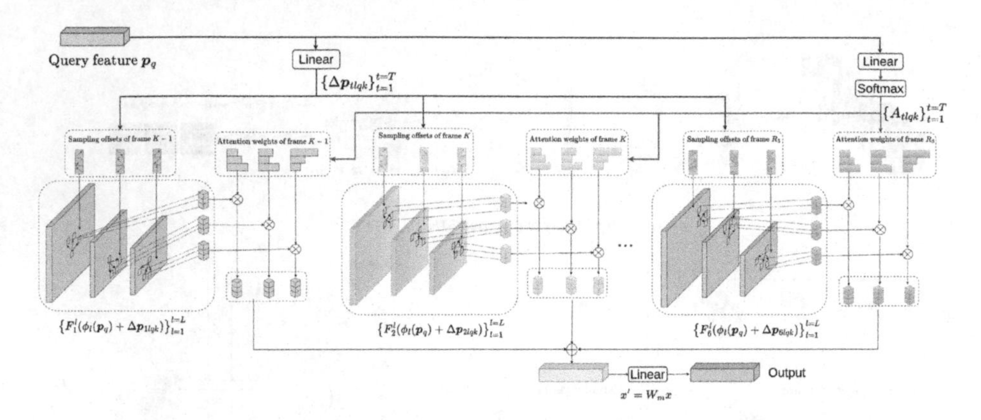

Fig. 2. Structure of our proposed Spatial-temporal deformable attention (STDA). Given a query feature and reference point, our STDA not only fuses multi-scale features within a frame, but also aggregates multi-scale features between different frames.

proposed STDA. Let $F_t = \left\{F_t^l\right\}_{l=1}^{L}$ represent the set of multi-scale feature maps at frame t, where $F_t^l \in \mathbb{R}^{C \times H_l \times W_l}$ is the feature map at level l. Given the query features $\boldsymbol{p}_q$ and corresponding reference points $\boldsymbol{z}_q$, the spatio-temporal multi-scale attention is given as:

$$\mathrm{STDA}\left(\boldsymbol{z}_q, \boldsymbol{p}_q, \{F_t\}_{t=0}^{T}\right) = \sum_{m=1}^{M} W_m \sum_{t=1}^{T} \sum_{l=1}^{L} \sum_{k=1}^{K} A_{tlqk} F_t^l(\phi_l(\boldsymbol{p}_q) + \Delta \boldsymbol{p}_{tlqk}), \quad (1)$$

where m represents multi-head index and k is sampling point index. W_m is a linear layer, A_{tlqk} indicates attention weight of sampling point, and $\Delta \boldsymbol{p}_{tlqk}$ indicates sample offset of sampling point. ϕ_l normalizes the coordinates $\boldsymbol{p}_q$ by the scale of feature map F_t^l. The sampling offset $\Delta \boldsymbol{p}_{tlqk}$ is predicted by the query feature $\boldsymbol{z}_q$ with a linear layer. The attention weight A_{tlqk} is predicted by feeding query feature $\boldsymbol{z}_q$ to a linear layer and a softmax layer. As a result, the sum of attention weights is equal to one as

$$\sum_{t=1}^{T} \sum_{l=1}^{L} \sum_{k=1}^{K} A_{tlqk} = 1. \quad (2)$$

Compared to the standard deformable attention, the proposed spatial-temporal deformable attention fully exploits spatial information within frame and temporal information across frames.

2.2 Spatial-Temporal Deformable Attention Based Encoder and Decoder

Here, we integrate the proposed spatial-temporal deformable attention (STDA) into encoder and decoder (called ST-Encoder and ST-Decoder). As shown in

Fig. 1(b), ST-Encoder takes deep multi-scale feature maps F_{k-1}, F_k, F_{k+1}, F_{r1}, F_{r2}, F_{r3} as inputs. Afterwards, we employ STDA to perform spatial and temporal fusion and generate the fused multi-scale feature maps F'_{k-1}, F'_k, F'_{k+1}, F'_{r1}, F'_{r2}, F'_{r3}, where the query corresponds to each pixel in multi-scale feature maps. Then, the fused feature map goes through a feed-forward network (FFN) to generate the output feature maps E_{k-1}, E_k, E_{k+1}, E_{r1}, E_{r2}, E_{r3}. Similar to the original deformable DETR, we adopt cascade structure to stack six STDA and FFN layers in ST-Encoder.

The ST-Decoder takes the output feature maps E_{k-1}, E_k, E_{k+1}, E_{r1}, E_{r2}, E_{r3} and a set of learnable queries $Q \in \mathbb{R}^{N \times C}$ as inputs. The learnable queries first go through a self-attention layer. Afterwards, STDA performs cross-attention operation between these feature maps and the queries, where the key elements are these output feature maps of ST-Encoder. Then, we employ a FFN layer to generate the prediction features $P_k \in \mathbb{R}^{N \times C}$. We also stack six self-attention, STDA, and FFN layers in ST-Decoder for deep feature extraction.

2.3 Multi-frame Prediction with Encoder Feature Shuffle

As discussed above, the proposed STNet adopts six frames to predict the results of one frame. Although STNet fully exploits temporal information for improved breast lesion detection, it becomes time-consuming for multi-frame prediction. To accelerate the detection speed, we introduce multi-frame prediction with encoder feature shuffle during inference. Instead of going through the entire network several times, we first share deep multi-scale feature maps before encoder and second perform the decoder several times for multi-frame prediction. To perform multi-frame prediction only in the decoder, we propose the encoder feature shuffle operation shown in Fig. 1(d). By exchanging the order of neighboring frame I_{k-1}, I_k, I_{k+1}, the decoder can predict the results of three neighboring frames, respectively. Compared to the original STNet, the proposed encoder feature shuffle strategy only employs decoder forward three frames and accelerates the inference speed.

3 Experiments

3.1 Dataset and Implementation Details

Dataset. We conduct the experiments on the public BLUVD-186 dataset [9], comprising 186 videos including 112 malignant and 74 benign cases. The dataset has totally 25,458 ultrasound frames, where the number of frames in a video ranges from 28 to 413. The videos encompass a comprehensive tumor scan, from its initial appearance to its largest section and eventual disappearance. All videos were captured using PHILIPS TIS L9-3 and LOGIQ-E9. The grounding-truths in a frame, including breast lesion bounding-boxes and corresponding categories, are labeled by two pathologists, which have eight years of professional background in the field of breast pathology. We adopt the same dataset splits as in

Table 1. State-of-the-art quantitative comparison of our approach with existing methods in literature on the BLUVD-186 dataset. Our approach achieves a superior performance on three different metrics. Compared to the recent CVA-Net [9], our approach obtains a gain of 3.9% in terms of overall AP. We show the best results in bold.

Method	Type	Backbone	AP	AP_{50}	AP_{75}
GFL [7]	image	ResNet-50	23.4	46.3	22.2
Cascade RPN [14]	image	ResNet-50	24.8	42.4	27.3
Faster R-CNN [12]	image	ResNet-50	25.2	49.2	22.3
VFNet [20]	image	ResNet-50	28.0	47.1	31.0
RetinaNet [8]	image	ResNet-50	29.5	50.4	32.4
DFF [24]	video	ResNet-50	25.8	48.5	25.1
FGFA [23]	video	ResNet-50	26.1	49.7	27.0
SELSA [15]	video	ResNet-50	26.4	45.6	29.6
Temporal ROI Align [5]	video	ResNet-50	29.0	49.9	33.1
MEGA [2]	video	ResNet-50	32.3	57.2	35.7
CVA-Net [9]	video	ResNet-50	36.1	65.1	38.5
STNet (Ours)	video	ResNet-50	**40.0**	**70.3**	**43.3**

the previous work CVA-Net [9], to guarantee a fair comparison. Specifically, the testing set comprises 38 videos randomly selected from all 186 videos, while the rest of the videos are used as the training set.

Evaluation Metrics. Three commonly-used metrics are employed for performance evaluation of breast lesion detection methods on the ultrasound videos, namely average precision (AP), AP_{50}, and AP_{75}.

Implementation Details. We employ the ResNet-50 [6] pre-trained on ImageNet [3], and use Xavier [4] to initialize the remaining network parameters. To enhance the diversity of training data, all videos are randomly subjected to horizontal flipping, cropping, and resizing. Similar to that of CVA-Net, we employ a two-phase training strategy to achieve better convergence. In the first phase, we employ Adam optimizer to train the model for 8 epochs. We then fine-tune the model for another 20 epochs with the SGD optimizer. Throughout both phases of training, we adopt the consistent hyper-parameters, where the learning rate is 5×10^{-5} and the weight decay is 1×10^{-4}. We train the model on a single NVIDIA A100 GPU and set the batch size as 1.

3.2 State-of-the-Art Comparison

Our proposed approach is compared with eleven state-of-the-art methods, comprising image-based and video-based methods. We report the detection performance of these state-of-the-art methods generated by CVA-Net [9]. Specifically, CVA-Net acquires the detection performance of these methods by utilizing their publicly available codes or re-implementing them if no publicly available codes.

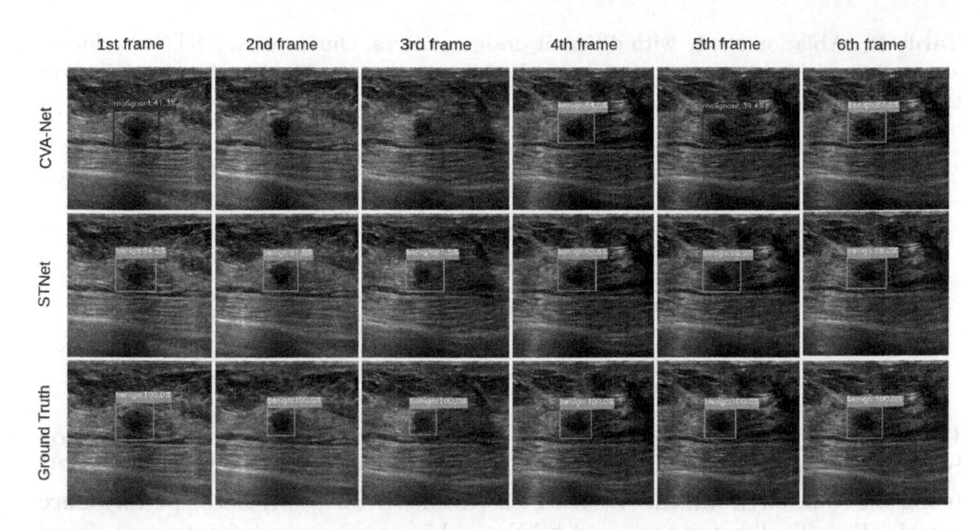

Fig. 3. Qualitative breast lesion detection comparison on example ultrasound video frames between the recent CVA-Net [9] and our proposed STNet. We also show the ground truth as reference. Our STNet achieves improved detection performance, compared to CVA-Net. Best viewed zoomed in. (Color figure online)

Quantitative Comparisons. Table 1 presents the state-of-the-art quantitative comparison of our approach with the eleven existing breast lesion video detection methods in literature. As a general trend, video-based methods tend to yield higher average precision (AP), AP50, and AP75 scores compared to image-based breast lesion detection methods. Among the eleven existing methods, the recent CVA-Net [9] achieves the best overall AP score of 36.1, AP50 score of 65.1, and AP75 score of 38.5. Our proposed STNet method consistently outperforms CVA-Net [9] on all three metrics (AP, AP50, and AP75). Specifically, our STNet achieves a significant improvement in the overall AP score from 36.1 to 40.0, the AP50 score from 65.1 to 70.3, and the AP75 score from 38.5 to 43.3. The significant improvement demonstrates the efficacy of our approach for detecting breast lesions in ultrasound videos.

Qualitative Comparisons. Figure 3 presents the qualitative breast lesion detection comparison between CVA-Net and our proposed approach on an ultrasound video containing the benign breast lesions. Moreover, we show the ground truth of each frame on the third row for reference. The first row of the figure shows that CVA-Net struggles to identify the breast lesions in the second and third frames. Further, although CVA-Net manages to identify the breast lesions in the first and fifth frames, the classification results are inaccurate (as highlighted by the blue rectangle in Fig. 3). In contrast, our STNet method in the second row of Fig. 3 accurately detects the breast lesions in all video frames and achieves accurate classification performance for each frame.

Table 2. Ablation study with different design choices. Our proposed STNet achieves a superior performance compared to the baseline and some different designs. We show the est results in bold.

	AP	AP_{50}	AP_{75}
Baseline + Single-frame	30.2	55.0	31.7
Baseline + Multi-frame	35.1	61.6	37.4
ST-Encoder + DA-Decoder	34.9	59.8	37.7
DA-Encoder + ST-Decoder	35.8	60.4	38.0
STNet (Ours)	**40.0**	**70.3**	**43.3**

Inference Speed Comparison. We present the inference speed comparison between our proposed STNet and CVA-Net on an NVIDIA RTX 3090 GPU using the same environment. We use FPS (frames per second) as the performance metric. Specifically, our proposed STNet achieves an averaged inference speed of 21.84 FPS, while CVA-Net achieves an averaged speed of 12.17 FPS. Our model operates around two times faster than CVA-Net, which we attribute to the ability of our model to predict three frames simultaneously.

3.3 Ablation Study

Effectiveness of STDA: To show the efficacy of our proposed STDA, we perform different ablation studies. The first baseline network, referred as "Baseline + Single-frame", uses the original deformable DETR and takes a single frame as input. The second baseline network, referred as "Baseline + Multi-frame", uses modified deformable DETR with multi-head attention module to fuse six input frames. For the third study, labeled "ST-Encoder + DA-Decoder", we retain the encoder with STDA in our model but replace the STDA in the decoder with the conventional deformable attention. Similarly, in the fourth study, labeled "DA-Encoder + ST-Decoder", we retain the decoder with STDA in our model but replace the STDA in the encoder with the conventional deformable attention. As shown in Table 2, the results show that "ST-Encoder + DA-Decoder" and "DA-Encoder + ST-Decoder" improve the AP by 4.7 and 5.6, respectively, compared to "Baseline + Single-frame". This demonstrates that STDA can effectively perform intra-frame and inter-frame multi-scale feature fusion, even when only partially adopted in the encoder or decoder. Furthermore, our proposed STNet improves the AP by 5.1 and 4.2 compared to "ST-Encoder + DA-Decoder" and "DA-Encoder + ST-Decoder", respectively, indicating that the integration of STDA in both the encoder and decoder is crucial for achieving superior detection performance.

4 Conclusion

We propose a novel breast lesion detection approach for ultrasound videos, termed as STNet, which performs local spatial-temporal feature fusion and deep feature aggregation in each stage of both encoder and decoder using our spatial-temporal deformable attention module. Additionally, we introduce the encoder feature shuffle strategy that enables multi-frame prediction during inference, thereby enabling us to accelerate the inference speed while maintaining better detection performance. The experiments conducted on a public breast lesion ultrasound video dataset show the efficacy of our STNet, resulting in a superior detection performance while operating at a fast inference speed. We believe STNet presents a promising solution and will help further promote future research in the direction of efficient and accurate breast lesion detection in videos.

Acknowledgment. This research is supported by the National Research Foundation, Singapore under its AI Singapore Programme (AISG Award No: AISG2-TC-2021-003) and Agency for Science, Technology and Research (A*STAR) Central Research Fund (CRF).

References

1. Chen, S., et al.: Semi-supervised breast lesion detection in ultrasound video based on temporal coherence. arXiv:1907.06941 (2019)
2. Chen, Y., Cao, Y., Hu, H., Wang, L.: Memory enhanced global-local aggregation for video object detection. In: Proceedings of the IEEE/CVF Conference on Computer Vision and Pattern Recognition, pp. 10337–10346 (2020)
3. Deng, J., Dong, W., Socher, R., Li, L.J., Li, K., Fei-Fei, L.: ImageNet: a large-scale hierarchical image database. In: 2009 IEEE Conference on Computer Vision and Pattern Recognition, pp. 248–255. IEEE (2009)
4. Glorot, X., Bengio, Y.: Understanding the difficulty of training deep feedforward neural networks. In: Proceedings of the Thirteenth International Conference on Artificial Intelligence and Statistics, pp. 249–256. JMLR Workshop and Conference Proceedings (2010)
5. Gong, T., et al.: Temporal RoI align for video object recognition. In: Proceedings of the AAAI Conference on Artificial Intelligence, vol. 35, pp. 1442–1450 (2021)
6. He, K., Zhang, X., Ren, S., Sun, J.: Deep residual learning for image recognition. In: IEEE Conference on Computer Vision and Pattern Recognition (2016)
7. Li, X., et al.: Generalized focal loss: learning qualified and distributed bounding boxes for dense object detection. Adv. Neural. Inf. Process. Syst. **33**, 21002–21012 (2020)
8. Lin, T.Y., Goyal, P., Girshick, R., He, K., Dollár, P.: Focal loss for dense object detection. In: Proceedings of the IEEE International Conference on Computer Vision, pp. 2980–2988 (2017)
9. Lin, Z., Lin, J., Zhu, L., Fu, H., Qin, J., Wang, L.: A new dataset and a baseline model for breast lesion detection in ultrasound videos. In: Wang, L., Dou, Q., Fletcher, P.T., Speidel, S., Li, S. (eds.) Medical Image Computing and Computer Assisted Intervention – MICCAI 2022. MICCAI 2022. LNCS, vol. 13433. Springer, Cham (2022). https://doi.org/10.1007/978-3-031-16437-8_59
10. Movahedi, M.M., Zamani, A., Parsaei, H., Tavakoli Golpaygani, A., Haghighi Poya, M.R.: Automated analysis of ultrasound videos for detection of breast lesions. Middle East J. Cancer **11**(1), 80–90 (2020)
11. Qi, X., et al.: Automated diagnosis of breast ultrasonography images using deep neural networks. Med. Image Anal. **52**, 185–198 (2019)

12. Ren, S., He, K., Girshick, R., Sun, J.: Faster R-CNN: towards real-time object detection with region proposal networks. arXiv preprint arXiv:1506.01497 (2015)
13. Vaswani, A., et al.: Attention is all you need. In: Advances in Neural Information Processing Systems (2017)
14. Vu, T., Jang, H., Pham, T.X., Yoo, C.D.: Cascade RPN: delving into high-quality region proposal network with adaptive convolution. arXiv preprint arXiv:1909.06720 (2019)
15. Wu, H., Chen, Y., Wang, N., Zhang, Z.: Sequence level semantics aggregation for video object detection. In: Proceedings of the IEEE/CVF International Conference on Computer Vision, pp. 9217–9225 (2019)
16. Xue, C., et al.: Global guidance network for breast lesion segmentation in ultrasound images. Med. Image Anal. **70**, 101989 (2021)
17. Yang, Z., Gong, X., Guo, Y., Liu, W.: A temporal sequence dual-branch network for classifying hybrid ultrasound data of breast cancer. IEEE Access **8**, 82688–82699 (2020)
18. Yap, M.H., et al.: Automated breast ultrasound lesions detection using convolutional neural networks. IEEE J. Biomed. Health Inform. **22**(4), 1218–1226 (2017)
19. Zhang, E., Seiler, S., Chen, M., Lu, W., Gu, X.: BIRADS features-oriented semi-supervised deep learning for breast ultrasound computer-aided diagnosis. Phys. Med. Biol. **65**(12), 125005 (2020)
20. Zhang, H., Wang, Y., Dayoub, F., Sunderhauf, N.: VarifocalNet: an IoU-aware dense object detector. In: IEEE/CVF Conference on Computer Vision and Pattern Recognition (CVPR), pp. 8510–8519 (2021)
21. Zhu, L., et al.: A second-order subregion pooling network for breast lesion segmentation in ultrasound. In: Martel, A.L., et al. (eds.) MICCAI 2020. LNCS, vol. 12266, pp. 160–170. Springer, Cham (2020). https://doi.org/10.1007/978-3-030-59725-2_16
22. Zhu, X., Su, W., Lu, L., Li, B., Wang, X., Dai, J.: Deformable DETR: deformable transformers for end-to-end object detection. In: International Conference on Learning Representations (2021)
23. Zhu, X., Wang, Y., Dai, J., Yuan, L., Wei, Y.: Flow-guided feature aggregation for video object detection. In: Proceedings of the IEEE International Conference on Computer Vision, pp. 408–417 (2017)
24. Zhu, X., Xiong, Y., Dai, J., Yuan, L., Wei, Y.: Deep feature flow for video recognition. In: Proceedings of the IEEE Conference on Computer Vision and Pattern Recognition, pp. 2349–2358 (2017)

A Flexible Framework for Simulating and Evaluating Biases in Deep Learning-Based Medical Image Analysis

Emma A. M. Stanley[1,2,3,4](✉), Matthias Wilms[3,4,5,6], and Nils D. Forkert[1,2,3,4,7]

[1] Department of Biomedical Engineering, University of Calgary, Calgary, Canada
emma.stanley@ucalgary.ca
[2] Department of Radiology, University of Calgary, Calgary, Canada
[3] Hotchkiss Brain Institute, University of Calgary, Calgary, Canada
[4] Alberta Children's Hospital Research Institute, University of Calgary, Calgary, Canada
[5] Department of Pediatrics, University of Calgary, Calgary, Canada
[6] Department of Community Health Sciences, University of Calgary, Calgary, Canada
[7] Department of Clinical Neurosciences, University of Calgary, Calgary, Canada

Abstract. Despite the remarkable advances in deep learning for medical image analysis, it has become evident that biases in datasets used for training such models pose considerable challenges for a clinical deployment, including fairness and domain generalization issues. Although the development of bias mitigation techniques has become ubiquitous, the nature of inherent and unknown biases in real-world medical image data prevents a comprehensive understanding of algorithmic bias when developing deep learning models and bias mitigation methods. To address this challenge, we propose a modular and customizable framework for bias simulation in synthetic but realistic medical imaging data. Our framework provides complete control and flexibility for simulating a range of bias scenarios that can lead to undesired model performance and shortcut learning. In this work, we demonstrate how this framework can be used to simulate morphological biases in neuroimaging data for disease classification with a convolutional neural network as a first feasibility analysis. Using this case example, we show how the proportion of bias in the disease class and proximity between disease and bias regions can affect model performance and explainability results. The proposed framework provides the opportunity to objectively and comprehensively study how biases in medical image data affect deep learning pipelines, which will facilitate a better understanding of how to responsibly develop models and bias mitigation methods for clinical use. Code is available at github.com/estanley16/SimBA.

M. Wilms and N.D. Forkert—Shared last authorship.

Supplementary Information The online version contains supplementary material available at https://doi.org/10.1007/978-3-031-43895-0_46.

H. Greenspan et al. (Eds.): MICCAI 2023, LNCS 14221, pp. 489–499, 2023.
https://doi.org/10.1007/978-3-031-43895-0_46

1 Introduction

Deep learning for medical image analysis is a key tool to facilitate precision medicine and support clinical decision making. However, it has become increasingly evident that biases in the training data can lead to obstacles for clinical implementation. In this work, we define bias as a property of the data (*e.g.*, class/attribute imbalance, spurious correlations) used for training a model that can lead to shortcut learning and/or failure to adequately represent data subgroups, which may lead to reduced generalizability and/or fairness when applied in real-world scenarios. For instance, such biases have been shown to lead to poor generalization capabilities of models evaluated on cohorts with sociodemographic population statistics different to those that it was trained on [9], which can lead to systematic misdiagnosis of subpopulations [16,18]. Moreover, image acquisition biases can act as spurious correlations to the target (shortcut learning) [8,23].

Due to these problems, a plethora of research has recently gone towards bias mitigation [13,14,25,26] and data harmonization [2,5,23]. However, the utility of real-world medical images to assess and address bias-related challenges is often limited and may not be a comprehensive or sustainable solution. This is because all real-world datasets inherently suffer from biases that can be related to cohort selection, varying scanners and protocols, biases in "ground truth" labels, or any other (un-)known confounding factors associated with the data or the labels. Additionally, many medical imaging datasets do not contain suitable sociodemographic information or representation to adequately investigate the full range of bias scenarios that could be encountered in practice, especially when considering intersectional analyses [22]. Moreover, limitations in the seminal work on underdiagnosis disparities in deep learning models for chest X-ray analysis [18] have been identified since the various sources of bias present in the dataset could not be effectively distinguished from algorithmic bias [3] and known confounding factors were not rigorously accounted for [15]. However, even when known confounders such as disease prevalence between groups are considered, it may not be possible to adequately correct for them and unknown confounders and associated spurious correlations may still exist that go unaccounted for, such as annotation bias in labels used for training. Thus, it is very challenging to understand how biases in medical image data affect deep learning pipelines, especially if it is unknown what biases are present in the dataset, their magnitude and frequency, and how to correct them. As noted by various researchers, "understanding the root cause of bias [...] is a key step towards eliminating that bias" [19]. Therefore, there is a need for a resource that enables researchers to objectively study how biases in medical images affect deep learning models, without the limitations associated with real-world datasets. As a first step towards addressing this need, we propose a flexible framework for generating synthetic neuroimaging data with controlled simulation of realistic biases.

Current methods that have been proposed for fully controlled simulation of features in deep learning datasets, where generating factors can be fully disentangled and are well known in advance, are largely limited to toy problems or

MNIST-like scenarios [4]. On the other hand, a considerable amount of recent research has gone into the supervised and unsupervised disentanglement of generating factors of medical images with generative models that can subsequently be used to synthesize data with specific factor variations [7,11]. However, in such setups, unknown biases could still exist, the true generating factors of real-world data are usually unknown, and it is often impossible to spatially localize an effect. Therefore, we believe that such standard generative models do not offer the flexibility and control of the data generation mechanism that is required to fully analyze the effect of data biases on deep learning models. With our proposed framework, we aim to provide a method for synthesizing realistic image data with a fidelity similar to standard generative models, while still providing a high level of flexibility and control over the type, scale, and proportions of simulated bias features that is usually only available in MNIST-like setups.

In this work, we simulate brain magnetic resonance (MR) images with region-specific morphology variations representing disease and bias effects. We also introduce global morphological variation representing distinct synthetic subjects. This option facilitates bridging the relationship between understanding the impacts of biases alone, and how biases combine with real-world variation when training deep learning models. We utilize neuroimaging data as an initial use case and focus on morphological biases in this work. However, the proposed modular framework is not limited to neuroimaging problems and could be modified to introduce other bias effects, such as gray value effects caused by acquisition parameters or pathologies. Ultimately, this framework can serve as a tool for generating datasets to facilitate analysis of how deep learning models handle various sources of bias. With complete control over the number of samples, types of bias, number of subgroups with different biases, intersections of biased subgroups, and strength and proportion of bias in target classes, datasets generated with this framework can be used as a tool for evaluating how proposed or state-of-the-art models are affected by biases in terms of performance, explainable AI, uncertainty, *etc.*, as well as for benchmarking bias mitigation and data harmonization strategies on a wide range of realistic, controlled scenarios.

The contributions of this paper are summarized as follows: (1) We propose a flexible framework for simulating brain MR datasets, which contain variable morphological disease and bias effects, as a first step towards the controlled and systematic study of how biases in medical imaging data affect deep learning pipelines. (2) We show how this modular framework can be customized to facilitate the investigation of a vast range of data cases that can lead to biased deep learning models. (3) We provide empirical evidence that data generated using this framework can be used to mimic realistic morphological biases in neuroimaging that lead to undesirable performance in a convolutional neural network, and show how these biases can be investigated with explainable AI methods.

2 Methods

The purpose of the proposed framework is to generate a dataset for a multi-class classification problem consisting of synthetic T1-weighted MR images, with N images I_i and associated labels corresponding to m disease classes. For simplicity, in this description of the methods, we focus on the binary classification task (m=2) with disease labels corresponding to disease (D) and no disease (ND). All images are derived by applying non-linear diffeomorphic transformations to a template image I_T, which represents an average brain morphology. More specifically, we consider three types of transformations: (1) φ_S, a subject morphology, (2) φ_D, a disease (target) effect, and (3) φ_B, a bias effect. φ_S is a global non-linear transformation that deforms I_T into a (simulated) subject morphology. In contrast, φ_D and φ_B are spatially localized deformations that only modify I_T locally to introduce an effect (φ_D) that can be used to differentiate disease classes, and a bias effect (φ_B). In our setup, each synthetic image is generated by sampling the transformations φ_S, φ_D, and optionally φ_B from dedicated generative models (Fig. 1A and Suppl. Mat. Fig. 1). Moreover, we assume that all diffeomorphic transformations are parameterized via stationary velocity fields—*e.g.*, $\varphi_S = \exp(v_S)$, where v_S denotes the velocity field and $\exp(\cdot)$ is the group exponential map from the Log-Euclidean framework, which can be efficiently computed via the scaling-and-squaring algorithm; see [1]. The resulting dataset is defined by the user-specified sample size, number of target disease classes, number of subgroups within the dataset containing bias effects, whether inter-subject variability effects are introduced to the datasets, types and degree of each respective effect, and proportions of each respective class and bias group.

Principal Component Analysis-Based Generative Models for Simulating Effects/Variability. To apply anatomically realistic morphological deformations to our template neuroimaging dataset in this work, we fit a principal component analysis (PCA) to the velocity fields of real T1-weighted MR images of different healthy subjects, which were non-linearly registered to the template image I_T. We treat the resulting low-dimensional affine subspace model as a generative model and sample velocity fields representing a range of real anatomical variation from it. For region-specific effects (φ_D and φ_B), the real T1-weighted MR image velocity fields within the regions defined by a label atlas are masked prior to PCA fitting, whereas the full brain is used in the PCA model for simulating subject morphology (φ_S). Thus, by sampling velocity fields v_D, v_B, and v_S from the latent space of the respective subspace models, we can model disease, bias, and subject morphology as variations within an expected extent of human neuroanatomy.

Disease and Bias Effects. We model disease (φ_D) and bias (φ_B) effects as morphological deformations localized to specific brain regions. We also assume that datasets belonging to each disease class have these localized effects sampled from respective distributions in a bimodal Gaussian mixture model within the PCA subspace of the disease effect model. We assume that bias effects are introduced as an additional morphological deformation in a separate brain region, and that these effects are sampled from a Gaussian distribution within the PCA subspace of the bias effect model. In general, an arbitrary number of target classes and bias groups can be introduced to the datasets in a similar sampling procedure.

Subject Morphology. To better emulate anatomical variation in clinical data and warrant the use of deep learning models, global morphological variation representing distinct subjects (φ_S) are applied to the entire anatomy within I_T. These are also sampled from a Gaussian distribution within the PCA subspace of the dedicated subject morphology model.

Introducing Effects to the Template Image. The sparsely defined velocity fields for the disease and bias effects, v_D and v_B, are densified using the scattered grid B-spline method [10] to produce a dense velocity field that includes both effects (if present). If inter-subject variability is desired, the conjugate action mechanism [12] is used to transport the deformation field to the "subject" space, where the "subject" is generated using the sampled v_S/φ_S from the subject morphology PCA model.

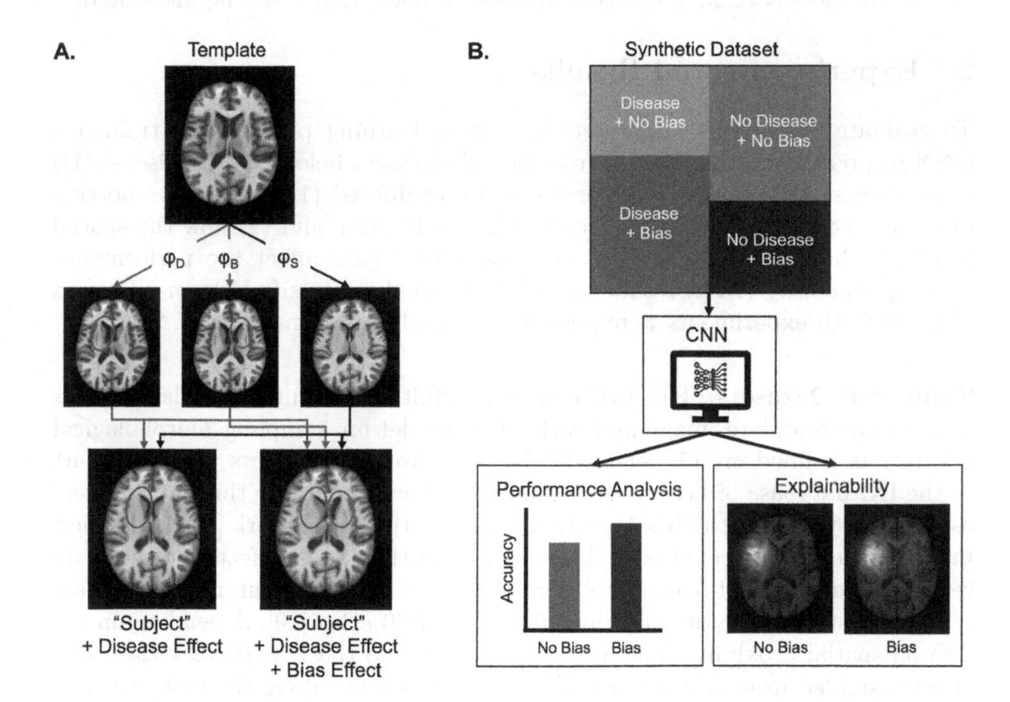

Fig. 1. A) Schematic representing how displacement fields for disease effects (φ_D), bias effects (φ_B), and subject morphology (φ_S) are introduced to a template image I_T to generate custom datasets. B) Synthetic dataset evaluation pipeline used in this paper. A convolutional neural network (CNN) is trained to classify the disease class from a dataset with subgroups containing bias features and evaluated with subgroup performance and explainability.

Framework Customization. For this initial work, we utilize velocity fields from real-world datasets to simulate anatomically realistic effects representing disease features, bias features, and subject morphology via different PCA-based generative models. Although real-world datasets do contain biases, the way in which we propose introducing these effects into the synthetic dataset is highly controlled in such a way where it is known exactly which and how many regions represent either disease or bias effects. Thus, this approach enables a controlled study of bias while benefiting from the utilization of 3D medical images that are representative of real-world clinical data. Moreover, due to the modularity of the proposed framework, such effects can also be introduced through a variety of other methods for generating deformation fields, ranging from highly precise but simple (*e.g.*, single displacement vectors) to more realistic but increasingly complex approaches (*e.g.*, generative models). Furthermore, in this work, we simulate morphological changes in brain images via diffeomorphic transformations as a use case, but the framework can be adapted to other disease or bias effects that would alter the topology (*e.g.*, gray value changes or lesions). Moreover, other imaging modalities or body regions (*e.g.*, cardiac MRI) as well as other generative models (*e.g.*, generative adversarial networks) could be integrated.

3 Experiments and Results

To evaluate our synthetic datasets in a deep learning pipeline, we trained a CNN to predict whether images from biased datasets belong to the disease (D) or no disease (ND) class. More precisely, we evaluated (1) how the proportion of datasets containing bias features within the D class, and (2) how the spatial proximity between the disease region and bias region affect the performance and explainability (XAI) results of a CNN trained to classify D from ND cases (Fig. 1B). All experiments were performed with Keras/Tensorflow v. 2.10.

Simulated Datasets. The SRI24 Normal Adult Brain Anatomy atlas [17] was used as the template image and each PCA model for sampling morphological effects was trained on T1-weighted MRI data from 50 subjects who were part of the IXI database of healthy individuals[1]. Velocity fields for this dataset were estimated by utilizing ITK's VariationalRegistration framework [6,24]. The left insular cortex was selected as the brain region for the disease effect, and the brain regions used to model bias effects were either the left putamen, right putamen, or right postcentral gyrus as defined by the LPB40 atlas [20], depending on the desired spatial proximity. The datasets belonging to the D and ND classes had effects sampled from $\mathcal{N}(0,1)$ and $\mathcal{N}(2,1)$, respectively, along the first principal component of the generative model for the disease region, and the datasets with bias features had effects sampled from $\mathcal{N}(2,1)$ along the first principal components of the models for the respective bias regions. Inter-subject variability effects were sampled from a Gaussian distribution of $\mathcal{N}(0,1)$ along the first 10 principal components of the subject morphology generative model.

[1] https://brain-development.org/ixi-dataset/.

Experiments. To evaluate the effect on model performance and XAI in relation to the proportion of datasets containing bias features within the disease class, the generated datasets had either 60% or 80% of the simulated images from the D class containing the bias feature, with 30% of the simulated images from the ND class containing the bias feature for all experiments. To evaluate the effect of proximity between disease and bias regions, the distances between regions were defined as either near, middle, or far, for the left putamen, right putamen, and right postcentral gyrus, respectively (see Fig. 2B). Each simulated dataset contained a balanced representation of D and ND labels. The proximity experiments were performed under both 60% and 80% conditions defined by the proportion experiments. Model performance with the biased datasets was compared against a baseline experiment in which the datasets do not contain any simulated bias features but only the disease effects.

Model and Training. A CNN was used as a model for predicting whether datasets belonged to the D or ND class. The model consisted of 5 blocks each containing a convolutional layer with (3×3×3) kernel, batch normalization, sigmoid activation, and (2×2×2) max pooling. The convolutional filter sizes were 32, 64, 128, 256, and 512 for each respective block. The sixth block contained average pooling, dropout (rate=0.2), and a dense classification layer with softmax activation. Binary cross entropy loss, Adam optimizer (learning rate = $1e^{-4}$), and batch size 4 with early stopping based on validation loss (patience=30) were used for training. Each experiment simulated and used 500 datasets of voxel dimensions (173×211×155) with a 55%/15%/30% train/validation/test split, stratified by disease and bias labels.

Evaluation. Model performance was evaluated using accuracy, sensitivity, and specificity computed for the aggregate test set, as well as separately for the bias (B) and no bias (NB) groups. Results are reported as the mean ± standard deviation of the models with 5 different weight initialization seeds on the same train/validation/test splits, following [18]. The SmoothGrad (SG) [21] method was used for XAI evaluation. Average SG maps were computed with 25 individual SG maps (5 from each seed) for the datasets in the test set with the bias feature, which were correctly identified as being in the disease class.

Results and Discussion. The results of our evaluation are summarized in Fig. 2, with full quantitative results shown in Tables 1 and 2 in the Supplementary Material. As seen in Fig. 2A, for all conditions with simulated dataset bias, the sensitivity is higher and specificity is lower within the B group, while the opposite was found for the NB group. Due to the higher representation of biased datasets in the D class, it seems reasonable to assume that the model uses the presence of bias features as a shortcut for predicting the disease state, and thus predicts the D class more often for the B group, resulting in fewer true negatives. Within the NB group, the absence of bias features seems to be also used as a shortcut for predicting the ND class, resulting in a higher number of ND class predictions and consequently fewer true positives. While these shortcuts are apparent in all experiments utilizing biased datasets, there is a stronger relationship between disease–bias region proximity and the degree of the shortcuts (measured by lower specificity in the bias group and lower sensitivity in the NB group) when the dataset has 80% of the D class containing the bias effect compared to 60%. In these 80% conditions, it was observed that the sensitivity in the NB group decreases as a function of region proximity, indicating that the model uses the absence of the bias effect as a shortcut for predicting the ND class more often when regions are further away. Likewise, specificity in the B group increases as a function of region proximity, indicating that the model uses the presence of the bias as a shortcut for predicting the D class label less often when regions are further away. A potential explanation for this may be that the CNN used has a spatially localized receptive field. Thus, when the bias and disease regions are near to each other, the network learns to associate them more closely and predicts the D class label more often for images with the bias effect. When the regions are farther apart from each other, the CNN becomes more tuned to recognize bias effects separately from disease effects. Thus, when the bias effect is not present, the model assumes the image belongs to the ND class.

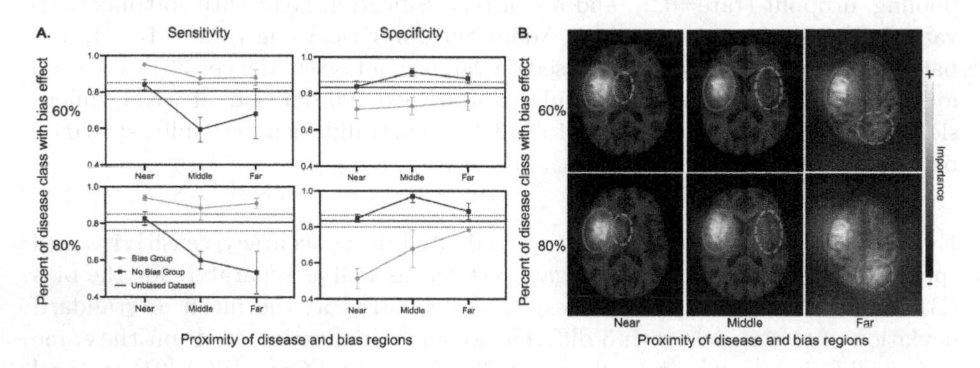

Fig. 2. Results of experiments investigating effect of the proportion of bias effect in the disease class and proximity between disease and bias regions. A) Mean sensitivity (left column) and specificity (right column) for test datasets with bias (green circle) and without bias (blue square). Error bars represent standard deviation over 5 different model initialization seeds with the same train/validation/test split. Black lines represent mean sensitivity and specificity with 95% confidence interval over 5 different model seeds with the same train/validation/test split of dataset containing no bias effects. B) Average SmoothGrad saliency maps with disease region circled in solid/magenta and bias region circled in dashed/cyan. (Color figure online)

Furthermore, as seen in Fig. 2B, with 60% of the D class containing the bias feature, the SG maps show minor activation in the region with the bias effect, particularly in the far proximity condition. However, when 80% of the D class contains the bias feature, the SG maps highlight the bias regions considerably more intensely for all proximities analyzed. Even though the model still uses prediction shortcuts, which affect performance of the B and NB groups when 60% of the images from the D class exhibit the bias feature, the regions associated with the bias are less clearly identifiable in the group-averaged SG saliency maps, suggesting that XAI may not always be a reliable tool to uncover sources of bias in medical image data.

4 Conclusion

In this work, we presented a flexible and modular framework for simulating bias in medical imaging datasets using realistic morphological effects in neuroimaging as a use case. By sampling brain region-specific morphological variation representing the disease state and bias features from generative models in a controlled manner, we can generate synthetic datasets of arbitrary size and composition, which enables the investigation of a vast range of dataset bias scenarios and corresponding impacts on deep learning pipelines. Directions for future work with this framework are extensive and could include the analysis of more variations of bias proportions and proximities on alternate model architectures (*e.g.*, vision transformers), evaluation of state-of-the-art bias mitigation strategies on various dataset compositions, as well as assessing other potential limitations of explainability methods as a tool for investigating bias. We believe that our work provides a strong foundation for advancing understanding of bias in deep learning for medical image analysis and consequently developing responsible models and methods for clinical use.

Acknowledgements. This work was supported by Alberta Innovates, the Natural Sciences and Engineering Research Council of Canada, the River Fund at Calgary Foundation, Canada Research Chairs Program, and the Canadian Institutes of Health Research.

References

1. Arsigny, V., Commowick, O., Pennec, X., Ayache, N.: A log-Euclidean framework for statistics on diffeomorphisms. In: Larsen, R., Nielsen, M., Sporring, J. (eds.) MICCAI 2006. LNCS, vol. 4190, pp. 924–931. Springer, Heidelberg (2006). https://doi.org/10.1007/11866565_113
2. Bashyam, V.M., et al.: The iSTAGING and PHENOM consortia: deep generative medical image harmonization for improving cross-site generalization in deep learning predictors. J. Magn. Reson. Imaging **55**(3), 908–916 (2022)
3. Bernhardt, M., Jones, C., Glocker, B.: Potential sources of dataset bias complicate investigation of underdiagnosis by machine learning algorithms. Nat. Med. **28**(6), 1157–1158 (2022)
4. Castro, D.C., Tan, J., Kainz, B., Konukoglu, E., Glocker, B.: Morpho-MNIST: quantitative assessment and diagnostics for representation learning. J. Mach. Learn. Res. **20**, 1–29 (2019)

5. Dinsdale, N.K., Jenkinson, M., Namburete, A.I.L.: Deep learning-based unlearning of dataset bias for MRI harmonisation and confound removal. Neuroimage **228**, 117689 (2021)
6. Ehrhardt, J., Schmidt-Richberg, A., Werner, R., Handels, H.: Variational registration. In: Handels, H., Deserno, T.M., Meinzer, H.-P., Tolxdorff, T. (eds.) Bildverarbeitung für die Medizin 2015. I, pp. 209–214. Springer, Heidelberg (2015). https://doi.org/10.1007/978-3-662-46224-9_37
7. Fragemann, J., Ardizzone, L., Egger, J., Kleesiek, J.: Review of disentanglement approaches for medical applications - towards solving the gordian knot of generative models in healthcare (2022). arXiv:2203.11132 [cs]
8. Glocker, B., Robinson, R., Castro, D.C., Dou, Q., Konukoglu, E.: Machine learning with multi-site imaging data: an empirical study on the impact of scanner effects (2019). arXiv:1910.04597 [cs, eess, q-bio]
9. Larrazabal, A.J., Nieto, N., Peterson, V., Milone, D.H., Ferrante, E.: Gender imbalance in medical imaging datasets produces biased classifiers for computer-aided diagnosis. Proc. Natl. Acad. Sci. **117**, 12592–12594 (2020)
10. Lee, S., Wolberg, G., Shin, S.: Scattered data interpolation with multilevel B-splines. IEEE Trans. Visual Comput. Graphics **3**(3), 228–244 (1997)
11. Liu, X., Sanchez, P., Thermos, S., O'Neil, A.Q., Tsaftaris, S.A.: Learning disentangled representations in the imaging domain. Med. Image Anal. **80**, 102516 (2022)
12. Lorenzi, M., Pennec, X.: Geodesics, parallel transport & one-parameter subgroups for diffeomorphic image registration. Int. J. Comput. Vision **105**(2), 111–127 (2013)
13. Luo, L., Xu, D., Chen, H., Wong, T.T., Heng, P.A.: Pseudo bias-balanced learning for debiased chest X-ray classification. In: Wang, L., Dou, Q., Fletcher, P.T., Speidel, S., Li, S. (eds.) Medical Image Computing and Computer Assisted Intervention – MICCAI 2022, pp. 621–631. LNCS, vol. 13438. Springer, Cham (2022). https://doi.org/10.1007/978-3-031-16452-1_59
14. Marcinkevics, R., Ozkan, E., Vogt, J.E.: Debiasing deep chest X-ray classifiers using intra- and post-processing methods. In: Proceedings of the 7th Machine Learning for Healthcare Conference, pp. 504–536. PMLR (2022)
15. Mukherjee, P., Shen, T.C., Liu, J., Mathai, T., Shafaat, O., Summers, R.M.: Confounding factors need to be accounted for in assessing bias by machine learning algorithms. Nat. Med. **28**(6), 1159–1160 (2022)
16. Puyol-Antón, E., et al.: Fairness in cardiac MR image analysis: an investigation of bias due to data imbalance in deep learning based segmentation. In: de Bruijne, M., et al. (eds.) Medical Image Computing and Computer Assisted Intervention - MICCAI 2021, pp. 413–423. LNCS, vol. 12903. Springer, Cham (2021). https://doi.org/10.1007/978-3-030-87199-4_39
17. Rohlfing, T., Zahr, N.M., Sullivan, E.V., Pfefferbaum, A.: The SRI24 multichannel atlas of normal adult human brain structure. Hum. Brain Mapp. **31**(5), 798–819 (2010)
18. Seyyed-Kalantari, L., Zhang, H., McDermott, M.B.A., Chen, I.Y., Ghassemi, M.: Underdiagnosis bias of artificial intelligence algorithms applied to chest radiographs in under-served patient populations. Nat. Med. **27**(12), 2176–2182 (2021)
19. Seyyed-Kalantari, L., Zhang, H., McDermott, M.B.A., Chen, I.Y., Ghassemi, M.: Reply to: 'Potential sources of dataset bias complicate investigation of underdiagnosis by machine learning algorithms' and 'Confounding factors need to be accounted for in assessing bias by machine learning algorithms'. Nat. Med. **28**(6), 1161–1162 (2022)
20. Shattuck, D.W., et al.: Construction of a 3D probabilistic atlas of human cortical structures. Neuroimage **39**(3), 1064–1080 (2008)

21. Smilkov, D., Thorat, N., Kim, B., Viégas, F., Wattenberg, M.: SmoothGrad: removing noise by adding noise (2017). arXiv: 1706.03825
22. Stanley, E.A.M., Wilms, M., Forkert, N.D.: Disproportionate subgroup impacts and other challenges of fairness in artificial intelligence for medical image analysis. In: Baxter, J.S.H., et al. (eds.) Ethical and Philosophical Issues in Medical Imaging, Multimodal Learning and Fusion Across Scales for Clinical Decision Support, and Topological Data Analysis for Biomedical Imaging, pp. 14–25. LNCS, vol. 13755. Springer, Cham (2022). https://doi.org/10.1007/978-3-031-23223-7_2
23. Wachinger, C., Rieckmann, A., Pölsterl, S.: Detect and correct bias in multi-site neuroimaging datasets. Med. Image Anal. **67**, 101879 (2021)
24. Werner, R., Schmidt-Richberg, A., Handels, H., Ehrhardt, J.: Estimation of lung motion fields in 4D CT data by variational non-linear intensity-based registration: a comparison and evaluation study. Phys. Med. Biol. **59**(15), 4247 (2014)
25. Zare, S., Nguyen, H.V.: Removal of confounders via invariant risk minimization for medical diagnosis. In: Wang, L., Dou, Q., Fletcher, P.T., Speidel, S., Li, S. (eds.) Medical Image Computing and Computer Assisted Intervention - MICCAI 2022, pp. 578–587. LNCS, vol. 13438. Springer, Cham (2022). https://doi.org/10.1007/978-3-031-16452-1_55
26. Zhao, Q., Adeli, E., Pohl, K.M.: Training confounder-free deep learning models for medical applications. Nat. Commun. **11**(11), 6010 (2020)

Client-Level Differential Privacy via Adaptive Intermediary in Federated Medical Imaging

Meirui Jiang[1], Yuan Zhong[1], Anjie Le[1], Xiaoxiao Li[2], and Qi Dou[1(✉)]

[1] Department of Computer Science and Engineering, The Chinese University of Hong Kong, Hong Kong, China
qidou@cuhk.edu.hk

[2] Department of Electrical and Computer Engineering, The University of British Columbia, Vancouver, Canada

Abstract. Despite recent progress in enhancing the privacy of federated learning (FL) via differential privacy (DP), the trade-off of DP between privacy protection and performance is still underexplored for real-world medical scenario. In this paper, we propose to optimize the trade-off under the context of client-level DP, which focuses on privacy during communications. However, FL for medical imaging involves typically much fewer participants (hospitals) than other domains (e.g., mobile devices), thus ensuring clients be differentially private is much more challenging. To tackle this problem, we propose an adaptive intermediary strategy to improve performance without harming privacy. Specifically, we theoretically find splitting clients into sub-clients, which serve as intermediaries between hospitals and the server, can mitigate the noises introduced by DP without harming privacy. Our proposed approach is empirically evaluated on both classification and segmentation tasks using two public datasets, and its effectiveness is demonstrated with significant performance improvements and comprehensive analytical studies. Code is available at: https://github.com/med-air/Client-DP-FL.

Keywords: Federated Learning · Client-level Differential Privacy · Medical Image Analysis

1 Introduction

Differential privacy (DP) has emerged as a promising technique to safeguard the privacy of sensitive data in federated learning (FL) [2,12,13,29,34], offering privacy guarantees in a mathematical format [7,25,33]. However, introducing noise to ensure DP often comes at the cost of performance. Some recent studies have noticed that the noise added to the gradient impedes optimization [6,14,27].

Supplementary Information The online version contains supplementary material available at https://doi.org/10.1007/978-3-031-43895-0_47.

H. Greenspan et al. (Eds.): MICCAI 2023, LNCS 14221, pp. 500–510, 2023.
https://doi.org/10.1007/978-3-031-43895-0_47

For critical medical applications requiring low error tolerance, such performance degradation makes the rigorous privacy guarantee diminish [5,22]. Therefore, it is imperative to maintain high performance while enhancing privacy, i.e., optimizing the privacy-performance trade-off. Unfortunately, despite its significance, such trade-off optimization in FL has not been sufficiently investigated to date.

Several studies have examined the trade-off in the centralized scenario. For instance, Li et al. [18] proposed enhancing utility by leveraging public data or data statistics to estimate gradient geometry. Amid et al. [3] utilized the loss on public data as a mirror map to improve performance. Li et al. [17] suggested constructing less noisy preconditioners using historical gradients. In contrast to these studies, we concentrate on promoting the trade-off in FL, where public dataset is limited and sharing side information may not be feasible [12,29]. Specifically, we aim to ensure that clients are differentially private. Our objective is not to protect a single data point, but rather to achieve that a learned model does not reveal whether a client participated in decentralized training. This ensures that a client's entire dataset is safeguarded against differential attacks from third parties. This is particularly crucial in medical imaging, where sensitive patient information is typically kept within each hospital. Nevertheless, in medical imaging, the number of participants (silos) is usually much smaller than in other domains, such as mobile devices [12]. This cross-silo situation necessitates adding a considerable amount of noise to protect client privacy, making the optimization of the trade-off uniquely challenging [20].

To improve the trade-off of privacy protection and performance, the key point is to mitigate the noise added to the client during gradient updates. Our idea is inspired by the observation in DP-FedAvg [24], which suggests that the utility of DP can be improved by utilizing a sufficiently large dataset with numerous users. Through an analysis of the DP accountant, we identified that the noise is closely related to the gradient clip bound and the number of participants. In this regard, we propose to split the original client into disjoint sub-clients, which act as intermediaries for exchanging information between the hospital and the server. This strategy increases the number of client updates against queries, thereby consequently reducing the magnitude of noise. However, finding an optimal splitting is not straightforward due to the non-identical nature of data samples. Splitting a client into more sub-clients may increase the diversity of FL training, which can adversely harm the final performance. Thus, there is a trade-off between noise level and training diversity. Our objective is to explore the relationships among clients, noise effects, and training diversities to identify a balance point that maximizes the trade-off between privacy and performance.

In this paper, we present a novel adaptive intermediary method to optimize the privacy-performance trade-off. Our approach is based on the interplay relationships among noise levels, training diversities, and the number of clients. Specifically, we observe a reciprocal correlation between the noise level and the number of intermediaries, as well as a linear correlation between the training diversity and the intermediary number. To determine the optimal number of intermediaries, we introduce a new term called intermediary ratio, which quantifies the ratio of noise level and training diversity. Our theoretical analysis

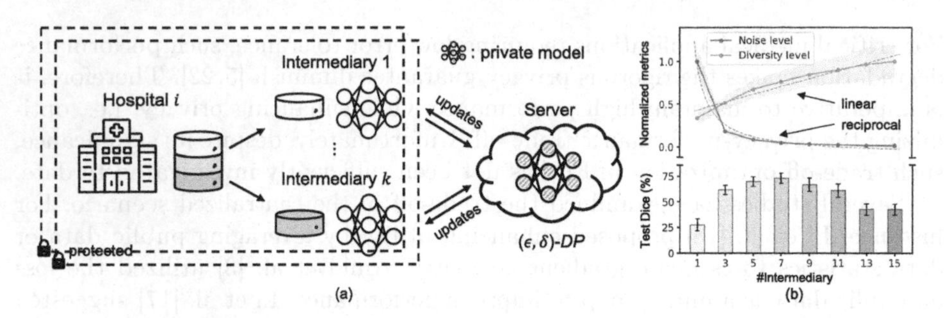

Fig. 1. (a) Overview of our intermediary strategy, which protects participating hospitals with superior privacy-performance trade-offs. The server aggregates local models from non-overlapping intermediaries with DP guarantees. (b) Continuously splitting intermediaries may not continuously improve performance, as it reduces gradient noises in a reciprocal manner, but also increases gradient diversity or heterogeneity linearly.

demonstrates that splitting the original clients into more intermediaries achieves DP with the same privacy budget and DP failure probability. Furthermore, we show that when sample-level DP and client-level DP have equivalent noise levels, the variance of the difference between noisy and original model diverges exponentially with more training steps, leading to poor performance. We evaluate our method on both classification and segmentation tasks, including the intracranial hemorrhage diagnosis with 25,000 CT slices, and the prostate MRI segmentation with heterogeneous data from different hospitals. Our method consistently outperforms various DP optimization methods on both tasks and can serve as a lightweight add-on with good compatibility. In addition, we conduct comprehensive analytical studies to demonstrate the effectiveness of our method.

2 Method

2.1 Preliminaries

In this work, we consider client-level differential privacy. We first introduce the definition of DP as follows:

Definition 1. *((ϵ, δ)-Differential Privacy [7,8]) For a randomized learning mechanism $M\colon \mathcal{X} \rightarrow \mathcal{R}$, where $\mathcal{X}$ is the collection of datasets it can be trained on, and $\mathcal{Y}$ is the collection of model it can generate, it is (ϵ, δ)-DP if:*

$$(\forall S \subseteq \mathcal{R})(\forall D, D' \in \mathcal{X}, D \sim D') \Pr[M(D) \in S] \leq \exp(\varepsilon) \cdot \Pr\left[M\left(D'\right) \in S\right] + \delta,$$

where ϵ denotes the privacy budget, and δ represents the probability that ϵ-DP fails in this mechanism. Note that the smaller the ϵ value is, the more private the mechanism is. Our aim of applying DP is to protect the collection of "datasets" $\mathcal{X}$, which are client model updates in every communication round in the context of FL. The protection can be done by incorporating a DP-preserving randomization mechanism into the learning process. One commonly used method is the Gaussian mechanism, which involves bounding the contribution (l_2-norm) of each client update followed by adding Gaussian noise proportional to that bound onto the aggregate [24]. Specifically, suppose there are N clients, denote the gradients of each client as Δ_i, the server model θ_{t+1} at round $t+1$ is updated by adding the Gaussian mechanism approximating the sum of updates as follows:

$$\theta_{t+1} \leftarrow \theta_t + \frac{1}{N}\left(\sum\nolimits_{i\in[N]} \Delta_i / \max(\frac{\|\Delta_i\|_2}{C}, 1) + \mathcal{N}(\mathbf{0}, z^2C^2\mathbf{I})\right), \tag{1}$$

where C is the gradient clipping threshold, and z is the noise multiplier determined by the privacy accountant with given ϵ, δ, and training steps. The noise multiplier z indicates the amount of noise required to reach a particular privacy budget. To privatize the participation of clients in FL, the noise added for client-level DP typically correlates with the number of clients. This incurs a large magnitude of noise in cross-silo FL in the medical field, which can significantly deteriorate the final server model performance.

2.2 Adaptive Intermediary for Improving Client-Level DP

The key to optimizing the privacy-performance trade-off lies in mitigating the effects of noise without compromising privacy protection. Based on the noise calculation in Eq. (1), we propose to study the final effects of noise on the server model, which can be denoted as $\zeta \sim \mathcal{N}(0, \sigma^2\mathbf{I})$, where $\sigma = {}^{zC}/_{N}$. Note that the final noise (ζ) is determined by σ, which relates to the noise multiplier z, clip threshold C, and the number of clients N. In DP, the clip threshold and the noise multiplier are usually pre-assigned. Therefore, the noise level can be reduced by increasing the number of clients N. To this end, we propose to reduce the noise by splitting the original clients into non-overlapping sub-clients, which serve as intermediaries to communicate with the server (see Fig. 1 (a)). We validate our hypothesis by studying the feasibility and analyzing the relationships between the intermediary number, noise, and performance.

Feasibility. We demonstrate the feasibility by showing the use of intermediary preserves privacy. For $\mathcal{X}$ the collection of possible datasets from extant clients, denote $D_i \in \mathcal{X}$ the dataset of client i, we randomly split D_i into v disjoint subsets $D_{i,1}, ..., D_{i,v}$, so that $\sqcup_j D_{i,j} = D_i$. We define the dataset $D_{i,j}$ of client i as the intermediary j. Then we show that partitioning extant clients into multiple intermediaries is capable of maintaining DP. Denote the collection of all possible datasets formed by the intermediaries as $\mathcal{Y}$, and note that $\mathcal{X} \subseteq \mathcal{Y}$. We have:

Theorem 1. *If a randomized learning mechanism $\mathcal{M} : \mathcal{X} \to \mathcal{R}$ is $(\epsilon, \delta) - DP$, then its induced mechanism $\tilde{\mathcal{M}} : \mathcal{Y} \to \mathcal{R}$ is also (ϵ, δ)-DP.*

This indicates that partitioning the original client into intermediaries keeps the same DP regime. The proof can be found in Appendix C. We also analyze the reverse relation in the appendix section to complete the overall relationship.

Privacy-Performance Trade-Off Analysis. With the above basis, we further investigate the privacy-performance trade-off by varying the number of intermediaries. According to the noise calculation of $\sigma = {}^{zC}/_{N}$, we can reduce noise by splitting clients into intermediaries to increase N. However, increasing the number of intermediaries causes each intermediary to hold fewer samples. This may affect the aggregation direction and harms final performance consequently. There is a trade-off behind intermediary splitting. To investigate the trade-off, we design and study two highly related metrics, i.e., noise level ξ and client update diversity level φ. Denoting clipped gradients as $\hat{\Delta}_i$, we define the noise level and diversity level as:

$$\xi = \frac{\|\zeta\|}{\|\sum_{i \in N} \hat{\Delta}_i\|_2}, \quad \varphi = \frac{\sum_{i \in N} \|\Delta_i\|_2}{\|\sum_{i \in N} \hat{\Delta}_i\|_2}. \tag{2}$$

By varying the number of intermediaries, we obtain different values for noise levels and diversities (see Fig. 1 (b)). By fitting the relations between noise level (client update diversity) and the number of intermediaries for each client (denoted as v), we surprisingly find the relations that:

$$\xi_v = v^{-1} \cdot \xi, \quad \varphi_v = v \cdot \varphi, \tag{3}$$

where ξ_v and φ_v denote the value when each client is split into v intermediaries. By defining the intermediary ratio as $\lambda = {}^{\xi}/_{\varphi}$, we can use this ratio to quantify the relations between noise level and diversity, which helps identify the optimal number of intermediaries to generate.

Adaptive Intermediary Generation. We can generate the intermediary based on the defined intermediary ratio λ. We experimentally investigated the relationships between the final performance and the number of intermediaries and found the optimal ratio lies in the range of ${}^{1}/_{N}$. Therefore, for each client, the number of intermediaries is $v = \sqrt[2]{N \cdot {}^{\xi}/_{\varphi}}$. Considering the extreme case of $\lim_{N \to \infty} \xi = 0$, we can also infer the ratio $\lambda = 0$, which further validates the rationality and consistency with our empirical findings. For the practical application, we can initialize the number of intermediaries via the first round results. Then, for each round, we will re-calculate the ratio using ξ and φ from the last round, and then adaptively split clients to make sure the new ratio lies around ${}^{1}/_{N}$.

2.3 Cumulation of Sample-Level DP to Client-Level

We further investigate the relationships between client-level DP and sample-level DP, by cumulating sample-level DP mechanism to a client level. In DP-SGD [1], denote the standard deviation of Gaussian noise as $\sigma = z(\epsilon, \delta)c/K$ with K being the batch size, c being the sample-level gradient clip bound and z being the noise multiplier determined by privacy accountant with (ϵ, δ). Noise is added to each batch gradient before taking a descent, so that each step is (ϵ, δ)-DP.

Note that z can take different forms, the form provided by moment accountant [1] is $z(\epsilon, \delta) = \mathcal{O}(\sqrt{\ln(1/\delta)/\epsilon^2})$. Through the use of the moment accountant and sensitivity cumulation, we can calculate the standard deviation of cumulated noise in $\mathcal{T}$ steps as $\sigma_{\mathcal{T}} = z'(\epsilon_{\mathcal{T}}, \delta_{\mathcal{T}})\mathcal{S}_{\mathcal{T}}$, where $\epsilon_{\mathcal{T}} = \mathcal{O}(\sqrt{\mathcal{T}}\epsilon)$, $\delta_{\mathcal{T}} = \mathcal{O}(\delta)$, and $\mathcal{S}_{\mathcal{T}} = \mathcal{O}(\mathcal{T}c)$. It follows that $z' = \mathcal{O}(1/\epsilon_{\mathcal{T}}) = \mathcal{O}(z/\sqrt{\mathcal{T}})$, and $\sigma'_s = \mathcal{O}(\sqrt{\mathcal{T}}\sigma_s)$. This indicates that the noise scale cumulates at a rate of $\mathcal{O}(\sqrt{\mathcal{T}})$. With regards to performance, we prove in Appendix C that the variance of the difference between the noisy model and the original model diverges with a rate of $\mathcal{O}((1-2\eta\beta+\eta^2\mu^2)^{\mathcal{T}})$ for μ-convex, β-smooth loss functions. This shows that increasing $\mathcal{T}$ also increases the probability of obtaining a model which diverges further from the original model, resulting in poorer performance.

On the Client-Level. For client-level noise, we can compute the standard deviation as $\sigma_c = z(\epsilon_c, \delta_c)C$, where C is the clip bound of client update. The clip bound is typically set to the median among $l2$-norms of all client updates. Assuming an identical distribution across clients and samples, we have $C = \mathcal{O}(\mathcal{T}c)$. As a result, we have $z_c = \mathcal{O}(z_{\mathcal{T}})$, indicating that the cumulation of sample-level noise in DP-SGD gives the same DP level up to a constant, which is equivalent to adding noise directly to the client level through the moment accountant. Regarding the performance, we note that by leveraging the noisy models from several clients that hold identically distributed datasets, we can reduce the probability of getting a significantly drifted model without additional privacy leakage.

3 Experiment

3.1 Experimental Setup

Datasets. We evaluate our method on two tasks: 1) intracranial hemorrhage (ICH) classification, and 2) prostate MRI segmentation. For ICH classification, we use the RSNA-ICH dataset [9] and follow [15] to relieve the class imbalance across ICH subtypes and perform the binary diseased-or-healthy classification. We randomly sample 25,000 slices and split them into 20 clients, where each client data is split into 60%, 20%, and 20% for training, validation, and testing. We resize images to 224×224 and perform data augmentation with random affine and horizontal flip. For prostate segmentation, we adopt a multi-site T2-weighted MRI dataset [21] which contains 6 different data sources from 3 public datasets [16,19,26]. We regard each data source as one client, resize images to 256×256, and use 50%, 25%, and 25% for for training, validation and testing.

Table 1. Performance comparison of different DP optimization methods and ours. We report mean and standard deviation across three independent runs with different seeds.

Intracranial Hemorrhage Diagnosis ($N = 20$)								
Method	No Privacy		$z = 0.5$		$z = 1.0$		$z = 1.5$	
	AUC ↑	Acc ↑	AUC ↑	Acc ↑	AUC ↑	Acc ↑	AUC ↑	Acc ↑
DP-FedAvg [24]	$90.88_{\pm 0.15}$	$82.85_{\pm 0.26}$	$70.38_{\pm 0.61}$	$64.94_{\pm 0.28}$	$68.00_{\pm 1.43}$	$63.55_{\pm 1.12}$	$66.77_{\pm 0.12}$	$62.01_{\pm 0.58}$
+Ours	-	-	$82.42_{\pm 0.29}$	$74.87_{\pm 0.43}$	$80.84_{\pm 0.78}$	$73.37_{\pm 0.92}$	$80.77_{\pm 0.80}$	$72.95_{\pm 0.47}$
DP-FedAdam [28]	$91.85_{\pm 0.30}$	$84.16_{\pm 0.56}$	$75.91_{\pm 0.28}$	$68.75_{\pm 0.16}$	$70.75_{\pm 2.18}$	$65.34_{\pm 0.64}$	$70.89_{\pm 2.05}$	$63.73_{\pm 1.16}$
+Ours	-	-	$82.86_{\pm 0.47}$	$75.20_{\pm 0.27}$	$81.63_{\pm 0.38}$	$73.99_{\pm 0.79}$	$80.55_{\pm 0.60}$	$73.06_{\pm 0.68}$
DP-FedNova [31]	$90.89_{\pm 0.25}$	$83.00_{\pm 0.17}$	$71.84_{\pm 1.51}$	$66.25_{\pm 0.91}$	$69.26_{\pm 1.76}$	$63.75_{\pm 1.49}$	$68.45_{\pm 0.93}$	$63.21_{\pm 1.13}$
+Ours	-	-	$82.73_{\pm 0.38}$	$75.35_{\pm 0.27}$	$80.64_{\pm 0.57}$	$73.55_{\pm 0.78}$	$79.39_{\pm 0.41}$	$71.70_{\pm 0.35}$
DP^2-RMSProp [17]	$88.89_{\pm 0.02}$	$80.77_{\pm 0.25}$	$70.59_{\pm 1.16}$	$64.92_{\pm 1.19}$	$67.43_{\pm 0.60}$	$62.05_{\pm 0.23}$	$65.91_{\pm 1.40}$	$61.77_{\pm 1.13}$
+Ours	-	-	$81.60_{\pm 0.68}$	$74.47_{\pm 0.95}$	$80.23_{\pm 0.22}$	$73.15_{\pm 0.44}$	$81.32_{\pm 0.50}$	$74.21_{\pm 0.85}$
Prostate MRI Segmentation ($N = 6$)								
Method	No Privacy		$z = 0.3$		$z = 0.5$		$z = 0.7$	
	Dice ↑	IoU ↑	Dice ↑	IoU ↑	Dice ↑	IoU ↑	Dice ↑	IoU ↑
DP-FedAvg [24]	$87.69_{\pm 0.12}$	$79.62_{\pm 0.13}$	$41.43_{\pm 3.89}$	$29.28_{\pm 3.40}$	$22.45_{\pm 3.15}$	$13.50_{\pm 2.24}$	$13.59_{\pm 0.96}$	$7.41_{\pm 0.59}$
+Ours	-	-	$70.59_{\pm 1.55}$	$67.72_{\pm 0.47}$	$63.28_{\pm 4.69}$	$61.12_{\pm 0.50}$	$58.14_{\pm 4.71}$	$56.18_{\pm 7.21}$
DP-FedAdam [28]	$87.63_{\pm 0.16}$	$79.65_{\pm 0.20}$	$38.24_{\pm 2.86}$	$38.24_{\pm 1.38}$	$16.50_{\pm 1.82}$	$15.03_{\pm 2.28}$	$9.15_{\pm 2.19}$	$5.49_{\pm 2.22}$
+Ours	-	-	$69.68_{\pm 1.45}$	$61.31_{\pm 0.71}$	$57.11_{\pm 6.30}$	$57.23_{\pm 1.17}$	$43.99_{\pm 9.04}$	$47.49_{\pm 7.81}$
DP-FedNova [31]	$87.44_{\pm 0.35}$	$79.49_{\pm 0.29}$	$41.91_{\pm 6.34}$	$29.33_{\pm 5.78}$	$17.10_{\pm 7.45}$	$9.96_{\pm 4.81}$	$11.41_{\pm 0.64}$	$6.06_{\pm 0.38}$
+Ours	-	-	$70.80_{\pm 1.28}$	$66.64_{\pm 1.42}$	$68.63_{\pm 2.17}$	$63.99_{\pm 0.96}$	$58.99_{\pm 4.43}$	$59.14_{\pm 2.03}$
DP^2-RMSProp [17]	$87.46_{\pm 0.08}$	$80.00_{\pm 0.09}$	$38.33_{\pm 2.44}$	$24.73_{\pm 3.80}$	$16.74_{\pm 0.83}$	$10.75_{\pm 0.99}$	$7.77_{\pm 0.41}$	$4.00_{\pm 0.23}$
+Ours	-	-	$63.05_{\pm 2.60}$	$63.05_{\pm 2.60}$	$53.53_{\pm 4.97}$	$60.84_{\pm 5.81}$	$47.82_{\pm 2.01}$	$59.66_{\pm 2.76}$

Privacy Setup. We use the Opacus' [32] implementation of privacy loss random variables (PRVs) accountant [10] for the Gaussian mechanism for our privacy accounting. We restrict the total number of training rounds and then account for any privacy overheads with various privacy levels controlled by the noise multiplier z, where a higher z indicates a higher privacy regime ϵ. Adaptive clipping [4] is employed to bound each client's contribution in the federation. Following [33], we report the results by exploring effects of different noise multiplier z values. We set z in the range of $\{0.5, 1.0, 1.5\}$ for ICH diagnosis, and $\{0.3, 0.5, 0.7\}$ for prostate segmentation, which induces privacy budgets of $\{245.6, 72.4, 36.9\}$ and $\{597.3, 224.7, 119.4\}$, respectively. We set $\delta = 10^{-k}$ where k is the smallest integer that satisfies $10^{-k} \leq 1/n$ for the client number n as suggested by [17].

Implementation Details. We use Adam optimize, set the local update epoch to 1, and set total communication rounds to 100. We use DenseNet121 [11] for classification, the batch size is 16 and the learning rate is 3×10^{-4}. We use UNet [30] for segmentation, the batch size of 8, and the learning rate is 10^{-3}.

3.2 Empirical Evaluation

First, we present experimental results using different global optimizers on the server with client-level DP. Then, we demonstrate how our adaptive intermediary strategy benefits privacy-performance trade-offs. We consider four popular private server optimizers: DP-FedAvg [24] which adds client-level privacy protection to FedAvg [23], DP-FedAdam which is a differentially private version of the optimizer FedAdam [28], DP-FedNova which we equip the global solver FedNova [31] for client-level DP, and DP²-RMSProp [17] which is a very recent private optimization framework and we deploy it as the global optimizer in FL.

We perform validation with different noise multiplier values. Non-private FL is also provided as a performance upper bound. Note that our method has the same performance ascompared methods in non-private settings, because there are no noises to harmonize. As can be observed from Table 1, severe performance degradation occurs in the private cross-silo FL setting, especially for high-privacy regimes (e.g., $z = 0.7$ for prostate segmentation). There are no significant differences among different global optimizers, which shows that the optimizers carefully designed for non-private FL are unable to address the noisy gradient issue in DP settings. However, our method relieves the gradient corruption and consistently and substantially boosts performance even with large noises (e.g., 44.55% Dice boost on prostate segmentation with $z = 0.7$). We also identify that the influences on performance introduced by DP may vary across different tasks and client numbers. For example, the segmentation task with fewer clients is more seriously damaged compared with the classification task with more clients.

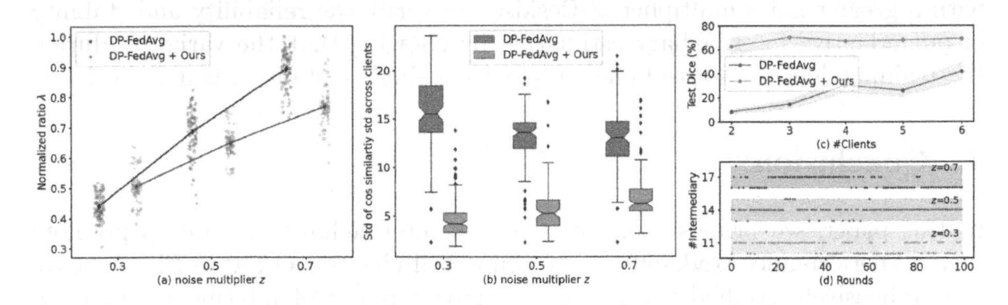

Fig. 2. Analytical studies on prostate segmentation. (a) The distribution of normalized ratio λ across communication rounds under different privacy levels. (b) The std of cosine similarities between Δ_i and the aggregated gradients in each round under different privacy levels. (c) Performance with different client numbers. (d) Intermediary variations across training rounds under different privacy regimes.

3.3 Analytical Studies

Effects of Optimizing Privacy-Performance Trade-Offs. We present the dynamic behavior of our method regarding variations of the intermediary ratio λ across different rounds in Fig. 2 (a). Compared with DP-FedAvg [24], where λ shows a significant increase with the rise of noise multiplier z, our method harmonizes this trend with more centralized distributions by the adaptive intermediary for better privacy-performance trade-offs. In Fig. 2 (b), we also study the standard deviation of similarities, which is another metric for quantifying gradient diversity between local and global gradients. Our method shows more stable optimization directions with less variance among clients. Moreover, we observe a decline in gradient diversities as the privacy regime rises for DP-FedAvg [24]. To interpret, we speculate that local optimization may be dominated by greater noises for more common gradient de-corruption.

Client Scalability Analysis. As the noise level is highly dependent on client numbers (see Eq. (1) and Table 1), we investigate the scalability of DP-FedAvg [24] and our method by varying number of clients. Figure 2 (c) presents the results on prostate segmentation with different training clients ($z = 0.3$). Notably, we keep test data unchanged for fair comparisons. We observe a dramatic drop in performance of DP-FedAvg [24] due to excessive noise when the number of clients shrinks. However, our method performs stably even under extreme conditions, e.g., the federation only has two participants.

Stability of Adaptive Intermediary Estimation. Finally, we analyze the historical variation of our adaptive intermediary strategy in Fig. 2 (d), where we present the intermediary numbers during the training progress. We expect that more intermediaries are required to balance the privacy-performance trade-off with a greater noise multiplier z. Besides, we verify the reliability and stability of our adaptive intermediary estimation by showing that the variation during the training does not exceed one, except for a single instance when $z = 0.7$.

4 Conclusion

In this paper, we propose a novel adaptive intermediary method to promote privacy-performance trade-offs in the context of client-level DP in FL. We have comprehensively studied the relations among number of intermediaries, noise levels and training diversities in our work. We also investigate relations between sample-level and client-level DP. Our proposed method outperforms compared methods on both medical image diagnosis and segmentation tasks and shows good compatibility with existing DP optimizers. For future work, it is promising to investigate our method for clients with imbalanced class distributions, where the intermediary may not have all labels.

Acknowledgement. This work was supported in part by Shenzhen Portion of Shenzhen-Hong Kong Science and Technology Innovation Cooperation Zone under HZQB-KCZYB-20200089, in part by National Natural Science Foundation of China (Project No. 62201485), in part by Hong Kong Innovation and Technology Commission Project No. ITS/238/21, in part by Science, Technology and Innovation Commission of Shenzhen Municipality Project No. SGDX20220530111201008, in part by Hong Kong Research Grants Council Project No. T45-401/22-N, and in part by NSERC Discovery Grant (DGECR-2022-00430).

References

1. Abadi, M., et al.: Deep learning with differential privacy. In: Proceedings of the 2016 Conference on Computer and Communications Security. ACM (2016)
2. Adnan, M., Kalra, S., Cresswell, J.C., et al.: Federated learning and differential privacy for medical image analysis. Sci. Rep. **12**(1), 1953 (2022)
3. Amid, E., et al.: Public data-assisted mirror descent for private model training. In: ICML, pp. 517–535. PMLR (2022)
4. Andrew, G., Thakkar, O., McMahan, B., Ramaswamy, S.: Differentially private learning with adaptive clipping. NeurIPS **34**, 17455–17466 (2021)
5. Dayan, I., et al.: Federated learning for predicting clinical outcomes in patients with COVID-19. Nat. Med. **27**(10), 1735–1743 (2021)
6. De, S., Berrada, L., Hayes, J., et al.: Unlocking high-accuracy differentially private image classification through scale. arXiv preprint arXiv:2204.13650 (2022)
7. Dwork, C., McSherry, F., Nissim, K., Smith, A.: Calibrating noise to sensitivity in private data analysis. In: Halevi, S., Rabin, T. (eds.) TCC 2006. LNCS, vol. 3876, pp. 265–284. Springer, Heidelberg (2006). https://doi.org/10.1007/11681878_14
8. Dwork, C., Roth, A., et al.: The algorithmic foundations of differential privacy. Found. Trends Theoret. Comput. Sci. **9**, 211–407 (2014)
9. Flanders, A.E., et al.: Construction of a machine learning dataset through collaboration: the RSNA 2019 brain CT hemorrhage challenge. Radiol. Artif. Intell. **2**(3), e190211 (2020)
10. Gopi, S., Lee, Y.T., Wutschitz, L.: Numerical composition of differential privacy. NeurIPS **34**, 11631–11642 (2021)
11. Huang, G., Liu, Z., Van Der Maaten, L., Weinberger, K.Q.: Densely connected convolutional networks. In: CVPR, pp. 4700–4708 (2017)
12. Kairouz, P., et al.: Advances and open problems in federated learning. Found. Trends Mach. Learn. **14**(1–2), 1–210 (2021)
13. Kaissis, G., et al.: End-to-end privacy preserving deep learning on multi-institutional medical imaging. Nat. Mach. Intell. **3**(6), 473–484 (2021)
14. Kim, M., Günlü, O., et al.: Federated learning with local differential privacy: trade-offs between privacy, utility, and communication. In: IEEE International Conference on Acoustics, Speech and Signal Processing, pp. 2650–2654 (2021)
15. Kyung, S., et al.: Improved performance and robustness of multi-task representation learning with consistency loss between pretexts for intracranial hemorrhage identification in head CT. Med. Image Anal. **81**, 102489 (2022)
16. Lemaître, G., Martí, R., et al.: Computer-aided detection and diagnosis for prostate cancer based on mono and multi-parametric MRI: a review. Comput. Biol. Med. **60**, 8–31 (2015)
17. Li, T., et al.: Differentially private adaptive optimization with delayed preconditioners. In: ICLR (2023)
18. Li, T., Zaheer, M., Reddi, S., Smith, V.: Private adaptive optimization with side information. In: ICML, pp. 13086–13105. PMLR (2022)
19. Litjens, G., et al.: Evaluation of prostate segmentation algorithms for MRI: the promise12 challenge. Med. Image Anal. **18**(2), 359–373 (2014)
20. Liu, K., Hu, S., Wu, S., Smith, V.: On privacy and personalization in cross-silo federated learning. In: NeurIPS (2022)
21. Liu, Q., Dou, Q., Yu, L., Heng, P.A.: MS-Net: multi-site network for improving prostate segmentation with heterogeneous MRI data. IEEE TMI **39**(9), 2713–2724 (2020)

22. Liu, X., Glocker, B., McCradden, M.M., Ghassemi, M., Denniston, A.K., Oakden-Rayner, L.: the medical algorithmic audit. Lancet Digital Health **4**(5), e384–e397 (2022)
23. McMahan, B., et al.: Communication-efficient learning of deep networks from decentralized data. In: AISTATS, pp. 1273–1282 (2017)
24. McMahan, H.B., Ramage, D., Talwar, K., Zhang, L.: Learning differentially private recurrent language models. In: ICLR (2018)
25. Mironov, I.: Rényi differential privacy. In: 2017 IEEE 30th Computer Security Foundations Symposium (CSF), pp. 263–275. IEEE (2017)
26. Nicholas, B., Anant, M., et al.: NCI-Proceedings of the IEEE-ISBI conference 2013 challenge: automated segmentation of prostate structures. The Cancer Imaging Archive (2015)
27. Papernot, N., et al.: Tempered sigmoid activations for deep learning with differential privacy. In: AAAI, vol. 35, pp. 9312–9321 (2021)
28. Reddi, S.J., et al.: Adaptive federated optimization. In: ICLR (2021)
29. Rieke, N., et al.: The future of digital health with federated learning. NPJ Digit. Med. **3**(1), 1–7 (2020)
30. Ronneberger, O., Fischer, P., Brox, T.: U-Net: convolutional networks for biomedical image segmentation. In: Navab, N., Hornegger, J., Wells, W.M., Frangi, A.F. (eds.) MICCAI 2015. LNCS, vol. 9351, pp. 234–241. Springer, Cham (2015). https://doi.org/10.1007/978-3-319-24574-4_28
31. Wang, J., et al.: Tackling the objective inconsistency problem in heterogeneous federated optimization. In: NeurIPS (2020)
32. Yousefpour, A., et al.: Opacus: user-friendly differential privacy library in PyTorch. arXiv preprint arXiv:2109.12298 (2021)
33. Zheng, Q., Chen, S., Long, Q., Su, W.: Federated f-differential privacy. In: AISTATS, pp. 2251–2259. PMLR (2021)
34. Ziller, A., et al.: Differentially private federated deep learning for multi-site medical image segmentation. arXiv preprint arXiv:2107.02586 (2021)

Inflated 3D Convolution-Transformer for Weakly-Supervised Carotid Stenosis Grading with Ultrasound Videos

Xinrui Zhou[1,2,3], Yuhao Huang[1,2,3], Wufeng Xue[1,2,3], Xin Yang[1,2,3], Yuxin Zou[1,2,3], Qilong Ying[1,2,3], Yuanji Zhang[1,2,3,4], Jia Liu[5], Jie Ren[5], and Dong Ni[1,2,3(✉)]

[1] National-Regional Key Technology Engineering Laboratory for Medical Ultrasound, School of Biomedical Engineering, Health Science Center, Shenzhen University, Shenzhen, China
nidong@szu.edu.cn

[2] Medical Ultrasound Image Computing (MUSIC) Lab, Shenzhen University, Shenzhen, China

[3] Marshall Laboratory of Biomedical Engineering, Shenzhen University, Shenzhen, China

[4] Shenzhen RayShape Medical Technology Co. Ltd., Shenzhen, China

[5] The Third Affiliated Hospital, Sun Yat-sen University, Guangzhou, China

Abstract. Localization of the narrowest position of the vessel and corresponding vessel and remnant vessel delineation in carotid ultrasound (US) are essential for carotid stenosis grading (CSG) in clinical practice. However, the pipeline is time-consuming and tough due to the ambiguous boundaries of plaque and temporal variation. To automatize this procedure, a large number of manual delineations are usually required, which is not only laborious but also not reliable given the annotation difficulty. In this study, we present the first video classification framework for automatic CSG. Our contribution is three-fold. First, to avoid the requirement of laborious and unreliable annotation, we propose a novel and effective video classification network for weakly-supervised CSG. Second, to ease the model training, we adopt an inflation strategy for the network, where pre-trained 2D convolution weights can be adapted into the 3D counterpart in our network for an effective warm start. Third, to enhance the feature discrimination of the video, we propose a novel attention-guided multi-dimension fusion (AMDF) transformer encoder to model and integrate global dependencies within and across spatial and temporal dimensions, where two lightweight cross-dimensional attention mechanisms are designed. Our approach is extensively validated on a large clinically collected carotid US video dataset, demonstrating state-of-the-art performance compared with strong competitors.

X. Zhou and Y. Huang—Contribute equally to this work.

Supplementary Information The online version contains supplementary material available at https://doi.org/10.1007/978-3-031-43895-0_48.

H. Greenspan et al. (Eds.): MICCAI 2023, LNCS 14221, pp. 511–520, 2023.
https://doi.org/10.1007/978-3-031-43895-0_48

Keywords: Ultrasound video · Carotid stenosis grading · Classification

1 Introduction

Carotid stenosis grading (CSG) represents the severity of carotid atherosclerosis, which is highly related to stroke risk [11]. In clinical practice, sonographers need to first visually locate the frame with the largest degree of vascular stenosis (i.e., minimal area of remnant vessels) in a dynamic plaque video clip based on B-mode ultrasound (US), then manually delineate the contours of both vessels and remnant vessels on it to perform CSG. However, the two-stage pipeline is time-consuming and the diagnostic results heavily rely on operator experience and expertise due to ambiguous plaque boundaries and temporal variation (see Fig. 1). Fully-supervised segmentation models can automatize this procedure, but require numerous pixel-level masks laboriously annotated by sonographers and face the risk of training failure due to unreliable annotation. Hence, tackling this task via weak supervision, i.e., video classification, is desired to avoid the requirement of tedious and unreliable annotation.

Achieving accurate automatic CSG with US videos is challenging. First, the plaque clips often have extremely high intra-class variation due to changeable plaque echo intensity, shapes, sizes, and positions (Fig. 1(d-e)). The second challenge lies in the inter-class similarity of important measurement indicators (i.e., diameter and area stenosis rate) for CSG among cases with borderlines of mild and severe, which makes designing automatic algorithms difficult (Fig. 1(c-d)).

A typical approach for this video classification task is CNN-LSTM [7]. Whereas, such *2D + 1D* paradigm lacks interaction with temporal semantics

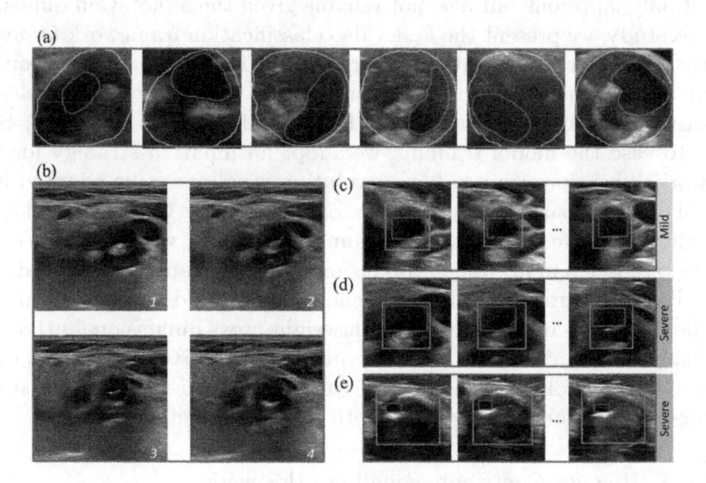

Fig. 1. (a) Carotid plaque images in US with annotated vessel (green contour) and remnant vessel (red contour). (b) An original video with vessels and plaques dynamically displayed. (c-e): Cropped plaque clips with annotated vessel (green box), remnant vessel (red box), and corresponding label. (Color figure online)

of input frames in the early stage. Instead, a more efficient way is to build 3D networks that handle spatial and temporal (ST) information simultaneously [21].

There are several types of 3D networks that have been widely used in visual tasks: (1) Pure 3D convolution neural networks (3D CNNs) refer to capturing local ST features using convolution operations [4,10,21]. However, most current 3D CNNs suffer from the lack of good initialization and capacity for extracting global representations [16]. (2) Pure 3D transformer networks (3D Trans) aim to exploit global ST features by applying self-attention mechanisms [3,9,15, 20]. However, their ability in extracting local ST information is weaker than 3D CNNs. Moreover, such designs have not deeply explored lightweight cross-dimensional attention mechanisms to gain refined fused features for classification.

Recently, Wang et al. [18] first introduced the self-attention mechanism in 3D CNN for video classification. [16,19] then proposed Convolution-Transformer hybrid networks for image classification. Li et al. [12] further extended such hybrid design to 3D for video recognition by seamlessly integrating 3D convolution and self-attention. Thanks to both operations, such networks can fully exploit and integrate local and global features, and thus achieve state-of-the-art results. However, current limited 3D hybrid frameworks are designed in cascade, which may lead to semantic misalignment between CNN- and Transformer-style features and thus degrade accuracy of video classification.

In this study, we present the first video classification framework based on 3D Convolution-Transformer design for CSG (named CSG-3DCT). Our contribution is three-fold. First, we propose a novel and effective video classification network for weakly-supervised CSG, which can avoid the need of laborious and unreliable mask annotation. Second, we adopt an inflation strategy to ease the model training, where pre-trained 2D convolution weights can be adapted into the 3D counterpart. In this case, our network can implicitly gain the pre-trained weights of existing large models to achieve an effective warm start. Third, we propose a novel play-and-plug attention-guided multi-dimension fusion (AMDF) transformer encoder to integrate global dependencies within and across ST dimensions. Two lightweight cross-dimensional attention mechanisms are devised in AMDF to model ST interactions, which merely use class (CLS) token [8] as Query. Extensive experiments show that CSG-3DCT achieve state-of-the-art performance in CSG task.

2 Methodology

Figure 2(a) shows the pipeline of our proposed framework. Note that the proposed CSG-3DCT is inflated from the 2D architecture. Thus, it can implicitly gain the pre-trained weights of current large model for effective initialization. In CSG-3DCT, given a video clip, vessel regions are first detected by the pre-trained detection model [14] for reducing redundant background information. Then, the cropped regions are concatenated to form a volumetric vessel and input to the 3D CNN and Transformer encoders. Specifically, to better model the global knowledge, ST features are decoupled and fused by the proposed AMDF transformer

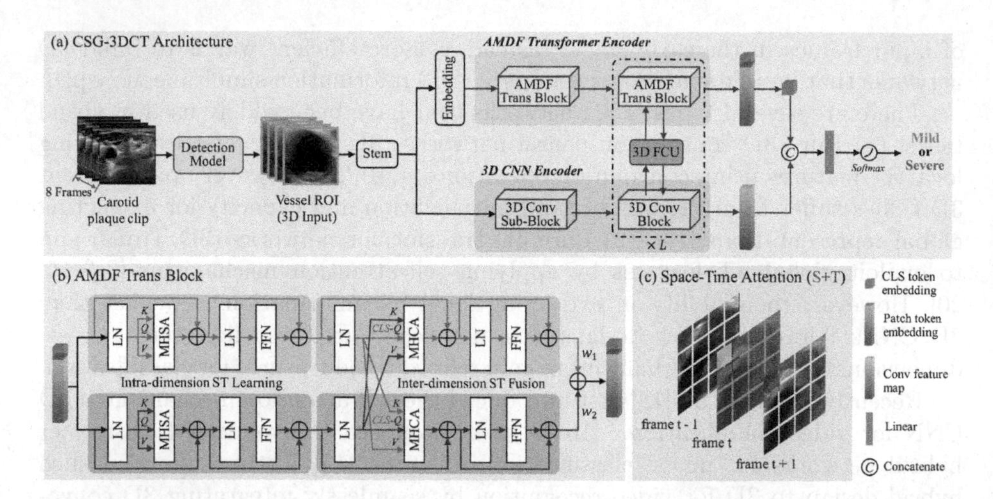

Fig. 2. (a) Overview of CSG-3DCT. It contains *L+1* and *L* repeated AMDF Trans and 3D Conv Blocks, respectively. A 3D Conv Block consists of two sub-blocks. (b) Pipeline of the AMDF Trans Block. (c) Visualization of the space-time attention used in the intra-dimension ST learning module. Yellow, blue and red patches indicate query, and attention separately adopted along temporal and spatial dimensions, respectively. (Color figure online)

encoder. CNN- and Transformer-style features are integrated by the 3D feature coupling unit (3D FCU) [16] orderly. Finally, by combining the CNN features and the CLS token, the model will output the label prediction.

3D Mix-Architecture for Video Classification. CNN and Transformer have been validated that they specialize in extracting local and global features, respectively. Besides, compared to the traditional 2D video classifiers, 3D systems have shown the potential to improve classification accuracy due to their powerful capacity of encoding multi-dimensional information. Thus, in CSG-3DCT, we propose to leverage the advantages of both CNN and Transformer and extend the whole framework to a 3D version.

The meta-architecture of our proposed CSG-3DCT follows the 2D Convolution-Transformer (Conformer) model [16]. It mainly has 5 stages (termed *c1*-*c5*). Extending it to 3D represents that both CNN and Transformer should be modified to adapt the 3D input. In specific, we tend to inflate the 2D $k \times k$ convolution kernels to 3D ones with the size of $t \times k^2$ by adding a temporal dimension, which is similar to [4]. Such kernels can be implicitly pre-trained on ImageNet through bootstrapping operation [4]. While translating the 2D transformer only requires adjusting the token number according to the input dimension.

Inflation Strategy for 3D CNN Encoder. We devise an inflation strategy for the 3D CNN encoder to relieve the model training and enhance the representation ability. For achieving 2D-to-3D inflation, a feasible scheme is to expand all the 2D convolution kernels at temporal dimension with *t>1* [4]. However, multi-temporal (*t>1*) 3D convolutions are computationally complex and hard to train. Thus, we only select part of the convolution kernels for inflating their temporal dimension larger than 1, while others restrict the temporal dimension to 1. By adapting pre-trained 2D convolution weights into the 3D counterpart, our network can achieve good initialization from existing large model. Moreover, we notice that performing convolutions at a temporal level in early layers may degrade accuracy due to the over-neglect of spatial learning [10]. Therefore, instead of taking a whole-stage temporal convolution, we only perform it on 3D Conv blocks of the last three stages (i.e., *c3-c5*). We highlight that our temporal convolutions are length-invariant, which indicates that we will not down-sample at the temporal dimension. It can benefit the maintenance of both video fidelity and time-series knowledge, especially for short videos. See supplementary material for more details.

Transformer Encoder with Play-and-plug AMDF Design. Simply translating the 2D transformer encoder into the 3D standard version mainly has two limitations: (1) It blindly compares the similarity of all ST tokens by self-attention, which tends to inaccurate predictions. Moreover, such video-based computation handles $t\times$ tokens simultaneously compared to image-based methods, leading to much computational cost. (2) It also has no ability to decide which information is more important during different learning stages. Thus, we propose to enhance the decoupled ST features and their interactions using different attention manners. The proposed encoder can improve computational efficiency, and can be flexibly integrated into 2D or 3D transformer-based networks.

Before the transformer encoder, we first decompose the feature maps X produced by the stem module into $t \times n^2$ embeddings without overlap. A CLS token $X_{cls} \in \mathbb{R}^d$ is then added in the start position of X to obtain merged embeddings $Z \in \mathbb{R}^{d\times(t\times n^2+1)}$. n^2 and d denote the number of spatial patch tokens and hidden dimensions, respectively. Then, the multiple AMDF Trans blocks in the transformer encoder drive Z to produce multi-dimensional enhanced representations. Specifically, the AMDF block has the following main components.

1) Intra-dimension ST Learning Module. Different from the cascade structure in [3], CSG-3DCT constructs two parallel branches to learn global ST features, respectively. As shown in Fig. 2(b), the proposed module is following ViT [8], which consists of a multi-head self-attention (MHSA) module and a feed-forward network (FFN). Query-Key-Value (QKV) projection after LayerNorms [2] is conducted before each MHSA module. Besides, the residual connections are performed in MHSA module and FFN. Taking token embeddings as input, the two branches can extract the ST features well by parallel spatial and temporal attention (see Fig. 2(c) for visualization of computation process).

2) Inter-dimension ST Fusion Module. To boost interactions between S and T dimensions, we build the inter-dimension fusion module after the intra-dimension learning module. The only difference between the two types of modules is the calculation mode of attention. As shown in Fig. 3, we consider the following

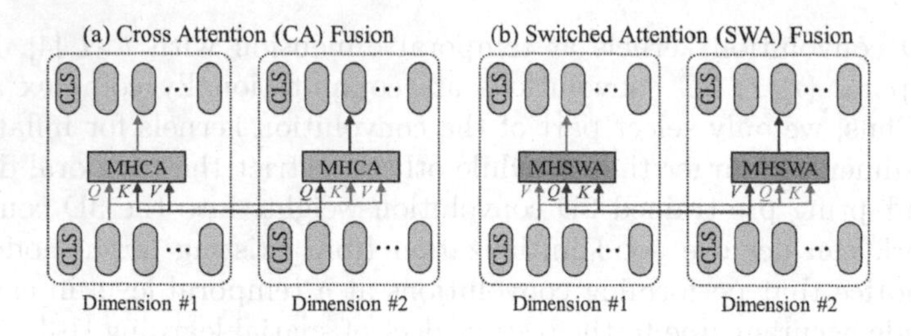

Fig. 3. An illustration of our proposed multi-dimension fusion methods.

two methods to interact the ST features: (i) *Switched Attention (SWA) Fusion* and (ii) *Cross Attention (CA) Fusion*. Here, we define one branch as the target dimension and the other branch as the complementary dimension. For example, when the temporal features are flowing to the spatial features, the spatial branch is the target and the temporal branch is the complementary one. SWA is an intuitive way for information interaction. It uses the attention weights (i.e., generated by **Q** and **K**) as the bridge to directly swap the information. For computing the target features in CA, **K** and **V** are from the complementary one, and **Q** is from its own. Intuition behind CA is that the target branch can *query* the useful information from the given **K** and **V** [6]. Thus, the *querying* process in CA can better encourage the knowledge flowing.

To improve computing efficiency in CA, Chen et al. [5] proposed to adopt the CLS token of the target branch to compute the CLS-**Q** to replace the common **Q** from token embeddings. Then, they transferred the target CLS token to the complementary branch to obtain the **K** and **V** and perform CA. However, such a design may lead to overfitting due to the query-queried feature dependency. Motivated by [5], we introduce a simple yet efficient attention strategy in inter-dimension ST fusion module. Specifically, the target dimension adopts its CLS token as a *query* to mine rich information, and this CLS token will not be inserted into the complementary dimension. Besides, using one token only can reduce the computation time quadratically compared to all tokens attention.

3) Learnable Mechanism for Adaptive Updating. Multi-dimensional features commonly have distinct degrees of contribution for prediction. For example, supposing the size of carotid plaque does not vary significantly in a dynamic segment, the spatial information may play a dominant role in making the final diagnosis. Thus, we introduce a learnable parameter to make the network adaptively adjust the weights of different branches and learn the more important features (see Fig. 2(b)). We highlight that this idea is easy to implement and general to be equipped with any existing feature-fusion modules.

3 Experimental Results

Dataset and Implementations. We validated the CSG-3DCT on a large in-house carotid transverse US video dataset. Approved by the local IRB, a total

Table 1. Quantitative results of methods. "MTV(B/2+S/8)" means to use the larger "B" model to encode shorter temporal information (2 frames), and the smaller "S" model to encode longer temporal information (8 frames) [20]. † denotes random initialization. ∗ indicates removing the learnable mechanism from AMDF encoder.

Methods	Accuracy	F1-score	Precision	Recall
I3D [4]	78.8%	78.8%	81.0%	80.8%
SlowFast [10]	70.3%	70.2%	70.3%	70.8%
TPN [21]	78.0%	77.8%	77.8%	78.5%
TimeSformer [3]	77.1%	76.2%	76.8%	75.9%
Vivit [1]	70.3%	70.2%	70.3%	70.8%
MTV (B/2+S/8) [20]	72.0%	72.0%	73.0%	73.4%
NL I3D [18]	78.8%	78.8%	81.0%	80.8%
UniFormer [12]	75.4%	75.4%	78.1%	77.6%
CSG-3DCT-Base	80.5%	80.2%	80.0%	80.4%
CSG-3DCT-Base†	76.3%	75.4%	75.8%	75.2%
CSG-3DCT-Base-16	79.8%	78.0%	79.3%	77.4%
CSG-3DCT-SWA*	82.2%	81.4%	82.5%	80.9%
CSG-3DCT-CA*	82.2%	82.1%	82.2%	83.0%
CSG-3DCT	83.1%	82.5%	82.8%	82.4%

of 200 videos (63225 images with size 560×560 and 380×380) were collected from 169 patients with carotid plaque. In clinic, sonographers often focus on a relatively narrow short plaque video clip instead of the long video. Thus, we remade the dataset by using the key plaque video clips instead of original long videos. Specifically, sonographers with 7-year experience manually annotated 8/16 frames for a plaque clip and labeled the corresponding stenosis grading (mild/severe) using the Pair annotation software package [13]. The final dataset was split randomly into 318, 23, and 118 plaque clips with 8 frames or into 278, 23, and 109 ones with 16 frames for training, validation, and independent testing set at the patient level with no overlap.

In this study, we implemented CSG-3DCT in *Pytorch*, using an NVIDIA A40 GPU. Unless specified, we trained our model using 8-frame input plaque clips. All frames were resized to 256 × 256. The learnable weights of QKV projection and LayerNorm weights in spatial dimension branch of intra-dimension ST learning module were initialized with those from transformer branch in Conformer [16], while other parameters in AMDF transformer encoder performed random initialization. We trained CSG-3DCT using Adam optimizer with the learning rate (*lr*) of 1e-4 and weight decay of 1e-4 for 100 epochs. Batch size was set as 4. Inspired by [18], CSG-3DCT with 16-frame inputs was initialized with 8-frame model and fine-tuned using an initial *lr* of 0.0025 for 40 epochs.

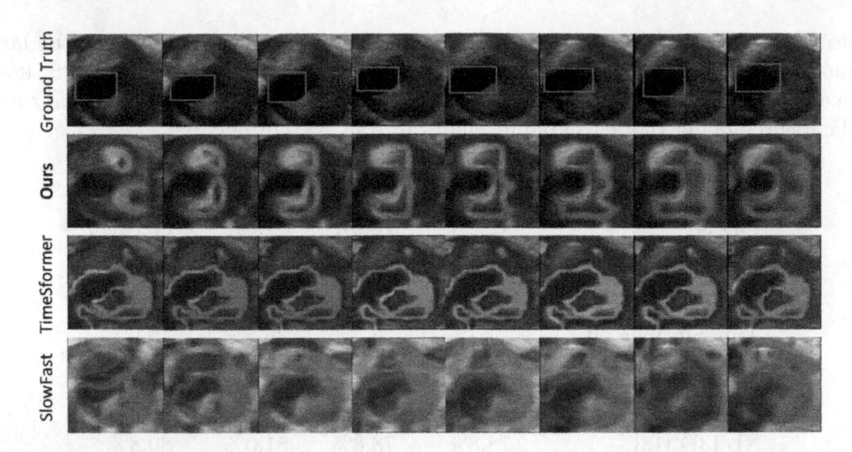

Fig. 4. Attention maps of one carotid severe stenosis testing case (shown in cropped volumetric vessel). Red box denotes remnant vessel annotated by sonographers. (Color figure online)

Quantitative and Qualitative Analysis. We conducted extensive experiments to evaluate CSG-3DCT. Accuracy, F1-score, precision and recall were evaluation metrics. Table 1 compares CSG-3DCT with other 8 strong competitors, including 3D CNNs, 3D Trans, and 3D Mix-architecture. Note that "-Base" is directly inflated from Conformer [16]. Among all the competitors, -Base achieves the best results on accuracy and f1-score. It can also be observed that our proposed CSG-3DCT achieves state-of-the-art results (at least **4.3%** improvement in accuracy).

Figure 4 visualizes feature maps of different typical networks using Grad-CAM [17]. We use models without temporal downsampling (i.e., TimeSformer [3] and the "fast" branch of SlowFast [10]) to observe attention changes along temporal dimension. Both models ignore capturing equally important local and global ST features simultaneously, resulting in imprecise and coarse attention to the key object, i.e., the plaque area. Compared to both cases, CSG-3DCT can progressively learn the ST contexts in an interactive fashion. As a result, the attention area is more accurate and complete, indicating the stronger discriminative ability of the learned features by CSG-3DCT, which proves the efficacy of our framework.

Ablation Study. We performed ablation experiments in the last 6 rows of Table 1. "-SWA*" uses SWA in AMDF transformer encoder, while "-CA*" uses CA instead. "-Base-16" denotes our "-Base" model with 16-frame inputs.

1) Effects of Different Key Components of Our Model Design. We compared CSG-3DCT with three variants (i.e., -Base, -SWA*, and -CA*) to analyze the effects of different key components. Compared with -Base, each of our proposed modules and their combination can help improve the accuracy. We adopt CA in our final model for its good performance.

2) Effects of Plaque Clip Length. We only investigated the effects of our model on 8-frame and 16-frame input clips due to limited GPU memory. We can find in Table 1 that longer input clips slightly degrade the performance. This is reasonable since the frame-extracting method has been applied in the original videos, causing the covered range of plaque from a longer plaque clip is relatively wider, which is not beneficial to stenosis grading.

3) Effectiveness of Initialization with ImageNet. We evaluated the value of training models starting from ImageNet-pretrained weights compared with scratch. It can be seen in Table 1 that model with pretraining significantly boosts +**4.2**% Acc., demonstrating the efficacy of good initialization.

4 Conclusion

We propose a novel and effective video classification network for automatic weakly-supervised CSG. To the best of our knowledge, this is the first work to tackle this task. By adopting an inflation strategy, our network can achieve effective warm start and make more accurate predictions. Moreover, we develop a novel AMDF Transformer encoder to enhance the feature discrimination of the video with reduced computational complexity. Experiments on our large in-house dataset demonstrate the superiority of our method. In the future, we will explore to validate the generalization capability of CSG-3DCT on more large datasets and extend two-grade classification to four-grade of carotid stenosis.

Acknowledgements. This work was supported by the grant from National Natural Science Foundation of China (Nos. 62171290, 62101343), Shenzhen-Hong Kong Joint Research Program (No. SGDX20201103095613036), Shenzhen Science and Technology Innovations Committee (No. 20200812143441001), Shenzhen College Stable Support Plan (Nos. 20220810145705001, 20200812162245001), and National Natural Science Foundation of China (No 81971632).

References

1. Arnab, A., Dehghani, M., Heigold, G., Sun, C., Lučić, M., Schmid, C.: ViViT: a video vision transformer. In: Proceedings of the IEEE/CVF International Conference On Computer Vision, pp. 6836–6846 (2021)
2. Ba, J.L., Kiros, J.R., Hinton, G.E.: Layer normalization. arXiv preprint arXiv:1607.06450 (2016)
3. Bertasius, G., Wang, H., Torresani, L.: Is space-time attention all you need for video understanding? In: ICML, vol. 2, p. 4 (2021)
4. Carreira, J., Zisserman, A.: Quo vadis, action recognition? A new model and the kinetics dataset. In: proceedings of the IEEE Conference on Computer Vision and Pattern Recognition, pp. 6299–6308 (2017)
5. Chen, C.F.R., Fan, Q., Panda, R.: CrossViT: cross-attention multi-scale vision transformer for image classification. In: Proceedings of the IEEE/CVF International Conference on Computer Vision, pp. 357–366 (2021)

6. Curto, D., et al.: Dyadformer: a multi-modal transformer for long-range modeling of dyadic interactions. In: Proceedings of the IEEE/CVF International Conference on Computer Vision, pp. 2177–2188 (2021)
7. Donahue, J., et al.: Long-term recurrent convolutional networks for visual recognition and description. In: Proceedings of the IEEE Conference on Computer Vision and Pattern Recognition, pp. 2625–2634 (2015)
8. Dosovitskiy, A., et al.: An image is worth 16x16 words: transformers for image recognition at scale. In: International Conference on Learning Representations (2021)
9. Fan, H., et al.: Multiscale vision transformers. In: Proceedings of the IEEE/CVF International Conference on Computer Vision, pp. 6824–6835 (2021)
10. Feichtenhofer, C., Fan, H., Malik, J., He, K.: SlowFast networks for video recognition. In: Proceedings of the IEEE/CVF International Conference on Computer Vision, pp. 6202–6211 (2019)
11. Howard, D.P., Gaziano, L., Rothwell, P.M.: Risk of stroke in relation to degree of asymptomatic carotid stenosis: a population-based cohort study, systematic review, and meta-analysis. Lancet Neurol. **20**(3), 193–202 (2021)
12. Li, K., et al.: UniFormer: unified transformer for efficient spatial-temporal representation learning. In: International Conference on Learning Representations (2022)
13. Liang, J., et al.: Sketch guided and progressive growing GAN for realistic and editable ultrasound image synthesis. Med. Image Anal. **79**, 102461 (2022)
14. Liu, J., et al.: Deep learning based on carotid transverse B-mode scan videos for the diagnosis of carotid plaque: a prospective multicenter study. Eur. Radiol. **32**, 1–10 (2022)
15. Liu, Z., et al.: Video swin transformer. In: Proceedings of the IEEE/CVF Conference on Computer Vision and Pattern Recognition, pp. 3202–3211 (2022)
16. Peng, Z., et al.: Conformer: local features coupling global representations for visual recognition. In: Proceedings of the IEEE/CVF International Conference on Computer Vision, pp. 367–376 (2021)
17. Selvaraju, R.R., Cogswell, M., Das, A., Vedantam, R., Parikh, D., Batra, D.: Grad-CAM: visual explanations from deep networks via gradient-based localization. In: Proceedings of the IEEE International Conference on Computer Vision, pp. 618–626 (2017)
18. Wang, X., Girshick, R., Gupta, A., He, K.: Non-local neural networks. In: Proceedings of the IEEE Conference on Computer Vision and Pattern Recognition, pp. 7794–7803 (2018)
19. Xu, Y., Zhang, Q., Zhang, J., Tao, D.: ViTAE: vision transformer advanced by exploring intrinsic inductive bias. Adv. Neural. Inf. Process. Syst. **34**, 28522–28535 (2021)
20. Yan, S., et al.: Multiview transformers for video recognition. In: Proceedings of the IEEE/CVF Conference on Computer Vision and Pattern Recognition, pp. 3333–3343 (2022)
21. Yang, C., Xu, Y., Shi, J., Dai, B., Zhou, B.: Temporal pyramid network for action recognition. In: Proceedings of the IEEE/CVF Conference on Computer Vision and Pattern Recognition, pp. 591–600 (2020)

One-Shot Federated Learning on Medical Data Using Knowledge Distillation with Image Synthesis and Client Model Adaptation

Myeongkyun Kang[1,3], Philip Chikontwe[1], Soopil Kim[1,3], Kyong Hwan Jin[2], Ehsan Adeli[3], Kilian M. Pohl[3], and Sang Hyun Park[1(✉)]

[1] Robotics and Mechatronics Engineering, Daegu Gyeongbuk Institute of Science and Technology (DGIST), Daegu, Korea
{mkkang,shpark13135}@dgist.ac.kr

[2] Electrical Engineering and Computer Science, Daegu Gyeongbuk Institute of Science and Technology (DGIST), Daegu, Korea

[3] Stanford University, Stanford, CA 94305, USA

Abstract. One-shot federated learning (FL) has emerged as a promising solution in scenarios where multiple communication rounds are not practical. Notably, as feature distributions in medical data are less discriminative than those of natural images, robust global model training with FL is non-trivial and can lead to overfitting. To address this issue, we propose a novel one-shot FL framework leveraging Image Synthesis and Client model Adaptation (FedISCA) with knowledge distillation (KD). To prevent overfitting, we generate diverse synthetic images ranging from random noise to realistic images. This approach (i) alleviates data privacy concerns and (ii) facilitates robust global model training using KD with decentralized client models. To mitigate domain disparity in the early stages of synthesis, we design noise-adapted client models where batch normalization statistics on random noise (synthetic images) are updated to enhance KD. Lastly, the global model is trained with both the original and noise-adapted client models via KD and synthetic images. This process is repeated till global model convergence. Extensive evaluation of this design on five small- and three large-scale medical image classification datasets reveals superior accuracy over prior methods. Code is available at https://github.com/myeongkyunkang/FedISCA.

Keywords: One-Shot Federated Learning · Knowledge Distillation · Noise · Image Synthesis · Client Model Adaptation

1 Introduction

One-shot federated learning (FL) allows a global model to be trained through a single communication round without sharing data between clients [6,8,15,33,35].

Supplementary Information The online version contains supplementary material available at https://doi.org/10.1007/978-3-031-43895-0_49.

H. Greenspan et al. (Eds.): MICCAI 2023, LNCS 14221, pp. 521–531, 2023.
https://doi.org/10.1007/978-3-031-43895-0_49

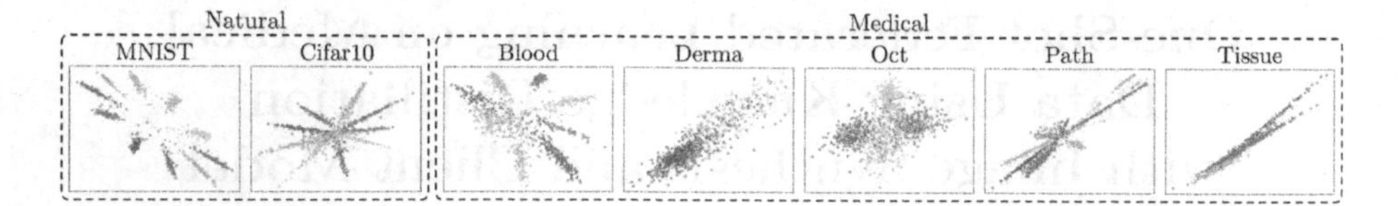

Fig. 1. Feature visualization on natural (MNIST and Cifar10) and medical (Blood, Derma, Oct, Path, and Tissue) images. For visualization, we placed a bottleneck layer before the class prediction layer, reducing the feature dimension to 2. Each color represents a classification label. Notably, the feature distribution in medical data is more complex.

This approach significantly reduces the risk of attack and communication costs compared to FL [21] and allows for decentralized training under extreme conditions. For instance, one-shot FL has emerged as a viable solution for reducing significant transmission costs in scenarios where patient data is only accessible within an isolated network requiring in-person transfer of client models. Since one-shot FL can only access clients' models once during training, recent one-shot FL suggests generating images and using them to transfer knowledge from multiple client models for global model training using knowledge distillation (KD) [33]. However, the lack of diversity in the generated images often leads to overfitting, posing a significant challenge for one-shot FL. To address this issue, [22,33] propose to enhance the transferability of client models by generating diverse natural images near the decision boundary. Compared to natural images, the decision boundaries in medical data are often more complex (*e.g.*, less discriminative as shown in Fig. 1), which limits the applicability of existing one-shot FL approaches to this application. Note, while the challenges in medical data and client heterogeneity can be mitigated through multiple communication rounds [12,18,23,36], the one-shot scenario presents a unique difficulty. Through this study, we reveal the inherent drawbacks of existing one-shot FL methods for medical data (see Table 1), and suggest a more suitable approach to address existing challenges *e.g.*, overfitting.

To prevent global model overfitting, we attempt to leverage random noise as a training source for KD (see Fig. 2). Baradad *et al.* [1] employs diverse types of structured noise for training in order to account for the difference between real images and random noise. However, due to the diversity of medical data [3,13,14], seeking a common noise space is more challenging than in natural images. Hence, we exploit DeepInversion [30], which synthesizes structured proxy noise specific to a task and thus ensures that generated noise matches the properties of medical data. Specifically, we first gather client models on the central server, where each client model is trained on its own dataset. Next, we synthesize images from random noise and store all intermediate samples in memory. Also, as images in the early stages of synthesis (*i.e.*, close to random noise) are different from real images, we design noise-adapted client models that employ adaptive batch normalization (AdaBN) [16]. AdaBN is based on the assumption that domain-related knowledge is represented by the statistics of the batch normalization

(BN) [11] and label-related knowledge is stored in the weight matrix of each layer, ultimately enhancing the KD signal for random noise. Lastly, we train a global model through KD with both the original- and noise-adapted client models using memory-stored images, repeating until global model convergences.

The contributions are as follows: (i) We propose one-shot FL leveraging image synthesis with client model adaptation. This allows to transfer knowledge from client models to the global model with synthesized images ranging from random noise to realistic images and contributes to preventing overfitting. (ii) We employ noise-adapted client models using AdaBN to produce a better KD signal for random noise. (iii) Comprehensive experiments on five small- and three large-scale medical image classification datasets consisting of microscopy, dermatoscopy, oct, histology, x-ray, and retinal images reveal that our method outperforms state-of-the-art one-shot FL methods.

Related Work. Due to the challenges of **one-shot FL**, prior methods were trained on public data [8,15], applying dataset distillation [35], or sharing additional information [6]. However, these assumptions may not hold for several real world scenarios, posing a challenge for their practical application. Recently, Zhang et al. [33] proposed the one-shot FL DENSE, which transfers knowledge from an ensemble of client models using KD and generated images. To enhance the transferability of client models, DENSE generates diverse images near the decision boundary to improve its accuracy. However, DENSE does not perform well in one-shot FL for medical data due to the complexity of decision boundaries. While DENSE diversifies generation using a generator, we propose to avoid overfitting by using synthesized images ranging from random noise to realistic images. For **data-free KD** [19], DeepInversion [30] synthesizes images by optimizing RGB pixels with cross-entropy and regularization losses and improves synthesis quality by minimizing feature statistics in BN layers. DAFL [2] uses a generator for image synthesis with a teacher model as a discriminator. To prevent student model overfitting, ZSKT [22] synthesizes images that exhibit mismatch between the student and teacher models. Unlike the methods that choose the best image as a training source for KD, our approach utilizes all intermediate synthesized images to prevent overfitting. Also, while Raikward et al. [24] proposed a method for KD that uses random noise as a training source, it requires real images during training and needs to adjust BN layer statistics multiple times iteratively. In contrast, our method performs one-shot FL without requiring real images during training.

2 Method

The overall training processes are shown in Fig. 2 and Algorithm 1. Given K client models $W^c = \{W_1^c, \dots, W_k^c\}$ with corresponding BN statistics μ_k and σ_k^2 with respect to data D_k, the objective of FL is to train a global model W^g, which represents all data $D = \{D_1, \dots, D_k\}$. Motivated by [17,33,34], KD enables the transfer of knowledge from client models W^c to the global model W^g. Due to restricted access of D, prior works [2,30,33] use synthetic images

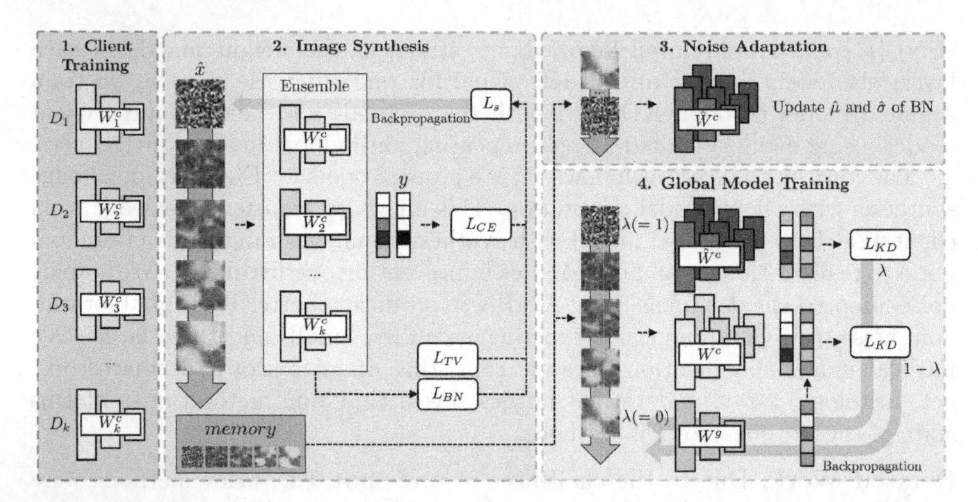

Fig. 2. Illustration of our proposed method. W_k^c denotes a client model with respect to data D_k and W^g denotes a global model. W^c denotes original client models and $\hat{W}^c$ denotes noise-adapted client models. $\hat{x}$ indicates random noise and λ indicates noise level. $\hat{x}$ is optimized to have a property of all D_k using L_{CE}, L_{BN}, and L_{TV}. Afterward, it is used as a training source for KD in global model training.

$\hat{x}$ as a training source for KD. However, since $\hat{x}$ may be monotonous for robust training, overfitting is a significant challenge in one-shot FL. To address this, we employ random Gaussian noise $\mathcal{N}(0,1)$ as a training source for KD [1]. However, in contrast with [1], $\mathcal{N}(0,1)$ does not capture common medical properties. Hence, we employ DeepInversion [30] to ensure random noise retains characteristics of D. Details regarding image synthesis with DeepInversion are described in the following section.

Image Synthesis. Given random noise $\hat{x} \in \mathbb{R}^{H\times W\times C}$ initialized from $\mathcal{N}(0,1)$, where H, W, and C denote height, width, and channels; the objective of image synthesis is to ensure $\hat{x}$ possesses a certain property of D. To achieve this, we optimize RGB pixels of $\hat{x}$ to synthesize a class-conditioned image with respect to a specific label y for I iterations. Formally,

$$L_s(\hat{x}, y; W^c) = L_{CE}(\hat{x}, y; W^c) + \lambda_{BN} L_{BN}(\hat{x}; W^c) + \lambda_{TV} L_{TV}(\hat{x}; W^c), \quad (1)$$

where L_{CE}, L_{BN}, and L_{TV} are cross-entropy, BN, and total variation losses [20]. Hyper-parameters λ_{BN} and λ_{TV} are used to balance the losses. Cross-entropy loss enables the synthesis of an image with respect to the label y, and total variation loss encourages image synthesis consistency. Additionally, $L_{BN}(\hat{x}) = \sum(\|\mu(\hat{x}) - \mu\| + \|\sigma^2(\hat{x}) - \sigma^2\|)$, where $\mu(\hat{x})$ & $\sigma^2(\hat{x})$ are the batch-wise mean & variance features of $\hat{x}$ and μ & σ^2 of the stored statistics of the BN layer. Since BN enforces feature similarity at all levels, this improves the quality of image synthesis significantly.

Recall that our method employs random noise $\hat{x}$ that has D's characteristics for training. In contrast to DeepInversion which selects the best image as a

Algorithm 1. Training process of our proposed method.

Input: Client models W^c with corresponding μ and σ^2, a global model W^g, a iteration I, a learning rate of image synthesis η_s, a learning rate of KD η_d, a momentum α.
$\hat{W}^c \leftarrow W^c, \hat{\mu} \leftarrow \mu, \hat{\sigma}^2 \leftarrow \sigma^2$ `// Initialize noise-adapted client models`
Repeat
  Initialize a batch of random noise $\hat{x}$ and arbitrary labels y.
  $memory \leftarrow [\,]$
  for $i = 1, \cdots, I$ **do**
    $\hat{x} \leftarrow \hat{x} - \eta_s \nabla L_s(\hat{x}, y; W^c)$ `// Synthesize image`
    $memory.append((\hat{x},\ 1 - i/I))$
  end for
  for $i = 1, \cdots, I$ **do**
    $\hat{x},\ \lambda \leftarrow memory[I - i]$
    $\hat{\mu} \leftarrow \alpha\hat{\mu} + (1-\alpha)\mu(\hat{x}),\ \hat{\sigma}^2 \leftarrow \alpha\hat{\sigma}^2 + (1-\alpha)\sigma^2(\hat{x})$ `// Adapt noise for` $\hat{W}^c$
  end for
  for $i = 1, \cdots, I$ **do**
    $\hat{x},\ \lambda \leftarrow memory[i]$
    $W^g \leftarrow W^g - \eta_d \nabla L_d(\hat{x}, \lambda; W^c, \hat{W}^c, W^g)$ `// Train global model`
  end for
until convergence.
Output: Trained global model W^g.

training source, our method employs all intermediate synthesized samples for KD. Thus we store all intermediate samples and the corresponding noise level λ (*e.g.*, $1 - i/I$ for i steps) in *memory* during I iterations. Due to the visual difference between $\mathcal{N}(0,1)$ and D, we design noise-adapted client models using AdaBN [16] to provide better KD signals for $\hat{x}$. The following section will describe more details regarding noise-adapted client models.

Noise Adaptation. BN [11] was proposed to mitigate internal covariate shifts, allowing to provide consistent input distributions to subsequent layers. Due to the existing discrepancy between $\mathcal{N}(0,1)$ and D, there is no guarantee BN will provide consistent input to subsequent parameters and may lead to poor model predictions. Thus we adapt $\mathcal{N}(0,1)$ by iteratively adjusting the running statistics of BN using AdaBN [16], producing better logit signals for KD. Formally,

$$\hat{\mu} = \alpha\hat{\mu} + (1-\alpha)\mu(\hat{x}),\ \hat{\sigma}^2 = \alpha\hat{\sigma}^2 + (1-\alpha)\sigma^2(\hat{x}), \tag{2}$$

where α represents momentum and $\hat{x}$ is a sample stored in *memory*. Initially, $\hat{\mu}$ and $\hat{\sigma}^2$ are set to μ and σ^2. The samples in *memory* ranging from characteristic images for D to $\mathcal{N}(0,1)$ by gradually adjusting $\hat{\mu}$ and $\hat{\sigma}^2$ towards $\mathcal{N}(0,1)$ through Eq. 2 for I steps. With this in mind, we now describe how to train the global model.

Global Model Training. KD allows to train a global model with multiple client models [17,33,34]. We denote W^c with original μ and σ^2 as W^c, and denote W^c with $\hat{\mu}$ and $\hat{\sigma}^2$ as $\hat{W}^c$. Since $\hat{x}$, W^c, and $\hat{W}^c$ are used for KD, this enables the model to avoid overfitting without being negatively impacted during global model training. Formally,

$$L_d(\hat{x}, \lambda; W^c, \hat{W}^c, W^g) = \lambda L_{KD}(\hat{x}; \hat{W}^c, W^g) + (1-\lambda) L_{KD}(\hat{x}; W^c, W^g), \tag{3}$$

where λ denotes a noise level stored in *memory*. $L_{KD}(\hat{x}; W^c, W^g)$ denotes the Kullback-Leibler divergence between $p(\hat{x}; W^c)$ and $p(\hat{x}; W^g)$ where $p(\cdot)$ is an ensemble (averaging) prediction of given models with a temperature on softmax inputs [10]. Overall, W^g is trained for I steps. To clarify, random noise contributes to avoiding overfitting, while noise-adapted client models help to produce a better KD signal for random noise, improving robust global model training. These processes *i.e.*, Image Synthesis, Noise Adaptation, and Global Model Training are repeated until the global model W^g converges.

3 Experiments

Datasets. For evaluation, we use five small-scale (28×28) medical image classification datasets *i.e.*, Blood, Derma, Oct, Path, and Tissue from MedMNIST [29]. Additionally, we use three large-scale (224×224) datasets *i.e.*, RSNA, Diabetic, and ISIC from RSNA Pneumonia Detection [25], Diabetic Retinopathy Detection [7], and ISIC2019-HAM-BCN20000 [4,5,28].

Experimental Settings. We explore three scenarios *i.e.*, (i) data heterogeneity levels, (ii) impact on large-scale datasets, and (iii) model heterogeneity *i.e.*, each client has different architectures. In (i), Blood, Derma, Oct, Path, and Tissue datasets are used with Independent and Identically Distributed (IID) clients and Dirichlet distributed [31] clients with $\alpha = 0.6$ and $\alpha = 0.3$. For (ii), RSNA, Diabetic, and ISIC datasets are used with IID clients, including ISIC$'$ where each client has a different image acquisition system [27]. For (iii), client models used either ResNet18 [9], ResNet34 [9], WRN-16-2 [32], VGG16(with BN) [26], and VGG8(with BN) [26], respectively.

Comparison Methods. We employ three one-shot FL methods: FedAvg [21] with single communication, DAFL [2], and DENSE [33], each evaluated using global model accuracy obtained on test data. For the upper bound, we report the FedAvg with 100 communications. For ablations, we evaluate (a) without image synthesis (w/o IS), (b) without image synthesis and noise adaptation (w/o IS&Ada) with only $\mathcal{N}(0,1)$ used for training, (c) without noise adaptation (w/o Ada), and (d) without intermediate random noise (w/o $\mathcal{N}$), this is equivalent to DeepInversion [30] in a one-shot FL scenario. For w/o $\mathcal{N}$, we synthesize all images and perform KD. For a fair comparison, we follow each method's original implementation and matched all training/parameter settings. For DAFL, an ensemble of client models was used as the teacher model following [17,33,34] with KD used for global model training. On large-scale datasets, an ImageNet pre-trained model was used with balanced classification accuracy reported for evaluation as in [27].

Implementation Details. We used ResNet18 [9] for our experiments with five clients by default. Client models were trained for 100 epochs with SGD optimizer using learning rate (LR) 1e-3 and batch size 128. For image synthesis, we used Adam optimizer with LR 5e-2 for 100 epochs with 500 and 1,000 synthesis iterations (*i.e.*, I) for small- and large-scale datasets, with batch sizes 256 and 50, respectively. Following [30], $\lambda_{TV} = 0.000025$ and $\lambda_{BN} = 10$, with KD temperature $T = 20$ and momentum $\alpha = 0.9$.

Table 1. Classification accuracy on five datasets with different heterogeneity levels. The first and second sub-rows show the accuracy of the upper bound and one-shot FL methods. The third sub-row shows ablation performance with IS, Ada, and $\mathcal{N}$ denoting w/o image synthesis, noise adaptation, and random noise, respectively. **Bold** indicates the best accuracy among one-shot FL methods.

	IID					Dirichlet ($\alpha = 0.6$)					Dirichlet ($\alpha = 0.3$)				
	Blood	Derma	Oct	Path	Tissue	Blood	Derma	Oct	Path	Tissue	Blood	Derma	Oct	Path	Tissue
FedAvg [21]	93.51	74.61	75.60	84.54	63.64	93.60	72.72	76.50	81.48	55.61	87.49	69.88	73.50	77.52	53.26
FedAvg(1)	13.74	66.88	25.00	5.86	32.07	18.24	66.88	25.00	5.86	32.07	16.92	10.97	25.00	5.86	32.07
DAFL [2]	7.13	66.43	25.00	7.63	11.55	7.13	66.88	34.40	14.97	39.15	7.13	13.62	25.00	18.64	45.00
DENSE [33]	39.37	66.93	33.80	21.89	21.35	34.52	67.78	39.40	30.31	9.47	30.78	12.77	25.80	19.87	9.33
FedISCA	**87.99**	**70.12**	70.20	**84.18**	**61.90**	**82.90**	**69.83**	**68.60**	**82.92**	**53.04**	**46.59**	**15.91**	**60.50**	**79.25**	**51.00**
w/o IS	9.09	66.88	25.20	24.69	23.70	9.09	66.88	26.10	22.41	9.31	23.27	11.02	27.10	18.70	9.31
w/o IS&Ada	7.13	11.12	35.80	4.72	7.13	7.13	66.88	25.00	14.15	32.07	7.13	11.12	25.00	4.72	7.13
w/o Ada	81.61	68.33	**70.30**	82.08	59.34	63.67	68.18	61.90	78.61	51.99	29.73	14.36	54.30	77.69	50.40
w/o $\mathcal{N}$ [30]	87.02	68.73	60.20	77.90	57.86	80.62	69.58	60.30	75.54	49.06	45.69	13.87	49.20	70.53	46.73

Table 2. Balanced classification accuracy on large-scale datasets.

	FedAvg [21]	FedAvg(1)	DAFL [2]	DENSE [33]	FedISCA	w/o IS	w/o IS&Ada	w/o Ada	w/o $\mathcal{N}$ [30]
RSNA	88.16	78.65	50.55	55.06	**85.34**	50.00	50.00	81.56	50.61
Diabetic	49.04	35.60	22.63	23.51	40.08	20.07	20.02	**40.91**	28.30
ISIC	62.88	38.05	14.51	13.69	**48.39**	12.50	12.50	47.21	25.61
ISIC′	57.15	18.08	18.37	16.46	**22.47**	11.29	12.52	21.72	14.80

3.1 Main Results

Table 1 shows the accuracy on five datasets with different heterogeneity levels. FedISCA outperforms all one-shot FL methods across all datasets regardless of the level of heterogeneity. In Table 2, FedISCA also reports improved performance against the compared methods, validating the viability of our approach on real-world large-scale data. On the contrary, DAFL and DENSE performed poorly on medical data since significant accuracy gaps exist between the upper bound and each competitor (except Derma). Additionally, though FedAvg reports higher accuracy for multiple communication rounds, it shows significantly lower accuracy for single communication (FedAvg(1)). To better explain this phenomenon, we analyzed the accuracy of FedAvg(1) by comparing the variance between client model parameters *i.e.*, client models with high variance *e.g.*, Path IID(=36.10), yield lower accuracy compared to those with low variance *e.g.*, Derma IID(=0.01). This suggests that the variance of client models is correlated with the accuracy of FedAvg(1).

In Fig. 3, we show the synthesized images of FedISCA, DAFL [2], and DENSE [33] on eight datasets. FedISCA generates more realistic images compared to the competitors. Note that DENSE aims to generate a diverse image (*e.g.*, generating highly transferable samples) distributed near the decision boundary, which may not be realistic. Although these methods have achieved higher accuracy on natural data, our experiments reveal that this assumption does not hold in the medical domain. In addition, DENSE outperforms FedAvg(1) on small-scale

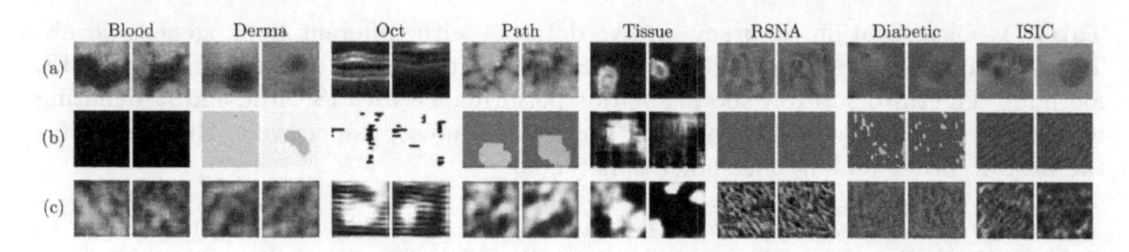

Fig. 3. The synthesized images of (a) FedISCA, (b) DAFL [2], and (c) DENSE [33] on eight datasets. Overall, FedISCA synthesizes more realistic images.

Table 3. Classification accuracy on five datasets with model heterogeneity.

	IID					Dirichlet ($\alpha = 0.6$)					Dirichlet ($\alpha = 0.3$)				
	Blood	Derma	Oct	Path	Tissue	Blood	Derma	Oct	Path	Tissue	Blood	Derma	Oct	Path	Tissue
DAFL [2]	7.13	65.69	25.00	15.72	35.66	7.13	67.23	37.10	28.15	39.54	7.13	13.47	45.30	29.68	19.54
DENSE [33]	46.86	66.88	44.00	33.08	38.28	23.47	67.93	40.70	28.68	36.70	34.67	13.42	44.00	39.37	38.37
FedISCA	**87.96**	**71.17**	**70.00**	**83.02**	**61.74**	**73.43**	**69.23**	**64.80**	**82.73**	**51.95**	**44.20**	**16.61**	**62.00**	**72.26**	**43.80**
w/o $\mathcal{N}$ [30]	87.55	69.93	51.20	74.05	57.90	68.78	68.93	61.00	72.79	46.91	43.85	15.61	51.50	64.65	39.89

datasets (except Tissue), but its accuracy is lower than FedAvg(1) on large-scale datasets. This suggests a difficulty in large-scale image generation *i.e.*, the generator in DENSE deteriorates global model training and leads to lower accuracy, while FedAvg(1) achieves high accuracy due to the low client model variance *e.g.*, RSNA(=0.61), Diabetic(=0.04), and ISIC(=0.09).

Ablations. We report ablation results in Table 1 and 2. In the medical field, generating realistic images is crucial for one-shot FL, as the accuracy of w/o IS, and w/o IS&Ada is significantly lower compared to FedISCA; this validates the need for image synthesis. However, relying on image synthesis alone is not enough to achieve high accuracy, as neither w/o Ada nor w/o $\mathcal{N}$ achieve the best accuracy across all datasets. w/o $\mathcal{N}$ performs worse than w/o Ada in most datasets (except Blood and Derma), showing that solely relying on the best image is not sufficient for robust training. On the contrary, the accuracy of FedISCA suggests that noise-adapted client models alleviate the negative effects of random noise, resulting in high accuracy. Overall, the experimental results support the idea that both components play an essential role in medical one-shot FL. Additionally, we also evaluate the variance in BN statistics between the original and noise-adapted client models. Here, we found that high variance (*e.g.*, RSNA(=0.0018)), yields improved accuracy compared to those with lower variance (*e.g.*, Diabetic(=0.0008)). Finally, Table 3 shows the accuracy of a global model trained on client models with model heterogeneity. The proposed method reports the best accuracy among all competitors, equally demonstrating the effectiveness of our method in one-shot FL with diverse types of model architectures.

4 Conclusion

We present a novel one-shot FL framework that uses image synthesis and client model adaptation with KD. We demonstrate that (i) random noise significantly reduces the risk of overfitting, resulting in robust global model training; (ii) noise-adapted client models enhance the KD signal leading to high accuracy; and (iii) through experiments on eight datasets, our method outperforms the state-of-the-art one-shot FL methods on medical data. Further investigation into severe heterogeneity in clients will be a topic of future research.

Acknowledgments. This work was supported by funding from the DGIST R&D program of the Ministry of Science and ICT of KOREA (22-KUJoint-02) and the framework of international cooperation program managed by the National Research Foundation of Korea (NRF-2022K2A9A1A01097840) and the NRF grant funded by the Korean Government (MSIT)(No. 2019R1C1C1008727) and the National Institute of Health (MH113406, DA057567, AA021697) and by the Stanford HAI Google Cloud Credit.

References

1. Baradad Jurjo, M., Wulff, J., Wang, T., Isola, P., Torralba, A.: Learning to see by looking at noise. Adv. Neural. Inf. Process. Syst. **34**, 2556–2569 (2021)
2. Chen, H., et al.: Data-free learning of student networks. In: International Conference on Computer Vision, pp. 3514–3522 (2019)
3. Chikontwe, P., Nam, S.J., Go, H., Kim, M., Sung, H.J., Park, S.H.: Feature re-calibration based multiple instance learning for whole slide image classification. In: International Conference on Medical Image Computing and Computer-Assisted Intervention. pp. 420–430. Springer (2022). https://doi.org/10.1007/978-3-031-16434-7_41
4. Codella, N.C., et al.: Skin lesion analysis toward melanoma detection: a challenge at the 2017 international symposium on biomedical imaging (isbi), hosted by the international skin imaging collaboration (isic). In: 2018 IEEE 15th International Symposium on Biomedical Imaging (ISBI 2018), pp. 168–172. IEEE (2018)
5. Combalia, M., et al.: Bcn20000: dermoscopic lesions in the wild. arXiv preprint arXiv:1908.02288 (2019)
6. Dennis, D.K., Li, T., Smith, V.: Heterogeneity for the win: one-shot federated clustering. In: International Conference on Machine Learning, pp. 2611–2620. PMLR (2021)
7. EyePACS: Diabetic retinopathy detection (2015)
8. Guha, N., Talwalkar, A., Smith, V.: One-shot federated learning. arXiv preprint arXiv:1902.11175 (2019)
9. He, K., Zhang, X., Ren, S., Sun, J.: Deep residual learning for image recognition. In: Computer Vision and Pattern Recognition, pp. 770–778 (2016)
10. Hinton, G., Vinyals, O., Dean, J.: Distilling the knowledge in a neural network. arXiv preprint arXiv:1503.02531 (2015)
11. Ioffe, S., Szegedy, C.: Batch normalization: accelerating deep network training by reducing internal covariate shift. In: International Conference on Machine Learning, pp. 448–456. PMLR (2015)

12. Jiang, M., Yang, H., Li, X., Liu, Q., Heng, P.A., Dou, Q.: Dynamic bank learning for semi-supervised federated image diagnosis with class imbalance. In: Medical Image Computing and Computer Assisted Intervention. pp. 196–206. Springer (2022). https://doi.org/10.1007/978-3-031-16437-8_19
13. Jung, E., Luna, M., Park, S.H.: Conditional gan with 3d discriminator for MRI generation of Alzheimer's disease progression. Pattern Recogn. **133**, 109061 (2023)
14. Kim, S., An, S., Chikontwe, P., Park, S.H.: Bidirectional rnn-based few shot learning for 3d medical image segmentation. In: Proceedings of the AAAI Conference on Artificial Intelligence, vol. 35, pp. 1808–1816 (2021)
15. Li, Q., He, B., Song, D.: Practical one-shot federated learning for cross-silo setting. In: International Joint Conference on Artificial Intelligence (2020)
16. Li, Y., Wang, N., Shi, J., Liu, J., Hou, X.: Adaptive batch normalization for practical domain adaptation. Pattern Recogn. **80**, 109–117 (2018)
17. Lin, T., Kong, L., Stich, S.U., Jaggi, M.: Ensemble distillation for robust model fusion in federated learning. Adv. Neural. Inf. Process. Syst. **33**, 2351–2363 (2020)
18. Liu, X., Li, W., Yuan, Y.: Intervention & interaction federated abnormality detection with noisy clients. In: Medical Image Computing and Computer Assisted Intervention, pp. 309–319. Springer (2022). https://doi.org/10.1007/978-3-031-16452-1_30
19. Liu, Y., Zhang, W., Wang, J., Wang, J.: Data-free knowledge transfer: a survey. arXiv preprint arXiv:2112.15278 (2021a
20. Mahendran, A., Vedaldi, A.: Understanding deep image representations by inverting them. In: Computer Vision and Pattern Recognition, pp. 5188–5196 (2015)
21. McMahan, B., Moore, E., Ramage, D., Hampson, S., y Arcas, B.A.: Communication-efficient learning of deep networks from decentralized data. In: Artificial Intelligence and Statistics, pp. 1273–1282. PMLR (2017)
22. Micaelli, P., Storkey, A.J.: Zero-shot knowledge transfer via adversarial belief matching. In: Advances in Neural Information Processing Systems 32 (2019)
23. Qi, X., Yang, G., He, Y., Liu, W., Islam, A., Li, S.: Contrastive re-localization and history distillation in federated cmr segmentation. In: Medical Image Computing and Computer Assisted Intervention, pp. 256–265. Springer (2022). https://doi.org/10.1007/978-3-031-16443-9_25
24. Raikwar, P., Mishra, D.: Discovering and overcoming limitations of noise-engineered data-free knowledge distillation. In: Advances in Neural Information Processing Systems (2022)
25. RSNA: Rsna pneumonia detection challenge (2018)
26. Simonyan, K., Zisserman, A.: Very deep convolutional networks for large-scale image recognition. In: International Conference on Learning Representations (2014)
27. Ogier du Terrail, J., et al.: Datasets and benchmarks for cross-silo federated learning in realistic healthcare settings. In: Advances in Neural Information Processing Systems (2022)
28. Tschandl, P., Rosendahl, C., Kittler, H.: The ham10000 dataset, a large collection of multi-source dermatoscopic images of common pigmented skin lesions. Scientific Data **5**(1), 1–9 (2018)
29. Yang, J., et al.: Medmnist v2: A large-scale lightweight benchmark for 2d and 3d biomedical image classification. arXiv preprint arXiv:2110.14795 (2021)
30. Yin, H., et al.: Dreaming to distill: data-free knowledge transfer via deepinversion. In: Computer Vision and Pattern Recognition, pp. 8715–8724 (2020)

31. Yurochkin, M., Agarwal, M., Ghosh, S., Greenewald, K., Hoang, N., Khazaeni, Y.: Bayesian nonparametric federated learning of neural networks. In: International Conference on Machine Learning, pp. 7252–7261. PMLR (2019)
32. Zagoruyko, S., Komodakis, N.: Wide residual networks. In: British Machine Vision Conference (BMVC) (2016)
33. Zhang, J., et al.: Dense: data-free one-shot federated learning. In: Advances in Neural Information Processing Systems (2022)
34. Zhang, S., Liu, M., Yan, J.: The diversified ensemble neural network. Adv. Neural. Inf. Process. Syst. **33**, 16001–16011 (2020)
35. Zhou, Y., Pu, G., Ma, X., Li, X., Wu, D.: Distilled one-shot federated learning. arXiv preprint arXiv:2009.07999 (2020)
36. Zhu, W., Luo, J.: Federated medical image analysis with virtual sample synthesis. In: Medical Image Computing and Computer Assisted Intervention, pp. 728–738. Springer (2022). https://doi.org/10.1007/978-3-031-16437-8_70

Multi-objective Point Cloud Autoencoders for Explainable Myocardial Infarction Prediction

Marcel Beetz[1(✉)], Abhirup Banerjee[1,2], and Vicente Grau[1]

[1] Institute of Biomedical Engineering, Department of Engineering Science, University of Oxford, Oxford OX3 7DQ, UK
marcel.beetz@eng.ox.ac.uk

[2] Division of Cardiovascular Medicine, Radcliffe Department of Medicine, University of Oxford, Oxford OX3 9DU, UK

Abstract. Myocardial infarction (MI) is one of the most common causes of death in the world. Image-based biomarkers commonly used in the clinic, such as ejection fraction, fail to capture more complex patterns in the heart's 3D anatomy and thus limit diagnostic accuracy. In this work, we present the multi-objective point cloud autoencoder as a novel geometric deep learning approach for explainable infarction prediction, based on multi-class 3D point cloud representations of cardiac anatomy and function. Its architecture consists of multiple task-specific branches connected by a low-dimensional latent space to allow for effective multi-objective learning of both reconstruction and MI prediction, while capturing pathology-specific 3D shape information in an interpretable latent space. Furthermore, its hierarchical branch design with point cloud-based deep learning operations enables efficient multi-scale feature learning directly on high-resolution anatomy point clouds. In our experiments on a large UK Biobank dataset, the multi-objective point cloud autoencoder is able to accurately reconstruct multi-temporal 3D shapes with Chamfer distances between predicted and input anatomies below the underlying images' pixel resolution. Our method outperforms multiple machine learning and deep learning benchmarks for the task of incident MI prediction by 19% in terms of Area Under the Receiver Operating Characteristic curve. In addition, its task-specific compact latent space exhibits easily separable control and MI clusters with clinically plausible associations between subject encodings and corresponding 3D shapes, thus demonstrating the explainability of the prediction.

Keywords: Myocardial infarction · Clinical outcome classification · 3D cardiac shape analysis · Multi-task learning · Geometric deep learning · Cardiac MRI · Cardiac function modeling

---

Supplementary Information The online version contains supplementary material available at https://doi.org/10.1007/978-3-031-43895-0_50.

H. Greenspan et al. (Eds.): MICCAI 2023, LNCS 14221, pp. 532–542, 2023.
https://doi.org/10.1007/978-3-031-43895-0_50

1 Introduction

Myocardial infarction (MI) is the deadliest cardiovascular disease in the developed world [22]. Consequently, an ability to predict future MI events is of immense importance on both an individual and population health level, as it would allow for improved risk stratification, preventative care, and treatment planning. In current clinical practice, MI prediction is typically based on volumetric biomarkers, such as ejection fraction. These can be derived from cardiac cine magnetic resonance imaging (MRI), which is considered the gold standard modality for cardiac function assessment [28]. While such metrics are relatively easy to calculate and interpret, they only approximate the complex 3D morphology and physiology of the heart with a single value, which hinders further improvements in predictive accuracy. Consequently, considerable research efforts have been dedicated to developing new methods capable of extracting novel biomarkers from images or segmentation masks using machine learning and deep learning techniques [1,16,17,21,23,29,30,34]. However, their focus on 2D data still limits the discovery of more intricate biomarkers whose important role for MI prediction and cardiac function assessment has previously been shown [10,12,20,29]. In order to efficiently process true 3D anatomical shape information, geometric deep learning methods for point clouds have recently been increasingly used for various cardiac image-based tasks [5,7,8,18,19,32,35].

In this work, we propose the multi-objective point cloud autoencoder as a novel geometric deep learning approach for interpretable MI prediction, based on 3D cardiac shape information. Its specialized multi-branch architecture allows for the direct and efficient processing of high resolution 3D point cloud representations of the multi-class cardiac anatomy at multiple time points of the cardiac cycle, while simultaneously predicting future MI events. Crucially, a low-dimensional latent space vector captures task-specific 3D shape information as an orderly multivariate probability distribution, offering pathology-specific separability and allowing for a straightforward visual analysis of associations between 3D structure and latent encodings. The resulting high explainability considerably boosts the method's clinical applicability and sets it apart from previous black-box deep learning approaches for MI classification [11,14,18]. To the best of our knowledge, this is the first point cloud deep learning approach to combine full 3D shape processing and multi-objective learning with an explicit focus on method interpretability for MI prediction.

2 Methods

2.1 Dataset and Preprocessing

We select the cine MRI acquisitions of 470 subjects of the UK Biobank study as our dataset in this work [24]. All images were acquired with a voxel resolution of $1.8 \times 1.8 \times 8.0$ mm^3 for short-axis and $1.8 \times 1.8 \times 6.0$ mm^3 for long-axis slices using a balanced steady-state free precession (bSSFP) protocol [25]. Half of the subjects in our dataset experienced an MI event after the image acquisition date

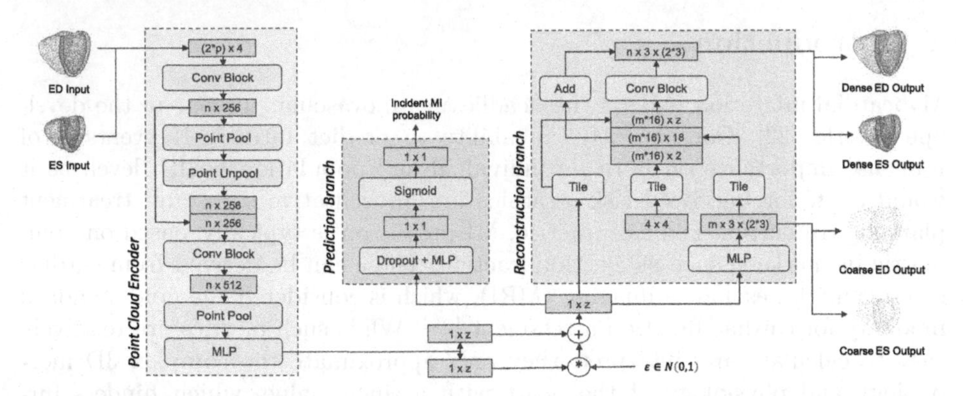

Fig. 1. Network architecture of the proposed multi-objective point cloud autoencoder. First, a point cloud deep learning-based encoder branch captures multi-scale shape information from multi-class and multi-temporal input anatomies in a low-dimensional latent space vector. Then, the resulting encodings are used in a reconstruction branch to recreate the original input shapes and in a prediction branch to output a clinical outcome probability (in this case for incident MI events).

(incident MI) as indicated by UK Biobank field IDs 42001 and 42000. The other 50% of subjects are considered as normal control cases. They were chosen to be free of any cardiovascular disease and other pathologies frequently observed in the UK Biobank study, following a similar selection as previous works [2,6,9] (see Table 1 of the Supplementary Material). For each subject, we reconstruct 3D multi-class point cloud representations of their biventricular anatomy from the corresponding raw cine MR images at both the end-diastolic (ED) and end-systolic (ES) phases of the cardiac cycle with the fully automatic multi-step process proposed in [3,4,13], and use them as inputs for our networks.

2.2 Network Architecture

The architecture of the multi-objective point cloud autoencoder consists of three task-specific branches, namely an encoder, a reconstruction, and a prediction branch, which are connected by a low-dimensional latent space vector (Fig. 1).

Concatenated multi-class point clouds at the ED and ES phases of the cardiac cycle with shape $(2*p)\times 4$ are first fed into the encoder branch as network inputs where $(2*p)$ represents the number of points p in the ED and ES point clouds and 4 are the x, y, z coordinate values in 3D space and a class label to encode the three cardiac substructures, namely left ventricular (LV) endocardium, LV epicardium, and right ventricular (RV) endocardium. The inputs are then passed through the point cloud-specific encoder, which is composed of two connected PointNet-style [26,27] blocks and a multi-layer perceptron (MLP), before outputting both a mean and standard deviation (SD) vector of size $1 \times z$. Next, the reparameterization trick is applied to these two vectors, and the resulting latent space vector is used as an input to both the reconstruction and prediction branches. This ensures that the latent space is influenced by both tasks during training and thus encourages an interpretable distribution that is both discriminative enough for the prediction task and also descriptive enough to allow accurate reconstruction. The reconstruction branch [33] starts with a MLP to produce an intermediate coarse point cloud output, which assures that the final fine point cloud preserves the global shape. It is then followed by a FoldingNet-style [31] layer to obtain the final dense output point cloud with both a local and global shape focus. The preliminary coarse and the dense output point cloud are represented as $m \times 3 \times (2 * 3)$ and $n \times 3 \times (2 * 3)$ tensors respectively, where m and n refer to the number of points with $n >> m$, the 3 to the spatial 3D coordinates, and the $(2 * 3)$ to the three cardiac substructures at ED and ES. In this work, we use the same total number of points to represent both the input and dense output point clouds. The prediction branch combines a Dropout layer, a MLP, and a Sigmoid activation function.

2.3 Loss and Training

The loss function of the multi-objective point cloud autoencoder consists of the sum of three subloss terms, each representing a different training objective in the multi-task setting, and weighted by two parameters β and γ.

$$L_{total} = L_{reconstruction} + \beta * L_{KL} + \gamma * L_{CE}. \tag{1}$$

The first loss term, $L_{reconstruction}$, encourages the network to accurately reconstruct input anatomies and thereby capture important shape information. It contains two subloss terms and a weighting parameter α.

$$L_{reconstruction} = \sum_{i=1}^{T} \sum_{j=1}^{C} \left(L_{coarse,i,j} + \alpha * L_{dense,i,j}\right). \tag{2}$$

Here, C and T refer to the number of cardiac substructures and phases respectively. We use $C = 3$ and $T = 2$ in this work. The L_{coarse} and L_{dense} loss terms compare the respective coarse and dense output predictions of the network with the same input point cloud using the symmetric Chamfer distance (CD). The weighting parameter α is increased stepwise from smaller (0.01) to larger (2.0) values during training in a monotonic annealing schedule to encourage the network to first focus on a good global reconstruction and gradually put more emphasis on a high local accuracy as training progresses. The second loss term in Eq. (1), L_{KL}, calculates the Kullback-Leibler divergence between the network's latent space and a multivariate standard normal distribution, which encourages high latent space quality and improves regularization. The third loss term, L_{CE}, refers to the binary cross entropy loss between the network's outcome prediction and the gold standard encoding. We again use a monotonic annealing schedule for the weighting parameter β to balance latent space quality and output accuracy and for γ to gradually put more focus on improving prediction performance. Hereby, we choose stepwise increases from 0.001 to 0.01 for β and from 1.0 to 5.0 for γ, based on empirical findings.

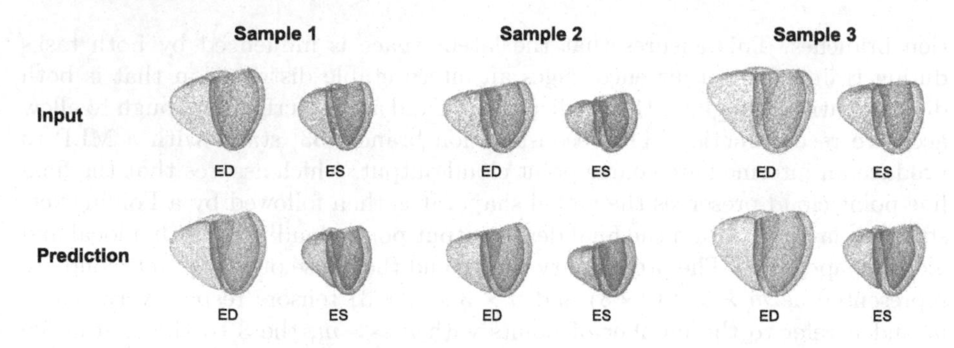

Fig. 2. Qualitative reconstruction results of three sample cases.

We randomly split the dataset into 70% training, 5% validation, and 25% test data. We train the network with the Adam optimizer and a mini-batch size of 8 for ~80,000 steps, since no improvement on the validation data was achieved during the 10,000 prior steps. The method is implemented using the TensorFlow library and has a post-training run time of ~15 ms. All experiments are performed on a GeForce RTX 2070 Graphics Card with 8 GB memory.

3 Experiments and Results

3.1 Input Shape Reconstruction

In our first experiment, we evaluate whether the multi-objective point cloud autoencoder is able to accurately reconstruct the ED and ES input anatomies. To this end, we pass all anatomies of the test dataset through the trained network and visualize both the input and corresponding predicted point clouds of three sample cases in Fig. 2. We observe good local and global shape alignment between the input and predicted anatomies in all cases. Relationships between cardiac substructures and between ED and ES phases are accurately retained.

Next, we quantify the reconstruction performance by calculating the symmetric Chamfer distances between the respective input and reconstructed point clouds of all subjects in the test dataset separately for each cardiac substructure and phase (Table 1). We find mean Chamfer distance values below the underlying acquisition's pixel resolution for both phases and all cardiac substructures.

Table 1. Reconstruction results of the proposed method.

Metric	Phase	LV endocardium	LV epicardium	RV endocardium
CD (mm)	ED	1.57 (±0.35)	1.53 (±0.23)	1.71 (±0.27)
	ES	1.26 (±0.30)	1.47 (±0.29)	1.67 (±0.32)

Values represent mean (±SD). CD = Chamfer distance.

3.2 Myocardial Infarction Prediction

We next evaluate the performance of the network for incident MI prediction as its second task. To this end, we first obtain both the gold standard MI outcomes and the MI predictions of our pre-trained network for all cases in the test dataset and quantify its performance using five common binary classification metrics (Table 2). To compare with clinical benchmarks, we select LV ejection fraction (EF) and the combination of LV and RV EF as widely used metrics and use each of them as input features for two separate logistic regression models. In addition, we choose a hierarchical convolutional neural network (CNN) and a standard PointNet [26] with 2D segmentation masks and 3D anatomy point clouds at ED and ES as respective inputs, as additional benchmarks (Table 2).

Table 2. Comparison of MI prediction results by multiple methods.

Input	Method	AUROC	Accuracy	Precision	Recall	F1-Score
LV EF	Regression	0.622	0.570	0.591	0.504	0.533
LV+RV EF	Regression	0.611	0.571	0.499	0.516	0.540
2D shapes	CNN	0.641	0.608	0.603	0.633	0.617
3D shape	PointNet	0.646	0.652	0.666	0.610	0.637
3D shape	Proposed	**0.767**	**0.694**	**0.706**	**0.683**	**0.695**

We find that the proposed multi-objective point cloud autoencoder outperforms all other approaches with improvements of 19% in terms of Area Under the Receiver Operating Characteristic (AUROC) curve.

3.3 Task-Specific Latent Space Analysis

In addition to validating the reconstruction and prediction performance of our network, we also investigate the ability of its latent space to store high-resolution 3D shape data in an interpretable and pathology-specific manner. To this end, we first pass the anatomy point clouds of both normal and MI cases through the encoder branch of the pre-trained network to obtain their respective latent space encodings. We then apply the Laplacian eigenmap [15] algorithm to the encodings as a non-linear dimensionality reduction technique and visualize the resulting 2D eigenmap of the latent space distribution in Fig. 3. In addition, in order to study associations between the latent subject encodings and their 3D anatomical shapes, we select 6 cases encoded at salient locations in the eigenmap and plot their pertinent 3D anatomies (Fig. 3).

We observe a clear differentiation between the encoded normal and MI cases in the eigenmap. Furthermore, the sample anatomies positioned in the area of normal subjects typically exhibit noticeably different shape patterns to the ones located in the MI cluster. For example, the left middle MI subject shows much smaller volume and myocardial thickness changes between ED and ES anatomies than the three normal cases.

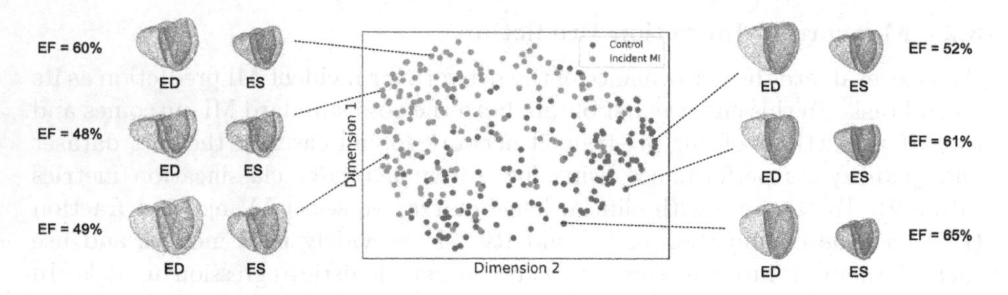

Fig. 3. Laplacian eigenmap of latent space encodings of both normal (blue) and MI (orange) subjects. ED and ES anatomies with LV ejection fraction (EF) values are shown for 6 cases located in salient map regions. (Color figure online)

3.4 Ablation Study

In order to assess the contributions of important parts of our network architecture, we next ablate multiple key components and study their effects on prediction performance. More specifically, we individually remove the dropout layer, the KL loss term, and the reconstruction branch, retrain each of the three ablated networks, and report their MI prediction results on the test dataset in Table 3. In addition, we investigate the importance of the multi-objective setting by first training the point cloud autoencoder without a prediction branch with a single reconstruction objective and then applying a logistic regression model for MI prediction to the learned general-purpose latent space representation (Table 3).

Table 3. Effects of architecture ablations on prediction performance.

Method	AUROC	Accuracy	Precision	Recall	F1-Score
Proposed	**0.767**	**0.694**	0.706	**0.683**	**0.695**
W/o dropout	0.739	0.652	0.655	0.667	0.661
W/o KL loss term	0.731	0.678	0.689	0.667	0.677
W/o reconstruction branch	0.755	0.686	**0.744**	0.583	0.654
W/o multi-objective training	0.717	0.655	0.659	0.648	0.648

All components contributed positively to the overall prediction performance with multi-objective training having the largest effect. The network without a reconstruction branch achieved the second-best AUROC, highest precision, and lowest recall score.

4 Discussion and Conclusion

In this paper, we have presented the multi-objective point cloud autoencoder as a novel geometric deep learning approach for interpretable MI prediction. The network is able to reconstruct input point clouds with high accuracy and only small localized smoothness artifacts despite the difficult multi-task setting. This shows the suitability of its architecture for efficient multi-scale feature extraction and its ability to effectively capture important 3D shape information in its latent space. Furthermore, the network can simultaneously process all three cardiac substructures at both ED and ES, indicating high flexibility and a potential for further extensions to the full cardiac cycle or other cardiac substructures. In addition, it also allows for more complex 3D shape-based biomarkers to be learned based on inter-temporal and inter-anatomical relationships. All these results are achieved directly on point cloud data, which offers a considerably more efficient storage of anatomical surface information than widely used voxelgrid-based deep learning approaches. Furthermore, the method is fast, fully automatic, and can be readily incorporated into a 3D shape analysis pipeline with cine MRI inputs.

The network also outperforms both machine learning techniques based on widely used clinical biomarkers as well as other deep learning approaches for MI prediction. On the one hand, this corroborates previous findings on the increased utility of full 3D shape information compared to single-valued or 2D biomarkers for MI assessment [14,20,29]. On the other hand, it shows the higher capacity of the proposed architecture and training process to extract important novel 3D biomarkers relevant for MI prediction. While we only study MI classification as a sample use case in this work, we believe that the proposed approach can be easily applied to other 3D shape-related pathologies or risk factors.

The network achieves these results based on a highly interpretable latent space with a clear differentiation between normal and MI subject encodings. Furthermore, the observed associations between encodings and 3D shapes demonstrate that the latent space is not only discriminative but also that the differentiation is based on clinically plausible 3D shape differences, such as reduced myocardial thinning between ED and ES in MI subjects which is indicative of impaired contraction ability of the heart. This greatly improves the explainability and applicability of the approach, as new subject phenotypes can be quickly and easily compared to other ones with similar encodings. Furthermore, the latent map not only shows well known associations of EF and MI but also a clear differentiation between some normal and MI cases with similar EF values. This indicates that the network is able to capture more intricate biomarkers that go beyond ejection fraction and to successfully utilize them in its MI prediction task while retaining high interpretability.

Finally, we show in our ablation studies that all major components of the architecture improve predictive accuracy. We hypothesize that the dropout layer, KL divergence term, and reconstruction branch introduce useful constraints, which have a positive regularizing effect and aid generalization. The multi-objective training procedure accounts for the largest performance gain. This is likely due to the exploited synergies of multiple tasks, which we also believe to be the primary reason for the high separability in the latent space.

Acknowledgment. This research has been conducted using the UK Biobank Resource under Application Number '40161'. The authors express no conflict of interest. The work of M. Beetz was supported by the Stiftung der Deutschen Wirtschaft

(Foundation of German Business). A. Banerjee is a Royal Society University Research Fellow and is supported by the Royal Society Grant No. URF\R1\221314. The work of A. Banerjee was partially supported by the British Heart Foundation (BHF) Project under Grant PG/20/21/35082. The work of V. Grau was supported by the CompBioMed 2 Centre of Excellence in Computational Biomedicine (European Commission Horizon 2020 research and innovation programme, grant agreement No. 823712).

References

1. Avard, E., et al.: Non-contrast cine cardiac magnetic resonance image radiomics features and machine learning algorithms for myocardial infarction detection. Comput. Biol. Med. **141**, 105145 (2022)
2. Bai, W., et al.: A population-based phenome-wide association study of cardiac and aortic structure and function. Nat. Med. **26**(10), 1654–1662 (2020)
3. Banerjee, A., et al.: A completely automated pipeline for 3D reconstruction of human heart from 2D cine magnetic resonance slices. Philosophical Trans. Royal Soc. A: Math. Phys. Eng. Sci. **379**(2212), 20200257 (2021)
4. Beetz, M., Banerjee, A., Grau, V.: Biventricular surface reconstruction from cine MRI contours using point completion networks. In: 2021 IEEE 18th International Symposium on Biomedical Imaging (ISBI), pp. 105–109. IEEE (2021)
5. Beetz, M., Banerjee, A., Grau, V.: Generating subpopulation-specific biventricular anatomy models using conditional point cloud variational autoencoders. In: Puyol Antón, E., et al. (eds.) STACOM 2021. LNCS, vol. 13131, pp. 75–83. Springer, Cham (2022). https://doi.org/10.1007/978-3-030-93722-5_9
6. Beetz, M., Banerjee, A., Grau, V.: Multi-domain variational autoencoders for combined modeling of MRI-based biventricular anatomy and ECG-based cardiac electrophysiology. In: Frontiers in Physiology, p. 991 (2022)
7. Beetz, M., Banerjee, A., Grau, V.: Point2Mesh-Net: combining point cloud and mesh-based deep learning for cardiac shape reconstruction. In: International Workshop on Statistical Atlases and Computational Models of the Heart, pp. 280–290. Springer (2023). https://doi.org/10.1007/978-3-031-23443-9_26
8. Beetz, M., Ossenberg-Engels, J., Banerjee, A., Grau, V.: Predicting 3D cardiac deformations with point cloud autoencoders. In: Puyol Antón, E., et al. (eds.) STACOM 2021. LNCS, vol. 13131, pp. 219–228. Springer, Cham (2022). https://doi.org/10.1007/978-3-030-93722-5_24
9. Beetz, M., et al.: Combined generation of electrocardiogram and cardiac anatomy models using multi-modal variational autoencoders. In: 2022 IEEE 19th International Symposium on Biomedical Imaging (ISBI), pp. 1–4 (2022)
10. Beetz, M., et al.: Interpretable cardiac anatomy modeling using variational mesh autoencoders. In: Frontiers in Cardiovascular Medicine, p. 3258 (2022)
11. Beetz, M., et al.: 3D shape-based myocardial infarction prediction using point cloud classification networks. arXiv preprint arXiv:2307.07298 (2023)
12. Beetz, M., et al.: Mesh U-Nets for 3D cardiac deformation modeling. In: International Workshop on Statistical Atlases and Computational Models of the Heart, pp. 245–257. Springer (2023). https://doi.org/10.1007/978-3-031-23443-9_23
13. Beetz, M., et al.: Multi-class point cloud completion networks for 3D cardiac anatomy reconstruction from cine magnetic resonance images. arXiv preprint arXiv:2307.08535 (2023)

14. Beetz, M., et al.: Post-infarction risk prediction with mesh classification networks. In: International Workshop on Statistical Atlases and Computational Models of the Heart. pp. 291–301. Springer (2023). https://doi.org/10.1007/978-3-031-23443-9_27
15. Belkin, M., Niyogi, P.: Laplacian eigenmaps and spectral techniques for embedding and clustering. In: Advances in Neural Information Processing Systems 14 (2001)
16. Bernard, O., et al.: Deep learning techniques for automatic MRI cardiac multi-structures segmentation and diagnosis: is the problem solved? IEEE Trans. Med. Imaging **37**(11), 2514–2525 (2018)
17. Cetin, I., et al.: A radiomics approach to computer-aided diagnosis with cardiac cine-MRI. In: Pop, M., et al. (eds.) STACOM 2017. LNCS, vol. 10663, pp. 82–90. Springer, Cham (2018). https://doi.org/10.1007/978-3-319-75541-0_9
18. Chang, Y., Jung, C.: Automatic cardiac MRI segmentation and permutation-invariant pathology classification using deep neural networks and point clouds. Neurocomputing **418**, 270–279 (2020)
19. Chen, X., et al.: Shape registration with learned deformations for 3D shape reconstruction from sparse and incomplete point clouds. Med. Image Anal. **74**, 102228 (2021)
20. Corral Acero, J., et al.: Understanding and improving risk assessment after myocardial infarction using automated left ventricular shape analysis. JACC: Cardiovascular Imaging (2022)
21. Isensee, F., Jaeger, P.F., Full, P.M., Wolf, I., Engelhardt, S., Maier-Hein, K.H.: Automatic cardiac disease assessment on cine-mri via time-series segmentation and domain specific features. In: Pop, M., et al. (eds.) STACOM 2017. LNCS, vol. 10663, pp. 120–129. Springer, Cham (2018). https://doi.org/10.1007/978-3-319-75541-0_13
22. Khan, M.A., et al.: Global epidemiology of ischemic heart disease: results from the global burden of disease study. Cureus **12**(7) (2020)
23. Khened, M., Alex, V., Krishnamurthi, G.: Densely connected fully convolutional network for short-axis cardiac cine MR image segmentation and heart diagnosis using random forest. In: Pop, M., et al. (eds.) STACOM 2017. LNCS, vol. 10663, pp. 140–151. Springer, Cham (2018). https://doi.org/10.1007/978-3-319-75541-0_15
24. Petersen, S.E., et al.: Imaging in population science: cardiovascular magnetic resonance in 100,000 participants of UK Biobank - rationale, challenges and approaches. J. Cardiovasc. Magn. Reson. **15**(46), 1–10 (2013)
25. Petersen, S.E., et al.: UK Biobank's cardiovascular magnetic resonance protocol. J. Cardiovasc. Magn. Reson. **18**(8), 1–7 (2016)
26. Qi, C.R., et al.: Pointnet: deep learning on point sets for 3D classification and segmentation. In: Proceedings of the IEEE Conference on Computer Vision and Pattern Recognition, pp. 652–660 (2017)
27. Qi, C.R., et al.: Pointnet++: deep hierarchical feature learning on point sets in a metric space. In: Advances in Neural Information Processing Systems, pp. 5099–5108 (2017)
28. Reindl, M., et al.: Role of cardiac magnetic resonance to improve risk prediction following acute ST-elevation myocardial infarction. J. Clin. Med. **9**(4), 1041 (2020)
29. Suinesiaputra, A., et al.: Statistical shape modeling of the left ventricle: myocardial infarct classification challenge. IEEE J. Biomed. Health Inform. **22**(2), 503–515 (2017)

30. Wolterink, J.M., Leiner, T., Viergever, M.A., Išgum, I.: Automatic segmentation and disease classification using cardiac cine MR images. In: Pop, M., et al. (eds.) STACOM 2017. LNCS, vol. 10663, pp. 101–110. Springer, Cham (2018). https://doi.org/10.1007/978-3-319-75541-0_11
31. Yang, Y., et al.: Foldingnet: interpretable unsupervised learning on 3D point clouds. arXiv preprint arXiv:1712.07262 (2017)
32. Ye, M., et al.: PC-U net: learning to jointly reconstruct and segment the cardiac walls in 3D from CT data. In: Puyol Anton, E., et al. (eds.) STACOM 2020. LNCS, vol. 12592, pp. 117–126. Springer, Cham (2021). https://doi.org/10.1007/978-3-030-68107-4_12
33. Yuan, W., et al.: PCN: point completion network. In: 2018 International Conference on 3D Vision (3DV), pp. 728–737 (2018)
34. Zhang, N., et al.: Deep learning for diagnosis of chronic myocardial infarction on nonenhanced cardiac cine MRI. Radiology **291**(3), 606–617 (2019)
35. Zhou, X.-Y., Wang, Z.-Y., Li, P., Zheng, J.-Q., Yang, G.-Z.: One-stage shape instantiation from a single 2D image to 3D point cloud. In: Shen, D., et al. (eds.) MICCAI 2019. LNCS, vol. 11767, pp. 30–38. Springer, Cham (2019). https://doi.org/10.1007/978-3-030-32251-9_4

Aneurysm Pose Estimation with Deep Learning

Youssef Assis[1(✉)], Liang Liao[1,2,3], Fabien Pierre[1], René Anxionnat[2,3], and Erwan Kerrien[1]

[1] Université de Lorraine,CNRS, Inria, LORIA, 54000 Nancy, France
youssef.assis@loria.fr
[2] Department of Diagnostic and Therapeutic Interventional Neuroradiology, Université de Lorraine, CHRU-Nancy, 54000 Nancy, France
[3] Université de Lorraine, Inserm, IADI, 54000 Nancy, France

Abstract. The diagnosis of unruptured intracranial aneurysms from time-of-flight Magnetic Resonance Angiography (TOF-MRA) images is a challenging clinical problem that is extremely difficult to automate. We propose to go beyond the mere detection of each aneurysm and also estimate its size and the orientation of its main axis for an immediate visualization in appropriate reformatted cut planes. To address this issue, and inspired by the idea behind YOLO architecture, a novel one-stage deep learning approach is described to simultaneously estimate the localization, size and orientation of each aneurysm in 3D images. It combines fast and approximate annotation, data sampling and generation to tackle the class imbalance problem, and a cosine similarity loss to optimize the orientation. We evaluate our approach on two large datasets containing 416 patients with 317 aneurysms using a 5-fold cross-validation scheme. Our method achieves a median localization error of 0.48 mm and a median 3D orientation error of 12.27 °C, demonstrating an accurate localization of aneurysms and an orientation estimation that comply with clinical practice. Further evaluation is performed in a more classical detection setting to compare with state-of-the-art nnDetecton and nnUNet methods. Competitive performance is reported with an average precision of 76.60%, a sensitivity score of 82.93%, and 0.44 false positives per case. Code and annotations are publicly available at https://gitlab.inria.fr/yassis/DeepAnePose.

Keywords: Object Pose Estimation · 3D YOLO · Intracranial Aneurysms

1 Introduction

Intracranial aneurysms are abnormal focal dilations of cerebral blood vessels. Their rupturing accounts for 85% of Subarachnoid Hemorrhages (SAH), and is associated with high morbidity and mortality rates [23]. Early detection and monitoring of Unruptured Intracranial Aneurysms (UIA) has become a problem of increasing clinical importance. Due to its non-invasive nature, 3D time-of-flight Magnetic Resonance Angiography (TOF-MRA) is the most suitable imaging technique for screening. However, detecting aneurysms in TOF-MRA volumes is a

H. Greenspan et al. (Eds.): MICCAI 2023, LNCS 14221, pp. 543–553, 2023.
https://doi.org/10.1007/978-3-031-43895-0_51

costly process that requires radiologists to scroll through different cut planes [7]. Therefore, an automated method to detect aneurysms and provide immediate appropriate visualization would be a valuable tool to assist radiologists in their clinical routine. We envision a dynamic browsing of cut planes rotating around the main axis of the aneurysm to facilitate the analysis of the aneurysm and the surrounding angioarchitecture. This requires estimating the location and orientation, i.e. the pose, of the aneurysm. This pose has been related to the risk of rupture [13] and could also be used for image registration [17].

Automated methods for detecting UIAs range from traditional Computer-Aided Detection (CAD) systems using image filtering techniques [1,27], to advanced deep learning methods based on Convolutional Neural Networks (CNNs). Although 2D and 2.5D methods have been proposed [19,24,26], most recent methods are fully 3D patch-based approaches. In 2020, the Aneurysm Detection and SegMentation (ADAM) challenge [25] compared various detection methods using TOF-MRA data. The class imbalance problem caused by the scarcity of aneurysm voxels inside an image volume, was addressed through loss functions and/or data augmentation. The top-performing method was nnDetection [3], which relies on a 3D bounding box representation. nnUNet [9], ranked third, uses the UNet [6] semantic segmentation architecture. Both methods consider large patches as input, which requires significant computing power and large databases for reliable sample modeling. Detection provides localization, but aneurysm orientation is challenging to estimate, due to the noise/artifacts in medical images, annotation burden, with inter- and intra-observer variability, and small size and shape diversity which imply more uncertainty than for larger objects.

Estimating the pose of organs has been investigated in the literature as a slice positioning problem. A set of slices must be optimally selected relative to the pose of the knee [5,14,28,29], shoulder [29], or brain [4,10,12]. These organs are single instances of large objects with specific shapes, standard positions and orientations within the images. On the contrary, aneurysms are very small pathologies with unspecific shapes, undefined locations and numbers. These algorithms can be categorized into registration-based and learning-based methods. Registration-based methods [5,12] rely on rigid transformations, which limits their application to (quasi-)rigid body parts. Learning-based methods [10,20,28,29] typically consist of a two-stage pipeline. The first stage detects the Regions-of-Interest (ROIs), while the second stage regresses/estimates the object orientation for each ROI. For instance, faster-RCNN [22] and V-Net [18] architectures were employed in [29] to localize and segment the orientation plane, defined by a center location and two unit vectors. However, existing methods are mainly intended for low-resolution images like MR scout scans, and their computational demands increase with high-resolution images. One-stage approaches, such as YOLO [21], demonstrate promising performance in object detection with greater flexibility than two-stage approaches.

In this paper, we introduce a novel one-stage method to simultaneously localize, and estimate the size and the orientation of aneurysms from 3D TOF-MRA images. A fast and approximate annotation is used. To address the class imbalance problem, a small patch approach is combined with dedicated data sampling and generation strategies. We follow a landmark approach to estimate the aneurysm pose, while avoiding rotation discontinuity problems associated with Euler angles and quaternions [30]. Furthermore, we propose a 3D extension of YOLO architecture, using a cosine similarity loss for the orientation.

2 Materiels and Methods

2.1 Datasets and Data Annotation

In this work, two TOF-MRA aneurysm datasets were used. The first dataset includes 132 exams (75 female, 57 male) collected at our medical institution between 2015 and 2021 according to the following inclusion criteria: diagnosed unruptured saccular aneurysms smaller than 20 mm, no pre-treated aneurysm or fusiform aneurysm. A single exam was included per patient (i.e. no follow-up exams). All images were acquired using a 3T scanner (GE Discovery MR750w). Acquisition parameters included TR = 28 ms, TE = 3.4 ms, slice thickness= 0.8 mm, and 4 slabs (54 slices/slab), resulting in $512 \times 512 \times 254$ volumes with a $0.47 \times 0.47 \times 0.4\text{mm}^3$ voxel size. Each DICOM data was anonymized on the clinical site before processing. As per the charter of our university hospital, the anonymous use of imaging data acquired in clinical practice is authorized for research purposes, in accordance with the principle of non-opposition of the patient. Each image contained from one (84/132) to five aneurysms (4 subjects), totaling 206 aneurysms with a mean diameter of 3.97 ± 2.32 mm (range: 0.96–19.63 mm). Most aneurysms were small, with 81 below 3 mm and 77 between 3–5 mm.

The second dataset is the public aneurysm dataset [7], which comprises 412 images. After applying the same inclusion criteria as the in-house dataset, 270 images were selected for analysis. Two expert neuroradiologists reviewed the dataset, identifying 7 additional aneurysms and removing 5 aneurysms as they were simple irregularities on the vessel surface. The resulting images contains 164 aneurysms with similar statistics to the first dataset: mean diameter of 3.74 ± 2.17 mm (range: 1.37–13.64 mm), 66 below 3 mm and 72 between 3–5 mm. Each image contained from 0 (130 healthy subjects) to 3 (3 subjects) aneurysms.

Previous works on aneurysm detection and segmentation relied on voxel-wise labeling, which is time-consuming and susceptible to intra- and inter-rater variability. To address these limitations, weak annotations using spheres have been recently investigated [2,7]. Similar to [2], our annotation involves labeling each aneurysm using two points: the center of the neck (i.e. ostium) and another point along its main axis (i.e. dome). This method provides information about aneurysms location, size, and their orientation (see Fig. 1). To simplify the placement of the two points in the volume rendering view, we developed a Python extension for the 3D Slicer software [8], which provides a real-time visualization of the sphere in the canonical cut planes.

2.2 Data Sampling and Generation

Accurate modeling of aneurysm and background properties is crucial for pose estimation tasks. We use small $96 \times 96 \times 96$ voxel patches with an isotropic voxel

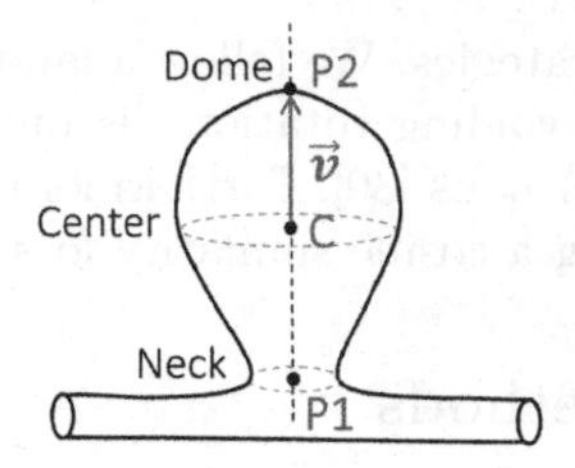

Fig. 1. Fast aneurysm annotation: 2 points ($P1$, $P2$) define a sphere. The ground truth pose is inferred as the center $C = (P1 + P2)/2$ and axis vector $\vec{v} = P2 - C$.

size of 0.4 mm, resulting in 38.4 mm side length patches. This approach is computationally efficient compared to larger patch methods, such as nnDetection [3] and nnUNet [9]. It also allows for the extraction of multiple non-intersecting negative (aneurysm-free) patches from each image for more training data and reliable background modeling. However, this approach introduces a class imbalance problem, as there is only a single positive patch for each aneurysm. To overcome this, we used adapted data sampling strategies. Our first strategy duplicates each positive patch 50 times and applies random distortions at each epoch to synthesize a variety of aneurysm shapes: each control point on a $3 \times 3 \times 3$ lattice enclosing the patch, except the central point, is moved randomly by 3 mm in all 3 space directions, and the distortion field is interpolated using cubic spline interpolation. To guide the model to discriminate between healthy vessels and aneurysms, our second strategy pre-selects 40 non-intersecting negative patches per image, 30 of which are centered on blood vessels by iteratively choosing the brightest voxels as patch centers. Each training epoch used a set composed of all positive patches, completed with random negative patches equally drawn among images (15% of the training set). Random rotations (0 to 180°), shifts (0 to 10 mm) and horizontal flips were applied as data augmentation.

2.3 Neural Network Architecture

Inspired by YOLO [21], we present a one-stage neural network architecture for aneurysm pose estimation in 3D images. As shown in Fig. 2, our architecture follows a grid-based approach and divides the input 3D patch ($96 \times 96 \times 96$ voxels) into $12 \times 12 \times 12 = 1728$ cells of $8 \times 8 \times 8$ voxels.

To encode the input patch into feature maps, we use residual convolutional blocks and down-sampling operations. The *Localization and Orientation Head* splits the encoded feature maps into a grid of 1728 cells using two consecutive convolutional blocks followed by three parallel convolutions. The first convolution generates a confidence probability score indicating whether the cell contains an aneurysm center. For positive cells (i.e. containing an aneurysm center), the second convolution, followed by sigmoid function, predicts the aneurysm center coordinates $C = (C_x, C_y, C_z)$ relative to the cell size, while the third convolution estimates the aneurysm size and its orientation by calculating the axis vector

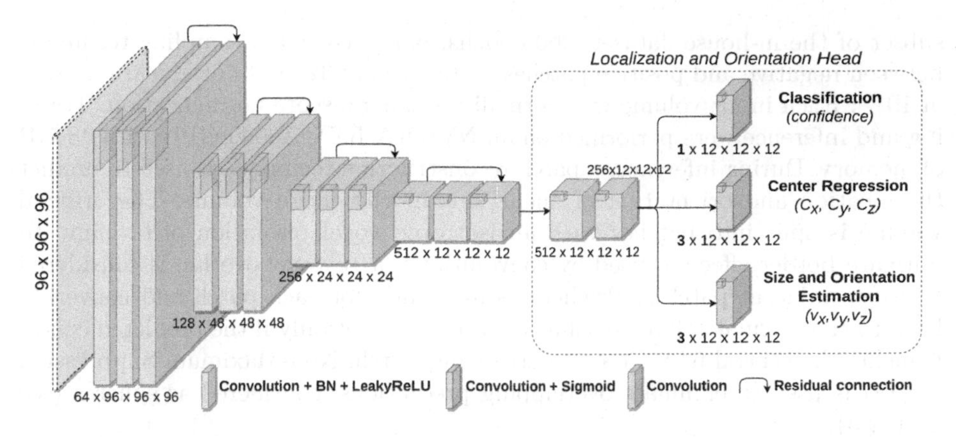

Fig. 2. Our aneurysm pose estimation architecture. 2D feature maps are used only for visualization purposes; their actual sizes in 3D are displayed.

$\vec{v} = (v_x, v_y, v_z)$. The aneurysm radius is given by $r = |\vec{v}|$, and its orientation by $\vec{v}/|\vec{v}|$. Each prediction can also be represented as a sphere (center C, radius r).

2.4 Loss Function

Due to the grid-based nature of our architecture, there is a high imbalance between the number of negative cells and a very small number of positive cells. Inspired by [21], a weighted loss function was employed, which encompasses the sum of terms pertaining to confidence, localization, and orientation estimation.

To optimize the detection confidence, we used the binary cross-entropy (BCE) loss function for both positive (BCE_P) and negative (BCE_N) cells. To prioritize identifying aneurysms over background, we weighted the negative cell term by half the number of positive cells ($\#P$) in the batch (Eq. 1). Aneurysm localization and dimensions are assessed using mean squared error (MSE) (Eq. 2). Orientation estimation is enforced through the cosine similarity of $\vec{v}$ (Eq. 3). These last two terms are only computed on positive cells with a weight of 5 to account for the limited number of such cells.

$$\begin{cases} \text{Confidence} = \text{BCE}_P + 0.5 \times \#P \times \text{BCE}_N & (1) \\ \text{Localization} = 5 \times \text{MSE}(C_x, C_y, C_z, v_x, v_y, v_z) \quad (0 \text{ for negative cells}) & (2) \\ \text{Orientation} = 5 \times (1 - \text{Cosine Similarity}(\vec{v})) \quad (0 \text{ for negative cells}) & (3) \end{cases}$$

2.5 Implementation Details

We implemented our method using PyTorch framework (1.10.0). The model has approximately 28 million parameters, that were optimized using the stochastic gradient descent algorithm. The hyper-parameters were determined using a subset of the in-house dataset: 200 epochs, balanced batch sampling technique between negative and positive patches, batch size of 32, and initial learning rate of 10^{-2}. Each input volume was normalized using z-score normalization. Training and inference were performed on an NVIDIA RTX A6000 GPU with 48 GB of memory. During inference, a patch reconstruction technique is used to predict the location and orientation of aneurysms in the entire volume. The original volume is split into patches with an isotropic voxel resolution of 0.4 mm. To mitigate border effects caused by convolutions, a 16 voxel overlap is considered between adjacent patches. Predictions are made for each patch and converted back to the original volume resolution: a pose is kept only if the predicted center is inside the central $64 \times 64 \times 64$ part of the patch. Non-Maximum Suppression (NMS) is used to eliminate overlapping predictions, considered as spheres (see Sect. 2.3).

2.6 Evaluation Metrics

For the pose estimation task, our method was evaluated based on two standard metrics. First, the Euclidean distance (in mm) was measured between the predicted aneurysm center (C) and its corresponding ground truth (GT). The second metric computes the angular difference (in degrees) between the predicted aneurysm orientation vector ($\vec{v}$) and its corresponding GT.

For the detection task, our evaluation was based on the Intersection-over-Union (IoU) between the predicted and GT spheres at a threshold of 10% [3,16]. A GT sphere with an IoU score above 10% was tagged as a true positive (TP), else it was a false negative (FN). A false positive (FP) was counted for each predicted sphere with no IoU score above 10%. We report the Average Precision metric ($\mathrm{AP}_{0.1}$), as well as the sensitivity score ($\mathrm{Sensitivity}_{0.5}$) and the number of false positives per case ($\mathrm{FPs/case}_{0.5}$), both at a default 50% confidence threshold.

3 Experiments and Results

3.1 Pose Estimation

We conducted 5-fold cross-validation separately on two large datasets (see Sect. 2.1) to evaluate the performance of our method for aneurysm pose estimation. Each dataset was randomly split into five subsets, with 25 or 26 patients per subset for the in-house dataset and 54 patients per subset for dataset [7]. The number of aneurysms and mean aneurysm size for each subset were as follows: (In-house) aneurysms: 47, 32, 45, 38, and 44; size: 4.15 mm, 3.61 mm, 3.88 mm, 3.85 mm, and 4.25 mm; (Dataset [7]) aneurysms: 32, 33, 43, 28, and 28; size: 3.56 mm, 3.19 mm, 4.05 mm, 4.21 mm, and 3.70 mm. We trained five models for each dataset, using four subsets for training and one subset for testing. This resulted in, for each fold, around 9655 training patches for the in-house dataset and 7595 training patches for dataset [7].

The results on both datasets are shown in Table 1. In the in-house dataset, the median (mean $\pm$ std) errors were 0.49 mm (0.54 mm$\pm$0.32) for the aneurysm

Table 1. Pose estimation performance evaluation of our method.

Datasets	Center error (mm)			Orientation error (°)		
	Mean ± std	**Median**	**Range**	**Mean ± std**	**Median**	**Range**
In-house	0.54 ± 0.32	0.49	0.05–1.74	15.26 ± 10.92	11.91	0.21 - 68.35
Dataset [7]	0.51 ± 0.26	0.48	0.05–1.43	14.58 ± 10.53	12.27	1.05–68.30

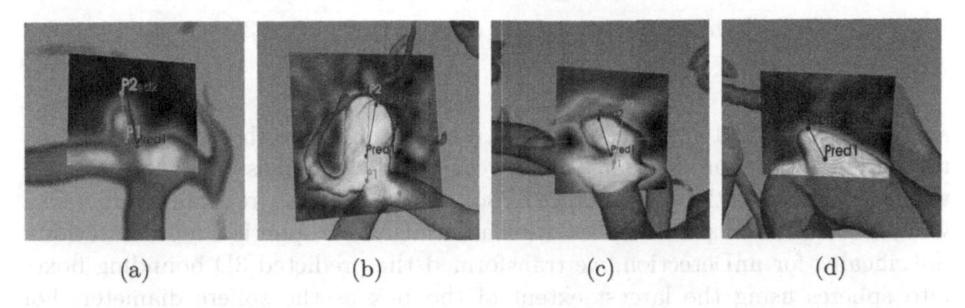

Fig. 3. Qualitative results on dataset [7]: predicted (blue) and GT (green) landmarks. Each reformatted cut plane was determined by rotating around the aneurysm axis passing through the predicted landmarks. The orientation error is (a) 8.2°, (b) 10.62°, (c) 41.54°, (d) unlabelled aneurysm detected by our method. (Color figure online)

center location; and 11.91° (15.26°±10.92) for its orientation. In dataset [7], the median errors were 0.48 mm (0.51 mm±0.26) for the center location; and 12.27° (14.58°±10.53), for the orientation.

Figure 3 illustrates that the pose computed by our method is sufficiently accurate for clinical use. It was used to display a cut plane through aneurysms with diverse shapes and sizes. Figure 3a reports on the case of a small aneurysm (size 1.97 mm). The pose was estimated with 8.20° orientation error and 0.82 mm center location error. This accuracy, especially on the location, makes it possible to infer a cut plane through the aneurysm that is fit for immediate clinical analysis. Similarly, the case of a larger, spherical-shaped aneurysm (size 7.69 mm) is shown in Fig. 3b. Our method estimated the pose with an orientation error of 10.62° and center location error of 0.72 mm. Larger orientation errors occurred in rare cases like the aneurysm in Fig. 3c (size 3.52 mm). We related such errors (here 41.54°) to the complex shape of the aneurysm, that implied annotation uncertainty for the axis orientation. Besides, our method was able to detect some aneurysms that were missed in the initial annotation by radiologists. Figure 3d shows such an aneurysm detection (size 3.50 mm).

3.2 Object Detection

We also evaluated the effectiveness of our method on the classical detection task by comparing it with two public and fully-automated state-of-the-art baselines, nnDetection [3] and nnUNet [9]. nnDetection is based on an improved RetinaNet architecture, which has demonstrated superior performance compared to SSD

Table 2. The results (mean ± std) of aneurysm detection task using dataset [7]. The results of our method on the in-house dataset are added for comparison.

Methods	$AP_{0.1}$ (%)	$\text{Sensitivity}_{0.5}$ (%)	$\text{FPs/case}_{0.5}$
nnDetection [3]	73.68 ± 6.38	84.76 ± 4.72	0.67 ± 0.12
nnUNet [9]	72.46 ± 4.74	71.95 ± 9.11	0.13 ± 0.06
Ours	76.60 ± 5.24	82.93 ± 5.92	0.44 ± 0.04
Ours (In-house)	82.48 ± 6.66	83.01 ± 6.30	0.34 ± 0.11

and Faster RCNN [15]. We used 5-fold cross-validation on the public dataset [7] to guarantee the reproducibility of the results. The 5 models trained in Sect. 3.1 were used to assess the performance of our method. To ensure a fair comparison, we converted the output of nnDetection and nnUNet to spherical representations. Specifically, for nnDetection, we transformed the predicted 3D bounding boxes into spheres using the largest extent of the box as the sphere diameter. For nnUNet, we fitted one sphere on each connected component from the segmented voxel image. The diameter was computed as the maximum distance between two voxel locations, and the confidence score as the maximum predicted voxel value.

As shown in Table 2, our method exhibited competitive performance compared to the two baselines achieving an $AP_{0.1}$ score of 76.60% (nnDetection: 73.68% and nnUNet: 72.46%). Additionally, our method demonstrated a good trade-off between sensitivity and FP/case, with a $\text{Sensitivity}_{0.5}$ score of 82.93% associated with 0.44 $\text{FPs/case}_{0.5}$. In comparison, based on Free-response Receiver Operating Characteristic (FROC) curves, nnUNet achieves a maximum sensitivity of 81.90% with a higher FP/case of 1.04, while nnDetection achieves the same sensitivity of 82.93% but with a higher FP/case of 0.51.

4 Conclusion

In this paper, we proposed a novel one-stage deep learning approach for aneurysm pose estimation from TOF-MRA images, which can also be used for the classical detection task. It was evaluated using two large datasets, including a public one [7]. The results demonstrate the effectiveness of our proposed method in both tasks.

In the pose estimation task, our method achieved good and similar performance on both datasets, accurately estimating the pose of aneurysms with diverse shapes and sizes. Rare errors in orientation were primarily due to small aneurysms and sometimes complex aneurysm shapes, leading to weak and uncertain GT annotations. Specifically, on the public dataset [7], the median orientation error was 14.79° for small aneurysms (<3 mm), 11.49° for medium-sized aneurysms (3–5 mm), and 10.69° for large aneurysms. A current work consists in giving a better clinical definition of the aneurysm axis to reduce this error.

In the aneurysm detection task, our proposed method exhibited promising performance compared to two state-of-the-art baselines, nnDetection [3] and nnUNet [9], with an average precision score of 76.60% and a good balance between sensitivity and FPs/case scores. Besides, these baselines are more computationally demanding compared to our method, which is based on small non-intersecting patches. Out of the 164 aneurysms in dataset [7], half of the 28 FN aneurysms had a size below 3 mm. Part of these misses are related to the annotation uncertainty on such aneurysms, which are difficult to diagnose in TOF-MRA [11]. Nevertheless, our future work will address this specific class of aneurysms, including the management of multiple annotators for finer uncertainty modeling.

Our method represents a promising step towards automated aneurysm pose estimation and detection, offering several advantages over existing approaches. It demonstrated multi-task learning capabilities by simultaneously localizing, and estimating the size and the orientation of aneurysms in a single forward pass. Preliminary qualitative tests are hopeful indicators for its clinical utility.

Acknowledgment. The authors would like to acknowledge the financial support provided by the Grand-Est Region and the University Hospital (CHRU) of Nancy, France.

References

1. Arimura, H., Li, Q., Korogi, Y., et al.: Automated computerized scheme for detection of unruptured intracranial aneurysms in three-dimensional magnetic resonance angiography1. Acad. Radiol. **11**(10), 1093–1104 (2004)
2. Assis, Y., Liao, L., Pierre, F., Anxionnat, R., Kerrien, E.: An efficient data strategy for the detection of brain aneurysms from mra with deep learning. In: Engelhardt, S., et al. (eds.) DGM4MICCAI/DALI -2021. LNCS, vol. 13003, pp. 226–234. Springer, Cham (2021). https://doi.org/10.1007/978-3-030-88210-5_22
3. Baumgartner, M., Jäger, P.F., Isensee, F., Maier-Hein, K.H.: nnDetection: a self-configuring method for medical object detection. In: de Bruijne, M., et al. (eds.) MICCAI 2021. LNCS, vol. 12905, pp. 530–539. Springer, Cham (2021). https://doi.org/10.1007/978-3-030-87240-3_51
4. Benner, T., Wisco, J.J., van der Kouwe, A.J., Fischl, B., et al.: Comparison of manual and automatic section positioning of brain MR images. Radiology **239**(1), 246–254 (2006)
5. Bystrov, D., Pekar, V., Young, S., Dries, S.P., Heese, H.S., van Muiswinkel, A.M.: Automated planning of MRI scans of knee joints. In: Medical Imaging 2007: Visualization and Image-Guided Procedures, vol. 6509, pp. 1023–1031. SPIE (2007)
6. Çiçek, Ö., Abdulkadir, A., Lienkamp, S.S., Brox, T., Ronneberger, O.: 3D U-Net: learning dense volumetric segmentation from sparse annotation. In: Ourselin, S., Joskowicz, L., Sabuncu, M.R., Unal, G., Wells, W. (eds.) MICCAI 2016. LNCS, vol. 9901, pp. 424–432. Springer, Cham (2016). https://doi.org/10.1007/978-3-319-46723-8_49
7. Di Noto, T., Marie, G., Tourbier, S., et al.: Towards automated brain aneurysm detection in TOF-MRA: open data, weak labels, and anatomical knowledge. In: Neuroinformatics pp. 1–14 (2022)
8. Fedorov, A., Beichel, R., Kalpathy-Cramer, J., et al.: 3D slicer as an image computing platform for the quantitative imaging network. Magnetic Resonance Imaging **30**(9), 1323–1341 (2012). https://slicer.org, pMID: 22770690

9. Isensee, F., Jaeger, P.F., Kohl, S.A., Petersen, J., Maier-Hein, K.H.: nnU-Net: a self-configuring method for deep learning-based biomedical image segmentation. Nat. Methods **18**(2), 203–211 (2021)
10. Iskurt, A., Becerikli, Y., Mahmutyazicioglu, K.: Automatic identification of landmarks for standard slice positioning in brain MRI. J. Magn. Reson. Imaging **34**(3), 499–510 (2011)
11. Jang, M., Kim, J., Park, J., et al.: Features of "false positive" unruptured intracranial aneurysms on screening magnetic resonance angiography. PloS one **15**(9), e0238597 (2020)
12. van der Kouwe, A.J., et al.: On-line automatic slice positioning for brain MR imaging. Neuroimage **27**(1), 222–230 (2005)
13. Lall, R., Eddleman, C.S., Bendok, B.R., et al.: Unruptured intracranial aneurysms and the assessment of rupture risk based on anatomical and morphological factors: sifting through the sands of data. Neurosurg. Focus **26**(5), E2 (2009)
14. Lecouvet, F.E., Claus, J., Schmitz, P., Denolin, V., Bos, C., Vande Berg, B.C.: Clinical evaluation of automated scan prescription of knee MR images. J. Magnetic Reson. Imaging Official J. Inter. Soc. Magnetic Reson. Med. **29**(1), 141–145 (2009)
15. Lin, T.Y., Goyal, P., Girshick, R., He, K., Dollár, P.: Focal loss for dense object detection. In: Proceedings of the IEEE International Conference on Computer Vision, pp. 2980–2988 (2017)
16. Maier-Hein, L., et al.: Metrics reloaded: Pitfalls and recommendations for image analysis validation. arXiv preprint arXiv:2206.01653 (2022)
17. Miao, S., Lucas, J., Liao, R.: Automatic pose initialization for accurate 2D/3D registration applied to abdominal aortic aneurysm endovascular repair. In: Medical Imaging 2012: Image-Guided Procedures, Robotic Interventions, and Modeling, vol. 8316, pp. 243–250. SPIE (2012)
18. Milletari, F., Navab, N., Ahmadi, S.A.: V-net: fully convolutional neural networks for volumetric medical image segmentation. In: 2016 Fourth International Conference on 3D Vision (3DV), pp. 565–571. Ieee (2016)
19. Nakao, T., Hanaoka, S., Nomura, Y., el al.: Deep neural network-based computer-assisted detection of cerebral aneurysms in MR angiography. Journal of Magnetic Resonance Imaging **47**(4), 948–953 (2018)
20. Reda, F.A., Zhan, Y., Zhou, X.S.: A Steering engine: learning 3-D anatomy orientation using regression forests. In: Navab, N., Hornegger, J., Wells, W.M., Frangi, A.F. (eds.) MICCAI 2015. LNCS, vol. 9351, pp. 612–619. Springer, Cham (2015). https://doi.org/10.1007/978-3-319-24574-4_73
21. Redmon, J., Farhadi, A.: Yolov3: an incremental improvement. arXiv preprint arXiv:1804.02767 (2018)
22. Ren, S., He, K., Girshick, R., Sun, J.: Faster R-CNN: towards real-time object detection with region proposal networks. In: Advances in Neural Information Processing Systems 28 (2015)
23. Sichtermann, T., Faron, A., Sijben, R., et al.: Deep learning-based detection of intracranial aneurysms in 3D TOF-MRA. Am. J. Neuroradiol. **40**(1), 25–32 (2019)
24. Stember, J.N., Chang, P., Stember, D.M., et al.: Convolutional neural networks for the detection and measurement of cerebral aneurysms on magnetic resonance angiography. J. Digit. Imaging **32**(5), 808–815 (2019)
25. Timmins, K.M., van der Schaaf, I.C., Bennink, E., et al.: Comparing methods of detecting and segmenting unruptured intracranial aneurysms on TOF-MRAs: the ADAM challenge. Neuroimage **238**, 118216 (2021)
26. Ueda, D., Yamamoto, A., Nishimori, M., et al.: Deep learning for MR angiography: automated detection of cerebral aneurysms. Radiology **290**(1), 187–194 (2019)

27. Yang, X., Blezek, D.J., Cheng, L.T., et al.: Computer-aided detection of intracranial aneurysms in MR angiography. J. Digit. Imaging **24**(1), 86–95 (2011)
28. Zhan, Y., Dewan, M., Harder, M., Krishnan, A., Zhou, X.S.: Robust automatic knee MR slice positioning through redundant and hierarchical anatomy detection. IEEE Trans. Med. Imaging **30**(12), 2087–2100 (2011)
29. Zeng, K., Zhao, Y., Zhao, Y., et al.: Deep learning solution for medical image localization and orientation detection. Med. Image Anal. **81**, 102529 (2022)
30. Zhou, Y., Barnes, C., Lu, J., Yang, J., Li, H.: On the continuity of rotation representations in neural networks. In: Proceedings of the IEEE/CVF Conference on Computer Vision and Pattern Recognition, pp. 5745–5753 (2019)

Joint Optimization of a β-VAE for ECG Task-Specific Feature Extraction

Viktor van der Valk[1](✉), Douwe Atsma[3], Roderick Scherptong[3], and Marius Staring[2]

[1] TECObiosciences GmbH, Landshut, Germany
viktorvandervalk@gmail.com

[2] Department of Radiology, Leiden University Medical Center, Leiden, The Netherlands

[3] Department of Cardiology, Leiden University Medical Center, Leiden, The Netherlands

Abstract. Electrocardiography is the most common method to investigate the condition of the heart through the observation of cardiac rhythm and electrical activity, for both diagnosis and monitoring purposes. Analysis of electrocardiograms (ECGs) is commonly performed through the investigation of specific patterns, which are visually recognizable by trained physicians and are known to reflect cardiac (dis)function. In this work we study the use of β-variational autoencoders (VAEs) as an explainable feature extractor, and improve on its predictive capacities by jointly optimizing signal reconstruction and cardiac function prediction. The extracted features are then used for cardiac function prediction using logistic regression. The method is trained and tested on data from 7255 patients, who were treated for acute coronary syndrome at the Leiden University Medical Center between 2010 and 2021. The results show that our method significantly improved prediction and explainability compared to a vanilla β-VAE, while still yielding similar reconstruction performance.

Keywords: Explainable AI · ECG · β-VAE · feature extraction · LVF prediction

1 Introduction

The electrocardiogram (ECG), is one of the most widely used methods to analyze cardiac morphology and function, by measuring the electrical signal from the heart with multiple electrodes. ECG data is used by clinicians for both diagnostic and monitoring purposes in various cardiac syndromes. A 12-lead ECG is routinely obtained in patients to diagnose and monitor disease development. However, for the interpretation of the ECG signal, the knowledge of an expert is required. Physicians usually analyze the ECG through the recognition of specific patterns, known to be associated with disease. This however requires substantial expertise, and potentially additional relevant information exists in a 12-lead ECG

H. Greenspan et al. (Eds.): MICCAI 2023, LNCS 14221, pp. 554–563, 2023.
https://doi.org/10.1007/978-3-031-43895-0_52

missed by human interpretation. Deep learning has already proven its usefulness in the interpretation of the ECG signal in multiple classification challenges [1,3] and more recently also in feature discovery by means of explainable AI algorithms [2,7,9,10,15]. The explainablity of AI algorithms is especially valued in medical settings, where trusting a black box AI algorithm is undesirable [1].

VAEs and in particular β-VAEs have been used as unsupervised explainable ECG feature generators in the explainable AI algorithms mentioned above [6]. It was shown that a β-VAE, trained on reconstruction of the ECG signal, is able to extract features from the ECG signal that can be made more interpretable by visualization of reconstructed latent space samples with the decoder of the β-VAE [10]. This is a first step towards an explainable deep learning pipeline for ECG analysis. However, the features generated by a β-VAE when only trained to minimized reconstruction loss, are likely not optimal for task specific predictions.

The aim of this paper is to explore further improvement of the latent features by improving their explainability and prediction performance. This is clinically relevant but unexplored for the post myocardial infarction setting. We propose to improve explainability by reducing the dimension of the latent space to a level more manageable for human assessment, while encouraging outcome specific information to be captured in a small part of the latent space, and while maintaining ECG reconstruction performance for visual assessment. To achieve this, we propose a novel method to jointly optimize the β-VAE with a combination of a task specific prediction loss for a subset of the latent space, and KL-divergence and reconstruction loss for the entire latent space. The task chosen to optimize here is left ventricular function (LVF), one of the most important determinants of prognosis in patients with cardiac disease. Current assessment of LVF requires advanced imaging methods and interpretation by a trained professional. The ECG, on the other hand, can be obtained by a patient at home. In combination with automated analysis this would facilitate remote monitoring of LVF in patients.

2 Methods

2.1 Data

To train the models for both reconstruction and LVF prediction, two datasets were used: i) A non-labeled dataset consisting of 119,886 raw 10 s 12-lead ECG signals taken at 500 Hz from 7255 patients diagnosed with acute coronary syndrome between 2010 and 2021 at the Leiden University Medical Center, the Netherlands; ii) A labeled dataset of 33,610 ECGs from 2736 patients of the same cohort. This dataset was labeled by visual assessment of an echocardiogram performed within 3 days before or after the ECG. The label categories, normal, mild, moderate and severe impairment were binarized for model training. When the ECG was taken within two weeks after cardiac intervention a 1-day margin was used. If a cardiac intervention was performed between ECG and echocardiography, the case was excluded. 11.5% of the ECGs were labeled with a moderate to severe impaired LVF. The institutional review board approved the study protocol (nWMODIV2_2022006) and waived the obligation to obtain informed consent.

2.2 Data Preprocessing

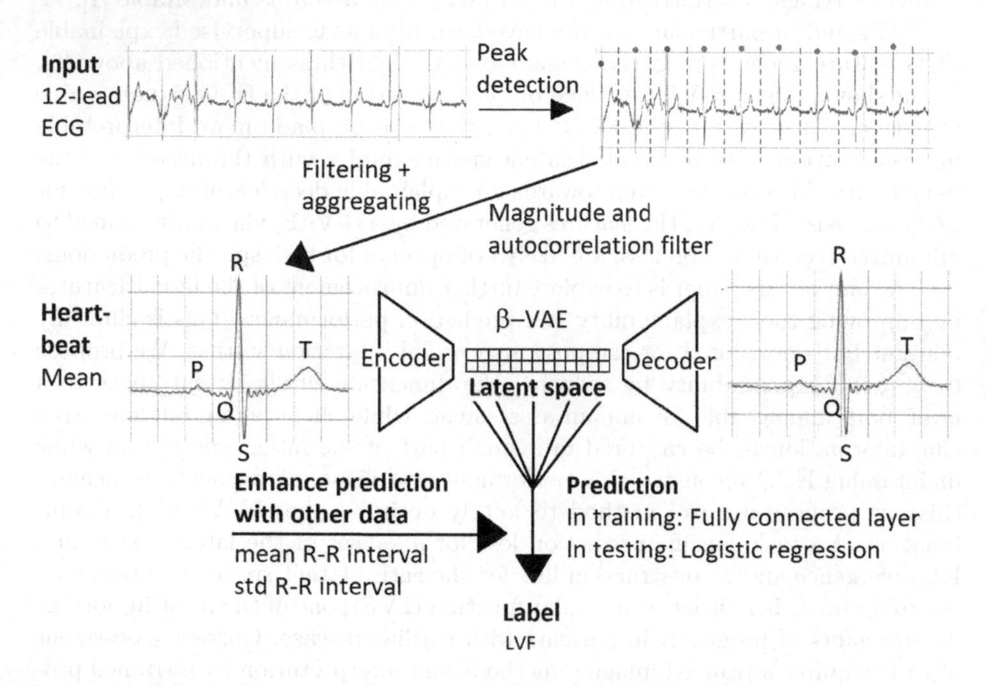

Fig. 1. Preprocessing, feature extraction and prediction pipeline.

The raw ECG signals were first split into separate heartbeats (400ms before and after the R-peak, the largest peak in the ECG, that represents depolarization of the ventricles) with a peak detection method inspired by RPNet, a U-Net structured CNN with inception blocks, that was trained on manually labeled peak locations [16]. The heartbeats were then filtered with a magnitude and an autocorrelation filter. The magnitude filter removed heartbeats with an average magnitude below a set threshold. The autocorrelation filter removed signals where both the mean and maximum autocorrelation between the heartbeats were below a set threshold. These two criteria were used, since ECG signals showing multiple rhythms can result in low mean autocorrelation, but, if not noisy, will not result in low maximum autocorrelation. The remaining heartbeats were then averaged per ECG lead. The μ and σ of the intervals of the subsequent R peaks were used as an additional feature for LVF prediction.

2.3 Model Overview

To investigate a general improvement to the VAE feature extraction pipeline [7,9,10,15], the proposed method was tested with two architectures: i) A small

VAE with 300k parameters consisting of an encoder and a mirrored decoder. Both parts contained 7 2D convolutional layers, of which 3 were residual layers, with respective channel sizes of [8,16,32,64,64,64,64] and a kernel size of 5; ii) A second larger VAE from the FactorECG pipeline as proposed by Van de Leur *et al.* (2022) [10] with 50M parameters. The VAEs were both extended at the bottleneck (the latent space, of size L), with a single fully connected layer for output prediction, in this case the LVF label, see Fig. 1. The μ and σ of the RR intervals (time between two subsequent R peaks), were added to the input of the prediction layer, since the information represented by these features is lost in averaging the heartbeats. To maintain explainability of the extracted features, only one fully connected layer is used, as otherwise the features will become weighted combinations of the latent space values, which makes visualization with the decoder and subsequent interpretation complex. However, for pure prediction performance, additional fully connected layers may have been helpful. The extracted features, again with the μ and σ of the RR interval, were subsequently analyzed with logistic regression using regularized l1 and l2 penalties on the LVF prediction task, ignoring the output of the prediction layer in the VAE. The VAEs were build and trained in the PyTorch 1.12 framework and trained on a Quadro RTX 6000 GPU with CUDA 11.4 [12,13], while for logistic regression we used the Scikit-learn toolbox [14]. The implementation of our models will be made publicly available via GitHub at https://github.com/ViktorvdValk/Task-Specific-VAE.

2.4 Model Training

The β-VAE was first pretrained in a self-supervised manner with the mean heartbeats of all filtered ECG signals, minimizing i) the mean squared reconstruction error (MSE) between the input and output ECG, and ii) the KL-divergence between the output of the encoder and the standard normal distribution. The KL-divergence loss was weighted with a β factor, like in the original paper [6]. This pretrained VAE was then fine-tuned in two-steps, first the encoder and decoder were fixed, and only the prediction layer was trained, then all layers were trained end-to-end. This training scheme was used to ensure more stable training. For these fine-tuning steps, the loss function was complemented with a binary cross-entropy loss, which was weighted with a γ factor. The *task naive* VAE resulting from pretraining was compared to the *task specific* VAE resulting from both fine-tuning steps. For pretraining, both datasets were combined and split in a training (85%) and a test set (15%). 5-fold cross validation was done with the training set with again an 85%:15% ratio between training and validation set. For fine-tuning, the same procedure was used on just the labeled dataset, making sure labeled ECGs were in the same set in both cases. All data splits were grouped by patient and stratified by label in case of the labeled data splits. Both pretraining and fine-tuning were done until convergence, i.e. until the loss on the validation set stopped improving for 25 epochs. This was done to prevent the advantage of additional training of the *task specific* network. To prevent overfitting, balanced sampling and regularization by means of drop out layers and the Adam optimizer with weight decay were used, this was especially necessary in the fine-tuning phase. To prevent gradient explosion, gradient clipping and He initialization were used [5].

2.5 Feature Evaluation

The differences between the features from the *task naive* and *task specific* VAEs, were compared w.r.t. reconstruction and prediction. For reconstruction, both MSE and correlation between input and output ECG, and for prediction the Area Under the Receiver Operator Characteristic Curve (AUROC) and the macro-averaged F1 score were used. Significant difference between AUROC scores was calculated as proposed in Hanley & McNeil (1983) [4]. For visualization of the representation of a latent space feature f in a so called factor traversal, all features except f were sampled at their mean, while f was sampled in a range between $\mu - 3\sigma$ and $\mu + 3\sigma$. Using these samples as input for the decoder, creates a representation of that feature, which can give insight in ECG features that are important for LVF prediction.

2.6 Baseline Methods

As a baseline method, a principal component analysis (PCA) was performed on the preprocessed ECGs, to extract features. PCA can be considered an ordered task naive linear feature extractor that focuses on the axis of the largest variance, in contrast to the VAEs which are non-ordered non-linear feature extractors, that are optimized for reconstruction. A logistic regression predictor with just sex and age as input was used as an additional baseline.

3 Experiments and Results

3.1 Experiments

The proposed pipeline contains several hyper-parameters, of which the latent space size L was optimized in this study. The influence of the β parameter was also briefly addressed. L was optimized for its importance in the explainability and the reconstruction and prediction quality of the model. A higher L increases the complexity of the model, and consequently decreases its explainability. An L that is too low, on the other hand, restricts the capacity of the model for reconstruction and prediction. The PCA baseline method was considered to give an upper bound of L, since the number of principal components, the PCA analog for L, indicates how many values would be needed to capture sufficient information.

3.2 Hyperparameter Optimization

The influences of γ on prediction and reconstruction performance was small and was therefore fixed to 500. The influence of L on prediction quality can be seen in Fig. 2. The PCA baseline performs more or less equal to the *task naive* networks for all L. For the *task specific* networks, the F1 scores are higher than their *task naive* counterparts and the PCA baseline, especially for lower L. The *task specific* VAEs already reached their best prediction performance starting at $L = 2$, as compared to the *task naive* VAEs and the PCA baseline that reach their best prediction performance from $L = 30$. The influence of L on reconstruction can be seen in Fig. 2a and b. All networks perform equal to the PCA baseline for low latent dimensions. The reconstruction for the small VAE and the FactorECG VAE does not seems to improve any further for respectively $L > 20$ and $L > 15$, where the PCA baseline reconstruction keeps improving with L. However, setting β to 0 and thereby ablating the variational nature of the VAEs prevents this stagnation of reconstruction performance. The *task specific* networks perform equally well as their *task naive* counterparts, which suggests that the additional joint optimization does not have a major negative impact on reconstruction. The optimization shows that the relevant information for LVF prediction in the ECG signal can be captured in just two features by both VAEs. Reconstruction, on the other hand, requires at least 10/15 features for the VAEs to reach maximum performance. Therefore, in another experiment, a *split task* VAE was trained, in which 8 of the latent space features where only optimized for reconstruction and only 2 also for prediction.

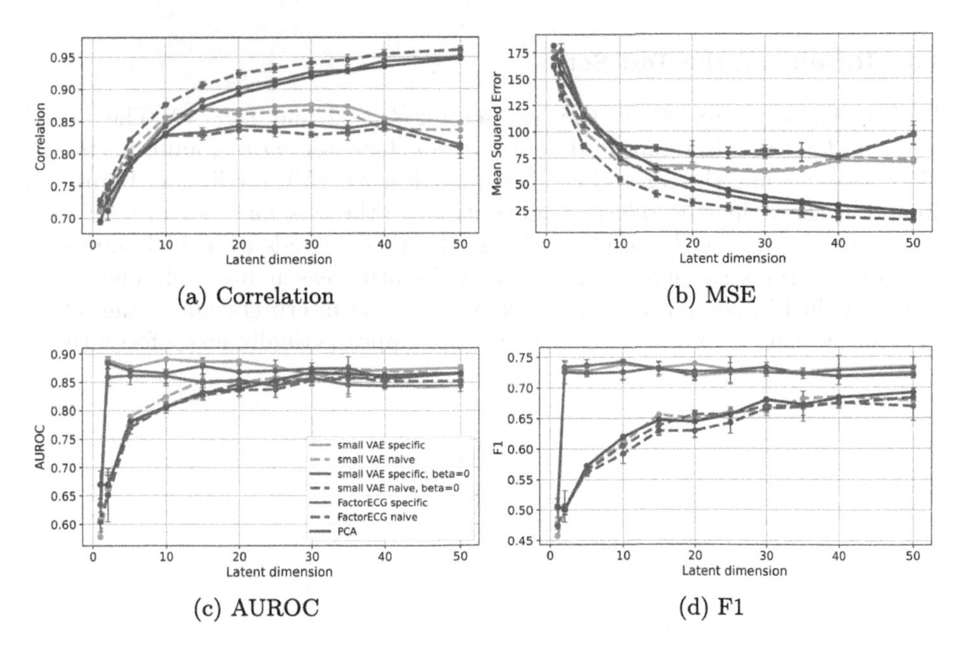

(a) Correlation (b) MSE

(c) AUROC (d) F1

Fig. 2. Influence of the latent dimension L on reconstruction quality: (a) correlation and (b) MSE, and on prediction quality: (c) AUROC and (d) F1-score, for various models. Plotted are the mean and standard deviation of 5-fold cross-validation on the validation set.

Table 1. Reconstruction and LVF prediction comparison on the test set for the *task naive* and *task specific* architectures. The results show the μ of 5-fold cross-validation. AUROC is shown with its 95% confidence interval.* p-value < 0.01 between AUROC of task naive and specific method for all folds.

Architecture	L	MSE	Correlation	AUROC	F1
Sex and age	2	–	–	0.556 (0.520–592)	0.474
PCA	2	147	0.724	0.656 (0,624–0.688)	0.496
Small VAE task naive	2	133	0.739	0.686 (0.655–0.716)	0.503
Small VAE task specific	2	164	0.711	0.842* (0.822–0.861)	0.682
Small VAE split task	2 (10)	76.5	0.838	0.839 (0.819–0.859)	0.695
Small VAE split task $\beta = 0$	2 (10)	73.6	0.838	0.846 (0.823–0.862)	0.674
FactorECG [10] task naive	2	139	0.735	0.685 (0.654–0.715)	0.507
FactorECG [10] task specific	2	161	0.724	0.770* (0.745–0.796)	0.695
PCA	10	77.2	0.826	0.761 (0.735–0.787)	0.580
Small VAE task naive	10	63.1	0.854	0.803 (0.781–0.826)	0.586
Small VAE task specific	10	70.6	0.847	0.853* (0.834–0.871)	0.679
FactorECG [10] task naive	10	84.6	0.820	0.770 (0.745–0.796)	0.579
FactorECG [10] task specific	10	87.2	0.822	0.833* (0.813–0.854)	0.707

3.3 Results on the Test Set

Table 1 shows the results on the test set for $L = 2$ and $L = 10$. The *(split) task specific* networks significantly outperform their *task naive* counterparts, the PCA baseline, and the sex and age benchmark w.r.t. LVF prediction. Figure 3a, 3b and 3c show that the *split task*, in contrast to the *task naive* small VAE with $\beta = 0$ and $\beta = 4$ can be used to encode the ECG signals to a landscape that visually separates the signals based on LVF status reasonably well. The factor traversals in Fig. 3e and d show an example of the interpretation of the latent features. Setting β to 0, creates features that appear visually less informative.

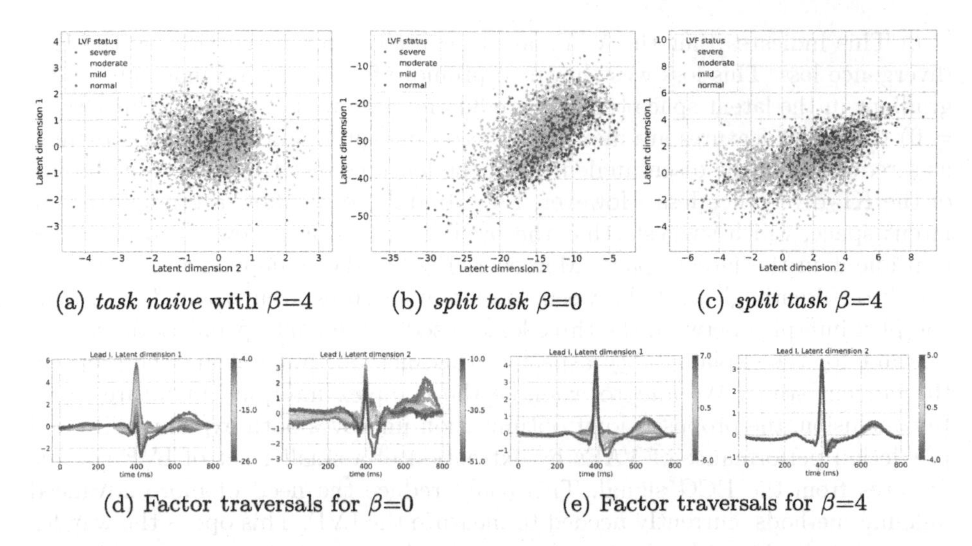

(a) ***task naive*** **with β=4** (b) ***split task*** **β=0** (c) ***split task*** **β=4**

(d) **Factor traversals for β=0** (e) **Factor traversals for β=4**

Fig. 3. Comparison of the latent space for different values of β, for the small VAE. For *task specific* methods, the scatter plots show the two dimensions of the latent space that are optimized for prediction: (a) *task naive* ($\beta = 4$); (b) *split task* ($\beta = 0$); (c) *split task* ($\beta = 4$). The latent space factor traversals (d) and (e) show the visual representation of the features for Lead I of the 12-lead ECG signal: (d) $\beta = 0$; (e) $\beta = 4$.

4 Discussion

Joint optimization of a β-VAE successfully generated features that contain more information about LVF, without hampering reconstruction of the ECG signal. We hypothesize that the β-VAEs have multiple optima for ECG reconstruction of which only some generate features that are relevant for LVF prediction. This study shows that joint optimization will favor this desired subset of optima, and that this is true for different architectures. In addition, we showed that jointly optimizing only a subset of the latent space features for prediction, results in aggregation of the predictive information, thereby improving explainability.

The AUROC score of the FactorECG VAE prediction is similar when compared to van der Leur *et al.* (2022) [10] (AUROC≈0.9 for $L = 36$). However, the proposed small VAE achieved equal if not better reconstruction and prediction performance with less than 1% of the parameters as shown in Fig. 2.

The F1 score is considered more robust than the AUROC score with data imbalance, which is the case here [8]. From Fig. 2d we can therefore conclude that the *task specific* networks outperform the *task naive* networks for any L. The differences between the *task specific* networks and their *task naive* versions in prediction, at similar reconstruction, indicate that the ECG signal can be summarized with a set of latent features of which only a subset is important for LVF prediction. The joint optimization promotes the extraction of this subset especially when L is small. Figure 2a and b show that the PCA baseline outperforms both VAEs in reconstruction for $L > 20$ when $\beta = 4$, but not for β

= 0. This indicates that the VAEs are restricted in reconstruction by the KL-divergence loss. This loss was shown to promote feature disentanglement and a gradient in the latent space [11]. Figure 3d and e show that without this loss (β = 0) the latent features are more complex to interpret. This could be explained as a reduction of the disentanglement of the features resulting from the absence of the KL-divergence loss. However, Fig. 3b and c both show a gradient in the latent space, which suggests that the prediction loss on its own also promotes a gradient in the latent space. Moreover, Fig. 3c shows dependence, and thus a lack of disentanglement, between the latent features even when $\beta = 4$. This complex interplay between the three losses used in the joint optimization, is very relevant for the explainability aspect of this method, but beyond the scope of the current study. We aim to examine the complex interplay in future work. In conclusion, the proposed joint optimization improves both explainability and prediction performance of VAEs by extraction of a smaller set of LVF specific features from the ECG signal. This could reduce the need of more advanced imaging methods, currently needed to measure the LVF. This opens the way for remote monitoring of left ventricular function in patients.

Acknowledgment. This project has received funding from the European Union's Horizon 2020 research and innovation programme under the Marie Sklodowska-Curie grant agreement No 860173.

References

1. Alday, E.A.P., et al.: Classification of 12-lead ECGs: the PhysioNet/computing in cardiology challenge 2020. Physiol. Meas. **41**(12), 124003 (2020)
2. Basu, S., Wagstyl, K., Zandifar, A., Collins, L., Romero, A., Precup, D.: Early prediction of Alzheimer's disease progression using variational autoencoders. In: Shen, D., et al. (eds.) MICCAI 2019. LNCS, vol. 11767, pp. 205–213. Springer, Cham (2019). https://doi.org/10.1007/978-3-030-32251-9_23
3. Clifford, G.D., et al.: AF classification from a short single lead ECG recording: the PhysioNet/computing in cardiology challenge 2017. In: Computing in Cardiology (CinC), pp. 1–4 (2017)
4. Hanley, J.A., McNeil, B.J.: A method of comparing the areas under receiver operating characteristic curves derived from the same cases. Radiology **148**(3), 839–843 (1983)
5. He, K., Zhang, X., Ren, S., Sun, J.: Delving deep into rectifiers: surpassing human-level performance on ImageNet classification. In: Proceedings of the IEEE International Conference on Computer Vision, pp. 1026–1034 (2015)
6. Higgins, I., et al.: beta-VAE: learning basic visual concepts with a constrained variational framework. In: International Conference on Learning Representations (2017). http://openreview.net/forum?id=Sy2fzU9gl
7. Jang, J.H., Kim, T.Y., Lim, H.S., Yoon, D.: Unsupervised feature learning for electrocardiogram data using the convolutional variational autoencoder. PLoS ONE **16**(12), 1–16 (2021)
8. Jeni, L.A., Cohn, J.F., De La Torre, F.: Facing imbalanced data-recommendations for the use of performance metrics. In: 2013 Humaine Association Conference on Affective Computing and Intelligent Interaction, pp. 245–251. IEEE (2013)

9. Kuznetsov, V.V., Moskalenko, V.A., Zolotykh, N.Y.: Electrocardiogram generation and feature extraction using a variational autoencoder. arXiv, pp. 1–6 (2020). http://arxiv.org/abs/2002.00254
10. van de Leur, R.R., et al.: Improving explainability of deep neural network-based electrocardiogram interpretation using variational auto-encoders. Eur. Heart J.-Digit. Health **3**(3), 390–404 (2022)
11. Mathieu, E., Rainforth, T., Siddharth, N., Teh, Y.W.: Disentangling disentanglement in variational autoencoders. In: International Conference on Machine Learning, pp. 4402–4412. PMLR (2019)
12. NVIDIA: Vingelmann, P., Fitzek, F.H.: CUDA, release: 10.2.89 (2020). http://developer.nvidia.com/cuda-toolkit
13. Paszke, A., et al.: PyTorch: an imperative style, high-performance deep learning library. In: Advances in Neural Information Processing Systems, pp. 8024–8035 (2019). http://papers.neurips.cc/paper/9015-pytorch-an-imperative-style-high-performance-deep-learning-library.pdf
14. Pedregosa, F., et al.: Scikit-learn: machine learning in python. J. Mach. Learn. Res. **12**, 2825–2830 (2011)
15. Van Steenkiste, T., Deschrijver, D., Dhaene, T.: Interpretable ECG beat embedding using disentangled variational auto-encoders. In: IEEE International Symposium on Computer-Based Medical Systems (CBMS), pp. 373–378 (2019)
16. Vijayarangan, S., Vignesh, R., Murugesan, B., Preejith, S., Joseph, J., Sivaprakasam, M.: RPnet: a deep learning approach for robust R peak detection in noisy ECG. In: International Conference of the IEEE Engineering in Medicine & Biology Society (EMBC), pp. 345–348 (2020)

Adaptive Multi-scale Online Likelihood Network for AI-Assisted Interactive Segmentation

Muhammad Asad[1(✉)], Helena Williams[2], Indrajeet Mandal[3], Sarim Ather[3], Jan Deprest[2], Jan D'hooge[4], and Tom Vercauteren[1]

[1] School of Biomedical Engineering and Imaging Sciences, King's College London, London, UK
muhammad.asad@kcl.ac.uk
[2] Department of Development and Regeneration, KU Leuven, Belgium
[3] Radiology Department, Oxford University Hospitals NHS Foundation Trust, Oxford, UK
[4] Department of Cardiovascular Sciences, KU Leuven, Belgium

Abstract. Existing interactive segmentation methods leverage automatic segmentation and user interactions for label refinement, significantly reducing the annotation workload compared to manual annotation. However, these methods lack quick adaptability to ambiguous and noisy data, which is a challenge in CT volumes containing lung lesions from COVID-19 patients. In this work, we propose an adaptive multi-scale online likelihood network (MONet) that adaptively learns in a data-efficient online setting from both an initial automatic segmentation and user interactions providing corrections. We achieve adaptive learning by proposing an adaptive loss that extends the influence of user-provided interaction to neighboring regions with similar features. In addition, we propose a data-efficient probability-guided pruning method that discards uncertain and redundant labels in the initial segmentation to enable efficient online training and inference. Our proposed method was evaluated by an expert in a blinded comparative study on COVID-19 lung lesion annotation task in CT. Our approach achieved 5.86% higher Dice score with 24.67% less perceived NASA-TLX workload score than the state-of-the-art. Source code is available at: https://github.com/masadcv/MONet-MONAILabel.

1 Introduction

Deep learning methods for automatic lung lesion segmentation from CT volumes have the potential to alleviate the burden on clinicians in assessing lung damage and disease progression in COVID-19 patients [20–22]. However, these

---

Supplementary Information The online version contains supplementary material available at https://doi.org/10.1007/978-3-031-43895-0_53.

H. Greenspan et al. (Eds.): MICCAI 2023, LNCS 14221, pp. 564–574, 2023.
https://doi.org/10.1007/978-3-031-43895-0_53

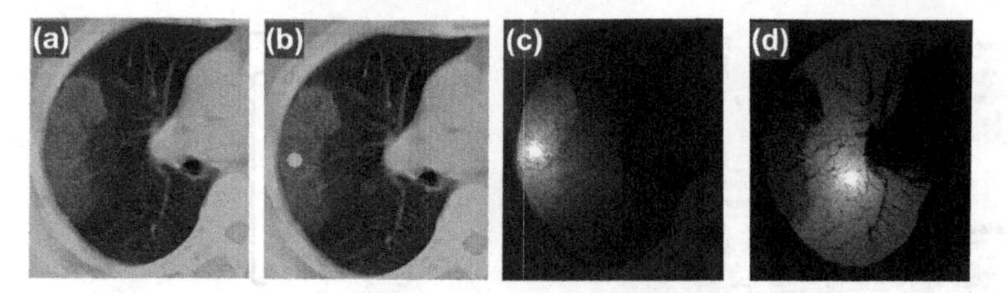

Fig. 1. Adaptive online training weights: (a) input, (b) foreground / background scribbles, (c) foreground and (d) background weights using $\tau = 0.2$ in Eq. (2). (Color figure online)

methods require large amounts of manually labeled data to achieve the level of robustness required for their clinical application [5,8,23,27]. Manual labeling of CT volumes is time-consuming and may increase the workload of clinicians. Additionally, applying deep learning-based segmentation models to data from new unseen sources can result in suboptimal lesion segmentation due to unseen acquisition devices/parameters, variations in patient pathology, or future coronavirus variants resulting in new appearance characteristics or new lesion pathologies [16]. To address this challenge, interactive segmentation methods that can quickly adapt to such changing settings are needed. These can be used either by end-users or algorithm developers to quickly expand existing annotated datasets and enable agile retraining of automatic segmentation models [4].

Related Work: Interactive segmentation methods for Artificial Intelligence (AI) assisted annotation have shown promising applications in the existing literature [14,18,25,26]. BIFSeg [26] utilizes a bounding box and scribbles with convolutional neural network (CNN) image-specific fine-tuning to segment potentially *unseen* objects of interest. MIDeepSeg [14] incorporates user-clicks with the input image using exponential geodesic distance. However, BIFSeg, MIDeepSeg, and similar deep learning-based methods exploit large networks that do not adapt rapidly to new data examples in an online setting due to the elevated computational requirements.

Due to their quick adaptability and efficiency, a number of existing online likelihood methods have been applied as interactive segmentation methods [2,3,24]. DybaORF [24] utilizes hand-crafted features with dynamically changing weights based on interactive labels' distribution to train a Random Forest classifier. ECONet [2] improves online learning with a shallow CNN that jointly learns both features and classifier to outperform previous online likelihood inference methods. While ECONet is, to the best of our knowledge, the only online learning method that addresses COVID-19 lung lesion segmentation, it is limited to learning from user scribbles only. This means that it requires a significant amount of user interaction to achieve expert-level accuracy. Additionally, the model uses a single convolution for feature extraction, limiting its accuracy to a specific scale of pathologies. For each CT volume, the model is trained from scratch, resulting in lack of prior knowledge about lesions.

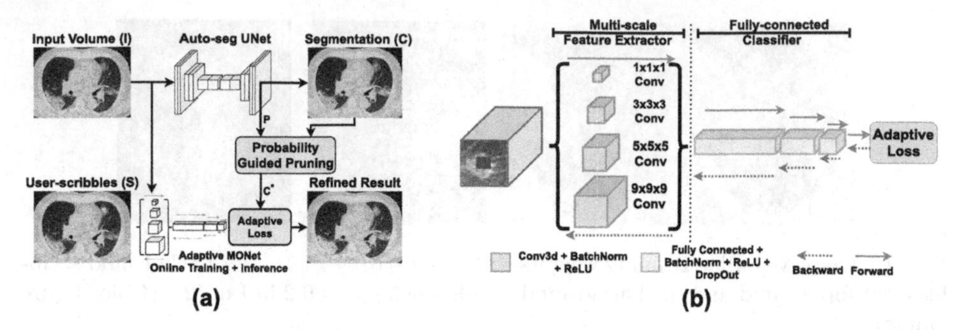

Fig. 2. Adaptive learning for interactive segmentation: (a) training and inference of MONet using adaptive loss and probability-guided pruning; (b) architecture of our multi-scale online likelihood network (MONet).

Contributions: To overcome limitations of existing techniques, we propose adaptive multi-scale online likelihood network (MONet) for AI-assisted interactive segmentation of lung lesions in CT volumes from COVID-19 patients. Our contributions are three-fold, we propose:

i. Multi-scale online likelihood network (MONet), consisting of a multi-scale feature extractor, which enables relevant features extraction at different scales for improved accuracy;
ii. Adaptive online loss that uses weights from a scaled negative exponential geodesic distance from user-scribbles, enabling adaptive learning from both initial segmentation and user-provided corrections (Fig. 1);
iii. Probability-guided pruning approach where uncertainty from initial segmentation model is used for pruning ambiguous online training data.

MONet enables human-in-the-loop online learning to perform AI-assisted annotations and should not be mistaken for an end-to-end segmentation model.

We perform expert evaluation which shows that adaptively learned MONet outperforms existing state-of-the-art, achieving 5.86% higher Dice score with 24.67% less perceived NASA-TLX workload score evaluated.

2 Method

Given an input image volume, I, a pre-trained CNN segmentation model generates an automatic segmentation C with associated probabilities P. When using data from a new domain, the automated network may fail to properly segment foreground/background objects. To improve this, the user provides scribbles-based interaction indicating corrected class labels for a subset of voxels in the image I. Let $\mathcal{S} = \mathcal{S}^f \cup \mathcal{S}^b$ represent these set of scribbles, where $\mathcal{S}^f$ and $\mathcal{S}^b$ denote the foreground and background scribbles, respectively, and $\mathcal{S}^f \cap \mathcal{S}^b = \emptyset$. Figure 2 (a) shows scribbles $\mathcal{S}$, along with the initial segmentation C and probabilities P.

2.1 Multi-scale Online Likelihood Network

Our proposed multi-scale online likelihood network (MONet), shown in Fig. 2 (b), uses a multi-scale feature extractor that applies a 3D convolution at various kernel sizes to capture spatial information at different scales. The output of each scale is concatenated and fed to a fully-connected classifier, which infers the likelihood for background/foreground classification of the central voxel in the input patch. Each layer in MONet is followed by batch normalization and ReLU activation.

2.2 Adaptive Loss for Online Learning

The scribbles $\mathcal{S}$ only provide sparse information for online learning. However, these corrections are likely also applicable to neighboring voxels with similar appearance features, thereby providing an extended source of training information. Concurrently, the initial automated segmentation C will often provide reliable results away from the scribbles. To extend the influence of the scribbles $\mathcal{S}$ while preserving the quality of the initial segmentation C, we propose a spatially-varying adaptive online loss:

$$\mathcal{L} = -\sum_{i} \left[(1 - W_i)\mathcal{L}_i^C + W_i \mathcal{L}_i^S\right], \tag{1}$$

where i is a voxel index, $\mathcal{L}^C$ and $\mathcal{L}^S$ are individual loss terms for learning from the automated segmentation C and the user-provided correction scribbles $\mathcal{S}$ respectively. W are spatially-varying interaction-based weights defined using the geodesic distance D between voxel i and the scribbles $\mathcal{S}$:

$$W_i = \exp\left(\frac{-D(i, S, I)}{\tau}\right), \tag{2}$$

where the temperature term τ controls the influence of W in I. The geodesic distance to the scribbles is defined as $D(i, \mathcal{S}, I) = \min_{j \in \mathcal{S}} d(i, j, I)$ where $d(i, j, I) = \min_{p \in \mathcal{P}_{i,j}} \int_0^1 \|\nabla I(p(x)) \cdot \mathbf{u}(x)\| \; dx$ and $\mathcal{P}_{i,j}$ is the set of all possible differentiable paths in I between voxels i and j. A feasible path p is parameterized by $x \in [0, 1]$. We denote $\mathbf{u}(x) = p'(x)/\|p'(x)\|$ the unit vector tangent to the direction of the path p. We further let $D = \infty$ for $\mathcal{S} = \emptyset$.

Dynamic Label-Balanced Cross-Entropy Loss: User-scribbles for online interactive segmentation suffer from dynamically changing class imbalance [2]. Moreover, lung lesions in CT volumes usually occupy a small subset of all voxels, introducing additional label imbalance and hence reducing their impact on imbalanced online training. To address these challenges, we utilize a label-balanced cross-entropy loss [2,10,11], with dynamically changing class weights from segmentations and scribbles distribution. Given an online model with parameters

θ, the foreground likelihood from this model is $p_i = P(s_i = 1|I, \theta)$. Then, the segmentations-balanced and scribbles-balanced cross-entropy terms are:

$$\mathcal{L}_i^C = \alpha^f y_i^C \log p_i + \alpha^b(1 - y_i^C)\log(1 - p_i), \tag{3}$$

$$\mathcal{L}_i^S = \beta^f y_i^S \log p_i + \beta^b(1 - y_i^S)\log(1 - p_i), \tag{4}$$

where α and β are class weights for labels C and scribbles $\mathcal{S}$ that are defined by labels and scribbles distributions during online interaction as: $\alpha^f = |\mathcal{T}|/|C^f|$, $\alpha^b = |\mathcal{T}|/|C^b|$, $\beta^f = |\mathcal{T}|/|\mathcal{S}^f|$, $\beta^b = |\mathcal{T}|/|\mathcal{S}^b|$ and $|\mathcal{T}| = |C| + |\mathcal{S}|$. y_i^C and y_i^S represent labels in C and $\mathcal{S}$, respectively.

The patch-based training approach from [2] is used to first extract K×K×K patches from I centered around each voxel in $\mathcal{S}$ and C and train MONet using Eq. (1). Once learned, efficient online inference from MONet is achieved by applying it to the whole input CT volumes as a fully convolutional network [12].

2.3 Improving Efficiency with Probability-Guided Pruning

MONet is applied as an online likelihood learning method, where the online training happens with an expert human annotator in the loop, which makes online training efficiency critical. We observe that the automatic segmentation models provide dense labels C which may significantly impact online training and inference performance. C may contain ambiguous predictions for new data, and a number of voxels in C may provide redundant labels. To improve online efficiency while preserving accuracy during training, we prune labels as $C^* = \mathcal{M} \odot C$ where: $\mathcal{M}_i$ is set to 1 if $P_i \geq \zeta$ and $U_i \geq \eta$ and 0 otherwise. $\zeta \in [0, 1]$ is the minimum confidence to preserve a label, $U_i \in [0, 1]$ is a uniformly distributed random variable, and $\eta \in [0, 1]$ is the fraction of samples to prune.

3 Experimental Validation

Table 1 outlines the different state-of-the-art interactive segmentation methods and their extended variants that we introduce for fair comparison. We compare our proposed MONet with ECONet [2] and MIDeepSeg [14]. As our proposed Eq. (2) is inspired by the exponential geodesic distance from MIDeepSeg [14], we introduce MIDeepSegTuned, which utilizes our proposed addition of a temperature term τ. Moreover, to show the importance of multi-scale features, we include MONet-NoMS which uses features from a single 3D convolution layer. We utilize MONAI Label to implement all online likelihood methods [7]. For methods requiring an initial segmentation, we train a 3D UNet [6] using MONAI [17] with features $[32, 32, 64, 128, 256, 32]$. Output from each method is regularized using GraphCut optimization. We also compare against a baseline interactive GraphCut (IntGraphCut) implementation, that updates UNet output with scribbles based on [3] and then performs GraphCut optimization. The proposed method is targeted for online training and inference, where quick adaptability with minimal

Table 1. State-of-the-art evaluated comparison methods, showing improvement in accuracy (Dice and ASSD) when using different features. Features in blue text are proposed in this paper. Key: OL - online learning, PP - post-processing.

Method	Technique	Initial Seg.	Multi Scale	Adaptive Loss	Temp. (τ)	Dice (%)	ASSD
MONet (proposed)	**OL**	✓	✓	✓	✓	**77.77**	**11.82**
MONet-NoMS	**OL**	✓	✗	✓	✓	77.06	13.01
ECONet [2]	**OL**	✗	✗	✗	✗	77.02	20.19
MIDeepSegTuned [14]	**PP**	✓	✗	✗	✓	76.00	20.16
MIDeepSeg [14]	**PP**	✓	✗	✗	✗	56.85	33.25
IntGraphCut	**PP**	✓	✗	✗	✗	68.58	28.64

latency is required. Note that incorporating more advanced deep learning methods in this context would result in a considerable decrease in online efficiency, rendering the method impractical for online applications [2]. We utilize a GPU-based implementation of geodesic distance transform [1] in Eq. (2), whereas MIDeepSeg uses a CPU-based implementation. We use NVIDIA Tesla V100 GPU with 32 GB memory for all our experiments. Comparison of accuracy for each method is made using Dice similarity (Dice) and average symmetric surface distance (ASSD) metrics against ground truth annotations [2,14]. Moreover, we compare performance using execution time (Time), including online training and inference time, average full annotation time (FA-Time), and number of voxels with scribbles (S) needed for a given accuracy.

Data: To simulate a scenario where the automatic segmentation model is trained on data from a different source than it is tested on, we utilize two different COVID-19 CT datasets. The dataset from the COVID-19 CT lesion segmentation challenge [21] is used for training and validation of 3D UNet for automatic segmentation task and patch-based pre-training of MONet/MONet-NoMS/ECONet. This dataset contains binary lung lesions segmentation labels for 199 CT volumes (160 training, 39 validation). We use UESTC-COVID-19 [27], a dataset from a different source, for the experimental evaluation of interactive segmentation methods (test set). This dataset contains 120 CT volumes with lesion labels, from which 50 are by expert annotators and 70 are by non-expert annotators. To compare robustness of the proposed method against expert annotators, we only use the 50 expert labelled CT volumes.

Training Parameters: Training of 3D UNet utilized a learning rate (lr) of $1e^{-4}$ for 1000 epochs and MONet/MONet-NoMS offline pre-training used 50 epochs, and lr $= 1e^{-3}$ dropped by 0.1 at the 35th and 45th epoch. Online training for MONet, MONet-NoMS and ECONet [2] used 200 epochs with lr $= 1e^{-2}$ set using cosine annealing scheduler [13]. Dropout of 0.3 was used for all fully-connected layers in online models. Each layer size in ECONet and MONet-NoMS was selected by repeating line search experiments from [2]: (i) input patch/3D

Table 2. Quantitative comparison of interactive segmentation methods using synthetic scribbler shows mean and standard deviation of Dice, ASSD, Time and Synthetic Scribbles Voxels.

Method	Dice (%)	ASSD	Time (s)	Scribbles
MONet (proposed)	**77.77 ± 6.84**	**11.82 ± 12.83**	6.18 ± 2.42	**20 ± 24**
MONet-NoMS	77.06 ± 7.27	13.01 ± 15.29	7.76 ± 8.16	20 ± 24
ECONet [2]	77.02 ± 6.94	20.19 ± 14.71	1.46 ± 1.22	2283 ± 2709
MIDeepSegTuned [14]	76.00 ± 7.37	20.16 ± 22.57	7.97 ± 2.47	23 ± 17
MIDeepSeg [14]	56.85 ± 14.25	33.25 ± 25.26	6.26 ± 1.46	436 ± 332
IntGraphCut	68.58 ± 9.09	28.64 ± 27.36	**0.11 ± 0.04**	480 ± 359

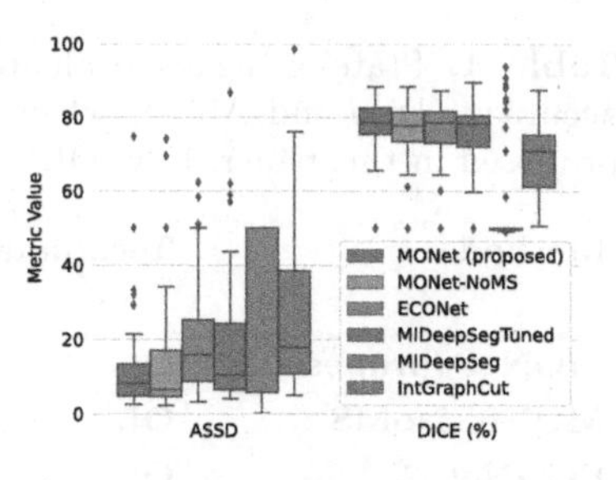

Fig. 3. Validation accuracy using synthetic scribbles.

convolution kernel size of $K = 9$, (ii) 128 input 3D convolution filters and (iii) fully-connected sizes of $32\times16\times2$. For MONet, we utilize four input 3D convolution with multi-scale kernel sizes $K = [1, 3, 5, 9]$ with each containing 32 filters (i.e., a total of 128 filters, same as (ii)). We utilize the same fully-connected sizes as in (iii) above. Parameters $\zeta = 0.8$ and $\eta = 0.98$ are selected empirically. We utilize $\tau = 0.3$ for MONet, MIDeepSegTuned and MONet-NoMS. We use GraphCut regularization, where $\lambda = 2.5$ and $\sigma = 0.15$ [3]. Search experiments used for selecting τ, λ, σ are shown in Fig. 1 and 2 in supplementary material.

3.1 Quantitative Comparison Using Synthetic Scribbler

We employ the synthetic scribbler method from [2,25] where mis-segmented regions in the inferred segmentations are identified by comparison to the ground truth segmentations. Table 2 and Fig. 3 present quantitative comparison of methods using synthetic scribbler. They show that MONet outperforms all existing state-of-the-art in terms of accuracy with the least number of synthetic scribbled voxels. In particular, MONet outperforms both MIDeepSeg [14] and MIDeepSeg-Tuned, where adaptive online learning enables it to quickly adapt and refine segmentations. In terms of efficiency, online training and inference of the proposed MONet takes around 6.18 s combined, which is 22.4% faster as compared to 7.97 s for MIDeepSeg. However, it is slower than ECONet and ISeg. MIDeepSeg performs the worst as it is unable to adapt to large variations and ambiguity within lung lesions from COVID-19 patients, whereas by utilizing our proposed Eq. (2) in MIDeepSegTuned, we improve its accuracy. When comparing to online learning methods, MONet outperforms MONet-NoMS, where the accuracy is improved due to MONet's ability to extract multi-scale features. Existing state-of-the-art online method ECONet [2] requires significantly more scribbled voxels as it only relies on user-scribbles for online learning.

3.2 Performance and Workload Validation by Expert User

This experiment aims to compare the performance and *perceived* subjective workload of the proposed MONet with the best performing comparison method

Table 3. Workload validation by expert user, shows Dice (%), ASSD, full annotation time, FA-Time (s), overall NASA-TLX perceived workload score and the % of data successfully annotated by expert.

	MONet (proposed)	MIDeepSeg-Tuned [14]
Finished	**100%**	33.33%
NASA-TLX	**52.33**	77.00
Dice (%)	$\mathbf{88.53 \pm 2.27}$	82.67 ± 12.36
ASSD	$\mathbf{2.91 \pm 1.58}$	11.30 ± 20.09
FA-Time (s)	**507.11**	567.43

Table 4. NASA-TLX perceived workload by expert user, shows total workload and individual sub-scale scores. The method with low score requires less effort, frustration, mental, temporal and physical demands with high perceived performance.

NASA-TLX weighted scores	MONet (proposed)	MIDeepSeg-Tuned [14]
Effort	**14.67**	21.33
Frustration	**7.67**	16.67
Mental Demand	**13.33**	15.00
Performance	**10.00**	17.00
Physical Demand	**4.67**	7.00
Temporal Demand	2.00	**0.00**
Total workload	**52.33**	77.00

MIDeepSegTuned based on [14]. We asked an expert, with 2 years of experience in lung lesion CT from Radiology Department, Oxford University Hospitals NHS Foundation Trust, to utilize each method for labelling the following pathologies as lung lesions in 10 CT volumes from UESTC-COVID-19 expert set [27]: ground glass opacity, consolidation, crazy-paving, linear opacities. One CT volume is used by the expert to practice usage of our tool. The remaining 9 CT volumes were presented in a random order, where the perceived workload was evaluated by the expert at half way (after 5 segmentations) and at the end. We use the National Aeronautics and Space Administration Task Load Index (NASA-TLX) [9] as per previous interactive segmentation studies [15,19,28]. The NASA-TLX asks the expert to rate the task based on six factors, being performance, frustration, effort, mental, physical and temporal demand. The weighted NASA-TLX score is then recorded as the expert answers 15 pair-wise questions rating factors based on importance. In addition, we also recorded accuracy metrics (Dice and ASSD) against ground truth labels in [27], time taken to complete annotation and whether the expert was able to successfully complete their task within 10 min allocated for each volume.

Table 3 presents an overview for this experiment, where using the proposed MONet, the expert was able to complete 100% of the labelling task, whereas using MIDeepSegTuned they only completed 33.33% within the allocated time. In addition, MONet achieves better accuracy with lower time for complete annotation and less overall perceived workload with NASA-TLX of 52.33% as compared to 77.00% for MIDeepSegTuned. Table 4 shows the individual scores that contribute to overall perceived workload. It shows that using the proposed MONet, the expert perceived reduced workload in all sub-scale scores except temporal demand. We believe this is due to the additional online train-

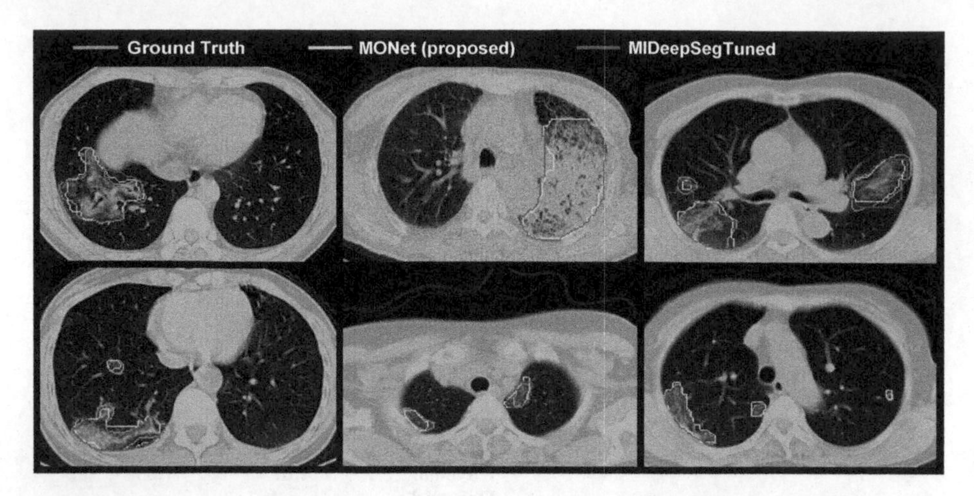

Fig. 4. Visual comparison of interactive segmentation results from Sect. 3.2. Segmentations are shown with contours on axial plane slices from different cases.

ing/inference overhead for MONet application. Figure 4 visually compares these results where MONet results in more accurate segmentation as compared to MIDeepSegTuned. We also note that MONet's ability to apply learned knowledge on the whole volume enables it to also infer small isolated lesions, which MIDeepSegTuned fails to identify.

4 Conclusion

We proposed a multi-scale online likelihood network (MONet) for scribbles-based AI-assisted interactive segmentation of lung lesions in CT volumes from COVID-19 patients. MONet consisted of a multi-scale feature extractor that enabled extraction of relevant features at different scales for improved accuracy. We proposed an adaptive online loss that utilized adaptive weights based on user-provided scribbles that enabled adaptive learning from both an initial automated segmentation and user-provided label corrections. Additionally, we proposed a dynamic label-balanced cross-entropy loss that addressed dynamic class imbalance, an inherent challenge for online interactive segmentation methods. Experimental validation showed that the proposed MONet outperformed the existing state-of-the-art on the task of annotating lung lesions in COVID-19 patients. Validation by an expert showed that the proposed MONet achieved on average 5.86% higher Dice while achieving 24.67% less perceived NASA-TLX workload score than the MIDeepSegTuned method [14].

Acknowledgment. This project has received funding from the European Union's Horizon 2020 research and innovation programme under grant agreement No 101016131 (icovid project). This work was also supported by core and project funding from the Wellcome/EPSRC [WT203148/Z/16/Z; NS/A000049/1; WT101957; NS/A000027/1].

This project utilized scribbles-based interactive segmentation tools from opensource project MONAI Label (https://github.com/Project-MONAI/MONAILabel) [7].

References

1. Asad, M., Dorent, R., Vercauteren, T.: Fastgeodis: Fast generalised geodesic distance transform. arXiv preprint arXiv:2208.00001 (2022)
2. Asad, M., Fidon, L., Vercauteren, T.: ECONet: Efficient convolutional online likelihood network for scribble-based interactive segmentation. In: Medical Imaging with Deep Learning (2022)
3. Boykov, Y.Y., Jolly, M.-P.: Interactive graph cuts for optimal boundary and region segmentation of objects in ND images. In: Proceedings Eighth IEEE International Conference on Computer Vision. ICCV 2001, pp. 105–112 (2001)
4. Budd, S., Robinson, E.C., Kainz, B.: A survey on active learning and human-in the-loop deep learning for medical image analysis. Med. Image Anal. **71**, 102062 (2021)
5. Chassagnon, G., et al.: AI-Driven CT-based quantification, staging and short-term outcome prediction of COVID-19 pneumonia. arXiv preprint arXiv:2004.12852 (2020)
6. Çiçek, Ö., Abdulkadir, A., Lienkamp, S.S., Brox, T., Ronneberger, O.: 3D U-Net: learning dense volumetric segmentation from sparse annotation. In: Ourselin, S., Joskowicz, L., Sabuncu, M.R., Unal, G., Wells, W. (eds.) MICCAI 2016. LNCS, vol. 9901, pp. 424–432. Springer, Cham (2016). https://doi.org/10.1007/978-3-319-46723-8_49
7. Diaz-Pinto, A., et al.: Monai label: A framework for AI-assisted interactive labeling of 3D medical images. arXiv preprint arXiv:2203.12362 (2022)
8. Gonzalez, C., Gotkowski, K., Bucher, A., Fischbach, R., Kaltenborn, I., Mukhopadhyay, A.: Detecting when pre-trained nnU-net models fail silently for Covid-19 lung lesion segmentation. In: de Bruijne, M., et al. (eds.) MICCAI 2021. LNCS, vol. 12907, pp. 304–314. Springer, Cham (2021). https://doi.org/10.1007/978-3-030-87234-2_29
9. Hart, S.G.: NASA-task load index (NASA-TLX); 20 years later. In: Proceedings of the Human Factors And Ergonomics Society Annual Meeting, pp. 904–908 (2006)
10. Ho, Y., Wookey, S.: The real-world-weight cross-entropy loss function: modeling the costs of mislabeling. IEEE Access **8**, 4806–4813 (2019)
11. Kukar, M., Kononenko, I., et al.: Cost-sensitive learning with neural networks. In: ECAI, pp. 8–94 (1998)
12. Long, J., Shelhamer, E., Darrell, T.: Fully convolutional networks for semantic segmentation. In: Proceedings of the IEEE Conference on Computer Vision And Pattern Recognition, pp. 3431–3440(2015)
13. Loshchilov, I., Hutter, F.: Sgdr: Stochastic gradient descent with warm restarts. arXiv preprint arXiv:1608.03983 (2016)
14. Luo, X., et al.: MIDeepSeg: minimally interactive segmentation of unseen objects from medical images using deep learning. Med. Image Anal. **72**, 102102 (2021)
15. McGrath, H., et al.: Manual segmentation versus semi-automated segmentation for quantifying vestibular schwannoma volume on MRI. Int. J. Comput. Assist. Radiol. Surg. **15**, 1445–1455 (2020)
16. McLaren, T.A., Gruden, J.F., Green, D.B.: The bullseye sign: a variant of the reverse halo sign in COVID-19 pneumonia. Clin. Imaging **68**, 191–96 (2020)

17. MONAI Consortium, MONAI: Medical Open Network for AI. (2020). https://github.com/Project-MONAI/MONAI
18. Rajchl, M. et al.: Deepcut: Object segmentation from bounding box annotations using convolutional neural networks. IEEE Trans. Med. Imaging **36**(2), 674–683 (2016)
19. Ramkumar, A., et al.: Using GOMS and NASA-TLX to to evaluate human-computer interaction process in interactive segmentation. Int. J. Human-Computer Interact. **33**(2), 123–34 (2017)
20. Revel, M.-P., et al.: Study of thoracic CT in COVID-19: the STOIC project. Radiology **301**(1), E361–E370 (2021)
21. Roth, H., et al.: Rapid Artificial Intelligence Solutions in a Pandemic-The COVID-19-20 Lung CT Lesion Segmentation Challenge (2021)
22. Rubin, G.D., et al.: The role of chest imaging in patient management during the COVID-19 pandemic: a multinational consensus statement from the Fleischner Society. Radiology **296**(1), 172–80 (2020)
23. Tilborghs, S., et al.: Comparative study of deep learning methods for the automatic segmentation of lung, lesion and lesion type in CT scans of COVID-19 patients. arXiv preprint arXiv:2007.15546 (2020)
24. Wang, G., et al.: Dynamically balanced online random forests for interactive scribble based segmentation. In: International Conference on Medical Image Computing and Computer-Assisted Intervention, pp. 35–360 (2016)
25. Wang, G., et al.: DeepIGeoS: a deep interactive geodesic framework for medical image segmentation. IEEE Trans. Pattern Anal. Mach. Intell. **41**(7), 155–1572 (2018)
26. Wang, G., et al.: Interactive medical image segmentation using deep learning with imag-specific fine tuning. IEEE transactions on medical imaging **37**(7), 1562–1573 (2018)
27. Wang, G., et al.: A noise-robust framework for automatic segmentation of COVID-19 pneumonia lesions from CT images. IEEE Trans. Med. Imaging **39**(8), 2653–2663 (2020)
28. Williams, H., et al.: Interactive segmentation via deep learning and b-spline explicit active surfaces. In: de Bruijne, M., et al. (eds.) MICCAI 2021. LNCS, vol. 12901, pp. 315–325. Springer, Cham (2021). https://doi.org/10.1007/978-3-030-87193-2_30

Explainable Image Classification with Improved Trustworthiness for Tissue Characterisation

Alfie Roddan[1(✉)], Chi Xu[1], Serine Ajlouni[2], Irini Kakaletri[3], Patra Charalampaki[2,4], and Stamatia Giannarou[1]

[1] The Hamlyn Centre for Robotic Surgery, Department of Surgery and Cancer, Imperial College London, London, UK
{a.roddan21,chi.xu20,stamatia.giannarou}@imperial.ac.uk

[2] Medical Faculty, University Witten Herdecke, Witten , Germany

[3] Medical Faculty, Rheinische Friedrich Wilhelms University of Bonn, Bonn, Germany

[4] Department of Neurosurgery, Cologne Medical Center, Cologne, Germany

Abstract. The deployment of Machine Learning models intraoperatively for tissue characterisation can assist decision making and guide safe tumour resections. For the surgeon to trust the model, explainability of the generated predictions needs to be provided. For image classification models, pixel attribution (PA) and risk estimation are popular methods to infer explainability. However, the former method lacks trustworthiness while the latter can not provide visual explanation of the model's attention. In this paper, we propose the first approach which incorporates risk estimation into a PA method for improved and more trustworthy image classification explainability. The proposed method iteratively applies a classification model with a PA method to create a volume of PA maps. We introduce a method to generate an enhanced PA map by estimating the expectation values of the pixel-wise distributions. In addition, the coefficient of variation (CV) is used to estimate pixel-wise risk of this enhanced PA map. Hence, the proposed method not only provides an improved PA map but also produces an estimation of risk on the output PA values. Performance evaluation on probe-based Confocal Laser Endomicroscopy (pCLE) data verifies that our improved explainability method outperforms the state-of-the-art.

Keywords: Explainability · Uncertainty · MC Dropout

1 Introduction

When using a Machine Learning (ML) model during intraoperative tissue characterisation, it is vital that the surgeon is able to assess how reliable a model's prediction is [8]. For the surgeon to trust the output predictions of the model, the model must be able to explain itself reliably in a clinical scenario [2]. To assess an

H. Greenspan et al. (Eds.): MICCAI 2023, LNCS 14221, pp. 575–585, 2023.
https://doi.org/10.1007/978-3-031-43895-0_54

explainability method we consider five metrics of performance: speed, usability, generalisability, trustworthiness and ability to localise semantic features. The explanation of a model's predictions is trustworthy if small perturbations in the input or model parameters, results in a similar output explanation. One form of explainability in the image classification domain is pixel attribution (PA) mapping. PA maps aim to highlight the "most important" pixels to the classification. PA maps can be used to visually highlight whether a model is poorly extracting semantic features [32] and/or that the model is misinformed due to spurious correlations within the data that it was trained on [16]. To efficiently process image data, these methods mainly rely on Convolutional Neural Networks (CNNs) and achieve state-of-the-art (SOTA) performance. One of the first PA methods proposed for CNNs was class activation maps (CAM) [33]. CAM uses one forward pass of the model to find the channels in the last convolutional layer that contributed most to the prediction. One of CAM's limitations is its reliance on global average pooling (GAP) [21] after the last convolutional layer as it dramatically reduces the number of architectures that can use CAM. To improve on this, Grad-CAM [30] generalises to all CNN architectures which are differentiable from the output logit layer to the chosen convolutional layer. However, Grad-CAM often lacks sharpness in object localisation, as noted and improved on in Grad-CAM++ [6] and SmoothGrad-CAM++ [24]. These extensions of Grad-CAM have good semantic feature localisation but they are unable to be deployed for use in surgery [5]. Both Score-CAM [31] and Recipro-CAM [5] also generalise to all CNN architectures but are deployable. Score-CAM improves on object localisation within the visual PA map without losing the class specific capabilities of Grad-CAM by masking out regions of the image and measuring the change in the output score. This is similar to perturbation methods like RISE [26], LIME [28] and other perturbation techniques [3,32]. On the other hand, Recipro-CAM focuses on the speed of PA map computation whilst maintaining comparable SOTA performance. By utilising the CNN's receptive field, Recipro-CAM generates a number of spatial masks and then measures the effect on the output score much like Score-CAM.

Despite being speedy, easy to deploy and able to localise semantic features, the above methods lack trustworthiness due to the training strategy of their underlying model. Deep learning (DL) models trained with empirical risk minimisation (ERM) are overconfident in prediction [12] and vulnerable to adversarial attacks [13]. Bayesian Neural Networks (BNNs) [23] bring improved regularisation and output uncertainty estimates. Unfortunately, the non-linearity and number of variables within NNs make Bayesian inference a computationally intensive task. For this reason, variational methods [15,18] are used to approximate Bayesian inference. More recently, the variational method Bayes by Backprop [4] used Dropout [19] to approximate Bayesian inference. Dropout is a regularisation technique which has also been noted to improve salient feature extraction. Although Bayes by Backprop is trustworthy, it often fails to scale to the complex architectures of SOTA models. To improve on this lack of generalisability, another variational method called Monte Carlo (MC) Dropout [12]

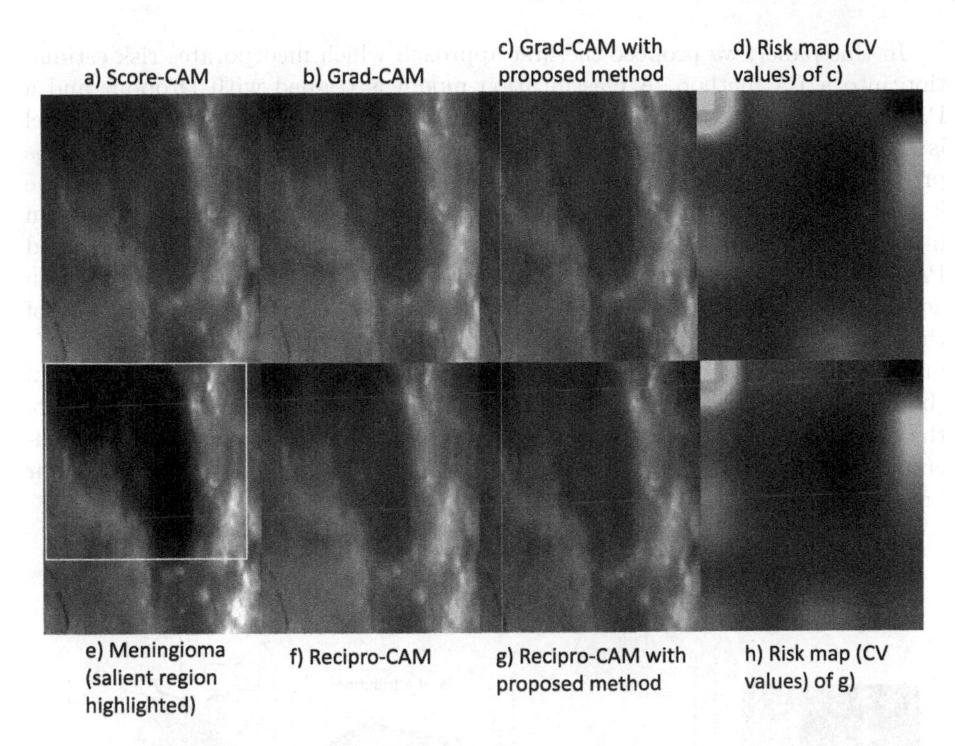

Fig. 1. PA maps generated using ResNet18 on meningioma pCLE data. a) Score-CAM PA map b) Grad-CAM PA map c) Grad-CAM PA map with our method applied d) Risk map (CV values) of c) e) Meningioma with the salient region highlighted with red bounding box f) Recipro-CAM PA map g) Recipro-CAM PA map with our method applied h) Risk map (CV values) of f). Yellow represents the highest PA value and black the lowest. (Color figure online)

proposes that a model trained with Dropout is equivalent to a probabilistic deep Gaussian process [7,11]. With this assumption, an estimated output distribution is computed after a number of forward passes with Dropout have been applied. This output distribution is used in practice to indicate risk in the model's predictions. Surgeons in practice can use this risk during diagnosis to trust the model for decision making [14]. Using Dropout to perturb a model is a computationally cheap method of model averaging [19]. It is worth noting though that this method's validity as a Bayesian Inference approximation was later questioned [10]. However, this does not affect the use of this method for risk estimation. So far, model explainability and risk estimation have mostly been used separately to assess models' suitability for surgical applications. DistDeepSHAP [20] computed the uncertainty of Shapley values to show uncertainty in explainability maps. However, DistDeepSHAP is a model-agnostic interpretability method that shows the global effect of perturbing inputs, instead of providing an insight to the model's learned representations. The aim of this paper is to show that the fusion of MC Dropout and PA methods leads to improved explainability.

In this paper, we propose the first approach which incorporates risk estimation into a PA method. A classification model is trained with Dropout and a PA method is used to generate a PA map. At test time, the classification model is employed with the Dropout enabled. In this work, we propose to repeat this process for a number of iterations creating a volume of PA maps. This volume is used to generate a pixel-wise distribution of PA values from which we can infer risk. More specifically, we introduce a method to generate an enhanced PA map by estimating the expectation values of the pixel-wise distributions. In addition, the coefficient of variation (CV) is used to estimate pixel-wise risk of this enhanced PA map. This provides an improved explanation of the model's prediction by clearly presenting to the surgeon which salient areas to trust in the model's enhanced PA map. In this work, we focus on the explainability of the classification of brain tumours using probe-based Confocal Laser Endomicroscopy (pCLE) data. Performance evaluation on pCLE data shows that our improved explainability method outperforms the SOTA.

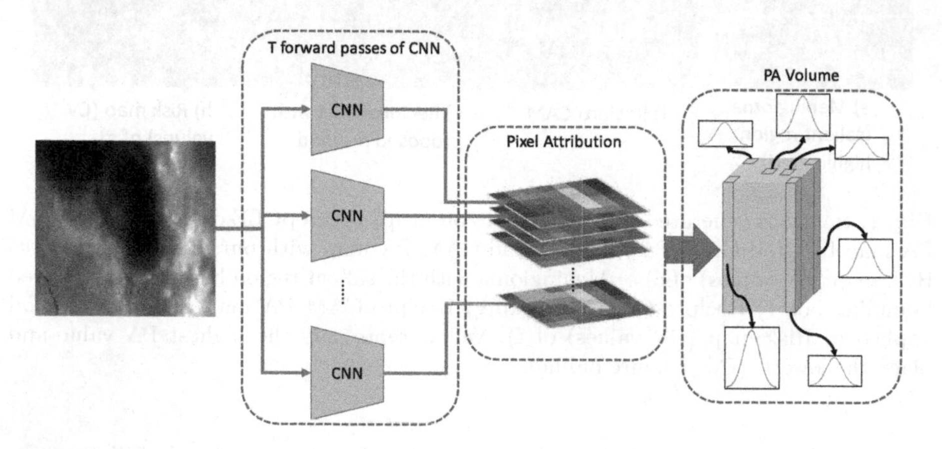

Fig. 2. Outline of the proposed method. A PA volume is generated using T forward passes of a CNN model with Dropout applied.

2 Methodology

The aim of the proposed method is to produce an improved PA map of a classification model, while providing risk estimation of the model's explainability to enhance trustworthiness in decision making during intraoperative tissue characterisation.

In our method, any CNN classification model trained with Dropout can be used. Let $\hat{\boldsymbol{Y}}$ be the output logits of the CNN model, where Dropout is enabled at test time, with input image $\boldsymbol{X} \in \mathbb{R}^{height \times width \times channels}$. Any PA method can be used to generate a PA map using the output logits $S = f_s(\hat{\boldsymbol{Y}}) \in \mathbb{R}^{height \times width}$ where $f_s(.)$ is the PA method. We propose to repeat the above process for T iterations to create a volume of PA maps $\boldsymbol{S} = \{S_1, ..., S_T\} \in \mathbb{R}^{height \times width \times T}$. A visual representation of how the volume is generated is show in Fig. 2. The aim is to use this volume to generate a pixel-wise distribution of PA values from which we can infer risk. To achieve this, we compute the expectation and variance values of the volume along the third dimension as:

$$\begin{aligned} \mathbb{E}(\boldsymbol{S}_{i,j}) &\approx \frac{1}{T}\sum_{t=1}^{T} f_s(\hat{\boldsymbol{Y}}_t)_{i,j} \\ Var(\boldsymbol{S}_{i,j}) &\approx \frac{1}{T}\sum_{t=1}^{T} f_s(\hat{\boldsymbol{Y}}_t)_{i,j}^T f_s(\hat{\boldsymbol{Y}}_t)_{i,j} - \mathbb{E}(\boldsymbol{S}_{i,j})^T\mathbb{E}(\boldsymbol{S}_{i,j}), \end{aligned} \tag{1}$$

where, i, j represent the pixel's row and column coordinates, respectively. The expectation $\mathbb{E}(\boldsymbol{S}_{i,j})$ of each pixel (i, j) is used to generate an enhanced PA map of size $height \times width$. The intuition is that the above distribution of PA values can produce less noisy and risky estimations of a pixel's contribution to the final explainability map compared to a single estimate.

As well as advancing SOTA PA methods, our method also estimates the trustworthiness of the enhanced PA map generated above. For risk estimation, it is important to consider that different pixels in the PA map correspond to different semantic features which contribute differently to the output logits. This makes the pixel-wise distributions have different scales. For this reason, the coefficient of variation (CV) is used to estimate pixel-wise risk, as it allows us to compare pixel-wise variances despite their different scales. This is mathematically defined as:

$$S_{i,j}^{cv} = \frac{\sqrt{Var(\boldsymbol{S}_{i,j})}}{\mathbb{E}(\boldsymbol{S}_{i,j})} = \frac{std(\boldsymbol{S}_{i,j})}{\mathbb{E}(\boldsymbol{S}_{i,j})}. \tag{2}$$

Our proposed method improves trustworthiness of explainability as it allows visualisation of both the explainability of the classification model (provided by the enhanced PA map) together with the pixel-wise risk of this map (provided by the CV map). For instance, salient areas on the PA map should not be trusted unless the CV values are low. An example of the enhanced PA and risk maps generated with the proposed method are shown in Fig. 3. This shows that the proposed method not only improves explainability but also provides associated risk information which improves trustworthiness.

3 Experiments and Analysis

Dataset. The developed explainability framework has been validated on an in vivo and ex vivo pCLE dataset of meningioma, glioblastoma and metastases of an invasive ductal carcinoma (IDC). All studies on human subjects were performed according to the requirements of the local ethic committee and in agreement with the Declaration of Helsinki (No. CLE-001 Nr: 2014480). The Cellvizio©by Mauna Kea Technologies, Paris, France has been used in combination with the mini laser probe CystoFlex©UHD-R. The distinguishing characteristic of the meningioma is the psammoma body with concentric circles that show various

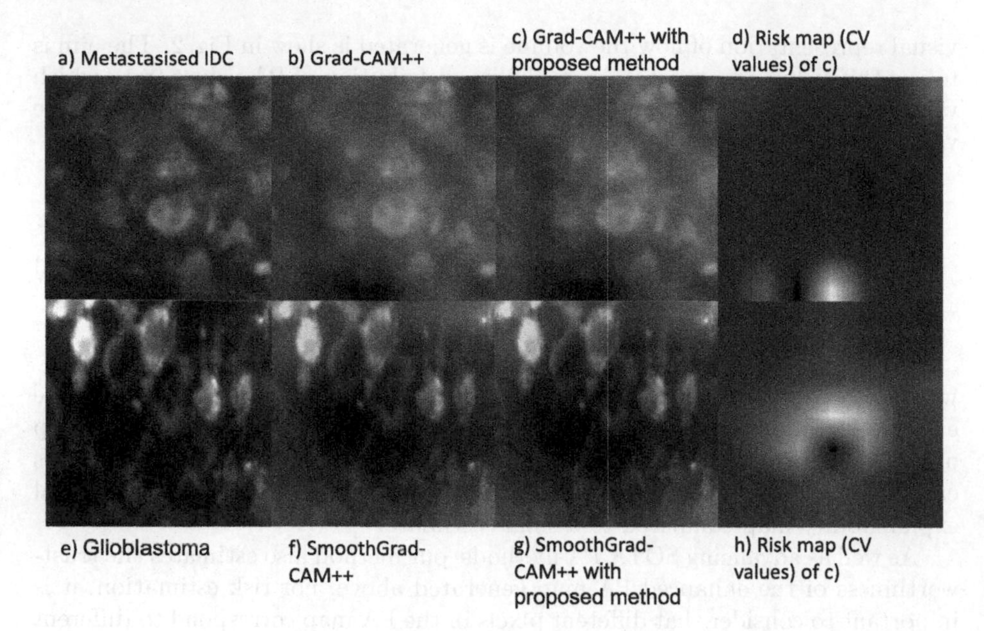

Fig. 3. PA maps generated using ResNet18 on pCLE data. a) Metastasised IDC b) Grad-CAM++ PA map on a) c) Grad-CAM++ PA map with our method applied on a) d) Risk map (CV values) of c) e) Glioblastoma f) SmoothGrad-CAM++ PA map on e) g) SmoothGrad-CAM++ PA map using our proposed method on e) h) Risk map (CV values) of g). Yellow represents the highest PA value and black the lowest. (Color figure online)

degrees of calcification. Regarding glioblastomas, the pCLE images allow for the visualization of the characteristic hypercellularity, evidence of irregular nuclei with mitotic activities or multinuclear appearance with irregular cell shape. When examining metastases of an IDC, the tumor presents as egg-shaped cells with uniform evenly spaced nuclei. Our dataset includes 38 meningioma videos, 24 glioblastoma and 6 IDC. Each pCLE video represents one tumour type and corresponds to a different patient. The data has been curated to remove noisy images and similar frames. This resulted in a training dataset of 2500 frames per class (7500 frames in total) and a testing dataset of the same size. The dataset is split into a training and testing subset, with the division done on the patient level.

Implementation. To implement the DL models we use the open-source framework PyTorch [25] and a NVIDIA Geforce RTX 3090 graphics card for parallel computation. To show our method generalises we trained two lightweight models: ResNet-18 [17] with a learning rate of 0.01 and MobileNetV2 [29] with a learning rate of 0.001. Both were trained using the Adam-W [22] optimiser with a weight decay of 0.01 and Dropout probability 0.1. We report the model's Top-1 accuracy for Resnet18 as 94.0% and for MobileNet as 86.6%. At test time, we set $T = 100$ to create a fair distribution of PA maps. PA methods were implemented with the help of TorchCAM [9] and ReciproCAM was implemented using the authors' source code.

Evaluation Metrics. Evaluating a PA method is not a trivial task because a PA map may not need to be inline with what a human deems "reasonable" [1]. Segmentation scores like intersection over union (IoU) may be used with caution to compare thresholded PA maps to ground truth maps with annotated salient regions. By doing so, we can measure how informed the model is about a particular class. To quantify how misinformed a model is, we can estimate at its average drop [6]:

$$AverageDrop(f_s, \hat{\boldsymbol{Y}}, \boldsymbol{X}) = \frac{max(0, \hat{\boldsymbol{Y}}(\boldsymbol{X}) - \hat{\boldsymbol{Y}}(\hat{\boldsymbol{X}}))}{\hat{\boldsymbol{Y}}(\boldsymbol{X})}, \tag{3}$$

where, $\hat{\boldsymbol{X}} = \boldsymbol{X} \odot f_s(\hat{\boldsymbol{Y}}(\boldsymbol{X})$. The above equation measures the effect on the output score of the classification model if we only include the pixels which the PA method scored highly. A minimum average drop is desired.

As average drop was found to not be sufficient on its own, the unified method ADCC [27] was introduced which is the harmonic mean of average drop, coherency and complexity, defined as:

$$\begin{aligned} ADCC(f_s(\hat{\boldsymbol{Y}})) =& 3(\frac{1}{Coherency(f_s(\hat{\boldsymbol{Y}}))} \\ &+ \frac{1}{1 - Complexity(f_s(\hat{\boldsymbol{Y}}))} \\ &+ \frac{1}{1 - AverageDrop(f_s, \hat{\boldsymbol{Y}}, \boldsymbol{X})})^{-1}. \end{aligned} \tag{4}$$

Coherency is the Pearson Correlation Coefficient which ensures that the remaining pixels after dropping are still important, defined as:

$$Coherency(f_s(\hat{\boldsymbol{Y}})) = \frac{Cov(f_s(\hat{\boldsymbol{Y}}(\hat{\boldsymbol{X}})), f_s(\hat{\boldsymbol{Y}}))}{\sigma(f_s(\hat{\boldsymbol{Y}}(\hat{\boldsymbol{X}}))\sigma(f_s(\hat{\boldsymbol{Y}}))}, \tag{5}$$

where, $Cov(.,.)$ is the covariance and σ is the standard deviation. A higher coherency is better. Complexity is the L1 norm of the output PA map.

$$Complexity(f_s(\hat{\boldsymbol{Y}}))) = ||f_s(\hat{\boldsymbol{Y}}))||_1. \tag{6}$$

Complexity is used to measure how cluttered a PA map is. For a good PA map, complexity should be a minimum. As it has been shown in the literature, the metrics in Eqs. (3), (5) and (6), can not be used individually to evaluate a PA method [27]. ADCC combined with computation time gives us a reliable overall metric of how a PA method is performing.

Table 1. Performance evaluation study based on the ADCC and time metrics. Coh is Coherence, Comp is Complexity, AD is average drop and each of these are reported for completeness. Time(s) is the average time to compute one PA map using a batch size of one. All metrics are run over the validation set and averaged. ScoreCAM with the proposed method takes >5 s per batch so was omitted due to resource constraints.

Architecture	PA method	Coh ↑	Comp ↓	AD ↓	ADCC ↑	Time(s) ↓
ResNet18	Standard - single iteration					
	Grad-CAM	90.1	32.7	10.1	76.6	**0.006**
	Grad-CAM++	90.6	33.1	10.6	76.2	**0.006**
	SmoothGradCAM++	88.3	27.6	14.3	74.8	**0.065**
	Score-CAM	90.0	32.3	5.9	**80.5**	**0.124**
	Recipro-CAM	91.0	41.2	10.0	72.8	**0.007**
	Proposed method					
	Grad-CAM	97.0	34.2	11.8	**77.7**	0.079
	Grad-CAM++	93.4	32.6	12.5	**78.2**	0.081
	SmoothGradCAM++	92.4	30.7	17.0	**75.8**	0.463
	Score-CAM	–	–	–	–	–
	Recipro-CAM	92.2	37.9	11.3	**76.1**	0.420
MoblieNetV2	Standard – single iteration					
	Grad-CAM	82.9	21.3	73.8	29.3	**0.010**
	Grad-CAM++	86.2	30.0	66.9	37.8	**0.010**
	SmoothGradCAM++	77.7	18.1	76.2	24.5	**0.072**
	Score-CAM	62.5	33.9	56.3	**43.9**	**0.324**
	Recipro-CAM	85.8	32.3	67.1	35.8	**0.008**
	Proposed method					
	Grad-CAM	90.0	27.0	59.4	**48.0**	0.103
	Grad-CAM++	91.4	35.9	41.5	**59.7**	0.105
	SmoothGradCAM++	89.8	22.1	71.0	**37.5**	0.322
	Score-CAM	–	–	–	–	–
	Recipro-CAM	90.6	33.7	48.3	**55.9**	0.674

Performance Evaluation. The proposed method has been compared to combinations of ResNet18 and MobileNetV2 with SOTA PA methods. At test time, Dropout is not enabled for these standard methods, it is only enabled for our method. In Table 1, we show that our method outperforms all the compared CNN-PA method combinations on ADCC. The Dropout version of ScoreCAM is too computationally expensive and therefore is not included in our comparison. We believe that the better performance of our method is because of the random dropping of features taking place during Dropout at test time which helps to suppress noise in the estimated enhanced PA map. The combination of Recipro-CAM with our proposed method improves performance (increases

ADCC) at the expense of increasing the computational complexity. We believe that this could be reduced using a batched implementation of Recipro-CAM. We attribute slow down in SmoothGradCAM++ when Dropout is applied during test time to the perturbations it adds on top of the PA method. Our validation study shows that Grad-CAM, Grad-CAM++ and Recipro-CAM are often leading in terms of speed as expected from the literature. In Fig. 1, we can see our proposed method reduces noise in the PA map around the salient region. The distinguishing characteristic of the meningioma is the psammoma body which is highlighted by all the PA methods. Risk estimations from Eq. (2) are also displayed and provide an added visualisation for a surgeon to trust the model. As it can be seen, areas of low CV match the areas of high PA values which verifies the trustworthiness of our method. We believe that the proposed explainability method could be used to support the surgeon intraoperatively in diagnosis and decision making during tumour resection. The enhanced PA map extracted with our method highlights the areas which were the most important to the model's prediction. When these areas correlate with clinically relevant areas, it shows that the model has learned to robustly classify the different tissue classes. Hence, it can be trusted by the surgeon for diagnosis.

4 Conclusion

In this work we have introduced the first combination of risk in an explainability method. Using our proposed framework we not only improve on all the tested SOTA PA method's ADCC performances but also produce an estimation of risk on the output PA values. The proposed method can clearly present to the surgeon areas of the explainability map that are more trustworthy. From this work we hope to encourage trust between the surgeon and DL models. For future work, we plan to reducing the computation time of our method and deploy the proposed framework for use in surgery.

Acknowledgement. This work was supported by the Engineering and Physical Sciences Research Council (EP/T51780X/1) and Intel R&D UK. Dr Giannarou is supported by the Royal Society (URF\R\201014).

References

1. Adebayo, J., et al.: Sanity Checks for Saliency Maps
2. Amann, J., Blasimme, A., Vayena, E., Frey, D., Madai, V.I.: Explainability for artificial intelligence in healthcare: a multidisciplinary perspective. BMC **20**(1) (2020)
3. Ancona, M., Ceolini, E., Öztireli, C., Gross, M.: Towards better understanding of gradient-based attribution methods for Deep Neural Networks (Dec 2017)
4. Blundell, C., Cornebise, J., Kavukcuoglu, K., Wierstra, D.: Weight Uncertainty in Neural Networks (May 2015)
5. Byun, S.Y., Lee, W.: Recipro-CAM: gradient-free reciprocal class activation map (Sep 2022)

6. Chattopadhyay, A., Sarkar, A., Howlader, P., Balasubramanian, V.N.: Grad-CAM++: improved Visual Explanations for Deep Convolutional Networks (Oct 2017)
7. Damianou, A.C., Lawrence, N.D.: Deep Gaussian Processes (Nov 2012)
8. Diprose, W.K., Buist, N., Hua, N., Thurier, Q., Shand, G., Robinson, R.: Physician understanding, explainability, and trust in a hypothetical machine learning risk calculator. J. Am. Med. Inform. Association **27**(4) (2020)
9. Fernandez, F.G.: TorchCAM: class activation explorer (2020)
10. Folgoc, L.L., et al.: Is MC Dropout Bayesian? (Oct 2021)
11. Gal, Y., Ghahramani, Z.: Dropout as a Bayesian Approximation: Appendix (June 2015)
12. Gal, Y., Ghahramani, Z.: Dropout as a Bayesian Approximation: Representing Model Uncertainty in Deep Learning (June 2015)
13. Goodfellow, I.J., Shlens, J., Szegedy, C.: Explaining and Harnessing Adversarial Examples (Dec 2014)
14. Gordon, L., Grantcharov, T., Rudzicz, F.: Explainable artificial intelligence for safe intraoperative decision support. JAMA Surg. **154**(11), 1064–1065 (2019)
15. Graves, A.: Practical Variational Inference for Neural Networks
16. Hagos, M.T., Curran, K.M., Mac Namee, B.: Identifying Spurious Correlations and Correcting them with an Explanation-based Learning (Nov 2022)
17. He, K., Zhang, X., Ren, S., Sun, J.: Deep Residual Learning for Image Recognition (Dec 2015)
18. Hinton, G.E., van Camp, D.: Keeping neural networks simple by minimizing the description length of the weights, pp. 5–13 (1993)
19. Hinton, G.E., Srivastava, N., Krizhevsky, A., Sutskever, I., Salakhutdinov, R.R.: Improving neural networks by preventing co-adaptation of feature detectors (July 2012)
20. Li, X., Zhou, Y., Dvornek, N.C., Gu, Y., Ventola, P., Duncan, J.S.: Efficient Shapley Explanation for Features Importance Estimation Under Uncertainty (2020)
21. Lin, M., Chen, Q., Yan, S.: Network In Network (Dec 2013)
22. Loshchilov, I., Hutter, F.: Decoupled Weight Decay Regularization (Nov 2017)
23. Neal, R.M.: Bayesian Learning for Neural Networks, vol. 118 (1996)
24. Omeiza, D., Speakman, S., Cintas, C., Weldermariam, K.: Smooth Grad-CAM++: An Enhanced Inference Level Visualization Technique for Deep Convolutional Neural Network Models (Aug 2019)
25. Paszke, A., etal.: PyTorch: an imperative style, high-performance deep learning library. In: Advances in Neural Information Processing Systems 32, pp. 8024–8035. Curran Associates, Inc. (2019)
26. Petsiuk, V., Das, A., Saenko, K.: RISE: Randomized Input Sampling for Explanation of Black-box Models (June 2018)
27. Poppi, S., Cornia, M., Baraldi, L., Cucchiara, R.: Revisiting The Evaluation of Class Activation Mapping for Explainability: A Novel Metric and Experimental Analysis (April 2021)
28. Ribeiro, M.T., Singh, S., Guestrin, C.: Why Should I Trust You?Explaining the Predictions of Any Classifier (Feb 2016)
29. Sandler, M., Howard, A., Zhu, M., Zhmoginov, A., Chen, L.C.: MobileNetV2: Inverted Residuals and Linear Bottlenecks (Jan 2018)
30. Selvaraju, R.R., Cogswell, M., Das, A., Vedantam, R., Parikh, D., Batra, D.: Grad-CAM: visual explanations from deep networks via gradient-based localization. Proceedings of the IEEE International Conference on Computer Vision 2017-October, pp. 618–626 (Dec 2017)

31. Wang, H., et al.: Score-CAM: Score-Weighted Visual Explanations for Convolutional Neural Networks (Oct 2019)
32. Zeiler, M.D., Fergus, R.: Visualizing and Understanding Convolutional Networks (Nov 2013)
33. Zhou, B., Khosla, A., Lapedriza, A., Oliva, A., Torralba, A.: Learning Deep Features for Discriminative Localization

A Video-Based End-to-end Pipeline for Non-nutritive Sucking Action Recognition and Segmentation in Young Infants

Shaotong Zhu[1], Michael Wan[1,2], Elaheh Hatamimajoumerd[1], Kashish Jain[1], Samuel Zlota[1], Cholpady Vikram Kamath[1], Cassandra B. Rowan[5], Emma C. Grace[5], Matthew S. Goodwin[3,4], Marie J. Hayes[5], Rebecca A. Schwartz-Mette[5], Emily Zimmerman[4], and Sarah Ostadabbas[1](✉)

[1] Augmented Cognition Lab, Department of Electrical and Computer Engineering, Northeastern University, Boston, MA, USA
ostadabbas@ece.neu.edu
[2] Roux Institute, Northeastern University, Portland, ME, USA
[3] Khoury College of Computer Sciences, Northeastern University, Boston, MA, USA
[4] Bouvé College of Health Sciences, Northeastern University, Boston, MA, USA
[5] Psychology Department, University of Maine, Orono, ME, USA

Abstract. We present an end-to-end computer vision pipeline to detect non-nutritive sucking (NNS)—an infant sucking pattern with no nutrition delivered—as a potential biomarker for developmental delays, using off-the-shelf baby monitor video footage. One barrier to clinical (or algorithmic) assessment of NNS stems from its sparsity, requiring experts to wade through hours of footage to find minutes of the relevant activity. Our NNS activity segmentation algorithm tackles this problem by identifying periods of NNS with high certainty—up to 94.0% average precision and 84.9% average recall across 30 heterogeneous 60 s clips, drawn from our manually annotated NNS clinical in-crib dataset of 183 h of overnight baby monitor footage from 19 infants. Our method is based on an underlying NNS action recognition algorithm, which uses spatiotemporal deep learning networks and infant-specific pose estimation, achieving 94.9% accuracy in binary classification of 960 2.5 s balanced NNS vs. non-NNS clips. Tested on our second, independent, and public NNS in-the-wild dataset, NNS recognition classification reaches 92.3% accuracy, and NNS segmentation achieves 90.8% precision and 84.2% recall. Our code and the manually annotated NNS in-the-wild dataset can be found at https://github.com/ostadabbas/NNS-Detection-and-Segmentation. Supported by MathWorks and NSF-CAREER Grant #2143882.

Keywords: Non-nutritive sucking · Action recognition · Action segmentation · Optical flow · Temporal convolution

Supplementary Information The online version contains supplementary material available at https://doi.org/10.1007/978-3-031-43895-0_55.

H. Greenspan et al. (Eds.): MICCAI 2023, LNCS 14221, pp. 586–595, 2023.
https://doi.org/10.1007/978-3-031-43895-0_55

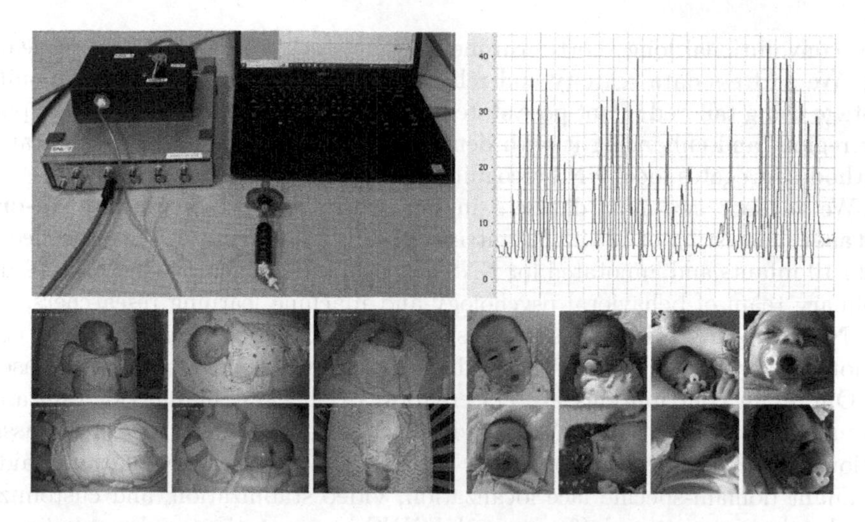

Fig. 1. *Top:* Illustration of non-nutritive sucking (NNS) signal extracted from a pressure transducer pacifier device [15]. Our computer vision-based NNS recognition and segmentation algorithms enable algorithmic identification of the relatively rare periods of high NNS activity from long videos, facilitating subsequent clinical expert evaluation. *Bottom:* Still frames from our NNS clinical in-crib dataset *(left)* and our public NNS in-the-wild dataset *(right)*.

1 Introduction

Non-nutritive sucking (NNS) is an infant oral sucking pattern characterized by the absence of nutrient delivery [11]. NNS reflects neural and motor development in early life [16] and may reduce the risk of SIDS [18,24], the leading cause of death for US infants aged 1–12 months [2]. However, studying the relationship between NNS patterns and breathing, feeding, and arousal during sleep has been challenging due to the difficulty of measuring the NNS signal.

NNS occurs in bursts of 6–12 sucks at 2 Hz per suck, with bursts happening a few times per minute during high activity periods [25]. However, active periods are sporadic, representing only a few minutes per hour, creating a burden for researchers studying characteristics of NNS. Current transducer-based approaches (see Fig. 1) are effective, but expensive, limited to research use, and may affect the sucking behavior [26]. This motivates our development of an end-to-end computer vision system to recognize and segment NNS actions from lengthy videos, enabling applications in automatic screening and telehealth, with a focus on high precision to enable periods of sucking activity to be reliably extracted for analysis by human experts.

Our contributions address the fine-grained NNS action recognition problem of classifying 2.5 s video clips, and the NNS action segmentation problem of detecting NNS activity in minute-long video clips. The action recognition method uses convolutional long short-term memory networks for spatiotemporal learning. We address data scarcity and reliability issues in real-world baby monitor footage using tailored infant pose state estimation, focusing on the face and pacifier region, and enhancing it with dense optical flow. The action segmentation method aggregates local NNS recognition signals from sliding windows.

We present two new datasets in our work: the **NNS clinical in-crib dataset**, consisting of 183 h of nighttime in-crib baby monitor footage collected from 19 infants and annotated for NNS activity and pacifier use by our interdisciplinary team of behavioral psychology and machine learning researchers, and the **NNS in-the-wild dataset**, consisting of 10 naturalistic infant video clips annotated for NNS activity. Figure 1 displays sample frames from both datasets.

Our main contributions are (1) creation of the first infant video datasets manually annotated with NNS activity; (2) development of an NNS classification system using a convolutional long short-term memory network, aided by infant domain-specific face localization, video stabilization, and customized signal enhancement, and (3) successful NNS segmentation on longer clips by aggregating local NNS recognition results across sliding windows.

2 Related Work

Current methods for measuring the NNS signal are limited to the pressured transducer approach [25], and a video-based approach that uses facial landmark detection to extract the jaw movement signal [9]. The latter relies on a 3D morphable face model [10] learned from adult face data, limiting its accuracy, given the domain gap between infant and adult faces [22]; its output also does not directly address NNS classification or segmentation. Our approach offers an efficient, end-to-end solution for both tasks and is freely available.

Action *recognition* is the task of identifying the action label of a short video clip from a set of predetermined classes. In our case, we wish to classify short infant clips based on the presence or absence of NNS. As with many action recognition algorithms, our core model is based on extending 2D convolutional neural networks to the temporal dimension for spatiotemporal data processing.

In particular, we make use of sequential networks (such as long short-term memory (LSTM) networks) after frame-wise convolution to enhance medium-range temporal dependencies [23].

Action *segmentation* is the task of identifying the periods of time during which specified events occur, often from longer untrimmed videos containing mixed activities. We follow an approach to segmentation common in limited-data contexts, patching together signals from a local low-level layer—our NNS action recognition—to obtain a global segmentation result [5].

3 Infant NNS Action Recognition and Segmentation

Figure 2 illustrates our NNS action segmentation pipeline, which predicts NNS event timestamps in long-form videos of infants using pacifiers. The process

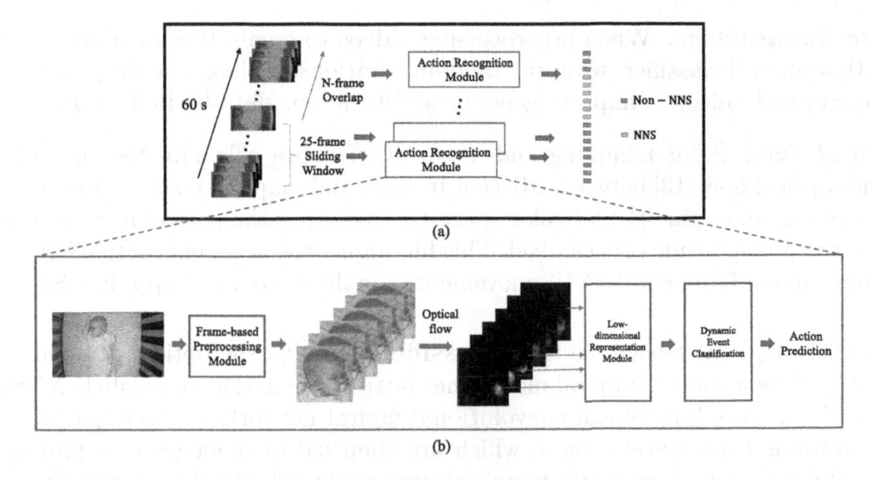

Fig. 2. *(a):* Proposed NNS segmentation pipeline: Aggregates local NNS action recognition results from sliding windows. *(b):* Proposed NNS action recognition pipeline: Utilizes dense optical flow on preprocessed frames, followed by a convolutional layer and a temporal layer to predict actions based on spatiotemporal information.

involves dividing the video into short sliding windows, applying our NNS action recognition module to classify NNS vs. non-NNS (or obtain confidence scores), and aggregating the output classes (or scores) to generate a segmentation result with predicted start and end timestamps.

3.1 NNS Action Recognition

The core of our model is the NNS action recognition system shown in Fig. 2b. It consists of a frame-based preprocessing module and a spatiotemporal classifier. The preprocessing module utilizes a pre-trained model, while only the spatiotemporal classifier is trained with our data.

Preprocessing Module. Our frame-based preprocessing module applies the following transformations in sequence. All three steps are used to produce training data for the subsequent spatiotemporal classifier, but during inference, the data augmentation step is not applicable and is omitted.

Smooth Facial Crop. The RetinaFace face detector [4] is applied to frames in each clip until a face bounding box is found and propagated to earlier and later frames using the minimum output sum of squared error (MOSSE) tracker [1]. To smooth the facial bounding box sequence and address temporal discontinuity, saliency corners [19] are detected from the initial frame and tracked to the next frame using the Lucas-Kanade optical flow algorithm [14]. The trajectory is smoothed using a moving average filter and applied to each bounding box to stabilize the facial area. The raw input video is then cropped to this smoothed bounding box, resulting in a video featuring the face alone.

Data Augmentation. When preprocessing videos to create training data for the spatiotemporal classifier, we apply random rotations, scaling, and flipping to the face-cropped video, to improve generalizability in our data-limited setting.

Optical Flow. After trimming and augmentation, we calculate the short-time dense optical flow [13] between adjacent frames, and map the results into the hue, saturation, and value (HSV) color space by cascading the optical flow direction vector and magnitude of each pixel. This highlights the apparent motion between frames, magnifying subtle NNS movements (as illustrated in Supp. Fig. S3.)[1].

Spatiotemporal-Based Action Classifier. Finally, the optical flow video is processed by a spatiotemporal model that outputs an action class label (NNS or non-NNS). Two-dimensional convolutional neural networks extract spatial representations from static images, which are then fed in sequence to a temporal convolution network for spatiotemporal processing. The final classification outcome is the output of the last temporal convolution network unit.

3.2 NNS Action Segmentation

To segment NNS actions in mixed videos with transitions between NNS and non-NNS activity, we applied NNS recognition in 2.5 s sliding windows and aggregated results to predict start and end timestamps. This window length provides fine-grained resolution for segmentation while being long enough (26 frames at a 10 Hz frame rate) for consistent human and machine detection of NNS behavior. To address concerns about the coarseness of this resolution, we tested the following window-aggregation configurations, the latter two of which have finer 0.5 s effective resolutions:

Tiled: 2.5 s windows precisely tile the length of the video with no overlaps, and the classification outcome for each window is taken directly to be the segmentation outcome for that window.

Sliding: 2.5 s windows are slid across with 0.5 s overlaps, and the classification outcome for each window is assigned to its (unique) middle-fifth 0.5 s segment as the segmentation outcome.

Smoothed: 2.5 s windows are slid across with 0.5 s overlaps, the classification *confidence score* for each window is assigned to its middle-fifth 0.5 s segment, a 2.5 s moving average of these confidence scores are taken, then the averaged confidence scores are thresholded for the final segmentation outcome.

4 Experiments, Results, and Ablation Study

4.1 NNS Dataset Creation

Our primary dataset is the **NNS clinical in-crib dataset**, consisting of 183 h of baby monitor footage collected from 19 infants during overnight sleep sessions by our clinical neurodevelopment team, with Institutional Review Board

[1] Informal qualitative tests determined the superiority of dense optical flow over other implementations, such as Farneback [6], TV-L1 [17], and RAFT [20].

(IRB #17-08-19) approval. Videos were shot in-crib with the baby monitors set up by caregivers, under low-light triggering the monochromatic infrared mode. Tens of thousands of timestamps for NNS and pacifier activity were placed, by two trained behavioral coders per video. For NNS, the definition of an event segment was taken to be an NNS *burst*: a sequence of sucks with <1 s gaps between. We restrict our subsequent study to NNS during pacifier use, which was annotated more consistently. Cohen κ annotator agreement of NNS events during pacifier use (among 10 pacifier-using infants) averaged 0.83 in 10 s incidence windows, indicating strong agreement by behavioral coding standards, but we performed further manual selection to increase precision for machine learning use, as detailed below[2]. We also created a smaller but publicly available **NNS in-the-wild dataset** of 14 YouTube videos featuring infants in natural conditions, with lengths ranging from 1 to 30 min, and similar annotations.

From each of these two datasets, we extracted 2.5 s clips for the classification task and 60 s clips for the segmentation task. In the NNS clinical in-crib dataset, we restricted our attention to six infant videos containing enough NNS activity during pacifier use for meaningful clip extraction. From each of these, we randomly drew up to 80 2.5 s clips consisting entirely of NNS activity and 80 2.5 s clips containing non-NNS activity for classification, for a total of 960; and five 60 s clips featuring transitions between NNS and non-NNS activity for segmentation, for a total of 30; redrawing if available when annotations were not sufficiently accurate. In the NNS in-the-wild dataset, we restricted to five infants exhibiting sufficient NNS activity during pacifier use, from which we drew 38 2.5 s clips each of NNS and no NNS activity for classification, for a total of 76; and from two to 26 60 s clips of mixed activity from each infant for segmentation, for a total of 39; again redrawing in cases of poor annotations. The 2.5 s clips for classification are equally balanced for NNS and non-NNS activity to support machine learning training; the 60 s mixed clips intended for segmentation intentionally over-represent NNS compared to its natural incidence rate (see Supp. Table S1), to enable meaningful statistical conclusions.

4.2 NNS Recognition Implementation and Results

For the spatiotemporal core of our NNS action recognition, we experimented with four configurations of 2D convolutional networks, a 1-layer CNN, ResNet18, ResNet50, and ResNet101 [8] (all ResNet are pre-trained using ImageNet dataset and we finetune their last fully connected layer on our data); and three configurations of sequential networks, an LSTM, a bi-directional LSTM, and a transformer model [21][3]. The models were trained for 20 epochs under a learning rate of 0.0001, and the best model was chosen based on a held-out validation set.

[2] See Supp. Fig. S1 and Supp. Fig. S2 for more on the creation of the NNS clinical in-crib dataset, and Supp. Table S1 for full Cohen κ scores, biographical data, and NNS and pacifier event statistics.

[3] Informal tests showed that the popular I3D [3] and X3D [7] models were not able to learn from the data, due possibly to the limited dataset size or subtleness of NNS movements. Formal quantitative tests will be included in forthcoming work.

We trained and tested this method with NNS clinical in-crib data from six infant subjects under a subject-wise leave-one-out cross-validation paradigm. Action recognition accuracies are reported on the top left of Table 1. The ResNet18-LSTM configuration performed best, achieving 94.9% average accuracy over six infants using optical flow input. The strong performance (≥85.2%) across all configurations indicates the viability of the overall method. We also evaluated a model trained on all six infants from the clinical in-crib dataset on the independent in-the-wild dataset. Results on the bottom left of Table 1 again show strong cross-configuration performance (≥79.5%), with ResNet101-Transformer reaching 92.3%, demonstrating strong generalizability of the method.

Table 1. Classification accuracy of our NNS action recognition model, under various convolutional and temporal configurations and two image modalities. We test on the NNS clinical in-crib data under subject-wise leave-one-out cross-validation, and on the NNS in-the-wild data directly, both with balanced classes. Strongest results in bold.

			Optical Flow				RGB			
		Convolutional	1-lr. CNN	ResNet18	ResNet50	ResNet101	1-lr. CNN	ResNet18	ResNet50	ResNet101
Dataset	Sequential	# Tr. Params.	333K	154K	614K	614K	333K	154K	614K	614K
Clinical	Transformer	393K	90.9	89.4	88.5	89.2	**63.5**	53.5	56.4	47.3
	LSTM	418K	90.7	**94.9**	87.9	85.2	52.9	52.1	57.5	46.8
	Bi-LSTM	535K	86.5	94.5	90.6	91.4	56.2	46.3	53.5	50.4
In-the-wild	Transformer	393K	83.6	79.5	81.4	**92.3**	54.0	53.3	48.9	**59.4**
	LSTM	418K	84.5	80.8	84.6	82.7	50.5	55.0	50.2	50.2
	Bi-LSTM	535K	87.2	85.2	87.5	87.2	54.4	51.7	50.2	49.8

As expected, models trained on the clinical in-crib data test worse on the independent in-the-wild data. But interestingly, models with the smaller ResNet18 network suffered steep drop-offs in performance when tested on the in-the-wild data, while models based on the complex ResNet101 fared better under the domain shift. Beyond this, it is hard to identify clear trends between configurations or capacities and performance.

Optical Flow Ablation. Performance of all models with raw RGB input replacing optical flow frames can be found on the right side of Table 1. The results are weak and close to random guessing, demonstrating the critical role played by optical flow in detecting the subtle NNS signal. This can also be seen clearly in the sample optical flow frames visualized in Supp. Fig. S3.

4.3 NNS Segmentation Results

Adopting the best ResNet18-LSTM recognition model, we tested the three configurations of the derived segmentation method on the 60 s mixed activity clips, under the same leave-one-out cross-validation paradigm on the six infants. In addition to the default classifier threshold of 0.5 used by our recognition model, we tested a 0.9 threshold to coax higher precision, as motivated in Sect. 1. We use the standard evaluation metrics of average precision AP_t and average recall AR_t based on hits and misses defined by an intersection-over-union (IoU) with threshold t, across common thresholds $t \in \{0.1, 0.3, 0.5\}$[4]. Averages are taken with subjects given equal weight, and results tabulated in Table 2.

[4] We follow definitions from [12], with tiebreaks decided by IoU instead of confidence.

The metrics reveal strong performance from all methods and both confidence thresholds on both test sets. Generally, as expected, setting a higher confidence threshold or employing the more tempered tiled or smoothed aggregation methods favours precision, while lowering the confidence threshold or employing the more responsive sliding aggregation method favours recall. The results are excellent at the IoU threshold of 0.1 but degrade as the threshold is raised, suggesting that while these methods can readily perceive NNS behavior, they are still limited by the underlying ground truth annotator accuracy. The consistency of the performance of the model across both cross-validation testing in the clinical in-crib dataset and the independent testing on the NNS in-the-wild dataset suggests strong generalizability. Figure 3 visualizes predictions (and underlying confidence scores) of the sliding model configuration with a confidence threshold of 0.9, highlighting the excellent precision characteristics and illustrating the overall challenges of the detection problem.

Table 2. Average precision AP_t and average recall AR_t performance for various IoU thresholds t of our NNS segmentation model. We test three local classification aggregation methods and two different classifier confidence thresholds. Precision-recall pairs with the highest precision in each threshold configuration in bold.

		Classifier Confidence Threshold = 0.9						Classifier Confidence Threshold = 0.5					
Dataset	Method	$AP_{0.1}$	$AR_{0.1}$	$AP_{0.3}$	$AR_{0.3}$	$AP_{0.5}$	$AR_{0.5}$	$AP_{0.1}$	$AR_{0.1}$	$AP_{0.3}$	$AR_{0.3}$	$AP_{0.5}$	$AR_{0.5}$
Clinical	Tiled	**94.0**	**84.9**	**80.6**	**74.9**	**56.4**	**53.0**	86.9	91.0	74.0	72.9	42.6	44.8
	Sliding	84.8	86.1	70.3	72.1	44.3	47.7	78.3	92.7	70.3	82.5	45.4	53.1
	Smoothed	91.4	74.9	70.4	60.2	39.6	36.5	**90.3**	**91.5**	**77.8**	**76.6**	**51.0**	**50.8**
In-the-wild	Tiled	90.7	81.5	**76.3**	**68.3**	56.7	49.7	**90.8**	**84.2**	**80.5**	**74.4**	67.9	63.5
	Sliding	82.7	80.5	70.6	64.7	**60.9**	**54.7**	79.0	85.1	67.2	72.7	62.8	66.5
	Smoothed	**90.8**	**72.4**	73.3	56.4	53.1	41.9	90.0	78.7	77.0	67.5	**72.2**	**62.6**

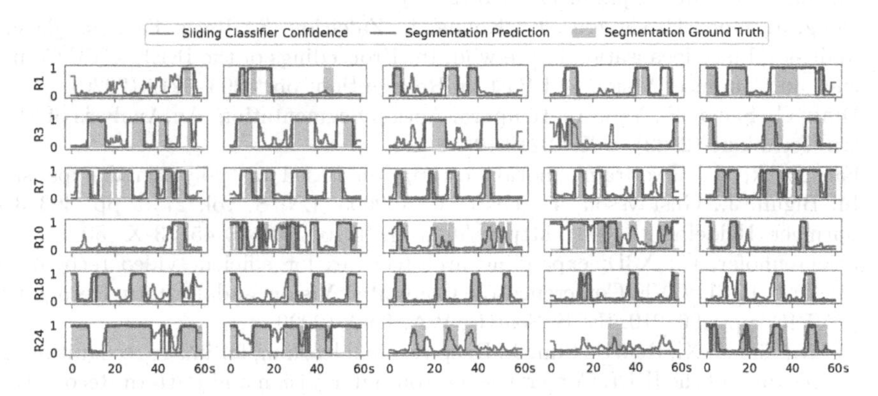

Fig. 3. Segmentation predictions and ground truth for each 60 s mixed clip from the NNS clinical in-bed dataset, under the sliding window aggregation model configuration and with a confidence threshold of 0.9, boosting precision at the cost of recall.

5 Conclusion

We present our novel computer vision method for the detection of non-nutritive sucking from videos, with a spatiotemporal action recognition model for classifying short video clips and a segmentation model for determining event timestamps in longer videos. Our work is grounded in our methodological collection and annotation of infant video data from varied settings. We use domain-specific techniques such as dense optical flow and infant state tracking to detect subtle sucking movements and ameliorate a relative scarcity of data. Future work could improve the robustness these methods in challenging examples of NNS activity, such as more ambiguous sucking or sucking while moving. This would require more precisely and reliably annotated data to train and evaluate, which in our experience could be difficult to obtain. An alternative approach would be to aim for more robust but less exacting, split-second results. Beyond improvements to the core NNS detection algorithms, algorithmic extraction of NNS signal characteristics, such as individual suck frequency, strength, duration, and temporal pattern, could further NNS research and one day aid in clinical care.

References

1. Bolme, D.S., Beveridge, J.R., Draper, B.A., Lui, Y.M.: Visual object tracking using adaptive correlation filters. In: 2010 IEEE Computer Society Conference on Computer Vision and Pattern Recognition, pp. 2544–2550 (2010)
2. Carlin, R.F., Moon, R.Y.: Risk factors, protective factors, and current recommendations to reduce sudden infant death syndrome: a review. JAMA Pediatr. **171**(2), 175–180 (2017)
3. Carreira, J., Zisserman, A.: Quo vadis, action recognition? A new model and the kinetics dataset. In: Proceedings of the IEEE Conference on Computer Vision and Pattern Recognition, pp. 6299–6308 (2017)
4. Deng, J., Guo, J., Ververas, E., Kotsia, I., Zafeiriou, S.: RetinaFace: single-shot multi-level face localisation in the wild. In: Proceedings of the IEEE/CVF Conference on Computer Vision and Pattern Recognition, pp. 5203–5212 (2020)
5. Ding, G., Sener, F., Yao, A.: Temporal Action Segmentation: An Analysis of Modern Technique (2022). arXiv:2210.10352 [cs]
6. Farnebäck, G.: Two-frame motion estimation based on polynomial expansion. In: Bigun, J., Gustavsson, T. (eds.) SCIA 2003. LNCS, vol. 2749, pp. 363–370. Springer, Heidelberg (2003). https://doi.org/10.1007/3-540-45103-X_50
7. Feichtenhofer, C.: X3D: expanding architectures for efficient video recognition. In: 2020 IEEE/CVF Conference on Computer Vision and Pattern Recognition (CVPR), pp. 200–210. IEEE, Seattle, WA, USA (2020)
8. He, K., Zhang, X., Ren, S., Sun, J.: Deep residual learning for image recognition. In: Proceedings of the IEEE Conference on Computer Vision and Pattern Recognition, pp. 770–778 (2016)
9. Huang, X., Martens, A., Zimmerman, E., Ostadabbas, S.: Infant contact-less non-nutritive sucking pattern quantification via facial gesture analysis. In: CVPR Workshops (2019)

10. Huber, P., et al.: A multiresolution 3D morphable face model and fitting framework. In: Proceedings of the 11th International Joint Conference on Computer Vision, Imaging and Computer Graphics Theory and Applications. University of Surrey (2016)
11. Humphrey, T.: The Development of Human Fetal Activity and its Relation to Postnatal Behavior, Advances in Child Development and Behavior, vol. 5. Academic Press, New York (1970)
12. Idrees, H., et al.: The THUMOS challenge on action recognition for videos "in the wild". Comput. Vis. Image Underst. **155**, 1–23 (2017)
13. Liu, C., et al.: Beyond pixels: exploring new representations and applications for motion analysis. Ph.D. thesis, Massachusetts Institute of Technology (2009)
14. Lucas, B.D., Kanade, T., et al.: An iterative image registration technique with an application to stereo vision. vol. 81. Vancouver (1981)
15. Martens, A., Hines, M., Zimmerman, E.: Changes in non-nutritive suck between 3 and 12 months. Early Hum. Dev. **149**, 105141 (2020)
16. Medoff-Cooper, B., Ray, W.: Neonatal sucking behaviors. Image: J. Nurs. Sch. **27**(3), 195–200 (1995)
17. Pock, T., Urschler, M., Zach, C., Beichel, R., Bischof, H.: A Duality Based Algorithm for TV-L ^{1}-Optical-Flow Image Registration. In: Ayache, N., Ourselin, S., Maeder, A. (eds.) MICCAI 2007. LNCS, vol. 4792, pp. 511–518. Springer, Heidelberg (2007). https://doi.org/10.1007/978-3-540-75759-7_62
18. Psaila, K., Foster, J.P., Pulbrook, N., Jeffery, H.E.: Infant pacifiers for reduction in risk of sudden infant death syndrome. Cochrane Database of Syst. Rev. **4**(4), CD011147 (2017)
19. Shi, J., et al.: Good features to track. In: 1994 Proceedings of IEEE Conference on Computer Vision and Pattern Recognition, pp. 593–600. IEEE (1994)
20. Teed, Z., Deng, J.: RAFT: recurrent all-pairs field transforms for optical flow. In: Vedaldi, A., Bischof, H., Brox, T., Frahm, J.-M. (eds.) ECCV 2020. LNCS, vol. 12347, pp. 402–419. Springer, Cham (2020). https://doi.org/10.1007/978-3-030-58536-5_24
21. Vaswani, A., et al.: Attention is all you need. In: Advances in Neural Information Processing Systems. vol. 30 (2017)
22. Wan, M., et al.: InfAnFace: bridging the infant-adult domain gap in facial landmark estimation in the wild. In: 26th International Conference on Pattern Recognition (ICPR) (2022)
23. Yue-Hei Ng, J., Hausknecht, M., Vijayanarasimhan, S., Vinyals, O., Monga, R., Toderici, G.: Beyond short snippets: deep networks for video classification. In: Proceedings of the IEEE Conference on Computer Vision and Pattern Recognition, pp. 4694–4702 (2015)
24. Zavala Abed, B., Oneto, S., Abreu, A.R., Chediak, A.D.: How might non nutritional sucking protect from sudden infant death syndrome. Med. Hypotheses **143**, 109868 (2020)
25. Zimmerman, E., Carpenito, T., Martens, A.: Changes in infant non-nutritive sucking throughout a suck sample at 3-months of age. PLOS ONE **15**(7), e0235741 (2020)
26. Zimmerman, E., Foran, M.: Patterned auditory stimulation and suck dynamics in full-term infants. Acta Paediatr. **106**(5), 727–732 (2017)

Reveal to Revise: An Explainable AI Life Cycle for Iterative Bias Correction of Deep Models

Frederik Pahde[1], Maximilian Dreyer[1], Wojciech Samek[1,2,3](✉), and Sebastian Lapuschkin[1](✉)

[1] Fraunhofer Heinrich-Hertz-Institute, Berlin, Germany
{wojciech.samek,sebastian.lapuschkin}@hhi.fraunhofer.de

[2] Technische Universität Berlin, Berlin, Germany

[3] BIFOLD – Berlin Institute for the Foundations of Learning and Data, Berlin, Germany

Abstract. State-of-the-art machine learning models often learn spurious correlations embedded in the training data. This poses risks when deploying these models for high-stake decision-making, such as in medical applications like skin cancer detection. To tackle this problem, we propose Reveal to Revise (R2R), a framework entailing the entire eXplainable Artificial Intelligence (XAI) life cycle, enabling practitioners to iteratively identify, mitigate, and (re-)evaluate spurious model behavior with a minimal amount of human interaction. In the first step (1), R2R *reveals* model weaknesses by finding outliers in attributions or through inspection of latent concepts learned by the model. Secondly (2), the responsible artifacts are *detected* and spatially *localized* in the input data, which is then leveraged to (3) *revise* the model behavior. Concretely, we apply the methods of RRR, CDEP and ClArC for model correction, and (4) (re-)evaluate the model's performance and remaining sensitivity towards the artifact. Using two medical benchmark datasets for Melanoma detection and bone age estimation, we apply our R2R framework to VGG, ResNet and EfficientNet architectures and thereby reveal and correct real dataset-intrinsic artifacts, as well as synthetic variants in a controlled setting. Completing the XAI life cycle, we demonstrate multiple R2R iterations to mitigate different biases. Code is available on https://github.com/maxdreyer/Reveal2Revise.

Keywords: XAI Life Cycle · Bias Identification · Model Correction

1 Introduction

Deep Neural Networks (DNNs) have successfully been applied in research and industry for a multitude of complex tasks. This includes various medical

F. Pahde and M. Dreyer—Contributed equally.

Supplementary Information The online version contains supplementary material available at https://doi.org/10.1007/978-3-031-43895-0_56.

H. Greenspan et al. (Eds.): MICCAI 2023, LNCS 14221, pp. 596–606, 2023.
https://doi.org/10.1007/978-3-031-43895-0_56

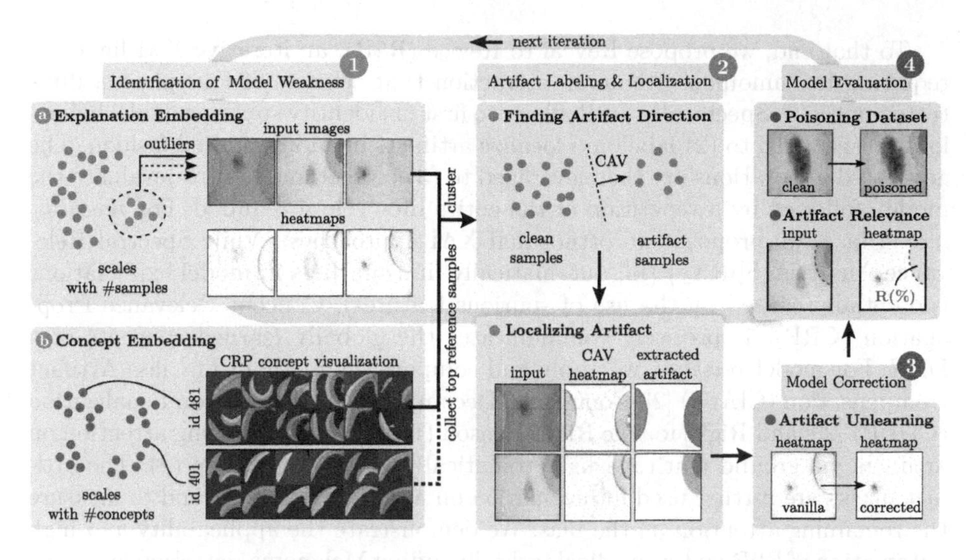

Fig. 1. Our R2R life cycle for *revealing* and *revising* spurious behavior of any pre-trained DNN. Firstly, we identify model weaknesses by finding either outliers in explanations using SpRAy (1a) or suspicious concepts using zoomed-in CRP concept visualizations (1b). Secondly (2), SpRAy clusters or collecting the top reference samples allows us to label artifactual samples and to compute an artifact CAV, which we use to model and localize the artifact in latent and input space, respectively. At this point, the artifact localization can be leveraged for (3) model correction, and (4) to evaluate the model's performance on a poisoned test set and measure its remaining attention on the artifact.

applications for which DNNs have even shown to be superior to medical experts, such as with Melanoma detection [5]. However, the reasoning of these highly complex and non-linear models is generally not transparent [23,24], and as such, their decisions may be biased towards unintended or undesired features, potentially caused by shortcut learning [2,9,14,27]. Particularly in high-stake decision processes, such as medical applications, unreliable or poorly understood model behavior may pose severe security risks.

The field of XAI brings light into the black boxes of DNNs and provides a better understanding of their decision processes. As such, local XAI methods reveal (input) features that are most relevant to a model, which, for image data, can be presented as heatmaps. In contrast, global XAI methods (e.g., [12,14]) reveal general prediction strategies employed or features encoded by a model, which is necessary for the identification and understanding of systematic (mis-)behavior. Acting on the insights from explanations, various methods have been introduced to correct for undesired model behavior [31]. While multiple approaches exist for either *revealing* or *revising* model biases, only few combine both steps, to be applicable as a framework. Such frameworks, however, either rely heavily on human feedback [25,29], are limited to specific bias types [2], or require labor-intensive annotations for both model evaluation and correction [13,25].

To that end, we propose Reveal to Revise (R2R), an iterative XAI life cycle requiring low amounts of human interaction that consists of four phases, illustrated in Fig. 1. Specifically, R2R allows to first (1) identify spurious model behavior and secondly, to (2) label and localize artifacts in an automated fashion. The generated annotations are then leveraged to (3) correct and (4) (re-)evaluate the model, followed by a repetition of the entire life cycle if required. For *revealing* model bias, we propose two orthogonal XAI approaches: While Spectral Relevance Analysis (SpRAy) [14] automatically finds outliers in model explanations (potentially caused by the use of spurious features), Concept Relevance Propagation (CRP) [1] precisely communicates the globally learned concepts of a DNN. For model *revision*, we apply and compare the methods of Class Artifact Compensation (ClArC) [2], Contextual Decomposition Explanation Penalization (CDEP) [20] and Right for the Right Reason (RRR) [22], penalizing attention on artifacts via ground truth masks automatically generated in step (2). The artifact masks are further used for evaluation on a poisoned test set and to measure the remaining attention on the bias. We demonstrate the applicability and high automation of R2R on two medical tasks, including Melanoma detection and bone age estimation, using the VGG-16, ResNet-18 and EfficientNet-B0 DNN architectures. In our experiments, we correct model behavior w.r.t. dataset-intrinsic, as well as synthetic artifacts in a controlled setting. Lastly, we showcase the R2R life cycle through multiple iterations, unveiling and unlearning different biases.

2 Related Work

Among other methods, e.g., leveraging auxiliary information [15,18,19,21], or training on de-biased representations [4,16], shortcut unlearning is often approached with XAI. The majority of related works introduce methods to either identify spurious behavior [1,14], or to align the model behavior with pre-defined priors [20,22], with only a few combining both, such as the eXplanatory Interactive Learning (XIL) framework [29] or the approach introduced by Anders et al. [2]. The former is based on presenting individual local explanations to a human, who, if necessary, provides feedback used for model correction [25,29]. However, studying individual predictions is slow and labor-extensive, limiting its practicability. In contrast, the authors of [2] use SpRAy [14] for the detection of spurious model behavior and labeling of artifactual samples. In addition to SpRAy, we suggest to study latent features of the model via CRP concept visualizations [1] as a tool for more fine-grained model inspection, catching systematic misbehavior which would not be visible through SpRAy clusters.

Most model correction methods require dense annotations, such as labels for artifactual samples or artifact localization masks, which are either crafted heuristically or by hand [13,20]. In our R2R framework, we automate the annotation by following [2] for data labeling through SpRAy outlier clusters, or by collecting the most representative samples of bias concepts according to CRP. The spatial artifact localization is further automated by computing artifact heatmaps as outlined in Sect. 3.1, thereby considerably easing the step from bias identification to correction.

Existing works for model correction measure the performance on the original or clean test set, with corrected models often showing an improved generalization [13,20]. A more targeted approach for measuring the artifact's influence is the evaluation on poisoned data [25], for which R2R is well suited by using its localization scheme to first extract artifacts and to then poison clean test samples. By precisely localizing artifacts, R2R further allows to measure the model's attention on an artifact through attribution heatmaps.

3 Reveal to Revise Framework

Our *Reveal to Revise* (R2R) framework comprises the entire XAI life cycle, including methods for (1) the identification of model bias, (2) artifact labeling and localization, (3) the correction of detected misbehavior, and (4) the evaluation of the improved model. To that end, we now describe the methods used for R2R.

3.1 Data Artifact Identification and Localization

The identification of spurious data artifacts using CRP concept visualizations or SpRAy clusters is firstly described, followed by our artifact localization approach.

CRP Concept Visualizations. CRP [1] combines global concept visualization techniques with local feature attribution methods. This provides an understanding of the relevance of latent concepts for a prediction and their localization in the input. In this work, we use Layer-wise Relevance Propagation (LRP) [3] for feature attribution under CRP and for heatmaps in general, however, other local XAI methods can be used as well. Jointly with Relevance Maximization [1], CRP is well suited for the identification of spurious concepts by precisely narrowing down the input parts that have been most relevant for model inference, as shown in Fig. 1 (*bottom left*) for band-aid concepts, where irrelevant background is overlaid with black semi-transparent color. The collection of top-ranked reference samples for spurious concepts allows us to label artifactual data.

Explanation Outliers Through SpRAy. Alternatively, SpRAy [14] is a strategy to find outliers in local explanations, which are likely to stem from spurious model behavior, such as the use of a Clever Hans features, *i.e.*, features correlating with a certain class that are unrelated to the actual task. Following [2,14], we apply SpRAy by clustering latent attributions computed through LRP. The SpRAy clusters then naturally allow us to label data containing the bias.

Artifact Localization. We automate artifact localization by training a Concept Activation Vector (CAV) $\mathbf{h}_l$ to model the artifact in latent space of a layer l, representing the direction from artifactual to non-artifactual samples obtained from a linear classifier. The artifact localization is given by a modified backward pass on the biased model with LRP for an artifact sample $\mathbf{x}$, where we initialize the relevances $\mathbf{R}_l(\mathbf{x})$ at layer l as

$$\mathbf{R}_l(\mathbf{x}) = \mathbf{a}_l(\mathbf{x}) \circ \mathbf{h}_l \tag{1}$$

with activations $\mathbf{a}_l$ and element-wise multiplication operator $\circ$. This is equivalent to explaining the output from the linear classifier given as $\mathbf{a}_l(\mathbf{x}) \cdot \mathbf{h}_l$. Using a threshold, the resulting CAV heatmap can be further processed into a binary mask to crop out the artifact from any corrupted sample, as illustrated in Fig. 1 (*bottom center*).

3.2 Methods for Model Correction

In the following, we present the methods used for mitigating model biases.

ClArC for Latent Space Correction. Methods from the ClArC framework correct model (mis-)behavior w.r.t. an artifact by modeling its direction $\mathbf{h}$ in latent space using CAVs [12]. The framework consists of two methods, namely Augmentive ClArC (a-ClArC) and Projective ClArC (p-ClArC). While a-ClArC adds $\mathbf{h}_l$ to the activations $\mathbf{a}_l$ of layer l for all samples in a fine-tuning phase, hence teaching the model to be invariant towards that direction, p-ClArC suppresses the artifact direction during the test phase and does not require any fine-tuning. More precisely, the perturbed activations $\mathbf{a}'_l$ are given by

$$\mathbf{a}'_l(\mathbf{x}) = \mathbf{a}_l(\mathbf{x}) + \gamma(\mathbf{x})\mathbf{h}_l \tag{2}$$

with perturbation strength $\gamma(\mathbf{x})$ dependent on input $\mathbf{x}$. Parameter $\gamma(\mathbf{x})$ is chosen such that the activation in direction of the CAV is as high as the average value over non-artifactual or artifactual samples for p-ClArC or a-ClArC, respectively.

RRR and CDEP for Correction Through Prior Knowledge. Model correction using RRR [22] or CDEP [20] is based on an additional λ-weighted loss term (besides the cross-entropy loss $\mathcal{L}_{\text{CE}}$) for neural network training that aligns the use of features by the model f_θ, described by an explanation $\exp_\theta$, to a defined prior explanation $\exp_{\text{prior}}$. The authors of RRR propose to penalize the model's attention on unfavorable artifacts using the input gradient w.r.t. the cross-entropy loss, leading to

$$\mathcal{L}_{\text{RRR}}\left(\exp_\theta(\mathbf{x}), \exp_{\text{prior}}(\mathbf{x})\right) = \left\|\nabla_{\mathbf{x}}\mathcal{L}_{\text{CE}}\left(f_\theta(\mathbf{x}), y_{\text{true}}\right) \circ \mathbf{M}_{\text{prior}}(\mathbf{x})\right\|_2{}^2 \tag{3}$$

with a binary mask $\mathbf{M}_{\text{prior}}(\mathbf{x})$ localizing an artifact and class label y_{true}.

Alternatively, CDEP [20] proposes to use CD [17] importance scores $\beta(\mathbf{x}_s)$ for a feature subset $\mathbf{x}_s$ based on the forward pass instead of gradient to align the model's attention. Penalizing artifact features via masked input $\mathbf{x}_M$ results in

$$\mathcal{L}_{\text{CDEP}}\left(\exp_\theta(\mathbf{x}), \exp_{\text{prior}}(\mathbf{x})\right) = \left\|\frac{e^{\beta(\mathbf{x}_M)}}{e^{\beta(\mathbf{x}_M)} + e^{\beta(\mathbf{x}-\mathbf{x}_M)}}\right\|_1. \tag{4}$$

4 Experiments

The experimental section is divided into the two parts of (1) identification, mitigation and evaluation of spurious model behavior with various correction methods and (2) showcasing the whole R2R framework in an iterative fashion.

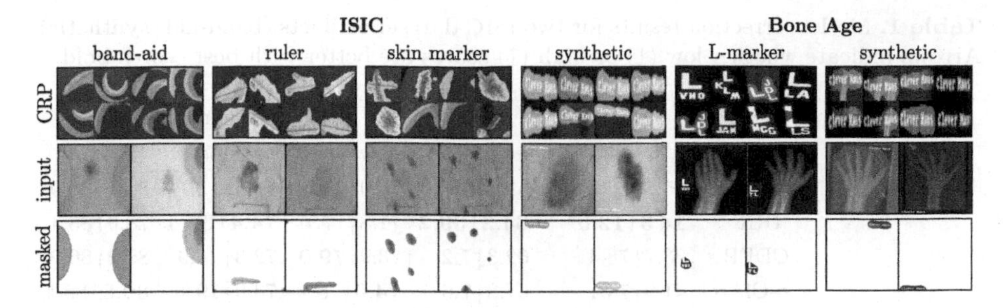

Fig. 2. Overview of artifacts with CRP visualizations of corresponding concepts zoomed-in using receptive field information (*top*), input samples (*middle*), and cropped out artifacts (*bottom*) using our artifact localization method. Shown are band-aid, ruler, skin marker, and synthetic artifacts for the ISIC dataset, as well as "L"-marker and synthetic artifacts for the Bone Age dataset.

4.1 Experimental Setup

We train VGG-16 [26], ResNet-18 [11] and EfficientNet-B0 [28] models on the ISIC 2019 dataset [7,8,30] for skin lesion classification and Pediatric Bone Age dataset [10] for bone age estimation based on hand radiographs. Besides evaluating our methodology on data-intrinsic artifacts occurring in these datasets, we artificially insert an artifact into data samples in a controlled setting. Specifically, we insert a "Clever Hans" text (shown in Fig. 2) into a subset of training samples of one specific class. See Appendix A.1 for additional experiment details.

4.2 Revealing and Revising Spurious Model Behavior

Revealing Bias: In the first step of the R2R life cycle, we can reveal the use of several artifacts by the examined models, including the well-known band-aid, ruler and skin marker [6] and our synthetic Clever Hans for the ISIC dataset, as shown in Fig. 2 for VGG-16. Here, we show concept visualizations and cropped out artifacts based on our automatic artifact localization scheme described in Sect. 3.1. The "band-aid" use can be further identified via SpRAy, as illustrated in Fig. 3 (*right*). Besides the synthetic Clever Hans for bone age classification, we encountered the use of "L" markings, resulting from physical lead markers placed by radiologist to specify the anatomical side. Interestingly, the "L" markings are larger for hands of younger children, as all hands are scaled to similar size [10], offering the model to learn a shortcut by estimating the bone age based on the relative size of the "L" markings, instead of valid features. While we revealed the "L" marking bias using CRP, we did not find corresponding SpRAy clusters, underlining the importance of both approaches for model investigation.

Revising Model Behavior: Having revealed spurious behavior, we now revise the models, beginning with model correction. Specifically, we correct for the band-aid, "L" markings as well as synthetic artifacts. The skin marker and ruler

Table 1. Model correction results for two ISIC dataset artifacts (band-aid | synthetic). Arrows indicate whether low (↓) or high (↑) scores are better with best scores bold.

architecture	method	↓ artifact relevance (%)	↑ F1 (%) *poisoned*	↑ F1 (%) *original*	↑ accuracy (%) *poisoned*	↑ accuracy (%) *original*
VGG-16	*Vanilla*	45.5 | 76.3	59.7 | 7.7	73.9 | 79.0	71.5 | 19.1	80.1 | 86.9
	RRR	**14.3** | **12.0**	**64.2** | **39.2**	71.8 | 77.7	**74.4** | **32.4**	78.0 | 85.4
	CDEP	23.7 | 78.4	62.8 | 7.2	73.9 | **79.0**	72.3 | 18.9	80.2 | 86.9
	p-ClArC	41.9 | 76.1	61.8 | 7.6	**74.0** | 78.1	73.0 | 19.1	**80.3** | 85.4
	a-ClArC	42.8 | 75.5	62.4 | 12.5	70.3 | 76.5	73.1 | 21.0	78.4 | **88.9**
ResNet-18	*Vanilla*	33.1 | 37.6	68.2 | 39.0	79.1 | 82.1	76.8 | 35.6	83.3 | 89.5
	RRR	30.3 | **16.9**	70.4 | **70.4**	**79.7** | 79.1	77.1 | **75.7**	**83.4** | 84.8
	CDEP	**25.4** | 22.2	**71.5** | 60.9	75.9 | 81.6	**77.5** | 59.4	81.5 | 87.9
	p-ClArC	32.0 | 33.6	69.2 | 38.9	78.3 | **81.8**	75.9 | 34.4	82.5 | **89.1**
	a-ClArC	32.9 | 38.4	70.1 | 52.9	78.3 | 80.5	76.2 | 45.3	81.1 | 88.9
Efficient-Net-B0	*Vanilla*	45.6 | 63.9	72.2 | 38.8	81.8 | 84.7	80.1 | 30.2	85.4 | 90.8
	RRR	**34.5** | **24.6**	**74.0** | 65.8	81.3 | 83.3	80.1 | 65.9	84.6 | 89.8
	p-ClArC	41.3 | 62.5	73.1 | 38.7	**82.0** | **84.4**	**80.4** | 29.8	**85.5** | **90.5**
	a-ClArC	45.7 | 65.6	72.7 | **72.4**	81.8 | 81.4	80.1 | **79.4**	84.9 | 87.3

artifacts are corrected for in iterative fashion in Sect. 4.3. For all methods (RRR, CDEP[1] and ClArC), including a *Vanilla* model without correction, we fine-tune the models' last dense layers for 10 epochs. Note that both RRR and CDEP require artifact masks to unlearn the undesired behavior. As part of R2R, we propose measures to automate this step by using the artifact localization strategy described in Sect. 3.1. Further note, that once generated, artifact localizations can be used for *all* investigated models. See Appendix A.1 for additional fine-tuning details.

We evaluate the effectiveness of model corrections based on two metrics: the attributed fraction of relevance to artifacts and prediction performance on both the original and a poisoned test set (in terms of F1-score and accuracy). Whereas in the synthetic case, we simply insert the artifact into all samples to poison the test set, data-intrinsic artifacts are cropped from random artifactual samples using our artifact localization strategy. Note that artifacts might overlap clinically informative features in poisoned samples, limiting the comparability of *poisoned* and *original* test performance. As shown in Tab. 1 (ISIC 2019) and Appendix A.2 (Bone Age), we are generally able to improve model behavior with all methods. The only exception is the synthetic artifact for VGG-16, where only RRR mitigates the bias to a certain extent, indicating that the artifact signal is too strong for the model. Here, fine-tuning only the last layer is not sufficient to learn alternative prediction strategies. Interestingly, despite successfully decreasing the models' output sensitivity towards artifacts,

[1] CDEP is not applied to EfficientNets, as existing implementations are incompatible.

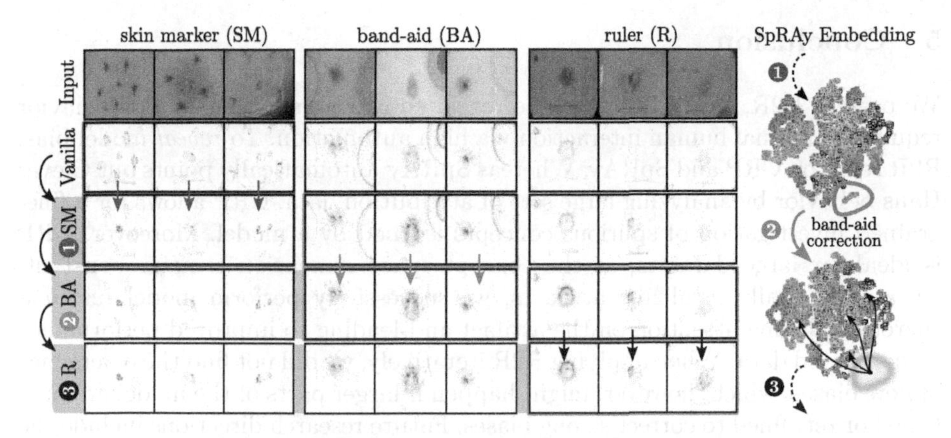

Fig. 3. The effect of iterative model correction on relevances attributed to artifacts for each iteration (*left*) and the band-aid artifact cluster from SpRAy, which dissipates after its correction step (*right*). See Appendix A.2 for quantitative results.

applying a-ClArC barely decreases the relevance attributed to artifacts in input space. This might result from ClArC methods not directly penalizing the use of artifacts, but instead encouraging the model to develop alternative prediction strategies. Overall, RRR yields the most consistent results, constantly reducing the artifact relevance while increasing the model performance on poisoned test sets. Both observations are underlined by heatmaps for revised models in Fig. A.1 (Appendix A.2), where RRR and CDEP visibly reduce the model attention on the artifacts.

4.3 Iterative Model Correction with R2R

Showcasing the full R2R life cycle (as shown in Fig. 1), we now perform multiple R2R iterations, revealing and revising undesired model behavior step by step. Specifically, we successively correct the VGG-16 model w.r.t. the skin marker, band-aid, and ruler artifacts discovered in Sect. 4.2 using RRR. In order to prevent the model from re-learning previously unlearned artifacts, we keep the previous artifact-specific RRR losses intact. Thus, we are able to correct for all artifacts, with evaluation results given in Appendix A.2, applying the same metrics as in Sect. 4.2. In Fig. 3, we show exemplary attribution heatmaps for all artifacts after each iteration. While there are large amounts of relevance on all artifacts initially, it can successfully be reduced in the according iterations to correct the model behavior w.r.t. skin marker (SM), band-aids (BA), and rulers (R). It is to note, that correcting for the skin marker also (slightly) improved the model w.r.t. other artifacts, which might result from corresponding latent features that are not independent, as shown by CRP visualizations in Fig. 2 for skin marker. Moreover, we show the SpRAy embedding of training samples after the first iteration in Fig. 3 (*right*), revealing an isolated cluster with samples containing the band-aid artifact, which dissipates after the correction step.

5 Conclusion

We present R2R, an XAI life cycle to reveal and revise spurious model behavior requiring minimal human interaction via high automation. To *reveal* model bias, R2R relies on CRP and SpRAy. Whereas SpRAy automatically points out Clever Hans behavior by analyzing large sets of attribution data, CRP allows for a fine-grained investigation of spurious concepts learned by a model. Moreover, CRP is ideal for large datasets, as the concept space dimension remains constant. By automatically localizing artifacts, we successfully perform model *revision*, thereby reducing attention on the artifact and leading to improved performance on corrupted data. When applying R2R iteratively, we did not find the emergence of new biases, which, however, might happen if larger parts of the model are fine-tuned or retrained to correct strong biases. Future research directions include the application to non-localizable artifacts, and addressing fairness issues in DNNs.

Acknowledgements. This work was supported by the Federal Ministry of Education and Research (BMBF) as grants [SyReal (01IS21069B), BIFOLD (01IS18025A, 01IS18037I)]; the European Union's Horizon 2020 research and innovation programme (EU Horizon 2020) as grant [iToBoS (965221)]; the European Union's Horizon 2022 research and innovation programme (EU Horizon Europe) as grant [TEMA (101093003)]; the state of Berlin within the innovation support program ProFIT (IBB) as grant [BerDiBa (10174498)]; and the German Research Foundation [DFG KI-FOR 5363].

References

1. Achtibat, R., et al.: From "where" to "what": towards human-understandable explanations through concept relevance propagation. arXiv preprint arXiv:2206.03208 (2022)
2. Anders, C.J., Weber, L., Neumann, D., Samek, W., Müller, K.R., Lapuschkin, S.: Finding and removing clever hans: using explanation methods to debug and improve deep models. Inf. Fusion **77**, 261–295 (2022)
3. Bach, S., Binder, A., Montavon, G., Klauschen, F., Müller, K.R., Samek, W.: On pixel-wise explanations for non-linear classifier decisions by layer-wise relevance propagation. PLoS ONE **10**(7), e0130140 (2015)
4. Bahng, H., Chun, S., Yun, S., Choo, J., Oh, S.J.: Learning de-biased representations with biased representations. In: ICML, pp. 528–539. PMLR (2020)
5. Brinker, T.J., et al.: Deep learning outperformed 136 of 157 dermatologists in a head-to-head dermoscopic melanoma image classification task. Eur. J. Cancer **113**, 47–54 (2019)
6. Cassidy, B., Kendrick, C., Brodzicki, A., Jaworek-Korjakowska, J., Yap, M.H.: Analysis of the ISIC image datasets: usage, benchmarks and recommendations. Med. Image Anal. **75**, 102305 (2022)
7. Codella, N.C., et al.: Skin lesion analysis toward melanoma detection: a challenge at the 2017 international symposium on biomedical imaging (ISBI), hosted by the international skin imaging collaboration (ISIC). In: 2018 IEEE 15th International Symposium on Biomedical Imaging (ISBI 2018), pp. 168–172. IEEE (2018)

8. Combalia, M., et al.: BCN20000: dermoscopic lesions in the wild. arXiv preprint arXiv:1908.02288 (2019)
9. Geirhos, R., et al.: Shortcut learning in deep neural networks. Nat. Mach. Intell. **2**(11), 665–673 (2020)
10. Halabi, S.S., et al.: The RSNA pediatric bone age machine learning challenge. Radiology **290**(2), 498–503 (2019)
11. He, K., Zhang, X., Ren, S., Sun, J.: Deep residual learning for image recognition. In: CVPR, pp. 770–778 (2016)
12. Kim, B., Wattenberg, M., Gilmer, J., Cai, C., Wexler, J., Viegas, F., et al.: Interpretability beyond feature attribution: quantitative testing with concept activation vectors (TCAV). In: ICML, pp. 2668–2677. PMLR (2018)
13. Kim, B., Kim, H., Kim, K., Kim, S., Kim, J.: Learning not to learn: training deep neural networks with biased data. In: CVPR, pp. 9012–9020 (2019)
14. Lapuschkin, S., Wäldchen, S., Binder, A., Montavon, G., Samek, W., Müller, K.R.: Unmasking clever hans predictors and assessing what machines really learn. Nat. Commun. **10**(1), 1096 (2019)
15. Makar, M., Packer, B., Moldovan, D., Blalock, D., Halpern, Y., D'Amour, A.: Causally motivated shortcut removal using auxiliary labels. In: International Conference on Artificial Intelligence and Statistics, pp. 739–766. PMLR (2022)
16. Mehrabi, N., Morstatter, F., Saxena, N., Lerman, K., Galstyan, A.: A survey on bias and fairness in machine learning. ACM Comput. Surv. (CSUR) **54**(6), 1–35 (2021)
17. Murdoch, W.J., Liu, P.J., Yu, B.: Beyond word importance: contextual decomposition to extract interactions from lstms. arXiv preprint arXiv:1801.05453 (2018)
18. Nauta, M., Walsh, R., Dubowski, A., Seifert, C.: Uncovering and correcting shortcut learning in machine learning models for skin cancer diagnosis. Diagnostics **12**(1), 40 (2021)
19. Puli, A., Zhang, L.H., Oermann, E.K., Ranganath, R.: Out-of-distribution generalization in the presence of nuisance-induced spurious correlations. arXiv preprint arXiv:2107.00520 (2021)
20. Rieger, L., Singh, C., Murdoch, W., Yu, B.: Interpretations are useful: penalizing explanations to align neural networks with prior knowledge. In: International Conference on Machine Learning, pp. 8116–8126. PMLR (2020)
21. Robinson, J., Sun, L., Yu, K., Batmanghelich, K., Jegelka, S., Sra, S.: Can contrastive learning avoid shortcut solutions? Adv. Neural. Inf. Process. Syst. **34**, 4974–4986 (2021)
22. Ross, A.S., Hughes, M.C., Doshi-Velez, F.: Right for the right reasons: training differentiable models by constraining their explanations. arXiv preprint arXiv:1703.03717 (2017)
23. Rudin, C.: Stop explaining black box machine learning models for high stakes decisions and use interpretable models instead. Nat. Mach. Intell. **1**(5), 206–215 (2019)
24. Samek, W., Montavon, G., Lapuschkin, S., Anders, C.J., Müller, K.R.: Explaining deep neural networks and beyond: a review of methods and applications. Proc. IEEE **109**(3), 247–278 (2021)
25. Schramowski, P., et al.: Making deep neural networks right for the right scientific reasons by interacting with their explanations. Nat. Mach. Intell. **2**(8), 476–486 (2020)
26. Simonyan, K., Zisserman, A.: Very deep convolutional networks for large-scale image recognition. arXiv preprint arXiv:1409.1556 (2014)

27. Stock, P., Cisse, M.: Convnets and imagenet beyond accuracy: understanding mistakes and uncovering biases. In: Proceedings of the European Conference on Computer Vision (ECCV), pp. 498–512 (2018)
28. Tan, M., Le, Q.: Efficientnet: rethinking model scaling for convolutional neural networks. In: International Conference on Machine Learning, pp. 6105–6114. PMLR (2019)
29. Teso, S., Kersting, K.: Explanatory interactive machine learning. In: Proceedings of the 2019 AAAI/ACM Conference on AI, Ethics, and Society, pp. 239–245 (2019)
30. Tschandl, P., Rosendahl, C., Kittler, H.: The HAM10000 dataset, a large collection of multi-source dermatoscopic images of common pigmented skin lesions. Sci. Data **5**(1), 1–9 (2018)
31. Weber, L., Lapuschkin, S., Binder, A., Samek, W.: Beyond explaining: opportunities and challenges of XAI-based model improvement. Inf. Fusion (2022)

Faithful Synthesis of Low-Dose Contrast-Enhanced Brain MRI Scans Using Noise-Preserving Conditional GANs

Thomas Pinetz[1(✉)], Erich Kobler[2], Robert Haase[2], Katerina Deike-Hofmann[2,3], Alexander Radbruch[2,3], and Alexander Effland[1]

[1] Institute of Applied Mathematics, Rheinische Friedrich-Wilhelms-Universität Bonn, Bonn, Germany
{pinetz,effland}@iam.uni-bonn.de
[2] Department of Neuroradiology, University Medical Center Bonn, Bonn, Germany
[3] German Center for Neurodegenerative Diseases (DZNE), Helmholtz Association of German Research Centers, Bonn, Germany

Abstract. Today Gadolinium-based contrast agents (GBCA) are indispensable in Magnetic Resonance Imaging (MRI) for diagnosing various diseases. However, GBCAs are expensive and may accumulate in patients with potential side effects, thus dose-reduction is recommended. Still, it is unclear to which extent the GBCA dose can be reduced while preserving the diagnostic value – especially in pathological regions. To address this issue, we collected brain MRI scans at numerous non-standard GBCA dosages and developed a conditional GAN model for synthesizing corresponding images at fractional dose levels. Along with the adversarial loss, we advocate a novel content loss function based on the Wasserstein distance of locally paired patch statistics for the faithful preservation of noise. Our numerical experiments show that conditional GANs are suitable for generating images at different GBCA dose levels and can be used to augment datasets for virtual contrast models. Moreover, our model can be transferred to openly available datasets such as BraTS, where non-standard GBCA dosage images do not exist.

Keywords: MRI · GANs · Optimal Transport · Noise Modelling

1 Introduction

Magnetic Resonance Imaging (MRI) of the brain is an essential imaging modality to accurately diagnose various neurological diseases ranging from inflammatory

---

T. Pinetz and A. Effland—are funded the German Research Foundation under Germany's Excellence Strategy - EXC-2047/1 - 390685813 and - EXC2151 - 390873048 and R. Haase is funded by a research grant (BONFOR; O-194.0002.1).
T. Pinetz and E. Kobler—contributed equally to this work.

Supplementary Information The online version contains supplementary material available at https://doi.org/10.1007/978-3-031-43895-0_57.

H. Greenspan et al. (Eds.): MICCAI 2023, LNCS 14221, pp. 607–617, 2023.
https://doi.org/10.1007/978-3-031-43895-0_57

lesions to brain tumors and metastases. For accurate depictions of said pathologies, gadolinium-based contrast agents (GBCA) are injected intravenously to highlight brain-blood barrier dysfunctions. However, these contrast agents are expensive and may cause nephrogenic systemic fibrosis in patients with severely reduced kidney function [31]. Moreover, [17] reported that Gadolinium accumulates inside patients with unclear health consequences, especially after repeated application. The American College of Radiology recommends administering the lowest GBCA dose to obtain the needed clinical information [1].

Driven by this recommendation, several research groups have recently published dose-reduction techniques focusing on maintaining image quality. Complementary to the development of higher relaxivity contrast agents [28], virtual contrast [3,8] – replacing a large fraction of the GBCA dose by deep learning – has been proposed. These approaches typically acquire a contrast-enhanced (CE) scan with a lower GBCA dose along with non-CE scans, e.g., T1w, T2w, FLAIR, or ADC. These input images are then processed by a deep neural network (DNN) to replicate the corresponding standard-dose scan. While promising, virtual contrast techniques have not been integrated into clinical practice yet due to false-positive signals or missed small lesions [3,23]. As with all deep learning-based approaches, the availability of large datasets is essential, which is problematic in the considered case since the additional CE low-dose scan is not acquired in clinical routine exams. Hence, there are no public datasets to easily benchmark and compare different algorithms or evaluate their performance. In general, the enhancement behavior of pathological tissues at various GBCA dosages has barely been researched due to a lack of data [12].

In recent years, generative models have been used to overcome data scarcity in the computer vision and medical imaging community. Frequently, generative adversarial networks (GANs) [9] are applied as state-of-the-art in image generation [30] or semantic translation/interpolation [5,18,21]. In a nutshell, the GAN framework trains two competing DNNs – the generator and the discriminator. The generator learns a non-linear transformation of a predefined noise distribution to fit the distribution of a target dataset, while the discriminator provides feedback by simultaneously approximating a distance or divergence between the generated and the target distribution. The choice of this distance leads to the well-known different GAN algorithms, e.g., Wasserstein GANs [4,10], Least Squares GANs [24], or Non-saturating GANs [9]. However, Lucic et al. [22] showed that this choice has only a minor impact on the performance.

Learning conditional distributions between images can be accomplished by additionally feeding a condition (additional scans, dose level, etc.) into both the generator and discriminator. In particular, for image-to-image translation tasks, these conditional GANs have been successfully applied using paired [14,25,27] and unpaired training data [35]. Within these methods, an additional content (cycle) loss typically penalizes pixel-wise deviations (e.g., ℓ_1) from a corresponding reference to enforce structural similarity, whereas a local adversarial loss (discriminator with local receptive field) controls textural similarity. In addition, embeddings have been used to inject metadata [7,18].

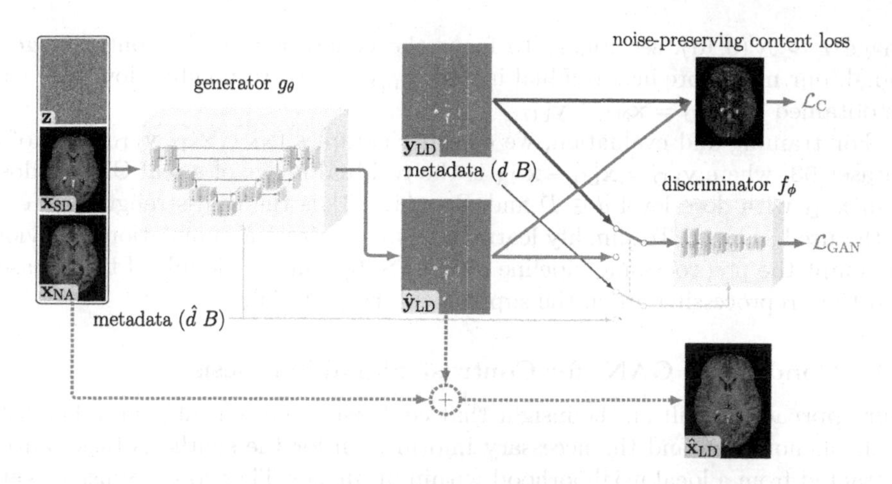

Fig. 1. Low-dose synthesis using a conditional GAN. The generator predicts a residual low-dose image $\hat{\mathbf{y}}_{LD}$ from a noise sample $\mathbf{z}$ conditioned on the native $\mathbf{x}_{NA}$ and standard-dose $\mathbf{x}_{SD}$ images as well as the field strength B and the artificial dose $\hat{d}$. Along with the discriminator, a novel noise-preserving loss – penalizing the Wasserstein distance of paired patches – is used for training. At inference, the generated residual $\hat{\mathbf{y}}_{LD}$ is added to the native image $\mathbf{x}_{NA}$ to yield the corresponding synthetic low-dose $\hat{\mathbf{x}}_{LD}$.

To study the GBCA accumulation behavior, we collected 453 CE scans with non-standard GBCA doses in the set of $\{10\%, 20\%, 33\%\}$ along with the corresponding standard-dose (0.1 mmol/kg) scan after applying the remaining contrast agent. Using this dataset, we aim at the semantic interpolation of the GBCA signal at various fractional dose levels. To this end, we use GANs to learn the contrast enhancement behavior from the dataset collective and thereby enable the synthesis of contrast signals at various dose levels for individual cases. Further, to minimize the smoothing effect [19] of typical content losses (e.g., ℓ_1 or perceptual [16]), we develop a noise-preserving content loss function based on the Wasserstein distance between paired image patches calculated using a Sinkhorn-style algorithm. This novel loss enables a faithful generation of noise, which is important for the identification of enhancing pathologies and their usability as additional training data.

With this in mind, the contributions of this work are as follows:

- synthesis of GBCA behavior at various doses using conditional GANs,
- loss enabling interpolation of dose levels present in training data,
- noise-preserving content loss function to generate realistic synthetic images.

2 Methodology

Given a *native* image $\mathbf{x}_{NA}$ (i.e. without any contrast agent injection) and a CE *standard-dose* image $\mathbf{x}_{SD}$, our conditional GAN approach synthesizes CE *low-dose* images $\hat{\mathbf{x}}_{LD}$ for selected dose levels $\hat{d} \in \mathcal{D} \subset [0, 1]$ from a uniform noise image $\mathbf{z} \sim \mathcal{N}(0, \text{Id})$, see Fig. 1. To focus the generation on the contrast agent signal, our model predicts residual images $\hat{\mathbf{y}}_{\text{LD}}$; the corresponding low-dose can be obtained by $\hat{\mathbf{x}}_{\text{LD}} = \mathbf{x}_{\text{NA}} + \hat{\mathbf{y}}_{\text{LD}}$.

For training and evaluation, we consider samples $(\mathbf{x}_{\text{NA}}, \mathbf{x}_{\text{SD}}, \mathbf{y}_{\text{LD}}, d, B)$ of a dataset $\texttt{DS}$, where $\mathbf{y}_{\text{LD}} = \mathbf{x}_{\text{LD}} - \mathbf{x}_{\text{NA}}$ is the residual image of a real CE low-dose scan $\mathbf{x}_{\text{LD}}$ with dose level $d \in \mathcal{D}$ and $B \in \{1.5, 3\}$ is the field-strength in Tesla of the used scanner. To simplify learning of the contrast accumulation behavior, we adapt the preprocessing pipeline of BraTS [6]. Further details of the dataset and the preprocessing are in the supplementary material.

2.1 Conditional GANs for Contrast Signal Synthesis

Our approach is built on the insight that contrast enhancement is an inherently local phenomenon and the necessary information for the synthesis task can be extracted from a local neighborhood within an image. Therefore, we use as generator g_θ a convolutional neural network (CNN) that is based on the U-Net [29] along with a local attention mechanism. The architecture design and the implementation details can be found in the supplementary material. As illustrated in Fig. 1, the generator uses a 3D noise sample $\mathbf{z} \sim \mathcal{N}(0, \text{Id})$ along with the native and SD images $(\mathbf{x}_{\text{NA}}, \mathbf{x}_{\text{SD}})$ as input. The synthesis is guided by the metadata $(\hat{d}\ B)^\top$, containing the artificial dose level $\hat{d} \in \mathcal{D}$ as well as the field strength of the corresponding scanner $B \in \{1.5, 3\}$. In particular, the metadata is injected into every residual block of the generator using an embedding, motivated by the recent success of diffusion-based models [13].

To learn this generator, a convolutional discriminator f_ϕ is used, which is in turn trained to distinguish the generated residual images $\hat{\mathbf{y}}_{\text{LD}}$ with random dose level $\hat{d}$ from the real residual images $\mathbf{y}_{\text{LD}}$ with the associated real dose level d. To make this a non-trivial task, label smoothing on the metadata is used, i.e., the real dose is augmented by zero-mean additive Gaussian noise with standard deviation 0.05. The discriminator architecture essentially implements the encoding side of the generator, however, no local attention layers are used as suggested by [20]. Like the generator, the discriminator is conditioned on the metadata using an embedding, which is not shared between both networks.

For training of the generator θ and discriminator ϕ, we consider the loss

$$\min_\theta \max_\phi \{\mathcal{L}_{\text{GAN}}(\theta, \phi) + \lambda_{\text{GP}} \mathcal{L}_{\text{GP}}(\phi) + \lambda_{\text{C}} \mathcal{L}_{\text{C}}(\theta)\}, \tag{1}$$

which consists of a Wasserstein GAN loss $\mathcal{L}_{\text{GAN}}$, a gradient penalty loss $\mathcal{L}_{\text{GP}}$, and a content loss $\mathcal{L}_{\text{C}}$ that are relatively weighted by scalar non-negative factors λ_{GP} and λ_{C}. In detail, the Wasserstein GAN loss reads as

$$\mathcal{L}_{\text{GAN}}(\theta, \phi) :=$$
$$\mathbb{E}_{(\mathbf{x}_{\text{NA}}, \mathbf{x}_{\text{SD}}, \mathbf{y}_{\text{LD}}, d, B) \sim \mathcal{U}(\texttt{DS})} \left\{ f_\phi(\mathbf{y}_{\text{LD}}, c) - \mathbb{E}_{\mathbf{z} \sim \mathcal{N}(0, \text{Id}), \hat{d} \sim \mathcal{U}(\mathcal{D})} \{ f_\phi(g_\theta(\mathbf{z}, \hat{c}), \hat{c}) \} \right\}$$

using condition tuples $c = (\mathbf{x}_{\text{NA}}, \mathbf{x}_{\text{SD}}, (d\ B)^\top)$ and $\hat{c} = (\mathbf{x}_{\text{NA}}, \mathbf{x}_{\text{SD}}, (\hat{d}\ B)^\top)$ to simplify notation. $\mathcal{U}(\mathcal{S})$ denotes a uniform distribution over a set $\mathcal{S}$. We highlight that the artificial dose levels $\hat{d}$ for the generated images are uniformly

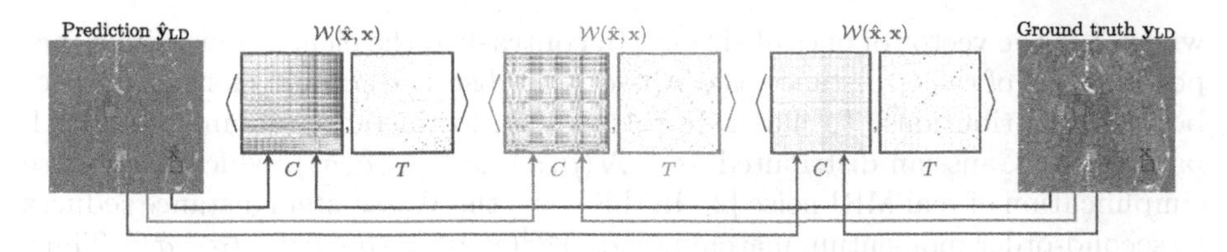

Fig. 2. Illustration of our patch-wise noise-preserving content loss. For each patch pair $((\hat{\mathbf{x}}, \mathbf{x}), (\hat{\mathbf{x}}, \mathbf{x}), (\hat{\mathbf{x}}, \mathbf{x}))$ extracted at the same position, the loss accounts for the Wasserstein distance $\mathcal{W}$ of the associated empirical distributions. In the center, the corresponding cost matrices C (pixel-wise absolute difference) along with the optimal transport maps T are shown, which are obtained by solving (2). The final loss is the sum of the element-wise multiplication of all C and T for every non-overlapping patch. (Color figure online)

sampled from $\mathcal{D} = [0.05, 0.5]$, which enables an interpolation around the dose levels present in the dataset DS. This is necessary since only a few distinct dose levels $\{0.1, 0.2, 0.33\}$ have been acquired. For regularizing the discriminator f_ϕ, we include the gradient penalty loss

$$\mathcal{L}_{\mathrm{GP}}(\phi) := \mathbb{E}_{\substack{(\mathbf{x}_{\mathrm{NA}}, \mathbf{x}_{\mathrm{SD}}, \mathbf{y}_{\mathrm{LD}}, d, B) \sim \mathcal{U}(\mathrm{DS}) \\ \mathbf{z} \sim \mathcal{N}(0, \mathrm{Id}), \hat{d} \sim \mathcal{U}(\mathcal{D}), \alpha \sim \mathcal{U}(0,1)}} \left\{ (\|\nabla f_\phi(h(\alpha, \mathbf{y}_{\mathrm{LD}}, \hat{\mathbf{y}}_{\mathrm{LD}}), h(\alpha, c, \hat{c}))\|_2 - 1)^2 \right\}$$

using $h(\alpha, \mathbf{y}, \hat{\mathbf{y}}) = \alpha\hat{\mathbf{y}} + (1 - \alpha)\mathbf{y}$. A penalty term is introduced, if f_ϕ is not Lipschitz continuous with factor 1 in its arguments as required by Wasserstein GANs [10]. Here, $\hat{\mathbf{y}}_{\mathrm{LD}} = g_\theta(\mathbf{z}, \hat{c})$ and the Lipschitz penalty is evaluated at convex combinations of real and synthetic images and condition tuples (essentially dose levels). Finally, using a distance ℓ_C, the content loss

$$\mathcal{L}_{\mathrm{C}}(\theta) := E_{(\mathbf{x}_{\mathrm{NA}}, \mathbf{x}_{\mathrm{SD}}, \mathbf{y}_{\mathrm{LD}}, d, B) \sim \mathcal{U}(\mathrm{DS}), \mathbf{z} \sim \mathcal{N}(0, \mathrm{Id})} \left\{ \ell_C\big(g_\theta(\mathbf{z}, c), \mathbf{y}_{\mathrm{LD}}\big) \right\}$$

guides the generator g_θ towards residual images in the dataset. Thus, it teaches the generator the principles of contrast enhancement. Typically, the ℓ_1-norm is used as a distance function, which leads to smooth results since it also penalizes deviations from the noise in y_{LD}.

2.2 Noise-Preserving Content Loss

To generate realistic CE images, it is also important to retain the original noise characteristics. Therefore, we introduce a novel loss that accounts for deviations in local statistics using optimal transport between empirical distributions of paired patches, as illustrated in Fig. 2.

Let $\mathbf{x}, \hat{\mathbf{x}} \in \mathbb{R}^{n^3}$ be patches of size $n \times n \times n$ extracted from the same location of a real and synthetic image, respectively. The Wasserstein distance of the associated empirical distributions using a transport plan $T \in \mathbb{R}_+^{n^3 \times n^3}$ and cost matrix $C \in \mathbb{R}_+^{n^3 \times n^3}$ given by $(C_{ij} = |\hat{x}_i - x_j|)$ is defined as

$$\mathcal{W}(\hat{\mathbf{x}}, \mathbf{x}) = \min_{T \in \mathbb{R}_+^{n^3 \times n^3}} \langle C, T \rangle_F \qquad \text{s.t.} \quad T\mathbf{1} = \mathbf{1}\tfrac{1}{n^3},\ T^\top \mathbf{1} = \mathbf{1}\tfrac{1}{n^3}, \tag{2}$$

where $\mathbf{1}$ is the vector of ones of size n^3. In contrast to the element-wise difference penalization of the ℓ_1-distance, the Wasserstein distance accounts for mismatches between distributions. To illustrate this, let us, for instance, assume that both patches are Gaussian distributed ($x \sim \mathcal{N}(\mu, \sigma)$, $\hat{x} \sim \mathcal{N}(\hat{\mu}, \hat{\sigma})$), which is a coarse simplification of real MRI noise [2]. In this case, the Wasserstein distance reduces to second-order momentum matching, i.e., $\mathcal{W}^2(\hat{\mathbf{x}}, \mathbf{x}) = (\mu - \hat{\mu})^2 + (\sigma - \hat{\sigma})^2$. Thus, the Wasserstein distance generalizes this distributional loss to any distribution within paired patches.

To efficiently solve problem (2), we use the inexact proximal point algorithm [34]. This algorithm is parallelized and applied to all non-overlapping patch pairs, to obtain our noise-preserving content loss

$$\ell_{\mathrm{NP}}(\hat{\mathbf{y}}, \mathbf{y}) = \mathbb{E}_{\mathbf{o} \sim \mathcal{U}(\mathcal{O})} \Big\{ \sum_{\mathbf{p} \in \mathcal{P}} \mathcal{W}(P_{\mathbf{p}+\mathbf{o}} \hat{\mathbf{y}}, P_{\mathbf{p}+\mathbf{o}} \mathbf{y}) \Big\},$$

where $P_{\mathbf{p}}$ extracts a local $n \times n \times n$ patch at location $\mathbf{p} \in \mathcal{P} = \{0, n, 2n, \dots\}^3$ using periodic boundary conditions. Note that we compute the expectation over offsets $\mathbf{o} \in \mathcal{O} = \{0, 1, \dots, \lceil \frac{n}{2} \rceil\}^3$ to avoid patching artifacts. In the numerical implementation, only a single offset is sampled for time and memory constraints.

3 Numerical Results

In this section, we evaluate the proposed conditional GAN approach with a particular focus on different content loss distance functions. All synthesis models were trained on 250 samples acquired on 1.5T and 3T Philips Achieva scanners and evaluated on 193 test cases, all collected at site ①. Further details of the dataset, model and training can be found in the supplementary. In our experiments, we observed that the choice of the content loss distance function $\ell_C(\hat{\mathbf{y}}, \mathbf{y})$ strongly influences the performance. Thus, we consider the different cases:

$$\ell_1: \quad \|\hat{\mathbf{y}} - \mathbf{y}\|_1 \qquad \text{VGG:} \quad \|h(\hat{\mathbf{y}}) - h(\mathbf{y})\|_1 \qquad \text{NP:} \quad \ell_{\mathrm{NP}}(\hat{\mathbf{y}}, \mathbf{y})$$

Following Johnsen et al. [16], $h(\mathbf{x})$ is the VGG-16 model [32] up to `relu3_3`.

A qualitative comparison of the different distance functions ℓ_C is visualized in Fig. 3. The first column depicts synthesized images using the ℓ_1-norm as the distance function. These images depict a plausible contrast signal, however, suffer from unrealistic smooth homogeneous regions. An improvement thereof is shown by the perceptual content loss (VGG). The NP-loss leads to a further improvement not only in the contrast signal behavior but also in the realism of the noise texture, cf. zoom regions in the lower corners.

To highlight the generalization capabilities, we depict in the bottom row of Fig. 3 a sample from site ②, which was acquired using a Philips Ingenia scanner. Moreover, the GBCA gadoterate was used, while our training data only consists of scans using the GBCA gadobutrol. Nevertheless, all models present realistically synthesized LD images. Comparing the zooms of the LD images, we observe that our NP-loss leads to a better synthesis of noise and thereby to

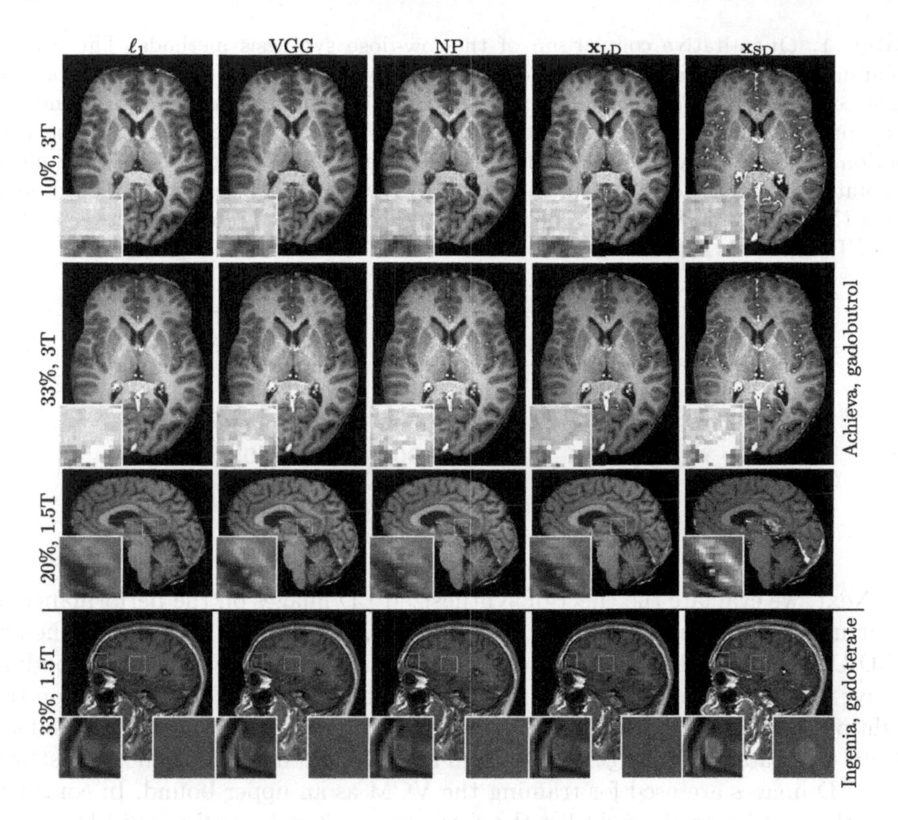

Fig. 3. Qualitative comparison of synthesized images using different loss functions to the corresponding reference $\mathbf{x}_{\mathrm{LD}}$. While the ℓ_1 loss yields smooth low-dose images, the noise pattern is preserved to some extent using the VGG loss; our loss helps to further retain the noise characteristics.

more realistic LD images. In the ℓ_1 and VGG columns, the noise is not faithfully synthesized, thus it is visually easy to spot the enhancing pathological regions.

For completeness, a quantitative ablation of the considered distance functions on the test images of site ① is shown in Table 1. Although neither maximizing PSNR nor SSIM [33] is our objective, we observe on-par performances of the perceptual (VGG) and our proposed content loss (NP) with the standard ℓ_1 distance function. Using the SD image, we define CE pixels as those pixels at which the intensity increases by at least 10% compared to the native scan. An example of these CE regions is illustrated in the supplementary. Thus, the mean absolute error for CE pixels ($\mathrm{MAE}_{\mathrm{CE}}$) quantifies the enhancement behavior. Further, we estimate the standard deviation of the non-CE pixels and report the MAE to the ground truth standard deviation (MAE_{σ}). As shown in Table 1, our loss outperforms the other content losses to a large extent on both metrics, proving its effectiveness for faithful contrast enhancement and noise generation. Further statistical analyses are presented in the supplementary.

Table 1. Quantitative comparison of the low-dose synthesis methods. The central columns present metrics evaluated on the synthesized low-dose images, whereas the right columns evaluate the effect of purely synthesized data for training the standard-dose prediction model [26]. Note, that the PSNR/SSIM of the standard dose prediction model was always evaluated on real LD images. The definitions of the mean absolute error on the contrast enhancement (MAE_{CE}) and on the noise standard deviation (MAE_{σ}) are in Sect. 3. A * denotes if a Wilcoxon signed rank test between VGG and $NP_{(our)}$ row is significant.

	low-dose synthesis				standard-dose prediction	
	PSNR	SSIM	MAE_{CE}	MAE_{σ}	PSNR	SSIM
ℓ_1	**38.34**	**0.978**	0.022	0.012	33.83	0.922
VGG	37.84	*0.976*	*0.019*	*0.009*	*36.33*	*0.958*
$NP_{(our)}$	*38.05**	*0.976*	**0.011***	**0.004***	**37.15***	**0.960***
$\mathbf{x}_{LD\,(real)}$					39.07	0.974

Next, we evaluate the effect of synthesized LD images on the performance of a virtual contrast model (VCM). In particular, we consider the state-of-the-art 2.5D U-Net model [11,23,26], which predicts an SD image given a corresponding native and LD image, see supplementary for further details. The columns on the right of Table 1 list the average PSNR and SSIM score on the *real* 33% LD subset of our test data from site①. The bottom row depicts the performance if just *real* 33% LD images are used for training the VCM as an upper bound. In contrast, the other entries on the right list the performance if *only synthesized* LD images are used for training. Both metrics show that the samples synthesized using our NP-loss model are superior to both ℓ_1 and VGG.

To determine the effectiveness of the LD synthesis models at different settings, we acquired 160 data samples from 1.5T and 3T Philips Ingenia scanners at site ②. This site used the GBCA gadoterate, which has a lower relaxivity compared to gadobutrol used at site ① [15]. For 80 samples real LD images were acquired, which are used for testing. Using the VCM solely trained on the real 33% LD data of site ① yields an average PSNR and MAE_{CE} on the test samples of site ② of 40.04 and 0.092, respectively. Extending the training data for the VCM by synthesized LD images from our model with NP-loss, we get a significantly improvemed ($p < 0.001$) PSNR score of 40.37 and MAE_{CE} of 0.075.

Finally, Fig. 4 visualizes synthesized LD images on the BraTS dataset [6] along with the associated VCM outputs. Comparing the predicted SD images $\hat{\mathbf{x}}_{SD}$ using 10% and 33% synthesized LD images $\hat{\mathbf{x}}_{LD}$, we observe that the weakly enhancing tumor at the bottom zoom is not preserved in the case of 10%, enabling evaluation of dose reduction methods on known pathological regions.

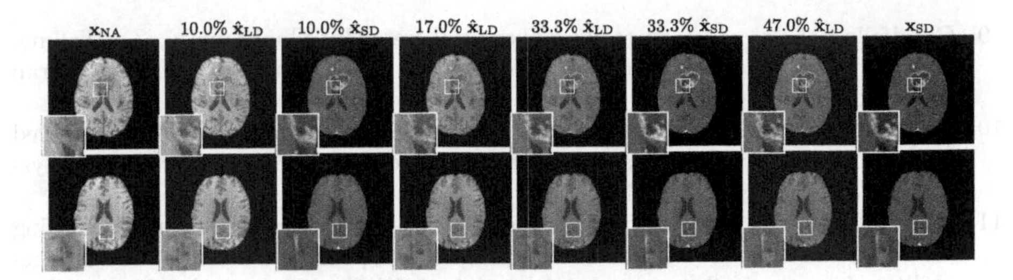

Fig. 4. Comparison of synthesized LD images $\hat{\mathbf{x}}_{LD}$ and corresponding predicted SD images $\hat{\mathbf{x}}_{SD}$ for different dose levels on BraTS [6] along with the native (left) and real SD image (right). We also included non-fractional dosage levels (17% and 47%) to showcase the wide applicability of our algorithm. Top: the tumor is well contrasted in all $\hat{\mathbf{x}}_{SD}$ even for 10%. Bottom: the subtle enhancement of the tumor cannot be recovered from the 10% LD image.

4 Conclusions

In this work, we used conditional GANs to synthesize contrast-enhanced images using non-standard GBCA doses. To this end, we introduced a novel noise-preserving content loss motivated by optimal transport theory. Numerous numerical experiments showed that our content loss improves the faithful synthesis of low-dose images. Further, the performance of virtual contrast models increases if training data is extended by synthesized images from our GAN model trained by the noise-preserving content loss.

References

1. ACR Manual on Contrast Media. American College of Radiology (2022)
2. Aja-Fernández, S., Vegas-Sánchez-Ferrero, G.: Statistical Analysis of Noise in MRI. Springer, Cham (2016). https://doi.org/10.1007/978-3-319-39934-8
3. Ammari, S., et al.: Can deep learning replace gadolinium in neuro-oncology?: A reader study. Invest. Radiol. **57**(2), 99–107 (2022)
4. Arjovsky, M., Chintala, S., Bottou, L.: Wasserstein generative adversarial networks. In: Proceedings of the 34th International Conference on Machine Learning (ICML), pp. 214–223 (2017)
5. Armanious, K., et al.: MedGAN: medical image translation using GANs. Comput. Med. Imaging Graph. **79**, 101684 (2020)
6. Baid, U., et al.: The RSNA-ASNR-MICCAI BraTS 2021 benchmark on brain tumor segmentation and radiogenomic classification. arXiv preprint arXiv:2107.02314 (2021)
7. Choi, Y., Uh, Y., Yoo, J., Ha, J.W.: StarGAN v2: diverse image synthesis for multiple domains. In: Proceedings of the IEEE/CVF Conference on Computer Vision and Pattern Recognition (CVPR), pp. 8188–8197 (2020)
8. Gong, E., Pauly, J.M., Wintermark, M., Zaharchuk, G.: Deep learning enables reduced gadolinium dose for contrast-enhanced brain MRI. J. Magn. Reson. Imaging **48**(2), 330–340 (2018)

9. Goodfellow, I., et al.: Generative adversarial nets. In: Ghahramani, Z., Welling, M., Cortes, C., Lawrence, N., Weinberger, K. (eds.) NeurIPS. vol. 27. Curran Associates, Inc. (2014)
10. Gulrajani, I., Ahmed, F., Arjovsky, M., Dumoulin, V., Courville, A.C.: Improved training of wasserstein GANs. In: Advances in Neural Information Processing Systems (NeurIPS). vol. 30 (2017)
11. Haase, R., et al.: Reduction of gadolinium-based contrast agents in MRI using convolutional neural networks and different input protocols: limited interchangeability of synthesized sequences with original full-dose images despite excellent quantitative performance. Invest. Radiol. **58**(6), 420–430 (2023)
12. Haase, R., et al.: Artificial contrast: Deep learning for reducing gadolinium-based contrast agents in neuroradiology. Invest. Radiol. **58**(8), 539–547 (2023)
13. Ho, J., Jain, A., Abbeel, P.: Denoising diffusion probabilistic models. NeurIPS **33**, 6840–6851 (2020)
14. Isola, P., Zhu, J.Y., Zhou, T., Efros, A.A.: Image-to-image translation with conditional adversarial networks. In: Proceedings of the IEEE Conference on Computer Vision and Pattern Recognition (CVPR), pp. 1125–1134 (2017)
15. Jacques, V., Dumas, S., Sun, W.C., Troughton, J.S., Greenfield, M.T., Caravan, P.: High relaxivity MRI contrast agents part 2: optimization of inner-and second-sphere relaxivity. Invest. Radiol. **45**(10), 613 (2010)
16. Johnson, J., Alahi, A., Fei-Fei, L.: Perceptual losses for real-time style transfer and super-resolution. In: Leibe, B., Matas, J., Sebe, N., Welling, M. (eds.) ECCV 2016. LNCS, vol. 9906, pp. 694–711. Springer, Cham (2016). https://doi.org/10.1007/978-3-319-46475-6_43
17. Kanda, T., Ishii, K., Kawaguchi, H., Kitajima, K., Takenaka, D.: High signal intensity in the dentate nucleus and globus pallidus on unenhanced t1-weighted MR images: relationship with increasing cumulative dose of a gadolinium-based contrast material. Radiology **270**(3), 834–841 (2014)
18. Karras, T., Laine, S., Aila, T.: A style-based generator architecture for generative adversarial networks. In: Proceedings of the IEEE/CVF Conference on Computer Vision and Pattern Recognition (CVPR), pp. 4401–4410 (2019)
19. Larsen, A.B.L., Sønderby, S.K., Larochelle, H., Winther, O.: Autoencoding beyond pixels using a learned similarity metric. In: International Conference on Machine Learning (ICML), pp. 1558–1566 (2016)
20. Lee, K., Chang, H., Jiang, L., Zhang, H., Tu, Z., Liu, C.: ViTGAN: training GANs with vision transformers. In: International Conference on Learning Representations (ICLR) (2022)
21. Liu, J., Pasumarthi, S., Duffy, B., Gong, E., Zaharchuk, G., Datta, K.: One model to synthesize them all: Multi-contrast multi-scale transformer for missing data imputation. arXiv preprint arXiv:2204.13738 (2022)
22. Lucic, M., Kurach, K., Michalski, M., Gelly, S., Bousquet, O.: Are GANs created equal? A large-scale study. In: Neural Information Processing Systems (NeurIPS) (2018)
23. Luo, H., et al.: Deep learning-based methods may minimize GBCA dosage in brain MRI. Eur. Radiol. **31**(9), 6419–6428 (2021)
24. Mao, X., Li, Q., Xie, H., Lau, R.Y., Wang, Z., Paul Smolley, S.: Least squares generative adversarial networks. In: Proceedings of the IEEE International Conference on Computer Vision (ICCV), pp. 2794–2802 (2017)

25. Nie, D., et al.: Medical image synthesis with context-aware generative adversarial networks. In: Descoteaux, M., Maier-Hein, L., Franz, A., Jannin, P., Collins, D.L., Duchesne, S. (eds.) MICCAI 2017. LNCS, vol. 10435, pp. 417–425. Springer, Cham (2017). https://doi.org/10.1007/978-3-319-66179-7_48
26. Pasumarthi, S., Tamir, J.I., Christensen, S., Zaharchuk, G., Zhang, T., Gong, E.: A generic deep learning model for reduced gadolinium dose in contrast-enhanced brain MRI. Magn. Reson. Med. **86**(3), 1687–1700 (2021)
27. Preetha, C.J., et al.: Deep-learning-based synthesis of post-contrast t1-weighted MRI for tumour response assessment in neuro-oncology: a multicentre, retrospective cohort study. Lancet Digital Health **3**(12), e784–e794 (2021)
28. Robic, C., et al.: Physicochemical and pharmacokinetic profiles of Gadopiclenol: a new macrocyclic gadolinium chelate with high t1 relaxivity. Invest. Radiol. **54**(8), 475 (2019)
29. Ronneberger, O., Fischer, P., Brox, T.: U-Net: convolutional networks for biomedical image segmentation. In: Navab, N., Hornegger, J., Wells, W.M., Frangi, A.F. (eds.) MICCAI 2015. LNCS, vol. 9351, pp. 234–241. Springer, Cham (2015). https://doi.org/10.1007/978-3-319-24574-4_28
30. Sauer, A., Schwarz, K., Geiger, A.: StyleGAN-XL: Scaling styleGAN to large diverse datasets. In: ACM SIGGRAPH, pp. 1–10 (2022)
31. Schieda, N., et al.: Gadolinium-based contrast agents in kidney disease: a comprehensive review and clinical practice guideline issued by the Canadian association of radiologists. Can. J. Kidney Health Dis. **5**, 136–150 (2018)
32. Simonyan, K., Zisserman, A.: Very deep convolutional networks for large-scale image recognition. In: International Conference on Learning Representations (ICLR), pp. 1–14 (2015)
33. Wang, Z., Bovik, A.C., Sheikh, H.R., Simoncelli, E.P.: Image quality assessment: from error visibility to structural similarity. Trans. Image Process. **13**(4), 600–612 (2004)
34. Xie, Y., Wang, X., Wang, R., Zha, H.: A fast proximal point method for computing exact wasserstein distance. In: Uncertainty in Artificial Intelligence, pp. 433–453 (2020)
35. Zhu, J.Y., Park, T., Isola, P., Efros, A.A.: Unpaired image-to-image translation using cycle-consistent adversarial networks. In: IEEE International Conference on Computer Vision (ICCV), pp. 2223–2232 (2017)

Prediction of Infant Cognitive Development with Cortical Surface-Based Multimodal Learning

Jiale Cheng[1,2], Xin Zhang[1,3](✉), Fenqiang Zhao[2], Zhengwang Wu[2], Xinrui Yuan[2], Li Wang[2], Weili Lin[2], and Gang Li[2](✉)

[1] School of Electronic and Information Engineering, South China University of Technology, Guangzhou, Guangdong, China
eexinzhang@scut.edu.cn

[2] Department of Radiology and Biomedical Research Imaging Center, University of North Carolina at Chapel Hill, Chapel Hill, NC, USA
gang_li@med.unc.edu

[3] Pazhou Laboratory, Guangzhou, Guangdong, China

Abstract. Exploring the relationship between the cognitive ability and infant cortical structural and functional development is critically important to advance our understanding of early brain development, which, however, is very challenging due to the complex and dynamic brain development in early postnatal stages. Conventional approaches typically use either the structural MRI or resting-state functional MRI and rely on the region-level features or inter-region connectivity features after cortical parcellation for predicting cognitive scores. However, these methods have two major issues: 1) *spatial information loss*, which discards the critical fine-grained spatial patterns containing rich information related to cognitive development; 2) *modality information loss*, which ignores the complementary information and the interaction between the structural and functional images. To address these issues, we unprecedentedly invent a novel framework, namely cortical surface-based multimodal learning framework (CSML), to leverage fine-grained multimodal features for cognition development prediction. First, we introduce the fine-grained surface-based data representation to capture spatially detailed structural and functional information. Then, a dual-branch network is proposed to extract the discriminative features for each modality respectively and further captures the modality-shared and complementary information with a disentanglement strategy. Finally, an age-guided cognition prediction module is developed based on the prior that the cognition develops along with age. We validate our method on an infant multimodal MRI dataset with 318 scans. Compared to state-of-the-art methods, our method consistently achieves superior performances, and for the first time suggests crucial regions and features for cognition development hidden in the fine-grained spatial details of cortical structure and function.

Keywords: Cognition Prediction · Multimodality · rs-fMRI · sMRI

H. Greenspan et al. (Eds.): MICCAI 2023, LNCS 14221, pp. 618–627, 2023.
https://doi.org/10.1007/978-3-031-43895-0_58

1 Introduction

Predictive modeling of the individual-level cognitive development during infancy is of great importance in advancing our understanding of the subject-specific relationship between the cognitive ability and early brain structural and functional development and their underlying neural mechanisms. It is also critical for early identifying cognitive delays and developing more effectively and timely personalized therapeutic interventions for at-risk infants. However, this is a very challenging task due to the complex and rapid development of brain structure, function and cognition during the first years of life [1, 2].

Recently, a few methods have been explored for predicting infant cognition using either resting-state functional MRI (rs-fMRI) [2–4] or structural MRI (sMRI) [5–7]. Although encouraging preliminary results have been achieved, two unaddressed major issues hinder the precise prediction of the individual-level cognitive development during infancy. 1) *Spatial information loss*: Previous works [3–8] typically rely on region-level features or inter-region connectivity features after parcellation of the brain cortex into a set of regions. Consequently, these features largely ignore fine-grained spatial patterns on cortical surfaces, which encode subject-specific rich information critical for cognitive prediction. 2) *Modality-information loss*: Previous methods use either functional features or structural features, and thus the complementary information between them and their underlying relationship are not leveraged for cognition development. Indeed, it is believed that the spontaneous neuronal activity is related to the intrinsic human brain functional organizations supported by the underlying structural substrates [2], which gives emphasis to understanding the underlying individual structure-functional profile during infancy. Therefore, an effective unified framework that can automatically learn the complementary and spatially fine-grained information from structural and functional data for cognition development prediction is critically desired.

To address the above issues, we propose a novel cortical surface-based multimodal learning framework (**CSML**), to enable learning of the fine-grained spatial patterns and complementary information from structural and functional MRI data for precise prediction of the individual-level cognitive development. Specifically, 1) to learn detailed spatial patterns of both functional connectivity and structural information, we propose to leverage the strong feature learning and representation ability of spherical surface networks [9] to automatically extract task-related features on cortical surfaces. 2) To effectively fuse structural and functional information, we propose a dual-branch surface network to simultaneously extract structural morphologic features and functional connectivity features on cortical surfaces, and further fuse their complementary information in a feature disentanglement module. 3) To enable precise prediction of cognitive outcome, we leverage the prior knowledge that the cognition function develops with age growing by jointly predicting age and cognition scales. To our best knowledge, this is the first work to leverage the multimodal, fine-grained spatial information on cortical surface explicitly for cognition development prediction. The experimental results based on a longitudinal infant dataset not only validate the superiority of our proposed model but also imply the tight association between the individual cognition development and the fine-grained cortical information.

2 Method

In this section, we present the details of CSML (Fig. 1), including three steps: 1) surface-based fine-grained information representation (Fig. 1(a)); 2) modality-specific information learning (Fig. 1(b)); and 3) multi-modality information fusion (Fig. 1(c)).

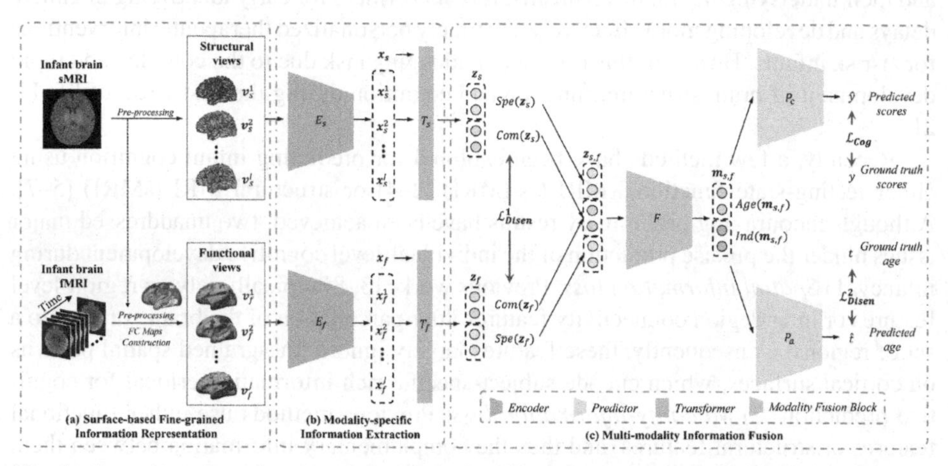

Fig. 1. Overview of our framework for cortical surface-based multimodal fine-grained information learning. Given the sMRI and fMRI of an infant, its structural and functional feature representations z_s and z_f are first extracted. Then, the modality shared $(Com(z))$ and specific $(Spe(z))$ information are disentangled and further fused by a modality fusion block F. After that, we constrain the fused latent variable $\boldsymbol{m}_{s,f}$ to be age-irrelevant by the age predictor P_a, and finally obtain the predicted cognitive scores from the predictor P_c.

2.1 Surface-Based Fine-Grained Information Representation

The input of the network framework consists of two branches for encoding cortical structural information and functional connectivity information, respectively. To integrate multi-modal MRI data together for cognition development prediction, we map all modality data to a common space, i.e., the cortical surface registered to UNC 4D infant surface atlas [10] and further resampled with 40,962 vertices, following the well-established pipelines [11–15]. To capture the spatially fine-grained information in structural MRI, the *structural branch* contains a set of surface maps of biologically meaningful cortical properties, including cortical thickness, surface area, cortical volume, sulcal depth, mean curvature, and average convexity. To preserve the fine-grained spatial patterns of functional connectivity, we leverage an infant-dedicated cortical functional parcellation map [15]. Specifically, for each parcel, we first calculate the Pearson's correlation coefficient between the averaged functional time series of all vertices within this parcel and the functional time series of each cortical vertex to build the parcel-specific cortical functional connectivity (FC) map and then perform Fisher's r-to-z transformation. Finally, we use these cortical FC maps from all parcels, which characterize rich spatially detailed FC information, as the input of the *functional branch*.

2.2 Modality-Specific Encoder

For the multi-modality input, we employ two modality-specific encoders E_s and E_f to describe its feature representation, respectively. To be specific, we regard each modality comprised of multiple feature channels, while each channel could be interpreted as an observation of the data from a certain view. Therefore, we process each view separately as $\boldsymbol{x}_s^i = E_s(\boldsymbol{v}_s^i), \boldsymbol{x}_f^j = E_f\left(\boldsymbol{v}_f^j\right)$, where $i\epsilon[1, I], I$ is the number of morphological features we used; $j\epsilon[1, J], J$ is the number of parcels we used in building FC maps. We implement E_s and E_f as the Spherical Res-Net [9, 16, 17], which is composed of stacks of spherical convolutional layers and spherical pooling layers to extract the fine-grained spatial patterns and generates the view-related feature representations of $\boldsymbol{v}_s^i$ and $\boldsymbol{v}_f^j$. Considering the different number of views in each modality, two Transformer layers [18] T_s and T_f are then adopted to fuse the multi-view feature representations for each modality separately. Herein, following the previous work [19], we prepend two learnable embeddings $\boldsymbol{x}_s$ and $\boldsymbol{x}_f$ as the first token for the sequences of view-related feature representations $\{\boldsymbol{v}_s^i | i\epsilon[1, I]\}$ and $\{\boldsymbol{v}_f^j | j\epsilon[1, J]\}$, respectively. Within the Transformer layers, for the structural-related features, $\boldsymbol{x}_s$ interact with and fuse the view-related feature representation $\{\boldsymbol{v}_s^i | i\epsilon[1, I]\}$ through the self-attention mechanism as follows,

$$A_s^i = Q_s(\boldsymbol{x}_s)K_s\left(\boldsymbol{v}_s^i\right)^T/const, \tag{1}$$

$$\widetilde{\boldsymbol{x}}_s = \boldsymbol{x}_s + \sum_{i=1}^{I} softmax\left(A_s^i\right)U_s(\boldsymbol{v}_s^i), \tag{2}$$

$$z_s = \widetilde{\boldsymbol{x}}_s + W_s(\widetilde{\boldsymbol{x}}_s), \tag{3}$$

where z_s is the aggregated representation for the structural data, $Q_s(\cdot)$, $K_s(\cdot)$, $U_s(\cdot)$, and $W_s(\cdot)$ are four multi-layer perceptrons (*MLP*), *const* is a constant for normalization. Similarly, we can obtain the functional-related variable z_f by feeding $\boldsymbol{x}_f$ and the functional view-related representations $\{\boldsymbol{v}_f^j | j\epsilon[1, J]\}$ into T_f.

2.3 Modality-Fusion Block

To better learn the complementary information between the two modalities, we further decompose the modality-specific latent variables z_s and z_f into two parts: $Com(z_n)$ and $Spe(z_n)$, where $n\epsilon\{s, f\}$, standing for the structure (s) and function (f) related variables, respectively. $Com(z_n)$ is the common code representing the shared information among modalities, while $Spe(z_n)$ is the specific code representing the complementary information that differentiates one modality from the other. The basic requirements of this disentanglement are: (1) The concatenation of $Com(z_n)$ and $Spe(z_n)$ equals z_n; (2) $Com(z_s)$ and $Com(z_f)$ should be as similar as possible; (3) $Spe(z_s)$ differs from $Spe(z_f)$ as much as possible. Accordingly, $\mathcal{L}_{Disen}^1$ is defined as:

$$\mathcal{L}_{Disen}^1 = \mathcal{L}_{Disen}^{Com}/\mathcal{L}_{Disen}^{Spe}, \tag{4}$$

$$\mathcal{L}_{Disen}^{Com} = ||Com(z_s) - Com(z_f)||_2, \tag{5}$$

$$\mathcal{L}_{Disen}^{Spe} = ||Spe(z_s) - Spe(z_f)||_2. \tag{6}$$

Since the latent variable of each modality has been disentangled into $Com(z_n)$ and $Spe(z_n)$, the combined information is formed as the concatenation of the common code and specific codes as follows: $z_{s,f} = \big(Spe(z_s), Common, Spe(z_f)\big)$, where $Common = 0.5(Com(z_s) + Com(z_f))$.

2.4 Cognitive Scores Prediction

Given the combined multimodal information $z_{s,f}$, it is intuitive to regress the cognitive scores directly. However, considering that cognitive functions develop rapidly during the first years of life [1], the regressor would be prone to learn the age-related information instead and thus cannot differentiate the individualized development discrepancy between subjects within the same age group. Therefore, we fuse the combined multimodal information through a MLP F as follows, $\boldsymbol{m}_{s,f} = F(z_{s,f})$, and further disentangle the age-related variance $Age(\boldsymbol{m}_{s,f})$ and the individual-related invariance $Ind(\boldsymbol{m}_{s,f})$ from $\boldsymbol{m}_{s,f}$ to precisely evaluate the cognition development level. The basic requirements of this disentanglement are: (1) The concatenation of $Age(\boldsymbol{m}_{s,f})$ and $Ind(\boldsymbol{m}_{s,f})$ equals $\boldsymbol{m}_{s,f}$; (2) $Age(\boldsymbol{m}_{s,f})$ is capable of age estimation through an age predictor P_a; (3) $Ind(\boldsymbol{m}_{s,f})$ is incapable of age estimation through P_a. Accordingly, $\mathcal{L}_{Disen}^2$ is defined as:

$$\mathcal{L}_{Disen}^2 = \mathcal{L}_{Disen}^{Age} - \mathcal{L}_{Disen}^{Ind}, \tag{7}$$

$$\mathcal{L}_{Disen}^{Age} = |t - P_a(Age(\boldsymbol{m}_{s,f}))|, \tag{8}$$

$$\mathcal{L}_{Disen}^{Ind} = |t - P_a(Ind(\boldsymbol{m}_{s,f}))|, \tag{9}$$

where t is the ground truth of age. Then, we can use the identity-related features $Ind(\boldsymbol{m}_{s,f})$ containing subject-specific structure-function profile to predict the cognitive scores through a cognitive score predictor P_c under the guidance of the corresponding age feature $Age(\boldsymbol{m}_{s,f})$. The loss function to train P_c is defined as:

$$\mathcal{L}_{Cog} = |y - P_c(Ind\big(\boldsymbol{m}_{s,f}\big), Age(\boldsymbol{m}_{s,f}))|, \tag{10}$$

where y is the ground truth of cognitive scores. Specifically, we implement P_a and P_c as two sets of *MLP*. Finally, the overall objective function to optimize the neural network is written as:

$$L = \lambda_1 L_{Disen}^1 + \lambda_2 L_{Disen}^2 + L_{Cog}, \tag{11}$$

where λ_1 and λ_2 are trade-off parameters to balance the multiple losses.

3 Experiments

3.1 Dataset

We verified the effectiveness of the proposed CSML model for infant cognition development prediction on a public high-resolution dataset including 318 pairs of sMRI and rs-fMRI scans acquired at different ages ranging from 88 to 1040 days in the UNC/UMN Baby Connectome Project [20]. All structural and functional MR images were preprocessed following state-of-the-art infant-tailored pipelines [11–15]. Cortical surfaces were reconstructed and aligned onto the public UNC 4D infant surface atlas [10, 11]. For each cortical vertex on the middle cortical surface, its representative fMRI time-series was extracted [13–15]. An infant-dedicated fine-grained functional parcellation map [15] with 432 cortical ROIs per hemisphere in UNC 4D infant surface atlas was warped onto each individual cortical surface.

To quantify the cognition development level of each participant, four Mullen cognitive scores [21] were collected at their corresponding scan ages, i.e., Visual Receptive Scale (VRS), Fine Motor Scale (FMS), Receptive Language Scale (RLS), and Expressive Language Scale (ELS). These four cognitive scales were respectively normalized into the [0, 1] range using the minimum and maximum values for the training efficiency.

3.2 Experimental Settings

In order to validate our methods, a 5-fold cross-validation strategy is employed, and each fold consists of 190 training samples, 64 validating samples, and 64 testing samples. To quantitatively evaluate the performance, the Pearson's correlation coefficient (PCC) and root mean square error (RMSE) between the ground truth and predicted values were calculated. In the testing phase, the mean and standard deviation of the 5-fold results were reported.

The encoders E_s and E_f in CSML constitutes 5 Res-blocks with the dimensions of {32, 32, 64, 64, 128}, respectively. The modality fusion block F, age predictor P_a, and cognitive score predictor P_c were designed as two-layer MLP with the ReLU activation function and the dimension of {192, 128}, {64, 1}, and {128, 1}, respectively. We implemented the model with PyTorch and used Adam as optimizer with the weight decay of 10^{-4} and the learning rate cyclically tuned within $[10^{-6}, 10^{-3}]$. The batch size was set to 1. The maximum training epoch is 500. After comparison, we empirically set the hyperparameters as λ_1=0.05 and λ_2=0.01.

3.3 Results

We first show the results of some ablated models of our method in Table 1, where *w/o Structure* and *w/o Function* denote for the variants using functional and structural features only. *w/o Age* denotes the variant with single task of cognition prediction. It can be observed that, the overall performance on four cognitive tasks has been extensively improved by jointly leveraging the structural and functional information. The disentanglement mechanism successfully separates the shared and complementary information amongst modalities and further removes the redundancy with the loss $\mathcal{L}^1_{Disen}$. Moreover, the joint age prediction and cognitive estimation also brings further improvement by differentiating the age-related and identity-related variables with $\mathcal{L}^2_{Disen}$. The scatter plots of predicted cognitive scores in five testing folds are depicted in Fig. 2(a), demonstrating that the scores are well predicted.

We also comprehensively compared with various traditional and state-of-the-art functional connectivity-based methods, including KNN, random forest (RF), SVR, gaussian process regression (GPR), GCN [22], GAT [23], and UniMP [24]. As shown in Table 2, our algorithm outperforms the previous methods by a large margin. Of note, the proposed method demonstrates better performance even with the functional information only, which highlights the importance to preserve the fined-grained FC information.

Table 1. The impact of each component of CSML on the prediction performance (in terms of PCC). * indicates statistically significantly better results than other methods with p-value < 0.05.

Components	VRS	FMS	RLS	ELS	Average
w/o Age	0.768 ± 0.034	0.819 ± 0.023	0.789 ± 0.032	0.745 ± 0.034	0.780
w/o Structure	0.748 ± 0.050	0.814 ± 0.037	0.790 ± 0.042	0.711 ± 0.039	0.757
w/o Function	0.771 ± 0.038	0.831 ± 0.013	0.793 ± 0.028	0.737 ± 0.052	0.782
w/o $\mathcal{L}^1_{Disen}$	0.791 ± 0.015	0.837 ± 0.018	0.814 ± 0.018	0.789 ± 0.021	0.808
w/o $\mathcal{L}^2_{Disen}$	0.820 ± 0.045	0.858 ± 0.016	0.848 ± 0.016	0.824 ± 0.019	0.838
Proposed	**0.855 ± 0.026***	**0.874 ± 0.030***	**0.873 ± 0.008***	**0.852 ± 0.009***	**0.864***

Table 2. Performance comparison of different methods (in terms of RMSE and PCC). The averaged values among four tasks were provided. * indicates statistically significantly better results than other methods with p-value < 0.05.

Methods		RMSE	PCC
Machine Learning-based Methods	*KNN*	0.1421 ± 0.0144	0.5698 ± 0.0751
	RF	0.1353 ± 0.0060	0.6705 ± 0.1354
	GPR	0.1175 ± 0.0127	0.7320 ± 0.0596
	SVR	0.1284 ± 0.0069	0.7355 ± 0.0192
Graph Convolution-based Methods	*GCN*	0.1382 ± 0.0175	0.6234 ± 0.0581
	GAT	0.1208 ± 0.0228	0.7001 ± 0.1327
	UniMP	0.1246 ± 0.0149	0.7073 ± 0.0249
Proposed		**0.0915 ± 0.0091***	**0.8635 ± 0.0066***

Additionally, based on our proposed model CSML, the prediction accuracy of infant cognition development is over 0.85 on average, suggesting that the model may observe plausible biomarkers for cognition development during infancy. Based on the well-trained models, we explored the explainability and interpretability of the proposed method by investigating the weights of the Transformers. Since the Transformer layers T_s and T_f fuse the multi-view representations v_s^i and v_f^j into z_s and z_f for further cognitive prediction, by analyzing the attention value A_n^i of each view v_n^i in the Transformer, we can infer which regions for functional data and which morphological features for structural data are more important for cognition prediction. The results are shown in Fig. 2 (b) and Fig. 2 (c), in line with the reports in related studies to some extent [25–28], demonstrating the scientific value of our method. For example, the left lateral prefrontal cortex involved in higher executive functions [25, 26] demonstrates high importance in functional data. Moreover, previous researchers [27, 28] have observed the close relevance between the visual cortex and the early cognitive process, which also confirms the result of our method.

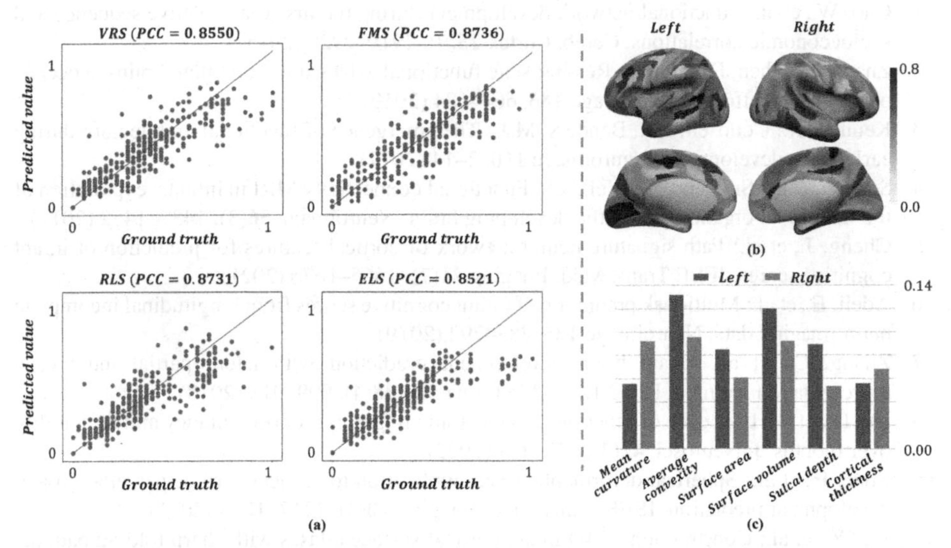

Fig. 2. The illustrations of (a) the predicted values distribution for four cognitive tasks, and the importance distribution of (b) each cortical regions and (c) each morphological feature on both hemispheres.

4 Conclusion

In this work, we develop an innovative cortical surface-based multimodal learning framework (CSML) to address the infant cognition prediction problem. Specifically, we unprecedentedly propose to explicitly leverage the surface-based feature representations to preserve the fine-grained, spatially detailed multimodal information for cognition prediction. In addition, by disentangling the modality-shared and complementary information, our model successfully captures the individualized cognition development patterns underlying the dramatic brain development. With its superior performance compared to state-of-the-art methods, our proposed CSML suggests that the informative clues for brain-cognitive relationship are hidden in the multimodal fine-grained details and validates itself as a potentially powerful framework for simultaneously learning effective representations from sMRI and rs-fMRI data.

Acknowledgements. The work of Gang Li was supported in part by NIH grants (MH116225, MH117943, MH127544, and MH123202). The work of Li Wang was supported by NIH grant (MH117943). This work also utilizes approaches developed by an NIH grant (1U01MH110274) and the efforts of the UNC/UMN Baby Connectome Project Consortium.

References

1. Gao, W., et al.: Functional network development during the first year: relative sequence and socioeconomic correlations. Cereb. Cortex **25**(9), 2919–2928 (2015)
2. Zhang, H., Shen, D., Lin, W.: Resting-state functional MRI studies on infant brains: a decade of gap-filling efforts. Neuroimage **185**, 664–684 (2019)
3. Keunen, K., Counsell, S.J., Benders, M.J.: The emergence of functional architecture during early brain development. Neuroimage **160**, 2–14 (2017)
4. Smyser, C.D., Snyder, A.Z., Neil, J.J.: Functional connectivity MRI in infants: exploration of the functional organization of the developing brain. Neuroimage **56**(3), 1437–1452 (2011)
5. Cheng, J., et al.: Path signature neural network of cortical features for prediction of infant cognitive scores. IEEE Trans. Med. Imaging **41**(7), 1665–1676 (2021)
6. Adeli, E., et al.: Multi-task prediction of infant cognitive scores from longitudinal incomplete neuroimaging data. Neuroimage **185**, 783–792 (2019)
7. Zhang, C., et al.: Infant brain development prediction with latent partial multi-view representation learning. IEEE Trans. Med. Imaging **38**(4), 909–918 (2018)
8. Hu, D., et al.: Existence of functional connectome fingerprint during infancy and its stability over months. J. Neurosci. **42**(3), 377–389 (2022)
9. Zhao, F., et al.: Spherical deformable u-net: application to cortical surface parcellation and development prediction. IEEE Trans. Med. Imaging **40**(4), 1217–1228 (2021)
10. Wu, Z., et al.: Construction of 4D infant cortical surface atlases with sharp folding patterns via spherical patch-based group-wise sparse representation. Hum. Brain Mapp. **40**(13), 3860–3880 (2019)
11. Li, G., et al.: Construction of 4D high-definition cortical surface atlases of infants: Methods and applications. Med. Image Anal. **25**(1), 22–36 (2015)
12. Li, G., et al.: Computational neuroanatomy of baby brains: a review. Neuroimage **185**, 906–925 (2019)
13. Li, G., et al.: Measuring the dynamic longitudinal cortex development in infants by reconstruction of temporally consistent cortical surfaces. Neuroimage **90**, 266–279 (2014)
14. Wang, L., Wu, Z., Chen, L., Sun, Y., Lin, W., Li, G.: iBEAT V2. 0: a multisite-applicable, deep learning-based pipeline for infant cerebral cortical surface reconstruction. Nat. Protoc. **18**(5), 1488–1509 (2023)
15. Wang, F., et al.: Fine-grained functional parcellation maps of the infant cerebral cortex. eLife (2023)
16. He, K., et al.: Deep residual learning for image recognition. In: Proceedings of the IEEE Conference on Computer Vision and Pattern Recognition, pp. 770–778 (2016)
17. Zhao, F., et al.: Harmonization of infant cortical thickness using surface-to-surface cycle-consistent adversarial networks. In: Shen, Dinggang, et al. (eds.) MICCAI 2019. LNCS, vol. 11767, pp. 475–483. Springer, Cham (2019). https://doi.org/10.1007/978-3-030-32251-9_52

18. Vaswani, A., et al.: Attention is all you need. In: Advances in Neural Information Processing Systems, vol. 30, pp. 6000-6010 (2017)
19. Dosovitskiy, A., et al.: An image is worth 16x16 words: transformers for image recognition at scale. In International Conference on Learning Representation (2021)
20. Howell, B.R., et al.: The UNC/UMN baby connectome project (BCP): an overview of the study design and protocol development. Neuroimage **185**, 891–905 (2019)
21. Mullen, E.M.: Mullen scales of early learning. AGS Circle Pines, MN (1995)
22. Kipf, T.N., Welling, M.: Semi-supervised classification with graph convolutional networks. arXiv preprint arXiv:1609.02907 (2016)
23. Veličković, P., et al.: Graph attention networks. arXiv preprint arXiv:1710.10903 (2017)
24. Shi, Y., et al.: Masked label prediction: Unified message passing model for semi-supervised classification. arXiv preprint arXiv:2009.03509 (2020)
25. Fuster, J.M.: Frontal lobe and cognitive development. J. Neurocytol. **31**(3), 373–385 (2002)
26. Kolk, S.M., Rakic, P.: Development of prefrontal cortex. Neuropsychopharmacology **47**(1), 41–57 (2022)
27. Roelfsema, P.R., de Lange, F.P.: Early visual cortex as a multiscale cognitive blackboard. Ann. Rev. Vis. Sci. **2**, 131–151 (2016)
28. Albers, A.M., et al.: Shared representations for working memory and mental imagery in early visual cortex. Curr. Biol. **23**(15), 1427–1431 (2013)

Distilling BlackBox to Interpretable Models for Efficient Transfer Learning

Shantanu Ghosh[1(✉)], Ke Yu[2], and Kayhan Batmanghelich[1]

[1] Department of Electrical and Computer Engineering, Boston University, Boston, MA, USA
shawn24@bu.edu

[2] Intelligent Systems Program, University of Pittsburgh, Pittsburgh, PA, USA

Abstract. Building generalizable AI models is one of the primary challenges in the healthcare domain. While radiologists rely on generalizable descriptive rules of abnormality, Neural Network (NN) models suffer even with a slight shift in input distribution (*e.g.,* scanner type). Fine-tuning a model to transfer knowledge from one domain to another requires a significant amount of labeled data in the target domain. In this paper, we develop an interpretable model that can be efficiently fine-tuned to an unseen target domain with minimal computational cost. We assume the interpretable component of NN to be approximately domain-invariant. However, interpretable models typically underperform compared to their Blackbox (BB) variants. We start with a BB in the source domain and distill it into a *mixture* of shallow interpretable models using human-understandable concepts. As each interpretable model covers a subset of data, a mixture of interpretable models achieves comparable performance as BB. Further, we use the pseudo-labeling technique from semi-supervised learning (SSL) to learn the concept classifier in the target domain, followed by fine-tuning the interpretable models in the target domain. We evaluate our model using a real-life large-scale chest-X-ray (CXR) classification dataset. The code is available at: https://github.com/batmanlab/MICCAI-2023-Route-interpret-repeat-CXRs.

Keywords: Explainable-AI · Interpretable models · Transfer learning

1 Introduction

Model generalizability is one of the main challenges of AI, especially in high stake applications such as healthcare. While NN models achieve state-of-the-art (SOTA) performance in disease classification [9,17,24], they are brittle to small shifts in the data distribution [7] caused by a change in acquisition protocol or scanner type [22]. Fine-tuning all or some layers of a NN model on the target domain can alleviate this problem [2], but it requires a substantial amount of labeled data and be computationally expensive [12,21]. In contrast, radiologists

---

Supplementary Information The online version contains supplementary material available at https://doi.org/10.1007/978-3-031-43895-0_59.

H. Greenspan et al. (Eds.): MICCAI 2023, LNCS 14221, pp. 628–638, 2023.
https://doi.org/10.1007/978-3-031-43895-0_59

follow fairly generalizable and comprehensible rules. Specifically, they search for patterns of changes in anatomy to read abnormality from an image and apply logical rules for specific diagnoses. This approach is transparent and closer to an interpretable-by-design approach in AI. We develop a method to extract a mixture of interpretable models based on clinical concepts, similar to radiologists' rules, from a pre-trained NN. Such a model is more data- and computation-efficient than the original NN for fine-tuning to a new distribution.

Standard interpretable by design method [18] finds an interpretable function (*e.g.*, linear regression or rule-based) between human-interpretable concepts and final output [14]. A concept classifier [19,26] detects the presence or absence of concepts in an image. In medical images, previous research uses TCAV scores [13] to quantify the role of a concept on the final prediction [3,6,23], but the concept-based interpretable models have been mostly unexplored. Recently Posthoc Concept Bottleneck models (PCBMs) [25] identify concepts from the embeddings of BB. However, the common design choice amongst those methods relies on a single interpretable classifier to explain the entire dataset, cannot capture the diverse sample-specific explanations, and performs poorly than their BB variants.

Our Contributions. This paper proposes a novel data-efficient interpretable method that can be transferred to an unseen domain. Our interpretable model is built upon human-interpretable concepts and can provide sample-specific explanations for diverse disease subtypes and pathological patterns. Beginning with a BB in the source domain, we progressively extract a mixture of interpretable models from BB. Our method includes a set of selectors routing the explainable samples through the interpretable models. The interpretable models provide First-order-logic (FOL) explanations for the samples they cover. The remaining unexplained samples are routed through the residuals until they are covered by a successive interpretable model. We repeat the process until we cover a desired fraction of data. Due to class imbalance in large CXR datasets, early interpretable models tend to cover all samples with disease present while ignoring disease subgroups and pathological heterogeneity. We address this problem by estimating the class-stratified coverage from the total data coverage. We then finetune the interpretable models in the target domain. The target domain lacks concept-level annotation since they are expensive. Hence, we learn a concept detector in the target domain with a pseudo labeling approach [15] and finetune the interpretable models. Our work is the first to apply concept-based methods to CXRs and transfer them between domains.

2 Methodology

Notation. Assume $f^0 : \mathcal{X} \rightarrow \mathcal{Y}$ is a BB, trained on a dataset $\mathcal{X} \times \mathcal{Y} \times \mathcal{C}$, with $\mathcal{X}$, $\mathcal{Y}$, and $\mathcal{C}$ being the images, classes, and concepts, respectively; $f^0 = h^0 \circ \Phi$, where Φ and h^0 is the feature extractor and the classifier respectively. Also, m is the number of class labels. This paper focuses on binary classification (having or not having a disease), so $m = 2$ and $\mathcal{Y} \in \{0, 1\}$. Yet, it can be extended to multiclass problems easily. Given a learnable projection [4,5], $t : \Phi \rightarrow \mathcal{C}$, our method learns

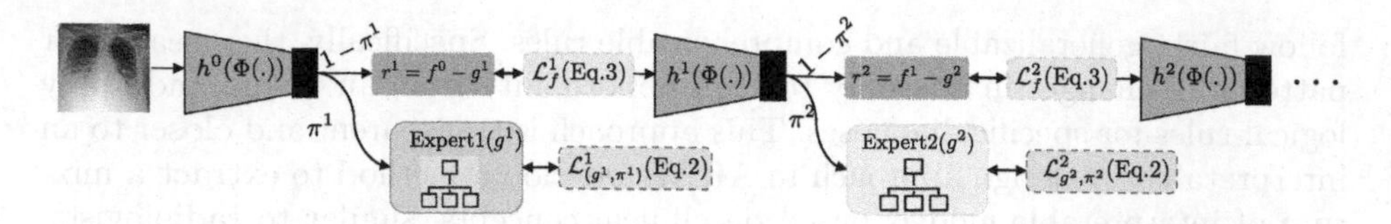

Fig. 1. Schematic view of our method. Note that $f^k(.) = h^k(\Phi(.))$. At iteration k, the selector *routes* each sample either towards the expert g^k with probability $\pi^k(.)$ or the residual $r^k = f^{k-1} - g^k$ with probability $1-\pi^k(.)$. g^k generates FOL-based explanations for the samples it covers. Note Φ is fixed across iterations.

three functions: (1) a set of selectors ($\pi : \mathcal{C} \rightarrow \{0, 1\}$) routing samples to an interpretable model or residual, (2) a set of interpretable models ($g : \mathcal{C} \rightarrow \mathcal{Y}$), and (3) the residuals. The interpretable models are called "experts" since they specialize in a distinct subset of data defined by that iteration's coverage τ as shown in SelectiveNet [16]. Figure 1 illustrates our method.

2.1 Distilling BB to the Mixture of Interpretable Models

Handling Class Imbalance. For an iteration k, we first split the given coverage τ^k to stratified coverages per class as $\{\tau_m^k = w_m \cdot \tau^k; w_m = N_m/N, \forall m\}$, where w_m denotes the fraction of samples belonging to the mth class; N_m and N are the samples of mth class and total samples, respectively.

Learning the Selectors. At iteration k, the selector π^k *routes* i^{th} sample to the expert (g^k) or residual (r^k) with probability $\pi^k(\boldsymbol{c}_i)$ and $1 - \pi^k(\boldsymbol{c}_i)$ respectively. For coverages $\{\tau_m^k, \forall m\}$, we learn g^k and π^k jointly by solving the loss:

$$\theta^*_{s^k}, \theta^*_{g^k} = \underset{\theta_{s^k}, \theta_{g^k}}{\arg\min}\, \mathcal{R}^k\Big(\pi^k(.;\theta_{s^k}), g^k(.;\theta_{g^k})\Big) \;\; \text{s.t.}\; \zeta_m\big(\pi^k(.;\theta_{s^k})\big) \geq \tau_m^k \;\; \forall m, \quad (1)$$

where $\theta^*_{s^k}, \theta^*_{g^k}$ are the optimal parameters for π^k and g^k, respectively. $\mathcal{R}^k$ is the overall selective risk, defined as, $\mathcal{R}^k(\pi^k, g^k) = \sum_m \frac{\frac{1}{N_m}\sum_{i=1}^{N_m} \mathcal{L}^k_{(g^k,\pi^k)}(\boldsymbol{x}_i, \boldsymbol{c}_i)}{\zeta_m(\pi^k)}$, where $\zeta_m(\pi^k) = \frac{1}{N_m}\sum_{i=1}^{N_m} \pi^k(\boldsymbol{c}_i)$ is the empirical mean of samples of mth class selected by the selector for the associated expert g^k. We define $\mathcal{L}^k_{(g^k,\pi^k)}$ in the next section. The selectors are neural networks with sigmoid activation. At inference time, π^k routes a sample to g^k if and only if $\pi^k(.) \geq 0.5$.

Learning the Experts. For iteration k, the loss $\mathcal{L}^k_{(g^k,\pi^k)}$ distills the expert g^k from f^{k-1}, BB of the previous iteration by solving the following loss:

$$\mathcal{L}^k_{(g^k,\pi^k)}(\boldsymbol{x}_i, \boldsymbol{c}_i) = \underbrace{\ell\Big(f^{k-1}(\boldsymbol{x}_i), g^k(\boldsymbol{c}_i)\Big)\pi^k(c_i)}_{\substack{\text{trainable component} \\ \text{for current iteration } k}} \underbrace{\prod_{j=1}^{k-1}\big(1 - \pi^j(\boldsymbol{c}_i)\big)}_{\substack{\text{fixed component trained} \\ \text{in the previous iterations}}}, \quad (2)$$

where $\pi^k(\boldsymbol{c}_i)\prod_{j=1}^{k-1}\left(1-\pi^j(\boldsymbol{c}_i)\right)$ is the cumulative probability of the sample covered by the residuals for all the previous iterations from $1,\cdots,k-1$ (*i.e.*, $\prod_{j=1}^{k-1}\left(1-\pi^j(\boldsymbol{c}_i)\right)$) and the expert g^k at iteration k (*i.e.*, $\pi^k(\boldsymbol{c}_i)$).

Learning the Residuals. After learning g^k, we calculate the residual as, $r^k(x_i, c_i) = f^{k-1}(x_i) - g^k(c_i)$ (difference of logits). We fix Φ and optimize the following loss to update h^k to specialize on those samples not covered by g^k, effectively creating a new BB f^k for the next iteration $(k+1)$:

$$\mathcal{L}_f^k(\boldsymbol{x}_j, \boldsymbol{c}_j) = \underbrace{\ell\left(r^k(\boldsymbol{x}_j, \boldsymbol{c}_j), f^k(\boldsymbol{x}_j)\right)}_{\substack{\text{trainable component} \\ \text{for iteration } k}} \underbrace{\prod_{i=1}^{k}\left(1-\pi^i(\boldsymbol{c}_j)\right)}_{\substack{\text{non-trainable component} \\ \text{for iteration } k}} \tag{3}$$

We refer to all the experts as the Mixture of Interpretable Experts (MoIE-CXR). We denote the models, including the final residual, as MoIE-CXR+R. Each expert in MoIE-CXR constructs sample-specific FOLs using the optimization strategy and algorithm discussed in [4].

2.2 Finetuning to an Unseen Domain

We assume the MoIE-CXR-identified concepts to be generalizable to an unseen domain. So, we learn the projection t_t for the target domain and compute the pseudo concepts using SSL [15]. Next, we transfer the selectors, experts, and final residual ($\{\pi_s^k, g_s^k\}_{k=1}^K$ and f_s^K) from the source to a target domain with limited labeled data and computational cost. Algorithm 1 details the procedure.

Algorithm 1. Finetuning to an unseen domain.

1: **Input:** Learned selectors, experts, and final residual from source domain: $\{\pi_s^k, g_s^k\}_{k=1}^K$ and f_s^K respectively, with K as the number of experts to transfer. BB of the source domain: $f_s^0 = h_s^0(\Phi_s)$. Source data: $\mathcal{D}_s = \{\mathcal{X}_s, \mathcal{C}_s, \mathcal{Y}_s\}$. Target data: $\mathcal{D}_t = \{\mathcal{X}_t, \mathcal{Y}_t\}$. Target coverages $\{\tau_k\}_{k=1}^K$.
2: **Output:** Experts $\{\pi_t^k, g_t^k\}_{k=1}^K$ and final residual f_t^K of the target domain.
3: Randomly select $n_t \ll N_t$ samples out of $N_t = |\mathcal{D}_t|$.
4: Compute the pseudo concepts for the correctly classified samples in the target domain using f_s^0, as, $\boldsymbol{c}_t^i = t_s\left(\Phi_s(\boldsymbol{x}_s^i)\right)$ *s.t.*, $y_t^i = f_s^0(\boldsymbol{x}_t^i)$, $i = 1\cdots n_t$
5: Learn the projection function t_t for target domain semi-supervisedly [15] using the pseudo labeled samples $\{\boldsymbol{x}_t^i, \boldsymbol{c}_t^i\}_{i=1}^{n_t}$ and unlabeled samples $\{\boldsymbol{x}_t^i\}_{i=1}^{N_t-n_t}$.
6: Complete the triplet for the target domain $\{\mathcal{X}_t, \mathcal{C}_t, \mathcal{Y}_t\}$, where $\boldsymbol{c}_t^i = t_t(\Phi_s(\boldsymbol{x}_t^i))$, $i = 1\cdots N_t$.
7: Finetune $\{\pi_s^k, g_s^k\}_{k=1}^K$ and f_s^K to obtain $\{\pi_t^k, g_t^k\}_{k=1}^K$ and f_t^K using equations 1, 2 and 3 respectively for 5 epochs. $\{\pi_t^k, g_t^k\}_{k=1}^K$ and $\left\{\{\pi_t^k, g_t^k\}_{k=1}^K, f_t^K\right\}$ represents MoIE-CXR and MoIE-CXR + R for the target domain.

---

3 Experiments

We perform experiments to show that MoIE-CXR 1) captures a diverse set of concepts, 2) does not compromise BB's performance, 3) covers "harder" instances with the residuals in later iterations resulting in their drop in performance, 4) is finetuned well to an unseen domain with minimal computation.

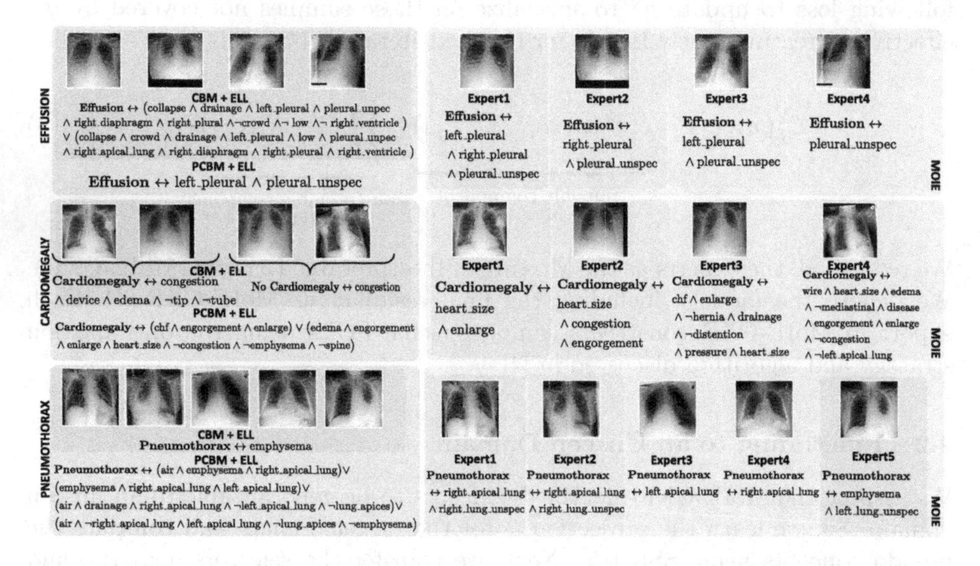

Fig. 2. Qualitative comparison of MoIE-CXR discovered concepts with the baselines.

Experimental Details. We evaluate our method using 220,763 frontal images from the MIMIC-CXR dataset [11]. We use Densenet121 [8] as BB (f^0) to classify cardiomegaly, effusion, edema, pneumonia, and pneumothorax, considering each to be a separate binary classification problem. We obtain 107 anatomical and observation concepts from the RadGraph's inference dataset [10], automatically generated by DYGIE++ [20]. We train BB following [24]. To retrieve the concepts, we utilize until the 4th Densenet block as feature extractor Φ and flatten the features to learn t. We use an 80%-10%-10% train-validation-test split with no patient shared across splits. We use 4, 4, 5, 5, and 5 experts for cardiomegaly, pneumonia, effusion, pneumothorax, and edema. We employ ELL [1] as g. Further, we only include concepts as input to g if their validation auroc exceeds 0.7. Refer to Table 1 in the supplementary material for the hyperparameters. We stop until all the experts cover at least 90% of the data cumulatively.

Baseline. We compare our method with 1) end-to-end CEM [26], 2) sequential CBM [14], and 3) PCBM [25] baselines, comprising of two parts: a) concept predictor $\Phi : \mathcal{X} \rightarrow \mathcal{C}$, predicting concepts from images, with all the convolution blocks; and b) label predictor, $g : \mathcal{C} \rightarrow \mathcal{Y}$, predicting labels from the concepts. We create CBM + ELL and PCBM + ELL by replacing the standard classifier with the identical g of MOIE-CXR to generate FOLs [1] for the baseline.

MoIE-CXR Captures Diverse Explanations. Figure 2 illustrates the FOL explanations. Recall that the experts (g) in MoIE-CXR and the baselines are ELLs [1], attributing attention weights to each concept. A concept with high attention weight indicates its high predictive significance. With a single g, the baselines rank the concepts in accordance with the identical order of attention weights for all the samples in a class, yielding a generic FOL for that class. In Fig. 2, the baseline PCBM + ELL uses *left_pleural* and *pleural_unspec* to identify effusion for all four samples. MoIE-CXR deploys multiple experts, learning to specialize in distinct subsets of a class. So different interpretable models in MoIE assign different attention weights to capture instance-specific concepts unique to each subset. In Fig. 2 expert2 relies on *right_pleural* and *pleural_unspec*, but expert4 relies only on *pleural_unspec* to classify effusion. The results show that the learned experts can provide more precise explanations at the subject level using the concepts, increasing confidence and trust in clinical use.

Table 1. MoIE-CXR does not compromize the performance of BB. We provide the mean and standard errors of AUROC over five random seeds. For MoIE-CXR, we also report the percentage of test set samples covered by all experts as "*Coverage*". We boldfaced our results and BB.

Model	Effusion	Cardiomegaly	Edema	Pneumonia	Pneumothorax
Blackbox (BB)	**0.92**	**0.84**	**0.89**	**0.79**	**0.91**
INTERPRETABLE BY DESIGN					
CEM [26]	$0.83_{\pm 1e-4}$	$0.75_{\pm 1e-4}$	$0.77_{\pm 2e-4}$	$0.62_{\pm 4e-4}$	$0.76_{\pm 3e-4}$
CBM (Sequential) [14]	$0.78_{\pm 1e-4}$	$0.72_{\pm 1e-4}$	$0.77_{\pm 5e-4}$	$0.60_{\pm 1e-3}$	$0.75_{\pm 6e-4}$
CBM + ELL [1,14]	$0.81_{\pm 1e-4}$	$0.72_{\pm 1e-4}$	$0.79_{\pm 5e-4}$	$0.62_{\pm 8e-4}$	$0.75_{\pm 6e-4}$
POSTHOC					
PCBM [25]	$0.88_{\pm 1e-4}$	$0.81_{\pm 1e-4}$	$0.82_{\pm 1e-4}$	$0.72_{\pm 1e-4}$	$0.85_{\pm 7e-4}$
PCBM-h [25]	$0.90_{\pm 1e-4}$	$0.83_{\pm 1e-4}$	$0.85_{\pm 1e-4}$	$0.77_{\pm 1e-4}$	$0.89_{\pm 7e-4}$
PCBM + ELL [1,25]	$0.90_{\pm 1e-4}$	$0.82_{\pm 1e-4}$	$0.85_{\pm 1e-4}$	$0.75_{\pm 1e-4}$	$0.85_{\pm 6e-4}$
PCBM-h + ELL [1,25]	$0.91_{\pm 1e-4}$	$0.83_{\pm 1e-4}$	$0.87_{\pm 1e-4}$	$0.77_{\pm 1e-4}$	$0.90_{\pm 1e-4}$
OURS					
MoIE-CXR $^{(\text{Coverage})}$	$\mathbf{0.93}^{(0.90)}_{\pm 1e-4}$	$\mathbf{0.85}^{(0.96)}_{\pm 1e-4}$	$\mathbf{0.91}^{(0.92)}_{\pm 1e-4}$	$\mathbf{0.80}^{(0.97)}_{\pm 1e-4}$	$\mathbf{0.91}^{(0.93)}_{\pm 2e-4}$
MoIE-CXR+R	$\mathbf{0.91}_{\pm 1e-4}$	$\mathbf{0.82}_{\pm 1e-4}$	$\mathbf{0.88}_{\pm 1e-4}$	$\mathbf{0.78}_{\pm 1e-4}$	$\mathbf{0.90}_{\pm 2e-4}$

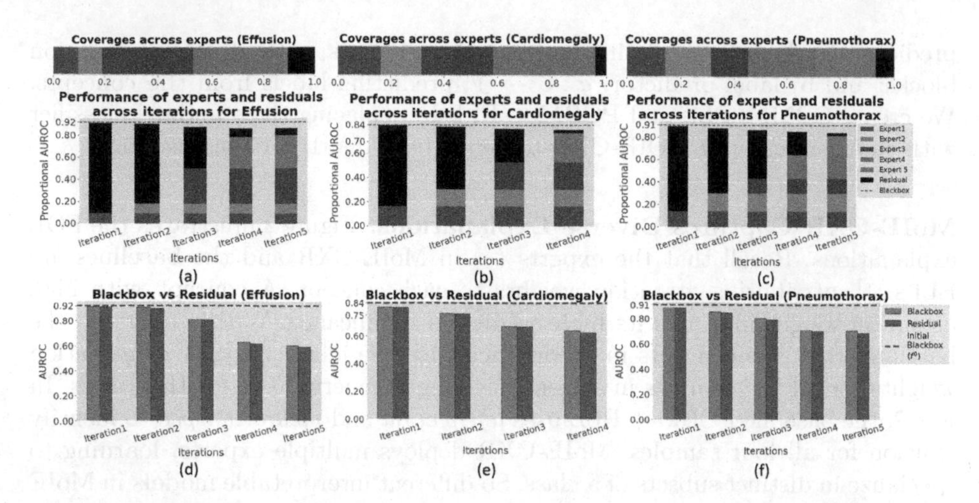

Fig. 3. Performance of experts and residuals across iterations. **(a-c):** Coverage and proportional AUROC of the experts and residuals. **(d-f):** Routing the samples covered by MoIE-CXR to the initial f^0, we compare the performance of the residuals with f^0.

MoIE-CXR does not Compromise BB's Performance. Analysing MoIE-CXR: Table 1 shows that MoIE-CXR outperforms other models, including BB. Recall that MoIE-CXR refers to the mixture of all interpretable experts, excluding any residuals. As MoIE-CXR specializes in various subsets of data, it effectively discovers sample-specific classifying concepts and achieves superior performance. In general, MoIE-CXR exceeds the interpretable-by-design baselines (CEM, CBM, and CBM + ELL) by a fair margin (on average, at least $\sim$ 10% $\uparrow$), especially for pneumonia and pneumothorax where the number of samples with the disease is significantly less ($\sim$ 750/24000 in the testset).

Analysing MoIE-CXR+R: To compare the performance on the entire dataset, we additionally report MoIE-CXR+R, the mixture of interpretable experts with the final residual in Table 1. MoIE-CXR+R outperforms the interpretable-by-design models and yields comparable performance as BB. The residualized PCBM baseline, *i.e.,* PCBM-h, performs similarly to MoIE-CXR+R. PCBM-h rectifies the interpretable PCBM's mistakes by learning the residual with the complete dataset to resemble BB's performance. However, the experts and the final residual approximate the interpretable and uninterpretable fractions of BB, respectively. In each iteration, the residual focuses on the samples not covered by the respective expert to create BB for the next iteration and likewise. As a result, the final residual in MoIE-CXR+R covers the "hardest" examples, reducing its overall performance relative to MoIE-CXR.

Identification of Harder Samples by Successive Residuals. Figure 3 (a–c) reports the proportional AUROC of the experts and the residuals per iteration.

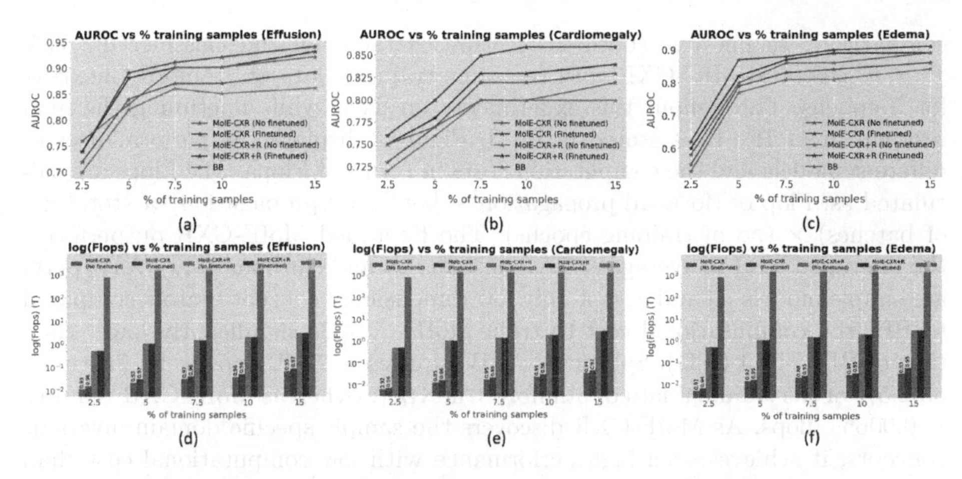

Fig. 4. Transferring the first 3 experts of MoIE-CXR trained on MIMIC-CXR to Stanford-CXR. With varying % of training samples of Stanford CXR, **(a-c):** reports AUROC of the test sets, **(d-g)** reports computation costs in terms of log (Flops) (T). We report the coverages in Stanford-CXR on top of the "finetuned" and "No finetuned" variants of MoIE-CXR (red and blue bars) in **(d-g)**. (Color figure online)

The proportional AUROC is the AUROC of that model times the empirical coverage, ζ^k, the mean of the samples routed to the model by the respective selector (π^k). According to Fig. 3a in iteration 1, the residual (black bar) contributes more to the proportional AUROC than the expert1 (blue bar) for effusion with both achieving a cumulative proportional AUROC ~ 0.92. All the final experts collectively extract the entire interpretable component from BB f^0 in the final iteration, resulting in their more significant contribution to the cumulative performance. In subsequent iterations, the proportional AUROC decreases as the experts are distilled from the BB of the previous iteration. The BB is derived from the residual that performs progressively worse with each iteration. The residual of the final iteration covers the "hardest" samples. Tracing these samples back to the original BB f^0, f^0 underperforms on these samples (Fig. 3 (d-f)) as the residual.

Applying MoIE-CXR to the Unseen Domain. In this experiment, we utilize Algorithm 1 to transfer MoIE-CXR trained on MIMIC-CXR dataset to Stanford Chexpert [9] dataset for the diseases – effusion, cardiomegaly and edema. Using 2.5%, 5%, 7.5%, 10%, and 15 % of training data from the Stanford Chexpert dataset, we employ two variants of MoIE-CXR where we (1) train only the selectors (π) without finetuning the experts (g) ("No finetuned" variant of MoIE-CXR in Fig. 4), and (2) finetune π and g jointly for only 5 epochs ("Finetuned" variant of MoIE-CXR and MoIE-CXR + R in Fig. 4). Finetuning π is essential to route the samples of the target domain to the appropriate expert. As later experts cover the "harder" samples of MIMIC-CXR, we only transfer the experts of the first three iterations (refer to Fig. 3). To ensure a fair comparison, we finetune (both the feature extractor Φ and classifier h^0) BB: $f^0 = h^0 \circ \Phi$ of MIMIC-CXR with the same training data of Stanford Chexpert for 5 epochs. Throughout this experiment, we fix Φ while finetuning the final residual in MoIE+R as stated in Eq. 3. Figure 4 displays the performances of different models and the computation costs in terms of Flops. The Flops are calculated as, Flop of (forward propagation + backward propagation) $\times$ (total no. of batches) $\times$ (no of training epochs). The finetuned MoIE-CXR outperforms the finetuned BB (on average $\sim 5\% \uparrow$ for effusion and cardiomegaly). As experts are simple models [1] and accept only low dimensional concept vectors compared to BB, the computational cost to train MoIE-CXR is significantly lower than that of BB (Fig. 4 (d-f)). Specifically, BB requires $\sim$ 776T flops to be finetuned on 2.5% of the training data of Stanford CheXpert, whereas MoIE-CXR requires $\sim$ 0.0065T flops. As MoIE-CXR discovers the sample-specific domain-invariant concepts, it achieves such high performance with low computational cost than BB.

4 Conclusion

This paper proposes a novel iterative interpretable method that identifies instance-specific concepts without losing the performance of the BB and is effectively fine-tuned in an unseen target domain with no concept annotation, limited labeled data, and minimal computation cost. Also, as in the prior work, MoIE-captured concepts may not showcase a causal effect that can be explored in the future.

Acknowledgement. This work was partially supported by NIH Award Number 1R01HL141813-01 and the Pennsylvania Department of Health. We are grateful for the computational resources from Pittsburgh Super Computing grant number TG-ASC170024.

References

1. Barbiero, P., Ciravegna, G., Giannini, F., Lió, P., Gori, M., Melacci, S.: Entropy-based logic explanations of neural networks. In: Proceedings of the AAAI Conference on Artificial Intelligence. vol. 36, pp. 6046–6054 (2022)
2. Chu, B., Madhavan, V., Beijbom, O., Hoffman, J., Darrell, T.: Best practices for fine-tuning visual classifiers to new domains. In: Hua, G., Jégou, H. (eds.) ECCV 2016. LNCS, vol. 9915, pp. 435–442. Springer, Cham (2016). https://doi.org/10.1007/978-3-319-49409-8_34
3. Clough, J.R., Oksuz, I., Puyol-Antón, E., Ruijsink, B., King, A.P., Schnabel, J.A.: Global and local interpretability for cardiac MRI classification. In: Shen, D., et al. (eds.) MICCAI 2019. LNCS, vol. 11767, pp. 656–664. Springer, Cham (2019). https://doi.org/10.1007/978-3-030-32251-9_72

4. Ghosh, S., Yu, K., Arabshahi, F., Batmanghelich, K.: Dividing and conquering a BlackBox to a mixture of interpretable models: route, interpret, repeat. In: Krause, A., Brunskill, E., Cho, K., Engelhardt, B., Sabato, S., Scarlett, J. (eds.) Proceedings of the 40th International Conference on Machine Learning. Proceedings of Machine Learning Research. vol. 202, pp. 11360–11397. PMLR (2023). https://proceedings.mlr.press/v202/ghosh23c.html
5. Ghosh, S., Yu, K., Arabshahi, F., Batmanghelich, K.: Tackling shortcut learning in deep neural networks: An iterative approach with interpretable models (2023)
6. Graziani, M., Andrearczyk, V., Marchand-Maillet, S., Müller, H.: Concept attribution: explaining CNN decisions to physicians. Comput. Biol. Med. **123**, 103865 (2020)
7. Guan, H., Liu, M.: Domain adaptation for medical image analysis: a survey. IEEE Trans. Biomed. Eng. **69**(3), 1173–1185 (2021)
8. Huang, G., Liu, Z., Van Der Maaten, L., Weinberger, K.Q.: Densely connected convolutional networks. In: Proceedings of the IEEE Conference on Computer Vision and Pattern Recognition, pp. 4700–4708 (2017)
9. Irvin, J., et al.: CheXpert: a large chest radiograph dataset with uncertainty labels and expert comparison. In: Proceedings of the AAAI Conference on Artificial Intelligence. vol. 33, pp. 590–597 (2019)
10. Jain, S., et al.: RadGraph: Extracting clinical entities and relations from radiology reports. arXiv preprint arXiv:2106.14463 (2021)
11. Johnson, A., et al.: MIMIC-CXR-JPG-chest radiographs with structured labels
12. Kandel, I., Castelli, M.: How deeply to fine-tune a convolutional neural network: a case study using a histopathology dataset. Appl. Sci. **10**(10), 3359 (2020)
13. Kim, B., et al.: Interpretability beyond feature attribution: Quantitative testing with concept activation vectors (tcav) (2017). arXiv preprint arXiv:1711.11279 (2017)
14. Koh, P.W., et al.: Concept bottleneck models. In: International Conference on Machine Learning, pp. 5338–5348. PMLR (2020)
15. Lee, D.H., et al.: Pseudo-label: the simple and efficient semi-supervised learning method for deep neural networks. In: Workshop on Challenges in Representation Learning, ICML. vol. 3, p. 896 (2013)
16. Rabanser, S., Thudi, A., Hamidieh, K., Dziedzic, A., Papernot, N.: Selective classification via neural network training dynamics. arXiv preprint arXiv:2205.13532 (2022)
17. Rajpurkar, P., et al.: CheXNet: Radiologist-level pneumonia detection on chest X-rays with deep learning. arXiv preprint arXiv:1711.05225 (2017)
18. Rudin, C., Chen, C., Chen, Z., Huang, H., Semenova, L., Zhong, C.: Interpretable machine learning: fundamental principles and 10 grand challenges. Stat. Surv. **16**, 1–85 (2022)
19. Sarkar, A., Vijaykeerthy, D., Sarkar, A., Balasubramanian, V.N.: Inducing semantic grouping of latent concepts for explanations: An ante-hoc approach. arXiv preprint arXiv:2108.11761 (2021)
20. Wadden, D., Wennberg, U., Luan, Y., Hajishirzi, H.: Entity, relation, and event extraction with contextualized span representations. In: Proceedings of the 2019 Conference on Empirical Methods in Natural Language Processing and the 9th International Joint Conference on Natural Language Processing (EMNLP-IJCNLP). pp. 5784–5789. Association for Computational Linguistics, Hong Kong, China (2019). https://doi.org/10.18653/v1/D19-1585, https://aclanthology.org/D19-1585

21. Wang, Y.X., Ramanan, D., Hebert, M.: Growing a brain: fine-tuning by increasing model capacity. In: Proceedings of the IEEE Conference on Computer Vision and Pattern Recognition, pp. 2471–2480 (2017)
22. Yan, W., et al.: MRI manufacturer shift and adaptation: increasing the generalizability of deep learning segmentation for MR images acquired with different scanners. Radiol. Artif. Intell. **2**(4), e190195 (2020)
23. Yeche, H., Harrison, J., Berthier, T.: UBS: a dimension-agnostic metric for concept vector interpretability applied to radiomics. In: Suzuki, K. (ed.) ML-CDS/IMIMIC -2019. LNCS, vol. 11797, pp. 12–20. Springer, Cham (2019). https://doi.org/10.1007/978-3-030-33850-3_2
24. Yu, K., Ghosh, S., Liu, Z., Deible, C., Batmanghelich, K.: Anatomy-Guided Weakly-Supervised Abnormality Localization in Chest X-rays. In: Wang, L., Dou, Q., Fletcher, P.T., Speidel, S., Li, S. (eds.) Medical Image Computing and Computer Assisted Intervention-MICCAI 2022. MICCAI 2022. Lecture Notes in Computer Science. vol. 13435. Springer, Cham (2022). https://doi.org/10.1007/978-3-031-16443-9_63
25. Yuksekgonul, M., Wang, M., Zou, J.: Post-hoc concept bottleneck models. arXiv preprint arXiv:2205.15480 (2022)
26. Zarlenga, M.E., et al.: Concept embedding models. arXiv preprint arXiv:2209.09056 (2022)

Gadolinium-Free Cardiac MRI Myocardial Scar Detection by 4D Convolution Factorization

Amine Amyar[1], Shiro Nakamori[1], Manuel Morales[1], Siyeop Yoon[1], Jennifer Rodriguez[1], Jiwon Kim[2], Robert M. Judd[3], Jonathan W. Weinsaft[2], and Reza Nezafat[1(✉)]

[1] Department of Medicine (Cardiovascular Division), Beth Israel Deaconess Medical Center and Harvard Medical School, Boston, MA, USA
rnezafat@bidmc.harvard.edu

[2] Division of Cardiology, Weill Cornell Medicine, New York, NY, USA

[3] Department of Medicine (Cardiology Division), Duke University, Durham, NC, USA

Abstract. Gadolinium-based contrast agents are commonly used in cardiac magnetic resonance (CMR) imaging to characterize myocardial scar tissue. Recent works using deep learning have shown the promise of contrast-free short-axis cine images to detect scars based on wall motion abnormalities (WMA) in ischemic patients. However, WMA can occur in patients without a scar. Moreover, the presence of a scar may not always be accompanied by WMA, particularly in non-ischemic heart disease, posing a significant challenge in detecting scars in such cases. To overcome this limitation, we propose a novel deep spatiotemporal residual attention network (ST-RAN) that leverages temporal and spatial information at different scales to detect scars in both ischemic and non-ischemic heart diseases. Our model comprises three primary components. First, we develop a novel factorized 4D (3D+time) convolutional layer that extracts 3D spatial features of the heart and a deep 1D kernel in the temporal direction to extract heart motion. Secondly, we enhance the power of the 4D (3D+time) layer with spatiotemporal attention to extract rich whole-heart features while tracking the long-range temporal relationship between the frames. Lastly, we introduce a residual attention block that extracts spatial and temporal features at different scales to obtain global and local motion features and to detect subtle changes in contrast related to scar. We train and validate our model on a large dataset of 3000 patients who underwent clinical CMR with various indications and different field strengths (1.5T, 3T) from multiple vendors (GE, Siemens) to demonstrate the generalizability and robustness of our model. We show that our model works on both ischemic and non-ischemic heart diseases outperforming state-of-the-art methods. Our code is available at https://github.com/HMS-CardiacMR/Myocardial_Scar_Detection.

Keywords: Contrast-Free MRI · Spatiotemporal Neural Network · 4D Convolution Factorization · Myocardial Scar Detection

H. Greenspan et al. (Eds.): MICCAI 2023, LNCS 14221, pp. 639–648, 2023.
https://doi.org/10.1007/978-3-031-43895-0_60

1 Introduction

Cardiovascular diseases continue to be the primary cause of death worldwide. Imaging of myocardial fibrosis/scar provides both diagnostic and prognostic information. Cardiac magnetic resonance (CMR) late gadolinium enhancement (LGE) is the gold standard for myocardial scar evaluation in ischemic and non-ischemic heart disease [7,8]. In LGE, imaging is performed 10–15 minutes after infusion of 0.1–0.2 mmol/kg of gadolinium-based contrast agent. However, many patients who undergo clinical CMR do not have any scars on LGE. While traditionally, gadolinium was considered safe; recent data show deposition of gadolinium in many organs, which is directly associated with the total dose of gadolinium [4,11,12]. Considering patients receiving multiple MRI scans throughout their life, it is important to minimize gadolinium use in imaging protocols with high field. Beyond patient safety, there is significant costs associated with gadolinium. Furthermore, concerns have arisen regarding environmental contamination due to excessive gadolinium use [5,15].

Recently, deep learning (DL) based methods have been proposed to limit gadolinium use by creating virtual LGE-like images using short-axis (sax) cine [22] or combined with native T_1 images [24]. The relationship between the motion field and myocardial infarction has also been explored to detect scar areas by learning temporal (dynamic) representations from cine images [20,21,23]. Another approach used radiomics alone [1,10,13] or combined with DL for myocardial scar screening [3]. Although promising, previous methods have several limitations. Changes in the mechanical properties of the myocardium caused by infarction can lead to regional wall motion abnormalities (WMA) [9]. However, WMA can occur in patients without scars [2]. Furthermore, the presence of scar may not always be accompanied by WMA, especially in non-ischemic heart disease [2], posing a significant challenge in detecting scars in such cases. The changes in contrast alone in cine images do not provide sufficient information to detect a scar [3]. Moreover, existing methods often overlook the 4D nature of the data (3D+time) and treat it as a 3D (2D+time) instead. Additionally, they require the detection of the left ventricle and were trained on 2D cine images that match LGE images, while in clinical practice, such information is not available beforehand. Finally, one of the major limitations is the lack of studies on large heterogeneous datasets [9].

Contribution: In this study, we develop a novel end-to-end deep spatiotemporal residual attention neural network (ST-RAN) for scar detection using whole heart imaging in ischemic and non-ischemic heart diseases. The proposed model leverages spatial information to capture changes in contrast and temporal information to capture WMA to detect scars, in a large heterogeneous dataset. To achieve this, we propose a novel efficient Conv3Plus1D layer that deploys a factorized 4D (3D+time) receptive field, to simultaneously extract hierarchical spatial features and deep temporal features (comprehensive spatiotemporal features), distinguishing between patients with and without a scar. We introduce a multi-scale residual attention block that learns global and local motions to detect significant and subtle changes, the latter more present in patients with small scar sizes and nearly preserved wall motion. We validate our proposed model on a large cohort of patients with and without scars, showing the robustness of the model, outperforming state-of-the-art methods.

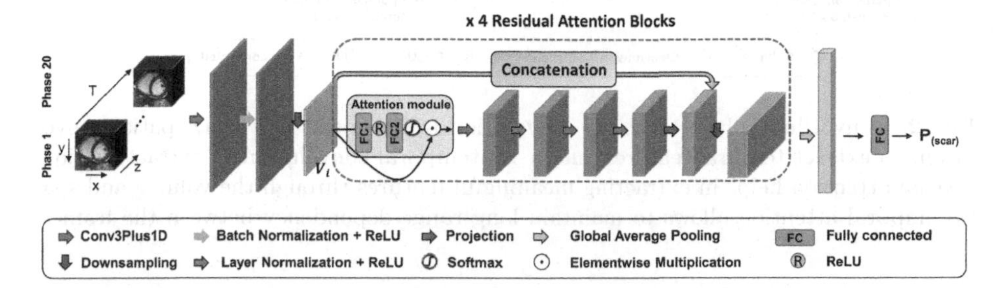

Fig. 1. Architecture overview. Our model takes an input a set of short-axis cine images of the whole heart, consisting of 20 phases, which are fed to a novel Conv3Plus1D layer, to extract spatial and temporal features. After batch normalization and non-linear transformation, the feature maps are fed to a series of residual attention blocks (RAB) at different scales to extract global and local features, subtle to changes due to myocardial scar. After the RAB, a global average pooling followed by a fully connected are used to predict presence of a scar.

2 Methods

As illustrated in Fig. 1, given a time series of a 3D volume of the heart, our goal is to predict the presence of a scar (Pscar). Inspired by recent work in image processing, we rely on residual attention network [18] to learn heart motion at different scales. In contrast to Zhang et al. [23] where they used a recurrent neural network to learn local motion features and an optical flow module to learn global motion features, our model enhances attention at different scales. This approach allows our model to capture local and global motion features within a single end-to-end network, reducing the complexity of using a two-stream model. The multi-scale temporal kernel allows to detect global motion, more likely to be related to a large scar size, and local motion to better estimate WMA at each segment. To address the heterogeneity of scar distribution, we have incorporated a spatiotemporal module to control the contribution of spatial features at different scales.

2.1 Spatiotemporal Decomposition Using 4D(3D+time) Layers

In sax cine, the data is in 4-dimension consisting of a 3D volume (stack of sax slices) with a temporal dimension. Therefore, effective representation of

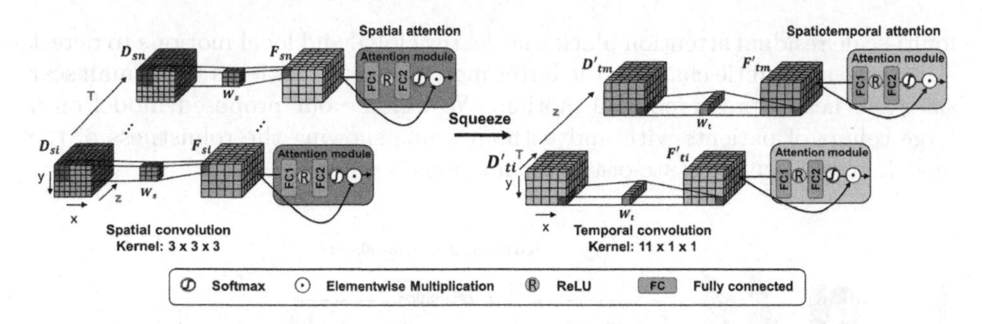

Fig. 2. Conv3Plus1D layer. The 4D convolution is factorized into a 3D spatial convolution to extract textural features, and a 1D temporal convolution to extract motion. Spatial attention helps in extracting meaningful features through the volume and spatiotemporal attention allows to maintain long range-dependency between the frames.

spatiotemporal features is crucial for accurate analysis. Inspired by Squeeze & Excitation network [6,14] and spatiotemporal network [17], we develop a novel efficient Conv3Plus1D layer that deploys a factorized 4D (3D+time) receptive field by applying a 3D spatial convolution of 3×3×3 to extract hierarchical spatial features, followed by a 1D temporal convolution of 11×1×1 to extract deep temporal features, as shown in Fig. 2. The large temporal filter allows maintain of the long-range dependency between the 20 frames. The input to the Conv3Plus1D layer is a 4D tensor $\mathcal{D} \in \mathbb{R}^{X \times Y \times Z \times T}$ where X is width, Y is high, Z is depth and T is time encoded in the channel-wise direction. The spatial convolution $\mathcal{W}_s$ is applied to input volume $\mathcal{D}_{si}$ across $X \times Y \times Z$ and its feature map output is $\mathcal{F}_{si} = W_s \times D_{si}$, where $i = 1....n$ and n is the number of feature maps in the T direction. The spatial attention module is trained to assign an attention score $\mathbf{a}_{fk}$ for each feature map F_{si}, and patch K in $X \times Y \times Z$ direction given as:

$$\mathbf{a}_{\mathrm{F_{si}k}} = \frac{e^{ReLU(\mathcal{F}_{si} * W_{c1}) * W_{c2}}}{\sum_{j=1}^{k} e^{ReLU(\mathcal{F}_{f|} * W_{c1}) * W_{c2}}} \tag{1}$$

where W_{c1} and W_{c2} are the weights for the fully connected layers 1 and 2 in the spatial attention module. The temporal convolution $\mathcal{W}_t$ is applied across a volume input $\mathcal{D'}_{ti}$ across $X \times Y \times T$ where the feature map output is $\mathcal{F}_{ti} = W_t \times D'_{ti}$, where $i = 1....m$ and m is the number of feature maps in the Z direction. The spatiotemporal attention module is trained to assign an attention score $\mathbf{a}_{F'k}$ for each feature map F_{ti} and patch K in $X \times Y \times T$ direction given as:

$$\mathbf{a}_{\mathrm{F'_{ti}k}} = \frac{e^{ReLU(\mathcal{F'}_{\mathrm{ti}} * W_{\mathrm{c3}}) * W_{\mathrm{c4}}}}{\sum_{j=1}^{k} e^{ReLU(\mathcal{F'}_{\mathrm{tj}}} * W_{\mathrm{c3}}) * W_{\mathrm{c4}}} \tag{2}$$

where W_{c3} and W_{c4} are the weights for the fully connected layers 1 and 2 in the spatiotemporal attention module.

To simplify the Eqs. (1) and (2), we have excluded the bias parameters. Following the application of softmax activation, the attention scores indicate the significance of each region in space and time, in determining the existence of a scar. Regions that are highly relevant to scar detection have scores near 1, while scar-free regions have scores near 0. Our proposed layer enables the learning of a more complete representation in spatial and temporal directions, surpassing a simple feature combination approach. Factorizing spatial and temporal kernels allows for reducing the model's parameters from 46M to 1.4M while learning rich spatiotemporal data representation.

2.2 Residual Attention Blocks

The motion patterns of the heart can evolve over time and scale. The residual attention network builds a stack of attention modules that generate attention-aware features. As the layers deepen, attention-aware features scale adaptively, enabling the detection of spatial and temporal subtle changes to be enhanced. This enhancement is crucial in accurately detecting small scar sizes. By aggregating information from tissues and motion across multiple scales, the attention module is able to learn and assign relative importance to each region with regard to the presence of a scar. The feature maps at a scale i where $i = 1....4$, are input to two fully-connected layers to encode spatial-wise and temporal-wise dependencies defined as $\mathcal{G} = W_{\mathrm{Ri1}} * ReLU(V_i * W_{\mathrm{Ri2}})$, with W_{Ri1} being the weights for the first fully connected layer at scale i, and W_{Ri2} being the weights for the second fully connected layer at scale i. The output $\mathcal{G}$ is then passed through the softmax activation to obtain the spatiotemporal residual weights, which will be applied to the input map V_i to extract the spatiotemporal features at scale i.

2.3 Network Details

The proposed network applies first a Conv3Plus1D layer followed by a batch normalization to mitigate internal covariate shift in the data, applies a small regularization effect, and a non-linear activation function ReLU. The output feature map is then resized through a customized resize frames layer that downsamples the size of the input by a factor of 2. This helps in increasing the batch size (×2 folds) and accelerates training (×12 folds) and testing (×4 folds) while maintaining the same high performance. This is followed by four attention residual blocks. Each block consists of residual attention applied to the input feature map and two sets of Conv3Plus1D layers, each followed by a layer normalization and ReLU. When layer normalization is located inside the residual blocks, it allows to speed-up convergence without a need for a learning rate warm-up stage [19]. Then a projection layer is applied to match the input data's last dimension and the residual block's output. When different-sized filters are used, and the input data is downsampled, the output may have a different number of filters than the input. The projection layer is used to project the last dimension of the input data to match the new filter size so that the input and output can be added together to form the residual connection. The concatenation helps tackle the vanishing gradient problem due to the 4D nature of the input data with a large 3D volume and long-range temporal dependency. By using all these layers, the network can learn from the input data more effectively, handle different-sized inputs and outputs, and learn to recognize patterns in the temporal domain. Finally, to allow the model to learn scar-specific feature representations, a global 3D average pooling is applied to enforce correspondences between feature maps and the probability of a scar, the latter estimated through one fully connected neuron with a sigmoid.

3 Materials and Implementation Details

Data Aacquisition. Cine images were collected breath-hold electrocardiogram-gated balanced steady-state free precession sequence of 10 sax slices. The data were acquired from institution anonymous from 2016 to 2020 using multivendor (GE Healthcare, Siemens Healthineers) and different field strengths (1.5 T, 3 T). The institutional review board approved the use of CMR data for research with a consent waiver. Patient information was handled in compliance with the Health Insurance Portability and Accountability Act. Patients were referred for a clinical CMR for different cardiac indications, resulting in a heterogeneous patient cohort, necessary for better evaluation of the model performance. In total, 3000 patients (1697 males, 54 ± 18 years) were used for training and evaluation. The data were split into training (n=2000, 762 scar+), validation (n=500, 169 scar+), and testing (n=500, 199 scar+). All images were cropped at the center to a size of 128×128 and normalized to a fixed intensity range (from 0 to 1).

Implementation Details. The model's optimization was performed using a mini-batch stochastic gradient descent of 64 with an initial learning rate of 0.001 and a weight decay of 0.0001 when the validation loss plateaus. The model was trained for a maximum of 500 epochs with an early stopping of 70. The binary cross-entropy loss function and binary accuracy metric for both training and validation were monitored to avoid overfitting and underfitting. All models were implemented using TensorFlow version 2.4.1 and trained on an NVIDIA DGX-1 system equipped with 8 T V100 graphics processing units (each with 32 GB memory and 5120 cores). All selected hyperparameters were optimized experimentally. DeLong's test was used to compare the AUC of the different models. All tests were two-sided with a significance level = 0.05.

4 Experiments and Results

Ablation Study on the Impact of Different Components Design. We first perform an ablation study to evaluate each component's contribution to our proposed model. We test the effect of having spatial kernels only (S-CNN), temporal kernels only (T-CNN), and a combination of both (ST-CNN). We test the impact of spatial and temporal attention (AST-CNN) and residual attention (ST-RAN) on the performance of our network. To this end, we train five variants

Table 1. Effectiveness of different components of our proposed model (Ablation) and comparison with state-of-the-art methods (SOTA).

Study	Method	AUC	Sensitivity	Specificity	F_1-score	Nb. parameters
Ablation	S-CNN	0.74	0.74	0.60	0.63	0.90M
	T-CNN	0.81	0.77	0.70	0.70	0.38M
	ST-CNN	0.82	0.83	0.68	0.71	1.38M
	AST-CNN	0.83	0.79	**0.72**	0.71	1.42M
	ST-RAN	**0.84**	**0.90**	0.60	**0.72**	1.43M
SOTA	3D-STCNN [17]	0.77	0.84	0.52	0.66	0.44M
	3D-CNN [16]	0.80	0.81	**0.63**	0.68	46.5M
	CNN-LSTM [23]	0.79	0.81	0.59	0.67	0.33M
	ST-RAN	**0.84**	**0.90**	0.60	**0.72**	1.43M

of our model. We can observe from Table 1 that temporal information significantly outperforms spatial information alone (AUC$=0.81$ vs. 0.74, $P = 0.004$). By combining both spatial and temporal information, we further increase the model's sensitivity (0.83 vs. 0.74, $P < 0.001$). Combining both spatial and temporal kernels with residual attention block yields the best performance on all dataset with an AUC of 0.84 and a sensitivity of 0.90. For both ischemic and non-ischemic heart diseases, ST-RAN showed higher sensitivity while having the lowest false-negative compared to others (Table 2).

Comparison with State-of-the-Art Methods. We then compare our model with state-of-the-art methods trained and tested on the same dataset for myocardial scar detection, including 3D (2D + time) spatiotemporal CNN (3D-STCNN) [17], 3D-CNN [16], and CNN with a long short-term memory network (CNN-LSTM) [23]. Our proposed model yields the best performance with a sensitivity of 0.90 and an F_1-score of 0.72, significantly outperforming all other methods (all $P < 0.05$), as shown in Table 1. We also compare different models on ischemic and non-ischemic patients, as shown in Table 2. We can notice the superiority of our model on both ischemic and non-ischemic patients outperforming other methods based on sensitivity, true positive, and false negative.

5 Discussion

Recent works using deep learning have shown the promise of contrast-free short-axis cine images to detect scars based on WMA in ischemic patients. However, these methods have limitations in detecting scar in non-ischemic heart diseases. Moreover, the large heterogeneous number of patients without scar in the dataset and with WMA has degraded these models' performance in detecting scar in ischemic patients. In contrast, our approach utilizes both spatial and temporal information to detect scar. Our model demonstrates superior performance over

Table 2. Ablation study and comparison with state-of-the-art methods for ischemic and non-ischemic heart diseases.

Study	Method	Sensitivity	True-positive	False-positive
Ischemic	S-CNN	0.74	79	28
	T-CNN	0.79	84	23
	ST-CNN	0.81	87	20
	AST-CNN	0.83	89	18
	3D-STCNN [17]	0.83	89	18
	3D-CNN [16]	0.82	88	19
	CNN-LSTM [23]	0.85	91	16
	ST-RAN	**0.92**	**98**	**9**
Non-Ischemic	S-CNN	0.74	68	24
	T-CNN	0.76	70	22
	ST-CNN	0.85	78	14
	AST-CNN	0.75	69	23
	3D-STCNN [17]	0.86	79	13
	3D-CNN [16]	0.79	73	19
	CNN-LSTM [23]	0.77	71	21
	ST-RAN	**0.89**	**82**	**10**

other methods in both ischemic and non-ischemic heart diseases. The inclusion of a multi-scale residual attention mechanism allows for the learning of global and local motions in an end-to-end network without the added complexity of a secondary component optical flow to extract global motion. Approximately half of non-ischemic cardiomyopathy patients who Gd-based imaging cardiac MRI exhibit no myocardial scars. Identifying these patients prior to Gd injection can significantly improve cost-effectiveness, scan efficiency, and safety by reducing unnecessary Gd administration.

To overcome the complexity of 4D convolution, we propose an effective training and inference strategy based on spatiotemporal factorization 4D (3D+time). This approach allows for a reduction in model parameters by a factor of 32 while maintaining high performance. The proposed layer extracts both spatial and temporal features, while enhancing attention on features in both directions, to detect subtle differences in left ventricle myocardial texture, as well as in cardiac motion.

In future work, we will investigate multimodality learning, incorporating other sequences such as T_1 maps, to enhance the model's precision to an even greater degree.

6 Conclusion

We propose a spatiotemporal residual attention neural network for myocardial scar detection, and we tackled the challenging non-ischemic patients. We showed that our model works on ischemic heart disease as well. Our results demonstrate the potential of our model in unmasking hidden information in native sax cine images, and allows for better detection of patients with a high likelihood of having a myocardial scar. These results indicate the potential of our proposed model in screening patients with and without a scar, thus, saving patients from unnecessary gadolinium based contrast agent administration, reducing costs, and environmental pollution. Finally, our proposed network has potential applications in various clinical contexts that require 4D processing.

References

1. Baessler, B., Mannil, M., Oebel, S., Maintz, D., Alkadhi, H., Manka, R.: Subacute and chronic left ventricular myocardial scar: accuracy of texture analysis on nonenhanced cine MR images. Radiology **286**(1), 103–112 (2018)
2. Csecs, I., et al.: Association between left ventricular mechanical deformation and myocardial fibrosis in Nonischemic cardiomyopathy. J. Am. Heart Assoc. **9**(19), e016797 (2020)
3. Fahmy, A.S., Rowin, E.J., Arafati, A., Al-Otaibi, T., Maron, M.S., Nezafat, R.: Radiomics and deep learning for myocardial scar screening in hypertrophic cardiomyopathy. J. Cardiovasc. Magn. Reson. **24**(1), 1–12 (2022)
4. Gulani, V., Calamante, F., Shellock, F.G., Kanal, E., Reeder, S.B., et al.: Gadolinium deposition in the brain: summary of evidence and recommendations. Lancet Neurol. **16**(7), 564–570 (2017)
5. Hatje, V., Bruland, K.W., Flegal, A.R.: Increases in anthropogenic gadolinium anomalies and rare earth element concentrations in san Francisco bay over a 20 year record. Environ. Sci. Technol. **50**(8), 4159–4168 (2016)
6. Hu, J., Shen, L., Sun, G.: Squeeze-and-excitation networks. In: Proceedings of the IEEE Conference on Computer Vision and Pattern Recognition, pp. 7132–7141 (2018)
7. Kim, R.J., et al.: Relationship of MRI delayed contrast enhancement to irreversible injury, infarct age, and contractile function. Circulation **100**(19), 1992–2002 (1999)
8. Kim, R.J., et al.: The use of contrast-enhanced magnetic resonance imaging to identify reversible myocardial dysfunction. N. Engl. J. Med. **343**(20), 1445–1453 (2000)
9. Leiner, T.: Deep learning for detection of myocardial scar tissue: Goodbye to gadolinium? (2019)
10. Mancio, J., et al.: Machine learning phenotyping of scarred myocardium from cine in hypertrophic cardiomyopathy. Eur. Heart J. Cardiovasc. Imaging **23**(4), 532–542 (2022)
11. McDonald, R.J., et al.: Gadolinium retention: a research roadmap from the 2018 NIH/ACR/RSNA workshop on gadolinium chelates. Radiology **289**(2), 517–534 (2018)
12. McDonald, R.J., et al.: Intracranial gadolinium deposition after contrast-enhanced MR imaging. Radiology **275**(3), 772–782 (2015)

13. Neisius, U., et al.: Texture signatures of native myocardial T1 as novel imaging markers for identification of hypertrophic cardiomyopathy patients without scar. J. Magn. Reson. Imaging **52**(3), 906–919 (2020)
14. Roy, A.G., Navab, N., Wachinger, C.: Concurrent spatial and channel 'Squeeze & Excitation' in fully convolutional networks. In: Frangi, A.F., Schnabel, J.A., Davatzikos, C., Alberola-López, C., Fichtinger, G. (eds.) MICCAI 2018. LNCS, vol. 11070, pp. 421–429. Springer, Cham (2018). https://doi.org/10.1007/978-3-030-00928-1_48
15. Schmidt, K., Bau, M., Merschel, G., Tepe, N.: Anthropogenic gadolinium in tap water and in tap water-based beverages from fast-food franchises in six major cities in Germany. Sci. Total Environ. **687**, 1401–1408 (2019)
16. Tran, D., Bourdev, L., Fergus, R., Torresani, L., Paluri, M.: Learning spatiotemporal features with 3d convolutional networks. In: Proceedings of the IEEE International Conference on Computer Vision, pp. 4489–4497 (2015)
17. Tran, D., Wang, H., Torresani, L., Ray, J., LeCun, Y., Paluri, M.: A closer look at spatiotemporal convolutions for action recognition. In: Proceedings of the IEEE Conference on Computer Vision and Pattern Recognition, pp. 6450–6459 (2018)
18. Wang, F., et al.: Residual attention network for image classification. In: Proceedings of the IEEE Conference on Computer Vision and Pattern Recognition, pp. 3156–3164 (2017)
19. Xiong, R., et al.: On layer normalization in the transformer architecture. In: International Conference on Machine Learning, pp. 10524–10533. PMLR (2020)
20. Xu, C., Howey, J., Ohorodnyk, P., Roth, M., Zhang, H., Li, S.: Segmentation and quantification of infarction without contrast agents via spatiotemporal generative adversarial learning. Med. Image Anal. **59**, 101568 (2020)
21. Xu, C., et al.: Direct detection of pixel-level myocardial infarction areas via a deep-learning algorithm. In: Descoteaux, M., Maier-Hein, L., Franz, A., Jannin, P., Collins, D.L., Duchesne, S. (eds.) MICCAI 2017. LNCS, vol. 10435, pp. 240–249. Springer, Cham (2017). https://doi.org/10.1007/978-3-319-66179-7_28
22. Xu, C., Xu, L., Ohorodnyk, P., Roth, M., Chen, B., Li, S.: Contrast agent-free synthesis and segmentation of ischemic heart disease images using progressive sequential causal gans. Med. Image Anal. **62**, 101668 (2020)
23. Zhang, N., et al.: Deep learning for diagnosis of chronic myocardial infarction on nonenhanced cardiac cine MRI. Radiology **291**(3), 606–617 (2019)
24. Zhang, Q., et al.: Toward replacing late gadolinium enhancement with artificial intelligence virtual native enhancement for gadolinium-free cardiovascular magnetic resonance tissue characterization in hypertrophic cardiomyopathy. Circulation **144**(8), 589–599 (2021)

Longitudinal Multimodal Transformer Integrating Imaging and Latent Clinical Signatures from Routine EHRs for Pulmonary Nodule Classification

Thomas Z. Li[1(✉)], John M. Still[2], Kaiwen Xu[3], Ho Hin Lee[3], Leon Y. Cai[1], Aravind R. Krishnan[4], Riqiang Gao[5], Mirza S. Khan[6], Sanja Antic[7], Michael Kammer[7], Kim L. Sandler[8], Fabien Maldonado[7], Bennett A. Landman[1,3,4,8], and Thomas A. Lasko[2,3]

[1] Biomedical Engineering, Vanderbilt University, Nashville, TN 37212, USA
thomas.z.li@vanderbilt.edu
[2] Biomedical Informatics, Vanderbilt University, Nashville, TN 37212, USA
[3] Computer Science, Vanderbilt University, Nashville, TN 37212, USA
[4] Electrical and Computer Engineering, Vanderbilt University, Nashville, TN 37212, USA
[5] Digital Technology and Innovation, Siemens Healthineers, Princeton, NJ 08540, USA
[6] Saint Luke's Mid America Heart Institute, Kansas City, MO 64111, USA
[7] Medicine, Vanderbilt University Medical Center, Nashville, TN 37235, USA
[8] Radiology, Vanderbilt University Medical Center, Nashville, TN 37235, USA

Abstract. The accuracy of predictive models for solitary pulmonary nodule (SPN) diagnosis can be greatly increased by incorporating repeat imaging and medical context, such as electronic health records (EHRs). However, clinically routine modalities such as imaging and diagnostic codes can be asynchronous and irregularly sampled over different time scales which are obstacles to longitudinal multimodal learning. In this work, we propose a transformer-based multimodal strategy to integrate repeat imaging with longitudinal clinical signatures from routinely collected EHRs for SPN classification. We perform unsupervised disentanglement of latent clinical signatures and leverage time-distance scaled self-attention to jointly learn from clinical signatures expressions and chest computed tomography (CT) scans. Our classifier is pretrained on 2,668 scans from a public dataset and 1,149 subjects with longitudinal chest CTs, billing codes, medications, and laboratory tests from EHRs of our home institution. Evaluation on 227 subjects with challenging SPNs revealed a significant AUC improvement over a longitudinal multimodal baseline (0.824 vs 0.752 AUC), as well as improvements over a single cross-section multimodal scenario (0.809 AUC) and a longitudinal imaging-only scenario (0.741 AUC). This work demonstrates significant advantages with a novel approach for co-learning longitudinal

Supplementary Information The online version contains supplementary material available at https://doi.org/10.1007/978-3-031-43895-0_61.

H. Greenspan et al. (Eds.): MICCAI 2023, LNCS 14221, pp. 649–659, 2023.
https://doi.org/10.1007/978-3-031-43895-0_61

imaging and non-imaging phenotypes with transformers. Code available at https://github.com/MASILab/lmsignatures.

Keywords: Multimodal Transformers · Latent Clinical Signatures · Pulmonary Nodule Classification

1 Introduction

The absence of highly accurate and noninvasive diagnostics for risk-stratifying benign vs malignant solitary pulmonary nodules (SPNs) leads to increased anxiety, costs, complications, and mortality [22,26]. The use of noninvasive methods to discriminate malignant from benign SPNs is a high-priority public health initiative [8,9]. Deep learning approaches have shown promise in classifying SPNs from longitudinal chest computed tomography (CT) [1,5,12,21], but approaches that only consider imaging are fundamentally limited. Multimodal models generally outperform single modality models in disease diagnosis and prediction [24], and this is especially true in lung cancer which is heavily contextualized through non-imaging risk factors [6,23,30]. Taken together, these findings suggest that learning across both time and multiple modalities is important in biomedical predictive modeling, especially SPN diagnosis. However, such an approach that scales across longitudinal multimodal data from comprehensive representations of the clinical routine has yet to be demonstrated [24].

Related Work. Directly learning from routinely collected electronic health records (EHRs) is challenging because observations within and between modalities can be sparse and irregularly sampled. Previous studies overcome these challenges by aggregating over visits and binning time series within a Bidirectional Encoder Representations from Transformers (BERT) architecture [2,14,20,25], limiting their scope to data collected on similar time scales, such as ICU measurements, [11,29], or leveraging graph guided transformers to handle asynchrony [33]. Self-attention [31] has become the dominant technique for learning powerful representations of EHRs with trade-offs in interpretability and quadratic scaling with the number of visits or bins, which can be inefficient with data spanning multiple years. In contrast, others address the episodic nature of EHRs by converting non-imaging variables to continuous longitudinal curves that provide the instantaneous value of categorical variables as intensity functions [17] or continuous variables as latent functions [16]. Operating with the hypothesis that distinct disease mechanisms manifest independently of one another in a probabilistic manner, one can learn a transformation that disentangles latent sources, or clinical signatures, from these longitudinal curves. Clinical signatures learned in this way are expert-interpretable and have been well-validated to reflect known pathophysiology across many diseases [15,18]. Given that several clinical risk factors have been shown to independently contribute to lung cancer risk, these signatures are well poised for this predictive task. Despite the wealth of studies seeking to learn comprehensive representations of routine EHRs, these techniques have not been combined with longitudinal imaging.

Present Work. In this work, we jointly learn from longitudinal medical imaging, demographics, billing codes, medications, and lab values to classify SPNs. We converted 9195 non-imaging event streams from the EHR to longitudinal curves to impute cross-sections and synchronize across modalities. We use Independent Component Analyses (ICA) to disentangle latent clinical signatures from these curves, with the hypothesis that the disease mechanisms known to be important for SPN classification can also be captured with probabilistic independence. We leverage a transformer-based encoder to fuse features from both longitudinal imaging and clinical signature expressions sampled at intervals ranging from weeks to up to five years. Due to the importance of time dynamics in SPN classification, we use the time interval between samples to scale self-attention with the intuition that recent observations are more important to attend to than older observations. Compared with imaging-only and a baseline that aggregates longitudinal data into bins, our approach allowed us to incorporate additional modalities from routinely collected EHRs, which led to improved SPN classification.

2 Methods

Latent Clinical Signatures via Probabilistic Independence. We obtained event streams for billing codes, medications, and laboratory tests across the full record of each subject in our EHR cohorts (up to 22 years). After removing variables with less than 1000 events and mapping billing codes to the SNOMED-CT ontology [7], we arrived at 9195 unique variables. We transformed each variable to a longitudinal curve at daily resolution, estimating the variable's instantaneous value for each day [18]. We used smooth interpolation for continuous variables [4] or a continuous estimate of event density per time for event data. Previous work used Gaussian process inference to compute both types of curves [16,17]. For this work we traded approximation for computational efficiency. To encode a limited memory into the curve values, each curve was smoothed using a rolling uniform mean of the past 365 d (Fig. 1, left).

We use an ICA model to estimate a linear decomposition of the observed curves from the EHR-Pulmonary cohort to independent latent sources, or clinical signatures. Formally, we have dataset $D_{\text{EHR-Pulmonary}} = \{L_k \mid k = 1, \dots, n\}$ with longitudinal curves denoted as $L_k = \{l_i | i = 1, \dots, 9195\}$. We randomly sample l_i $\forall i \in [1, 9195]$ at a three-year resolution and concatenate samples across all subjects as $x_i \in R^m$. For $D_{\text{EHR-Pulmonary}}$, m was empirically found to be 630037. We make a simplifying assumption that x_i is a linear mixture of c latent sources, s, with longitudinal expression levels $e \in R^m$:

$$x_i = e_1 s_{i,1} + e_2 s_{i,2} + \dots + e_c s_{i,c} \tag{1}$$

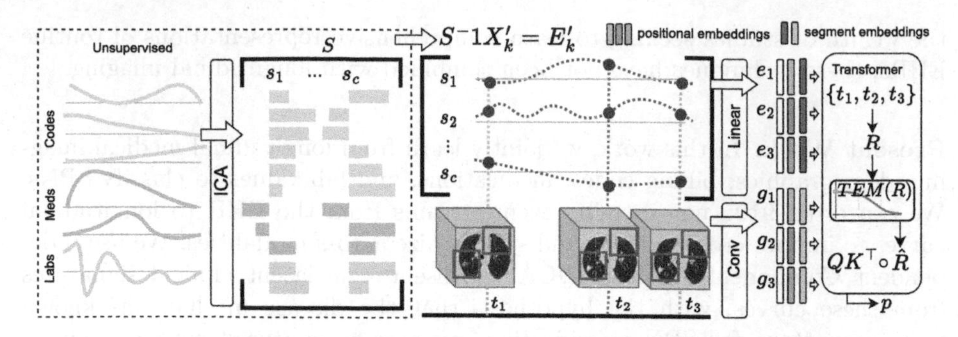

Fig. 1. Left: Event streams for non-imaging variables are transformed into longitudinal curves. ICA learns independent latent signatures, S, in an unsupervised manner on a large non-imaging cohort. Right: Subject k's expressions of the signatures, E'_k, are sampled at scan dates. Input embeddings are the sum of 1) token embedding derived from signatures or imaging, 2) a fixed positional embedding indicating the token's position in the sequence, and 3) a learnable segment embedding indicating imaging or non-imaging modality. The time interval between scans is used to compute a time-distance scaled self-attention. This is a flexible approach that handles asynchronous modalities, incompleteness over varying sequence lengths, and irregular time intervals.

The linear mixture is then $X = SE$ with x_i forming the rows of X, $S \in R^{9195\times c}$ denoting the independent latent sources and $E \in R^{c\times m}$ denoting the expression matrix. We set $c = 2000$ and estimated S in an unsupervised manner using FastICA [13]. Given longitudinal curves for another cohort, for instance $D_{\text{Image-EHR}} = \{X'_k \mid k = 1, \ldots, n\}$, we obtain expressions of clinical signatures for subject k via $E'_k = S^{-1}X'_k$ (Fig. 1, left).

Longitudinal Multimodal Transformer (TDSig). We represent our multimodal datasets $D_{\text{Image-EHR}}$ and $D_{\text{Image-EHR-SPN}} = \{(E_k, G_k) \mid k = 1, \ldots, n\}$ as a sequence of clinical expressions $E_k = \{e_{k,1}, \ldots, e_{k,T}\}$ sampled at the same dates as images $G_k = \{g_{k,1}, \ldots, g_{k,T}\}$, where T is the maximum sequence length. We set $T = 3$ and added a fixed padding embedding to represent missing items in the sequence. Embeddings that incorporate positional and segment information are computed for each item in the sequence (Fig. 1, right). Token embeddings for images are a convolutional embeddings of five concatenated 3D patches proposed by a pretrained SPN detection model [21]. We use a 16-layer ResNet [10] to compute this embedding. Likewise, token embeddings for clinical signature expressions are linear transformations to the same dimension as imaging token embeddings. The sequence of embeddings are then passed through a multi-headed Transformer. All embeddings except the nodule detection model are co-optimized with the Transformer. We will refer to this approach as TDSig.

Time-Distance Self-attention. Following [5,19,32], we intuit that if medical data is sampled as a cross-sectional manifestation of a continuously progressing phenotype, we can use a temporal emphasis model (TEM) emphasize the importance of recent observations over older ones. Additionally, self-attention is masked for padded embeddings, allowing our approach to scale with varying sequence lengths across subjects. Formally, if subject k has a sequence of T images at relative acquisition days $t_1 \dots t_T$, we construct a matrix R of relative times with entries $R_{i,j} = |t_T - t_i|$ where t_i is the acquisition day of tokens $\hat{e}_{k,i}$ and $\hat{g}_{k,i}$, or 0 if they are padded embeddings. We map the relative times in R to a [0,1] value in $\hat{R}$ using a TEM of the form:

$$\hat{R}_{i,j} = \text{TEM}(R_{i,j}) = \frac{1}{1 + exp(bR_{i,j} - c)} \tag{2}$$

This is a flipped sigmoid function that monotonically decreases with the relative time from the most recent observation. Its slope of decline and decline offset are governed by learnable non-negative parameters b and c respectively. A separate TEM is instantiated for each attention head, with the rationale that separate attention heads can learn to condition on time differently. The transformer encoder computes query, key, and value matrices as linear transformations of input embedding $H = \{\hat{E} \parallel \hat{G}\}$ at attention head p

$$Q_p = H_p W_p^Q \qquad K_p = H_p W_p^K \qquad V_p = H_p W_p^V$$

TEM-scaled self-attention is computed via element-wise multiplication of the query-key product and $\hat{R}$:

$$\text{softmax}\left(\frac{\text{ReLU}(Q_p K_p^\top + M) \circ \hat{R}}{\sqrt{d}}\right) V_p \tag{3}$$

where M is the padding mask [31] and d is the dimension of the query and key matrices. ReLU gating of the query-key product allows the TEM to adjust the attention weights in an unsigned direction.

Baselines. We compared against a popular multimodal strategy that aggregates event streams into a sequence of bins as opposed to our method of extracting instantaneous cross-sectional representations. For each scan, we computed a TF-IDF [27] weighted vector from all billing codes occurring up to one year before the scan acquisition date. We passed this through a published Word2Vec-based medical concept embedding [3] to compute a contextual representation $\in R^{100}$. This, along with the subject's scans, formed a sequence that was used as input to a model we call TDCode2vec. Our search for contextual embeddings for medications and laboratory values did not yield any robust published models that were compatible with our EHR's nomenclature, so these were not included in TDCode2vec. We also performed experiments using only image sequences as input, which we call TDImage. Finally, we implemented single cross-sectional versions of TDImage, TDCode2vec, and TDSig, CSImage, CSCode2vec, and CSSig respectively, using the scan date closest to the lung malignancy diagnosis for cases or SPN date for controls. All baselines except CSImage, which

Table 1. Breakdown of modalities, size, and longitudinality of each dataset.

	Modalities					Counts (cases/controls)	
	Demo	Img	Code	Med	Lab	Subjects	Scans
EHR-Pulmonary	✓	–	✓	✓	✓	288,428	–
NLST	✓	✓	–	–	–	533/801	1066/1602
Image-EHR	✓	✓	✓	✓	✓	257/665	641/1624
Image-EHR-SPN	✓	✓	✓	✓	✓	58/169	76/405

Demo: Demographics, Img: Chest CTs, Code: ICD billing codes,
Med: Medications, Lab: Laboratory tests.

employed a multi-layer perceptron directly after the convolutional embedding, used the same architecture and time-distance self-attention as TDSig. The transformer encoders in this study were standardized to 4 heads, 4 blocks, input token size of 320, multi-layer perception size of 124, self-attention weights of size 64. This work was supported by Pytorch 1.13.1, CUDA 11.7.

3 Experimental Setup

Datasets. This study used an imaging-only cohort from the NLST [28] and three multimodal cohorts from our home institution with IRB approval (Table 1). For the **NLST** cohort (https://cdas.cancer.gov/nlst/), we identified cases who had a biopsy-confirmed diagnosis of lung malignancy and controls who had a positive screening result for an SPN but no lung malignancy. We randomly sampled from the control group to obtain a 4:6 case control ratio. Next, **EHR-Pulmonary** was the unlabeled dataset used to learn clinical signatures in an unsupervised manner. We searched all records in our EHR archives for patients who had billing codes from a broad set of pulmonary conditions, intending to capture pulmonary conditions beyond just malignancy. Additionally, **Image-EHR** was a labeled dataset with paired imaging and EHRs. We searched our institution's imaging archive for patients with three chest CTs within five years. In the EHR-Image cohort, malignant cases were labeled as those with a billing code for lung malignancy and no cancer of any type prior. Importantly, this case criteria includes metastasis from cancer in non-lung locations. Benign controls were those who did not meet this criterion. Finally, **Image-EHR-SPN** was a subset of Image-EHR with the inclusion criteria that subjects had a billing code for an SPN and no cancer of any type prior to the SPN. We labeled malignant cases as those with a lung malignancy billing code occurring within three years after any scan and only used data collected before the lung malignancy code. All data within the five-year period were used for controls. We removed all billing codes relating to lung malignancy. A description of the billing codes used to define SPN and lung cancer events are provided in Supplementary 1.2.

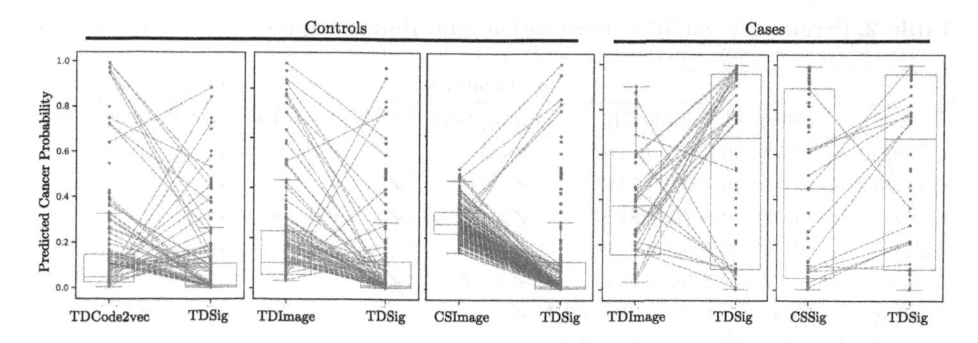

Fig. 2. A comparison of median and interquartile range of predicted probabilities reveals that TDSig is more correctly confident than baselines. Blue and red indicate subjects that were correctly and incorrectly reclassified by TDSig respectively. When compared to these baselines, TDSig is more often reclassifying correctly than not. (Color figure online)

Training and Validation. All models were pretrained with the NLST cohort after which we froze the convolutional embedding layer. While this was the only pretraining step for image-only models (CSImage and TDimage), the multimodal models underwent another stage of pretraining using the Image-EHR cohort with subjects from Image-EHR-SPN subtracted. In this stage, we randomly selected one scan and the corresponding clinical signature expressions for each subject and each training epoch. Models were trained until the running mean over 100 global steps of the validation loss increased by more than 0.2. For evaluation, we performed five-fold cross-validation with Image-EHR-SPN, using up to three of the most recent scans in the longitudinal models. We report the mean AUC and 95% confidence interval from 1000 bootstrapped samples, sampling with replacement from the pooled predictions across all test folds. A two-sided Wilcoxon signed-rank test was used to test if differences in mean AUC between models were significant.

Reclassification Analysis. We performed a reclassification analysis of low, medium, and high-risk tiers separated by thresholds of 0.05 and 0.65, which are the cutoffs used to guide clinical management. Given a baseline comparison, our approach reclassifies a subject correctly if it predicts a higher risk tier than the baseline in cases, or a lower risk tier than the baseline in controls (Fig. 2).

4 Results

The significant improvement with TDSig over CSSig demonstrates the advantage of longitudinally in the context of combining images and clinical signatures (Table 2). There were large performance gaps between TDSig and TDCode2vec, as well as between CSSig and CSCode2vec, demonstrating the advantage of

Table 2. Performance on SPN classification using different approaches and modalities.

		Modalities					Pretrain	
	Mean AUC [95% CI]	Img	Demo	Code	Med	Lab	NLST	Image-EHR
CSImage	0.7392 [0.7367, 0.7416]	✓	–	–	–	–	✓	–
CSCode2vec	0.7422 [0.7398, 0.7447]	✓	✓	✓	–	–	✓	✓
CSSig	0.8097 [0.8075, 0.8120]	✓	✓	✓	✓	✓	✓	✓
TDImage	0.7406 [0.7381, 0.7432]	✓	–	–	–	–	✓	–
TDCode2vec	0.7524 [0.7499, 0.7550]	✓	✓	✓	–	–	✓	✓
TDSig	**0.8238 [0.8216, 0.8260]***	✓	✓	✓	✓	✓	✓	✓

*: $p < 0.01$ against all other methods.

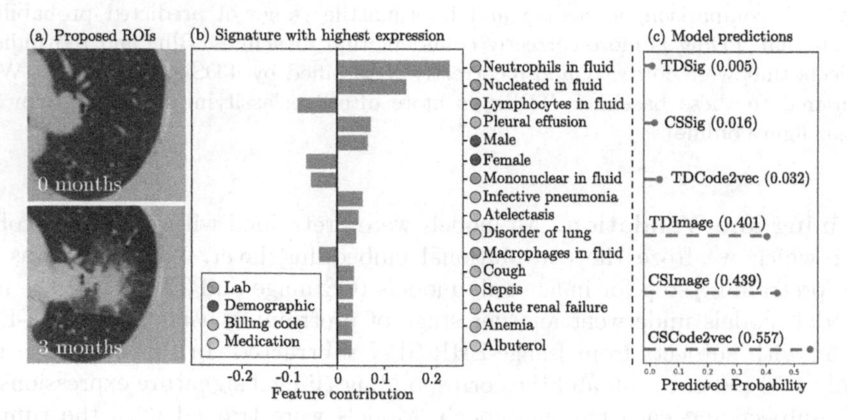

Fig. 3. This is a control subject who developed a lesion over 3 months (a), to which the imaging-only approaches assigned a cancer probability of 0.4 (c). However, the subject's highest expressed clinical signature at the 3-month mark was a new pattern of bacterial pneumonia (b), offering to the model a benign explanation of an image that it would otherwise be less correctly confident in.

clinical signatures over a binned embedding strategy. Cross-sectional embedded billing codes did not significantly improve performance over images alone (CSCode2vec vs CSImage, $p = 0.56$), but adding clinical signatures did (CSSig vs CSImage, $p < 0.01$; TDSig vs TDImage, $p < 0.01$) and the greatest improvement in longitudinal data over single cross sections occurred when clinical signatures were included.

For control subjects, TDSig correctly/incorrectly reclassified 40/18 from TDCode2vec, 54/8 from TDImage, 12/18 from CSSig, 104/7 from CSCode2vec, and 125/5 from CSImage. For case subjects, TDSig correctly/incorrectly reclassified 13/10 from TDCode2vec, 17/8 from TDImage, 12/2 from CSSig, 23/16 from CSCode2vec, and 29/16 from CSImage (Fig. 2). Full reclassification matrices are reported in Supplementary 6.1. On qualitative inspection of a control subject, clinical signatures likely added clarity to benign imaging findings that were difficult for baseline approaches to classify (Fig. 3).

5 Discussion and Conclusion

This work presents a novel transformer-based strategy for integrating longitudinal imaging with interpretable clinical signatures learned from comprehensive multimodal EHRs. We demonstrated large performance gains in SPN classification compared with baselines, although calibration of our models is needed to assess clinical utility. We evaluated on clinically-billed SPNs, meaning that clinicians likely found these lesions difficult enough to conduct a clinical workup. In this setting, we found that adding clinical context increased the performance gap between longitudinal data and single cross-sections. Our clinical signatures incorporated longitudinality and additional modalities to build a better representation of clinical context than binned embeddings. We release our implementation at https://github.com/MASILab/lmsignatures.

The lack of longitudinal multimodal datasets has long been a limiting factor [24] in conducting studies such as ours. One of our contributions is demonstrating training strategies in a small-dataset, incomplete-data regime. We were able to overcome our small cohort size (Image-EHR-SPN) by leveraging unsupervised learning on datasets without imaging (EHR-Pulmonary), pretraining on public datasets without EHRs (NLST), and pretraining on paired multimodal data with noisy labels (Image-EHR) within a flexible transformer architecture.

Our approach of sampling cross-sections where clinical decisions are likely to be made scales well with long, multi-year observation windows, which may not be true for BERT-based embeddings [20,25]. We did not compare against these contextual embeddings because none have been publically released, but integrating these with longitudinal imaging is an area of future investigation.

Acknowledgements. This research was funded by the NIH through R01CA253923-02 and in part by NSF CAREER 1452485 and NSF 2040462. This research is also supported by ViSE through T32EB021937-07 and the Vanderbilt Institute for Clinical and Translational Research through UL1TR002243-06. We thank the National Cancer Institute for providing data collected through the NLST.

References

1. Ardila, D., et al.: End-to-end lung cancer screening with three-dimensional deep learning on low-dose chest computed tomography (2019). https://www.nature.com/articles/s41591-019-0447-x
2. Choi, E., Bahadori, M.T., Sun, J., Kulas, J., Schuetz, A., Stewart, W.: RETAIN: an interpretable predictive model for healthcare using reverse time attention mechanism. In: Advances in Neural Information Processing Systems. vol. 29 (2016)
3. Finch, A., et al.: Exploiting hierarchy in medical concept embedding. JAMIA Open **4**(1), ooab022 (2021)
4. Fritsch, F.N., Butland, J.: A method for constructing local monotone piecewise cubic interpolants. SIAM J. Sci. Stat. Comput. **5**(2), 300–304 (1984)
5. Gao, R., et al.: Time-distanced gates in long short-term memory networks. Med. Image Anal. **65**(101785), 101785 (2020)

6. Gao, R., et al.: Deep multi-path network integrating incomplete biomarker and chest CT data for evaluating lung cancer risk. In: Medical Imaging 2021: Image Processing. vol. 11596, pp. 387–393. SPIE (2021)
7. Gaudet-Blavignac, C., Foufi, V., Bjelogrlic, M., Lovis, C.: Use of the Systematized Nomenclature of Medicine Clinical Terms (SNOMED CT) for Processing Free Text in Health Care: Systematic Scoping Review. J. Med. Internet Res. **23**(1), e24594 (2021)
8. Gómez-Sáez, N., et al.: Prevalence and variables associated with solitary pulmonary nodules in a routine clinic-based population: a cross-sectional study. Eur. Radiol. **24**(9), 2174–2182 (2014)
9. Gould, M.K., et al.: Recent trends in the identification of incidental pulmonary nodules. Am. J. Respir. Crit. Care Med. **192**(10), 1208–1214 (2015)
10. He, K., Zhang, X., Ren, S., Sun, J.: Deep residual learning for image recognition (2015)
11. Huang, K., Altosaar, J., Ranganath, R.: Clinicalbert: Modeling clinical notes and predicting hospital readmission. arXiv:1904.05342 (2019)
12. Huang, P., et al.: Prediction of lung cancer risk at follow-up screening with low-dose CT: a training and validation study of a deep learning method. Lancet Dig. Health **1**(7), e353–e362 (2019)
13. Hyvarinen, A.: Fast and robust fixed-point algorithms for independent component analysis. IEEE Trans. Neural Netw. **10**(3), 626–634 (1999)
14. Labach, A., Pokhrel, A., Yi, S.E., Zuberi, S., Volkovs, M., Krishnan, R.G.: Effective self-supervised transformers for sparse time series data (2023). https://openreview.net/forum?id=HUCgU5EQluN
15. Lasko, T., et al.: EHR-driven machine-learning model to distinguish benign from malignant pulmonary nodules (2023)
16. Lasko, T.: Nonstationary gaussian process regression for evaluating repeated clinical laboratory tests. In: Proceedings of the AAAI Conference on Artificial Intelligence. vol. 29 (2015)
17. Lasko, T.A.: Efficient inference of gaussian-process-modulated renewal processes with application to medical event data. In: Uncertainty in Artificial Intelligence: Proceedings of the Conference. Conference on Uncertainty in Artificial Intelligence. vol. 2014, p. 469. NIH Public Access (2014)
18. Lasko, T.A., Mesa, D.A.: Computational phenotype discovery via probabilistic independence. arXiv preprint arXiv:1907.11051 (2019)
19. Li, T.Z., et al.: Time-distance vision transformers in lung cancer diagnosis from longitudinal computed tomography. arXiv preprint arXiv:2209.01676 (2022)
20. Li, Y., et al.: BEHRT: transformer for electronic health records. Sci. Rep. **10**(1), 1–12 (2020)
21. Liao, F., Liang, M., Li, Z., Hu, X., Song, S.: Evaluate the malignancy of pulmonary nodules using the 3D deep leaky noisy-or network. IEEE Trans. Neural Netw. Learn. Syst. **30**(11), 3484–3495 (2019)
22. Massion, P.P., Walker, R.C.: Indeterminate pulmonary nodules: risk for having or for developing lung cancer? Cancer Prev. Res. (Phila) **7**(12), 1173–1178 (2014)
23. McWilliams, A., et al.: Probability of cancer in pulmonary nodules detected on first screening CT. N. Engl. J. Med. **369**(10), 910–919 (2013)
24. Mohsen, F., Ali, H., El Hajj, N., Shah, Z.: Artificial intelligence-based methods for fusion of electronic health records and imaging data. Sci. Rep. **12**(1), 17981 (2022)
25. Rasmy, L., Xiang, Y., Xie, Z., Tao, C., Zhi, D.: Med-BERT: pretrained contextualized embeddings on large-scale structured electronic health records for disease prediction. NPJ Dig. Med. **4**(1), 86 (2021)

26. Rivera, M.P., Mehta, A.C., Wahidi, M.M.: Establishing the diagnosis of lung cancer: diagnosis and management of lung cancer, 3rd ed: American college of chest physicians evidence-based clinical practice guidelines. Chest **143**(5 Suppl), e142S-e165S (2013)
27. Schütze, H., Manning, C.D., Raghavan, P.: Introduction to information retrieval. vol. 39. Cambridge University Press, Cambridge (2008)
28. Team, N.L.S.T.R.: Reduced lung-cancer mortality with low-dose computed tomographic screening. N. Engl. J. Med. **365**(5), 395–409 (2011)
29. Tipirneni, S., Reddy, C.K.: Self-supervised transformer for sparse and irregularly sampled multivariate clinical time-series. ACM Trans. Knowl. Discov. Data (TKDD) **16**(6), 1–17 (2022)
30. Vanguri, R.S., et al.: Multimodal integration of radiology, pathology and genomics for prediction of response to PD-(L) 1 blockade in patients with non-small cell lung cancer. Nat. Cancer **3**(10), 1151–1164 (2022)
31. Vaswani, A., et al.: Attention is all you need. In: Advances in Neural Information Processing Systems. vol. 30 (2017)
32. Wu, C., Wu, F., Huang, Y.: Da-transformer: Distance-aware transformer. arXiv preprint arXiv:2010.06925 (2020)
33. Zhang, X., Zeman, M., Tsiligkaridis, T., Zitnik, M.: Graph-guided network for irregularly sampled multivariate time series. arXiv preprint arXiv:2110.05357 (2021)

FedContrast-GPA: Heterogeneous Federated Optimization via Local Contrastive Learning and Global Process-Aware Aggregation

Qin Zhou and Guoyan Zheng(✉)

Institute of Medical Robotics, School of Biomedical Engineering,
Shanghai Jiao Tong University, No. 800, Dongchuan Road, Shanghai 200240, China
guoyan.zheng@sjtu.edu.cn

Abstract. Federated learning is a promising strategy for performing privacy-preserving, distributed learning for medical image segmentation. However, the data-level heterogeneity as well as system-level heterogeneity makes it challenging to optimize. In this paper, we propose to improve **Fed**erated optimization via local **Contrast**ive learning and **G**lobal **P**rocess-aware **A**ggregation (referred as FedContrast-GPA), aiming to jointly address both data-level and system-level heterogeneity issues. In specific, To address data-level heterogeneity, we propose to learn a unified latent feature space via an intra-client and inter-client local prototype based contrastive learning scheme. Among which, intra-client contrastive learning is adopted to improve the discriminative ability of learned feature embedding at each client, while inter-client contrastive learning is introduced to achieve cross-client distribution perception and alignment in a privacy preserving manner. To address system-level heterogeneity, we further propose a simple yet effective process-aware aggregation scheme to achieve effective straggler mitigation. Experimental results on six prostate segmentation datasets demonstrate large performance boost over existing state-of-the-art methods.

Keywords: Heterogeneous federated learning · Process-aware aggregation · Local prototype learning · Contrastive learning

1 Introduction

Recently, federated learning has emerged as a promising strategy for performing privacy-preserving, distributed learning for medical image segmentation. Among various methods, FedAvg [1] has been the de facto approach for federated learning, where the server maintains a global model which is dispatched to each client for updating locally on their own private data. After that, the updated local models

This study was partially supported by the Natural Science Foundation of China via project U20A20199 and 62201341.

H. Greenspan et al. (Eds.): MICCAI 2023, LNCS 14221, pp. 660–670, 2023.
https://doi.org/10.1007/978-3-031-43895-0_62

are collected and averaged to produce a global model for the training of the next round. For FedAvg and its variants, a well-known issue is "client drift" caused by non-IID data distribution across different clients (i.e., data-level heterogeneity). To address this issue, a group of methods resort to designing proximal terms or re-parametrization strategies [2,3] to restrain the client drift from the global model. However, these regularization terms inherently limit the local convergence potential, Other methods try to improve the local models' generalization ability without strict proximal restrictions on model parameters [4,5]. In [27], the authors proposed to learn compact local representations on each device and a global model across all devices, reducing both the intra-client and inter-client data variance. However, these methods perform local updates blindly, totally ignoring the feature distributions of other clients. In medical image segmentation, FedDG [6] was proposed to improve the local models' generalization ability via exchanging amplitude spectrum to transmit the distribution information across clients. However, the distribution perception step was processed offline, which was fixed upon finished, limiting its potential adaptability to various subsequent tasks.

Different from the above-mentioned methods, in this paper, we aim to tackle the "client drift" problem by exploring a unified latent feature space for different clients in a privacy-preserving manner and by enhancing the feature discriminability of each client. Concretely, we propose to extract local prototypes to represent the feature distribution at each client. Since local prototypes are statistical characteristics, we can share them among different clients without the concern of privacy issues. Then performing cross-client pixel to local prototype matching can help not only to perceive the global feature distribution but also to explicitly align the cross-client features, leading to a more unified latent feature space. Besides, by performing pixel to local prototype matching at each client, we can directly shape and enhance the discriminability of the learned feature space at each client.

Another well-acknowledged concern of federated networks is the "straggler" problem caused by system-level heterogeneity. FedAvg and some of its variants [2,3] directly average the local models weighted by their data amount ratio, which may lead to unexpected deterioration due to asynchronous learning process of local models. Based on the intuition that well-trained local models should contribute more to the global model, in this paper, we propose a simple yet effective process-aware model aggregation scheme, which is demonstrated to effectively suppress the influence of "stragglers".

The contributions of our method can be summarized as follows:

- We propose a novel FedContrast-GPA framework to simultaneously alleviate both data-level and system-level heterogeneity issues in federated optimization.
- We propose an intra-client and inter-client local prototype based contrastive learning scheme, which not only enhances the feature discriminability of each client, but also explicitly performs cross-client feature distribution perception and alignment in a privacy-preserving manner.
- We introduce a simple yet effective process-aware weighting scheme to suppress the influence of "stragglers" in global model aggregation.

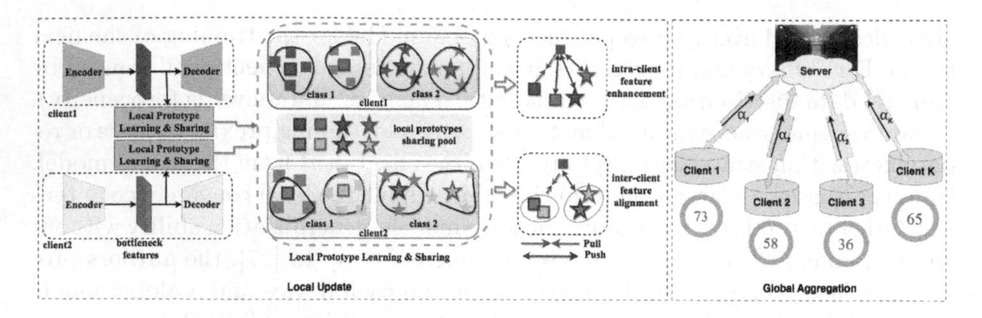

Fig. 1. The workflow of the proposed FedContrast-GPA framework. Please note in local update, same shapes mean the features belong to the same semantic class while same shapes of the same color figure online form a sub-cluster of the related semantic class. The local prototypes are marked with border lines. In global aggregation, the numbers indicate the training process. (Color figure online)

2 Method

A typical federated learning process consists of two stages: local update at each client and global aggregation at the server side. In this paper, we propose the FedContrast-GPA framework (as shown in Fig. 1), which consists of the intra- and inter-client local prototype based contrastive learning scheme (during local update) and the process-aware aggregation scheme (during global aggregation).

Assuming there exist K clients in the federated network, we denote client k as S_k. Then private data set on the k-th client can be denoted as $\{\boldsymbol{I}_k^n, \boldsymbol{Y}_k^n\}, n \in \{1, \cdots, N_k\}$, where $\boldsymbol{I}_k^n, \boldsymbol{Y}_k^n$ are the image and the corresponding segmentation map for the n-th instance in S_k, and N_k is the number of instances in S_k. In this paper, we adopt U-Net as the backbone architecture for segmentation. Denote $\varphi_k = f_k^e \circ f_k^d$ as the mapping function for S_k, where f_k^e and f_k^d are the encoder and decoder, and $\circ$ means sequentially executing f_k^e and f_k^d. Denote $\boldsymbol{w}_k$ as the parameters of local model on S_k, then the goal of federated learning is to find the optimal global model $\boldsymbol{w} = GA(\{\boldsymbol{w}_k\}, k \in \{1, \cdots, K\})$ that generalizes well across different clients, where $GA(\cdot)$ refers to a certain strategy for model aggregation.

2.1 Intra- and Inter-client Local-Prototype based Contrastive Learning

Local Prototype Learning. Denote the bottleneck features of class c on S_k as $\boldsymbol{F}_k^c = \{\boldsymbol{f}_k^{c,i}, i \in \{1, \cdots, N_k^c\}\}$, where N_k^c represents the number of pixels belonging to class c in the intermediate feature maps. In order to model the feature distribution of S_k from a statistical view, we propose to generate the class-specific local prototypes to capture semantic-aware feature distribution. Considering that the spatial coverage and visual changes may vary dramatically across different classes, we extend the method introduced in [8] to allow learning different number of sub-clusters for different semantic classes. For detailed derivation of local prototype learning, please refer to [8]. Denote the learned local prototypes for class c as $\boldsymbol{P}_k^c = \{\boldsymbol{p}_k^{c,t}, t \in \{1, \cdots, T_c\}\}$ (where T_c refers to the number of sub-clusters for class c), and the pixel-to-local-prototype mapping as $\boldsymbol{M}_k^c = \{m_k^{c,i}, i = \{1, \cdots, N_k^c\}\}$, where $m_k^{c,i} \in \{1, \cdots, T_c\}$ represents the assigned sub-cluster index of pixel i in class c. Then the learned local prototypes are utilized to perform intra-client feature enhancement and inter-client feature alignment. Please note, T_c may vary for different semantic class to flexibly adapt to its visual characteristics.

Intra-client Local-Prototype Based Contrastive Learning (Intra-LPCL) for Feature Enhancement. The motivation of Intra-LPCL is to enhance the discriminability of local features. Specifically, given the bottleneck features $\boldsymbol{F}_k^c$, and the learned local prototypes $\{\boldsymbol{P}_k^c = \{\boldsymbol{p}_k^{c,t}, t \in \{1, \cdots, T_c\}\}, c \in \{1, \cdots, C\}\}$ from all the semantic classes, where C is the total class number. Then contrastive learning is introduced to enforce compactness within a sub-cluster and separation among different sub-clusters. Specifically, the intra-client pixel-to-local-prototype contrastive loss is calculated as,

$$L_k^c = \frac{1}{Z} \sum_{c=1}^{C} \sum_{i=1}^{N_k^c} -log \frac{e^{s_k^{c,m_k^{c,i}}}}{\sum_{c'=1}^{C} \sum_{t=1}^{T_{c'}} e^{s_k^{c',t}}}, \tag{1}$$

where $s_k^{c,m_k^{c,i}}$ denotes the similarity between the i-th local feature of class c (i.e., $\boldsymbol{f}_k^{c,i}$) and the local prototype from the sub-cluster that it belongs to (i.e., $\boldsymbol{p}_k^{c,m_k^{c,i}}$), and $s_k^{c',t}$ represents the similarity score between $\boldsymbol{f}_k^{c,i}$ and the local prototype from the t-th sub-cluster of class c', where $T_{c'}$ denotes the number of sub-clusters for class c', and Z is the normalization factor to average over all the pixels within a mini-batch. In our method, we adopt cosine similarity to get the similarity score,

$$s_k^{c',t} =< \boldsymbol{f}_k^{c,i}, \boldsymbol{p}_k^{c',t} >, \tag{2}$$

where $<,>$ denotes the cosine similarity function. Please note, visual compactness is only imposed at the sub-cluster granularity, which means the local features should distribute faraway from not only sub-clusters of the other semantic classes, but also other sub-clusters of the same semantic class. Apart from the contrastive loss term, in Intra-LPCL, we also explicitly maximize the feature similarities between local features and their assigned local prototypes as,

$$L_k^d = \frac{1}{Z} \sum_{c=1}^{C} \sum_{i=1}^{N_k^c} 1.0 - < \boldsymbol{f}_k^{c,i}, \boldsymbol{p}_k^{c,m_k^{c,i}} >, \tag{3}$$

Then the final Intra-LPCL loss is calculated as,

$$L_k^{intra} = L_k^c + L_k^d, \tag{4}$$

Inter-client Local-Prototype Based Contrastive Learning (Inter-LPCL) for Feature Alignment. The aim of Inter-LPCL is to perform distributed feature alignment across different clients in a privacy-preserving manner, such that the aggregated global model can generalize well across clients. Given the i-th local feature of class c from S_k (i.e., $\boldsymbol{f}_k^{c,i}$), and the prototypes pool $\{\boldsymbol{P}_{k'}^{c'} = \{\boldsymbol{p}_{k'}^{c',t}, t \in \{1, \cdots, T_{c'}\}\}, c' \in \{1, \cdots, C\}\}$ from $S_{k'}$ ($k' \neq k$), we don't know the cross-client pixel-to-prototype assignments. Thus, instead of imposing strict restrictions on sub-cluster compactness as done in Intra-LPCL, we loosen the alignment restrictions to category level. Specifically, the local features from S_k are supposed to distribute closer to one of the sub-clusters belonging to the same class in $S_{k'}$, and faraway from sub-clusters of the other semantic classes. Mathematically, the inter-LPCL loss is calcualted as,

$$L_{k,k'}^{inter} = \frac{1}{Z}\sum_{c=1}^{C}\sum_{i=1}^{N_k^c} -log\frac{e^{max(\{s_{k,k'}^{c,t},\, t\in\{1,\cdots,T_c\}\})}}{\sum_{c'=1}^{C} e^{max(\{s_{k,k'}^{c',t}, t\in\{1,\cdots,T_{c'}\}\})}}, \tag{5}$$

where $max(\cdot)$ returns the maximum value in the set, $\{s_{k,k'}^{c',t}, t \in \{1, \cdots, T_{c'}\}\}$ denotes the similarity set calculated between $\boldsymbol{f}_k^{c,i}$ and all the local prototypes from class c' in $S_{k'}$, which is formulated as,

$$s_{k,k'}^{c',t} =< \boldsymbol{f}_k^{c,i}, \boldsymbol{p}_{k'}^{c',t} >, \tag{6}$$

The final Inter-LPCL loss of S_k is then calculated by averaging over k' in $L_{k,k'}^{inter}$, which is formally defined as,

$$L_k^{inter} = \frac{1}{K-1}\sum_{k'\neq k} L_{k,k'}^{inter}, \tag{7}$$

Overall Objective for Local Update. The overall loss function for updating local model from S_k is formulated as,

$$L_k = L_k^{seg} + \lambda_1 L_k^{intra} + \lambda_2 L_k^{inter}, \tag{8}$$

where λ_1, λ_2 are the hyper-parameters, L_k^{seg} is the segmentation loss,

$$L_k^{seg} = \frac{1}{Z}\sum_n CE(\varphi_k(\boldsymbol{I}_k^n), \boldsymbol{Y}_k^n), \tag{9}$$

where $CE(\cdot)$ denotes the cross entropy loss.

2.2 Process-Aware Global Model Aggregation

During each federated communication, FedAvg updates the global model as weighted average over local models,

$$\boldsymbol{w} = \alpha_k \boldsymbol{w}_k, k \in \{1, \cdots, K\}, \tag{10}$$

Table 1. Dice similarity coefficients (DSC) of different settings to demonstrate effectiveness of each component in our method.

Intra-LPCL	Inter-LPCL	GA_p	S_1	S_2	S_3	S_4	S_5	S_6	Average
			87.5	83.0	88.9	80.1	90.0	76.3	84.3
✓			86.3	85.5	87.1	81.3	90.8	82.5	85.6
	✓		88.8	86.0	87.8	80.0	91.4	80.2	85.7
		✓	88.2	**86.3**	86.9	82.8	91.1	80.5	86.0
✓	✓	✓	**89.7**	85.0	**89.5**	**83.4**	**91.5**	**87.3**	**87.7**

where α_k is the aggregation weight for S_k, which is commonly set as $\frac{N_k}{\sum_k N_k}$ (N_k is the number of images in S_k). Instead of weighting the local models by its data amount ratio, in this paper, we argue that the aggregation weights should reflect the training process of each local model (i.e., well-trained model that generates good segmentation results should contribute more during aggregation). Specifically, denote the mean Dice Similarity Coefficient obtained on the training and validation data of S_k as DSC_k, then the normalized weights in our method are calculated as,

$$\alpha_k = \frac{DSC_k}{\sum_k DSC_k}. \tag{11}$$

By introducing the process-aware aggregation scheme, we can effectively detect the straggler, improving the robustness of aggregated global model.

3 Experiments and Results

Datasets and Implementation Details. We validate our method on the challenging task of prostate segmentation from 3D MR images. T2-weighted MRI images used in our study are collected from 6 different data sources [24–26], where each source is treated as a client in our study. We follow [28] to preprocess the data. For data augmentation, both geometric transformations (including elastic deformation, translation, rotation and scaling) and intensity augmentations (including contrast and gaussian noise) are employed in our method. The local model is trained using Adam optimizer with a batch size of 64 and Adam momentum of 0.9 and 0.999. The learning rate is initialzed as 0.001 and multiplied by 0.9 after each round of federated communication. The local epoch in each federated round is empirically set as 1. The hyper-parameters λ_1, λ_2 are empirically set as 0.03 and 0.001 respectively. The number of local prototypes for prostate and background are chosen by grid search, and empirically set as 3 and 6, respectively.

Ablation Study on the Effectiveness of Each Component: We denote the process-aware global aggregation as GA_p, then the detailed analysis on component effectiveness is presented in Table 1. One can see from this table that

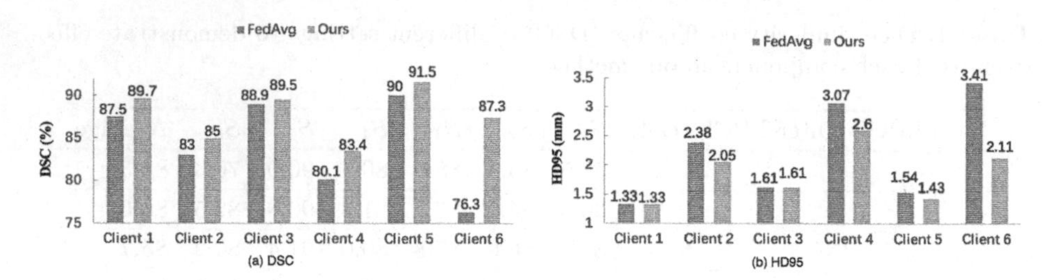

Fig. 2. Analysis on the straggler mitigation effect in terms of both DSC and 95% Hausdorff Distance (HD95) metrics.

incorporating the "Intra-LPCL" and "Inter-LPCL" terms can bring +1.3% and +1.4% overall DSC performance gains respectively, validating the effectiveness of intra-client discriminability enhancement and inter-client feature perception and alignment. Albeit its simplicity, our process-aware global aggregation scheme can also boost the DSC segmentation performance by a large margin (i.e., a 1.7% increase over the baseline). Combined together, the proposed FedContrast-GPA framework can witness a gain of 3.4% on the overall DSC performance.

To further demonstrate the advantage of our federated learning strategy, we conduct experiments to analyze the performance of centralized training and separate training, where centralized training is trained by updating the global model sequentially using private data from each client, while separate training is trained by updating each local client with only private data and no global communication. The average Dice performance for each client in centralized training is 73%, 77%, 84%, 72%, 86%, and 76%, respectively, while the average Dice performance for each client in separate training is 85%, 79%, 86%, 73%, 91%, and 27%, respectively. We can see that directly putting data together in centralized training does not bring performance gain due to data heterogeneity. Besides, in separate training, we can see a severe performance drop in some clients without enough learning data and knowledge from others.

Analysis on the Straggler Mitigation Effect. To demonstrate the effectiveness of our method in straggler mitigation, we compare the client-specific DSC and HD95 performance between the baseline (FedAvg) and Ours. For clarity, we first define the "stragglers" in a federated learning network as follows: the clients whose performance from the baseline (FedAvg) rank among the worst half of all the clients. Note that the "stragglers" are recognized according to the performance of FedAvg, since our method aims to address the "straggler" problem in FedAvg.

As shown in Fig. 2, for the prostate segmentation task, "Client 2", "Client 4" and "Client 6" are the "stragglers". Compared to the "non-stragglers", we can observe large performance gains on the "stragglers". In specific, the DSC performance gains on "Client 2", "Client 4" and "Client 6" are 2.0%, 3.3% and +11.0% (on average a 5.4% DSC gain), respectively. The HD95 performance

Table 2. Comparison with state-of-the-art methods in terms of DSC. Please note that larger DSC numbers indicate better performance. Best results are marked in bold.

Methods	S_1	S_2	S_3	S_4	S_5	S_6	Average
FedAvg	87.5	83.0	88.9	80.1	90.0	76.3	84.3
FedAvg-LG	83.8	82.4	85.3	75.9	89.9	56.3	78.9
FedDG	85.7	83.4	84.2	80.6	89.4	81.4	84.1
FedProx	88.1	84.3	86.2	83.1	90.6	83.1	85.9
MOON	87.0	84.9	87.7	82.7	90.3	83.0	85.9
Ours	**89.7**	**85.0**	**89.5**	**83.4**	**91.5**	**87.3**	**87.7**

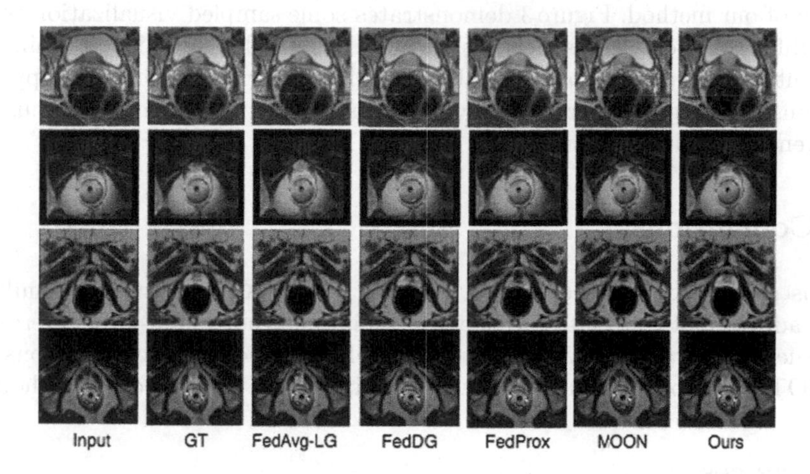

Fig. 3. Qualitative comparison results between our method and other state-of-the-arts on prostate segmentation.

gains are –0.33 mm, –0.47 mm, and –1.3 mm, respectively (on average –0.7 mm gain in terms of HD95). Meanwhile, for the "non-stragglers", the improvements are respectively 2.2%, 0.6% and 1.5% (on average 1.4%) in terms of DSC and –0.0 mm, –0.0 mm, and –0.11 mm (on average –0.04 mm) in terms of HD95. From above analysis, we can see that the proposed method can achieve effective straggler mitigation by bringing larger performance gains over stragglers, and slightly boost the performance over the 'non-stragglers".

Comparison with State-of-the-Art Methods. We compare the performance of our method with five state-of-the-art (SOTA) methods, including FedAvg [1], FedAvg-LG [27], FedDG [6], FedProx [2] and MOON [4]. For fair comparison, all the SOTA methods were trained/tested on our own dataset splits. The base parameter settings are kept the same as ours, other hyper-parameters are chosen by grid-search (In FedAvg-LG, the number of layers for global aggregation is set as 13, the hyper-parameters in FedDG are the same as the original paper, the weight for proxy term in FedProx is 2.5e-4, and the model contrastive coefficient in MOON is 0.01). In the following, we conduct both quantitative and qualitative comparisons with SOTA methods.

To analyse the performance of our proposed FedContrast-GPA framework, we report the DSCs for all the distributed clients (i.e., $S_1 - S_6$ in Table 2). For a straightforward comparison with the SOTA methods, we also record the average DSC across all the clients. Detailed comparison results are illustrated in Table 2. As shown, overall, our proposed FedContrast-GPA framework achieves superior performance than the listed SOTA methods. Specifically, FedContrast-GPA outperforms the second best method by 1.8% in terms of average DSC, and generates the best DSC performance at each client, demonstrating favorable generalization ability of our method. Figure 3 demonstrates some sampled visualization results from different clients. As shown, listed SOTA methods may fail to obtain good segmentation results on some samples from different clients, while our approach can consistently generate reasonably good results, demonstrating the robustness and generalizability of our method.

4 Conclusions

In this paper, we proposed a novel FedContrast-GPA framework to simultaneously address both the data-level heterogeneity and the system-level heterogeneity issues in federated networks. Extensive ablation studies and comparisons with the SOTA methods demonstrated the effectiveness of the proposed method.

References

1. McMahan, B., Moore, E., Ramage, D., Hampson, S., y Arcas, B.A.: Communication-efficient learning of deep networks from decentralized data. In: Artificial Intelligence and Statistics, pp. 1273–1282. PMLR (2017)
2. Li, T., Sahu, A.K., Zaheer, M., Sanjabi, M., Talwalkar, A., Smith, V.: Federated optimization in heterogeneous networks. Proc. Mach. Learn. Syst. **2**, 429–450 (2020)
3. Karimireddy, S.P., Kale, S., Mohri, M., Reddi, S., Stich, S., Suresh, A.T.: SCAFFOLD: stochastic controlled averaging for federated learning. In: International Conference on Machine Learning, pp. 5132–5143. PMLR (2020)
4. Li, Q., He, B., Song, D.: Model-contrastive federated learning. In: Proceedings of the IEEE/CVF Conference on Computer Vision and Pattern Recognition, pp. 10713–10722) (2021)
5. Mendieta, M., Yang, T., Wang, P., Lee, M., Ding, Z., Chen, C.: Local learning matters: rethinking data-level heterogeneity in federated learning. In: Proceedings of the IEEE/CVF Conference on Computer Vision and Pattern Recognition, pp. 8397–8406 (2022)
6. Liu, Q., Chen, C., Qin, J., Dou, Q., Heng, P.A.: FedDG: federated domain generalization on medical image segmentation via episodic learning in continuous frequency space. In: Proceedings of the IEEE/CVF Conference on Computer Vision and Pattern Recognition, pp. 1013–1023 (2021)

7. Ouyang, C., Biffi, C., Chen, C., Kart, T., Qiu, H., Rueckert, D.: Self-supervision with Superpixels: training few-shot medical image segmentation without annotation. In: Vedaldi, A., Bischof, H., Brox, T., Frahm, J.-M. (eds.) ECCV 2020. LNCS, vol. 12374, pp. 762–780. Springer, Cham (2020). https://doi.org/10.1007/978-3-030-58526-6_45
8. Zhou, T., Wang, W., Konukoglu, E., Van Gool, L.: Rethinking semantic segmentation: a prototype view. In: Proceedings of the IEEE/CVF Conference on Computer Vision and Pattern Recognition, pp. 2582–2593 (2022)
9. Chen, T., Kornblith, S., Norouzi, M., Hinton, G.: A simple framework for contrastive learning of visual representations. In: International Conference on Machine Learning, pp. 1597–1607. PMLR (2020)
10. Wang, H., Yurochkin, M., Sun, Y., Papailiopoulos, D., Khazaeni, Y.: Federated learning with matched averaging. arXiv preprint arXiv:2002.06440 (2020)
11. Hsu, T.M.H., Qi, H., Brown, M.: Measuring the effects of non-identical data distribution for federated visual classification. arXiv preprint arXiv:1909.06335 (2019)
12. Wang, J., Liu, Q., Liang, H., Joshi, G., Poor, H.V.: Tackling the objective inconsistency problem in heterogeneous federated optimization. Adv. Neural Inf. Process. Syst. **33**, 7611–7623 (2020)
13. Wu, Y., Zeng, D., Wang, Z., Shi, Y., Hu, J.: Federated contrastive learning for volumetric medical image segmentation. In: de Bruijne, M., et al. (eds.) MICCAI 2021. LNCS, vol. 12903, pp. 367–377. Springer, Cham (2021). https://doi.org/10.1007/978-3-030-87199-4_35
14. Dong, N., Xing, E.P.: Few-shot semantic segmentation with prototype learning. In: British Machine Vision Conference (BMVC). vol. 3, no. 4 (2018)
15. Liu, J., Qin, Y.: Prototype refinement network for few-shot segmentation. arXiv preprint arXiv:2002.03579 (2020)
16. Liu, Y., Zhang, X., Zhang, S., He, X.: Part-aware prototype network for few-shot semantic segmentation. In: Vedaldi, A., Bischof, H., Brox, T., Frahm, J.-M. (eds.) ECCV 2020. LNCS, vol. 12354, pp. 142–158. Springer, Cham (2020). https://doi.org/10.1007/978-3-030-58545-7_9
17. Yu, Q., Dang, K., Tajbakhsh, N., Terzopoulos, D., Ding, X.: A location-sensitive local prototype network for few-shot medical image segmentation. In: IEEE 18th International Symposium on Biomedical Imaging (ISBI), pp. 262–266. IEEE (2021)
18. Jaiswal, A., Babu, A.R., Zadeh, M.Z., Banerjee, D., Makedon, F.: A survey on contrastive self-supervised learning. Technologies **9**(1), 2 (2020)
19. Liu, W., Wu, Z., Ding, H., Liu, F., Lin, J. and Lin, G.: Few-shot segmentation with global and local contrastive learning. arXiv preprint arXiv:2108.05293 (2021)
20. Chaitanya, K., Erdil, E., Karani, N., Konukoglu, E.: Contrastive learning of global and local features for medical image segmentation with limited annotations. Adv. Neural Inf. Process. Syst. **33**, 12546–12558 (2020)
21. Zeng, D., et al.: Positional contrastive learning for volumetric medical image segmentation. In: International Conference on Medical Image Computing and Computer-Assisted Intervention, pp. 221–230. Springer (2021)
22. Wu, Y., Zeng, D., Wang, Z., Shi, Y., Hu, J.: Distributed contrastive learning for medical image segmentation. Med. Image Anal. **81**, 102564 (2022)
23. Ronneberger, O., Fischer, P., Brox, T.: U-Net: convolutional networks for biomedical image segmentation. In: Navab, N., Hornegger, J., Wells, W.M., Frangi, A.F. (eds.) MICCAI 2015. LNCS, vol. 9351, pp. 234–241. Springer, Cham (2015). https://doi.org/10.1007/978-3-319-24574-4_28
24. Bloch, N., Madabhushi, A., Huisman, H., Freymann, J., et al.: NCI-ISBI 2013 Challenge: Automated Segmentation of Prostate Structures (2015)

25. Lemaitre, G., Marti, R., Freixenet, J., Vilanova. J. C., et al.: Computer-aided detection and diagnosis for prostate cancer based on mono and multi-parametric MRI: a review. In: Computers in Biology and Medicine. vol. 60, pp. 8–31 (2015)
26. Litjens, G., Toth, R., Ven, W., Hoeks, C., et al.: Evaluation of prostate segmentation algorithms for MRI: the promise12 challenge. In: Medical Image Analysis. vol. 18, pp. 359–373 (2014)
27. Liang, P.P., et al.: Think locally, act globally: federated learning with local and global representations. In: Workshop on Federated Learning at Advances in Neural Information Processing Systems. vol. 32 (2019)
28. Liu, Q., Dou, Q., Yu, L., Heng, P.A.: MS-Net: multi-site network for improving prostate segmentation with heterogeneous MRI data. IEEE Trans. Med. Imaging **39**(9), 2713–2724 (2020)
29. Patro, S., Sahu, K.K.: Normalization: A preprocessing stage. arXiv preprint arXiv:1503.06462 (2015)

Partially Supervised Multi-organ Segmentation via Affinity-Aware Consistency Learning and Cross Site Feature Alignment

Qin Zhou, Peng Liu, and Guoyan Zheng(✉)

Institute of Medical Robotics, School of Biomedical Engineering, Shanghai Jiao Tong University, No. 800, Dongchuan Road, Shanghai 200240, China
guoyan.zheng@sjtu.edu.cn

Abstract. Partially Supervised Multi-Organ Segmentation (PSMOS) has attracted increasing attention. However, facing with challenges from lacking sufficiently labeled data and cross-site data discrepancy, PSMOS remains largely an unsolved problem. In this paper, to fully take advantage of the unlabeled data, we propose to incorporate voxel-to-organ affinity in embedding space into a consistency learning framework, ensuring consistency in both label space and latent feature space. Furthermore, to mitigate the cross-site data discrepancy, we propose to propagate the organ-specific feature centers and inter-organ affinity relationships across different sites, calibrating the multi-site feature distribution from a statistical perspective. Extensive experiments manifest that our method generates favorable results compared with other state-of-the-art methods, especially on hard organs with relatively smaller sizes.

Keywords: Multi-organ Segmentation · Partially Supervised · Affinity Relationship · Consistency Learning

1 Introduction

Automatic multi-organ segmentation (MOS) plays a vital role in computer-aided diagnosis and treatment planning. Recently, deep learning based methods have made remarkable progress in solving MOS tasks. However, they typically require a large amount of expert-level accurate, densely-annotated data for training, which is laborious and time consuming to collect. Therefore, existing fully labeled datasets (termed as FLDs) are very few and often low in sample size [1]. While there exist many publicly available partially labeled datasets (PLDs) [2,3], each with one or a few out of the many organs annotated. This has motivated the development of various Partially-Supervised Multi-Organ Segmentation (PSMOS) methods that aim to learn a unified model from a union of such datasets. For example, Dmitriev and Kaufman proposed the conditional

This study was partially supported by the National Natural Science Foundation of China via projects U20A20199 and 62201341.

H. Greenspan et al. (Eds.): MICCAI 2023, LNCS 14221, pp. 671–680, 2023.
https://doi.org/10.1007/978-3-031-43895-0_63

U-Net to enable PSMOS using a single unified network [4]. Co-training between two models with consistency constraints on soft pseudo labels [6], and multi-scale features learned in a pyramid-input and pyramid-output network [7] were both explored for PSMOS. Other researchers resorted to prior knowledge to guide the training process. In PaNN [8], the average organ size distributions on the PLDs were constrained to resemble the prior statistics obtained from the FLD. Another method was introduced in [9] where the non-overlapping characteristics between different organs were exploited to design the exclusion loss.

Although witnessed great progress in PSMOS, existing methods are faced with the following challenges: 1) Shortage in sufficiently labeled samples for supervised learning, since voxel-level labels are only available for a subset of organs in PLDs; 2) Significant cross-site appearance variations caused by different imaging protocols or subject cohorts. Different from existing methods, we propose a novel framework to explicitly tackle the above-mentioned challenges.

To handle the label-scarcity problem in PLDs, we propose a novel Affinity-aware Consistency Learning (ACL) scheme to incorporate voxel-to-organ affinity in the embedding space into consistency learning. Although consistency learning is frequently used for leveraging unlabeled data in label-efficient learning [10–12], it is mostly deployed in the label space [13–15], while little attention has been paid to exploring consistency in the latent feature space. Zheng et al. [16] proposed to adopt auxiliary student-teacher networks to utilize the features for consistency learning, which introduced more parameters, thus were computationally expensive. By incorporating voxel-to-organ affinity in the embedding space into consistency learning, our ACL scheme is plug-and-play and can capture rich context information in the embedding space.

To tackle the data discrepancy problem [17], based on the assumption that a well trained joint model should generate consistent feature distributions across different sites, we propose a novel Cross-Site Feature Alignment (CSFA) module, where two terms are introduced to attend to both the organ-specific and inter-organ statistics in the latent feature space. Concretely, for each PLD, we restrain the organ-specific prototypes calculated in each mini-batch to be close to the corresponding prototypes generated on the small-sized FLD. To further reduce the data discrepancy problem, we constrain the affinity relationships across different organ-specific prototypes to be consistent among different sites. By doing this, we transfer not only the single-class centroid, but also the inter-organ affinity learned from the small-sized FLD to PLDs, allowing for knowledge propagation at multiple granularity levels. Our contributions can be summarized as follows:

- We propose a novel affinity-aware consistency learning scheme to incorporate voxel-to-organ affinity in the embedding space into a consistency learning framework, which can capture semantic context in the latent feature space.
- We design a novel cross site feature alignment module to calibrate feature distributions of PLDs with distribution priors learned from a small-sized FLD, alleviating the cross-site data discrepancy.
- We demonstrate on five datasets collected from different sites that our method can effectively learn a unified MOS model from multi-source datasets, achieving superior performance over the state-of-the-art (SOTA) methods.

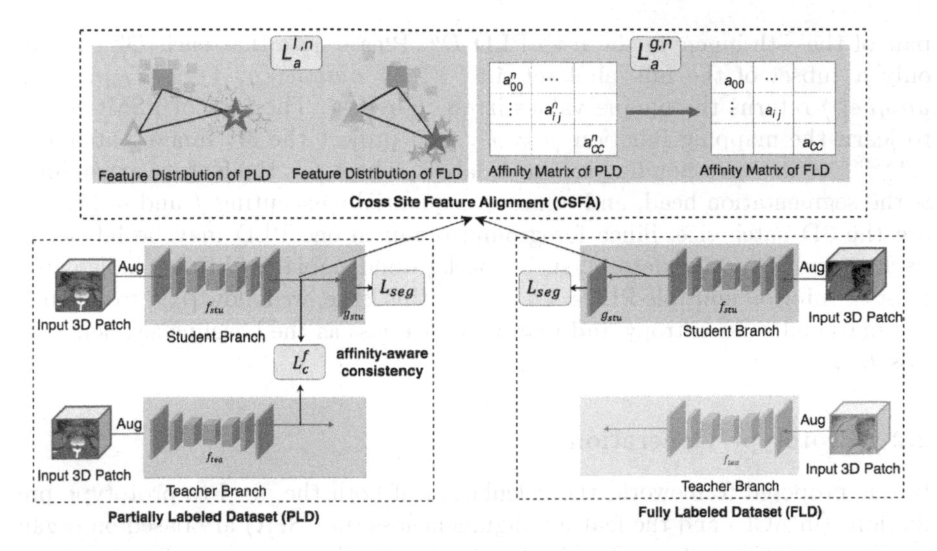

Fig. 1. A schematic illustration of our framework. "Aug" refers to perturbations with data augmentations. In the CSFA module, hollow shapes refer to the features belonging to unlabeled organs in the PLDs, while solid ones refer to labeled organs. The affinity matrix is calculated according to Eq. 10 and Eq. 11. L_{seg} is the segmentation loss.

2 Methodology

To learn a unified model from a small-sized FLD and a number of PLDs, we propose a novel framework to address the issues of label-scarcity and cross-site data discrepancy. The overall workflow of our method is presented in Fig. 1. During training, in each batch, we sample 3D patches from both the FLD and one of the PLDs, where the teacher-student scheme [14] is adopted to impose consistency constraints on the unlabeled voxels of the PLD. In our method, apart from the label space consistency, we introduce the ACL scheme to explore consistency in the embedding space. We further leverage the CSFA module to perform feature alignment between the FLD and the PLD. Please note that consistency constraints are only imposed on the unlabeled voxels of PLDs. The label space consistency loss is omitted in Fig. 1 for brevity.

2.1 Preliminaries

Denote Y_{full} as the full label set, i.e., $Y_{full} = \{0, 1, 2, \cdots, C\}$, where 0 refers to the background class, and $\{1, \cdots, C\}$ are one-to-one mappings to the target organs, C is the number of target organs. Given a small-sized FLD D_f and a number of PLDs $D_p = \{D_p^n, n \in \{1, \cdots, N\}\}$, where N is the number of PLDs. Each dataset can then be formally defined as either $D_f = \{I_{j,i}^f, y_{j,i}^f\}$ or $D_p^n = \{I_{j,i}^n, y_{j,i}^n\}$, where $I_{j,i}^f$ is the i-th pixel of the j-th image in the FLD D_f, and $y_{j,i}^f$ is its corresponding label. Similarly, $(I_{j,i}^n, y_{j,i}^n)$ is the i-th pixel-label pair of the j-th image in the n-th PLD D_p^n. Please note that each D_p^n contains only a subset of the full label set, i.e., $Y_p^n = unique(\{y_{j,i}^n\}) \subsetneqq Y_{full}$, where $unique(\cdot)$ returns the unique values in the label set. The task of PSMOS aims to learn the mapping function $\varphi = f \circ g$ to project the 3D image patch $I_j \in \mathbb{R}^{h \times w \times z}$ to its corresponding semantic labels, where f is the feature extractor, g is the segmentation head, and $\circ$ means sequentially executing f and g, (h, w, z) are the 3D patch size. Since foreground organ in one PLD may be labeled as background in another dataset, such a background ambiguity brings challenges to joint training on multiple PLDs. To address this issue, we follow [7,9] to calculate the marginal cross entropy and marginal Dice loss as the baseline segmentation loss L_{seg}.

2.2 Prototype Generation

In our proposed framework, the calculation of both the pixel-to-prototype predictions (in ACL) and the feature alignment loss (in CSFA) are based on organ-specific prototypes. In each mini-batch, denote the organ-specific prototypes for the FLD as $\{\boldsymbol{q}_c\}, c \in \{0, \cdots, C\}$ and prototypes for the n-th PLD as $\{\boldsymbol{q}_c^n\}, c \in \{0, \cdots, C\}$, then they are generated as follows. On the FLD, we generate the prototypes in an exponential moving average scheme. Specifically, the feature prototype of the t-th iteration is calculated as (for brevity, we omit the iteration superscript t),

$$\boldsymbol{q}_c = \alpha \boldsymbol{q}_c + (1 - \alpha) \boldsymbol{q}_c^{update}, c \in \{0, \cdots, C\}, \tag{1}$$

where $\boldsymbol{q}_c^{update}$ is the average feature of the c-th class in current mini-batch of the FLD and α is the weighting coefficient. Given the feature maps $\boldsymbol{F} = \{\boldsymbol{f}_i\}$ and their related labels $\{y_i\}$, where $\boldsymbol{f}_i$ represents the i-th pixel in the feature maps of current mini-batch, the feature center of the c-th class is then calculated as,

$$\boldsymbol{q}_c^{update} = \frac{1}{Z_c} \sum_{i, y_i = c} \boldsymbol{f}_i, c \in \{0, \cdots, C\}, \tag{2}$$

where Z_c is the number of pixels belonging to the c-th class in current mini-batch.

On the n-th PLD, we directly adopt the feature centers calculated in each mini-batch as the organ-specific prototypes. In specific, for the labeled organs, the prototypes $\{\boldsymbol{q}_c^n\}, c \in Y_p^n$ are calculated according to Eq. 2, with feature maps generated on 3D patches from the n-th PLD. While on the unlabeled organs, only reliable features are used for calculating the pseudo feature centers as,

$$\boldsymbol{q}_c^n = \frac{1}{Z_c} \sum_{i, y_i = c} \mathbb{1}[p_i^n > \tau] \boldsymbol{f}_i, c \notin Y_p^n, \tag{3}$$

where p_i^n is the normalized prediction score generated from the teacher model, y_i denotes the corresponding pseudo label, τ is the confidence threshold, Z_c is the number of reliable predictions in class c, and $\mathbb{1}[\cdot]$ returns 1 if the inside condition is True, otherwise, returns 0.

2.3 Affinity-Aware Consistency Learning

In this paper, we propose to incorporate the voxel-to-organ affinity into consistency learning. Specifically, instance-to-prototype matching is calculated to capture the voxel-to-organ affinity. The affinities are then transformed into normalized scores for calculating the consistency constraint on two perturbed inputs. We adopt the teacher-student scheme [14] for consistency learning on the unlabeled data. Formally, denote I_t, I_s as the perturbed versions of the same sampled 3D patch for the teacher branch and the student branch respectively. In the teacher branch, denote $\phi_i = f_{tea}(I_{t,i}) \in \mathbb{R}^d$ as the extracted feature for the i-th pixel of 3D image patch I_t. Given the prototypes generated on the FLD $\{\boldsymbol{q}_c\}, c \in \{0, \cdots, C\}$, then the pixel-to-prototype classification logit $\boldsymbol{p}_{t,i} = \{p_{t,i}^c\}$ is calculated as,

$$p_{t,i}^c =< \phi_i, \boldsymbol{q}_c >, \ c \in \{0, \cdots, C\} \tag{4}$$

where $< \cdot, \cdot >$ calculates the cosine similarity between the two terms.

Similarly, in the student branch, denote ψ_i as the i-th feature $\psi_i = f_{stu}(I_{s,i}) \in \mathbb{R}^d$, then prototype based predictions $\boldsymbol{p}_{s,i} = \{p_{s,i}^c\}$ can be obtained as,

$$p_{s,i}^c =< \psi_i, \boldsymbol{q}_c >, c \in \{0, \cdots, C\}, \tag{5}$$

Since $\boldsymbol{p}_{t,i}, \boldsymbol{p}_{s,i}$ model the voxel-to-organ affinities in the embedding space, constraining consistency on them introduces rich context information for training on the unlabeled data, which is formulated as,

$$L_c^f = \frac{1}{Z_c^f} \sum_i KL(\boldsymbol{p}_{s,i}, \boldsymbol{p}_{t,i}), \tag{6}$$

where $\frac{1}{Z_c^f}$ is the normalization factor to get the mean KL-Divergence in the feature embedding space. Denote $\varphi_{tea} = f_{tea} \circ g_{tea}$, $\varphi_{stu} = f_{stu} \circ g_{stu}$ as the teacher and student segmentation model respectively, the logits from the student and the teacher branch can be calculated as $\boldsymbol{l}_{s,i} = \varphi_{stu}(I_{s,i}), \boldsymbol{l}_{t,i} = \varphi_{tea}(I_{t,i})$. Then the consistency loss in the label space is calculated as,

$$L_c^l = \frac{1}{Z_c^l} \sum_i KL(\boldsymbol{l}_{s,i}, \boldsymbol{l}_{t,i}), \tag{7}$$

where $\frac{1}{Z_c^l}$ is the normalization factor. The overall affinity-aware consistency loss is finally formulated as,

$$L_c = L_c^f + L_c^l, \tag{8}$$

2.4 Cross-Site Feature Alignment (CSFA) Module

The CSFA module is proposed to calibrate feature distributions across different sites. Specifically, given the learned prototypes from current mini-batch of the n-th PLD ($\{\boldsymbol{q}_c^n\}, c \in \{0, \cdots, C\}$), they can be regarded as the organ-specific cluster centers in the embedding space. Then, compactness loss is introduced to calibrate D_p^n with the cluster centers learned from the FLD as,

$$L_a^{l,n} = \frac{1}{|Y_p^n|} \sum_{c \in Y_p^n} ||\boldsymbol{q}_c^n - \boldsymbol{q}_c||_2^2, \tag{9}$$

where $|Y_p^n|$ returns the number of labeled organs in D_p^n.

To further take into consideration the inter-organ affinity relationships during feature distribution alignment, we first model inter-organ affinity relationships on the FLD by calculating the affinity matrix $\boldsymbol{A} = \{a_{ij}\} \in \mathbb{R}^{(C+1)\times(C+1)}$ as shown in Fig. 1,

$$a_{ij} =< \boldsymbol{q}_i, \boldsymbol{q}_j >, i \in \{0, \cdots, C\},\ j \in \{0, \cdots, C\}, \tag{10}$$

Similarly, we can obtain the affinity matrix $\boldsymbol{A}_p^n = \{a_{ij}^n\} \in \mathbb{R}^{(C+1)\times(C+1)}$ on partially labeled dataset D_p^n as,

$$a_{ij}^n =< \boldsymbol{q}_i^n, \boldsymbol{q}_j^n >, i \in \{0, \cdots, C\},\ j \in \{0, \cdots, C\}, \tag{11}$$

Then the affinity relationship aware feature alignment loss is calculated as,

$$L_a^{g,n} = \frac{1}{C+1} \sum_c KL(\boldsymbol{a}_c, \boldsymbol{a}_{p,c}^n), \tag{12}$$

where $\boldsymbol{a}_c$, $\boldsymbol{a}_{p,c}^n$ refer to the c-th row of $\boldsymbol{A}$ and $\boldsymbol{A}_p^n$ respectively.

The overall cross-site alignment loss is then calculated as the sum of the compactness loss and the affinity relationship aware calibration loss,

$$L_a = L_a^{l,n} + L_a^{g,n}, \tag{13}$$

The overall training objective is finally formulated as,

$$L = L_{seg} + L_c + \lambda_a L_a. \tag{14}$$

where λ_a is the tradeoff parameter.

3 Experiments and Results

Datasets and Implementation Details. We use five abdominal CT datasets (MALBCVWC [1], Decathlon Spleen [3], KiTS [2], Decathlon Liver [3] and Decathlon Pancreas [3] datasets respectively) to evaluate the effectiveness of our method [1–3]. The spatial resolution of all these datasets are resampled to $(1 \times 1 \times 3)mm^3$. We randomly split each dataset into training (60%), validation (20%) and testing (20%). We adopt 3D U-Net [18] as our backbone model. The patch size (h, w, z) is set to $(160, 160, 96)$. The hyper-parameters α and τ are empirically set to 0.9, and 0.8, respectively. λ_a is initialized as 0.01 and linearly decreased to $1e-3$ at 20000 iterations. We use SGD optimizer to train the model and the initial learning rate is set to 0.01. We adopt Dice similarity coefficient (DSC) as metric to evaluate the performance of different methods.

Table 1. Results of the ablation study on the effectiveness of each component in our method (Metric: DSC (%)).

Settings	Liver	Spleen	Pancreas	RK	LK	Overall
baseline	94.7	91.9	77.5	94.1	93.5	90.3
baseline + ACL	94.5	**94.1**	77.4	95.5	95.3	91.4
baseline + ACL + CSFA	**95.2**	93.8	**79.0**	**96.0**	**95.5**	**91.9**

Table 2. Analysis on the effectiveness of CSFA in mitigating cross-site data discrepancy. Please note D_0 - D_4 refer to the MALBCVWC [1], Decathlon-Spleen [3], KiTS [2], Decathlon-Liver [3] and Decathlon-Pancreas [3] datasets respectively, where D_0 is the FLD, and others are PLDs. Please note small MMD indicates small data discrepancy.

Settings	D_0 vs D_1	D_0 vs D_2	D_0 vs D_3	D_0 vs D_4	Overall
Ours wo/CSFA	0.3030	0.3187	0.2818	0.3577	0.3153
Ours w/CSFA	0.1589	0.2036	0.1358	0.2925	0.1977

Ablation Study. In this subsection, we carry out experiments to investigate effectiveness of each component in the proposed framework. Concretely, the baseline results are trained with only the L_{seg} loss. In Table 1, the "baseline+ACL" setting reports the results with our proposed affinity-aware consistency learning scheme. Comparing to the baseline, it brings a 1.1% performance gain in terms of DSC. By introducing the CSFA module, the "baseline+ACL+CSFA" setting can further boost the performance by 0.5% in terms of DSC.

We further study the effectiveness of the CSFA module in alleviating cross-site data discrepancy. Concretely, we measure the feature distribution discrepancy between the FLD and each PLD by calculating the Maximum Mean Discrepancy (MMD) using gaussian kernel [19], which was designed to quantify domain discrepancy. We conduct "full vs partial" MMD analysis on the following two settings: "Ours w/CSFA" and "Ours wo/CSFA", where "Ours w/CSFA" is the proposed framework, while "Ours wo/CSFA" setting refers to removing the CSFA module from our framework. In the MMD calculation, for each dataset, we first generate features from the penultimate layer. Then we randomly select 2000 features in each class for MMD calculation. Please note, for each PLD, we adopt the pseudo labels for feature selection. Detailed comparison results are illustrated in Table 2. As shown, by introducing the CSFA module, the feature distribution discrepancy in terms of MMD can be effectively alleviated across all the "full vs partial" dataset pairs.

Comparison with the State-of-the-Art (SOTA) Methods. We compare with four SOTA methods, including PaNN [8], PIPO [7], Marginal Loss [9], and DoDNet [5]. For fair comparison, all the SOTA methods were trained/tested on our own dataset splits. We also implemented our method taking the nnUNet as the backbone to compare with Marginal Loss [9] and PaNN [8]. We reported the

Table 3. Comparison with state-of-the-art methods in terms of DSC. "RK", "LK" refer to "Right Kidney" and "Left Kidney" respectively.

Methods	backbone	Liver	Spleen	Pancreas	RK	LK	Overall	Avg_{hard}
PIPO [7]	3D-UNet	93.01	93.63	76.51	93.50	89.98	89.3	86.7
DoDNet [5]	3D-UNet	95.41	95.09	70.01	94.06	92.00	89.3	85.4
Marginal Loss [9]	nnUNet	95.45	94.88	77.91	94.14	91.52	90.8	87.9
PaNN [8]	nnUNet	95.13	95.14	78.88	96.21	91.02	91.3	88.7
Ours	3D-UNet	95.2	93.82	79.03	96.04	95.49	91.9	90.2
Ours	nnUNet	95.8	95.0	83.1	94.7	93.8	92.5	90.5

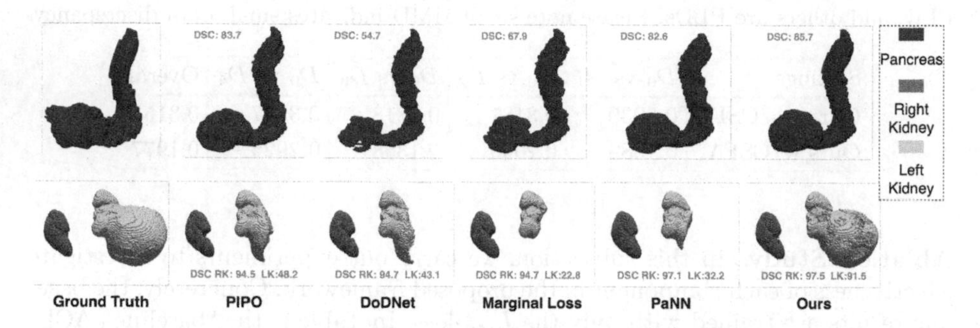

Fig. 2. 3D visualized results of some hard samples.

DSC values for each organ across test sets from all the datasets. For a straightforward comparison with the SOTA, we also recorded the average DSC over all the organs. Detailed results are illustrated in Table 3. As shown, our method achieves the best performance. Specifically, our method outperforms the second-best method PaNN [8] with a 1.2% DSC gain using the same nnUNet backbone. And our method when taking 3D-UNet as the backbone also outperforms the listed SOTA methods. We further conduct paired t-test to compare the difference between ours and other SOTA methods, the p-values are 2E-8 (PIPO), 2E-5 (DoDNet), 2E-4 (Marginal Loss), 0.037 (PaNN), respectively. As all p-values are smaller than 0.05, the differences between ours and other SOTA methods are statistically significant.

In practice, some organs are much harder to be well-segmented than others due to their relatively small organ sizes. Therefore, we pay more attention to the performance on those hard organs (in our datasets, Pancreas and Kidneys are deemed to be more difficult due to their relatively small sizes). From the last column of Table 3, we can see that the segmentation performance gains of our method are more pronounced on hard organs (on average a 1.8% DSC gain). Figure 2 demonstrates the qualitative visualization results on some hard samples. As shown in this figure, our method can generate better segmentation results than other SOTA methods. Besides, the reasonable performance on segmenting kidney with tumors (row 2 in Fig. 2) makes our method promising in clinical practice.

4 Conclusion

In this paper, we designed a novel Affinity-aware Consistency Learning scheme (ACL) to model voxel-to-organ affinity context in the feature embedding space into consistency learning. Meanwhile, the CSFA module was designed to perform feature distribution alignment across different sites, where both organ-specific cluster centers and the inter-organ affinity relationships were propagated from the small-sized FLD to PLDs for cross-site feature alignment. Extensive ablation studies validated effectiveness of each component in our method. Quantitative and Qualitative comparison results with other SOTA methods demonstrated superior performance of our method.

References

1. Bennett, L., Xu, Z., Igelsias, J., Styner, M., Langerak, T., Klein, A.: Miccai multi-atlas labeling beyond the cranial vault-workshop and challenge. In: Proceedings of MICCAI Multi-Atlas Labeling Beyond Cranial Vault-Workshop Challenge. vol. 5, pp. 12 (2015)
2. Nicholas, H., et al.: The kits19 challenge data: 300 kidney tumor cases with clinical context, CT semantic segmentations, and surgical outcomes. arXiv preprint arXiv:1904.00445 (2019)
3. Amber L.S., et al.: A large annotated medical image dataset for the development and evaluation of segmentation algorithms. arXiv preprint arXiv:1902.09063 (2019)
4. Konstantin, D., Kaufman, A.E.: Learning multi-class segmentations from single-class datasets. In: Proceedings of the IEEE/CVF Conference on Computer Vision and Pattern Recognition, pp. 9501–9511 (2019)
5. Zhang, J., Xie, Y., Xia, Y., Shen, C.: DoDNet: learning to segment multi-organ and tumors from multiple partially labeled datasets. In: Proceedings of the IEEE/CVF Conference on Computer Vision and Pattern Recognition, pp. 1195–1204 (2021)
6. Huang, R., Zheng, Y., Hu, Z., Zhang, S., Li, H.: Multi-organ segmentation via co-training weight-averaged models from few-organ datasets. In: Martel, A.L., et al. (eds.) MICCAI 2020. LNCS, vol. 12264, pp. 146–155. Springer, Cham (2020). https://doi.org/10.1007/978-3-030-59719-1_15
7. Xi, F., Yan, P.: Multi-organ segmentation over partially labeled datasets with multi-scale feature abstraction. IEEE Trans. Med. Imaging **39**(11), 3619–3629 (2020)
8. Zhou, Y., et al.: Prior-aware neural network for partially-supervised multi-organ segmentation. In: Proceedings of the IEEE/CVF International Conference on Computer Vision, pp. 10672–10681 (2019)
9. Shi, G., Xiao, L., Chen, Y., Zhou, S.K.: Marginal loss and exclusion loss for partially supervised multi-organ segmentation. Med. Image Anal. **70**, 101979 (2021)
10. Jisoo, J., Lee, S., Kim, J., Kwak, N.: Consistency-based semi-supervised learning for object detection. In: Advances in Neural Information Processing Systems. vol. 32 (2019)

11. Abuduweili, A., Li, X., Shi, H., Xu, C.Z., Dou, D.: Adaptive consistency regularization for semi-supervised transfer learning. In: Proceedings of the IEEE/CVF Conference on Computer Vision and Pattern Recognition, pp. 6923–6932 (2021)
12. Ouali, Y., Hudelot, C., Tami, M. Semi-supervised semantic segmentation with cross-consistency training. In: Proceedings of the IEEE/CVF Conference on Computer Vision and Pattern Recognition, pp. 12674–12684 (2020)
13. Samuli, L., Aila, T.: Temporal ensembling for semi-supervised learning. In: International Conference on Learning Representations (ICLR) (2017)
14. Antti, T., Valpola, H.: Mean teachers are better role models: weight-averaged consistency targets improve semi-supervised deep learning results. In: Advances in Neural Information Processing Systems. vol. 30 (2017)
15. French, G., Laine, S., Aila, T., Mackiewicz, M., Finlayson, G.: Semi-supervised semantic segmentation needs strong, varied perturbations. In: British Machine Vision Conference (2020)
16. Zheng, K., Xu, J., Wei, J.: Double noise mean teacher self-Ensembling model for semi-supervised tumor segmentation. In: IEEE International Conference on Acoustics, Speech and Signal Processing (ICASSP), pp. 1446–1450 (2022)
17. Bento, M., Fantini, I., Park, J., Rittner, L., Frayne, R.: Deep learning in large and multi-site structural brain MR imaging datasets. Front. Neuroinformatics **15**(82), 805669 (2022)
18. Çiçek, Ö., Abdulkadir, A., Lienkamp, S. S., Brox, T., Ronneberger, O.: 3D-UNet: learning dense volumetric segmentation from sparse annotation. In: Medical Image Computing and Computer-Assisted Intervention-MICCAI, pp. 424–432 (2016)
19. Gretton, A., Borgwardt, K.M., Rasch, M.J., Schölkopf, B., Smola, A.: A kernel two-sample test. J. Mach. Learn. Res. **13**(1), 723–773 (2012)

Attentive Deep Canonical Correlation Analysis for Diagnosing Alzheimer's Disease Using Multimodal Imaging Genetics

Rong Zhou[1], Houliang Zhou[1], Brian Y. Chen[1], Li Shen[2], Yu Zhang[3], and Lifang He[1(✉)]

[1] Department of Computer Science and Engineering, Lehigh University, Bethlehem, PA, USA
lih319@lehigh.edu

[2] Department of Biostatistics, Epidemiology and Informatics, University of Pennsylvania, Philadelphia, PA, USA

[3] Department of Bioengineering, Lehigh University, Bethlehem, PA, USA

Abstract. Integration of imaging genetics data provides unprecedented opportunities for revealing biological mechanisms underpinning diseases and certain phenotypes. In this paper, a new model called attentive deep canonical correlation analysis (ADCCA) is proposed for the diagnosis of Alzheimer's disease using multimodal brain imaging genetics data. ADCCA combines the strengths of deep neural networks, attention mechanisms, and canonical correlation analysis to integrate and exploit the complementary information from multiple data modalities. This leads to improved interpretability and strong multimodal feature learning ability. The ADCCA model is evaluated using the ADNI database with three imaging modalities (VBM-MRI, FDG-PET, and AV45-PET) and genetic SNP data. The results indicate that this approach can achieve outstanding performance and identify meaningful biomarkers for Alzheimer's disease diagnosis. To promote reproducibility, the code has been made publicly available at https://github.com/rongzhou7/ADCCA.

Keywords: Brain imaging genetics · Canonical correlation analysis · Self-attention · Alzheimer's disease

1 Introduction

Alzheimer's disease (AD) is an irreversible neurodegenerative disorder that affects millions of people worldwide [5]. In recent years, brain imaging genetics has emerged as a promising field for the diagnosis and prediction of AD and its prodromal stage – mild cognitive impairment (MCI). This approach

Supplementary Information The online version contains supplementary material available at https://doi.org/10.1007/978-3-031-43895-0_64.

H. Greenspan et al. (Eds.): MICCAI 2023, LNCS 14221, pp. 681–691, 2023.
https://doi.org/10.1007/978-3-031-43895-0_64

largely focuses on using neuroimaging techniques, such as MRI and PET, to identify brain regions that are associated with specific genetic variants such as single nucleotide polymorphisms (SNPs). Such analyses have produced a wealth of research findings [23,26,28] that have demonstrated significant associations between imaging characteristics and genetics in AD, and have the great potential to identify new multimodal biomarkers affecting specific brain systems and provide an enormous impetus for drug discovery.

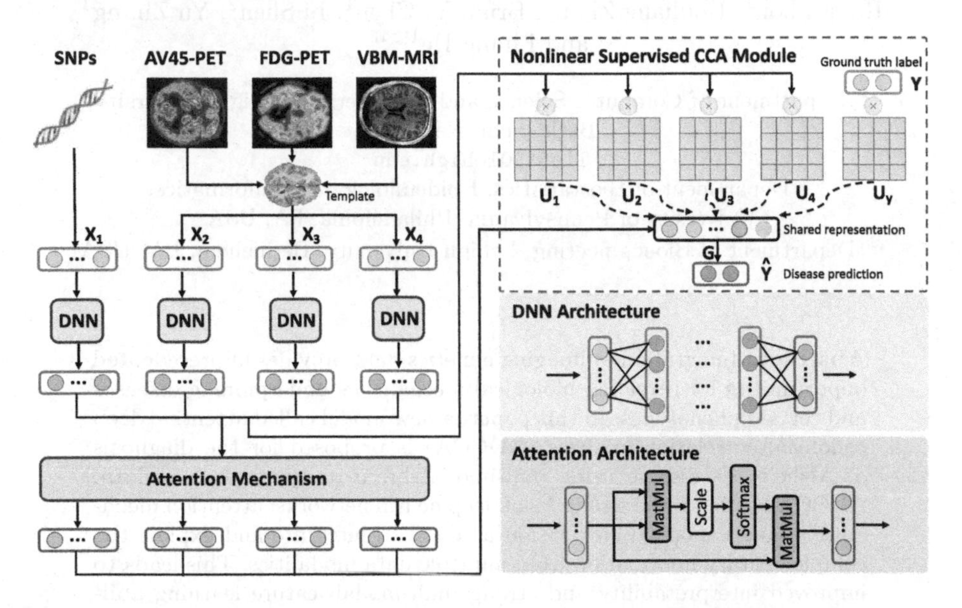

Fig. 1. An overview of the proposed framework. $\mathbf{X}_1, \cdots, \mathbf{X}_4$ are input modality data, and $\mathbf{Y}$ is the label information. DNNs first operate on each modality, generating hidden representations for each modality. These hidden representations go through a self-attention mechanism generating improved self-attention representations. At the same time, the hidden representations and label $\mathbf{Y}$ are multiplied by individual projection matrices $\mathbf{U}_1, \ldots, \mathbf{U}_4, \mathbf{U}_y$ based on CCA, thus mapping them to a shared representation $\mathbf{G}$. Finally, the disease prediction is calculated by self-attention representations with projection matrices and shared representation $\mathbf{G}$.

In the literature, various methods have been proposed to brain imaging genetics analysis [3,9–11,13,19,27,29]. In particular, canonical correlation analysis (CCA) [12] is a powerful multivariate statistical technique for quantifying the associations between different sets of data. CCA and its variations have been widely applied in imaging genetics studies because of its advantages in biological interpretation. For example, Du et al. [8] proposed a joint multitask sparse canonical correlation analysis and classification (MTSCCALR) for identifying imaging genetics biomarkers of AD. Kim et al. [16] introduced a multi-task learning-based structured sparse canonical correlation analysis (MTS2CCA) for identifying brain imaging genetics related to sleep. Moon et al. [20] proposed a supervised deep generalized canonical correlation analysis (SDGCCA) for improving the classification of phenotypes and revealing biomarkers associated with phenotypes in the context of AD. Despite much progress made in this area, CCA-based traditional shallow models assume that the relationships between genetic and imaging data are linear. However, this may not always be the case, and nonlinear relationships may exist in brain imaging genetics data, leading to biased results. On the other hand, the existing CCA-based deep models do not provide a direct interpretation of the underlying biological mechanisms driving the observed associations between genetic and imaging data. Most of them explored post-hoc explanations as justifications for model predictions. This can limit the ability to translate findings into clinically relevant insights.

In this paper, we propose a novel attentive deep canonical correlation analysis (ADCCA) model for diagnosing AD disease and discovering biomarkers using multimodal brain imaging genetics data. As illustrated in Fig. 1, the proposed framework comprises three key components: (i) deep neural network (DNN) modeling for generating latent representations of each modality to capture intra-modality correlations; (ii) attention update mechanism for focusing on the most salient regions of input data; and (iii) nonlinear supervised CCA modeling for integrating multiple modalities to discriminate phenotypic groups. By combining the power of these techniques, the ADCCA approach effectively models nonlinear relationships among multimodal imaging genetics data and provides simultaneous predictions and interpretations. The model is trained end-to-end using a combination of classification and correlation losses.

Through extensive experiments on the real-world ADNI dataset with three imaging modalities (VBM-MRI, FDG-PET, and AV45-PET) and genetic SNP data, we show that our model achieves outstanding performance for classifying AD vs. HC, AD vs. MCI, and MCI vs. HC groups. Also, it is demonstrated that the model explanation can reveal disorder-specific biomarkers coinciding with neuroscience findings. Last, we show that the combination of classification and correlation models can boost disease prediction performance.

2 Method

Suppose that the problem includes N subjects with M modalities. Let $\mathbf{X}_m \in \mathbb{R}^{N \times d_m}$ denote the m-th modality data, where d_m represents the dimension of features in the m-th modality, $m = 1, 2, \cdots, M$. Let $\mathbf{Y} \in \mathbb{R}^N$ denote the label information of all subjects. In this work, we seek to learn a disease prediction model that estimates $\hat{\mathbf{Y}}$ from $\{\mathbf{X}_m\}_{m=1}^{M}$ by making full use of all M modalities, as well as identify disease-specific biomarkers for clinical interpretation.

The proposed ADCCA aims to combine the strengths of DNN, attention mechanism, and CCA to integrate and exploit the complementary information from multiple data modalities (Fig. 1). First, we use a separate DNN containing several fully-connected hidden layers to learn hidden representations for each modality, denoted as $f_m(\mathbf{X}_m) \in \mathbb{R}^{N \times l_m}$, where l_m represents the dimension of last layer of DNN corresponding to the m-th modality. Second, we employ the attention mechanism [25] on the basis of the DNN model. With the help of the attention mechanism, our method can explicitly capture the important features hidden in the input data. Specifically, we use self-attention, sometimes called intra-attention, which is regarded as an improvement in attention that focuses on internal links of features and reduces external data dependency to compute its representation. Suppose there are three linear transformation matrices for the m-th modality: $\mathbf{W}_Q^m, \mathbf{W}_K^m, \mathbf{W}_V^m$. Mathematically, the self-attention representation of $f_m(\mathbf{X}_m)$ can be calculated as:

$$\mathrm{Att}(f_m(\mathbf{X}_m)) = \mathrm{Softmax}\Big(\frac{f_m(\mathbf{X}_m)\mathbf{W}_Q^m (f_m(\mathbf{X}_m)\mathbf{W}_K^m)^\top}{\sqrt{l_m}}\Big) f_m(\mathbf{X}_m)\mathbf{W}_V^m. \quad (1)$$

Third, following [20], we learn cross-modality features and incorporate the label information of samples for supervised learning based on CCA. The correlation loss function is defined as follows:

$$L_{cor} = \left\|\mathbf{G} - \mathbf{U}_y^\top \mathbf{Y}\right\|_F^2 + \sum_{m=1}^{M} \left\|\mathbf{G} - \mathbf{U}_m^\top f_m(\mathbf{X}_m)\right\|_F^2, \quad s.t.\ \mathbf{G}\mathbf{G}^\top = \mathbf{I}. \quad (2)$$

where $\mathbf{U}_1, \cdots, \mathbf{U}_4, \mathbf{U}_y$ are projection matrices for each modality and label information, respectively. $\mathbf{I}$ denotes the identity matrix.

According to Eq. (2), we have $\mathbf{G} \approx \mathbf{U}_m^\top f_m(\mathbf{X}_m) \approx \mathbf{U}_y^\top \mathbf{Y}$. Thus, the label $\mathbf{Y}$ can be approximated as follows: $\mathbf{Y} \approx (\mathbf{U}_y^\top)^\dagger \mathbf{U}_m^\top f_m(\mathbf{X}_m)$, where $\mathbf{U}_y^\dagger$ denotes the pseudo-inverse of $\mathbf{U}_y$. Then, we substitute self-attention representations that are more representative of each modality into the above equation and let $\hat{\mathbf{Y}}_m = (\mathbf{U}_y^\top)^\dagger \mathbf{U}_m^\top \mathrm{Att}(f_m(\mathbf{X}_m))$. Further, the conventional supervised cross-entropy loss [7] is used to enable the propagation of label information directly to the DNN of each modality.

$$L_{cls} = \sum_{m=1}^{M} \mathrm{CrossEntropy}\Big(\mathbf{Y}, \mathrm{Softmax}(\hat{\mathbf{Y}}_m)\Big). \quad (3)$$

The final label prediction of ADCCA can be obtained using the following soft voting of the label presentation of each modality: $\hat{\mathbf{Y}} = \mathrm{Softmax}((\sum_{m=1}^{M} \hat{\mathbf{Y}}_m)/M)$. Overall, our final training objective can be defined as:

$$L = L_{cls} + \lambda L_{cor}, \quad (4)$$

where L_{cls} is the supervised cross-entropy disease prediction loss, L_{cor} is the correlation loss, and λ is a tunable hyperparameter that scales the numerical value of each loss item to the same order of magnitude to balance their influence. The solution on loss function L is similar to the SGDCCA method except for substituting the outputs of DNN models to their self-attention representations.

3 Experiments and Results

3.1 Data Acquisition and Preprocessing

Brain imaging genetic data used in this study were obtained from the public ADNI database [22]. There is a total of 597 participants with both genotype and brain imaging data, including 104 AD, 305 MCI, and 188 healthy control (HC) subjects. The image data consisted of three modalities including structural Magnetic Resonance Imaging (VBM-MRI), 18 F-fluorodeoxyglucose Positron Emission Tomography (FDG-PET), and 18 F-florbetapir PET (AV45-PET). These three imaging modalities allowed us to examine brain structure, glucose metabolism, and amyloid plaque deposition, respectively.

Following the previous studies [2,30], we preprocessed neuroimaging data to extract ROI-based features. Specifically, the multi-modality imaging scans were aligned to each participant's same visit. All imaging scans were aligned to a T1-weighted template image, and segmented into gray matter (GM), white matter (WM) and cerebrospinal fluid (CSF) maps. They were normalized to the standard Montreal Neurological Institute (MNI) space as $2 \times 2 \times 2$ mm^3 voxels, being smoothed with an 8 mm FWHM kernel. We preprocessed the structural MRI scans with voxel-based morphometry (VBM) by using the SPM software [1], and registered the FDG-PET and AV45-PET scans to the MNI space by SPM. We subsampled the whole brain imaging and contained 90 ROIs (excluding the cerebellum and vermis) based on the AAL-90 atlas [24]. ROI-level measures were calculated by averaging all the voxel-level measures within each ROI.

For genetic SNP data, according to the AlzGene database[1], only the SNPs belonging to top AD gene candidates were selected. The genetic data were genotyped by the Human 610-Quad or OmniExpress Array platform (Illumina, Inc., San Diego, CA, USA), and preprocessed following standard quality control and imputation procedures. There were 54 SNPs included which were collected from the neighbor of AD risk gene APOE according to the ANNOVAR annotation.

3.2 Evaluation of Disease Classification Performance

In our experiments, the whole data were separated into three groups, including AD vs. HC, AD vs. MCI, and MCI vs. HC. To quantitatively evaluate the performance of different methods, we considered four commonly-used evaluation metrics: accuracy (ACC), F1-score (F1), area under receiver operating characteristic curve (AUC), and Matthews correlation coefficient (MCC) [6]. Since the number of subjects was limited, we calculated the mean and standard deviation of all metrics using 5-fold cross-validation (CV). Many researchers have successfully adopted multimodal brain imaging data into CCA. We carefully choose five related methods for comparison: 1) vanilla DNN [18], 2) generalized CCA (GCCA) [15], 3) deep generalized CCA (DGCCA) [4], 4) MTSCCALR [8], and 5) SDGCCA [20]. Note that GCCA and DGCCA are unsupervised learning methods, and the others are supervised learning methods. The proposed model includes four DNNs, one for each modality, with three fully-connected layers and a Tanh activation function, which is trained with Adam optimizer with the learning rate set to 0.0001 and weight decay set to 0.001.

Table 1 presents the classification results, where ± represents the standard deviation of evaluation scores across the 5 folds. From the results, it can be

[1] www.alzgene.org.

observed that the proposed ADCCA method significantly outperforms all other methods in terms of all four metrics. The higher AUC and MCC scores indicate that our method is able to accurately identify both positive and negative cases of AD. The smaller standard deviations of ADCCA illustrated the overall stability and reproducibility of the experiment.

Table 1. Classification performance comparison. The best results are in bold.

Task	Measures	DNN	GCCA	DGCCA	MTSCCALR	SDGCCA	ADCCA
AD vs. HC	ACC	.866 ± .037	.812 ± .037	.837 ± .028	.828 ± .047	.914 ± .029	**.932 ± .010**
	F1	.873 ± .049	.811 ± .054	.833 ± .041	.862 ± .046	.883 ± .034	**.901 ± .025**
	AUC	.943 ± .030	.930 ± .015	.939 ± .013	.893 ± .051	.978 ± .013	**.979 ± .015**
	MCC	.720 ± .080	.652 ± .079	.688 ± .060	.629 ± .087	.822 ± .057	**.895 ± .043**
AD vs. MCI	ACC	.689 ± .035	.618 ± .059	.638 ± .017	.746 ± .049	.812 ± .063	**.825 ± .011**
	F1	.579 ± .032	.583 ± .038	.535 ± .037	.679 ± .041	.683 ± .079	**.823 ± .032**
	AUC	.811 ± .025	.726 ± .050	.756 ± .022	.836 ± .039	.880 ± .043	**.925 ± .050**
	MCC	.413 ± .046	.256 ± .054	.281 ± .048	.482 ± .104	.569 ± .110	**.625 ± .024**
MCI vs. HC	ACC	.523 ± .026	.499 ± .024	.519 ± .044	.594 ± .029	.647 ± .058	**.758 ± .033**
	F1	.529 ± .031	.543 ± .084	.513 ± .044	.513 ± .025	.702 ± .058	**.799 ± .030**
	AUC	.570 ± .030	.540 ± .032	.574 ± .054	.637 ± .022	.796 ± .074	**.816 ± .051**
	MCC	.103 ± .058	.105 ± .075	.109 ± .100	.172 ± .045	.273 ± .110	**.407 ± .073**

3.3 The Most Discriminative Brain Regions and SNPs

Identifying the most discriminative brain regions (*i.e.*, ROIs) and SNPs is crucial for AD diagnosis. Here, we employed the integrated gradients interface provided by Captum [17] to assign importance scores to each feature of different modalities by analyzing the pre-trained model, which can provide a comprehensive explanation of how the input features of a deep learning model contribute to the model's output. The reason why not using the self-attention weights is that we use the self-attention to assign attention scores to hidden representations instead of the original features, thus it may not fully capture the importance of the original features in the input data. Figure 2(a-c) shows the top 20 discriminative ROIs identified by the proposed method from each individual brain imaging modality. Figure 2(d) shows the top 20 discriminative ROIs selected by the average importance scores of ROIs from the three modalities. We found that the hippocampal, amygdala, uncus, and gyrus regions are only identified by using the three modalities together. These selected regions are known to be highly related to AD and MCI in previous studies [21]. Besides, the result shows that the selected ROIs exhibited differences across different classification groups, indicating that our model can effectively differentiate the important ROIs for specific diseases. Figure 3 shows the most frequently selected SNPs with importance scores. The result indicates that rs6448453, rs3865444, and rs2718058 are the most discriminative SNPs which is consistent with previous evidence [14].

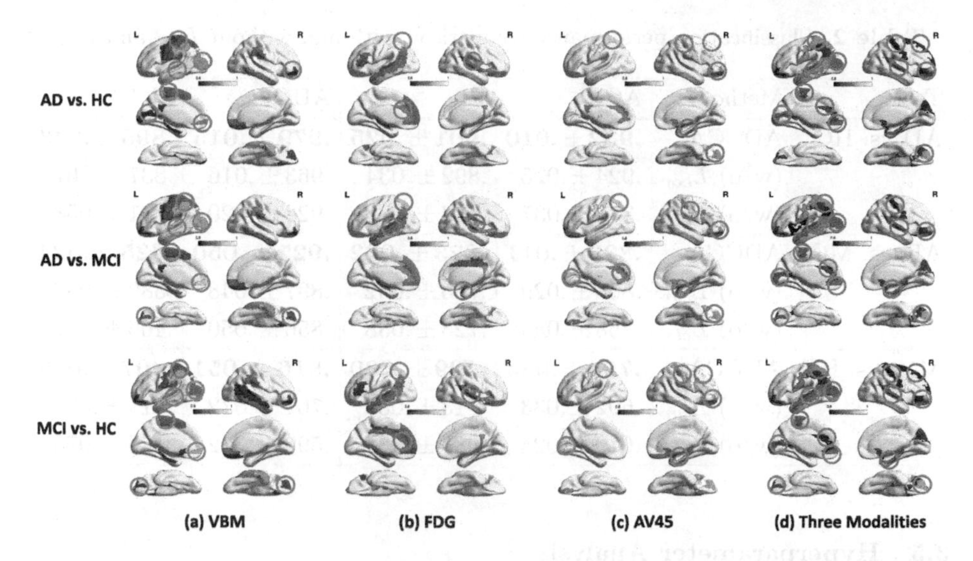

Fig. 2. Top 20 discriminative ROIs identified by ADCCA from three brain imaging modalities for three different classification groups in lateral, medial, and ventral view. The color bar indicates the importance score. The commonly selected ROIs across different modalities are circled in blue. (Color figure online)

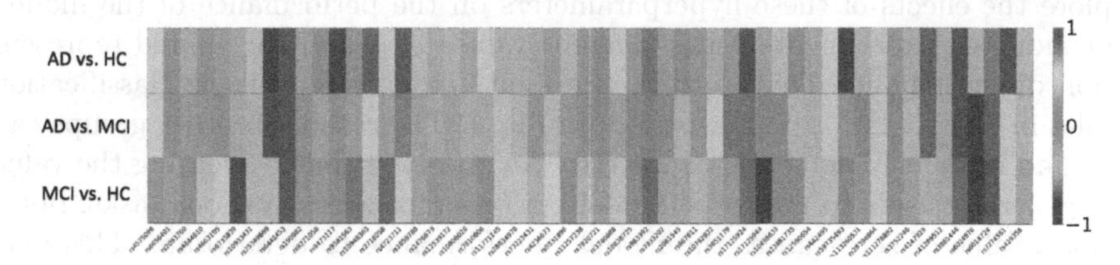

Fig. 3. The importance scores of SNPs. The red color indicates a high score. (Color figure online)

3.4 Ablation Study

The proposed ADCCA is trained using both correlation and classification losses. To understand the impact of each loss on classification, we conducted ablation studies by evaluating the performance of two additional models: the ADCCA model trained without the correlation loss (w/o L_{cor}) and without the classification loss (w/o L_{cls}). The results presented in Table 2 indicate that ADCCA outperforms the other two models for all evaluation metrics on all three classification tasks, suggesting that both correlation and classification losses contribute to ADCCA's improved performance. Removing either loss leads to decreased performance, and the impact will be particularly significant if the classification loss is eliminated.

Table 2. Classification performance comparison with and without L_{cor} and L_{cls}

Task	Method	ACC	F1	AUC	MCC
AD vs. HC	ADCCA	**.932 ± .010**	**.901 ± .025**	**.979 ± .015**	**.895 ± .043**
	(w/o) L_{cor}	.924 ± .025	.892 ± .034	.963 ± .016	.837 ± .067
	(w/o) L_{cls}	.876 ± .037	.853 ± .029	.928 ± .020	.791 ± .058
AD vs. MCI	ADCCA	**.825 ± .011**	**.823 ± .032**	**.925 ± .050**	**.625 ± .024**
	(w/o) L_{cor}	.806 ± .029	.795 ± .032	.897 ± .048	.589 ± .033
	(w/o) L_{cls}	.758 ± .059	.723 ± .038	.856 ± .050	.466 ± .054
MCI vs. HC	ADCCA	**.758 ± .033**	**.799 ± .030**	**.816 ± .051**	**.407 ± .073**
	(w/o) L_{cor}	.692 ± .033	.713 ± .056	.761 ± .072	.317 ± .092
	(w/o) L_{cls}	.619 ± .024	.683 ± .084	.599 ± .032	.176 ± .075

3.5 Hyperparameter Analysis

We investigated the impact of two important hyperparameters in the ADCCA model: λ, which appears in the loss function to balance the classification and correlation losses, and the dimension of the shared representation $\mathbf{G}$. In order to explore the effects of these hyperparameters on the performance of the model, we conducted experiments using different values of λ and the shared representation dimensionality. Due to the space limit, we only report the classification results in AD vs. HC group, as shown in Fig. 4. The results in other groups can be found in the supplementary material. We observed that decreasing the value of λ generally leads to improved model performance across various tasks, but a lambda value of zero causes the model's performance to deteriorate. This may indicate that for the ADCCA model, L_{cls} is more important than L_{cor}. Furthermore, combining these two loss functions to jointly guide the model can lead to improved model performance. We also found that for the AD vs. HC group, the model achieves good performance even with a low-dimensional shared representation. However, for other groups, the impact of the shared representation dimension on the model's performance seems not significant. One explanation for this could be that the AD vs. HC group exhibits distinct feature differences, allowing the original features to be well represented even when mapped into a low-dimensional shared representation.

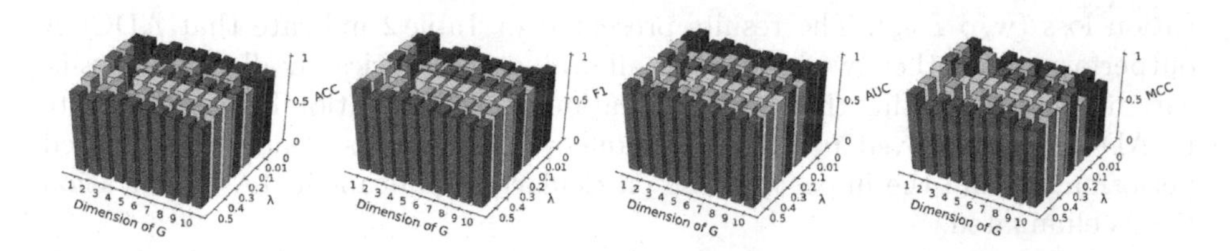

Fig. 4. Sensitivity analysis of hyperparameters on AD vs. HC

4 Conclusion

In this work, we propose a novel deep canonical correlation analysis method for multimodal Alzheimer's disease diagnosis that leverages attention mechanisms to enhance interpretability and multimodal feature learning. Experimental results on the real-world imaging-genetics dataset demonstrate that our approach achieves better classification performance than the existing state-of-the-art methods in terms of both classification accuracy and correlation between the modalities. In an exploratory analysis, we further show that the biomarkers identified by our model are closely associated with Alzheimer's disease. Our proposed approach is applicable to other diseases with multimodal data available. However, the limited size of medical datasets may restrict the effectiveness and generalization ability of such deep learning models. To address this issue, a potential future direction is to employ pre-training and transfer learning techniques that facilitate learning across datasets.

Acknowledgements. This work is partially supported by the National Science Foundation (MRI-2215789 and IIS-1909879), National Institutes of Health (U01AG068057, U01AG-066833, R01LM013463, R01MH129694, and R21MH130956), Alzheimer's Association grant (AARG-22-972541), and Lehigh's grants under Accelerator (S00010293), CORE (001250), and FIG (FIGAWD35).

References

1. Ashburner, J., Friston, K.J.: Voxel-based morphometry-the methods. Neuroimage **11**(6), 805–821 (2000)
2. Barshan, E., Fieguth, P.: Stage-wise training: An improved feature learning strategy for deep models. In: Feature extraction: modern questions and challenges, pp. 49–59. PMLR (2015)
3. Batmanghelich, N.K., Dalca, A., Quon, G., Sabuncu, M., Golland, P.: Probabilistic modeling of imaging, genetics and diagnosis. IEEE Trans. Med. Imaging **35**(7), 1765–1779 (2016)
4. Benton, A., Khayrallah, H., Gujral, B., Reisinger, D.A., Zhang, S., Arora, R.: Deep generalized canonical correlation analysis. In: Proceedings of the 4th Workshop on Representation Learning for NLP (RepL4NLP-2019), pp. 1–6 (2019)
5. Catania, M., et al.: A novel bio-inspired strategy to prevent amyloidogenesis and synaptic damage in Alzheimer's disease. Mol. Psych. 1–8 (2022)
6. Chicco, D., Jurman, G.: The advantages of the Matthews correlation coefficient (mcc) over f1 score and accuracy in binary classification evaluation. BMC Genomics **21**, 1–13 (2020)
7. De Boer, P.T., Kroese, D.P., Mannor, S., Rubinstein, R.Y.: A tutorial on the cross-entropy method. Ann. Oper. Res. **134**, 19–67 (2005)

8. Du, L., et al.: Identifying diagnosis-specific genotype-phenotype associations via joint multitask sparse canonical correlation analysis and classification. Bioinformatics **36**, i371–i379 (2020)
9. Du, L., et al.: Detecting genetic associations with brain imaging phenotypes in Alzheimer's disease via a novel structured SCCA approach. Med. Image Anal. **61**, 101656 (2020)
10. Ghosal, S., et al.: Bridging imaging, genetics, and diagnosis in a coupled low-dimensional framework. In: Shen, D., et al. (eds.) Medical Image Computing and Computer Assisted Intervention – MICCAI 2019: 22nd International Conference, Shenzhen, China, October 13–17, 2019, Proceedings, Part IV, pp. 647–655. Springer, Cham (2019). https://doi.org/10.1007/978-3-030-32251-9_71
11. Ghosal, S., et al.: A biologically interpretable graph convolutional network to link genetic risk pathways and imaging phenotypes of disease. In: ICLR (2022)
12. Hotelling, H.: Relations between two sets of variates. Biometrika **28**(3/4), 321–377 (1936)
13. Hu, W., et al.: Adaptive sparse multiple canonical correlation analysis with application to imaging (epi) genomics study of schizophrenia. IEEE Trans. Biomed. Eng. **65**(2), 390–399 (2017)
14. Jansen, I.E., et al.: Genome-wide meta-analysis identifies new loci and functional pathways influencing Alzheimer's disease risk. Nat. Genet. **51**(3), 404–413 (2019)
15. Kettenring, J.R.: Canonical analysis of several sets of variables. Biometrika **58**(3), 433–451 (1971)
16. Kim, M., et al.: Multi-task learning based structured sparse canonical correlation analysis for brain imaging genetics. Med. Image Anal. **76**, 102297 (2022)
17. Kokhlikyan, N., et al.: Captum: A unified and generic model interpretability library for pytorch. arXiv preprint arXiv:2009.07896 (2020)
18. LeCun, Y., Bengio, Y., Hinton, G.: Deep learning. Nature **521**(7553), 436–444 (2015)
19. Liu, J., Calhoun, V.D.: A review of multivariate analyses in imaging genetics. Front. Neuroinform. **8**, 29 (2014)
20. Moon, S., Hwang, J., Lee, H.: SDGCCA: supervised deep generalized canonical correlation analysis for multi-omics integration. J. Comput. Biol. **29**(8), 892–907 (2022)
21. Mu, Y., Gage, F.H.: Adult hippocampal neurogenesis and its role in Alzheimer's disease. Mol. Neurodegener. **6**(1), 1–9 (2011)
22. Muller, S.G., et al.: The Alzheimer's disease neuroimaging initiative. Neuroimaging Clin. **15**(4), 869–877 (2005)
23. Shen, L., Thompson, P.M.: Brain imaging genetics: integrated analysis and machine learning. In: IEEE International Conference on Bioinformatics and Biomedicine (BIBM), pp. 1–1. IEEE Computer Society (2021)
24. Tzourio-Mazoyer, N., et al.: Automated anatomical labeling of activations in SPM using a macroscopic anatomical parcellation of the MNI MRI single-subject brain. Neuroimage **15**(1), 273–289 (2002)
25. Vaswani, A., et al.: Attention is all you need. Adv. Neural Inform. Process. Syst. **30** (2017)
26. Viding, E., Williamson, D.E., Forbes, E.E., Hariri, A.R.: The integration of neuroimaging and molecular genetics in the study of developmental cognitive neuroscience. MIT press (2008)
27. Wang, M.L., Shao, W., Hao, X.K., Zhang, D.Q.: Machine learning for brain imaging genomics methods: a review. Mach. Intell. Res. **20**(1), 57–78 (2023)

28. Xin, Y., Sheng, J., Miao, M., Wang, L., Yang, Z., Huang, H.: A review of imaging genetics in Alzheimer's disease. J. Clin. Neurosci. **100**, 155–163 (2022)
29. Zhou, H., Zhang, Yu., Chen, B.Y., Shen, L., He, L.: Sparse interpretation of graph convolutional networks for multi-modal diagnosis of Alzheimer's disease. In: Wang, L., Dou, Q., Fletcher, P.T., Speidel, S., Li, S. (eds.) Medical Image Computing and Computer Assisted Intervention – MICCAI 2022: 25th International Conference, Singapore, September 18–22, 2022, Proceedings, Part VIII, pp. 469–478. Springer, Cham (2022). https://doi.org/10.1007/978-3-031-16452-1_45
30. Zhu, Y., et al.: Graphene and graphene oxide: synthesis, properties, and applications. Adv. Mater. **22**(35), 3906–3924 (2010)

FedIIC: Towards Robust Federated Learning for Class-Imbalanced Medical Image Classification

Nannan Wu[1], Li Yu[1], Xin Yang[1], Kwang-Ting Cheng[2], and Zengqiang Yan[1](✉)

[1] School of Electronic Information and Communications, Huazhong University of Science and Technology, Wuhan, China
{wnn2000,hustlyu,xinyang2014,z_yan}@hust.edu.cn

[2] School of Engineering, Hong Kong University of Science and Technology, Hong Kong, China
timcheng@ust.hk

Abstract. Federated learning (FL), training deep models from decentralized data without privacy leakage, has shown great potential in medical image computing recently. However, considering the ubiquitous class imbalance in medical data, FL can exhibit performance degradation, especially for minority classes (*e.g.* rare diseases). Existing methods towards this problem mainly focus on training a balanced classifier to eliminate class prior bias among classes, but neglect to explore better representation to facilitate classification performance. In this paper, we present a privacy-preserving FL method named FedIIC to combat class imbalance from two perspectives: feature learning and classifier learning. In feature learning, two levels of contrastive learning are designed to extract better class-specific features with imbalanced data in FL. In classifier learning, per-class margins are dynamically set according to real-time difficulty and class priors, which helps the model learn classes equally. Experimental results on publicly-available datasets demonstrate the superior performance of FedIIC in dealing with both real-world and simulated multi-source medical imaging data under class imbalance. Code is available at https://github.com/wnn2000/FedIIC.

Keywords: Federated learning · Class imbalance · Contrastive learning · Classification

1 Introduction

Federated learning (FL), allowing decentralized data sources to train a unified deep learning model collaboratively without data sharing, has drawn great attention in medical imaging due to its privacy-preserving properties [13,22,25,40]. Existing studies of FL mainly focus on data heterogeneity across clients [19,20,31], while ignoring the widely-existed class imbalance problem in

---

Supplementary Information The online version contains supplementary material available at https://doi.org/10.1007/978-3-031-43895-0_65.

H. Greenspan et al. (Eds.): MICCAI 2023, LNCS 14221, pp. 692–702, 2023.
https://doi.org/10.1007/978-3-031-43895-0_65

medical scenarios. In clinical practice, the number of samples for different diseases may vary greatly due to varying incidence rates in the population. When conducting FL on cooperative medical institutions with global class-imbalanced data, the global model may suffer from significant performance degradation, which typically manifests as the recognition accuracy of minority classes (*e.g.* rare diseases) being lower than that of majority classes (*e.g.* common diseases) [34]. Deploying such a biased global/federated model is fatal, especially for misdiagnosing a rare disease [15,42]. Therefore, addressing class imbalance in federated learning is of great value.

Several FL frameworks have been proposed to tackle imbalanced data [9,41]. Following re-weighting [7], Wang *et al.* [39] presented a weighted form of cross entropy loss named ratio loss depending on a balanced auxiliary dataset for the server to calculate weights. Sarkar *et al.* [33] introduced focal loss [24] to up-weight hard samples. CLIMB [35] assigned larger weights to clients more likely to own minority classes via a meta-algorithm. Inspired by decoupling [17], CReFF [34] retrained a new classifier with balanced synthetic features in the server. All these methods aim to balance classes from the classifier perspective without exploring better representations with class-imbalanced data for performance improvement.

In this paper, we formulate the effect of class imbalance in FL into the attribute bias and the class bias [37]. The attribute bias means minority classes have more imbalanced background attributes in their class-specific attributes compared to majority classes, making them less distinguishable. The class bias represents the difference in prior probabilities across classes, resulting in biased predictions toward majority classes. To handle the two biases, we present a new class-balancing FL method named **FedIIC** from two perspectives: feature learning and classifier learning. The key idea of FedIIC is to alleviate the two biases through the calibration of the feature extractor and the classifier. Specifically, two-level supervised contrastive learning [18], *i.e.* intra- and inter-client contrastive learning, is built to calibrate the feature extractor for better feature learning. For classifier learning, difficulty-aware logit adjustment is adopted to calibrate the classifier dynamically for better decision boundaries. Extensive comparison experiments on both real-world and simulated multi-source data validate FedIIC's effectiveness.

The main contributions are summarized as follows. (1) A new viewpoint of realistic medical FL scenarios where global training data is class-imbalanced. (2) A novel privacy-preserving framework FedIIC for balanced federated learning. (3) Superior performance in dealing with class imbalance under both real-world and simulated multi-source decentralized settings.

2 Methodology

2.1 Preliminaries and Overview

Considering a typical FL scenario for multi-class image classification with K participants, each participant is assumed to own a private dataset $D_k = \{(x_i, y_i)\}_{i=1}^{N_k}$, $k \in [K]$, where N_k is the data amount of D_k, and denote each image-label pair as $(x_i \in \mathcal{X} \subseteq \mathbb{R}^d, y_i \in \mathcal{Y} = [L])$. The goal of FL is to

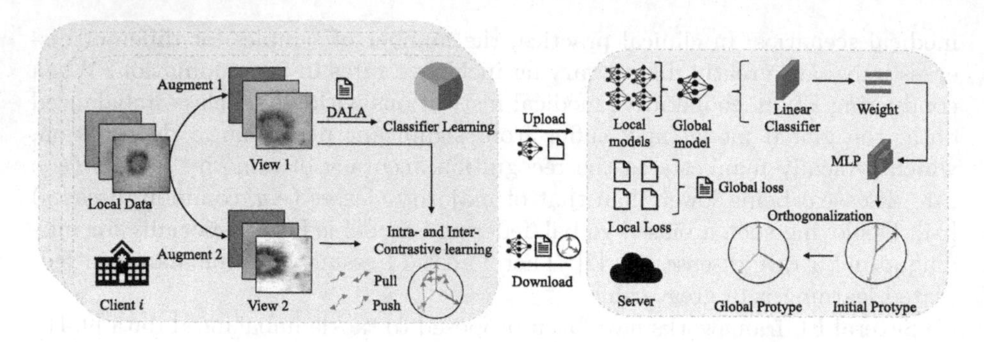

Fig. 1. Overview of the proposed FedIIC.

train a global model $f(g(\cdot))$ with the union of all cooperative data sources $D := \cup_{k\in[K]} D_k$ without privacy leakage, where $f(\cdot)$ and $g(\cdot)$ represent the linear classifier and the feature extractor respectively. Note that D is set as class-imbalanced in this paper.

Assuming each image has two kinds of latent attributes, *i.e.* $\mathcal{Z}_c$ and $\mathcal{Z}_a$, representing the class-specific attributes (determining the category of the image, *e.g.* texture, color, *etc.*) and the variant background attributes (*e.g.* brightness, contrast, *etc.*) respectively [37], based on the Bayes theorem, the posterior probability of classification can be formulated as

$$P(y \mid x) = P(y \mid \mathcal{Z}_c, \mathcal{Z}_a) = \frac{p(\mathcal{Z}_c \mid y)}{p(\mathcal{Z}_c)} \cdot \frac{p(\mathcal{Z}_a \mid y, \mathcal{Z}_c)}{p(\mathcal{Z}_a \mid \mathcal{Z}_c)} \cdot p(y), \quad (1)$$

where the last two items represent the attribute bias and the class bias respectively, which widely exist in class-imbalanced data and affect the posterior probability. For robust FL with class-imbalanced data, the key idea is to alleviate the two biases simultaneously, instead of focusing on the latter as [34]. Hence, we propose FedIIC to address class imbalance from the two perspectives as illustrated in Fig. 1. Details are presented in the following.

2.2 Intra-Client Contrastive Learning

Limited local data affects data diversity (*i.e.*, limited $(\mathcal{Z}_c, \mathcal{Z}_a)$ combinations), especially for minority classes, making $\mathcal{Z}_c$ less distinguishable. To emphasize more on the learning of $\mathcal{Z}_c$, supervised contrastive learning (SCL), proven to be effective for representation learning [16,21,27,45], is introduced in local training. The basic loss function of SCL can be formulated as

$$\mathcal{L}_{SCL} = \sum_{i\in I} \frac{-1}{|P(i)|} \sum_{j\in P(i)} \log \frac{\exp(z_i \cdot z_j/\tau)}{\sum_{a\in A(i)} \exp(z_i \cdot z_a/\tau)}, \quad (2)$$

where I denotes the index set of the multi-view batch generated by different augmentations (*e.g.* the two views in Fig. 1), $|\cdot|$ measures the number of elements in a set, $A(i) = I \backslash \{i\}, P(i) = \{s \in A(i) | y_s = y_i\}$, τ represents the temperature, and z denotes the l_2-normalized embedding of a sample x. Note that in this paper, we use a 2-layer MLP $h(\cdot)$ to obtain z before it is normalized as [3], *i.e.* $z = \frac{h(g(x))}{\|h(g(x))\|_2}$. In the multi-view batch, $\mathcal{L}_{SCL}$ keeps the embeddings of the same class closer while pushing the embeddings of different classes further away, which helps the model learn better $\mathcal{Z}_c$ of each class due to richer $\mathcal{Z}_a$. However, SCL can not perfectly address class imbalance as the majority classes would benefit more from Eq. 2 following traditional training losses (*e.g.* the cross entropy loss). To overcome this problem, we propose to employ a dynamic temperature $\tau' := P\tau = (p^i p^j)^t \tau$ in Eq. 2 inspired by [16,45], where p^i is the prior probability of class i in the local dataset and t is a parameter set as 0.5 by default. Hence, the loss function is rewritten as

$$\mathcal{L}_{Intra} = \sum_{i \in I} \frac{-1}{|P(i)|} \sum_{j \in P(i)} \log \frac{\exp(z_i \cdot z_j / \tau')}{\sum_{a \in A(i)} \exp(z_i \cdot z_a / \tau')}, \tag{3}$$

named intra-client contrastive learning. Through P, sample pairs of the minority classes are up-weighted compared to those of the majority classes, leading to better balance.

2.3 Inter-client Contrastive Learning

Given limited local data under FL, the effectiveness of intra-client contrastive learning may be bounded. How to better utilize cross-client data from the global perspective is crucial for further performance improvement. Inspired by learning from prototypes [4,12,31], we propose inter-client contrastive learning. Assuming a set of shared class-wise prototypes $V = \{v^1, v^2, ..., v^L\}$ across clients, the local model can be trained by

$$\mathcal{L}_{Inter} = \sum_{i \in I} \frac{-1}{|P(i)|} \log \frac{\exp(z_i \cdot v^{y_i} / \tau)}{\sum_{j=1}^{L} \exp(z_i \cdot v^j / \tau)}, \tag{4}$$

where y_i is the label of sample i. When minimizing $\mathcal{L}_{Inter}$, the embedding of each sample will get closer to the prototype of the same class while farther from the prototypes of different classes, encouraging local models to learn common attributes (*i.e.* class-specific attributes) for samples with the same classes.

To this end, how to produce high-quality prototypes is the key to inter-client contrastive learning. In previous studies, one common method to generate prototypes is uploading and aggregating local information. For example, Mu *et al.* [31] and Chen *et al.* [4] uploaded features to the server directly to generate prototypes. However, it may cause privacy leakage under well-designed attacks and will introduce extra communication costs. Different from these methods, in FedIIC, we propose a new method to generate global prototypes without uploading extra information. Considering that the essence of linear classification is similarity calculation based on vector inner product, the weights of a well-trained linear classifier are nearly co-linear with the feature vectors of different classes [11,32,45]. Therefore, the weights of a linear classifier denoted as $W = \{w^1, w^2, ..., w^L\}$, can represent the corresponding features of L classes learned by the feature extractor $g(\cdot)$ to some extent. Specifically, given a global model $[f_g(\cdot), g_g(\cdot), h_g(\cdot)]$ after model aggregation in the server, the weights of $g_g(\cdot)$ are fed to $h_g(\cdot)$ to calculate the initial prototypes $\widetilde{V} = \{\widetilde{v}^1, \widetilde{v}^2, ..., \widetilde{v}^L\}$ as shown in Fig. 1. Considering that features of different classes should have low inter-class similarity, we further fine-tune $\widetilde{V}$ via gradient descent by

$$\widetilde{V} \leftarrow \widetilde{V} - \nabla \sum_{i \in Y} \max_{j \in Y, j \neq i} \left(\frac{\widetilde{v}^i}{\|\widetilde{v}^i\|_2} \cdot \frac{\widetilde{v}^j}{\|\widetilde{v}^j\|_2}\right). \tag{5}$$

In this way, the cosine similarity of any $(\widetilde{v}^i, \widetilde{v}^j)$ pair in $\widetilde{V}$ is minimized to be equal, resulting in $\widetilde{V}$ with lower inter-class similarity. This operation is called orthogonalization. Finally, the class-wise prototypes V are defined as the element-wise l_2-normalization of $\widetilde{V}$ and are sent to clients for inter-client contrastive learning.

2.4 Difficulty-Aware Logit Adjustment

After calibrating the feature extractor $g(\cdot)$, one common method to calibrate the linear classifier $f(\cdot)$ is logit adjustment (LA) [2,30] to alleviate the impact of class imbalance in local training. Specifically, Zhang *et al.* [43] proposed to add per-class margins to logits and re-compute the cross entropy (CE) loss by

$$\mathcal{L}_{LA} = \sum_{i \in I} - \log \frac{\exp(f(g(x_i))_{y_i} - \delta_{y_i})}{\sum_{y' \in \mathcal{Y}} \exp(f(g(x_i))_{y'} - \delta_{y'})}, \tag{6}$$

where δ_y denotes the positive per-class margin and is inversely proportional to the local class frequency $p(y)$. In this way, during local training, the logits of minority classes will increase to compensate for the item, which in turn trains the model to emphasize more on minority classes. However, the frequency-dependent margin may not be appropriate for medical data. For instance, some disease types/classes may have large intra-class variations and are difficult to diagnose even with a large amount of data, which may result in even smaller per-class margins. To address this, in FedIIC, the per-class margin is calculated based on not only the class frequency but also difficulties inspired by [44]. Specifically, we define $\delta_y := \log([\bar{l}_{ce}(y)]^q / p(y))$, where $\bar{l}_{ce}(y)$ is the average CE loss of all samples belonging to class y in any round and q is a hyper-parameter set as 0.25 by default. $\bar{l}_{ce}(y)$ is calculated as follows. At any round r, the total sample number of class y, denoted as N_r^y, belonging to clients of communication is first calculated. After receiving the global model from the server and before local training, each client i uploads $l_{ce}^i(y)$, *i.e.* the total loss of class y, to the server. Finally, $\bar{l}_{ce}(y)$ is calculated as $\frac{1}{N_r^y} \sum_i l_{ce}^i(y)$. This process to calculate average loss value can be privacy-preserving under the existing secure multi-party computation framework based on homomorphic encryption [35]. Based on the newly defined δ_y, Eq. 6 is renamed as $\mathcal{L}_{DALA}$. Note that the calculation of $\mathcal{L}_{DALA}$ does not rely on the multi-view batch like $\mathcal{L}_{Intra}$ and $\mathcal{L}_{Inter}$. For a fair

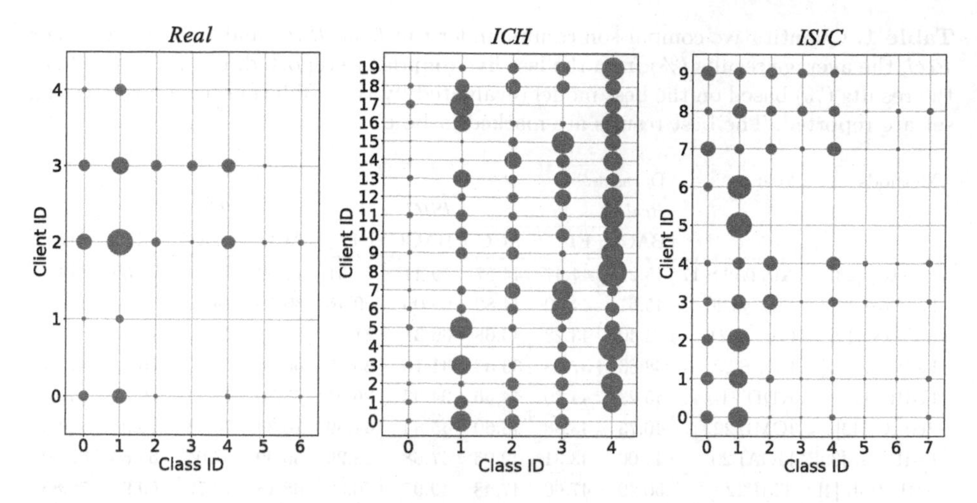

Fig. 2. Illustration of imbalanced data distributions. The radius of each solid circle represents each client's data amount of a specific class.

comparison with other methods trained by the CE loss, only one view of the multi-view batch is used to calculate $\mathcal{L}_{DALA}$. The overall loss function in local training is written as

$$\mathcal{L} = \mathcal{L}_{DALA} + k_1 \mathcal{L}_{Intra} + k_2 \mathcal{L}_{Inter}, \tag{7}$$

where k_1 and k_2 are trade-off hyper-parameters. After minimizing $\mathcal{L}$ during the local training phase of each client, the global model is updated by FedAvg [28].

3 Experiments

Datasets. Three FL scenarios with class-imbalanced global data are used for evaluation, which are described as follows:

1. Real Multi-Source Dermoscopic Image Datasets (denoted as ***Real***) consisting of five data sources from three datasets, including PH^2 [29], Atlas [1], and HAM10000 [38] where each source is treated as an individual client. For evaluation, we construct a separate test set by randomly sampling from the training set of ISIC 2019 [5,38] and ensure that the test set has no overlap with the above five data sources.
2. Intracranial Hemorrhage Classification (denoted as ***ICH***). The RNSA ICH dataset [10], containing five ICH subtypes, is adopted for experiments. The same pre-processing strategies in [14,26] are adopted, and images with only one single hemorrhage type are selected. Following [14,26], data is split according to 7:1:2 for training, validation, and testing respectively. To simulate heterogeneous multi-source data, following [34], Dirichlet distribution, *i.e.* $Dir(\alpha = 1.0)$, is used to divide the training set to 20 clients.

Table 1. Quantitative comparison results under the *Real*, *ISIC*, and *ICH* settings. For *Real*, the average results (%) from the last five rounds are reported. For *ISIC* and *ICH*, the results (%) based on the best model (evaluated by the validation set) on the testing set are reported. The best results are marked in bold.

Methods	Year	Datasets								
		Real			*ISIC*			*ICH*		
		BACC	F1	ACC	BACC	F1	ACC	BACC	F1	ACC
FedAvg [28]	AISTATS'17	45.21	44.47	44.57	49.41	54.31	72.50	73.75	77.35	84.83
FedProx [20]	MLSys'20	45.61	44.90	44.89	69.00	69.46	80.50	79.62	82.45	86.78
MOON [19]	CVPR'21	44.40	43.28	43.68	66.31	71.27	81.38	77.05	78.81	84.87
FedProc [31]	FGCS'23	38.83	37.98	39.36	31.16	35.45	66.88	73.29	76.23	84.89
FedRS [23]	KDD'21	45.23	44.50	44.46	24.93	26.01	61.39	72.44	76.51	84.13
FedLC [43]	ICML'22	46.73	45.88	45.60	45.84	41.89	70.33	76.53	78.96	84.92
FedFocal [33]	IJCAI'20	44.00	43.31	42.96	47.68	38.29	56.99	63.04	54.80	52.30
PRR-Imb [4]	TMI'22	50.49	47.60	47.48	49.97	46.52	68.18	71.72	69.98	78.85
CLIMB [35]	ICLR'22	46.07	45.91	45.86	49.70	52.32	71.65	72.64	76.08	84.73
CReFF [34]	IJCAI'22	51.13	48.56	49.46	71.52	57.83	72.92	82.21	74.64	81.63
FedIIC (ours)	-	**55.12**	**51.57**	**51.67**	**78.84**	**78.05**	**85.71**	**84.22**	**84.73**	**87.77**

3. Skin Lesion Classification (denoted as ***ISIC***).The training data of ISIC 2019 [5,38], containing eight classes, is used for evaluation. Following [14,26], we split the dataset by 7:1:2 for training, validation, and testing respectively. Similarly, Dirichlet distribution, *i.e.* $Dir(\alpha = 1.0)$, is used to generate highly heterogeneous data partitions of 10 clients.

Data distributions of the three training settings are illustrated in Fig. 2, and imbalance ratios are 35.43, 19.59 and 57.60, respectively.

Implementation Details. EfficientNet-B0 [36], pre-trained by ImageNet [8], is adopted as the backbone trained by an Adam optimizer with betas as 0.9 and 0.999, a weight decay as 5e-4, constant learning rates of 1e-4 for *Real* and 3e-4 for both *ICH* and *ISIC*, and a batch size of 32. For *ICH*, the multi-view batch for contrastive learning is generated by following [14,26]. For both *Real* and *ISIC*, the multi-view batch is generated by 1) RandAug [6] and 2) SimAugment [3]. The hyper-parameters k_1 and k_2 in Eq. 7 are set as 2.0. For federated training, the local training epoch is set as 1 and the global training round is set as 200 for *ICH* and *ISIC* and 30 for *Real*. At each round, all clients (*i.e.*, 100%) are included for model aggregation.

3.1 Comparison with State-of-the-Art Methods

Ten related approaches are included for comprehensive comparison, including FedAvg [28], FedProx [20] addressing data heterogeneity, MOON [19] and FedProc [31] utilizing contrastive learning in FL, FedFocal [33] utilizing focal loss

Table 2. Component-wise study.

FedAvg	DALA	Intra	Inter	*ISIC*		*ICH*	
				BACC	F1	BACC	F1
✓				49.41	54.31	73.75	77.35
✓	✓			50.65	42.12	81.26	76.47
✓	✓	✓		51.30	43.84	81.96	82.68
✓	✓		✓	75.78	76.55	83.81	82.39
✓	✓	✓	✓	**78.84**	**78.05**	**84.22**	**84.73**

Table 3. Parameter-wise study.

Param		BACC	F1
t	0.0	76.33	75.43
	0.5	**78.84**	**78.05**
orth.	w/o	72.64	75.34
	w	**78.84**	**78.05**
d	0.0	77.32	**78.41**
	0.25	**78.84**	78.05

[24] for balancing, FedRS [23] addressing the class-missing problem, FedLC [43] applying frequency-dependent logits adjustment in FL, PRR-Imb [4] training personalized models with heterogeneous and imbalanced data, and CLIMB [35] and CReFF [34] addressing class-imbalance global data in FL. All the methods share the same experimental details described above for a fair comparison. *More implementation details and visualization results can be found in supplemental materials.*

Following the ISIC 2019 competition, balanced accuracy (BACC) is used as the primary metric for class-imbalanced testing sets. Two key metrics in classification, *i.e.* F1 score (F1) and accuracy (ACC) are also employed for evaluation. Comparison results are summarized in Table 1. As can see, FedIIC achieves the best performance against all previous methods across the three metrics, outperforming the second-best approach (CReFF) by 3.99%, 7.32%, and 2.01% in BACC on *Real*, *ISIC*, and *ICH* respectively.

3.2 Ablation Study

To validate the effectiveness of each component in FedIIC, a series of ablation studies are conducted on *ISIC* and *ICH* following the same experimental details described in Sect. 3. Quantitative results are summarized in Table 2. Under severe global imbalance, FedAvg is struggling. With the introduction of DALA, the performance is improved in BACC but degraded in F1. It is consistent with the quantitative results between CReFF and FedAvg on *ICH* in Table 1, indicating the limitation of only eliminating class bias through classifier calibration while ignoring attribute bias. The above results validate the necessity of addressing the imbalance in feature learning for performance improvement. Therefore, introducing either intra- or inter-client contrastive learning for better representation learning under class imbalance is beneficial in both BACC and F1. By combining all the components, FedIIC achieves the best overall performance, outperforming FedAvg with large margins.

Ablation studies of hyper-parameters in FedIIC are conducted on *ISIC* as stated in Table 3. Setting $t = 0$ encounters noticeable performance degradation, indicating the necessity of dynamic temperatures based on class priors in intra-client contrastive learning. Meanwhile, the performance gap between the initial prototypes $\widetilde{V}$ with and without orthogonalization validates the effectiveness of reducing inter-class similarity in prototypes. When introducing difficulty to logit adjustment (*i.e.*, $d = 0.25$), we observe an increase in BACC and a decrease in F1, which is consistent with the above analysis in Table 1 (*i.e.*, CReFF vs. FedAvg).

4 Conclusion

This paper discusses a more realistic federated learning (FL) setting in medical scenarios where global data is class-imbalanced and presents a novel framework FedIIC. The key idea behind FedIIC is to calibrate both the feature extractor and the classification head to simultaneously eliminate attribute biases and class biases. Specifically, both intra- and inter-client contrastive learning are introduced for balanced feature learning, and difficulty-aware logit adjustment is deployed to balance decision boundaries across classes. Experimental results on both real-world and simulated medical FL scenarios demonstrate FedIIC's superiority against the state-of-the-art FL approaches. We believe that this study is helpful to build real-world FL systems for clinical applications.

Acknowledgement. This work was supported in part by the National Natural Science Foundation of China under Grants 62202179 and 62271220, in part by the Natural Science Foundation of Hubei Province of China under Grant 2022CFB585, and in part by the Research Grants Council GRF Grant 16203319. The computation is supported by the HPC Platform of HUST.

References

1. Argenziano, G., et al.: Interactive atlas of dermoscopy (2000)
2. Cao, K., Wei, C., Gaidon, A., Arechiga, N., Ma, T.: Learning imbalanced datasets with label-distribution-aware margin loss. In: NeurIPS, vol. 32 (2019)
3. Chen, T., Kornblith, S., Norouzi, M., Hinton, G.: A simple framework for contrastive learning of visual representations. In: ICML, pp. 1597–1607 (2020)
4. Chen, Z., Yang, C., Zhu, M., Peng, Z., Yuan, Y.: Personalized retrogress-resilient federated learning toward imbalanced medical data. IEEE Trans. Med. Imaging **41**(12), 3663–3674 (2022)
5. Combalia, M., et al.: BCN20000: dermoscopic lesions in the wild. arXiv:1908.02288 (2019)
6. Cubuk, E.D., Zoph, B., Shlens, J., Le, Q.V.: RandAugment: practical automated data augmentation with a reduced search space. In: NeurIPS (2020)
7. Cui, Y., Jia, M., Lin, T.Y., Song, Y., Belongie, S.: Class-balanced loss based on effective number of samples. In: CVPR, pp. 9268–9277 (2019)
8. Deng, J., et al.: ImageNet: a large-scale hierarchical image database. In: CVPR, pp. 248–255 (2009)
9. Duan, M., et al.: Self-balancing federated learning with global imbalanced data in mobile systems. IEEE Trans. Parallel Distrib. Syst. **32**(1), 59–71 (2020)
10. Flanders, A.E., et al.: Construction of a machine learning dataset through collaboration: the RSNA 2019 brain CT hemorrhage challenge. Radiol. Artif. Intel. **2**(3), e190211 (2020)
11. Graf, F., Hofer, C., Niethammer, M., Kwitt, R.: Dissecting supervised constrastive learning. In: ICML, pp. 3821–3830 (2021)

12. Guo, Q., Qi, Y., Qi, S., Wu, D.: Dual class-aware contrastive federated semi-supervised learning. arXiv:2211.08914 (2022)
13. Jiang, M., Wang, Z., Dou, Q.: HarmoFL: harmonizing local and global drifts in federated learning on heterogeneous medical images. In: AAAI, pp. 1087–1095 (2022)
14. Jiang, M., et al.: Dynamic bank learning for semi-supervised federated image diagnosis with class imbalance. In: Wang, L., Dou, Q., Fletcher, P.T., Speidel, S., Li, S. (eds.) Medical Image Computing and Computer Assisted Intervention – MICCAI 2022: 25th International Conference, Singapore, September 18–22, 2022, Proceedings, Part III, pp. 196–206. Springer, Cham (2022). https://doi.org/10.1007/978-3-031-16437-8_19
15. Ju, L., et al.: Flexible sampling for long-tailed skin lesion classification. In: Wang, L., Dou, Q., Fletcher, P.T., Speidel, S., Li, S. (eds.) Medical Image Computing and Computer Assisted Intervention – MICCAI 2022: 25th International Conference, Singapore, September 18–22, 2022, Proceedings, Part III, pp. 462–471. Springer, Cham (2022). https://doi.org/10.1007/978-3-031-16437-8_44
16. Kang, B., Li, Y., Xie, S., Yuan, Z., Feng, J.: Exploring balanced feature spaces for representation learning. In: ICLR (2021)
17. Kang, B., et al.: Decoupling representation and classifier for long-tailed recognition. In: ICLR (2020)
18. Khosla, P., et al.: Supervised contrastive learning. In: NeurIPS, vol. 33, pp. 18661–18673 (2020)
19. Li, Q., He, B., Song, D.: Model-contrastive federated learning. In: CVPR, pp. 10713–10722 (2021)
20. Li, T., et al.: Federated optimization in heterogeneous networks. Proc. Mach. Learn. Syst. **2**, 429–450 (2020)
21. Li, T., et al.: Targeted supervised contrastive learning for long-tailed recognition. In: CVPR, pp. 6918–6928 (2022)
22. Li, X., Jiang, M., Zhang, X., Kamp, M., Dou, Q.: FedBN: federated learning on non-IID features via local batch normalization. In: ICLR (2021)
23. Li, X.C., Zhan, D.C.: FedRS: federated learning with restricted softmax for label distribution non-IID data. In: KDD, pp. 995–1005 (2021)
24. Lin, T.Y., Goyal, P., Girshick, R., He, K., Dollár, P.: Focal loss for dense object detection. In: CVPR, pp. 2980–2988 (2017)
25. Liu, Q., Chen, C., Qin, J., Dou, Q., Heng, P.A.: FedDG: federated domain generalization on medical image segmentation via episodic learning in continuous frequency space. In: CVPR, pp. 1013–1023 (2021)
26. Liu, Q., Yang, H., Dou, Q., Heng, P.-A.: Federated semi-supervised medical image classification via inter-client relation matching. In: de Bruijne, M., et al. (eds.) Medical Image Computing and Computer Assisted Intervention – MICCAI 2021: 24th International Conference, Strasbourg, France, September 27–October 1, 2021, Proceedings, Part III, pp. 325–335. Springer, Cham (2021). https://doi.org/10.1007/978-3-030-87199-4_31
27. Marrakchi, Y., Makansi, O., Brox, T.: Fighting class imbalance with contrastive learning. In: de Bruijne, M., et al. (eds.) Medical Image Computing and Computer Assisted Intervention – MICCAI 2021: 24th International Conference, Strasbourg, France, September 27–October 1, 2021, Proceedings, Part III, pp. 466–476. Springer, Cham (2021). https://doi.org/10.1007/978-3-030-87199-4_44
28. McMahan, B., Moore, E., Ramage, D., Hampson, S., y Arcas, B.A.: Communication-efficient learning of deep networks from decentralized data. In: AISTATS, pp. 1273–1282 (2017)

29. Mendonça, T., Ferreira, P.M., Marques, J.S., Marcal, A.R., Rozeira, J.: PH2-A dermoscopic image database for research and benchmarking. In: EMBC, pp. 5437–5440 (2013)
30. Menon, A.K., et al.: Long-tail learning via logit adjustment. In: ICLR (2021)
31. Mu, X., et al.: FedProc: prototypical contrastive federated learning on non-IID data. Future Gener. Comput. Syst. **143**, 93–104 (2023). https://doi.org/10.1016/j.future.2023.01.019
32. Papyan, V., Han, X., Donoho, D.L.: Prevalence of neural collapse during the terminal phase of deep learning training. Proc. Natl. Acad. Sci. U.S.A. **117**(40), 24652–24663 (2020)
33. Sarkar, D., Narang, A., Rai, S.: Fed-Focal loss for imbalanced data classification in federated learning. In: IJCAI (2020)
34. Shang, X., Lu, Y., Huang, G., Wang, H.: Federated learning on heterogeneous and long-tailed data via classifier re-training with federated features. In: IJCAI (2022)
35. Shen, Z., Cervino, J., Hassani, H., Ribeiro, A.: An agnostic approach to federated learning with class imbalance. In: ICLR (2022)
36. Tan, M., Le, Q.: EfficientNet: rethinking model scaling for convolutional neural networks. In: ICML, pp. 6105–6114 (2019)
37. Tang, K., Tao, M., Qi, J., Liu, Z., Zhang, H.: Invariant feature learning for generalized long-tailed classification. In: Avidan, S., Brostow, G., Cissé, M., Farinella, G.M., Hassner, T. (eds.) Computer Vision – ECCV 2022: 17th European Conference, Tel Aviv, Israel, October 23–27, 2022, Proceedings, Part XXIV, pp. 709–726. Springer, Cham (2022). https://doi.org/10.1007/978-3-031-20053-3_41
38. Tschandl, P., Rosendahl, C., Kittler, H.: The HAM10000 dataset, a large collection of multi-source dermatoscopic images of common pigmented skin lesions. Sci. Data **5**(1), 1–9 (2018)
39. Wang, L., Xu, S., Wang, X., Zhu, Q.: Addressing class imbalance in federated learning. Proc. AAAI Conf. Artif. Intell. **35**(11), 10165–10173 (2021). https://doi.org/10.1609/aaai.v35i11.17219
40. Yan, Z., Wicaksana, J., Wang, Z., Yang, X., Cheng, K.T.: Variation-aware federated learning with multi-source decentralized medical image data. IEEE J. Biomed. Health Inform. **25**(7), 2615–2628 (2020)
41. Yang, M., Wang, X., Zhu, H., Wang, H., Qian, H.: Federated learning with class imbalance reduction. In: EUSIPCO, pp. 2174–2178 (2021)
42. Yang, Z., et al.: ProCo: prototype-aware contrastive learning for long-tailed medical image classification. In: Wang, L., Dou, Q., Fletcher, P.T., Speidel, S., Li, S. (eds.) Medical Image Computing and Computer Assisted Intervention – MICCAI 2022: 25th International Conference, Singapore, September 18–22, 2022, Proceedings, Part VIII, pp. 173–182. Springer, Cham (2022). https://doi.org/10.1007/978-3-031-16452-1_17
43. Zhang, J., et al.: Federated learning with label distribution skew via logits calibration. In: ICML, pp. 26311–26329 (2022)
44. Zhao, Y., Chen, W., Tan, X., Huang, K., Zhu, J.: Adaptive logit adjustment loss for long-tailed visual recognition. Proc. AAAI Conf. Artif. Intell. **36**(3), 3472–3480 (2022). https://doi.org/10.1609/aaai.v36i3.20258
45. Zhu, J., Wang, Z., Chen, J., Chen, Y.P.P., Jiang, Y.G.: Balanced contrastive learning for long-tailed visual recognition. In: CVPR, pp. 6908–6917 (2022)

Transferability-Guided Multi-source Model Adaptation for Medical Image Segmentation

Chen Yang[1], Yifan Liu[2], and Yixuan Yuan[2](✉)

[1] Department of Electrical Engineering, City University of Hong Kong, Hong Kong, SAR, China
[2] Department of Electronic Engineering, The Chinese University of Hong Kong, Hong Kong, SAR, China
yxyuan@ee.cuhk.edu.hk

Abstract. Unsupervised domain adaptation has drawn sustained attentions in medical image segmentation by transferring knowledge from labeled source data to unlabeled target domain. However, most existing approaches assume the source data are collected from a single client, which cannot be successfully applied to explore complementary transferable knowledge from multiple source domains with large distribution discrepancy. Moreover, they require access to source data during training, which is inefficient and unpractical due to privacy preservation and memory storage. To address these challenges, we study a novel and practical problem, named multi-source model adaptation (MSMA), which aims to transfer multiple source models to the unlabeled target domain without any source data. Since no target label and source data is provided to evaluate the transferability of each source model or domain gap between the source and the target domain, we may encounter negative transfer by those less related source domains, thus hurting target performance. To solve this problem, we propose a transferability-guided model adaptation (TGMA) framework to eliminate negative transfer. Specifically, 1) A label-free transferability metric (LFTM) is designed to evaluate transferability of source models without target annotations for the first time. 2) Based on the designed metric, we compute instance-level transferability matrix (ITM) for target pseudo label correction and domain-level transferability matrix (DTM) to achieve model selection for better target model initialization. Extensive experiments on multi-site prostate segmentation dataset demonstrate the superiority of our framework.

Keywords: Source-free Domain Adaptation · Multi-source · Label-free transferability metric

1 Introduction

Deep neural networks have greatly advanced medical image analysis in recent years [12]. However, a large amount of annotated data is required for training,

H. Greenspan et al. (Eds.): MICCAI 2023, LNCS 14221, pp. 703–712, 2023.
https://doi.org/10.1007/978-3-031-43895-0_66

which is time-consuming and error-prone, especially in medical image segmentation task that needs pixel-wise annotations. Moreover, a segmentation model trained on one clinical centre (source domain) often fails to generalize well when deployed in a new centre (target domain) due to the discrepancy in the data distribution [2,9,16]. Unsupervised domain adaptation (UDA) [5,14,17] seeks to tackle this dilemma by transferring the knowledge from label-rich source domain to label-rare target domain. However, the source data may become inaccessible due to storage and privacy concerns in medical settings, which hinders the wide applications of domain adaptation. Towards this obstacle, great interests have been invoked to explore source-free domain adaptation (SFDA) [2,6,8,9,16], where a model pre-trained on the labeled source data are adapted to the unlabeled target domain without accessing source data. Though great successes, how to achieve adaptation to the unlabeled target domain with the knowledge from multiple source domains under privacy protection is still an open question to be solved.

To this end, we study a practical and challenging domain adaptation problem which explores transferable knowledge from multiple source domains to target domain with only pre-trained source models rather than the source data, namely *multi-source model adaptation* (MSMA). Although MSMA methods [1,3,7] have made great progress for natural object recognition, there is still a blank in the multi-source-free domain adaptive medical image segmentation. Directly applying existing MSMA methods on medical image segmentation by optimizing all source segmentation models are time-consuming and inefficient due to larger model capacity of segmentation model than classification model. Another trivial solutions to tackle MSMA via SFDA methods [2,6,16,18] are to adapt each source model individually and simply take an average prediction of adapted models. However, this strategy does not take into account the varying contributions of different source models to the target domain, which can result in negative transfer from less related source domains. To rank pre-trained models, transferability metrics [10,13,20] have been widely applied to measure the domain relevance or task relevance for transfer learning, but all of them need target annotations, which is not accessible for multi-source model adaptation. Automatically select an optimal subset of the source models without requiring source data and target annotations in an unsupervised fashion is of far-reaching significance for MSMA.

To address this problem, we develop a novel Transferability-Guided Model Adaptation (TGMA) model, which represents the first attempt to solve MSMA in medical image segmentation. Specifically, a label-free transferability metric (LFTM) is designed to evaluate the relevance between source and target domain without access to the source data. Based on the designed LFTM, we can compute instance-level transferability matrix (ITM) to achieve pseudo-label correction for precise supervision, and domain-level transferability matrix (DTM) to accomplish model selection for better target initialization. To this end, we can achieve adaptation to unlabeled target domain with clean pseudo label and proper model initialization. The main contributions are summarized as:

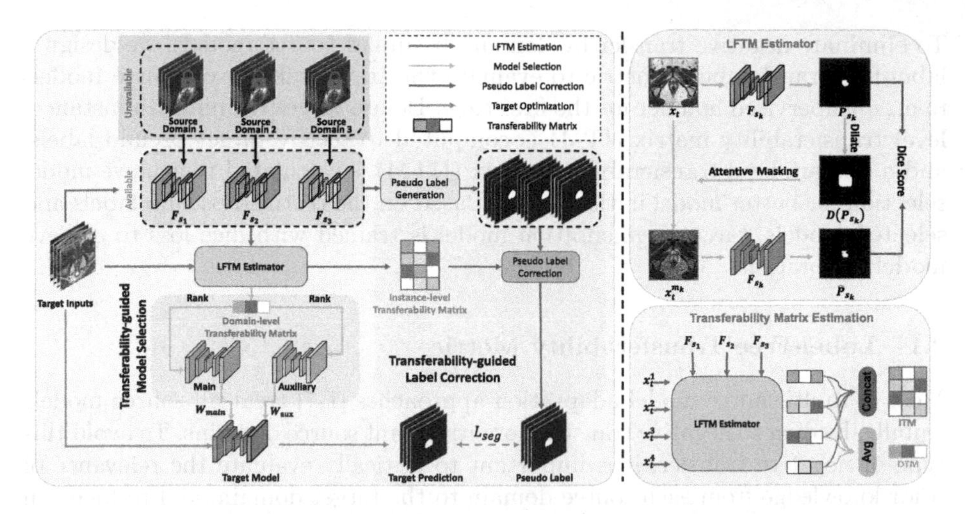

Fig. 1. Illustration of Transferability-Guided multi-source Model Adaptation (TGMA) framework, including (a) label-free transferability metric (LFTM) estimator, (b) transferability-guided model selection and (c) transferability-guided label correction.

- We present the first work that studies the practical domain adaptation problem of transferring knowledge from multiple source segmentation models rather than the source data to unlabeled target domain.
- We design a novel label-free transferability metric (LFTM) based on attentive masking consistency to evaluate the domain relevance for the first time.
- Based on the LFTM, we propose a transferability-guided model adaptation (TGMA) framework including pseudo-label correction by instance-level transferability matrix (ITM) and model selection by domain-level transferability matrix (DTM).
- Extensive experiments on the multi-site prostate segmentation dataset demonstrate the superiority of our TGMA compared with state-of-the-art domain adaptation methods.

2 Method

In MSMA scenario, we address the problem of jointly adapting multiple segmentation models, trained on a variety domains, to a new unlabeled target domain. Formally, let us consider we have a set of source models $\{F_{s_j}\}_{j=1}^{M}$, where the jth model $\{F_{s_j}\}$ is a segmentation model learned using the source dataset $\mathcal{D}_s^j = \{x_{s_j}^i, y_{s_j}^i\}_{i=1}^{N_j}$, with N_j data points, where $x_{s_j}^i$ and $y_{s_j}^i$ denote the i-th source image and the corresponding segmentation label respectively. Now, given a target unlabeled dataset $\mathcal{D}_t = \{x_{t^i}\}_{i=1}^{N_t}$, the problem is to learn a segmentation model F_t, using only the learned source models, without any access to the source dataset. Figure 1 gives an overview of our proposed TGMA framework.

To eliminate negative transfer by domain-dissimilar source models, we design a label-free transferability metric to evaluate the transferability of source models in an unsupervised manner for the first time. Before target training, an instance-level transferability matrix (ITM) is computed to rectify target pseudo labels, and a domain-level transferability matrix (DTM) is calculated to achieve model selection for better model initialization. Based on the rectified pseudo labels and selected models, target segmentation model is trained with dice loss to achieve model adaptation.

2.1 Label-Free Transferability Metric

Most of multi-source model adaptation approaches [1,7] treat all source models equally, leading to negative transfer from irrelevant source domains. To avoid this type of negative transfer, it is important to critically evaluate the relevance of prior knowledge from each source domain to the target domain, and to focus on the most relevant source domains for learning in the target domain. However, it's challenging to evaluate the domain relevance in the absence of source data and target ground truths. To identify the transferability of source models, we develop a label-free transferability metric (LFTM) on the basis of attentive masking consistency to prevent negative transfer for the first time. Our metric is designed based on two assumptions: 1) *Sample relevance*: similar samples should hold identical predictions; 2) *Model stability*: if a source model makes accurate decision on this sample, little permutation on irrelevant regions will not influence the prediction. We follow these two assumptions to construct augmented sample by attentive masking, and compute the consistency as the transferability.

Given unlabeled target data $\mathcal{D}_t = \{x_t^i\}_{i=1}^N$ and a pre-trained source segmentation model F_{s_k}, we import them to the LFTM estimator and compute the transferability metric $LFTM(x_t, F_{s_k})$ with only twice forwards as shown in Fig. 1. In the first forward process, the original target sample x_t is passed into the source model F_{s_k} to generate segmentation map $P_{s_k} = F_{s_k}(x_t)$. Based on the assumption that masking the normal regions from the diseased image will not affect the lesion regions, we preserve the segmentation region of the original image and randomly mask the other regions to generate masked image $x_t^{m_k}$. Since the segmentation results may be affected by receptive field, we enlarge the segmentation map P_{s_k} to $D(P_{s_k})$ by dilation. Then masked image $x_t^{m_t}$ is generated by combination of enlarged lesion regions and masked normal regions:

$$x_t^{m_k} = M(x_t) * (1 - D(P_{s_k})) + x_t * D(P_{s_k}), \tag{1}$$

where $M(x_t)$ is the masking operation to randomly remove pixels. In the second forward process, the masked target sample $x_t^{m_k}$ is passed into the source model F_{s_k} to generate segmentation map $\hat{P}_{s_k} = F_{s_k}(x_t^{m_k})$. Then we calculate the dice score between these two predictions as transferability metric:

$$LFTM(x_t, F_{s_k}) = 2 * \frac{P_{s_k} \cap \hat{P}_{s_k}}{P_{s_k} + \hat{P}_{s_k}}. \tag{2}$$

The larger the LFTM is, the more stable the source model is on the target sample. With M source models and N_t target samples, we can compute the instance-level transferability matrix (ITM) $T_{instance} \in \mathbb{R}^{M \times N_t}$, which can be utilized to correct target pseudo labels. Averaging $T_{instance}$ on the domain-space can generate domain-level transferability matrix (DTM) $T_{domain} \in \mathbb{R}^{M \times 1}$, which represents the contribution of each source model to the target domain. The detailed process is illustrated in Transferability Matrix Estimation of Fig. 1.

2.2 Transferability-Guided Model Adaptation

The basic pipeline for target training needs accurate pseudo labels and suitable model initialization. While there are multiple pseudo labels and source models, simply averaging them as target supervision and model initialization is trivial solution, which ignores the contribution differences of these source domains. To tackle this problem, we propose a transferability-guided model adaptation (TGMA) framework on the basis of LFTM, which consists of two modules: Label Correction and Model Selection. Based on the instance-level transferability matrix $T_{instance}$, we re-weight the pseudo labels generated by multiple source models to achieve pseudo label correction. With the domain-level transferability matrix T_{domain}, we select the most portable source model as the main model initialization and make full use of other source models by weighted optimization strategy.

Transferability-Guided Label Correction. In MSMA, we generate pseudo labels as supervision because no target ground truth is available. However, with multiple pseudo labels predicted by source models for a target sample, prior works [1,7] typically average these labels equally to obtain the final pseudo label. However, negative source models that are poorly suited to the target domain may generate inaccurate pseudo labels, resulting in noisy or unreliable training data. To eliminate negative transfer and improve pseudo-label correction, we can re-weight model predictions from all source models using the calculated instance-level transferability matrix $T_{instance}$.

Taking a target sample x_t for example, we pass this sample to source models $\{F_{s_1}, F_{s_2}, ..., F_{s_M}\}$ to obtain corresponding predictions $\{P_{s_1}, P_{s_2}, ..., P_{s_M}\}$. We take argmax operation on these predictions to generate one-hot pseudo labels $\{y_{s_1}, y_{s_2}, ..., y_{s_M}\}$, where $y = argmax(P)$. The instance-level transferability matrix $T_{instance}$ is applied on these pseudo labels to achieve noise correction by contribution re-weighting:

$$y_t = argmax(\sum_{i=1}^{M} LFTM(x_t, F_{s_i}) * y_{s_i}), \tag{3}$$

where each pseudo label is weighted by the corresponding LFTM score for better combination. This strategy largely prevents the negative transfer problem caused by noisy labels of those domain-irrelevant source models.

Transferability-Guided Model Selection. Previous MSMA methods [1,7] usually treat all models equally and optimize all source models parameters to achieve adaptation to the target domain. On the one hand, they ignore the negative transfer problem led by some less related domains. On the other hand, optimizing all source parameters is time-consuming and inefficient. To better make full use of the source models, we utilize the calculated domain-level transferability matrix T_{domain} to rank all source models.

With T_{domain} representing the transferability of source models, we choose the best source model as main network F_{main} and the second best model as auxiliary network F_{aux}. Only initialing the target model from F_{main} may ignore complementary knowledge of other source models, while optimizing all source models are inefficient. To obtain a compromise solution, we take the second model as auxiliary parameter knowledge. Then a weighted optimization strategy is utilized on the best model and the auxiliary model with weight W_{main} and W_{aul} respectively:

$$F_t = \min_{F \cap W} \mathcal{L}_{dice}(y_t, W_{main} * F_{main}(x_t) + W_{aux} * F_{aux}(x_t)), \tag{4}$$

where $\mathcal{L}_{dice}$ is calculated on the combined target prediction and corresponding pseudo label. This loss optimizes model parameter $W_{main} * F_{main} + W_{aux} * F_{aux}$. The model selection strategy choose optimal source model while makes full use of those sub-optimal source models for better model initialization, thus avoiding the negative transfer by those domain-irrelevant domains.

3 Experiment

3.1 Dataset

Extensive experiments are conducted to verify the effectiveness of our proposed framework on Prostate MR (**PMR**) dataset which is collected and labeled from six different public data sources for prostate segmentation [15]. All of the MRI images have been re-sampled to the same spacing and center-cropped with the size of 384×384. We divide them into six sites, each of which contains $\{261, 384, 158, 468, 421, 175\}$ slices. We denote these six sites as $\{A, B, C, D, E, F\}$ for convenience. At each adaptation process, five sites are selected as source domains and the rest one is set as the target domain. We conduct leave-one-domain-out experiments by selecting one domain to hold out as the target. For example, $\rightarrow A$ denotes adapting source models from $\{B, C, D, E, F\}$ to unlabeled images of A.

3.2 Implementation Details and Evaluation Metrics

The framework is implemented with Pytorch 1.7.0 using an NVIDIA RTX 2080Ti GPU. Following [15], we adopt UNet as our segmentation backbone. We train the target model for 200 epochs with the batch size of 6. Adam optimizer is adopted with the momentum of 0.9 and 0.999, and the learning rate is set to 0.001. We adopt the well-known metrics Dice score for segmentation evaluation.

Table 1. Comparison with state-of-the-art domain adaptation approaches on PMR dataset, measured by dice score.

Type	Method	Source Data	→ A	→ B	→ C	→ D	→ E	→ F	Average
SFDA	SHOT (20') [6]	✗	28.76	34.32	50.57	31.55	33.60	28.70	34.58
	NRC (21') [18]	✗	32.44	38.05	59.39	28.03	40.47	27.35	37.62
	FSM (22') [16]	✗	33.57	38.56	70.72	29.40	36.76	32.52	40.25
MSDA	KD3A (21') [4]	✓	38.05	55.78	64.16	31.77	37.69	49.36	46.13
	CWAN (21') [19]	✓	51.18	61.96	71.62	75.45	55.88	60.11	62.69
	PTMDA (22') [11]	✓	65.16	68.37	79.05	77.23	61.58	69.42	70.13
MSMA	Source only	✗	31.26	39.80	66.95	9.86	14.93	32.77	32.59
	DECISION (22') [1]	✗	48.03	60.72	69.85	71.34	52.94	63.16	61.01
	DINE (22') [7]	✗	54.20	62.82	74.11	72.59	53.46	64.78	63.66
	TGMA (Ours)	✗	62.76	65.73	76.14	75.10	58.59	65.63	67.32
	Ours w/o ITM	✗	56.41	60.65	72.83	71.29	54.42	62.04	62.94
	Ours w/o DTM	✗	59.48	61.56	73.16	73.97	56.85	63.24	64.71

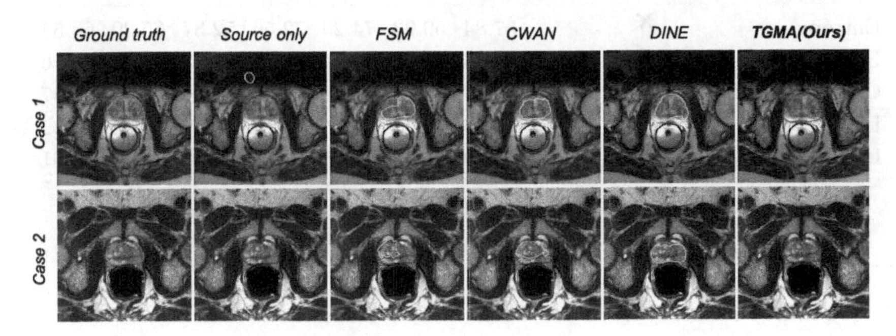

Fig. 2. Qualitative comparison on the PMR dataset of different DA methods.

3.3 Comparison with State-of-the-Arts

We compare our methods to several domain adaptation frameworks, including the single source-free domain adaptation (SFDA) [6,16,18], multi-source domain adaptation (MSDA) [4,11,21] and multi-source model adaptation (MSMA) [1,7] methods. For implementation, as most of these methods are originally designed for the image classification task, we try out best to keep their design principle and adapt them to our image segmentation task. Specifically, SFDA methods are performed on each source model and averaging the adapted model predictions as the final results. The results on prostate segmentation is listed in Table 1. As observed, MSDA methods shows superior performance than MSMA approaches due to access to the source data. Notably, compared with SFDA and MSMA approaches, our TGMA achieves higher performance on nearly all metrics with 67.32% on Average Dice. These clear improvements benefit from our LFTM metric which considers the different contributions of each source model, and largely eliminate negative transfer from the perspective of pseudo label generation and model initialization. Without the rectification by instance-level transferability matrix (Ours w/o ITM), the pseudo labels are simply generated by average combination of predictions from source models. The significant decrease in performance by 4.38% on Average Dice highlights the criticality of weighting pseudo labels with scores that reflect the relevance of the source domains. Without the model selection by domain-level transferability matrix (Ours w/o DTM), the target models are initialized from each source pre-trained network and trained separately, leading to 2.61% performance drop on Average Dice. It demonstrates that model initialization is also essential to the transfer learning. Moreover, Fig. 2 shows the segmentation results of different methods on two typical cases. We observe that our model with transferability guidance can well eliminate the negative transfer interference by some domain-irrelevant domains.

Table 2. Comparison with different unsupervised metrics on PMR dataset.

Method	Source Data	→ A	→ B	→ C	→ D	→ E	→ F	Average
Entropy	✗	57.84	59.03	74.29	72.56	52.37	65.20	63.54
Rotation	✗	60.46	61.35	73.74	71.82	53.92	62.13	63.90
Cropping	✗	61.14	62.37	74.59	73.83	55.44	63.05	65.07
LFTM	✗	62.76	65.73	76.14	75.10	58.59	65.63	67.32
LFTM w/o Dilation	✗	61.69	63.08	73.95	74.17	56.22	64.54	65.61

3.4 Ablation Analysis

The performance improvement mainly comes from our designed LFTM to detect negative transfer. There are some other unsupervised metrics that can evaluate model stability, such as entropy, rotation-consistency and crop-consistency. To better evaluate the effectiveness of LFTM, we apply these unsupervised metrics to estimate ITM and DTM for label correction and model selection. The comparison results are shown in Table 2. It's obvious that our proposed LFTM outperforms other unsupervised metrics with a large margin. Entropy may make overconfident decisions on model predictions, thus leading to high transferability on those domain-irrelevant source models. Rotation and Cropping are simple data augmentation methods, which can only evaluate the model stability. Our proposed LFTM makes full use of the segmentation mask to construct feature-nearest sample, thus applying sample relevance to evaluate model transferability. Removing dilation operation leads to 1.71% performance degradation on Average Dice, revealing the effect of receptive field.

4 Conclusion

In this paper, we study a practical domain adaptation problem, named multi-source model adaptation where only multiple pre-trained source segmentation models rather than the source data are provided for adaptation to unlabeled target domain. To eliminate the negative transfer by domain-dissimilar source models, we design a label-free transferability metric based on the attentive masking consistency to evaluate the transferability of each source segmentation model with only target images. Using this metric, we calculate two types of transferability matrices: an instance-level matrix to adjust the target pseudo label, and a domain-level matrix to choose an optimal subset for improved model initialization.

Acknowledgements. This work was supported by National Natural Science Foundation of China 62001410, Hong Kong Research Grants Council (RGC) Early Career Scheme grant 21207420, General Research Fund 11211221.

References

1. Ahmed, S.M., Raychaudhuri, D.S., Paul, S., Oymak, S., Roy-Chowdhury, A.K.: Unsupervised multi-source domain adaptation without access to source data. In: Proceedings of the IEEE/CVF Conference on Computer Vision and Pattern Recognition, pp. 10103–10112 (2021)
2. Bateson, M., Kervadec, H., Dolz, J., Lombaert, H., Ayed, I.B.: Source-free domain adaptation for image segmentation. Med. Image Anal. **82**, 102617 (2022)
3. Dong, J., Fang, Z., Liu, A., Sun, G., Liu, T.: Confident anchor-induced multi-source free domain adaptation. Adv. Neural. Inf. Process. Syst. **34**, 2848–2860 (2021)
4. Feng, H., et al.: KD3A: Unsupervised multi-source decentralized domain adaptation via knowledge distillation. In: ICML, pp. 3274–3283 (2021)
5. Ganin, Y., et al.: Domain-adversarial training of neural networks. J. Mach. Learn. Res. **17**(1), 2030–2096 (2016)
6. Liang, J., Hu, D., Feng, J.: Do we really need to access the source data? source hypothesis transfer for unsupervised domain adaptation. In: International Conference on Machine Learning, pp. 6028–6039. PMLR (2020)
7. Liang, J., Hu, D., Feng, J., He, R.: Dine: Domain adaptation from single and multiple black-box predictors. In: Proceedings of the IEEE/CVF Conference on Computer Vision and Pattern Recognition, pp. 8003–8013 (2022)
8. Liu, X., Yuan, Y.: A source-free domain adaptive polyp detection framework with style diversification flow. IEEE Trans. Med. Imaging **41**(7), 1897–1908 (2022)
9. Liu, Y., Zhang, W., Wang, J.: Source-free domain adaptation for semantic segmentation. In: Proceedings of the IEEE/CVF Conference on Computer Vision and Pattern Recognition, pp. 1215–1224 (2021)
10. Nguyen, C., Hassner, T., Seeger, M., Archambeau, C.: Leep: A new measure to evaluate transferability of learned representations. In: International Conference on Machine Learning, pp. 7294–7305. PMLR (2020)
11. Ren, C.X., Liu, Y.H., Zhang, X.W., Huang, K.K.: Multi-source unsupervised domain adaptation via pseudo target domain. IEEE Trans. Image Process. **31**, 2122–2135 (2022)
12. Ronneberger, O., Fischer, P., Brox, T.: U-Net: convolutional networks for biomedical image segmentation. In: Navab, N., Hornegger, J., Wells, W.M., Frangi, A.F. (eds.) MICCAI 2015. LNCS, vol. 9351, pp. 234–241. Springer, Cham (2015). https://doi.org/10.1007/978-3-319-24574-4_28

13. Tran, A.T., Nguyen, C.V., Hassner, T.: Transferability and hardness of supervised classification tasks. In: Proceedings of the IEEE/CVF International Conference on Computer Vision, pp. 1395–1405 (2019)
14. Tzeng, E., Hoffman, J., Saenko, K., Darrell, T.: Adversarial discriminative domain adaptation. In: CVPR, pp. 7167–7176 (2017)
15. Wang, J., Jin, Y., Wang, L.: Personalizing federated medical image segmentation via local calibration. In: Computer Vision-ECCV 2022: 17th European Conference, Tel Aviv, Israel, October 23–27, 2022, Proceedings, Part XXI. pp. 456–472. Springer (2022). https://doi.org/10.1007/978-3-031-19803-8_27
16. Yang, C., Guo, X., Chen, Z., Yuan, Y.: Source free domain adaptation for medical image segmentation with Fourier style mining. Med. Image Anal. **79**, 102457 (2022)
17. Yang, C., Guo, X., Zhu, M., Ibragimov, B., Yuan, Y.: Mutual-prototype adaptation for cross-domain polyp segmentation. IEEE J. Biomed. Health Inform. **25**(10), 3886–3897 (2021). https://doi.org/10.1109/JBHI.2021.3077271
18. Yang, S., van de Weijer, J., Herranz, L., Jui, S., et al.: Exploiting the intrinsic neighborhood structure for source-free domain adaptation. Adv. Neural. Inf. Process. Syst. **34**, 29393–29405 (2021)
19. Yao, Y., Li, X., Zhang, Y., Ye, Y.: Multisource heterogeneous domain adaptation with conditional weighting adversarial network. IEEE Trans. Neural Netw. Learn. Syst. (2021)
20. You, K., Liu, Y., Wang, J., Long, M.: Logme: practical assessment of pre-trained models for transfer learning. In: International Conference on Machine Learning, pp. 12133–12143. PMLR (2021)
21. Zhao, S., et al.: Multi-source distilling domain adaptation. In: Proceedings of the AAAI Conference on Artificial Intelligence. vol. 34, pp. 12975–12983 (2020)

Explaining Massive-Training Artificial Neural Networks in Medical Image Analysis Task Through Visualizing Functions Within the Models

Ze Jin, Maolin Pang, Yuqiao Yang, Fahad Parvez Mahdi, Tianyi Qu, Ren Sasage, and Kenji Suzuki(✉)

Biomedical Artificial Intelligence Research Unit, Institute of Innovative Research, Tokyo Institute of Technology, Kanagawa, Japan
{jin.z.ab,suzuki.k.di}@m.titech.ac.jp

Abstract. In this study, we proposed a novel explainable artificial intelligence (XAI) technique to explain massive-training artificial neural networks (MTANNs). Firstly, we optimized the structure of an MTANN to find a compact model that performs equivalently well to the original one. This enables to "condense" functions in a smaller number of hidden units in the network by removing "redundant" units. Then, we applied an unsupervised hierarchical clustering algorithm to the function maps in the hidden layers with the single-linkage method. From the clustering and visualization results, we were able to group the hidden units into those with similar functions together and reveal the behaviors and functions of the trained MTANN models. We applied this XAI technique to explain the MTANN model trained to segment liver tumors in CT. The original MTANN model with 80 hidden units (F1 = 0.6894, Dice = 0.7142) was optimized to the one with nine hidden units (F1 = 0.6918, Dice = 0.7005) with almost equivalent performance. The nine hidden units were clustered into three groups, and we found the following three functions: 1) enhancing liver area, 2) suppressing non-tumor area, and 3) suppressing the liver boundary and false enhancement. The results shed light on the "black-box" problem with deep learning (DL) models; and we demonstrated that our proposed XAI technique was able to make MTANN models "transparent".

Keywords: Deep Learning · Explainable AI (XAI) · Visualizing Functions · Liver Tumor Segmentation · Unsupervised Hierarchical Clustering

1 Introduction

Artificial intelligence (AI) research has evolved rapidly, and unprecedented breakthroughs have been made in many fields. Applications of AI products can be witnessed in our daily life, such as autonomous driving, computer-aided diagnosis, automatic voice customer service, etc. The development of AI is undoubtedly a revolution in the course of human history.

H. Greenspan et al. (Eds.): MICCAI 2023, LNCS 14221, pp. 713–722, 2023.
https://doi.org/10.1007/978-3-031-43895-0_67

The most effective and commonly used AI model is the one based on deep neural networks [1]. However, with continuous research being held in methodologies, DL models are becoming more and more complicated. Researchers found that the deeper and more complex DL models are, the better the performance they could achieve for the tasks that traditional AI algorithms could not work well. The complexity of DL models reduces interpretability and transparency substantially; therefore, the current DL models are "black-box" [2]. It is difficult to find how the model works in a way that humans can understand. Because of that, what researchers can do is only to prepare enough data and spend time training a model to obtain a high performance. Therefore, researchers or users can hardly find the reason why a DL model made a wrong decision.

XAI is an old area in AI research, but was named relatively recently [3, 4], focusing now on the explainability of DL models. The final goal of XAI is to develop methods for revealing a basis for the decision made by a DL model and how the decision was made by the model to let users understand and trust the decision and model. Many XAI methods have been proposed to explain a trained DL model (i.e., post-hoc methods). Representative XAI methods include class activation mapping (CAM) [5], grad-CAM, layer-wise relevance propagation (LRP) [6], DL important features (DeepLIFT) [7], local interpretable model-agnostic explanations (LIME) [8], and SHapley additive explanations (SHAP) [9]. These XAI methods offer post-hoc explanations that indicate which areas in a given input image the trained model focuses on and identify which areas in the image have a positive or negative impact on the model decision. In other words, those XAI methods are "instance-based" and limited to the visual explanation of model's attentions in a given input image (i.e., an instance). However, they do not offer explanations of the learned functions of the network.

In this study, we developed and presented an original XAI approach that can reveal the learned functions of groups of neurons in a neural network, which we call "functional explanations" and define as explanations of the model behavior by a combination of functions, as opposed to the visualization of a pattern to which a neuron responds. To our knowledge, there is no XAI method that offers functional explanations. Thus, our method is a post-hoc method that offers both instance-based and model-based functional explanations. We applied our XAI method to an MTANN model to emphasize the explainability and trustability of the MTANN, so that users can trust the MTANN.

2 Method

2.1 MTANN Deep Learning

In the field of image processing, supervised nonlinear filters and edge enhancers based on an artificial neural network (ANN) [10] have been investigated for the reduction of the quantum noise in angiograms and supervised semantic segmentation of the left ventricles in angiography [11], which are called neural filters and neural edge enhancers, respectively. By extending the neural filter and edge enhancer, massive-training artificial neural networks (MTANNs) have been developed to reduce false positives in the computerized detection of lung nodules in computed tomography (CT) [12]. The MTANNs have also shown promising performance in pattern recognition and classification tasks [13, 14].

An MTANN is a deep learning model consisting of linear-output artificial neural network regression model that directly operates on pixels in an input image, as shown in Fig. 1. A large number of patches are extracted from input images; and corresponding pixels at the same positions in desired output images, named as teaching images, are extracted for the MTANN to learn. This patch-based training leads to the fact that the MTANN can be trained with only a small number of input and teaching images.

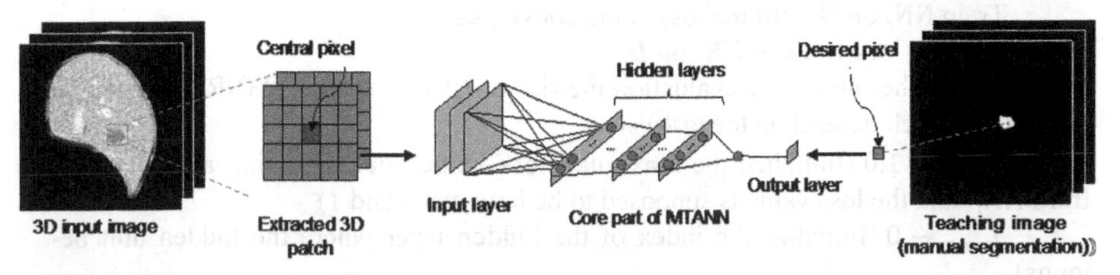

Fig. 1. Illustration of the structure of MTANN, extracting a patch from an input image and a desired pixel from a teaching image.

2.2 Sensitivity-Based Structure Optimization

The numbers of hidden layers and their units in an MTANN model are adjustable hyper-parameters. A relatively large structure is used to ensure that the model performs well on a specific task. A trained large model, however, may contain redundant units, and functions of neurons for the task would be "distributed and diluted" in many neurons in the model. This makes the analysis of the functions of neurons very difficult [15].

To address this issue, we applied our sensitivity-based structure optimization algorithm [16] to a trained large MTANN model to "consolidate" the diluted functions of neurons in the MTANN model. With this algorithm, redundant hidden units of the model are gradually removed; and a compact model with equivalent performance is obtained. The algorithm is described as the following steps:

Algorithm 1: Structure optimization for the MTANN.

Require: $\boldsymbol{D} = \{(x_i,y_i)|1\leq i\leq N\}$: The training data
Require: $\boldsymbol{H} = \{h_i|1\leq i\leq n\}$: The numbers of units in each hidden layer
$t \leftarrow 0$ (Initialize timestamp)
Initialize the weights in the model NN_i
While $\exists h_i>1(h_i\in H)$ **do**
$t\leftarrow t+1$
Train NN_t on $\boldsymbol{D}$ until the loss value converges
$l_t \leftarrow$ the loss value of NN_t on $\boldsymbol{D}$
$m_t \leftarrow$ other necessary evaluation metrics of NN_t on $\boldsymbol{D}$ (like PSNR, dice coefficient, etc., which depend on the task)
$l_{max} \leftarrow 1.0$ (Initialize the maximum loss value after removing a hidden unit from NN_t, and the loss value is supposed to be between 0 and 1)
$i_{max} \leftarrow 0$ (Initialize the index of the hidden layer where the hidden unit belongs)
$j_{max} \leftarrow 0$ (Initialize the index of the hidden layer until in the i_{max}-th hidden layer)
for i in $\{1...n\}$ **do** (Go through each hidden layer)
if $h_i = 1$ **do** (This layer has only one unit which cannot be removed)
Skip to the next iteration
for j in $\{1...h_i\}$ **do** (Go through each hidden unit in the i-th hidden layer)
Remove the j-th hidden unit in the i-th hidden layer from NN_t temporarily
$l_0 \leftarrow$ the loss value of NN_t on $\boldsymbol{D}$
if $l_0 < l_{max}$ **do**
$l_{max} \leftarrow l_0$
$i_{max} \leftarrow i$
$j_{max} \leftarrow j$
Put the j-th hidden unit in the i-th hidden layer back to NN_t
$NN_{t+1} \leftarrow NN_t$ (Copy current model's weights and structure)
Remove the j-th hidden unit in the i-th hidden layer from NN_{t+1} permanently
return $\{(NN_i,l_i,m_i)|1\leq i\leq t\}$

With the proposed optimization algorithm, the hidden units of MTANN could be gradually removed until the performance drops greatly when any of the rest unit is deleted.

2.3 Calculation of Weighted Function Maps

After applying the structure optimization algorithm, every hidden unit in the compact model is expected to have an essential function for the target task. To understand the functions of the hidden units, function maps were obtained by performing the MTANN convolution of a hidden unit over a given input image. For better discrimination between enhancement and suppression, the function maps were normalized and then multiplied by the sign of the weight between the hidden units and the output unit. Weighted function maps were finally generated by shifting the range of the function map by 0.5. Namely, for a given hidden unit, in the weighted function map, a pixel value >0.5 means enhancement of patterns in the input image, whereas a pixel value <0.5 means suppression.

2.4 Unsupervised Hierarchical Clustering

To group similar functions of the hidden units of the MTANN, we applied an unsupervised hierarchical clustering algorithm [17] to the weighted functional visualization maps. With this algorithm, the hidden units were automatically divided into several groups based on the following distance function between the weighted function maps of the hidden units:

$$distance(x, y) = \alpha(1 - SSIM(x, y)) + NRMSE(x, y) \tag{1}$$

where SSIM is the structural similarity index, and NMRSE is the normalized root mean square error. With the unsupervised hierarchical clustering algorithm, we visualize the function maps of the hidden units group by group to explain the behavior of each group of the hidden units.

3 Experiments

3.1 Dynamic Contrast-Enhanced Liver CT

Our XAI technique was applied to explain the MTANN model's decision in a liver tumor segmentation task [20]. Dynamic contrast-enhanced liver CT scans consisting of 42 patients with 194 liver tumors in the portal venous phase from the LiTS database [21] were used in this study. Each slice of the CT volumes in the dataset has a matrix size of 512 × 512 pixels, with in-plane pixel sizes of 0.60–1.00 mm and thicknesses of 0.20–0.70 mm. The dataset consists of the original hepatic CT image with the liver mask and the "gold-standard" liver tumor region manually segmented by a radiologist, as illustrated in Fig. 2.

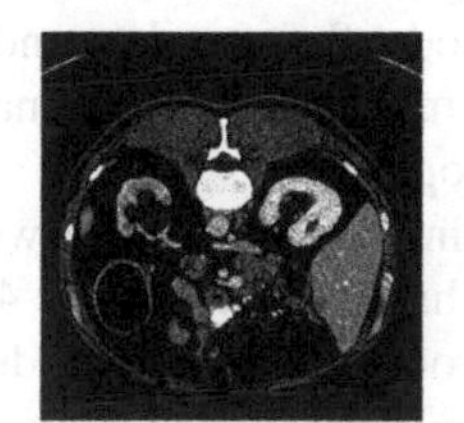
(a) Original Hepatic CT Image

(b) Manually Segmented Liver Mask

(c) Gold-standard Manual Segmentation of Tumor

Fig. 2. An example from the dynamic contrast-enhanced liver CT dataset.

Firstly, to have the same physical scale on spatial coordinates, bicubic interpolation was applied on the original hepatic CT images together with the corresponding liver mask and "gold-standard" tumor segmentation to obtain isotropic images with a voxel size of $0.60 \times 0.60 \times 0.60$ mm^3. Then, to unify the image size into the same size, the isotropic image was cropped to obtain the liver region volume of interest (VOI) with an in-plane matrix size of 512×512. An anisotropic diffusion filter was applied to reduce the quantum noise, which could substantially reduce the noise while major structures such as tumors and vessels maintained [22]. Finally, a Z-score normalization was applied to unify complex histograms of tumors in different cases. The final pre-processed CT images were used as the input images.

In addition, since most liver tumors' shape is ellipsoidal, the liver tumors can also be enhanced by the Hessian-based method and utilized in the model to improve the performance [23, 24]. Hence, the model consisted of these two input channels: segmented liver CT image and its Hessian-enhanced image. Also, the patches were extracted from input images from both channels: a $5 \times 5 \times 5$ sized patch in the same spatial position was extracted to form a training patch with a size of $2 \times 5 \times 5 \times 5$ pixels.

Seven cases and 24 cases in the dynamic contrast-enhanced CT scans dataset were used for training and testing, respectively. 10,000 patches were randomly selected from the liver mask region in each case, summing up to a total of 70,000 training samples for training. The number of input units in the MTANN model with one hidden layer was 250. The structure optimization process started with 80 hidden units in the hidden layer. The binary cross-entropy (BCE) loss function was used to train the model. The MTANN model classified the input patches into tumor or non-tumor classes, and the output pixels represented the probability of being a tumor class. During the structure optimization process, the F1 score on the training patches and the Dice coefficient on the training images were also calculated as the reference to select a suitable compact model that performed equivalently to the original large model.

As observed in the four evaluation metric curves in Fig. 3, as the number of hidden units was reduced from 80 to 9, the performance of the model fluctuated up and down, and after it was reduced below 9, the performance of the model dropped dramatically. Therefore, we chose a number of hidden units of 9 as the optimized structure.

Then, we applied the unsupervised hierarchical clustering algorithm to the weighted function maps from the optimized compact model with 9 hidden units. Figure 4 shows that the 9 hidden units are clearly divided into 3 different groups. We denote hidden units 3, 4, and 7 as group A, hidden units 2, 6, 1, and 8 as group B, and hidden units 0 and 5 as group C. The hidden units in the same group should have a similar function, and the function maps from each group should show the function of the group.

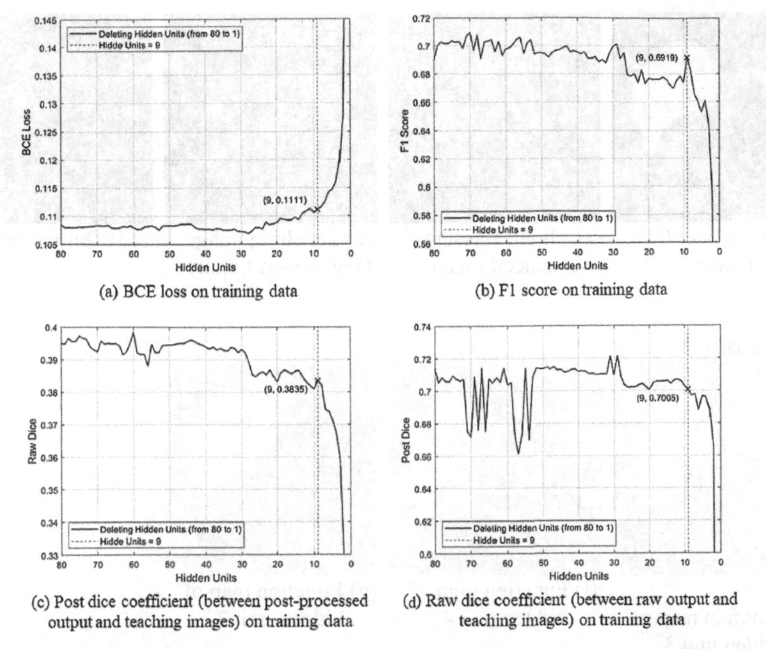

(a) BCE loss on training data

(b) F1 score on training data

(c) Post dice coefficient (between post-processed output and teaching images) on training data

(d) Raw dice coefficient (between raw output and teaching images) on training data

Fig. 3. Performance change of an MTANN segmentation scheme (in terms of BCE loss, F1 score, raw dice, and post dice) in the structure optimization process.

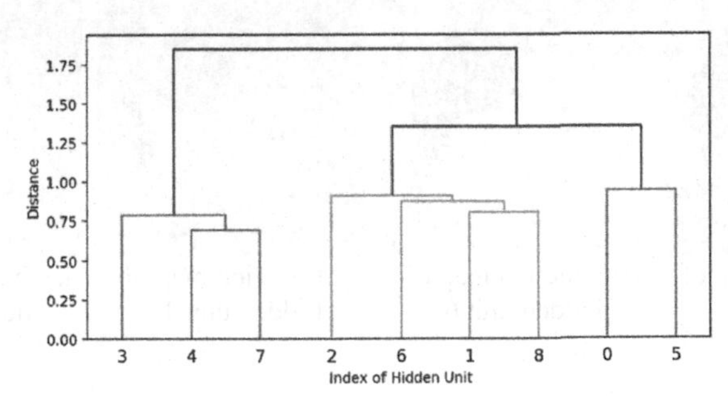

Fig. 4. Result of the unsupervised hierarchical clustering process for the function visualization maps for 9 hidden units.

As illustrated in Fig. 5, the low-intensity areas in the function maps of hidden units 0 and 5 in group C match the high-intensity areas in the Hessian-enhanced input image, which means they suppress the high-intensity areas. Likewise, group A enhances the liver area, and group B suppresses the non-tumor area. We also understood that groups A and B worked together to enhance the tumor area, and group C suppressed the liver's boundary as well as reduced the false enhancements inside the liver. Thus, our XAI method was able to reveal the learned functions of groups of neurons in the neural network, which we call "functional explanations" and define as the explanations of the

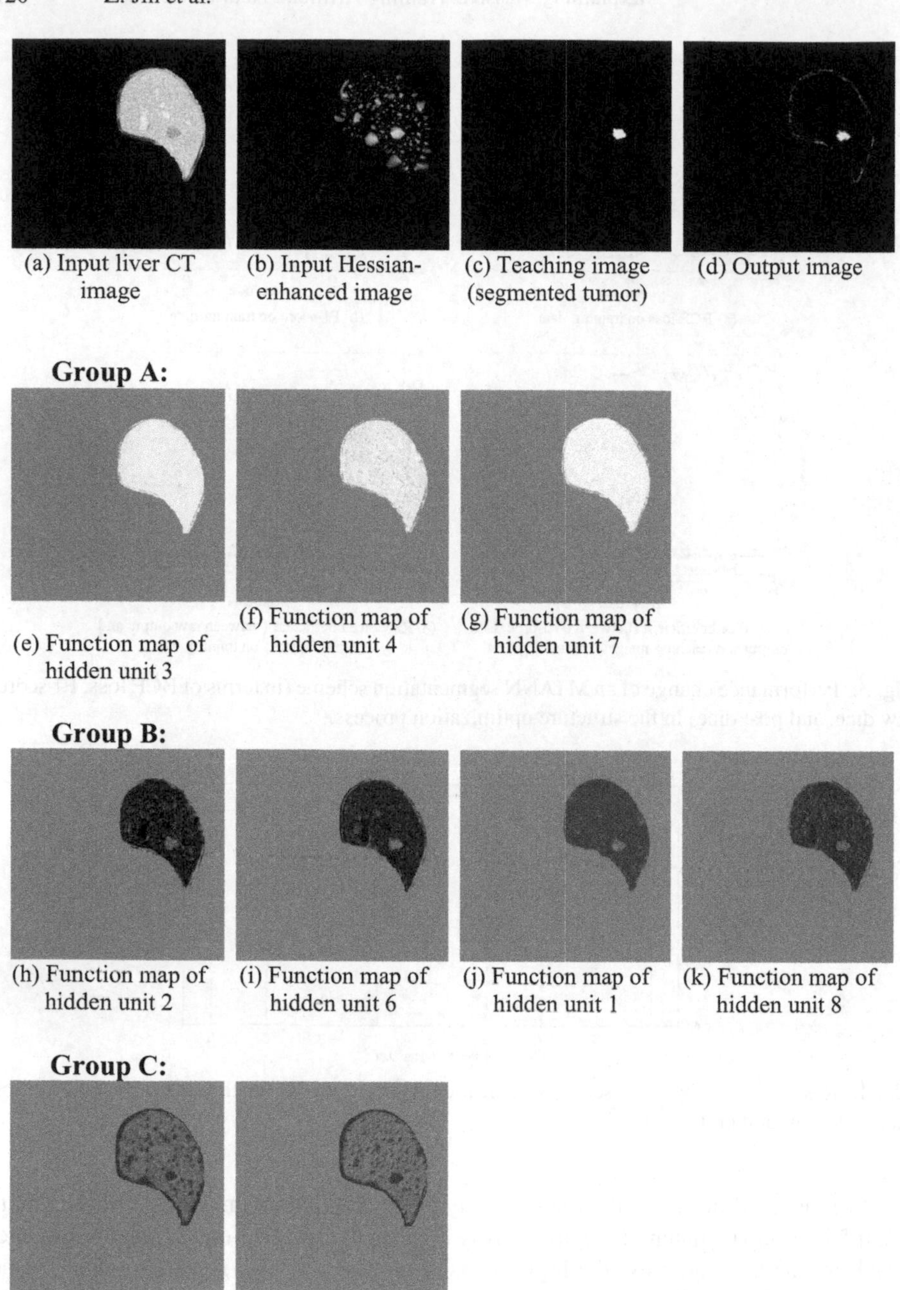

Fig. 5. Functional visualization maps for the 9 hidden units in groups A, B, and C obtained by using our XAI method, and the comparison with the input, teaching, and output images.

model behavior by a combination of functions. Our method is a post-hoc method that offers both instance-based and model-based functional explanations.

4 Conclusion

In this study, we proposed a novel XAI approach to explain the functions and behavior of an MTANN model for semantic segmentation of liver tumors in CT. Our structure optimization algorithm refined the structure and made every hidden unit in the model have a clear, meaningful function by removing redundant hidden units and "condensing" the functions into fewer hidden units, which solved the issue of unstable XAI results with conventional XAI methods. The unsupervised hierarchical clustering algorithm in our XAI approach grouped the hidden units with a similar function into one group so as to explain their functions by group. Through the experiments, we successfully proved that the MTANN model was explainable by functions.

Acknowledgment. This paper is based on results obtained from a project commissioned by the New Energy and Industrial Technology Development Organization (NEDO).

References

1. LeCun, Y., Bengio, Y., Hinton, G.: Deep learning. Nature **521**(7553), 436 (2015)
2. Castelvecchi, D.: Can we open the black box of AI? Nat. News **538**(7623), 20 (2016)
3. Gunning, D., Aha, D.: DARPA's explainable artificial intelligence (XAI) program. AI Mag. **40**(2), 44–58 (2019)
4. Adabi, A., Berrada, M.: Peeking inside the black-box: a survey on explainable artificial intelligence (XAI). IEEE Access **6**, 52138–52160 (2018)
5. Zhou, B., Khosla, A., Lapedriza, A., Oliva, A., Torralba, A.: Learning deep features for discriminative localization. In: Proceedings of the IEEE Conference on Computer Vision and Pattern Recognition, pp. 2921–2929 (2016)
6. Bach, S., Binder, A., Montavon, G., Klauschen, F., Müller, K.R., Samek, W.: On pixel-wise explanations for non-linear classifier decisions by layer-wise relevance propagation. PLoS ONE **10**, 1–46 (2015)
7. Shrikumar, A., Greenside, P., Kundaje, A.: Learning important features through propagating activation differences. In: Proceedings International Conference Machine Learning, pp. 3145–3153 (2017)
8. Ribeiro, M.T., Singh, S., Guestrin, C.: Why should i trust you? Explaining the predictions of any classifier. In: Proceedings of the 22nd ACM SIGKDD International Conference on Knowledge Discovery and Data Mining, pp. 1135–1144 (2016)
9. Lundberg, S.M., Lee, S.-I.: A unified approach to interpreting model predictions. In: Advances in Neural Information Processing Systems, vol. 30 (2017)
10. Suzuki, K., Horiba, I., Sugie, N.: Neural edge enhancer for supervised edge enhancement from noisy images. IEEE Trans. Pattern Anal. Mach. Intell. **25**(12), 1582–1596 (2003)
11. Suzuki, K., Horiba, I., Sugie, N., et al.: Neural filter with selection of input features and its application to image quality improvement of medical image sequences. IEICE Trans. Inf. Syst. **85**(10), 1710–1718 (2002)
12. Suzuki, K., et al.: Extraction of left ventricular contours from left ventriculograms by means of a neural edge detector. IEEE Trans. Med. Imaging **23**(3), 330–339 (2004)

13. Suzuki, K., Li, F., Sone, S., Doi, K.: Computer-aided diagnostic scheme for distinction between benign and malignant nodules in thoracic low-dose CT by use of massive training artificial neural network. IEEE Trans. Med. Imaging **24**(9), 1138–1150 (2009)
14. Suzuki, K., Rockey, D.C., Dachman, A.H.: CT colonography: advanced computer-aided detection scheme utilizing MTANNs for detection of 'missed' polyps in a multicenter clinical trial. Med. Phys **37**(1), 12–21 (2010)
15. Weigend, A.: On overfitting and the effective number of hidden units. In: Proceedings of the 1993 Connectionist Models Summer School, vol. 1 (1994)
16. Suzuki, K., Horiba, I., Sugie, N.: A simple neural network pruning algorithm with application to filter synthesis. Neural Process. Lett **13**(1), 43–53 (2001). https://doi.org/10.1023/A:1009639214138
17. Bar-Joseph, Z., Gifford, D.K., Jaakkola, T.S.: Fast optimal leaf ordering for hierarchical clustering. Bioinformatics **17**(1), 22–29 (2001)
18. Bauer, E., Kohavi, R.: An empirical comparison of voting classification algorithms: bagging, boosting, and variants. Mach. Learn. **36**, 105–139 (1999). https://doi.org/10.1023/A:1007515423169
19. Wang, Z., Bovik, A.C., Sheikh, H.R., Simoncelli, E.P., Simoncelli, E.P.: Image quality assessment: from error visibility to structural similarity. IEEE Trans Image Process. **13**(4), 600–612 (2004)
20. Sato, M., Jin, Z., Suzuki, K.: Semantic segmentation of liver tumor in contrast-enhanced hepatic CT by using deep learning with hessian-based enhancer with small training dataset size. In: 2021 IEEE 18th International Symposium on Biomedical Imaging (ISBI), pp. 34–37 (2021)
21. Simpson, A.L., Antonelli, M., Bakas, S., et al.: A large annotated medical image dataset for the development and evaluation of segmentation algorithms. ArXiv Prepr. ArXiv190209063 (2019)
22. Huynh, H.T., Le-Trong, N., Bao, P.T., Oto, A., Suzuki, K.: Fully automated MR liver volumetry using watershed segmentation coupled with active contouring. Int. J. Comput. Assist. Radiol. Surg. **12**(2), 235–243 (2017). https://doi.org/10.1007/s11548-016-1498-9
23. Sato, Y., et al.: Tissue classification based on 3D local intensity structures for volume rendering. IEEE Trans. Vis. Comput. Graph. **6**(2), 160–180 (2000)
24. Jin, Z., Arimura, H., Kakeda, S., Yamashita, F., Sasaki, M., Korogi, Y.: An ellipsoid convex enhancement filter for detection of asymptomatic intracranial aneurysm candidates in CAD frameworks. Med. Phys. **43**(2), 951–960 (2016)

An Explainable Geometric-Weighted Graph Attention Network for Identifying Functional Networks Associated with Gait Impairment

Favour Nerrise[1], Qingyu Zhao[2], Kathleen L. Poston[3], Kilian M. Pohl[2], and Ehsan Adeli[2(✉)]

[1] Department of Electrical Engineering, Stanford University, Stanford, CA, USA
fnerrise@stanford.edu

[2] Department of Psychiatry and Behavioral Sciences, Stanford University, Stanford, CA, USA
eadeli@stanford.edu

[3] Department of Neurology and Neurological Sciences, Stanford University, Stanford, CA, USA

Abstract. One of the hallmark symptoms of Parkinson's Disease (PD) is the progressive loss of postural reflexes, which eventually leads to gait difficulties and balance problems. Identifying disruptions in brain function associated with gait impairment could be crucial in better understanding PD motor progression, thus advancing the development of more effective and personalized therapeutics. In this work, we present an explainable, geometric, weighted-graph attention neural network (**xGW-GAT**) to identify functional networks predictive of the progression of gait difficulties in individuals with PD. **xGW-GAT** predicts the multi-class gait impairment on the MDS-Unified PD Rating Scale (MDS-UPDRS). Our computational- and data-efficient model represents functional connectomes as symmetric positive definite (SPD) matrices on a Riemannian manifold to explicitly encode pairwise interactions of entire connectomes, based on which we learn an attention mask yielding individual- and group-level explainability. Applied to our resting-state functional MRI (rs-fMRI) dataset of individuals with PD, **xGW-GAT** identifies functional connectivity patterns associated with gait impairment in PD and offers interpretable explanations of functional subnetworks associated with motor impairment. Our model successfully outperforms several existing methods while simultaneously revealing clinically-relevant connectivity patterns. The source code is available at https://github.com/favour-nerrise/xGW-GAT.

Keywords: Resting-state fMRI · Geometric learning · Attention mechanism · Gait impairment · Explainability · Neuroimaging biomarkers

---

Supplementary Information The online version contains supplementary material available at https://doi.org/10.1007/978-3-031-43895-0_68.

H. Greenspan et al. (Eds.): MICCAI 2023, LNCS 14221, pp. 723–733, 2023.
https://doi.org/10.1007/978-3-031-43895-0_68

1 Introduction

Parkinson's Disease (PD) is an age-related neurodegenerative disease with complex symptomology that significantly impacts the quality of life, with nearly 90,000 people diagnosed each year in North America [29]. Recent research has shown that gait difficulty and postural impairment symptoms of PD are highly correlated with alterations in various brain networks, including the motor, cerebellar, and cognitive control networks [25]. Understanding brain functional networks associated with an individual's gait impairment severity is essential for developing targeted interventions, such as physical therapy or brain stimulation techniques. However, most prior works have *either* focused only on a binary diagnosis (PD vs. Control) [16] (ignoring the progression and heterogeneity of the disease symptoms) *or* only used sensor- and vision-based technologies [8,18] to quantify PD symptoms (abstaining from identifying brain networks associated with gait impairment severity).

Graph Neural Networks (GNNs) have been highly successful in inferring neural activity patterns in resting-state fMRI (rs-fMRI) [23]. These models represent functional connectivity matrices as weighted graphs, where each node is a brain region of interest (ROI), and the edges between them capture the magnitude of connectivity, i.e., interactions, as weights. Changes in the connectivity strengths can reflect intrinsic representations in a high-dimensional space that correlate with symptom or disease severity. Assuming that edges with higher weights exert greater functional connectivity (and vice versa), GNNs can encode how ROIs and their neighbors across various individuals can possess similar attributes. GAT [26] is a well-known GNN model that encodes pairwise interactions (edges) into an attention mechanism and uses eigenvectors and eigenvalues of each node as positional embeddings for local structures. However, since each node or ROI in a brain network has the same degree and connects to every other node, standard graph representations are limited in modeling functional connectivity differences in a high-dimensional space that can be used for inter-subject functional covariance comparison. Riemannian geometry [14] is another robust, mathematical framework for rs-fMRI analysis that projects a functional, connectivity matrix in a manifold of symmetric positive-definite (SPD) matrices, making it possible to model high-dimensional, edge interactions and dependencies. It has been applied to analyzing gait patterns [20] in Parkinson's disease and to functional brain network analysis in other neurological disorders (e.g., Mild Cognitive Impairment [7] and autism [30]).

Addressing the problem of identifying brain functional network alterations related to the severity of gait impairments presents several challenges: (**i**) clinical datasets are often sparse or highly imbalanced, especially for severely impaired disease states; (**ii**) although substantial progress has been made in modeling functional connectomes using graph theory, few studies exist that capture the individual variability in disease progression and they often fall short of generating clinically relevant explanations that are symptom-specific.

In this work, we propose a novel, explainable, geometric weighted-graph attention network (**xGW-GAT**) that embeds functional connectomes in a learnable, graph structure that encodes discriminative edge attributes used for attention-based, transductive classification tasks. We train the model to predict a gait impairment rating score (MDS-UPDRS Part 3.10) for each PD participant. To mitigate limited clinical data across all different classes of gait impairment and data imbalance (challenge **i**), we propose a stratified, learning-based sample selection method that leverages non-Euclidean, centrality features of connectomes to sub-select training samples with the highest predictive power. To provide clinical interpretability (challenge **ii**), **xGW-GAT** innovatively produces individual and global attention-based, explanation masks per gait category and soft assigns nodes to functional, resting-state brain networks. We apply the proposed framework on our dataset of 35 clinical participants and compare it with existing methods. We observe significant improvements in classification accuracy while enabling adequate clinical interpretability.

In summary, our contributions are: (1) we propose a novel, geometric attention-based model, **xGW-GAT**, that uses edge-weights to depict neighborhood influence from local node embeddings during dynamic, attention-based learning; (2) we develop a multi-classification pipeline that mitigates sparse and imbalanced sampling with stratified, learning-based sample selection during training on real-world clinical datasets; (3) we provide an explanation generator to interpret attention-based, edge explanations that highlight salient brain network interactions for gait impairment severity states; (4) we establish a new benchmark for PD gait impairment assessment using brain functional connectivities.

2 XGW-GAT: Explainable, Geometric-Weighted GAT

Problem definition. Assume a set of functional connectomes, $\mathcal{G}_n \in \mathbb{R}^{d\times d}, \mathcal{G}_2, \ldots, \mathcal{G}_N$ are given, where N is the number of samples and d is the number of ROIs. Each connectome is represented by a weighted, undirected graph $\mathcal{G} = (\mathcal{V}, \mathcal{E}, \mathbf{W})$, where $\mathcal{V} = \{v_i\}_{i=1}^{d}$ is the set of nodes, $\mathcal{E} \subseteq \mathcal{V} \times \mathcal{V}$ is the edge set, and $\mathbf{W} \in \mathbb{R}^{|\mathcal{V}|\times|\mathcal{V}|}$ denotes the matrix of edge weights. The weight w_{ij} of an edge $e_{ij} \in \mathcal{E}$ represents the strength of the functional connection between nodes v_i and v_j, i.e., the Pearson correlation coefficient of the time series of the pair of the nodes. Each $\mathcal{G}_n$ contains node attributes X_n and edge attributes H_n. We develop a model that predicts a gait impairment score, $\mathcal{Y}_n$ and outputs an individual explanation mask $\mathbf{M}_c \in \mathbb{R}^{d\times d}$ per class c to assign ROIs to functional brain networks.

2.1 Connectomes in a Riemannian Manifold

Functional connectivity matrices belong to the manifold of symmetric positive-definite (SPD) matrices [31]. We leverage Riemannian geometry to perform principled comparisons between different connectomes, such as prior work [24]. To highlight connections between adjacent nodes, each weight matrix $\mathbf{W}_n \in \mathbb{R}^{d\times d}$ can be represented as a symmetric, adjacency matrix with zero, non-negative

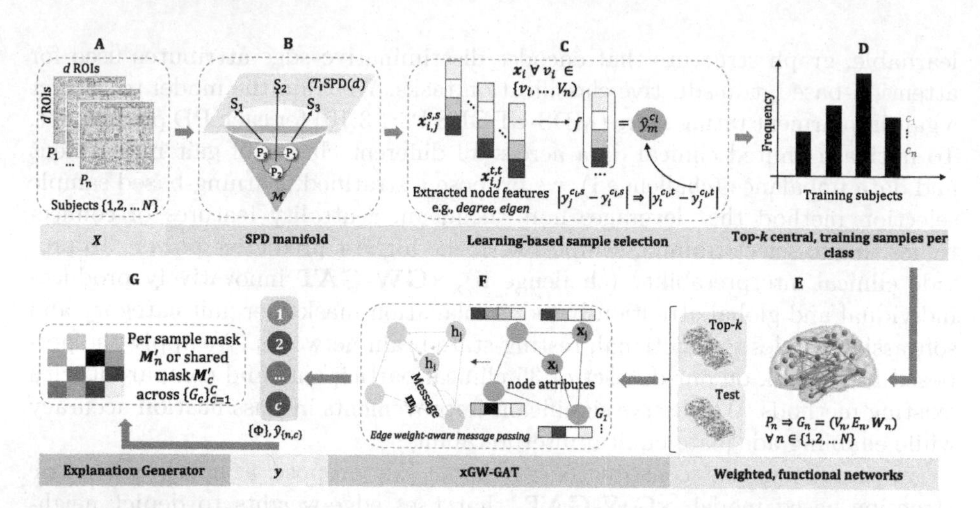

Fig. 1. xGW-GAT. (A) Input: functional connectomes. (B) Extract pairwise tangent matrices in SPD(d). (C) Compress tangent matrices into weighted graphs (connectomes) (D) Use linear regression to train a mapping, f, on training samples to learn pairwise differences between target and record scores. (E) Group top-k samples per class across N-fold cross-validation runs with the lowest predicted difference and oversample for imbalance. (F) Represent samples as weighted, graphs and use edge weight-aware attention to encode and propagate learning; predict gait score. (G) Produce explanation masks for each class or individual participants within functional brain networks.

eigenvalues, where each element of the adjacency matrix is the edge weight, w_{ij} between nodes i and j. We then consider $\mathbf{W}_n$ to be a point, $\mathbf{S}_n$, in the manifold of SPD matrices Sym_d^+ that locally looks like a topological Euclidean space. However, Sym_d^+ does not form a vector space; thus, we project each SPD matrix $\mathbf{S}_n$ onto a common tangent space using parallel transport. Given a reference point $\mathbf{S}_i \in Sym_d^+$, we transport a tangent vector $v \in T_{\mathbf{S}_j}$ from $\mathbf{S}_j$ to $\mathbf{S}_i$ along the geodesic connecting $\mathbf{S}_j$ and $\mathbf{S}_i$ (see Fig. 1-B). This process is performed for each subject $n = 1, 2, \ldots, N$, yielding a set of tangent vectors in a common tangent space that can be analyzed using traditional Euclidean methods.

To calculate the geodesic distance between two SPD matrices $\mathbf{S}_i$ and $\mathbf{S}_j \in Sym_d^+$, we adopt the Log-Euclidean Riemannian Metric (LERM) [1] distance, d_{LERM} as follows:

$$d_{\mathrm{LERM}}(\mathbf{S}_i, \mathbf{S}_j) = \|log(\mathbf{S}_i) - log(\mathbf{S}_j)\|_F^2 \tag{1}$$

where $\|\cdot\|_F^2$ is the Frobenius norm. LERM is invariant to similarity transformations (scaling and orthogonality) and is computationally efficient for high-dimensional data. See the Supplementary Material for results with other distance metrics.

2.2 Stratified Learning-Based Sample Selection

Data availability and dataset imbalance are re-occurring challenges with real-world clinical datasets, often leading to bias and overfitting during model training. We address this by expanding a learning-based sample selection method [11] to weight per-class distributions. We assume that similar brain connectivity networks are correlated with disease severity whereas connectomes that vary in topological patterns might elicit different gait impairment scores. Our subsampling technique selects training samples containing the highest representative power, i.e., contributing the least amount of pairwise differences for predicting a gait score. We divide our training samples into subgroups: train-in, n_s, and holdout, n_t using N-fold cross-validation. For each pair of symmetric d-by-d tangent matrices, $\mathbf{S}_{i,j}^{s,s} \in T_{\mathbf{I}}\text{SPD}(d)$, we encode the pairwise differences between the connectomes from the train-in, n_s, to obtain a set of $n_s(n_s-1)/2$ tangent matrices in $T_{\mathbf{I}}\text{SPD}(d)$. Each tangent matrix represents the "difference" between two connectomes. We affix a threshold of k samples to be selected from each class c to identify l central training samples with the highest expected predictive power, i.e., the *lowest average difference* in target gait impairment scores per class, $\hat{y}_c$, between samples j from the train-in and holdout group. We select degree, closeness, and eigenvector centrality as our topological features that encode information on *changes* in node connectivity. We train a linear regression mapping f on the Riemannian geometric distances $\mathcal{D}_{le}(\mathbf{S}_i^s, \mathbf{S}_j^s)$ between the connectomes from n_s using the vectorized upper triangular portion (including the diagonal) of the tangent matrices. The *absolute difference in target score, per class*, between samples i and j from the train-in group n_s is denoted by $|\hat{y}_{c,j}^s - \hat{y}_{c,i}^s|$ (see Fig. 1-C). The top-k samples per class with the highest predictive power are sub-selected from the total training set, oversampled for class imbalance with RandomOverSampler [15], and used for training xGW-GAT layers (see Fig. 1-D).

2.3 Dynamic Graph Attention Layers

Attention. We employ Graph Attention Network version 2 (GATv2) [2], a GAT [26] variant to perform dynamic, multi-head, edge-weight attention message passing for classifying each $\mathbf{S}_n$. We assume that every node $i \in \mathcal{V}$ has an initial representation $\mathbf{h}_i^{(0)} \in \mathbb{R}^{d_0}$. GATv2 updates each node representation, $\mathbf{h}$ based on the features of neighboring nodes and the edge weights between nodes by computing attention scores α_{ij} for every edge (i,j) by normalizing attention coefficients $e(\mathbf{h}_i, \mathbf{h}_j)$. α_{ij} measures the importance of node j's features to node i's at layer l by performing a weighted sum over the neighboring nodes $j \in \mathcal{N}_i$:

$$e(\mathbf{h}_i, \mathbf{h}_j) := \text{LeakyReLU}(\mathbf{a}^{(l)\top}[\mathbf{\Theta}^{(l)}\mathbf{h}_i^{(l-1)} \| \mathbf{\Theta}^{(l)}\mathbf{h}_j^{(l-1)}]) \quad (2)$$

$$\alpha_{ij} := softmax_j(e(\mathbf{h}_i, \mathbf{h}_j)) \quad (3)$$

$$\mathbf{h}_i^{(l+1)} := \sigma\left(\sum_{j \in \mathcal{N}_i} \alpha_{ij}^{(l)} \mathbf{h}_j^{(l-1)}\right). \quad (4)$$

where $\mathbf{a}^{(l)} \in \mathbb{R}^{2F}$ and $\mathbf{\Theta}^{(l)}$ are trainable parameters and learned, $\mathbf{h}_i^{(l)} \in \mathbb{R}^F$ is the embedding for node i, σ represents a non-linearity activation function, and $\|$ denotes vector concatenation. As conventional graph attention mechanisms for transductive tasks typically do *not* incorporate edge attributes, we introduce an attention-based, message-passing mechanism incorporating edge weights, similar to [6]. The algorithm uses a message vector $\mathbf{m}_{ij} \in \mathbb{R}^F$ by concatenating node features of neighboring nodes i, j, and edge weight $\mathbf{W}_{i,j}$:

$$\mathbf{m}_{ij}^{(l)} = \text{MLP}_1([\mathbf{h}_i^{(l)}; \mathbf{h}_j^{(l)}; \mathbf{W}_{ij}]), \tag{5}$$

where MLP_1 is a Multi-Layer Perceptron. Accordingly, an update of each ROI representation is influenced by its neighboring regions weighted by their connectivity strength. After stacking L layers, a readout function summarizing all node embeddings is employed to obtain a graph-level embedding $\mathbf{g}$:

$$\mathbf{z} = \sum_{i \in V} \mathbf{h}_i^{(L)}, \mathbf{g} = \text{MLP}_2(\mathbf{z}) + \mathbf{z}. \tag{6}$$

Loss Function. xGW-GAT layers (Fig. 1-F) are trained with a supervised, weighted negative log-likelihood loss function to mitigate class imbalance across classes, C, defined as:

$$\mathcal{L}_{\text{NLL}} := \frac{-1}{N} \sum_{p=1}^{N} \sum_{q=1}^{C} r_q y_{pq} log(\hat{y}_{pq}), \tag{7}$$

where r_q is the rescaling weight for the q-th class, y_{pq} is the q-th element of the true label vector y_p for the p-th sample, and $\hat{y}_{pq}$ is the predicted label vector.

2.4 Individual- And Global-Level Explanations

We define an attention explanation mask for each sample, $n \in 1, 2, \ldots, N$ and for each class $c \in 1, 2, \ldots, C$ that identifies the most important node/ROI connections contributing to the classification of subjects. We return a set of attention coefficients $\alpha^n = [\alpha_1^n, \alpha_2^n, \ldots, \alpha_S^n]$ for each sample n, where S is the number of attention heads. We aggregate trained, attention coefficients per sample used for predicting each $\hat{y}$ using a *max* operation that returns $\alpha_{\max}^n \in \mathbb{R}^{d \times d}$. An explanation mask per class, M_c, or per sample, M_n, can be derived using the max attention coefficients, $\alpha_{\max}$ (Fig. 1-G):

$$\mathbf{M}_c = \frac{1}{N} \sum_{n=1}^{N} \alpha_{\max}; \quad \mathbf{M}_n = \frac{1}{C} \sum_{c=1}^{C} \alpha_{\max}. \tag{8}$$

$\mathbf{M}$ can be soft-thresholded to retain the top-L most positively attributed attention weights to the mask as follows:

$$\mathbf{M}'[i] = \begin{cases} \mathbf{M}[i] & \text{if } \mathbf{M}[i] \in \text{Top-L}(\mathbf{M}) \\ 0, & \text{otherwise,} \end{cases} \tag{9}$$

where Top-L($\mathbf{M}$) represents the set of top-L elements in $\mathbf{M}$.

3 Experiments

Dataset. We obtained data from a private dataset ($n = 35$, mean age 69 ± 7.9) defined in [19], which contains MDS-UPDRS exams from all participants. Following previously published protocols [21], all participants are recorded during the off-medication state. Participants were evaluated by a board-certified movement disorders specialist on a scale from 0 to 4 based on MDS-UPDRS Sect. 3.10 [10]. The dataset includes 22 participants with a score 1, 4 participants with a score 2, 4 participants with a score of 3, 4 participants with a score 4, and 1 participant with a score 0 on MDS-UPDRS item 3.10. The single score-0 participant (normal) was combined with the score-1 participants (minor gait impairment) to adjust for severe class imbalance. We pre-processed functional connectivity matrices and corrected them for possible motion artifacts using the CONN toolbox [28]. The FC matrices were obtained using a combined Harvard-Oxford and AAL parcellation atlas [28] with 165 ROIs, where each entry in row i and column j in the matrix is the Pearson correlation between the average rs-fMRI signal measured in ROI i and ROI j. We imputed any missing ROI network scores with the mean score per column and Z-transformed FC matrices $[\mu = 0, \sigma = 1]$. This dataset (like other clinical datasets in practice) poses highly imbalanced distributions for classes with severe impairment, which makes it useful to demonstrate our method's capability in an imbalanced and limited-data scenario. In addition, most existing studies focus on differentiating participants from controls, while the severity of specific impairments is understudied (our focus).

Software. All experiments were implemented in Python 3.10 and ran on Nvidia A100 GPU runtimes. We used PyTorch Geometric [9], PyTorch, and Scikit-learn for machine learning methods. We used the SPD class from the Morphometrics package to compute Riemannian geometrics and NetworkX to extract the topological features of the graphs from the tangent matrices. Hyper-parameters are tuned automatically with the open-source AutoML toolkit NNI (https://github.com/microsoft/nni).

Setup. We used the mean, connectivity profile, i.e., **W** [5], as the node feature for xGW-GAT layers, a weighted ADAM optimizer, a learning rate of $1e-4$, a batch size of 2 for training and 1 for test, and 100 training epochs. We used 2 GATv2 layers, dropout rate = 0.1, hidden_dim = 8, heads= 2, and a global mean pooling layer. We used 4-fold cross-validation to partition training and holdout sets and selected $k = 4$ as the optimal number of selected training samples between 2 and 15. We report weighted, macro average scores for F_1, area under the ROC curve (AUC), precision (Pre), and recall (Rec) over 100 trials.

3.1 Results

We perform a multi-class classification task of *Slight(1), Mild(2), Moderate(3), Severe(4)* gait impairment severity. To benchmark our method, we compare our

Table 1. Comparison with baseline and ablated methods. * indicates statistical difference by Wilcoxon signed rank test at ($p < 0.05$) compared with our method.

Method	Pre	Rec	F_1	AUC
GCN* [13]	0.46	0.48	0.47	0.54
PNA* [4]	0.52	0.54	0.53	0.56
BrainNetCNN*[12]	0.62	0.71	0.66	0.57
BrainGNN*[17]	0.66	0.53	0.59	0.62
GAT* [26]	0.70	0.58	0.64	0.71
xGW-GAT (dc)*	0.61	0.65	0.63	0.51
xGW-GAT (ec)*	0.64	0.62	0.63	0.72
xGW-GAT (cc)*	0.61	0.53	0.57	0.57
xGW-GAT (!ss)*	0.55	0.47	0.51	0.54
xGW-GAT (ss)	**0.75**	**0.77**	**0.76**	**0.83**

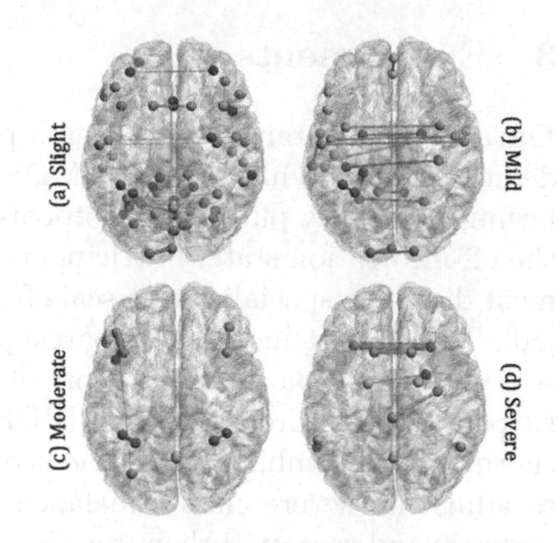

Fig. 2. Salient ROI connections on explanation brain networks across the four classes of gait impairment (DMN, SMN, VN, SN, DAN, FPN, LN, CN, and BLN).

results with several state-of-the-art classifiers: GAT [26], GCN [13], PNA [4], and two state-of-the-art deep models design for brain networks: BrainNetCNN [12] and BrainGNN [17]. We also perform an ablation study on sample selection and the type of topological features used in training our method: (!ss) no [stratified, learning-based] sample selection, (dc) node degree centrality, (cc) node closeness centrality, and (ec) eigenvector centrality. Results for the highest-performing settings of **xGW-GAT** are displayed in Table 1 and node feature descriptions are included in the Supplementary Material.

The results (Table 1) show that **xGW-GAT** yields significant improvement in performance over SOTA graph-based models, including models designed for brain network analysis. xGW-GAT with our stratified, learning-based selection method combined with the RandomOverSampler technique to temper the effects of class imbalance outperforms a standard xGW-GAT by 42%. Compared with SOTA deep models like GCN and PNA, our model also outperforms them by large margins, with up to 29% improvement for AUC. These predictions are promising for an explainable analysis of PD gait impairment while also minimizing random uncertainties introduced in individual participant graphs.

4 Discussion

Brain Networks Mapping. As shown in Fig. 2, we aid interpretability for clinical relevance by partitioning the ROIs into nine "networks" based on their functional roles: Default Mode Network (DMN), SensoriMotor Network (SMN), Visual Network (VN), Salience Network (SN), Dorsal Attention Network (DAN), FrontoParietal Network (FPN), Language Network (LN), Cerebellar Network (CN), and Bilateral Limbic Network (BLN) are colored accordingly, while edges across different systems are colored gray. Edge widths here are the attention weights.

Salient ROIs. We provide per-class and individual-level interpretations for understanding how ROIs contribute to predicting gait impairment scores. We build the node and edge files with the thresholded attention explanation masks, $\mathbb{M}'$ per PD participant or per class and plot glass brains using BrainNet Viewer ((https://www.nitrc.org/projects/bnv/)).

We observe that rich interactions decrease significantly for the *Mild* class, Fig. 2(b), within the CN, primarily associated with coordinating voluntary movement, the SN, responsible for thought, cognition, and planning behavior, and the VN, the center for visual processing during resting and task states. These observations are consistent with existing neuroimaging findings, which support that PD is positively associated with the severity of cognitive deficits and neuromotor control for inter-network and intra-network interactions within the salience network, cerebellar lobules, and visual network [22,32]. Similarly, there are significantly lower connections within CN and VN and sparser connections within the SMN for the *Moderate* and *Severe* classes, Fig. 2(c)-(d). Existing studies show functional connectivity losses within the sensorimotor network (SMN) [3] are correlated with disruptions in regional and global topological organization for SMN areas for people with PD, resulting in loss of motor control. For *Mild*, *Moderate*, and *Severe* PD participants, abrupt connectivity is also observed for the frontoparietal network, FPN, known for coordinating behavior and associated with connectivity alterations correlated with motor deterioration [27].

5 Conclusion

This study showcases a novel benchmark for using an explainable, geometric-weighted graph attention network to discover patterns associated with gait impairment. The framework innovatively integrates edge-weighted attention encoding and explanations to represent neighborhood interactions in functional brain connectomes, providing interpretable functional network clustering for neurological analysis. Despite a small sample size and imbalanced settings, the lightweight model offers stable results for quick inference on categorical PD neuromotor states. Future work includes new experiments, an expanded, multi-modal dataset, and sensitivity and specificity analysis to discover subtypes associated with the severity of PD gait impairment.

Acknowledgements. This work was partially supported by NIH grants (AA010723, NS115114, P30AG066515), Stanford School of Medicine Department of Psychiatry and Behavioral Sciences Jaswa Innovator Award, UST (a Stanford AI Lab alliance member), and the Stanford Institute for Human-Centered AI (HAI) Google Cloud credits. FN is funded by the Stanford Graduate Fellowship and the Stanford NeuroTech Training Program Fellowship.

References

1. Arsigny, V., Fillard, P., Pennec, X., Ayache, N.: Geometric means in a novel vector space structure on symmetric positive-definite matrices. SIAM J. Matrix Anal. Appl. **29**(1), 328–347 (2007)

2. Brody, S., Alon, U., Yahav, E.: How attentive are graph attention networks? arXiv preprint arXiv:2105.14491 (2021)
3. Caspers, J., et al.: Within-and across-network alterations of the sensorimotor network in Parkinson's disease. Neuroradiology **63**(12), 2073–2085 (2021)
4. Corso, G., Cavalleri, L., Beaini, D., Liò, P., Veličković, P.: Principal neighbourhood aggregation for graph nets. NeurIPS **33**, 13260–13271 (2020)
5. Cui, H., et al.: Braingb: a benchmark for brain network analysis with graph neural networks. IEEE TMI 2022 (2022)
6. Cui, H., Dai, W., Zhu, Y., Li, X., He, L., Yang, C.: Interpretable graph neural networks for connectome-based brain disorder analysis. In: MICCAI 2022, pp. 375–385. Springer (2022). https://doi.org/10.1007/978-3-031-16452-1_36
7. Dodero, L., Minh, H.Q., Biagio, M.S., Murino, V., Sona, D.: Kernel-based classification for brain connectivity graphs on the riemannian manifold of positive definite matrices. In: 2015 IEEE ISBI, pp. 42–45 (2015)
8. Endo, M., Poston, K.L., Sullivan, E.V., Fei-Fei, L., Pohl, K.M., Adeli, E.: GaitForeMer: Self-Supervised Pre-Training of Transformers via Human Motion Forecasting for Few-Shot Gait Impairment Severity Estimation. MICCAI, pp. 130–139 (2022). https://doi.org/10.1007/978-3-031-16452-1_13
9. Fey, M., Lenssen, J.E.: Fast graph representation learning with pytorch geometric. arXiv preprint arXiv:1903.02428 (2019)
10. Goetz, C.G., et al.: The MDS-sponsored revision of the unified Parkinson's disease rating scale. Official MDS Dutch Translation (2019)
11. Hanik, M., Demirtaş, M.A., Gharsallaoui, M.A., Rekik, I.: Predicting cognitive scores with graph neural networks through sample selection learning. Brain Imaging Behav. **16**(3), 1123–1138 (2022)
12. Kawahara, J., et al.: Convolutional neural networks for brain net-works; towards predicting neurodevelopment. Neu-roImage (2017)
13. Kipf, T.N., Welling, M.: Semi-supervised classification with graph convolutional networks. arXiv preprint arXiv:1609.02907 (2016)
14. Klingenberg, W.: Contributions to Riemannian geometry in the large. Ann. Math. **69**(3), 654–666 (1959)
15. Lemaître, G., Nogueira, F., Aridas, C.K.: Imbalanced-learn: a python toolbox to tackle the curse of imbalanced datasets in machine learning. J. Mach. Learn. Res. **18**(17), 1–5 (2017)
16. Li, K., Su, W., Li, S.H., Jin, Y., Chen, H.B.: Resting state fMRI: a valuable tool for studying cognitive dysfunction in pd. Parkinson's Disease 2018 (2018)
17. Li, x, et al.: Braingnn: interpretable brain graph neural network for fMRI analysis. Med. Image Anal. **74**, 102233 (2021)
18. Lu, M., et al.: Vision-based estimation of MDS-UPDRS gait scores for assessing Parkinson's disease motor severity. MICCAI **2020**(12263), 637–647 (2020)
19. Lu, M.: Quantifying Parkinson's disease motor severity under uncertainty using MDS-UPDRS videos. Med. Image Anal. **73**, 102179 (2021)
20. Olmos, J., Galvis, J., Martínez, F.: Gait patterns coded as Riemannian mean covariances to support Parkinson's disease diagnosis. In: IBERAMIA, pp. 3–14 (2023)
21. Poston, K.L., et al.: Compensatory neural mechanisms in cognitively unimpaired Parkinson disease. Ann. Neurol. **79**(3), 448–463 (2016)
22. Ruan, X., et al.: Impaired topographical organization of functional brain networks in parkinson's disease patients with freezing of gait. Front. Aging Neurosci. **12**, 580564 (2020)

23. Rubinov, M., Sporns, O.: Complex network measures of brain connectivity: uses and interpretations. Neuroimage **52**(3), 1059–1069 (2010)
24. Shahbazi, M., Shirali, A., Aghajan, H., Nili, H.: Using distance on the Riemannian manifold to compare representations in brain and in models. Neuroimage **239**, 118271 (2021)
25. Togo, H., Nakamura, T., Wakasugi, N., Takahashi, Y., Hanakawa, T.: Interactions across emotional, cognitive and subcortical motor networks underlying freezing of gait. NeuroImage: Clin. **37**, 103342 (2023)
26. Velickovic, P., Cucurull, G., Casanova, A., Romero, A., Lio, P., Bengio, Y., et al.: Graph attention networks. stat **1050**(20), 10–48550 (2017)
27. Vervoot, G., et al.: Functional connectivity alterations in the motor and fronto-parietal network relate to behavioral heterogeneity in parkinson's disease. Parkinsonism Related Disorders **24**, 48–55 (2016)
28. Whitfield-Gabrieli, S., Nieto-Castanon, A.: Conn: a functional connectivity toolbox for correlated and anticorrelated brain networks. Brain connectivity **2**(3), 125–141 (2012)
29. Willis, A., et al.: Incidence of Pakinson disease in north America. NPJ Parkinson's Disease **8**(1), 170 (2022)
30. Wong, E., Anderson, J.S., Zielinski, B.A., Fletcher, P.T.: Riemannian regression and classification models of brain networks applied to autism. In: CNI 2018, Held in Conjunction with MICCAI 2018,pp. 78–87 (2018)
31. You, K., Park, H.J.: Re-visiting Riemannian geometry of symmetric positive definite matrices for the analysis of functional connectivity. Neuroimage **225**, 117464 (2021)
32. Zhu, H., et al.: Abnormal dynamic functional connectivity associated with subcortical networks in Parkinson's disease: a temporal variability perspective. Front. Neurosci. **13**, 80 (2019)

An Interpretable and Attention-Based Method for Gaze Estimation Using Electroencephalography

Nina Weng[1(✉)], Martyna Plomecka[2], Manuel Kaufmann[3], Ard Kastrati[3], Roger Wattenhofer[3], and Nicolas Langer[2]

[1] Technical University of Denmark, Kongens Lyngby, Denmark
ninwe@dtu.dk
[2] University of Zurich, Zurich, Switzerland
martyna.plomecka@uzh.ch, n.langer@psychologie.uzh.ch
[3] ETH Zurich, Zurich, Switzerland
{kamanuel,akastrati,wattenhofer}@ethz.ch

Abstract. Eye movements can reveal valuable insights into various aspects of human mental processes, physical well-being, and actions. Recently, several datasets have been made available that simultaneously record EEG activity and eye movements. This has triggered the development of various methods to predict gaze direction based on brain activity. However, most of these methods lack interpretability, which limits their technology acceptance. In this paper, we leverage a large data set of simultaneously measured Electroencephalography (EEG) and Eye tracking, proposing an interpretable model for gaze estimation from EEG data. More specifically, we present a novel attention-based deep learning framework for EEG signal analysis, which allows the network to focus on the most relevant information in the signal and discard problematic channels. Additionally, we provide a comprehensive evaluation of the presented framework, demonstrating its superiority over current methods in terms of accuracy and robustness. Finally, the study presents visualizations that explain the results of the analysis and highlights the potential of attention mechanism for improving the efficiency and effectiveness of EEG data analysis in a variety of applications.

Keywords: EEG · Interpretable model · Attention Mechanism

1 Introduction

Gaze information is a widely used behavioral measure to study attentional focus [7], cognitive control [19], memory traces [23] and decision making [28]. The most commonly used gaze estimation technique in laboratory settings is the infrared

Supplementary Information The online version contains supplementary material available at https://doi.org/10.1007/978-3-031-43895-0_69.

H. Greenspan et al. (Eds.): MICCAI 2023, LNCS 14221, pp. 734–743, 2023.
https://doi.org/10.1007/978-3-031-43895-0_69

eye tracker, which detects gaze position by emitting invisible near-infrared light and then capturing the reflection from the cornea [6]. While infrared eye tracker still remains the most accurate and reliable solution for the gaze estimation, these systems have several limitations, including individual differences in the contrast of the pupil and iris and the need for time-consuming setup and calibration before each scanning session [3,11].

Recently, Electroencephalogram (EEG) has been explored as an alternative method to estimate eye movements by recording electrical activity from the brain non-invasively with high temporal resolution [16]. The growing body of literature has shown that Deep Learning architectures could be significantly effective for many EEG-based tasks [4,26]. Nevertheless, with the advantages that Deep Learning brings, new challenges arise. Most of these models applied to electroencephalography (EEG) data tend to lack *interpretability*, making it difficult to understand the underlying reasons for their predictions, which subsequently leads to a decrease in the acceptability of advanced technology in neuroscience [25]. However, a potential solution already exists, in the form of the attention mechanism [29]. The attention mechanism has the potential to provide a more transparent and understandable way of analyzing EEG data, enabling us to comprehend the relationships between different brain signals better and make more informed decisions based on the results. With the development and implementation of these techniques, we can look forward to a future where EEG data can be utilized more effectively and efficiently in various applications.

Attention mechanisms have recently emerged as a powerful tool for processing sequential data, including time-series data in various fields such as natural language processing, speech recognition, and computer vision [5,24,29]. In the context of EEG signal analysis, attention mechanism has shown promising results in various applications, including sleep stage classification, seizure detection, and event-related potential analysis [8,13,17]. Since different electrodes record the brain activity from the different brain areas and functions, the information density from each electrode can vary for different tasks [15].

In this study, we introduce a new deep learning framework for analyzing EEG signals applying attention mechanisms. For the method evaluation, we used the EEGEyeNet dataset and benchmark [16], which includes concurrent EEG and infrared eye-tracking recordings, with eye tracking data serving as a ground truth. Our method incorporates attention modules to assign weights to individual electrodes based on their importance, allowing the network to prioritize relevant information in the signal. Specifically, we demonstrate the ability of our framework to accurately predict gaze position and saccade direction, achieving superior performance compared to previously benchmarked methods. Furthermore, we provide visualizations of model's interpretability through case studies.

2 Model

2.1 Motivation

In this study, our primary goal was to build a model sensitive to different electrodes. The motivation for this goal is two-fold. Firstly, with regards to interpreting the model, the electrodes can be considered the smallest entity as they record signals from specific regions of the brain. Therefore, the electrode-based explanation is a reasonable approach considering human understanding. Second, in the context of model learning, incorporating adaptive weighting of electrodes within a neural network can potentially enhance the accuracy and reliability of gaze estimation systems. This is because electrodes are functionally connected to cognitive behaviors. Specifically, in tasks such as gaze estimation, electrodes positioned near the eyes can capture electrical signals from the orbicularis oculi muscles [2], thereby making the pre-frontal brain areas more crucial for precise estimation [15]. Additionally, the noise of EEG recordings could be induced by broken wire contacts, too much or dried gel, or loose electrodes [27], the influence of such electrodes should be reduced in the network under ideal circumstances.

As shown in Fig. 1, our model design focuses on enhancing an existing deep learning architecture with an electrode-sensitive component. This component first extracts electrode-related information, and then utilizes this information for two purposes: (1) emphasizing the reliable electrodes and diminishing the influence of suspicious electrodes, while simultaneously (2) providing explanations for each prediction.

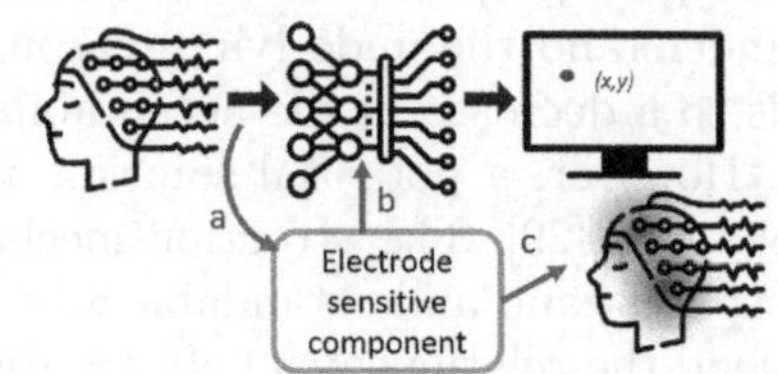

Fig. 1. We augment an electrode-sensitive component to a deep learning model, which works as follows: a) extract electrode-wise information from input data, b) control the predictions, and c) provide explanations.

2.2 Attention-CNN

Following the idea from the previous section, we propose the Attention-CNN model, where the attention blocks are used as the electrode-sensitive component. As shown in Fig. 2, the Attention-CNN model is structured by adding an attention block after each convolution block in every layer and an additional single attention block before the final prediction block (the blocks in blue). A

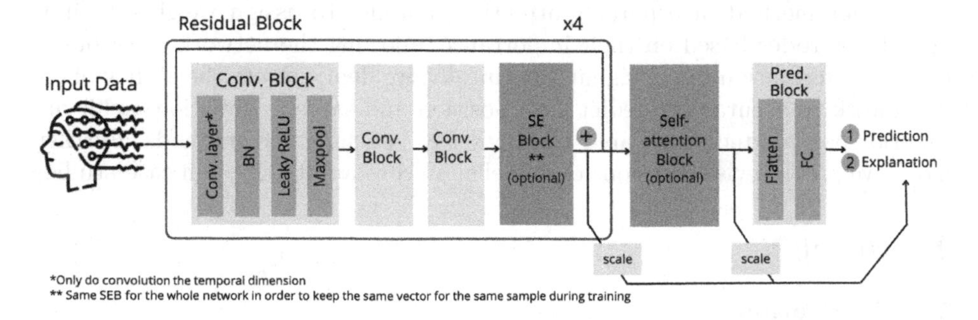

Fig. 2. The Architecture of the Attention-CNN model. (color figure online)

convolution block contains a convolution layer, a batch-norm layer [14], a leaky ReLU [18] and a max-pooling layer. In addition, the residual [10] techniques are applied in the CNN framework. The convolution layer operates only in the time dimension. The attention blocks, acting as an electrode-sensitive component, can be carried out by Squeeze-and-Excitation Block (SE Block) [12] and/or Self-Attention Block (SA Block) [29]. In the attention blocks, the retrieved electrode importance is used to weigh the features in each layer. Additionally, the same weights can provide explanations for the predictions of the model. In the prediction block, the features are flattened and then fed into the fully connected layer to finally obtain the predictions. While the SA Block is only required once in the process, the SE Blocks are added in every residual block. In order to keep the same scale for the same sample, the parameters of the SE Blocks are shared for the whole process. All building blocks are trained end-to-end, including the weights for the electrode importance used in the attention blocks.

Squeeze and Excitation Block: the SE block involves two principle operations. The **Squeeze** operation compresses features $u \in \mathbb{R}^{T' \times J}$ into electrode-wise vectors $z \in \mathbb{R}^J$ by using global average pooling. Here, T' denotes the feature size, and J is the number of electrodes. More precisely, the j-th element of z is calculated by $z_j = \mathbf{F}_{sq}(\mathbf{u}_j) = \frac{1}{T'} \sum_{i=1}^{T'} u_j(i)$. The **Excitation** operation first computes activation s by employing the gating mechanism with sigmoid activation: $s = \mathbf{F}_{ex}(\mathbf{z}, \mathbf{W}) = \sigma(\mathbf{W}_2 \delta(\mathbf{W}_1 \mathbf{z}))$, where σ refers to the sigmoid function, δ represents the ReLU [20] function, and $\mathbf{W}$ are learnable weights. The final output of SE block weigh each channel adaptively by re-scaling U with s: $\tilde{\mathbf{x}_j} = \mathbf{F}_{scale}(\mathbf{u}_j, s_j) = s_j \cdot \mathbf{u}_j$. In contrast to the original implementation [12] which deals with 3-dimensional data, the input data in our setup has only 2 dimensions (electrodes and time).

Self Attention Block: The *self-attention* mechanism [22] was first used in the field of Natural language processing (NLP), aiming at catching the attention of/between different words in a sentence or paragraph. The attention is obtained by letting the input data interact with *themselves* and determining which features are more important. This was implemented by introducing the *Query, Key, Value* technique, which is defined as $\mathbf{Q} = \phi_Q(\mathbf{U}, \mathbf{W}_Q)$, $\mathbf{K} = \phi_K(\mathbf{U}, \mathbf{W}_K)$, $\mathbf{V} = \phi_V(\mathbf{U}, \mathbf{W}_V)$, where U denotes the input of self-attention block and $\phi(\cdot, \cdot)$ represents linear transformation.

Then, *Attention Weights* are computed using Query and Key:

$$\mathbf{M}_{att} = softmax(\frac{\mathbf{Q} \cdot \mathbf{K}^T}{\sqrt{d_k}})$$

where d_k stands for the dimensions of the Key, and $\sqrt{d_k}$ works as a scaling factor. The softmax function was applied to adjust the range of the value in attention weights ($\mathbf{M_{att}}$) to $[0, 1]$.

Unlike the transformer model, the attention weights are first compressed into a one-dimensional vector by a layer of global average pooling (ψ) and normalized by a sigmoid function. More precisely, we compute $\mathbf{Z}_{att} = sigmoid(\psi(\mathbf{M}_{att}))$. Finally, the output of SA Block $\mathbf{X}$ is computed by : $\mathbf{X} = \kappa(\mathbf{Z}_{att}, V)$, where κ denotes the electrode-wise production.

3 Experiments and Results

3.1 Materials and Experimental Settings

EEGEyeNet Dataset: For our experiments, we utilized the EEGEyeNet dataset [16], which includes synchronized EEG and Eye-tracking data. The EEG signals were collected using a high-density, 128-channel EEG Geodesic Hydrocel system sampled at a frequency of 500 Hz. Eye-tracking data, including eye position and pupil size, were gathered using an infrared video-based eye tracker (EyeLink 1000 Plus, SR Research), also operating at a sampling rate of 500 Hz. The recorded EEG and eye-tracking information was pre-processed, synchronized and segmented into 1-second clips based on eye movements. The infrared eye tracking recordings were used as ground truth. In this paper, the processed dataset we utilized contains two parts: the *Position Task* and *Direction Task*, which correspond to two types of eye movements: *fixation*, i.e., the maintaining of the gaze on a single location, and *saccade*, i.e. the rapid eye movements that shift the centre of gaze from one point to another. While *Position Task* estimates the absolute position from fixation, *Direction Task* estimates the relative changes during saccades, involving two sub-tasks, i.e., the prediction of amplitude and angle. The statistics and primary labels of these two parts are shown in Table 1.

Table 1. Dataset Description

Task	#Subjects	#Samples	Primary labels
Position	72	50264	`subject_id`: the identical ID of the participant `pos`: the fixation position in the form of (x, y)
Direction	72	41783	`subject_id`: the identical ID of the participant `amplitude`:the distance in pixels during the saccade `angle`: the saccade direction in radians

To ensure data integrity and prevent data leakage, the dataset was split into training, validation, and test sets across subjects, with 70 % of the subjects used for training, and 15% each for validation and testing. This procedure ensures that no data from the same subject appears in both the training and validation/testing phases, thereby avoiding potential subject-related patterns from being learned by the model during training and tested on in validation/testing. For more details of this dataset, please refer to [16].

Implementation Details: The experiments are implemented with PyTorch [21]. When training the Attention-CNN model, the batch size is set to 32, the number of epochs is 50, and the learning rate is $1e^{-4}$. There are 12 convolution blocks, and the residual operation repeats every three convolution blocks. The feature length of the hidden layer is set as 64, and the kernel size is 64. The number of convolutional layers, kernel size and hidden feature length, are selected based on validation performance. We conducted experiments with three configurations: the SE Block and the SA Block together, only one of the attention blocks, or no attention blocks at all. For the angle prediction in Direction Task, we use angle loss $l_{angle} = |(atan(sin(p-t), cos(p-t))|$, where p denotes the predicted results, and t denotes the targets. For Position Task and Amplitude prediction in the Direction Task, the loss function is set to smooth-L1 [9].

Evaluation: For Position task, Euclidean distance is applied as the evaluation metric in both pixels and visual angles. Compared to pixel distance, visual angles depend on both object size on the screen and the viewing distance, thus enabling the comparison across varied settings. The performance of Direction Task is measured by the square root of the mean squared error (RMSE) for the angle (in radians) and the amplitude (in pixels) of saccades. In order to avoid the error caused by the repeatedness of angles in the plane (i.e. 2π and 0 rad represents the same direction), $atan(sin(\alpha), cos(\alpha))$ is applied, just like in angle loss.

3.2 Performance of the Attention-CNN

Table 2 shows the quantitative performance of the Attention-CNN in this work. For the Position Task, CNN with SE block has an average performance with the RMSE of 109.58 pixels. Likewise, the CNN model with both SE block and the SA block has a similar performance (110.05 pixels). Similar to Position Task, in amplitude prediction of Direction Task, the attention blocks aid the prediction evidently, heightening the performance by 5 pixels. Here, the model with both attention blocks has a lower variance. For angle prediction, the CNN model with both SE block and SA block has the best performance among all with the RMSE of 0.1707 rad.

We can conclude that the CNN model with both attention blocks consistently outperforms the CNN model alone by 5 to 10 percent across all tasks, indicating that electrode-wise attention assists in the learning process of the models.

Table 2. The performance of the Attention-CNN on Direction and Position Task.

Models	*Angle/Amplitude*		*Abs. Position*
	Angle RMSE	**Amp. RMSE**	**Euclidean Distance (Visual Angle)**
CNN	0.1947 ± 0.021	57.4486 ± 2.053	115.0143 ± 0.648 (2.39 ± 0.010)
CNN + SE	0.1754 ± 0.007	55.1656 ± 3.513	**109.5816 ± 0.238 (2.27 ± 0.004)**
CNN + SA	0.1786 ± 0.010	**52.1583 ± 1.943**	112.3823 ± 0.851 (2.33 ± 0.013)
CNN + both	**0.1707 ± 0.011**	**52.2782 ± 1.169**	**110.0523 ± 0.670 (2.28 ± 0.010)**

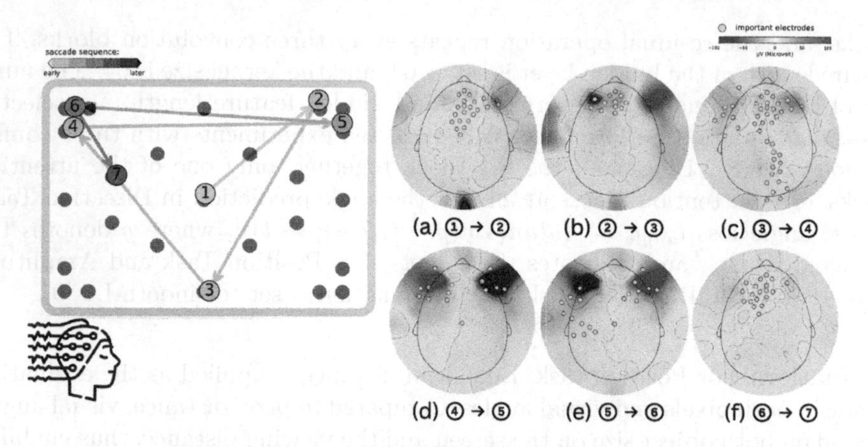

Fig. 3. Visualization of signal intensity across scalp and electrode importance from our models. Left: the track of a continuous sequence of saccades. Right: the corresponding brain activities (red: positive electrical signal, blue: negative electrical signal) and the important electrodes detected by the attention-based model (denoted as yellow nodes, the threshold is set as the mean value of all electrodes during the sequence). The model used here is the CNN with SA block. (Color figure online)

3.3 Model Interpretability by Case Studies

To provide a more detailed analysis of the interpretability of our proposed Attention-CNN model, as well as to further investigate the underlying reasons for the observed accuracy improvement, we conducted a visual analysis of the model performance, with a particular focus on the role of the attention block. Our analysis yielded two key findings, which are as follows:

Firstly, the attention blocks were able to detect the electrical difference between the right and left pre-frontal area in case of longer saccades, i.e. rapid eye movements from one side of the screen to the other; see the saccades (d) and (e) in Fig. 3. We present the sequence of saccades and observed the EEG signals as well as the electrode importance from proposed models in Fig. 3. The attention block effectively captured this phenomenon by highlighting the electrodes surrounding the prominent signals (saccades (d) and (e) in Fig. 3). Conversely, in cases where the saccade was of a shorter distance (other saccades in Fig. 3), attention was more widely distributed across the scalp rather than being concentrated in specific regions. This is justifiable as the neural network aims to integrate a more comprehensive set of information from all EEG channels.

Additionally, the attention block effectively learned to circumvent the interference caused by noisy electrodes and redirected attention towards the frontal region. Figure 4 illustrates a scenario where problematic electrodes were situated around both ears, exhibiting abnormal amplitudes ($\pm 100\ \mu$V). Using Layer-wise Relevance Propagation [1] to elucidate the CNN model's predictions, the result depicted in Fig. 4b revealed that the most significant electrodes were located over the left ear, coinciding with the noisy electrodes. In contrast, as shown in

Fig. 4c, the Attention-CNN model effectively excluded the unreliable electrodes and allocated greater attention to the frontal region of the brain.

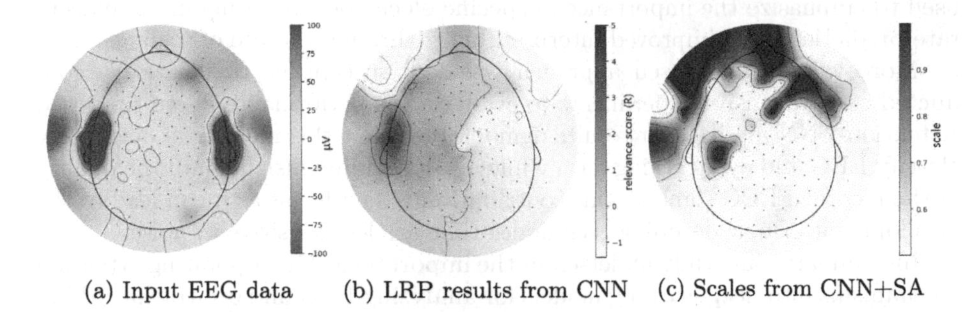

(a) Input EEG data (b) LRP results from CNN (c) Scales from CNN+SA

Fig. 4. One example of test samples containing problematic electrodes is the Position Task. As shown in (a), the dark red areas around the ears represent intense electrical signals with abnormal amplitudes (>100 V). In (b), the Layer-wise Relevance Propagation (LRP) results from the CNN model reveal that the electrodes around the left ear still play a crucial role in the prediction process. Conversely, the Attention-CNN model's results (c), indicate that it bypasses the ear area and allocates more emphasis to the pre-frontal region. As a result, the error in Euclidean Distance improved by 200.85 pixels for this specific sample (from 265.18 to 64.33). (Color figure online)

3.4 Explainability Quantification

We further examine the validity in explainability of the proposed method by comparing the distribution of learned attention of noisy and non-noisy electrodes in the Direction Task. The attention block's effectiveness is demonstrated by its ability to assign lower weights to these noisy electrodes in contrast to the non-noisy ones. Within all samples in the Direction Task that feature at least one noisy electrode, only 19% of the non-noisy electrodes had normalized attention weights below 0.05. In contrast, 42% of the noisy electrodes exhibited this trait, implying the attention block's ability to reduce weights of abnormal electrodes. We direct readers to the Supplementary materials for a distribution plot showcasing the difference between noisy and non-noisy electrodes, along with additional details. It's important to note that quantifying explainability methods for signal-format data, such as EEG, presents a significant challenge and has limited existing research. Therefore, additional investigations in this field are anticipated in future studies.

4 Conclusion

In this study, we aimed to address the issue of the lack of interpretability in deep learning models for EEG-based tasks. Our approach was to leverage the fact that EEG signal noise or artifacts are often localized to specific electrodes. We accomplished this by incorporating attention modules as electrode-sensitive components within a neural network architecture. These attention blocks were used to emphasize the importance of specific electrodes, resulting in more accurate predictions and improved interpretability through the use of scaling.

Moreover, our proposed approach was less susceptible to noise. We conducted comprehensive experiments to evaluate the performance of our proposed Attention-CNN model. Our results demonstrate that this model can accurately classify EEG and eye-tracking data while also providing insights into the quality of the recorded EEG signals. This contribution is significant as it can lead to the development of new decoding techniques that are less sensitive to noise.

In summary, our study underscores the importance of incorporating attention mechanisms into deep learning models for analyzing EEG and eye-tracking data. This approach opens up new avenues for future research in this area and has the potential to provide valuable insights into the neural basis of cognitive processes.

References

1. Bach, S., Binder, A., Montavon, G., Klauschen, F., Müller, K.R., Samek, W.: On pixel-wise explanations for non-linear classifier decisions by layer-wise relevance propagation. PLoS ONE **10**(7), e0130140 (2015)
2. Bulling, A., Ward, J.A., Gellersen, H., Tröster, G.: Eye movement analysis for activity recognition using electrooculography. IEEE Trans. Pattern Anal. Mach. Intell. **33**(4), 741–753 (2010)
3. Carter, B.T., Luke, S.G.: Best practices in eye tracking research. Int. J. Psychophysiol. **155**, 49–62 (2020)
4. Craik, A., He, Y., Contreras-Vidal, J.L.: Deep learning for electroencephalogram (EEG) classification tasks: a review. J. Neural Eng. **16**(3), 031001 (2019)
5. Devlin, J., Chang, M.W., Lee, K., Toutanova, K.: Bert: Pre-training of deep bidirectional transformers for language understanding. arXiv preprint arXiv:1810.04805 (2018)
6. Duchowski, A., Duchowski, A.: Eye tracking techniques. eye tracking methodology: Theory Pract. 51–59 (2007)
7. Eckstein, M.K., Guerra-Carrillo, B., Singley, A.T.M., Bunge, S.A.: Beyond eye gaze: what else can eyetracking reveal about cognition and cognitive development? Dev. Cogn. Neurosci. **25**, 69–91 (2017)
8. Feng, L.X., et al.: Automatic sleep staging algorithm based on time attention mechanism. Front. Hum. Neurosci. **15**, 692054 (2021)
9. Girshick, R.: Fast R-CNN. In: Proceedings of the IEEE International Conference on Computer Vision, pp. 1440–1448 (2015)
10. He, K., Zhang, X., Ren, S., Sun, J.: Deep residual learning for image recognition. In: Proceedings of the IEEE Conference on Computer Vision and Pattern Recognition, pp. 770–778 (2016)
11. Holmqvist, K., Nyström, M., Mulvey, F.: Eye tracker data quality: what it is and how to measure it. In: Proceedings of the Symposium on Eye Tracking Research and Applications, pp. 45–52 (2012)
12. Hu, J., Shen, L., Sun, G.: Squeeze-and-excitation networks. In: Proceedings of the IEEE Conference on Computer Vision and Pattern Recognition, pp. 7132–7141 (2018)

13. Hu, Z., Chen, L., Luo, Y., Zhou, J.: EEG-based emotion recognition using convolutional recurrent neural network with multi-head self-attention. Appl. Sci. **12**(21), 11255 (2022)
14. Ioffe, S., Szegedy, C.: Batch normalization: accelerating deep network training by reducing internal covariate shift. In: International Conference on Machine Learning, pp. 448–456. PMLR (2015)
15. Kastrati, A., Plomecka, M.B., Küchler, J., Langer, N., Wattenhofer, R.: Electrode clustering and bandpass analysis of eeg data for gaze estimation. arXiv preprint arXiv:2302.12710 (2023)
16. Kastrati, A., et al.: Eegeyenet: a simultaneous electroencephalography and eye-tracking dataset and benchmark for eye movement prediction. arXiv preprint arXiv:2111.05100 (2021)
17. Lee, Y.E., Lee, S.H.: EEG-transformer: Self-attention from transformer architecture for decoding eeg of imagined speech. In: 2022 10th International Winter Conference on Brain-Computer Interface (BCI), pp. 1–4. IEEE (2022)
18. Maas, A.L., Hannun, A.Y., Ng, A.Y., et al.: Rectifier nonlinearities improve neural network acoustic models. In: Proceedings of ICML. vol. 30, p. 3. Atlanta, Georgia, USA (2013)
19. Munoz, D.P., Everling, S.: Look away: the anti-saccade task and the voluntary control of eye movement. Nat. Rev. Neurosci. **5**(3), 218–228 (2004)
20. Nair, V., Hinton, G.E.: Rectified linear units improve restricted boltzmann machines. In: ICML (2010)
21. Paszke, A., et al.: Pytorch: an imperative style, high-performance deep learning library. Adv. Neural Inform. Process. Syst. **32** (2019)
22. Ribeiro, M.T., Singh, S., Guestrin, C.: "why should i trust you?" explaining the predictions of any classifier. In: Proceedings of the 22nd ACM Sigkdd International Conference on Knowledge Discovery and Data Mining, pp. 1135–1144 (2016)
23. Ryan, J.D., Riggs, L., McQuiggan, D.A.: Eye movement monitoring of memory. JoVE (J. Visualized Exp.) (**42**), e2108 (2010)
24. Shaw, P., Uszkoreit, J., Vaswani, A.: Self-attention with relative position representations. arXiv preprint arXiv:1803.02155 (2018)
25. Sturm, I., Lapuschkin, S., Samek, W., Müller, K.R.: Interpretable deep neural networks for single-trial EEG classification. J. Neurosci. Methods **274**, 141–145 (2016)
26. Tabar, Y.R., Halici, U.: A novel deep learning approach for classification of EEG motor imagery signals. J. Neural Eng. **14**(1), 016003 (2016)
27. Teplan, M., et al.: Fundamentals of EEG measurement. Measure. Scie. Rev. **2**(2), 1–11 (2002)
28. Vachon, F., Tremblay, S.: What eye tracking can reveal about dynamic decision-making. Adv. Cogn. Eng. Neuroergonom. **11**, 157–165 (2014)
29. Vaswani, A., et al.: Attention is all you need. Adv. Neural Inform. Process. Syst. **30** (2017)

On the Relevance of Temporal Features for Medical Ultrasound Video Recognition

D. Hudson Smith[1(✉)], John Paul Lineberger[1], and George H. Baker[2]

[1] Clemson University, Clemson, SC 29634, USA
{dane2,jplineb}@clemson.edu
[2] Medical University of South Carolina, Charleston, SC 29425, USA
baker@musc.edu

Abstract. Many medical ultrasound video recognition tasks involve identifying key anatomical features regardless of when they appear in the video suggesting that modeling such tasks may not benefit from temporal features. Correspondingly, model architectures that exclude temporal features may have better sample efficiency. We propose a novel multi-head attention architecture that incorporates these hypotheses as inductive priors to achieve better sample efficiency on common ultrasound tasks. We compare the performance of our architecture to an efficient 3D CNN video recognition model in two settings: one where we expect not to require temporal features and one where we do. In the former setting, our model outperforms the 3D CNN - especially when we artificially limit the training data. In the latter, the outcome reverses. These results suggest that expressive time-independent models may be more effective than state-of-the-art video recognition models for some common ultrasound tasks in the low-data regime. Code is available at https://github.com/MedAI-Clemson/pda_detection.

Keywords: Ultrasound · Video · Sample Efficiency · Attention

1 Introduction and Related Work

Ultrasound (US) is one of the most common imaging techniques in medical practice, with applications to fetal imaging, cardiac imaging, sports medicine, and more. With the rise of US for routine clinical care, there is a growing interest in applying computer vision techniques to automate or enhance the analysis of US imagery [13]. Many US examinations involve the collection of video clips showing different anatomical regions. The medical imaging community is in the early stages of applying techniques from the video recognition community to US recognition tasks. These applications face several challenges arising from the

---

Supplementary Information The online version contains supplementary material available at https://doi.org/10.1007/978-3-031-43895-0_70.

H. Greenspan et al. (Eds.): MICCAI 2023, LNCS 14221, pp. 744–753, 2023.
https://doi.org/10.1007/978-3-031-43895-0_70

nature of US as an imaging modality, differences between US imagery and natural imagery, and the lack of large representative datasets. To make matters worse, the collection of large medical datasets is often unethical or prohibitively costly. There is, therefore, a significant need for efficient methods that can produce high levels of performance using the minimum number of samples. In this work, we propose an efficient US video recognition architecture that takes advantage the nature of common US recognition tasks.

To design an efficient US recognition architecture, it is necessary to consider the space of US recognition tasks and evaluate the algorithmic structures needed to efficiently capture the semantics in those settings. We posit that many of these tasks amount to the identification of specific visual characteristics at key moments in the clip. The identification of the *standard plane* in fetal head US depends on recognizing key structures in fetal brain tissue [3,19]; the quality assessment of FAST clips [24] relies on the ability to recognize that key organs and other structures have been visualized in the clip; view identification relies on recognizing orientation of the anatomical structures in relation to one another [8,11]; and the quantification of heart function requires measurement of ventricular volumes at two key moments in the cardiac cycle [22]. Based on these observations, we propose a novel *US Video Network* (USVN) that treats frames as independent and unordered. USVN constructs expressive video representations by combining information from multiple frames using a novel multi-head attention mechanism. We demonstrate a setting in which USVN yields better performance and far better sample efficiency than a competing model that includes temporal features. We also demonstrate that, in a setting where temporal dependence is important, USVN lags behind the competing model. These contrasting outcomes demonstrate the importance of tailoring the model architecture to the structure of the US recognition task in data-constrained settings.

A large body of work has addressed video recognition tasks, including object tracking [14], temporal action localization [28], captioning [1], action recognition [30], and many others. Driven by the availability of large human action datasets, the field of action recognition has focused on the need to capture expressive spatiotemporal features. This has led to the development of two-stream networks using optical flow [21], the use of 3D convolutional networks [10,25], and, of course, the use of transformer-based architectures [15,18]. Our main point of departure with these methods is the importance placed upon temporal features. We posit that temporal features are not relevant in some common US tasks and that excluding these features leads to better sample efficiency. To explore this idea, we assume temporal independence *a priori*, placing our problem formulation in the format of a Multi-instance Learning (MIL) task.

Multi-instance learning (MIL) describes the situation where labels apply to bags of instances rather than to individual instances. Instances within a bag are assumed to be unordered and, conditional on the bag label, independent from one another [2]. Under our assumption that all video frames can be treated independently, video recognition can be viewed as MIL where the bag is the video, and the instances are the frames. MIL has a long history of applications to video recognition that predates deep learning [5,6,23,29]. In the classical

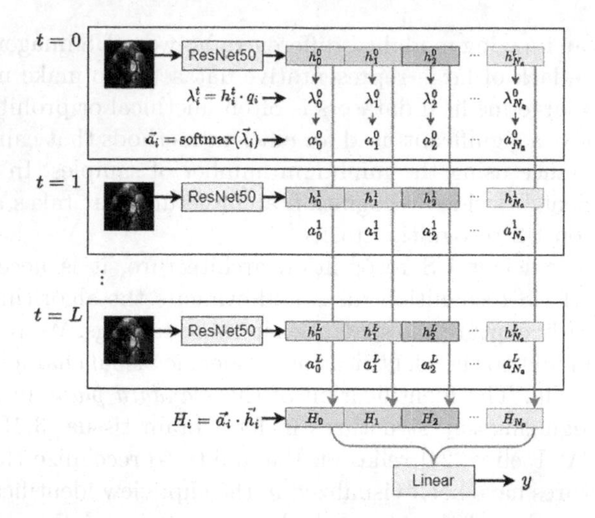

Fig. 1. Proposed video-recognition architecture. Frame representations from ResNet50 are partitioned into N_a equal-sized vectors, h_i^t, represented by the colored boxes at each time step. These are compared by dot product with global query vectors q_i to compute attention weights a_i^t. The video-level representation, H_i, is the attention-weighted sum of the partitions across frames. y is the video-level prediction.

formulation of MIL it is assumed that instances have unobserved labels, and the task is to extract these as latent variables and aggregate them to predict the bag-level label. In their paper *Attention-based deep multiple instance learning* Ilse, Tomczak, and Welling [9] depart from this classical perspective by aggregating embeddings rather than instance labels. We take a similar approach. Unlike their work, however, we use multiple attention heads focused on different subspaces of the image-level embeddings, with their work as a special case of ours. To our knowledge, we are the first to introduce a MIL formalism using multiple attention heads in this way.

There is growing interest in applying action recognition techniques to medical US video with applications to fetal [3,19,20], abdominal [11,24], and cardiac [4,8,17,22] US. Most existing applications make MIL assumptions but only apply a fixed pooling function to frame-level labels. Howard et al. [8] apply a range of techniques, including average pooling, two-stream networks, and 3D convolutions to identifying cardiac views. They conclude that two-stream networks yield the best performance. The authors do not test any methods that adaptively pool frame information in a time-independent manner. Lei et al. [12] specifically consider the detection of Patent Ductus Arteriosus (PDA). They make MIL assumptions by applying the video-level label to the individual frames and training a 2D CNN to estimate these noisy labels. Video-level labels are generated by applying a decision threshold to the frame-level predictions and then voting with equal weight across frames. Ouyang et al. [16] use 3D convolutions, specifically the R(2+1)D architecture [25], to predict ejection fraction from cardiac

US obtaining human-level performance. They do not assess the performance of any time-independent methods. Among these examples, we see a divide between methods that have no ability to adaptively weight different frames and those that can express arbitrary spatiotemporal features. We fill this gap by proposing a time-independent method that adaptively pools information from different moments in time.

2 Proposed Method

2.1 USVN

Architecture. Our video recognition architecture, shown in Fig. 1, pools information across frames using a multi-head attention mechanism. Like the attention mechanism in the transformer architecture [26], we compute attentions over subspaces of the frame-level representations. We hypothesize that US video recognition requires the detection of distinct visual features that may appear at different points of time in the video. The individual attention heads can function as detectors of these features. Unlike ordinary multi-head attention, the subspaces are not compared with other frames in the sequence but with a set of global query vectors inferred during training. The use of global query vectors arises from our inductive prior that the recognition task amounts to locating key pieces of information at any point in the sequence, and the inferred query vectors are representations of that key information.

Frames are first embedded into 2048-dimensional vectors using a CNN encoder. This encoder is initialized via ImageNet pretraining and fine-tuned during training. Rather than learn N_a projections from scratch for the attention weighting, we simply partition the frame representations into N_a vectors h_i^t each of size $d_a = 2048/N_a$ and rely on the final convolutional layers of the CNN to adapt. We then compute the un-normalized attention scores via dot product with the global query vectors: $\lambda_i^t = h_i^t \cdot q_i$. The resulting scores are normalized resulting in N_a attention vectors, $\boldsymbol{a_i} = \text{softmax}(\boldsymbol{\lambda_i})$, where the arrow notation represents vectorization in time. The video-level representation from the ith head is then simply $H_i = \boldsymbol{a_i} \cdot \boldsymbol{h_i}$, and the full video representation is the concatenation $H = \text{concat}([H_1, H_2, \ldots, H_{N_a}])$. The video-level prediction can then be computed using a shallow fully-connected network, $y = f(H)$.

Augmentation by Frame Sampling. Because USVN treats all frames independently, it is not necessary to use contiguous spans of frames during training. Instead, we randomly sample fixed-size sets of frames from each video. This can have a regularizing effect by using novel frames for each training epoch. During evaluation we use all video frames. We accommodate the varying numbers of frames in each video by zero padding and masked attention.

Model Interpretability. We identify prototype frames for each attention head. These prototypes produce embedding subspace vectors h_i^t that are closely aligned with the corresponding query vector q_i. These prototype images can then be qualitatively evaluated by the clinical specialist (see Supplemental Material).

2.2 Benchmark Implementations

A simple and common approach for video recognition is to use fixed pooling functions to aggregate the frame-level representations across time, treating each element of the representation as a channel. We evaluate this approach using max and average pooling functions. Our attention-based method can implement average pooling by assigning equal weight to all frames for each attention head. Neglecting potential optimization challenges, this suggests that attention-based pooling should be at least as good as average pooling. On the other hand, our model can only approximate max pooling in the $N_a = 2048$ case by assigning very large, positive values to the single-element query vectors causing the attentions to become sharply concentrated at one time step. However, this solution pushes the softmax over time into regions with very small gradients. We conclude that max pooling can learn video representations that cannot be expressed by USVN (and vice versa).

R(2+1)D is a 3D CNN video recognition architecture that decomposes the spatial and temporal convolution into two successive steps [25]. First, a 2D convolution is applied over space then a 1D convolution is applied over time. Compared to its 3D ResNet counterparts on Sports-1M and Kinetics datasets, R(2+1)D is a very capable model that can learn complex features while having the same number of parameters in a more data-efficient way. We choose to benchmark against this architecture due to its efficiency and because this is the architecture used by Ouyang et al. to achieve human-level performance on the EchoNet-Dynamic US dataset [16].

3 Experimental Results

3.1 Datasets

Patent Ductus Arteriosus (PDA). PDA is an opening between the aorta and pulmonary artery that, in severe cases, can cause heart failure shortly after birth. Ultrasound imaging is the primary diagnostic tool for detecting and characterizing PDA. Specifically, doppler US imaging can visualize the motion of the blood through the PDA opening. This motion appears as a characteristic blob of color in the region of the PDA. Physicians are trained to recognize the color and shape of the blob as well as where it appears in relation to other visible anatomy. Superficially, this recognition task makes no reference to the dynamics of the video. We therefore expect that temporal features are not required for accurate PDA recognition. For this dataset we train USVN to predict whether or not an image indicates the presence of PDA. The model output, y, is therefore a single number interpreted as the log-odds of PDA.

We retrospectively collected a set of 1,145 doppler US clips from 165 distinct examinations involving 66 distinct patients. Each clip was labeled to indicate the presence (661 clips) or absence (484 clips) of PDA. Patients were divided into training (44), validation (11), and test (11) sets with stratification on the presence of PDA. These sets contained 755, 118, and 272 videos, respectively.

The large variation in the number of videos in the validation and test sets results from the fact that patients have a variable number of examinations ranging from 1 to 10.

Table 1. Model performance comparison. EchoNet benefits from modeling temporal features; PDA does not. Performance is measured on the test set.

Model	PDA (ROC AUC)	EchoNet (r^2)
R(2+1)D	0.816	**0.822**
Average Pool	0.837	0.679
Max Pool	0.835	0.657
USVN (Ours)	**0.855**	0.765

EchoNet-Dynamic. The Echonet Dynamic dataset consists of 10,030 apical-4 chamber echocardiograms downsampled to 112×112. Each study has clinical measurements: ejection fraction (EF), end systolic volume (ESV), and end diastolic volume (EDV). EF is commonly used to assess cardiac function and is computed from ESV and EDV as

$$\mathrm{EF} = 1 - \mathrm{ESV}/\mathrm{EDV}. \tag{1}$$

The echocardiograms were obtained by registered sonographers and level 3 echocardiographers. For each of these videos, a masking and cropping transformation was performed to remove text and instrument information from the scanning area.

For this dataset, we train USVN to predict ejection fraction. Rather than predict EF directly, we output a tuple of real numbers (y_1, y_2) and insert them in place of ESV and EDV in Eq. (1). This choice is motivated by the knowledge that ESV and EDV are determined from different phases of the cardiac cycle. We speculate that decomposing EF into ESV and EDV effectively linearizes the estimation of EF as a function of the video representation H with different attention heads responsible for estimating ESV and EDV.

3.2 Results

Model Performance. Table 1 summarizes the performance of USVN and our benchmark implementations on the PDA and EchoNet tasks. For PDA classification, we evaluate using the area under the ROC curve (ROC AUC). For EchoNet, we use the percent of variance explained (r^2). USVN results are based on $N_a = 16$ and $N_a = 128$ for PDA and EchoNet, respectively, based on a hyperparameter search (see Supplemental Material). For the PDA dataset, we expected that temporal features are not beneficial and, indeed, we see that R(2+1)D performs worse than all other methods, likely due to the unneeded capacity in the temporal convolutions and the relatively small size of the PDA dataset. USVN leads to a small benefit over average and max pooling for this task. The EchoNet task does benefit from modeling temporal features as indicated by R(2+1)D obtaining the highest score. However, USVN significantly outperforms the fixed pooling methods and is surprisingly close to R(2+1)D. This suggests that temporal features play a relatively small part in explaining the variability in the EchoNet dataset.

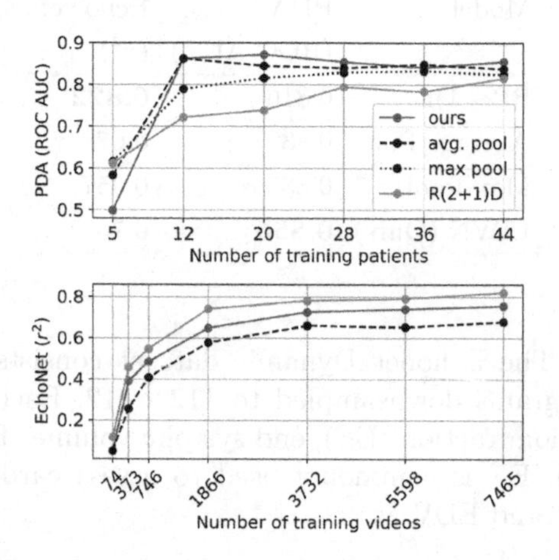

Fig. 2. Dependence on number of patients in training set for PDA classification (*top*) and EchoNet ejection fraction prediction (*bottom*). For PDA, we show patients, rather than videos along the x-axis due to the non-independence of videos from the same patient. For EchoNet, we omit the "max pool" variant because it failed to obtain positive r^2 values for several points along the x-axis. Performance is measured on the test set.

Sample Efficiency. In Fig. 2 we evaluate the sample efficiency of USVN by artificially limiting the amount of training data. In the case of PDA, we downsample the number of patients because videos from a single patient are correlated with one another. For EchoNet, we downsample the number of videos. In both cases, we use the full validation and test sets to better isolate variation due to limited training data from variation due to model selection and evaluation.

For PDA, R(2+1)D underperforms the time-independent methods, and the gap is larger for smaller numbers of training patients (see Fig. 2, top panel). Surprisingly, USVN and average pooling have very similar performance across samples and saturate for a small subset of the available patients. R(2+1)D needs all available patients to approach a similar level of performance. This result aligns with our expectation that the inductive prior of time independence can yield sample efficiency benefits when applied to the appropriate task.

R(2+1)D outperforms the time-independent models across all samples for the EchoNet task (see Fig. 2, bottom panel). Despite being a much simpler architecture than R(2+1)D and approaching similar levels of performance, USVN does not exhibit any sample efficiency benefits in the low-data regime for the EchoNet task. Solving the EchoNet task with spatial features alone may require more adaptation of the pretrained encoder than is required when solving with temporal features. For instance, it may be possible through extensive adaptation of the encoder network to recognize the visual characteristics associated with the end of diastole. However, the end of diastole may also manifest as, for example, an extremum in time of some visual characteristic. A model with access to temporal features such as R(2+1)D may be able to capture such an extremum with relatively little adaptation of the pretrained network.

3.3 Implementation Details

For the fixed pooling methods and USVN, we use an ImageNet-pretrained ResNet50 image encoder provided through the `timm` library [27]. We train using the `timm` implementation of the AdamP optimizer [7] with $\beta_{1,2} = 0.9, 0.999$, weight decay of 0.001, batch size of 20 clips, and initial learning rates of $3 \cdot 10^{-5}$ and 0.001 for PDA and EchoNet, respectively. We sample 32 frames per clip during training. We reduce the learning rate by a factor of 10 after 3 epochs with no improvement of the validation loss, and we terminate training after ten consecutive epochs of no improvement. We use 50% dropout on the inputs to the linear layer for each dataset.

To reproduce the results of R(2+1)D on Echonet Dynamic Dataset by Ouyang et al. [16], we cloned their github repo and re-ran their experiments with their best found hyperparameters. Our training runs show similar, if not better, results than stated in the original work. To adapt the model for PDA classification, we modified their data loader, training script, and the R(2+1)D model to allow PDA images. We also removed the manual bias term initialization, left over from predicting ejection fraction on the fully connected linear layer, and initialize it randomly instead. Finally, we replaced MSE loss with binary cross entropy with logits in the training loop. Every run was done for 45 epochs with a batch size of 20 for Echonet Dynamic dataset and 10 for PDA dataset. Model saving occurred for every epoch that showed improvement to the validation loss.

4 Conclusions and Discussion

The field of video recognition has been driven by large human action recognition datasets. Unlike videos of human actions, the accurate recognition of medical ultrasound images often only requires identifying key pieces of information at any point in the video and does not make reference to the sequence of events. The contrast between results for the PDA task (where USVN excels) and the EchoNet task (where USVN suffers) demonstrates the importance of tailoring the model architecture to the task at hand in data-constrained settings. Our results suggest that models developed for human action recognition are not optimal in some practical scenarios involving medical ultrasound and that models that assume temporal independence have better sample efficiency. We introduce an architecture, USVN, that is tailored to the medical ultrasound context and demonstrate a situation where the inductive prior of time independence leads to significant sample efficiency benefits. We also present a situation where temporal features are relevant and show that, even for very small datasets, USVN produces no efficiency benefits. Practitioners of deep learning who work with medical ultrasound in the low-data regime should take care to match the architecture choice to the nature of the recognition task.

Acknowledgement. We thank Clemson University for their generous allotment of compute time on the Palmetto Cluster.

References

1. Amirian, S., Rasheed, K., Taha, T.R., Arabnia, H.R.: Automatic image and video caption generation with deep learning: a concise review and algorithmic overlap. IEEE Access **8**, 218386–218400 (2020)
2. Carbonneau, M.A., Cheplygina, V., Granger, E., Gagnon, G.: Multiple instance learning: a survey of problem characteristics and applications. Pattern Recogn. **77**, 329–353 (2018)
3. Chen, H., et al.: automatic fetal ultrasound standard plane detection using knowledge transferred recurrent neural networks. In: Navab, N., Hornegger, J., Wells, W.M., Frangi, A.F. (eds.) MICCAI 2015. LNCS, vol. 9349, pp. 507–514. Springer, Cham (2015). https://doi.org/10.1007/978-3-319-24553-9_62
4. Dezaki, F.T., et al.: Deep residual recurrent neural networks for characterisation of cardiac cycle phase from echocardiograms. In: Cardoso, M.J., et al. (eds.) DLMIA/ML-CDS -2017. LNCS, vol. 10553, pp. 100–108. Springer, Cham (2017). https://doi.org/10.1007/978-3-319-67558-9_12
5. Ding, X., Li, B., Hu, W., Xiong, W., Wang, Z.: Horror video scene recognition based on multi-view multi-instance learning. In: Lee, K.M., Matsushita, Y., Rehg, J.M., Hu, Z. (eds.) ACCV 2012. LNCS, vol. 7726, pp. 599–610. Springer, Heidelberg (2013). https://doi.org/10.1007/978-3-642-37431-9_46
6. Gu, Z., Mei, T., Hua, X.S., Tang, J., Wu, X.: Multi-layer multi-instance learning for video concept detection. IEEE Trans. Multimedia **10**(8), 1605–1616 (2008)
7. Heo, B., et al.: Adamp: slowing down the slowdown for momentum optimizers on scale-invariant weights. arXiv preprint arXiv:2006.08217 (2020)
8. Howard, J.P., et al.: Improving ultrasound video classification: an evaluation of novel deep learning methods in echocardiography. J. Med. Artif. Intell. **3** (2020)
9. Ilse, M., Tomczak, J., Welling, M.: Attention-based deep multiple instance learning. In: International Conference on Machine Learning, pp. 2127–2136. PMLR (2018)
10. Ji, S., Xu, W., Yang, M., Yu, K.: 3D convolutional neural networks for human action recognition. IEEE Trans. Pattern Anal. Mach. Intell. **35**(1), 221–231 (2012)
11. Kornblith, A.E., et al.: Development and validation of a deep learning strategy for automated view classification of pediatric focused assessment with sonography for trauma. J. Ultrasound Med. **41**(8), 1915–1924 (2022)
12. Lei, H., Ashrafi, A., Chang, P., Chang, A., Lai, W.: Patent ductus arteriosus (PDA) detection in echocardiograms using deep learning. Intell.-Based Med. **6**, 100054 (2022)

13. Liu, S., et al.: Deep learning in medical ultrasound analysis: a review. Engineering **5**(2), 261–275 (2019)
14. Luo, W., Xing, J., Milan, A., Zhang, X., Liu, W., Kim, T.K.: Multiple object tracking: a literature review. Artif. Intell. **293**, 103448 (2021)
15. Mazzia, V., Angarano, S., Salvetti, F., Angelini, F., Chiaberge, M.: Action transformer: a self-attention model for short-time pose-based human action recognition. Pattern Recogn. **124**, 108487 (2022)
16. Ouyang, D., et al.: Video-based AI for beat-to-beat assessment of cardiac function. Nature **580**(7802), 252–256 (2020)
17. Patra, A., Huang, W., Noble, J.A.: Learning spatio-temporal aggregation for fetal heart analysis in ultrasound video. In: Cardoso, M.J., et al. (eds.) DLMIA/ML-CDS -2017. LNCS, vol. 10553, pp. 276–284. Springer, Cham (2017). https://doi.org/10.1007/978-3-319-67558-9_32
18. Plizzari, C., Cannici, M., Matteucci, M.: Spatial temporal transformer network for skeleton-based action recognition. In: Del Bimbo, A., et al. (eds.) ICPR 2021. LNCS, vol. 12663, pp. 694–701. Springer, Cham (2021). https://doi.org/10.1007/978-3-030-68796-0_50
19. Pu, B., Li, K., Li, S., Zhu, N.: Automatic fetal ultrasound standard plane recognition based on deep learning and IIoT. IEEE Trans. Industr. Inf. **17**(11), 7771–7780 (2021)
20. Rasheed, K., Junejo, F., Malik, A., Saqib, M.: Automated fetal head classification and segmentation using ultrasound video. IEEE Access **9**, 160249–160267 (2021)
21. Simonyan, K., Zisserman, A.: Two-stream convolutional networks for action recognition in videos. Adv. Neural Inform. Process. Syst. **27** (2014)
22. Sofka, M., Milletari, F., Jia, J., Rothberg, A.: Fully convolutional regression network for accurate detection of measurement points. In: Cardoso, M.J., et al. (eds.) DLMIA/ML-CDS -2017. LNCS, vol. 10553, pp. 258–266. Springer, Cham (2017). https://doi.org/10.1007/978-3-319-67558-9_30
23. Stikic, M., Schiele, B.: Activity recognition from sparsely labeled data using multi-instance learning. In: Choudhury, T., Quigley, A., Strang, T., Suginuma, K. (eds.) LoCA 2009. LNCS, vol. 5561, pp. 156–173. Springer, Heidelberg (2009). https://doi.org/10.1007/978-3-642-01721-6_10
24. Taye, M., Morrow, D., Cull, J., Smith, D.H., Hagan, M.: Deep learning for fast quality assessment. J. Ultrasound Med. **42**(1), 71–79 (2022)
25. Tran, D., Wang, H., Torresani, L., Ray, J., LeCun, Y., Paluri, M.: A closer look at spatiotemporal convolutions for action recognition. In: Proceedings of the IEEE conference on Computer Vision and Pattern Recognition, pp. 6450–6459 (2018)
26. Vaswani, A., et al.: Attention is all you need. Adv. Neural Inform. Process. Syst. **30** (2017)
27. Wightman, R.: Pytorch image models. https://github.com/rwightman/pytorch-image-models (2019). https://doi.org/10.5281/zenodo.4414861
28. Xia, H., Zhan, Y.: A survey on temporal action localization. IEEE Access **8**, 70477–70487 (2020)
29. Yang, J., Yan, R., Hauptmann, A.G.: Multiple instance learning for labeling faces in broadcasting news video. In: Proceedings of the 13th Annual ACM International Conference on Multimedia, pp. 31–40 (2005)
30. Zhang, H.B., et al.: A comprehensive survey of vision-based human action recognition methods. Sensors **19**(5), 1005 (2019)

Synthetic Augmentation with Large-Scale Unconditional Pre-training

Jiarong Ye[1], Haomiao Ni[1], Peng Jin[1], Sharon X. Huang[1], and Yuan Xue[2,3](✉)

[1] The Pennsylvania State University, University Park, Pennsylvania, USA
[2] Johns Hopkins University, Baltimore, MD, USA
[3] The Ohio State University, Columbus, OH, USA
Yuan.Xue@osumc.edu

Abstract. Deep learning based medical image recognition systems often require a substantial amount of training data with expert annotations, which can be expensive and time-consuming to obtain. Recently, synthetic augmentation techniques have been proposed to mitigate the issue by generating realistic images conditioned on class labels. However, the effectiveness of these methods heavily depends on the representation capability of the trained generative model, which cannot be guaranteed without sufficient labeled training data. To further reduce the dependency on annotated data, we propose a synthetic augmentation method called HistoDiffusion, which can be pre-trained on large-scale unlabeled datasets and later applied to a small-scale labeled dataset for augmented training. In particular, we train a latent diffusion model (LDM) on diverse unlabeled datasets to learn common features and generate realistic images without conditional inputs. Then, we fine-tune the model with classifier guidance in latent space on an unseen labeled dataset so that the model can synthesize images of specific categories. Additionally, we adopt a selective mechanism to only add synthetic samples with high confidence of matching to target labels. We evaluate our proposed method by pre-training on three histopathology datasets and testing on a histopathology dataset of colorectal cancer (CRC) excluded from the pre-training datasets. With HistoDiffusion augmentation, the classification accuracy of a backbone classifier is remarkably improved by 6.4% using a small set of the original labels. Our code is available at https://github.com/karenyyy/HistoDiffAug.

1 Introduction

The recent advancements in medical image recognition systems have greatly benefited from deep learning techniques [15,28]. Large-scale well-annotated datasets

J. Ye and H. Ni—These authors contributed equally to this work.

Supplementary Information The online version contains supplementary material available at https://doi.org/10.1007/978-3-031-43895-0_71.

H. Greenspan et al. (Eds.): MICCAI 2023, LNCS 14221, pp. 754–764, 2023.
https://doi.org/10.1007/978-3-031-43895-0_71

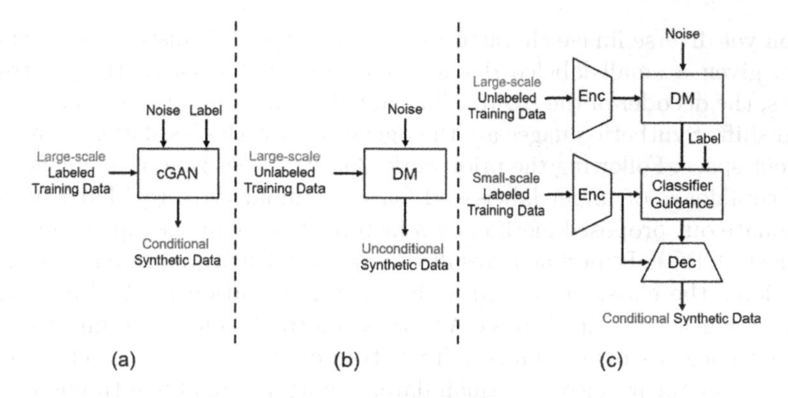

Fig. 1. Comparison between different deep generative models for synthetic augmentation. (a) cGAN-based method which requires relatively large-scale annotated training data; (b) Diffusion model (DM) which cannot take conditional input; (c) Our proposed HistoDiffusion model that can be pretrained on large-scale unannotated data and later applied to *unseen* small-scale annotated data for augmentation.

are one of the key components for training deep learning models to achieve satisfactory results [3,17]. However, unlike natural images in computer vision, the number of medical images with expert annotations is often limited by the high labeling cost and privacy concerns. To overcome this challenge, a natural choice is to employ data augmentation to increase the number of training samples. Although conventional augmentation techniques [23] such as flipping and cropping can be directly applied to medical images, they merely improve the diversity of datasets, thus leading to marginal performance gains [1]. Another group of studies employ conditional generative adversarial networks (cGANs) [10] to synthesize visually appealing medical images that closely resemble those in the original datasets [36,37]. While existing works have proven effective in improving the performance of downstream models to some extent, a sufficient amount of labeled data is still required to adequately train models to generate decent-quality images. More recently, diffusion models have become popular for natural image generation due to their impressive results and training stability [4,13,31]. A few studies have also demonstrated the potential of diffusion models for medical image synthesis [19,24].

Although annotated data is typically hard to acquire for medical images, unannotated data is often more accessible. To mitigate the issue existed in current cGAN-based synthetic augmentation methods [8,36–38], in this work, we propose to leverage the diffusion model with unlabeled pre-training to reduce the dependency on the amount of labeled data (see comparisons in Fig. 1). We propose a novel synthetic augmentation method, named HistoDiffusion, which can be pre-trained on large-scale unannotated datasets and adapted to small-scale annotated datasets for augmented training. Specifically, we first employ a latent diffusion model (LDM) and train it on a collection of unlabeled datasets from multiple sources. This large-scale pre-training enables the model to learn common yet diverse image characteristics and generate realistic medical images. Second, given a small labeled dataset that does not exist in the pre-training datasets, the decoder of the LDM is fine-tuned using annotations to adapt to the domain shift. Synthetic images are then generated with classifier guidance [4] in the latent space. Following the prior work [36], we select generated images based on the confidence of target labels and feature similarity to real labeled images. We evaluate our proposed method on a histopathology image dataset of colorectal cancer (CRC). Experiment results show that when presented with limited annotations, the classifier trained with our augmentation method outperforms the ones trained with the prior cGAN-based methods. Our experimental results show that once HistoDiffusion is well pre-trained using large datasets, it can be applied to any future incoming small dataset with minimal fine-tuning and may substantially improve the flexibility and efficacy of synthetic augmentation.

2 Methodology

Figure 2 illustrates the overall architecture of our proposed method. First, we train an LDM on a large-scale set of unlabeled datasets collected from multiple sources. We then fine-tune the decoder of this pretrained LDM on a small labeled dataset. To enable conditional image synthesis, we also train a latent classifier on the same labeled dataset to guide the diffusion model in LDM. Once the classifier is trained, we apply the fine-tuned LDM to generate a pool of candidate images conditioned on the target class labels. These candidate images are then passed through the image selection module to filter out any low-quality results. Finally, we can train downstream classification models on the expanded training data, which includes the selected images, and then use them to perform inference on test data. In this section, we will first introduce the background of diffusion models and then present details about the HistoDiffusion model.

2.1 Diffusion Models

Diffusion models (DM) [13,30,32] are probabilistic models that are designed to learn a data distribution. Given a sample from the data distribution $z_0 \sim q(z_0)$, the DM *forward* process produces a Markov chain $z_1, \ldots, z_T$ by gradually adding Gaussian noise to z_0 based on a variance schedule $\beta_1, \ldots, \beta_T$, that is:

$$q(z_t|z_{t-1}) = \mathcal{N}(z_t; \sqrt{1-\beta_t} z_{t-1}, \beta_t \mathbf{I}) \ , \tag{1}$$

where variances β_t are constants. If β_t are small, the posterior $q(z_{t-1}|z_t)$ can be well approximated by diagonal Gaussian [21,30]. Furthermore, when the T of the chain is large enough, z_T can be well approximated by standard Gaussian distribution $\mathcal{N}(\mathbf{0}, \mathbf{I})$. These suggest that the true posterior $q(z_{t-1}|z_t)$ can be estimated by $p_\theta(z_{t-1}|z_t)$ defined as [22]:

$$p_\theta(z_{t-1}|z_t) = \mathcal{N}(z_{t-1}; \mu_\theta(z_t), \Sigma_\theta(z_t)) \ . \tag{2}$$

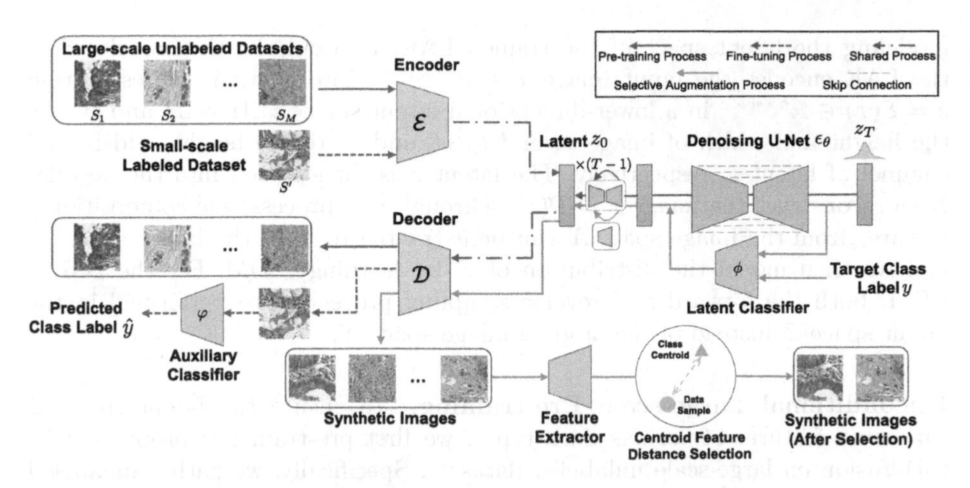

Fig. 2. The architecture of our proposed HistoDiffusion, which consists of a pre-training process (blue solid lines), a fine-tuning process (blue dashed lines), and a selective augmentation process (orange lines). During pre-training, a latent autoencoder (LAE) and a diffusion model (DM) are trained on large-scale unlabeled datasets for unconditional image synthesis. HistoDiffusion is then fine-tuned on a small-scale dataset for conditional image synthesis under the guidance of a trained latent classifier. During selective augmentation, given a target class label, the synthetic images generated by the fine-tuned model are selected and added to the training set based on their distances to the class centroids in the feature space. (Color figure online)

The DM *reverse* process (also known as *sampling*) then generates samples $z_0 \sim p_\theta(z_0)$ by initiating a Markov chain with Gaussian noise $z_T \sim \mathcal{N}(\mathbf{0}, \mathbf{I})$ and progressively decreasing noise in the chain of $z_{T-1}, z_{T-2}, \cdots, z_0$ using the learnt $p_\theta(z_{t-1}|z_t)$. To learn $p_\theta(z_{t-1}|z_t)$, Gaussian noise ϵ is added to z_0 to generate samples $z_t \sim q(z_t|z_0)$, then a model ϵ_θ is trained to predict ϵ using the following mean-squared error loss:

$$L_{\mathrm{DM}} = \mathbb{E}_{t \sim \mathcal{U}(1,T), z_0 \sim q(z_0), \epsilon \sim \mathcal{N}(\mathbf{0},\mathbf{I})} [||\epsilon - \epsilon_\theta(z_t, t)||^2] \ , \tag{3}$$

where time step t is uniformly sampled from $\{1, \dots, T\}$. Then $\mu_\theta(z_t)$ and $\Sigma_\theta(z_t)$ in Eq. 2 can be derived from $\epsilon_\theta(z_t, t)$ to model $p_\theta(z_{t-1}|z_t)$ [13,22]. The denoising model ϵ_θ is typically implemented using a time-conditioned U-Net [27] with residual blocks [11] and self-attention layers [35]. Sinusoidal position embedding [35] is also usually used to specify the time step t to ϵ_θ.

2.2 HistoDiffusion

Model Architecture. Our proposed HistoDiffusion is built on Latent Diffusion Models (LDM) [26], which requires fewer computational resources without degradation in performance, compared to prior works [4,15,28]. LDM first trains a latent autoencoder (LAE) [16] to encode images as lower-dimensional latent representations and then learns a diffusion model (DM) for image synthesis by modeling the latent space of the trained LAE. Particularly, the encoder $\mathcal{E}$ of the LAE encodes the input image $x \in \mathbb{R}^{H \times W \times 3}$ into a latent representation $z = \mathcal{E}(x) \in \mathbb{R}^{h \times w \times c}$ in a lower-dimensional latent space $\mathcal{Z}$. Here H and W are the height and width of image x, and h, w, and c are the height, width, and channel of latent z, respectively. The latent z is then passed into the decoder $\mathcal{D}$ to reconstruct the image $\hat{x} = \mathcal{D}(z)$. Through this process, the compositional features from the image space $\mathcal{X}$ can be extracted to form the latent space $\mathcal{Z}$, and we then model the distribution of $\mathcal{Z}$ by learning a DM. For the DM in LDM, both the forward and reverse sampling processes are performed in the latent space $\mathcal{Z}$ instead of the original image space $\mathcal{X}$.

Unconditional Large-scale Pre-training. To ensure the latent space $\mathcal{Z}$ can cover features of various data types, we first pre-train our proposed HistoDiffusion on large-scale unlabeled datasets. Specifically, we gather unlabeled images from M different sources to construct a large-scale set of datasets $\mathcal{S} = \{S_1, S_2, \ldots, S_M\}$. We then train an LAE using the data from $\mathcal{S}$ with the following self-reconstruction loss to learn a powerful latent space $\mathcal{Z}$ that can describe diverse features:

$$L_{\text{LAE}} = \mathcal{L}_{\text{rec}}(\hat{x}, x) + \lambda_{\text{KL}} D_{\text{KL}}(q(z)||\mathcal{N}(\mathbf{0}, \mathbf{I})) \ , \tag{4}$$

where $\mathcal{L}_{\text{rec}}$ is the loss measuring the difference between the output reconstructed image $\hat{x}$ and the input ground truth image x. Here we implement $\mathcal{L}_{\text{rec}}$ with a combination of a pixel-wise L_1 loss, a perceptual loss [39], and a patch-base adversarial loss [6,7]. To avoid arbitrarily high-variance latent spaces, we also add a KL regularization term D_{KL} [16,26] to constrain the variance of the latent space $\mathcal{Z}$ with a slight KL-penalty.

After training the LAE, we fixed the trained encoder $\mathcal{E}$ and then train a DM with the loss L_{DM} in Eq. 3 to model $\mathcal{E}$'s latent space $\mathcal{Z}$. Here $z_0 = \mathcal{E}(x)$ in Eq. 3. Once the DM is trained, we can use denoising model ϵ_θ in the DM reverse sampling process to synthesize a novel latent $\tilde{z}_0 \in \mathbb{R}^{h \times w \times c}$ and employ the trained decoder $\mathcal{D}$ to generate a new image $\tilde{x} = \mathcal{D}(\tilde{z}_0)$, which should satisfy the similar distribution as the data in $\mathcal{S}$.

Conditional Small-scale Fine-tuning. Using the LAE and DM pretrained on $\mathcal{S}$, we can only generate the new image $\tilde{x}$ following the similar distribution in $\mathcal{S}$. To generalize our HistoDiffusion to the small-scale labeled dataset S' collected from a different source (*i.e.*, $S' \not\subset \mathcal{S}$), we further fine-tune HistoDiffusion using the labeled data from S'. Let y be the label of image x in S'. To minimize the training cost, we fix both the trained encoder $\mathcal{E}$ and trained DM model ϵ_θ to keep latent space $\mathcal{Z}$ unchanged. Then we only fine-tune the decoder $\mathcal{D}$ using labeled data (x, y) from S' with the following loss function:

$$L_{\mathcal{D}} = \mathcal{L}_{\text{rec}}(\hat{x}, x) + \lambda_{\text{CE}} \mathcal{L}_{\text{CE}}(\varphi(\hat{x}), y) \ , \tag{5}$$

where $\mathcal{L}_{\text{rec}}(\hat{x}, x)$ is the self-reconstruction loss between the output reconstructed image $\hat{x} = \mathcal{D}(\mathcal{E}(x))$ and the input ground truth image x. To enhance the correlation between the decoder output $\hat{x}$ and label y, we also add an auxiliary image classifier φ trained with (x, y) on the top of $\mathcal{D}$ and impose the cross-entropy classification loss $\mathcal{L}_{\text{CE}}$ when fine-tuning $\mathcal{D}$. λ_{CE} is the balancing parameter. We annotate this fine-tuned decoder as $\mathcal{D}'$ for differentiation.

Classifier-guided Conditional Synthesis. To enable conditional image generation with our HistoDiffusion, we further apply the classifier-guided diffusion sampling proposed in [4,29,30,33] using the labeled data (x, y) from small-scale labeled dataset S'. We first utilize the trained encoder $\mathcal{E}$ to encode the data x from S' as latent z_0. Then we train a time-dependent latent classifier ϕ with paired (z_t, y) using the following loss function:

$$L_\phi = \mathcal{L}_{\text{CE}}(\phi(z_t), y) \ , \tag{6}$$

where $z_t \sim q(z_t|z_0)$ is the noisy version of z_0 at the time step t during the DM forward process, and $\mathcal{L}_{\text{CE}}$ is the cross-entropy classification loss. Based on the trained unconditional diffusion model ϵ_θ, and a classifier ϕ trained on noisy input z_t, we enable conditional diffusion sampling by perturbing the reverse-process mean with the gradient of the log probability $p_\phi(y|z_t)$ of a target class y predicted by the classifier ϕ as follows:

$$\hat{\mu}_\theta(z_t|y) = \mu_\theta(z_t) + g \cdot \Sigma_\theta(z_t) \nabla_{z_t} \log p_\phi(y|z_t) \ , \tag{7}$$

where g is the guidance scale. Then the DM reverse process in HistoDiffusion can finally generate a novel latent $\tilde{z}_0$ satisfying the class condition y through a Markov chain starting with a standard Gaussian noise $z_T \sim \mathcal{N}(\mathbf{0}, \mathbf{I})$ using $p_{\theta,\phi}(z_{t-1}|z_t, y)$ defined as follows:

$$p_{\theta,\phi}(z_{t-1}|z_t, y) = \mathcal{N}(z_{t-1}; \hat{\mu}_\theta(z_t|y), \Sigma_\theta(z_t)) \ . \tag{8}$$

The final image $\tilde{x}$ of class y can be generated by applying the fine-tuned decoder $\mathcal{D}'$, *i.e.*, $\tilde{x} = \mathcal{D}'(\tilde{z}_0)$.

Selective Augmentation. To further improve the efficacy of synthetic augmentation, we follow [36] to selectively add synthetic images to the original labeled training data based on centroid feature distance. The augmentation ratio is defined as the ratio between the selected synthetic images and the original training images. More results are demonstrated later in Table 1.

3 Experiments

Datasets. We employ three public datasets of histopathology images during the large-scale pre-training procedure. The first one is the H&E breast cancer dataset [2], containing 312,320 patches extracted from the hematoxylin & eosin (H&E) stained human breast cancer tissue micro-array (TMA) images [18]. Each patch has a resolution of 224 × 224. The second dataset is PanNuke [9], a pan-cancer histology dataset for nuclei instance segmentation and classification. The PanNuke dataset includes 7,901 patches of 19 types of H&E stained

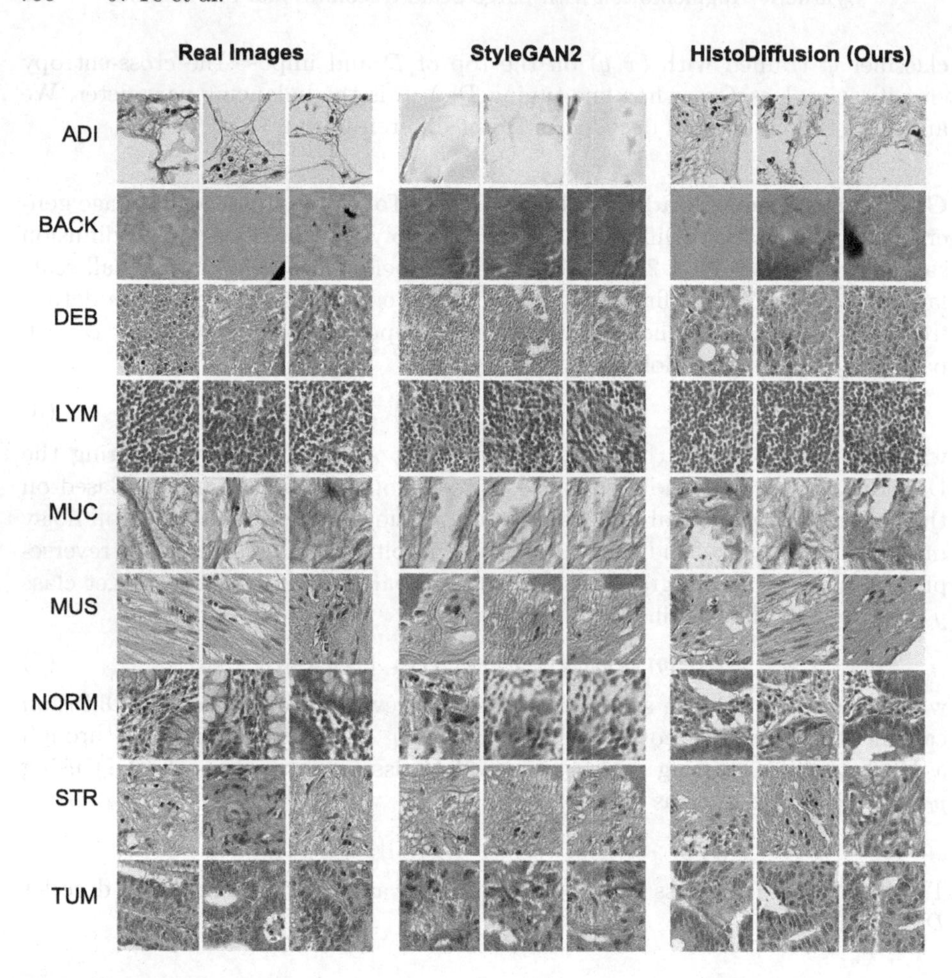

Fig. 3. Comparison of real images from training subset, synthesized images generated by StyleGAN2 [14] and our proposed HistoDiffusion (zoom in for clear observation). Qualitatively, our synthesized results contain more realistic and diagnosable patterns than results synthesized from StyleGAN2.

tissues obtained from multiple data sources, and each patch has a unified size of 256×256 pixels. The third dataset is TCGA-BRCA-A2/E2 [34], a subset derived from the TCGA-BRCA breast cancer histology dataset [20]. The subset consists of 482,958 patches with a resolution of 256×256. Overall, there are 803,179 patches used for pre-training. As for fine-tuning and evaluation, we employ the NCT-CRC-HE-100K dataset that contains 100,000 patches from H&E stained histological images of human colorectal cancer (CRC) and normal tissue. The patches have been divided into 9 classes: Adipose (ADI), background (BACK), debris (DEB), lymphocytes (LYM), mucus (MUC), smooth muscle (MUS), normal colon mucosa (NORM), cancer-associated stroma (STR), colorectal adenocarcinoma epithelium (TUM). The resolution of each patch is 224×224.

To replicate a scenario where only a small annotated dataset is available for training, we have opted to utilize a subset of 5,000 (5%) samples for fine-tuning. This subset has been carefully selected through an even sampling without replacement from each tissue type present in the train set. It is worth noting that the labels for these samples have been kept, which allows the fine-tuning process to be guided by labeled data, leading to better predictions on the specific task or domain being trained. By ensuring that the fine-tuning process is representative of the entire dataset through even sampling from each tissue type, we can eliminate bias towards any particular tissue type. We evaluate the fine-tuned model on the official test set. The related data use declaration and acknowledgment can be found in our supplementary materials.

Evaluation Metrics. We employ Fréchet Inception Distance (FID) score [12] to assess the image quality of the synthetic samples. We further compute the accuracy, F1-score, sensitivity, and specificity of the downstream classifiers to evaluate the performance gain from different augmentation methods.

Model Implementation. All the patches are resized to $256 \times 256 \times 3$ before being passed into the models. Our implementation of HistoDiffusion basically follows the LDM-4 [26] architecture, where the input is downsampled by a factor of 4, resulting in a latent representation with dimensions of $64 \times 64 \times 3$. We use 1000 timesteps ($T = 1000$) for the training of diffusion model and sample with classifier-free guidance scale $g = 1.0$ and 200 DDIM steps. The latent classifier ϕ is constructed using the encoder architecture of the LAE and an additional attention pooling layer [25] added before the output layer.

We use the same architecture for the auxiliary image classifier φ. For downstream evaluation, we implement the classifier using the ViT-B/16 architecture [5] in all experiments to ensure fair comparisons. The default hyper-parameter settings provided in their officially released codebases are followed.

Comparison to State-of-the-Art. We compare our proposed HistoDiffusion with the current state-of-the-art cGAN-based method [36]. We employ StyleGAN2 [14] as the backbone generative model for cGAN-based synthesis. To ensure a fair comparison, all images synthesized by StyleGAN2 and HistoDiffusion model are further selected based on feature centroid distances [36]. More implementation details of our proposed HistoDiffusion, StyleGAN2, and baseline classifier can also be found in our supplementary materials.

Result Analysis. As shown in Table 1, under the same synthetic augmentation setting, HistoDiffusion shows better FID scores and outperforms the state-of-the-art cGAN model StyleGAN2 in all classification metrics. A qualitative comparison between synthetic images by HistoDiffusion and StyleGAN2 can be

Table 1. Quantitative comparison results of synthetic image quality and augmented classification. "Random" refers to directly augmenting the training dataset with synthesized images without any image selections while "selective" indicates applying selective module [36] to filter out low-quality images. The number ($X\%$) suggests that the number of the synthesized images is $X\%$ of the original training set.

	FID↓	Accuracy↑	F1 Score↑	Sensitivity↑	Specificity↑
Baseline (5% real images)	/	0.855	0.850	0.855	0.983
StyleGAN2 [14]					
+ random 50%	5.714	0.860	0.856	0.860	0.980
+ selective [36] 50%	5.088	0.868	0.861	0.867	0.978
100%	5.927	0.879	0.876	0.879	0.982
200%	7.550	0.895	0.888	0.895	0.983
300%	10.643	0.898	0.896	0.898	0.987
HistoDiffusion (Ours)					
+ random 50%	4.921	0.870	0.869	0.870	0.982
+ selective [36] 50%	4.544	0.891	0.888	0.891	0.983
100%	**3.874**	0.903	0.902	0.903	0.991
200%	4.583	**0.919**	**0.916**	**0.919**	**0.992**
300%	8.326	0.910	0.912	0.910	0.988

found in Fig. 3, where HistoDiffusion consistently generates more realistic images matching the given class conditions than SytleGAN2, especially for classes ADI and BACK.

When augmenting the training dataset with different numbers of images synthesized from HistoDiffusion and StyleGAN2, one can observe that when increasing the ratio of synthesized data to 100%, the FID score of StyleGAN2 increases quickly and can become even worse than the one without using image selection strategy. In contrast, HistoDiffusion can keep synthesizing high-quality images until the augmentation ratio reaches 300%. Regarding classification performance improvement of the baseline classifier, the accuracy and F1 score of using HistoDiffusion augmentation are increased by up to **6.4**% and **6.6**%, respectively. Even when not using the image selection module to filter out the low-quality results (*i.e.*, +random 50%), our HistoDiffusion can still improve the accuracy by 1.5%. The robustness and effectiveness of HistoDiffusion can be attributed to the unconditional large-scale pre-training, our specially-designed conditional fine-tuning, and classifier-guided generation, among others.

4 Conclusions

In this study, we have introduced a novel synthetic augmentation technique, termed HistoDiffusion, to enhance the performance of medical image recognition systems. HistoDiffusion leverages multiple unlabeled datasets for large-scale, unconditional pre-training, while employing a labeled dataset for small-scale conditional fine-tuning. Experiment results on a histopathology image dataset excluded from the pre-training demonstrate that given limited labels, HistoDiffusion with image selection remarkably enhances the classification performance of the baseline model, and can potentially handle any future incoming small dataset for augmented training using the same pre-trained model.

References

1. Chen, Y., et al.: Generative adversarial networks in medical image augmentation: a review. Comput. Biol. Med. **144** 105382 (2022)
2. Claudio Quiros, A., Murray-Smith, R., Yuan, K.: Pathologygan: learning deep representations of cancer tissue. MELBA **2021**(4), 1–48 (2021)
3. Deng, J., Dong, W., Socher, R., Li, L.J., Li, K., Fei-Fei, L.: ImageNet: A large-scale hierarchical image database. In: CVPR, pp. 248–255. Ieee (2009)
4. Dhariwal, P., Nichol, A.: Diffusion models beat GANs on image synthesis. NeurIPS **34**, 8780–8794 (2021)
5. Dosovitskiy, A., et al.: An image is worth 16x16 words: Transformers for image recognition at scale. ICLR (2021)
6. Dosovitskiy, A., Brox, T.: Generating images with perceptual similarity metrics based on deep networks. In: NeurIPS, vol. 29 (2016)
7. Esser, P., Rombach, R., Ommer, B.: Taming transformers for high-resolution image synthesis. In: CVPR, pp. 12873–12883 (2021)
8. Frid-Adar, M., Diamant, I., Klang, E., Amitai, M., Goldberger, J., Greenspan, H.: Gan-based synthetic medical image augmentation for increased CNN performance in liver lesion classification. Neurocomputing **321**, 321–331 (2018)
9. Gamper, J., Alemi Koohbanani, N., Benet, K., Khuram, A., Rajpoot, N.: PanNuke: an open pan-cancer histology dataset for nuclei instance segmentation and classification. In: Reyes-Aldasoro, C.C., Janowczyk, A., Veta, M., Bankhead, P., Sirinukunwattana, K. (eds.) ECDP 2019. LNCS, vol. 11435, pp. 11–19. Springer, Cham (2019). https://doi.org/10.1007/978-3-030-23937-4_2
10. Goodfellow, I., et al.: Generative adversarial networks. Commun. ACM **63**(11), 139–144 (2020)
11. He, K., Zhang, X., Ren, S., Sun, J.: Deep residual learning for image recognition. In: CVPR, pp. 770–778 (2016)
12. Heusel, M., Ramsauer, H., Unterthiner, T., Nessler, B., Hochreiter, S.: GANs trained by a two time-scale update rule converge to a local nash equilibrium. In: NeurIPS, vol. 30 (2017)
13. Ho, J., Jain, A., Abbeel, P.: Denoising diffusion probabilistic models. NeurIPS **33**, 6840–6851 (2020)
14. Karras, T., Laine, S., Aittala, M., Hellsten, J., Lehtinen, J., Aila, T.: Analyzing and improving the image quality of stylegan. In: CVPR, pp. 8110–8119 (2020)
15. Ker, J., Wang, L., Rao, J., Lim, T.: Deep learning applications in medical image analysis. IEEE Access **6**, 9375–9389 (2017)
16. Kingma, D.P., Welling, M.: Auto-encoding variational bayes. arXiv (2013)
17. Lin, T.-Y., et al.: Microsoft COCO: common objects in context. In: Fleet, D., Pajdla, T., Schiele, B., Tuytelaars, T. (eds.) ECCV 2014. LNCS, vol. 8693, pp. 740–755. Springer, Cham (2014). https://doi.org/10.1007/978-3-319-10602-1_48

18. Marinelli, R.J., et al.: The stanford tissue microarray database. Nucleic Acids Research **36**(suppl_1), D871–D877 (2007)
19. Moghadam, P.A., et al.: A morphology focused diffusion probabilistic model for synthesis of histopathology images. In: WACV, pp. 2000–2009 (2023)
20. Network, T.C.G.A.: Comprehensive molecular portraits of human breast Tumours. Nature **490**(7418), 61–70 (2012)
21. Nichol, A., et al.: Glide: Towards photorealistic image generation and editing with text-guided diffusion models. arXiv (2021)
22. Nichol, A.Q., Dhariwal, P.: Improved denoising diffusion probabilistic models. In: ICML, pp. 8162–8171. PMLR (2021)
23. Perez, L., Wang, J.: The effectiveness of data augmentation in image classification using deep learning. arXiv (2017)
24. Pinaya, W.H., et al.: Brain imaging generation with latent diffusion models. In: DGM4MICCAI, pp. 117–126. Springer (2022). https://doi.org/10.1007/978-3-031-18576-2_12
25. Radford, A., et al.: Learning transferable visual models from natural language supervision. In: ICML, pp. 8748–8763. PMLR (2021)
26. Rombach, R., Blattmann, A., Lorenz, D., Esser, P., Ommer, B.: High-resolution image synthesis with latent diffusion models. In: CVPR, pp. 10684–10695 (2022)
27. Ronneberger, O., Fischer, P., Brox, T.: U-Net: convolutional networks for biomedical image segmentation. In: Navab, N., Hornegger, J., Wells, W.M., Frangi, A.F. (eds.) MICCAI 2015. LNCS, vol. 9351, pp. 234–241. Springer, Cham (2015). https://doi.org/10.1007/978-3-319-24574-4_28
28. Shen, D., Wu, G., Suk, H.I.: Deep learning in medical image analysis. Annu. Rev. Biomed. Eng. **19**, 221–248 (2017)
29. Shi, C., Ni, H., Li, K., Han, S., Liang, M., Min, M.R.: Exploring compositional visual generation with latent classifier guidance. In: CVPR, pp. 853–862 (2023)
30. Sohl-Dickstein, J., Weiss, E., Maheswaranathan, N., Ganguli, S.: Deep unsupervised learning using nonequilibrium thermodynamics. In: ICML, pp. 2256–2265 (2015)
31. Song, J., Meng, C., Ermon, S.: Denoising diffusion implicit models. arXiv (2020)
32. Song, Y., Ermon, S.: Generative modeling by estimating gradients of the data distribution. NeurIPS **32** (2019)
33. Song, Y., Sohl-Dickstein, J., Kingma, D.P., Kumar, A., Ermon, S., Poole, B.: Score-based generative modeling through stochastic differential equations. arXiv (2020)
34. van Treeck, M., et al.: Deepmed: A unified, modular pipeline for end-to-end deep learning in computational pathology. BioRxiv 2021–12 (2021)
35. Vaswani, A., et al.: Attention is all you need. In: NeurIPS, vol. 30 (2017)
36. Xue, Y., et al.: Selective synthetic augmentation with histogan for improved histopathology image classification. Med. Image Anal. **67**, 101816 (2021)
37. Xue, Y., et al.: Synthetic augmentation and feature-based filtering for improved cervical histopathology image classification. In: Shen, D., et al. (eds.) MICCAI 2019. LNCS, vol. 11764, pp. 387–396. Springer, Cham (2019). https://doi.org/10.1007/978-3-030-32239-7_43
38. Ye, J., et al.: Synthetic sample selection via reinforcement learning. In: Martel, A.L., et al. (eds.) MICCAI 2020. LNCS, vol. 12261, pp. 53–63. Springer, Cham (2020). https://doi.org/10.1007/978-3-030-59710-8_6
39. Zhang, R., Isola, P., Efros, A.A., Shechtman, E., Wang, O.: The unreasonable effectiveness of deep features as a perceptual metric. In: CVPR, pp. 586–595 (2018)

DeDA: Deep Directed Accumulator

Hang Zhang[1(✉)], Rongguang Wang[2], Renjiu Hu[1], Jinwei Zhang[1], and Jiahao Li[1]

[1] Cornell University, Ithaca, USA
hz459@cornell.edu
[2] University of Pennsylvania, Philadelphia, USA

Abstract. Chronic active multiple sclerosis lesions, also referred to as rim+ lesions, are characterized by a hyperintense rim observed at the lesion's edge on quantitative susceptibility maps. Despite their geometrically simple structure, characterized by radially oriented gradients at the lesion edge with a greater gradient magnitude compared to non-rim+ (rim-) lesions, recent studies indicate that the identification performance for these lesions is subpar due to limited data and significant class imbalance. In this paper, we propose a simple yet effective image processing operation, deep directed accumulator (DeDA), which provides a new perspective for injecting domain-specific inductive biases (priors) into neural networks for rim+ lesion identification. Given a feature map and a set of sampling grids, DeDA creates and quantizes an accumulator space into finite intervals and accumulates corresponding feature values. This DeDA operation can be regarded as a symmetric operation to the grid sampling within the forward-backward neural network framework, the process of which is order-agnostic, and can be efficiently implemented with the native CUDA programming. Experimental results on a dataset with 177 rim+ and 3986 rim- lesions show that 10.1% of improvement in a partial (false positive rate < 0.1) area under the receiver operating characteristic curve (pROC AUC) and 10.2% of improvement in an area under the precision recall curve (PR AUC) can be achieved respectively comparing to other state-of-the-art methods. The source code is available online at https://github.com/tinymilky/DeDA.

Keywords: Directed accumulator · Neural networks · Multiple sclerosis · Quantitative susceptibility mapping

1 Introduction

Over the past decade, we have observed the substantial success of Convolutional Neural Networks (CNNs) in a multitude of grid-based medical imaging applications, such as magnetic resonance imaging (MRI) reconstruction [20,27] and lesion segmentation [13,25]. Despite the effectiveness of general inductive biases like translation equivariance [15] and locality [16], the diverse nature of

Supplementary Information The online version contains supplementary material available at https://doi.org/10.1007/978-3-031-43895-0_72.

H. Greenspan et al. (Eds.): MICCAI 2023, LNCS 14221, pp. 765–775, 2023.
https://doi.org/10.1007/978-3-031-43895-0_72

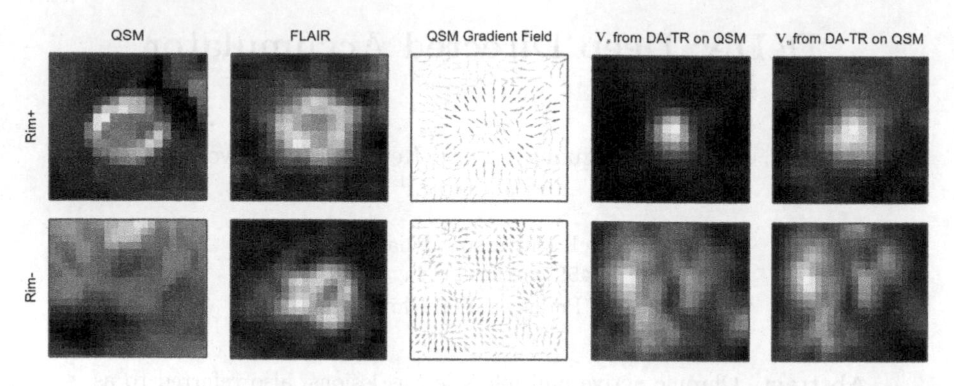

Fig. 1. Visual illustration of the difference between a rim+ and a rim- lesion. QSM image patches represent the lesion magnetic susceptibility, while Fluid Attenuated Inversion Recovery (FLAIR) image patches pinpoint the exact location of the lesions. The gradient field map of the QSM images presents normalized gradient vectors (the darker the blue, the larger the gradient vector's magnitude). The two rightmost columns display gradient magnitude maps $\mathbf{V}_s$ and QSM value maps $\mathbf{V}_u$ processed by DA-TR (see Sect. 2.2). Notably, the rim+ lesion exhibits structured patterns in the accumulator space by aggregating feature values along gradients, a characteristic absent in the rim- lesion. (Color figure online)

diseases represented in medical images necessitates highly domain-specific knowledge. Consequently, the question of how to incorporate domain-specific inductive biases, or priors, beyond general ones into neural networks for medical image processing remains an open challenge.

In this study, we strive to answer this question by addressing the identification problem associated with a specific type of multiple sclerosis (MS) lesion, referred to as a chronic active lesion, or rim+ lesion. Histopathology studies characterize rim+ lesions by an iron-rich rim of activated macrophages and microglia [2,6,9,14]. These lesions are visible with in-vivo quantitative susceptibility mapping (QSM) [7,22] and phase imaging techniques [1,2]. Notably, they display a paramagnetic hyperintense rim at the edge (see Fig. 1). Despite several efforts to tackle the issue [4,18,24], a clinically reliable solution remains elusive.

Given the limited amount of data and high class imbalance, it's more advantageous to explicitly incorporate domain knowledge into the network as priors. As illustrated in Fig. 1, rim+ lesions distinguish themselves from rim- lesions in three primary aspects. Firstly, rim+ lesions exhibit a hyperintense ring-like structure at the lesion's edge on QSM. Secondly, a higher magnitude of gradients is observed near the edge of rim+ lesions, a feature not present in rim- lesions. Lastly, rim+ lesions are characterized by radially oriented gradients at the edge, whereas rim- lesions lack such structured orientations.

In this work, we introduce the **De**ep **D**irected **A**ccumulator (DeDA), a novel image processing operation. DeDA, symmetric to grid sampling within a neural network's forward-backward framework, explicitly encodes the aforementioned prior information. Given a feature map and sampling grids, DeDA creates an accumulator space, quantizes it into finite intervals, and accumulates feature values. DeDA can also be viewed as a generalized discrete Radon transform, as it accumulates values between two discrete feature spaces. Our contributions are twofold: Firstly, we present DeDA, a simple yet powerful method that augments neural networks' representation capacity by explicitly incorporating domain-specific priors. Secondly, our experimental results on rim+ lesion identification demonstrate a notable improvement of 10.1% in partial area under the receiver operating characteristic curve (pROC AUC) and a 10.2% improvement in area under the precision recall curve (PR AUC), outperforming existing state-of-the-art methods.

2 Methodology

Numerous signal processing techniques, including the Fourier transform, Radon transform, and Hough transform, map discrete signals from image space to another functional space. We call this new space accumulator space, where each cell's value in the new space constitutes a weighted sum of values from all cells in the original image space. For our purposes, an appealing feature of the accumulator space is that local convolutions within it, like those in Hough and sinogram spaces, result in global aggregation of structural features, such as lines, in the feature map space. This proves beneficial for incorporating geometric priors into neural networks. Differing from attention-based methods, this convolution in accumulator space explicitly captures long-range information through direct geometric prior parameterization.

2.1 Differentiable Directed Accumulation

The process of transforming an image to an accumulator space involves a critical step, directed accumulation (DA), in which a cell from the accumulator space is pointed by multiple cells from the image space. Figure 2, Eq. (1) and Eq. (3) have shown that this DA operation is a symmetric operation to the grid sampling [12] within the forward-backward learning framework, where the backward pass of DA possesses the same structure as the forward pass of grid sampling if only one sampling grid is given, and vice versa for the forward pass. In addition, DA is further generalized to allow multiple sampling grids to accumulate values from the source feature map. Here we briefly review the grid sampling method and then derive the proposed DeDA.

Grid Sampling: Given a source feature map $\mathbf{U} \in \mathbb{R}^{C \times H \times W}$, a sampling grid $\mathbf{G} \in \mathbb{R}^{2 \times H' \times W'} = (\mathbf{G}^x, \mathbf{G}^y)$ specifying pixel locations to read from $\mathbf{U}$, and a kernel function $\mathcal{K}()$ defining the image interpolation, then the output value of a particular position (i, j) at the target feature map $\mathbf{V} \in \mathbb{R}^{C \times H' \times W'}$ can be written as follows:

$$\mathbf{V}^c_{ij} = \sum_n^H \sum_m^W \mathbf{U}^c_{nm} \mathcal{K}(\mathbf{G}^x_{ij}, n) \mathcal{K}(\mathbf{G}^y_{ij}, m), \tag{1}$$

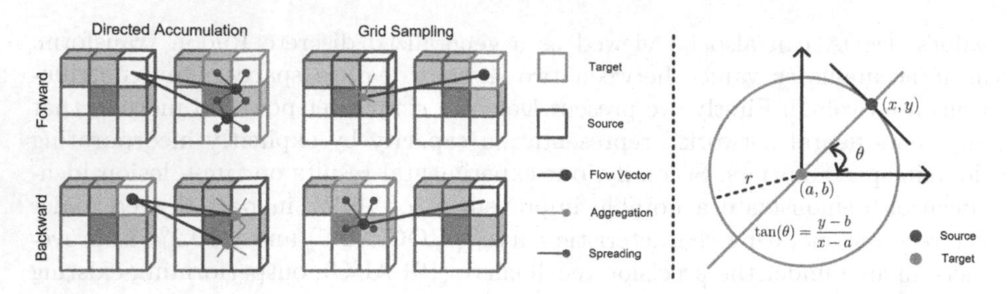

Fig. 2. Visual illustration of the proposed method. The left panel shows differences between the grid sampling and the proposed DeDA using bilinear sampling kernel. The right panel shows the schematic for rim parameterization, where the knowledge of a triple (x, y, θ) is mapped to a straight line (marked in orange) in the accumulator space. (Color figure online)

where the kernel function $\mathcal{K}()$ can be replaced with any other specified kernels, e.g. integer sampling kernel $\delta(\lfloor \mathbf{G}^x_{ij} + 0.5 \rfloor - n) \cdot \delta(\lfloor \mathbf{G}^y_{ij} + 0.5 \rfloor - m)$. Here $\lfloor x + 0.5 \rfloor$ rounds x to the nearest integer and $\delta()$ is the Kronecker delta function. The gradients with respect to $\mathbf{U}$ and $\mathbf{G}$ for back propagation can be defined accordingly [12].

DeDA: Given a source feature map $\mathbf{U} \in \mathbb{R}^{C \times H \times W}$, a target feature map $\mathbf{V} \in \mathbb{R}^{C \times H' \times W'}$, a set of sampling grids $\mathcal{G} = \{\mathbf{G}[k] \in \mathbb{R}^{2 \times H \times W} = (\mathbf{G}^x[k], \mathbf{G}^y[k]) \mid k \in \mathbb{Z}^+, 1 \leq k \leq N\}$ ($N \geq 1$ denotes the number of grids), and a kernel function $\mathcal{K}()$, the output value of a particular position (i, j) at the target feature map $\mathbf{V}$ can be written as follows:

$$\mathbf{V}^c_{ij} = \sum_k^N \sum_n^H \sum_m^W \mathbf{U}^c_{nm} \mathcal{K}(\mathbf{G}^x_{nm}[k], i) \mathcal{K}(\mathbf{G}^y_{nm}[k], j). \tag{2}$$

It is worth noting that the spatial dimension of the grid $\mathbf{G}[k]$ should be the same as that of $\mathbf{U}$, but the first dimension of $\mathbf{G}[k]$ can be an arbitrary number as long as it aligns with the number of spatial dimensions of $\mathbf{V}$, e.g. if given $\mathbf{U} \in \mathbb{R}^{H \times W}$ and $\mathbf{G}[k] \in \mathbb{R}^{3 \times H \times W}$, it is expected that $\mathbf{V} \in \mathbb{R}^{H' \times W' \times D'}$. Basically, the DeDA operation in Eq. (2) performs a tensor mapping by $\mathcal{D} : (\mathbf{U}, \mathcal{G}; \mathcal{K}) \rightarrow \mathbf{V}$, where $\mathcal{K}$ is the sampling kernel. For simplicity, function $\mathcal{D}()$ will be used to denote the DeDA forward for the rest of the paper.

To allow back propagation for training networks with DeDA, the gradients with respect to $\mathbf{U}$ are derived using the chain rule as follows:

$$\frac{\partial \mathcal{L}}{\partial \mathbf{V}^c_{nm}} \frac{\partial \mathbf{V}^c_{nm}}{\partial \mathbf{U}^c_{ij}} = \sum_k^N \sum_n^{H'} \sum_m^{W'} \mathbf{A}^c_{nm} \mathcal{K}(\mathbf{G}^x_{ij}[k], n) \mathcal{K}(\mathbf{G}^y_{ij}[k], m). \tag{3}$$

The gradient tensor with respect to $\mathbf{V}$ is $\mathbf{A}$. The structure of Eq. (3) reduces to Eq. (1) when $N = 1$, indicating DeDA's symmetry with grid sampling. Given identical transformations for each channel c in DeDA's forward and backward passes, we denote the feature map with spatial dimensions alone henceforth.

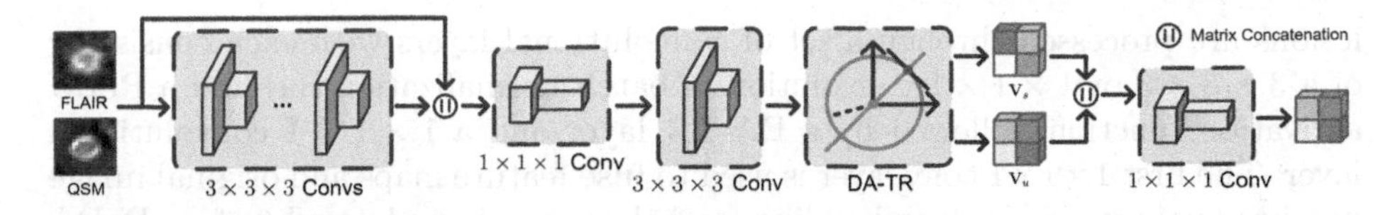

Fig. 3. Schematic of the network layer for DA-TR. Conv denotes a convolutional layer, and each of these layers consists a 3×3×3 or 1×1×1 convolution, a batch normalization, and a ReLU activation.

2.2 DeDA-Based Transformation Layer for Rim Parameterization

In this section, we derive DeDA-based transformation and its convolution layers for rim parameterization. As shown in Fig. 1, a rim+ lesion can be characterized by a hyperintense rim at the lesion edge on QSM and differs from a rim- lesion in both image intensities and gradients at the edge. To account for both image intensities and gradients, the rim is parameterized as $\tan(\theta) = \frac{y-b}{x-a}$, where (a, b) are parameters of coordinates for the rim center in the accumulator space and θ represents the gradient direction at (x, y) in the image space. As can be seen from the right panel of Fig. 2, mapping a single (x, y, θ) to the accumulator space produces a straight line, and thus coordinates of the rim center can be identified by the intersection of many of these lines.

DeDA Transformation of the Rim: Given a source feature map $\mathbf{U} \in \mathbb{R}^{H\times W}$, the magnitude of image gradients can be obtained as follows $\mathbf{S} = \sqrt{\mathbf{U}_x \odot \mathbf{U}_x + \mathbf{U}_y \odot \mathbf{U}_y}$, where $\odot$ denotes the Hadamard product, $\mathbf{U}_x = \frac{\partial \mathbf{U}}{\partial x}$, and $\mathbf{U}_y = \frac{\partial \mathbf{U}}{\partial y}$. The image gradient tensor $\mathbf{U}_x$ and $\mathbf{U}_y$ can be efficiently computed using convolution kernels such as the Sobel operator. Normalized gradients can be obtained by $\hat{\mathbf{U}}_x = \frac{\mathbf{U}_x}{\mathbf{S}+\epsilon}$ and $\hat{\mathbf{U}}_y = \frac{\mathbf{U}_y}{\mathbf{S}+\epsilon}$, where ϵ is a small real value to avoid zero denominator. The mesh grids of $\mathbf{U}$ are denoted as $\mathbf{M}_x$ (value range: $(0, H-1)$) and $\mathbf{M}_y$ (value range: $(0, W-1)$). We can then generate a set of sampling grids as follows:

$$\mathcal{G} = \{\mathbf{G}[k] = (\mathbf{G}^x[k], \mathbf{G}^y[k]) \mid k \in \mathbb{Z}^+, 1 \le k \le N\}, \tag{4}$$

where $\mathbf{G}[k] \in \mathbb{R}^{2\times H\times W}$, $\mathbf{G}^x[k] = k\hat{\mathbf{U}}_x + \mathbf{M}_x$, $\mathbf{G}^y[k] = k\hat{\mathbf{U}}_y + \mathbf{M}_y$, and $N = \max(H, W)$. Now the DeDA-based transformation of Rim (DA-TR) can be formulated as $\mathbf{V}_s = \mathcal{D}(\mathbf{S}, \mathcal{G}; \mathcal{K})$ and $\mathbf{V}_u = \mathcal{D}(\mathbf{U}, \mathcal{G}; \mathcal{K})$, where the integer sampling kernel is used. It is worth noting that feature and gradient magnitude values are accumulated separately due to differences of image intensity and gradients between rim+ and rim- lesions (see Fig. 1).

Network Layer for DA-TR: To gain more representation ability and capture long-range contextual information, DA-TR is applied to both intermediate feature maps and original images. As can be seen from Fig. 3, image patches of lesions are processed through a set of convolutional layers with each consisting of a $3 \times 3 \times 3$ or $1 \times 1 \times 1$ convolution, a batch normalization [11] and a ReLU activation function, followed by a DA-TR layer and a $1 \times 1 \times 1$ convolutional layer. The first $1 \times 1 \times 1$ conv layer is used to fuse feature maps and original image patches for better feature embedding, and the second one is used to fuse DeDA transformed gradient magnitude maps $\mathbf{V}_s$ and feature maps $\mathbf{V}_u$. It is worth noting that only in-plane rims are observed, and thus the DA-TR is performed on the 2D feature map slices along the axial direction.

3 Experiments and Results

For fair and consistent comparison, the dataset applied in the previous work [24] was asked for and used to demonstrate the performance of the proposed DeDA-based rim parameterization DA-TR. A total of 172 subjects were included in the dataset, and 177 lesions were identified as rim+ lesions and 3986 lesions were identified as rim- lesions, please refer to [24] for more details about the image acquisition and pre-processing.

3.1 Comparator Methods and Implementation Details

Comparator Methods: Three methods have been developed so far for rim+ lesion identification, of which APRL [18] and RimNet [4] are on phase imaging and QSMRim-Net [24] is on QSM. In comparison with these methods, we use QSM along with T2-FLAIR images as the network inputs for RimNet and QSMRimNet, and use the QSM image to extract first-order radiomic features for APRL. Furthermore, we applied residual networks (ResNet) [10], vision transformer (ViT) [8], Swin transformer [17], and Nested transformer [28] as backbone architecture for our application, and determined that ResNet with 18 convolution layers works the best. Transformer-based networks with fewer inductive biases rely heavily on the use of a large training dataset or depends strongly on the feature reuse [19], as a result, these networks as well as CNNs with deeper structures are prone to overfit small datasets.

Implementation Details: A stratified five-fold cross-validation procedure was applied to train and validate the performance, and all experiments including ablation study were carried out within this setting. Each lesion was cropped into patches with a fixed size of $32 \times 32 \times 8$ voxels. Random flipping, random affine transformation and random Gaussian blurring were used to augment our data. More details of the training procedure can be found out in the supplementary materials.

3.2 Results and Ablation Study

Lesion-wise Results: To evaluate the performance of each method and produce clinically relevant results, pROC curves with false positive rates (FPRs) in the range of $(0, 0.1)$ and PR curves of the different validation folds were interpolated using piece-wise constant interpolation and averaged to show the overall

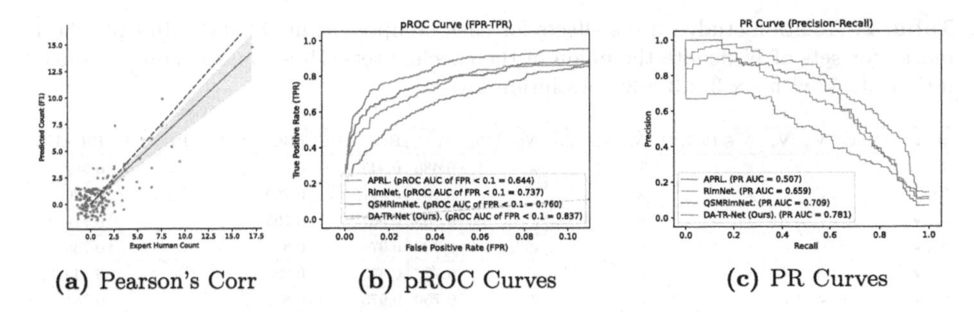

(a) Pearson's Corr **(b)** pROC Curves **(c)** PR Curves

Fig. 4. The predicted count of rim + lesions from DA-TR-Net versus the expert human count is shown in (a), where points in the plot have been jittered for better visualization. The pROC and PR curves for the proposed and other comparator methods are shown in (b) and (c), where AUC denotes the area under the curve.

performance at the lesion level. For each curve, AUC was computed directly from the interpolated and averaged curves. The binary indicators of rim+/rim- lesions were generated by thresholding the model probabilities to maximize the F_1 score, where $F_1 = 2 \cdot \frac{precision \cdot sensitivity}{precision + sensitivity}$. In addition, accuracy, F1 score, sensitivity, specificity, and precision were used to characterize the performance of each method. Table 1 and Fig. 4 show the lesion-wise performance metrics of the proposed methods in comparison with the other methods. DA-TR-Net outperformed the other competitors in all evaluation metrics. With a slightly higher overall accuracy and specificity with other methods, DA-TR-Net resulted in a 5.5%, 15.4% and 39.4% improvement in F_1 score, 10.1%, 13.6% and 30.0% improvement in pROC (FPR < 0.1) AUC, and 10.2%, 18.5% and 54.0% improvement in PR AUC compared to QSMRimNet, RimNet and APRL, respectively.

Subject-wise Results: We also evaluated the performance at the subject-level. Pearson's correlation coefficient was used to measure the correlation model predicted count and human expert count. Mean Squared Error (MSE) was also used to measure the averaged accuracy for the model predicted count. Figure 4a shows the scatter-plot for the predicted count v.s. the human expert count, along with the identity line, and the Pearson's correlation coefficient (ρ) for DA-TR-Net was $\rho = 0.93 (95\% CI : 0.90, 0.95)$ As can be seen from Table 1, the Pearson's correlations and MSE for the proposed DA-TR-Net was found higher

Table 1. Results of the proposed and other methods using a stratified five-fold cross-validation scheme. The best performing metric is bolded.

Method	Accuracy	F_1	Sensitivity	Specificity	Precision	ROC AUC	pROC AUC	PR AUC	ρ (95%CI)	MSE
APRL [18]	0.954	0.538	0.627	0.969	0.470	0.940	0.644	0.507	0.68 (0.59,0.75)	3.16
RimNet [4]	0.970	0.650	0.655	0.984	0.644	0.950	0.737	0.659	0.75 (0.67,0.81)	2.41
QSMRimNet [24]	0.977	0.711	0.667	0.991	0.761	0.939	0.760	0.709	0.89 (0.86,0.92)	1.00
DA-TR-Net (Ours)	**0.980**	**0.750**	**0.712**	**0.992**	**0.792**	**0.975**	**0.837**	**0.781**	**0.93(0.90,0.95)**	**0.69**

Table 2. Ablation study on the effects for each component in DA-TR. Multiple check marks for sets of N denote the union of the checked sets. Pre-Convs denotes a convolution block with six $3 \times 3 \times 3$ convolution layers.

#	Pre-Convs	$\mathbf{V}_u$	$\mathbf{V}_s$	$N \in \{5,7,9\}$	$N \in \{11,13\}$	$N \in \{15\}$	F_1	ROC AUC	pROC AUC (FPR< 0.1)	PR AUC
1	×	×	×	×	×	×	0.685	0.945	0.753	0.689
2	×	✓	✓	×	×	✓	0.701	0.971	0.790	0.720
3	✓	✓	×	×	×	✓	0.703	0.967	0.795	0.714
4	✓	✓	✓	×	×	✓	0.702	0.976	0.817	0.736
5	✓	✓	✓	×	✓	✓	0.727	0.975	0.825	0.743
6	✓	✓	✓	✓	✓	✓	0.750	0.975	0.837	0.781

than other competitors. This demonstrates that the performance of DA-TR-Net at the subject-level is statistically significantly higher than that of APRL, Rim-Net, and QSMRim-Net (Table 2).

Ablation Study: We conducted an ablation study to investigate the effects of each component accompanied with DA-TR. First, we examined the effects of applying the proposed DA-TR to the latent feature maps and raw images. Second, we examined the effects of using $\mathbf{V}_u$ and $\mathbf{V}_s$, because rim+ lesions differ from rim- lesions in both gradient magnitudes and values at the edge of the lesion. We then investigated how multi-radius rim parameterization can affect the results, as the size of rim+ lesions vary greatly with a radius from 5 to 15 among different subjects. Results from models #1, #2 and #4 show that the rim parametrization DA-TR is useful for rim+ identification, and DA-TR used in the latent feature map space performs even better. Comparing model #3 and #4, one can see that accumulating both gradient magnitudes and feature values is beneficial. The consistent performance improvement from model #4 to #5 and from model #5 to #6 has demonstrated the effectiveness of applying multi-radius rim parameterization. More results on backbone networks can be found in the appendix.

3.3 Discussions

Medical images often require processing of a primary target or region of interest (ROI), such as rims, left ventricles, or tumors. These ROIs frequently exhibit distinct geometric structures [26] or possess specific spatial relationships [25] with their surroundings. Capturing these characteristics poses a challenge for modern neural networks, especially given limited and imbalanced training data. While differentiable grid sampling [12] can tackle some of these issues within a certain scope, another major class involving transformations (e.g. Hough transform [3]) that necessitate directed accumulation is overlooked. Our proposed DeDA bridges this gap, enabling the use of image transformations with directed accumulation within a neural network. This allows for the parametrization of geometric shapes and the modeling of spatial correlations in a differentiable manner.

While the study focuses on rim+ lesion identification, the proposed DeDA can be extended to other applications. These include the utilization of polar transformation for skin lesion recognition/segmentation, symmetric circular transformation for cardiac image registration [23], parabola transformation for curvilinear structure segmentation [21], and high-dimensional bilateral filtering [5].

4 Conclusions

We present DeDA, an image processing operation that helps parameterize rim and effectively incorporates prior information into networks through a value accumulation process. The experimental results demonstrate that DeDA surpasses existing state-of-the-art methods in all evaluation metrics by a significant margin. Furthermore, DeDA's versatility extends beyond lesion identification and can be applied in other image processing applications such as Hough transform, bilateral grid, and Polar transform. We are excited about the potential of DeDA to advance numerous medical applications and other image processing tasks.

Acknowledgement. The database was approved by the local Institutional Review Board and written informed consent was obtained from all patients prior to their entry into the database. We would like to thank folks from Weill Cornell for sharing the data used in this paper.

References

1. Absinta, M., et al.: Seven-tesla phase imaging of acute multiple sclerosis lesions: a new window into the inflammatory process. Ann. Neurol. **74**(5), 669–678 (2013)
2. Absinta, M., et al.: Persistent 7-tesla phase rim predicts poor outcome in new multiple sclerosis patient lesions. J. Clin. Investig. **126**(7), 2597–2609 (2016)
3. Ballard, D.H.: Generalizing the Hough transform to detect arbitrary shapes. Pattern Recogn. **13**(2), 111–122 (1981)
4. Barquero, G., et al.: RimNet: a deep 3D multimodal MRI architecture for paramagnetic rim lesion assessment in multiple sclerosis. NeuroImage Clinical **28**, 102412 (2020)
5. Chen, Jiawen, Paris, Sylvain, Durand, Frédo.: Real-time edge-aware image processing with the bilateral grid. ACM Trans. Graph. **26**(3), 103 (2007). https://doi.org/10.1145/1276377.1276506
6. Dal-Bianco, A., et al.: Slow expansion of multiple sclerosis iron rim lesions: pathology and 7 t magnetic resonance imaging. Acta Neuropathol. **133**(1), 25–42 (2017)
7. De Rochefort, L., et al.: Quantitative susceptibility map reconstruction from MR phase data using Bayesian regularization: validation and application to brain imaging. Magn. Reson. Med. Official J. Int. Soc. Magn. Reson. Med. **63**(1), 194–206 (2010)

8. Dosovitskiy, A., et al.: An image is worth 16 × 16 words: transformers for image recognition at scale. In: International Conference on Learning Representations (2020)
9. Gillen, K.M., et al.: QSM is an imaging biomarker for chronic glial activation in multiple sclerosis lesions. Ann. Clin. Transl. Neurol. **8**(4), 877–886 (2021)
10. He, K., Zhang, X., Ren, S., Sun, J.: Deep residual learning for image recognition. In: Proceedings of the IEEE Conference on Computer Vision and Pattern Recognition, pp. 770–778 (2016)
11. Ioffe, S., Szegedy, C.: Batch normalization: accelerating deep network training by reducing internal covariate shift. In: International Conference on Machine Learning, pp. 448–456. PMLR (2015)
12. Jaderberg, M., Simonyan, K., Zisserman, A., et al.: Spatial transformer networks. In: Advances in Neural Information Processing Systems, vol. 28 (2015)
13. Kamnitsas, K., et al.: Efficient multi-scale 3D CNN with fully connected CRF for accurate brain lesion segmentation. Med. Image Anal. **36**, 61–78 (2017)
14. Kaunzner, U.W., et al.: Quantitative susceptibility mapping identifies inflammation in a subset of chronic multiple sclerosis lesions. Brain **142**(1), 133–145 (2019)
15. Kayhan, O.S., Gemert, J.C.V.: On translation invariance in CNNs: convolutional layers can exploit absolute spatial location. In: Proceedings of the IEEE/CVF Conference on Computer Vision and Pattern Recognition, pp. 14274–14285 (2020)
16. Lenc, K., Vedaldi, A.: Understanding image representations by measuring their equivariance and equivalence. In: Proceedings of the IEEE Conference On Computer Vision and Pattern Recognition, pp. 991–999 (2015)
17. Liu, Z., et al.: Swin transformer: hierarchical vision transformer using shifted windows. In: Proceedings of the IEEE/CVF International Conference on Computer Vision, pp. 10012–10022 (2021)
18. Lou, C., et al.: Fully automated detection of paramagnetic rims in multiple sclerosis lesions on 3t susceptibility-based MR imaging. NeuroImage Clin. **32**, 102796 (2021)
19. Matsoukas, C., Haslum, J.F., Sorkhei, M., Söderberg, M., Smith, K.: What makes transfer learning work for medical images: feature reuse and other factors. In: Proceedings of the IEEE/CVF Conference on Computer Vision and Pattern Recognition, pp. 9225–9234 (2022)
20. Muckley, M., et al.: Results of the 2020 fastMRI challenge for machine learning MR image reconstruction. IEEE Trans. Med. Imaging **40**(9), 2306–2317 (2021)
21. Shi, T., Boutry, N., Xu, Y., Géraud, T.: Local intensity order transformation for robust curvilinear object segmentation. IEEE Trans. Image Process. **31**, 2557–2569 (2022)
22. Wang, Y., Liu, T.: Quantitative susceptibility mapping (QSM): decoding MRI data for a tissue magnetic biomarker. Magn. Reson. Med. **73**(1), 82–101 (2015)
23. Zhang, H., Hu, R., Chen, X., Wang, R., Zhang, J., Li, J.: DAGrid: directed accumulator grid. arXiv preprint arXiv:2306.02589 (2023)
24. Zhang, H., et al.: QSMRim-Net: imbalance-aware learning for identification of chronic active multiple sclerosis lesions on quantitative susceptibility maps. NeuroImage Clin. **34**, 102979 (2022)
25. Zhang, H., et al.: ALL-Net: anatomical information lesion-wise loss function integrated into neural network for multiple sclerosis lesion segmentation. NeuroImage Clin. **32**, 102854 (2021)
26. Zhang, H., et al.: Geometric loss for deep multiple sclerosis lesion segmentation. In: 2021 IEEE 18th International Symposium on Biomedical Imaging (ISBI), pp. 24–28. IEEE (2021)

27. Zhang, H., et al.: Efficient folded attention for medical image reconstruction and segmentation. In: Proceedings of the AAAI Conference on Artificial Intelligence, vol. 35, pp. 10868–10876 (2021)
28. Zhang, Z., Zhang, H., Zhao, L., Chen, T., Arik, S.Ö., Pfister, T.: Nested hierarchical transformer: towards accurate, data-efficient and interpretable visual understanding. In: Proceedings of the AAAI Conference on Artificial Intelligence, vol. 36, pp. 3417–3425 (2022)

Mixing Temporal Graphs with MLP for Longitudinal Brain Connectome Analysis

Hyuna Cho[1], Guorong Wu[2], and Won Hwa Kim[1(✉)]

[1] Pohang University of Science and Technology (POSTECH), Pohang, South Korea
{hyunacho,wonhwa}@postech.ac.kr
[2] University of North Carolina, Chapel Hill, USA

Abstract. Analyses of longitudinal brain networks, i.e., graphs, are of significant interest to understand the dynamics of brain changes with respect to aging and neurodegenerative diseases. However, each subject has a graph of heterogeneous structure and time-points as the data are obtained over several years. Moreover, most existing datasets suffer from lack of samples as the images are expensive to acquire, which leads to overfitting with complex deep neural networks. To address these issues for characterizing progressively alternations of brain connectome and region-wise measures as early as possible, we develop Spatio-Temporal Graph Multi-Layer Perceptron (STGMLP) that mixes features over both graph and time spaces to classify sets of longitudinal human brain connectomes. The proposed model is made efficient and interpretable such that it can be easily adopted to medical imaging datasets and identify personalized features responsible for a specific diagnostic label. Extensive experiments show that our method achieves successful results in both performance and computational efficiency on Alzheimer's Disease Neuroimaging Initiative (ADNI) and Adolescence Brain Cognitive Development (ABCD) datasets independently.

1 Introduction

Consider a longitudinal brain connectome study where each participant goes through imaging protocol multiple times over the study period. Given a population of such subjects, analyzing them can be posed as a *spatio-temporal graph analysis* where each sample in a dataset is given as a set of longitudinal graphs of different cardinality. An exemplar sample corresponding to this task is shown in Fig. 1a that consists of graphs from T_m time points with multi-variate node features (denoted in different colors). The fundamental goal of such longitudinal studies is to characterize progressive change patterns of time-varying graphs due to certain factors such as aging [32] and neurodegenerative diseases [19,22].

There are several practical bottlenecks to extract meaningful results from the longitudinal brain connectome and region-wise imaging measures. The data are temporally sparse, i.e., the participants pay a different number of visits which can be very few. Also, each brain network has a different structure of white-matter fiber connections unlike regular lattice structure in images. Last but importantly,

H. Greenspan et al. (Eds.): MICCAI 2023, LNCS 14221, pp. 776–786, 2023.
https://doi.org/10.1007/978-3-031-43895-0_73

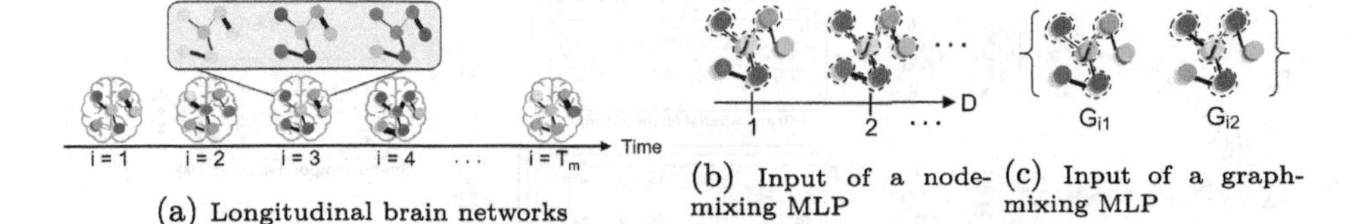

(a) Longitudinal brain networks (b) Input of a node-mixing MLP (c) Input of a graph-mixing MLP

Fig. 1. (a) A set of longitudinal graphs for T_m timepoints. The node colors (i.e., yellow, blue, and red) represent different node features, and the edge thickness stands for edge strength. (b) A node-mixing MLP in GSMs takes node features of a local graph centered at node j and its neighbors (in dotted lines) as an input. As the depth D of GSM increases, the range of the local graph is broadened. (c) A graph-mixing MLP in GTMs takes a graph pair p of different time-points (i.e., $p = \{i_1, i_2\}$). (Color figure online)

most neuroimaging datasets suffer from lack of samples as the data are expensive to acquire and process. These spatio-temporal heterogeneities and sample-size issue make longitudinal analyses of the brain network challenging, but it must be investigated to characterize the progressive disease-relevant variations.

Therefore, it is necessary to develop an efficient prediction model for a "set of longitudinal graphs" and corresponding regional measures (i.e., node features) over sparse time-points. Most graph neural networks are designed for a fixed template graph for predicting node values [18] or graph-level labels [33,35], where the graph topology is used as a domain and predictions heavily rely on node features. Moreover, recent spatio-temporal graph methods for a stream of images (e.g., video) often include complex architectures that require a large-scale dataset to train [2,3], which cannot be easily adopted for medical applications due to the limited sample-size. Notice that the sample-size is a much bigger issue for a longitudinal study as a single label is given for a "set" of graphs, as opposed to a cross-sectional study where the label is given for each graph.

We tackle the aforementioned issues by designing a flexible architecture with a "mix" of Multi-layer Perceptron (MLP) [30] to investigate time-varying graph structure and measurements on the nodes. We propose Spatio-Temporal Graph MLP (STGMLP) which integrates the following three mixing modules: 1) graph spatial mixer (GSM), 2) graph temporal mixer (GTM), and 3) spatio-temporal mixer (STM) that extract space, time, and spatio-temporal features, respectively. The features curated from the three components are fed into a downstream classifier to discriminate labels for the sets of longitudinal graphs. The core idea is to efficiently mix features along irregular space and time with simple MLPs: brain network structure guides spatial mixing as a graph, and temporal pooling extracts the most effective features from disjoint time-space across subjects.

Contributions of Our Work: our model 1) can be trained efficiently compared to existing spatio-temporal graph deep models with a significantly reduced number of parameters, 2) flexibly incorporates irregular space and time into prediction, 3) yields interpretable results that quantify the contribution of each node

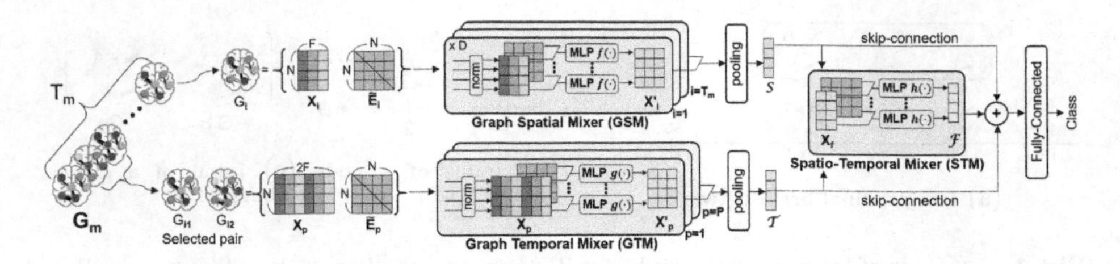

Fig. 2. Architecture of STGMLP. It consists of T_m Graph Spatial Mixers (GSM), P Graph Temporal Mixers (GTM), a Spatio-Temporal Mixer (STM), and a classifier head. GSMs and GTMs contain a node-mixing MLP and a graph-mixing MLP, respectively. STM mixes the jointly obtained features from each spatial and temporal aspects. If the GSM depth D is $D>1$, the GSM output X_i' is used as an input of the same GSM to deepen model layers and to widen an encoding range of neighbor nodes.

to classify different clinical labels. Extensive validation was performed on two independent public datasets, i.e., Alzheimer's Disease Neuroimaging Initiative (ADNI) and Adolescent Brain Cognitive Development (ABCD), for classifying pre-defined groups to demonstrate the efficiency and efficacy of our model.

2 STGMLP: Spatio-Temporal Graph MLP

Problem Definition. Consider a longitudinal graph set $\boldsymbol{G_m} = \{G_1, ..., G_{T_m}\}$ for $m = 1, 2, \cdots, M$ samples where the graphs for $i = 1, ..., T_m$ timepoints (different across samples, $T_m \geq 2$) are presented in chronological order. For each undirected graph $G_i = \{E_i, X_i\}$, $X_i \in \mathbb{R}^{N\times F}$ contains F node features for N nodes and $E_i \in \mathbb{R}^{N\times N}$ is a weighted adjacency matrix whose elements denote connection strength between two nodes. Given a population of $\boldsymbol{G_m}$ with C classes, STGMLP aims to classify the label of each $\boldsymbol{G_m}$ by leveraging both temporal and spatial variations of the graph set from different groups. Note that the label of each sample (i.e., longitudinal graph set) is consistent over time.

Overview of STGMLP. STGMLP mixes graph features across space and time with Graph Spatial Mixer (GSM) and Graph Temporal Mixer (GTM), respectively. GSM performs a per-graph operation (i.e., *node-mixing*) and GTM accounts for cross-temporal operation (i.e., *graph-mixing*) between multiple graphs in a $\boldsymbol{G_m}$. Figure 1b and 1c show inputs to node-mixing and graph-mixing MLPs. The node-mixing MLP projects node features from a local neighborhood of each node (i.e., local graph) onto a latent space. On the other hand, the graph-mixing MLP extracts hidden relationships between graphs across time by projecting spatially equivalent local graphs from multiple G_i's to the same latent space.

The overall structure of STGMLP is shown in Fig. 2, which integrates GSMs, GTMs, a Spatio-Temporal Mixer (STM), and a downstream classifier. Due to the heterogeneous number of timepoints T_m across samples, pooling operations are applied to the outputs of GSMs and GTMs to reduce them into a coherent dimension. Taking outcomes from the pooling layers, the STM fuses both spatial and temporal features. The fused feature is combined with its inputs through a skip-connection [14] to maximize the use of multi-level (i.e., space, time, and spatio-temporal) information extracted from the input. Finally, a downstream classifier takes the mixed features to predict labels for a given longitudinal graph set $\boldsymbol{G}_m$. The details of each module and variables are given below.

Graph Spatial Mixer. GSM encodes node features and a graph structure of a graph G_i with graph convolution. The node-mixing MLP ($\mathbb{R}^F \mapsto \mathbb{R}^F$) acts on rows of X_i, and it is shared across $\boldsymbol{G}_m$ for all $N \times T_m$ nodes. Let $f(\cdot)$ be an operation of node-mixing MLP which takes $\tilde{E}_i$ and X_i as inputs, where $\tilde{E}_i$ is a normalized E_i to ensure unbiased strength of the connectivity. It includes self-connections I_N (i.e., identity matrix) and computed as $\tilde{E}_i = \tilde{D}_i^{-\frac{1}{2}}(E_i+I_N)\tilde{D}_i^{-\frac{1}{2}}$, where $(\tilde{D}_i)_{jj} = \sum_{k=1}^{N}(E_i + I_N)_{jk}$ is a diagonal degree matrix of $E_i + I_N$ for $j = 1, ..., N$ nodes. Layer normalization [1] is applied across all features to prevent biased learning from unbalanced node feature distributions. Including two fully-connected (FC) layers and a GELU nonlinearity [15] $\sigma(\cdot)$, the MLP operates independently on each j-th node with two-layer graph convolutions [18] as

$$(X_i')_j = f(\tilde{E}_i, X_i) = \tilde{E}_i\, \sigma\big(\tilde{E}_i\, \mathrm{LN}(X_i)_j W^0\big) W^1, \tag{1}$$

where W^0 and W^1 are trainable weights and $\mathrm{LN}(\cdot)$ is a layernorm function. The output $(X_i')_j \in \mathbb{R}^F$ accounts for a latent vector of local graph structure at node j. Stacking $(X_i')_j$s up to the number of nodes N, an outcome $X_i' \in \mathbb{R}^{N\times F}$ is derived for an input G_i. In this way, a set of whole outputs from T_m GSMs is derived as $\{X_i'\}_{i=1}^{T_m} \in \mathbb{R}^{T_m \times N \times F}$. Notice that the GSM can be stacked D times by iteratively taking the X_i' as an updated input to encode a wider range of local graph structures. After performing max pooling on T_m and F dimensions of $\{X_i'\}_{i=1}^{T_m}$, the condensation of spatial features across $\boldsymbol{G}_m = \{G_1, ..., G_{T_m}\}$ is obtained as a N-dimensional vector $\mathcal{S}$.

Graph Temporal Mixer. GTM performs a cross-temporal operation on multiple "pairs" of graphs. This graph-mixing encodes the relations between graphs of different time-points. Given T_m graphs from a subject, P pairs of graphs, each pair as a set $\{G_{i_1}, G_{i_2}\}$, are selected where P is a user parameter. For each pair for $\{i_1, i_2\}$, an averaged connectivity $\tilde{E}_p{=}(\tilde{E}_{i_1}{+}\tilde{E}_{i_2})/2$ and $X_p \in \mathbb{R}^{N\times 2F}$ as a concatenation of X_{i_1} and X_{i_2} are inputted into the graph-mixing MLP $g(\cdot)$. In our work, we choose to input pairs of temporally adjacent graphs together with the first-and-last graph pair to encode a temporal sequence. The $g(\cdot)$ acts on rows of X_p, mapping $\mathbb{R}^{2F} \mapsto \mathbb{R}^F$. It transforms the features of node j (i.e., $(X_p)_j$) into F-dimensional latent vector, and the projection is performed across the whole node pairs in parallel. Similar to the node-mixing MLP, graph-mixing MLP contains two FC layers with weights W^2 and W^3 and a GELU $\sigma(\cdot)$ as

$$(X_p')_j = g(\tilde{E}_p, X_p) = \sigma\big(\tilde{E}_p\, \mathrm{LN}(X_p)_j W^2\big) W^3. \tag{2}$$

As in the GSM, each $(X_p')_j$ is stacked N times to be a X_p'. For P GTMs, an output $\{X_p'\}_{p=1}^{P} \in \mathbb{R}^{P\times N\times F}$ is obtained and reduced into $\mathcal{T} \in \mathbb{R}^N$ by max pooling on P pairs and F node features. Note that, unlike GSMs, $\tilde{E}_p$ is used only once in Eq. (2). Using $\tilde{E}_p$ multiple times causes encoding of a wider range of local graph structures, unnecessarily encompassing graph spatial relations (i.e., non-adjacent neighbors) for extracting temporal relations of j-th node pairs.

Spatio-Temporal Mixer. To capture comprehensive spatio-temporal relations across the whole graphs in $\boldsymbol{G}_m$, the spatial and temporal features, i.e., $\mathcal{S}$ and $\mathcal{T}$, are embedded into a latent vector $\mathcal{F} \in \mathbb{R}^N$ in STM. To do so, the $\mathcal{S}$ and $\mathcal{T}$ are stacked as $X_f = [\mathcal{S}, \mathcal{T}]$, where $(X_f)_j$ represents the spatial and temporal node features of the j-th node. A spatio-temporal mixing MLP $h(\cdot)$ is applied to $(X_f)_j$ for all j's in parallel with an averaged edge matrix $\tilde{E}_f$ across $\boldsymbol{G}_m$ as

$$\mathcal{F}_j = h(\tilde{E}_f, X_f) = \tilde{E}_f \, \sigma(\tilde{E}_f (X_f)_j W^4) W^5, \tag{3}$$

where W^4 and W^5 are weight matrices and $\sigma(\cdot)$ is a nonlinearity. With this STM, irregular space and time components can be flexibly integrated into a prediction.

Longitudinal Graph Classifier. To take a full advantage from the extracted features, $\mathcal{S}$ and $\mathcal{T}$ are combined together with $\mathcal{F}$ via a skip connection. These features collected from diverse branching paths contain both low and high-level information extracted from the graphs, and their integration provides strong ensemble-like results [31]. Using a FC layer and softmax, a set of predicted label probabilities $\hat{Y}$ is obtained for C classes as

$$\hat{Y}_c = \frac{\left((\mathcal{F} + \mathcal{S} + \mathcal{T})W^6\right)_c}{\sum_{c=1}^{C} \left((\mathcal{F} + \mathcal{S} + \mathcal{T})W^6\right)_c} \qquad \text{for } c = 1, \dots C, \tag{4}$$

where W^6 is a set of trainable weights of the FC layer for class prediction. Given the ground truth Y, the cross-entropy loss is defined with ℓ_2-regularization as

$$L = -\frac{1}{M} \sum_{m=1}^{M} \sum_{c=1}^{C} Y_{mc} \cdot \log(\hat{Y}_{mc}) + \frac{\lambda}{2M} ||\mathbf{W}||_{\ell_2}^2, \tag{5}$$

where $\mathbf{W}$ is a set of trainable parameters and λ controls a regularization strength.

3 Experiments

In this section, we evaluate STGMLP on two independent datasets, i.e., ADNI and ABCD, whose demographics are given in Table 1. We discuss the quantitative results, model behavior, and neuroscientific interpretations below.

3.1 Materials and Setup

ADNI Dataset. The ADNI is the largest public AD dataset providing longitudinal and multimodal images such as magnetic resonance imaging (MRI) and positron emission tomography (PET). As node features, cortical thickness from MRI, standardized uptake value ratio (SUVR) from FDG-PET and Amyloid-PET at all ROIs were measured. Structural brain networks were obtained by in-house probabilistic tractography on diffusion tensor images (DTI) on Destrieux

Table 1. Demographics of the ADNI and ABCD datasets.

Category	ADNI			ABCD	
	Preclinical	MCI	AD	BP	NP
# of Subject	45	61	25	835	734
# of Record	163	253	75	1,670	1,468
Gender (M/F)	18/27	38/23	14/11	439/396	399/335
Age (Mean ± std)	73.8 ± 5.6	72.5 ± 7.3	76.0 ± 7.5	9.9 ± 0.4	9.9 ± 0.4

atlas [8] with 148 ROIs. The number of visits (i.e., T_m) by participants varied from 2 to 7. Five labels were initially given: cognitively normal (CN), significant memory concern (SMC), early mild cognitive impairment (EMCI), late MCI (LMCI) and AD. They were redefined as Preclinical (CN and SMC), MCI (EMCI and LMCI) and AD groups to secure sufficient sample size.

ABCD Dataset. The ABCD dataset (v4.0) contains two timepoints with multivariate features: baseline data for children aged 8–10 and their 2-year follow-up measurements such as fractional anisotropy, mean diffusivity, and cortical thickness obtained via DTI and MRI. Morphometric similarity network [27] was used to construct a graph per subject. As in other works [5,9,24,34] studying the relationship between socioeconomic status (SES) and brain development on the ABCD, we categorized the longitudinal samples into Below-Poverty (BP) and Non-Poverty (NP) groups based on the annual household income. The poverty criterion from U.S. Census Bureau ($27,479) is used to set the BP group, and the NP group is set whose annual household income is $200k and greater.

Setup. As baselines, we adopt various graph convolutional networks (GCNs) for spatio-temporal graph analysis such as ST-GCN [11], IT-GNN [16], infoGCN [4], ShiftGCN [3], and CTRGCN [2]. Also, typical machine learning (ML) methods such as Linear SVM, Linear Regression (LR) and MLP are used for comparisons. Along with the all node and edge features, the maximum time difference $X_{T_m} - X_1$ is used to train these ML models. We applied early stopping via test loss with 5-fold cross validation (CV) for all methods including ours.

To implement STGMLP, the learning rate, weight decay (λ), dropout rate, and depth D of GSM were set to 1e-2, 5e-4, 5%, and 3, respectively. For GTMs on the ADNI, total $P{=}T_m$ pairs are selected: $T_m - 1$ pairs for adjacent graphs in time, and one pair for the first and last (i.e., end-to-end) timepoints. For the ABCD, P is set to 1 as $T_m{=}2$ for all samples. Note that, the combination of timepoints can be flexibly selected to include domain knowledge.

3.2 Evaluation and Discussions on the Results

Quantitative results (i.e., mean accuracy, precision, and recall) of all experiments and the number of trainable parameters are compared in Table 2. The results demonstrate that our model with a small computational cost outperformed baselines with vast parameters on both datasets. Also, our method showed no overfitting, as the mean training accuracies for ADNI and ABCD were 74.9% and

Table 2. Comparison of the number of trainable parameters and model performance with 5-fold cross validation from ADNI and ABCD experiments.

Method	# param	ADNI			ABCD		
		Accuracy (%)	Recall (%)	Precision (%)	Accuracy (%)	Recall (%)	Precision (%)
SVM	-	47.3 ± 4.2	42.9 ± 6.0	48.3 ± 3.2	64.4 ± 7.2	65.0 ± 5.7	70.5 ± 3.2
LR	68k	55.8 ± 1.9	45.4 ± 4.5	52.2 ± 14.0	67.8 ± 5.0	66.7 ± 5.3	68.7 ± 4.8
MLP	1,461k	55.7 ± 2.7	46.4 ± 6.6	50.8 ± 11.4	62.8 ± 6.6	61.2 ± 7.6	66.3 ± 5.1
ST-GCN [11]	62k	47.6 ± 2.2	46.0 ± 5.2	47.3 ± 1.7	54.1 ± 1.5	54.7 ± 6.1	46.2 ± 7.9
IT-GNN [16]	5k	50.4 ± 4.4	51.6 ± 5.0	50.8 ± 5.1	54.2 ± 1.5	51.4 ± 2.3	36.3 ± 6.4
ShiftGCN [3]	753k	58.9 ± 3.9	60.7 ± 6.8	57.2 ± 11.4	65.8 ± 3.3	66.6 ± 4.1	67.1 ± 3.5
CTRGCN [2]	683k	62.0 ± 2.9	57.7 ± 5.5	57.4 ± 14.8	67.4 ± 3.6	67.5 ± 3.7	67.9 ± 3.6
CTRGCN (w/ 1 layer) [2]	67k	63.5 ± 8.6	57.4 ± 11.0	55.8 ± 12.4	66.5 ± 3.3	66.6 ± 4.1	67.1 ± 3.5
infoGCN [4]	40,584k	57.3 ± 7.5	52.1 ± 10.1	43.2 ± 13.7	53.9 ± 2.9	51.1 ± 2.5	27.0 ± 0.9
infoGCN (w/ 1 layer) [4]	840k	58.9 ± 2.5	58.0 ± 5.4	51.2 ± 12.0	53.8 ± 1.3	51.5 ± 3.4	27.0 ± 0.9
STGMLP (Ours)	**1k**	$\mathbf{71.3 \pm 4.6}$	$\mathbf{66.4 \pm 3.3}$	$\mathbf{76.5 \pm 6.1}$	$\mathbf{72.3 \pm 2.9}$	$\mathbf{72.1 \pm 2.9}$	$\mathbf{72.4 \pm 3.0}$

Table 3. The nodal Grad-CAM for top 10 ROIs with the highest contribution to classify the AD class of ADNI dataset (Left) and the BP class of ABCD dataset (Right).

idx	ROI	Preclinical	MCI	AD
1	r g.front.sup	9.0e-3	8.5e-2	**1.8e-1**
2	l s.temporal.sup	1.6e-1	8.6e-2	**1.7e-1**
3	r g.temp.sup.g.t.transv	1.4e-1	**1.9e-1**	1.7e-1
4	r s.temporal.sup	7.5e-3	7.6e-2	**1.5e-1**
5	l g.temp.sup.lateral	6.0e-2	1.3e-1	**1.5e-1**
6	r g.insular.short	2.0e-1	**2.1e-1**	1.5e-1
7	r g.temp.sup.plan.polar	1.9e-1	**2.0e-1**	1.5e-1
8	r g.temp.sup.lateral	5.9e-2	1.3e-1	**1.4e-1**
9	l g.front.sup	1.1e-2	7.3e-2	**1.4e-1**
10	r g.orbital	3.8e-2	1.0e-1	**1.4e-1**

idx	ROI	BP	NP
1	l g.insular.short	**2.8e-1**	2.5e-1
2	r s.postcentral	**1.8e-1**	1.5e-1
3	l s.oc.middle.and.Lunatus	**1.6e-1**	1.4e-1
4	r g.occipital.sup	**1.6e-1**	1.2e-1
5	r s.oc.middle.and.Lunatus	**1.5e-1**	1.4e-1
6	l g.occipital.sup	**1.4e-1**	1.3e-1
7	r g.insular.short	**1.4e-1**	1.2e-1
8	r. g.Ins.lg.and.s.cent.ins	**1.4e-1**	1.0e-1
9	l g.temp.sup.Plan.polar	1.4 e-1	**1.7e-1**
10	l s.postcentral	**1.4e-1**	1.3e-1

73.6%. Moreover, it is worth noting that the improvement in performance comes from the effectiveness of our method, not solely from the reduction in the number of parameters. On ADNI, our model even showed a 20.9%p accuracy margin over IT-GNN, which has a similar parameter scale (5k) as ours (1k).

Preclinical vs. MCI vs. AD on ADNI Dataset. Here, STGMLP achieved 71.3% accuracy with 7.8∼24.0%p margin over baselines. Here, we provide clinically interpretable results by analyzing nodal contributions to classify each class via class-averaged Grad-CAM [28]. In Table 3, we reported the top 10 regions with the highest gradient activation for AD group classification, which are mostly distributed in *the temporal and frontal* regions. The ROI showing the highest activation is *the right superior frontal gyrus*, which is a majorly damaged area where atrophy of white matter is discovered in various AD studies [13,17,25]. Also, both sides of *the superior temporal gyri and sulci* were found, which are highly activated in auditory and verbal memory processing [20,36]. The visualization of averaged class-wise activations for all ROIs is shown in Fig 3.

Below-Poverty vs. Non-poverty on ABCD Dataset. As shown in Table 3, ROIs that played a decisive role to classify the BP class are mostly distributed

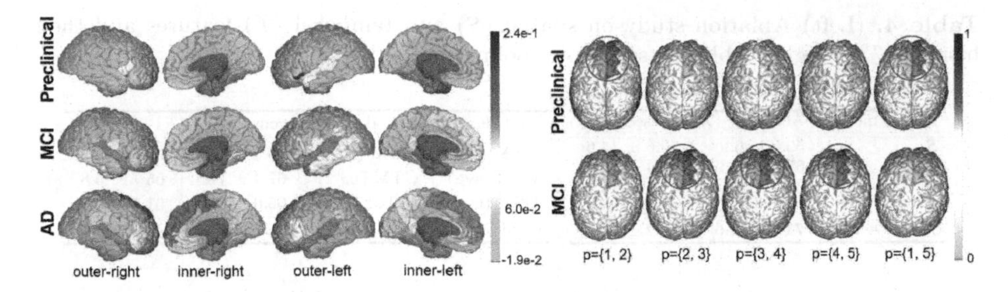

Fig. 3. (Left) Visualization of the nodal class-wise activations for Preclinical vs. MCI vs. AD group analysis. The activations are averaged on a per-class basis. Progressive variations in several ROIs are distinguishable for the labels. (Right) Averaged temporal feature maps extracted with GTM from subjects with T_m=5. For each timepoint pair p=$\{i_1, i_2\}$, higher values (circled in red) indicate higher activation from GELU.

along the *insular* and *occipital* regions. For example, insular subregions such as *the long insular gyrus and central sulcus of the insula* and both left and right sides of *the short insular gyri* were identified, which are implicated to social decision making [23,26] and emotional processing [6,12]. These ROIs responsible for somatosensory are thought to be impacted by environmental exposures such as SES and play a key role in overall cognition in children [7,10]. Also, developmental changes in occipital subregions such as *the superior occipital gyrus* and *the middle occipital sulcus* are closely related to the parental SES [21,29], which appear to be consistent with the result of our experiment.

3.3 Temporal Analysis on AD-Specific Activation

In Fig. 3, we also investigated pairwise temporal features from the ADNI experiment. While the Preclinical group shows strong activations in the initial and end-to-end pairs, features of later timepoints are highly activated in MCI group, showing consistency to the neurodegenerative dynamics in AD. We also observed that the use of pairwise information is sufficient to analyze the entire time series, as the signs of trained weights (i.e., W^2 in GTM) were totally opposite between adjacent timepoints, i.e., averaged weights were -7.7e-2 (std:0.56) vs. +4.4e-2 (std:0.29). In this way, our method investigates network alterations (around 1–2 year) in a pairwise manner and captures the whole temporal variation via pooling, rather than directly looking at the whole changes (over several years). These results confirm that GTM can capture the key temporal features for a given pair and label, and the following nonlinear function intensifies the difference.

3.4 Ablation Study on Hyperparamters

Ablation studies on each module (i.e., GSM, GTM, and STM) and pooling methods were performed on ADNI. The results, reported in Table 4, show that using max pooling for both GSM and GTM performed best. This suggests that there

Table 4. (Left) Ablation study on spatial ($\mathcal{S}$) and temporal ($\mathcal{T}$) features and their fusions ($\mathcal{F}$). (Right) Ablation study on pooling methods.

Fusion method	Accuracy (%)	F1-score (%)
$\mathcal{S}$	59.6 ± 6.9	50.6 ± 14.6
$\mathcal{T}$	67.4 ± 7.3	62.4 ± 7.0
$\mathcal{F}$	60.4 ± 7.4	51.2 ± 15.1
$\mathcal{S}+\mathcal{T}$	70.5 ± 6.4	**69.4** ± 7.2
$\mathcal{F}+\mathcal{S}+\mathcal{T}$	**71.3** ± 4.6	68.2 ± 2.9

Fusion Method	Accuracy (%)	F1-score (%)
GSM (avg) + GTM (avg)	65.8 ± 4.7	60.5 ± 5.9
GSM (avg) + GTM (max)	67.4 ± 7.1	65.7 ± 5.8
GSM (max) + GTM (avg)	65.0 ± 5.4	61.6 ± 7.1
GSM (max) + GTM (max)	**71.3**±4.6	**68.2**±2.9

exist particularly significant time-points (or time-points pairs) for classifying longitudinal brain networks and strongly reflecting these points in decision making is more useful than smoothing features for the entire time with average pooling.

4 Conclusion

In this work, we proposed a novel longitudinal graph mixer to investigate longitudinal variations of spatio-temporal graphs. The idea was driven by mixing features temporally and spatially along the topology of brain networks, and its structure was designed to deal with the heterogeneity of data with a significantly reduced number of parameters compared to deep graph convolutional models. Experiments validate the superiority of our framework, successfully identifying key ROIs in classifying different classes, suggesting a significant potential to be deployed for other longitudinal connectome analyses of various brain disorders.

Acknowledgement. This research was supported by NRF-2022R1A2C2092336 (50%), IITP-2022-0-00290 (20%), IITP-2019-0-01906 (AI Graduate Program at POSTECH, 10%) funded by MSIT, HU22C0171 (10%) and HU22C0168 (10%) funded by MOHW in South Korea, NIH R03AG070701 and Foundation of Hope in the U.S.

References

1. Ba, J.L., Kiros, J.R., et al.: Layer normalization. arXiv preprint arXiv:1607.06450 (2016)
2. Chen, Y., Zhang, Z., et al.: Channel-wise topology refinement graph convolution for skeleton-based action recognition. In: Proceedings of the IEEE/CVF International Conference on Computer Vision, pp. 13359–13368 (2021)
3. Cheng, K., Zhang, Y., et al.: Skeleton-based action recognition with shift graph convolutional network. In: Proceedings of the IEEE/CVF Conference on Computer Vision and Pattern Recognition, pp. 183–192 (2020)
4. Chi, H.g., Ha, M.H., et al.: Infogcn: Representation learning for human skeleton-based action recognition. In: Proceedings of the IEEE/CVF Conference on Computer Vision and Pattern Recognition, pp. 20186–20196 (2022)

5. Cho, H., Park, G., Isaiah, A., Kim, W.H.: Covariate correcting networks for identifying associations between socioeconomic factors and brain outcomes in children. In: de Bruijne, M., et al. (eds.) Medical Image Computing and Computer Assisted Intervention – MICCAI 2021: 24th International Conference, Strasbourg, France, September 27 – October 1, 2021, Proceedings, Part VII, pp. 421–431. Springer, Cham (2021). https://doi.org/10.1007/978-3-030-87234-2_40
6. Craig, A.D.: How do you feel-now? the anterior insula and human awareness. Nat. Rev. Neurosci. **10**(1), 59–70 (2009)
7. Craig, A.D., Chen, K., et al.: Thermosensory activation of insular cortex. Nat. Neurosci. **3**(2), 184–190 (2000)
8. Destrieux, C., Fischl, B., et al.: Automatic parcellation of human cortical gyri and sulci using standard anatomical nomenclature. Neuroimage **53**(1), 1–15 (2010)
9. Ellwood-Lowe, M., Irving, C., et al.: Exploring neural correlates of behavioral and academic resilience among children in poverty. Dev. Cogn. Neurosci. **54**, 101090 (2022)
10. Failla, M.D., Peters, B.R., et al.: Intrainsular connectivity and somatosensory responsiveness in young children with ASD. Molecular Autism **8**(1), 1–11 (2017)
11. Gadgil, S., Zhao, Q., Pfefferbaum, A., Sullivan, E.V., Adeli, E., Pohl, K.M.: Spatio-Temporal Graph Convolution for Resting-State fMRI Analysis. In: Martel, A.L., et al. (eds.) Medical Image Computing and Computer Assisted Intervention – MICCAI 2020: 23rd International Conference, Lima, Peru, October 4–8, 2020, Proceedings, Part VII, pp. 528–538. Springer, Cham (2020). https://doi.org/10.1007/978-3-030-59728-3_52
12. Gu, X., Hof, P.R., et al.: Anterior insular cortex and emotional awareness. J. Comp. Neurol. **521**(15), 3371–3388 (2013)
13. Guo, X., Wang, Z., et al.: Voxel-based assessment of gray and white matter volumes in Alzheimer's disease. Neurosci. Lett. **468**(2), 146–150 (2010)
14. He, K., Zhang, X., et al.: Deep residual learning for image recognition. In: Proceedings of the IEEE Conference on Computer Vision and Pattern Recognition, pp. 770–778 (2016)
15. Hendrycks, D., Gimpel, K.: Gaussian error linear units (GELUs). arXiv preprint arXiv:1606.08415 (2016)
16. Kim, M., Kim, J., et al.: Interpretable temporal graph neural network for prognostic prediction of Alzheimer's disease using longitudinal neuroimaging data. In: 2021 IEEE International Conference on Bioinformatics and Biomedicine, pp. 1381–1384. IEEE (2021)
17. Kim, W.H., Singh, V., Chung, M.K., Hinrichs, C., et al.: Multi-resolutional shape features via non-euclidean wavelets: applications to statistical analysis of cortical thickness. Neuroimage **93**, 107–123 (2014)
18. Kipf, T.N., Welling, M.: Semi-supervised classification with graph convolutional networks. arXiv preprint arXiv:1609.02907 (2016)
19. Kundu, S., Lukemire, J., et al.: A novel joint brain network analysis using longitudinal Alzheimer's disease data. Sci. Rep. **9**(1), 1–18 (2019)
20. Lenzi, D., Serra, L., Perri, R., et al.: Single domain amnestic mci: A multiple cognitive domains fMRI investigation. Neurobiol. Aging **32**(9), 1542–1557 (2011)
21. Lu, Y.C., Kapse, K., et al.: Association between socioeconomic status and in utero fetal brain development. JAMA Netw. Open **4**(3), e213526–e213526 (2021)
22. Olde Dubbelink, K.T., Hillebrand, A., et al.: Disrupted brain network topology in Parkinson's disease: a longitudinal magnetoencephalography study. Brain **137**(1), 197–207 (2014)

23. Quarto, T., Blasi, G., et al.: Association between ability emotional intelligence and left insula during social judgment of facial emotions. PLoS ONE **11**(2), e0148621 (2016)
24. Rakesh, D., Zalesky, A., et al.: Similar but distinct-effects of different socioeconomic indicators on resting state functional connectivity: findings from the adolescent brain cognitive development (ABCD) study. Dev. Cogn. Neurosci. **51**, 101005 (2021)
25. Ribeiro, L.G., Busatto Filho, G.: Voxel-based morphometry in Alzheimer's disease and mild cognitive impairment: systematic review of studies addressing the frontal lobe. Dementia & Neuropsychol. **10**, 104–112 (2016)
26. Rogers-Carter, M.M., Varela, J.A., et al.: Insular cortex mediates approach and avoidance responses to social affective stimuli. Nat. Neurosci. **21**(3), 404–414 (2018)
27. Seidlitz, J., Váša, F., et al.: Morphometric similarity networks detect microscale cortical organization and predict inter-individual cognitive variation. Neuron **97**(1), 231–247 (2018)
28. Selvaraju, R.R., Cogswell, M., et al.: Grad-cam: Visual explanations from deep networks via gradient-based localization. In: Proceedings of the IEEE/CVF International Conference on Computer Vision, pp. 618–626 (2017)
29. Spann, M.N., Bansal, R., et al.: Prenatal socioeconomic status and social support are associated with neonatal brain morphology, toddler language and psychiatric symptoms. Child Neuropsychol. **26**(2), 170–188 (2020)
30. Tolstikhin, I.O., Houlsby, N., et al.: Mlp-mixer: An all-mlp architecture for vision. In: Annual Conference on Neural Information Processing Systems, vol. 34 (2021)
31. Veit, A., Wilber, M.J., et al.: Residual networks behave like ensembles of relatively shallow networks. In: Annual Conference on Neural Information Processing Systems, vol. 29 (2016)
32. Wu, K., Taki, Y., et al.: A longitudinal study of structural brain network changes with normal aging. Front. Hum. Neurosci. **7**, 113 (2013)
33. Xu, K., Hu, W., et al.: How powerful are graph neural networks? arXiv preprint arXiv:1810.00826 (2018)
34. Yang, F., Isaiah, A., Kim, W.H.: COVLET: covariance-based wavelet-like transform for statistical analysis of brain characteristics in children. In: Martel, A.L., et al. (eds.) Medical Image Computing and Computer Assisted Intervention – MICCAI 2020: 23rd International Conference, Lima, Peru, October 4–8, 2020, Proceedings, Part VII, pp. 83–93. Springer, Cham (2020). https://doi.org/10.1007/978-3-030-59728-3_9
35. Yang, F., Meng, R., Cho, H., Wu, G., Kim, W.H.: Disentangled sequential graph autoencoder for preclinical Alzheimer's disease characterizations from ADNI Study. In: de Bruijne, M., et al. (eds.) Medical Image Computing and Computer Assisted Intervention – MICCAI 2021: 24th International Conference, Strasbourg, France, September 27–October 1, 2021, Proceedings, Part II, pp. 362–372. Springer, Cham (2021). https://doi.org/10.1007/978-3-030-87196-3_34
36. Zlatar, Z.Z., Bischoff-Grethe, A., et al.: Higher brain perfusion may not support memory functions in cognitively normal carriers of the apoe $\varepsilon 4$ allele compared to non-carriers. Front. Aging Neurosci. **8**, 151 (2016)

Correction to: COLosSAL: A Benchmark for Cold-Start Active Learning for 3D Medical Image Segmentation

Han Liu(✉), Hao Li, Xing Yao, Yubo Fan, Dewei Hu,
Benoit M. Dawant, Vishwesh Nath, Zhoubing Xu, and Ipek Oguz

Correction to:
Chapter "COLosSAL: A Benchmark for Cold-Start Active Learning for 3D Medical Image Segmentation" in:
H. Greenspan et al. (Eds.): ***Medical Image Computing and Computer Assisted Intervention – MICCAI 2023*****, LNCS 14221, https://doi.org/10.1007/978-3-031-43895-0_3**

In the originally published version of chapter 3, the second and third affiliation stated wrong locations. This has been corrected.

The updated original version of this chapter can be found at
https://doi.org/10.1007/978-3-031-43895-0_3

H. Greenspan et al. (Eds.): MICCAI 2023, LNCS 14221, p. C1, 2023.
https://doi.org/10.1007/978-3-031-43895-0_74

Author Index

H. Greenspan et al. (Eds.): MICCAI 2023, LNCS 14221, pp. 787–791, 2023.
https://doi.org/10.1007/978-3-031-43895-0